Transformation Groups and (Co)Homology

Vom Fachbereich Mathematik
der Technischen Universität Darmstadt
zur Erlangung des Grades eines
Doktors der Naturwissenschaften
(Dr. rer. nat.)
genehmigte Dissertation

von

Dipl.-Math. Martin Fuchssteiner
aus Darmstadt

Referent:	Prof. Dr. K.-H. Neeb
Koreferent:	Prof. Dr. L. Kramer
Tag der Einreichung:	07. Dezember 2009
Tag der mündlichen Pruefung:	23. Februar 2010

Darmstadt 2010
D 17

Bibliografische Information der Deutschen Nationalbibliothek

Die Deutsche Nationalbibliothek verzeichnet diese Publikation in der
Deutschen Nationalbibliografie; detaillierte bibliografische Daten sind
im Internet über http://dnb.d-nb.de abrufbar.

ISBN 978-3-8325-2524-8

Logos Verlag Berlin GmbH
Comeniushof, Gubener Str. 47,
10243 Berlin
Tel.: +49 (0)30 42 85 10 90
Fax: +49 (0)30 42 85 10 92
INTERNET: http://www.logos-verlag.de

Acknowledgements

I would like to express my gratitude to the people who supported the formation of this thesis. First and most of all, I thank my advisor Prof. Dr. KARL-HERMANN NEEB. He suggested the "inetgrability of local group cocycles" as a research problem. I thank him for giving me all the freedom one could wish for, which lead to research (and results) in unforeseen areas.

Furthermore, I would like to express my thanks to the members of the research group "Algebra, Geometry and Functional Analysis" at the mathematics department of the Technical Univerity Darmstadt. They provided a stimulating and pleasant working environment. For help in all administrative efforts during the work I express my gratitude to Gerlinde Gehring.

Contents

VIII   Contents

Zusammenfassung

Die vorliegende Arbeit befaßt sich hauptsächlich mit algebraischer Topologie und verwandten Gebieten wie z.B. der homologischen Algebra. Insbesondere werden die Grundlagen für eine Verallgemeinerung klassischer Spektralsequenzen zu solchen in der Kategorie **TopAb** der abelschen topologischen Gruppen oder der Kategorie **TopRMod** der topologischen Moduln über einem (kommutativen) topologischen Ring **R** gelegt. Diese sind nötig um z.B. Spektralsequenzen der Form

$$H_c^p(G; H_c^q(N; V)) \Rightarrow H_c^{p+q}(\hat{G}; V)$$

für stetige Koketten und Erweiterungen $N \to \hat{G} \twoheadrightarrow G$ topologischer Gruppen oder das Analogon mit glatten Kozykeln für Lie-Gruppen Erweiterungen konstruieren zu können. Da die benutzten Kategorien nicht abelsch sind, ist es notwendig die klassischen Ergebnisse in gutartigen nicht abelschen Kategorien neu zu beweisen. Dies erfordert neue Konzepte, da in den betrachteten Kategorien das "Schlangen Lemma", eines der wichtigsten Werkzeuge der homologischen Algebra, nicht gilt. Die notwendigen Betrachtungen hierzu werden in Kapitel 1 gemacht. Einige der wichtigsten der dort bewiesenen Verallgemeinerungen klassischer Ergebnisse für die betrachteten Objekte in nicht abelschen Kategorien sind

- das Erhalten von Monomorphismen, Epimorphismen, Einbettungen unter dem Normalisierungsfunktor sowie die Kostetigkeit und Exaktheit desselben,
- der Normalisierungssatz für die Kohomologie von Kokettenkomplexen
- das induzieren einer langen exakten Kohomologiesequenz durch eine kurze exakte Sequenz kosimplizialer Objekte,
- die Universalität des Kohomologiefunktors für kosimpliziale Objekte,

sowie Analoge Sätze für simpliziale Objekte und deren Homologie. Diese Ergebnisse werden dann vor allem in der Kategorie **TopRMod** der topologischen Moduln über einem topologischen Ring **R** angewandt. Dort wird zuerst der "freie topologische Modul"-Funktor betrachtet mit dessen Hilfe dann einige bekannte Eigenschaften des Tensorprodukts von Moduln auch für den Tensorproduktfunktor in **TopRMod** nachgewiesen werden. Für diesen wird z.B. bewiesen, daß

- binäre Produkte von Quotientenhomomorphismen wieder solche sind,
- der Endofunktor $M \otimes -$ für einen festen topologischen Modul M Kokerne erhält,
- und für jede freie topologische abelsche Gruppe A der Endofunktor $A \otimes -$ der Kategorie **TopAb** der topologischen abelschen Gruppen exakt bezüglich der Klasse aller lokal trivialen kurzen exakten Sequenzen ist.

Diese Ergebnisse werden daraufhin in Kapitel 2 angewandt. Dort werden topologisierte Versionen klassischer Homologie- und Kohomologietheorien betrachtet, wie z.B. eine topologisierte singuläre Homologie, eine topologisierte singuläre Kohomologie, äquivariante Kohomologie, Alexander-Spanier Kohomologie und äquivariante

Alexander-Spanier Kohomologie, alle mit topologischen Koeffizienten. Für diese Konstruktionen werden wiederum einige Eigenschaften der klassischen (nicht topologischen Versionen) nachgewiesen. Dies sind vor allem

- die Homotopieinvarianz,
- das induzieren einer langen exakten Sequenz durch eine kurze exakte Sequenz von Koeffizienten in der Kategorie **TopAb** der topologischen abelschen Gruppen (manchmal unter schwachen zusätzlichen Bedingungen),
- und die Existenz einer langen exakten Sequenz von topologischen Homologie- bzw. Kohomologiegruppen für Raumpaare (X, A) mit oft auftretenden Eigenschaften, wie z.B. der Abgeschlossenheit von A in X.

In den Kapiteln 3,4,5 und 6 werden geometrische Konstruktionen und Verfahren eingeführt, welche es später erlauben lokale äquivariante Kozykel auf topologischen Räumen mit Gruppenwirkung zu globalen äquivarianten Kozykeln fortzusetzen bzw. deren Fortsetzbarkeit zu globalen äquivarianten Kozykeln zu charakterisieren. Diese Konstruktionen werden vor allem in Kapitel 16 benötigt, um die Integrabilität äquivarianter Differentialformen zu lokal glatten globalen äquivarianten Kozykeln zu charakterisieren bzw. nachzuweisen. Eine erste Anwendung dieser Verfahren ist in Kapitel 7 zu finden. Dort wird die Erweiterbarkeit lokaler Gruppen-2-kozykel, bzw. deren Keime zu globalen Gruppen-2-Kozykeln charakterisiert. Im nächsten Kapitel werden die Koalgebra-Struktur der rationalen Homologie eingeführt und die kokommutativen Hopf-Algebren wegzusammenhängender H-Räume studiert. Die wichtigsten Ergebnisse dort sind

- eine Verallgemeinerung des Struktursatzes über die rationale Homologie wegzusammenhängender topologischer Gruppen endlichen homologietyps zu einem analogen Struktursatz ohne jegliche Voraussetzungen,
- und eine Verallgemeinerung desselben Struktursatzes zu einem analogen Struktursatz über die rationale Homologie wegzusammenhängender H-Räume unter der einzigen Voraussetzung daßdiese dem Trennungsaxiom T_1 genügen.

Die darauf folgenden Kapitel 9,10,11, und 12 befassen sich wieder mit algebraischer Topologie und geometrischen Methoden derselben. In den Kapiteln 9 und 10 wird die allgemeine Geometrische Realisierung simplizialer topologischer Räume untersucht. Hier lassen sich viele Eigenschaften der geometrischen Realisierung simplizialer Mengen verallgemeinern. Diese sind unter anderem

- Die Erhaltung von Monomorphismen, Epimorphismen, der Dichtheit des Bildes, Identifikationen, Proklusionen, Abgeschlossenheit von Morphismen in hausdorffsche simpliziale Räume,
- die Darstellung der geometrische Realisierung $|X|$ eines simplizialen topologischen Raumes X als Kolimes der Realisierungen der k-Skeleta,
- die Existenz eines rechts adjungierten Funktors und somit die Kostetigkeit,
- die Eigenschaft daß die geometrische Realisierung des topologisierten "singulären simplizialen Raum"-Funktors natürlich äquivalent zur Identität ist,
- die Erhaltung von Egalisatoren,
- und die Erhaltung von Homotopie.

Darüberhinaus wird nachgewiesen, daß der geometrische Realisierungsfunktor nicht alle endlichen Produkte erhält. Weiterhin wird bewiesen, daß

- die geometrische Realisierung simplizialer k_ω-Räumen wieder k_ω Räume sind,
- daß die geometrische Realisierung simplizialer Mengen endliche Produkte (in der Kategorie der topologischen Räume) erhält.
- und somit, daß die geometrische Realisierung einer (diskreten) simplizialen Algebra beliebigen Typs eine topologische Algebra desselben Typs ist.

In Kapitel 11 wird der geometrische Realisierungsfunktor benutzt um einen "einhüllender kontrahierbarer Raum"-Funktor E zu konstruieren, der natürlich kontrahierbar ist. (Dies entspricht nicht Milnors Konstruktion universeller Bündel.) Dieser wird daraufhin benutzt um universelle Bündel und klassifizierende Räume zu konstruieren. Mit Hilfe der in den Kapiteln 9 und 10 nachgewiesenen Eigenschaften wird sodann gezeigt, daß

- der "einhüllende kontrahierbare Raum"-Funktor E Egalisatoren erhält,
- der einhüllende Raum eines Hausdorff-Raumes wieder hausdorffsch ist,
- der einhüllende Raum eines k_ω-Raumes wieder ein k_ω-Raum ist,
- der "einhüllende kontrahierbare Raum"-Funktor E endliche Limiten von Hausdorff-Räumen erhält,
- der einhüllende Raum EA einer hausdorffschen k_ω-Algebra A beliebigen Typs wieder eine hausdorffsche k_ω-Algebra des gleichen Typs ist,
- und die Einschränkung von E auf hausdorffsche k_ω-Gruppen exakt ist.

Diese Eigenschaften des "einhüllenden kontrahierbaren Raum"-Funktors E erlauben dann die darauf folgende Konstruktion (numerierbarer) universeller Bündel lokal kontrahierbarer Hausdorff k_ω-Gruppen. Diese hat die besondere Eigenschaft, daß die Totalräume der Bündel topologische Gruppen sind und der klassifizierende Raum ein Nebenklassenraum ist. Sie läßt sich insbesondere auf reelle und komplexe Kac-Moody Gruppen und abzählbare Limiten endlich-dimensionaler lokal kompakter Gruppen anwenden, wie z.B. $\mathbf{O} = \lim O_n$, $\mathbf{SO} = \lim SO_n$, $\mathbf{U} = \lim U_n$, $\mathbf{PU} = \lim PU_n$ etc..

Zu Ende des 11. Kapitels werden Eilenberg-MacLane Räume und verwandte Konstruktionen untersucht. Die wichtigen Resultate sind die Erkenntnis, daß

- Zu jeder folge A_n (diskreter) abelscher Gruppen eine abelsche topologische Gruppe G existiert, welche ein CW Komplex ist und deren Homotopie-Gruppen $\pi_n(G)$ genau die abelschen Gruppen A_n sind,
- und somit für alle $n \in \mathbb{N}$ zu jeder abelschen Gruppe π ein Eilenberg-MacLane $K(\pi, n)$ existiert, welche ein CW-Komplex ist,
- die rationale singuläre Homologie eines Eilenberg-MacLane Raumes $K(\pi, n)$ entweder isomorph zur Polynomalgebra über $\pi \otimes \mathbb{Q}$ oder zur äußeren Algebra über $\pi \otimes \mathbb{Q}$ isomorph ist, je nachdem ob n gerade oder ungerade ist.

Hierbei sollte erwähnt werden, daß die letzten beiden Resultate für die Spezialfälle abzählbarer abelscher Gruppen π bzw. endlich-dimensionaler $\pi \otimes \mathbb{Q}$ schon bekannt sind, jedoch bis jetzt noch nicht ohne diese Voraussetzungen. Die Erkenntnisse über Eilenberg-MacLane Räume ermöglichen die Konstruktionen von Whitehead- und Postnikovtürmen, deren Faserungen Hauptfaserbündel sind. Diese Konstruktionen werden in Kapitel 12 vorgestellt.

In Kapitel 13 wird eine Leray-Serre Spektralfolge für die deRham Kohomologie unendlich-dimensionaler flacher Bündel konstruiert. Mit Hilfe dieser wird der Fluß-Homomorphismus der symplektischen Topologie verallgemeinert. Die allgemeine Version des Fluß-Homomorphismus wird zur Charakterisierung der Integrabilität äquivarianter Differentialformen zu lokal glatten äquivarianten globalen Kozykeln verwendet. Kapitel 14 enthält leichte Verallgemeinerungen der van Est Spektralfolgen für topologische Gruppen bzw. unendlich-dimensionale glatte Transformationsgruppen. Diese werden dann in den letzten beiden Kapiteln 14 und 16 zur Beschreibung der äquivarianten Fortsetzbarkeit lokaler äquivarianter Kozykel und zur Beschreibung der Integrabilität äquivarianter Differentialformen zu lokal glatten äquivarianten globalen Kozykeln benötigt. Das Ergebnis ist die Erkenntnis,

- daß eine äquivariante Differentialform auf dem ausreichend zusammenhängenden G-Raum M einer glatten Transformationsgruppe (G, M) zu einem lokal glatten globalen äquivarianten Kozyklus integrierbar ist, falls alle Fluß-Homomorphismen auf singulären Zykeln trivial sind.

Abstract/Introduction

Chapter 1: Homological Algebra in TopRMod

The first chapter is devoted to developing the basic notions of homological algebra in categories of topological algebras (such as the category **TopRMod** of topological modules over a topological ring R). Many algebras associated with geometric objects such as manifolds, (principal) fibre bundles, topological groups and transformation groups can naturally be endowed with a topology, turning them into topological algebras. Doing so provides a finer tool to differentiate these geometric objects as just considering the underlying (discrete) algebras. Examples are the topological singular (co)homology, continuous cohomology of topological groups (cf. [Seg70]) and topological equivariant cohomology. In the case of continuous group cohomology, we wish to drive a spectral sequence

$$H^p(B, H^q(N; A)) \Rightarrow H^{p+q}(G; A).$$

This requires the cochain complex $C^*(N; A)$ of continuous group cochains and its cohomology groups $H^q(N; A)$ to be Abelian topological groups. Furthermore one has to use homological algebra in the category **TopAb** of Abelian topological groups. This category – along with most categories of topological algebras – is not Abelian. For example a monomorphism $f : A \to A'$ in **TopAb** in general fails to be a kernel of its cokernel. The fact that most categories of topological algebras are not Abelian prevents an application of the abstract machinery of homological algebra developed for Abelian categories. Therefore it has to be carefully examined which constructions of homological algebra can be generalised and which results carry over to these categories. There are some constructions of homological algebra that can be generalised to many categories of topological algebras. Recent progress on this subject has been made in the articles [BC05] and [BC06]. The results presented there are the starting point for our work. We will especially concerned with the category **TopRMod** of topological modules over a topological ring R and the category **TopAb** of Abelian topological groups, which is a special case thereof. (It is the category of topological modules over the discrete ring $\mathbb{Z}$ of integers.) The results obtained here will have applications on Lie algebra cohomology, continuous cohomology of topological groups, topological de Rham- and Alexander-Spanier cohomology as well as for their equivariant versions.

In order to generalise the most basic constructions of homological algebra it has in particular to be verified that the homology functor has the usual properties, e.g. that every short exact sequence

$$0 \to M_* \to M'_* \to M''_* \to 0$$

of simplicial topological R-modules leads to a long exact sequence of homology modules. Furthermore it has to be proved that for any any module M the tensor product functor $M \otimes -$: **TopRMod** $\to$ **TopRMod** behaves sufficiently "nice". The first section of this chapter covers the basic notions of homological algebra in **TopRMod**. There we introduce normalised (co)chain complexes of (co)simplicial topological modules and verify that the normalisation functor is exact (cf. Theorem 1.1.37). We also show that the (co)homology of normalised (co)chain complexes associated to (co)simplicial modules is a universal (co)homological δ-functor (Theorems 1.1.19 and 1.1.48).

In the second section we derive a general normalisation theorem for (co)chain complexes in homological categories, which states that the (co)homology of chain complexes and the (co)homology of normalised (co)chain complexes of (co)simplicial objects in these categories are naturally isomorphic universal (co)homological δ-functors (Theorems 1.2.19 and 1.2.34). This includes many categories of (co)simplicial topological algebras. This in particular includes the category **TopRMod**, concerning which we show that the (co)homology of unnormalised (co)chain complexes associated to (co)simplicial modules is a universal (co)homological δ-functor which is naturally isomorphic to the (co)homology of the unnormalised associated (co)chain complexes (Theorems 1.2.22 and 1.2.37).

The third section contains a short digression into the realm of simplicial objects in homological categories (such as **TopRMod**), where we show that the chain complexes associated to simplicial objects are proper and that their homology is a coeffaceable and universal homological δ-functor.

The last section is devoted to more concrete computations and the examination of the free Abelian topological group functor and of binary tensor products in the category **TopRMod**. Here we observe that the free Abelian topological group functor preserves quotients and that the binary tensor product of quotients in **TopRMod** is again a quotient. Moreover tensoring with a fixed module M is shown to preserve cokernels. The further examination of the free Abelian topological group functor is largely concerned with P-embeddings, which play a crucial role for the exactness of short sequences (cf. Theorem 1.4.18). Here we derive a slight generalisation of the characterisation of P-embeddings (Theorem 1.4.36), which is needed for the the proof that tensoring with a free Abelian topological group is an exact functor w.r.t. large classes of short exact sequences, which in particular includes those group extensions which are locally trivial fibre bundles (Theorem 1.4.53).

Chapter 2: Topological Homology- and Cohomology Theories

It is often useful to equip homology groups $H_n(X)$ or cohomology groups $H^n(X;V)$ of a topological space X or equivariant cohomology (e.g. group cohomology) with a topology. Nevertheless no systematic approach to this subject has been made so far. In this chapter we introduce (generalised) topological homology theories or cohomology theories as functors **Top** $\to$ **TopAb** from the category **Top** of topological spaces to the category **TopAb** of Abelian topological groups. This includes the classical case, where the coefficients are given the discrete topology.

We begin with the definition of topological (co)homology theories and then introduce topologised versions of classical (co)chain complexes and (co)homologies. We then verify that these topologised versions share many of the important properties of their classical counterparts. This requires us to do homological algebra in the category **TopAb** of topological Abelian groups and relies on all the results derived in Chapter 1. In particular we introduce topologised singular homotopy, singular homology, singular cohomology, (equivariant) continuous Alexander-Spanier cohomology (all with topological coefficients) and show the homotopy invariance of these functors into the category **TopAb** of Abelian topological groups. The preservation of homotopy by the singular simplicial space functor $C(\Delta, -)$ and the the standard simplicial space functor is also proved. Moreover we show that large classes of short exact sequences of coefficients (e.g. short exact sequences which are locally trivial fibre bundles) induce long exact sequence of the (co)homology groups in **TopAb**. In addition it is shown that large classes of embeddings of subspaces (e.g. P-embeddings, retracts, closed neighbourhood retracts in normal spaces, closed subspaces of the homotopy type of countable CW-complexes, closed subspaces of metric or paracompact $\aleph$-spaces, closed submanifolds, fibres of bundles over completely regular spaces) induce long exact sequences of the (co)homology groups of various topologised (co)homology theories.

Chapter 3: Equivariant Systems of Singular Simplices

In this Chapter we define and construct systems of (smooth) singular simplices in various topological spaces, e.g. Lie groups and Riemannian manifolds. These are then used to construct an inverse map to the vertex transformation and, in conjunction with the subdivision introduced in Chapters 4, 5 and 6, to extend local equivariant cocycles to global equivariant cocycles on spaces with group actions. They are also needed for integrating equivariant differential forms to locally smooth global cocycles in Chapter 16.

Chapters 4 and 5: Relative Subdivision and Complete Relative Subdivision

In the third chapter we establish a way to subdivide singular singular simplices in a topological space X similar to the barycentric subdivision, but in such a way that faces of a singular simplex are not subdivided if they are "small enough". Here "small enough" means that they are $\mathfrak{U}$-small for a given open covering $\mathfrak{U}$ of X. Furthermore the special subdivision to be constructed has to intertwine the face and degeneracy maps of singular simplices.

In the fourth chapter we prove that repeated relative subdivision w.r.t. an open cover becomes stable after finitely many steps. This is used to construct a chain equivalence $\mathrm{Csd}_{\mathfrak{U}} : S_*(X) \to S_*(X, \mathfrak{U})$ for any given open covering $\mathfrak{U}$ of a topological space X. This equivalence maps any singular simplex σ to the 'minimal' $\mathfrak{U}$-relative subdivision $\mathrm{Sd}_{\mathfrak{U}}^n(\sigma)$, so that the chain $\mathrm{Sd}_{\mathfrak{U}}^n(\sigma)$ consists of $\mathfrak{U}$-small simplices only.

The main motivation is the extension of local equivariant cocycles to to global equivariant cocycles, for which this kind of subdivision is needed (see the procedure in Section 6.2 and Chapter 15). A very convenient special case thereof is the following observation: Suppose $f \in A_{eq}(X, \mathfrak{U}; V)$ is a local equivariant n-cocycle defined on the open neighbourhood $\mathrm{SS}(X, \mathfrak{U})([n])$ of $\mathrm{SS}(X)([n]) \cong X^{n+1}$ and $\hat{\sigma}_n$ is an equivariant system of simplices on $\mathrm{SS}(X, \mathfrak{U})([n])$. Then the function f extends to a group homomorphism $f : \mathbb{Z}^{(\mathrm{SS}(X, \mathfrak{U})([n]))} \to V$ and the function

$$F : \mathrm{SS}(X)([n]) \cong X^{n+1} :\to V, \quad F(\mathbf{x}) = f(\mathrm{Csd}_{\mathfrak{U}}(\hat{\sigma}_n(\mathbf{x}))),$$

where $\mathrm{Csd}_{\mathfrak{U}}$ denotes our special type of complete subdivision, is a global equivariant cocycle on X with the same germ as f at the diagonal (cf. Theorem 6.2.17).

Chapter 6: Relative Subdivision In Transformation Semi-Groups

In this chapter we examine how the relative subdivision can be applied to transformation semi-groups (G, X). This in particular includes the equivariance of the subdivision chain map $\mathrm{Sd}_\mathfrak{U} : S_*(X) \to S_*(X)$ and the complete relative subdivision chain map $\mathrm{Csd}_\mathfrak{U} : S_*(X) \to S_*(X, \mathfrak{U})$ for G-invariant open coverings $\mathfrak{U}$. It is especially fruitful in the case of highly connected spaces. Here the relative subdivision can be used to extend equivariant Alexander-Spanier cocycles to global equivariant cocycles (see section 6.2 below), a case which includes the extension of germs of (homogeneous) group cocycles.

After the study of the relative subdivision we introduce the "vertex transformation" which induces a natural morphism ${}^t H^*_{AS,c}(X; V) \to{}^t H^*(X; V)$ from the topologised continuous Alexander-Spanier cohomology ${}^t H^*_{AS,c}(X; V)$ into the topologised singular cohomology ${}^t H^*(X; V)$ of a topological space X. This is a transfer of the classical morphism to the topologised version of these cohomologies.

Chapter 7: Abelian Extensions of Topological Groups

Given a topological group G, an Abelian divisible topological group A and a local group cocycle f, under which circumstances is there an extension to a global cocycle F on G? For $n = 2$ this is equivalent to the existence of an extension

$$0 \to A \to A \times_F G \to G \to 1$$

of G by A which is locally given by f. Especially when G and A are Lie groups and f is known to be a local integral of a Lie algebra cocycle $\omega \in Z^2(\mathfrak{g}, \mathfrak{a})$ (which always exists if A is a quotient of a sequentially complete TVS by a discrete subgroup) then this equivalent to an integration of the extension

$$0 \to \mathfrak{a} \to \mathfrak{a} \oplus_\omega \mathfrak{g} \to \mathfrak{g} \to 0$$

of Lie algebras to an extension of Lie groups. Here the constructed extension would be a locally trivial fibration with base space G. For this case the question has been answered by [Nee04] (, where A is assumed to be a quotient of a sequentially complete topological vector space by a discrete subgroup). In this chapter we consider this problem for locally contractible connected topological groups and establish a necessary and sufficient condition for the germ of a local group 2-cocycle f to be extendible to a global group cocycle F by deriving an exact sequence

$$Z^2(G; V) \to Z^2_{AS,eq}(G; V)^G \to H^2_{AS}(G; V) \oplus \hom(\pi_1(M), H^1(G; V)).$$

This is the main result of this chapter. It is a generalisation of the result for locally smooth Lie group cocycles in [Nee04] to arbitrary connected topological groups G which are locally contractible. The proof relies on the existence of an equivariant system of singular 2-simplices derived in Section 3, the relative subdivision introduced in Chapter 4, the complete relative subdivision proven to exist in Chapter 5 and the observations concerning relative subdivision in transformation groups made in Chapter 6.

The considerations leading to the proof are of general nature for topological groups G and do not rely on the differential structure provided by Lie groups. The constructions made do rely on the existence of a universal simply connected covering group $\tilde{G}$. So the group G is assumed to be connected and semi-locally simply connected, which implies the existence of a universal covering group $\tilde{G}$. (These conditions are always satisfied by connected groups that are CW-complexes or locally contractible, e.g. Lie groups, loop groups thereof, Kac-Moody groups etc.).

Chapter 8: The Rational Singular Homology of Arc-Wise Connected H-Spaces

In this chapter we prove a general version of the "Cartan-Serre Theorem" and the "Milnor-Moore Theorem" on the rational singular homology $H(X;\mathbb{Q})$ of an arc-wise connected H-space X without any additional restrictions on the topological space X. The classical versions of these Theorems assures that for any H-space of finite type the rational Hurewicz homomorphism extends to an isomorphism

$$\text{hur} : \pi_*(X) \otimes \mathbb{Q} \xrightarrow{\cong} P_*(X;\mathbb{Q})$$

of Lie algebras onto the space of primitives in the singular Hopf algebra $H(X;\mathbb{Q})$ and this isomorphism extends to an isomorphism

$$U(\pi_*(X) \otimes Q) \xrightarrow{\cong} H(X;\mathbb{Q})$$

of Hopf algebras, where $U(\pi_*(X)\otimes\mathbb{Q})$ denotes the universal enveloping algebra of the graded rational Lie algebra $\pi_*(X)\otimes\mathbb{Q}$. Below we prove an analogous theorem without the assumption that $H(X;\mathbb{Q})$ is of finite type. First we recall the construction of the coalgebra and algebra structure of the rational singular homology $H(X;\mathbb{Q})$ and the Hopf algebra structure of the rational singular homology $H(\Omega X;\mathbb{Q})$ of the loop space ΩX of a simply connected topological space X. This has very useful consequence for the structure of the rational singular homology of loop spaces of simply connected H-spaces (Theorem 8.3.9) which in particular includes loop groups of simply connected topological groups e.g. simply connected Lie groups. This is then used to derive a "Cartan-Serre Theorem" for arc-wise connected topological groups (Theorem 8.4.10) and and then derive an even more general structure theorem on the rational singular homology of arc-wise connected H-spaces:

Theorem (The Structure of the Rational Singular Homology of Arc-wise Connected H-spaces). The rational Hurewicz homomorphism for an arc-wise connected H-space X satisfying the separation axion T_1 is a Lie algebra isomorphism of $\pi(X)\otimes\mathbb{Q}$ onto the graded Lie algebra $P(X;\mathbb{Q})$ of primitive elements in the Hopf algebra $H(X;\mathbb{Q})$ and extends to an isomorphism of graded Hopf algebras from the universal enveloping algebra $U(\pi(X)\otimes\mathbb{Q})$ of the rational graded Lie algebra $\pi(X)\otimes\mathbb{Q}$ onto the singular graded Hopf algebra $H(X;\mathbb{Q})$:

$$\begin{array}{ccc} \pi(G)\otimes\mathbb{Q} & \xrightarrow[\cong]{hur} & P(G;\mathbb{Q}) \\ \downarrow & & \downarrow \\ U(\pi(G)\otimes\mathbb{Q}) & \xrightarrow{\cong} & H(G;\mathbb{Q}) \end{array}$$

In particular the rational singular homology $H(X;\mathbb{Q})$ of an arc-wise connected H-space X is a primitively generated Hopf algebra. Its subspace $P(X;\mathbb{Q})$ of primitive elements is the image $hur(\pi(X)\otimes\mathbb{Q})$ and the multiplication in $H(X;\mathbb{Q})$ is induced by the multiplication in X.

This structure theorem is the main result of this chapter. It extends the classical "Cartan-Serre Theorem" to full generality without relying on such very restrictive requirements as the H-space X to be of finite type. It enables us to test the triviality of differential forms on an infinite dimensional Lie group G by integration over products of smooth spheres in G. This will particularly useful in characterising the triviality of flux homomorphisms in terms of period homomorphisms, both of which will be introduced in Chapter 13. The triviality of these homomorphisms will later turn out to be essential for the characterisation of the integrability of equivariant closed forms to smooth equivariant global global cocycles or the extensibility of equivariant Alexander-Spanier cocycles to global equivariant cocycles.

Chapter 9: Geometric Realisation

In this chapter some results concerning the classical geometric realisation functor $|-|$ as a functor $\mathbf{Set}^{\mathcal{O}^{op}} \to \mathbf{kTop}$ or $\mathbf{kTop}^{\mathcal{O}^{op}} \to \mathbf{kTop}$ (as defined for example in Mays book [May92] or the work [GZ67] of Gabriel and Zisman) are generalised to a functor $\mathbf{Top}^{\mathcal{O}^{op}} \to \mathbf{Top}$ (see Appendix F for a definition of the simplicial category $\mathcal{O}$ and simplicial objects). This includes in particular the preservation of certain finite limits and the existence of a right adjoint. These generalisations will be needed in the characterisation of the rational homology of Eilenberg-Mac Lane spaces and for the construction of Eilenberg-Mac Lane spaces which are Abelian topological groups in Chapter 11. The results obtained in this chapter are among the most important in this work. They lay the foundations for the results on classifying bundles and spaces in Chapter 11, the construction of Eilenberg-MacLane spaces which are Abelian topological groups in Chapter 11 and the derivation of the structure Theorem on the rational singular homology algebra of Eilenberg-MacLane spaces in Chapter 11. These result furthermore allow us to construct higher analogues of simply connected coverings and enables the construction of Whitehead and Postnikov towers consisting of principal bundles with Abelian structure group in Chapter 12.

In the first section we introduce the general geometric realisation functor $|-|$. Then the preservation of special kinds of morphisms by the functor $|-|$ is studied, where we in particular prove the preservation of *injectivity, surjectivity, denseness of images, proclusions, identifications* and *closedness of morphisms into Hausdorff (simplicial) spaces* by the general geometric realisation functor $|-| : \mathbf{Top}^{\mathcal{O}^{op}} \to \mathbf{Top}$. Afterwards we consider the restriction of the geometric realisation functor $|-|$ to important subcategories of $\mathbf{Top}$ such as the category $\mathbf{CGTop}^{\mathcal{O}^{op}}$ of compactly generated simplicial topological spaces and the category $\mathbf{kTop}^{\mathcal{O}^{op}}$ of compactly Hausdorff generated simplicial topological spaces. Here we observe that these restrictions corestrict to functors

$$\mathbf{CGTop}^{\mathcal{O}^{op}} \to \mathbf{CGTop} \quad \text{and} \quad \mathbf{kTop}^{\mathcal{O}^{op}} \to \mathbf{kTop}$$

respectively. Thereafter we show that the geometric realisation $|X|$ of a simplicial topological space is the colimit of the geometric realisations $|X_k|$ of its k-skeleta. All of these observations are generalisations from the classical geometric realisation of simplicial sets (regarded as discrete simplicial spaces) to the general setting.

In the second section we prove the cocontinuouity of the general geometric realisation functor $|-| : \mathbf{Top}^{\mathcal{O}^{op}} \to \mathbf{Top}$ by explicitly constructing a right adjoint. This right adjoint is the simplicial singular space functor

$$C(\Delta, -) : \mathbf{Top} \to \mathbf{Top}$$

where the function spaces are equipped with the compact open topology. This the first main result in this chapter:

Theorem. *The singular simplicial space functor $C(\Delta, -)$ is right adjoint to the geometric realisation functor $|-|$.*

This observation generalises the important well known fact, that the geometric realisation of simplicial sets regarded as a functor $\mathbf{Set}^{\mathcal{O}^{op}} \to \mathbf{kTop}$ is cocontinuous. Our result is far more general and is new even for discrete simplicial spaces, because we do not corestrict the geometric realisation functor to the category $\mathbf{kTop}$ of k-spaces. Last but not least the counit $\epsilon : |C(\Delta, -)| \to \mathrm{id}_{\mathbf{Top}}$ of this adjunction is shown to be a left inverse to the inclusion $i : \mathrm{id}_{\mathbf{Top}} \hookrightarrow |C(\Delta, -)|$ of topological spaces into the geometric realisation of their singular simplicial spaces:

Lemma. *The natural transformation i is a right inverse of ϵ, i.e. $\epsilon i = \mathrm{id}_{\mathbf{Top}}$ and the transformations fit into the commutative diagram*

$$\mathrm{id}_{\mathbf{Top}} \xrightarrow{\quad i \quad} |C(\Delta, -)| \xrightarrow{\quad \epsilon \quad} \mathrm{id}_{\mathbf{Top}} .$$
$$\underset{\mathrm{id}}{}$$

It will be shown later in Chapter 10 that the counit ϵ actually consists of homotopy equivalences.

In Section 3 the preservation of finite limits is examined. After showing that the general geometric realisation functor $|-| : \mathbf{Top}^{\mathcal{O}^{op}} \to \mathbf{Top}$ preserves equalisers we prove that it also preserves certain finite products. The classical theorem concerning the question of the preservation of finite products is [Mil57, Theorem 2] which was proven by Milnor. It states that the restriction $|-| : \mathbf{Set}^{\mathcal{O}^{op}} \to \mathbf{kHaus}$ preserves finite products of at most countable simplicial sets. In addition, if X and Y are simplicial sets and either CW-complex $|X|$ or $|Y|$ is locally finite, then the product $X \times Y$ is also preserved. In Section 2 below we show that the finiteness condition there can be weakened to a local countability condition:

Proposition. *If X and Y are discrete simplicial spaces and either X or Y is locally countable, then the natural map $|X \times Y| \to |X| \times |Y|$ is a homeomorphism.*

We then proceed to show that the geometric realisation of a k_ω-space is a k_ω-space, a fact that will be useful for the classifying space constructions considered in Chapter 11. Thereafter we generalise the above Proposition even more and observe that the geometric realisation functor

$$|-| : \mathbf{Set}^{\mathcal{O}^{op}} \to \mathbf{kHaus}$$

preserves finite limits. All this culminates in the following theorem:

Theorem. *Finite products of polytopes are polytopes.*

These results enable us to show that the geometric realisation of discrete simplicial algebras of arbitrary type are topological algebras of the same type (Theorem 9.3.52). In particular the geometric realisation of an (Abelian) discrete simplicial group is an (Abelian) topological group. Concerning the geometric realisation of discrete simplicial algebras we then observe:

Theorem. *The corestricted functor $|-| : \mathbf{Alg}_{(\Omega, \mathrm{E})}^{\mathcal{O}^{op}} \to \mathbf{TopAlg}_{(\Omega, \mathrm{E})}$ preserves finite limits. If the category $\mathbf{Alg}_{(\Omega, \mathrm{E})}$ contains a zero object, then the restriction $|-| : \mathbf{Alg}_{(\Omega, \mathrm{E})}^{\mathcal{O}^{op}} \to \mathbf{TopAlg}_{(\Omega, \mathrm{E})}$ of the geometric realisation functor preserves kernels.*

We then return to the question whether or when the geometric realisation preserves products and prove that it does not preserve finite products in general but binary products with locally finite discrete simplicial spaces are always preserved. Last but not least we generalise a classical result concerning the geometric realisation of simplicial sets by proving:

Theorem. *The geometric realisation functor $|-| : \mathbf{Top}^{\mathcal{O}^{op}} \to \mathbf{Top}$ preserves homotopy, i.e. if f and g are homotopic maps of simplicial spaces then the induced maps $|f|$ and $|g|$ are homotopic as well.*

All these results are new in the more general context of geometric realisation of simplicial topological spaces.

Chapter 10: Preservation of Homotopy

In this section we show that the simplicial space functor $C(\Delta, -) : \mathbf{Top} \to \mathbf{Top}^{\mathcal{O}^{op}}$ preserve homotopies. In addition it is shown that the counit $\epsilon : |C(\Delta, -)| \to \mathrm{id}_{\mathbf{Top}}$ of the adjunction between the geometric realisation functor functor $|-|$ and the singular simplicial space functor $C(\Delta, -)$ provides a homotopy equivalence $|C(\Delta, -)| \simeq \mathrm{id}_{\mathbf{Top}}$. Therefore every topological space X is naturally equivalent to the geometric realisation $|C(\Delta, X)|$ of its singular simplicial space $C(\Delta, X)$, i.e. it is up to homotopy recovered by the geometric realisation $|C(\Delta, X)|$ of its singular simplicial space $C(\Delta, X)$. In particular application of the singular simplicial space functor $C(\Delta, -)$ does not loose any information on the homotopy type of topological spaces. Summarising we observe that all morphisms in the commutative diagram

$$\mathrm{id}_{\mathbf{Top}} \xrightarrow{\;\;i\;\;} |C(\Delta, -)| \xrightarrow{\;\;\epsilon\;\;} \mathrm{id}_{\mathbf{Top}} \;,$$
$$\mathrm{id}$$

are homotopy equivalences. This generalises the classical Theorem concerning the geometric realisation of simplicial sets $\mathbf{Set}^{\mathcal{O}^{op}} \to \mathbf{kHaus}$ which states that the geometric realisation $|\hom_{\mathbf{Top}}(\Delta, X)|$ of the discrete singular simplicial set $\hom(\Delta, X)$ of a CW-complex X is homotopy equivalent to X itself. This classical theorem is proved by abstract nonsense using the Whitehead-Serre Theorem, which is not available for arbitrary topological spaces. We prove the generalisation by explicitly constructing the homotopy equivalence and showing that it is natural in the topological space X.

Chapter 11: Abelian Hausdorff Eilenberg-Mac Lane Groups

In this chapter we exploit the new results on the geometric realisation of simplicial topological spaces obtained in the previous chapters. This is prepared in Section 1, where we derive preservation properties of the standard simplicial space functor $\mathrm{SS} : \mathbf{Top} \to \mathbf{Top}^{\mathcal{O}^{op}}$. These will be needed to prove the analogous properties of the enveloping space functor E to be constructed in Section 2. At first we observe that the standard simplicial space functor SS preserves Hausdorffness and restricts and corestricts to a functor

$$\mathbf{k}_\omega \mathbf{Top} \to \mathbf{k}_\omega \mathbf{Top}^{\mathcal{O}^{op}} \quad \text{and} \quad \mathbf{k}_\omega \mathbf{Haus} \to \mathbf{k}_\omega \mathbf{Haus}^{\mathcal{O}^{op}}.$$

The most important preservation properties derived are the preservation of *monomorphisms, epimorphisms, open proclusions, open identifications, (closed) embeddings* and the preservation of *limits*. The preservation of limits in particular implies that for each category $\mathbf{TopAlg}_{(\Omega, \mathrm{E})}$ of topological algebras of type (Ω, E) the standard simplicial space functor SS restricts and corestricts to functors

$$\mathrm{SS} : \mathbf{TopAlg}_{(\Omega, \mathrm{E})} \to \mathbf{TopAlg}_{(\Omega, \mathrm{E})}^{\mathcal{O}^{op}} \qquad \mathrm{SS} : \mathbf{k}_\omega \mathbf{TopAlg}_{(\Omega, \mathrm{E})} \to \mathbf{k}_\omega \mathbf{TopAlg}_{(\Omega, \mathrm{E})}^{\mathcal{O}^{op}}$$

$$\mathrm{SS} : \mathbf{k}_\omega \mathbf{HausAlg}_{(\Omega, \mathrm{E})} \to \mathbf{k}_\omega \mathbf{HausAlg}_{(\Omega, \mathrm{E})}^{\mathcal{O}^{op}}.$$

A kind of topological algebras we are particularly interested in are topological groups. concerning these we prove that the restriction and corestriction

$$\mathrm{SS} : \mathbf{TopGrp} \to \mathbf{TopGrp}^{\mathcal{O}^{op}}$$

of the standard simplicial space functor preserves *monomorphisms, epimorphisms, proclusions, kernels, cokernels,* exact sequences and extensions. In addition the standard simplicial space of an Abelian topological group is Abelian. Moreover the application of the standard simplicial space functor is compatible with (semi-)group actions, i.e. one obtains functor

$$\mathrm{SS} : \mathbf{TrGrp} \to \mathbf{TrGrp}^{\mathcal{O}^{op}} \quad \text{and} \quad \mathrm{SS} : \mathbf{TrSemGrp} \to \mathbf{TrSemGrp}^{\mathcal{O}^{op}}.$$

In particular the standard simplicial space $\mathrm{SS}(X)$ of the G-space X of a transformation (semi-)group (G, X) is a simplicial G-space. Where possible, these results are also be shown for the standard simplicial object functor in subcategories $\mathbf{Top}_\mathrm{C}$ of $\mathbf{Top}$. Here we are primarily interested in the category $\mathbf{CGTop}$ of compactly generated spaces and the category $\mathbf{kTop}$ of k-spaces as well as several 'Hausdorff' versions of them.

In the second section we use the preparations made in Section 1 to construct an enveloping space functor $E : \mathbf{Top} \to \mathbf{Top}$ which assigns to each topological space a contractible enveloping space EX. The enveloping space EX is the geometric realisation $|\mathrm{SS}(X)|$ of the standard simplicial space $\mathrm{SS}(X)$ of X. This construction has already been used in the category $\mathbf{kTop}$ of k-spaces (cf. [Seg68]). Here we verify that the result generalises to the category $\mathbf{Top}$ of topological spaces and that the contraction of EX is natural in X. Furthermore we show that the enveloping space functor $E : \mathbf{Top} \to \mathbf{Top}$ preserves *monomorphisms, epimorphisms, equalisers, denseness of images, open proclusions, Hausdorffness* and *closed embeddings into Hausdorff spaces*. In addition we prove that the enveloping space functor E restricts and corestricts to an endofunctor

$$E : \mathbf{k}_\omega \mathbf{Haus} \to \mathbf{k}_\omega \mathbf{Haus}$$

of the category $\mathbf{k}_\omega \mathbf{Haus}$ of k_ω Hausdorff spaces and this endofunctor also preserves *equalisers* and *finite limits*. In particular, for every type (Ω, E) of algebras the enveloping space functor restricts and corestricts to an endofunctor

$$E : \mathbf{k}_\omega \mathbf{HausAlg}_{(\Omega,\mathrm{E})} \to \mathbf{k}_\omega \mathbf{HausAlg}_{(\Omega,\mathrm{E})},$$

which includes the category $\mathbf{k}_\omega \mathbf{HausGrp}$ of k_ω Hausdorff groups. This category contains real and complex finite dimensional Lie groups and real and complex Kac-Moody groups (cf. [GGH06]) and is therefore of interest in itself. The endofunctor $E : \mathbf{k}_\omega \mathbf{HausGrp} \to \mathbf{k}_\omega \mathbf{HausGrp}$ so obtained is then shown to preserve *kernels, cokernels* and to be *exact*.

In Section 3 In this section we introduce a construction of universal bundles and classifying spaces for (G, A)-bundles, where G and A are Hausdorff k_ω topological groups. This construction relies on the enveloping space functor E introduced in the previous section. In addition to the functor E we introduce another endofunctor B of $\mathbf{Top}$ resp. $\mathbf{k}_\omega \mathbf{Haus}$ and a natural transformation $E \to B$ which consists of quotient maps. The main results of this section are the following theorem and its consequences:

Theorem. *If A is a locally contractible Hausdorff k_ω-group, then the principal bundle $EA \to BA$ is numerable, hence a universal A-bundle and BA is a classifying space for A.*

Thus for each locally contractible k_ω Hausdorff group A there exists a universal bundle $EA \to BA$ whose total space EA is a k_ω Hausdorff group containing the fibre A as a subgroup. This in particular is shown to hold for the direct limit groups $\mathbf{SO} = \lim SO_n \mathbf{SO} = \lim SO_n$, $\mathbf{U} = \lim U_n$, $\mathbf{SU} = \lim SU_n$, $\mathbf{PU} = \lim PU_n$ and real and complex Kac-Moody groups. Moreover, if the group A is Abelian, we prove that the classifying space BA is an Abelian k_ω Hasdorff group as well and the sequence

$$0 \hookrightarrow A \to EA \twoheadrightarrow BA \to 0$$

is exact and locally trivial. This allows the construction to be iterated as in [Seg70], which we do not carry out here.

In the fourth section we consider Eilenberg-Mac Lane spaces. Given a countable discrete Abelian group π and $n \in \mathbb{N}$ it is known that there exist Eilenberg Mac Lane spaces $K(\pi, n)$ which are Abelian topological groups. This has been proved by Ivanov in [Iva85]. He uses free Abelian topological groups constructed in [DT58]. This construction leads to the existence of the Eilenberg Mac Lane spaces $K(\pi, n)$ which are Abelian topological groups. Unfortunately his proof relies on the countability of the group π. Different constructions using classifying spaces have been found by [Mil57, Theorem 3] and [McC69]. These constructions also relied on the countability of the group π. In this chapter we give a general natural procedure to construct compactly generated locally contractible Hausdorff Eilenberg-Mac Lane spaces which are Abelian topological groups or even CW-complexes. This extends the result concerning countable Abelian groups to the full class of Abelian groups. The new result is then used to prove a generalised "Cartan-Serre" Theorem on the rational (co)homology of Eilenberg Mac Lane spaces. This theorem does not rely on the finite dimensionality of the rational homotopy or homology groups, which is the strong restriction of the classical result of Cartan and Serre. We derive an even more general result by presenting a functorial construction of Abelian topological groups with prescribed homotopy groups. The construction establishes a correspondence between the category **AbGrLieAlg** of Abelian graded lie algebras and the category CW-**TopAb** of Abelian topological groups of the homotopy type of CW-complexes. It requires no prerequisites and only relies on the preservation of finite products under the geometric realisation functor $|-| : \mathbf{Set}^{\mathcal{O}^{op}} \to \mathbf{Top}$ (Theorem 9.3.49). This in particular includes a general procedure to construct Abelian Eilenberg Mac Lane topological groups $K(\pi, n)$.

In the last section we use the results obtained in Section 4 to derive the following structure theorem on the rational singular homology of Eilenberg-Mac Lane spaces:

Theorem. *The rational homology $H(X; \mathbb{Q})$ of any Eilenberg-Mac Lane space $K(\pi, n)$ is the exterior algebra over $\pi \otimes \mathbb{Q}$ if n is odd and the polynomial algebra on $\pi \otimes \mathbb{Q}$ if n is even.*

Chapter 12: Towers of Principal Bundles

In this chapter we construct special versions of Whitehead and Postnikov towers. Each stage of these towers will consist of principal bundles with Abelian structure group. The special version of the Whitehead tower generalises the notion of universal coverings: Recall that to each semi-locally simply connected space X there exists a principal bundle

$$\pi_1(X) \longrightarrow \tilde{X} \longrightarrow X$$

whose total space $\tilde{X}$ is the universal covering space of X and whose fibre is the discrete fundamental group $\pi_1(X)$ of X. Moreover every arc-wise connected covering space Y of X with fundamental group $\pi_1(Y)$ is the orbit space of the action of a subgroup $N \cong \pi_1(Y)$ of $\pi_1(X)$ on $\tilde{X}$. Simple connectivity of topological spaces is a very useful property that makes many constructions possible. (The existence of universal covering groups for example ensures the existence of a left adjoint to the Lie algebra functor for classical Lie groups.) This makes covering spaces a very useful tool in general topology. Higher connectivity is even more desirable, so it would be very convenient to have a construction of principal bundles killing higher homotopy groups analogous to that of universal covering spaces. Unfortunately no such generalisation (using principal bundles) to higher dimensions is known. The Whitehead tower to be constructed generalises the notion of universal coverings and resolves this issue.

In contrast the special Postnikov tower to be constructed eases the computation of homology groups. It finally enables us to make fundamental observations concerning the rational homology and cohomology of CW-complexes and topological spaces with Abelian fundamental group.

Chapter 13: The Generalised Flux Homomorphism

The subject of this chapter is the de Rham complex for infinite dimensional flat smooth fibre bundles. We start with a reminder on the differential calculus of infinite dimensional manifolds proposed by BERTRAM, GLÖCKNER and NEEB in their exposition [BGN04], where the maps of class C^0 are the continuous ones. We assume that the reader is familiar with the notions of differential calculus, manifolds smooth bundles and Lie groups developed in this article and with the topological of spaces of C^k-functions with the C^k-compact open topology introduced in the follow up [Glö04]. At first we observe some topological properties of manifolds in the above sense and then we recall some fundamental theorems on function spaces, which we later generalise to a larger class of manifolds. To be precise we generalise the exponential law

$$\hom_{\mathbf{Mf}}(M \times N, V) \cong \hom_{\mathbf{Mf}}(M, C^\infty(N, V))$$

to the case of infinite dimensional manifolds M and N which have the property that arbitrary finite products $W^p \times W'^q$ of the model spaces W of M and W' of N are k-spaces. This extends the recent result [Glö04, Proposition 12.6 (b)] of Glöckner to a more general setting.

Following these general consideration on differential calculus we derive a decomposition of the topological de Rham spaces $^t\Omega^n(E; V)$ of the total space E of a flat smooth bundle. This is a straightforward generalisation from the classical finite dimensional context to the topological de Rham cohomology of infinite dimensional bundles, where we use the notion of infinite dimensional calculus introduced above. This decomposition is then used to construct a double complex $\Omega^{*,*}(E; V)$ consisting of spaces of differential forms. This double complex in turn is the necessary tool used in deriving a Leray-Serre spectral sequence

$$\Omega^p(B, {}^t\Omega^q(F; V)) \Rightarrow H^{p+q}_{dR}(E; V)$$

for flat infinite dimensional bundles with base manifold B and fibre F.

The double complex $\Omega^{*,*}(E; V)$ will also be necessary to construct a generalisation of the flux homomorphism of symplectic topology to infinite dimensional bundles. Here the diffeomorphism group of a (compact) manifold is replaced by an infinite dimensional Lie group G acting on a G-bundle of infinite dimensional manifolds. After constructing the flux homomorphisms we also define general period maps. These are necessary tools to be used to characterise the integrability of equivariant differential forms in Chapter 16 and the extensibility of equivariant Alexander-Spanier cocycles in Chapter 15.

Chapter 14: Spectral Sequences for Transformation Groups

Let G be a topological group, X be a G-space and V be a G-module. To compute the equivariant $\mathfrak{U}$-local cohomology $H^*_{eq}(X, \mathfrak{U}; V)$ for some open G-invariant cover of X we define a suitable spectral sequence. The first approach to this problem was described in two articles [vE62a], [vE62b] of W.T. van Est. He considered a transformation group (G, X) and a G-vector space V and imposed the restriction of local contractibility on the group G and finite dimensionality of the cohomology modules $H^*_{AS}(G; V)$. With these restrictions he constructed a spectral sequence

$$E_2^{p,q} = H_c^p(G; H_{AS}^q(G; V)) \Rightarrow H_{AS,eq}^{p+q}(G; V),$$

where $H_c^p(G, H_{AS}^q(G; V))$ denotes the group cohomology with continuous cochains with values in the vector space $H_{AS}^q(G; V)$. Another prominent example of a similar spectral sequence connects the smooth group cohomology and the Lie algebra cohomology of a finite dimensional Lie group G:

$$E_2^{p,q} = H_s^p(G; H_{dR}^q(G; V)) \Rightarrow H_c^{p+q}(\mathfrak{g}; V)$$

This spectral sequence straightforwardly generalises to infinite dimensional Lie groups whose de Rham complex splits when given the C^∞-compact-open topology. A common example of Lie groups of this type are diffeomorphism groups of compact manifolds (cf. [Beg87]). There also exists a version for properly discontinuous actions of a group G on a finite dimensional manifold M:

$$E_2^{p,q} = H^p(G; H_{dR}(M; V)) \Rightarrow H_{dR,eq}^{p+q}(M; V),$$

We will construct a slightly more general spectral sequence for arbitrary transformation groups (G, X), G-modules V and a version for smooth actions on (possibly infinite dimensional) manifolds. Instead of considering Vietoris cohomology $H^*(\Gamma_u; V)$ as can Est did, we work with the equivariant $\mathfrak{U}$-local cohomology here. Using an elementary approach similar to van Est, we prove that the above restrictions on G and the dimension of the modules $H_{dR}^q(M; V)$ are unnecessary, i.e. we neither have to restrict ourselves to finite dimensional Lie groups G, manifolds M and vector spaces V nor do we have to assume the manifolds G, M or V to be (smoothly) paracompact.

We begin with a smooth transformation group (G, M), a topological vector space V that is a G-module and construct a spectral sequence

$$E_0^{p,q} = A_{s,dR}^{p,q}(M, M; V)^G \Rightarrow H_{dR,eq}^{p+q}(M; V)$$

linking the smooth equivariant cohomology $H_{s,eq}^*(M; V)$ of M to the equivariant de Rham cohomology $H_{dR,eq}^*(M; V)$. The general setting goes back to van Est. In addition to the spectral sequence described there, we explicitly construct a row contraction h, which gives further information on the edge maps. The result we obtain is known only for finite dimensional Lie groups acting on themselves by left translation, but not in the generality obtained below. The spectral sequence so obtained enables us to use classical spectral sequence arguments for infinite dimensional smooth transformation groups (G, M), which has not been possible until now.

The smoothness of an equivariant cocycle is a major restriction. In general smooth cocycles are not the appropriate concept to describe fibre bundles or Lie group extensions. This is so because the global smoothness of cocycles amounts to the existence of global smooth sections. A globally smooth group 2-cocycle $f \in C^2(G; A)$ of a Lie group G for example describes an isomorphy class of group extensions

$$0 \to A \to A \times_f G \to G \to 1$$

that admit a global smooth section, i.e. $A \times_f G = A \times G$ as manifolds. Since we are interested in the more general class of locally trivial bundles (resp. group extensions that are principal fibre bundles) and not necessarily smooth extensions of local cocycles, we introduce a new double complex, that is appropriate for describing these objects. This double complex then gives rise to a spectral sequence

$$H^p_{eq}(X; H^q(X, \mathfrak{U}; V)) \Rightarrow H^{p+1}_{eq}(X, \mathfrak{U}; V)$$

linking the cohomology $H^*_{eq}(X; V)$ of global equivariant cochains on the G-space of a transformation group (G, X) to the equivariant $\mathfrak{U}$-local cohomology $H_{eq}(X, \mathfrak{U}; V)$ of X. This spectral sequence generalises the one for smooth transformation groups. It enables us to consider arbitrary equivariant $\mathfrak{U}$-local cocycles and examine their extensibility to global equivariant cocycles. This in particular includes the extensibility of (germs of) local group cocycles to global group cocycles.

The spectral sequences constructed in this chapter will be used to characterise the integrability of equivariant differential forms in Chapter 16 resp. the extensibility of equivariant $\mathfrak{U}$-local cocycles to global equivariant cocycles in Chapter 15.

Chapter 15: Equivariant Extension of Local Cocycles

The purpose of this chapter is to relate the cohomology $H_{eq}(M; V)$ of equivariant global cochains on the G-space X of a transformation group (G, X) of special type to the equivariant $\mathfrak{U}$-local cohomology $H_{eq}(X, \mathfrak{U}; V)$ for an open cover $\mathfrak{U}$ of X. This includes the relation of the group cohomology $H(G; V)$ of a topological group G and the cohomology of local group cochains as a special case. In contrast to the integrability of differential forms considered in the next chapter there are no difficulties arising from the non-existence of classical theorems in the infinite dimensional context. Also, we do not have any restrictions on the coefficient space V, which here is just an abstract Abelian group.

The cohomologies $H_{eq}(X; V)$ and $H_{eq}(X, \mathfrak{u}; V)$ of a transformation group (G, X) of special type are related by characterising the extendibility of equivariant $\mathfrak{U}$-local cocycles on X to global equivariant cocycles; we call an equivariant $\mathfrak{U}$-local cocycle on X *extendible* to a global equivariant cocycle if is the restriction of a global equivariant cocycle on X. Under different assumptions on the space X the necessary and sufficient condition for a G-equivariant $\mathfrak{U}$-local cocycle $[f] \in Z^n_{eq}(M, \mathfrak{U}; V)^G$ ($n > 0$) to be extendible to a global equivariant cocycle will be shown to be the triviality of its flux.

This characterisation of extensibility uses the flux and the period maps introduced in Section 13.6. The main results of this chapter are the characterisation of the extensibility of equivariant $\mathfrak{U}$-local cocycles to equivariant global cocycles under varying connectedness-conditions on the space X:

1. Concerning equivariant Alexander-Spanier 1-cocycles we observe the exactness of the sequence

$$Z^1(X; V)^G \to Z^1_{AS}(X; V)^G \xrightarrow{\mathrm{per}_\varrho(-)_1} \mathrm{hom}(\pi_1(X), V),$$

   which describes the extensibility of an equivariant Alexander-Spanier 1-cocycle $[f]$ on X in terms of the period map $\pi_1(X) \to V$, $[s] \mapsto \lambda^1([f])(s)$.
2. The extendibility of (germs of) $\mathfrak{U}$-local n-cocycles on $(n-1)$-connected spaces X is described by an exact sequence

$$Z^n(X; V)^G \to \mathrm{colim}_\mathfrak{U} Z^n(X, \mathfrak{U}; V)^G \xrightarrow{\mathrm{per}_{\varrho t}(-)_n} \mathrm{hom}(\pi_n(X); V^G)$$

   If the space X is not $(n-1)$-connected but only $(n-2)$-connected, we add another term and a flux homomorphism on the right to obtain an exact sequence

$$Z^n(X; V)^G \to \mathrm{colim}_\mathfrak{U} Z^n(X, \mathfrak{U}; V)^G \xrightarrow{\mathrm{per}_{\varrho t}(-)_n} \mathrm{hom}(Z_n(X), V^G) \oplus \mathrm{hom}(Z_{n-1}(X), Z^1(G; V)^G)$$

Both result generalise the characterisation of the integrability of continuous Lie algebra cocycles to locally smooth Lie group cocycles in [Nee02a] to general transformation groups. In contrast to the procedure there we do neither require any of the spaces X or V to be Hausdorff. Moreover we consider equivariant $\mathfrak{U}$-local cocycles in any dimension:

In any of the above cases (the germ of) an equivariant $\mathfrak{U}$-local n-cocycle is extendible to a global equivariant cocycle if and only if all relevant flux homomorphisms are trivial. If the coefficients V are divisible, then one can w.l.o.g. switch to the rational version of the flux homomorphisms.

It should be remarked that the above characterisations of the extensibility of equivariant $\mathfrak{U}$-local cocycles do not require any condition on the G-space to be satisfied. They hold in full generality.

The proof of the above statements on the extensibility of equivariant Alexander-Spanier cocycles relies on the spectral sequence $_IE^{p,q} \Rightarrow H_{eq}^{p+q}(X,\mathfrak{U};V)$ introduced in Section 14.3 and on the flux and period homomorphisms introduced in Chapter 13.

Chapter 16: Integrability of Differential Forms

The purpose of this chapter is to relate the cohomology $H_{ls,eq}(M;V)$ of locally smooth global equivariant cochains and $H_{s,eq}(M;V)$ smooth equivariant global cochains on the G-manifold M of a smooth transformation group (G,M) of special type to the equivariant de Rham cohomology $H_{dR,eq}(M;V)$ of M. This includes the relation of the locally smooth and smooth group cohomologies $H_{ls}(G;V)$ and $H_s(G;V)$ of a Lie group G and the continuous Lie algebra cohomology $H_c(\mathfrak{g};V)$ of its Lie algebra $\mathfrak{g}$ as a special case. A great difficulty arising in the general context of infinite dimensional manifolds M and Lie groups G is the the fact that neither of them needs to be smoothly paracompact, so de Rahm's Theorem is not available. Even Abelian Lie-groups that are Banach spaces (e.g. $C([0,1],\mathbb{R})$ or $l^1(\mathbb{N},\mathbb{R})$) need not be smoothly paracompact. Another difficulty arising in infinite dimensional analysis is the absence of the "exponential law" for spaces of smooth functions. Here we can rely on the observations concerning k-spaces (e.g. metrisable vector spaces) in Section 13.1 or ensure the exponential law of certain function spaces in different way. The coefficient space V is required to be Mackey complete to ensure the existence of integrals of V-valued differential forms over smooth singular simplices.

The cohomologies $H_{ls,eq}(M;V)$ resp. $H_{s,eq}(M;V)$ and $H_{dR,eq}(M;V)$ of a smooth transformation group (G,M) of special type are related by characterising the integrability of closed equivariant forms on M to smooth global equivariant cocycles; we call an equivariant close n-form ω on M *integrable* to a locally smooth equivariant global cocycle if it is in the image of the derivation homomorphism

$$Z_{ls}^n(M;V)^G \xrightarrow{\tau^n} Z_{dR}^n(M;V)^G.$$

Under additional assumptions on the manifold M, the necessary and sufficient condition for a G-equivariant closed n-form $\omega \in Z_{dR}^n(M;V)^G$ $(n>0)$ to be integrable to a (locally) smooth global cocycle will be shown to be the triviality of its flux. All (topological) vector spaces and manifolds considered in this chapter will be vector spaces or manifolds over the real or complex numbers.

The characterisation of integrability is done in in terms of the flux and the period maps introduced in Section 13.6. The main results of this chapter are the characterisation of the integrability of closed equivariant forms to locally smooth and smooth equivariant global cocycles under varying connectedness-conditions on the manifold M:

1. Concerning equivariant closed 1-forms we observe the exactness of the sequence

$$Z_s^1(M;V)^G \to Z_{dR}^1(M;V)^G \xrightarrow{\text{per}_{\varrho t}(-)_1} \hom(\pi_1(M), V^G),$$

which describes the integrability of a closed equivariant 1-form ω on M in terms of the period map $\pi_1(M) \to V^G$, $[s] \otimes 1 \mapsto \int_s \omega$.

2. For closed equivariant n-forms on $(n-1)$-connected manifolds M with additional structure (such as Lie groups) a similar sequence

$$Z_{ls}^n(M;V)^G \to Z_{dR}^n(M;V)^G \xrightarrow{\text{per}_{\varrho t}(-)_n} \hom(\pi_n(M), V)$$

turns out to be exact. If the manifold M is not $(n-1)$-connected but $(n-2)$-connected, we obtain an exact sequence

$$Z_{ls}^n(M;V)^G \to Z_{dR}^2(M;V)^G \to \hom(Z_n(M), V^G) \oplus \hom(Z_{n-1}(M), Z_c^1(\mathfrak{g};V))$$

Both result generalise the corresponding observations concerning smooth Lie group cocycles and continuous Lie algebra cocycles in [Nee02a] to more general transformation groups. In contrast to the procedure there we do neither require any of the spaces $\mathfrak{g}$ or V to be locally convex nor do we require M or G to be Hausdorff. Moreover we extend these results by characterising the integrability of equivariant closed n-forms on $\lfloor \frac{n-1}{2} \rfloor$-connected manifolds(with additional structure). Here a sequence of the form

3.

$$Z_{ls}^n(M;V)^G \to Z_{dR}^n(M;V)^G \to \bigoplus_{p \leq \lfloor \frac{n-1}{2} \rfloor + 1}^{n} \hom(Z_p(M), Z_c^{n-p}(\mathfrak{g};V))$$

is exact. In all cases the homomorphisms $Z_{dR}^n(M;V)^G \to \hom(Z_p(M), Z_c^{n-p}(\mathfrak{g};V))$ occurring in the above sequences are flux homomorphisms. Thus in any of the above cases an equivariant closed n-form is integrable to a locally smooth global equivariant n-cocycle if and only if all relevant flux homomorphisms are trivial. The last result (3) is completely new and was not even known for finite dimensional Lie groups.

Under the additional assumptions that the exponential law holds for the natural linear continuous maps $C^\infty(M^p \times T^{\oplus q}M, V) \to C^\infty(M^p, C^\infty(T^{\oplus q}M, V))$ and that the topological de Rham complex ${}^t\Omega^*(M;V)$ of M splits we can also prove the analogon of (2) for globally smooth cocycles, i.e. if the manifold M is $(n-2)$-connected, then the sequence

$$Z_{ls}^n(M;V)^G \to Z_{dR}^2(M;V)^G \to \hom(Z_n(M), V^G) \oplus \hom(Z_{n-1}(M), Z_c^1(\mathfrak{g};V))$$

is exact.

It should be mentioned that Hilbert manifolds satisfy any of the above conditions on the manifold M and its de Rham complex. This in particular includes all finite dimensional manifolds. Moreover, any diffeomorphism group $\text{Diff}(M)$ of a compact manifold M has split de Rham complex (cf. [Beg87, Theorem 7.5]) and the group $\Omega_s(M)$ of smooth loops in a separable Hilbert manifold M has split de Rham complex. This is also true for the space $C^\infty(M,N)$ of smooth functions from a compact manifold M into a separable metrisable Hilbert manifold N (cf. [Beg87, Theorem 7.6]).

The proof of the above statements on the integrability of closed equivariant forms relies on the spectral sequences introduced in Section 14.2 and on the flux and period homomorphisms introduced in chapter 13.

1

Homological Algebra in TopRMod

This chapter is devoted to developing the basic notions of homological algebra in categories of topological algebras (such as the category **TopRMod** of topological modules over a topological ring R). Many algebras associated with geometric objects such as manifolds, (principal) fibre bundles, topological groups and transformation groups can naturally be endowed with a topology, turning them into topological algebras. Doing so provides a finer tool to differentiate these geometric objects as just considering the underlying (discrete) algebras. Examples are the topological singular (co)homology, continuous cohomology of topological groups (cf. [Seg70]) and topological equivariant cohomology. In the case of continuous group cohomology, we wish to drive a spectral sequence

$$H^p(B, H^q(N; A)) \Rightarrow H^{p+q}(G; A).$$

This requires the cochain complex $C^*(N; A)$ of continuous group cochains and its cohomology groups $H^q(N; A)$ to be abelian topological groups. Furthermore one has to use homological algebra in the category **TopAb** of abelian topological groups. This category – along with most categories of topological algebras – is not abelian. For example a monomorphism $f : A \to A'$ in **TopAb** in general fails to be a kernel of its cokernel. The fact that most categories of topological algebras are not abelian prevents an application of the abstract machinery of homological algebra developed for abelian categories. Therefore it has to be carefully examined which constructions of homological algebra can be generalised and which results carry over to these categories. There are some constructions of homological algebra that can be generalised to many categories of topological algebras. Recent progress on this subject has been made in the articles [BC05] and [BC06]. The results presented there are the starting point for our work. We will especially concerned with the category **TopRMod** of topological modules over a topological ring R and the category **TopAb** of abelian topological groups, which is a special case thereof. (It is the category of topological modules over the discrete ring $\mathbb{Z}$ of integers.) The results obtained here will have applications on Lie algebra cohomology, continuous cohomology of topological groups, topological deRham- and Alexander-Spanier cohomology as well as for their equivariant versions.

In order to generalise the most basic constructions of homological algebra it has in particular to be verified that the homology functor has the usual properties, e.g. that every short exact sequence

$$0 \to M_* \to M'_* \to M''_* \to 0$$

of simplicial topological R-modules leads to a long exact sequence of homology modules. Furthermore it has to be proved that for any any module M the tensor product functor $M \otimes - : \mathbf{TopRMod} \to \mathbf{TopRMod}$ behaves sufficiently "nice".

The first section of this chapter covers the basic notions of homological algebra in **TopRMod**. There we introduce normalised (co)chain complexes of (co)simplicial topological modules and verify that the normalisation functor is exact (Theorem 1.1.37). We also show that the (co)homology of normalised (co)chain complexes associated to (co)simplicial modules is a universal (co)homological δ-functor (Theorems 1.1.19 and 1.1.48).

In the second section we derive a general normalisation theorem for (co)chain complexes in homological categories, which states that the (co)homology of chain complexes and the (co)homology of normalised (co)chain complexes of (co)simplicial objects in these categories are naturally isomorphic universal (co)homological δ-functors (Theorems 1.2.19 and 1.2.34). This includes many categories of (co)simplicial topological algebras. This in particular includes the category **TopRMod**, concerning which we show that the (co)homology of unnormalised (co)chain complexes associated to (co)simplicial modules is a universal (co)homological δ-functor which is naturally isomorphic to the (co)homology of the unnormalised associated (co)chain complexes (Theorems 1.2.22 and 1.2.37).

The third section contains a short digression into the realm of simplicial objects in homological categories (such as **TopRMod**), where we show that the chain complexes associated to simplicial objects are proper and that their homology is a coeffaceable and universal homological δ-functor.

The last section is devoted to more concrete computations and the examination of the free Abelian topological group functor and of binary tensor products in the category **TopRMod**. Here we observe that the free Abelian topological group functor preserves quotients and that the binary tensor product of quotients in **TopRMod** is again a quotient. Moreover tensoring with a fixed module M is shown to preserve cokernels. The further examination of the free Abelian topological group functor is largely concerned with P-embeddings, which play a crucial role for the exactness of short sequences (cf. Theorem 1.4.18). Here we derive a slight generalisation of the characterisation of P-embeddings (Theorem 1.4.36), which is needed for the the proof that tensoring with a free Abelian topological group is an exact functor w.r.t. large classes of short exact sequences, which in particular includes those group extensions which are locally trivial fibre bundles (Theorem 1.4.53).

1.1 Basic Notions of Homological Algebra in TopRMod

We start by recalling the basic definitions of homological algebra in the category **TopRMod** of topological modules over a topological ring R. *The kernel* of a homomorphism $f : M \rightarrow M'$ of topological R-modules is the inclusion of the submodule $K(f) := \{m \in M \mid f(m) = 0\}$. It is denoted by $\ker f$. In general there may exist many isomorphic kernels of a homomorphism f but when we speak of *the kernel* we understand it to be the inclusion $\ker f$ of the submodule $K(f)$ given above. *The coimage* of f is the quotient homomorphism onto the quotient module $CI(f) = M/K(f)$. This quotient homomorphism is denoted by $\mathrm{CoIm}\, f$. The inclusion $f(M) \hookrightarrow M'$ of the submodule $f(M)$ of M' is called the *image of* f and is denoted by $\mathrm{Im}\, f$. The quotient homomorphism $M' \twoheadrightarrow M'/f(M)$ is a cokernel of f. It is denoted by $\mathrm{coker}\, f$. The quotient module $M'/f(M)$ will also be denoted by $CoK(f)$. Analogously to kernels, there may in general exist many isomorphic cokernels, but when we speak of *the cokernel* of f we always think of the quotient homomorphism $\mathrm{coker}\, f : M' \rightarrow Cok(f)$. Each homomorphism $f : M \rightarrow M'$ in **TopRMod** factors through the coimage $\mathrm{CoIm}\, f : M \twoheadrightarrow M/K(f)$ and the image $\mathrm{Im}\, f : f(M) \hookrightarrow M'$:

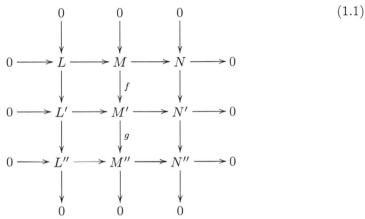

The induced (dotted) homomorphism $m_f : CI(f) \to M'$ is a monomorphism but not always an inclusion. This is a unique coimage-image factorisation in the regular category **TopRMod** (cf. A.1.34,A.1.35). This unique coimage-image factorisation enables us to use much of the homological algebra known from abelian categories. As a first step we observe that the "3x3-Lemma" (or "Nine Lemma") carries over to the category **TopRMod**. Consider a commutative diagram of the following form in **TopRMod**:

$$(1.1)$$

$$
\begin{array}{ccccc}
0 & & 0 & & 0 \\
\downarrow & & \downarrow & & \downarrow \\
0 \longrightarrow L \longrightarrow & & M \longrightarrow & & N \longrightarrow 0 \\
\downarrow & & \downarrow f & & \downarrow \\
0 \longrightarrow L' \longrightarrow & & M' \longrightarrow & & N' \longrightarrow 0 \\
\downarrow & & \downarrow g & & \downarrow \\
0 \longrightarrow L'' \longrightarrow & & M'' \longrightarrow & & N'' \longrightarrow 0 \\
\downarrow & & \downarrow & & \downarrow \\
0 & & 0 & & 0
\end{array}
$$

Lemma 1.1.1 (The 3×3 Lemma in TopRMod). *If all horizontal sequences in the above diagram 1.1 are exact and $g \circ f$ is the zero homomorphism then the exactness of any two of the vertical rows implies the exactness of the third row.*

Proof. This is Theorem A.6.5 for the homological category **TopRMod**. □

Next we observe that the "Noether Isomorphism Theorems" hold true for the category **TopRMod** of topological R-modules.

Theorem 1.1.2 (Noether Isomorphism Theorems). *Let K and L be submodules of a topological R-module M.*

1. *The restriction of the quotient homomorphism $M \to M/K$ to $L+K$ is a quotient homomorphism, i.e. $(L + K)/K$ is a submodule of M/K.*
2. *The inclusion $L \hookrightarrow M$ induces an isomorphism $L/(L \cap K) \cong (L + K)/K$.*
3. *The projections $(L+K)/(L \cap K) \twoheadrightarrow (L+K)/L$ and $(L+K)/(L \cap K) \twoheadrightarrow (L+K)/K$ induce an isomorphism $(L + K)/(L \cap K) \cong ((L + K)/L) \times ((L + K)/K)$.*

Proof. The category **TopRMod** is a homological category. Therefore the Noether Isomorphism Theorems follow from the general version [BB04, Theorem 4.3.10] and [BB04, Theorem 4.3.12] for homological categories. □

Having established the Noether Isomorphism Theorems we proceed with the homology of chain complexes. The homology of chain complexes in **TopRMod** is defined as usual via quotients of kernels by images. (See A.6.9 for the general definition in homological categories). It is a functor $H : \mathbf{Ch}(\mathbf{TopRMod}) \to \mathbf{TopRMod}$ from the category $\mathbf{Ch}(\mathbf{TopRMod})$ of chain complexes in **TopRMod** into the category **TopRMod** of topological R-modules. In order to see that the homology functor is a homological δ-functor, the Snake Lemma has to be proved for the category

TopRMod. This turns out to be possible only in a specialised version. Consider a commutative diagram

$$
\begin{array}{ccccccc}
L & \longrightarrow & M & \longrightarrow & N & \longrightarrow & 0 \;, \\
\downarrow{\scriptstyle f} & & \downarrow{\scriptstyle g} & & \downarrow{\scriptstyle h} & & \\
0 \longrightarrow & L' & \longrightarrow & M' & \longrightarrow & N' &
\end{array}
$$

where both horizontal sequences are exact. This commutative diagram can be enlarged by adjoining the kernels and cokernels of the homomorphisms f, g and h. All in all one obtains the following commutative diagram

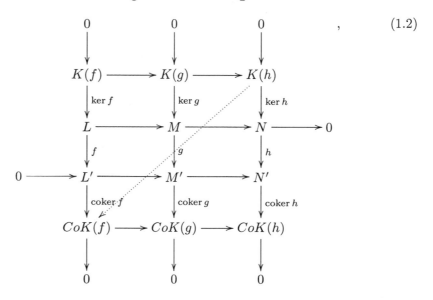

(1.2)

in which all rows and columns are exact and in which the dotted arrow indicates a homomorphism we wish to exist.

Lemma 1.1.3 (Snake Lemma in TopRMod). *If the homomorphisms f, g and h in in the above diagram 1.2 are quotients onto their image, then there exists a unique connecting homomorphism $\delta : \ker h \to \operatorname{coker} f$ filling in as the dotted arrow such that the sequence*

$$
K(f) \to K(g) \to K(h) \xrightarrow{\delta} Cok(f) \to CoK(g) \to CoK(h)
$$

is exact and the so enlarged diagram is commutative.

Proof. The homomorphisms which are quotients onto their image are the proper morphisms in the regular category **TopRMod**. The existence of the connecting homomorphism follows from the general Snake Lemma in homological categories (see Theorem A.6.6). The underlying homomorphism $U\delta$ of R-modules is a connecting homomorphism of the underlying diagram in the category **RMod** of R-modules. Thus the uniqueness of δ follows from the uniqueness of the connecting homomorphisms in the category **RMod**. □

Homomorphisms of topological R-modules which are quotients onto their image are regular epimorphisms whose image is a kernel. Recall that morphisms of this kind are called *proper* (cf. A.3.17). (This is not to be confused with proper maps between topological spaces.) Thus the definition of proper chain complexes adapts to the following form:

Definition 1.1.4. *A chain complex in* **TopRMod** *is called* proper, *if all differentials are quotients onto their image. The category of proper chain complexes in* **TopRMod** *is denoted by* **PCh(TopRMod)**.

The specialised version of the Snake Lemma can be used to prove that homology is a universal homological δ-functor on the category **PCh(TopRMod)** of proper chain complexes of topological R-modules.

Theorem 1.1.5 (The long exact homology sequence). *Every short exact sequence of proper chain complexes of topological R-modules*

$$0 \to M_* \to M'_* \to M''_* \to 0$$

induces a long exact sequence of homology modules

$$\cdots \xrightarrow{\delta_{n+2}} H_{n+1}(M_*) \to H_{n+1}(M'_*) \to H_{n+1}(M''_*) \xrightarrow{\delta_{n+1}} H_n(M_*) \to \cdots,$$

which depends naturally on the given short exact sequence.

Proof. The category **TopRMod** of topological R-modules is a pointed regular and protomodular category. A general proof for such categories can be found in [EVdL04, Proposition 2.4]. □

The restriction on the chain complexes M_*, M'_* and M''_* to be proper can be circumvented if these chain complexes come from a short exact sequence of simplicial topological R-modules. Recall that assigning to each simplicial module M its associated chain complex M_* and to each morphism $f : M \to M'$ of simplicial modules the induced morphism $f_* : M_* \to M'_*$ between the associated chain complexes is a functor $(-)_* : \mathbf{TopRMod}^{\mathcal{O}^{op}} \to \mathbf{Ch(TopRMod)}$ from the category $\mathbf{TopRMod}^{\mathcal{O}^{op}}$ of simplicial topological R-modules into the category **Ch(TopRMod)** of chain complexes in **TopRMod**.

Definition 1.1.6. *The functor* $(-)_* : \mathbf{TopRMod}^{\mathcal{O}^{op}} \to \mathbf{Ch(TopRMod)}$ *is called* the associated chain complex functor.

The components $f_n : M_n \to M'_n$ of a morphism $f_* : M_* \to M'_*$ of chain complexes associated to a morphism $f : M \to M'$ of simplicial topological modules coincide with the components $f([n]) : M([n]) \to M'([n])$ of f. So we note:

Lemma 1.1.7. *The associated chain complex functor* $(-)_*$ *preserves and creates limits and colimits.*

Proof. In functor categories (co)limits can be calculated point-wise, provided the point-wise limits exist (cf. [ML98, Theorem V.3.1]). Therefore limits and colimits in $\mathbf{TopRMod}^{\mathcal{O}^{op}}$ are taken point-wise, whence the components of (co)limits of simplicial modules coincide with those of the (co)limits of the associated chain complexes. Therefore the associated chain complex functor $(-)_*$ is continuous and cocontinuous. The completeness and cocompleteness of **TopRMod** guarantees the existence of all limits and colimits, hence the functor $(-)_*$ also creates limits and colimits. □

In case a cochain complex stems from a simplicial module it is possible to consider "normalised" subcomplex (to be defined below) instead, which can be shown to always be proper. For simplicial objects in pointed, exact and protomodular categories this has been done in [EVdL04, Theorem 3.6]. Since the categories **TopAb** and **TopRMod** are pointed and protomodular but not exact, this result needs to be proven for more general categories, which we will do in the following. As in the classical case, the tool to achieve this result is the Moore normalisation functor.

Definition 1.1.8. *The* Moore *(or normalised) chain complex* NM_* *of a simplicial topological R-module M is the chain complex given by*

$$NM_n := \bigcap_{i=0}^{n-1} K\left(M(\epsilon_i) : M([n]) \to M([n-1])\right)$$

and differential $d_n = (-1)^n M(\epsilon_n)$.

Remark 1.1.9. The index $i = n$ is the only one left out in this Intersection. It is also possible to leave out the index $i = 0$. In this case one uses a different Definition of the normalised chain complex but obtains the same result (the Normalisation Theorem 1.2.19) in the end. Indeed, in the literature different authors choose different Definitions, but these can always be interchanged, using the "front-to-back"-dualism of simplicial objects.

Example 1.1.10. The singular topological chain complex $S^t(X)$ of a topological space X (which is introduced in Section 2.3) has a normalised subcomplex which is generated by non-degenerate simplices.

Observe that the differential $(-1)^n M(\epsilon_n) : NM_n \to NM_{n-1}$ on the normalised chain complex NM_* of a simplicial topological R-module M is the restriction of the differential $\sum_{i=0}^{n}(-1)^i M(\epsilon_i)$ of the chain complex M_* to the submodule NM_n of M_n. Therefore the normalised chain complex NM_* of a simplicial topological module M is a subcomplex of the chain complex M_* associated to M. Every morphism $f : M \to M'$ of simplicial topological modules induces a morphism $f_* : M_* \to M'_*$ of the associated chain complexes which further restricts to morphism $Nf_* : NM_* \to NM'_*$ of the normalised subcomplexes. Thus assigning to each simplicial topological R-module M its normalised chain complex NM_* is the object part of a functor $N : \mathbf{TopRMod}^{\mathcal{O}^{op}} \to \mathbf{Ch}(\mathbf{TopRMod})$ from the category $\mathbf{TopRMod}^{\mathcal{O}^{op}}$ of simplicial topological R-modules into the category $\mathbf{Ch}(\mathbf{TopRMod})$ of chain complexes in $\mathbf{TopRMod}$ (cf. [EVdL04, Definition 3.1]).

This – regarding the normalisation as a functor into a category of chain complexes – is the classical point of view of the normalisation of simplicial objects. It is still used in recent publications such as (cf. [EVdL04]), whose results we also will use, but the view of the normalisation as an endofunctor of $\mathbf{TopRMod}^{\mathcal{O}^{op}}$ is more promising and enables us to use a functorial treatment of the "Aushängung" of simplicial objects. Therefore we observe:

Lemma 1.1.11. *For each simplicial topological module M the restriction of the face and degeneracy morphisms to the submodules NM_n also corestricts to the submodules NM_n, i.e. the modules NM_n form a simplicial submodule NM of M.*

Proof. Let M be a simplicial topological module. It is to show that the boundary and degeneracy morphisms of M map the submodules NM_n of M_n into each other. If $M(\epsilon_i) : M([n+1]) \to M([n])$ is a face map of M, then the simplicial identity $\epsilon_j \epsilon_i = \epsilon_i \epsilon_{j-1}$ for $i < j$ (or equivalently $\epsilon_j \epsilon_i = \epsilon_{i+1} \epsilon_j$ for $i \leq j$) implies that the face morphism $M(\epsilon_i)$ maps NM_{n+1} into the intersection

$$NM_n := \bigcap_{i=0}^{n-1} K\left(M(\epsilon_i) : M([n]) \to M([n-1])\right).$$

If $M(\eta_i) : M([n]) \to M([n+1])$ is a degeneracy morphism of M, then the simplicial identity $\eta_j \epsilon_i = \epsilon_i \eta_{j-1}$ for $i < j$ (or equivalently $\epsilon_i \eta_j = \eta_{i+1} \epsilon_i$ for $i \leq j$) and the simplicial identity $\eta_j \epsilon_i = \epsilon_{i-1} \eta_j$ for $i > j$ (or equivalently $\epsilon_i \eta_j = \eta_j \epsilon_{i+1}$ for $i > j$)

imply that the degeneracy morphism $M(\eta_i)$ maps the submodule NM_n of $M([n])$ into the intersection

$$NM_{n+1} := \bigcap_{i=0}^{n} K\left(M(\epsilon_i) : M([n]) \to M([n-1])\right).$$

Since he the boundary and degeneracy morphisms of M map the submodules NM_n of M_n into each other, hence the submodules NM_n of the modules $M([n])$ form a simplicial submodule of M. □

The simplicial submodule formed by the modules NM_n shall always be denoted by NM. Assigning to each simplicial module M its simplicial submodule NM is the object part of a functor N from the category $\mathbf{TopRMod}^{\mathcal{O}^{op}}$ into itself.

Definition 1.1.12. *The endofunctor* N : $\mathbf{TopRMod}^{\mathcal{O}^{op}} \to \mathbf{TopRMod}^{\mathcal{O}^{op}}$ *is called the* Moore normalisation functor. *For each simplicial module M the simplicial submodule NM of M is called its* normalised submodule.

The composition $(-)_* \circ N : \mathbf{TopRMod}^{\mathcal{O}^{op}} \to \mathbf{Ch}(\mathbf{TopRMod})$ of the Moore normalisation functor N and the associated chain complex functor $(-)_*$ then assigns to each simplicial topological module M its normalised chain complex NM_*. We sometimes abbreviate this composition by $(N-)_*$.

In the following we prove that the normalised chain complexes of simplicial topological modules are proper. To begin with, we need some preparatory observations.

Lemma 1.1.13. *Each face homomorphism $M(\epsilon_i)$ of a simplicial topological R-module M is a split epimorphism.*

Proof. This is a consequence of the simplicial identity $\eta_i \epsilon_i = \mathrm{id}$ (cf. F.1). □

Lemma 1.1.14. *Split epimorphisms in* **TopRMod** *are quotient homomorphisms.*

Proof. Let $e : M \to M'$ be an epimorphism in **TopRMod** and $s : M' \to M$ be a section of e. It suffices to show that e is an open map. Let $U \subset M$ be an open subset of M. The e-load $U + K(e)$ of U is open as well. The image $e(U)$ equals the inverse image $s^{-1}(U + K(e))$, which is open because s is continuous. Thus e is an continuous open surjection, hence a quotient homomorphism. □

Lemma 1.1.15. *Consider a commutative square in* **TopRMod** *with horizontal quotient homomorphisms:*

$$\begin{array}{ccc} N & \xrightarrow{\ f\ } & N' \\ {\scriptstyle v}\downarrow & & \downarrow{\scriptstyle w} \\ M & \xrightarrow[\ g\ ]{} & M' \end{array}$$

If w is a monomorphism and v an inclusion of a submodule then w is an inclusion as well.

Proof. Assume that the homomorphism v in the above diagram is an inclusion of a submodule and that the homomorphism w is a monomorphism. Because w is a monomorphism, the domain $K(f)$ of the kernel of f is the intersection of the domain $K(g)$ of the kernel of g with the submodule N of M:

$$K(f) = K(w \circ f) = K(g \circ v) = v^{-1}(K(g)) = N \cap K(g)$$

Because of this and the fact that the homomorphism f is a quotient map, the homomorphism w is induced by the inclusion v. That the induced homomorphism w is an isomorphism onto its image $w(N')$ follows from the Noether Isomorphism Theorem 1.1.2. □

Theorem 1.1.16. *The normalised chain complex NM_* of a simplicial topological R-module M is proper.*

Proof. The proof follows that of [EVdL04, Theorem 3.6] for simplicial objects in exact categories. The exactness of the category considered therein is not necessary. Let M be a simplicial topological R-module and consider the commutative square

$$
\begin{array}{ccc}
NM_n & \xrightarrow{\mathrm{CoIm}\,(-1)^n M(\epsilon_n)_{|NM_n}} & CI((-1)^n M(\epsilon_n)_{|NM_n}) \ , \\
\downarrow & & \downarrow \\
M_n & \xrightarrow[(-1)^n M(\epsilon_n)]{} & M_{n-1}
\end{array}
$$

where the upper horizontal arrow is a quotient homomorphism by definition and the lower horizontal arrow is a quotient homomorphism by Lemma 1.1.14. Both vertical arrows are monomorphisms and the left one is an inclusion. By Lemma 1.1.15 the right vertical arrow is an inclusion as well. Thus the homomorphism

$$
d_{n|NM_n} = (-1)^n M(\epsilon_n)_{|NM_n} : NM_n \to M_{n-1}
$$

is proper and so is its corestriction to NM_{n-1}. Since this is true for any boundary morphism of NM_*, the normalised chain complex NM_* is proper. □

Thus the composition $(-)_* \circ N : \mathbf{TopRMod}^{\mathcal{O}^{op}} \to \mathbf{Ch}(\mathbf{TopRMod})$ of the Moore normalisation functor N and the associated chain complex functor $(-)_*$ corestricts to a functor into the category $\mathbf{PCh}(\mathbf{TopRMod})$ of proper chain complexes of topological R-modules. Moreover, this functor preserves exact sequences:

Proposition 1.1.17. *The composition $(-)_* \circ N$ is an exact functor.*

Proof. This is an application of [EVdL04, Proposition 5.6] to the pointed protomodular category $\mathbf{TopRMod}$. □

Corollary 1.1.18. *The Moore normalisation functor N is exact.*

Proof. This follows from Proposition 1.1.17 and the fact that the associated chain complex functor $(-)_*$ preserves and creates limits and colimits by Lemma 1.1.7 □

Thus each short exact sequence of simplicial topological modules gives rise to a short exact sequence of normalised submodules and a short exact sequence of associated normalised chain complexes. The fact that normalised chain complexes are proper enables us to show that the homology of normalised chain complexes is a homological δ-functor:

Theorem 1.1.19. *The homology of normalised chain complexes $H_* \circ N$ is a universal homological δ-functor $\mathbf{TopRMod}^{\mathcal{O}^{op}} \to \mathbf{TopRMod}$, especially each short exact sequence*

$$
0 \to M \to M' \to M'' \to 0
$$

of simplicial topological R-modules induces a long exact sequence of homology modules

$$
\cdots \xrightarrow{\delta_{n+2}} H_{n+1}(NM_*) \to H_{n+1}(NM'_*) \to II_{n+1}(NM''_*) \xrightarrow{\delta_{n+1}} II_n(NM_*) \to \cdots \ ,
$$

which depends naturally on the given short exact sequence.

Proof. Let $0 \to M \to M' \to M'' \to 0$ be a short exact sequence of simplicial topological modules. The sequence $0 \to NM_* \to NM'_* \to NM''_* \to 0$ of normalised chain complexes is also exact by Proposition 1.1.17. Because the normalised chain complexes NM_*, NM'_* and NM''_* are proper by Theorem 1.1.16 an application of Theorem 1.1.5 yields that $H_* \circ N$ is a homological δ-functor. The universality follows from a modification of the classical proof of [Wei94, Theorem 2.4.7] as described in [EVdL04, 5.8,5.9]. □

Analogous considerations can be made for cochain complexes. Here we note that – similar to chain complexes – the definition of proper cochain complexes adapts to the following form:

Definition 1.1.20. *A cochain complex in* **TopRMod** *is called* proper, *if all differentials are quotients onto their image. The category of proper cochain complexes in* **TopRMod** *is denoted by* **PcoCh(TopRMod)**.

Analogously to simplicial objects, assigning to each cosimplicial module M its associated cochain complex M^* and to each morphism $f : M \to M'$ of cosimplicial modules the induced morphism $f^* : M^* \to M'^*$ between the associated cochain complexes is a functor $(-)^* : \mathbf{TopRMod}^{\mathcal{O}} \to \mathbf{coCh(TopRMod)}$ from the category $\mathbf{TopRMod}^{\mathcal{O}}$ of cosimplicial topological R-modules into the category **coCh(TopRMod)** of cochain complexes in **TopRMod**.

Definition 1.1.21. *The functor* $(-)^* : \mathbf{TopRMod}^{\mathcal{O}} \to \mathbf{coCh(TopRMod)}$ *is called the* associated cochain complex functor.

The components $f^n : M^n \to M'^n$ of a morphism $f^* : M^* \to M'^*$ of cochain complexes associated to a morphism $f : M \to M'$ of cosimplicial topological modules coincide with the components $f([n]) : M([n]) \to M'([n])$ of f. So we note:

Lemma 1.1.22. *The associated cochain complex functor* $(-)^*$ *preserves and creates limits and colimits.*

Proof. The proof is analogous to that of Lemma 1.1.7. □

Observe that the change $n \mapsto -n$ of indices turns a chain complex (M_n, d_n) into a cochain complex (M^n, d^n) and vice versa. In particular the cochain complex (M^n, d^n) is proper exactly if the corresponding chain complex (M_n, d_n) is proper. Furthermore the cohomology $H^*(M^*)$ of the cochain complex (M^n, d^n) (as a graded module) coincides with the homology $H_*(M_*)$ of the corresponding chain complex (M_n, d_n). As a consequence we obtain an analogue to the long exact homology sequence for proper chain complexes.

Theorem 1.1.23 (The long exact cohomology sequence). *Every short exact sequence of proper cochain complexes of topological R-modules*

$$0 \to M^* \to M'^* \to M''^* \to 0$$

induces a long exact sequence of cohomology modules

$$\cdots \xrightarrow{\delta_{n-1}} H^n(M^*) \to H^n(M'^*) \to H^n(M''^*) \xrightarrow{\delta_n} H^{n+1}(M^*) \to \cdots,$$

which depends naturally on the given short exact sequence.

Proof. This follows from Theorem 1.1.5 with reversed directions of arrows. □

Similar to chain complexes, the restriction that the cochain complexes are proper can be circumvented if the cochain complex comes from a cosimplicial topological R-module. This can be proven with a version of the Moore normalisation functor for cosimplicial R-modules. There are two ways to define such a functor. One possible way is to dualise the Definition for simplicial R-modules, i.e. to consider cosimplicial R-modules as simplicial objects in the opposite category **TopRMod**^{op}. This would turn equalisers into coequalisers, i.e. quotients. Thus the "normalised" cochain complex of a cosimplicial module M would be a quotient complex of the cochain complex M^* associated to M. We choose a different approach, where the normalised cochain complex is a subcomplex of M^* and which is a straightforward generalisation of the classical definition.

Definition 1.1.24. *The* Moore normalised cochain complex NM^* *of a cosimplicial R-module M is the chain complex given by*

$$NM^n := \bigcap_{i=0}^{n-1} K\left(M(\eta_i) : M([n]) \to M([n-1])\right)$$

and differential $d_n = \sum_i (-1)^i M(\epsilon_i)_{|NM_n}$.

Remark 1.1.25. In contrast to the normalised chain complex of simplicial modules, none of the epimorphisms $M(\eta_i) : M([n]) \to M([n-1])$ is left out in the intersection defining the normalised cochain complex. This will happen when defining the "degenerate cochain complex" (see Definition 1.1.50 below).

Example 1.1.26. The Alexander-Spanier topological cochain complex $A_{AS}^{t*}(X; V)$ of a topological space X (to be introduced in Section 2.5 has a normalised subcomplex consisting of germs of functions which are trivial on degenerate simplices. Similarly the singular topological cochain complex $S^{t*}(X; V)$ of a topological space X (to be introduced in Section 2.7 has a normalised subcomplex consisting of functions which are trivial on degenerate singular simplices.

As for simplicial objects, it is profitable to regard normalisation as an endofunctor of the category **TopRMod**$^{\mathcal{O}}$ of cosimplicial topological R-modules.

Lemma 1.1.27. *For each cosimplicial topological module M the modules NM^n form a cosimplicial submodule of M.*

Proof. Let M be a cosimplicial topological module. It is to show that the coface and codegeneracy morphisms of M map the submodules NM^n of M^n into each other. If $M(\epsilon_i) : M([n]) \to M([n+1])$ is a coface map of M, then the simplicial identities $\eta_j \epsilon_i = \epsilon_i \eta_{j-1}$ for $i < j$ (or equivalently $\epsilon_i \eta_j = \eta_{i+1} \epsilon_i$ for $i \leq j$) and $\eta_j \epsilon_i = \epsilon_{i-1} \eta_j$ for $i > j+1$ (or equivalently $\epsilon_i \eta_j = \eta_j \epsilon_{i+1}$ for $i > j$) imply that the coface morphism $M(\epsilon_i)$ maps the submodule NM^n of $M([n])$ into the intersection

$$NM^{n+1} := \bigcap_{i=0}^{n} K\left(M(\epsilon_i) : M([n]) \to M([n-1])\right).$$

The submodule NM^{n+1} is by definition contained in the kernel of each codegeneracy morphism $M(\eta_i) : M([n+1]) \to M([n])$. There fore every codegeneracy morphism $M(\eta_i) : M([n+1]) \to M([n])$ maps the submodule NM^{n+1} to zero, hence into the submodule NM^n of $M([n])$. $\square$

The cosimplicial submodule formed by the submodules NM^n shall always be denoted by NM. Assigning to each cosimplicial module M its cosimplicial submodule NM is the object part of a functor N from the category **TopRMod**$^{\mathcal{O}}$ of cosimplicial topological R-modules into itself.

Definition 1.1.28. *The endofunctor* $N : \textbf{TopRMod}^{\mathcal{O}} \to \textbf{TopRMod}^{\mathcal{O}}$ *is called the* Moore normalisation functor. *For each cosimplicial module* M *the cosimplicial submodule* NM *of* M *is called the* normalised submodule *of* M.

The composition $(-)^* \circ N : \textbf{TopRMod}^{\mathcal{O}} \to \textbf{coCh}(\textbf{TopRMod})$ of the Moore normalisation functor N and the associated cochain complex functor $(-)^*$ then assigns to each cosimplicial topological module M its normalised cochain complex NM^*. Since NM is a submodule of M, the normalised cochain complex NM^* is a subcomplex of the associated cochain complex M^* and the differential d^n on NM^n is the restriction of the differential $\sum_{i=0}^{n}(-1)^i M(\epsilon_i)$ of the cochain complex M^* to the submodule NM^n of M^n.

For further proceeding we will use the *right shift functor* $T : \mathcal{O} \to \mathcal{O}$ (cf. F.1.4). Recall that this endofunctor of the simplicial category $\mathcal{O}$ is defined via

$$T[n] = [n+1], \quad T(\alpha)(k) = \begin{cases} 0 & \text{if } k = 0 \\ 1 + \alpha(k-1) & \text{if } k > 0 \end{cases} \quad \text{for all } \alpha : [m] \to [n].$$

Composition of cosimplicial modules with the right shift functor T on $\mathcal{O}$ is an endofunctor $(-T) : \textbf{TopRMod}^{\mathcal{O}} \to \textbf{TopRMod}^{\mathcal{O}}$ of the category $\textbf{TopRMod}^{\mathcal{O}}$ of cosimplicial topological R-modules. This endofunctor maps a cosimplicial module M to the cosimplicial module MT. Furthermore the face morphisms $\epsilon_0 : [n] \hookrightarrow [n+1]$ for all the different $n \in \mathbb{N}$ form a natural transformation $\epsilon_0 : \text{id}_{\mathcal{O}} \to T$. This natural transformation induces a natural transformation $M(\epsilon_0) : M \to MT$ for each cosimplicial topological R-module M. The simplicial identity $\eta_0 \epsilon_0 = \text{id}$ implies that each face map $M(\epsilon_0) : M([n]) \to M([n+1])$ is a split monomorphism, therefore an embedding. So one obtains an exact sequence

$$0 \to M \xrightarrow{M(\epsilon_0)} MT \to (MT/M) \to 0$$

for each cosimplicial topological R-module M. Motivated by the proof for the Eilenberg-MacLane Theorem for simplicial modules found in [DP61, Chapter 3], we call the codomain MT/M of the cokernel of $M(\epsilon_0)$ the *Aushängung* of M. Since this construction is functorial in M, we define:

Definition 1.1.29. *The functor* $A = (-T)/(-) : \textbf{TopRMod}^{\mathcal{O}} \to \textbf{TopRMod}^{\mathcal{O}}$, *which assigns to each cosimplicial module* M *the quotient module* $AM = MT/M$ *is called the* "Aushängungs functor".

The degeneracy morphisms $\eta_0 : T([n]) = [n+1] \twoheadrightarrow [n]$ for different $n \in \mathbb{N}$ do not form a natural transformation $T \to \text{id}_{\mathcal{O}}$, but for each $n \in \mathbb{N}$ the degeneracy morphism $\eta_0 : [n+1] \to [n]$ is a left inverse of the face map $\epsilon_0 : [n] \to [n+1] = T([n])$. Thus for each cosimplicial topological R-module M and each $n \in \mathbb{N}$ the short exact sequence

$$0 \to M([n]) \xrightarrow{M(\epsilon_0)} MT([n]) = M([n+1]) \to AM([n]) \to 0$$

splits, and the quotient module $AM([n])$ is isomorphic to the kernel of the splitting map $M(\eta_0) : MT([n]) = M([n+1]) \twoheadrightarrow M([n])$. This causes the n-th component $MT([n])$ of the cosimplicial module MT to be isomorphic to the direct sum of the n-th component $AM([n])$ of the Aushängung AM of M and the n-th component of M itself:

$$MT([n]) \cong AM([n]) \oplus M([n]) \tag{1.3}$$

Under this isomorphism the embedding $M(\epsilon_0) : M([n]) \hookrightarrow M([n+1] = MT([n])$ corresponds to the inclusion of the second summand $M([n])$ and the splitting homomorphism $M(\eta_0) : MT([n]) \twoheadrightarrow M([n])$ corresponds to the projection onto this second summand. The simplicial identity $\eta_j\eta_i = \eta_i\eta_{j+1}$ for $i \leq j$ implies the identity $M(\eta_0)M(\eta_i) = M(\eta_{i-1})M(\eta_0)$ for all $i > 0$, hence the codegeneracy morphisms $MT(\eta_i) = M(\eta_{i+1}) : MT([n+1]) \to MT([n])$ of MT map the kernel of $M(\eta_0) : MT([n+1]) \to M([n+1])$ into the kernel of $M(\eta_0) : MT([n]) \to M([n])$. The coface morphisms do not have this property in general. Note that the splitting morphism $M(\eta_0) : M([n+1]) = MT([n]) \to M([n])$ is functorial in the cosimplicial module M, hence so is the decomposition of $MT([n])$ into the Aushängung $AM([n])$ and $M([n])$ in equation 1.3.

Proposition 1.1.30. *The right shift functor* $(-T) : \mathbf{TopRMod}^\mathcal{O} \to \mathbf{TopRMod}^\mathcal{O}$ *is continuous, cocontinuous and creates limits and colimits.*

Proof. In functor categories – such as $\mathbf{TopRMod}^\mathcal{O}$ – (co)limits can be calculated point-wise, provided the point-wise limits exist (cf. [ML98, Theorem V.3.1]). Therefore limits and colimits in $\mathbf{TopRMod}^\mathcal{O}$ are taken point-wise, whence the components of (co)limits of cosimplicial topological modules, shifted in dimension, coincide with those of the (co)limits of the right shifted cosimplicial modules. ☐

Corollary 1.1.31. *The right shift functor* $(-T) : \mathbf{TopRMod}^\mathcal{O} \to \mathbf{TopRMod}^\mathcal{O}$ *preserves and reflects kernels and cokernels.*

Recall that the kernels in the category $\mathbf{TopRMod}^\mathcal{O}$ of topological R-modules are the inclusions and the cokernels in $\mathbf{TopRMod}^\mathcal{O}$ are the quotient morphisms.

Lemma 1.1.32. *The right shift functor* $(-T) : \mathbf{TopRMod}^\mathcal{O} \to \mathbf{TopRMod}^\mathcal{O}$ *preserves and reflects monomorphisms and epimorphisms.*

Proof. Let $f : M \to M'$ be a morphism of cosimplicial topological R-modules. The morphism $fT : MT \to M'T$ of cosimplicial modules is a monomorphism or an epimorphism f is a monomorphism or epimorphism respectively. Conversely, if $fT : MT \to M'T$ is a monomorphism then for every $n > 0$ the morphism $f([n]) = fT([n-1])$ is a monomorphism. In addition, all arrows but the upper horizontal one in the commutative diagram

$$
\begin{array}{ccc}
M([0]) & \xrightarrow{\;\;f([0])\;\;} & M'([0]) \\
{\scriptstyle M(\epsilon_0)}\big\uparrow & & \big\downarrow{\scriptstyle M'(\epsilon_0)} \\
M([1]) = MT([0]) & \xrightarrow[\;fT([0])\;]{} & M'T([0]) = M'([1])
\end{array}
$$

are monomorphisms, which forces the morphism $f([0]) : M([0]) \to M'([0])$ to be a monomorphism as well. If $fT : MT \to M'T$ is an epimorphism then for every $n > 0$ the morphism $f([n]) = fT([n-1])$ is an epimorphism. In addition, all arrows but the lower horizontal one in the commutative diagram

$$
\begin{array}{ccc}
M([1]) = MT([0]) & \xrightarrow{\;fT([0])=f([1])\;} & M'T([0]) = M'([1]) \\
{\scriptstyle M(\eta_0)}\big\downarrow & & \big\downarrow{\scriptstyle M'(\eta_0)} \\
M([0]) & \xrightarrow[\;f([0])\;]{} & M'([0])
\end{array}
$$

are epimorphisms, which forces the morphism $f([0]) : M([0]) \to M'([0])$ to be an epimorphism as well. ☐

Lemma 1.1.33. *The Aushängungs functor* $A : \mathbf{TopRMod}^\mathcal{O} \to \mathbf{TopRMod}^\mathcal{O}$ *preserves and reflects monomorphisms, epimorphisms inclusions and quotient morphisms.*

Proof. Let $f : M \to M'$ be a morphism of cosimplicial topological R-modules. The morphism $fT : MT \to M'T$ of cosimplicial modules is a monomorphism, epimorphism, inclusion or quotient morphism if and only if f is a monomorphism, epimorphism, inclusion or quotient morphism respectively. In the decomposition $MT([n]) \cong AM([n]) \oplus M([n])$ (Equation 1.3), the morphism $fT([n])$ corresponds to the morphism $Af([n]) \oplus f([n])$, which thus is a monomorphism, epimorphism, inclusion or quotient morphism if and only if f is a monomorphism, epimorphism, inclusion or quotient morphism respectively. Therefore the morphism Af is a monomorphism, epimorphism, inclusion or quotient morphism if and only if f is a monomorphism, epimorphism, inclusion or quotient morphism respectively. $\square$

Recall that for each cosimplicial topological R-module M the components of the cosimplicial module MT are given by $MT([n]) = M([n+1])$. Furthermore the translation functor shifts the indices of the face and degeneracy morphisms (cf. Section F.1). Concerning the compositions $NA : \mathbf{TopRMod}^{\mathcal{O}} \to \mathbf{TopRMod}^{\mathcal{O}}$ and $N(-T) : \mathbf{TopRMod}^{\mathcal{O}} \to \mathbf{TopRMod}^{\mathcal{O}}$ this results in the following observation:

Lemma 1.1.34. *The endofunctors NA and $(N-)T$ of $\mathbf{TopRMod}^{\mathcal{O}}$ are naturally isomorphic.*

Proof. Let M be a cosimplicial topological R-module and consider its Aushängung AM. The n-th component $NAM([n])$ of the normalised Aushängung NAM of M is a submodule of the n-th component $AM([n])$ of the Aushängung AM of M, which in turn is a submodule of the n-th component $MT([n])$ of the cosimplicial module MT. The right shifted normalised cosimplicial module $(NM)T$ is a cosimplicial submodule of MT, because NM is a submodule of M and composition with the right shift functor T preserves inclusions. It remains to show that for each $n \in \mathbb{N}$ the submodules $(NAM)([n])$ and $(NM)T([n])$ of $MT([n])$ coincide. The n-th component $NAM([n])$ of the cosimplicial topological R-module NAM is given by

$$(NAM)([n]) = \bigcap_{i=0}^{n-1} K(AM(\eta_i : [n] \to [n-1]))$$

$$= \bigcap_{i=0}^{n-1} K(MT(\eta_i : [n] \to [n-1])_{|AM([n])}).$$

The morphism $MT(\eta_i : [n] \to [n-1])$ is the morphism $M(\eta_{i+1} : [n+1] \to [n])$ and the submodule $AM([n])$ of $MT([n])$ these homomorphisms are restricted to is the domain of the kernel of the coface homomorphism $M(\eta_0) : M([n+1]) \to M([n])$. So all in all one obtains the equality

$$NAM([n]) = \bigcap_{i=1}^{n} K(M(\eta_0) : M([n+1]) \to M([n])) \cap K(M(\eta_i) : [n+1] \to [n])$$

$$= \bigcap_{i=0}^{n} K(M(\eta_i) : M([n+1]) \to M([n]))) = (NM)T([n]).$$

Thus for each cosimplicial topological R-module M the cosimplicial submodules NAM and $(NM)T$ of MT coincide, so the functors NA and $(N-)T$ are naturally isomorphic. $\square$

Lemma 1.1.35. *The functor $N : \mathbf{TopRMod}^{\mathcal{O}} \to \mathbf{TopRMod}^{\mathcal{O}}$ preserves monomorphisms, epimorphisms and inclusions and kernels.*

Proof. The preservation of monomorphisms, inclusions and kernels follows directly from the definition. We prove that the normalisation functor preserves epimorphisms by induction on n. The equalities $NM([0]) = M([0])$ and $NM'([0]) = M'([0])$ for every cosimplicial topological module M imply that $Nf([0]) = f([0])$ is an epimorphism whenever f is an epimorphism. Now assume that for each epimorphism f of topological R-modules the homomorphisms $Nf([m])$ are epimorphisms for all $m \leq n$. This especially implies that the morphism $Af([n])$ is an epimorphism for all epimorphisms f in of cosimplicial topological R-modules. If f is such an epimorphism, then the isomorphism $(NM)T \cong NAM$ (Lemma 1.1.34) shows, that the homomorphism $(Nf)T([n])$ corresponds to the homomorphism $Af([n])$, hence is an epimorphism as well. So $Nf([n+1]) = (Nf)T([n])$ is an epimorphism, which completes the inductive step. $\square$

Proposition 1.1.36. *The functor* $N : \mathbf{TopRMod}^{\mathcal{O}} \to \mathbf{TopRMod}^{\mathcal{O}}$ *preserves quotients and cokernels.*

Proof. Let $f : M' \to M''$ be a quotient homomorphism in $\mathbf{TopRMod}^{\mathcal{O}}$. By the preceding Lemma, the corresponding morphism $Nf : NM' \to NM''$ of normalised cosimplicial modules is an epimorphism. It remains to show that NM'' carries the quotient topology. Let $K(f)$ be the kernel of f. Because the normalisation functor preserves kernels, the inclusion $NK(f) \hookrightarrow NM'$ is the kernel of Nf. There fore the epimorphism Nf factors through the quotient morphism $q : NM' \twoheadrightarrow NM'/NK(f)$ as $Nf = gq$, where $g : NM'/NK(f) \to NM''$ is a continuous bijection. Let i denote the inclusion $NM'' \hookrightarrow M''$ of NM'' in M''. The Noether Isomorphism Theorem 1.1.2 now implies that $NM'/NK(f)$ is a submodule of M'', the inclusion $NM'/NK(f) \hookrightarrow M''$ being induced by the inclusion $NM' \hookrightarrow M'$. Since the submodules $NM'/NK(f) \leq M''$ and $NM'' \leq M''$ coincide, the continuous bijection $g : NM'/NK(f) \to NM''$ has to be an isomorphism. Thus the Moore normalisation functor preserves quotients.

If $f : M' \to M''$ is the cokernel of a morphism $g : M \to M'$ of cosimplicial topological R-modules, then the inclusion $g(M) \hookrightarrow M'$ is the kernel of f. Since the Moore normalisation functor N preserves epimorphisms (by Lemma 1.1.35), we see that the morphism $Ng : NM \to N(g(M))$ is surjective, i.e. $Ng(NM) = N(g(M))$. Because N also preserves kernels, the inclusion $Ng(NM) = N(g(M)) \hookrightarrow NM'$ is the kernel of Nf. So f is a quotient morphism with kernel $Ng(NM) \hookrightarrow NM'$, hence a cokernel of Ng. This shows that the functor N also preserves cokernels. $\square$

Theorem 1.1.37. *The functor* $N : \mathbf{TopRMod}^{\mathcal{O}} \to \mathbf{TopRMod}^{\mathcal{O}}$ *is exact.*

Proof. The Moore normalisation functor N preserves kernels by Lemma 1.1.36 and cokernels by Proposition 1.1.35, hence it is an exact functor. $\square$

In the following we prove that not only the normalised cochain complex NM^* of a cosimplicial topological module M but also the cochain complex M^* associated to M is proper. Let M be a cosimplicial topological R-module. Being covariant, the functor M maps the face morphisms ϵ_i in $\mathcal{O}$ to monomorphisms $M(\epsilon_i)$ in $\mathbf{TopRMod}$. Therefore we can not proceed as in the case of simplicial modules, because neither does the differential $d^n = \sum (-1)^i M(\epsilon_i)$ restrict to a single coface morphism on the modules NM^n nor are these coface morphisms split epimorphism as required for the proof of Proposition 1.1.16. Instead we are going to prove the properness of the (normalised) cochain complex using the Noether Isomorphism Theorem. Here fore we first identify the summation

$$\sum_{i=0}^{n} p_i : \bigoplus_{i=0}^{n} M([n]) \to M([n]), \quad (m_0, \ldots, m_n) \mapsto \sum_{i=0}^{n} m_i$$

as a quotient homomorphism:

Lemma 1.1.38. *The summation $\sum_{i=0}^{n} p_i : \bigoplus_{i=0}^{n} M([n]) \to M([n])$ is a split epimorphism and a quotient homomorphism.*

Proof. The embedding $M([n]) \hookrightarrow \bigoplus_{i=0}^{n} M([n])$, $m \mapsto (m, 0, \dots, 0)$ is a right inverse for the summation, so the latter is a split epimorphism, hence a quotient homomorphism by Lemma 1.1.14. $\square$

Lemma 1.1.39. *Each coboundary homomorphism $M(\epsilon_i)$ of a cosimplicial topological R-module M is a split monomorphism.*

Proof. This is a consequence of the simplicial identity $\eta_i \epsilon_i = \mathrm{id}$ (cf. F.1). $\square$

Lemma 1.1.40. *Split monomorphisms in **TopRMod** are inclusions.*

Proof. Let $m : M \to M'$ be a split monomorphism with left inverse $e : M' \to M$. If $O \subset M$ is an open subset of M then the identity $m^{-1}e^{-1} = \mathrm{id}_M$ implies the equality $m^{-1}e^{-1}(O) = O$, hence $m(O) = e^{-1}(O) \cap m(M)$. Therefore m is a homeomorphism onto its image, i.e. an inclusion of a submodule. $\square$

Theorem 1.1.41. *The cochain complex M^* associated to a cosimplicial topological R-module M is proper, i.e. the associated cochain complex functor $(-)^*$ correstricts to a functor $(-)^* : \mathbf{TopRMod}^{\mathcal{O}} \to \mathbf{PcoCh}(\mathbf{TopRMod})$.*

Proof. Let M be a cosimplicial topological R-module and consider the coface homomorphisms $M(\epsilon_i) : M([n]) \to M([n+1])$. Each of these coface homomorphisms $M(\epsilon_i)$ has a left inverse (namely $M(\eta_i)$) and thus is an embedding. Therefore the homomorphism

$$j^n : M([n]) \to \bigoplus_{i=0}^{n+1} M([n+1]),$$
$$j^n(m) = (M(\epsilon_0)(m), -M(\epsilon_1)(m), \dots, (-1)^n M(\epsilon_n)(m))$$

also is a monomorphism. Furthermore it also has a left inverse (e.g. the projection onto the first factor followed by $M(\eta_0)$), hence it is a split monomorphism and an embedding as well. So the homomorphism j^n embeds $M([n])$ as a submodule of the direct sum $\bigoplus_{i=0}^{n+1} M([n+1])$. The boundary homomorphism on $M([n])$ is the composition

$$d^n = \left(\sum_{i=0}^{n+1} p_i \right) \circ j^n : M([n]) \to M([n+1])$$

of the summation $\sum_{i=0}^{n+1} p_i : \bigoplus_{i=0}^{n+1} M([n+1]) \to M([n+1])$, which is a split epimorphism, and the embedding j^n, which is an inclusion. So one obtains the commutative diagram

$$
\begin{array}{ccc}
M([n]) & \xrightarrow{\ \mathrm{CoIm}\, d^n\ } & CI(d^n) \\
{\scriptstyle j^n} \downarrow & & \downarrow {\scriptstyle w} \\
\bigoplus_{i=0}^{n+1} M([n]) & \xrightarrow[\sum_{i=0}^{n+1} p_i]{} & M([n+1])
\end{array}
$$

where the upper horizontal arrow is a quotient homomorphism by definition and the lower horizontal arrow is a split epimorphism, hence a quotient as well (by Lemma 1.1.14). Both vertical arrows are monomorphisms and the left one is an inclusion. By Lemma 1.1.15 the right vertical arrow is an inclusion as well. Thus the boundary homomorphism d^n is proper. Since this is true for any boundary morphism of M^*, the chain complex M^* associated to M is proper. $\square$

Theorem 1.1.42. *The cohomology of associated cochain complexes is a cohomological δ-functor $H \circ (-)^* : \mathbf{TopRMod}^{\mathcal{O}} \to \mathbf{TopRMod}$, i.e. each short exact sequence*

$$0 \to M \to M' \to M'' \to 0$$

of cosimplicial topological R-modules induces a long exact sequence of cohomology modules

$$\cdots \xrightarrow{\delta_{n-1}} H^n(M^*) \to H^n(M'^*) \to H^n(M''^*) \xrightarrow{\delta_n} H^{n+1}(M^*) \to \cdots,$$

which depends naturally on the given short exact sequence.

Proof. Let $0 \to M \to M' \to M'' \to 0$ be a short exact sequence of cosimplicial topological modules. Because the associated cochain complexes M^*, M'^* and M''^* are proper by Theorem 1.1.41 an application of Theorem 1.1.23 yields the desired result.   □

Lemma 1.1.43. *For each cosimplicial topological R-module M, the cohomology $H(MT^*)$ of the cochain complex MT^* associated to the cosimplicial module MT is trivial.*

Proof. Let M be a cosimplicial topological R-module and consider the cosimplicial topological module MT. The inclusions $[0] \hookrightarrow T([n]) = [n+1]$ form a morphism from the constant cosimplicial object $[0]$ in $\mathcal{O}$ to T, whose left inverse is the unique projection $p : T \to [0]$ (cf. Section F.6). The embedding i in $\mathcal{O}$ induces an embedding $M(i) : M^0 \hookrightarrow MT$ of the constant cosimplicial module M^0 into MT with right inverse the projection $M(p) : MT \twoheadrightarrow M^0$. By Theorem F.6.3 the morphism $M(i)M(p) = M(ip)$ and the identity morphism id_{MT} are homotopic. Therefore the induced endomorphisms $M(ip)^* : MT^* \to MT^*$ and $\mathrm{id} : MT^* \to MT^*$ of the associated cochain complex MT^* are homotopic as well. Since the cohomology of the constant cochain complex M^0 is trivial, this forces the cohomology of the cochain complex MT^* to be trivial as well.   □

Proposition 1.1.44. *The cohomology of associated cochain complexes is an effaceable functor $H \circ (-)^* : \mathbf{TopRMod}^{\mathcal{O}} \to \mathbf{TopRMod}$.*

Proof. It is to prove that for every cosimplicial topological R-module M there exists a monomorphism $i : M \to I$ into cosimplicial topological R-module I such that $H(i^*) = 0$. Setting $I = MT$ and $i : M \to MT$ to be the inclusion of M into MT, we see that the triviality of $H(I) = H(MT)$ implies $H(i^*) = 0$ for all $n \in \mathbb{N}$.   □

Theorem 1.1.45. *The cohomology of associated cochain complexes is a universal cohomological δ-functor $H \circ (-)^* : \mathbf{TopRMod}^{\mathcal{O}} \to \mathbf{TopRMod}$,*

Proof. The cohomology of associated cochain complexes functor is a cohomological δ-functor $H \circ (-)^* : \mathbf{TopRMod}^{\mathcal{O}^{op}} \to \mathbf{TopRMod}$. Effaceable (additive) cohomological δ-functors are universal. (The dual proof of [Wei94, Theorem 2.4.7] is valid for any homological category, if one replaces the embeddings in injective objects by the morphisms guaranteed to exist by the effaceability of $H \circ (-)^*$.) Thus the functor $H \circ (-)^*$ is a universal cohomological δ-functor.   □

The proof of the fact that associated cochain complexes are proper (Theorem 1.1.41) can be refined to also work for normalised cochain complexes. Herefore we first need a generalisation of Lemma 1.1.15:

Lemma 1.1.46. *Consider a commutative square in* **TopRMod** *with upper horizontal arrow f a quotient homomorphism:*

$$
\begin{array}{ccc}
N & \xrightarrow{\;f\;} & N' \\
\downarrow{\scriptstyle v} & & \downarrow{\scriptstyle w} \\
M & \xrightarrow{\;g\;} & M'
\end{array}
$$

If w is a monomorphism, v an inclusion of a submodule and g is a composition g = em of a split epimorphism e and a split monomorphism m then w is an inclusion as well.

Proof. Assume that the homomorphism v in the above diagram is an inclusion of a submodule, the homomorphism g is a composition $g = em$ of a split epimorphism $e : M'' \to M$ and a split monomorphism $m : M \to M''$ and that the homomorphism w is a monomorphism. The monomorphism m is a split homomorphism, hence an inclusion by Lemma 1.1.40. The composition mv of the inclusions v and m is an inclusion $mv : M \to M''$. So the above diagram can be rewritten as

$$
\begin{array}{ccc}
N & \xrightarrow{\;f\;} & N' \\
\downarrow{\scriptstyle v} & & \\
M & & w \\
\downarrow{\scriptstyle m} & & \\
M'' & \xrightarrow{\;e\;} & M'
\end{array}
$$

Because w is a monomorphism, the domain $K(f)$ of the kernel of f is the intersection of the kernel of e with the submodule N of M:

$$
K(f) = K(w \circ f) = K(g \circ v) = (vm)^{-1}(K(e)) = N \cap K(e)
$$

Because of this and the fact that the homomorphism f is a quotient map, the homomorphism w is induced by the inclusion mv. That the induced homomorphism w is an isomorphism onto its image $w(N')$ follows from the Noether Isomorphism Theorem 1.1.2. $\qquad\square$

Theorem 1.1.47. *The normalised cochain complex NM^* of a cosimplicial topological R-module M is proper, i.e. the normalised cochain complex functor $(N-)^*$ correstricts to a functor $(N-)^* : \mathbf{TopRMod}^{\mathcal{O}} \to \mathbf{PcoCh}(\mathbf{TopRMod})$.*

Proof. The proof is a refinement of that of Theorem 1.1.47. We use the notation introduced therein. Let M be a cosimplicial topological R-module and consider the decomposition

$$
d^n = \left(\sum_{i=0}^{n+1} p_i \right) \circ j^n : M([n]) \to M([n+1])
$$

of the coboundary homomorphism d^n into the inclusion $j^n : M([n])$ and the split epimorphism $\sum_{i=0}^{n+1} p_i$. Let $i : NM^* \to M^*$ denote the inclusion of the normalised subcomplex and consider the commutative diagram

$$
\begin{array}{ccc}
NM([n]) & \xrightarrow{\text{CoIm } d^n_{|NM^n}} & CI(d^n_{|NM^n}) \\
\downarrow{\scriptstyle i^n} & & \downarrow{\scriptstyle w} \\
M([n]) & \xrightarrow{\;d^n\;} & M([n+1])
\end{array}
$$

where the upper horizontal arrow is a quotient homomorphism by definition and the lower horizontal arrow is a composition of a split epimorphism with a split monomorphism. An application of Lemma 1.1.46 shows that w is an inclusion. Thus the homomorphism

$$d^n{}_{|NM^n} : NM^n \to M^{n+1}$$

is proper and so is its corestriction to NM^{n+1}. Since this is true for any boundary morphism of NM^*, the normalised chain complex NM^* is proper.   □

Since the Moore normalisation functor N is exact and the associated chain complex functor $(-)^* : \textbf{TopRMod}^{\mathcal{O}} \to \textbf{coCh(TopRMod)}$ also preserves exactness, the composition $N \circ (-)^* : \textbf{TopRMod}^{\mathcal{O}} \to \textbf{coCh(TopRMod)}$ corestricts to a functor into the category $\textbf{PcoCh(TopRMod)}$ of proper cochain complexes of topological R-modules. So each short exact sequence of cosimplicial topological modules gives rise to a short exact sequence of normalised cochain complexes. The fact that normalised cochain complexes are proper implies that the cohomology of normalised cochain complexes is a cohomological δ-functor:

Theorem 1.1.48. *The cohomology of normalised cochain complexes $H \circ (N-)^*$ is a cohomological δ-functor* $\textbf{TopRMod}^{\mathcal{O}} \to \textbf{TopRMod}$, *i.e. each short exact sequence*

$$0 \to M \to M' \to M'' \to 0$$

of cosimplicial topological R-modules induces a long exact sequence of cohomology modules

$$\cdots \xrightarrow{\delta_{n-1}} H^n(NM^*) \to H^n(NM'^*) \to H^n(NM''^*) \xrightarrow{\delta_n} H^{n+1}(NM^*) \to \cdots,$$

which depends naturally on the given short exact sequence.

Proof. Let $0 \to M \to M' \to M'' \to 0$ be a short exact sequence of cosimplicial topological modules. The sequence $0 \to NM^* \to NM'^* \to NM''^* \to 0$ of normalised cochain complexes is also exact by Proposition 1.1.37. Because the normalised cochain complexes NM^*, NM'^* and NM''^* are proper by Theorem 1.1.47 an application of Theorem 1.1.23 yields the desired result.   □

The cohomologies of the associated cochain complex M^* and the normalised cochain complex NM^* of a cosimplicial topological module M coincide, if the inclusion $NM^* \hookrightarrow M^*$ is a homotopy equivalence of cochain complexes. Classically this is proved with the help of another subcomplex of the cochain complex M^* – called the "degenerate" subcomplex – that serves as a complement of the normalised subcomplex NM^* in M^*. The definition of this subcomplex generalises to cosimplicial objects in **TopRMod**:

Definition 1.1.49. *The* degenerate cochain complex DM^* *of a cosimplicial topological R-module M is the cochain complex given by*

$$DM^n := \left\langle \bigcup_{i=0}^{n-1} M(\epsilon_i) \right\rangle$$

and differential $d_n = \sum_{i=0}^n (-1)^i M(\epsilon_i)$.

Note, that the degenerate cochain complex DM^* of a cosimplicial topological module M is a subcomplex of the cochain complex M^* associated to a M. Every morphism $f : M \to M'$ between cosimplicial topological modules M and M' induces a morphism $Df^* : DM^* \to DM'^*$ of the degenerate subcomplexes. Thus assigning to each cosimplicial topological R-module M its degenerate cochain complex DM^* is the object part of a functor $(D-)^* : \mathbf{TopRMod}^{\mathcal{O}} \to \mathbf{coCh}(\mathbf{TopRMod})$ from the category $\mathbf{TopRMod}^{\mathcal{O}}$ of cosimplicial topological R-modules into the category $\mathbf{coCh}(\mathbf{TopRMod})$ of cochain complexes in $\mathbf{TopRMod}$. The normalised cochain complex functor $(N-)^*$ and the degenerate cochain complex functor $(D-)^*$ are both subfunctors of the associated cochain complex functor

$$(-)^* : \mathbf{TopRMod}^{\mathcal{O}} \to \mathbf{coCh}(\mathbf{TopRMod}).$$

Similar to cosimplicial modules, the normalisation of a chain complex M_* associated to a simplicial topological R-module M can be avoided, if the inclusion $NM_* \hookrightarrow M_*$ of the normalised chain complex NM_* into the chain complex M_* is a weak equivalence. Classically this also is proved with the help of a "degenerate" subcomplex of the associated chain complex M_*. The definition thereof generalises to simplicial objects in $\mathbf{TopRMod}$:

Definition 1.1.50. *The* degenerate chain complex DM_* *of a simplicial topological R-module M is the chain complex given by*

$$DM_n := \left\langle \bigcup_{i=0}^{n} M(\eta_i) \right\rangle$$

and differential $d_n = \sum_{i=0}^{n}(-1)^i M(\epsilon_i)$.

Similar to cosimplicial modules, the degenerated chain complex DM_* of a simplicial module M is a subcomplex of the chain complex M_* associated to M. Also, assigning to each simplicial module M the degenerate subcomplex DM_* of M_* is the object part of a functor $(D-)_* : \mathbf{TopRMod}^{\mathcal{O}^{op}} \to \mathbf{Ch}(\mathbf{TopRMod})$ from the category $\mathbf{TopRMod}^{\mathcal{O}^{op}}$ of simplicial topological R-modules into the category $\mathbf{Ch}(\mathbf{TopRMod})$ of chain complexes in $\mathbf{TopRMod}$. The normalised chain complex functor $(N-)_*$ and the degenerate chain complex functor $(D-)_*$ are subfunctors of the associated chain complex functor

$$(-)_* : \mathbf{TopRMod}^{\mathcal{O}^{op}} \to \mathbf{Ch}(\mathbf{TopRMod}).$$

In the next section we will construct natural transformations $(-)_* \to (-)_*$ and $(-)^* \to (-)^*$, which both have image the normalised (co)chain complex and kernel the degenerate (co)chain complex respectively. This will be a generalisation of the normalisation of chain complexes in abelian categories and ultimately enable us to omit the Moore normalisation functor in Theorem 1.1.19, thus identifying the homology of chain complexes associated to simplicial topological modules as a universal homological δ-functor. This result will not rely any more on the properness of the chain complexes. We will also derive analogous results concerning the normalisation of cochain complexes associated to cosimplicial modules.

1.2 Normalisation

In this section we derive general normalisation theorems for simplicial and cosimplicial objects in pointed additive categories with images. These normalisation theorems will especially apply to simplicial and cosimplicial topological modules (e.g. the singular simplicial abelian topological group and the cosimplicial module of forms on a differentiable manifold, cochain complexes of forms on topological Lie algebras and similar objects). They are a generalisation of the classical results for abelian categories, containing them as a special case. The normalisation theorem for simplicial objects then implies that the normalised chain complex NM_* of a simplicial topological module M is weakly equivalent to the cochain complex M_* and that the inclusion $NM_* \hookrightarrow M_*$ induces an isomorphism in homology.

Consider an Ab-category $\mathbf{C}$. Analogously to the category of topological modules, there exists a functor $(-)_* : \mathbf{C}^{\mathcal{O}^{op}} \to \mathbf{Ch}(\mathbf{C})$ assigning to each simplicial object C in $\mathbf{C}$ its associated chain complex C_*.

Definition 1.2.1. *For any Ab-category $\mathbf{C}$ the functor $(-)_* : \mathbf{C}^{\mathcal{O}^{op}} \to \mathbf{Ch}(\mathbf{C})$ is called the* associated chain complex functor.

Lemma 1.2.2. *. The associated chain complex functor $(-)_*$ is continuous and co-continuous.*

Proof. In functor categories (co)limits can be calculated point-wise, provided the point-wise limits exist (cf. [ML98, Theorem V.3.1]). Therefore limits and colimits in $\mathbf{C}^{\mathcal{O}^{op}}$ are taken point-wise, whence the components of (co)limits of simplicial modules coincide with those of the (co)limits of the associated chain complexes.   □

As has been shown in [Eps66] normalising of associated chain complexes in abelian categories can be described by the use of an extension $\mathcal{O}^{+0}$ of the simplicial category $\mathcal{O}$. We take his approach because it can be generalised to arbitrary pointed Ab-categories with images such as **TopRMod** or **TopAb**. It can furthermore also be applied to cochain complexes associated to cosimplicial objects.

Definition 1.2.3. *The* Ab-hull $\mathcal{O}^+$ *of the simplicial category $\mathcal{O}$ is the category with the same set of objects $[n]$ and set of morphisms $\hom_{\mathcal{O}^+}([m],[n])$ the free abelian group $\mathbb{Z}^{(\hom_{\mathcal{O}}([m],[n]))}$ on the set $\hom_{\mathcal{O}}([m],[n])$ of morphisms from $[m]$ to $[n]$ in $\mathcal{O}$. Composition of morphisms is given by the unique bilinear extension of the composition of morphisms in $\mathcal{O}$.*

The Ab-hull $\mathcal{O}^+$ is an Ab-category by definition, containing the simplicial category $\mathcal{O}$ as a subcategory. The embedding $\mathcal{O} \hookrightarrow \mathcal{O}^+$ of $\mathcal{O}$ in its Ab-hull $\mathcal{O}^+$ has a universal property:

Lemma 1.2.4. *Every functor $C : \mathcal{O} \to \mathbf{C}$ into an Ab-category $\mathbf{C}$ can be uniquely be extended to an additive functor $C^+ : \mathcal{O}^+ \to \mathbf{C}$.*

Proof. If $C : \mathcal{O} \to \mathbf{C}$ is a functor into an Ab-category $\mathbf{C}$ then the extension C^+ is defined via $C^+([n]) = C([n])$ and $C^+(\sum z_i \varphi_i) = \sum z_i C(\varphi_i)$ for any finite $\mathbb{Z}$-linear combination of morphisms $\varphi_i : [n] \to [m]$. The uniqueness follows from the requirement that C^+ is additive.   □

Note that this holds for both co- and contravariant functors $C : \mathcal{O} \to \mathbf{C}$. To put it differently, each simplicial object in an Ab-category $\mathbf{C}$ can be described by a contravariant additive functor $C^+ : \mathcal{O}^+ \to \mathbf{C}$ and vice versa. Similarly, every cosimplicial object C in an Ab-category $\mathbf{C}$ can uniquely be extended to an additive covariant functor $C^+ : \mathcal{O}^+ \to \mathbf{C}$ and this correspondence is bijective. The Ab-hull $\mathcal{O}^+$ of $\mathcal{O}$ can be further extended by adjoining a zero object:

Definition 1.2.5. *For every category $\mathbf{C}$, the category $\mathbf{C}^0$ is the category obtained from $\mathbf{C}$ by adjoining a zero object 0.*

Thus the category $\mathcal{O}^{+0}$ is a pointed Ab-category containing the Ab-hull $\mathcal{O}^+$ as a full subcategory. The usefulness of the category $\mathcal{O}^{+0}$ stems from the fact that any simplicial object $C : \mathcal{O}^{op} \to \mathbf{C}$ in a pointed Ab-category $\mathbf{C}$ can uniquely be extended to a contravariant additive functor $C^{+0} : \mathcal{O}^{+0} \to \mathbf{C}$ preserving the zero object and any cosimplicial object $C : \mathcal{O} \to \mathbf{C}$ in a pointed Ab-category $\mathbf{C}$ can uniquely be extended to a covariant additive functor $C^{+0} : \mathcal{O}^{+0} \to \mathbf{C}$ preserving the zero object:

Lemma 1.2.6. *Every functor $C : \mathcal{O} \to \mathbf{C}$ into a pointed Ab-category $\mathbf{C}$ can be uniquely be extended to an additive functor $C^{+0} : \mathcal{O}^+ \to \mathbf{C}$ preserving the zero object.*

Proof. Each functor $C : \mathcal{O} \to \mathbf{C}$ into a pointed category $\mathbf{C}$ can be uniquely extended to an additive functor $C^+ : \mathcal{O}^+ \to \mathbf{C}$. The requirement that C^{+0} is to preserve the zero object implies that there exists a unique extension C^{+0} of C^+ preserving the zero object. $\qquad\square$

This Lemma also holds for both co- and contravariant functors. So every simplicial object C in a pointed Ab-category $\mathbf{C}$ can be identified with its unique contravariant additive extension $C^{+0} : \mathcal{O}^{+0} \to \mathbf{C}$ and this correspondence is bijective. Similarly, every cosimplicial object C in a pointed Ab-category $\mathbf{C}$ can be identified with its unique (covariant) additive extension $C^{+0} : \mathcal{O}^{+0} \to \mathbf{C}$ and this correspondence is bijective as well.

Since the category $\mathcal{O}^{+0}$ is a pointed Ab-category, we can consider cochain complexes in $\mathcal{O}^{+0}$. The unique additive extension C^{+0} of a simplicial object C in a pointed Ab-category $\mathbf{C}$ transforms cochain complexes and morphisms of cochain complexes in chain complexes and morphisms of chain complexes in $\mathbf{C}$. In this way one obtains a functor $C_*^{+0} : \mathbf{coCh}(\mathcal{O}^{+0}) \to \mathbf{Ch}(\mathbf{C})$ from the category $\mathbf{coCh}(\mathcal{O}^{+0})$ of cochain complexes in $\mathcal{O}^{+0}$ into the category $\mathbf{Ch}(\mathbf{C})$ of chain complexes in $\mathbf{C}$. In particular, we can consider the objects $[n]$ to form a cochain complex with differentials

$$d_n : [n] \to [n+1], \quad d_n = \sum_{i=0}^{n+1} (-1)^i \epsilon_i.$$

Every chain complex C_* associated to a simplicial object C in a pointed Ab-category $\mathbf{C}$ is the image of this cochain complex $([n], d_n)$ under the functor C_*^{+0}. We also denote this chain complex by C_*^{+0} when using the extension of the functor C to the category $\mathcal{O}^{+0}$. This provides an abstract framework for the normalisation of chain complexes associated to simplicial objects. The normalisation of such a chain complex C_* can be modelled by an endomorphism α^* of the cochain complex $([n], d_n)$ in the category $\mathcal{O}^{+0}$. Here we define endomorphisms α^n of the objects $[n]$ via

$$\alpha^n : [n] \to [n], \quad \alpha^n = (\mathrm{id}_{[n]} - \epsilon_0\eta_0)(\mathrm{id}_{[n]} - \epsilon_2\eta_2)\cdots(\mathrm{id}_{[n]} - \epsilon_{n-1}\eta_{n-1})$$

Regarding the cochain complex $([n], d_n)$ in the pointed Ab-category $\mathcal{O}^{+0}$ we find:

Proposition 1.2.7. *The endomorphisms $\alpha^n : [n] \to [n]$ form an idempotent endomorphism α^* of the cochain complex $([n], d_n)$ and this endomorphism α^* is homotopic to the identity.*

Proof. The proof is analogous to that of [Eps66, Proposition 2.3], where the analogous statement form for the front-to-back dual in the opposite category $\mathcal{O}^{+0op}$ is proved. Thus the statement follows from [Eps66, Proposition 2.3] by dualising and taking the front-to-back dual. $\qquad\square$

Corollary 1.2.8. *The endomorphisms $\mathrm{id}_{[n]} - \alpha^n : [n] \to [n]$ form an idempotent endomorphism $\mathrm{id} - \alpha^*$ of the cochain complex $([n], d_n)$ and this endomorphism $\mathrm{id} - \alpha^*$ is homotopic to the zero endomorphism.*

The endomorphisms α^* and $\mathrm{id} - \alpha^*$ of the cochain complex $([n], d_n)$ induce endomorphisms of chain complexes associated to simplicial objects in pointed Ab-categories: For every simplicial object C in a pointed Ab-category $\mathbf{C}$ the morphisms $C^{+0}(\alpha^n) : C_n \to C_n$ form an idempotent endomorphism $C_*^{+0}(\alpha^*) : C_* \to C_*$ of the chain complex C_* associated to C. Likewise, the morphisms $C^{+0}(\mathrm{id}_{[n]} - \alpha^n)$ form an endomorphism $C_*^{+0}(\mathrm{id} - \alpha^*) = \mathrm{id} - C_*^{+0}(\alpha^*) : C_* \to C_*$ of the chain complex C_*. Here the above Proposition translates as:

Corollary 1.2.9. *The endomorphism $C_*^{+0}(\alpha^*)$ of the chain complex C_* associated to a simplicial object C in a pointed Ab-category* **C** *is idempotent and chain homotopic to the identity.*

Corollary 1.2.10. *The endomorphism $\mathrm{id} - C_*^{+0}(\alpha^*)$ of the chain complex C_* associated to a simplicial object C in a pointed Ab-category* **C** *is idempotent and chain homotopic to the zero endomorphism.*

If C is a simplicial object in a pointed Ab-category **C** which admits images, then the endomorphism $C_*^{+0}(\alpha^*) : C_* \to C_*$ of the chain complex C_* has a a a unique image factorisation

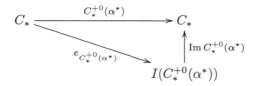

as an epimorphism $e_{C_*^{+0}(\alpha^*)}$ followed by the image morphism $\operatorname{Im} C_*^{+0}(\alpha^*)$ (cf. Lemma A.1.29). The endomorphism $\mathrm{id} - C_*^{+0}(\alpha^*)$ of C_* has an analogous factorisation. The complex $I(C_*^{+0}(\alpha^*))$ is a subcomplex of the chain complex C_* by definition. Likewise the domain $I(\mathrm{id} - C_*^{+0}(\alpha^*))$ of the image of $\mathrm{id} - C_*^{+0}(\alpha^*)$ is a subcomplex of C_*.

Example 1.2.11. If **C** = **TopAlg**$_{(\Omega,\mathrm{E})}$ is a pointed Ab-category of topological algebras of certain type (Ω, E), then the image $\operatorname{Im} C_n^{+0}(\alpha^n)$ is the inclusion of the topological subalgebra $C^{+0}(\alpha^n)((C[n]))$ of $C([n])$ and the image $\operatorname{Im} C_*^{+0}(\alpha^*)$ is the inclusion of the subcomplex $C_*^{+0}(\alpha^*)(C_*)$ of topological subalgebras of C_*.

If we assume further that the category **C** admits binary products (or binary coproducts), then it is an additive category and admits arbitrary finite coproducts. (Recall that such finite coproducts in additive categories are called direct sums and are denoted by the symbol "$\oplus$".) In this case the image-factorisations of the endomorphisms $C_*^{+0}(\alpha^*)$ and $\mathrm{id} - C_*^{+0}(\alpha^*)$ of the chain complex C_* lead to a decomposition of the chain complex C_*:

Proposition 1.2.12. *For any simplicial object C in an additive category* **C** *with images the morphism*

$$\operatorname{Im} C_*^{+0}(\alpha^*) \circ \mathrm{pr}_1 + \operatorname{Im}(\mathrm{id} - C_*^{+0}(\alpha^*)) \circ \mathrm{pr}_2 : I(C_*^{+0}(\alpha^*)) \oplus I(\mathrm{id} - C_*^{+0}(\alpha^*)) \to C_*$$

of chain complexes is an isomorphism with inverse $(e_{C_*^{+0}(\alpha^*)}, e_{\mathrm{id}-C_*^{+0}(\alpha^*)})$.

Proof. This is an application of Theorem A.2.39.   □

Corollary 1.2.13. *If C is a simplicial object in an additive category* **C** *with images then the image $\operatorname{Im} C_*^{+0}(\alpha^*)$ is a kernel of $\mathrm{id} - C_*^{+0}(\alpha^*)$ and the kernel $\ker C_*^{+0}(\alpha)$ is an image of $\mathrm{id} - C_*^{+0}(\alpha^*)$.*

Example 1.2.14. If **C** $-$ **TopAlg**$_{(\Omega,\mathrm{E})}$ is a pointed additive category of topological algebras of a given type (Ω, E) then the chain complex C_* associated to a simplicial algebra C in **C** is isomorphic to the direct sum $C_*^{+0}(\alpha^*)(C_*) \oplus (\mathrm{id} - C_*^{+0}(\alpha^*))(C_*)$. The second summand is isomorphic to the domain $K(C_*(\alpha^*))$ of the kernel of $C_*^{+0}(\alpha^*)$, so all in all one obtains an isomorphism $C_* \cong C_*^{+0}(\alpha^*)(C_*) \oplus K(C_*^{+0}(\alpha^*))$.

Example 1.2.15. The chain complex M_* associated to a a simplicial topological module M over a topological ring R is isomorphic to the direct sum of its subcomplexes $M_*^{+0}(\alpha^*)(M_*)$ and $(\mathrm{id} - M_*^{+0}(\alpha^*))(M_*)$. The second summand is isomorphic to the domain $K(M_*(\alpha^*))$ of the kernel $\ker M_*^{+0}(\alpha^*)$, so all in all one obtains an isomorphism $M_* \cong M_*^{+0}(\alpha^*)(M_*) \oplus K(M_*^{+0}(\alpha^*))$.

The preceding proposition shows that the chain complex C_* of a simplicial object C in a sufficiently "nice" category contains the image and the kernel of the morphism $C_*^{+0}(\alpha^*)$ as direct summands. This decomposition is functorial in the simplicial objects considered. To make this precise, we denote the functor $\mathbf{C}^{\mathcal{O}^{op}} \to \mathbf{Ch}(\mathbf{C})$, which assigns to each simplicial object C in a pointed Ab-category $\mathbf{C}$ its associated chain complex C_* by $(-)_*$. The endomorphism α^* of the cochain complex $([n], d_n)$ in $\mathcal{O}^{+0}$ induces a natural transformation $(-)_*^{+0}(\alpha^*) : (-)_* \to (-)_*$. Provided that there exists an image functor for the category $\mathbf{C}$ considered, this natural transformation $(-)_*^{+0}(\alpha^*)$ factors through its image:

$$
\begin{array}{ccc}
(-)_* & \xrightarrow{\;(-)_*^{+0}(\alpha^*)\;} & (-)_* \\
& \searrow{\scriptstyle e_{(-)_*^{+0}(\alpha^*)}} & \uparrow{\scriptstyle \mathrm{Im}(-)_*^{+0}(\alpha^*)} \\
& & I(-)^{+0}(\alpha^*)
\end{array}
$$

In this case the natural transformation $\mathrm{id} - C_*^{+0}(\alpha^*)$ has an analogous image factorisation. This can be used to define a generalisation of the normalisation of chain complexes in abelian categories:

Definition 1.2.16. *For any additive category* $\mathbf{C}$ *with functorial images the functor*

$$(N-)_* := I(-)_*^{+0}(\alpha^*) : \mathbf{C}^{\mathcal{O}^{op}} \to \mathbf{Ch}(\mathbf{C})$$

is called the normalised chain complex functor. *The image complex* NC_* *of a simplicial object* C *in* $\mathbf{C}$ *is called the* normalised subcomplex of C_*.

We will show below that this recovers the Moore normalisation functor.

Definition 1.2.17. *For any additive category* $\mathbf{C}$ *with functorial images the functor*

$$(D-)_* := I(-)_*^{+0}(\mathrm{id} - \alpha^*) : \mathbf{C}^{\mathcal{O}^{op}} \to \mathbf{Ch}(\mathbf{C})$$

is called the degenerate chain complex functor. *The image complex* DC_* *for a simplicial object* C *in* $\mathbf{C}$ *is called the* degenerate subcomplex of C_*.

The abstract framework provided above enables us to prove a general normalisation theorem, which especially holds in the category **TopRMod** of topological modules over a topological ring R. Here fore we note that the endomorphism α^* of the cochain complex $([n], d_n)$ in $\mathcal{O}^{+0}$ has the following additional properties:

Proposition 1.2.18. *The morphism* $\alpha^0 : [0] \to [0]$ *is the identity and the morphisms* α^n *satisfy the identities*

$$\alpha^n \epsilon_i = 0 \text{ for } i < n \quad \text{and} \quad \eta_i \alpha^n = 0 \text{ for all } 0 \le i \le n - 1.$$

In addition there exist morphisms $\beta_i^n : [n-1] \to [n]$, $0 \le i \le n-1$ *such that* $\mathrm{id}_{[n]} - \alpha^n = \beta_0^n \eta_0 + \cdots + \beta_{n-1}^n \eta_{n-1}$.

Proof. This is the front to back dual of [Eps66, Proposition 2.3]. (Be aware that the author there uses the opposite category $\mathcal{O}^{op}$.) $\qquad\square$

The normalised chain complex functor $(N-)_*$ and the degenerate chain complex functor $(D-)_*$ then are subfunctors of the functor $(-)_*$ and the morphism α^* of cochain complexes provides a natural transformation $(-)_*^{+0}(\alpha^*) : (-)_*^{+0} \to (-)_*^{+0}$. Recall that the natural transformation $e_{(-)_*^{+0}(\alpha^*)}$ consists of epimorphisms. This natural transformation can be used to decompose the natural transformation $(-)_*^{+0}(\alpha^*)$:

Theorem 1.2.19 (Normalisation of chain complexes). *For every additive category* **C** *with functorial images the natural transformation* $(-)_*^{+0}(\alpha^*)$ *is homotopic to the identity and has kernel* $(D-)_*$ *and image* $(N-)_*$. *Furthermore the sequence*

$$ 0 \to (D-)_* \to (-)_* \xrightarrow{e_{(-)_*^{+0}(\alpha^*)}} (N-)_* \to 0 $$

of natural transformations is split exact. Splittings are given by the the image $\mathrm{Im}(-)_*^{+0}(\alpha^*) : (N-)_* \to (-)_*$ *and the morphism* $\mathrm{id} - (-)_*^{+0}(\alpha^*) : (-)_* \to (D-)_*$. *For every simplicial object* C *in* **C** *the objects* NC_n *of the normalised chain complex* NC_* *are isomorphic to the equalisers*

$$ NC_n \cong \mathrm{Eq}\{0, \partial_0, \dots, \partial_{n-1}\}, $$

where ∂_i *stands for the face morphism* $C(\epsilon_i) : C([n]) \to C([n-1])$. *If* **C** *is a category* **TopAlg**$_{(\Omega, \mathrm{E})}$ *of topological algebras, then the normalised chain complex is given by the Moore normalisation functor. Likewise, the degenerate chain complex is given by*

$$ DC_n \cong K(\mathrm{Coeq}\{\mathrm{Im}\, s_0, \dots, \mathrm{Im}\, s_{n-1}\}) $$

where s_i *stands for the degeneracy morphism* $C(\eta_i) : C([n-1]) \to C([n])$. *If* **C** *is a category* **TopAlg**$_{(\Omega, \mathrm{E})}$ *of topological algebras, then the degenerate chain complex is given by the subcomplex generated by the images of the degeneracy maps.*

Proof. The proofs of [Eps66, Proposition 2.3] and [Eps66, Corollary 2.4] carry over the the general case. The only prerequisite needed is that the category **C** considered is a pointed Ab-category and has all kernels and images (so the normalised and degenerate cochain complexes exist). □

Thus the abstract Definitions 1.2.16 and 1.2.17 of the normalisation transformation and the degenerate chain complex transformation recover in the definitions 1.1.8 and 1.1.50 of the Moore normalised chain complex NM_* and the degenerate chain complex DM_* associated to a simplicial topological module M. But the Normalisation Theorem 1.2.19 furthermore applies to all simplicial objects in sufficiently "nice" categories.

Example 1.2.20. If M is a simplicial topological module over a topological ring R then the sequence

$$ 0 \to DM_* \hookrightarrow M_* \xrightarrow{e_{M_*^{+0}(\alpha^*)}} NM_* \to 0 $$

is split exact. This especially applies to the singular simplicial abelian topological group $S^t(X)$ of a topological space X, which is examined in the next chapter.

As a consequence of the Normalisation Theorem 1.2.19 the homology of the chain complex C_* associated to a simplicial object C in an additive category with functorial images and kernels and the homology of the normalised chain complex are isomorphic:

Corollary 1.2.21. *For every additive category* **C** *with functorial images and kernels the functors* $H \circ (-)_* : \mathcal{O}^{op} \to \mathbf{Ch}(\mathbf{C})$ *and* $H \circ (N-)_* : \mathcal{O}^{op} \to \mathbf{Ch}(\mathbf{C})$ *are naturally isomorphic.*

The fact that the homologies of the proper normalised chain complexes are isomorphic to the corresponding homologies of the unnormalised chain complexes enables us to identify the homology of chain complexes associated to simplicial modules as a homological δ-functor:

Theorem 1.2.22. *The homology $H \circ (N-)_*$ of normalised chain complexes and the homology $H \circ (-)_*$ of associated chain complexes are naturally isomorphic universal homological δ-functors* $\mathbf{TopRMod}^{\mathcal{O}^{op}} \to \mathbf{TopRMod}$, *i.e. each short exact sequence*

$$0 \to M \to M' \to M'' \to 0$$

of simplicial topological R-modules induces a long exact sequence of homology modules

$$\cdots \xrightarrow{\delta_{n+2}} H_{n+1}(M_*) \to H_{n+1}(M'_*) \to H_{n+1}(M''_*) \xrightarrow{\delta_{n+1}} H_n(M_*) \to \cdots ,$$

which depends naturally on the given short exact sequence.

Proof. This is a consequence of the natural isomorphism $H \circ (N-)_* \cong H \circ (-)_*$ and Theorem 1.1.19. □

All the results obtained so far will also be needed in their analogous form for cosimplicial topological R-modules and cochain complexes of topological R-modules. These analogous forms are derived below. First, observe that – similar to the category $\mathbf{TopRMod}$ of topological modules over a topological ring R – for any Ab-category $\mathbf{C}$ there exists a functor $(-)^* : \mathbf{C}^{\mathcal{O}} \to \mathbf{coCh}(\mathbf{C})$ assigning to each cosimplicial object C in $\mathbf{C}$ its associated cochain complex C^*.

Definition 1.2.23. *For any Ab-category $\mathbf{C}$ the functor $(\ )^* : \mathbf{C}^{\mathcal{O}} \to \mathbf{coCh}(\mathbf{C})$ is called the* associated cochain complex functor.

Lemma 1.2.24. *The associated cochain complex functor $(-)^* : \mathbf{C}^{\mathcal{O}} \to \mathbf{coCh}(\mathbf{C})$ is continuous and cocontinuous.*

Proof. The proof is analogous to that of Lemma 1.2.2. □

As in the previous section, that dealt with the category $\mathbf{TopRMod}$, we choose an approach, where the normalised cochain complex NC^* of a cosimplicial object C is a subcomplex of the cochain complex C^* associated to C. This approach is given by applying the abstract framework provided by the cochain transformation α^* in $\mathcal{O}^{+0}$ to covariant functors instead of contravariant functors, but still using images factorisations (instead of switching to coimage factorisations).

Recall that any cosimplicial object C in a pointed Ab-category $\mathbf{C}$ has a unique additive extension $C^{+0} : \mathcal{O}^{+0} \to \mathbf{C}$ preserving the zero object. This extension preserves cochain complexes and thus gives rise to a functor

$$C^{+0*} : \mathbf{coCh}(\mathcal{O}^{+0*}) \to \mathbf{coCh}(\mathbf{C})$$

from the category $\mathbf{coCh}(\mathcal{O}^{+0*})$ of cochain complexes in $\mathcal{O}^{+0}$ to the category $\mathbf{coCh}(\mathbf{C})$ of cochain complexes in $\mathbf{C}$. The cochain transformation α^* so induces a cochain transformation $C^{+0*}(\alpha^*)$ of the cochain complex C^* associated to C. As a consequence of Proposition 1.2.7 we observe:

Corollary 1.2.25. *The endomorphism $C^{+0*}(\alpha^*)$ of the cochain complex C^* associated to a cosimplicial object C in a pointed Ab-category $\mathbf{C}$ is idempotent and chain homotopic to the identity.*

Corollary 1.2.26. *The endomorphism* $\mathrm{id} - C^{+0*}(\alpha^*)$ *of the chain complex* C^* *associated to a cosimplicial object* C *in a pointed Ab-category* $\mathbf{C}$ *is idempotent and chain homotopic to the zero endomorphism.*

If C is a cosimplicial object in a category $\mathbf{C}$ admitting images then the cochain transformation $C^{+0*}(\alpha^*)$ has an image factorisation

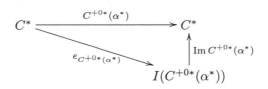

as an epimorphism $e_{C^{+0*}(\alpha^*)}$ followed by the image morphism $\mathrm{Im}\, C^{+0*}(\alpha^*)$ (cf. Lemma A.1.29). The endomorphism $\mathrm{id} - C^{+0*}(\alpha^*)$ of the cochain complex C^* has an analogous factorisation. The cochain complex $I(C^{+0*}(\alpha^*))$ is a subcomplex of the cochain complex C^* by definition. Likewise the domain $I(\mathrm{id} - C^{+0*}(\alpha^*))$ of the image of $\mathrm{id} - C^{+0*}(\alpha^*)$ is a subcomplex of C^*.

Example 1.2.27. If $\mathbf{C} = \mathbf{TopAlg}_{(\Omega, \mathrm{E})}$ is a pointed Ab-category of topological algebras of certain type (Ω, E), then the image $\mathrm{Im}\, C^{+0n}(\alpha^n)$ is the inclusion of the topological subalgebra $C^{+0}(\alpha^n)((C[n]))$ of $C([n])$ and the image $\mathrm{Im}\, C^{+0*}(\alpha^*)$ is the inclusion of the subcomplex $C^{+0*}(\alpha^*)(C^*)$ of topological subalgebras of C^*.

Assuming further that the category $\mathbf{C}$ admits binary coproducts, the image-factorisations of the endomorphisms $C^{+0*}(\alpha^*)$ and $\mathrm{id} - C^{+0*}(\alpha^*)$ of the cochain complex C^* lead to a decomposition of the cochain complex C^*:

Proposition 1.2.28. *For any cosimplicial object* C *in an additive category* $\mathbf{C}$ *with functorial images the morphism*

$$\mathrm{Im}\, C^{+0*}(\alpha^*) \circ \mathrm{pr}_1 + \mathrm{Im}(\mathrm{id} - C^{+0*}(\alpha^*)) \circ \mathrm{pr}_2 : I(C^{+0*}(\alpha^*)) \oplus I(\mathrm{id} - C^{+0*}(\alpha^*)) \to C^*$$

of chain complexes is an isomorphism with inverse $(e_{C^{+0*}(\alpha^*)}, e_{\mathrm{id} - C^{+0*}(\alpha^*)})$.

Proof. This is an application of Theorem A.2.39. □

Corollary 1.2.29. *If* C *is a cosimplicial object in an additive category* $\mathbf{C}$ *with images then the image* $\mathrm{Im}\, C^{+0*}(\alpha^*)$ *is a kernel of* $\mathrm{id} - C^{+0*}(\alpha^*)$ *and the kernel* $\ker C^{+0*}(\alpha)$ *is an image of* $\mathrm{id} - C^{+0*}(\alpha^*)$.

Example 1.2.30. If $\mathbf{C} = \mathbf{TopAlg}_{(\Omega, \mathrm{E})}$ is an additive category of topological algebras of type (Ω, E) then the cochain complex C^* associated to a cosimplicial algebra C of type (Ω, E) is isomorphic to the direct sum $C^{+0*}(\alpha^*)(C^*) \oplus (\mathrm{id} - C^{+0*}(\alpha^*))(C_*)$.

Example 1.2.31. The cochain complex M^* associated to a cosimplicial topological module M over a topological ring R is isomorphic to the direct sum of its subcomplexes $M^{+0*}(\alpha^*)(M^*)$ and $(\mathrm{id} - M^{+0*}(\alpha^*))(M^*)$. The second summand is isomorphic to the domain $K(M^*(\alpha^*))$ of the kernel $\ker M^{+0*}(\alpha^*)$, so all in all one obtains an isomorphism $M^* \cong M^{+0*}(\alpha^*)(M^*) \oplus K(M^{+0*}(\alpha^*))$.

The preceding proposition shows that – similar to chain complexes associated to simplicial objects – the cochain complex C^* associated to a cosimplicial object C in a sufficiently "nice" category contains the image and the kernel of the morphism $C^{+0*}(\alpha^*)$ as direct summands. This decomposition also is functorial in the cosimplicial objects considered: To make this precise, we denote the functor $\mathbf{C}^{\mathcal{O}} \to \mathbf{coCh}(\mathbf{C})$, which assigns to each cosimplicial object C in a pointed Ab-category $\mathbf{C}$ its associated cochain complex C^* by $(-)^*$. The endomorphism α^* of the cochain complex $([n], d_n)$ in $\mathcal{O}^{+0}$ induces a natural transformation $(-)^{+0*}(\alpha^*) : (-)^* \to (-)^*$. Provided that images exist in the category $\mathbf{C}$ considered, this natural transformation $(-)^{+0*}(\alpha^*)$ factors through its image:

$$
\begin{array}{ccc}
(-)^* & \xrightarrow{\;(-)^{+0*}(\alpha^*)\;} & (-)^* \\
& e_{(-)^{+0*}(\alpha^*)} \searrow & \big\uparrow \mathrm{Im}(-)^{+0*}(\alpha^*) \\
& & I(-)^{+0*}(\alpha^*)
\end{array}
$$

In this case the natural transformation $\mathrm{id} - C^{+0*}(\alpha^*)$ has an analogous image factorisation. This leads to a generalisation of the normalisation of cochain complexes in abelian categories:

Definition 1.2.32. *For any additive category $\mathbf{C}$ with functorial images and the functor*

$$
N := I(-)^{+0*}(\alpha^*) : \mathbf{C}^{\mathcal{O}} \to \mathbf{coCh}(\mathbf{C})
$$

is called the normalised cochain complex functor. The image complex NC^* of a simplicial object C in $\mathbf{C}$ is called the normalised subcomplex of C^*.

We will show below that this recovers the Moore normalisation functor for cosimplicial topological modules.

Definition 1.2.33. *For any additive category $\mathbf{C}$ with functorial images the functor*

$$
D := I(-)^{+0*}(\mathrm{id} - \alpha^*) : \mathbf{C}^{\mathcal{O}} \to \mathbf{coCh}(\mathbf{C})
$$

is called the degenerate cochain complex functor. The image complex DC^* is called the degenerate subcomplex of C^*.

The abstract framework provided above enables us to prove a general normalisation theorem for cochain complexes, which especially holds in the category $\mathbf{TopRMod}$ of topological modules over a topological ring R. The normalised cochain complex functor $(N-)^*$ and the degenerate cochain complex functor $(D-)^*$ are subfunctors of the functor $(-)^*$ and the natural transformation $(-)^{+0*}(\alpha^*) : (-)^{+0*} \to (-)^{+0*}$ can be used to decompose the functor $(-)^{+0*}(\alpha^*)$ into $(N-)^*$ and $(D-)^*$:

Theorem 1.2.34 (Normalisation of cochain complexes). *For every additive category $\mathbf{C}$ with functorial images the natural transformation $(-)^{+0*}(\alpha^*)$ is homotopic to the identity and has kernel $(D-)^*$ and image $(N-)^*$. Furthermore the sequence*

$$
0 \to (D-)^* \to (-)^* \xrightarrow{\;e_{(-)^{+0*}(\alpha^*)}\;} (N-)^* \to 0
$$

of natural transformations is split exact. Splittings are given by the the image $\mathrm{Im}(-)^{+0*}(\alpha^*) : (N-)^* \to (-)^*$ and the morphism $\mathrm{id} - (-)^{+0*}(\alpha^*) : (-)^* \to (D-)^*$. For every cosimplicial object C in $\mathbf{C}$ the objects NC^n of the normalised chain complex NC^* are isomorphic to domains of the equalisers

$$
NC^n \cong \mathrm{dom}(\mathrm{Eq}\{0, s_0, \ldots, s_{n-1}\}),
$$

where s_i stands for the codegeneracy morphism $C(\eta_i) : C([n]) \to C([n-1])$. If $\mathbf{C}$ is a category $\mathbf{TopAlg}_{(\Omega, \mathrm{E})}$ of topological algebras, then the normalised cochain complex is given by the Moore normalisation functor. Likewise, the degenerate chain complex is given by

$$DC^n \cong K(\mathrm{Coeq}\{\mathrm{Im}\,\partial_0, \ldots, \mathrm{Im}\,\partial_{n-1}\})$$

where ∂_i stands for the coface morphism $C(\epsilon_i) : C([n-1]) \to C([n])$. If $\mathbf{C}$ is a category $\mathbf{TopAlg}_{(\Omega, \mathrm{E})}$ of topological algebras, then the degenerate chain complex is given by the subcomplex generated by the images of the degeneracy maps.

Proof. The proofs of [Eps66, Proposition 2.3] and [Eps66, Corollary 2.4] also work for cosimplicial objects. The only prerequisite needed is that the category $\mathbf{C}$ considered is an additive category and has all kernels and images (so the normalised and degenerate chain complexes exist).   □

Thus the abstract Definitions 1.2.32 and 1.2.33 of the normalisation transformation and the degenerate chain complex transformation recover in the definitions 1.1.28 and 1.1.49 of the Moore normalised chain complex NM^* and the degenerate chain complex DM_* associated to a simplicial topological module M. But the Normalisation Theorem 1.2.34 moreover applies to all cosimplicial objects in sufficiently "nice" categories.

Example 1.2.35. If M is a cosimplicial topological module over a topological ring R then the sequence

$$0 \to DM^* \hookrightarrow M^* \xrightarrow{e_{M+0*}(\alpha*)} NM^* \to 0$$

of cochain complexes is split exact. This is especially useful, because it enables one to w.l.o.g. work with normalised Alexander-Spanier- or group cochain complexes only.

As a consequence of the Normalisation Theorem for cochain complexes the cohomology of the cochain complex M^* associated to a cosimplicial topological module M and the cohomology of the normalised cochain complex are isomorphic:

Corollary 1.2.36. *For every additive category $\mathbf{C}$ with functorial images and kernels the functors $H \circ (-)^* : \mathcal{O} \to \mathbf{coCh}(\mathbf{C})$ and $H \circ (N-)^* : \mathcal{O} \to \mathbf{coCh}(\mathbf{C})$ are naturally isomorphic.*

The additional fact that normalised cochain complexes of topological modules are proper enables us to identify the cohomology of cochain complexes associated to cosimplicial modules as a cohomological δ-functor:

Theorem 1.2.37. *The cohomology of normalised cochain complexes $H \circ (N-)^*$ and the cohomology of associated cochain complexes $H \circ (-)^*$ are naturally isomorphic universal cohomological δ-functors $\mathbf{TopRMod}^{\mathcal{O}} \to \mathbf{TopRMod}$, i.e. each short exact sequence*

$$0 \to M \to M' \to M'' \to 0$$

of cosimplicial topological R-modules induces a long exact sequence of homology modules

$$\cdots \xrightarrow{\delta_{n-1}} H^n(M^*) \to H^n(M'^*) \to H_n(M''^*) \xrightarrow{\delta_n} H^{n+1}(M^*) \to \cdots,$$

which depends naturally on the given short exact sequence.

Proof. This is a consequence of the natural isomorphism $H \circ (N-)^* \cong H \circ (-)^*$ and the Normalisation Theorem 1.2.34.   □

The Normalisation Theorems 1.2.19 and 1.2.34 allow us to treat (co)chain complexes associated to (co)simplicial objects as if they were proper, replacing them by their normalised subcomplexes. This is essential in verifying the properties of many topological (co)homology theories and will be used extensively in Chapter 2.

1.3 Simplicial Objects in Homological Categories

In this section we show that (co)chain complexes associated to (co)simplicial objects in homological categories (such as **TopRMod**) are proper. This includes the normalised and degenerate subcomplexes as special cases and extends the result in [EVdL04, Theorem 3.6] to unnormalised (co)chain complexes in even more general categories. It furthermore implies that the normalisation of (co)chain complexes associated to (co)simplicial objects in homological categories is not necessary to obtain a (co)homological δ-functor.

The normalised chain complex associated to a simplicial object is interesting in its own right. It simplifies many calculations by enabling one to neglect non-normalised chains (cf. Theorem 1.2.19). However, for homological categories, it is even possible to show that all chain complexes associated to simplicial objects are proper. Here the normalisation of the chain complexes is not necessary to obtain a homological δ-functor. We begin the proof thereof by some preparatory observations on split epimorphisms.

Lemma 1.3.1. *Products and coproducts of split epimorphisms are split epimorphisms.*

Proof. Let e_i, $i \in I$ be split epimorphisms with right inverse m_i respectively. Then the product $\prod_{i \in I} m_i$ is a right inverse to the product $\prod_{i \in I} e_i$, hence the latter one is a split epimorphism. Similarly the coproduct $\coprod_{i \in I} m_i$ is a right inverse to the coproduct $\coprod_{i \in I} e_i$, which therefore is a split epimorphism. $\square$

Example 1.3.2. The face morphisms $C(\epsilon_i) : C([n+1]) \to C([n])$ of a simplicial object C in a category $\mathbf{C}$ are split epimorphisms, because the simplicial identity $\eta_i \epsilon_i = \mathrm{id}_{[n]}$ implies the equalities $C(\epsilon_i)C(\eta_i) = C(\eta_i \epsilon_i) = C(\mathrm{id}_{[n]}) = \mathrm{id}_{C([n])}$. Therefore their product and their coproduct

$$\prod_{i=0}^{n+1} C(\epsilon_i) : \prod_{i=0}^{n+1} C([n+1]) \to \prod_{i=0}^{n+1} C([n]),$$

$$\coprod_{i=0}^{n+1} C(\epsilon_i) : \coprod_{i=0}^{n+1} C([n+1]) \to \coprod_{i=0}^{n+1} C([n])$$

also are a split epimorphisms. Similarly the codegeneracy morphisms of a cosimplicial object in $\mathbf{C}$ are split epimorphisms and so are their products and coproducts.

Proposition 1.3.3. *Split epimorphisms are regular epimorphisms.*

Proof. See [BB04, Proposition A.4.11] for a proof. $\square$

Corollary 1.3.4. *Products and coproducts of split epimorphisms are regular epimorphisms.*

Example 1.3.5. If C is a simplicial object in an Ab-category $\mathbf{C}$, then all the morphisms $C(\epsilon_i)$ are split epimorphisms and so are the morphisms $(-1)^i C(\epsilon_i)$. Furthermore finite products coincide with finite coproducts and are called finite direct sums (cf. A.2.9, A.2.11). Therefore their direct sums (if they exist)

$$\bigoplus_{i=0}^{n+1} (-1)^i C(\epsilon_i) : \bigoplus_{i=0}^{n+1} C([n+1]) \to \bigoplus_{i=0}^{n+1} C([n])$$

also are a split epimorphisms, which are regular by the preceding Corollary. Moreover in additive categories (such as **TopRMod**), these finite direct sums always exist.

Since additive categories have all finite products (and thus direct sums) there exists an $(n+1)$-fold direct sum $\bigoplus_{i=0}^{n} C$ for each object C in an additive category $\mathbf{C}$. Let $i_0, \ldots, i_n$ denote the inclusions of the $n+1$ summands and $p_0, \ldots, p_n$ denote the projections onto these summands respectively. By the definition of the direct sum, the sum $\sum_{i=0}^{n} p_i : \bigoplus_{i=0}^{n} C \to C$ of the projections is the unique morphism which restricts to the identity on every summand C of the direct product $\bigoplus_{i=0}^{n} C$, i.e. it is the unique morphism $\bigoplus_{i=0}^{n} C \to C$ making the following diagram commutative:

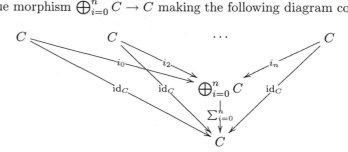

This morphism $\sum_{i=0}^{n} p_i$ is called the *summation on* $\bigoplus_{i=0}^{n} C$. For each $n \in \mathbb{N}$ the summations on $(n+1)$-fold products for different objects form a natural transformation

$$S_n := \sum_{i=0}^{n} p_i : \left(\bigoplus_{i=0}^{n} \right) \circ D_{n+1} \to \mathrm{id}_{\mathbf{C}},$$

where D_{n+1} denotes the diagonal transformation $\mathbf{C} \to \mathbf{C}^{n+1}$. The inclusions $i_0, \ldots, i_n$ into the direct sum $\bigoplus_{i=0}^{n} C$ are also natural in the objects C. Moreover each of the morphisms $i_0, \ldots, i_n$ in the cocone of the direct sum is a right inverse of the summation on $\bigoplus_{i=0}^{n} C$, so we note:

Lemma 1.3.6. *The natural transformations S_n are split epimorphisms in the functor category $\mathbf{C}^{\mathbf{C}}$.*

These observations suffice to decompose the boundary morphisms of simplicial objects in homological Ab-categories. Let C be a simplicial object in an additive category $\mathbf{C}$ and consider the diagonal morphism $D_{n+1,C} : C([n]) \to \bigoplus_{i=0}^{n} C([n])$. The boundary morphism $d_{n+1} : C([n+1]) \to C([n])$ of the cochain complex associated to C can be decomposed into the monomorphism D_{n+2} followed by a direct sum of face morphisms $C(\epsilon_i)$ and a summation morphism:

$$d_{n+1} = \sum_{i=0}^{n+1} (-1)^i C(\epsilon_i) = \left(\sum_{i=0}^{n+1} p_i \right) \circ \underbrace{\left(\bigoplus_{i=0}^{n+1} (-1)^i C(\epsilon_i) \right)}_{=:e_{n+1,C}} \circ D_{n+2,C([n+1])} \quad (1.4)$$

$$= S_{n+1,C([n])} e_{n+1,C} D_{n+2,C([n+1])}$$

In particular the last two morphisms $S_{n+1,C([n])}$ and $e_{n+1,C}$ are split epimorphisms by Lemma 1.3.6 and Lemma 1.3.1 respectively. Because compositions of split epimorphisms are split epimorphisms, the boundary morphism d_{n+1} is a composition of the monomorphism $D_{n+1,C}$ and a split epimorphism. Each of the projections p_i from $\bigoplus_{i=0}^{n+1} C$ onto C is a left inverse to the diagonal morphism $D_{n+1,C}$, so the latter is a split monomorphism. So every boundary morphism is a composition of a split monomorphism and a split epimorphism. We would like to observe that the split monomorphism $D_{n+1,C}$ embeds $C([n+1])$ as a proper subobject of $\bigoplus_{i=0}^{n+1} C([n+1])$, so we can apply the Noether Isomorphism Theorem. Here fore we need to show that $D_{n+1;C([n+1])}$ is a proper monomorphism. We start by noting that the identity $\mathrm{id}_{C([n+1])}$ of $C([n+1])$ is a coimage of $D_{n+1,C([n+1])}$:

Lemma 1.3.7. *For every monomorphism $m : C \to C'$ the identity $\mathrm{id}_C : C \to C$ is a coimage of m.*

Proof. Let $m : C \to C'$ be a monomorphism in an arbitrary category **C**. By the definition of monomorphisms, the equality $mf = mg$ for any pair of morphisms $f, g : C'' \to C$ implies $f = g$. Therefore the commutative diagram

$$
\begin{array}{ccc}
C & \xrightarrow{\ \mathrm{id}_C\ } & C \\
{\scriptstyle \mathrm{id}_C}\downarrow & & \downarrow{\scriptstyle f} \\
C & \xrightarrow[f]{} & C'
\end{array}
$$

is a pullback square, i.e. the fibre product $C \times_m C = C$ exists, is isomorphic to C and $(\mathrm{id}_C, \mathrm{id}_C)$ is a kernel pair of f. Thus the identity id_C is a coequaliser of the kernel pair $(\mathrm{id}_C, \mathrm{id}_C)$, hence a coimage of m. □

Lemma 1.3.8. *Split monomorphisms in pointed Ab-categories are proper.*

Proof. Let $m : C \to C'$ be a split monomorphism in an Ab-category **C**. By the definition of split monomorphisms, there exists a left inverse $e : C' \to C$ of m. Furthermore the identity of C is a coimage of m, i.e. $m = m \circ \mathrm{id}_C$ is a coimage factorisation of m. We assert that m also is a kernel, namely the kernel of the endomorphism $(me - \mathrm{id}_{C'})$ of C'. We prove this by showing that m has the required universal property. If $f : C'' \to C$ is a morphism satisfying $(me - \mathrm{id}_{C'})f = 0$ then this implies $f = (me)f = m(ef)$, so the morphism f factors through m. Since m is a monomorphism, this factorisation is unique. Being a kernel, the monomorphism m has to be proper. □

This especially implies that the diagonal morphism $D_{n+1,C([n])}$ in the factorisation $d_{n+1} = S_{n+1,C([n])}e_{n+1,C}D_{n+1,C([n+1])}$ of the boundary morphism d_{n+1} is a proper monomorphism. The decomposition of the boundary morphisms into a regular epimorphism followed by proper monomorphism allows us to apply the Noether Isomorphism Theorems:

Theorem 1.3.9. *The chain complex associated to a simplicial object in a homological Ab-category is proper.*

Proof. Let C be a simplicial object in an additive homological category **C**. For each $n \in \mathbb{N}$ the boundary morphism $d_{n+1} : C([n+1]) \to C([n])$ is the composition of the proper split monomorphism $D_{n+1,C([n+1])}$ and the regular split epimorphism

$$
S_{n+1,C([n])}e_{n+1,C} = \left(\sum_{i=0}^{n+1} p_i \right) \circ \left(\bigoplus_{i=0}^{n+1} (-1)^i C(\epsilon_i) \right) : \bigoplus_{i=0}^{n+1} C([n+1]) \to C([n])
$$

Since every morphisms in a homological category has a kernel, one can consider the kernel $\ker e_{n+1,C} : K(e_{n+1,C}) \to \bigoplus_{i=0}^{n+1} C([n+1])$ of the split epimorphism $e_{n+1,C}$ and the kernel $\ker d_{n+1} : K(d_{n+1}) \to C([n+1])$ of the boundary morphism d_{n+1}. All in all one obtains a commutative diagram

$$
\begin{array}{ccccc}
K(d_{n+1}) & \xrightarrow{\ \ker d_{n+1}\ } & C([n+1]) & \xrightarrow{\ \mathrm{CoIm}\, d_n\ } & CI(d_n) \\
\downarrow & & {\scriptstyle D_{n+1,C([n+1])}}\Big\downarrow & & \Big\downarrow{\scriptstyle m_{d_{n+1}}} \\
0 \longrightarrow K(e_{n+1}) & \xrightarrow[\ \ker e_{n+1,C}\ ]{} & \bigoplus_{i=0}^{n+1} C([n+1]) & \xrightarrow[e_{n+1,C}]{} & C([n]) \longrightarrow 0
\end{array} \quad ,
$$

where the lower row is exact and the left hand square is a pullback square (by Lemma A.2.24). Because the monomorphism $D_{n+1,C([n+1])}$ is proper, the Noether Isomorphism Theorem implies that the monomorphism $m_{d_{n+1}}$ embeds $CI(d_{n+1})$ as a proper subobject of $C([n])$, hence it is proper as well. Thus the boundary morphism d_{n+1} is proper. □

Corollary 1.3.10. *The chain complex M_* associated to a simplicial topological module M over a topological ring R is proper.*

Proof. The category **TopRMod** is homological. □

Theorem 1.3.11. *The homology of associated chain complexes $H \circ (-)_*$ in a homological category $\mathbf{C}$ is a homological δ-functor $\mathbf{C}^{\mathcal{O}^{op}} \to \mathbf{C}$.*

Corollary 1.3.12. *The homology of associated chain complexes in **TopRMod** is a homological δ-functor $\mathbf{TopRMod}^{\mathcal{O}^{op}} \to \mathbf{TopRMod}$.*

It remains to observe that the homological δ-functor $H \circ (-)_* : \mathbf{C}^{\mathcal{O}^{op}} \to \mathbf{C}$ is a universal one. This will follow at once, if it is coeffaceable.

Lemma 1.3.13. *For any homological Ab-category $\mathbf{C}$ the functor $H \circ (-)_*$ on $\mathbf{C}^{\mathcal{O}^{op}}$ is coeffaceable.*

Proof. It is to prove that to each simplicial object C in a homological Ab-category there exists an epimorphism $e : P \to C$ such that $H^*(e) = 0$. For each simplicial object C in a homological Ab-category $\mathbf{C}$ one can consider the right shifted simplicial object CT, where T is the right shift functor on $\mathcal{O}$ (cf. F.1.4). The simplicial object C is the codomain of the epimorphism $C(\epsilon_0) : CT \to C$. Since the right shifted simplicial object is homotopic to the constant simplicial object C_0, the homology of its associated chain complex is trivial. Thus setting $P = CT$, $e = C(\epsilon_0) : P \to C$ shows that $H^*(-)^*$ is coeffaceable. □

Theorem 1.3.14. *The homology of associated chain complexes $H \circ (-)_*$ for a homological category $\mathbf{C}$ is a universal homological δ-functor $\mathbf{C}^{\mathcal{O}^{op}} \to \mathbf{C}$.*

Proof. The homology of associated chain complexes functor is a homological δ-functor $H(-)_* : \mathbf{C}^{\mathcal{O}^{op}} \to \mathbf{C}$. Effaceable (additive) homological δ-functors are universal. (The proof of [Wei94, Theorem 2.4.7] is valid for any homological category, if one replaces the epimorphisms from projective objects by the epimorphisms guaranteed to exist by the coeffaceability of $H(-)_*$.) Thus the functor $H_*(-)_*$ is a universal homological δ-functor. □

Theorem 1.3.15. *The homology of associated chain complexes in **TopRMod** is a universal homological δ-functor $\mathbf{TopRMod}^{\mathcal{O}^{op}} \to \mathbf{TopRMod}$.*

Remark 1.3.16. It is also possible to prove an analogous result for cosimplicial objects in homological categories. This is a generalisation of the Theorems 1.1.41 and 1.1.45 for cosimplicial topological modules. The proof is completely analogous to the derivation of those Theorems. All the properties used therein are those of homological categories.

1.4 Tensor Products in TopRMod

The last section of this chapter covers the properties of (binary) tensor products in the category **TopRMod**. Many constructions in **TopRMod** given below are topologised versions of the underlying algebraic constructions. Therefore we will frequently use forgetful functors; these forgetful functors will always be denoted by U. If the domain of such forgetful functors is not evident from the context, they will be specified. The categories **TopRMod** and **TopAb** are regarded as subcategories of the category **Top** of topological spaces and the forgetful functors **TopRMod** $\to$ **Top** and **TopAb** $\to$ **Top** are omitted from notation.

We start by recalling the construction of binary tensor products in the category **TopRMod** of left modules over a topological ring R.

Definition 1.4.1. *A tensor product of two R-modules M_1 and M_2 is a biadditive continuous map $\otimes : M_1 \times M_2 \to M_1 \otimes M_2$ into an R-module $M_1 \otimes M_2$ such that for each biadditive continuous map $\omega : M_1 \times M_2 \to M$ into an R-module M there exists a unique homomorphism $f : M_1 \otimes M_2 \to M$ of topological R-modules satisfying $\omega = f \circ \otimes$.*

A tensor product of two topological modules M_1 and M_2 always exists. It can be constructed analogously to the discrete case: Let M_1 and M_2 be topological R-modules. Consider free topological R-module $\mathrm{F}_{\textbf{TopRMod}}(M_1 \times M_2)$ on the product space $M_1 \times M_2$ and let N be the submodule of $\mathrm{F}_{\textbf{TopRMod}}(M_1 \times M_2)$ generated by all elements of the form $(g + g', h) - (g, h) - (g', h)$, $(g, h + h') - (g, h) - (g, h')$, $r(g, h) - (rg, h)$ and $r(g, h) - (g, rh)$. The continuous bilinear map

$$\otimes : M_1 \times M_2 \to M_1 \otimes M_2 := \mathrm{F}_{\textbf{TopRMod}}(M_1 \times M_2)/N, \quad \otimes(g, h) = (g, h) + N$$

satisfies the universal property of a tensor product of R-modules. In particular, the underlying module $U(M_1 \otimes M_2)$ of the tensor product coincides with the tensor product $UM_1 \otimes UM_2$ in the category **RMod** of modules over the ring UR underlying R. As usual we denote an element $(g, h) + N$ by $g \otimes h$. All in all we have observed:

Lemma 1.4.2. *The map $\otimes : M_1 \times M_2 \to M_1 \otimes M_2$ is a tensor product in the category **TopRMod** of topological R-modules. Its underlying map $U\otimes$ is the tensor product of the underlying modules UM_1 and UM_2 in **RMod**.*

Theorem 1.4.3. *The free topological R-module functor $\mathrm{F}_{\textbf{TopRMod}}$ preserves quotients, i.e. for every quotient map $q : X \twoheadrightarrow Y$ of topological spaces the induced homomorphism $\mathrm{F}_{\textbf{TopRMod}}(q) : \mathrm{F}_{\textbf{TopRMod}}X \to \mathrm{F}_{\textbf{TopRMod}}Y$ is a quotient homomorphism.*

Proof. The proof uses the facts that quotient maps in **Top** are colimits and that the free topological R-module functor $\mathrm{F}_{\textbf{TopRMod}}$ preserves colimits. Let $q : X \twoheadrightarrow Y$ be a quotient map in **Top**. Consider the singleton spaces $\{y\}$ for all $y \in Y$ and the continuous maps

$$h_{x,y} : \{y\} \to X, \quad h_{x,y}(y) = x$$

for all pairs (x, y) such that x is contained in the fibre $q^{-1}(y)$ of y. These maps $h_{x,y}$ form a diagram, which we denote by D. (Strictly speaking D is a functor from a suitable category into **Top**). The quotient map $q : X \twoheadrightarrow Y$ identifies two points x and x' exactly if there exists a singleton space $\{y\}$ and two maps $h_{x,y}$ and $h_{x',x}$ such that $h_{x,y}(y) = x$ and $h_{x',y}(y) = x'$. Therefore the quotient space Y is the colimit of the diagram D and the quotient map $q : X \twoheadrightarrow Y$ is a map of the colimiting cocone under the diagram D. Since the free topological R-module functor $\mathrm{F}_{\textbf{TopRMod}}$ preserves colimits, the free topological R-module $\mathrm{F}_{\textbf{TopRMod}}Y$ is the colimit of the diagram $\mathrm{F}_{\textbf{TopRMod}} \circ D$. Every free topological R-module $\mathrm{F}_{\textbf{TopRMod}}\{y\}$ on a singleton space $\{y\}$ is isomorphic to the topological ring R regarded as a left R-module. The colimit topological R-module $\mathrm{F}_{\textbf{TopRMod}}Y$ is the topological R-module obtained by identifying for each $x \in Y$ and each $r \in R$ all points $\mathrm{F}_{\textbf{TopRMod}}(h_{x,y})(r)$ for all maps $h_{x,y}$ in the diagram D. This is to say it is the factor module obtained by factoring out the submodule generated by all elements of the form $F_{\mathbf{TopRMod}}(h_{x,y})(r) - F_{\mathbf{TopRMod}}(h_{x',y})(r)$. So $F_{\mathbf{TopRMod}}Y$ is a quotient of the free topological R-module $F_{\mathbf{TopRMod}}X$ and the map $F_{\mathbf{TopRMod}}(q)$ is the quotient map.   □

Corollary 1.4.4. *The free abelian topological group functor* $F_{\mathbf{TopAb}}$ *preserves quotients, i.e. for every quotient map* $q : X \twoheadrightarrow Y$ *of topological spaces the induced homomorphism* $F_{\mathbf{TopAb}}(q) : F_{\mathbf{TopAb}}X \to F_{\mathbf{TopAb}}Y$ *is a quotient homomorphism.*

Proof. This is the version of Theorem 1.4.3 for the discrete ring $\mathbb{Z}$ of integers.   □

Recall that quotient homomorphisms of topological groups are open maps. Any finite product of quotient homomorphisms of topological groups is a product of surjective open maps, hence open and surjective and thus a quotient homomorphism as well. This fact is needed in the next proof:

Proposition 1.4.5. *For any topological module* M *over a topological ring* R, *the endofunctor* $M \otimes -$: **TopRMod** $\to$ **TopRMod** *preserves quotients, i.e. if the homomorphism* $q : M' \to M''$ *is a quotient map then* $\mathrm{id}_M \otimes q$ *is a quotient homomorphism.*

Proof. Let M, M' and M'' be topological modules over a topological unital ring R and $q : M' \to M''$ be a quotient homomorphism. The product homomorphism $\mathrm{id}_M \times q$ then also is a quotient homomorphism. Consider the commutative diagram

$$
\begin{array}{ccc}
F_{\mathbf{TopRMod}}(M \times M') & \xrightarrow{\ F_{\mathbf{TopRMod}}(\mathrm{id}_M \times q)\ } & F_{\mathbf{TopRMod}}(M \times M'') \\
\downarrow & & \downarrow \\
M \otimes M' & \xrightarrow{\ \ \ \ \ \ \ \ \ \mathrm{id}_M \otimes q \ \ \ \ \ \ \ \ \ } & M \otimes M''
\end{array}
$$

of topological R-modules. The homomorphism $\mathrm{id}_M \times q$ is a quotient homomorphism of R-modules. Therefore the upper horizontal arrow $F_{\mathbf{TopRMod}}(\mathrm{id}_M \times q)$ is a quotient map by Theorem 1.4.3. The vertical arrows are the quotient homomorphisms from the construction of tensor products as quotients of free topological R-modules. Therefore the lower horizontal homomorphism $\mathrm{id}_M \otimes q$ is a quotient map as well.   □

Corollary 1.4.6. *Binary tensor products of quotient homomorphisms are quotient homomorphisms.*

Proof. Let $q_1 : M_1 \to M_1'$ and $q_2 : M_2 \to M_2'$ quotient homomorphisms of topological R-modules. By Proposition 1.4.5 the tensor product maps $\mathrm{id}_{M_1} \otimes q_2$ and $q_1 \otimes \mathrm{id}_{M_2'}$ both are quotient homomorphisms. The composition $q_1 \otimes q_2 = (q_1 \otimes \mathrm{id}_{M_2'})(\mathrm{id}_{M_1} \otimes q_2)$ is a composition of quotient homomorphisms, hence a quotient homomorphism.   □

Lemma 1.4.7. *For any topological module* M *over a topological ring* R *the endofunctor* $M \otimes -$: **TopRMod** $\to$ **TopRMod** *preserves cokernels.*

Proof. Let M be topological R-module, $f : N \to N'$ be a homomorphism of topological R-modules and let $U :$ **TopRMod** $\to$ **RMod** denote the forgetful functor. The cokernel of f is the quotient of N' by the image $f(N)$ of f. It is a quotient map $q : N' \to N''$ such that the sequence

$$
0 \to f(N) \xrightarrow{\ i\ } N' \xrightarrow{\ q\ } N'' \to 0
$$

of topological R-modules is exact. Here i denotes the inclusion of the image $f(N)$ as a submodule in N'. Tensoring the above sequence of topological R-modules with the topological R-module M yields another sequence

$$M \otimes f(N) \xrightarrow{\mathrm{id}_M \otimes i} M \otimes N' \xrightarrow{\mathrm{id}_M \otimes q} A \otimes N'' \to 0$$

of topological modules, which might not be exact. The images of the homomorphisms $\mathrm{id}_M \otimes i$ and $\mathrm{id}_M \otimes f$ coincide, so if $\mathrm{id}_M \otimes q$ is a cokernel of $\mathrm{id}_M \otimes i$ it also is a cokernel of $\mathrm{id}_M \otimes f$. The homomorphism $\mathrm{id}_M \otimes q$ is a quotient map by Proposition 1.4.5. Since the underlying sequence of modules over the ring UR is exact at $U(M \times N')$, (the functor $UM \otimes - : (\mathbf{U}\mathbf{R})\mathbf{Mod} \to (\mathbf{U}\mathbf{R})\mathbf{Mod}$ is right exact), the homomorphism $\mathrm{id}_M \otimes q$ is the quotient of $M \otimes N'$ by the image of $\mathrm{id}_M \otimes i$. Thus it is a cokernel of $\mathrm{id}_M \otimes i$ and a cokernel of $\mathrm{id}_M \otimes f$. □

For the discrete ring $R = \mathbb{Z}$ of integers this implies that any endofunctor of **TopAb** of the form $A \otimes - :$ **TopAb** $\to$ **TopAb** preserves cokernels. All in all these observations can be summarised as follows:

In the abelian category **RMod** of modules over a ring R tensoring with a free module M is an exact endofunctor $M \otimes - :$ **RMod** $\to$ **RMod**. The reason for this lies in the fact that free modules are projective, hence flat. This cannot be generalised to the non-abelian categories **TopRMod** or **TopAb**. Analogously to the discrete case one first observes:

Lemma 1.4.8. *A projective topological module M in* **TopRMod** *is a direct summand of a free topological R-module.*

Proof. The classical proof for the category **RMod** of modules carries over. □

Unfortunately the converse implication is in general neither true in the category **TopRMod** of topological modules over a unital topological ring R in general nor in the category **TopAb** of abelian topological groups. Here the situation is as follows: Let $M = \mathrm{F}_{\mathbf{TopAb}}X$ be a free abelian topological group on a completely regular topological space X. Suppose $f : M \to N'$ is a homomorphism of abelian topological groups and $g : N_d \to N'$ is a homomorphism from a discrete abelian topological group N_d onto the abelian topological group N':

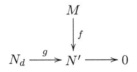

The restriction $f_{|X} : X \to N'$ of f to the subspace X of $\mathrm{F}_{\mathbf{TopAb}}X$ is a continuous function into the topological space N'. If there exists a homomorphism $h : M \to N_d$ satisfying $f = g \circ h$ then the restriction $h_{|X}$ of h to the subspace X of $\mathrm{F}_{\mathbf{TopAb}}X$ is a lift of the continuous function $f_{|X}$. Moreover such lifts of continuous functions need to always exist for $\mathrm{F}_{\mathbf{TopAb}}X$ to be projective:

Lemma 1.4.9. *A free abelian topological abelian group $\mathrm{F}_{\mathbf{TopAb}}X$ on a topological space X is projective if and only if it is discrete.*

Proof. Let $\mathrm{F}_{\mathbf{TopAb}}X$ be a free abelian topological abelian group on a topological space X. Denote the underlying abelian group equipped with the discrete topology by N. Then the natural continuous bijective homomorphism $N \to \mathrm{F}_{\mathbf{TopAb}}X$ is an epimorphism. If the abelian topological group $\mathrm{F}_{\mathbf{TopAb}}X$ is projective, the the identity of $\mathrm{F}_{\mathbf{TopAb}}X$ can be lifted to N. Thus the group $\mathrm{F}_{\mathbf{TopAb}}X$ itself has to be discrete. Conversely if $\mathrm{F}_{\mathbf{TopAb}}X$ is discrete then it certainly has the universal property, because any homomorphism $\mathrm{F}_{\mathbf{TopAb}}N$ of abelian groups into an abelian topological group N is continuous. □

Theorem 1.4.10. *A free abelian topological abelian group $\mathrm{F}_{\mathbf{TopAb}}X$ on a topological space X is projective if and only if X is discrete.*

Proof. By the preceding Lemma a free abelian topological abelian group $F_{\mathbf{TopAb}}X$ on a topological space X is projective if and only if it is discrete. The free abelian topological group $F_{\mathbf{TopAb}}X$ contains the regularisation $\mathrm{Cr}X$ (the image of X under the reflector $\mathrm{Cr} : \mathbf{Top} \to \mathbf{Cr}$) as a subspace. The topology of the regularisation $\mathrm{Cr}X$ is coarser than that of X. Therefore, if $F_{\mathbf{TopAb}}X$ is discrete, then so is its subspace $\mathrm{Cr}X$ and thus space X itself. Conversely the free abelian group on a discrete space X is discrete, hence projective. □

This line of reasoning can be generalised to the category **TopRMod** of topological modules over a topological ring R. Consider a free topological R-nodule $M = F_{\mathbf{TopRMod}}X$ on a completely regular space X. If the base ring R is not discrete (like $\mathbb{Z}$) then the discrete topology on M is not an R-module topology. Therefore a replacement for the discrete abelian group N_d in the above arguments is needed. Let X_d denote the discrete topological space on the set of points of X. The free topological module $F_{\mathbf{TopRMod}}X_d$ will serve as a substitute for the discrete abelian group N_d.

Theorem 1.4.11. *A free topological module* $F_{\mathbf{TopRMod}}X$ *on a topological space* X *is projective if and only if* X *is discrete.*

Proof. Let $F_{\mathbf{TopRMod}}X$ be a free topological module on a topological space X and let $i : X \to F_{\mathbf{TopRMod}}X$ be the natural injection of X into $F_{\mathbf{TopRMod}}X$ as a generating space. Denote the underlying set of X equipped with the discrete topology by X_d. The natural continuous bijection $X_d \to X$ induces an epimorphism $f : F_{\mathbf{TopRMod}}X_d \to F_{\mathbf{TopRMod}}X$ of topological modules. This epimorphism restricts to a continuous bijection $f_{|X_d} : X_d \to i(X)$. If the topological module $F_{\mathbf{TopRMod}}X$ is projective, the identity of $F_{\mathbf{TopRMod}}X$ can be lifted to $F_{\mathbf{TopRMod}}X_d$, i.e. there exists a homomorphism $g : F_{\mathbf{TopRMod}}X \to F_{\mathbf{TopRMod}}X_d$ which is a right inverse of f. The right inverse g then restricts to a right inverse $g_{|i(X)} : i(X) \to X_d$ of the continuous bijection $f_{|X_d} : X \to i(X)$. The discreteness of the space X_d then implies the discreteness of $i(X)$. So the topological space X itself has to be discrete. Conversely if X is discrete then the free topological R-module $F_{\mathbf{TopRMod}}X$ certainly has the universal property, because any homomorphism $F_{\mathbf{TopRMod}}X \to N$ of R-modules into a topological R-module N is continuous. □

So many free topological modules M are not projective. Nevertheless the endofunctor $M \otimes - : \mathbf{TopRMod} \to \mathbf{TopRMod}$ may preserve the exactness of exact sequences. A necessary condition here fore is that an inclusion $N \hookrightarrow N'$ of a submodule N of N' leads to an inclusion $M \otimes N \hookrightarrow M \otimes N'$ of a topological submodule. While the underlying map of (discrete) modules is injective (tensoring with a free module is an exact endofunctor of $(\mathbf{UR})\mathbf{Mod}$), the homomorphism $M \otimes N \hookrightarrow M \otimes N'$ may not be an embedding. For $\mathbb{Z}$-modules (i.e. abelian topological groups) this can be guaranteed by requiring the homomorphism $F_{\mathbf{TopAb}}(M \times N) \to F_{\mathbf{TopAb}}(M \times N')$ to be an embedding.

Lemma 1.4.12. *If* A *is an abelian topological group and* $V \xrightarrow{i} V'$ *an inclusion of a subgroup* V *of an abelian topological group* V' *such that the induced homomorphism* $F_{\mathbf{TopAb}}(\mathrm{id}_A \times i) : F_{\mathbf{TopAb}}(A \times V) \to F_{\mathbf{TopAb}}(A \times V')$ *is an embedding then the homomorphism* $\mathrm{id}_A \otimes i : A \otimes V \to A \otimes V'$ *is an embedding as well.*

Proof. Let A be a free abelian topological group, $V \xrightarrow{i} V'$ be an inclusion of a subgroup V of an abelian topological group V' and let U denote the forgetful functor $\mathbf{TopAb} \to \mathbf{Ab}$. The underlying group UA of the free abelian topological group A is a free abelian group (cf. E.1). Because tensoring with the free abelian group UA is an exact endofunctor $\mathbf{Ab} \to \mathbf{Ab}$, the sequence

$$0 \to U(A \otimes N) \xrightarrow{U(\mathrm{id}_A \otimes i)} U(A \otimes N')$$

of abelian groups is exact. In particular $U(\mathrm{id}_A \otimes i)$, hence $\mathrm{id}_A \otimes i$ is injective. It remains to show that the injective homomorphism $\mathrm{id}_A \otimes i$ of topological groups is an embedding. To see that $\mathrm{id}_A \otimes i$ is an inclusion consider the commutative diagram

$$
\begin{array}{ccc}
\mathrm{F_{TopAb}}(A \times V) & \xrightarrow{\mathrm{F_{TopAb}}(i)} & \mathrm{F_{TopAb}}(A \times V') \\
\downarrow{\scriptstyle q_{A \times V}} & & \downarrow{\scriptstyle q_{A \times V'}} \\
A \otimes V & \xrightarrow{\quad \mathrm{id}_A \otimes i \quad} & A \otimes V'
\end{array}
$$

of abelian topological groups, where the vertical arrows $q_{A \times V}$ and $q_{A \times V'}$ are quotient maps. If U is an open subset of $A \otimes V$ then its inverse image $\tilde{U} = q_{A \times V}^{-1}(U)$ is open in $\mathrm{F_{TopAb}}(A \times V)$. The homomorphism $\mathrm{F_{TopAb}}(i)$ is an embedding by assumption, so there exists an open $q_{A \times V'}$-saturated set $\tilde{W}$ in $\mathrm{F_{TopAb}}(A \times V')$ such that $\mathrm{F_{TopAb}}(i)^{-1}(\tilde{W}) = \tilde{U}$. Since the quotient homomorphism $q_{A \times V'}$ is an open map, the image $V = q_{A \times V'}(\tilde{W})$ of $\tilde{W}$ is an open set in $A \otimes V'$. The commutativity of the above diagram now implies

$$
\begin{aligned}
(\mathrm{id}_A \otimes i)^{-1}(W) &= q_{A \times V'}[q_{A \times V'}^{-1}(\mathrm{id}_A \otimes i)^{-1}(W)] \\
&= q_{A \times V'}[q_{A \times V'}^{-1}(\tilde{W})] = q_{A \times V'}(\tilde{U}) = U,
\end{aligned}
$$

i.e. the open set U is the inverse image of the open set W in $A \otimes V'$. So every open set in $A \otimes V$ is the inverse image of an open set in $A \otimes V'$ under the continuous injection $\mathrm{id}_A \otimes i$, hence the homomorphism $\mathrm{id}_A \otimes V$ is an embedding. $\qquad\square$

Unfortunately not for all embeddings $A \times V \hookrightarrow A \times V'$ of topological spaces the induced homomorphism $\mathrm{F_{TopAb}}(A \times V) \to \mathrm{F_{TopAb}}(A \times V')$ is an embedding. In the following we derive the necessary and sufficient condition for this to be the case.

Lemma 1.4.13. *An embedding $X \hookrightarrow Y$ of completely regular spaces induces an embedding $\mathrm{F_{TopAb}}X \to \mathrm{F_{TopAb}}Y$ of free abelian topological groups if every continuous bounded pseudometric on X can be extended to a pseudometric on Y.*

Proof. Let $i : X \hookrightarrow Y$ be the inclusion of a completely regular subspace X of a completely regular topological space Y. If X is the empty space, then $\mathrm{F_{TopAb}}X$ is the trivial abelian group $\{0\}$. The inclusion $\{0\} \hookrightarrow \mathrm{F_{TopAb}}Y$ of the trivial subgroup $\{0\}$ into the abelian topological group $\mathrm{F_{TopAb}}Y$ is always an embedding. If X is not empty choose a common base point $*$ of X and Y, such that i is an inclusion of based spaces. Let $U : \mathbf{Top}_* \to \mathbf{Top}$ denote the forgetful functor and $P : \mathbf{Top} \to \mathbf{Top}_*$ its left adjoint, assigning to every topological space X the based topological space $X \dot\cup \{*\}$. The functors $\mathrm{F_{TopAb}}$ and $\mathrm{F_{TopAb}}P$ are naturally isomorphic (cf. Theorem E.1.14). In particular the topological group $\mathrm{F_{TopAb}}X$ is naturally isomorphic to the topological group $\mathrm{F_{TopAb}^*}X \oplus \mathbb{Z}$ (by Corollary E.1.15). So one obtains a commutative diagram

$$
\begin{array}{ccc}
\mathrm{F_{TopAb}}X & \xrightarrow{\mathrm{F_{TopAb}}(i)} & \mathrm{F_{TopAb}}Y \\
{\scriptstyle \cong}\downarrow & & \downarrow{\scriptstyle \cong} \\
\mathrm{F_{TopAb}^*}X \oplus \mathbb{Z} & \xrightarrow{\mathrm{F_{TopAb}^*}(i) \oplus \mathrm{id}_{\mathbb{Z}}} & \mathrm{F_{TopAb}^*}Y \oplus \mathbb{Z}
\end{array}
$$

of homomorphisms of abelian topological groups. The vertical arrows are natural isomorphisms. If every continuous pseudometric on X can be extended to a continuous pseudometric on X then the lower horizontal homomorphism $\mathrm{F_{TopAb}^*}(i) \oplus \mathrm{id}_{\mathbb{Z}}$ is an embedding by Theorem E.1.21. This forces the homomorphism $\mathrm{F_{TopAb}}(i)$ to be an embedding as well. $\qquad\square$

Example 1.4.14. If X is a completely regular topological space, then the natural functions $X \to F_{\mathbf{TopGrp}}X$ and $X \to F_{\mathbf{TopAb}}X$ are embeddings (by Lemma E.1.9). Every continuous pseudometric ϱ on X can be extended to a maximal invariant pseudometric d_ϱ on the free topological group $F_{\mathbf{TopGrp}}X$ resp. the free abelian topological group $F_{\mathbf{TopAb}}X$, called the Graev-extension of ϱ. (In fact the group topology on $F_{\mathbf{TopGrp}}X$ resp. $F_{\mathbf{TopAb}}X$ is given by such Graev-extensions of pseudometrics, see [Sip03, 1.6,1.9].) Thus the inclusion $X \hookrightarrow F_{\mathbf{TopAb}}X$ induces an embedding $F_{\mathbf{TopAb}}X \hookrightarrow F_{\mathbf{TopAb}}^2 X$.

Lemma 1.4.13 can be generalised to arbitrary topological spaces. Let $\mathbf{Cr}$ denote the full subcategory of **Top** with objects all completely regular spaces. Recall that $\mathbf{Cr}$ is a reflective subcategory of **Top** (by Theorem B.2.10) and let $Cr : \mathbf{Top} \to \mathbf{Cr}$ be the reflector. Consider the embedding $I : \mathbf{Cr} \hookrightarrow \mathbf{Top}$ of the subcategory $\mathbf{Cr}$ into **Top** and let $\eta' : \mathrm{id}_{\mathbf{Top}} \to ICr$ be the counit of the adjunction between the reflector Cr and the embedding I. The natural transformation η' induces a natural transformation $F_{\mathbf{TopAb}} \to F_{\mathbf{TopAb}}ICr$. The counit $\eta : \mathrm{id}_{\mathbf{Top}} \to UF_{\mathbf{TopAb}}$ of the adjunction between the free abelian topological group functor $F_{\mathbf{TopAb}} : \mathbf{Top} \to \mathbf{TopAb}$ and the forgetful functor $U : \mathbf{TopAb} \to \mathbf{Top}$ intertwines the natural transformations η' and $F_{\mathbf{TopAb}}(\eta')$:

$$\begin{array}{ccc} \mathrm{id}_{\mathbf{Top}} & \xrightarrow{\ \eta_X\ } & UF_{\mathbf{TopAb}} \\ \eta'_X \downarrow & & \downarrow F_{\mathbf{TopAb}}(\eta') \\ ICrX & \xrightarrow{\ \eta_{ICrX}\ } & UF_{\mathbf{TopAb}}ICr \end{array} \qquad (1.5)$$

The maps η_X for all topological spaces X factor through the maps η'_X in a natural way:

Lemma 1.4.15. *The counit $\eta : \mathrm{id}_{\mathbf{Top}} \to UF_{\mathbf{TopAb}}$ of the adjunction between $\mathrm{id}_{\mathbf{Top}}$ and $UF_{\mathbf{TopAb}}$ factors through the counit $\eta' : \mathrm{id}_{\mathbf{Top}} \to ICr$, i.e. there exits a natural transformation $f : ICr \to UF_{\mathbf{TopAb}}$ satisfying $\eta = f \circ \eta'$.*

Proof. Because the underlying space of every topological group is completely regular, the functor $UF_{\mathbf{TopAb}}$ takes values in the category $\mathbf{Cr}$ of completely regular spaces. Therefore the counit $\mathrm{id}_{\mathbf{Top}} \to UF_{\mathbf{TopAb}}$ of the adjunction between $F_{\mathbf{TopAb}}$ and U factors through the counit $\eta' : \mathrm{id}_{\mathbf{Top}} \to ICr$, i.e. there exists a natural transformation $f : ICr \to UF_{\mathbf{TopAb}}$ satisfying $\eta = f\eta'$. $\qquad \square$

The natural transformation $f : ICr \to UF_{\mathbf{TopAb}}$ can be used to construct a left inverse to the natural transformation $F_{\mathbf{TopAb}}(\eta') : UF_{\mathbf{TopAb}} \to UF_{\mathbf{TopAb}}ICr$ on the right in Diagram 1.5. Moreover, we assert that the natural transformation $F_{\mathbf{TopAb}}(\eta')$ is a natural isomorphism:

Proposition 1.4.16. *The counit $\eta' : \mathrm{id}_{\mathbf{Top}} \to ICr$ of the adjunction between $\mathrm{id}_{\mathbf{Top}}$ and Cr induces a natural isomorphism $F_{\mathbf{TopAb}}(\eta') : F_{\mathbf{TopAb}} \cong F_{\mathbf{TopAb}}Cr$.*

Proof. If X is a topological space then the natural continuous bijection $X \to ICrX$ induces a continuous bijective homomorphism $F_{\mathbf{TopAb}}X \to F_{\mathbf{TopAb}}ICrX$ of free abelian topological groups. Consider the commutative diagram

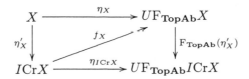

of natural functions. The natural continuous function $f_X : I\mathrm{Cr}X \to U\mathrm{F}_{\mathbf{TopAb}}X$ induces a unique homomorphism $\varphi_X : \mathrm{F}_{\mathbf{TopAb}}I\mathrm{Cr}X \to \mathrm{F}_{\mathbf{TopAb}}X$ of abelian topological groups satisfying $U\varphi \circ \eta_{I\mathrm{Cr}X} = f_X$. As a consequence one obtains the equality

$$\eta_X = f_X \circ \eta'_X = \varphi_X \eta_{I\mathrm{Cr}X} \eta'_X = \varphi_X \circ U\mathrm{F}_{\mathbf{TopAb}}(\eta'_X) \circ \eta_X$$

of continuous functions. So the homomorphism $\varphi_X \mathrm{F}_{\mathbf{TopAb}}(\eta'_X)$ of free abelian topological groups restricts to the identity on the generating subspace $\eta_X(X)$ of $U\mathrm{F}_{\mathbf{TopAb}}X$. This forces $\varphi_X \mathrm{F}_{\mathbf{TopAb}}(\eta'_X)$ to be the identity of $\mathrm{F}_{\mathbf{TopAb}}X$, hence $\mathrm{F}_{\mathbf{TopAb}}(\eta'_X)$ is an isomorphism. □

This allows us to identify the free abelian topological group $\mathrm{F}_{\mathbf{TopAb}}X$ on a topological space X with the free abelian topological group $\mathrm{F}_{\mathbf{TopAb}}I\mathrm{Cr}X$ on the regularisation $I\mathrm{Cr}X$ of X.

Corollary 1.4.17. *The image of the counit* $\eta : \mathrm{id}_{\mathbf{Top}} \to U\mathrm{F}_{\mathbf{TopAb}}$ *is naturally isomorphic to the counit* $\eta' : \mathbf{Top} \to \mathbf{Top}$.

Proof. The reflection $\mathrm{Cr}X$ of any topological space X is completely regular, so the function $\eta_{\mathrm{Cr}X} : I\mathrm{Cr}X \to \mathrm{F}_{\mathbf{TopAb}}I\mathrm{Cr}X$ is always an embedding (by Lemma E.1.9). Let $\varphi_X : \mathrm{F}_{\mathbf{TopAb}}I\mathrm{Cr}X \to \mathrm{F}_{\mathbf{TopAb}}X$ be the unique homomorphism of of abelian topological groups satisfying $f = \varphi_X \circ \eta_{\mathrm{Cr}X}$. From the proof of the preceding Proposition we infer that the homomorphism φ_X is a left inverse to the isomorphism $\mathrm{F}_{\mathbf{TopAb}}(\eta')$, hence an isomorphism as well. Thus the function $f_X = \varphi_X \eta_{I\mathrm{Cr}X}$ is an embedding. □

The preceding observations enable us to generalise Theorem 1.4.13 to arbitrary topological spaces.

Theorem 1.4.18. *An embedding* $X \hookrightarrow Y$ *of a topological subspace X of Y induces an embedding* $\mathrm{F}_{\mathbf{TopAb}}X \to \mathrm{F}_{\mathbf{TopAb}}Y$ *of free abelian topological groups if and only if every continuous bounded pseudometric on X can be extended to a continuous pseudometric on Y.*

Proof. Let $i : X \hookrightarrow Y$ be the inclusion of a subspace X of a topological space Y. The embedding i induces a continuous function $\mathrm{Cr}(i) : \mathrm{Cr}X \to \mathrm{Cr}Y$ and a homomorphism $\mathrm{F}_{\mathbf{TopAb}}(i) : \mathrm{F}_{\mathbf{TopAb}}X \to \mathrm{F}_{\mathbf{TopAb}}Y$ of abelian topological groups. Because the image of the counit $\eta : \mathrm{id}_{\mathbf{Top}} \to U\mathrm{F}_{\mathbf{TopAb}}$ is naturally isomorphic to the counit $\eta' : \mathbf{Top} \to \mathbf{Top}$, all the functions fit in the commutative diagram

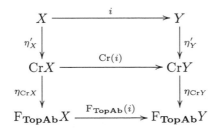

where we have identified the topological groups $\mathrm{F}_{\mathbf{TopAb}}X$ and $\mathrm{F}_{\mathbf{TopAb}}\mathrm{Cr}X$ and the groups $\mathrm{F}_{\mathbf{TopAb}}Y$ and $\mathrm{F}_{\mathbf{TopAb}}\mathrm{Cr}Y$. The condition that every pseudometric on X can be extended to a pseudometric on Y implies that the reflection $\mathrm{Cr}X$ is a subspace of $\mathrm{Cr}Y$. This in turn implies that the homomorphism $\mathrm{F}_{\mathbf{TopAb}}(\mathrm{Cr}i)$ is an embedding by Lemma 1.4.13. □

The condition that continuous pseudometrics on a subspace X of a topological space Y can be extended to the space Y has been thoroughly investigated. There are several characterisations of subspaces where this is the case:

Theorem 1.4.19 (Shapiro). *For a subspace X of a topological space Y the following are equivalent:*

1. *Any bounded continuous pseudometric on X can be extended to a bounded continuous pseudometric on Y.*
2. *Any bounded continuous pseudometric on X can be extended to a continuous pseudometric on Y.*
3. *Any continuous pseudometric on X can be extended to a continuous pseudometric on Y.*
4. *Every normal locally finite cozero-set cover of X has a refinement that is the restriction of a normal open cover of Y.*

Proof. The equivalence of these statements is proved in [Sha66, Theorem 2.1]. □

Theorem 1.4.20 (Arens). *For a closed subspace X of a topological space Y the following are equivalent:*

1. *Any continuous pseudometric on X can be extended to a continuous pseudometric on Y.*
2. *Every continuous map $f : X \to S$ into a metrisable subset S of a locally convex topological vector space can be extended to a function $f' : Y \to S$ if either S is complete in one of its metrics or X is the support of a continuous real valued function on Y.*

Proof. This is Theorem 2.4 of the article [Are53]. The proof presented there uses [Are52, Theorem 4.1]. □

Definition 1.4.21. *A subspace X of a topological space Y is called P-embedded in Y if every continuous pseudometric on X can be extended to a continuous pseudometric on Y.*

Example 1.4.22. Every singleton subspace $\{*\} \subseteq Y$ of a topological space Y is P-embedded in Y.

Example 1.4.23. The topologies of the free pointed topological group $F^*_{\mathbf{TopGrp}}X$ and the pointed free abelian topological group $F^*_{\mathbf{TopAb}}X$ on a pointed topological space $(X, *)$ are constructed by extending all continuous pseudometrics on X to pseudometrics on $F^*_{\mathbf{TopGrp}}X$ resp. $F^*_{\mathbf{TopAb}}X$ and then endowing $F^*_{\mathbf{TopGrp}}X$ resp. $F^*_{\mathbf{TopAb}}X$ with the initial topology w.r.t. all these pseudometrics. (See the construction in [Gra51].) Therefore the inclusions $X \hookrightarrow F^*_{\mathbf{TopGrp}}X$ and $X \hookrightarrow F^*_{\mathbf{TopAb}}X$ of a pointed completely regular space $(X, *)$ into the topological groups $F^*_{\mathbf{TopGrp}}X$ and $F^*_{\mathbf{TopAb}}X$ are P-embeddings.

Recall the natural isomorphism $F_{\mathbf{TopAb}}X \cong F^*_{\mathbf{TopAb}}X \oplus \mathbb{Z}$ (cf. Corollary E.1.15) for pointed spaces X. This isomorphism shows that the topology on the free abelian topological group $F_{\mathbf{TopAb}}X$ on a topological space X is constructed in a similar way to the topology on $F^*_{\mathbf{TopAb}}X$ (after choosing a base point $*$ in X). This indicates that completely regular spaces are also P-embedded in the free abelian topological group $F_{\mathbf{TopAb}}X$. There exists a short proof of this fact without the use of the actual construction in [Gra51]:

Lemma 1.4.24. *The natural inclusion $X \hookrightarrow F_{\mathbf{TopAb}}X$ of a completely regular space X into the free abelian topological group $F_{\mathbf{TopAb}}X$ on X is a P-embedding.*

Proof. Let X be a completely regular topological space. The universal uniformity $\mathcal{U}$ on X is the restriction of the left uniformity $\mathcal{V}_l$ on the abelian topological group $F_{\mathbf{TopAb}}X$ to the subspace X (cf. [Pes82]). Every continuous pseudometric ϱ on

X is uniformly continuous w.r.t. the universal uniformity $\mathcal{U}$. Uniformly continuous pseudometrics on uniform subspaces of uniform spaces are extendible to continuous pseudometrics on the whole space ([Gan68, Corollary 3.5]). Thus every pseudometric ϱ on X is extendible to a continuous pseudometric on $F_{\mathbf{TopAb}}X$, i.e. the subspace X is P-embedded in $F_{\mathbf{TopAb}}X$. $\qquad\square$

Remark 1.4.25. This also shows that the topology on a free abelian group $F_{\mathbf{TopAb}}X$ on a topological space X is actually defined by extending all continuous pseudometrics on X to semi-norms on $F_{\mathbf{TopAb}}X$ and then endowing $F_{\mathbf{TopAb}}X$ with the initial topology w.r.t. all these pseudometrics.

With the notion of P-embeddings Theorem 1.4.18 can now be reformulated as:

Theorem 1.4.26. *An embedding $X \hookrightarrow Y$ of topological spaces induces an embedding $F_{\mathbf{TopAb}}X \to F_{\mathbf{TopAb}}Y$ of free abelian topological groups if and only if X is P-embedded in Y.*

Corollary 1.4.27. *Every singleton subspace $\{*\} \subseteq X$ of a topological space Y generates a discrete subgroup in $F_{\mathbf{TopAb}}Y$ which is isomorphic to $\mathbb{Z}$.*

Proof. The singleton subspace $\{*\}$ of Y is P-embedded in Y. Therefore its inclusion into Y induces an inclusion $\mathbb{Z} \cong F_{\mathbf{TopAb}}\{*\} \hookrightarrow F_{\mathbf{TopAb}}Y$. $\qquad\square$

Not only singleton subsets but every P-embedded subset A of a topological space X generates a free abelian subgroup $F_{\mathbf{TopAb}}A$ of the free abelian topological group $F_{\mathbf{TopAb}}X$. It has been shown in [Mor70a, Theorem 1.10] that the subgroup $F_{\mathbf{TopAb}}A$ generated by A is closed if X is a T_3-space and A is compact. Below we show that the compactness assumption on A is unnecessary.

Lemma 1.4.28. *If a free abelian topological group $F_{\mathbf{TopAb}}X$ on a topological space X is Hausdorff, then the discrete subgroup $F_{\mathbf{TopAb}}\{*\}$ generated by a point $* \in X$ is closed.*

Proof. If the free abelian topological group $F_{\mathbf{TopAb}}X$ on a topological space X is Hausdorff then its points are closed. In this case any discrete subgroup of $F_{\mathbf{TopAb}}X$ is locally closed in $F_{\mathbf{TopAb}}X$, hence closed. $\qquad\square$

Lemma 1.4.29. *A P-embedded closed subset A of a T_3-space X generates a closed subgroup $F_{\mathbf{TopAb}}A$ in the free abelian topological group $F_{\mathbf{TopAb}}X$.*

Proof. Let $A \hookrightarrow X$ be a closed P-embedding into a T_3-space. Let $q : X \twoheadrightarrow X/A$ denote the quotient map onto the factor space X/A obtained by identifying all points of A. Because X is a T_3-space and A is closed, the factor space X/A is Hausdorff. The point $* = A$ of this factor space generates a discrete closed subgroup $F_{\mathbf{TopAb}}\{*\}$ of $F_{\mathbf{TopAb}}(X/A)$. The subgroup $F_{\mathbf{TopAb}}A$ of the free abelian topological group $F_{\mathbf{TopAb}}X$ is the inverse image of the closed subgroup $F_{\mathbf{TopAb}}\{*\}$ under the homomorphism $F_{\mathbf{TopAb}}(q) : F_{\mathbf{TopAb}}X \to F_{\mathbf{TopAb}}(X/A)$ induced by q. Being an inverse image of a closed set under a continuous map it is closed as well. $\qquad\square$

The characterisation of P-embeddings found by Arens (Theorem 1.4.20) could be useful without the restrictive condition on the subspace X to be closed. Fortunately, this restrictions turns out to be unnecessary. To prove this, we need some preparatory observations on paracompactness:

Lemma 1.4.30. *A topological space X carrying the initial topology w.r.t. a surjection $f : X \twoheadrightarrow Y$ onto a paracompact space Y is paracompact.*

Proof. Let X be a topological space equipped with the initial topology w.r.t. a surjection $f : X \twoheadrightarrow Y$ onto a paracompact topological space Y. (This is to say that the topology on X is the inverse image of the topology on Y under the surjection f.) If $\mathfrak{U} = \{U_i \mid i \in I\}$ is an open covering of X then each open set U_i is the inverse image $f^{-1}(V_i)$ of an open set V_i in Y. We assert that $\mathfrak{V} = \{V_i \mid i \in I\}$ is an open covering of Y. Every point $y \in Y$ is the image $f(x)$ of a point $x \in X$. The point x is contained in some open set U_i. There fore the image $y = f(x)$ is contained in $f(U_i) = f(f^{-1}(V_i)) \subset V_i$. Thus $\mathfrak{V}$ is an open covering of Y. Because Y is paracompact, the open covering $\mathfrak{V}$ has an open neighbourhood-finite refinement $\mathfrak{V}'$. The inverse image $\mathfrak{U}' = f^{-1}(\mathfrak{V}')$ then is an open neighbourhood-finite refinement of the covering $\mathfrak{U}$ of X. □

Lemma 1.4.31. *Pseudometric topological spaces are paracompact.*

Proof. Let (X, ϱ) be a pseudometric topological space and consider the equivalence relation $\sim$ on X given by $x \sim y \Leftrightarrow \varrho(x, y) = 0$. The quotient space $X/\sim$ is a metric space and X carries the initial topology w.r.t. the quotient map $q : X \twoheadrightarrow X/\sim$. Since metric spaces are paracompact (cf. [Dug89, Theorem IX.5.3]) Lemma 1.4.30 implies that the pseudometric space X also is paracompact. □

Proposition 1.4.32. *If (X, ϱ) is a pseudometric space and A a subset of X satisfying $A = \varrho(-, A)^{-1}(0)$ and V an affine space V of type m, then there exists a linear map $e : C(A, V) \to C(X, V)$ of (non-topological) vector spaces which assigns to every continuous function $f : A \to V$ a continuous extension $e(f) : X \to V$ of f satisfying $e(f)(X) \subset \mathrm{conv}(f(A))$. If the topology on V is induced by a semi-norm, then the linear map e can be chosen to preserve the supremum semi-norm of each bounded function $f : A \to X$.*

Proof. This is a generalisation of [Dug89, Theorem IX.6.1] and the proof is a modification of the proof thereof. Let X be a pseudometric space and A a subset of X satisfying $A = \varrho(-, A)^{-1}(0)$. For each $x \in X \setminus A$ let B_x be the open Ball with radius $\frac{1}{2}\varrho(x, A)$ centred at x. Because the pseudometric space $X \setminus A$ is paracompact by Lemma 1.4.31, the open covering $\{B_x \mid x \in X \setminus A\}$ of $X \setminus A$ has an open neighbourhood refinement $\mathfrak{U}$. Form here on, the proof follows that of [Dug89, Theorem IX.6.1 and Remark after IX.6.2]. □

Theorem 1.4.33. *If an inclusion $A \hookrightarrow X$ is a P-embedding then every continuous function $f : A \to V$ into a pseudometric topological vector space V has an extension $e(f) : X \to V$ satisfying $e(f)(X) \subset \mathrm{conv}(\overline{f(A)})$ provided that either $f(A)$ is contained in some complete subspace of V or A is a cozero set in X. If the pseudometric V is induced by a semi-norm and f is bounded, then the extension $e(f)$ can be chosen to have the same supremum semi-norm.*

Proof. Let $A \hookrightarrow X$ be a P-embedding and let $f : A \to V$ be a continuous function into a pseudometric topological vector space V with pseudometric d. Suppose that the image $f(A)$ is contained in some complete subspace Y of V. Let $\varrho = d \circ (f \times f)$ be the pullback of the pseudometric d to the subspace A of X. Because $A \hookrightarrow X$ is a P-embedding, the pseudometric ϱ has an extension to X, which we also denote by ϱ. This pseudometric ϱ induces a topology on X that is coarser than the original one. Therefore it suffices to show that there exists an extension of f to X which is continuous w.r.t. the topology induced by the pseudometric ϱ, i.e. one can w.l.o.g. assume that the topology on X is induced by ϱ, which we will do from now on. The function f is uniformly continuous w.r.t. the pseudometrics ϱ on A and d on V. Therefore there exists a unique extension $\overline{f} : \overline{A} \to Y \subset V$ of f to the closure $\overline{A}$ of A and this extension $\overline{f}$ is uniformly continuous as well (cf. [Bou71, Théorème 6.2, Chap. II, §3]). Moreover the image $\overline{f}(\overline{A})$ of $\overline{A}$ is contained in the closure $\overline{f(A)}$ of the image of A. If d is induced by a semi-norm and f is bounded, then the supremum semi-norm $\|\overline{f}\|$ coincides with the supremum semi-norm of f. The closure $\overline{A}$ in the pseudometric space (X, ϱ) is the set $\{x \in X \mid \varrho(x, A) = 0\}$, so by Proposition 1.4.32 there exists a continuous extension $e(f)$ of $\overline{f}$ to X. The function $e(f)$ then is an extension of f by construction. If d is induced by a semi-norm and f is bounded, then $\overline{f}$ is bounded as well and (by Proposition 1.4.32) $e(f)$ can be chosen to have the same supremum semi-norm as f. If A is a cozero set, then there exists a continuous function $g : X \to \mathbb{R}$ such that $A = g^{-1}(0)$. In this case one can replace the pseudometric ϱ on X by the pseudometric given by $\varrho'(x, x') = \varrho(x, x') + |g(x) - g(y)|$. Thus $A = \varrho'(-, A)^{-1}(0)$ and, as before, we can apply Proposition 1.4.32 to obtain a continuous extension $e(f)$ of f having the desired properties.   □

This can especially by applied to function spaces that are Banach spaces. Recall that for each topological space A and Banach space V the space $C^b(A, V)_{\|\cdot\|_\infty}$ of bounded continuous functions $A \to V$ equipped with the supremum norm is again a Banach space.

Corollary 1.4.34. *If $A \hookrightarrow X$ is a P-embedding and V a Banach space then there exists an continuous function $e : C^b(A, V)_{\|\cdot\|_\infty} \to C^b(X, V)_{\|\cdot\|_\infty}$, which is an isometry onto its image, assigning to every bounded continuous function $f : A \to V$ an extension $e(f) : X \to V$ such that $e(f)(X) \subset \mathrm{conv}(\overline{f(A)})$*

Proof. This follows from the fact that the linear map e in Theorem 1.4.33 can be chosen to preserve the supremum norm, so it is a continuous isometry onto its image.   □

The converse implication also holds. Here it suffices to demand that every bounded Banach space valued function on the subspace A of X has a continuous extension to the space X:

Proposition 1.4.35. *If $i : A \hookrightarrow X$ is an inclusion of a subspace such that every continuous bounded Banach space valued function f on A has an extension to X, then i is a P-embedding.*

Proof. Let $i : A \hookrightarrow X$ be an inclusion with the property that each bounded Banach space valued function $f : A \to X$ has an extension to X. In order to show that i is a P-embedding, it suffices to show that every continuous bounded pseudometric on A has an extension to a continuous bounded pseudometric on X (cf. Theorem 1.4.19). Consider the Banach space $C^b(A, \mathbb{R})_{\|\cdot\|_\infty}$ of bounded continuous real valued functions on A equipped with the supremum norm. For every bounded continuous pseudometric $\varrho : A \times A \to \mathbb{R}$ define a continuous function f_ϱ from A into the Banach space $C^b(A, \mathbb{R})_{\|\cdot\|_\infty}$ via

$$f_\varrho : A \to C^b(A, \mathbb{R})_{\|\cdot\|_\infty}, \quad f_\varrho(x)(a) = \varrho(x, a).$$

By assumption, the continuous function f_ϱ can be extended to a continuous function $F_\varrho : X \to C^b(A, \mathbb{R})_{\|\cdot\|_\infty}$. Setting $d_\varrho(x, x') = \|F_\varrho(x) - F_\varrho(x')\|_\infty$ defines a pseudometric d_ϱ on X. We assert that the pseudometric d_ϱ on X is an extension of the pseudometric ϱ on A. The triangle inequality for the pseudometric ϱ implies the inequality

$$|[f_\varrho(x) - f_\varrho(x')](a)| = |\varrho(x, a) - \varrho(x', a)| \leq \varrho(x, x')$$

for all x, x' in A. So for all $x, x' \in X$ the continuous function $f_\varrho(x) - f_\varrho(x')$ on A is bounded by $\varrho(x, x')$, i.e. $d_\varrho(x, x') = \|F_\varrho(x) - F_\varrho(x')\|_\infty \leq \varrho(x, x')$. Evaluation of the continuous function $F_\varrho(x) - F_\varrho(x')$ at $x \in A$ shows that the upper bound $\varrho(x, x')$ is attained:

$$|[f_\varrho(x) - f_\varrho(x')](x)| = |\varrho(x,x) - \varrho(x,x')| = |0 - \varrho(x,x')| = \varrho(x,x')$$

This implies the equality $d_\varrho(x,x') = \varrho(x,x')$ for all $x, x' \in A$, i.e. the pseudometric d_ϱ is an extension of ϱ to X. Since this is true for all bounded pseudometrics ϱ on A, the inclusion $i : A \hookrightarrow X$ is a P-embedding. $\qquad\square$

Summarising Corollary 1.4.34 and Proposition 1.4.35 yields the desired generalisation of Arens characterisation of P-embeddings:

Theorem 1.4.36. *An inclusion $i : A \hookrightarrow X$ of a subspace is a P-embedding if and only if every continuous bounded Banach space valued function f on A has an extension to X.*

So a subspace X of Y is P-embedded in Y if any of the conditions of Theorem 1.4.19 or Theorem 1.4.36 is satisfied. One possibility to ensure the existence of extensions of pseudometrics on X to pseudometrics on Y is to require the subspace X to be a retract of Y:

Lemma 1.4.37. *If X is a retract of the topological space Y then X is P-embedded in Y.*

Proof. Let Y be a topological space and X be a subspace that is a retract of Y. Let $r : Y \to X$ be a retraction of Y onto X. The retraction r restricts to the identity on Y. If $f : X \to V$ is a continuous function from X into a topological vector space V then $f \circ r : Y \to V$ is an extension of f to Y. By Theorem 1.4.36 X is P-embedded in Y. $\qquad\square$

Example 1.4.38. Any base space B of a vector bundle $p : E \to B$ is a retract of E. Therefore all base spaces of vector bundles are P-embedded into the total spaces. In particular any tangent bundle TM of a (possibly infinite dimensional) manifold M contains M as a P-embedded subspace.

Another way to ensure that every pseudometric on a subspace X of a topological space Y can be extended to Y is to require X to be closed and the space Y to be paracompact.

Lemma 1.4.39. *Every closed subspace X of a paracompact topological space Y is P-embedded in Y.*

Proof. Let X be a closed subspace of a paracompact topological space Y and $\mathfrak{U}$ a normal locally finite cozero-set covering of X. The open covering $\mathfrak{U}$ is the restriction of a covering $\mathfrak{U}'$ of X by open sets in Y. Since the complement $Y \setminus X$ of X is open in Y, the set $\mathfrak{V} = \mathfrak{U} \cup \{Y \setminus X\}$ is an open covering of Y that restricts to $\mathfrak{U}$ on X. Because all open coverings in paracompact spaces are normal, the covering $\mathfrak{V}$ is normal. So every normal locally finite cozero-set covering of X is a restriction of a normal open covering of Y; thus (by Theorem 1.4.19) every continuous pseudometric on X can be extended to a continuous pseudometric on Y. $\qquad\square$

Example 1.4.40. If X is a discrete subspace of a paracompact Hausdorff space Y then every pseudometric on X is the restriction of a continuous pseudometric on Y. In particular discrete subgroups of finite dimensional Lie groups are P-embedded.

Example 1.4.41. Finite dimensional manifolds are paracompact. Therefore every continuous pseudometric on a submanifold M of a finite dimensional manifold M' can be extended to a continuous pseudometric on the whole manifold M'. In particular closed subgroups of finite dimensional Lie groups are P-embedded.

Example 1.4.42. CW-complexes are paracompact. Therefore every continuous pseudometric on a sub-CW-complex K of a CW-complex L can be extended to a continuous pseudometric on the whole CW-complex L.

If the spaces considered are the fibre F and the total space E of a bundle $p : E \to B$, then the existence of extensions of pseudometrics can be accomplished by assuming the bundle to be locally trivial and the base space B to be completely regular:

Lemma 1.4.43. *If $F \subset E$ is the a fibre of a locally trivial bundle $p : E \to B$ over a completely regular base space B then F is P-embedded in E.*

Proof. Let $p : E \to B$ be a locally trivial bundle over a completely regular base space B and $F = p^{-1}(*)$ be the fibre of a point $* \in B$. Because the bundle $p : E \to B$ is locally trivial there exists a a local section on a neighbourhood U of $*$. This implies that the inverse image $p^{-1}(U)$ of U is homeomorphic to the space $F \times U$ and the quotient map p corresponds to the projection onto the second factor under this homeomorphism. Identifying the spaces $p^{-1}(U)$ and $F \times U$ we consider $F \times U$ as a subspace of E. Let $\varphi : U \to I$ be a continuous function with support in U satisfying $\varphi(*) = 1$. (Such a function always exists because the base space B was assumed to be completely regular.) If $f : F \to V$ is a continuous function into a real vector space V then define a function $f' : E \to V$ via

$$f'_{|F \times U} = (f \circ \mathrm{pr}_1) \cdot (\varphi \circ \mathrm{pr}_2) \quad f'_{|p^{-1}(B \setminus \mathrm{supp}\varphi)} = 0$$

where the spaces $F \times U$ and $p^{-1}(U)$ have been identified. The so defined function f' then is an extension of f to the total space E. By Theorem 1.4.36 the subspace F is P-embedded in E. $\square$

Example 1.4.44. All Hurewicz G-fibrations over locally contractible spaces are locally trivial. Therefore the fibre G of a Hurewicz principal G-fibration $p : E \to B$ over a locally contractible completely regular space B is P-embedded in E.

Example 1.4.45. The fibre F of a locally trivial fibre bundle $p : E \to M$ over a completely regular manifold M is P-embedded in the total space E.

Example 1.4.46. All topological groups are completely regular. Thus the fibre F of a locally trivial fibre bundle $p : E \to G$ over a topological group G is P-embedded in the total space E.

Example 1.4.47. The fibre G of a Hurewicz G-fibration $p : E \to G''$ over a locally contractible topological group G'' is P-embedded in E. In particular the fibre G of a Hurewicz G-fibration $p : E \to G''$ over a Lie group G'' is P-embedded in E.

Example 1.4.48. The subgroup G of a group extension $0 \to G \to G' \to G'' \to 0$ of a topological group G'', which admits local sections, is P-embedded in G'. In particular all normal subgroups occurring in such extensions of Lie groups are P-embedded subgroups.

In all these examples there exists a "tubular neighbourhood" V of the subspace F and a continuous function $f : V \to \mathbb{R}$ with $f_{|F} = 1$ and support in V. Particular examples are group extensions and coset spaces of topological groups, where the quotient map is a locally trivial bundle. We give such inclusions a name:

Definition 1.4.49. *An inclusion $G \hookrightarrow G'$ of a subgroup G into a topological group G' is called* locally trivial *if there exists a right G-invariant open neighbourhood V of G such that the bundle $V \to V/G$ is trivial.*

This equivalent to the bundle $G' \to G'/G$ being locally trivial. Topological modules are abelian topological groups, so an inclusion $M \hookrightarrow M'$ of topological modules over a topological ring R is said to be locally trivial if the inclusion of the underlying abelian topological groups is locally trivial.

Definition 1.4.50. *A short exact sequence* $0 \to G \to G' \to G'' \to 0$ *of topological groups is called* locally trivial *if the bundle* $G' \to G''$ *is locally trivial. Analogously a short exact sequence* $0 \to M \to M' \to M'' \to 0$ *of topological modules is called* locally trivial *if the bundle* $M' \to M''$ *is locally trivial.*

Observe that the local triviality of embeddings of subgroups is preserved under finite products. In particular taking a binary product with a fixed group A preserves local triviality of short exact sequences.

Theorem 1.4.51. *If an inclusion* $V \xrightarrow{i} V'$ *of a subgroup of a topological group V' is locally trivial, then the induced homomorphism* $\mathrm{F}_{\mathbf{TopAb}}(i) : \mathrm{F}_{\mathbf{TopAb}}V \to \mathrm{F}_{\mathbf{TopAb}}V'$ *of free abelian topological groups is an embedding.*

Proof. Let $V \xrightarrow{i} V'$ be a locally trivial inclusion of a subgroup V of a topological group V'. By Lemma 1.4.43 V is P-embedded into V'. Therefore the induced homomorphism $\mathrm{F}_{\mathbf{TopAb}}(i) : \mathrm{F}_{\mathbf{TopAb}}V \to \mathrm{F}_{\mathbf{TopAb}}V'$ is an embedding by Theorem 1.4.18. $\qquad\square$

Definition 1.4.52. *A functor* $F : \mathbf{C} \to \mathbf{D}$ *between pointed categories is called* exact *with respect to a class $\mathcal{E}$ of short exact sequences if it preserves the exactness of all short exact sequences in $\mathcal{E}$.*

Theorem 1.4.53. *For any free abelian topological group A the tensor product functor* $A \otimes - : \mathbf{TopAb} \to \mathbf{TopAb}$ *is exact w.r.t. the class $\mathcal{E}$ of locally trivial short exact sequences.*

Proof. Let A be a free abelian topological group and $0 \to V \xrightarrow{i} V' \to V'' \to 0$ be a locally trivial short exact sequence in **TopAb**. The endofunctor $A \otimes -$ of **TopAb** preserves cokernels by Proposition 1.4.7. The homomorphism $\mathrm{id}_A \otimes i$ is injective, because the underlying homomorphism of abelian groups is injective. (Tensoring with a free abelian group is an exact endofunctor of **Ab**.) It remains to prove that the homomorphism $\mathrm{id}_A \otimes i : A \otimes V \to A \otimes V'$ of topological groups is an embedding. Because the inclusion $i : V \hookrightarrow V'$ was assumed to be locally trivial, the inclusion $\mathrm{id}_A \times i : A \times V \hookrightarrow A \times V'$ is locally trivial as well. Since any topological group $A \times V''$ is completely regular and the sequence $0 \to A \times V \to A \times V' \to A \times V'' \to 0$ is locally trivial, the fibre $A \times V$ is P-embedded in $A \times V'$ by Lemma 1.4.43. Therefore the induced homomorphism $\mathrm{F}_{\mathbf{TopAb}}(\mathrm{id}_A \times i) : \mathrm{F}_{\mathbf{TopAb}}(A \times V) \to \mathrm{F}_{\mathbf{TopAb}}(A \times V')$ is an embedding by Theorem 1.4.18. The homomorphism $\mathrm{id}_A \otimes i : A \otimes V \to A \otimes V'$ then is an embedding by Lemma 1.4.12. $\qquad\square$

Topological Homology- and Cohomology Theories

It is often useful to equip homology groups $H_n(X)$ or cohomology groups $H^n(X;V)$ of a topological space X or equivariant cohomology (e.g. group cohomology) with a topology. Nevertheless no systematic approach to this subject has been made so far. In this chapter we introduce (generalised) topological homology theories or cohomology theories as functors $\mathbf{Top} \to \mathbf{TopAb}$ from the category $\mathbf{Top}$ of topological spaces to the category $\mathbf{TopAb}$ of abelian topological groups. This includes the classical case, where the coefficients are given the discrete topology.

We begin with the definition of topological (co)homology theories and then introduce topologised versions of classical (co)chain complexes and (co)homologies. We then verify that these topologised versions share many of the important properties of their classical counterparts. This requires us to do homological algebra in the category $\mathbf{TopAb}$ of topological Abelian groups and relies on all the results derived in Chapter 1. In particular we introduce topologised singular homotopy, singular homology, singular cohomology, (equivariant) continuous Alexander-Spanier cohomology (all with topological coefficients) and show the homotopy invariance of these functors into the category $\mathbf{TopAb}$ of Abelian topological groups. The preservation of homotopy by the singular simplicial space functor $C(\Delta, -)$ and the the standard simplicial space functor is also proved. Moreover we show that large classes of short exact sequences of coefficients (e.g. short exact sequences which are locally trivial fibre bundles) induce long exact sequence of the (co)homology groups in $\mathbf{TopAb}$. In addition it is shown that large classes of embeddings of subspaces (e.g. P-embeddings, retracts, closed neighbourhood retracts in normal spaces, closed subspaces of the homotopy type of countable CW-complexes, closed subspaces of metric or paracompact $\aleph$-spaces, closed submanifolds, fibres of bundles over completely regular spaces) induce long exact sequences of the (co)homology groups of various topologised (co)homology theories.

After the study of topologised (co)homology theories we introduce the "vertex transformation" which induces a natural morphism ${}^tH^*_{AS,c}(X;V) \to {}^t H^*(X;V)$ from the topologised continuous Alexander-Spanier cohomology ${}^tH^*_{AS,c}(X;V)$ into the topologised singular cohomology ${}^tH^*(X;V)$ of a topological space X. This is a transfer of the classical morphism to the topologised version of these cohomologies.

At last we define and construct systems of (smooth) singular simplices in various topological spaces, e.g. Lie groups and Riemannian manifolds. These are used later to construct an inverse map to the vertex transformation and, in conjunction with the subdivision introduced in Chapters 4, 5 and 6, to extend equivariant Alexander-Spanier cocycles to global equivariant cocycles on spaces with group actions.

2.1 Definition of Topological Homology Theories

Like the classical homology and cohomology theories the topologised versions will be defined on a suitable subcategory of the category $\mathbf{Top}(2)$ of pairs of topological spaces. We recall that the objects of the category $\mathbf{Top}(2)$ are the pairs (X, A) of topological spaces, where A is a subspace of X and the morphisms $(X, A) \to (Y, B)$ are the pairs (f, f') of continuous functions such that f maps the subspace A of X into the subspace B of Y and F' is the restriction and corestriction of f. The categories $\mathbf{Top}(3)$ of triples of topological spaces and $\mathbf{Top}(n)$ of n-ads are defined analogously. Particular examples of subcategories of $\mathbf{Top}(2)$ are:

1. The full subcategory of all pairs (G, N) of (possibly infinite dimensional) Lie groups G with a subgroup N.
2. The full subcategory of all pairs (M, N) of (possibly infinite dimensional) manifolds M with a submanifold N.
3. The full subcategory of all pairs (K, L) of a CW-complex K with a subcomplex L.

If one restricts oneself to morphisms compatible with the structure of the objects in the above examples one obtains the subcategory $\lg(2)$ of all pairs (G, N) of possibly infinite dimensional) Lie groups G with a subgroup N and pairs of morphisms of Lie groups and the subcategory $\mathbf{Mf}(2)$ of all pairs (M, N) of possibly infinite dimensional) manifolds M with a submanifold N and pairs of smooth mappings.

As usual the category $\mathbf{Top}$ is regarded as a full subcategory of $\mathbf{Top}(2)$ via the inclusion $X \mapsto (X, \emptyset)$. We use the endofunctor

$$\kappa_l : \mathbf{Top}(2) \to \mathbf{Top}(2), \quad \kappa_l(X, A) = (A, \emptyset), \quad \kappa_l((f, f')) = (f', \emptyset)$$

of $\mathbf{Top}(2)$ to define homology and cohomology theories. A homotopy H from (f, f') to (g, g') in $\mathbf{Top}(2)$ is a homotopy H from f to g which restricts and corestricts to a homotopy from f' to g'. The category whose objects are the topological spaces and whose morphisms from X to X' are the homotopy classes of continuous functions $X \to X'$ is denoted by $\mathbf{Toph}$. The hom-sets $\hom_{\mathbf{Toph}}(X, X')$ are also abbreviated by $[X, X']$. The category whose objects are the pairs of topological spaces and whose morphisms the homotopy classes of morphisms in $\mathbf{Top}(2)$ is denoted by $\mathbf{Top}(2)\mathbf{h}$. The hom-sets $\hom_{\mathbf{Top}(2)\mathbf{h}}((X, A), (X', A'))$ of this category are also written as $[(X, A), (X', A')]$.

Definition 2.1.1. *A homology theory on* $\mathbf{Top}(2)$ *with values in a pointed regular category* $\mathbf{D}$ *is a series of functors* $h_n : \mathbf{Top}(2) \to \mathbf{D}$ *with natural transformations* $\partial_n : h_n \to h_{n-1} \circ \kappa$ *satisfying the following conditions:*

1. *(Homotopy invariance) Each functor* h_n *descends to the category* $\mathbf{Top}(2)\mathbf{h}$, *i.e.* h_n *maps homotopic pairs of functions* $(f, f') \simeq (g, g')$ *to the same morphism* $h_n(f, f') = h_n(g, g')$.
2. *(Exact homology sequence) For each pair* (X, A) *of topological spaces in* $\mathbf{Top}(2)$ *the sequence*

$$\cdots \to h_{n+1}(X, A) \xrightarrow{\partial_{n+1}} h_n(A, \emptyset) \to h_n(X, \emptyset) \to h_n(X, A) \xrightarrow{\partial_n} \cdots$$

   is exact.

A homology theory on a subcategory $\mathbf{C}$ *of* $\mathbf{Top}(2)$ *with values in a pointed regular category* $\mathbf{D}$ *is a series of functors* $h_n : \mathbf{C} \to \mathbf{D}$ *together with natural transformations* $\partial_n : h_n \to h_{n-1} \circ \kappa$ *satisfying the above conditions1 and 2 for all objects* (X, A) *in* $\mathbf{C}$. *A homology theory is called* proper *if it satisfies the following condition:*

3. (Dimension) For every singleton space $*$ and all $n > 0$ the object $h_n(*)$ is a zero object in $\mathbf{D}$.

Remark 2.1.2. We have dropped the excision axiom in the above definition.

Using the abelian category $\mathbf{Ab}$ of abelian groups as codomain $\mathbf{D}$ in the above definition of proper homology theories recovers the definition of homology theories due to Eilenberg and Steenrod except the excision axiom. We are interested in topologised versions of these theories. So we define:

Definition 2.1.3. *A topological homology theory on a subcategory $\mathbf{C}$ of $\mathbf{Top}(2)$ is a homology theory with values in the category $\mathbf{TopAb}$ of abelian topological groups or the category $\mathbf{TopRMod}$ of topological modules over a topological ring R.*

Example 2.1.4. All homology theories with values in the category $\mathbf{Ab}$ of abelian groups can be given the discrete topology, turning them into topological homology theories.

Example 2.1.5. The Čech and singular homology groups of finite dimensional manifolds with real coefficients are finite dimensional vector spaces. These can uniquely be equipped with a Hausdorff topology, turning them into real topological vector spaces. The connecting homomorphisms are linear maps of finite dimensional vector spaces, hence continuous. Therefore these homology theories can be turned into topological homology theories (with values in $\mathbf{Top\mathbb{R}mod}$).

Example 2.1.6. All homology groups of homology theories with coefficients in the category $\mathbf{Kmod}$ modules over a topological field $\mathbf{K}$ can be given the free $\mathbf{K}$-vector space topology on a discrete basis. This turns these theories into topological homology theories with values in $\mathbf{TopKmod}$. If $\mathbf{K}$ contains the real numbers $\mathbb{R}$ as a subfield, one can also use the free locally convex vector space topology.

2.2 Homotopy Groups

In this section we equip the homotopy groups $\pi_n(X, *)$ of a pointed topological space $(X, *)$ with a topology turning them into semi-topological groups. For a very special class of spaces (arc-wise connected locally semi-locally simply connected $\omega - LC_1$ spaces with barycentric refinements) this has already been done by J. Dugundji in [Dug50]. Fundamental groups ought to have been shown to be topological groups in [Bis02]; unfortunately the proof of the continuity of the group multiplication has a conceptual error. Furthermore the approach chosen there does not generalise to higher homotopy groups. We here give a general procedure to turn the functor π_* into a functor with values in the category of semi-topological groups. We begin with some considerations of attaching spaces and their function spaces. All function spaces occuring in this section are equipped with the compact open topology.

Let A be a subspace of the topological space X and $f : A \to Y$ be a continuous function. The quotient space $X \cup_f Y$ of the disjoint union $X \dot\cup Y$ by the equivalence relation generated by $a \sim f(a)$ for all points $a \in A$ is the pushout of f and the inclusion $A \hookrightarrow X$ (see [Dug89, Theorem VI.6.5]). So the diagram

$$
\begin{array}{ccc}
A & \longrightarrow & X \\
{\scriptstyle f}\downarrow & & \downarrow \\
Y & \xrightarrow[j]{} & X \cup_f Y
\end{array}
$$

is a pushout diagram. If the space A is a closed subspace of X then the function $j : Y \to X \cup_f Y$ embeds Y as a closed subspace (cf. [Dug89, Theorem VI.6.3]). We are interested in the cases where the image of X in $X \cup_f Y$ also is a closed subspace. Since the space $X \cup_f Y$ carries the quotient topology of $X \dot\cup Y$, the image of X in the quotient is closed if and only if the image $f(A)$ of f is closed in Y.

Lemma 2.2.1. *If A is a subspace of X and $f : A \to Y$ is a continuous closed injection, then the inclusion $X \hookrightarrow X \dot\cup Y$ induces a closed embedding $X \hookrightarrow X \cup_f Y$.*

Proof. Let A be a subspace of X and $f : A \to Y$ be a continuous closed injection. We consider the quotient map $q : X \dot\cup Y \twoheadrightarrow X \cup_f Y$. Because f is injective, the restriction $q_{|X} : X \to X \cup_f Y$ is injective. It remains to prove that it is closed. Let B be a closed subspace of X. The image $q_{|X}(B)$ of B is closed in $X \cup_f Y$ if and only if its inverse image

$$q^{-1} q_{|X}(B) = B \dot\cup f(B \cap A)$$

under the quotient map q is closed in $X \dot\cup Y$. The intersection $B \cap A$ is closed in A and since f was assumed to be closed, the image $f(B \cap A)$ then is closed in Y. This in turn implies that the image $q_{|X}(B)$ is closed in $X \cup_f Y$. Thus the image of every closed subset of X under $q_{|X}$ is closed and $q_{|X}$ is a closed embedding. $\square$

Proposition 2.2.2. *If A is a closed subspace of X and $f : A \to Y$ is a continuous closed injection, then the inclusions $X \hookrightarrow X \dot\cup Y$ induce $Y \hookrightarrow X \dot\cup Y$ induce closed embeddings $X \hookrightarrow X \cup_f Y$ and $Y \hookrightarrow X \cup_f Y$ respectively.*

Thus we can regard X and Y as closed subspaces of $X \cup_f Y$ provided that the attaching map is a closed injection and its domain is closed in X. In this case the intersections of every compact subset $K \subset X \cup_f Y$ with the closed subspaces X and Y are closed in K, hence compact. So every compact subset $K \subset X \cup_f Y$ is the union of the two compact subsets $K \cap X$ and $K \cap Y$. So we note:

Corollary 2.2.3. *If A is a subspace of X and $f : A \to Y$ is a continuous closed injection, then every compact subset $K \subset X \cup_f Y$ is the union of the compact subsets $K \cap X$ and $K \cap Y$ of the closed subspaces X and Y respectively.*

Common examples are the attaching of cells and the coproduct $X \vee Y$ of two pointed spaces X and Y in the category $\mathbf{Top}_*$ of pointed topological spaces. In the latter case the base points of X and Y are identified. If the base points of X and Y are closed in X and Y respectively, then the attaching map $f : \{*\} \to Y$ is a closed injection with closed domain in X. So we observe:

Lemma 2.2.4. *If X and Y are pointed topological spaces with closed base points, then the natural maps $X \to X \vee Y$ and $Y \to X \vee Y$ are closed embeddings.*

Thus we can regard X and Y as closed subspaces of $X \vee Y$ provided that either one of the base points is closed. In this case every compact subset $K \subset X \vee Y$ is the union of the two compact subsets $K \cap X$ and $K \cap Y$. This has implications on the compact open topology of function spaces. Let X, Y and Z be topological spaces, A be a subspace of X and $f : A \to Y$ be an attaching map. The surjective quotient map $q : X \dot\cup Y \twoheadrightarrow X \cup_f Y$ induces an injective continuous function

$$C(q, Z) : C(X \cup_f Y, Z) \to C(X \dot\cup Y, Z)$$

between function spaces (equipped with the compact open topology). We assert that this map is an embedding under the above imposed conditions on A and f.

Theorem 2.2.5. *If A is a closed subspace of X and $f : A \to Y$ is a continuous closed injection, then the quotient map $X \dot\cup Y \twoheadrightarrow X \cup_f Y$ induces an embedding $C(q, Z) : C(X \cup_f Y, Z) \hookrightarrow C(X \dot\cup Y, Z)$ of function spaces.*

Proof. Let A be a closed subspace of X and $f : A \to Y$ be a continuous closed injection. It suffices to show that $C(q, Z)$ is open onto its image. If (K, U) is a subbasic open set of $C(X \cup_f Y, Z)$ then the compact set K is the union of the compact subsets $K \cap X$ and $K \cap Y$ of $X \cup_f Y$ by Corollary 2.2.3. Therefore the open set (K, U) is the intersection of the subbasic open sets $(K \cap X, U)$ and $(K \cap Y, U)$:

$$(K, U) = ((K \cap X) \cup (K \cap Y), U) = (K \cap X, U) \cap (K \cap Y, U)$$

Let $K_1 = K \cap X$ denote the compact subset of X and (K_1, U) denote the subbasic open set in the function space $C(X, Z)$. The image of the set $(K \cap X, U)$ under $C(q, Z)$ is the set $(K_1, U) \cap \operatorname{Im} C(q, Z)$. Analogously, if $K_2 = K \cap Y$ is regarded as a compact subset of Y and (K_2, U) is a subbasic open set in the function space $C(Y, Z)$, the image of the set $(K \cap Y, U)$ under $C(q, Z)$ is the set $(K_1, U) \cap \operatorname{Im} C(q, Z)$. Since the map $C(q, Z)$ is injective, it preserves intersections. So the image of the subbasic open set (K, U) under $C(q, Z)$ is given by

$$
\begin{aligned}
C(q, Z)((K, U)) &= C(q, Z)((K \cap X, U) \cap (K \cap Y, U)) \\
&= (K_1, U) \cap \operatorname{Im} C(q, Z) \cap (K_2, U) \cap \operatorname{Im} C(q, Z) \\
&= (K_1, U) \cap (K_2, U) \cap \operatorname{Im} C(q, Z),
\end{aligned}
$$

i.e. it is open in the image $\operatorname{Im} C(q, Z)$ of $C(q, Z)$. Thus the function $C(q, Z)$ is a homeomorphism onto its image. $\qquad\square$

This carries over to function spaces of pairs of topological spaces in the following way: Let B be a subspace of the topological space Z, so that (X, A) and (Z, B) are pairs of topological spaces (where A is understood to be a subspace of X etc.). We denote with $C((X, A), (Z, B))$ the subspace of $C(X, Z)$ consisting of all functions that map the subspace A of X into the subspace B of Z. The induced function $C(q, Z)$ maps all functions in the subspace $C((X \dot\cup Y, A \dot\cup f(A)), (Z, B))$ into the subspace $C((X \dot\cup Y, A \dot\cup f(A)), (Z, B))$:

$$
\begin{array}{ccc}
C((X \dot\cup Y, A \dot\cup f(A)), (Z, B)) & \xrightarrow{\ C(q,Z)_{|C((X \dot\cup Y, A \dot\cup f(A)),(Z,B))}\ } & C((X \dot\cup Y, A \dot\cup f(A)), (Z, B)) \\
\downarrow & & \downarrow \\
C(X \cup_f Y, Z) & \xrightarrow[\ \ C(q,Z)\ \ ]{} & C(X \dot\cup Y, Z)
\end{array}
$$

Corollary 2.2.6. *If A is a closed subspace of X and $f : A \to Y$ is a closed injection, then the embedding $C(q, Z)$ restricts to an embedding of $C((X \cup_f Y, A), (Z, B))$ into the space $C((X \dot\cup Y, A \dot\cup f(A)), (Z, B))$.*

The inclusions $i_X : X \hookrightarrow X \dot\cup Y$ and $i_Y : Y \hookrightarrow X \dot\cup Y$ of X resp. Y into the disjoint union $X \dot\cup Y$ induce projections $p_X = C(i_X, Z) : C(X \dot\cup Y, Z) \twoheadrightarrow C(X, Z)$ and $p_Y = C(i_X, Z) : C(X \dot\cup Y, Z) \twoheadrightarrow C(Y, Z)$ of function spaces. These projections assemble to a natural isomorphism

$$C(X \dot\cup Y, Z) \xrightarrow[\cong]{(p_X, p_Y)} C(X, Z) \times C(Y, Z)$$

of function spaces.

Theorem 2.2.7. *If A is a closed subspace of X, B a subspace of Z and $f : A \to Y$ a continuous closed injection, then (p_X, p_Y) embeds the subspace $C((X \cup_f Y, A), (Z, B))$ of $C(X \cup_f Y, Z)$ into the subspace $C((X, A), (Z, B)) \times C((Y, f(A)), (Z, B))$ of $C(X, Z) \times C(Y, Z)$.*

Proof. Let A and B be subspaces of X and Z respectively and suppose that A is closed in X and $f : A \to Y$ is a continuous closed injection. Consider the following commutative diagram

$$
\begin{array}{ccc}
C((X \cup_f Y, A), (Z, B)) & \xrightarrow{(p_X, p_Y)} & C((X, A), (Z, B)) \times C((Y, f(A)), (Z, B)) \ , \\
\downarrow & & \downarrow \\
C(X \cup_f Y, Z) & \longrightarrow & C(X, Z) \times C(Y, Z)
\end{array}
$$

where the lower horizontal function is an embedding. The vertical arrows are inclusions of subspaces. Thus the restriction of (i_X, i_Y) to $C((Y, f(A)), (Z, B))$ is an embedding. $\qquad\square$

A special case of this are the function spaces in the category $\mathbf{Top}_*$ of pointed topological spaces. If X, Y and Z are based spaces and the base points in X and Y are closed, one can actually strengthen the result of Theorem 2.2.7:

Theorem 2.2.8. *If X and Y are pointed topological spaces with closed base points, then the natural map $C_*(X \vee Y, Z) \to C_*(X, Z) \times C_*(Y, Z)$ is a homeomorphism.*

Proof. The morphisms of based spaces in $C(X \vee Y, Z)$ correspond bijectively to the pairs of morphisms of based spaces in $C(X, Z) \times C(Y, Z)$. So if the base points of X and Y are closed, Theorem 2.2.7 ensures that $C_*(X \vee Y, Z) \to C_*(X, Z) \times C_*(Y, Z)$ is a homeomorphism. $\qquad\square$

Let $X = Y = \mathbb{S}^1$ be the one-sphere. We identify the one-sphere with the quotient space $\mathbb{R}/\mathbb{Z}$ and regard the one point union $\mathbb{S}^1 \vee \mathbb{S}^1$ as a subspace of $\mathbb{S}^1 \times \mathbb{S}^1$. The function

$$
\gamma : \mathbb{S}^1 \to \mathbb{S}^1 \times \mathbb{S}^1, \quad \gamma([t]) = \begin{cases} ([2t], *) & \text{for } t \leq \tfrac{1}{2} \\ (*, [2t]) & \text{for } t \geq \tfrac{1}{2} \end{cases}
$$

is a coproduct in $\mathbb{S}^1$. This coproduct describes the concatenation of paths in the following way: Let i_1 and i_2 be the identification of $\mathbb{S}^1$ with the first and second factor of $\mathbb{S}^1 \vee \mathbb{S}^1$ respectively. In the the sequence

$$
C_*(\mathbb{S}^1, Z) \times C_*(\mathbb{S}^1, Z) \xrightarrow[\cong]{(i_1, i_2)^{-1}} C(\mathbb{S}^1 \vee \mathbb{S}^1, Z) \xrightarrow{C_*(\gamma, Z)} C_*(\mathbb{S}^1, Z)
$$

each pair (f, g) of morphisms $\mathbb{S}^1 \to Z$ of based spaces is mapped to the concatenation of the loops f and g. This description allows to topologise the fundamental groups of based spaces, turning them into semi-topological groups.

Lemma 2.2.9. *The quotient topology of $C_*(\mathbb{S}^1, Z)$ turns the fundamental group $\pi_1(Z)$ of a based space Z into a semi-topological group.*

Proof. Let Z be a based space and equip the group $\pi_1(Z)$ with the quotient topology. It is to show that left and right translations in $\pi_1(Z)$ are continuous. Let $[\alpha]$ be a class in $\pi_1(Z)$ represented by the map $\alpha \in C_*(\mathbb{S}^1, Z)$. We use the slice

$$
s_{1,\alpha} : C_*(\mathbb{S}^1, Z) \hookrightarrow C_*(\mathbb{S}^1, Z) \times C_*(\mathbb{S}^1, Z), \quad \beta \mapsto (\alpha, \beta)
$$

through $(\alpha, *)$ to describe the left translation $\lambda_{[\alpha]}$ on $\pi_1(Z)$. The left translation $\lambda_{[\alpha]}$ on $\pi_1(Z)$ is induced by the continuous function

$$C_*(\mathbb{S}^1, Z) \xrightarrow{s_{1,\alpha}} C_*(\mathbb{S}^1, Z) \times C_*(\mathbb{S}^1, Z) \xrightarrow[\cong]{(i_1, i_2)^{-1}} C(\mathbb{S}^1 \vee \mathbb{S}^1, Z) \xrightarrow{C_*(\gamma, Z)} C_*(\mathbb{S}^1, Z).$$

This continuous function is compatible with the quotient maps $C(\mathbb{S}^1, Z) \twoheadrightarrow \pi_1(Z)$. Thus the induced map $\lambda_{[\alpha]}$ on the quotient space $\pi_1(Z)$ is continuous. The continuity of the right translation follows in a similar way. $\qquad\square$

Definition 2.2.10. *The topologised fundamental group $\pi_1^t(X)$ of a based topological space X is the space of path components of $C_*(\mathbb{S}^1, X)$ endowed with the quotient topology.*

Remark 2.2.11. It is also possible to define the semi-topological fundamental group $\pi_1^t(X)$ of a based space X using the function space $C((I, \partial I), (X, *))$. Since the function spaces $C_*(\mathbb{S}^1, X)$ and $C((I, \partial I), (X, *))$ are homeomorphic, this leads to the same result.

In an analogous fashion (presented below) it can be shown that the quotient topology turns all homotopy groups $\pi_n(Z)$ of a based topological space Z into semi-topological groups. We fix a homeomorphism $\mathbb{S}^n \cong \mathbb{S}^1 \wedge \mathbb{S}^{n-1}$ for each $n \geq 1$. Recall that for compactly generated Hausdorff spaces the smash-product distributes over the one point union (cf. [Whi78, III.2.2]). Therefore the function $\gamma : \mathbb{S}^1 \to \mathbb{S}^1 \vee \mathbb{S}^1$ introduced above induces a function

$$\gamma_n : \mathbb{S}^n \cong \mathbb{S}^1 \wedge \mathbb{S}^{n-1} \xrightarrow{\gamma \wedge \mathrm{id}_{\mathbb{S}^{n-1}}} (\mathbb{S}^1 \vee \mathbb{S}^1) \wedge \mathbb{S}^{n-1} \cong (\mathbb{S}^1 \wedge \mathbb{S}^{n-1}) \vee (\mathbb{S}^1 \wedge \mathbb{S}^{n-1}) \cong \mathbb{S}^n \vee \mathbb{S}^n.$$

These functions describe the addition in higher homotopy groups: Let i_1 and i_2 be the identification of $\mathbb{S}^n$ with the first and second factor of $\mathbb{S}^n \vee \mathbb{S}^n$ respectively. If Z is a based topological space then the composition

$$C_*(\mathbb{S}^n, Z) \times C_*(\mathbb{S}^n, Z) \xrightarrow[\cong]{(i_1, i_2)^{-1}} C_*(\mathbb{S}^n \vee \mathbb{S}^n, Z) \xrightarrow{C(\gamma_n, Z)} C_*(\mathbb{S}^n, Z)$$

induces the addition in $\pi_n(Z)$. Using the functions γ_n one can generalise Lemma 2.2.9 to arbitrary homotopy groups.

Theorem 2.2.12. *Every homotopy group $\pi_n(Z)$ of a based topological space Z is a semi-topological group when equipped with the quotient topology of $C_*(\mathbb{S}^n, Z)$.*

Proof. The proof is analogous to that of Lemma 2.2.9. Let Z be a based space and $n \geq 1$. The underlying set of $\pi_n(Z)$ is known to be a group. It remains to show that left and right translations in $\pi_n(Z)$ are continuous. Let $[\alpha]$ be a class in $\pi_n(Z)$ represented by the map $\alpha \in C_*(\mathbb{S}^n, Z)$. We use the slice

$$s_{1,\alpha} : C_*(\mathbb{S}^n, Z) \hookrightarrow C_*(\mathbb{S}^n, Z) \times C_*(\mathbb{S}^n, Z), \quad \beta \mapsto (\alpha, \beta)$$

through $(\alpha, *)$ to describe the left translation $\lambda_{[\alpha]}$ on $\pi_n(Z)$. The left translation $\lambda_{[\alpha]}$ on $\pi_n(Z)$ is induced by the continuous function

$$C_*(\mathbb{S}^n, Z) \xrightarrow{s_{1,\alpha}} C_*(\mathbb{S}^n, Z) \times C_*(\mathbb{S}^1 n, Z) \xrightarrow[\cong]{(i_1, i_2)^{-1}} C(\mathbb{S}^n \vee \mathbb{S}^n, Z) \xrightarrow{C_*(\gamma_n, Z)} C_*(\mathbb{S}^n, Z),$$

This continuous function is compatible with the quotient maps $C(\mathbb{S}^1, Z) \twoheadrightarrow \pi_n(Z)$. Thus the induced map $\lambda_{[\alpha]}$ on the quotient space $\pi_n(Z)$ is continuous. The continuity of the right translation follows in a similar way. $\qquad\square$

Definition 2.2.13. *The* topologised n-th homotopy group $\pi_n^t(Z,*)$ *of a based space* $(Z,*)$ *is the space of path components of* $C_*(\mathbb{S}^n, Z)$ *equipped with the quotient topology.*

For every map $f : Z \to Z'$ of based spaces the induced function $C(\mathbb{S}^n, f)$ on the function spaces descends to a continuous function on the homotopy groups. So the homotopy group functors π_n on **Top** generalise to functors π_n^t into the category of semi-topological groups. We assert that these functors π_n^t satisfy the homotopy invariance condition of a topological homology theory:

Lemma 2.2.14. *Homotopic functions* $f \simeq g : X \to Y$ *between topological spaces* X *and* Y *induce identical morphisms* $\pi_n^t(f), \pi_n^t(g) : \pi_n^t(X) \to \pi_n^t(Y)$.

Proof. This follows from the fact that the underlying morphisms $\pi_n(f)$ and $\pi_n(g)$ between the discrete homotopy groups $\pi_n(X)$ and $\pi_n(Y)$ coincide.   □

It remains to show that the long sequence of semi-topological homotopy groups for certain pairs of spaces is exact.

2.3 Singular Topological Homology

In this section we define a generalisation of singular homology to a topological homology theory on various subcategories of **Top**(2). This requires the coefficients to be an Abelian topological group. The construction is analogous to the one of classical singular homology, but the functors used here have codomain the category of topological spaces or Abelian topological groups. Consider the function space functors $C(\Delta_n, -)$, where for each topological space X the spaces $C(\Delta_n, X)$ of all singular n-simplices are equipped with the compact-open topology. These functors assemble to a functor $C(\Delta, -) : \textbf{Top} \to \textbf{Top}^{\mathcal{O}^{op}}$ (where $\mathcal{O}$ denotes the simplicial category cf. F.1, F.4.2). This functor assigns to each topological space X the simplicial topological space $C(\Delta, X)$. The underlying simplicial set $\hom_{\textbf{Top}}(\Delta, X)$ is used to define the classical singular homology of X. Below we derive a topological version of the classical construction.

Definition 2.3.1. *The functor* $C(\Delta, -) : \textbf{Top} \to \textbf{Top}^{\mathcal{O}^{op}}$ *is called the* singular simplicial space functor. *For any topological space* X, *the functor* $C(\Delta, X) : \mathcal{O}^{op} \to \textbf{Top}$ *is called the* singular simplicial space *associated to the topological space* X.

Remark 2.3.2. Since the zeroth standard simplex Δ_0 is a singleton space, the part $C(\Delta_0, X)$ of the singular simplicial space of a topological space X is homeomorphic to X itself. The part $C(\Delta_1, X)$ is the underlying space of the path groupoid of X.

Remark 2.3.3. The singular simplicial space functor $C(\Delta, -)$ is a presheaf of simplicial topological spaces by definition.

Each part $C(\Delta_n, X)$ of the singular simplicial space of a topological space X contains the space X as a the subspace of constant functions. In this way one can regard the constant simplicial space X as a simplicial subspace of the singular simplicial space $C(\Delta, X)$. Since function spaces with the compact open topology inherit some of the separation properties of the codomain space we obtain:

Lemma 2.3.4. *The singular simplicial space* $C(\Delta, X)$ *of a topological space* X *is* $T_1, T_2,$ *regular, completely regular or* T_3 *if and only if* X *is* $T_1, T_2,$ *regular, completely regular or* T_3 *respectively.*

Proof. Let X be a topological space. Since each function space $C(\Delta_n, X)$ contains X as a subspace, X has to satisfy every of the above separation axioms that is satisfied by the singular simplicial space $C(\Delta, X)$. For the converse implications see [Dug89, XII.1.3] for a proof in the case of Hausdorffness or regularity. The proof for the separation axiom T_1 is analogous to that of the Hausdorffness. It remains to prove that $C(\Delta, X)$ is completely regular if X is completely regular. Here it suffices to show that each point in $C(\Delta_n, X)$ has a neighbourhood basis of supports of continuous real valued functions. Let X be completely regular and (K, U) a subbasic open neighbourhood of a function $f : \Delta_n \to X$, i.e. K is compact, $U \subset X$ is open and the inverse image $f^{-1}(U)$ contains K. We construct a continuous function separating f and the complement of (K, U). The image $f(K) \subset U$ is compact, thus there exists a continuous function $\varphi : X \to \mathbb{R}$ with support in U satisfying $f(K) \subset \varphi^{-1}((0, 1))$ (cf. Lemma B.2.5). We assert that the composition of the functions

$$C(\Delta_n, X) \xrightarrow{g \mapsto g|_K} C(K, X) \xrightarrow{g \mapsto \varphi \circ g} C(K, \mathbb{R}) \xrightarrow{\min} \mathbb{R}$$

is a continuous function separating f and the complement of (K, U). To show continuity we note that each of the functions above is continuous: The restriction of functions to a compact subspace is continuous, composition with a fixed function is continuous (cf. B.5.1) and taking the minimum of real valued functions on compact spaces is also continuous. Therefore the assignment

$$\tilde{\varphi} : C(\Delta_n, X) \to \mathbb{R}, \quad \tilde{\varphi}(g) = \min_{k \in K}\{f(k)\}.$$

defines a continuous function on $C(\Delta_n, X)$. The value of $\tilde{\varphi}$ at f is not zero because the image $f(K)$ is contained in $\varphi^{-1}((0, 1))$. If $h : \Delta_n \to X$ is not contained in (K, U), then there exists a point $p \in K$ such that $h(p) \notin U$. This implies $\varphi(h(p)) = 0$ because the support of φ is contained in U. Therefore the continuous function $\tilde{\varphi}$ has support in the open set (K, U). □

Example 2.3.5. Topological groups are completely regular. Thus the singular simplicial space $C(\Delta, G)$ of any topological group G is completely regular. It is Hausdorff iff X is Hausdorff.

Example 2.3.6. Topological manifolds which satisfy the separation axiom T_0 are completely regular. Thus the singular simplicial space $C(\Delta, M)$ of any T_0-manifold M is completely regular. It is Hausdorff iff M is Hausdorff.

We expect the singular simplicial space functor $C(\Delta, -)$ to preserve homotopy, i.e. it should translate homotopic functions $f \simeq g : X \to Y$ between topological spaces X and Y into (simplicially) homotopic morphisms of simplicial topological spaces. If $f \simeq g : X \to Y$ are such homotopic functions and $H : X \times I \to Y$ is a homotopy from f to g then the singular simplicial space functor $C(\Delta, -)$ maps the homotopy H to a morphism

$$C(\Delta, H) : C(\Delta, X \times I) \to C(\Delta, Y)$$

of simplicial topological spaces. The standard simplices Δ_n are locally compact, so the singular simplicial space functor is continuous by Corollary B.5.13. Therefore the simplicial spaces $C(\Delta, X) \times C(\Delta, I)$ and $C(\Delta, X \times I)$ are isomorphic. The second factor $C(\Delta, I)$ of the former product space contains the discrete simplicial space $\mathrm{hom}(-, [1])$ as a subspace. The latter can be embedded into the singular simplicial space $C(\Delta, I)$ via the map

$$i : \mathrm{hom}(-, [1]) \to C(\Delta, I), \quad \alpha \mapsto \langle \alpha(0), \dots, \alpha(n) \rangle,$$

which assigns to each morphism $\alpha : [m] \to [n]$ in $\mathcal{O}$ the affine n-simplex with vertices $\alpha(0), \ldots, \alpha(n)$ in I. Using this embedding one obtains the following chain of morphisms of simplicial topological spaces:

$$C(\Delta, X) \times \hom(-, [1]) \xrightarrow{\mathrm{id}_{C(\Delta,X)} \times i} C(\Delta, X) \times C(\Delta, I) \cong C(\Delta, X \times I) \xrightarrow{C(\Delta, H)} C(\Delta, Y)$$

The natural isomorphism $C(\Delta, X) \times C(\Delta, I) \cong C(\Delta, X \times I)$ of simplicial spaces in this chain of morphisms is the inverse of the morphism $(C(\Delta, \mathrm{pr}_1), C(\Delta, \mathrm{pr}_2))$, where pr_1 and pr_2 denote the projections of $X \times I$ onto the first and second factor respectively. The composition of the morphisms in the above chain is a candidate for a simplicial homotopy. We assign it a special symbol:

Definition 2.3.7. *For any homotopy* $H : X \times I \to Y$ *in* **Top** *the composition* $C(\Delta, H) \circ (C(\Delta, \mathrm{pr}_1), C(\Delta, \mathrm{pr}_2))^{-1} \circ (\mathrm{id}_{C(\Delta,Y)} \times i)$ *is denoted by* $h(H)$.

Let $f, g : X \to Y$ be continuous functions as above. We verify that for each homotopy $H : X \times I \to Y$ from f to g the morphism $h(H)$ indeed provides a simplicial homotopy from $C(\Delta, f)$ to $C(\Delta, g)$:

Lemma 2.3.8. *For any homotopy* $H : X \times I \to Y$ *from* f *to* g, *the morphism* $h(H)$ *provides a homotopy from* $C(\Delta, f)$ *to* $C(\Delta, g)$.

Proof. Let $f, g : X \to Y$ be continuous maps between topological spaces X and Y and $H : X \times I \to Y$ be a homotopy from g to f. The zero morphisms in the components $\hom([n], [1])$ of $\hom(-, [1])$ form a singleton simplicial subspace of $\hom(-, [1])$. We denote the slice $C(\Delta, X) \hookrightarrow C(\Delta, X) \times \hom(-, [1])$ through this singleton subspace by $j_{C(\Delta,Y),1}$. (The index 1 stems from the fact that the slice is induced by the inclusion $\hom(-, \epsilon_1) : \hom(-, [0]) \to \hom(-, [1])$.) Analogously, the constant 1-morphisms in $\hom(-, [1])$ form a singleton simplicial subspace of $\hom(-, [1])$. We denote the slice $C(\Delta, X) \hookrightarrow C(\Delta, X) \times \hom(-, [1])$ through this singleton subspace by $j_{C(\Delta,Y),0}$. By Theorem F.6.7 it suffices to prove that the compositions of $h(H)$ with the embeddings $j_{C(\Delta,Y),0}$ and $j_{C(\Delta,Y),1}$ are the morphisms $C(\Delta, f)$ and $C(\Delta, g)$ respectively, i.e. that the following diagram is commutative:

$$C(\Delta, X) \xrightarrow{j_{C(\Delta,Y),0}} C(\Delta, Y) \times \hom(-, [1]) \xleftarrow{j_{C(\Delta,Y),1}} C(\Delta, X)$$

with $C(\Delta, f)$ and $C(\Delta, g)$ going to $C(\Delta, Y')$ via h.

Let $\sigma \in C(\Delta_n, Y)$ be a singular n-simplex in Y and let 1 denote the constant morphism 1 in $\hom([n], [1])$. Straight from the definitions one derives:

$$h(H) \circ j_{C(\Delta,Y),0}(\sigma) = h(H)(\sigma, 1) = C(\Delta, H)(\sigma, \langle 1, \ldots, 1 \rangle)$$
$$= H \circ (\sigma, \langle 1, \ldots, 1 \rangle) = H(-, 1) \circ \sigma = f \circ \sigma = C(\Delta, f)(\sigma)$$

Because this is true for arbitrary $n \in \mathbb{N}$ and singular n-simplices $\sigma \in C(\Delta_n, Y)$ the morphisms $h(H) \circ j_{C(\Delta,Y),0}$ and $C(\Delta, f)$ are equal. In the same fashion one concludes the equality of the morphisms $h(H) \circ j_{C(\Delta,Y),1}$ and $C(\Delta, g)$. Thus the morphism $h(H)$ provides a homotopy from $C(\Delta, f)$ to $C(\Delta, g)$. $\square$

All in all we can summarise these observations in the following theorem:

Theorem 2.3.9. *The singular simplicial space functor* $C(\Delta, -)$ *preserves homotopy equivalence of morphisms.*

Corollary 2.3.10. *The singular simplicial space functor $C(\Delta, -)$ preserves the homotopy equivalence of spaces.*

Proof. Let $X \simeq Y$ be homotopy equivalent spaces. Then there exist continuous functions $f : X \to Y$ and $g : Y \to X$ such that $fg \simeq \mathrm{id}_Y$ and $gf \simeq \mathrm{id}_X$. By Theorem 2.3.9 there exist simplicial homotopies h' from $C(\Delta, fg)$ to $\mathrm{id}_{C(\Delta,Y)}$ and h from $C(\Delta, gf)$ to $\mathrm{id}_{C(\Delta,X)}$. Thus the simplicial spaces $C(\Delta, X)$ and $C(\Delta, Y)$ are homotopy equivalent. $\qquad\square$

Having verified the preservation of homotopy we proceed with constructing a topological version of singular homology. Recall the free Abelian topological group functor $F_{\mathbf{TopAb}} : \mathbf{Top} \to \mathbf{TopAb}$ from the category $\mathbf{Top}$ of topological spaces into the category $\mathbf{TopAb}$ of Abelian topological groups. It assigns to each topological space X the free Abelian topological group $F_{\mathbf{TopAb}}X$ on X. The functor $F_{\mathbf{TopAb}}$ is left adjoint to the forgetful functor $U : \mathbf{TopAb} \to \mathbf{Top}$ and thus preserves colimits. We compose the functor $F_{\mathbf{TopAb}}$ with the singular simplicial space functor $C(\Delta, -)$ to obtain a topologised version of the singular simplicial group of a topological space:

Definition 2.3.11. *The functor $S^t := F_{\mathbf{TopAb}} \circ C(\Delta, -) : \mathbf{Top} \to \mathbf{TopAb}^{\mathcal{O}^{op}}$ is called the* singular simplicial topological group functor. *For any topological space X, the functor $S^t(X) : \mathcal{O}^{op} \to \mathbf{TopAb}$ is called the* singular simplicial topological group *associated to X.*

Since any topological group is completely regular, one infers that for non completely regular spaces X the natural map $X \to UF_{\mathbf{TopAb}}X$ is not an embedding. However, it is always injective, because the abstract group underlying the topological group $F_{\mathbf{TopAb}}X$ is the free group on the set of points of X. (cf. E.1). Therefore the counit $\mathrm{id}_{\mathbf{Top}} \to UF_{\mathbf{TopAb}}$ of the adjunction between the forgetful functor $U : \mathbf{TopAb} \to \mathbf{Top}$ and the free Abelian topological group functor $F_{\mathbf{TopAb}}$ consists of continuous injections.

Lemma 2.3.12. *If a topological space X is completely regular then the natural map $C(\Delta, X) \to S^t(X)$ is an embedding. If X is Hausdorff in addition, then the singular simplicial topological group $S^t(X)$ is Hausdorff and the embedding $C(\Delta, X) \hookrightarrow S^t(X)$ is closed.*

Proof. If the topological space X is completely regular then the natural map $C(\Delta, X) \to S^t(X)$ is an embedding by Lemma E.1.9. If X also is Hausdorff, then the simplicial space $C(\Delta, X)$ is also completely regular and Hausdorff by Lemma 2.3.4. Therefore the free Abelian topological group $S^t(X)$ on $C(\Delta, X)$ is Hausdorff and the inclusion $C(\Delta, X) \hookrightarrow S^t(X)$ is a closed embedding by Lemma E.1.18. $\quad\square$

Example 2.3.13. The singular simplicial space $C(\Delta, G)$ of any Hausdorff topological group G is completely regular and Hausdorff. Thus the singular simplicial topological group of Hausdorff topological groups are Hausdorff.

Example 2.3.14. The singular simplicial space $C(\Delta, G)$ of any Hausdorff topological manifold is completely regular and Hausdorff. Thus the singular simplicial topological group $S^t(M)$ of any Hausdorff topological manifold M is Hausdorff.

Lemma 2.3.15. *The singular simplicial topological group functor S^t preserves homotopy.*

Proof. Let $f \simeq g : X \to Y$ be homotopic continuous functions. The singular simplicial space functor $C(\Delta, -)$ preserves homotopy by theorem 2.3.9, so a homotopy $H : X \times I \to Y$ from f to g yields a homotopy from the morphism $C(\Delta, f)$ to the morphism $C(\Delta, g)$. Applying the free Abelian topological group functor $F_{\mathbf{TopAb}}$ to a homotopy from $C(\Delta, f)$ to $C(\Delta, g)$ yields a homotopy from $S^t(f) = F_{\mathbf{TopAb}}C(\Delta, f)$ to $S^t(g) = F_{\mathbf{TopAb}}C(\Delta, g)$. □

Considering the chain complex $S_*^t(X)$ associated to the singular simplicial topological group $S^t(X)$ associated to a topological space X yields another functor $S_*^t := (-)_* \circ S^t : \mathbf{Top} \rightarrow \mathbf{Ch}(\mathbf{TopAb})$ into the category $\mathbf{Ch}(\mathbf{TopAb})$ of chain complexes of Abelian topological groups.

Definition 2.3.16. *The functor $S_*^t : \mathbf{Top} \rightarrow \mathbf{Ch}(\mathbf{TopAb})$ is called the* singular topological chain complex functor. *The chain complex $S_*^t(X)$ associated to a topological space X is called the* singular topological chain complex *associated to the space X.*

Lemma 2.3.17. *The singular topological chain complex functor S_*^t preserves homotopy.*

Proof. The singular simplicial space functor S^t preserves homotopy by Lemma 2.3.15. Simplicially homotopic morphisms between simplicial Abelian topological groups induce homotopic morphisms between the associated chain complexes by Lemma F.6.6, so S_*^t preserves homotopy as well. □

Definition 2.3.18. *The functor $H_*^t := H_* \circ S_*^t$ is called the* singular topological homology functor. *The graded Abelian topological group $H_*^t(X)$ associated to a space X is called its* topological singular homology.

Lemma 2.3.19. *The singular topological homology functor H_*^t is homotopy invariant, i.e. if $f \simeq g : X \rightarrow Y$ are homotopic continuous functions then $H_*^t(f) = H_*^t(g)$.*

Proof. This is a direct consequence of Lemma 2.3.17. □

Definition 2.3.20. *Let $\mathfrak{U}$ be a family of subsets of a topological space X. A map $f : Y \rightarrow X$ is called* $\mathfrak{U}$-small *if its image $f(Y)$ is contained in one of the sets $U \in \mathfrak{U}$.*

Given a family $\mathfrak{U}$ of subsets of a topological space X, a singular n-simplex σ in $C(\Delta_n, X)$ is then called $\mathfrak{U}$-small if it is a $\mathfrak{U}$-small map $\sigma : \Delta_n \rightarrow X$. Like in classical singular homology, we can consider only $\mathfrak{U}$-small simplices.

Definition 2.3.21. *For any family $\mathfrak{U}$ in a topological space X the subspace of $C(\Delta_n, X)$ consisting of $\mathfrak{U}$-small simplices is denoted by $C(\Delta_n, X, \mathfrak{U})$.*

In case there is no restriction on the singular n-simplices (e.g. if the family $\mathfrak{U}$ contains the space X,) we simply write $C(\Delta_n, X)$ for the space of singular n-simplices. Since boundaries and degeneracies of $\mathfrak{U}$-small singular simplices are $\mathfrak{U}$-small as well, the spaces $C(\Delta_n, X, \mathfrak{U})$ form a simplicial subspace of the singular simplicial space $C(\Delta, X)$.

Definition 2.3.22. *The singular space $C(\Delta, X, \mathfrak{U})$ is called the* $\mathfrak{U}$-small singular simplicial space *of X.*

These simplicial spaces can best be described by using pairs $(X, \mathfrak{U})$ of topological spaces X with a family of subspaces $\mathfrak{U}$. These pairs are the objects of a category:

Definition 2.3.23. *The category $\mathbf{TF}$ of topological spaces with families of subsets is given by the following data:*

  1. *Its objects are the pairs $(X, \mathfrak{U})$ of a topological space X and a family $\mathfrak{U}$ of subsets of X.*

2. *The morphisms* $(X, \mathfrak{U}) \to (Y, \mathfrak{V})$ *are the continuous functions* $f : X \to Y$ *satisfying* $f(\mathfrak{U}) \prec \mathfrak{V}$, *i.e.* $f(\mathfrak{U})$ *refines* $\mathfrak{V}$.

To each topological space X one can assign the trivial covering $\mathfrak{U} = \{X\}$ consisting only of the space X itself. In this way one can embed the category **Top** as a full subcategory of **TF**. As the above observations show, the singular simplicial space functor on the subcategory **Top** of **TF** extends to a functor

$$C(\Delta, -, -) : \mathbf{TF} \to \mathbf{Top}^{\mathcal{O}^{op}}.$$

Definition 2.3.24. *The extension* $C(\Delta, -, -)$ *is called the* singular simplicial space functor *on* **TF**. *The functor* $S^t := F_{\mathbf{TopAb}} \circ C(\Delta, -, -) : \mathbf{TF} \to \mathbf{TopAb}^{\mathcal{O}^{op}}$ *is called the* singular simplicial topological group functor on **TF**. *For any object* $(X, \mathfrak{U})$ *of* **TF** *the functor* $S^t(X, \mathfrak{U}) : \mathcal{O}^{op} \to \mathbf{TopAb}$ *is called the* $\mathfrak{U}$-small singular simplicial topological group *associated to* X.

Definition 2.3.25. *The chain complex* $S_*^t(X, \mathfrak{U})$ *associated to* $S^t(X, \mathfrak{U})$ *is called the* $\mathfrak{U}$-small singular topological chain complex *(of* X*)*.

If $\mathfrak{U}$ and $\mathfrak{V}$ are families of subsets of a topological space X and $\mathfrak{U}$ is a refinement of V then $C(\Delta, X, \mathfrak{U})$ is a subspace of $C(\Delta, X, \mathfrak{V})$. If $\mathfrak{U}$ is a family of open subsets then for any set $U \in \mathfrak{U}$ the inclusion $i_U : U \hookrightarrow X$ induces an embedding of the space $C(\Delta_n, U)$ as an open subspace in $C(\Delta_n, X)$. In this case the space $C(\Delta_n, X, \mathfrak{U})$ of $\mathfrak{U}$-small n-simplices is the union of open subspaces

$$C(\Delta_n, X, \mathfrak{U}) = \bigcup_{U \in \mathfrak{U}} C(\Delta_n, U),$$

hence is itself an open subspace of the space $C(\Delta_n, X)$ of all singular n-simplices. So we note:

Lemma 2.3.26. *If* $\mathfrak{U}$ *is a family of open subsets of the topological space* X *then* $C(\Delta, X, \mathfrak{U})$ *is an open simplicial subspace of* $C(\Delta, X)$.

Unfortunately this result does not extend to the free Abelian groups $S^t(X, \mathfrak{U})$ and $S^t(X)$, even if they are completely regular (see Lemma 1.4.13).

Definition 2.3.27. *The homology of the complex* $S_*^t(X, \mathfrak{U})$ *in* **TopAb** *is called the* $\mathfrak{U}$-small singular topological homology of X. *It is denoted by* $H_*^t(X; \mathfrak{U})$. *The extension* $H_*^t := H_* \circ S_*^t \circ F_{\mathbf{TopAb}} \circ C(\Delta, -, -)$ *of the singular homology functor to the category* **TF** *is called the* singular topological homology functor on **TF**.

If one composes the forgetful functor $U : \mathbf{TopAb} \to \mathbf{Ab}$ with $S^t(X; \mathfrak{U})$ one obtains the underlying simplicial group $S(X, \mathfrak{U})$ defining ordinary singular homology. As a consequence one obtains:

Lemma 2.3.28. *The composition of the forgetful functor* $U : \mathbf{TopAb} \to \mathbf{Ab}$ *with the topological homology* H_*^t *is the classical singular homology:* $H_* = U \circ H_*^t$.

Proof. The forgetful functor $U : \mathbf{TopAb} \to \mathbf{Ab}$ preserves kernels and cokernels. □

Similar to the classical singular homology one can generalise the above constructions using an Abelian group V of "coefficients". Here the coefficient group V is required to be an Abelian topological group. If V is such an Abelian topological group then one can compose the functor

$$- \otimes V : \mathbf{TopAb} \to \mathbf{TopAb}$$

(where the tensor product is taken in the category **TopAb** of Abelian topological groups) with the singular simplicial Abelian topological group functor S^t. The composition $(- \otimes V) \circ S^t$ is abbreviated by $S^t \otimes V$.

Definition 2.3.29. *The functor* $S^t \otimes V : \mathbf{Top} \to \mathbf{TopAb}^{\mathcal{O}^{op}}$ *is called the* singular simplicial topological group functor with coefficients V. *For any topological space* X, *the functor* $S^t(X) \otimes V : \mathcal{O}^{op} \to \mathbf{TopAb}$ *is called the* singular simplicial topological group with coefficients V associated to X.

As before, considering the chain complex associated with the singular simplicial topological group $S^t(X) \otimes V$ with coefficients V of a space X yields a functor $S^t_* \otimes V : \mathbf{Top} \to \mathbf{Ch}(\mathbf{TopAb})$ from the category $\mathbf{Top}$ of topological spaces into the category $\mathbf{Ch}(\mathbf{TopAb})$ of chain complexes of Abelian topological groups. The chain complex $S^t_*(X) \otimes V$ obtained for a topological space X is the tensor product of the singular topological chain complex $S_*(X)$ with the Abelian topological group V.

Definition 2.3.30. *The functor* $S^t_* \otimes V : \mathbf{Top} \to \mathbf{Ch}(\mathbf{TopAb})$ *is called the* singular topological chain complex functor with coefficients V. *The chain complex* $S^t_*(X) \otimes V$ *associated to a topological space* X *is called the* singular topological chain complex with coefficients V associated to the space X.

Lemma 2.3.31. *The functors* $S^t \otimes V$ *and* $S^t_* \otimes V$ *preserve homotopy.*

Proof. If $f \simeq g : X \to Y$ are homotopic continuous functions between topological spaces X and Y then the morphisms $S^t(f)$ and $S^t(g)$ of simplicial Abelian topological groups are homotopic by Lemma 2.3.15. Tensoring a simplicial homotopy from $S^t(f)$ to $S^t(g)$ with the identity id_V of V yields a simplicial homotopy from $(S^t \otimes V)(f) = S^t(f) \otimes \mathrm{id}_V$ to $(S^t \otimes V)(g) = S^t(g) \otimes \mathrm{id}_V$, so the functor $S^t \otimes V$ preserves homotopy. Since simplicially homotopic morphisms between simplicial Abelian topological groups induce homotopic morphisms between the associated chain complexes (by Lemma F.6.6), the singular topological chain complex functor $S^t_* \otimes V$ also preserves homotopy. $\square$

Definition 2.3.32. *The functor* $H^t_*(-; V) := H_* \circ (S^t_* \otimes V)$ *is called the* singular topological homology functor with coefficients V. *The graded Abelian topological group* $H^t_*(X; V)$ *associated to a space* X *is called its* topological singular homology with coefficients V.

Lemma 2.3.33. *The singular topological homology functor* $H^t_*(-; V)$ *with coefficients* V *is homotopy invariant.*

Proof. This is a direct consequence of Lemma 2.3.31. $\square$

As before, these functors can be adapted to work only with $\mathfrak{U}$-small singular simplices, where $\mathfrak{U}$ is a family of subsets of the topological space X considered. All that is to be done here fore is to tensor all Abelian topological groups that occur with the group V of coefficients. We use the extension of S^t to the category $\mathbf{TF}$ of topological spaces with families of subsets introduced before:

Definition 2.3.34. *The functor* $S^t \otimes V := \mathbf{TF} \to \mathbf{TopAb}^{\mathcal{O}^{op}}$ *is called the* singular simplicial topological group functor with coefficients V on $\mathbf{TF}$. *For any object* $(X, \mathfrak{U})$ *of* $\mathbf{TF}$ *the functor* $S^t(X, \mathfrak{U}) \otimes V : \mathcal{O}^{op} \to \mathbf{TopAb}$ *is called the* $\mathfrak{U}$-small singular simplicial topological group with coefficients V associated to X.

Definition 2.3.35. *The chain complex* $S^t_*(X, \mathfrak{U}) \otimes V$ *associated to* $S^t(X, \mathfrak{U})$ *is called the* $\mathfrak{U}$-small singular topological chain complex with coefficients V (of X).

Definition 2.3.36. *The homology of the complex* $S^t_*(X, \mathfrak{U}) \otimes$ *in* $\mathbf{TopAb}$ *is called the* $\mathfrak{U}$-small topological singular homology with coefficients V of X. *It is denoted by* $H^t_*(X; \mathfrak{U}; V)$. *The extension* $H^t_*(-; V) := H_* \circ S^t_* \otimes V$ *of the singular homology functor to the category* $\mathbf{TF}$ *is called the* singular topological homology functor with coefficients on $\mathbf{TF}$.

If one composes the forgetful functor $U : \textbf{TopAb} \to \textbf{Ab}$ with $S^t(X; \mathfrak{U}) \otimes V$, one obtains the underlying simplicial group defining ordinary singular homology with coefficients UV. As a consequence one obtains:

Lemma 2.3.37. *The composition of the forgetful functor $U : \textbf{TopAb} \to \textbf{Ab}$ with the topological homology functor $H_*^t(-; V)$ (with coefficients V) is the classical singular homology functor $H_*(-; UV)$ with coefficients UV.*

Proof. The forgetful functor $U : \textbf{TopAb} \to \textbf{Ab}$ preserves kernels and cokernels. $\square$

This can be used to show that, like ordinary singular homology, the topological singular homology is actually independent of the family $\mathfrak{U}$ chosen, provided it is an open covering.

Theorem 2.3.38. *If $\mathfrak{U}$ is an open covering of X and V a group of coefficients, then the inclusion $S^t(X, \mathfrak{U}) \otimes V \hookrightarrow S^t(X) \otimes V$ induces an isomorphism in homology.*

Proof. Let $\mathfrak{U}$ be an open covering of the topological space X and V be an Abelian topological group. Let $Z_n^t(X, \mathfrak{U}; V)$ denote the subspace of $\mathfrak{U}$-small cycles in $S^t(X, \mathfrak{U})([n]) \otimes V$ and $B_n^t(X, \mathfrak{U}; V)$ denote the subspace of boundaries of $\mathfrak{U}$-small singular $n+1$ chains. Similarly we denote by $Z_n^t(X; V)$ the subspace of cycles in $S^t(X)([n]) \otimes V$ and by $B_n^t(X; V)$ the subspace of boundaries of singular $n+1$-chains. The inclusion $S(X, \mathfrak{U}) \otimes V \hookrightarrow S(X) \otimes V$ of the underlying singular simplicial groups induces an isomorphism $H_*(X, \mathfrak{U}; V) \cong H_*(X : V)$ in classical singular homology. This implies the equality

$$B_n^t(X, \mathfrak{U}; V) = Z_n^t(X, \mathfrak{U}; V) \cap B_n^t(X; V).$$

Applying the Noether Isomorphism Theorem 1.1.2 now shows that the inclusion $Z_n^t(X, \mathfrak{U}; V) \hookrightarrow Z_n^t(X; V)$ induces an inclusion $H_n^t(X, \mathfrak{U}; V) \hookrightarrow H_*^t(X; V)$. Since this inclusion is bijective, it is an isomorphism. $\square$

All the functors $S^t \otimes V$, $S_*^t \otimes V$ and $H_*^t(-; V)$ are natural in the topological group V, i.e. one can consider these as parts of bifunctors:

$$S^t \otimes - : \textbf{Top} \times \textbf{TopAb} \to \textbf{TopAb}^{\mathcal{O}^{op}},$$
$$S_*^t \otimes - : \textbf{Top} \times \textbf{TopAb} \to \textbf{Ch}(\textbf{TopAb}),$$

and

$$H_*^t(-; -) : \textbf{Top} \times \textbf{TopAb} \to \textbf{TopAb}.$$

Thus a homomorphism $\varphi : V \to V'$ between Abelian topological groups V and V' induces natural transformations $S^t \otimes \varphi : S^t \otimes V \to S^t \otimes V'$, $S_*^t \otimes \varphi : S_*^t \otimes V \to S_*^t \otimes V'$ and $H_*^t(-; \varphi) : H_*^t(-; V) \to H_*^t(-; V')$. Especially if $V \hookrightarrow V' \twoheadrightarrow V''$ is an exact sequence of Abelian topological groups, one obtains sequences

$$S^t \otimes V \to \ S^t \otimes V' \ \to S^t \otimes V''$$
$$S_*^t \otimes V \to \ S_*^t \otimes V' \ \to S_*^t \otimes V''$$
$$H^t(-; V) \to H^t(-; V') \to H^t(-; V'')$$

of natural transformations. The sequences of the underlying groups lead to a long exact sequence in classical singular homology. Unfortunately the derivation of a long exact sequence for topological homology is delicate and needs some additional requirements on the exact sequence $V \hookrightarrow V' \twoheadrightarrow V''$. The reason here fore is that the free Abelian topological group functor $F_{\textbf{TopAb}}$ does not preserve embeddings in general. (See Section 1.4 for a discussion of this topic.) So tensoring the singular topological chain complex $S_*^t(X)$ of a topological space X with the exact sequence $V \hookrightarrow V' \twoheadrightarrow V''$ does in general not yield a short exact sequence of chain complexes. This can be resolved by requiring that the short exact sequence $V \hookrightarrow V' \to V''$ is a locally trivial bundle.

Recall that such short exact sequences are called *locally trivial* (see Definition 1.4.50). The local triviality of short exact sequences ensures that tensoring with a free Abelian topological group A preserves the exactness of this sequence:

Theorem 2.3.39. *A locally trivial short exact sequence* $0 \to V \to V' \to V'' \to 0$ *induces a short exact sequence* $0 \to S^t \times V \to S^t \otimes V' \to S^t \otimes V'' \to 0$ *of singular simplicial topological group functors.*

Proof. Let X be a topological space and $0 \to V \to V' \to V'' \to 0$ be a locally trivial short exact sequence of Abelian topological groups. Each group $S^t(X)([n])$ is a free Abelian topological group. Therefore all the sequences

$$0 \to S^t(X)([n]) \otimes V \to S^t(X)([n]) \otimes V' \to S^t(X)([n]) \otimes V'' \to 0$$

are exact by Theorem 1.4.53. So the sequence $0 \to S^t \times V \to S^t \otimes V' \to S^t \otimes V'' \to 0$ of singular simplicial topological group functors is exact.   □

Corollary 2.3.40. *A locally trivial short exact sequence* $0 \to V \to V' \to V'' \to 0$ *induces a short exact sequence* $0 \to S_*^t \otimes V \to S_*^t \otimes V' \to S_*^t \otimes V'' \to 0$ *of singular simplicial topological chain complex functors.*

We would like to derive a long exact sequence in homology from the short exact sequence $0 \to S_*^t \otimes V \to S_*^t \otimes V' \to S_*^t \otimes V'' \to 0$ above. The usual way is to apply the Snake Lemma, which requires the differentials of the complexes to be quotients onto their images. Fortunately, the fact that the singular topological cochain complexes (with coefficients) arise as associated chain complexes of simplicial Abelian topological groups in conjunction with Theorem 1.3.9 (resp. Corollary 1.3.10) ensures that these chain complexes are proper.

Theorem 2.3.41. *A locally trivial short exact sequence* $0 \to V \to V' \to V'' \to 0$ *of Abelian topological groups induces a long exact sequence*

$$\cdots \to H_{n+1}^t(-;V'') \xrightarrow{\delta_{n+1}} H_n^t(-;V) \to H_n^t(-;V') \to H_n^t(-;V'') \xrightarrow{\delta_n} \cdots$$

of singular topological homology functors.

Proof. If $0 \to V \to V' \to V'' \to 0$ is a locally trivial short exact sequence of Abelian topological groups, then the sequence $0 \to S^t \otimes V \to S^t \otimes V' \to S^t \otimes V'' \to 0$ of singular simplicial topological group functors is exact by Theorem 2.3.39. By Corollary 1.3.15 every exact short sequence of simplicial Abelian topological groups induces a long exact sequence in homology.   □

Example 2.3.42. If V is an open subgroup of an Abelian topological group V, then the factor group $V'' = V'/V$ is discrete and the exact sequence $0 \to V \to V' \to V'' \to 0$ is locally trivial. Therefore the long sequence

$$\cdots \to H_{n+1}^t(\quad;V'') \xrightarrow{\delta_{n+1}} H_n^t(-;V) \to H_n^t(-;V') \to H_n^t(-;V'') \xrightarrow{\delta_n} \cdots$$

of homology functors is exact.

Example 2.3.43. If V is a discrete subgroup of an Abelian topological group V, then the quotient homomorphism $V' \twoheadrightarrow V'' = V'/V$ is a covering map and the short exact sequence $0 \to V \to V' \to V'' \to 0$ is locally trivial. Therefore the long sequence

$$\cdots \to H_{n+1}^t(-;V'') \xrightarrow{\delta_{n+1}} H_n^t(-;V) \to H_n^t(-;V') \to H_n^t(-;V'') \xrightarrow{\delta_n} \cdots$$

of homology functors is exact.

Each pair (X,A) of spaces in **Top**(2) gives rise to a long exact sequence of classical singular homology groups. This result does not generalise to singular topological homology. The reason here fore is – similar to topological singular homology with coefficients – that the free Abelian group functor $F_{\mathbf{TopAb}}$ does not preserve embeddings. As a consequence the simplicial free Abelian topological group $S^t(A)$ need not be a subgroup of the free Abelian topological group $S^t(X)$. We derive some sufficient conditions for this to be the case.

The long exact sequence of classical singular homology groups stems from a short exact sequence $S(A) \to S(X) \to S(X,A)$ of Abelian groups. Here the quotient morphism $S(X) \twoheadrightarrow S(X,A)$ is the cokernel of the morphism $S(i_A) : S(A) \to S(X)$ induced by the inclusion $i_A : A \hookrightarrow X$ of the subspace A of X. Proceeding analogously for topological homology we define:

Definition 2.3.44. *The codomain of the cokernel of the morphism $S^t(i)$ induced by an embedding $i : A \hookrightarrow X$ of topological spaces is denoted by $S^t(X,A)$. It is called the* singular simplicial topological group *of the pair (X,A).*

This construction is functorial in the pairs (X,A) of topological spaces, hence it defines a functor $S^t(-,-) : \mathbf{Top}(2) \to \mathbf{TopAb}^{\mathcal{O}^{op}}$. The singular simplicial topological group $S^t(X)$ of a topological space X can be identified with the simplicial topological group $S^t(X,\emptyset)$, which is the quotient of $S^t(X)$ by the trivial subgroup $\{0\} = F_{\mathbf{TopAb}}\emptyset$. Similar to the singular simplicial topological group functor $S^t : \mathbf{Top} \to \mathbf{TopAb}^{\mathcal{O}^{op}}$ we observe:

Lemma 2.3.45. *The functor $S^t(-,-) : \mathbf{Top}(2) \to \mathbf{TopAb}^{\mathcal{O}^{op}}$ preserves homotopy.*

Proof. Let $(f,f_{|A}),(g,g_{|A}) : (X,A) \to (Y,B)$ be morphisms in **Top**(2) and let i_A resp. i_B denote the injections $S^t(A) \to S^t(X)$ resp. $S^t(B) \to S^t(Y)$. A homotopy H from $(f,f_{|A})$ to $(g,g_{|A})$ is a homotopy $H : X \times I \to Y$ from f to g that restricts and corestricts to a homotopy $H_{|A \times I} : A \times I \to B$ from $f_{|A}$ to $g_{|A}$. Therefore the homotopy from $C(\Delta,f)$ to $C(\Delta,g)$ obtained from a homotopy $H : X \times I \to Y$ restricts to a homotopy from $C(\Delta,f_{|A})$ to $C(\Delta,g_{|A})$. As a consequence the induced homotopy from $S^t(f)$ to $S^t(g)$ maps the simplicial subgroup $S^t(i_A)(S^t(A))$ of $S^t(X)$ into the simplicial subgroup $S^t(i_B)(S^t(B))$, and thus descends to the quotient spaces, i.e. to a homotopy from $S^t(f,f_{|A})$ to $S^t(g,g_{|A})$. $\square$

We denote the chain complex associated to the singular simplicial topological group $S^t(X,A)$ of a pair (X,A) of topological spaces by $S_*^t(X,A)$. Recall that the the associated chain complex functor $\mathbf{TopAb}^{\mathcal{O}^{op}} \to \mathbf{Ch}(\mathbf{TopAb})$ is cocontinuous (by Lemma 1.1.7) and thus preserves cokernels. Therefore the singular topological chain complex $S_*^t(X,A)$ of the pair (X,A) is the cokernel of the morphism $S_*^t(A) \to S_*^t(X)$ of singular topological chain complexes.

Definition 2.3.46. *The chain complex $S_*^t(X,A)$ associated to the singular simplicial topological group $S^t(X,A)$ of a pair (X,A) of topological spaces is called the* singular topological chain complex *of the pair (X,A). The homology thereof is denoted by $H^t(X,A)$ and called the* singular topological homology *of the pair (X,A).*

Lemma 2.3.47. *The singular topological chain complex functor* $S_*^t(-,-)$ *on* **Top**(2) *preserves homotopy.*

Proof. The singular simplicial topological group functor $S^t(-,-)$ preserves homotopy by Lemma 2.3.45. Because simplicially homotopic morphisms of simplicial Abelian topological groups induce homotopic morphisms of the associated chain complexes (Lemma F.6.6), the functor $S_*^t(-,-)$ preserves homotopy as well.   □

Lemma 2.3.48. *The singular topological homology functor* $H_*^t(-,-)$ *on* **Top**(2) *is homotopy invariant.*

Proof. This is a direct consequence of the fact that the singular topological chain complex functor $S_*^t(-,-)$ on **Top**(2) preserves homotopy.   □

So far for every pair (X,A) of topological spaces one obtains a sequence

$$S^t(A) \to S^t(X) \twoheadrightarrow S^t(X,A) \to 0$$

of singular simplicial topological groups. The sequence is exact at $S^t(X,A)$ and the projection onto $S^t(X,A)$ is the cokernel of the morphism $S^t(i) : S^t(A) \to S^t(X)$ by definition, but the latter morphism might not be an embedding. It is an embedding if and only if the subspace A is P-embedded into X (cf. Theorem 1.4.18). In this case the sequence $0 \to S^t(A) \to S^t(X) \twoheadrightarrow S^t(X,A) \to 0$ of singular simplicial topological groups is exact and so is the sequence of associated chain complexes.

Proposition 2.3.49. *Every P-embedding $A \hookrightarrow X$ of a subspace A of a topological space X induces a long exact sequence*

$$\cdots \to H_{n+1}^t(X,A) \xrightarrow{\delta_{n+1}} H_n^t(A) \to H_n^t(X) \to H_n^t(X,A) \xrightarrow{\delta_n} \cdots \qquad (2.1)$$

of singular homology groups.

Proof. If $A \hookrightarrow X$ is a P-embedding of a subspace A of a topological space X then the morphism $S^t(A) \to S^t(X)$ is an inclusion by Theorem 1.4.26. As a consequence the sequence $0 \to S^t(A) \to S^t(X) \twoheadrightarrow S^t(X,A) \to 0$ of singular simplicial topological groups is exact. Because the homology of chain complexes associated to simplicial Abelian topological groups is a homological δ-functor by Theorem 1.3.14 (resp. Corollary 1.3.15), this exact sequence leads a long exact sequence in homology.   □

Theorem 2.3.50. *The singular topological homology functor* $H_*^t(-)$ *is a homology theory on the subcategory of* **Top**(2) *consisting of pairs* (X,A) *of topological spaces X and P-embedded subspaces $A \subseteq X$.*

Proof. The singular topological homology functor $H_*^t(-,-)$ on **Top**(2) is homotopy invariant by Lemma 2.3.48. In addition, for each pair (X,A) of a topological space X with a P-embedded subspace A the long sequence of homology groups in Equation 2.1 is exact by Proposition 2.3.49, so $H_*^t(-,-)$ is a homology theory on the subcategory of **Top**(2) consisting of pairs (X,A) of topological spaces X and P-embedded subspaces $A \subseteq X$.   □

Lemma 2.3.51. *If A is a retract of X then the inclusion $A \hookrightarrow X$ induces an embedding $S^t(A) \hookrightarrow S^t(X)$ of singular simplicial topological groups.*

Proof. if A is a retract of X then A is P-embedded into X by Lemma 1.4.37. Thus the simplicial free Abelian topological group $S^t(A)$ is a subgroup of the simplicial free Abelian topological group $S^t(X)$ by Theorem 1.4.18.   □

If the inclusion $S^t(A) \hookrightarrow S^t(X)$ is a P-embedding, then it is the kernel of the projection onto $S^t(X, Y)$ and the sequence $0 \to S^t(A) \to S^t(X) \twoheadrightarrow S^t(X, A) \to 0$ of singular simplicial topological groups is exact. So the sequence of associated chain complexes is exact as well. Analogously to short exact sequences of coefficient groups, the Snake Lemma in **TopAb** can be applied to such short exact sequences of singular topological chain complexes, because these are chain complexes associated to simplicial topological groups and thus are always proper by Theorem 1.3.15.

Lemma 2.3.52. *If A is a retract of the topological space X then the short exact sequence $0 \to S^t(A) \to S^t(X) \twoheadrightarrow S^t(X, A) \to 0$ splits and the inclusion $A \hookrightarrow X$ induces an inclusion of $H_*^t(A)$ as a direct summand: $H_*^t(X) \cong H_*^t(A) \oplus H_*^t(X, A)$.*

Proof. Let A be a retract of X and $r : X \to A$ be a retraction. Denote the inclusion $A \hookrightarrow X$ by i. The morphism $S^t(r) : S^t(X) \to S^t(A)$ of singular simplicial topological groups satisfies $S^t(r) \circ S^t(i) = S^t(r \circ i) = S^t(\mathrm{id}_A) = \mathrm{id}_{S^t(A)}$ and thus provides a splitting of the short exact sequence $0 \to S^t(A) \to S^t(X) \twoheadrightarrow S^t(X, A) \to 0$. Consequently the induced long exact sequence in homology decomposes into split short exact sequences, hence the homomorphism $H_*^t(i) : H_*^t(A) \to H_*^t(X)$ is the inclusion of a direct summand and $H_*^t(X) \cong H_*^t(A) \oplus H_*^t(X, A)$.   □

Example 2.3.53. Any base space B of a vector bundle $p : E \to B$ with fibre F is a retract of E. Therefore the zero section of a vector bundle $p : E \to B$ induces an embedding $S^t(B) \hookrightarrow S^t(E)$ of singular simplicial topological groups, hence the homology of the total space E decomposes as $H_*^t(E) \cong H_*^t(B) \oplus H_*^t(F) \cong H_*^t(B)$. In particular any tangent bundle TM of a (possibly infinite dimensional) manifold M with model space V contains M as a retract and the zero section $M \to TM$ induces an embedding $S^t(M) \hookrightarrow S^t(TM)$ of singular simplicial topological groups, which implies the equality $H_*^t(TM) \cong H_*^t(M) \oplus H_*^t(V) \cong H_*^t(M)$.

This readily generalises to singular simplicial topological groups with coefficients. Recall that tensoring with an Abelian topological group V preserves quotients by Theorem 1.4.5 and cokernels by Lemma 1.4.7. Thus for each pair (X, A) of topological spaces and coefficient group V the quotient morphism $S^t(X) \otimes V \twoheadrightarrow S^t(X, A) \otimes V$ is the cokernel of the morphism $S^t(A) \otimes V \twoheadrightarrow S^t(X) \otimes V$. Tensoring with a topological group V might not preserve P-embeddings, but a retraction of X onto the subspace A ensures that the morphism $S^t(A) \otimes V \twoheadrightarrow S^t(X) \otimes V$ is an inclusion and that the sequence of singular simplicial topological groups with coefficients V splits:

Lemma 2.3.54. *If A is a retract of the topological space X and V an Abelian topological group of coefficients, then the sequence*

$$0 \to S^t(A) \otimes V \to S^t(X) \otimes V \twoheadrightarrow S^t(X, A) \otimes V \to 0$$

is split exact and the inclusion $A \hookrightarrow X$ induces an inclusion of $H_*^t(A; V)$ as a direct summand: $H_*^t(X; V) \cong H_*^t(A; V) \oplus H_*^t(X, A; V)$.

Proof. Let A be a retract of X and $r : X \to A$ be a retraction. Denote the inclusion $A \hookrightarrow X$ by i. The morphism $S^t(r) \otimes \mathrm{id}_V : S^t(X) \otimes V \to S^t(A) \otimes V$ of singular simplicial topological groups with coefficients V satisfies the equations

$$\left(S^t(r) \otimes \mathrm{id}_V\right) \circ \left(S^t(i) \otimes \mathrm{id}_V\right) = (S^t(r) \circ S^t(i)) \otimes \mathrm{id}_V = S^t(r \circ i) \otimes \mathrm{id}_V$$
$$= S^t(\mathrm{id}_A) \otimes \mathrm{id}_V = \mathrm{id}_{S^t(A)} \otimes \mathrm{id}_V = \mathrm{id}_{S^t(A) \otimes V},$$

i.e. it is a left inverse of $S^t \otimes V(i)$. Therefore the morphism $S^t(i) \otimes \mathrm{id}_V$ is an inclusion and the sequence $0 \to S^t(A) \otimes V \to S^t(X) \otimes V \twoheadrightarrow S^t(X, A) \otimes V \to 0$ is split exact. Consequently the induced long exact sequence in homology decomposes into split short exact sequences, hence the homomorphism $H_*^t(i; V) : H_*^t(A; V) \to H_*^t(X; V)$ is the inclusion of a direct summand and $H_*^t(X; V) \cong H_*^t(A; V) \oplus H_*^t(X, A; V)$.   □

Example 2.3.55. (cf. Example 2.3.53) Any base space B of a vector bundle $E \to B$ with fibre F is a retract of E. Therefore the topological singular homology $H_*^t(E; V)$ of the total space E decomposes as $H_*^t(E; V) \cong H_*^t(B; V) \oplus H_*^t(F; V) \cong H_*^t(B; V)$. In particular any tangent bundle TM of a manifold M with model space W contains M as a retract and this implies $H_*^t(TM; V) \cong H_*^t(M; V) \oplus H_*^t(W; V) = H_*^t(M; V)$.

Another important example of P-embeddings is given by closed subspaces of metrisable spaces. Here we note:

Lemma 2.3.56. *If X is a metrisable topological space then the singular simplicial space $C(\Delta, X)$ (endowed with the compact open topology) is metrisable.*

Proof. See [Dug89, XII.8.2 (3)] for a proof.   □

Example 2.3.57. All finite dimensional manifolds are metrisable. Thus the singular simplicial space of any finite dimensional manifold is metrisable. This includes in particular all finite dimensional Lie groups.

Example 2.3.58. All paracompact manifolds with metrisable model spaces are metrisable. This includes all paracompact Hilbert-manifolds, Banach manifolds, Fréchet-manifolds, and in particular all Hilbert-,Banach, Fréchet-, or just metrisable Lie groups.

The metrisability of the singular simplicial space $C(\Delta, X)$ turns out to be sufficient to guarantee that the morphism $S^t(A) \to S^t(X)$ of singular simplicial topological groups is an embedding for closed subspaces A.

Proposition 2.3.59. *If A is a closed subspace of a metrisable topological space X then the singular simplicial topological group $S^t(A)$ is a closed subgroup of the singular simplicial topological group $S^t(X)$.*

Proof. If A is a closed subspace of a metrisable space X, then the singular simplicial space $C(\Delta, A)$ is a closed subspace of the metrisable simplicial space $C(\Delta, X)$. Metrisable spaces are paracompact. Therefore the singular simplicial space $C(\Delta, A)$ is P-embedded in the singular simplicial space $C(\Delta, X)$ by Lemma 1.4.39. By Theorem 1.4.26 the homomorphism $S^t(A) \to S^t(X)$ is an embedding. Because metrisable spaces are T_3-spaces, the subgroup $S^t(A)$ of $S^t(X)$ is closed by Lemma 1.4.29.   □

Corollary 2.3.60. *If A is a closed subspace of a metrisable space X then the sequence*

$$0 \to S^t(A) \to S^t(X) \to S^t(X, A) \to 0$$

of singular simplicial topological groups is exact and the singular simplicial group $S^t(X, A)$ is Hausdorff.

Example 2.3.61. Finite dimensional manifolds are metrisable. Therefore each pair (M, N) of a finite dimensional manifold M and a closed submanifold N leads to a short exact sequence

$$0 \to S^t(N) \to S^t(M) \to S^t(M, N) \to 0$$

of singular simplicial topological groups. More generally this holds for all closed submanifolds N of a metrisable (possibly infinite dimensional) manifold M. This includes in particular all pairs (G, H) of a metric Lie group G and a closed subgroup H of G.

In order to derive a long exact sequence in homology, we once again have to use the fact that the singular topological cochain complexes stem from simplicial Abelian topological groups. Doing so we observe:

Theorem 2.3.62. *Topological singular homology is a topological homology theory on the full subcategory of* **Top**(2) *whose objects are the pairs of metrisable spaces and closed subspaces.*

Proof. Each inclusion $A \hookrightarrow A$ of a closed subspace into a metrisable space X induces an inclusion $S^t(A) \hookrightarrow S^t(X)$ of singular simplicial topological groups and gives rise to a short exact sequence

$$0 \to S^t(A) \to S^t(X) \to S^t(X, A) \to 0$$

of singular simplicial topological groups. The homology of associated chain complexes is a universal homological δ-functor $H \circ (-)_* : \mathbf{TopAb}^{\mathcal{O}^{op}} \to \mathbf{TopAb}$ by Theorem 1.3.14 (resp. Corollary 1.3.15). Therefore the above short exact sequence of singular simplicial topological groups induces a long exact sequence of topological homology groups, which depends naturally on the pair (X, A). Furthermore the functor $H_*^t(-, -)$ on **Top**(2) is homotopy invariant by Lemma 2.3.48, so it restricts to a homotopy invariant functor on the subcategory of **Top**(2) whose objects are the pairs of metrisable spaces and closed subspaces. $\qquad\square$

The metrisability of the space X was sufficient but not necessary to ensure that the morphism $S^t(A) \to S^t(X)$ is an embedding. The crucial point is the paracompactness of the function spaces $C(\Delta_n, X)$. This can also be accomplished by requiring the space X to be a $\aleph$-space, relaxing the condition of metrisability.

Lemma 2.3.63. *If X is a paracompact $\aleph$-space then the singular simplicial space $C(\Delta, X)$ is a paracompact $\aleph$-space as well.*

Proof. The domain of each function space $C(\Delta_n, X)$ is a separable metric space, hence an $\aleph$-space. Function spaces with domain an $\aleph$-space and codomain a paracompact $\aleph$-space are paracompact $\aleph$-spaces by [O'M71, Theorem 1]. Thus the singular simplicial space $C(\Delta, X)$ of a paracompact $\aleph$-space X is paracompact. $\qquad\square$

Example 2.3.64. All separable metrisable spaces are paracompact $\aleph_0$-spaces. This includes in particular all finite dimensional manifolds and all countable CW-complexes.

Example 2.3.65. All first countable topological groups are metrisable. Therefore all separable first countable topological groups are $\aleph$-spaces. This includes in particular all Lie groups with separable model space.

Proposition 2.3.66. *If A is a closed subspace of a paracompact $\aleph$-space X, then the singular simplicial topological group $S^t(A)$ is a subgroup of the singular simplicial topological group $S^t(X)$.*

Proof. Let A be a closed subspace of a paracompact $\aleph$-space X. The singular simplicial space $C(\Delta, X)$ is paracompact by Lemma 2.3.63. The singular simplicial space $C(\Delta, A)$ then is a closed subspace of the simplicial paracompact space $C(\Delta, X)$, hence P-embedded in $C(\Delta, X)$ by Lemma 1.4.39. Thus the simplicial free Abelian topological group $S^t(A)$ is a simplicial subgroup of the simplicial group $S^t(X)$ by Theorem 1.4.26. $\qquad\square$

Having obtained that morphism $S^t(A) \to S^t(X)$ of singular simplicial topological groups is an embedding one can proceed as for metrisable spaces.

Theorem 2.3.67. *Topological singular homology is a topological homology theory on the full subcategory of* **Top**(2) *whose objects are the pairs of paracompact $\aleph$-spaces and closed subspaces.*

Proof. The proof is analogous to that of Theorem 2.3.62.   □

Lemma 2.3.68. *If F is the fibre of a locally trivial fibre bundle $p : E \to B$ over a completely regular base space B, then the inclusion $F \hookrightarrow E$ induces an embedding $S^t(F) \hookrightarrow S^t(E)$ of singular simplicial topological groups. If the total space E is a T_3-space and the fibre F is closed in E, then the subgroup $S^t(F)$ of $S^t(E)$ is closed.*

Proof. If F is the fibre of a locally trivial fibre bundle $p : E \to B$ over a completely regular base space B, then F is P-embedded into E by Lemma 1.4.43. Thus the morphism $S^t(F) \to S^t(E)$ is an embedding by Theorem 1.4.26. If E is a T_3-space and the fibre F is closed in E, then the subgroup $S^t(F)$ of $S^t(E)$ is closed by Lemma 1.4.29   □

Example 2.3.69. All fibre bundles $p : E \to B$ over locally contractible base spaces B are locally trivial. Therefore the inclusion $F \hookrightarrow E$ of the fibre F into the total space E of a bundle $p : E \to B$ over a locally contractible completely regular space B induces an embedding $S^t(F) \to S^t(E)$ of singular simplicial topological groups. If the total space is E a T_3-space and the fibre F is closed in E, then the subgroup $S^t(F)$ of $S^t(E)$ is closed.

Example 2.3.70. All Hausdorff manifolds are locally contractible and completely regular. Thus the inclusion $F \hookrightarrow E$ of the fibre F into the total space E of a fibre bundle $p : E \to M$ over a Hausdorff manifold M induces an embedding $S^t(F) \to S^t(E)$ of singular simplicial topological groups. If the total space is E a T_3-space and the fibre F is closed in E, then the subgroup $S^t(F)$ of $S^t(E)$ is closed.

Example 2.3.71. All topological groups are completely regular. Thus the inclusion of the fibre F into the total space E of a locally trivial fibre bundle $p : E \to G$ over a topological group G induces an embedding $S^t(F) \to S^t(E)$ of singular simplicial topological groups. If the total space is E a T_3-space and the fibre F is closed in E, then the subgroup $S^t(F)$ of $S^t(E)$ is closed.

Example 2.3.72. The fibre F of a fibre bundle $p : E \to G$ over a locally contractible topological group G is P-embedded in E. In particular the fibre F of a bundle $p : E \to G$ over a Lie group G is P-embedded in E. If the total space is E a T_3-space and the fibre F is closed in E, then the subgroup $S^t(F)$ of $S^t(E)$ is closed.

Example 2.3.73. Classifying spaces of locally contractible topological groups are locally contractible (cf. [Seg70, Proposition A.1]). Thus the inclusion $G \hookrightarrow EG$ of a locally contractible group G into the total space EG of its classifying bundle induces an embedding $S^t(G) \hookrightarrow S^t(EG)$ and a long exact sequence of topological homology groups.

Example 2.3.74. The inclusion $G \hookrightarrow G'$ of a subgroup G of a topological group G' in a group extension $0 \to G \to G' \to G'' \to 0$ of a locally contractible topological group G'' induces an embedding $S^t(G) \hookrightarrow S^t(G)$ of singular simplicial topological groups. In particular all normal subgroups occurring in extensions of Lie groups induce embeddings of singular simplicial topological groups.

Summarising the preceding observations the singular topological homology functor restricts to a topological homology theory on various subcategories of **Top**(2):

Theorem 2.3.75. *The singular topological homology H_*^t is a topological homology theory on the full subcategory of **Top**(2) whose objects are the pairs (X, A) of a topological space X and a P-embedded subspace A of X; this includes the full subcategories whose objects are the pairs (X, A) where*

1. *A is a retract of X, or*
2. *A is a closed subspace of a metrisable space X, or*
3. *A is a closed subspace of a paracompact ℵ-space, or*
4. *A is a closed subgroup of a first countable topological group X, or*
5. *A is a closed subspace of the homotopy type of a countable CW-complex, or*
6. *A is a closed submanifold of a manifold X.*

2.4 Definition of Topological Cohomology Theories

The definition of topological cohomology theories is dual to the definition of topological homology theories:

Definition 2.4.1. *A cohomology theory on* **Top**(2) *with values in a pointed regular category* **D** *is a series of contravariant functors* $h^n :$ **Top**$(2) \to$ **D** *with natural transformations* $\partial^n : h^n \circ \kappa_l \to h^{n+1}$ *satisfying the following conditions:*

1. *(Homotopy invariance) Each functor* h^n *descends to the category* **Top**(2)**h**, *i.e.* h^n *maps homotopic morphisms* $(f, f') \simeq (g, g')$ *to the same morphism* $h^n(f, f') = h^n(g, g')$.
2. *(Exact cohomology sequence) For each pair* (A, B) *of topological spaces in* **Top**(2) *the sequence*

$$\cdots \to h^{n-1}(A, \emptyset) \xrightarrow{\partial^{n-1}} h^n(X, A) \to h_n(X, \emptyset) \to h_n(A, \emptyset) \xrightarrow{\partial^n} \cdots$$

is exact.

A cohomology theory on a subcategory **C** *of* **Top**(2) *with values in a pointed regular category* **D** *is a series of contravariant functors* $h^n :$ **C** $\to$ **D** *together with natural transformations* $\partial^n : h^n \to h_{n+1} \circ \kappa$ *satisfying the above conditions 1 and 2 for all objects* (X, A) *in* **C**. *A cohomology theory is called* proper *if it satisfies the following condition:*

3. *(Dimension) For every singleton space* $*$ *and all* $n > 0$ *the object* $h^n(*)$ *is a zero object in* **D**.

Remark 2.4.2. We have dropped the excision axiom in the above definition.

Using the Abelian category **Ab** of Abelian groups as codomain **D** in the above definition of proper cohomology theories recovers the definition of cohomology theories due to Eilenberg and Steenrod without the excision axiom. We are interested in topologised versions of these theories. So we define:

Definition 2.4.3. *A topological cohomology theory on a subcategory* **C** *of* **Top**(2) *is a cohomology theory with values in the category* **TopAb** *of Abelian topological groups or the category* **TopRMod** *of topological modules over a topological ring* R.

Example 2.4.4. All cohomology theories with values in the category **Ab** of Abelian groups can be given the discrete topology, turning them into topological cohomology theories.

Example 2.4.5. The Čech and singular cohomology groups of finite dimensional manifolds with real coefficients are finite dimensional vector spaces. These can uniquely be equipped with a Hausdorff topology, turning them into real topological vector spaces. The connecting homomorphisms are linear maps of finite dimensional vector spaces, hence continuous. Therefore these cohomology theories can be turned into topological cohomology theories (with values in **TopℝMod**).

Example 2.4.6. All cohomology groups of cohomology theories with coefficients in the category **Kmod** modules over a topological field **K** can be given the free **K**-vector space topology on a discrete basis. This turns these theories into topological cohomology theories with values in **TopKmod**. If **K** contains the real numbers $\mathbb{R}$ as a subfield, one can also use the free locally convex vector space topology.

2.5 Continuous Alexander-Spanier Cohomology

In this section we define (equivariant) topological Alexander-Spanier cohomology. This will be a topological cohomology theory on various subcategories of **Top**(2). The procedure given below relies on the homological algebra in **TopRMod** presented in chapter 1 and requires the coefficients to be an Abelian topological group (which may be discrete or indiscrete).

The simplicial category $\mathcal{O}$ can be considered to be a subcategory of the category **Top** of topological spaces by endowing each set $[n] = \{0, \ldots, n\}$ with the discrete topology. We will use this identification without further reference. Let X be a topological space. Since each of the topological spaces $[n]$ is finite and discrete the compact open topology on the function space $C([n], X)$ coincides with the topology of point-wise convergence, i.e. $C([n], X) \cong X^{n+1}$. We also write **x** for a point $(x_0, \ldots, x_n)$ in this function space. This is our starting point for the construction of simplicial topological spaces out of topological spaces:

Definition 2.5.1. *For any topological space X, the* standard simplicial space $SS(X)$ *is the simplicial space $SS(X) : \mathcal{O}^{op} \to$ **Top** which is given by*

$$SS(X)([n]) := C([n], X), \quad \text{and} \quad SS(X)(\alpha) = C(\alpha, X)$$

for any morphism $\alpha : [m] \to [n]$ in $\mathcal{O}$. The standard simplicial space functor *is the functor* SS $:$ **Top** $\to$ **Top**$^{\mathcal{O}^{op}}$ *which additionally maps a continuous function $f : X \to Y$ to the morphism $C(-, f) : C(-, X) \to C(-, Y)$ of simplicial spaces. A point $(x_0, \ldots, x_n) \in SS(X)([n])$ is called an n-simplex.*

More explicitly, the standard simplicial space $SS(X)$ of a topological space X is given by $SS(X)([n]) \cong X^{n+1}$ and $SS(X)(\alpha)(x_0, \ldots, x_n) = (x_{\alpha(0)}, \ldots, x_{\alpha(m)})$ for any morphism $\alpha : [m] \to [n]$ in $\mathcal{O}$ and point $(x_0, \ldots, x_n)$ of X^{n+1}. Using this identification the standard simplicial space functor SS then additionally maps a continuous function $f : X \to Y$ to the morphisms

$$SS(f)([n]) = \prod_{i=0}^{n} f : \prod_{i=0}^{n} X = SS(X)([n]) \to \prod_{i=0}^{n} Y = SS(Y)([n])$$

for each $n \in \mathbb{N}$. Any product functor on **Top** is right adjoint to the diagonal functor and thus continuous. Since the standard simplicial space functor SS $:$ **Top** $\to$ **Top**$^{\mathcal{O}^{op}}$ is simply built out of product functors, it preserves many additional structures a topological space X can have.

Example 2.5.2. If the standard simplicial space functor SS is applied to a manifold M, then the resultant simplicial space $SS(M)$ is a simplicial manifold.

Example 2.5.3. The standard simplicial space $SS(G)$ of a topological group G is a simplicial topological group. This simplicial topological group is Abelian if and only if G is Abelian.

Example 2.5.4. The standard simplicial space $SS(G)$ of a topological semi-group G is a simplicial topological semi-group. This simplicial topological semi-group is Abelian if and only if G is Abelian.

More generally the standard simplicial space $\mathrm{SS}(A)$ of a topological algebra of type (Ω, E) is a simplicial topological algebra of type (Ω, E). This not only applies to total algebras but also to topological groupoids:

Example 2.5.5. If $G \rightrightarrows M$ is a topological groupoid with source map $\alpha : G \to M$ and target map $\beta : G \to M$, then $\mathrm{SS}(G) \rightrightarrows \mathrm{SS}(M)$ is a simplicial topological groupoid with source map $\mathrm{SS}(\alpha) : \mathrm{SS}(G) \to \mathrm{SS}(M)$ and target map $\mathrm{SS}(\beta) : \mathrm{SS}(G) \to (M)$.

Another useful property of the standard simplicial space functor SS is the preservation of homotopy:

Lemma 2.5.6. *The standard simplicial space functor* $\mathrm{SS} : \mathbf{Top} \to \mathbf{Top}^{\mathcal{O}^{op}}$ *preserves homotopy.*

Proof. Let $f, g : X \to Y$ be continuous functions between topological spaces X and Y and $H \times I \to Y$ be a homotopy from g to f. The homotopy H induces a morphism $\mathrm{SS}(H) : \mathrm{SS}(X \times I) \to \mathrm{SS}(Y)$ of standard simplicial spaces. The simplicial topological space $\mathrm{SS}(X \times I)$ is naturally isomorphic to the simplicial space $\mathrm{SS}(X) \times \mathrm{SS}(I)$; the isomorphism is given by the pair $(\mathrm{SS}(\mathrm{pr}_1), \mathrm{SS}(\mathrm{pr}_2))$ of morphisms induced by the projections pr_1 and pr_2 of $X \times I$ onto the first and second factor respectively. The simplicial topological space $\mathrm{SS}(I)$ contains the discrete simplicial topological space $\hom(-, [1])$ as a simplicial subspace in the following way: The embedding $i : \hom(-, [1]) \to \mathrm{SS}(I)$ maps a morphism $\alpha \in \hom([n], [1])$ to the point $(\alpha(0), \dots, \alpha(n))$ in $\mathrm{SS}(I)$. All in all one obtains a chain

$$\mathrm{SS}(X) \times \hom(-, [1]) \xrightarrow{\mathrm{id}_{\mathrm{SS}(X)} \times i} \mathrm{SS}(X) \times \mathrm{SS}(I) \cong \mathrm{SS}(X \times I) \xrightarrow{\mathrm{SS}(H)} \mathrm{SS}(Y)$$

of morphisms of simplicial spaces. We denote the composition of all the morphisms in this chain by $\tilde{h}(H) : \mathrm{SS}(X) \times \hom(-, [1]) \to \mathrm{SS}(Y)$. Recall that the simplicial space $\hom(-, [0])$ is a singleton simplicial space. The face morphism $\epsilon_0 : [0] \hookrightarrow [1]$ induces an embedding $\hom(-, \epsilon_0) : \hom(-, [0]) \to \hom(-, [1])$, which maps the singleton simplicial space $\hom(-, [0])$ onto the simplicial subspace of $\hom(-, [1])$ consisting of the constant morphisms 1. This embedding in turn induces an inclusion

$$j_{X,0} : \mathrm{SS}(X) \cong \mathrm{SS}(X) \times \hom(-, [0]) \hookrightarrow \mathrm{SS}(X) \times \hom(-, [1])$$

of simplicial topological spaces. Analogously the face morphism $\epsilon_1 : [0] \hookrightarrow [1]$ induces an embedding $\hom(-, \epsilon_1) : \hom(-, [0]) \to \hom(-, [1])$, which maps the singleton simplicial space $\hom(-, [0])$ onto the simplicial subspace of $\hom(-, [1])$ consisting of the zero morphisms. This embedding in turn induces another inclusion

$$j_{X,1} : \mathrm{SS}(X) \cong \mathrm{SS}(X) \times \hom(-, [0]) \hookrightarrow \mathrm{SS}(X) \times \hom(-, [1])$$

of simplicial topological spaces. We use these embeddings to prove that there exists a simplicial homotopy from $\mathrm{SS}(f)$ to $\mathrm{SS}(g)$. In view of Theorem F.6.7 it suffices to show that the morphisms $\mathrm{SS}(f)$ and $\mathrm{SS}(g)$ coincide with the compositions $\tilde{H}(H)j_{X,0}$ resp. $\tilde{h}(H)j_{X,1}$, i.e. that the following diagram is commutative:

$$
\begin{array}{ccccc}
\mathrm{SS}(X) & \xrightarrow{\;\;j_{X,0}\;\;} & \mathrm{SS}(X) \times \hom(-, [1]) & \xleftarrow{\;\;j_{X,1}\;\;} & \mathrm{SS}(X) \\
& \searrow_{\mathrm{SS}(f)} & \downarrow{\tilde{h}(H)} & \swarrow_{\mathrm{SS}(g)} & \\
& & \mathrm{SS}(Y) & &
\end{array}
$$

The evaluation of the composition $\tilde{h}(H)j_{X,0}$ at a point $\mathbf{x} = (x_0, \dots, x_n)$ in $\mathrm{SS}(X)([n])$ is given by

$$[\tilde{h}(H)([n])][j_{X,0}([n])](\mathbf{x}) = [\tilde{h}(H)([n])]((x_0,\ldots,x_n),(1,\ldots,1))$$
$$= [\mathrm{SS}(H)([n])]((x_0,1),\ldots,(x_n,1))$$
$$= (H(x_0,1),\ldots,H(x_n,1)) = (f(x_0),\ldots,f(x_n))$$
$$= [\mathrm{SS}(f)([n])](\mathbf{x}).$$

So the morphisms $\mathrm{SS}(f)$ and $\tilde{h}(H)j_{X,0}$ coincide. The equality of the morphisms $\mathrm{SS}(g)$ and $\tilde{h}(H)j_{X,1}$ is proven similarly. $\qquad\square$

Let X be a topological space and V be an Abelian topological group. A (trivial) topologised cohomology of X with coefficient group V can be obtained with the help of the function space functor $C(-;V)$. As usual all function spaces are equipped with the compact-open topology. Applying the contravariant function space functor $C(-,V)$ to the standard simplicial space $\mathrm{SS}(X)$ yields the cosimplicial Abelian topological group $C(\mathrm{SS}(X),V)$.

Definition 2.5.7. *The cosimplicial group $C(\mathrm{SS}(X),V)$ is denoted by $A(X;V)$. It is called the* standard cosimplicial Abelian group with coefficients V *of X. The bifunctor $A(-;-) = C(\mathrm{SS}(-),-) : \mathbf{Top} \times \mathbf{TopAb} \to \mathbf{TopAb}^{\mathcal{O}}$ is called the* standard cosimplicial Abelian group functor. *The elements of $A(X;V)([n])$ are called* n-cochains.

We note that the restriction of $A(-;V)$ to the category of open subsets of a topological space X (whose morphisms are the inclusions) is a presheaf of cosimplicial Abelian topological groups on X by construction.

Lemma 2.5.8. *Every standard cosimplicial group $A(X;V)$ is completely regular. It is Hausdorff if and only if the coefficient group V is Hausdorff.*

Proof. Let X be any topological space. Each of the function spaces $C(\mathrm{SS}(X)([n]),V)$ is a topological group and therefore completely regular. Furthermore each function space $C(\mathrm{SS}(X)([n]),V)$ contains the coefficient group V as a subspace, so V has to be Hausdorff if $C(C(\Delta_n,X),V)$ is Hausdorff. Conversely, if the coefficient group V is Hausdorff, then so is every function space of the form $C(Y,V)$. (See [Dug89, XII.1.3] for a proof.) Thus the cosimplicial group $C(\mathrm{SS}(X),V)$ is Hausdorff if and only if V is Hausdorff. $\qquad\square$

Passing to the cochain complex associated to the cosimplicial Abelian topological group $A(X;V)$ yields another functor $A^*(-;V) : \mathbf{Top} \to \mathbf{coCh}(\mathbf{TopAb})$ into the category $\mathbf{coCh}(\mathbf{TopAb})$ of cochain complexes of Abelian topological groups. By Theorem 1.1.41 these cochain complexes are proper, i.e. the differentials are quotient maps onto their image.

Definition 2.5.9. *The functor $A^*(-;V) : \mathbf{Top} \to \mathbf{coCh}(\mathbf{TopAb})$ is called the* standard topological cochain complex functor with coefficients V *and the cochain complex $A^*(X;V)$ associated to a topological space X is called the* standard topological cochain complex (with coefficients V) *associated to the space X. The bifunctor $A^*(-;-) : \mathbf{Top} \times \mathbf{TopAb} \to \mathbf{coCh}(\mathbf{TopAb})$ is called the* standard topological cochain complex functor.

Recall that the category $\mathbf{Top}$ of topological spaces contains the subcategory $\mathbf{Top}_*$ of based spaces (cf. Example A.4.4). Analogously to the discrete versions of the standard topological cochain complexes we observe:

Lemma 2.5.10. *Every standard cochain complex $A^*(X;V)$ in $\mathbf{coCh}(\mathbf{TopAb})$ associated to a topological space X is split exact and this splitting is functorial on the subcategory $\mathbf{Top}_* \times \mathbf{TopAb}$ of $\mathbf{Top} \times \mathbf{TopAb}$.*

Proof. Let X be a topological space and V be an Abelian topological group. If X is not a based space pick a base point $*$ in X and consider the embeddings

$$\tilde{h}_n : \mathrm{SS}(X)([n]) \to \mathrm{SS}(X)([n+1]), \quad (x_0, \ldots, x_n) \mapsto (*, x_0, \ldots, x_n).$$

of topological spaces. These embeddings induce morphisms $h_n = C(\tilde{h}_n, V)$ of Abelian topological groups. We assert that the morphisms h_n form a contracting homotopy of the cochain complex $A^*(X; V)$. Let $f \in A^n(X; V)$ be a continuous function. The evaluation of $(h_{n-1}d_n + d_{n+1}h_n)(f)$ at a point $\mathbf{x} = (x_0, \ldots, x_n)$ computes to

$$
\begin{aligned}
[(h_{n+1}d_n + d_{n-1}h_n)(f)](\mathbf{x}) &= [h_{n+1}(d_n f)](\mathbf{x}) + [d_{n-1}(f\tilde{h}_n)](\mathbf{x}) \\
&= (d_n f)(\tilde{h}_n(\mathbf{x})) + \sum_{i=0}^{n}(-1)^i (f\tilde{h}_n)(x_0, \ldots, \hat{x}_i, \ldots, x_n) \\
&= f(x_0, \ldots, x_n) + \sum_{i=1}^{n+1}(-1)^i f(*, x_0, \ldots, \hat{x}_{i-1}, \ldots, x_n) \\
&\quad + \sum_{i=0}^{n}(-1)^i f(*, x_0, \ldots, \hat{x}_i, \ldots, x_n) \\
&= f(x_0, \ldots, x_n).
\end{aligned}
$$

So the morphism $h_{n-1}d_n + d_{n+1}h_n$ is the identity $A^n(X : V)$. Since this is true for arbitrary $n \in \mathbb{N}$ the morphisms h_n form a contracting homotopy of $A^*(X; V)$. The contracting homotopy is functorial in the coefficients and functorial in based spaces by construction (because morphisms between based spaces map the base points onto each other). $\qquad\square$

Analogously to singular simplices one can restrict oneself to subsets of simplices $(x_0, \ldots, x_n)$ in the spaces $\mathrm{SS}(X)([n])$, which 'fit' in families of subsets of X:

Definition 2.5.11. *Let $\mathfrak{U}$ be a family of subsets of a topological space X. An n-simplex $(x_0, \ldots, x_n)$ in $\mathrm{SS}(X)([n])$ is called $\mathfrak{U}$-small if there exists a set $U \in \mathfrak{U}$ such that the product space U^{n+1} contains $(x_0, \ldots, x_n)$.*

Definition 2.5.12. *The subspace of $\mathrm{SS}(X)([n])$ consisting only of $\mathfrak{U}$-small simplices is denoted by $\mathrm{SS}(X; \mathfrak{U})([n])$.*

In case there is no restriction on the n-simplices (e.g. if the family $\mathfrak{U}$ contains the space X,) we simply write $\mathrm{SS}(X)([n])$ for the space of $\mathfrak{U}$-small n-simplices. Since boundaries and degeneracies of $\mathfrak{U}$-small simplices are $\mathfrak{U}$-small as well, the spaces $\mathrm{SS}(X, \mathfrak{U})([n])$ form a simplicial subspace of the standard simplicial space $\mathrm{SS}(X)$ of the topological space X.

Definition 2.5.13. *The simplicial space $\mathrm{SS}(X, \mathfrak{U})$ is called the $\mathfrak{U}$-small standard simplicial space of X.*

Definition 2.5.14. *The cosimplicial Abelian topological group $C(\mathrm{SS}(X, \mathfrak{U}), V)$ is denoted by $A(X, \mathfrak{U}; V)$. It is called the $\mathfrak{U}$-small standard cosimplicial Abelian topological group of X.*

Lemma 2.5.15. *The $\mathfrak{U}$-small standard cosimplicial group $A(X, \mathfrak{U}; V)$ of every topological space X is completely regular. It is Hausdorff if and only if the coefficient group V is Hausdorff.*

Proof. Let X be a topological space. Each of the function spaces $C(\mathrm{SS}(X,\mathfrak{U})([n]),V)$ is a topological group and therefore completely regular. Furthermore each function space $C(\mathrm{SS}(X,\mathfrak{U})([n]),V)$ contains the coefficient group V as a subspace, so V has to be Hausdorff if $C(C(\Delta_n,X),V)$ is Hausdorff. Conversely, if the coefficient group V is Hausdorff, then so is every function space of the form $C(Y,V)$. (See [Dug89, XII.1.3] for a proof.) Thus the cosimplicial group $C(\mathrm{SS}(X,\mathfrak{U}),V)$ is Hausdorff if and only if V is Hausdorff. $\square$

Let $\mathfrak{U}$ and $\mathfrak{V}$ be families of subsets of a topological space X. We write $\mathfrak{V} \prec \mathfrak{U}$ if $\mathfrak{V}$ is a refinement of $\mathfrak{U}$. In this case every $\mathfrak{V}$-small simplex also is $\mathfrak{U}$-small, so the simplicial space $\mathrm{SS}(X,\mathfrak{V})$ is a simplicial subspace of $\mathrm{SS}(X,\mathfrak{U})$. The inclusion as a simplicial subspace induces a restriction morphism $i_{\mathfrak{V},\mathfrak{U}} : A(X,\mathfrak{U};V) \to A(X,\mathfrak{V};V)$ of cosimplicial Abelian topological groups. Thus the $\mathfrak{U}$-small standard cosimplicial Abelian topological groups $A(X,\mathfrak{U};V)$ for different families $\mathfrak{U}$ of subsets of X (with the relation "is refined by") form a directed set of cosimplicial Abelian topological groups. The same is true if one restricts oneself to open coverings of X.

Definition 2.5.16. *The colimit Abelian topological group* $\mathrm{colim}_{\mathfrak{U}}\, A(X,\mathfrak{U};V)$ *where* $\mathfrak{U}$ *ranges over all open coverings of* X *is denoted by* $A^t_{AS}(X;V)$. *It is called the Alexander-Spanier cosimplicial Abelian topological group with coefficients* V *of* X. *The elements of* $A^t_{AS}(X;V)([n])$ *are called continuous Alexander-Spanier n-cochains.*

This is the cosimplicial topological group of germs of continuous functions defined on a neighbourhood of the diagonal. In case the topological space X is completely regular and paracompact, the Alexander-Spanier cosimplicial Abelian topological group $A^t_{AS}(X;V)$ of X can also be obtained from the standard cosimplicial Abelian topological group $A(X;V)$ by factoring out the cosimplicial subgroup formed by the subgroups

$$A_0(X;V)([n]) := \{ f \in A(X;V)([n]) \mid \exists \text{ open covering } \mathfrak{U} \text{ of } X : f_{|\mathrm{SS}(X,\mathfrak{U})([n])} = 0 \}$$

of functions that vanish on a neighbourhood of the diagonal. The morphisms forming the colimit cocone under the directed set of cosimplicial topological groups $A(X,\mathfrak{U};V)$ will be denoted by $\varrho_{\mathfrak{U}} : A(X,\mathfrak{U};V) \to A^t_{AS}(X;V)$.

Let X and Y be topological spaces and $f : X \to Y$ be a continuous function. For each open covering $\mathfrak{U}$ of Y the pullback family $f^{-1}(\mathfrak{U})$ is an open covering of X. Because the push forward family $ff^{-1}(\mathfrak{U})$ refines $\mathfrak{U}$, the morphism $\mathrm{SS}(f)$ of simplicial spaces restricts and corestricts to a morphism

$$\mathrm{SS}(f)|^{\mathrm{SS}(Y,\mathfrak{U})}_{|\mathrm{SS}(X,f^{-1}(\mathfrak{U}))} : \mathrm{SS}(X, f^{-1}(\mathfrak{U})) \to \mathrm{SS}(Y,\mathfrak{U})$$

of simplicial spaces. An application of the contravariant functor $C(-,V)$ yields a morphism $m_{f,\mathfrak{U}} : A(Y,\mathfrak{U};V) \to A(X, f^{-1}(\mathfrak{U});V)$ of cosimplicial Abelian topological groups. The morphisms $m_{f,\mathfrak{U}}$ for different open coverings $\mathfrak{U}$ of Y can be composed with the colimit morphisms $\varrho_{f^{-1}(\mathfrak{U})} : A(X, f^{-1}(\mathfrak{U});V) \to A^t_{AS}(X;V)$. In this way one obtains a morphism $\varrho_{f^{-1}(\mathfrak{U})} m_{f,\mathfrak{U}} : A(Y,\mathfrak{U};V) \to A^t_{AS}(X;V)$ for every open covering $\mathfrak{U}$ of Y. We assert that these morphisms of cosimplicial groups are compatible with the morphisms $i_{\mathfrak{U},\mathfrak{V}} : A(X,\mathfrak{V};V) \to A(X,\mathfrak{U};V)$ for open coverings $\mathfrak{U} \prec \mathfrak{V}$ of Y.

Proposition 2.5.17. *For every continuous function* $f : X \to Y$ *the collection of morphisms* $\varrho_{f^{-1}(\mathfrak{U})} m_{f,\mathfrak{U}}$ *for open coverings* $\mathfrak{U}$ *of* Y *form a cocone under the diagram of topological groups* $A(Y,\mathfrak{U};V)$, *i.e. the equations* $\varrho_{f^{-1}(\mathfrak{U})} m_{f,\mathfrak{U}} = \varrho_{f^{-1}(\mathfrak{V})} m_{f,\mathfrak{V}} i_{\mathfrak{V},\mathfrak{U}}$ *are satisfied for all open coverings* $\mathfrak{V} \prec \mathfrak{U}$ *of* Y.

Proof. Let $f : X \to Y$ be a continuous functions and $\mathfrak{V} \prec \mathfrak{U}$ be open coverings of Y. The open covering $f^{-1}(\mathfrak{V})$ of X is a refinement of the open covering $f^{-1}(\mathfrak{U})$ of X. Therefore $\mathrm{SS}(X, f^{-1}(\mathfrak{V}))$ is a simplicial subspace of $\mathrm{SS}(X, f^{-1}(\mathfrak{U}))$ and the diagram

$$
\begin{array}{ccc}
\mathrm{SS}(X, f^{-1}(\mathfrak{V})) & \xrightarrow{\;\mathrm{SS}(f)|^{\mathrm{SS}(X,\mathfrak{V})}_{|\mathrm{SS}(X,f^{-1}(\mathfrak{V}))}\;} & \mathrm{SS}(Y, \mathfrak{V}) \\
\downarrow & & \downarrow \\
\mathrm{SS}(X, f^{-1}(\mathfrak{U})) & \xrightarrow{\;\mathrm{SS}(f)|^{\mathrm{SS}(X,\mathfrak{U})}_{|\mathrm{SS}(X,f^{-1}(\mathfrak{U}))}\;} & \mathrm{SS}(Y, \mathfrak{U})
\end{array}
$$

is commutative. The vertical arrows are inclusions of simplicial subspaces. These inclusions induce the morphisms $i_{f^{-1}(\mathfrak{V}),f^{-1}(\mathfrak{U})} : A(X, f^{-1}(\mathfrak{U}); V) \to A(X, f^{-1}(\mathfrak{V}); V)$ and $i_{\mathfrak{V},\mathfrak{U}} : A(Y, \mathfrak{U}; V) \to A(Y, \mathfrak{V}; V)$ respectively. So application of the contravariant function space functor $C(-, V)$ to the above diagram yields the equality

$$
m_{f,\mathfrak{V}}i_{\mathfrak{V},\mathfrak{U}} = i_{f^{-1}(\mathfrak{V}),f^{-1}(\mathfrak{U})}m_{f,\mathfrak{U}} : A(X, \mathfrak{U}; V) \to A(X, f^{-1}(\mathfrak{V}); V)
$$

of morphisms of Abelian topological groups. Composing both sides of the equality with the colimit morphism $\varrho_{f^{-1}(\mathfrak{V})} : A(X, f^{-1}(\mathfrak{V}); V) \to A^t_{AS}(X; V)$ one obtains:

$$
\varrho_{f^{-1}(\mathfrak{V})}m_{f,\mathfrak{V}}i_{\mathfrak{V},\mathfrak{U}} = \varrho_{f^{-1}(\mathfrak{V})}i_{f^{-1}(\mathfrak{V}),f^{-1}(\mathfrak{U})}m_{f,\mathfrak{U}} = \varrho_{f^{-1}(\mathfrak{U})}m_{f,\mathfrak{U}}
$$

The last equality follows from the fact that the morphisms $\varrho_{\mathfrak{W}}$ for open coverings $\mathfrak{W}$ of X form a cocone under the diagram of cosimplicial groups $A(X, \mathfrak{W}; V)$. $\square$

Corollary 2.5.18. *A continuous function $f : X \to Y$ between topological spaces X and Y induces a unique morphism $A^t_{AS}(f; V) : A^t_{AS}(Y; V) \to A^t_{AS}(X; V)$ of cosimplicial Abelian topological groups satisfying $\varrho_{f^{-1}(\mathfrak{U})}m_{f,\mathfrak{U}} = A^t_{AS}(f; V)\varrho_{\mathfrak{U}}$ for all open coverings $\mathfrak{U}$ of Y.*

The identity function $\mathrm{id}_X : X \to X$ of any topological space induces the identity morphism $A^t_{AS}(\mathrm{id}_X; V) = \mathrm{id}_{A^t_{AS}(X;V)}$. In addition the assignment $f \mapsto A^t_{AS}(f; V)$ preserves composition of morphisms by construction, but reverses their order. So we note:

Lemma 2.5.19. *The assignments $X \mapsto A^t_{AS}(X; V)$ and $f \mapsto A^t_{AS}(f; V)$ form a contravariant functor $A^t_{AS}(-; V) : \mathbf{Top} \to \mathbf{TopAb}^{\mathcal{O}}$.*

Moreover all the constructions made are natural in the coefficient group V. Thus the contravariant functors $A^t_{AS}(-; V)$ for all Abelian topological groups V are part of a bifunctor $A^t_{AS}(-; -) : \mathbf{Top} \times \mathbf{TopAb} \to \mathbf{TopAb}^{\mathcal{O}}$.

Definition 2.5.20. *The bifunctor $A^t_{AS}(-; -) : \mathbf{Top} \times \mathbf{TopAb} \to \mathbf{TopAb}^{\mathcal{O}}$ is called the Alexander-Spanier cosimplicial Abelian topological group functor. For each coefficient group V the functor $A^t_{AS}(-; V) : \mathbf{Top} \to \mathbf{TopAb}^{\mathcal{O}}$ is called the Alexander-Spanier cosimplicial Abelian topological group functor with coefficients V.*

We expect the Alexander-Spanier cosimplicial Abelian topological group functor $A^t_{AS}(-; -)$ to preserve homotopy. Consider two continuous functions $f, g : X \to Y$ and a homotopy $H \times I \to Y$ from g to f. The homotopy H induces a morphism $\mathrm{SS}(H) : \mathrm{SS}(X \times I) \to \mathrm{SS}(Y)$ of standard simplicial spaces. The simplicial topological space $\mathrm{SS}(X \times I)$ is naturally isomorphic to the simplicial space $\mathrm{SS}(X) \times \mathrm{SS}(I)$. The second factor of the latter product contains the discrete simplicial topological space $\mathrm{hom}(-, [1])$ as a simplicial subspace. Composing the inclusion of this subspace with the natural isomorphism $\mathrm{SS}(X) \times \mathrm{SS}(I) \cong \mathrm{SS}(X \times I)$ and the morphism $\mathrm{SS}(H)$ of simplicial topological spaces yields a chain

$$\mathrm{SS}(X) \times \hom(-, [1]) \hookrightarrow \mathrm{SS}(X) \times \mathrm{SS}(I) \cong \mathrm{SS}(X \times I) \xrightarrow{\mathrm{SS}(H)} \mathrm{SS}(Y)$$

of morphisms of simplicial topological spaces. We denote the composition of all the morphisms in this chain by $\tilde{h} : \mathrm{SS}(X) \times \hom(-, [1]) \to \mathrm{SS}(Y)$. For each $n \in \mathbb{N}$ the topological spaces $\hom([n], [1])$ are discrete, so the subspaces $\mathrm{SS}(X)([n]) \times \{\alpha\}$ are open and closed subspaces of $\mathrm{SS}(X)([n]) \times \hom([n], [1])$ for all $\alpha \in \hom([n], [1])$. We denote the restriction of a continuous function f on $\mathrm{SS}(X)([n]) \times \hom([n], [1])$ to the subspace $\mathrm{SS}(X)([n]) \times \{\alpha\}$ by f_α. The subspaces $\mathrm{SS}(X)([n]) \times \{\alpha\}$ are all isomorphic to $\mathrm{SS}(x)([n]) \cong X^{n+1}$. As a consequence, for each Abelian topological group V the function space of continuous functions from $\mathrm{SS}(X)([n]) \times \hom([n], [1])$ to V (which is an Abelian topological group) decomposes into a direct sum:

$$C(\mathrm{SS}(X)([n]) \times \hom([n], [1]); V) \xrightarrow{\cong} \bigoplus_{\alpha \in \hom([n], [1])} C(\mathrm{SS}(X)([n]), V)$$

$$f \mapsto \sum_{\alpha \in \hom([n], [1])} f_\alpha.$$

Note that the direct sum on the right is finite, so it can also be written as a finite product of Abelian topological groups. Because the function space functor $C(-, V)$ is contravariant in the first argument, these direct sums for different $n \in \mathbb{N}$ form a cosimplicial Abelian topological group. A morphism $\beta : [m] \to [n]$ in $\mathcal{O}$ induces a morphism of this cosimplicial group which maps a sum $\sum_{\alpha \in \hom([m], [1])} f_\alpha$ to the sum $\sum_{\alpha' \in \hom([n], [1])} f_{\alpha' \beta}$. Because finite direct sums in **TopAb** are (isomorphic to) finite products, this establishes an isomorphism

$$C(\mathrm{SS}(X) \times \hom(-, [1]); V) \xrightarrow{\cong} C(\mathrm{SS}(X); V)^{\hom(-, [1])}$$

$$C(\mathrm{SS}(X)([n]) \times \hom([n], [1]); V) \ni f \mapsto \prod_{\alpha \in \hom([n], [1])} f_\alpha$$

of cosimplicial Abelian topological groups. All in all the homotopy $H : X \times I \to Y$ induces a morphism $C(\mathrm{SS}(Y), V) \to C(\mathrm{SS}(X), V)^{\hom(-, [1])}$. We assert that there exists a similar morphism between the cosimplicial groups of germs of continuous functions.

Lemma 2.5.21. *For each open covering $\mathfrak{U}$ of Y there exists an open covering $\mathfrak{V}$ of X such that $\tilde{h}(H)$ maps $\mathrm{SS}(X, \mathfrak{V}) \times \hom(-, [1])$ into $\mathrm{SS}(Y, \mathfrak{U})$.*

Proof. Let $\mathfrak{U} = \{U_i \mid i \in I\}$ be an open covering of Y and consider the pullback coverings $\mathfrak{U}_0 = H(-, 0)^{-1}(\mathfrak{U})$ and $\mathfrak{U}_1 = H(-, 1)^{-1}(\mathfrak{U})$ of X. Their common refinement $\mathfrak{U}_0 \wedge \mathfrak{U}_1 := \{H(-, 0)^{-1}(U_i) \cap H(-, 1)^{-1}(U_j) \mid i, j \in I\}$ has the desired properties. $\square$

Let $\mathfrak{V}$ be an open covering of X. Analogously to the above decomposition of function spaces into directs sums resp. finite products the cosimplicial Abelian topological group $C(\mathrm{SS}(X, \mathfrak{V}) \times \hom(-, [1]); V)$ is isomorphic to the cosimplicial Abelian topological group $C(\mathrm{SS}(X); V)^{\hom(-, [1])} = A(X, \mathfrak{V}; V)^{\hom(-, [1])}$. By Lemma 2.5.21, if $\mathfrak{U}$ is an open covering of Y then there exists an open covering $\mathfrak{V}$ of X such that the morphism $\tilde{h}(H) : \mathrm{SS}(X) \times \hom(-, [1]) \to \mathrm{SS}(Y)$ of simplicial topological space restricts and corestricts to a morphism $\mathrm{SS}(X, \mathfrak{V}; V) \times (-, [1]) \to \mathrm{SS}(Y, \mathfrak{U}; V)$. This morphism induces a morphism $A(Y, \mathfrak{U}; V) \to A(X, \mathfrak{V}; V)^{\hom(-, [1])}$ of cosimplicial Abelian topological groups. Passing to the colimit (first in the codomain and then in the domain) yields a morphism $h(H) : A^t_{AS}(Y; V) \to A^t_{AS}(X; V)^{\hom(-, [1])}$ of cosimplicial Abelian topological groups. Collecting the above observations we can now prove:

Theorem 2.5.22. *The Alexander-Spanier cosimplicial Abelian topological group functor $A^t_{AS}(-; -)$ preserves homotopy.*

Proof. Let $f, g : X \to Y$ be continuous function between the topological spaces X and Y, V be a topological group of coefficients, and let $H \times I \to Y$ be a homotopy from g to f. The homotopy H induces a morphism

$$\tilde{h}(H) : \mathrm{SS}(X) \times \hom(-, [1]) \to \mathrm{SS}(Y),$$

which in turn induces a morphism $h(H) : A_{AS}^t(Y; V) \to A_{AS}^t(X; V)^{\hom(-,[1])}$ of cosimplicial Abelian topological groups. We use the characterisation of cosimplicial homotopies in Theorem F.6.9. The embedding $\hom(-, \epsilon_0)$ of the singleton simplicial set $\hom(-, [0])$ into $\hom(-, [1])$ induces a projection

$$p_0 : A_{AS}^t(X; V)^{\hom(-,[1])} \to A_{AS}^t(X; V)^{\hom(-,[0])} \cong A_{AS}^t(X; V),$$

which projects each Abelian topological group $A_{AS}^t(X; V)([n])^{\hom([n]),[1])}$ onto the last factor of the product $A_{AS}^t(X; V)([n])^{\hom([n],[1])}$. Analogously the embedding $\hom(-, \epsilon_1)$ of the zero morphisms into $\hom(-, [1])$ induces a projection

$$p_1 : A_{AS}^t(X; V)^{\hom(-,[1])} \to A_{AS}^t(X; V)^{\hom(-,[0])} \cong A_{AS}^t(X; V),$$

which projects each group $A_{AS}^t(X; V)^{\hom([n]),[1])}$ onto the first factor of the product $A_{AS}^t(X; V)([n])^{\hom([n],[1])}$. In view of Theorem F.6.9 it suffices to prove the equalities $p_0 h(H) = A_{AS}^t(f; V)$ and $p_1 h(H) = A_{AS}^t(g; V)$ to show that $A_{AS}^t(f; V)$ is homotopic to $A_{AS}^t(g; V)$. The morphism $p_0 h(H)$ is induced by the morphism

$$\mathrm{SS}(X) \cong \mathrm{SS}(X) \times \hom(-, 0) \xrightarrow{\mathrm{id}_{\mathrm{SS}(X)} \times \hom(-,\epsilon_0)} \mathrm{SS}(X) \times \hom(-, [1]) \xrightarrow{\tilde{h}(H)} \mathrm{SS}(Y)$$

of simplicial topological spaces. The morphism $\mathrm{id}_{\mathrm{SS}(X)} \times \hom(-, \epsilon_0)$ maps an n-simplex $(x_0, \ldots, x_n)$ in $\mathrm{SS}(X)([n])$ to the n-simplex $((x_0, \ldots, x_n), 1)$ where the last entry denotes the constant map 1 in $\hom([n], [1])$. The morphism $\tilde{h}(H)$ of simplicial spaces maps maps this n-simplex to the n-simplex

$$\begin{aligned}
\tilde{h}(H)((x_0, \ldots, x_n), 1) &= \mathrm{SS}(H)((x_0, 1), \ldots, (x_m, 1)) \\
&= (H(x_0, 1), \ldots, H(x_n, 1)) \\
&= (f(x_0), \ldots, f(x_n)) \\
&= \mathrm{SS}(f)(x_0, \ldots, x_n),
\end{aligned}$$

i.e. the morphism $\tilde{h}(H)(\mathrm{id}_{\mathrm{SS}(X)} \times \hom(-, [1]))$ of simplicial topological spaces coincides with the morphism $\mathrm{SS}(f)$. Therefore the morphism $p_0 h(H)$ coincides with the morphism $A_{AS}^t(f; V)$. Analogously one shows that the morphisms $p_1 h(H)$ and $A_{AS}^t(g; V)$ of cosimplicial Abelian topological groups coincide. So by Theorem F.6.9 the morphism $h(H)$ provides a homotopy from $A_{AS}^t(f; V)$ to $A_{AS}^t(g; V)$. The morphism $h(H)$ and thus the resulting homotopy from $A_{AS}^t(f; V)$ to $A_{AS}^t(g; V)$ is functorial in the coefficients V, so all in all one obtains a homotopy from $A_{AS}^t(f; -)$ to $A_{AS}^t(g; -)$. $\square$

Considering the cochain complexes the associated to the Alexander-Spanier cosimplicial Abelian topological groups $A_{AS}^t(X; V)$ of topological spaces X yields another functor $A_{AS}^{t*}(-; -) := (-)^* \circ A_{AS}^t(-, -) : \mathbf{Top} \times \mathbf{TopAb} \to \mathbf{coCh}(\mathbf{TopAb})$ into the category $\mathbf{coCh}(\mathbf{TopAb})$ of cochain complexes of Abelian topological groups. Theorem 1.1.41 guarantees that the cochain complexes so obtained are proper.

Definition 2.5.23. *The bifunctor* $A_{AS}^{t*}(-, -) : \mathbf{Top} \times \mathbf{TopAb} \to \mathbf{coCh}(\mathbf{TopAb})$ *is called the* Alexander-Spanier topological cochain complex functor. *For each topological space X and coefficient group V the cochain complex $A_{AS}^{t*}(X; V)$ is called the* Alexander-Spanier topological cochain complex with coefficients V of the space X.

Lemma 2.5.24. *The bifunctor $A_{AS}^{t*}(-;-)$ preserves homotopy.*

Proof. This follows from the fact that cosimplicially homotopic morphisms of cosimplicial Abelian topological groups induce cochain homotopic morphisms between the associated cochain complexes (cf. Lemma F.6.6). □

Taking the cohomology of the topological Alexander-Spanier cochain complexes yields a functor $\mathbf{Top} \times \mathbf{TopAb} \to \mathbf{TopAb}$.

Definition 2.5.25. *The bifunctor $H_{AS}^t(-;-) := H \circ A_{AS}^{t*}(-;-)$ is called the* Alexander-Spanier topological cohomology functor. *For each coefficient group V the functor $H_{AS}^t(-;V)$ is called the* Alexander-Spanier topological cohomology functor *with coefficients V. The graded Abelian topological group $H_{AS}^t(X;V)$ associated to a space X is called its* topological Alexander-Spanier cohomology *of X with coefficients V.*

Lemma 2.5.26. *The Alexander-Spanier topological cohomology functor is homotopy invariant.*

Proof. This is a direct consequence of Lemma 2.5.24. □

If one composes the forgetful functor $U : \mathbf{TopAb} \to \mathbf{Ab}$ with $A_{AS}^t(X;V)$ one obtains the underlying cosimplicial group defining ordinary Alexander-Spanier cohomology with coefficients UV. As a consequence one obtains:

Lemma 2.5.27. *The composition of the forgetful functor $U : \mathbf{TopAb} \to \mathbf{Ab}$ with the topological cohomology functor $H_{AS}^t(-;V)$ (with coefficients V) is the classical Alexander-Spanier cohomology functor $H_{AS}(-;UV)$ with coefficients UV.*

Proof. The forgetful functor $U : \mathbf{TopAb} \to \mathbf{Ab}$ preserves kernels and cokernels. □

Analogously to the classical Alexander-Spanier cochain complexes one can consider a "normalised" version of the Alexander-Spanier topological cochain complex $A_{AS}^{t*}(X;V)$ of a topological space X. It is obtained by composing the Moore normalisation functor $N : \mathbf{TopAb}^{\mathcal{O}} \to \mathbf{TopAb}^{\mathcal{O}}$ (see Definitions 1.1.24 and 1.1.28) with the Alexander-Spanier cosimplicial topological group functor $A_{AS}^t(-;-)$ and then taking the associated cochain complex:

Definition 2.5.28. *The bifunctor $NA_{AS}^t(-;-) : \mathbf{Top} \times \mathbf{TopAb} \to \mathbf{TopAb}^{\mathcal{O}}$ is called the* normalised Alexander-Spanier cosimplicial topological group functor. *The bifunctor $NA_{AS}^{t*}(-;-) : \mathbf{Top} \times \mathbf{TopAb} \to \mathbf{coCh}(\mathbf{TopAb})$ is called the* normalised Alexander-Spanier topological cochain complex functor. *For each coefficient group V the functor $NA_{AS}^{t*}(-;V)$ is called the* normalised Alexander-Spanier topological cochain complex *(with coefficients V) functor. The cochain complex $NA_{AS}^{t*}(X;V)$ associated to a topological space X is called the* normalised Alexander-Spanier topological cochain complex *with coefficients V of X.*

Recall that for each topological space X and coefficient group V the components $NA_{AS}^t(X;V)([n])$ of the normalised Alexander-Spanier cosimplicial topological group $NA_{AS}^t(X;V)$ are the subgroups

$$NA_{AS}^t(X;V)([n]) = \bigcap_{i=0}^{n-1} K\left(A_{AS}^t(X;V)(\eta_i : [n] \to [n-1])\right).$$

of the groups $A_{AS}^t(X;V)([n])$ (cf. Definition 1.1.24). These also are the objects of the normalised topological Alexander-Spanier cochain complex $NA_{AS}^{t*}(X;V)$. Similarly to before one observes:

Lemma 2.5.29. *The functors $NA^t_{AS}(-;-)$ and $NA^{t*}_{AS}(-;-)$ preserve homotopy.*

Proof. Let $f, g : X \to Y$ be continuous functions between topological spaces X and Y and $H : X \times I \to Y$ be a homotopy from g to f. The homotopy constructed in the proof of Theorem 2.5.22 maps normal germs to normal germs and thus restricts to a homotopy from $NA^t_{AS}(f;-)$ to $NA^t_{AS}(g;-)$. As a consequence the induced morphisms $NA^{t*}_{AS}(f;-)$ and $NA^{t*}_{AS}(g;-)$ of cochain complexes are homotopic as well (by Lemma F.6.6). $\square$

One can also form the degenerate subcomplex $DA^{t*}_{AS}(X;V)$ of the topological Alexander-Spanier cochain complex $A^{t*}_{AS}(X;V)$ (cf. Definition 1.1.49). Straight from the general Normalisation Theorem 1.2.34 one then derives:

Theorem 2.5.30. *The inclusion $DA^{t*}_{AS}(-;-) \hookrightarrow A^{t*}_{AS}(-;-)$ of the subfunctor $DA^{t*}_{AS}(-;-)$ of $A^{t*}_{AS}(-;-)$ is cochain homotopic to the identity. Furthermore there exists a projection $p : A^t_{AS}(-;-) \twoheadrightarrow NA^t_{AS}(-;-)$ onto the subfunctor $NA^t_{AS}(-;-)$ of $A^t_{AS}(-;-)$ such that the sequence*

$$0 \to DA^{t*}_{AS}(-;-) \hookrightarrow A^{t*}_{AS}(-;-) \xrightarrow{p} NA^{t*}_{AS}(-;-) \to 0$$

of bifunctors $\mathbf{Top} \times \mathbf{TopAb} \to \mathbf{coCh(TopAb)}$ *is split exact.*

Corollary 2.5.31. *The inclusion $NA^{t*}_{AS}(-;-) \hookrightarrow A^{t*}_{AS}(-;-)$ induces an isomorphism $H(NA^{t*}_{AS}(-;-)) \cong H^t_{AS}(-;-)$ of cohomology functors. In particular the Alexander-Spanier topological cohomology $H^{t*}_{AS}(X;V)$ with coefficients V of a topological space X can be computed by using normalised cochains only.*

Lemma 2.5.32. *For each topological space X and coefficient group V the normalised cosimplicial subgroup $NA^t_{AS}(X;V)$ of the Alexander-Spanier cosimplicial topological group $A^t_{AS}(X;V)$ is the cosimplicial subgroup of germs of functions which vanish on degenerate simplices.*

Proof. Let X be a topological space and V be a topological group of coefficients. Any codegeneracy morphism $A^t_{AS}(X;V)(\eta_i) : A^t_{AS}(X;V)([n]) \to A^t_{AS}(X;V)([n-1])$ of the cosimplicial group $A^t_{AS}(X;V)$ maps the germ of a function $f \in A(X;\mathfrak{U};V)([n])$ to the germ of the function $f \circ \mathrm{SS}(X,\mathfrak{U})(\eta_i)$, where $\mathrm{SS}(X,\mathfrak{U})(\eta_i)$ is the degeneracy morphism

$$\mathrm{SS}(X,\mathfrak{U})(\eta_i)(x_0,\ldots,x_{n-1}) = (x_0,\ldots,x_i,x_i,\ldots,x_n),$$

of the $\mathfrak{U}$-small standard simplicial space $\mathrm{SS}(X,\mathfrak{U})$. If f vanishes on all degenerate simplices, then the function $A(X,\mathfrak{U})(\eta_i)(f) = f \circ \mathrm{SS}(X,\mathfrak{U})(\eta_i)$ is trivial and so is the germ of $f \circ \mathrm{SS}(X,\mathfrak{U})(\eta_i)$ in $A^t_{AS}(X;V)([n-1])$. Therefore all germs of continuous functions that vanish on degenerate simplices in $A^t_{AS}(X;V)([n])$ are contained in all the domains $K(A^t_{AS}(X;V)(\eta_i))$ of the kernels $\ker A^t_{AS}(X;V)(\eta_i)$ of the codegeneracy morphisms $A^t_{AS}(X;V)(\eta_i) : A^t_{AS}(X;V)([n]) \to A^t_{AS}(X;V)([n-1])$. Thus they are contained in the intersection

$$NA^t_{AS}(X;V)([n]) = \bigcap_{i=0}^{n-1} K(A^t_{AS}(X;V)(\eta_i : [n] \to [n-1])).$$

Conversely assume that the germ of a function $f \in A^t_{AS}(X,\mathfrak{U};V)([n])$ is contained in the subgroup $NA^t_{AS}(X;V)([n])$. Then the germ of $f \circ \mathrm{SS}(X,\mathfrak{U})(\eta_i)$ is trivial for all degeneracy morphisms $\eta_i : [n] \to [n-1]$. Thus there exists an open refinement $\mathfrak{U}'$ of the open cover $\mathfrak{U}$ such that the restriction of $f \circ \mathrm{SS}(X,\mathfrak{U})(\eta_i)$ to $\mathrm{SS}(X,\mathfrak{U}')([n-1])$ is trivial for all degeneracy morphisms $\eta_i : [n] \to [n-1]$. Since all degenerate simplices in $\mathrm{SS}(X,\mathfrak{U}')([n])$ are of the form $\mathrm{SS}(X,\mathfrak{U}')(\eta_i)(\mathbf{x})$ for some degeneracy morphism

$\eta_i : [n] \to [n-1]$ and some simplex $\mathbf{x} \in \mathrm{SS}(X, \mathfrak{U}')([n-1])$, this implies that the restriction of f to $\mathrm{SS}(X, \mathfrak{U}')([n])$ vanishes on all degenerate simplices. So the germ of f in $A^t_{AS}(X; V)$ is the germ of a function which vanishes on all degenerate simplices.

□

The fact that normal Alexander-Spanier cochains vanish on degenerate simplices especially implies that all normal Alexander-Spanier cochains in $NA^t_{AS}(X; V)([n])$ vanish on the diagonal of $\mathrm{SS}(X)([n]) \cong X^{n+1}$. This enables us to prove the following property:

Lemma 2.5.33. *If X is a topological space and $0 \to V \to V' \xrightarrow{q} V'' \to 0$ is a locally trivial short exact sequence of Abelian topological groups then the morphism $NA^t_{AS}(X; q) : NA^t_{AS}(X; V') \to NA^t_{AS}(X, \mathfrak{U}; V'')$ of cosimplicial topological groups is an epimorphism.*

Proof. Let $0 \to V \to V' \xrightarrow{q} V'' \to 0$ be a locally trivial short exact sequence of Abelian topological groups. Then there exists a local section $s : O \to V'$ defined on a neighbourhood O of the identity 0 in V'' which maps this identity to $0 \in V'$. Let X be a topological space and the normal n-cochain $[f] \in NA^n_{AS}(X; V'')$ be the germ of a continuous function $f \in NA^n(X, \mathfrak{U}; V'')$ defined on a neighbourhood $\mathrm{SS}(X, \mathfrak{U})([n])$ of the diagonal. Since f is normal, it vanishes on the diagonal; so the inverse image $f^{-1}(O)$ of the identity neighbourhood O in V'' is a neighbourhood of the diagonal in $\mathrm{SS}(X, \mathfrak{U})([n])$. The restriction of f to the neighbourhood $\mathrm{SS}(X, \mathfrak{U}) \cap f^{-1}(O)$ of the diagonal takes values in the identity neighbourhood O only. Consider the open covering

$$\mathfrak{U}' = \{ U' \mid (U' \text{ is open}) \wedge (\exists U \in \mathfrak{U} : U' \subseteq U) \wedge (U'^{n+1} \subseteq f^{-1}(O)) \}$$

of X. By definition this covering is a refinement of $\mathfrak{U}$. Furthermore the restriction $\varrho_{\mathfrak{U}',\mathfrak{U}} f \in NA^t_{AS}(X, \mathfrak{U}'; V'')([n])$ of f takes values in the identity neighbourhood O of V'' only. Therefore this restriction can be composed with the section $s : O \to V'$ to obtain a function

$$g := s \circ (\varrho_{\mathfrak{U}',\mathfrak{U}} f) \in NA^t_{AS}(X; V')([n]).$$

(This function is a normal cochain by construction.) The germ $[g] \in NA^t_{AS}(X; V')$ of g then is a lift of f so $[f]$ is contained in the image of $NA^t_{AS}(X; q)([n])$.   □

We proceed to show that for many topological spaces X and locally trivial short exact sequences $0 \to V \to V' \xrightarrow{q} V'' \to 0$ of Abelian topological groups one in fact obtains a quotient homomorphism $NA^t_{AS}(X; V') \twoheadrightarrow NA^t_{AS}(X; V'')$.

Proposition 2.5.34. *If every open covering $\mathfrak{U}$ of a topological space X has a refinement $\mathfrak{U}'$ such that the diagonals in all the spaces $\mathrm{SS}(X, \mathfrak{U}')([n])$ are deformation retracts and the quotient homomorphism $q : V' \twoheadrightarrow V''$ of Abelian topological groups is a Hurewicz fibration, then each function $NA^t_{AS}(X; V')([n]) \to NA^t_{AS}(X; V'')([n])$ has a continuous section.*

Proof. If X is a topological space for which every open covering $\mathfrak{U}$ has a refinement $\mathfrak{U}'$ such that the diagonals $D_{n+1}X$ in all the spaces $\mathrm{SS}(X, \mathfrak{U}')([n])$ are deformation retracts (DR), then one can use the latter open coverings only to define the normalised Alexander-Spanier topological groups:

$$NA^t_{AS}(X; V'')([n]) = \operatorname*{colim}_{\mathfrak{U}' : D_{n+1}X \text{ is a DR of } \mathrm{SS}(X,\mathfrak{U}')([n])} NA(X, \mathfrak{U}'; V'')$$

For each $n \in \mathbb{N}$ and covering $\mathfrak{U}'$ of X such that diagonal $D_{n+1}X$ in $\mathrm{SS}(X, \mathfrak{U}')([n])$ is a deformation retract, let $H : \mathrm{SS}(X, \mathfrak{U}')([n]) \times I \to \mathrm{SS}(X, \mathfrak{U}')([n])$ be a deformation of $\mathrm{SS}(X, \mathfrak{U}')([n])$ to the diagonal $D_{n+1}X$. The function $H(-, 0)$ then is the identity on $SS(X, \mathfrak{U}')([n])$. For each continuous function $f : SS(X, \mathfrak{U}')([n]) \to V''$ in $NA(X; \mathfrak{U}'; V'')([n])$ the composition fH is a homotopy $SS(X, \mathfrak{U}')([n]) \times I \to V''$, which satisfies the equations $fH(-, 0) = f$ and $fH(-, 1) = 0$. Because the quotient homomorphism $V' \twoheadrightarrow V''$ was assumed to be a Hurewicz fibration, the homotopy fH can be lifted to a homotopy $\tilde{H}_f$ into V' by setting

$$\tilde{H}_f : SS(X, \mathfrak{U}')([n]) \times I \to V', \quad \tilde{H}_f(\mathbf{x}, t) := [\Lambda(fH(\mathbf{x}, 1 - (-)), 0)](t),$$

where Λ is a lifting function for the Hurewicz fibration $V' \to V''$ (cf. Theorem D.2.7.) The function $\tilde{H}_f(-, 0)$ then is a lift of f to V' by construction. We assert that the assignment $f \mapsto \tilde{H}_f(-, 0)$ is a continuous function from the function space $NA(X, \mathfrak{U}'; V'')([n])$ to $NA(X, \mathfrak{U}'; V')([n])$. Let H' denote the homotopy obtained from H by setting $H'(\mathbf{x}, t) = H(\mathbf{x}, 1 - t)$. Composition with H' is a continuous function from $C(SS(X, \mathfrak{U}')([n]), V'')$ to $C(SS(X, \mathfrak{U}')([n]) \times I, V'')$ (cf. Lemma B.5.1). Because the unit interval I is compact, the latter function space is naturally isomorphic to the function space $C(SS(X, \mathfrak{U}')([n]), C(I, V''))$. The isomorphism maps a function fH' to its adjoint $\widehat{fH'}$. Since all the functions fH satisfy the equations $fH'(-, 0) = 0$ and $fH'(-, 1) = f$ the adjoint function $\widehat{fH'}$ map any point $\mathbf{x}$ in $SS(X, \mathfrak{U}')([n])$ into the subspace PV'' of of $C(I, V'')$ consisting of paths in V'' starting at 0. Summarising we have obtained a continuous function

$$A(X, \mathfrak{U}'; V'') \to C(SS(X, \mathfrak{U}')([n]), PV''), \quad f \mapsto \widehat{fH'}.$$

Let Λ be the above used lifting function of the Hurewicz fibration $V' \to V''$. Paths in PV'' can be lifted to V'' by composing them with the function $\Lambda(-, 0)$ (cf. Theorem D.2.7). Thus the application of $C(SS(X, \mathfrak{U}')([n]), \Lambda(-, 0))$ lifts the functions in $C(SS(X, \mathfrak{U}')([n]), PV'')$ to $C(SS(X, \mathfrak{U}')([n]), PV')$. The latter function space is a subspace of the function space $C(SS(X, \mathfrak{U}')([n]), C(I, V''))$ which in turn is naturally isomorphic to the function space $C(SS(X, \mathfrak{U}')([n]) \times I, V')$. Restriction of functions is continuous, so all in we obtain a chain

$$NA(X, \mathfrak{U}'; V'') \xrightarrow{f \mapsto \widehat{fH'}} C(SS(X, \mathfrak{U}')([n]), PV'') \xrightarrow{C(SS(X, \mathfrak{U}')([n]), \Lambda(-, 0))}$$

$$\longrightarrow C(SS(X, \mathfrak{U}')([n]), PV') \hookrightarrow C(SS(X, \mathfrak{U}')([n]), C(I, V')) \xrightarrow{\cong}$$

$$\xrightarrow{\cong} C(SS(X, \mathfrak{U}')([n]) \times I, V') \to C(SS(X, \mathfrak{U}')([n]) \times \{0\}, V') \xrightarrow{\cong}$$

$$\xrightarrow{\cong} C(SS(X, \mathfrak{U}')([n]), V') = A(X, \mathfrak{U}'; V')([n])$$

of continuous functions. The composition of all these continuous functions maps a function $f \in A(X, \mathfrak{U}'; V'')$ to $\tilde{H}_f$ by construction (and thus can be corestricted to $NA(X, \mathfrak{U}'; V')([n])$). Summarising we see that the assignment $f \mapsto \tilde{H}_f(-, 0)$ is a continuous function from $NA(X, \mathfrak{U}'; V'')([n])$ to $NA(X, \mathfrak{U}'; V')([n])$ that lifts the continuous functions in $NA(X, \mathfrak{U}'; V'')([n])$ all together. So all the morphisms $NA(X, \mathfrak{U}'; q)([n])$ for G-invariant open coverings $\mathfrak{U}'$ of X have a section and thus are quotient homomorphisms. Because direct limits of identification maps are identification maps (cf. [Dug89, Appendix 2, Theorem 1.5]) the homomorphism $NA_{AS}^t(X; q)([n])$ is a quotient map as well. $\square$

Theorem 2.5.35. *If every open covering $\mathfrak{U}$ of a topological space X has a refinement $\mathfrak{U}'$ such that the diagonals in all the spaces $SS(X, \mathfrak{U}')([n])$ are deformation retracts, then every short exact sequence $0 \to V \to V' \xrightarrow{q} V'' \to 0$ Abelian topological groups which is a Hurewicz fibration induces a long exact sequence in the Alexander-Spanier topological cohomology of X.*

Proof. Let X be a topological space for which every open covering $\mathfrak{U}$ has a refinement $\mathfrak{U}'$ such that the diagonals $D_{n+1}X$ in all the spaces $SS(X, \mathfrak{U}')([n])$ are deformation retracts. If $0 \to V \to V' \xrightarrow{q} V'' \to 0$ is a short exact sequence of coefficients which is a Hurewicz fibration, then each of the homomorphisms $NA_{AS}^{t}(X; q)([n])$ is a quotient homomorphism by Proposition 2.5.34. Therefore the morphism $NA_{AS}^{t}(X; q) : NA_{AS}^{t}(X; V') \to NA_{AS}^{t}(X; V'')$ of normalised Alexander-Spanier cosimplicial topological groups is a quotient morphism. In this case the sequence

$$0 \to NA_{AS}^{t}(X; V) \to NA_{AS}^{t}(X; V') \to NA_{AS}^{t}(X; V'') \to 0$$

of normalised Alexander-Spanier cosimplicial topological groups is exact and induces a long exact sequence in cohomology.   □

Corollary 2.5.36. *If every open covering $\mathfrak{U}$ of a topological space X has a refinement $\mathfrak{U}'$ such that the diagonals in all the spaces $\mathrm{SS}(X, \mathfrak{U}')([n])$ are deformation retracts then every locally trivial short exact sequence $0 \to V \to V' \to V'' \to 0$ of coefficients with paracompact group V'' induces a long exact sequence in the Alexander-Spanier topological cohomology of X.*

Proof. If $0 \to V \to V' \xrightarrow{q} V'' \to 0$ is a locally trivial short exact sequence of coefficients and the group V'' is paracompact, then the bundle $V \to V''$ is a Hurewicz fibration (cf. [Dug89, Theorem XX.4.2]).   □

Corollary 2.5.37. *If every open covering $\mathfrak{U}$ of a topological space X has a refinement $\mathfrak{U}'$ such that the diagonals in all the spaces $\mathrm{SS}(X, \mathfrak{U}')([n])$ are deformation retracts then every short exact sequence $0 \to V \to V' \to V'' \to 0$ of coefficients with locally contractible group V'' induces a long exact sequence in the Alexander-Spanier topological cohomology of X.*

Proof. If $0 \to V \to V' \xrightarrow{q} V'' \to 0$ is a short exact sequence of coefficients and the group V'' is locally contractible, then there exists a contractible identity neighbourhood U and a partition of unity subordinate to the trivialising cover $v.U$, $v \in V''$ (cf. [,]). Therefore the bundle $V \to V''$ is a Hurewicz fibration (cf. [Dug89, Theorem XX.4.2]).   □

The conditions on the topological space X may sound a bit technical, but are always satisfied by Lie groups, as the following example shows:

Example 2.5.38. If G is a Lie group then there exists a chart $\varphi : U \to U'$ from some identity neighbourhood U onto a disc-like neighbourhood U' in a topological vector space V (the model space of the Lie group G). The neighbourhood U' can be contracted to the identity via the homotopy $H : V \times I \to V$, $H(v, t) = (1 - t)v$. Choose a neighbourhood base U_i', $i \in I$ of the identity in U' consisting of disc-like neighbourhoods. The contraction H restricts to a contraction $H_i : U_i' \times I \to U_i'$ of each of the neighbourhoods U_i'. For every open covering $\mathfrak{U}$ of G there exists a refinement $\mathfrak{V}$ consisting of open subsets of the form $g\varphi^{-1}(U_i')$ for elements g of G. For each $n \in \mathbb{N}$ the function

$$H : \mathrm{SS}(G, \mathfrak{V}) \times I \to \mathrm{SS}(G, \mathfrak{V})$$
$$H((g_0, \ldots, g_n), t) = g_0 \cdot \varphi^{-1}\left(1, H(\varphi(g_0^{-1}g_1), \ldots, H(\varphi(g_0^{-1}g_n)))\right)$$

then is a deformation of $\mathrm{SS}(G, \mathfrak{V})([n])$ to the diagonal $D_{n+1}G$ in $\mathrm{SS}(G, \mathfrak{V})([n])$. Moreover this deformation is invariant under left translation by construction.

Corollary 2.5.39. *For every Lie group G and every locally trivial short exact sequence $0 \to V \to V' \to V'' \to 0$ of coefficients with paracompact group V'' there exist a corresponding long exact sequence in the Alexander-Spanier topological cohomology of G.*

Analogously to the classical case we also define Alexander-Spanier cosimplicial topological groups for pairs (X, A) of topological spaces in $\mathbf{Top}(2)$. If A is a subspace of a topological space X then the inclusion $i_A : A \hookrightarrow X$ of A into X induces an inclusion $\mathrm{SS}(i_A) : \mathrm{SS}(A) \hookrightarrow \mathrm{SS}(X)$ of standard simplicial spaces which in turn induces a morphism $A^t_{AS}(i_A; V) : A^t_{AS}(X; V) \to A^t_{AS}(A; V)$ of cosimplicial Abelian topological groups for each coefficient group V.

Definition 2.5.40. *The domain of the kernel of the morphism $A^t_{AS}(i_A; V)$ induced by an inclusion $i_A : A \hookrightarrow X$ of a subspace A of X is denoted by $A^t_{AS}(X, A; V)$. It is called the* Alexander-Spanier cosimplicial topological group (with coefficients V) *of the pair (X, A).*

For each $n \in \mathbb{N}$ and pair (X, A) of topological spaces the Alexander-Spanier topological group $A^t_{AS}(X, A; V)([n])$ of the pair (X, A) is the group of germs of continuous functions on the diagonal in $\mathrm{SS}(X)([n]) \cong X^{n+1}$ that vanish on a neighbourhood of the diagonal in $\mathrm{SS}(A)([n]) \cong A^{n+1}$. The Alexander-Spanier cosimplicial topological group $A^t_{AS}(X; V)$ of a topological space X can be identified with the Alexander-Spanier cosimplicial topological group $A^t_{AS}(X, \emptyset; V)$ of the pair $(X, \emptyset)$. Furthermore the construction is functorial in the coefficients, so it defines a functor $A^t_{AS}(-, -; -) : \mathbf{Top}(2) \times \mathbf{TopAb} \to \mathbf{TopAb}^{\mathcal{O}}$. Similar to the Alexander-Spanier cosimplicial group functor $A^t_{AS}(-; -)$ we observe:

Lemma 2.5.41. *The functor $A^t_{AS}(-, -; -) : \mathbf{Top}(2) \times \mathbf{TopAb} \to \mathbf{TopAb}^{\mathcal{O}}$ preserves homotopy.*

Proof. Let $(f, f_{|A}), (g, g_{|A}) : (X, A) \to (Y, B)$ be morphisms in $\mathbf{Top}(2)$. A homotopy H from $(g, g_{|A})$ to $(f, f_{|A})$ is a homotopy $H : X \times I \to Y$ from g to f that restricts and corestricts to a homotopy $H_{|A \times I} : A \times I \to B$ from $g_{|A}$ to $f_{|A}$. Therefore the morphism $\tilde{h}(H) : \mathrm{SS}(X) \times \hom(-, [1]) \to \mathrm{SS}(Y)$ obtained from a such a homotopy $H : X \times I \to Y$ restricts and corestricts to a morphism $\mathrm{SS}(A) \times \hom(-, [1]) \to \mathrm{SS}(B)$. As a consequence the induced homotopy from $A^t_{AS}(f; V)$ to $A^t_{AS}(g; V)$ maps germs of continuous functions in $A^t_{AS}(Y; V)([n])$ that vanish on a neighbourhood of the diagonal in $\mathrm{SS}(B)([n])$ to germs of functions in $A^t_{AS}(X; V)([n])$ that vanish on a neighbourhood of the diagonal in $\mathrm{SS}(B)([n])$, i.e. it maps the cosimplicial subgroup $A^t_{AS}(Y, B; V)$ of $A^t_{AS}(Y; V)$ into the cosimplicial subgroup $A^t_{AS}(X, A; V)$ of $A^t_{AS}(X; V)$. Therefore it restricts to a homotopy from $A^t_{AS}((f, f_{|A}); V)$ to $A^t_{AS}((g, g_{|A}); V)$. $\qquad\square$

We denote the cochain complex associated to an Alexander-Spanier cosimplicial topological group $A^t_{AS}(X, A; V)$ by $A^{t*}_{AS}(X, A; V)$. The associated cochain complex functor $(-)^* : \mathbf{TopAb}^{\mathcal{O}} \to \mathbf{coCh}(\mathbf{TopAb})$ is continuous (by Lemma 1.1.22) and thus preserves kernels. Therefore the Alexander-Spanier topological cochain complex $A^{t*}_{AS}(X, A)$ is the kernel of the morphism $A^{t*}_{AS}(X; V) \to A^{t*}_{AS}(A; V)$ of Alexander-Spanier topological cochain complexes.

Definition 2.5.42. *The cochain complex $A^{t*}_{AS}(X, A; V)$ associated to the Alexander-Spanier cosimplicial topological group $A^t_{AS}(X, A; V)$ of a pair (X, A) of topological spaces is called the* Alexander-Spanier topological cochain complex *of the pair (X, A) (with coefficients V). The cohomology thereof is denoted by $H^t_{AS}(X, A; V)$ and called the* Alexander-Spanier topological cohomology *of the pair (X, A) (with coefficients V).*

Lemma 2.5.43. *The functor $A^{t*}_{AS}(-, -; -) : \mathbf{Top}(2) \times \mathbf{TopAb} \to \mathbf{TopAb}$ preserves homotopy.*

Proof. The Alexander-Spanier topological cochain complex functor $A^{t*}_{AS}(-,-;-)$ preserves homotopy by Lemma 2.5.41. Because cosimplicially homotopic morphisms of cosimplicial Abelian topological groups induce homotopic morphisms of the associated cochain complexes (Lemma F.6.6), the functor $A^{t*}_{AS}(-,-;-)$ preserves homotopy as well. □

Lemma 2.5.44. *The cohomology functor $H^t_{AS}(-,-;-)$ is homotopy invariant.*

Proof. This is a direct consequence of the fact that the Alexander-Spanier topological cochain complex functor $A^{t*}_{AS}(-,-;-)$ preserves homotopy. □

Straight from the definition of the Alexander-Spanier cosimplicial topological group functors one infers that for every pair (X, A) of topological spaces and any group V of coefficients one obtains a sequence

$$0 \to A^t_{AS}(X, A; V) \xrightarrow{\ker A^t_{AS}(i_A; V)} A^t_{AS}(X; V) \to A^t_{AS}(A; V)$$

of cosimplicial topological groups, in which the inclusion of the cosimplicial group $A^t_{AS}(X, A; V)$ is the kernel of the morphism $A^t_{AS}(X; V) \to A^t_{AS}(A; V)$ by definition, but the latter morphism might not be the cokernel of this inclusion.

Proposition 2.5.45. *If A is a retract of the topological space X then the short sequence $0 \to A^t_{AS}(X, A; V) \to A^t_{AS}(X; V) \to A^t_{AS}(A; V) \to 0$ is split exact for any group V of coefficients.*

Proof. Let A be a retract of X and $r : X \to A$ be a retraction and denote the inclusion $A \hookrightarrow X$ by i_A. The morphism $A^t_{AS}(r; V) : A^t_{AS}(A; V) \to A^t_{AS}(XV)$ of Alexander-Spanier cosimplicial topological groups satisfies the equations

$$A^t_{AS}(i_A; V) \circ A^t_{AS}(r; V) = A^t_{AS}(r \circ i_A; V) = A^t_{AS}(\mathrm{id}_A; V) = \mathrm{id}_{A^t_{AS}(A;V)},$$

hence the morphism $A^t_{AS}(i_A; V) : A^t_{AS}(XV) \to A^t_{AS}(AV)$ is a split epimorphism. Split epimorphisms are quotient morphisms by Lemma 1.1.14, so the short sequence $0 \to A^t_{AS}(X, A; V) \to A^t_{AS}(X; V) \to A^t_{AS}(A; V) \to 0$ is split exact. □

Corollary 2.5.46. *If A is a retract of the topological space X then the inclusion $A^t_{AS}(X, A; V) \hookrightarrow A^t_{AS}(X; V)$ induces an inclusion of $H^t_*(X, A; V)$ as a direct summand: $H^t_*(X; V) \cong H^t_*(X, A; V) \oplus H^t_*(A; V)$.*

Example 2.5.47. Any base space B of a vector bundle $p : E \to B$ with fibre F is a retract of E. Therefore the zero section of a vector bundle $p : E \to B$ induces an embedding $A^t_{AS}(E, B; V) \hookrightarrow A^t_{AS}(E; V)$ of Alexander-Spanier cosimplicial topological groups, hence the cohomology of the total space E decomposes as $H^t_{AS}(E; V) \cong H^t_{AS}(B; V) \oplus H^t_{AS}(F; V) \cong HAS^t(B; V)$. In particular any tangent bundle TM of a (possibly infinite dimensional) manifold M with model space W contains M as a retract and the zero section $M \to TM$ induces an embedding $A^t_{AS}(TM, M; V) \hookrightarrow A^t_{AS}(TM; V)$ of cosimplicial topological groups, which implies the equality $H^t_{AS}(TM; V) \cong H^t_{AS}(M; V) \oplus H^t_{AS}(W; V) \cong H^t_{AS}(M)$.

If A is not a retract but only a neighbourhood retract (NR), then a weaker result can be obtained by requiring the space A to be a closed, X to be normal and V to be a real topological vector space.

Proposition 2.5.48. *If A is a closed NR in a normal space X then the short sequence $0 \to A^t_{AS}(X, A; V) \to A^t_{AS}(X; V) \to A^t_{AS}(A; V) \to 0$ is exact for every real topological vector space V.*

Proof. Let A be a closed NR in a normal space X, the coefficient group V be a real topological vector space and denote the inclusion $A \hookrightarrow X$ by i_A. It suffices to show that each homomorphism $A_{AS}^t(i_A; V)([n]) : A_{AS}^t(X; V)([n]) \to A_{AS}^t(A; V)([n])$ induced by the inclusion i_A is a quotient homomorphism. By assumption here exist a neighbourhood U of A and a retraction $r : U \to A$ of U onto A. Because the space X is normal, there exists a real valued continuous function $\varphi : X \to \mathbb{R}$ on X whose support lies in U and contains the subspace A. Let pr_i denote the projection of the spaces X^{n+1} resp. U^{n+1} onto the i-th factor. The product function

$$\varphi_n := \prod_{i=0}^{n} \varphi \circ \mathrm{pr}_i : X^{n+1} \to \mathbb{R}$$

has support in U^{n+1} and its restriction the subspace A^{n+1} the constant function 1 by construction. So for each open covering $\mathfrak{U}$ of U the multiplication with φ_n defines a (continuous) endomorphism $\phi_n : A(U; \mathfrak{U}; V)([n]) \to A(U; \mathfrak{U}; V)([n])$. All the functions $\phi_n(f)$ in the image of this endomorphism are trivial on the complement $\mathrm{SS}(U, \mathfrak{U})([n]) \setminus \mathrm{supp}(\varphi_n)$. Moreover, they can be continuously extended to a function $\Phi_{n,\mathfrak{U}}(f)$ on the neighbourhood $\mathrm{SS}(X, \mathfrak{U} \cup \{X \setminus A\})([n])$ of the diagonal in X^{n+1} by defining

$$\Phi_{n,\mathfrak{U}}(f)(\mathbf{x}) := \begin{cases} \phi_n(f)(\mathbf{x}) & \text{on } \mathrm{SS}(U, \mathfrak{U})([n]) \\ 0 & \text{on } \mathrm{SS}(X, \mathfrak{U} \cup \{X \setminus A\})([n]) \setminus \mathrm{supp}(\varphi_n) \end{cases}.$$

The function $\Phi_{n,\mathfrak{U}}$ is linear and continuous by construction. It descends to a continuous linear map $\phi_n : A_{AS}^t(U; V)([n]) \to A_{AS}^t(X; V)([n])$. In addition each function $\Phi_{n,\mathfrak{U}}(f)$ restricts to f on the subspace $\mathrm{SS}(A, \mathfrak{U}_{|A})([n])$, so Φ_n is a right inverse to the morphism $A_{AS}^t(i_A; V)([n]) : A_{AS}^t(X; V)([n]) \to A_{AS}^t(A; V)([n])$ induced by the inclusion i_A. Therefore $A_{AS}^t(i_A; V)([n])$ is a split epimorphism, hence a quotient homomorphism. □

Corollary 2.5.49. *For every pair (X, A) of a normal space X and a closed neighbourhood retract A in X and every real topological vector space V the long sequence*

$$\cdots \xrightarrow{\partial^{n-1}} H_{AS}^{tn}(X, A; V) \to H_{AS}^{tn}(X; V) \to H_{AS}^{tn}(A; V) \xrightarrow{\partial^n} \cdots$$

of Alexander-Spanier topological cohomology groups is exact.

Example 2.5.50. If $G \hookrightarrow G' \twoheadrightarrow G''$ is a locally trivial extension of Lie groups then the long sequence

$$\cdots \xrightarrow{\partial^{n-1}} H_{AS}^{tn}(G', G; V) \to H_{AS}^{tn}(G'; V) \to H_{AS}^{tn}(G; V) \xrightarrow{\partial^n} \cdots$$

of Alexander-Spanier topological cohomology groups is exact.

Theorem 2.5.51. *For each coefficient group V the Alexander-Spanier topological cohomology $H_{AS}^t(-, -; V)$ is a topological cohomology theory on the full subcategory of $\mathbf{Top}(2)$ with objects pairs (X, A) with a topological space X and a retract A of X. If V is a real topological vector space then the Alexander-Spanier topological cohomology $H_{AS}^t(-, -; V)$ is a topological cohomology theory on the full subcategory of $\mathbf{Top}(2)$ with objects pairs (X, A) of a closed NR A of a normal space X.*

2.6 Equivariant Continuous Alexander-Spanier Cohomology

In this section we define an equivariant version of topological Alexander-Spanier cohomology. This will be a topological cohomology theory on various subcategories of the category **TrSemGrp**(2) of pairs of transformation semi-groups. The procedure given below requires the coefficients to be a G-module, for some semi-group G. (The action is allowed to be trivial.)

We reconsider the standard simplicial space functor and examine its behaviour on transformation semi-groups. Recall that the standard simplicial space $SS(G)$ of a topological semi-group G is a simplicial topological semi-group. So for a transformation semi-group (G, X) one obtains a pair $(SS(G), SS(X))$ of simplicial spaces the first of which is a simplicial topological semi-group. Let $\varrho : G \times X \to X$ denote the action of G on X. The simplicial topological spaces $SS(G \times X)$ and $SS(G) \times SS(X)$ are naturally isomorphic. The natural isomorphism is the morphism $SS((\mathrm{pr}_G, \mathrm{pr}_X),$ where pr_G and pr_X denote the projections of $G \times X$ onto G and X respectively. The composition of $SS(\varrho)$ with the inverse of this natural isomorphism is a simplicial action

$$\widetilde{SS}(\varrho) := SS(\varrho) \circ SS((\mathrm{pr}_G, \mathrm{pr}_X)^{-1} : SS(X) \times SS(X) \to SS(X),$$
$$\widetilde{SS}(\varrho)([n])((g_0, \ldots, g_n)(x_0, \ldots, x_n)) = (g_0.x_0, \ldots g_n.x_n)$$

of the simplicial topological semi-group $SS(X)$ on the simplicial topological space $SS(X)$. Regarding the category **Top** of topological spaces as the full subcategory of the category **TrSemGrp** of transformation semi-groups (with the trivial group acting trivially on each topological space) we find:

Lemma 2.6.1. *The standard simplicial space functor* SS *on* **Top** *extends to a functor* SS : **TrSemGrp** $\to$ **TrSemGrp**$^{\mathcal{O}^{op}}$.

We can embed the group G as the constant simplicial space into $SS(G)$ via the diagonal embedding. In this way one obtains an action of the group G on the simplicial topological space $SS(X)$, hence we note:

Lemma 2.6.2. *For any topological semi-group G the standard simplicial space functor* SS *restrict to a functor* SS : **GTop** $\to$ **GTop**$^{\mathcal{O}^{op}}$.

Recall that the standard simplicial space functor SS : **Top** $\to$ **Top**$^{\mathcal{O}^{op}}$ on **Top** preserves homotopy by Lemma 2.5.6. The homotopies in the category **TrSemGrp** of transformation semi-groups are homotopies in **Top** which are morphisms of transformation groups if the unit interval I is equipped with the trivial action. Likewise homotopies between morphisms of G-spaces (where G is a topological semi-group) are those homotopies which are G-equivariant when I is endowed with the trivial action. Reinspecting the proof of Lemma 2.5.6 we observe:

Proposition 2.6.3. *The functor* SS : **TrSemGrp** $\to$ **TrSemGrp**$^{\mathcal{O}^{op}}$ *preserves homotopy.*

Proof. If $(\varphi_f, f), (\varphi_g, g) : (G, X) \to (G', Y)$ are morphisms of transformation semi-groups and (φ_H, H) is a homotopy from (φ_g, g) to (φ_f, f) then $(SS(\varphi_H), SS(H))$ is a morphism of simplicial transformation groups. Furthermore the natural isomorphism $SS(X \times I) \cong SS(X) \times SS(I)$ is compatible with the group action. Therefore the morphism $\tilde{h}(H)$ constructed in the proof of Lemma 2.5.6 is a morphism of transformation semi-groups. As a consequence the resultant homotopy from $SS(f)$ to $SS(f)$ is a homotopy of transformation semi-groups. So the standard simplicial space functor SS : **TrSemGrp** $\to$ **TrSemGrp**$^{\mathcal{O}^{op}}$ preserves homotopy. $\square$

Corollary 2.6.4. *The functor* SS : **TrSemGrp** $\to$ **TrSemGrp**$^{\mathcal{O}^{op}}$ *preserves homotopy.*

Proof. The category **GTop** is a subcategory of **TrSemGrp** and homotopies of transformation semi-groups in this subcategory are exactly the homotopies in **GTop**, hence the restriction SS : **GTop** → **GTop**$^{\mathcal{O}^{op}}$ preserves homotopies as well.   □

Let $G \times X \to X$ be an abstract (i.e. not necessarily continuous) action of a semi-group G on the topological space X and the topological coefficient group V be an abstract (possibly trivial) G-module. We consider the action of G on the simplicial space $SS(X)$. Similar to group actions we define:

Definition 2.6.5. *A function $f : SS(X)([n]) \to V$ is called G-equivariant if it satisfies the equation $f(g.\mathbf{x}) = g.f(\mathbf{x})$ for all $\mathbf{x} \in SS(X)([n])$ and all $g \in G$.*

If G is a group We can define a not necessarily continuous action of the G on the groups $A(X;V)([n])$ via

$$G \times A(X;V)([n]) \to A(X;V)([n]), \quad (g.f)(x_0,\ldots,x_n) = g.(f(g^{-1}x_0,\ldots,g^{-1}x_n))$$

The G-fixed point sets $A(X;V)([n])^G$ of these group actions on the topological groups $A(X;V)([n])$ form a cosimplicial subgroup $A_{eq}(X;V)$ of the standard cosimplicial topological group of X with coefficients V. The elements of the groups $A_{eq}(X;V)([n])$ are exactly the G-equivariant n-cochains.

Definition 2.6.6. *The cosimplicial topological subgroup of $A(X;V)$ consisting of all equivariant cochains is denoted by $A_{eq}(X;V)$ It is called the* equivariant standard cosimplicial Abelian group with coefficients V *of X.*

Lemma 2.6.7. *For every abstract action of a semi-group on a topological space X and the coefficient group V the cosimplicial topological group $A_{oq}(X;V)$ is completely regular. It is Hausdorff if and only if the coefficient group V is Hausdorff.*

Proof. This follows from Lemma 2.5.8 and the fact that $A_{eq}(X;V)$ is a cosimplicial subgroup of the cosimplicial topological group $A(X;V)$.   □

For a fixed topological semi-group G and G-module V the passage to the cochain complex associated to the cosimplicial Abelian topological group $A_{eq}(X;V)$ yields another functor $A^*_{eq}(-;V) : \mathbf{GTop} \to \mathbf{coCh(TopAb)}$ from the category **GTop** of G-spaces into the category **coCh(TopAb)** of cochain complexes of Abelian topological groups. By Theorem 1.1.41 these cochain complexes are proper, i.e. the differentials are quotient maps onto their image.

Definition 2.6.8. *The functor $A^*_{eq}(-;V) : \mathbf{GTop} \to \mathbf{coCh(TopAb)}$ is called the* equivariant standard topological cochain complex functor with coefficients V *and the cochain complex $A^*_{eq}(X;V)$ associated to a G-space X is called the* equivariant standard topological cochain complex (with coefficients V) *associated to the space X. The bifunctor $A^*_{eq}(-;-) : \mathbf{GTop} \times \mathbf{GTopAb} \to \mathbf{coCh(TopAb)}$ is called the* equivariant standard topological cochain complex functor.

Remark 2.6.9. If the semi-group G is endowed with the discrete topology then every action of G on a topological space is continuous. Thus it is no restriction to only consider transformation semi-groups (G, X) (where the action is continuous by definition).

Example 2.6.10. In the special case in which G is a topological group and the action on $X = G$ is the left translation the cochain complex $A^*_{eq}(G;V)$ is the complex of continuous homogeneous group cochains. Its cohomology is the continuous group cohomology of G with coefficients V.

Remark 2.6.11. As the preceding example shows, the equivariant standard topological cochain complex $A^*_{eq}(X; V)$ of a transformation semi-group (G, X) with coefficients a G-module V is not exact in general.

Let (G, X) be a transformation semi-group and the group V of coefficients be a G-module. Each element $g \in G$ acts on X an on the power set $\mathcal{P}(X)$ of X. The latter action maps a family of $\mathfrak{U}$ of subsets of X to the family $g.\mathfrak{U} = \{g.U \mid U \in \mathfrak{U}\}$ of subsets of X. Therefore the action of g on the standard simplicial space $SS(X)$ of X maps the simplicial subspace $SS(X, \mathfrak{U})$ to the simplicial subspace $SS(X, g.\mathfrak{U})$. If the family $\mathfrak{U}$ of subsets of X is G-invariant then these simplicial subspaces of the standard simplicial space $SS(X)$ always coincide, i.e. the $\mathfrak{U}$-small standard simplicial space $SS(X, \mathfrak{U})$ is a G-invariant simplicial subspace of the standard simplicial space $SS(X)$. In this case we can consider G-equivariant n-cochains on $SS(X, \mathfrak{U})([n])$.

Definition 2.6.12. *For every G-invariant family $\mathfrak{U}$ of subsets of a G-space X the cosimplicial subgroup of $A(X, \mathfrak{U}; V)$ consisting of all G-equivariant cochains is denoted by $A_{eq}(X, \mathfrak{U}; V)$. It is called the* equivariant $\mathfrak{U}$-small standard cosimplicial Abelian topological group of X.

If G is not only a semi-group but a group then we can define an action of G on the space $A(X, \mathfrak{U}; V)$ by setting

$$G \times A(X, \mathfrak{U}; V) \to A(X, \mathfrak{U}; V), \quad [g.f](\mathbf{x}) = g.f(g^{-1}.\mathbf{x}).$$

So we can consider G-fixed points in $A(X, \mathfrak{U}; V)$ if $\mathfrak{U}$ is a G-invariant family of subsets of X. these fixed points are exactly the G-equivariant cochains in $A(X, \mathfrak{U}; V)$.

Lemma 2.6.13. *Any equivariant $\mathfrak{U}$-small standard cosimplicial group $A_{eq}(X, \mathfrak{U}; V)$ is completely regular. If the coefficient group V is Hausdorff then $A_{eq}(X, \mathfrak{U}; V)$ is also Hausdorff.*

Proof. Let (G, X) be any transformation semi-group and V be any G-module. Each of the function spaces $A_{eq}(X, \mathfrak{U}; V)$ is a topological group and therefore completely regular. Furthermore, if the coefficient group V is Hausdorff, then so is every function space of the form $C(Y, V)$. (See [Dug89, XII.1.3] for a proof.) Thus the cosimplicial subgroup $A_{eq}(X, \mathfrak{U}; V)$ of $C(SS(X), V)$ is Hausdorff if V is Hausdorff.   □

Let (G, X) be a transformation semi-group and the group V of coefficients be a G-module. Recall that the action of an element $g \in G$ maps an open covering $\mathfrak{U}$ of X to the open covering $g.\mathfrak{U} = \{g.U \mid U \in \mathfrak{U}\}$ and furthermore the action of g on the standard simplicial space $SS(X)$ of X maps the simplicial subspace $SS(X, \mathfrak{U})$ to the simplicial subspace $SS(X, g.\mathfrak{U})$. The coverings $\mathfrak{U}$ and $g.\mathfrak{U}$ have a common refinement where it can be tested whether f intertwines the action of g on X and V. This gives rise to the equivariant version of the Alexander-Spanier cosimplicial topological groups.

Definition 2.6.14. *A germ $[f] \in A_{AS}(X; V)([n])$ of a function $f \in A(X, \mathfrak{U}; V)([n])$ is called* equivariant *if for every element $g \in G$ there exists a refinement $\mathfrak{V}$ of $\mathfrak{U}$ and $g.\mathfrak{U}$ such that f satisfies the equation $f(g.\mathbf{x}) = g.f(\mathbf{x})$ for all $\mathbf{x} \in SS(X, \mathfrak{V}([n]))$.*

if G is not only a semi-group but a group and $f : SS(X, \mathfrak{U})([n]) \to V$ is a continuous function then

$$g.f : SS(X, g.\mathfrak{U})([n]) \to V, \quad [g.f](\mathbf{x}) = g.f(g^{-1}.\mathbf{x})$$

is a continuous function on $SS(X, g.\mathfrak{U})$. Setting $g.[f] = [g.f]$ defines an action of G on the colimit group $A^t_{AS}(X; V)([n])$. Because the action of G intertwines the coboundary and codegeneracy maps one in this way obtains an action of G on the cosimplicial group $A_{AS}^t(X;V)$. (This action of G on $A_{AS}^t(X;V)$ is by construction compatible with th action of G on the standard cosimplicial group $A(X;V)$.) The G-fixed points are exactly the G-invariant cochains.

Definition 2.6.15. *The cosimplicial subgroup of $A_{AS}^t(X,\mathfrak{U};V)$ consisting of G-equivariant cochains is denoted by $A_{AS,eq}^t(X,\mathfrak{U};V)$ It is called the* equivariant *Alexander-Spanier cosimplicial Abelian topological group with coefficients V of X.*

If we consider morphisms $(\varphi_f, f) : (G,X) \to (G',Y)$ of transformation semi-groups where both semi-groups act on the coefficients then the induced morphisms between the Alexander-Spanier cosimplicial topological groups map equivariant cochains to equivariant cochains:

Lemma 2.6.16. *If $(\varphi_f, f) : (G,X) \to (G',Y)$ is a morphism of transformation semi-groups and the group V is a G-module and a G'-module then the morphism $A_{AS}^t(f;V)$ restricts to a morphism $A_{AS,eq}^t(f;V) : A_{AS,eq}^t(Y;V) \to A_{AS,eq}^t(X;V)$.*

Proof. Let $(\varphi_f, f) : (G,X) \to (G',Y)$ be a morphism of transformation semi-groups and V be a G-module as well as a G'-module. If $[h] \in A_{AS,eq}(Y;V)([n])$ is the germ of a G'-equivariant n-cochain h defined on a neighbourhood of the diagonal then the equivariance of f implies

$$A(f;V)([n])(h)(g.\mathbf{x}) = f((h(g.\mathbf{x})) = f(g.(h(\mathbf{x}))) = g.(fh(\mathbf{x})) = g.A(f;V)([n])(h)$$

for all $\mathbf{x}$ in a some sufficiently small neighbourhood $SS(X,\mathfrak{V})([n])$ of the diagonal. Thus the germ $A(f;V)([n])(h)$ of fh is equivariant. $\qquad\square$

Similarly to the standard cosimplicial topological group functors $A(-;V)$ for different coefficient groups V the contravariant functors $A_{AS,eq}^t(-;V)$ on **GTop** (for G-modules V) are part of a bifunctor $A_{AS,eq}^t(-;-) : \mathbf{GTop} \times \mathbf{GTopAb} \to \mathbf{TopAb}$.

Definition 2.6.17. *The bifunctor $A_{AS,eq}^t(-;-) : \mathbf{GTop} \times \mathbf{GTopAb} \to \mathbf{TopAb}^{\mathcal{O}}$ is called the* equivariant *Alexander-Spanier cosimplicial Abelian topological group functor. For each G-module V the functor $A_{AS,eq}^t(-;V) : \mathbf{GTop} \to \mathbf{TopAb}^{\mathcal{O}}$ is called the* equivariant *Alexander-Spanier cosimplicial Abelian topological group functor with coefficients V.*

Recall that the Alexander-Spanier cosimplicial Abelian topological group functor $A_{AS}^t(-;-)$ preserves homotopy. Concerning the equivariant Alexander-Spanier cosimplicial topological group functor $A_{AS,eq}^t(-;-) : \mathbf{GTop} \times \mathbf{GTopAb} \to \mathbf{TopAb}$ an analogous result can be derived. Here fore one only needs to notice that the morphism $\tilde{h}(H)$ constructed in the proof of Theorem 2.5.22 is equivariant if the homotopy H started with is a homotopy of G-spaces. So we note:

Lemma 2.6.18. *The equivariant Alexander-Spanier cosimplicial Abelian topological group functor $A_{AS,eq}^t(-;-) : \mathbf{GTop} \times \mathbf{GTopAb} \to \mathbf{TopAb}$ preserves homotopy.*

Considering the cochain complexes associated to equivariant Alexander-Spanier cosimplicial topological groups yields yet another functor $A_{AS,eq}^{t*}(-;-)$ from the category $\mathbf{GTop} \times \mathbf{GTopAb}$ into the category $\mathbf{coCh}(\mathbf{TopAb})$ of cochain complexes of Abelian topological groups. The cochain complexes so obtained are proper as well (cf. Theorem 1.1.41).

Definition 2.6.19. *The bifunctor $A_{AS,eq}^{t*}(-,-)$ on $\mathbf{GTop} \times \mathbf{GTopAb}$ is called the* equivariant *Alexander-Spanier topological cochain complex functor. For each G-space X and G-module V the cochain complex $A_{AS,eq}^{t*}(X;V)$ is called the* equivariant *Alexander-Spanier topological cochain complex with coefficients V of the space X.*

Lemma 2.6.20. *The bifunctor* $A_{AS,eq}^{t*}(-;-) : \mathbf{GTop} \times \mathbf{GTopAb} \to \mathbf{TopAb}$ *preserves homotopy.*

Proof. This also follows from the fact that cosimplicially homotopic morphisms of cosimplicial Abelian topological groups induce cochain homotopic morphisms between the associated cochain complexes (cf. Lemma F.6.6). □

The passage to the cohomology of the equivariant topological Alexander-Spanier cochain complexes yields another functor $\mathbf{GTop} \times \mathbf{GTopAb} \to \mathbf{TopAb}$.

Definition 2.6.21. *The bifunctor* $H_{AS,eq}^{t}(-;-) := H \circ A_{AS,eq}^{t*}(-;-)$ *is called the* equivariant Alexander-Spanier topological cohomology functor. *For each G-module V the functor $H_{AS,eq}(-;V)$ is called the* equivariant Alexander-Spanier topological cohomology functor with coefficients V. *The graded Abelian topological group $H_{AS,eq}(X;V)$ associated to a G-space X is called its* equivariant topological Alexander-Spanier cohomology of X with coefficients V.

Lemma 2.6.22. *The equivariant Alexander-Spanier topological cohomology functor is homotopy invariant.*

Proof. This is a direct consequence of Lemma 2.6.20. □

Analogously to the Alexander-Spanier topological cochain complexes one can consider a "normalised" version of the equivariant Alexander-Spanier topological cochain complex $A_{AS,eq}^{t*}(X;V)$ of a transformation semi-group (G,X) and a G-module V. This normalised version is also obtained by composition with the Moore normalisation functor $N : \mathbf{TopAb}^{\mathcal{O}} \to \mathbf{TopAb}^{\mathcal{O}}$ (see Definitions 1.1.24 and 1.1.28).

Definition 2.6.23. *The bifunctor* $NA_{AS}^{t}(-;-)^{G} : \mathbf{GTop} \times \mathbf{GTopAb} \to \mathbf{TopAb}^{\mathcal{O}}$ *is called the* normalised equivariant Alexander-Spanier cosimplicial topological group functor. *The bifunctor* $NA_{AS}^{t*}(-;-) : \mathbf{GTop} \times \mathbf{GTopAb} \to \mathbf{coCh}(\mathbf{TopAb})$ *is called the* normalised equivariant Alexander-Spanier topological cochain complex functor. *For each G-module V the functor $NA_{AS}^{t*}(-;V)$ is called the* normalised equivariant Alexander-Spanier topological cochain complex (with coefficients V) functor. *The cochain complex $NA_{AS}^{t*}(X;V)$ associated to a topological space X is called the* normalised Alexander-Spanier topological cochain complex with coefficients V of X.

The normalised equivariant Alexander-Spanier cosimplicial topological group functor $NA_{AS}^{t}(-;-)^{G}$ is a subfunctor of the normalised Alexander-Spanier cosimplicial topological group functor $NA_{AS}^{t}(-;-)$ by construction. Likewise, for each G-module V the normalised equivariant Alexander-Spanier topological cochain complex functor $NA_{AS}^{t*}(-;V)^{G}$ is a subfunctor of the normalised equivariant Alexander-Spanier topological cochain complex functor $NA_{AS}^{t*}(-;V)$. The elements of components $NA_{AS}^{t}(X;V)([n])^{G}$ of the normalised equivariant Alexander-Spanier cosimplicial topological group $NA_{AS}^{t}(X;V)$ are the normal cochains $f \in A_{AS}^{t}(X;V)([n])$ which are G-equivariant. Similarly to before one observes:

Lemma 2.6.24. *The functors $NA_{AS}^{t}(-;-)^{G}$ and $NA_{AS}^{t*}(-;-)^{G}$ preserve homotopy.*

Proof. Let $f,g : X \to Y$ be equivariant functions between G-spaces, V be a G-module and $H : X \times I \to Y$ be a G-homotopy from g to f. The homotopy constructed in the proof of Theorem 2.5.22 is equivariant because H is equivariant. In addition it and maps normal germs to normal germs and thus restricts to a homotopy from $NA_{AS,eq}^{t}(f;-)$ to $NA_{AS,eq}^{t}(g;-)$. As a consequence the induced morphisms $NA_{AS,eq}^{t*}(f;-)$ and $NA_{AS,eq}^{t*}(g;-)$ of cochain complexes are homotopic as well (by Lemma F.6.6). □

Like always one can also form the degenerate subcomplex $DA_{AS,eq}^{t*}(X;V)$ of the equivariant Alexander-Spanier topological cochain complex $A_{AS,eq}^{t*}(X;V)$ (cf. Definition 1.1.49). Straight from the general Normalisation Theorem 1.2.34 one then derives:

Theorem 2.6.25. *The inclusion $DA_{AS,eq}^{t*}(-;-) \hookrightarrow A_{AS,eq}^{t*}(-;-)$ of the subfunctor $DA_{AS,eq}^{t*}(-;-)$ of $A_{AS,eq}^{t*}(-;-)$ is cochain homotopic to the identity. Furthermore there exists a projection $p : A_{AS,eq}^{t}(-;-) \twoheadrightarrow NA_{AS,eq}^{t}(-;-)$ onto the subfunctor $NA_{AS,eq}^{t}(-;-)$ of $A_{AS,eq}^{t}(-;-)$ such that the sequence*

$$0 \to DA_{AS,eq}^{t*}(-;-) \hookrightarrow A_{AS,eq}^{t*}(-;-) \xrightarrow{p} NA_{AS,eq}^{t*}(-;-) \to 0$$

of bifunctors $\mathbf{GTop} \times \mathbf{GTopAb} \to \mathbf{coCh(TopAb)}$ *is split exact.*

Corollary 2.6.26. *The inclusion $NA_{AS,eq}^{t*}(-;-) \hookrightarrow A_{AS,eq}^{t*}(-;-)$ induces an isomorphism $H(NA_{AS,eq}^{t*}(-;-)) \cong H_{AS,eq}^{t}(-;-)$ of cohomology functors. In particular the equivariant Alexander-Spanier topological cohomology $H_{AS,eq}^{t*}(X;V)$ with coefficients V of a G-space X can be computed by using normalised cochains only.*

Because the normalised equivariant Alexander-Spanier cosimplicial topological group $NA_{AS,eq}^{t}(X;V)$ is a cosimplicial subgroup of the normalised Alexander-Spanier cosimplicial topological group $NA_{AS}^{t}(X;V)$ we note:

Lemma 2.6.27. *For each G-space X and G-module V the normalised cosimplicial subgroup $NA_{AS,eq}^{t}(X;V)$ of the equivariant Alexander-Spanier cosimplicial topological group $A_{AS,eq}^{t}(X;V)$ is the cosimplicial subgroup of germs of functions which vanish on degenerate simplices.*

Proof. This follows from the fact that all normal Alexander-Spanier cochains have this property (by Lemma 2.5.32). □

The fact that normal equivariant Alexander-Spanier cochains vanish on degenerate simplices especially implies that all normal equivariant Alexander-Spanier cochains in $NA_{AS,eq}^{t}(X;V)([n])$ vanish on the diagonal of $SS(X)([n]) \cong X^{n+1}$. Similar to the non-equivariant case this enables us to prove:

Lemma 2.6.28. *If X is a G-space and $0 \to V \to V' \xrightarrow{q} V'' \to 0$ is a short exact sequence of G-modules with G-equivariant local sections then the morphism $NA_{AS,eq}^{t}(X;q) : NA_{AS,eq}^{t}(X;V') \to NA_{AS,eq}^{t}(X,\mathfrak{U};V'')$ of cosimplicial topological groups is an epimorphism.*

Proof. The proof of Lemma 2.5.33 carries over to the equivariant case if one uses G-equivariant local sections. □

Analogously to the non-equivariant case we also define equivariant Alexander-Spanier cosimplicial topological groups for pairs (X, A) of G-spaces in $\mathbf{GTop}(2)$. If A is a G-invariant subspace of a G-space X then the inclusion $i_A : A \hookrightarrow X$ of A into X induces an inclusion $SS(i_A) : SS(A) \hookrightarrow SS(X)$ of standard simplicial G-spaces which in turn induces a morphism $A_{AS,eq}^{t}(i_A;V) : A_{AS,eq}^{t}(X;V) \to A_{AS,eq}^{t}(A;V)$ of cosimplicial Abelian topological groups for each G-module V.

Definition 2.6.29. *The domain of the kernel of the morphism $A_{AS,eq}^{t}(i_A;V)$ induced by an inclusion $i_A : A \hookrightarrow X$ of a G-subspace A of a G-space X is denoted by $A_{AS,eq}^{t}(X, A;V)$. It is called the* equivariant Alexander-Spanier cosimplicial topological group *(with coefficients V) of the pair (X, A).*

This is the cosimplicial subgroup group of the cosimplicial topological group $A^t_{AS}(X, A; V)$ consisting of equivariant cochains: For each $n \in \mathbb{N}$ and pair (X, A) of G-spaces the equivariant Alexander-Spanier topological group $A^t_{AS,eq}(X, A; V)([n])$ of the pair (X, A) is the group of equivariant germs of continuous functions on the diagonal in $\mathrm{SS}(X)([n]) \cong X^{n+1}$ that vanish on a neighbourhood of the diagonal in $\mathrm{SS}(A)([n]) \cong A^{n+1}$. The equivariant Alexander-Spanier cosimplicial topological group $A^t_{AS,eq}(X; V)$ of a G-space X can be identified with the equivariant Alexander-Spanier cosimplicial topological group $A^t_{AS,eq}(X, \emptyset; V)$ of the pair $(X, \emptyset)$. Furthermore the construction is functorial in the coefficients, so it defines a functor $A^t_{AS,eq}(-, -; -) : \mathbf{GTop}(2) \times \mathbf{GTopAb} \to \mathbf{TopAb}^{\mathcal{O}}$. Similar to the Alexander-Spanier cosimplicial group functor we observe:

Lemma 2.6.30. *The functor* $A^t_{AS,eq}(-, -; -) : \mathbf{GTop}(2) \times \mathbf{GTop} \to \mathbf{TopAb}^{\mathcal{O}}$ *preserves homotopy.*

Proof. Let $(f, f_{|A}), (g, g_{|A}) : (X, A) \to (Y, B)$ be two morphisms in $\mathbf{GTop}(2)$. A homotopy H from $(g, g_{|A})$ to $(f, f_{|A})$ in $\mathbf{GTop}$ is an equivariant homotopy $H : X \times I \to Y$ from g to f that restricts and corestricts to an equivariant homotopy $H_{|A \times I} : A \times I \to B$ from $g_{|A}$ to $f_{|A}$. Therefore the equivariant morphism $\tilde{h}(H) : \mathrm{SS}(X) \times \hom(-, [1]) \to \mathrm{SS}(Y)$ obtained from such a homotopy $H : X \times I \to Y$ restricts and corestricts to an equivariant morphism $\mathrm{SS}(A) \times \hom(-, [1]) \to \mathrm{SS}(B)$. As a consequence the induced homotopy from $A^t_{AS}(f; V)$ to $A^t_{AS}(g; V)$ is equivariant and maps equivariant germs of continuous functions in $A^t_{AS,eq}(Y; V)([n])$ that vanish on a neighbourhood of the diagonal in $\mathrm{SS}(B)([n])$ to equivariant germs of functions in $A^t_{AS,eq}(X; V)([n])$ that vanish on a neighbourhood of the diagonal in $\mathrm{SS}(B)([n])$, i.e. it maps the cosimplicial subgroup $A^t_{AS,eq}(Y, B; V)$ of $A^t_{AS}(Y; V)$ into the cosimplicial subgroup $A^t_{AS,eq}(X, A; V)$ of $A^t_{AS,eq}(X; V)$. Therefore it restricts to an equivariant homotopy from $A^t_{AS,eq}((f, f_{|A}); V)$ to $A^t_{AS,eq}((g, g_{|A}); V)$. $\square$

We denote the cochain complex associated to an equivariant Alexander-Spanier cosimplicial topological group $A^t_{AS,eq}(X, A; V)$ by $A^{t*}_{AS,eq}(X, A; V)$. The associated cochain complex functor $(-)^* : \mathbf{TopAb}^{\mathcal{O}} \to \mathbf{coCh}(\mathbf{TopAb})$ is continuous (by Lemma 1.1.22) and thus preserves kernels. Therefore the equivariant Alexander-Spanier topological cochain complex $A^{t*}_{AS,eq}(X, A; V)$ is the kernel of the morphism $A^{t*}_{AS,eq}(X; V) \to A^{t*}_{AS,eq}(A; V)$ of equivariant Alexander-Spanier topological cochain complexes.

Definition 2.6.31. *The cochain complex* $A^{t*}_{AS,eq}(X, A; V)$ *associated to the equivariant Alexander-Spanier cosimplicial topological group* $A^t_{AS,eq}(X, A; V)$ *of a pair* (X, A) *of topological spaces is called the* equivariant Alexander-Spanier topological cochain complex *of the pair* (X, A) *(with coefficients* V). *The cohomology thereof is denoted by* $H^t_{AS,eq}(X, A; V)$ *and called the* equivariant Alexander-Spanier topological cohomology *of the pair* (X, A) *(with coefficients* V).

The reader is reminded that the homotopies in the category $\mathbf{GTop}$ of G-spaces are morphisms of G-spaces where the unit interval I is a trivial G-space; thus they are equivariant by definition.

Lemma 2.6.32. *The functor* $A^{t*}_{AS,eq}(-, -; -) : \mathbf{GTop}(2) \times \mathbf{GTop} \to \mathbf{TopAb}$ *preserves homotopy.*

Proof. The Alexander-Spanier topological cochain complex functor $A^{t*}_{AS,eq}(-, -; -)$ preserves homotopy by Lemma 2.6.30. Because cosimplicially homotopic morphisms of cosimplicial Abelian topological groups induce homotopic morphisms of the associated cochain complexes (Lemma F.6.6), the functor $A^{t*}_{AS,eq}(-, -; -)$ preserves homotopy as well. $\square$

Lemma 2.6.33. *The cohomology functor $H^t_{AS,eq}(-,-;-)$ is homotopy invariant.*

Proof. This is a consequence of the fact that the equivariant Alexander-Spanier topological cochain complex functor $A^{t*}_{AS,eq}(-,-;-)$ preserves homotopy. $\square$

Straight from the definition of the equivariant Alexander-Spanier cosimplicial topological group functors one infers that for every pair (X,A) of G-spaces and any G-module V of coefficients one obtains a sequence

$$0 \to A^t_{AS,eq}(X,A;V) \xrightarrow{\ker A^t_{AS,eq}(i_A;V)} A^t_{AS,eq}(X;V) \to A^t_{AS,eq}(A;V)$$

of cosimplicial topological groups, in which the inclusion of the cosimplicial group $A^t_{AS,eq}(X,A;V)$ is the kernel of the morphism $A^t_{AS,eq}(X;V) \to A^t_{AS,eq}(A;V)$ by definition, but the latter morphism might not be the cokernel of this inclusion.

Proposition 2.6.34. *If the G-space A is a retract of the G-space X in **GTop** then the short sequence $0 \to A^t_{AS,eq}(X,A;V) \to A^t_{AS,eq}(X;V) \to A^t_{AS,eq}(A;V) \to 0$ is split exact for any G-module V of coefficients.*

Proof. Let A be a retract of X in **GTop** and $r : X \to A$ be an equivariant retraction. Denote the inclusion $A \hookrightarrow X$ by i_A. The morphism $A^t_{AS,eq}(r;V)$ of equivariant Alexander-Spanier cosimplicial topological groups satisfies the equations

$$A^t_{AS,eq}(i_A;V) \circ A^t_{AS,eq}(r;V) = A^t_{AS,eq}(r \circ i_A;V) = A^t_{AS,eq}(\mathrm{id}_A;V) = \mathrm{id}_{A^t_{AS,eq}(A;V)},$$

hence the morphism $A^t_{AS,eq}(i_A;V) : A^t_{AS,eq}(XV) \to A^t_{AS,eq}(AV)$ is a split epimorphism. Split epimorphisms are quotient morphisms by Lemma 1.1.14, so the short sequence $0 \to A^t_{AS,eq}(X,A;V) \to A^t_{AS,eq}(X;V) \to A^t_{AS,eq}(A;V) \to 0$ is exact and splits. $\square$

Corollary 2.6.35. *If a A is a retract of the G-space X in **GTop** then the inclusion $A^t_{AS,eq}(X,A;V) \hookrightarrow A^t_{AS,eq}(X;V)$ induces an inclusion of $H^{t*}_{AS,eq}(X,A;V)$ as a direct summand: $H^{t*}_{AS,eq}(X;V) \cong H^{t*}_{AS,eq}(X,A;V) \oplus H^{t*}_{AS,eq}(A;V)$.*

If A is not a retract in **GTop** but only a neighbourhood retract in **GTop** (abbreviated by G-NR), then a weaker result can be obtained by requiring the space A to be a closed, and X and V to satisfy additional conditions.

Definition 2.6.36. *A G-space is called G-normal if for any two G-invariant closed subsets A and B of X there exists a G-equivariant continuous function $f : X \to I$ satisfying $f_{|A} = 1$ and $f_{|B} = 0$.*

Proposition 2.6.37. *If A is a closed G-NR in a G-normal space X then the short sequence $0 \to A^t_{AS,eq}(X,A;V) \to A^t_{AS,eq}(X;V) \to A^t_{AS,eq}(A;V) \to 0$ is exact for every real topological vector space V.*

Proof. The proof is analogous to that of Proposition 2.5.48. The only difference is that all maps involved are equivariant. $\square$

Corollary 2.6.38. *For every pair (X,A) of a G-normal G-space X and a closed G-neighbourhood retract A in X and every G-module that is a real topological vector space the long sequence*

$$\cdots \xrightarrow{\partial^{n-1}} H^{tn}_{AS,eq}(X,A;V) \to H^{tn}_{AS,eq}(X;V) \to H^{tn}_{AS,eq}(A;V) \xrightarrow{\partial^n} \cdots$$

of equivariant Alexander-Spanier topological cohomology groups is exact.

Example 2.6.39. If $p : E \to B$ is a principal bundle over a normal base space B with fibre a topological semi-group G then E is G-normal. If the bundle is locally trivial and B is a T_1 space then the fibre G is a closed G-NR in E. As a consequence the long sequence

$$\cdots \xrightarrow{\partial^{n-1}} H^{tn}_{AS,eq}(E, G; V) \to H^{tn}_{AS,eq}(E; V) \to H^{tn}_{AS,eq}(G; V) \xrightarrow{\partial^n} \cdots$$

of equivariant Alexander-Spanier topological cohomology groups is exact. Special cases hereof are extensions of locally contractible T_1-groups.

Theorem 2.6.40. *For each G-module V the equivariant Alexander-Spanier topological cohomology $H^t_{AS,eq}(-, -; V)$ is a topological cohomology theory on the full subcategory of $\mathbf{GTop}(2)$ with objects pairs (X, A) with a G-space X and a G-retract A of X. If V is a real topological vector space then the equivariant Alexander-Spanier topological cohomology $H^t_{AS,eq}(-, -; V)$ is a topological cohomology theory on the full subcategory of $\mathbf{GTop}(2)$ with objects pairs (X, A) of a closed G-NR A of a G-normal space X.*

2.7 Singular Topological Cohomology

In analogy to singular topological homology one can define a topologised version of singular cohomology. Like before the coefficient group is required to be an Abelian topological group. In extreme cases this group can be discrete or indiscrete. There are two natural constructions to obtain a singular topological cohomology. One uses the singular simplicial space and one the singular simplicial topological group of a topological space. We choose the first one.

Let X be a topological space and V be an Abelian topological group. A topologised cohomology of X with coefficient group V can be obtained with the help of the function space functor $C(-, V)$. As usual all function spaces are equipped with the compact-open topology. Applying the contravariant function space functor $C(-, V)$ to the singular simplicial space $C(\Delta, X)$ yields the cosimplicial Abelian topological group $C(C(\Delta, X), V)$.

Definition 2.7.1. *The cosimplicial group $C(C(\Delta, X), V)$ is denoted by $S^t(X; V)$. It is called the* singular cosimplicial topological group (with coefficients V) *of X. The bifunctor $S^t(-; -) : \mathbf{Top} \times \mathbf{TopAb} \to \mathbf{TopAb}^{\mathcal{O}}$, $(X, V) \mapsto S^t(X; V)$ is called the* singular cosimplicial topological group functor.

Lemma 2.7.2. *The singular cosimplicial topological group $S^t(X; V)$ is always completely regular. It is Hausdorff if and only if the coefficient group V is Hausdorff.*

Proof. Let X be any topological space. Each of the function spaces $C(C(\Delta_n, X), V)$ is a topological group and therefore completely regular. Furthermore each function space $C(C(\Delta_n, X), V)$ contains the coefficient group V as a subspace, so V has to be Hausdorff if $C(C(\Delta_n, X), V)$ is Hausdorff. Conversely, if the coefficient group V is Hausdorff, then so is every function space of the form $C(Y, V)$. (See [Dug89, XII.1.3] for a proof.) Thus the cosimplicial group $C(C(\Delta, X), V)$ is Hausdorff if and only if V is Hausdorff. □

Example 2.7.3. If V is a Hausdorff vector space, e.g. a Hausdorff Lie algebra, then the topological singular cosimplicial group $S^t(X; V)$ of any space X is Hausdorff.

Lemma 2.7.4. *The singular cosimplicial topological group functor $S^t(-; -)$ preserves homotopy.*

Proof. Let V be a group of coefficients. The singular simplicial space functor $C(\Delta, -)$ preserves homotopy by Theorem 2.3.9. Applying the contravariant functor $C(-, V)$ transforms simplicial homotopies into cosimplicial homotopies.   □

Considering the cochain complex associated to the singular cosimplicial topological group $S^t(X; V)$ yields another functor $S^{t*}(-; -)$ from **Top** × **TopAb** into the category **coCh(TopAb)** of cochain complexes of Abelian topological groups.

Definition 2.7.5. *The functor* $S^{t*}(-; -)$: **Top** × **TopAb** → **coCh(TopAb)** *is called the* singular topological cochain complex functor. *The cochain complex* $S^{t*}(X; V)$ *associated to a topological space* X *is called the* singular topological cochain complex *associated to the space* X.

Lemma 2.7.6. *The singular topological cochain complex functor* $S^{t*}(-; -)$ *preserves homotopy.*

Proof. The singular simplicial space functor $S^t(-; -)$ preserves homotopy by Lemma 2.7.4. Cosimplicially homotopic morphisms between cosimplicial Abelian topological groups induce homotopic morphisms between the associated cochain complexes by Lemma F.6.6, so $S^{t*}(-; -)$ preserves homotopy as well.   □

Definition 2.7.7. *The functor* $H^{t*} := H^* \circ S^{t*}(-; -)$ *is called the* singular topological cohomology functor. *The graded Abelian topological group* $H^{t*}(X)$ *associated to a space* X *is called its* singular topological cohomology.

Lemma 2.7.8. *The singular topological cohomology functor* $H^{t*}(-; -)$ *is homotopy invariant.*

Proof. This is a direct consequence of Lemma 2.7.6.   □

Similar to singular topological homology, one can also define a version of singular topological cohomology using $\mathfrak{U}$-small singular simplices only, where $\mathfrak{U}$ is a family of subsets of the topological space X considered. Recall the category **TF** from section 2.3 and that the singular simplicial space functor $C(\Delta, -)$ extends to a functor **TF** → **Top**$^{\mathcal{O}^{op}}$ from **TF** into the category **Top**$^{\mathcal{O}^{op}}$ of simplicial topological spaces. Since the singular cosimplicial topological group (with coefficients V) is defined by composition with $C(-, V)$ we can straightforwardly generalise:

Definition 2.7.9. *For each family* $\mathfrak{U}$ *of subsets of a topological space* X, *the cosimplicial group* $C(C(\Delta, X, \mathfrak{U}), V)$ *is called the* $\mathfrak{U}$-small singular cosimplicial topological group (with coefficients V) *of* X *It is denoted by* $S^t(X, \mathfrak{U}; V)$. *The extension* $S^t(-, -; -)$: **TF** × **TopAb** → **TopAb**$^{\mathcal{O}}$, $(X, \mathfrak{U}, V) \mapsto S^t(X, \mathfrak{U}; V)$ *of* $S^t(-; -)$ *is also called the* singular cosimplicial topological group functor.

The extension $S^t(-, -; -)$ of the singular cosimplicial topological group functor has the same behaviour as $S^t(-; -)$ concerning Hausdorffness. As before one observes:

Lemma 2.7.10. *The singular cosimplicial topological group* $S^t(X, \mathfrak{U}; V)$ *is always completely regular. It is Hausdorff if and only if the coefficient group* V *is Hausdorff.*

Proof. The proof is analogous to the proof of Lemma 2.7.2.   □

Passing to the cochain complexes associated to $\mathfrak{U}$-small singular cosimplicial topological groups yields a functor from **TF** into the category **coCh(TopAb)** of cochain complexes of Abelian topological groups:

Definition 2.7.11. *The cochain complex associated to a $\mathfrak{U}$-small singular cosimplicial topological group $S^t(X, \mathfrak{U}; V)$ is denoted by $S^{t*}(X, \mathfrak{U}; V)$. It is called the $\mathfrak{U}$-small singular topological cochain complex of the topological space X. The extension $S^{t*}(-, -; -) : \mathbf{TF} \times \mathbf{TopAb} \to \mathbf{coCh(TopAb)}$, $(X, \mathfrak{U}, V) \mapsto S^{t*}(X, \mathfrak{U}; V)$ of $S^{t*}(-; -)$ is also called the* singular topological cochain complex functor.

Definition 2.7.12. *The cohomology of a $\mathfrak{U}$-small singular topological cochain complex $S^{t*}(X, \mathfrak{U}; V)$ is denoted by $H^{t*}(X, \mathfrak{U}; V)$. It is called the $\mathfrak{U}$-small singular topological cohomology of X. The extension of the singular topological cohomology functor $H^{t*}(-; -)$ to $\mathbf{TF} \times \mathbf{TopAb})$ is also called the* singular topological cohomology functor.

Theorem 2.7.13. *If $\mathfrak{U}$ is an open covering of X and V a group of coefficients then the inclusion $S^t(X, \mathfrak{U}; V) \hookrightarrow S^t(X; V)$ induces an isomorphism in cohomology.*

Proof. Let $\mathfrak{U}$ be an open covering of the topological space X and V be an Abelian topological group. Let $Z^{tn}(X, \mathfrak{U}; V)$ denote the subspace of $\mathfrak{U}$-small cocycles in $S^t(X, \mathfrak{U}; V)([n])$ and $B^{tn}(X, \mathfrak{U}; V)$ denote the subspace of coboundaries of $\mathfrak{U}$-small singular $(n-1)$ cochains. Similarly we denote by $Z^{tn}(X; V)$ the subspace of cocycles in $S^t(X; V)([n])$ and by $B^{tn}(X; V)$ the subspace of coboundaries of singular $(n-1)$-chains. The inclusion $S(X, \mathfrak{U}; V) \hookrightarrow S(X; V)$ of the underlying singular cosimplicial groups induces an isomorphism $H^*(X, \mathfrak{U}; V) \cong H^*(X : V)$ in classical singular cohomology. This implies the equality

$$B^{tn}(X, \mathfrak{U}; V) = Z^{tn}(X, \mathfrak{U}; V) \cap B^{tn}(X; V).$$

Applying the Noether Isomorphism Theorem 1.1.2 now shows that the inclusion $Z^{tn}(X, \mathfrak{U}; V) \hookrightarrow Z^{tn}(X; V)$ induces an inclusion $H^{tn}(X, \mathfrak{U}; V) \hookrightarrow H^{tn}(X; V)$. Since this inclusion is bijective, it is an isomorphism. $\qquad\square$

We wish to derive a long exact sequence in singular topological cohomology for a short exact sequence $0 \to V \to V' \to V'' \to 0$ of coefficient groups. An inclusion $V \hookrightarrow V'$ of a subgroup of an Abelian topological group induces an inclusion $S^t(X; V) \hookrightarrow S^t(X; V')$ of singular cosimplicial topological groups for every topological space X. But, because the functors $S^t(X; -)([n]) = C(C(\Delta_n, X), -)$ do not preserve quotients in general, not every short exact sequence $0 \to V \to V' \to V'' \to 0$ of coefficient groups gives rise to a short exact sequence of singular cosimplicial topological groups. So one can not expect to obtain a long exact sequence in singular topological cohomology unless some conditions on the topological space X or the coefficient groups are imposed. One possibility to resolve this issue is to require the quotient homomorphism $V' \twoheadrightarrow V''$ to be a regular Hurewicz fibration. The procedure is similar to the one used in Proposition 2.5.34. We begin with a preparatory observation:

Lemma 2.7.14. *The space of constant singular n-simplices in a topological space X is a deformation retract of the space $C(\Delta_n, X)$.*

Proof. Let X be a topological space. Every standard simplex Δ_n is contractible to the zeroth vertex $\mathbf{e}_0$ via the contraction

$$H_n : \Delta_n \times I \to \Delta_n, \quad H_n(\mathbf{t}, s) = s\mathbf{t}.$$

Applying the function space functor $C(-, X)$ to this contraction yields a continuous function $C(H_n, X)$ from $C(\Delta_n, X)$ to $C(\Delta_n \times I, X)$. The spaces Δ_n and I are compact Hausdorff spaces, so the latter function space is naturally isomorphic to the function space $C(I, C(\Delta_n, X))$ by Lemma B.5.18. Composing the continuous function $C(H_n, X)$ with this natural isomorphism yields a continuous function

$$H'_n : C(\Delta_n, X) \to C(I, C(\Delta_n, X)), \quad H'_n(\sigma) = (s \mapsto \sigma H_n(-, s)).$$

Because the unit interval is a compact Hausdorff space, the evaluation map for the function space $C(I, C(\Delta_n, X))$ is continuous. The composition $\mathrm{ev} \circ (H'_n \times \mathrm{id}_I)$ of the continuous evaluation map $\mathrm{ev} : C(I, C(\Delta_n, X)) \times I \to C(I, C(\Delta_n, X))$ with $H'_n \times \mathrm{id}_I$ is the continuous function

$$H''_n : C(\Delta_n, X) \times I \to C(\Delta_n, X), \quad (\sigma, s) \mapsto \sigma H_n(-, s).$$

The homotopy H_n satisfies the equation $H_n(-, 1) = \mathrm{id}_{\Delta_n}$ which implies the identity $H''_n(-, 1) = \mathrm{id}_{C(\Delta_n, X)}$. Similarly the identity $H_n(-, 0) = \mathbf{e}_0$ (where $\mathbf{e}_0$ denotes the constant map $\mathbf{e}_0$) implies $H''_n(\sigma, 0) = \sigma(\mathbf{e}_0)$. There fore the subspace of constant maps in $C(\Delta_n, X)$ is a deformation retract of the space $C(\Delta_n, X)$ of singular n-simplices in X. □

If a quotient morphism $V \twoheadrightarrow V''$ of coefficient groups is a regular Hurewicz fibration, then we expect that – similar to Proposition 2.5.34 – continuous functions $f : C(\Delta_n, X) \to V''$ can be simultaneously lifted to V' provided that they are trivial on constant simplices. Fortunately the Normalisation Theorem 1.2.19 allows us to restrict ourselves to normal singular n-cochains. Recall that for each topological space X and coefficient group V the components $NS^t(X; V)([n])$ of the normalised singular cosimplicial topological group $NS^t(X; V)$ are the subgroups

$$NAS^t(X; V)([n]) = \bigcap_{i=0}^{n-1} K\left(S^t(X; V)(\eta_i : [n] \to [n-1])\right)$$

of the groups $S^t(X; V)([n])$ (cf. Definition 1.1.24). These also are the objects of the normalised singular topological cochain complex $NS^*(X; V)$. The normal singular cochains are the functions $f \in S^t(X; V'')([n])$ which are trivial on degenerate simplices, such as the constant ones. Similarly to before one now observes:

Lemma 2.7.15. *The functors $NS^t(-; -)$ and $NS^*(-; -)$ preserve homotopy.*

Proof. Let $f, g : X \to Y$ be continuous functions between topological spaces X and Y and $H : X \times I \to Y$ be a homotopy from g to f. The homotopy between the morphisms $C(\Delta, f)$ and $C(\Delta, g)$ of singular simplicial spaces $C(\Delta)$ constructed in the proof of Theorem 2.3.9 maps degenerate simplices to degenerate simplices. So the simplicial homotopy from $S^t(f; V)$ to $S^t(g; V)$ obtained therefrom maps normal singular cochains to normal singular cochains and thus restricts to a homotopy from $NS^t(f; -)$ to $NS^t(g; -)$. As a consequence the induced morphisms $NS^{t*}(f; -)$ and $NS^{t*}(g; -)$ of cochain complexes are homotopic as well (by Lemma F.6.6). □

One can also form the degenerate subcomplex $DS^{t*}(X; V)$ of the singular topological cochain complex $AS^{t*}(X; V)$. (cf. Definition 1.1.49). Straight from the general Normalisation Theorem 1.2.34 one then derives:

Theorem 2.7.16. *The inclusion $DS^{t*}(-; -) \hookrightarrow S^{t*}(-; -)$ of the subfunctor of degenerate cochains $S^{t*}(-; -)$ is cochain homotopic to the identity. Furthermore there exists a projection $p : S^t(-; -) \twoheadrightarrow NS^t(-; -)$ onto the subfunctor $NS^t(-; -)$ of $S^t(-; -)$ such that the sequence*

$$0 \to DS^{t*}(-; -) \hookrightarrow S^{t*}(-; -) \xrightarrow{p} NS^{t*}(-; -) \to 0$$

of bifunctors $\mathbf{Top} \times \mathbf{TopAb} \to \mathbf{coCh(TopAb)}$ is split exact.

Corollary 2.7.17. *The inclusion $NS^{t*}(-;-) \hookrightarrow S^{t*}(-;-)$ induces an isomorphism $H(NS^{t*}(-;-)) \cong H^{t*}(-;-)$ of cohomology functors. In particular the singular topological cohomology $H^{t*}(X;V)$ with coefficients V of a topological space X can be computed by using normalised cochains only.*

Using normal singular cochains only, we can now prove that a regular Hurewicz fibration of coefficient groups induces a quotient morphism of singular cosimplicial topological groups. Here fore we first observe:

Proposition 2.7.18. *If a quotient homomorphism $V' \xrightarrow{q} V''$ of coefficient groups is a regular Hurewicz fibration, then each function $NS^t(X;q)([n])$ has a continuous section, in particular it is surjective.*

Proof. Let $q : V \twoheadrightarrow V''$ be a quotient homomorphism of Abelian topological groups which is a regular Hurewicz fibration and X be any topological space. The deformation H_n'' from $C(\Delta_n, X)$ onto the subspace of constant n-simplices constructed in Lemma 2.7.14 provides a homotopy

$$H : S^t(X;V)([n]) \times I \to S^t(X;V)([n]), \quad H(f,s) \mapsto (fH_n'')(-,s).$$

Let $DC(\Delta_n, X)$ denote the space of degenerate n-simplices in X. Because the homotopy H_n'' restricts to a homotopy $DC(\Delta_n, X) \times I \to DC(\Delta_n, X)$, the homotopy H restricts to a homotopy $NS^t(X;V'')([n]) \times I \to NS^t(X;V'')([n])$. Every normal singular cochain $f \in NS^t(X;V'')([n])$ is trivial on degenerate simplices. Therefore the function $H(-,0)$ maps normalised singular cochains to the constant function $0 \in V''$. This constant function can be lifted to the constant function $0 \in V'$. So for every normal singular n-cochain f one obtains a commutative diagram

$$
\begin{array}{ccc}
 & & V' \; . \\
 & \nearrow^{\;0} & \downarrow{q} \\
C(\Delta_n, X) \times I & \xrightarrow[fH_n'']{} & V''
\end{array}
$$

Because the quotient homomorphism q was assumed to be a regular Hurewicz fibration, there exists a regular lifting function Λ (cf. Theorem D.2.7). Proceeding analogously to Proposition 2.5.34, we see that the continuous function

$$L : NS^t(X;V'')([n]) \to S^t(X;V')([n]), \quad L(f)(\sigma) = [\Lambda((fH_n'')(\sigma,-),0)](1)$$

lifts all normal singular n-cochains simultaneously. It remains to prove that the lifted cochains $L(f)$ are normal. If f is a normal singular n-cochain and σ a degenerate singular n-simplex in X, then the path $s \mapsto fH_n''(\sigma,s)$ is the constant path 0 in V''. Because the lifting function Λ was assumed to be regular, the lifted path $\lambda(fH_n''(\sigma,-),0)$ is constant as well. Since this path in V' starts at 0, the endpoint $[\lambda(fH_n''(\sigma,-),0)](1)$ also is zero, which implies that the singular n-cochain $L(f)$ is trivial on the degenerate n-simplex σ. Because this is true for arbitrary normal n-cochains in $NS^t(X;V'')([n])$ and degenerate singular n-simplices, the function L corestricts to a continuous function $L : NS^t(X;V'')([n]) \to NS^t(X;V')([n])$, which lifts all normal singular n-cochains in $NS^t(X;V'')([n])$. $\square$

Theorem 2.7.19. *If $0 \to V \to V' \xrightarrow{q} V'' \to 0$ is a short exact sequence of coefficients for which the quotient homomorphism q is a regular Hurewicz fibration, then the short exact sequence of coefficients induces a long exact sequence in singular topological cohomology.*

Proof. If $0 \to V \to V' \xrightarrow{q} V'' \to 0$ is a short exact sequence of coefficients which is a regular Hurewicz fibration, then each of the homomorphisms $NS^t(X;q)([n])$ is the projection of a trivial bundle by Proposition 2.7.18, hence a quotient map. Therefore the morphism $NS^t(X;q) : NS^t(X;V') \to NS^t(X;V'')$ of normalised Alexander-Spanier cosimplicial topological groups is a quotient morphism. In this case the sequence

$$0 \to NS^t(X;V) \to NS^t(X;V') \to NS^t(X;V'') \to 0$$

of normalised singular cosimplicial topological groups is exact and induces a long exact sequence in cohomology. □

Corollary 2.7.20. *Any locally trivial short exact sequence* $0 \to V \to V' \to V'' \to 0$ *of coefficient groups with paracompact group* V'' *induces a long exact sequence in singular topological cohomology.*

Proof. If $0 \to V \to V' \xrightarrow{q} V'' \to 0$ is a locally trivial short exact sequence of coefficients and the group V'' is paracompact, then the bundle $V \to V''$ is a regular Hurewicz fibration (cf. [Dug89, Theorem XX.4.2]). □

Corollary 2.7.21. *Every short exact sequence* $0 \to V \to V' \to V'' \to 0$ *of coefficients with locally contractible group* V'' *induces a long exact sequence in the Alexander-Spanier topological cohomology of* X.

Proof. If $0 \to V \to V' \xrightarrow{q} V'' \to 0$ is a short exact sequence of coefficients and the group V'' is locally contractible, then there exists a contractible identity neighbourhood U in V'' and a partition of unity subordinate to the trivialising cover $v.U$, $v \in V''$ (cf. [,]). Therefore the bundle $V \to V''$ is a Hurewicz fibration (cf. [Dug89, Theorem XX.4.2]). □

Each pair (X, A) of spaces in **Top**(2) gives rise to a long exact sequence of classical singular cohomology groups. This result does not generalise to singular topological cohomology. The reason here fore is that the restriction of the contravariant functor $C(-, V)$ does in general neither turn embeddings into quotients nor quotients into inclusions. As a consequence the cosimplicial Abelian topological group $S^t(X;V)$ need not be a subgroup of the cosimplicial Abelian topological group $S^t(A;V)$. We derive some sufficient conditions for this to be the case.

The long exact sequence of classical singular cohomology groups stems from a short exact sequence $S(X, A; V) \hookrightarrow S(X;V) \twoheadrightarrow S(A;V)$ of cosimplicial Abelian groups. Here the inclusion $S(X, A; V) \hookrightarrow S(X;V)$ is the kernel of the morphism $S(i_A) : S(X;V) \to S(X;V)$ induced by the inclusion $i_A : A \hookrightarrow X$ of the subspace A of X. Proceeding analogously for singular topological cohomology we define:

Definition 2.7.22. *The domain of the kernel of the morphism* $S^t(i_A;V)$ *induced by an embedding* $i_A : A \hookrightarrow X$ *of topological spaces is denoted by* $S^t(X, A; V)$. *It is called the* singular cosimplicial topological group *of the pair* (X, A).

This construction is functorial in the pairs (X, A) of topological spaces, hence it defines a bifunctor $S^t(-, -; -) : \mathbf{Top}(2) \times \mathbf{TopAb} \to \mathbf{TopAb}^{\mathcal{O}^{op}}$. The singular cosimplicial topological group $S^t(X;V)$ of a topological space X can be identified with the cosimplicial topological group $S^t(X, \emptyset; V)$, which is the kernel of the morphism $S^t(X) \to \{0\} = S^t(\emptyset; V)$. Similar to the singular cosimplicial topological group functor $S^t(-; -) : \mathbf{Top} \to \mathbf{TopAb}^{\mathcal{O}^{op}}$ we observe:

Lemma 2.7.23. *The singular cosimplicial topological group* $S^t(X, A; V)$ *of a pair* (X, A) *is Hausdorff if and only if the coefficient group is Hausdorff. In this case it is closed in* $S^t(X;V)$.

Proof. The function spaces $S^t(X, A; V)([n]) \leq S^t(X; V)([n])$ are Hausdorff if and only if V is Hausdorff. Since $S^t(X, A; V)$ is the kernel of a morphism into $S^t(A; V)$, this kernel is closed if $S^t(A; V)$ is Hausdorff, i.e. if V is Hausdorff. □

Lemma 2.7.24. *The functor* $S^t(-, -; V) : \mathbf{Top}(2) \times \mathbf{TopAb} \to \mathbf{TopAb}^{\mathcal{O}^{op}}$ *preserves homotopy.*

Proof. Let $(f, f_{|A}), (g, g_{|A}) : (X, A) \to (Y, B)$ be morphisms in $\mathbf{Top}(2)$. A homotopy H from $(g, g_{|A})$ to $(f, f_{|A})$ is a homotopy $H : X \times I \to Y$ from g to f that restricts and corestricts to a homotopy $H_{|A \times I} : A \times I \to B$ from $g_{|A}$ to $f_{|A}$. Therefore the homotopy from $C(\Delta, f)$ to $C(\Delta, g)$ obtained from a homotopy $H : X \times I \to Y$ restricts to a homotopy from $C(\Delta, f_{|A})$ to $C(\Delta, g_{|A})$. As a consequence the induced homotopy from $S^t(f; V)$ to $S^t(g; V)$ maps the cosimplicial subgroup $S^t(Y, B; V)$ of $S^t(Y; V)$ into the cosimplicial subgroup $S^t(X, A)$ of $S^t(X; V)$ and thus restricts to a homotopy from $S^t((f, f_{|A}); V)$ to $S^t((g, g_{|A}); V)$ □

We denote the cochain complex associated to the singular cosimplicial topological group $S^t(X, A; V)$ of a pair (X, A) of topological spaces by $S^{t*}(X, A; V)$. Recall that the associated cochain complex functor $\mathbf{TopAb}^{\mathcal{O}} \to \mathbf{coCh}(\mathbf{TopAb})$ is continuous (by Lemma 1.1.22) and thus preserves kernels. Therefore the singular topological cochain complex $S^{t*}(X, A; V)$ of the pair (X, A) is the kernel of the morphism $S^{t*}(X; V) \to S^{t*}(A; V)$ of singular topological cochain complexes.

Definition 2.7.25. *The cochain complex* $S^{t*}(X, A; V)$ *associated to the singular cosimplicial topological group* $S^t(X, A; V)$ *of a pair* (X, A) *of topological spaces is called the* singular topological cochain complex with coefficients V *of the pair* (X, A). *The cohomology thereof is denoted by* $H^{t*}(X, A; V)$ *and called the* singular topological cohomology with coefficients V *of the pair* (X, A).

Lemma 2.7.26. *The singular topological cochain complex functor* $S^{t*}(-, -; V)$ *on* $\mathbf{Top}(2)$ *preserves homotopy.*

Proof. The singular cosimplicial topological group functor $S^t(-, -; V)$ preserves homotopy by Lemma 2.7.24. Because cosimplicially homotopic morphisms of cosimplicial Abelian topological groups induce homotopic morphisms of the associated cochain complexes (Lemma F.6.6), the functor $S^{t*}(-, -; V)$ preserves homotopy as well. □

Lemma 2.7.27. *The singular topological cohomology functor* $H^{t*}(-, -; V)$ *is homotopy invariant.*

Proof. This is a direct consequence of the fact that the singular topological cochain complex functor $S^{t*}(-, -; V)$ on $\mathbf{Top}(2)$ preserves homotopy. □

So far one obtains for every pair (X, A) of topological spaces and inclusion $i_A : A \hookrightarrow X$ of A into X a sequence

$$0 \to S^t(X, A; V) \hookrightarrow S^t(X; V) \xrightarrow{S^t(i_A; V)} S^t(A; V)$$

of singular cosimplicial topological groups. The sequence is exact at $S^t(X, A; V)$ and the inclusion of $S^t(X, A; V)$ into $S^t(X; V)$ is the kernel of the morphism $S^t(i_A; V) : S^t(X; V) \to S^t(A; V)$ by definition, but the latter morphism might not be a quotient. Similar to singular topological homology we observe:

Lemma 2.7.28. *If A is a retract of the topological space X, then the short sequence $0 \to S^t(X, A; v) \to S^t(X; V) \twoheadrightarrow S^t(A; V) \to 0$ is split exact for any group V of coefficients and the inclusion $A \hookrightarrow X$ induces an inclusion of $H^{t*}(A; V)$ as a direct summand:* $H^{t*}(X; V) \cong H^{t*}(A; V) \oplus H^{t*}(X, A; V)$.

Proof. Let A be a retract of X and $r : X \to A$ be a retraction. Let V be a group of coefficients and denote the inclusion $A \hookrightarrow X$ by i_A. The morphism $S^t(r;V)$: $S^t(A) \to S^t(X;V)$ of singular cosimplicial topological groups satisfies the equations

$$S^t(i;v) \circ S^t(r;V) = S^t(r \circ i;V) = S^t(\mathrm{id}_A;V) = \mathrm{id}_{S^t(A;V)}.$$

Therefore the morphism $S^t(I_A;V)$ is a split epimorphism, hence a quotient morphism and the sequence $0 \to S^t(X,A;V) \to S^t(X;V) \twoheadrightarrow S^t(X,A) \to 0$ is exact. Furthermore the morphism $S^t(r;V)$ provides a splitting of this short exact sequence Consequently the induced long exact sequence in homology decomposes into split short exact sequences, hence the homomorphism $H^{t*}(i_A) : H^{t*}(A;V) \to H^{t*}(X;V)$ is the inclusion of a direct summand and $H^{t*}(X) \cong H^{t*}(A) \oplus H^{t*}(X,A)$. □

Example 2.7.29. (cf. Example 2.3.53) The base space of a vector bundle $E \to B$ with fibre F is a retract of E. Therefore the topological singular cohomology $H^{t*}(E;V)$ of the total space E decomposes as $H^{t*}(E;V) \cong H^{t*}(B;V) \oplus H^{t*}(F;V) \cong H^{t*}(B;V)$. In particular any tangent bundle TM of a manifold M with model space W contains M as a retract and this implies $H^{t*}(TM;V) \cong H^{t*}(M;V) \oplus 0$

If A is not a retract but only a neighbourhood retract (NR), then a weaker result can be obtained by requiring the space A to be a closed, X to be normal and V to be a real topological vector space.

Proposition 2.7.30. *If A is a closed NR in a normal space X, then the short sequence $0 \to S^t(X,A;V) \to S^t(X;V) \to S^t(A;V) \to 0$ is exact for every real topological vector space V.*

Proof. Let A be a closed NR in a normal space X, the coefficient group V be a real topological vector space and denote the inclusion $A \hookrightarrow X$ by i_A. It suffices to show that each homomorphism $S^t(i_A;V)([n]) : S^t(X;V)([n]) \to S^t(A;V)([n])$ induced by the inclusion i_A is a quotient homomorphism. By assumption here exist a neighbourhood U of A and a retraction $r : U \to A$ of U onto A. Because the space X is normal, there exists a real valued continuous function $\varphi : X \to \mathbb{R}$ on X whose support lies in U and contains the subspace A. Let pr_i denote the projection of the spaces X^{n+1} resp. U^{n+1} onto the i-th factor. The product function

$$\varphi_n := \prod_{i=0}^{n} \varphi \circ \mathrm{pr}_i : X^{n+1} \to \mathbb{R}$$

has support in U^{n+1} and its restriction the subspace A^{n+1} the constant function 1 by construction. We use this function to show that each homomorphism $S^t(i_A;V)([n]) : S^t(X;V)([n]) \to S^t(A;V)([n])$ induced by the inclusion i_A is a quotient homomorphism. The simplicial category $\mathcal{O}$ can be regarded as a discrete cosimplicial space (the objects $[n]$ being discrete spaces). This discrete cosimplicial space can be included into the standard cosimplicial space Δ (cf. Definition F.2.8). The inclusion $i : \mathcal{O} \hookrightarrow \Delta$ maps a natural number $i \in [n]$ to the vertex $\mathbf{e}_i$ of the standard simplex Δ_n. This inclusion of cosimplicial spaces induces a morphism $C(i,X) : C(\Delta,X) \to C([-],X) = \mathrm{SS}(X)$ from the singular simplicial space of X to the standard simplicial space of X. For each $n \in \mathbb{N}$ the composition $\phi_n := \varphi_n \circ C(i([n]),X)$ is a continuous function on the n-th component $C(\Delta_n,X)$ of the singular simplicial space of X. Since φ_n is the constant function 1 on A^{n+1} the function ϕ_n is the constant function 1 on the subspace $C(\Delta_n,A)$ of $C(\Delta_n,X)$. Furthermore this function f_n is trivial on all singular n-simplices whose vertices are not all contained in $\mathrm{supp}\varphi \subset U$. Therefore we can define a continuous linear map $\Phi_n : S^t(A;V)([n]) \to S^t(X;V)([n])$ which simultaneously extends all continuous functions $f : C(\Delta,X) \to V$ by setting

$$\Phi_n(f)(\sigma) = \begin{cases} \phi_n(\sigma) \cdot S^t(r;V)([n])(f) & \text{if } \sigma \in C(\Delta_n, U) \\ 0 & \text{if } \sigma \notin \text{supp}\phi_n \end{cases}.$$

The linear map $\Phi_n : S^t(A;V)([n]) \to S^t(X;V)([n])$ so constructed is a right inverse to the homomorphism $S^t(i_A;V)([n]) : S^t(X;V)([n]) \to S^t(A;V)([n])$, hence the latter one is a quotient homomorphism. $\square$

Corollary 2.7.31. *For every pair (X, A) of a normal space X and a closed neighbourhood retract A in X and every real topological vector space V the long sequence*

$$\cdots \xrightarrow{\partial^{n-1}} H^{tn}(X, A; V) \to H^{tn}(X; V) \to H^{tn}(A; V) \xrightarrow{\partial^n} \cdots$$

of singular topological cohomology groups is exact.

Example 2.7.32. If $p : E \to B$ is a locally trivial bundle over a T_1 base space B and total space E of the bundle is normal, then the long sequence

$$\cdots \xrightarrow{\partial^{n-1}} H^{tn}(E, G; V) \to H^{tn}(E; V) \to H^{tn}(G; V) \xrightarrow{\partial^n} \cdots$$

of equivariant singular topological cohomology groups is exact. Special cases hereof are normal extensions of locally contractible T_1-groups.

Theorem 2.7.33. *For each coefficient group V the singular topological cohomology $H^{t*}(-, -; V)$ is a topological cohomology theory on the full subcategory of $\mathbf{Top}(2)$ with objects pairs (X, A) of a closed retract A of a topological space X. If V is a real topological vector space, then the singular topological cohomology $H^{t*}(-, -; V)$ is a topological cohomology theory on the full subcategory of $\mathbf{Top}(2)$ with objects pairs (X, A) of a closed NR A of a normal space X.*

3

Equivariant Systems of Singular Simplices

In this chapter we construct morphisms $SS(X, \mathfrak{U}) \to C(\Delta, X, \mathfrak{U})$ of simplicial topological spaces which assign to each $\mathfrak{U}$-small n-simplex $(x_0, \ldots, x_n)$ in X^{n+1} a $\mathfrak{U}$-small singular n-simplex with vertices $x_0, \ldots, x_n$. We first start with a non-continuous version and later consider continuous or even smooth maps. Let X be a topological space and $\varrho : G \times X \to X$ be an abstract action of a group G on X. As always we assume the unit interval I and the standard simplices Δ_n to be trivial G-spaces. The forgetful functor $\mathbf{Top} \to \mathbf{Set}$ is denoted by U. So the set $\hom_{\mathbf{Top}}(\Delta_n, X)$ of singular n-simplices in a topological space X can also be written as $UC(\Delta_n, X)$. Recall the vertex morphism $\lambda : C(\Delta, X) \to SS(X)$ which assigns to each singular n-simplex τ its (ordered set of) vertices $(\tau(\mathbf{e}_0), \ldots, \tau(\mathbf{e}_n)) \in X^{n+1}$.

Definition 3.0.1. *A right inverse* $\hat{\sigma}_n : UX^{n+1} \to \hom_{\mathbf{Top}}(\Delta_n, X)$ *to the vertex map* $U\lambda([n]) : \hom_{\mathbf{Top}}(\Delta_n, X) \to X^{n+1}$ *is called a* system of (singular) n-simplices *on* X.

Example 3.0.2. The standard 0-simplex Δ_0 is a singleton space. Therefore the set $\hom_{\mathbf{Top}}(\Delta_0, X)$ of singular 0-simplices in a topological space X can be identified with the underlying set UX of X. In particular the assignment

$$\hat{\sigma}_0 : UX \to \hom_{\mathbf{Top}}(\Delta_0, X), \quad \hat{\sigma}_0(x_0)(t_0) = x_0$$

is a system of singular 0-simplices on X. Moreover, this system of 0-simplices is necessarily unique.

Definition 3.0.3. *A system* $\hat{\sigma}_n : UX^{n+1} \to \hom_{\mathbf{Top}}(\Delta_n, X)$ *of singular n-simplices is called* equivariant *if it intertwines the action of G on UX and $\hom_{\mathbf{Top}}(\Delta_n, X)$, i.e. if it satisfies the equation* $\hat{\sigma}_n(g.x_0, \ldots, g.x_n) = g.\hat{\sigma}_n(x_0, \ldots, x_n)$ *for all $g \in G$ and all $(x_0, \ldots, x_n) \in X^{n+1}$.*

Example 3.0.4. The unique system system $\hat{\sigma}_0$ of singular 0-simplices on a G-space X is always equivariant.

Lemma 3.0.5. *If the action of G on X is free and X is arc-wise connected, then there exists an equivariant system* $\hat{\sigma}_1 : UX^2 \to \hom_{\mathbf{Top}}(\Delta_1, X)$ *of singular 1-simplices on X.*

Proof. Let the action of G on X be free and $\Sigma \subset X$ be a fundamental domain. Consider all points $(x_0, x_1) \in \Sigma \times X$ and choose a path $\hat{\sigma}_1(x_0, x_1)$ from x_0 to x_1 for every such point $(x_0, x_1) \in \Sigma \times X$. Then extend the map $\hat{\sigma}_1$ to $U(X \times X)$ by requiring the equation

$$\hat{\sigma}_n(g.x_0, \ldots, g.x_n) = g.\hat{\sigma}_n(x_0, \ldots, x_n)$$

to hold for all $g \in G$ and all $(x_0, \dots, x_n) \in UX^{n+1}$. This is well defined because the action of G is free. The resulting map $\hat{\sigma}_1$ is an equivariant system of singular 1-simplices on X. □

Example 3.0.6. If $H \leq G$ is a subgroup of an arc-wise connected topological group G, then there exists a system $\hat{\sigma}_1 : UG^2 \to \mathrm{hom}_{\mathbf{Top}}(\Delta_1, G)$ of singular 1-simplices on G, which is equivariant w.r.t. the action of H on G by left translation.

Example 3.0.7. If X is a connected manifold, then it is arc-wise connected. Thus for every manifold with group action $G \times M \to M$ there exists a G-equivariant system $\hat{\sigma}_1$ of singular 1-simplices on M.

Example 3.0.8. Connected Lie groups are always arc-wise connected. Thus for every subgroup H of a connected Lie group G there exists a system $\hat{\sigma}_1$ of singular 1-simplices on G, which is equivariant w.r.t. the action of H on G by left translation.

Definition 3.0.9. *A map $\hat{\sigma}_n : UV_n \to \mathrm{hom}_{\mathbf{Top}}(\Delta_n, X)$ defined for a neighbourhood V_n of the diagonal in X^{n+1} satisfying $\hat{\sigma}_n(x_0, \dots, x_n)(\mathbf{e}_i) = x_i$ for all $0 \leq i \leq n$ is called a* local system of (singular) n-simplices *on X.*

Definition 3.0.10. *A local system $\hat{\sigma}_n : UV_n \to \mathrm{hom}_{\mathbf{Top}}(\Delta_n, X)$ of singular n-simplices defined for a G-invariant neighbourhood V_n of the diagonal in X^{n+1} is called* equivariant *if it intertwines the action of G on UV_n and $\mathrm{hom}_{\mathbf{Top}}(\Delta_n, X)$, i.e. if it satisfies the equation $\hat{\sigma}_n(g.x_0, \dots, g.x_n) = g.\hat{\sigma}_n(x_0, \dots, x_n)$ for all $g \in G$ and all $(x_0, \dots, x_n) \in V_n$.*

Lemma 3.0.11. *If the action of G on X is free and X is locally arc-wise connected, then there exists a local equivariant system $\hat{\sigma}_1 : UV_1 \to \mathrm{hom}_{\mathbf{Top}}(\Delta_1, X)$ of singular 1-simplices on X defined for a neighbourhood V_1 of the diagonal in $X \times X$.*

Proof. Let the action of G on X be free and $\Sigma \subset X$ be a fundamental domain. Each point $x \in X$ has an arc-wise connected neighbourhood V_x. Consider the covering $\mathfrak{V} = \{V_x \mid x \in X\}$ of X, all $\mathfrak{V}$-small simplices $(x_0, x_1) \in \Sigma \times X$ and choose a path $\hat{\sigma}_1(x_0, x_1)$ from x_0 to x_1 for every such $\mathfrak{V}$-small simplex $(x_0, x_1) \in \Sigma \times X$. Let V_1 denote the G-invariant neighbourhood $\bigcup_{x \in \Sigma} G.(V_x \times V_x)$ of the diagonal in X^2 and extend the map $\hat{\sigma}_1$ to UV_1 by requiring the equation

$$\hat{\sigma}_n(g.x_0, g.x_1) = g.\hat{\sigma}_n(x_0, x_1)$$

to hold for all $g \in G$ and all $(x_0, x_1) \in V_1$. This is well defined because the action of G on X is free. The resulting map $\hat{\sigma}_1 : UV_1 \to \mathrm{hom}_{\mathbf{Top}}(\Delta_1, X)$ is a local equivariant system of singular 1-simplices defined for the neighbourhood V_1. □

Example 3.0.12. If $H \leq G$ is a subgroup of a locally arc-wise connected topological group G, then there exists a local system $\hat{\sigma}_1 : UV_2 \to \mathrm{hom}_{\mathbf{Top}}(\Delta_1, G)$ of singular 1-simplices on G, which is equivariant w.r.t. the action of H on G by left translation.

Example 3.0.13. Every manifold M is locally arc-wise connected. Thus for every manifold with group action $G \times M \to M$ there exists a local G-equivariant system $\hat{\sigma}_1$ of singular 1-simplices on M.

Example 3.0.14. Lie groups are always locally arc-wise connected. Thus for every subgroup H of a Lie group G there exists a local system $\hat{\sigma}_1$ of singular 1-simplices on G, which is equivariant w.r.t. the action of H on G by left translation.

Definition 3.0.15. *Two local systems of simplices $\hat{\sigma}_n : UV_p \to \mathrm{hom}_{\mathbf{Top}}(\Delta_p, X)$ and $\hat{\sigma}_q : UV_q \to \mathrm{hom}_{\mathbf{Top}}(\Delta_q, X)$ are called* compatible *if the domains UV_p and UV_q are components of a simplicial subset UV of $USS(X)$ and the systems $\hat{\sigma}_p$ and $\hat{\sigma}_q$ intertwine all maps induced by morphisms $\alpha : [p] \to [q]$ and $\beta : [q] \to [p]$ in the simplicial category $\mathcal{O}$.*

Consequently, a local system $\hat{\sigma}_p : UV_p \to \hom_{\mathbf{Top}}(\Delta_p, X)$ of singular n-simplices is called *compatible*, if it is compatible with itself. This especially implies that the all the faces of the singular n-simplices $\hat{\sigma}_n(\mathbf{x})$ are compatible in the sense that for simplices $\mathbf{x} \in UV_n$ and $\mathbf{x}' \in UV_n$ with common boundary $\partial_i \mathbf{x} = \partial_j \mathbf{x}'$ the corresponding faces $\partial_i \hat{\sigma}_n(\mathbf{x})$ and $\partial_j \hat{\sigma}_n(\mathbf{x}')$ of the singular simplices $\hat{\sigma}_n(\mathbf{x})$ and $\hat{\sigma}_n(\mathbf{x}')$ coincide:

Proposition 3.0.16. *If $\hat{\sigma}_n : UV_n \to \hom_{\mathbf{Top}}(\Delta_n, X)$, $n > 0$ is a compatible equivariant local system of singular n-simplices, then there exists a G-invariant neighbourhood V_{n-1} of the diagonal in X^n and a local equivariant system $\hat{\sigma}_{n-1} : UV_{n-1} \to \hom_{\mathbf{Top}}(\Delta_{n-1}, X)$ of singular $(n-1)$-simplices which is compatible to $\hat{\sigma}_n$. In the case $V_n = X^{n+1}$ the neighbourhood V_{n-1} can be chosen to be X^n.*

Proof. Let $\hat{\sigma}_n : UV_n \to \hom_{\mathbf{Top}}(\Delta_n, X)$, $n > 0$ be a compatible local equivariant system of singular n-simplices on X. Define a neighbourhood V_{n-1} of the diagonal in X^n by

$$V_{n-1} := \bigcup_{i=0}^{n} \partial_i V_n,$$

so that V_{n-1} consists of all faces $\partial_i \mathbf{x}$ of simplices $\mathbf{x} \in V_n$. Since the subspace V_n is G-invariant the so obtained neighbourhood V_{n-1} of the diagonal in X^n is G-invariant as well. For this G-invariant neighbourhood V_{n-1} of the diagonal in X^n we can define the system $\hat{\sigma}_{n-1}$ of singular $(n-1)$-simplices via

$$\hat{\sigma}_{n-1} : UV_{n-1} \to \hom \mathbf{Top}(\Delta_{n-1}, X),$$
$$\hat{\sigma}_{n-1}(\mathbf{x}) = \partial_n \hat{\sigma}_n(\mathbf{x}, x_{n-1}) = C(\Delta(\epsilon_{n-1}), X)\,(\hat{\sigma}_n(\mathbf{x}, x_{n-1})).$$

This is a local system $\hat{\sigma}_{n-1}$ of singular $(n-1)$-simplices which is equivariant by construction. We assert that this local system $\hat{\sigma}_{n-1}$ of singular $(n-1)$-simplices is compatible to the system $\hat{\sigma}_n$ of singular n-simplices. Let $\alpha : [n-1] \to [n]$ be a morphism in $\mathcal{O}$ and $\mathbf{x}$ be a point in the neighbourhood V_n of the diagonal in X^n. Using the simplicial identity $\eta_{n-1}\epsilon_{n-1} = \mathrm{id}_{[n-1]}$ we observe the equalities

$$
\begin{aligned}
\hat{\sigma}_{n-1}\left(SS(X)(\alpha)(\mathbf{x})\right) &= C(\Delta(\epsilon_{n-1}), X)\left[\hat{\sigma}_n\left(SS(X)(\eta_{n-1})SS(X)(\alpha)(\mathbf{x})\right)\right] \\
&= C(\Delta(\epsilon_{n-1}), X)\left[\hat{\sigma}_n\left(SS(X)(\alpha\eta_{n-1})(\mathbf{x})\right)\right] \\
&= C(\Delta(\epsilon_{n-1}), X) \circ C(\Delta(\alpha\eta_{n-1}), X)\,(\hat{\sigma}_n(\mathbf{x})) \\
&= C(\Delta(\alpha\eta_{n-1})\Delta(\epsilon_{n-1}), X)\,(\hat{\sigma}_n(\mathbf{x})) \\
&= C(\Delta(\alpha\eta_{n-1}\epsilon_{n-1}), X)\,(\hat{\sigma}_n(\mathbf{x})) \\
&= C(\Delta(\alpha), X)\,(\hat{\sigma}_n(\mathbf{x})).
\end{aligned}
$$

Similarly, if $\mathbf{x}$ is a point in V_{n-1} and $\alpha : [n] \to [n-1]$ a morphism in the simplicial category $\mathcal{O}$ we observe:

$$
\begin{aligned}
\hat{\sigma}_n\left(SS(X)(\alpha)(\mathbf{x})\right) &= \hat{\sigma}_n\left(SS(X)(\eta_{n-1}\epsilon_{n-1}\alpha)(\mathbf{x})\right) \\
&= \hat{\sigma}_n\left(SS(X)(\epsilon_{n-1}\alpha)SS(X)(\epsilon_{n-1})(\mathbf{x})\right) \\
&= C(\Delta(\epsilon_{n-1}\alpha), X)\left[\hat{\sigma}_n\left(SS(X)(\eta_{n-1})\mathbf{x}\right)\right] \text{ because } \hat{\sigma}_n \text{ is compatible} \\
&= C(\Delta(\alpha), X) \circ C(\Delta(\epsilon_{n-1}), X)\left[\hat{\sigma}_n(x_0, \ldots, x_{n-1}, x_{n-1})\right] \\
&= C(\Delta(\alpha), X)\left[\hat{\sigma}_{n-1}(x_0, \ldots, x_{n-1})\right]
\end{aligned}
$$

Thus the local equivariant system $\hat{\sigma}_{n-1}$ of singular $(n-1)$-simplices is compatible to the local equivariant system $\hat{\sigma}_n$ of singular n-simplices. $\square$

Corollary 3.0.17. *If $\hat{\sigma}_n : UV_n \to \hom_{\mathbf{Top}}(\Delta_n, X)$, $n > 0$ is a compatible equivariant local system of singular n-simplices, then there exists G-invariant neighbourhoods $V_0, \ldots, V_{n-1}$ of the diagonals in $X, \ldots, X^n$ resp. and local equivariant systems*

$\hat{\sigma}_k : UV_k \to \hom_{\mathbf{Top}}(\Delta_k, X)$ *of singular k-simplices for all $0 \le k \le n$ which are compatible. In the case $V_n = X^{n+1}$ the neighbourhoods V_k of the diagonal in X^{k+1} can be chosen to be X^{k+1}.*

Proof. This follows from repeated application of the preceding Proposition.   □

Lemma 3.0.18. *If $\hat{\sigma}_n : USS(X, \mathfrak{U})([n]) \to \hom_{\mathbf{Top}}(\Delta_n, X)$ a local equivariant system of singular n-simplices (where $\mathfrak{U}$ is a G-invariant covering of X), then there exists a local equivariant system $\hat{\sigma}_{n-1} : USS(X, \mathfrak{U})([n-1]) \to \hom_{\mathbf{Top}}(\Delta_{n-1}, X)$ of singular $(n-1)$-simplices on X which is compatible to $\hat{\sigma}_n$.*

Proof. This follows from the fact that the neighbourhood $\bigcup_{i=0}^{n} \partial_i SS(X, \mathfrak{U})([n])$ of the diagonal in X^n defined in the proof of Proposition 3.0.16 contains the neighbourhood $SS(X, \mathfrak{U})([n-1])$ and the spaces $SS(X, \mathfrak{U})([n])$ and $SS(X, \mathfrak{U})([n-1])$ are parts of a simplicial space.   □

Lemma 3.0.19. *If $\hat{\sigma}_n : USS(X, \mathfrak{U})([n]) \to \hom_{\mathbf{Top}}(\Delta_n, X)$ a local system of singular n-simplices (where $\mathfrak{U}$ is a G-invariant covering of X), then there exist compatible local equivariant systems $\hat{\sigma}_k : USS(X, \mathfrak{U})([k]) \to \hom_{\mathbf{Top}}(\Delta_k, X)$, $0 \le k \le n$ of singular k-simplices on X.*

Proof. This follows from repeated application of the preceding Lemma.   □

Let $\mathfrak{U}$ be an open covering of X. A local system $\hat{\sigma}_n : USS(X, \mathfrak{U})([n]) \to \hom_{\mathbf{Top}}(\Delta_n, X)$ of singular n-simplices induces a unique homomorphism

$$\mathrm{F}_{\mathbf{Ab}}(\hat{\sigma}_n) : \mathbb{Z}^{(USS(X,\mathfrak{U})([n]))} \to S_n(X)$$

from the free Abelian group $\mathbb{Z}^{(USS(X,\mathfrak{U})([n]))}$ on the set $USS(X, \mathfrak{U})([n])$ underlying the neighbourhood $SS(X, \mathfrak{U})([n])$ of the diagonal in X^{n+1} to the Abelian group $S_n(X)$ generated by the singular n-simplices in X. These groups are parts of the Abelian simplicial groups $\mathbb{Z}^{(USS(X,\mathfrak{U}))}$ and $S(X)$ respectively. If $\hat{\sigma}_k : USS(X, \mathfrak{U})([k]) \to \hom_{\mathbf{Top}}(\Delta_k, X)$, $0 < k \le n$ is a collection of compatible local systems of singular simplices, then the compatibility of the systems $\hat{\sigma}_k$ implies that the induced homomorphisms

$$\mathrm{F}_{\mathbf{Ab}} : \mathbb{Z}^{(USS(X,\mathfrak{U})([k]))} \to S_k(X)$$

intertwine all homomorphisms induced by morphisms $\alpha : [k] \to [k']$, $k, k' \le n$ in the simplicial category $\mathcal{O}$. Let V be any Abelian group of coefficients. The application of the contravariant functor $\hom_{\mathbf{Ab}}(-, V)$ to the induced homomorphisms $\mathrm{F}_{\mathbf{Ab}}(\hat{\sigma}_k)$, $k \le n$ yields homomorphisms

$$S^k(X; V) \to A^k(X, \mathfrak{U}; V), \quad f \mapsto f \circ \hat{\sigma}_k$$

of Abelian groups which assigns to each singular k-cochain an $\mathfrak{U}$-local cochain. By construction, these homomorphisms intertwine all homomorphisms induced by morphisms $\alpha : [k] \to [k']$, $k, k' \le n$ in the simplicial category $\mathcal{O}$. As a consequence the homomorphisms $f \mapsto f \hat{\sigma}_k$, $k < n$ intertwine the coboundary operators on the cochain complexes $S^*(X; V)$ on $A^*(X, \mathfrak{U}; V)$. The homomorphisms $f \mapsto f \hat{\sigma}_k$ of Abelian groups can be used to show that the natural homomorphisms $H^k(X, \mathfrak{U}; V) \to H^k(X; V)$, $k \le n$ from the $\mathfrak{U}$-local into the singular induced by the vertex transformation $C(\lambda_{X, \mathfrak{U}}, X)$ (cf. Section 5.5) are injective:

Lemma 3.0.20. *If $\hat{\sigma}_n : USS(X, \mathfrak{U})([n]) \to \hom_{\mathbf{Top}}(\Delta_n, X)$ is a local system of singular n-simplices then the natural map $H^n(X, \mathfrak{U}; V) \to H^n(X; V)$ is injective.*

Proof. Let $\mathfrak{U}$ be an open cover of X and $\hat{\sigma}_n : USS(X,\mathfrak{U})([n]) \to \hom_{\mathbf{Top}}(\Delta_n, X)$ be a system of singular n-simplices. We will use the fact that the singular cohomology $H(X;V)$ is naturally isomorphic to the cohomology $H_{sing}(X,\mathfrak{U};V)$ of $\mathfrak{U}$-small singular cochains. Consider an $\mathfrak{U}$-local n-cocycle $f \in A^n(X,\mathfrak{U};V)$. If the singular cohomology class $[f \circ \lambda_{X,\mathfrak{U},n}] \in H^n_{sing}(X,\mathfrak{U};V)$ is trivial, then there exists a singular $(n-1)$-cochain $c \in S_{n-1}(X,\mathfrak{U};V)$ with coboundary $dc = f \circ \lambda_{X,\mathfrak{U},n}$. Extend this $\mathfrak{U}$-small singular $(n-1)$-cochain c to a singular cochain $C \in S_{n-1}(X;V)$ defined on all singular $(n-1)$-simplices in X. By Lemma 3.0.16 there exists a local system $\hat{\sigma}_{n-1} : USS(X,\mathfrak{U})([n-1]) \to \hom_{\mathbf{Top}}(\Delta_{n-1}, X)$ of singular $(n-1)$-simplices which is compatible with $\hat{\sigma}_n$. With the use of this local system of singular $(n-1)$-simplices we define a $\mathfrak{U}$-local $(n-1)$-cochain h by setting

$$h : SS(X,\mathfrak{U})([n-1]) \to V, \quad h = C \circ \hat{\sigma}_{n-1}.$$

We assert that the coboundary of h is exactly the n-cocycle f. Let $\mathbf{x} \in SS(X,\mathfrak{U})([n])$ be a $\mathfrak{U}$-small n-simplex. The evaluation of the coboundary dh at $\mathbf{x}$ computes to

$$[dh](\mathbf{x}) = \sum_{i=0}^n (-1)^i h(x_0, \ldots, \hat{x}_i, \ldots, x_{n+1})$$
$$= \sum_{i=0}^n (-1)^i C\left(\hat{\sigma}_{n-1}(x_0, \ldots, \hat{x}_i, \ldots, x_{n+1})\right)$$
$$= C\left(\partial \hat{\sigma}_n(\mathbf{x})\right) = [dC]\left(\hat{\sigma}_n(\mathbf{x})\right) = [f\lambda_{X,\mathfrak{U},n}]\left(\hat{\sigma}_n(\mathbf{x})\right)$$
$$= [f\lambda_{X,\mathfrak{U},n}\hat{\sigma}_n](\mathbf{x}) = f(\mathbf{x}).$$

Therefore the $\mathfrak{U}$-local cocycle f is the coboundary of the $\mathfrak{U}$-local $(n-1)$-cochain h. Consequently its cohomology class $[f] \in H^n(X,\mathfrak{U};V)$ is trivial. $\square$

Proposition 3.0.21. *If the topological space X admits a compatible local system $\hat{\sigma}_n : USS(X,\mathfrak{B})([n]) \to \hom_{\mathbf{Top}}(\Delta_n, X)$ of singular n-simplices, then the natural maps $H^k(X,\mathfrak{U};V) \to H^k(X;V)$ are injective for all $k \leq n$.*

Proof. Let $\hat{\sigma}_n : USS(X,\mathfrak{U})([n]) \to \hom_{\mathbf{Top}}(\Delta_n, X)$, be a compatible equivariant local system of singular n-simplices. By Lemma 3.0.19 there exists a collection $\hat{\sigma}_k : SS(X,\mathfrak{U})([k]) \to \hom_{\mathbf{Top}}(\Delta_k, X), k \leq n$ of compatible equivariant local systems of singular n-simplices. Therefore, by Lemma 3.0.20, the natural homomorphisms $H^k(X,\mathfrak{U};V) \to H^k(X;V)$ are injective for all $k \leq n$. $\square$

Consider the colimit cosimplicial Abelian group $\operatorname{colim}_{\mathfrak{U}} A(X,\mathfrak{U};V)$, where $\mathfrak{U}$ ranges over all G-invariant open covers $\mathfrak{U}$ of X. If there exists a compatible local system $\hat{\sigma}_n : USS(X,\mathfrak{U})([n]) \to \hom_{\mathbf{Top}}(\Delta_n, X)$ of singular n-simplices, then the natural homomorphism $Z^n(X,\mathfrak{U};V) \to H^n(X;V)$ factors uniquely through the colimit $H^n(\operatorname{colim}_{\mathfrak{U}} A(X,\mathfrak{U};V))$. Because cohomology commutes with colimits, the latter colimit can also be written as $\operatorname{colim}_{\mathfrak{U}} H^n(X,\mathfrak{U};V)$, so the equivariant local system $\hat{\sigma}_n$ of singular n-simplices induces a homomorphism $\operatorname{colim}_{\mathfrak{U}} H^n(X,\mathfrak{U};V) \to H^n(X;V)$. Using this notation, the above result on the injectivity can now be generalised:

Lemma 3.0.22. *If the topological space X admits a compatible local equivariant system $\hat{\sigma}_n : USS(X,\mathfrak{U})([n]) \to \hom_{\mathbf{Top}}(\Delta_n, X)$ of singular n-simplices, then the natural homomorphism $\operatorname{colim}_{\mathfrak{U}} H^n(X,\mathfrak{U};V) \to H^n(X;V)$ is injective.*

Proof. Let $\hat{\sigma}_n : SS(X,\mathfrak{B})([n]) \to \hom_{\mathbf{Top}}(\Delta_n, X)$ be a compatible equivariant local system of singular n-simplices. Consider a cohomology class $[f]$ in the cohomology $H^n(\operatorname{colim}_{\mathfrak{U}} A(X,\mathfrak{U};V))$. This cohomology class is represented by a $\mathfrak{V}$-local n-cocycle $f \in A^n(X,\mathfrak{V};V)$, where we w.l.o.g. assume $\mathfrak{V} \prec \mathfrak{B}$. If the image of the cohomology class $[f] \in H^n(\mathrm{colim}_\mathfrak{U} A(X,\mathfrak{U};V))$ in the singular cohomology $H^n(X;V)$ is trivial, the commutativity of diagram

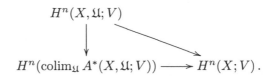

$$H^n(X,\mathfrak{U};V)$$

$$H^n(\mathrm{colim}_\mathfrak{U} A^*(X,\mathfrak{U};V)) \longrightarrow H^n(X;V).$$

shows that the cohomology class $[f] \in H^n(X,\mathfrak{U};V)$ is also trivial, because the downright homomorphism is injective. As a consequence the cohomology class $[f]$ in $\mathrm{colim}_\mathfrak{U} H^n(X,\mathfrak{U};V) \cong H^n(\mathrm{colim}_\mathfrak{U} A(X,\mathfrak{U};V))$ is trivial as well. This proves the injectivity of the natural homomorphism $\mathrm{colim}_\mathfrak{U} H^k(X,\mathfrak{U};V) \to H^k(X;V)$.   □

Proposition 3.0.23. *If the G-space X admits a compatible local equivariant system $\hat{\sigma}_n : USS(X,\mathfrak{U})([n]) \to \mathrm{hom}_{\mathbf{Top}}(\Delta_n, X)$ of singular n-simplices, then the natural homomorphisms $\mathrm{colim}_\mathfrak{U} H^k(X,\mathfrak{U};V) \to H^k(X;V)$ are injective for all $k \leq n$.*

Proof. If the topological space X admits a compatible local equivariant system $\hat{\sigma}_n : USS(X,\mathfrak{U})([n]) \to \mathrm{hom}_{\mathbf{Top}}(\Delta_n, X)$ of singular n-simplices, then there exists a collection $\hat{\sigma}_k : USS(X,\mathfrak{U})([k]) \to \mathrm{hom}_{\mathbf{Top}}(\Delta_k, X)$, $k \leq n$ of compatible local equivariant systems of singular k-simplices. The assertion follows now from the preceding Proposition.   □

Lemma 3.0.24. *If V_n is a G-invariant neighbourhood of the diagonal in X^{n+1}, then there exists a G-invariant open cover $\mathfrak{U}$ of X satisfying $SS(X,\mathfrak{U})([n]) \subset V_n$.*

Proof. Let V_n be a G-invariant neighbourhood of the diagonal in X^{n+1}. Because U_n is a neighbourhood of the diagonal, every point $x \in X$ has a neighbourhood V_x satisfying $V_x^{n+1} \subset V_n$. The collection $\mathfrak{U}' = \{V_x \mid x \in X\}$ is an open cover of X satisfying

$$SS(X,\mathfrak{U}')([n]) = \bigcup_{x \in X} V_x^{n+1} \subset V_n.$$

Consider the G-invariant open covering $\mathfrak{U} = G.\mathfrak{U}'$ of X. Since the neighbourhood U_n of the diagonal is G-invariant, it also contains the subspace $SS(X,\mathfrak{U})([n])$.   □

Theorem 3.0.25. *If X admits a compatible equivariant local system $\hat{\sigma}_n$ of singular n-simplices, then the natural homomorphisms $\mathrm{colim}_\mathfrak{U} H^k(X,\mathfrak{U};V) \to H^k(X;V)$ are injective for all $k \leq n$.*

Proof. Let $\hat{\sigma}_n : UV_n \to \mathrm{hom}_{\mathbf{Top}}(\Delta_n, X)$ be a compatible equivariant local system of singular n-simplices. By Lemma 3.0.24 there exists a compatible equivariant local system $\hat{\sigma}_n : USS(X,\mathfrak{U})([n]) \to \mathrm{hom}_{\mathbf{Top}}(\Delta_n, X)$ of singular n-simplices, where $\mathfrak{B}$ is a G-invariant open covering of X. By Lemma 3.0.23 the natural homomorphisms $H^k(X,\mathfrak{U};V) \to H^k(X;V)$ are injective for all $k \leq n$ in this case.   □

If the action of G on the space is trivial, then every open covering $\mathfrak{U}$ is G-invariant. In this case the colimit cosimplicial group $\mathrm{colim}_\mathfrak{U} A(X,\mathfrak{U};V)$ is the Alexander-Spanier cosimplicial Abelian group $A_{AS}(X;V)$. In addition every (local) system of singular n-simplices is equivariant if the action of G is trivial. So we observe:

Corollary 3.0.26. *If the space X admits a compatible local system $\hat{\sigma}_n$ of singular n-simplices, then the natural homomorphisms $H^k_{AS}(X,\mathfrak{U};V) \to H^k(X;V)$ are injective for all $k \leq n$.*

Theorem 3.0.27. *If the action of G on X is free and X is n-connected, then there exists a compatible equivariant system $\hat{\sigma}_n$ of singular $(n+1)$-simplices on X.*

Proof. The proof is by induction on n. For the case $n = 0$ of arc-wise connected spaces the assertion has already been proved in Lemma 3.0.5. Assume the assertion to be true for $k \leq n$ and let X be an $(n+1)$-connected G-space with free action $G \times X \to X$. By the inductive assumption there exists a compatible equivariant system $\hat{\sigma}_{n+1} : UX^{n+2} \to \hom_{\mathbf{Top}}(\Delta_{n+1}, X)$ of singular $(n+1)$-simplices on X. By Proposition 3.0.16 there exists an equivariant system $\hat{\sigma}_n$ of singular n-simplices on X which is compatible with $\hat{\sigma}_{n+1}$. The compatibility of both systems of singular simplices implies the equalities

$$\partial_j \hat{\sigma}_{n+1}(x_0, \ldots, \hat{x}_i, \ldots, x_{n+2}) = \hat{\sigma}_n(x_0, \ldots, \hat{x}_i, \ldots, \hat{x}_{j-1}, \ldots, x_{n+2})$$
$$= \partial_i \hat{\sigma}_{n+1}(x_0, \ldots, \hat{x}_{j-1}, \ldots, x_{n+2})$$

for all points $\mathbf{x} \in X^{n+2}$ and all $0 \leq i < j \leq n+1$. Each singular $(n+1)$-simplex $\hat{\sigma}_{n+1}(x_0, \ldots, \hat{x}_i, \ldots, x_{n+2})$ can be regarded as a continuous map $\Delta_{n+2}^i \to X$ from the i-th face of the standard simplex Δ_{n+2} into X. The above equations show that these maps coincide at the intersections of their domains, i.e. they assemble to a continuous map $f_{\mathbf{x}} : \partial\Delta_{n+2} \to X$ defined on the topological boundary $\partial\Delta_{n+2}$ of the standard $(n+1)$-simplex Δ_{n+2}. Choose a fundamental domain $\Sigma \subset X$ and consider all points $\mathbf{x} \in \Sigma \times X^{n+2}$ and the corresponding continuous maps $f_{\mathbf{x}} : \partial\Delta_{n+1} \to X$. Because the topological space X is $(n+1)$-connected, there exists a continuous map $\hat{\sigma}_{n+2}(\mathbf{x}) : \Delta_{n+2} \to X$ which restricts to $f_{\mathbf{x}}$ on the topological boundary of Δ_{n+2}. In this way we have obtained a map $U(\Sigma \times X^{n+2}) \to \hom_{\mathbf{Top}}(\Delta_{n+2}, X)$. We define the desired system $\hat{\sigma}_{n+2}$ of singular $(n+2)$-simplices by requiring the equation

$$\hat{\sigma}_{n+2}(g.x_0, \ldots, g.x_{n+2}) = g.\hat{\sigma}_{n+2}(x_0, \ldots, x_n)$$

to hold for all $g \in G$ and all $(x_0, \ldots, x_{n+2}) \in X^{n+3}$. This is well defined because the action of G is free. The resulting map $\hat{\sigma}_{n+2} : UX^{n+3} \to \hom_{\mathbf{Top}}(\Delta_{n+2}, X)$ is then an equivariant system of singular $(n+2)$-simplices on X. By construction, the i-th face of the singular $(n+2)$-simplex $\hat{\sigma}_{n+2}(\mathbf{x})$ is the singular $(n+1)$-simplex $\hat{\sigma}_{n+1}(x_0, \ldots, \hat{x}_i, \ldots, x_{n+2})$. Thus the maps $\hat{\sigma}_{n+2}(\mathbf{x})$ form a system $\hat{\sigma}_{n+2}$ of singular $(n+2)$-simplices which is compatible to $\hat{\sigma}_{n+1}$. This completes the inductive step. $\square$

Corollary 3.0.28. *If the topological space X is n-connected, then there exists a compatible system of singular $(n+1)$-simplices on X.*

A special case are the weakly connected topological spaces, which are n-connected for all $n \in \mathbb{N}$. They admit compatible systems of singular simplices in each dimension. This motivates the following definition:

Definition 3.0.29. *A morphism $\hat{\sigma} : USS(X) \to \hom_{\mathbf{Top}}(\Delta, X)$ of simplicial sets, whose component $\hat{\sigma}([n])$ in each dimension n is a system of singular n-simplices is called a* system of (singular) simplices *on X.*

Corollary 3.0.30. *If the action of G on X is free and X is weakly connected, then there exists an equivariant system $\hat{\sigma}_n : USS(X) \to \hom_{\mathbf{Top}}(\Delta, X)$ of singular simplices on X.*

Corollary 3.0.31. *If the topological space X is weakly connected, then there exists a system $\hat{\sigma}_n : USS(X) \to \hom_{\mathbf{Top}}(\Delta, X)$ of singular simplices on X.*

Example 3.0.32. If there exists a fixed point $*$ of the G-action, then the path space $PX = C_*((I, 0), (X, *))$ of all paths in X starting at $*$ is equivariantly contractible via

$$\Phi : PX \times I \to PX, \quad \Phi(\gamma)(s)(t) = \gamma(st).$$

Thus there exists an equivariant system $\hat{\sigma}_n : PX^{n+1} \to \hom_{\mathbf{Top}}(\Delta_n, PX)$ of singular n-simplices for each $n \in \mathbb{N}$.

Example 3.0.33. If G is an n-connected topological group acting on itself via left translations, then there exists a compatible equivariant system $\hat{\sigma}_n : G^{n+1} \to \hom_{\mathbf{Top}}(\Delta_n, X)$ of singular n-simplices on X. More generally, if the set underlying the topological space X admits a group structure, then there exists a compatible system $\hat{\sigma}_n : X^{n+1} \to \hom_{\mathbf{Top}}(\Delta_n, X)$ of singular n-simplices on X which equivariant w.r.t. the actin of X on itself via left translation.

Example 3.0.34. The path group PG of a topological group G acts on itself via left translation. Since this action is free and the space PG is contractible there exists a system $\hat{\sigma} : USS(PG) \to \hom_{\mathbf{Top}}(\Delta_n, PG)$ of singular simplices on PG.

Example 3.0.35. The total space EG of the universal bundle $EG \to BG$ of a topological group G is weakly contractible. Since the action of G on EG is free there exists a system $\hat{\sigma} : USS(EG) \to \hom_{\mathbf{Top}}(\Delta_n, EG)$ of singular simplices on EG.

In the following we will consider (local) systems $\hat{\sigma}_n$ of singular n-simplices where the singular n-simplex $\hat{\sigma}_n(\mathbf{x})$ depends continuously on its vertices $x_0, \ldots, x_n$. This property is obtained by considering continuous maps $X^{n+1} \to C(\Delta_n, X)$ instead of maps $(UX)^{n+1} \to \hom_{\mathbf{Top}}(\Delta_n, X)$ of the underlying sets. (As always we equip the function spaces $C(\Delta_n, X)$ with the compact open topology here.)

Definition 3.0.36. *A continuous right inverse $\hat{\sigma}_n : X^{n+1} \to C(\Delta_n, X)$ to the vertex map $\lambda([n]) : C(\Delta_n, X) \to X^{n+1}$ is called a continuous system of (singular) n-simplices on X.*

Remark 3.0.37. Since all standard simplices Δ_n are compact, the continuity of a system of n-simplices $\hat{\sigma}_n : X^{n+1} \to C(\Delta_n, X)$ is equivalent to the continuity of the adjoint map $\sigma_n : X^{n+1} \times \Delta_n \to X$.

Example 3.0.38. The unique system of singular 0-simplices

$$\hat{\sigma}_0 : X \to \hom_{\mathbf{Top}}(\Delta_0, X), \quad \hat{\sigma}_0(x_0)(t_0) = x_0$$

on the topological space X is always continuous.

Definition 3.0.39. *A continuous system $\hat{\sigma}_n : X^{n+1} \to C(\Delta_n, X)$ of singular n-simplices is called* equivariant *if it intertwines the action of G on X and $C(\Delta_n, X)$, i.e. if it satisfies the equation $\hat{\sigma}_n(g.x_0, \ldots, g.x_n) = g.\hat{\sigma}_n(x_0, \ldots, x_n)$ for all $g \in G$ and all $(x_0, \ldots, x_n) \in X^{n+1}$.*

Example 3.0.40. The unique continuous system system $\hat{\sigma}_0$ of singular 0-simplices on a G-space X is always equivariant.

Definition 3.0.41. *A continuous map $\hat{\sigma}_n : U_n \to C(\Delta_n, X)$ defined on a neighbourhood U_n of the diagonal in X^{n+1} satisfying $\hat{\sigma}_n(x_0, \ldots, x_n)(\mathbf{e}_i) = x_i$ for all $0 \leq i \leq n$ is called a continuous local system of (singular) n-simplices on X.*

Remark 3.0.42. Since all standard simplices Δ_n are compact, the continuity of a local system of n-simplices $\hat{\sigma}_n : U_n \to C(\Delta_n, X)$ is equivalent to the continuity of the adjoint map $\sigma_n : U_n \times \Delta_n \to X$.

Definition 3.0.43. *A continuous local system $\hat{\sigma}_n : U_n \to C(\Delta_n, X)$ of singular n-simplices is called* equivariant *if it intertwines the action of G on X and $C(\Delta_n, X)$, i.e. if it satisfies the equation $\hat{\sigma}_n(g.x_0, \ldots, g.x_n) = g.\hat{\sigma}_n(x_0, \ldots, x_n)$ for all $g \in G$ and all $(x_0, \ldots, x_n) \in X^{n+1}$.*

Remark 3.0.44. The equivariance of a continuous local system of singular n-simplices $\hat{\sigma}_n : U_n \to C(\Delta_n, X)$ is equivalent to the equivariance of the adjoint map $\sigma_n : U_n \times \Delta_n \to X$.

Definition 3.0.45. *Two continuous local systems of simplices* $\hat{\sigma}_p : U_p \to C(\Delta_p, X)$ *and* $\hat{\sigma}_q : U_q \to C(\Delta_q, X)$ *are called* compatible *if the domains* U_p *and* U_q *are components of a simplicial subspace* U *of* $\mathrm{SS}(X)$ *and the systems* $\hat{\sigma}_p$ *and* $\hat{\sigma}_q$ *intertwine all maps induced by morphisms* $\alpha : [p] \to [q]$ *and* $\beta : [q] \to [p]$ *in the simplicial category* $\mathcal{O}$.

In the following we will show that a large class of topological spaces (the strongly uniformly locally contractible spaces) admits continuous (equivariant) local systems of singular n-simplices. Recall that we identified the product $X \times X$ with the component $\mathrm{SS}(X)([1])$ of the standard simplicial space $\mathrm{SS}(X)$, hence the projection onto the first factor of the product $X \times X$ will be denoted by pr_0.

Definition 3.0.46. *A* strong uniform local contraction *of a topological space X is a continuous function $F : U_1 \times I \to X$ defined for a neighbourhood U_1 of the diagonal $D_2 X$ in $X \times X$ with the following properties:*

1. *The map* $(\mathrm{pr}_0, F) : U_1 \times I \to X \times X$ *is a deformation onto the diagonal $D_2 X$ in $X \times X$ within U_1, rel. $D_2 X$ (i.e. it leaves the diagonal fixed).*
2. *The map F satisfies the equality $F(x_0, F(x_0, x_1, s), t) = F(x_0, x_1, s + t - st)$ for all $(x_0, x_1) \in U_1$ and $s, t \in I$.*

A topological space X is called strongly uniformly locally contractible *(SULC) if there exists a strong uniform local contraction of X. A G-space X is called* strongly equivariantly universally contractible *(SEULC) if the neighbourhood U_1 of the diagonal can be chosen to be G-invariant and the function F can be chosen to be equivariant.*

Remark 3.0.47. This is a variation of the property "uniformly locally contractible" (ULC). Both properties are designed such that the function F is compatible with the face maps $\mathrm{SS}(X, \mathfrak{U})([1]) \to \mathrm{SS}(X, \mathfrak{U})([0])$ in a certain way.

Example 3.0.48. Disc-like identity neighbourhoods U in topological vector spaces V are *SULC*: Define a function $F : U_1 \times I \to U$ by $F(v_0, v_1, s) = sv_0 + (1 - s)v_1$. The function (pr_0, F) is a contraction of U_1 onto the diagonal in $U \times U$. Moreover it satisfies the condition

$$
\begin{aligned}
F(v_0, F(v_0, v_1, s), t) &= tv_0 + (1 - t)[sv_0 + (1 - s)v_1] \\
&= (s + t - st)v_0 + (1 - s)(1 - t)v_1 \\
&= (s + t - st)v_0 + 1 - (s + t - st)v_1 = F(v_0, v_1, s + t - st),
\end{aligned}
$$

which is the condition for U to be a *SULC* space.

This example generalises to arbitrary identity identity neighbourhoods of topological vector spaces:

Example 3.0.49. Identity neighbourhoods U in topological vector spaces V are *SULC*: Let U_1 be the neighbourhood of the diagonal consisting of all points $v_1, v_2 \in U \times U$ for which the line segment from v_0 to v_1 is contained in U. Define a function $F : U_1 \times I \to U$ by $F(v_0, v_1, s) = sv_0 + (1 - s)v_1$. The function (pr_0, F) is a contraction of U_1 onto the diagonal in $U \times U$. Moreover it satisfies the condition

$$
\begin{aligned}
F(v_0, F(v_0, v_1, s), t) &= tv_0 + (1 - t)[sv_0 + (1 - s)v_1] \\
&= (s + t - st)v_0 + (1 - s)(1 - t)v_1 \\
&= (s + t - st)v_0 + 1 - (s + t - st)v_1 = F(v_0, v_1, s + t - st),
\end{aligned}
$$

which is the condition for U to be a *SULC* space.

Let U be an identity neighbourhood of a topological group G. If the identity neighbourhood U is SULC and $F' : U_1 \times I \to U$ is a strong uniform local contraction defined for a neighbourhood U_1 of the diagonal in $U \times U$, then $G.U_1$ is a G-invariant neighbourhood of the diagonal in $G \times G$ and the function

$$F : G.U_1 \times I \to X, \quad F(g, g', t) = g.F'(g^{-1}g', t)$$

defines a strong uniform local contraction on G. Moreover the function F is left equivariant w.r.t. the action of G on itself by left translation. Analogously one can construct a strong uniform local contraction which is equivariant w.r.t. the action of G on itself by right translation, so we note:

Lemma 3.0.50. *Locally SULC topological groups are SEULC w.r.t. the actions by left and right translation.*

Lie groups have identity neighbourhoods which are isomorphic to disc-like identity neighbourhoods in real or complex vector spaces. These identity neighbourhoods are *SULC* (cf. Example 3.0.48). As a consequence we observe:

Lemma 3.0.51. *Real and complex Lie groups are SEULC w.r.t. the actions by left and right translation.*

Example 3.0.52. The path group $PG = C_*(I, G)$ of a topological group G (i.e. the paths starting at the identity) is contractible. A natural contraction is given by the assignment $\Phi(\gamma, s)(t) = \gamma((1-s)t)$ for all paths $\gamma \in PG$ and $s, t \in I$. The function

$$F : PG \times PG \to PG, \quad F(\gamma_0, \gamma_1, s) = \gamma_0 . \Phi(\gamma_0^{-1}\gamma_1, 1 - s)$$

is a strong uniform contraction of PG. Moreover it is equivariant w.r.t. the action of PG on itself by left translation. Thus the path group PG is a strongly uniformly (locally) contractible PG-space.

Definition 3.0.53. *A G-space is called* trivial, *if there exists a fundamental domain Σ such that the action $\varrho : G \times X \to X$ of G on X induces an homeomorphism $G \times \Sigma \cong X$. The fundamental domain Σ is then called* splitting.

Remark 3.0.54. The action of the group G on a trivial G-space X is necessarily free.

Lemma 3.0.55. *A G-space X is trivial if and only if there exists an equivariant continuous map $p : X \to G$.*

Proof. If X is a trivial G-space, then there exists a fundamental domain Σ such that the action of G on X induces an homeomorphism $\varrho_{G \times \Sigma} : G \times \Sigma \cong X$. The composition $\mathrm{pr}_G \varrho_{G \times \Sigma}^{-1} : X \to G$ is then a continuous equivariant map $X \to G$. Conversely, if there exists a continuous equivariant map $p : X \to G$, then the map

$$X \to G \times \Sigma, \quad x \mapsto (p(x), p(x)^{-1}.x)$$

is a continuous inverse of the restriction $\varrho_{G \times \Sigma}$. Thus the G-space X is trivial. □

Proposition 3.0.56. *If (G, X) is a trivial transformation group and there exists a strong uniform local contraction defined on a G-invariant neighbourhood U_1 of the diagonal, then X is SEULC.*

Proof. Let (G, X) be a transformation group and Σ be a fundamental domain such that the action induces an homeomorphism $h : G \times \Sigma \xrightarrow{\cong} X$. Assume that there exists a G-invariant open neighbourhood U_1 of the diagonal D_2X in $X \times X$ and a strong uniform local contraction

$$F : U_1 \times I \to X.$$

Let pr_G and pr_Σ denote the projections of $G \times \Sigma$ onto G and Σ respectively. The compositions $\mathrm{pr}_G h^{-1}$ and $\mathrm{pr}_\Sigma h^{-1}$ can be used to define a continuous function $\tilde{F}$ on $U_1 \times I$ via

$$\tilde{F} : U_1 \times I \to X,$$
$$\tilde{F}(x_0, x_1, t) = [\mathrm{pr}_G h^{-1}(x_0)].F\left(\mathrm{pr}_\Sigma h^{-1}(x_0), [\mathrm{pr}_G h^{-1}(x_0)]^{-1}.x_1, t\right).$$

The function $\tilde{F}$ is equivariant by construction and has the properties required for X to be strongly equivariantly uniformly contractible. $\qquad\square$

The property SULC can be used to construct continuous local systems of singular simplices. In the following (G, X) will denote a transformation group. We will consider equivariant constructions only. This is no restriction since the trivial group $\{1\}$ acts on every topological space and every SULC space is SEULC w.r.t. this group action.

Lemma 3.0.57. *Every equivariantly SEULC G-space X admits a continuous equivariant system $\hat{\sigma}_1$ of 1-simplices on a G-invariant neighbourhood U_1 of the diagonal $D_2 X$ in $X \times X$.*

Proof. Let $F : U_1 \times I \to U_1$ be a contraction rel $D_2 X$ of a G-invariant neighbourhood U_1 of the diagonal. Then $\hat{\sigma}_1(x_0, x_1, t_0, t_1) = F(x_0, x_1, t_0)$ defines a continuous equivariant system of 1-simplices on U_1. $\qquad\square$

Note that this system $\hat{\sigma}_1$ of singular 1-simplices maps the diagonal $D_2 X$ to constant simplices by construction, so the systems $\hat{\sigma}_0$ and $\hat{\sigma}_1$ of singular simplices intertwine the restrictions of the face maps $U_1 \cap \mathrm{SS}(X)([1]) \to X = \mathrm{SS}(X)([0])$ and the degeneracy map $\mathrm{SS}(X)([0]) = X \to U_1 \subset X \times X$, hence they are compatible.

From now on we assume the topological space X to be SEULC and fix a G-invariant open neighbourhood U_1 of the diagonal in $X \times X$ and a strong uniform local contraction $F : U_1 \times I \to X$. Furthermore the elements $(x_0, \ldots, x_n)$ of $\mathrm{SS}(X)([n])$ will be abbreviated by the symbol $\mathbf{x}$. Likewise, elements $(t_0, \ldots, t_n)$ of the standard simplex Δ_n will be written shortly as $\mathbf{t}$. The i-th vertex $(0, \ldots, 0, t_i = 1, 0, \ldots, 0)$ of the standard simplex Δ_n will be denoted by $\mathbf{e}_i$. The continuous equivariant system of 1-simplices

$$\hat{\sigma}_1 : U_1 \to C(\Delta_1, X), \quad \hat{\sigma}_1(x_0, x_1)(t_0.t_1) = F(x_0, x_1, t_0)$$

constructed above is always denoted by $\hat{\sigma}_1$ and the unique continuous equivariant system of 0-simplices on X will always be denoted by $\hat{\sigma}_0$. Recall that these two systems of singular simplices are compatible. We will now prove the existence of a continuous equivariant local system $\hat{\sigma}_2 : U_2 \to C(\Delta_2, X)$ of singular 2-simplices which is compatible to $\hat{\sigma}_0$ and $\hat{\sigma}_1$. We start by defining the neighbourhood U_2 of the diagonal in X^3 on which this system $\hat{\sigma}_2$ of singular 2-simplices will be defined.

Consider the continuous equivariant function $(\mathrm{pr}_0, \hat{\sigma}_1) : X \times U_1 \times \Delta_1 \to U_1$. It maps the subspace $D_3 X \times \Delta_1$ of $X \times U_1 \times \Delta_1$ onto the diagonal $D_2 X$. Therefore the inverse image $(\mathrm{pr}_0, \hat{\sigma}_1)^{-1}(U_1)$ of the G-invariant open set U_1 is a G-invariant open set containing $D_2 X \times \Delta_1$. We assert that the subset

$$U_2 := \{(x_0, x_1, x_2) \in X \times U_1 \mid \{x_0\} \times \hat{\sigma}_1(x_1, x_2)(\Delta_1) \subset U_1\}$$

of $\mathrm{SS}(X)([2])$ has the desired properties.

Lemma 3.0.58. *The set U_2 is an open G-invariant neighbourhood of the diagonal.*

Proof. Let $\mathbf{x}$ be a point in U_2. Because the inverse image $(\mathrm{pr}_0, \hat{\sigma}_1)^{-1}(U_1)$ of the G-invariant open set U_1 is open, every point $(\mathbf{x}, \mathbf{t})$ in $U_2 \times \Delta_1$ has an open neighbourhood of the form $V_\mathbf{t} \times W_\mathbf{t}$ in $\mathrm{SS}(X)([2]) \times I$. Since the standard simplex Δ_1 is compact, it is covered by finitely many such open sets $W_{\mathbf{t}_1}, \ldots, W_{\mathbf{t}_n}$. The intersection $\bigcap_{i=1}^n V_{\mathbf{t}_i}$ then is an open neighbourhood of $\mathbf{x}$ which is contained in U_2. Since this is true for all points $\mathbf{x}$ in U_2, the set U_2 is open. Moreover, if $\mathbf{x}$ is a point in U_2 then the equivariance of $(\mathrm{pr}_0, \hat{\sigma}_1)$ implies the equalities

$$(\mathrm{pr}_0, \hat{\sigma}_1)(\{g.\mathbf{x}\} \times \Delta_1) = (\mathrm{pr}_0, \hat{\sigma}_1)(g.(\{\mathbf{x}\} \times \Delta_1)) = g.(\mathrm{pr}_0, \hat{\sigma}_1)(\{\mathbf{x}\} \times \Delta_1) \subset g.U_1 \subset U_1$$

for all $g \in G$. This shows the G-invariance of U_2. All in all we have proved that U_2 is a G-invariant open subset of $\mathrm{SS}(X)([2])$. $\qquad \square$

Lemma 3.0.59. *The neighbourhood U_1 of the diagonal in $\mathrm{SS}(X)([1])$ contains the images of U_2 under all face maps $\mathrm{SS}(X)(\epsilon_i) : \mathrm{SS}(X)([2]) \to \mathrm{SS}(X)([1])$.*

Proof. Let $\mathbf{x} = (x_0, x_1, x_2) \in U_2$ be a point in U_2 and $\epsilon_i : [1] \to [2]$ be a face morphism. The set U_2 is defined in such a way that the neighbourhood U_1 contains the subset $\{x_0\} \times \hat{\sigma}_1(x_1, x_2)(\Delta_1)$. Since $\hat{\sigma}_1$ is a continuous system of 1-simplices the singular 1-simplex $\hat{\sigma}_1(x_1, x_2)$ has vertices x_1 and x_2. Therefore the neighbourhood U_1 of the diagonal contains the first face $(x_0, x_2) = (x_0, \hat{\sigma}_1(x_1, x_2)(0, 1))$ and the second face $(x_0, x_1) = (x_0, \hat{\sigma}_1(x_1, x_2)(1, 0))$ of $\mathbf{x}$. The zeroth face (x_1, x_2) of $\mathbf{x}$ is contained in U_1 because the neighbourhood U_2 of the diagonal is a subset of $X \times U_1$ by definition. $\qquad \square$

Lemma 3.0.60. *The neighbourhood U_2 of the diagonal contains all images of U_1 under degeneracy maps $\mathrm{SS}(X)(\eta_i) : \mathrm{SS}(X)([1]) \to \mathrm{SS}(X)([2])$.*

Proof. Let $\mathbf{x} = (x_0, x_1)$ be a point in U_1. The singular 1-simplex $\hat{\sigma}_1(x_1, x_1)$ is the constant function x_1, hence $\{x_0\} \times \hat{\sigma}_1(x_1, x_1)(\Delta_1) = \{x_0\} \times \{x_1\} \subset U_1$ and the 1-degeneracy (x_0, x_1, x_1) of $\mathbf{x}$ satisfies the defining condition of U_2. Concerning the 0-degeneracy we remember that the function $F : U_1 \times I \to X$ satisfies the additional property $(\mathrm{pr}_0, F)(U_1 \times I) \subset U_1$ by definition, which implies the inclusions

$$\{x_0\} \times \hat{\sigma}_1(x_0, x_1)(\Delta_1) = (\mathrm{pr}_0, F)(\{(x_0, x_1)\} \times I) \subset (\mathrm{pr}_0, F)U_1 \times I \subset U_1$$

i.e. the 0-degeneracy (x_0, x_0, x_1) of $\mathbf{x}$ is contained in U_2 as well. $\qquad \square$

So far we have observed that the set $U_2 \subset \mathrm{SS}(X)([2])$ generates a simplicial subspace of $\mathrm{SS}(X)$, whose components in dimension 0, 1 and 2 are X, U_1 and U_2 respectively. Moreover there exist continuous equivariant systems 0-simplices on X and of 1-simplices on U_1 intertwining the face and degeneracy maps between X and U_1. In analogy to the definition of $\hat{\sigma}_1$ we now define a not necessarily continuous function σ_2 via

$$\sigma_2 : U_2 \times \Delta_2 \to X$$

$$\sigma_2(x_0, x_1, x_2)(t_0, t_1, t_2) = \begin{cases} F\left(x_0, \hat{\sigma}_1(x_1, x_2)\left(\frac{t_1}{1-t_0}, \frac{t_2}{1-t_0}\right), t_0\right) & \text{if } t_0 \neq 1 \\ x_0 & \text{if } t_0 = 1. \end{cases}$$

Lemma 3.0.61. *The functions $\hat{\sigma}_2(\mathbf{x}) : \Delta_2 \to X$ are continuous for all $\mathbf{x} \in U_2$.*

Proof. Let $\mathbf{x}$ be a point in U_2. Since the function $\hat{\sigma}_2(\mathbf{x})$ is continuous on $(\Delta_2 \setminus \{\mathbf{e}_0\})$ by definition it remains to prove the continuity at $\mathbf{e}_0 = (1, 0, 0) \in \Delta_2$. Let U be an open neighbourhood of x_0 and let $K = \hat{\sigma}_1(x_1, x_2)(\Delta_1)$ denote the compact image of Δ_1 under the continuous function $\hat{\sigma}_1(x_1, x_2)$. Since F is continuous and maps a point $(x_0, k, 1) \in \{x_0\} \times K \times \{1\}$ to x_0 there exists a neighbourhood V_k of k in X and a neighbourhood W_k of 1 in I such that $V_k \times W_k$ is mapped into U under F. Because K is compact it is covered by finitely many open sets $V_{k_1}, \ldots, V_{k_n}$. The intersection $W = \bigcap_{i=1}^n W_{k_i}$ is a neighbourhood of 1 in I such that F maps $\{x_0\} \times K \times W$ into U. The set $W' = \{(t_0, t_1, t_2) \in \Delta_2 \mid t_0 \in W\}$ then is a neighbourhood of $\mathbf{e}_0 = (1, 0, 0)$ and the image

$$\hat{\sigma}_2(x_0, x_1, x_2)(W') \subseteq F(\{x_0\} \times \hat{\sigma}_1(x_1, x_2)(\Delta_1) \times W) \subseteq U$$

of W' under $\hat{\sigma}_2(x_0, x_1, x_2)$ is contained in U. This shows that $\hat{\sigma}_2(x_0, x_1, x_2)$ is a continuous function. $\qquad\square$

Lemma 3.0.62. *The function* $\sigma_2 : U_2 \times \Delta_2 \to X$ *is continuous.*

Proof. The function σ_2 is continuous on $U_2 \times (\Delta_2 \setminus \{\mathbf{e}_0\})$ by definition. Furthermore any point $((x_0, x_1, x_2), (1, 0, 0)) \in U_2 \times \Delta_2$ is mapped onto x_0 under σ_2. It remains to show that for any open neighbourhood U of x_0 there exists an open neighbourhood V of $((x_1, x_2, x_3), (1, 0, 0))$ which is mapped into U under σ_2. Let U be an open neighbourhood of x_0 and let $K = \hat{\sigma}_1(x_1, x_2)(\Delta_1)$ denote the compact image of Δ_1 under the continuous function $\hat{\sigma}_1(x_1, x_2)$. It has been shown in the proof of the preceding Lemma that there exists a neighbourhood W of $1 \in I$ such that F maps the set $\{x_0\} \times K \times W$ into U. We assume w.l.o.g. that W is compact. Since F is continuous and W' and the image K are compact, there exist open neighbourhoods U' of x_0 and V' of K such that F maps $U' \times V' \times W$ into U. Because σ_1 is continuous and Δ_1 is compact there exists a neighbourhood V'' of (x_1, x_2) such that $\sigma_1(V'' \times \Delta_1)$ is contained in V. The set $W' = \{(t_0, t_1, t_2) \in \Delta_2 \mid t_0 \in W\}$ then is a neighbourhood of $(1, 0, 0) \in \Delta_2$ and $U' \times V'' \times W'$ is a neighbourhood of $((x_0, x_1, x_2), (1, 0, 0))$. The image

$$\sigma_2(U' \times V'' \times W') = F(U \times \sigma_1(V'' \times \Delta_1) \times W') \subseteq F(U' \times V' \times W') \subseteq U$$

of the neighbourhood $U' \times V'' \times W'$ under σ_2 is contained in U. This shows that σ_2 is continuous. $\qquad\square$

Lemma 3.0.63. *The faces of a singular 2-simplex $\hat{\sigma}_2(x_0, x_1, x_3)$ are the singular 1-simplices associated to the corresponding faces of (x_0, x_1, x_2).*

Proof. Let (x_0, x_1, x_2) be a point in U_2. The zeroth face of $\hat{\sigma}_2(x_0, x_1, x_2)$ is given by

$$\hat{\sigma}_2(x_0, x_1, x_2)^0(t_0, t_1) = \hat{\sigma}_2(x_0, x_1, x_2)(0, t_0, t_1) = F(x_0, \hat{\sigma}_1(x_1, x_2)(t_0, t_1), 0)$$
$$= \hat{\sigma}_1(x_1, x_2)(t_0, t_1).$$

So the zeroth face of $\hat{\sigma}_2(x_0, x_1, x_2)$ is the singular simplex $\hat{\sigma}_1(x_1, x_2)$ associated to the zeroth face (x_1, x_2) of (x_1, x_2, x_3). The first face of $\hat{\sigma}_2(x_0, x_1, x_2)$ is given by

$$\hat{\sigma}_2(x_0, x_1, x_2)^1(t_0, t_1) = \hat{\sigma}_2(x_0, x_1, x_2)(t_0, 0, t_1) = F(x_0, \hat{\sigma}_1(x_1, x_2)(0, \frac{t_1}{1 - t_0}), t_0)$$
$$= F(x_0, \hat{\sigma}_1(x_1, x_2)(0, 1), t_0) = F(x_0, x_2, t_0)$$
$$= \hat{\sigma}_1(x_0, x_1)(t_0, t_1).$$

Thus the first face of $\hat{\sigma}_2(x_0, x_1, x_2)$ is the singular simplex $\hat{\sigma}_1(x_0, x_2)$ associated to the first face (x_0, x_2) of (x_1, x_2, x_3). The equality of the third face of $\hat{\sigma}_2(x_0, x_1, x_2)$ and $\hat{\sigma}_1(x_0, x_1)$ follows analogously. $\qquad\square$

Lemma 3.0.64. *The singular 2-simplex associated to a degeneracy of a point $(x_0, x_1) \in U_1$ is the corresponding degeneracy of the singular 1-simplex $\hat{\sigma}_1(x_0, x_1)$.*

Proof. Let $(x_0, x_1) \in U_1$ be a point in U_1. The zeroth degeneracy of (x_0, x_1) is the 2-simplex (x_0, x_0, x_1) in U_2. The zeroth degeneracy of the singular 1-simplex $\hat{\sigma}_1(x_0, x_1)$ is given by $(t_0, t_1, t_2) \mapsto \hat{\sigma}_1(x_0, x_1)(t_0 + t_1, t_2)$. The zeroth vertex of this singular simplex is the point x_0 which is also the zeroth vertex of the singular 2-simplex $\hat{\sigma}_2(x_0, x_0, x_1)$, so both singular simplices coincide at $\mathbf{e}_0 = (1, 0, 0) \in \Delta_2$. For $t_0 \neq 1$ the singular 2-simplex $\hat{\sigma}_2(x_0, x_0, x_1)$ is given by

$$
\begin{aligned}
\hat{\sigma}_2(x_0, x_0, x_1)(t_0, t_1, t_2) &= F\left(x_0, \hat{\sigma}_1(x_1, x_2)\left(\frac{t_1}{1-t_0}, \frac{t_2}{1-t_0}\right), t_0\right) \\
&= F\left(x_0, F\left(x_0, x_1, \frac{t_1}{1-t_0}\right), t_0\right) \\
&= F\left(x_0, x_1, t_0 + \frac{t_1}{1-t_0} - \frac{t_0 t_1}{1-t_0}\right) \\
&= F\left(x_0, x_1, \frac{t_0 - t_0^2 + t_1 - t_0 t_1}{1-t_0}\right) \\
&= F\left(x_0, x_1, \frac{(1-t_0)(t_0 + t_1)}{1-t_0}\right) \\
&= F(x_0, x_1, t_0 + t_1) = \hat{\sigma}_1(x_0, x_1)(t_0 + t_1, t_2),
\end{aligned}
$$

which shows that the singular simplex $\hat{\sigma}_2(x_0, x_0, x_1)$ coincides with the zeroth degeneracy of $\hat{\sigma}_1(x_0, x_1)$. The first degeneracy of $(x_0, x_1,)$ is the simplex (x_0, x_1, x_1) in U_2. The first degeneracy of the singular 1-simplex $\hat{\sigma}_1(x_0, x_1)$ is the continuous function $(t_0, t_1, t_2) \mapsto \hat{\sigma}_1(x_0, x_1)(t_0, t_1 + t_2)$. Both functions coincide at the point $\mathbf{e}_0 = (1, 0, 0) \in \Delta_2$. For $t_0 \neq 1$ the singular 2-simplex $\hat{\sigma}_2(x_0, x_1, x_1)$ is given by

$$
\begin{aligned}
\hat{\sigma}_2(x_0, x_1, x_1)(t_0, t_1, t_2) &= F\left(x_0, \hat{\sigma}_1(x_1, x_1)\left(\frac{t_1}{1-t_0}, \frac{t_2}{1-t_0}\right), t_0\right) \\
&= F(x_0, x_1, t_0) \qquad \text{because } \hat{\sigma}_1(x_1, x_1) \text{ is constant} \\
&= \hat{\sigma}_1(x_0, x_1)(t_0, t_1 + t_2)
\end{aligned}
$$

which shows that the singular simplex $\hat{\sigma}_2(x_0, x_1, x_1)$ coincides with the first degeneracy of $\hat{\sigma}_1(x_0, x_1)$.   □

Proceeding by induction one can now show the existence of G-invariant neighbourhoods U_n of the diagonal in $SS(X)([n])$ and continuous equivariant systems $\hat{\sigma}_n$ of n-simplices on these sets U_n with the following property:

Property 3.0.65.

1. The sets U_n are G-invariant and the maps $\hat{\sigma}_n$ are equivariant.
2. The neighbourhood U_n of the diagonal in $SS(X)([n])$ generates a simplicial subspace of $SS(X)$ whose first $n+1$ components are exactly the spaces $U_0, \ldots, U_n$.
3. In the case $n > 0$ the space U_n is the set

$$
U_n := \{\mathbf{x} \in X \times U_{n-1} \mid \{x_0\} \times \hat{\sigma}_{n-1}(x_1, \ldots, x_n)(\Delta_{n-1}) \subset U_1\}.
$$

4. The continuous systems of simplices $\hat{\sigma}_1, \ldots, \hat{\sigma}_n$ intertwine the face and degeneracy maps between the spaces $U_1, \ldots, U_n$.
5. In the case $n > 0$ the singular n-simplices $\hat{\sigma}_n(\mathbf{x})$, $n > 0$ are given by the formula

$$
\sigma_n(\mathbf{x})(\mathbf{t}) = \begin{cases} F\left(x_0, \hat{\sigma}_{n-1}(x_1, \ldots, x_n)\left(\left(\frac{1}{1-t_0}\right)(t_1, \ldots, t_n)\right), t_0\right) & \text{if } t_0 \neq 1 \\ x_0 & \text{if } t_0 = 1. \end{cases}
$$

For $n = 2$ the existence compatible continuous equivariant local systems $\hat{\sigma}_0, \hat{\sigma}_1, \hat{\sigma}_2$ defined on neighbourhoods U_0, U_1, U_2 of the diagonals in X, X^2 and X^3 respectively has already been shown. We now assume the neighbourhoods U_n of the diagonal in $\mathrm{SS}(X)([n])$ and continuous systems $\hat{\sigma}_n$ of n-simplices on U_n with Property 3.0.65 to be given. From this assumption we derive the existence of a neighbourhood U_{n+1} of the diagonal in $\mathrm{SS}(X)([n+1])$ and a continuous system $\hat{\sigma}_{n+1}$ of $(n+1)$-simplices on U_{n+1} satisfying Property 3.0.65. The procedure is similar to the construction of U_2 and $\hat{\sigma}_2$. Analogously to the definition of U_2 we define a subset of $\mathrm{SS}(X)([n])$ via

$$U_{n+1} := \{\mathbf{x} \in X \times U_n \mid \{x_0\} \times \hat{\sigma}_n(x_1, \ldots, x_n)(\Delta_n) \subset U_1\}.$$

Lemma 3.0.66. *The set U_{n+1} is an open G-invariant neighbourhood of the diagonal in $\mathrm{SS}(X)([n+1])$.*

Proof. Let $\mathbf{x}$ be a point in U_{n+1}. Because the inverse image $(\mathrm{pr}_0, \hat{\sigma}_n)^{-1}(U_1)$ of the G-invariant open set U_1 is open, every point $(\mathbf{x}, \mathbf{t})$ in $U_{n+1} \times \Delta_n$ has an open neighbourhood of the form $V_{\mathbf{t}} \times W_{\mathbf{t}}$ in $\mathrm{SS}(X)([n+1]) \times \Delta_n$. Since the standard simplex Δ_n is compact, it is covered by finitely many such open sets $W_{\mathbf{t}_1}, \ldots, W_{\mathbf{t}_n}$. The intersection $\bigcap_{i=1}^{n} V_{\mathbf{t}_i}$ then is an open neighbourhood of $\mathbf{x}$ which is contained in U_{n+1}. Since this is true for all points $\mathbf{x}$ in U_{n+1}, the set U_{n+1} is open. Moreover, if $\mathbf{x}$ is a point in U_{n+1} then the equivariance of $(\mathrm{pr}_0, \hat{\sigma}_n)$ implies the equalities

$$(\mathrm{pr}_0, \hat{\sigma}_n)(\{g.\mathbf{x}\} \times \Delta_n) = (\mathrm{pr}_0, \hat{\sigma}_n)(g.(\{\mathbf{x}\} \times \Delta_n))$$
$$= g.(\mathrm{pr}_0, \hat{\sigma}_n)(\{\mathbf{x}\} \times \Delta_n) \subset g.U_n \subset U_n$$

for all $g \in G$. This shows the G-invariance of U_{n+1}. All in all we have proved that U_{n+1} is a G-invariant open subset of $\mathrm{SS}(X)([n+1])$. $\qquad\square$

Lemma 3.0.67. *The neighbourhood U_n of the diagonal in $\mathrm{SS}(X)([n])$ contains the images of U_{n+1} under all face maps $\mathrm{SS}(X)(\epsilon_i) : \mathrm{SS}(X)([n+1]) \to \mathrm{SS}(X)([n])$.*

Proof. Let $\mathbf{x} \in U_{n+1}$ be a point in U_{n+1} and $\epsilon_i : [n] \to [n+1]$ be a face morphism. The neighbourhood U_{n+1} of the diagonal in $\mathrm{SS}(X)([n])$ is defined in such a way that neighbourhood U_1 contains the subset $\{x_0\} \times \hat{\sigma}_n(x_1, \ldots, x_{n+1})(\Delta_n)$. By the inductive assumption (Property 3.0.65) the faces of the singular n-simplex $\hat{\sigma}_n(x_1, \ldots, x_{n+1})$ are the singular $(n-1)$-simplices $\hat{\sigma}_{n-1}(x_1, \ldots, \hat{x}_i, \ldots, x_{n+1})$. Therefore the subset $\{x_0\} \times \hat{\sigma}_n(x_1, \ldots, x_{n+1})(\Delta_n)$ of U_1 especially contains all the subsets $\{x_0\} \times \hat{\sigma}_{n-1}(x_1, \ldots, \hat{x}_i, \ldots, x_{n+1})(\Delta_{n-1})$, $i > 0$. A look at the definition of U_n shows that the simplices $(x_0, \ldots, \hat{x}_i, \ldots, x_{n+1})$, $i > 0$ are all elements of U_n. The zeroth face $(x_1, \ldots, x_{n+1})$ of $\mathbf{x}$ is contained in U_n because the neighbourhood U_{n+1} of the diagonal in $\mathrm{SS}(X)([n])$ is a subset of $X \times U_n$ by definition. $\qquad\square$

As required for Property 3.0.65 we now define not necessarily continuous functions $\hat{\sigma}_{n+1}$ on U_{n+1} via

$$\sigma_{n+1}(\mathbf{x})(\mathbf{t}) = \begin{cases} F\left(x_0, \hat{\sigma}_n(x_1, \ldots, x_{n+1})\left(\left(\frac{1}{1-t_0}\right)(t_1, \ldots, t_{n+1})\right), t_0\right) & \text{if } t_0 \neq 1 \\ x_0 & \text{if } t_0 = 1. \end{cases}$$

Lemma 3.0.68. *The functions $\hat{\sigma}_{n+1}(\mathbf{x}) : \Delta_{n+1} \to X$ are continuous for all points $\mathbf{x} \in U_{n+1}$.*

Proof. Let $\mathbf{x}$ be a point in U_{n+1}. The function $\hat{\sigma}_{n+1}(\mathbf{x})$ is continuous on $(\Delta_{n+1} \setminus \{\mathbf{e}_0\})$ by definition, so it remains to prove the continuity at $\mathbf{e}_0 \in \Delta_{n+1}$. Let U be an open neighbourhood of x_0 and let $K = \hat{\sigma}_n(x_1, \ldots, x_{n+1})(\Delta_n)$ denote the compact image of Δ_n under the continuous function $\hat{\sigma}_n(x_1, \ldots, x_{n+1})$. Since F is continuous and maps a point $(x_0, k, 1) \in \{x_0\} \times K \times \{1\}$ to x_0 there exists a neighbourhood V_k of k

in X and a neighbourhood W_k of 1 in I such that $V_k \times W_k$ is mapped into U under F. Because K is compact it is covered by finitely many open sets $V_{k_1}, \ldots, V_{k_n}$. The intersection $W = \bigcap_{i=1}^{n} W_{k_i}$ is a neighbourhood of 1 in I such that F maps $\{x_0\} \times K \times W$ into U. The set $W' = \{\mathbf{t} \in \Delta_{n+1} \mid t_0 \in W\}$ then is a neighbourhood of $\mathbf{e}_0$ and the image

$$\hat{\sigma}_{n+1}(\mathbf{x})(W') \subseteq F(\{x_0\} \times \hat{\sigma}_n(x_1, \ldots, x_{n+1})(\Delta_n) \times W) \subseteq U$$

of W' under $\hat{\sigma}_{n+1}(\mathbf{x})$ is contained in U. This shows that $\hat{\sigma}_{n+1}(\mathbf{x})$ is a continuous function. $\qquad\square$

Lemma 3.0.69. *The function $\sigma_{n+1} : U_{n+1} \times \Delta_{n+1} \to X$ is continuous.*

Proof. The function σ_{n+1} is continuous on $U_{n+1} \times (\Delta_{n+1} \setminus \{\mathbf{e}_0\})$ by definition. Furthermore any point $(\mathbf{x}, \mathbf{e}_0) \in U_{n+1} \times \Delta_{n+1}$ is mapped onto x_0 under σ_{n+1}. It remains to show that for any open neighbourhood U of x_0 there exists an open neighbourhood V of $(\mathbf{x}, \mathbf{e}_0)$ which is mapped into U under σ_2. Let U be an open neighbourhood of x_0 and let $K = \hat{\sigma}_1(x_1, \ldots, x_{n+1})(\Delta_n)$ denote the compact image of Δ_n under the continuous function $\hat{\sigma}_n(x_1, \ldots, x_{n+1})$. It has been shown in the proof of the preceding Lemma that there exists a neighbourhood W of $1 \in I$ such that F maps the set $\{x_0\} \times K \times W$ into U. We assume w.l.o.g. that W is compact. Since F is continuous and W' and the image K are compact, there exist open neighbourhoods U' of x_0 and V' of K such that F maps $U' \times V' \times W$ into U. Because σ_n is continuous by the inductive assumption and Δ_n is compact there exists a neighbourhood V'' of $(x_1, \ldots, x_{n+1})$ such that $\sigma_n(V'' \times \Delta_n)$ is contained in V. The set $W' = \{\mathbf{t} \in \Delta_2 \mid t_0 \in W\}$ then is a neighbourhood of $\mathbf{e}_0 \in \Delta_2$ and $U' \times V'' \times W'$ is a neighbourhood of $(\mathbf{x}, \mathbf{e}_0)$. The image

$$\sigma_2(U' \times V'' \times W') = F(U \times \sigma_1(V'' \times \Delta_1) \times W') \subseteq F(U' \times V' \times W') \subseteq U$$

of the neighbourhood $U' \times V'' \times W'$ under σ_{n+2} is contained in U. This shows that σ_{n+1} is continuous. $\qquad\square$

Lemma 3.0.70. *The faces of a singular $(n+1)$-simplex $\hat{\sigma}_{n+1}(\mathbf{x})$ are the singular n-simplices associated to the corresponding faces of $\mathbf{x}$.*

Proof. Let $\mathbf{x}$ be a point in U_{n+1}. The zeroth face of the singular $(n+1)$-simplex $\hat{\sigma}_{n+1}(\mathbf{x})$ is given by

$$\hat{\sigma}_{n+1}(\mathbf{x})^0(\mathbf{t}) = \hat{\sigma}_{n+1}(\mathbf{x})(0, t_0, \ldots, t_n) = F(x_0, \hat{\sigma}_n(x_1, \ldots, x_{n+1})(\mathbf{t}), 0)$$
$$= \hat{\sigma}_n(x_1, \ldots, x_{n+1})(\mathbf{t}).$$

So the zeroth face of $\hat{\sigma}_{n+1}(\mathbf{x})$ is the singular simplex $\hat{\sigma}_n(x_1, \ldots, x_{n+1})$ associated to the zeroth face $(x_1, \ldots, x_{n+1})$ of $\mathbf{x}$. For $i > 0$ the i-th face of $\hat{\sigma}_{n+1}(\mathbf{x})$ is given by

$$\hat{\sigma}_{n+1}(\mathbf{x})^i(\mathbf{t}) = \hat{\sigma}_{n+1}(\mathbf{x})(t_0, \ldots, t_{i-1}, 0, t_i, \ldots t_n)$$
$$= F\left(x_0, \hat{\sigma}_n(x_1, \ldots, x_{n+1})\left(\frac{t_1}{1-t_0}, \ldots, \frac{t_{i-1}}{1-t_0}, 0, \frac{t_i}{1-t_0}, \ldots, \frac{t_n}{1-t_0}\right), t_0\right)$$
$$= F\left(x_0, \hat{\sigma}_n(x_1, \ldots, x_{n+1})^{i-1}\left(\frac{t_1}{1-t_0}, \ldots, \frac{t_n}{1-t_0}\right), t_0\right)$$
$$= F\left(x_0, \hat{\sigma}_{n-1}(x_1, \ldots, \hat{x}_i, \ldots, x_{n+1})\left(\frac{t_1}{1-t_0}, \ldots, \frac{t_n}{1-t_0}\right), t_0\right)$$
$$= \hat{\sigma}_n(x_0, \ldots, \hat{x}_i, \ldots, x_{n+1})(\mathbf{t}).$$

Therefore the i-th face of $\hat{\sigma}_{n+1}(\mathbf{x})$ is the singular n-simplex $\hat{\sigma}_n(x_0, \ldots, \hat{x}_i, \ldots, x_{n+1})$ associated to the i-th face $(x_0, \ldots, \hat{x}_i, \ldots, x_{n+1})$ of $\mathbf{x}$. $\qquad\square$

Lemma 3.0.71. *The singular $(n+1)$-simplex associated to a degeneracy of a point* $\mathbf{x} \in U_n$ *is the corresponding degeneracy of the singular n-simplex $\hat{\sigma}_n(\mathbf{x})$.*

Proof. Let $\mathbf{x} \in U_n$ be a point in U_n. The zeroth degeneracy of $\mathbf{x}$ is the $(n+1)$-simplex $(x_0, x_0, \ldots, x_n)$ in U_{n+1}. The zeroth degeneracy of the singular n-simplex $\hat{\sigma}_n(\mathbf{x})$ is given by $\mathbf{t} \mapsto \hat{\sigma}_1(\mathbf{x})(t_0 + t_1, t_2, \ldots, t_{n+1})$. The zeroth vertex of this singular simplex is the point x_0 which is also the zeroth vertex of the singular $(n+1)$-simplex $\hat{\sigma}_{n+1}(\mathbf{x})$, so both singular simplices coincide at $\mathbf{e}_0 \in \Delta_{n+1}$. For $t_0 \neq 1$ the singular $(n+1)$-simplex $\hat{\sigma}_{n+1}(x_0, x_0, \ldots, x_n)$ is given by

$$\hat{\sigma}_{n+1}(x_0, x_0, \ldots, x_n)(\mathbf{t}) = F\left(x_0, \hat{\sigma}_n(x_1, \ldots, x_{n+1})\left(\frac{t_1}{1-t_0}, \ldots, \frac{t_{n+1}}{1-t_0}\right), t_0\right)$$

$$= F\left(x_0, F\left(x_0, \hat{\sigma}_{n-1}(x_1, \ldots, x_n)\left(\frac{t_2}{1-(t_0+t_1)}, \ldots, \frac{t_{n+1}}{1-(t_0+t_1)}\right), \frac{t_1}{1-t_0}\right), t_0\right)$$

$$= F\left(x_0, \hat{\sigma}_{n-1}(x_1, \ldots, x_n)\left(\frac{t_2}{1-(t_0+t_1)}, \ldots, \frac{t_{n+1}}{1-(t_0+t_1)}\right), t_0 + \frac{t_1}{1-t_0} - \frac{t_0 t_1}{1-t_0}\right)$$

$$= F\left(x_0, \hat{\sigma}_{n-1}(x_1, \ldots, x_n)\left(\frac{t_2}{1-(t_0+t_1)}, \ldots, \frac{t_{n+1}}{1-(t_0+t_1)}\right), \frac{t_0 - t_0^2 + t_1 - t_0 t_1}{1-t_0}\right)$$

$$= F\left(x_0, \hat{\sigma}_{n-1}(x_1, \ldots, x_n)\left(\frac{t_2}{1-(t_0+t_1)}, \ldots, \frac{t_{n+1}}{1-(t_0+t_1)}\right), \frac{(1-t_0)(t_0+t_1)}{1-t_0}\right)$$

$$= F\left(x_0, \hat{\sigma}_{n-1}(x_1, \ldots, x_n)\left(\frac{t_2}{1-(t_0+t_1)}, \ldots, \frac{t_{n+1}}{1-(t_0+t_1)}\right), t_0 + t_1\right)$$

$$= \hat{\sigma}_n(\mathbf{x})(t_0 + t_1, \ldots, t_n),$$

which shows that the singular simplex $\hat{\sigma}_{n+1}(x_0, x_0, \ldots, x_n)$ coincides with the zeroth degeneracy of $\hat{\sigma}_n(\mathbf{x})$. The i-th degeneracy of $\mathbf{x}$ is the simplex $(x_0, \ldots, x_i, x_i, \ldots, x_n)$ in U_{n+1}. The i-th degeneracy of the singular 1-simplex $\hat{\sigma}_1(\mathbf{x})$ is the continuous function $\mathbf{t} \mapsto \hat{\sigma}_n(\mathbf{x})(t_0, \ldots, t_i + t_{i+1}, \ldots, t_{n+1})$. Both functions coincide at the point $\mathbf{e}_0$ in Δ_{n+1}. For $t_0 \neq 1$ and $i > 0$ the singular $(n+1)$-simplex $\hat{\sigma}_{n+1}(x_0, \ldots, x_i, x_i, \ldots, x_n)$ is given by

$$\hat{\sigma}_{n+1}(x_0, \ldots, x_i, x_i, \ldots, x_n)(\mathbf{t}) =$$

$$F\left(x_0, \hat{\sigma}_n(x_1, \ldots, x_i, x_i, \ldots, x_n)\left(\frac{t_1}{1-t_0}, \ldots, \frac{t_{n+1}}{1-t_0}\right), t_0\right)$$

$$= F\left(x_0, \hat{\sigma}_{n-1}(x_1, \ldots, x_n)\left(\frac{t_1}{1-t_0}, \ldots, \frac{t_i + t_{i+1}}{1-t_0}, \ldots, \frac{t_{n+1}}{1-t_0}\right), t_0\right)$$

$$= \hat{\sigma}_n(\mathbf{x})(t_0, \ldots, t_{i-1}, t_i + t_{i+1}, \ldots, t_{n+1})$$

which shows that the singular simplex $\hat{\sigma}_2(x_0, \ldots, x_i, x_i, \ldots, x_n)$ coincides with the i-th degeneracy of $\hat{\sigma}_n(\mathbf{x})$. $\qquad\square$

This completes the inductive step. The equivariant continuous systems $\hat{\sigma}_n$ of simplices on the neighbourhoods U_n of the diagonals in $\mathrm{SS}(X)([n])$ assemble to a morphism $\sigma : U \to C(\Delta, X)$ of simplicial spaces by construction. This motivates the following definition:

Definition 3.0.72. *An* equivariant continuous system of simplices *on a simplicial subspace U of $\mathrm{SS}(X)$ is an equivariant morphism $\hat{\sigma} : U \to C(\Delta, X)$ of simplicial spaces.*

All in all we can summarise the above constructions in the following observation:

Theorem 3.0.73. *For every SEULC space X there exists a continuous equivariant system of simplices defined on a semisimplicial subspace U of $\mathrm{SS}(X)$ consisting of neighbourhoods of the diagonals.*

If one considers the action of the trivial group $G = \{1\}$ on, then every SULC G-space X is SEULC. We note the consequence:

Corollary 3.0.74. *For every SULC space X there exists a continuous system of simplices defined on a semisimplicial subspace U of $\mathrm{SS}(X)$ consisting of neighbourhoods of the diagonals.*

Let V be an Abelian group of coefficients and recall (cf. Theorem 3.0.25) that the existence of an equivariant system $\hat{\sigma}_n$ of singular n-simplices implies the injectivity of the natural homomorphism $H^n(\mathrm{colim}_{\mathfrak{U}} A^*(X, \mathfrak{U}; V)) \to H^n(X; V)$ into the singular cohomology $H(X; V)$, where $\mathfrak{U}$ ranges over all G-invariant open coverings of X. So we observe:

Corollary 3.0.75. *If the topological space X is a SEULC space then the natural homomorphism $H(\mathrm{colim}_{\mathfrak{U}} A^*(X, \mathfrak{U}; V)) \to H(X; V)$ is injective.*

Corollary 3.0.76. *For every locally SULC topological group G acting on itself by left translation the natural homomorphism $H(\mathrm{colim}_{\mathfrak{U}} A^*(G, \mathfrak{U}; V)) \to H(G; V)$ is injective.*

Proof. By Lemma 3.0.50 Locally SULC topological groups are SEULC w.r.t. the action by left translation Thus for every locally SULC topological group G the natural homomorphism $H(\mathrm{colim}_{\mathfrak{U}} A^*(G, \mathfrak{U}; V)) \to H(G; V)$ is injective. $\square$

Example 3.0.77. Real and complex Lie groups are SEULC w.r.t. the action by left translation. Thus for every real or complex Lie group G the natural homomorphism $H(\mathrm{colim}_{\mathfrak{U}} A^*(G, \mathfrak{U}; V)) \to H(G; V)$ is injective.

Similarly, if there exists a local system $\hat{\sigma}_n$ of singular n-simplices on X, then the natural homomorphism $H^n_{AS}(X; V) \to H^n(X; V)$ is injective.

Corollary 3.0.78. *If the topological space X is a SULC space then the natural homomorphism $H_{AS}(X, \mathfrak{U}; V) \to H(X; V)$ is injective.*

Example 3.0.79. Locally SULC topological groups are SEULC w.r.t. the action by left translation. Thus for every locally SULC topological group G the natural homomorphism $H_{AS}(G, \mathfrak{U}; V) \to H(G; V)$ is injective.

Example 3.0.80. Real and complex Lie groups are SEULC w.r.t. the action by left translation. Thus for every real or complex Lie group G the natural homomorphism $H_{AS}(G, \mathfrak{U}; V)) \to H(G; V)$ is injective.

Sometimes a continuous local system $\hat{\sigma}_n : U_n \to C(\Delta_x, X)$ of singular n-simplices may not be extendable to a global continuous system of singular n-simplices, but the underlying map of sets can be extended. This motivates the next definition.

Definition 3.0.81. *A system $\hat{\sigma}_n : UV'_n \to \hom_{\mathbf{Top}}(\Delta_n, X)$ of singular n-simplices defined for a neighbourhood V_n of the diagonal in X^{n+1} is called* locally continuous, *if there exists a smaller neighbourhood $V_n \subset V'_n$ of the diagonal such that the map $\hat{\sigma}_n : V_n \to C(\Delta_n, X)$ is continuous.*

Theorem 3.0.82. *If the topological space X is $(n-1)$-connected and the action of G on X is free, then the underlying maps of any collection $\hat{\sigma}_k : V_k \to C(\Delta_k, X)$, $k \leq n$ of compatible continuous equivariant local systems of singular k-simplices can be extended to a collection $\hat{\sigma}'_k : UX^{k+1} \to \hom_{\mathbf{Top}}(\Delta_k, X)$, $k \leq n$ of compatible equivariant local systems of singular k-simplices.*

Proof. The construction is analogous to that of systems of piecewise smooth singular simplices presented in [Fuc03, Lemma 5.2.1]. $\square$

Corollary 3.0.83. *If the topological space X is $(n-1)$-connected and SEULC and the action of G on the space X is free, then there exist compatible equivariant systems $\hat{\sigma}_k : UX^{k+1} \to \hom_{\mathbf{Top}}(\Delta_k, X)$, $k \leq n$, which are locally continuous.*

Proof. This follows directly from Theorems 3.0.73 and 3.0.82.   □

A special case are locally contractible topological groups:

Theorem 3.0.84. *If the topological group G is locally contractible, then it admits a compatible local equivariant system $\hat{\sigma}$ of singular simplices.*

Proof. Let G be a locally contractible group and $\phi : U \to U$ be a contraction of an identity neighbourhood U rel $\{1\}$. The continuous system $\hat{\sigma}_1$ of singular 1-simplices is given by

$$\hat{\sigma}_1 : G.(\{1\} \times U) \to G, \quad \hat{\sigma}_1(g_0, g_1)(t_0, t_1) = g_0.\Phi(g_0^{-1}g_1)(t_0).$$

The systems $\hat{\sigma}_k$ of higher dimension can be inductively defined by the formula

$$\hat{\sigma}_{k+1}(g_0, \ldots, g_{k+1})(t_0, \ldots, t_{k+1}) =$$
$$g_0.\Phi\left(g_0^{-1}\hat{\sigma}_k(g_1, \ldots, g_{k+1})\left((\tfrac{1}{1-t_0})(t_1, \ldots, t_{k+1})\right), t_0\right)$$

cf. the previous inductive construction of systems of simplices in this section.   □

Corollary 3.0.85. *If a topological group G is locally contractible, then the natural homomorphism $H(\colim_{\mathfrak{U}} A^*(G, \mathfrak{U}; V)) \to H(G; V)$, where $\mathfrak{U}$ ranges over all left-invariant open covers of G, is injective.*

This generalises part of a result of van Est (cf. [vE02b, §11]), from connected locally contractible topological groups to all locally contractible topological groups.

Theorem 3.0.86. *If the topological group G is locally contractible and $(n-1)$-connected, then there exist compatible equivariant systems $\hat{\sigma}_0, \ldots, \hat{\sigma}_n$ of singular simplices defined on $X, \ldots, X^{n+1}$.*

Proof. This follows from the preceding theorem and Theorem 3.0.82.   □

If the space X is a smooth manifold M, then one can restrict oneself to smooth singular simplices only. We equip the simplicial space $C^\infty(\Delta, M)$ of smooth singular simplices in M with the C^∞-compact open topology and adapt our definitions:

Definition 3.0.87. *A smooth equivariant map $\hat{\sigma}_n : M^{n+1} \to C^\infty(\Delta_n, M)$ which is a right inverse to the vertex map (, i.e the vertices of each singular n-simplex $\hat{\sigma}_n(m_0, \ldots, m_n)$ are exactly the points $m_0, \ldots, m_n$) is called a* smooth system of *singular n-simplices.*

Example 3.0.88. The space $C^\infty(\Delta_0, X)$ of smooth singular 0-simplices in a manifold M is naturally isomorphic to the manifold M itself. Therefore the unique equivariant system $\hat{\sigma}_0$ of singular 0-simplices on M is smooth.

Definition 3.0.89. *A smooth equivariant map $\hat{\sigma}_n : U_n \to C^\infty(\Delta_n, X)$ defined on an open neighbourhood U_n of the diagonal in M^{n+1} which is a right inverse to the vertex map (, i.e the vertices of each singular n-simplex $\hat{\sigma}_n(m_0, \ldots, m_n)$ are exactly the points $m_0, \ldots, m_n$) is called a* smooth local system of *singular n-simplices. If U_n is G-invariant and $\hat{\sigma}_n$ is equivariant, then the smooth system of singular n-simplices is called* equivariant.

Remark 3.0.90. Since all standard simplices Δ_n are compact, the smoothness of a system of n-simplices $\hat{\sigma}_n : U_n \to C^\infty(\Delta_n, X)$ is equivalent to the smoothness of the adjoint map $\sigma_n : U_n \times \Delta_n \to X$.

Definition 3.0.91. *An* smooth equivariant local system of simplices *on an open simplicial subspace U of $\mathrm{SS}(X)$ is an equivariant morphism $\hat{\sigma} : U \to C^\infty(\Delta, M)$ of simplicial manifolds.*

If M is a smooth SEULC manifold and the strong uniform local contraction F is smooth, then the inductive construction of continuous local systems of singular simplices presented before can be adapted (where one has to use smoothing functions when defining $\hat{\sigma}_{n+1}$). In this case one proceed analogously and obtains a smooth local systems of singular simplices. So we note:

Theorem 3.0.92. *Every smooth manifold M with smooth strong uniform local contraction admits a smooth local system of singular simplices.*

A slightly different construction for Lie groups has been made in [Fuc03]. There the construction of the systems of simplices explicitly used the addition and multiplication in an identity neighbourhood of the Lie group. This construction has the advantage, that it is 'integrating' in the following sense:

Definition 3.0.93. *A smooth system $\hat{\sigma}_n : U_n \to C^\infty(\Delta_n, M)$ of singular n-simplices defined on an open neighbourhood U_n of the diagonal in M^{n+1} is called* integrating, *if it satisfies the identity*

$$\tau^n \left(\mathbf{m} \mapsto \int_{\sigma_n(\mathbf{m})} \omega \right) = \omega$$

for all differential n-forms omega on M, where $\tau^* : A_s^*(M, \mathfrak{U}; V) \to \Omega^*(M; V)$ is the derivation homomorphism.

Lemma 3.0.94. *Every Lie group admits integrating smooth equivariant systems of simplices in all dimensions.*

Proof. This has been shown in [Fuc03, Satz 4.3.12]. □

In fact, the inductive definition of the neighbourhoods U_n presented in this section and combined with the inductive definition of the smooth singular n-simplices in [Fuc03, Satz 4.3.12] yields the following result:

Theorem 3.0.95. *Every Lie group admits an integrating smooth equivariant local system of singular simplices.*

4

Relative Subdivision

In this chapter we establish a way to subdivide singular singular simplices in a topological space X similar to the barycentric subdivision, but in such a way that faces of a singular simplex are not subdivided if they are "small enough". Here "small enough" means that they are $\mathfrak{U}$-small for a given open covering $\mathfrak{U}$ of X. Furthermore the special subdivision to be constructed has to intertwine the face and degeneracy maps of singular simplices. The main motivation is the extension of local equivariant cocycles to to global equivariant cocycles, for which this kind of subdivision is needed (see the procedure in Section 6.2 and Chapter 15). A very convenient special case thereof is the following observation: Suppose $f \in A_{eq}(X, \mathfrak{U}; V)$ is a local equivariant n-cocycle defined on the open neighbourhood $SS(X, \mathfrak{U})([n])$ of $SS(X)([n]) \cong X^{n+1}$ and $\hat{\sigma}_n$ is an equivariant system of simplices on $SS(X, \mathfrak{U})([n])$. Then the function f extends to a group homomorphism $f : \mathbb{Z}^{(SS(X, \mathfrak{U})([n]))} \to V$ and the function

$$F : SS(X)([n]) \cong X^{n+1} \to V, \quad F(\mathbf{x}) = f(\mathrm{Csd}_{\mathfrak{U}}(\hat{\sigma}_n(\mathbf{x}))),$$

where $\mathrm{Csd}_{\mathfrak{U}}$ denotes our special type of complete subdivision, is a global equivariant cocycle on X with the same germ as f at the diagonal (cf. Theorem 6.2.17). We start with the subdivision of affine simplices in the standard simplex Δ_n.

4.1 Affine Simplices

Let $L_p(\Delta_q)$ denote the free abelian group on the set of all affine p-simplices in Δ_q. In analogy to the diameter of affine simplices and the mesh of chains in $L_p(\Delta_q)$ we define a $\mathfrak{U}$-diameter of affine simplices in Δ_q and a $\mathfrak{U}$-mesh of chains in $L_p(\Delta_q)$ for every family $\mathfrak{U}$ of subsets of Δ_q.

Definition 4.1.1. *Let σ be a p-simplex in the standard q-simplex Δ_q and $\mathfrak{U}$ a family of subsets of Δ_q. The $\mathfrak{U}$-diameter $\mathrm{diam}_{\mathfrak{U}}(\sigma)$ of σ is*

$$\mathrm{diam}_{\mathfrak{U}}(\sigma) := \begin{cases} 0 & \text{if } \sigma \text{ is } \mathfrak{U}\text{-small} \\ \mathrm{diam}(\sigma) & \text{if } \sigma \text{ is not } \mathfrak{U}\text{-small} \end{cases}$$

Every chain $c \in L_p(\Delta_q)$ of affine p-simplices in Δ_q can uniquely be expressed as $\mathbb{Z}$-linear combination $c = \sum z_\sigma \sigma$ of affine p-simplices σ in Δ_q. We denote the coefficient z_σ of an affine p-simplex σ in this representation by $c(\sigma)$.

Definition 4.1.2. *Let $\mathfrak{U}$ be a family of subsets of Δ_q. The $\mathfrak{U}$-mesh of a chain c of simplices in Δ_q is the maximum of the $\mathfrak{U}$-diameters of the simplices in c:*

$$\mathrm{mesh}_{\mathfrak{U}}(c) := \max\{\mathrm{diam}_{\mathfrak{U}}(\sigma) \mid c(\sigma) \neq 0\}$$

Using the notion of the $\mathfrak{U}$-diameter we now define a generalisation of the linear subdivision of affine simplices, with the property that an affine p-simplex σ in Δ_q is subdivided if its $\mathfrak{U}$-diameter $\text{diam}_{\mathfrak{U}}(\sigma)$ is non-zero and not subdivided otherwise. We recall that the i-th boundary of an affine simplex σ is denoted by σ^i.

Definition 4.1.3. *Let $\mathfrak{U}$ a family of subsets of Δ_q. The $\mathfrak{U}$-subdivision $\text{LSd}_{\mathfrak{U}}$ is the unique group endomorphism on $L_1(\Delta_q)$ that is given by*

$$\text{LSd}_{\mathfrak{U}}(\sigma) := \begin{cases} \sigma & \text{if } \sigma \text{ is } \mathfrak{U}\text{-small} \\ \mathbf{b}_\sigma \sigma^0 - \mathbf{b}_\sigma \sigma^1 & \text{if } \sigma \text{ is not } \mathfrak{U}\text{-small} \end{cases}$$

for any affine 1-simplex σ. The $\mathfrak{U}$-subdivision of affine simplices of higher dimension is defined by induction:

$$\text{LSd}_{\mathfrak{U}}(\sigma) := \begin{cases} \sigma & \text{if } \sigma \text{ is } \mathfrak{U}\text{-small} \\ \mathbf{b}_\sigma \text{LSd}_{\mathfrak{U}}(\partial \sigma) & \text{if } \sigma \text{ is not } \mathfrak{U}\text{-small} \end{cases}$$

for any affine $(p+1)$-simplex and $\text{LSd}_{\mathfrak{U}} : L_p(\Delta_q) \to L_p(\Delta_q)$ already defined. By linear extension to $(p+1)$-chains this defines an endomorphism of the abelian group $L_{n+1}(\Delta_q)$, also denoted by $\text{LSd}_{\mathfrak{U}}$. The relative subdivision of 0-simplices is the identity.

If the family $\mathfrak{U}$ is taken to be the family of all singletons in Δ_q, then the $\mathfrak{U}$-subdivision coincides with the barycentric subdivision. And, similar to the barycentric subdivision, the relative subdivision induces a chain map on the complex $L_*(\Delta_q)$:

Lemma 4.1.4. *For any family $\mathfrak{U}$ of subsets of Δ_q the $\mathfrak{U}$-relative subdivision is a chain map $L_*(\Delta_q) \to L_*(\Delta_q)$.*

Proof. We shall show $\text{LSd}_{\mathfrak{U}}(\partial\sigma) = \partial(\text{LSd}_{\mathfrak{U}}(\sigma))$ by induction on p where σ is an affine p-simplex. First observe that a $\mathfrak{U}$-small p-simplex has $\mathfrak{U}$-small faces. So for any $\mathfrak{U}$-small p-simplex σ we have $\text{LSd}_{\mathfrak{U}}(\partial\sigma) = \partial\sigma = \partial(\text{LSd}_{\mathfrak{U}}(\sigma))$. Therefore the formula remains to be proved only for simplices that are not $\mathfrak{U}$-small. If $p = 0$ then both sides vanish because the relative subdivision is the identity on 0-chains and $\partial(L_0(\Delta_q)) = 0$. If $p = 1$ and the simplex σ is not $\mathfrak{U}$-small, then $\text{LSd}_{\mathfrak{U}}(\partial\sigma) = \partial\sigma$ while

$$\partial(\text{LSd}_{\mathfrak{U}}(\sigma)) = \partial(\text{LSd}(\sigma)) = \text{LSd}(\partial\sigma) = \text{LSd}_{\mathfrak{U}}(\partial\sigma).$$

For $p > 1$ and assuming the formula to be true for chains of degree $< p$ we have

$$\text{LSd}_{\mathfrak{U}}(\partial\sigma) = \text{LSd}_{\mathfrak{U}}(\partial\sigma) - \mathbf{b}_\sigma(\partial\text{LSd}_{\mathfrak{U}}\partial\sigma) = \partial(\mathbf{b}_\sigma(\text{LSd}_{\mathfrak{U}}(\partial\sigma))) = \partial(\text{LSd}_{\mathfrak{U}}(\sigma)),$$

since $\partial\text{LSd}_{\mathfrak{U}}\partial\sigma = \text{LSd}_{\mathfrak{U}}\partial\partial\sigma = 0$ by the inductive assumption. $\square$

Lemma 4.1.5. *For any affine simplex σ and families $\mathfrak{U}$ and $\mathfrak{U}'$ in Δ_q such that $\mathfrak{U}_{|\sigma(\Delta_p)} = \mathfrak{U}'_{|\sigma(\Delta_p)}$, the $\mathfrak{U}$-relative and $\mathfrak{U}'$-relative subdivision of σ coincide.*

Proof. This is proved by induction on p, where σ is an affine p-simplex in Δ_q. Let $\mathfrak{U}$ and $\mathfrak{U}'$ be two families of subsets of Δ_q such that $\mathfrak{U}_{|\sigma(\Delta_p)} = \mathfrak{U}'_{|\sigma(\Delta_p)}$. Since both relative subdivisions are the identity on $L_0(\Delta_q)$ the two relative subdivisions coincide on all 0-simplices. Let σ be an affine p-simplex $(0 < p)$. Either the simplex σ is $\mathfrak{U}$- and $\mathfrak{U}'$-small, then $\text{LSd}_{\mathfrak{U}}(\sigma) = \sigma = \text{LSd}'_{\mathfrak{U}}(\sigma)$, or σ is not $\mathfrak{U}$-small. In this case, assuming the hypothesis to be true for all $p' < p$, we get

$$\text{LSd}_{\mathfrak{U}}(\sigma) - \mathbf{b}_v \text{LSd}_{\mathfrak{U}}(\partial\sigma)$$
$$= \mathbf{b}_\sigma \text{LSd}_{\mathfrak{U}'}(\partial\sigma)$$
$$= \text{LSd}_{\mathfrak{U}'}(\sigma),$$

since for any face σ^i of σ the families $\mathfrak{U}_{|\sigma^i(\Delta_{p-1})}$ and $\mathfrak{U}'_{|\sigma^i(\Delta_{p-1})}$ coincide as well. $\square$

Proposition 4.1.6. *For any affine map $f : \mathbb{R}^q \to \mathbb{R}^{q'}$ and any affine p-simplex σ in $\mathbb{R}^q$ the barycenter $\mathbf{b}_{f\sigma}$ of $f\sigma$ is $f(\mathbf{b}_\sigma)$.*

Proof. Let $f : \mathbb{R}^q \to \mathbb{R}^{q'}$ be an affine map and $\sigma = (v_0, \ldots, v_p)$ be an affine p-simplex in Δ_q. This simplex is mapped to the p-simplex $f\sigma = (f(v_0), \ldots, f(v_p))$ under f. Since the map $f - f(0)$ is a linear function we get

$$\mathbf{b}_{f\sigma} = \frac{1}{p+1} \sum_{i=0}^{p} f(v_i)$$

$$= \frac{1}{p+1} \sum_{i=0}^{p} [(f - f(0))(v_i) + f(0)]$$

$$= \left(\frac{1}{p+1} \sum_{i=0}^{p} (f - f(0))(v_i) \right) + f(0)$$

$$= (f - f(0)) \left(\frac{1}{p+1} \sum_{i=0}^{p} (v_i) \right) + f(0)$$

$$= (f - f(0))(\mathbf{b}_\sigma) + f(0) = f(\mathbf{b}_\sigma).$$

Thus the barycenters of simplices map to the barycenters of the images under affine mappings. □

Corollary 4.1.7. *For any affine q-simplex $f : \Delta_q \to \Delta_{q'}$ and any affine p-simplex σ in Δ_q the barycenter of $f\sigma$ is $f(\mathbf{b}_\sigma)$.*

Proposition 4.1.8. *For any affine q-simplex $f : \Delta_q \to \Delta_{q'}$ and family $\mathfrak{U}$ in $\Delta_{q'}$, the following square commutes:*

$$\begin{array}{ccc} L_p(\Delta_q) & \xrightarrow{\ L_p(f)\ } & L_p(\Delta_{q'}) \\ {\scriptstyle \mathrm{LSd}_{f^{-1}(\mathfrak{U})}} \downarrow & & \downarrow {\scriptstyle \mathrm{LSd}_{\mathfrak{U}}} \\ L_p(\Delta_q) & \xrightarrow{\ L_p(f)\ } & L_p(\Delta_{q'}) \end{array} \qquad (4.1)$$

Proof. We prove $L_p(f)\mathrm{LSd}_{f^{-1}(\mathfrak{U})}(\sigma) = \mathrm{LSd}_{\mathfrak{U}} L_p(f)(\sigma)$ by induction on p. For $p = 0$ all relative subdivisions are the identity, so the equality holds. Assume the hypothesis true for all $p' < p$ and let σ be an affine p-simplex in Δ_q. Either the simplex $f\sigma$ is $\mathfrak{U}$-small and σ is $f^{-1}(\mathfrak{U})$-small, then $\mathrm{LSd}_{\mathfrak{U}} L_p(f)(\sigma) = f\sigma = L_p(f)\mathrm{LSd}_{f^{-1}(\mathfrak{U})}(\sigma)$, or $f\sigma$ is not $\mathfrak{U}$-small and σ is not $f^{-1}(\mathfrak{U})$-small. Using the inductive definition of the relative subdivision in this case we get

$$L_p(f)\mathrm{LSd}_{f^{-1}(\mathfrak{U})} = L_p(f) \left(\mathbf{b}_\sigma \mathrm{LSd}_{f^{-1}(\mathfrak{U})}(\partial\sigma) \right)$$

$$= \mathbf{b}_{f\sigma} \left(L_{p-1}(f)\mathrm{LSd}_{f^{-1}(\mathfrak{U})}(\partial\sigma) \right)$$

$$= \mathbf{b}_{f\sigma} \left(\mathrm{LSd}_{\mathfrak{U}}(L_{p-1}\partial\sigma) \right) \quad \text{by the inductive assumption}$$

$$= \mathbf{b}_{f\sigma} \left(\mathrm{LSd}_{\mathfrak{U}} \partial L_p(f)\sigma \right) \quad \text{since } L_p \text{ (f) is a chain map}$$

$$= \mathrm{LSd}_{\mathfrak{U}}(f\sigma) = \mathrm{LSd}_{\mathfrak{U}} \circ L_p(f)(\sigma)$$

for any affine simplex σ in Δ_q. Thus the square 4.1 commutes. □

An immediate consequence can be obtained by setting $f = \mathrm{id}^i_{\Delta_q}$ and applying proposition 4.1.8 to the identity in Δ_q.

Corollary 4.1.9. *For any standard simplex Δ_p and family $\mathfrak{U}$ in Δ_p, the subdivision of the i-th face of id_{Δ_p} is equal to $\mathrm{LSd}_{(\mathrm{id}^i_{\Delta_p})^{-1}(\mathfrak{U})}(\mathrm{id}_{\Delta_{p-1}})$.*

Proof. Let $f = \mathrm{id}^i_{\Delta_{q'}}$ be the i-th face of the identity of $\Delta_{q'}$ (so $q = q' - 1$) and $\mathfrak{U}$ a family in $\Delta_{q'}$. Application of the operator equality of proposition 4.1.8 to the identity of $\mathrm{id}_{\Delta_{q'}}$ yields

$$L_p(\mathrm{id}_{\Delta^i_{q'}}) \circ \mathrm{LSd}_{(\mathrm{id}^i_{\Delta_{q'}})^{-1}(\mathfrak{U})}(\mathrm{id}_{\Delta_{q'-1}}) = \mathrm{LSd}_{\mathfrak{U}} \circ L_p(\mathrm{id}_{\Delta^i_{q'}})(\mathrm{id}_{\Delta_{q'-1}}).$$

Since the chain on the right side is just the $\mathfrak{U}$-relative subdivision of the i-th face $\mathrm{id}^i_{\Delta_{q'}}$ of $\mathrm{id}_{\Delta_{q'}}$, setting $p = q' - 1$ leads to the desired equality. $\square$

Theorem 4.1.10. *Any relative subdivision $\mathrm{LSd}_{\mathfrak{U}}$ on $L_*(\Delta_q)$ is chain homotopic to the identity.*

Proof. Since the chain $L_*(\Delta_q)$ is a free resolution of the integers $\mathbb{Z}$, all chain endomorphisms of $L_*(\Delta_q)$ that are the identity on $\mathbb{Z}$ (and $L_0(\Delta_q)$) have to be homotopy equivalent. Thus any relative subdivision $\mathrm{LSd}_{\mathfrak{U}}$ is chain homotopic to the identity. $\square$

Corollary 4.1.11. *Any power $\mathrm{LSd}^n_{\mathfrak{U}}$ of a relative subdivision $\mathrm{LSd}_{\mathfrak{U}}$ on $L_*(\Delta_q)$ is chain homotopic to the identity.*

Corollary 4.1.12. *For any family $\mathfrak{U}$ of subsets of Δ_q the relative subdivision $\mathrm{LSd}_{\mathfrak{U}}$ induces isomorphisms in the homology of $L_*(\Delta_q)$.*

4.2 Singular Simplices

In this section the notion of relative subdivision is generalised to singular simplices in arbitrary topological spaces. The procedure is analogous to the classical barycentric subdivision.

Let X be a topological space and $\sigma \in S_p(X)$ a singular p-simplex. Any family $\mathfrak{U}$ of subsets of X can be pulled back by σ to obtain a family $\sigma^{-1}(\mathfrak{U})$ of subsets of the standard p-simplex Δ_p, where the relative subdivision of affine simplices can be applied:

Definition 4.2.1. *Let X be a topological space and $\mathfrak{U}$ a family of subsets of X. The $\mathfrak{U}$-subdivision is the unique group endomorphism $\mathrm{Sd}_{\mathfrak{U}} : S_p(X) \to S_p(X)$ that is given by*

$$\mathrm{Sd}_{\mathfrak{U}}(\sigma) := S_p(\sigma) \circ \mathrm{LSd}_{\sigma^{-1}(\mathfrak{U})}(\mathrm{id}_{\Delta_p})$$

for any singular p-simplex σ in X.

Note that, like before, any relative subdivision is the identity on singular 0-simplices. Since the relative subdivision of a simplex σ depends only on the family $\sigma^{-1}(\mathfrak{U})$ we find:

Lemma 4.2.2. *For any singular simplex σ in a topological space X and families $\mathfrak{U}$ and $\mathfrak{U}'$ of subsets of X such that $\mathfrak{U}_{|\sigma(\Delta_p)} = \mathfrak{U}'_{|\sigma(\Delta_p)}$ the $\mathfrak{U}$-relative and $\mathfrak{U}'$-relative subdivision of σ coincide.*

Proof. If the the families $\mathfrak{U}_{|\sigma(\Delta_p)}$ and $\mathfrak{U}'_{|\sigma(\Delta_p)}$ coincide, then the families $\sigma^{-1}(\mathfrak{U})$ and $\sigma^{-1}(\mathfrak{U}')$ coincide as well. By definition 4.2.1 the two relative subdivisions are equal. $\square$

So the relative subdivision of a simplex σ only depends on the induced family $\mathfrak{U}_{\sigma(\Delta_p)}$.

Let $X \to Y$ be a continuous function between topological spaces X and Y. Then this map induces a chain map $S_*(f) : S_*(X) \to S_*(Y)$. Moreover any family $\mathfrak{U}$ of subsets of Y can be pulled back by f^{-1} to a family $f^{-1}(\mathfrak{U})$ of subsets of X. Here the $f^{-1}(\mathfrak{U})$-relative subdivision can be applied. One should expect that the relative subdivision commutes with $S_p(f)$.

Lemma 4.2.3. *If $f : X \to Y$ is a continuous map between topological spaces X and Y and $\mathfrak{U}$ a family of subsets of Y then $\mathrm{Sd}_{\mathfrak{U}} \circ S_*(f) = S_*(f) \circ \mathrm{Sd}_{f^{-1}(\mathfrak{U})}$.*

Proof. Let X, Y be topological spaces, $f : X \to Y$ a continuous map and $\mathfrak{U}$ be a family of subsets of Y. For any singular p-simplex σ in X the $f^{-1}(\mathfrak{U})$-relative subdivision of σ is given by

$$\mathrm{Sd}_{f^{-1}(\mathfrak{U})}(\sigma) = S_p(\sigma) \circ \mathrm{Sd}_{\sigma^{-1}f^{-1}(\mathfrak{U})}(\mathrm{id}_{\Delta_p}) = S_p(\sigma) \circ \mathrm{Sd}_{(f\sigma)^{-1}(\mathfrak{U})}(\mathrm{id}_{\Delta_p}).$$

Therefore we find:

$$\begin{aligned}
S_p(f) \circ \mathrm{Sd}_{f^{-1}(\mathfrak{U})}(\sigma) &= S_p(f) \circ S_p(\sigma) \circ \mathrm{Sd}_{(f\sigma)^{-1}(\mathfrak{U})}(\mathrm{id}_{\Delta_p}) \\
&= S_p(f\sigma) \, \mathrm{Sd}_{(f\sigma)^{-1}(\mathfrak{U})}(\mathrm{id}_{\Delta_p}) \\
&= \mathrm{Sd}_{\mathfrak{U}}(f\sigma),
\end{aligned}$$

by definition of the relative subdivision of singular simplices. Thus the maps $\mathrm{Sd}_{\mathfrak{U}} \circ S_*(f)$ and $S_*(f) \circ \mathrm{Sd}_{f^{-1}(\mathfrak{U})}$ are the same. $\square$

This can especially be applied to the inclusion $\iota_* : L_*(\Delta_q) \hookrightarrow S_*(\Delta_q)$ of the chain complex $L_*(\Delta_q)$ of affine simplices in Δ_q into the chain complex $S_*(\Delta_q)$ of singular simplices in Δ_q.

Lemma 4.2.4. *The chain map $\mathrm{Sd}_{\mathfrak{U}}$ commutes with the inclusion $L_*(\Delta_q) \hookrightarrow S_*(\Delta_q)$.*

Corollary 4.2.5. *For any affine simplex σ and families $\mathfrak{U}$ and $\mathfrak{U}'$ in Δ_q such that $\mathfrak{U}_{|\sigma(\Delta_p)} = \mathfrak{U}'_{|\sigma(\Delta_p)}$ the $\mathfrak{U}$-relative and $\mathfrak{U}'$-relative subdivision of σ coincide.*

Since the subgroup $L_p(\Delta_q)$ of $S_p(\Delta_q)$ is invariant under relative subdivisions the inclusion ι_* is in fact a chain map. The same is true for the relative subdivision on singular simplices:

Lemma 4.2.6. *For any family $\mathfrak{U}$ of subsets of a topological space X the $\mathfrak{U}$-subdivision is a chain map $\mathrm{Sd}_{\mathfrak{U}} : S_*(X) \to S_*(X)$.*

Proof. Let σ be an singular p-simplex. The boundary of the $\mathfrak{U}$-relative subdivision of σ is by definition given by

$$\begin{aligned}
\partial \, \mathrm{Sd}_{\mathfrak{U}}(\sigma) &= \partial \left[S_p(\sigma) \circ \mathrm{Sd}_{\sigma^{-1}(\mathfrak{U})}(\mathrm{id}_{\Delta_p}) \right] \\
&= S_p(\sigma) \circ \mathrm{Sd}_{\sigma^{-1}(\mathfrak{U})}(\partial \mathrm{id}_{\Delta_p}) \\
&= \sum_{i=0}^{p} (-1)^i \, S_p(\sigma) \circ \mathrm{Sd}_{\sigma^{-1}(\mathfrak{U})}(\mathrm{id}_{\Delta_p}^i)
\end{aligned}$$

since the maps $S_p(\sigma)$ and $\mathrm{Sd}_{\sigma^{-1}(\mathfrak{U})}$ are chain maps. The i-th face of the affine simplex id_{Δ_p} is equal to $S_{p-1}(\mathrm{id}_{\Delta_p}^i)(\mathrm{id}_{\Delta_{p-1}})$. Therefore the $\mathrm{Sd}_{\sigma^{-1}(\mathfrak{U})}$-relative subdivision of this face is given by

$$\begin{aligned}
\mathrm{Sd}_{\sigma^{-1}\mathfrak{U}}(\mathrm{id}_{\Delta_p}^i) &= S_{p-1}(\mathrm{id}_{\Delta_p}^i) \circ \mathrm{Sd}_{(\mathrm{id}_{\Delta_p}^i)^{-1}\sigma^{-1}\mathfrak{U}}(\mathrm{id}_{\Delta_{p-1}}) \\
&= S_{p-1}(\mathrm{id}_{\Delta_p}^i) \circ \mathrm{Sd}_{(\sigma^i)^{-1}\mathfrak{U}}(\mathrm{id}_{\Delta_{p-1}}).
\end{aligned}$$

So we get

$$\begin{aligned}
\mathrm{Sd}_{\sigma^{-1}\mathfrak{U}}(\mathrm{id}_{\Delta_p}^i) &= \sum_{i=0}^{p} (-1)^i \, S_{p-1}(\sigma) \circ S_{p-1}(\mathrm{id}_{\Delta_p}^i) \circ \mathrm{Sd}_{\sigma^i{}^{-1}\mathfrak{U}}(\mathrm{id}_{\Delta_{p-1}}) \\
&= \sum_{i=0}^{p} (-1)^i \, S_{p-1}(\sigma^i) \circ \mathrm{Sd}_{(\sigma^i)^{-1}(\mathfrak{U})}(\mathrm{id}_{\Delta_{p-1}}) \\
&= \sum_{i=0}^{p} (-1)^i \, \mathrm{Sd}_{\mathfrak{U}}(\sigma^i) = \mathrm{Sd}_{\mathfrak{U}}(\partial \sigma),
\end{aligned}$$

for any singular p-simplex σ. Thus the $\mathfrak{U}$-relative subdivision is a chain map. $\square$

Since the chains $S_*(\Delta_q)$ are free resolutions of the integers and $\mathrm{Sd}_{\mathfrak{U}}$ is the identity on $S_0(\Delta_q)$, the chain map $\mathrm{Sd}_{\mathfrak{U}}$ has to be chain homotopic to the identity. So theorem 4.1.10 straightforwardly generalises to singular simplices in Δ_q. But for future generalisation we prove it by explicitly constructing a chain homotopy.

Theorem 4.2.7. *Any relative subdivision* $\mathrm{Sd}_{\mathfrak{U}} *$ *on* $L_*(\Delta_q)$ *is chain homotopic to the identity.*

Proof. At first we define for every standard q-simplex Δ_q and family $\mathfrak{U}$ in Δ_q a chain homotopy $\tilde{h}_{\mathfrak{U}}$ from $\mathrm{Sd}_{\mathfrak{U}}$ to the identity. Let $\tilde{h}_{\mathfrak{U},0} : L_0(\Delta_q) \to L_1(\Delta_q)$ be the constant function 0 and $\tilde{h}_{\mathfrak{U},p} : L_p(\Delta_q) \to L_{p+1}(\Delta_q)$ be defined by induction on the formula

$$\tilde{h}_{\mathfrak{U},p}(\sigma) = \mathbf{b}_\sigma \left(\mathrm{Sd}_{\mathfrak{U}}(\sigma) - \sigma - \tilde{h}_{\mathfrak{U},p-1}(\partial \sigma) \right),$$

for affine simplices σ in Δ_q. We will prove $\partial \circ \tilde{h}_{\mathfrak{U}} + \tilde{h}_{\mathfrak{U}} \circ \partial = \mathrm{Sd}_{\mathfrak{U}} -\mathrm{id}$ by induction on p. For any 0-simplex σ we get $\partial \tilde{h}_{\mathfrak{U},0}(\sigma) + \tilde{h}_{\mathfrak{U},-1}\partial(\sigma) = 0 = (\mathrm{Sd}_{\mathfrak{U}} -\mathrm{id})(\sigma)$. Assuming the formula to be true for chains of degree $< p$ we observe

$$\partial \tilde{h}_{\mathfrak{U},p-1}\partial = (\mathrm{Sd}_{\mathfrak{U}} -\mathrm{id} - \tilde{h}_{\mathfrak{U},p-2} \circ \partial) \circ \partial = (\mathrm{Sd}_{\mathfrak{U}} -\mathrm{id}) \circ \partial,$$

by the inductive assumption. So that we can compute

$$\begin{aligned}
\partial(\tilde{h}_{\mathfrak{U},p}(\sigma)) &= \partial \left(\mathbf{b}_\sigma (\mathrm{Sd}_{\mathfrak{U}}(\sigma) - \sigma - \tilde{h}_{\mathfrak{U},p-1}(\partial \sigma)) \right) \\
&= \left(\mathrm{Sd}_{\mathfrak{U}}(\sigma) - \sigma - \tilde{h}_{\mathfrak{U},p-1}(\partial \sigma) \right) - \mathbf{b}_\sigma \left(\partial \mathrm{Sd}_{\mathfrak{U}}(\sigma) - \partial \sigma - \partial \tilde{h}_{\mathfrak{U},p-1}\partial(\sigma) \right) \\
&= \left(\mathrm{Sd}_{\mathfrak{U}}(\sigma) - \sigma - \tilde{h}_{\mathfrak{U},p-1}(\partial \sigma) \right) - \mathbf{b}_\sigma \left(\mathrm{Sd}_{\mathfrak{U}}(\partial \sigma) - \partial \sigma - \partial \tilde{h}_{\mathfrak{U},p-1}\partial(\sigma) \right) \\
&= (\mathrm{Sd}_{\mathfrak{U}} -\mathrm{id})(\sigma) - \tilde{h}_{\mathfrak{U},p-1}(\partial \sigma) - 0(\partial \sigma),
\end{aligned}$$

for any p-simplex σ. Hence $\partial \circ \tilde{h}_{\mathfrak{U}} + \tilde{h}_{\mathfrak{U}} \circ \partial = \mathrm{Sd}_{\mathfrak{U}} -\mathrm{id}$.   □

Corollary 4.2.8. *Any power* $\mathrm{Sd}_{\mathfrak{U}}^n *$ *of a relative subdivision* $\mathrm{Sd}_{\mathfrak{U}} *$ *on* $L_*(\Delta_q)$ *is chain homotopic to the identity.*

Proposition 4.2.9. *For any affine simplex* $f : \Delta_q \to \Delta_{q'}$ *and family* $\mathfrak{U}$ *in* $\Delta_{q'}$ *the following square commutes.*

$$\begin{CD}
L_p(\Delta_q) @>{L_p(f)}>> L_p(\Delta_{q'}) \\
@V{\tilde{h}_{f^{-1}(\mathfrak{U}),p}}VV @VV{\tilde{h}_{\mathfrak{U},p}}V \\
L_{p+1}(\Delta_q) @>{L_{p+1}(f)}>> L_{p+1}(\Delta_{q'})
\end{CD} \tag{4.2}$$

Proof. We prove $L_{p+1}(f) \circ \tilde{h}_{f^{-1}(\mathfrak{U}),0} = \tilde{h}_{\mathfrak{U},0} \circ L_p(f)$ by induction on p. For $p = 0$ the homomorphisms $\tilde{h}_{\mathfrak{U},0}$ and $\tilde{h}_{f^{-1}(\mathfrak{U}),0}$ vanish so equality holds in this case. Assume the hypothesis to be true for all $p' < p$ and let σ be an affine p-simplex in Δ_q. Using the inductive definition of the homotopies $\tilde{h}$ we obtain

$$\begin{aligned}
L_{p+1}(f)\tilde{h}_{f^{-1}(\mathfrak{U})}(\sigma) &= L_{p+1}(f) \left[\mathbf{b}_\sigma \left(\mathrm{Sd}_{f^{-1}(\mathfrak{U})}(\sigma) - \sigma - \tilde{h}_{f^{-1}(\mathfrak{U})}(\partial \sigma) \right) \right] \\
&= \mathbf{b}_{f\sigma} \left[L_p(f) \left(\mathrm{Sd}_{f^{-1}(\mathfrak{U})}(\sigma) - \sigma - \tilde{h}_{f^{-1}(\mathfrak{U})}(\partial \sigma) \right) \right] \\
&= \mathbf{b}_{f\sigma} \left(\mathrm{Sd}_{\mathfrak{U}}(L_p(f)(\sigma) - L_p(f)(\sigma) - \tilde{h}_{\mathfrak{U},p-1}(\partial L_p(f)\sigma) \right) \\
&= \tilde{h}_{\mathfrak{U},p}(f\sigma) = \tilde{h}_{\mathfrak{U},p} \circ L_p(f)(\sigma)
\end{aligned}$$

for any affine p-simplex σ in Δ_q. This completes the inductive step.   □

Corollary 4.2.10. *For any affine p-simplex σ and family $\mathfrak{U}$ in Δ_q the following equation holds:*

$$\tilde{h}_{\sigma^{-1}(\mathfrak{U}),p-1}(\mathrm{id}^i_{\Delta_p}) = L_p(\mathrm{id}^i_{\Delta_p}) \circ \tilde{h}_{(\sigma^i)^{-1}(\mathfrak{U}),p-1}(\mathrm{id}_{\Delta_{p-1}})$$

Proof. Let σ be an affine p-simplex in Δ_q and $\mathfrak{U}$ a family of subsets of Δ_q. The i-th face of σ is given by $\sigma^i = \sigma \circ \mathrm{id}^i_{\Delta_p}$. This implies $(\sigma^i)^{-1}(\mathfrak{U}) = (\mathrm{id}^i_{\Delta_p})^{-1}\sigma^{-1}(\mathfrak{U})$. We apply Proposition 4.2.9 to the affine simplex $\mathrm{id}^i_{\Delta_p}$ and the family $\sigma^{-1}(\mathfrak{U})$ in Δ_p to obtain

$$\tilde{h}_{\sigma^{-1}(\mathfrak{U}),p-1} \circ L_p(\mathrm{id}^i_{\Delta_p}) = L_p(\mathrm{id}^i_{\Delta_p}) \circ \tilde{h}_{(\mathrm{id}^i_{\Delta_p})^{-1}\sigma^{-1}(\mathfrak{U}),p} = L_p(\mathrm{id}^i_{\Delta_p}) \circ \tilde{h}_{(\sigma^i)^{-1}(\mathfrak{U}),p}.$$

Applying the operators on both sides to the affine simplex $\mathrm{id}_{\Delta_{p-1}}$ leads to the desired equality:

$$\tilde{h}_{\sigma^{-1}(\mathfrak{U}),p-1}(\mathrm{id}^i_{\Delta_p}) = \tilde{h}_{\sigma^{-1}(\mathfrak{U}),p-1} \circ L_{p-1}(\mathrm{id}^i_{\Delta_p})(\mathrm{id}_{\Delta_{p-1}})$$
$$= L_p(\mathrm{id}^i_{\Delta_p}) \circ \tilde{h}_{(\sigma^i)^{-1}(\mathfrak{U}),p}(\mathrm{id}_{\Delta_{p-1}})$$

$\square$

For arbitrary topological spaces the chain of singular simplices need not be even be exact, but the theorem still remains true.

Theorem 4.2.11. *Any relative subdivision on the chain singular simplices in a topological space X is chain homotopic to the identity.*

Proof. We utilise the chain homotopies $\tilde{h}_{\mathfrak{U}}$ for families $\mathfrak{U}$ in the standard simplices to define homomorphisms $h_p : S_p(X) \to S_{p+1}(X)$. Let h_p be the unique homomorphism that is given by

$$h_p(\sigma) = S_{p+1}(\sigma)\left(\tilde{h}_{\sigma^{-1}(\mathfrak{U}),p}(\mathrm{id}_{\Delta_p})\right)$$

for any singular p-simplex σ in X. We will prove that h is a chain homotopy from $\mathrm{Sd}_{\mathfrak{U}}$ to the identity. This is done by induction on p. For any 0-simplex σ we get

$$\partial h_0(\sigma) + h_{-1}\partial(\sigma) = \partial S_1(\sigma)\left(\tilde{h}_{\sigma^{-1}(\mathfrak{U}),0}(\mathrm{id}_{\Delta_p})\right) + S_0(\sigma)\tilde{h}_{\sigma^{-1}(\mathfrak{U}),-1}\partial(\mathrm{id}_{\Delta_p})$$
$$= 0 + 0 = (\mathrm{Sd}_{\mathfrak{U}} - \mathrm{id})(\sigma).$$

Assuming the formula to be true for chains of degree $< p$ we obtain

$$(\partial h_p + h_{p-1}\partial)(\sigma) = \partial S_{p+1}(\sigma)\tilde{h}_{\sigma^{-1}(\mathfrak{U}),p}(\mathrm{id}_{\Delta_p}) + \sum_{i=0}^{p}(-1)^i S_p(\sigma^i)\tilde{h}_{(\sigma^i)^{-1}(\mathfrak{U}),p-1}(\mathrm{id}_{\Delta_{p-1}})$$

But

$$S_p(\sigma^i)\tilde{h}_{(\sigma^i)^{-1}(\mathfrak{U}),p-1}(\mathrm{id}_{\Delta_{p-1}}) = S_p(\sigma) \circ S_p(\mathrm{id}^i_{\Delta_p}) \circ \tilde{h}_{(\sigma^i)^{-1}(\mathfrak{U}),p-1}(\mathrm{id}_{\Delta_{p-1}})$$
$$= S_p(\sigma)\tilde{h}_{\sigma^{-1}(\mathfrak{U}),p-1}(\mathrm{id}^i_{\Delta_p}) \quad \text{By Corollary 4.2.10}$$

So all in all we get

$$(\partial h_p + h_{p-1}\partial)(\sigma) = S_p(\sigma)\partial\tilde{h}_{\sigma^{-1}(\mathfrak{U}),p}(\mathrm{id}_{\Delta_p}) + \sum_{i=0}^{p}(-1)^i S_p(\sigma)\tilde{h}_{\sigma^{-1}(\mathfrak{U}),p-1}(\mathrm{id}^i_{\Delta_p})$$

$$= S_p(\sigma)\left(\partial\tilde{h}_{\sigma^{-1}(\mathfrak{U}),p}(\mathrm{id}_{\Delta_p}) + \sum_{i=0}^{p}(-1)^i\tilde{h}_{\sigma^{-1}(\mathfrak{U}),p-1}(\mathrm{id}^i_{\Delta_p})\right)$$

$$= S_p(\sigma)\left(\partial\tilde{h}_{\sigma^{-1}(\mathfrak{U}),p}(\mathrm{id}_{\Delta_p}) + \tilde{h}_{\sigma^{-1}(\mathfrak{U}),p-1}(\partial\mathrm{id}_{\Delta_p})\right)$$

$$= S_p(\sigma)\left(\mathrm{Sd}_{\sigma^{-1}(\mathfrak{U}),p}(\mathrm{id}_{\Delta_p}) - \mathrm{id}_{\Delta_p}\right)$$

$$= \mathrm{Sd}_{\mathfrak{U}}(\sigma) - \sigma$$

for any affine p-simplex σ. Thus h is a chain homotopy from $\mathrm{Sd}_{\mathfrak{U}}$ to id.   □

Corollary 4.2.12. *Any power* $\mathrm{Sd}_{\mathfrak{U}}^n *$ *of a relative subdivision* $\mathrm{Sd}_{\mathfrak{U}} *$ *is chain homotopic to the identity.*

Corollary 4.2.13. *For any family* $\mathfrak{U}$ *of subsets of a topological space the relative subdivision* $\mathrm{Sd}_{\mathfrak{U}} *$ *induces isomorphisms in singular homology.*

Corollary 4.2.14. *For any family* $\mathfrak{U}$ *of subsets of a topological space the relative subdivision* $\mathrm{Sd}_{\mathfrak{U}} *$ *induces isomorphisms in singular cohomology.*

Next we observe that the relative subdivision behaves well on $\mathfrak{B}$-small simplices, where $\mathfrak{B}$ is an arbitrary family in the topological space X.

Lemma 4.2.15. *For any family* $\mathfrak{B}$ *in* X *the subcomplex* $S_*(X, \mathfrak{B})$ *of* $S_*(X)$ *is invariant under relative subdivision.*

Proof. Let $\mathfrak{U}$ and $\mathfrak{B}$ be two families in X. Since for any $\mathfrak{B}$-small n-simplex σ the chain $\mathrm{Sd}_{\mathfrak{U}}(\sigma)$ consists of simplices τ whose image $\tau(\Delta_n)$ lies in the image $\sigma(\Delta_n)$ of σ, these simplices are $\mathfrak{B}$-small as well. So all the groups $S_n(X, \mathfrak{B})$ are mapped into itself under the $\mathfrak{U}$-relative subdivision.   □

As a consequence we immediately obtain that for any two families $\mathfrak{U}$ and $\mathfrak{B}$ in X the $\mathfrak{U}$-relative subdivision restricts to an endomorphism of $S_*(X, \mathfrak{B})$. Like before this endomorphism is chain homotopic to the identity. The proof consists of two steps. First we prove the statement for affine simplices in $L_*(\Delta_q)$ and then in the general case.

Theorem 4.2.16. *Any restriction of a relative subdivision to the complex* $L_*(\Delta_q, \mathfrak{B})$ *is chain homotopic to the identity.*

Proof. The proof is analogous to that of theorem 4.2.7. One only needs to verify, that the chain homotopy $\tilde{h}_{\mathfrak{U}}$ constructed in the proof of theorem 4.2.7 maps $\mathfrak{B}$-small simplices to $\mathfrak{B}$-small simplices. The map $\tilde{h}_{\mathfrak{U},0}$ is the constant function 0. So $\tilde{h}_{\mathfrak{U},0}$ maps $L_0(\Delta_q, \mathfrak{B})$ into $L_0(\Delta_q, \mathfrak{B})$. We will prove $\tilde{h}_{\mathfrak{U},p}(\sigma)(\Delta_p) \subset \sigma(\Delta_p)$ for affine p-simplices and $p > 1$ by induction on p. Let σ be an affine $\mathfrak{B}$-small 1-simplex in Δ_q. The chain $\tilde{h}_{\mathfrak{U},p}(\sigma)$ is given by

$$\tilde{h}_{\mathfrak{U},1}(\sigma) = \mathbf{b}_\sigma \left(\mathrm{Sd}_{\mathfrak{U}}(\sigma) - \sigma - \tilde{h}_{\mathfrak{U},0}(\partial\sigma) \right)$$

$$= \mathbf{b}_\sigma \left(\mathrm{Sd}_{\mathfrak{U}}(\sigma) - \sigma \right) \text{ because } \tilde{h}_{\mathfrak{U},0} \text{ is the 0-function.}$$

The barycenter $\mathbf{b}_\sigma$ as well as the images of the chain $\mathrm{Sd}_{\mathfrak{U}}(\sigma) - \sigma$ all lie in the image of σ. Since this image is convex and the simplex σ was assumed to be $\mathfrak{B}$-small, the chain $\tilde{h}_{\mathfrak{U},1}(\sigma)$ consists of $\mathfrak{B}$-small simplices only, whose images all lie in $\sigma(\Delta_q)$. Assuming $\tilde{h}_{\mathfrak{U},p'} L_{p'}(\Delta_q, \mathfrak{B}) \subset L_{p'+1}(\Delta_q, \mathfrak{B})$ to be true for all $p' < p$ a similar argument can be made. If σ is a $\mathfrak{B}$-small affine p-simplex in Δ_q, the the chain $\tilde{h}_{\mathfrak{U},p}(\sigma)$ is given by

$$\tilde{h}_{\mathfrak{U},p}(\sigma) = \mathbf{b}_\sigma \left(\mathrm{Sd}_{\mathfrak{U}}(\sigma) - \sigma - \tilde{h}_{\mathfrak{U},p-1}(\partial\sigma) \right).$$

The images of the chain $\tilde{h}_{\mathfrak{U},p-1}(\partial\sigma)$ lie in $\sigma(\Delta_q)$ by the inductive assumption. The barycenter $\mathbf{b}_\sigma$ and the images of the chain $\mathrm{Sd}_{\mathfrak{U}}(\sigma) - \sigma$ all lie in the image of σ as well. Since this image is convex and the simplex σ was assumed to be $\mathfrak{B}$-small, the chain $\tilde{h}_{\mathfrak{U},p}(\sigma)$ consists of $\mathfrak{B}$-small simplices only, whose images all lie in $\sigma(\Delta_q)$. This completes the inductive step.   □

Theorem 4.2.17. *Any restriction of a relative subdivision to the complex $S_*(X, \mathfrak{B})$ is chain homotopic to the identity.*

Proof. Let $\mathfrak{U}$ be a family in X. By theorem 4.2.11 the $\mathfrak{U}$-relative subdivision $\mathrm{Sd}_\mathfrak{U}$ is chain homotopic to the identity. We only need to prove, that the homotopy h used in the proof maps $\mathfrak{B}$-small simplices to $\mathfrak{B}$-small simplices. The homomorphisms $h_p : S_p(X) \to S_{p+1}(X)$ are given by

$$h_p(\sigma) = S_{p+1}(\sigma)\left(\tilde{h}_{\sigma^{-1}(\mathfrak{U}),p}(\mathrm{id}_{\Delta_p})\right).$$

If a p-simplex σ in X is $\mathfrak{B}$-small, then the affine simplex id_{Δ_p} is $\sigma^{-1}(\mathfrak{U})$-small. So the chain $\tilde{h}_{\sigma^{-1}(\mathfrak{U}),p}(\mathrm{id}_{\Delta_p})$ lies in $L_{p+1}(\Delta_p, \mathfrak{B})$ by theorem 4.2.16. Hence the chain $S_{p+1}(\sigma)\left(\tilde{h}_{\sigma^{-1}(\mathfrak{U}),p}(\mathrm{id}_{\Delta_p})\right)$ lies in $S_{p+1}(X, \mathfrak{B})$. Thus the homotopy h maps $\mathfrak{B}$-small simplices to $\mathfrak{B}$-small simplices. $\square$

Corollary 4.2.18. *Any restriction of a relative subdivision to the complex $S_*(X, \mathfrak{B})$ induces the identity in singular homology.*

Even though the relative subdivision might not be a continuous operation on the topological singular simplicial groups $S^t(X)$ of a topological space X, this result also implies:

Corollary 4.2.19. *Any restriction of a relative subdivision to the complex $S_*^t(X, \mathfrak{B})$ induces the identity in singular topological homology.*

5

Complete Relative Subdivision

In this chapter we prove that repeated relative subdivision w.r.t. an open cover becomes stable after finitely many steps. This is used to construct a chain equivalence $\mathrm{Csd}_{\mathfrak{U}} : S_*(X) \to S_*(X, \mathfrak{U})$ for any given open covering $\mathfrak{U}$ of a topological space X. This equivalence maps any singular simplex σ to the 'minimal' $\mathfrak{U}$-relative subdivision $\mathrm{Sd}_{\mathfrak{U}}^n(\sigma)$, so that the chain $\mathrm{Sd}_{\mathfrak{U}}^n(\sigma)$ consists of $\mathfrak{U}$-small simplices only. Unfortunately the usual short procedure for proving the existence of such a 'minimal' subdivision, i.e. the one used in the case of barycentric subdivision, cannot be generalised in a straight forward way. So for simplicity we first restrict ourselves to affine simplices in low dimensions.

5.1 Low Dimensions

If σ is an affine 1-simplex in the euclidean m-space and $\mathfrak{U}$ an open covering of σ, then the $\mathfrak{U}$-relative subdivision of σ is the barycentric subdivision, provided that σ is not $\mathfrak{U}$-small. In this case the diameter of the simplex σ is divided by 2. So we find:

Lemma 5.1.1. *For any affine 1-simplex σ in Δ_q and any family $\mathfrak{U}$ in Δ_q the $\mathfrak{U}$-mesh of $\mathrm{Sd}_{\mathfrak{U}}^n(\sigma)$ converges to 0.*

Proof. Let σ be an affine 1-simplex in Δ_q and $\mathfrak{U}$ a family of subsets of Δ_q. If σ is not $\mathfrak{U}$-small then $\mathrm{mesh}_{\mathfrak{U}}(\mathrm{Sd}_{\mathfrak{U}}(\sigma)) \leq \frac{1}{2}\mathrm{diam}_{\mathfrak{U}}(\sigma)$. If σ is $\mathfrak{U}$-small then $\mathrm{Sd}_{\mathfrak{U}}(\sigma) = \sigma$ and $\mathrm{mesh}_{\mathfrak{U}}(\mathrm{Sd}_{\mathfrak{U}}(\sigma)) = \mathrm{mesh}_{\mathfrak{U}}(\sigma) = 0 = \mathrm{diam}_{\mathfrak{U}}(\sigma)$. So in any case the $\mathfrak{U}$-mesh of a chain c in $L_1(\Delta_q)$ is divided by 2 under $\mathfrak{U}$-relative subdivision. $\square$

Corollary 5.1.2. *For any affine 1-simplex σ in Δ_q and any open covering $\mathfrak{U}$ of σ there exists a smallest natural number $s(\sigma)$ such that all simplices in $\mathrm{Sd}_{\mathfrak{U}}^{s(\sigma)}(\sigma)$ are $\mathfrak{U}$-small.*

Proof. Let σ be an affine 1-simplex in Δ_q and $\mathfrak{U}$ an open covering of σ. Since the image $\sigma(\Delta_1)$ is compact there exists a Lebesgue number ϵ to this covering. Let $n \in \mathbb{N}$ such that the $\mathfrak{U}$-mesh of $\mathrm{Sd}_{\mathfrak{U}}^n(\sigma)$ is less than $\frac{1}{2}\epsilon$. Any simplex τ in $\mathrm{Sd}_{\mathfrak{U}}^n(\sigma)$ has $\mathfrak{U}$-diameter $< \epsilon$. So either $\mathrm{diam}_{\mathfrak{U}}(\tau) = 0$ and τ is $\mathfrak{U}$-small or $0 < \mathrm{diam}_{\mathfrak{U}}(\tau) < \epsilon$. But then the diameter of τ is less than the Lebesgue number of $\mathfrak{U}$ and τ is $\mathfrak{U}$-small as well. Thus the chain $\mathrm{Sd}_{\mathfrak{U}}^n(\sigma)$ lies in $L_1(\Delta_q, \mathfrak{U})$. $\square$

Regarding the relative subdivision of simplices of higher dimension an analogous result is more difficult to prove. The reason for this is that the mesh of a chain might not change, even if all simplices are subdivided. This results from the fact that a simplex σ might be not $\mathfrak{U}$-small although all the faces of σ are $\mathfrak{U}$-small.

Definition 5.1.3. *The subgroup of $L_p(\Delta_q)$ generated by all affine p-simplices of diameter smaller or equal to r is denoted by $L_p(\Delta_q, r)$:*

$$L_p(\Delta_q, r) := \{c \in L_p(\Delta_q) \mid \mathrm{mesh}(c) \leq r\}$$

Similarly, the subgroup of $L_p(\Delta_q)$ generated by all affine p-simplices of $\mathfrak{U}$-diameter smaller or equal to r is denoted by $L_{p,\mathfrak{U}}(\Delta_q, r)$:

$$L_{p,\mathfrak{U}}(\Delta_q, r) := \{c \in L_p(\Delta_q) \mid \mathrm{mesh}_{\mathfrak{U}}(c) \leq r\}$$

Let σ be an affine 2-simplex that is not $\mathfrak{U}$-small. If all the faces of σ are not $\mathfrak{U}$-small either, then the $\mathfrak{U}$-relative subdivision of σ is the barycentric subdivision. In this case the mesh of $\mathrm{Sd}_{\mathfrak{U}}(\sigma)$ is bounded by $\frac{2}{3}\mathrm{diam}(\sigma)$. This is a classical result. It is a consequence of the following fact:

Lemma 5.1.4. *For any affine p-simplex σ in the euclidean q-space the distance of the barycenter $\mathbf{b}_\sigma$ to any of the vertices of σ is less than $\frac{p}{p+1}$ the diameter of σ.*

Proof. Let $\sigma = (v_0, \dots, v_p)$ be an affine p-simplex. The distance from $\mathbf{b}_\sigma$ to a vertex v_i computes to

$$\|\mathbf{b}_\sigma - v_i\| = \left\| \frac{1}{p+1} \sum_{j=0}^{p} v_j - v_i \right\| = \left\| \frac{1}{p+1} \sum_{j=0, j\neq i}^{p} -\frac{p}{p+1} v_i \right\|$$

$$\leq \frac{1}{p+1} \sum_{j=0, j\neq i}^{p} \|v_j - v_i\| \leq \frac{p}{p+1} \mathrm{diam}(\sigma),$$

what was to be proved. $\qquad\square$

If one of the faces of σ is $\mathfrak{U}$-small, this face is not subdivided, so the mesh of $\mathrm{Sd}_{\mathfrak{U}}(\sigma)$ might be equal to the diameter of σ:

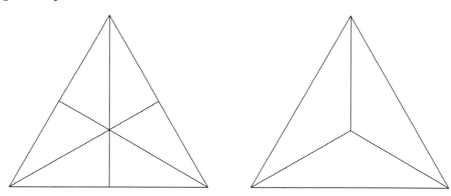

So we do not know per se if the $\mathfrak{U}$-mesh of $\mathrm{Sd}_{\mathfrak{U}}^n(\sigma)$ converges to 0. The difficulties arise from the simplices of the form $\mathbf{b}_\sigma \sigma^i$, where σ^i is a $\mathfrak{U}$-small face of σ^i. This gives rise to the following definition:

Definition 5.1.5. *Let $\sigma = (v_0, \dots, v_p)$ be an affine p-simplex in Δ_q. For any $l \leq p$ the affine l-simplex $(v_{p-l}, \dots, v_p)$ is called the l-end of σ. The barycenter of the l-end of σ is denoted by $\mathbf{b}_l(\sigma)$:*

$$\mathbf{b}_l(\sigma) := \frac{1}{l+1} \sum_{n-l \leq i \leq n} v_i$$

Remark 5.1.6. The l-end of any affine simplex is an affine l-simplex by definition.

Definition 5.1.7. *Let* $\sigma = (v_0, \ldots, v_p)$ *be an affine p-simplex in* Δ_q. *For any* $l \leq n$ *the affine l-simplex* $(v_0, \ldots, v_l)$ *is called the l-start of* σ.

So the simplices $\mathbf{b}_\sigma \sigma^i$ that form the obstructions have a $\mathfrak{U}$-small 1-end. By Lemma 5.1.4 the distance of the first vertex $\mathbf{b}_\sigma$ to the 1-end σ^i is smaller than $\frac{2}{3}\mathrm{diam}(\sigma)$. This leads to the next (slightly more general) definition.

Definition 5.1.8. *Let* $\sigma = (v_0, \ldots, v_p)$ *be an affine p-simplex in* Δ_q. *The l-diameter* $d_l(\sigma)$ *of* σ *is the maximal distance of a vertex in the* $(l$-$1)$-*start of* σ *to a vertex of* σ *if* σ *is not* $\mathfrak{U}$-*small and* 0 *if* σ *is* $\mathfrak{U}$-*small:*

$$d_l(\sigma) := \begin{cases} 0 & \text{if } \sigma \text{ is } \mathfrak{U}\text{-small} \\ \max_{0 \leq i < l, 0 \leq j \leq p} \|v_i - v_j\| & \text{if } \sigma \text{ is not } \mathfrak{U}\text{-small} \end{cases}$$

The 0-diameter of any simplex is defined to be 0.

Let r be the diameter of σ. So the simplices $\mathbf{b}_\sigma \sigma^i$ we want to study have 1-diameter $\leq \frac{2}{3}\mathrm{diam}(\sigma)$. All simplices of that kind generate a subgroup of $L_2(\Delta_q)$:

Definition 5.1.9. *The group* $A_2(1, r)$ *is the subgroup of* $L_2(\Delta_q, r)$ *generated by* $L_{2,\mathfrak{U}}(\Delta_q, \frac{2}{3}r)$ *and all simplices with* $\mathfrak{U}$-*small 1-end whose 1-diameter is at most* $\frac{2}{3}r$.

The exact purpose of this rather complicated definition will become apparent later. The group $A_2(1, r)$ is a subgroup of $L_{2,\mathfrak{U}}(\Delta_q, r)$ and contains the group $L_{2,\mathfrak{U}}(\Delta_q, \frac{2}{3}r)$. The chain $\mathrm{Sd}_\mathfrak{U}(\sigma)$ now splits in chains of two types of simplices: The simplices in $L_{2,\mathfrak{U}}(\Delta_q, \frac{2}{3}r)$ and the simplices of the form $\mathbf{b}_\sigma \sigma^i$ with $\mathfrak{U}$-small 1-end σ^i that lie in the group $A_2(1, r)$. We are going to show that the latter can be mapped into $L_{2,\mathfrak{U}}(\Delta_q, \frac{2}{3}r)$ by repeated $\mathfrak{U}$-relative subdivision. For this we examine the $\mathfrak{U}$-relative subdivision on $A_2(1, r)$ modulo $L_{2,\mathfrak{U}}(\Delta_q, \frac{2}{3}r)$.

Lemma 5.1.10. *For any affine simplex* τ *in* $A_2(1, r)$ *the* $\mathfrak{U}$-*relative subdivision of* τ *is equal to* $\mathbf{b}_\tau \mathrm{Sd}_\mathfrak{U}(\tau^0)$ *modulo* $L_{2,\mathfrak{U}}(\Delta_q, \frac{2}{3}r)$.

Proof. Let $\tau = (v_0, v_1, v_2)$ be an affine 2-simplex in $A_2(1, r)$ that is not $\mathfrak{U}$-small. The distance of the vertex v_0 to any of the vertices v_1 and v_2 is less than $\frac{2}{3}r$. So the faces τ^1 and τ^2 have diameter at most $\frac{2}{3}r$. Hence the mesh of $\mathrm{Sd}_\mathfrak{U}(\tau^1)$ and $\mathrm{Sd}_\mathfrak{U}(\tau^2)$ is also bounded by $\frac{2}{3}r$. Because the distance of the barycenter $\mathbf{b}_\tau$ to any vertex of τ is smaller than $\frac{2}{3}r$ the mesh of the chains $\mathbf{b}_\tau \mathrm{Sd}_\mathfrak{U}(\tau^1)$ and $\mathbf{b}_\tau \mathrm{Sd}_\mathfrak{U}(\tau^2)$ is also bounded by $\frac{2}{3}r$. Therefore these chains are contained in $L_2(\Delta_q, \frac{2}{3}r)$ and $\mathrm{Sd}_\mathfrak{U}(\tau) = \mathbf{b}_\tau \mathrm{Sd}_\mathfrak{U}(\tau^0)$ mod $L_{2,\mathfrak{U}}(\Delta_q, \frac{2}{3}r)$. $\square$

So in order to show that any simplex τ in $A_2(1, r)$ is mapped into $L_{2,\mathfrak{U}}(\Delta_q, \frac{2}{3}r)$ by repeated relative subdivision we only have to consider subdivisions of the above type. This motivates the following definitions.

Definition 5.1.11. *The endomorphism* $\mathrm{Sd}_{1,\mathfrak{U}} : A_2(1, r) \to A_2(1, r)$ *is the unique homomorphism that is given by*

$$\mathrm{Sd}_{1,\mathfrak{U}}(\tau) := \mathbf{b}_\tau \tau^0$$

for any affine simplex τ *in* $A_2(1, r)$.

As we have seen the distance of the barycenter $\mathbf{b}_\sigma$ to the vertices of σ is smaller than $\frac{2}{3}r$. Since the barycenter $\mathbf{b}_{\sigma^0}$ is an affine combination of these vertices the distance from $\mathbf{b}_\sigma$ to $\mathbf{b}_{\sigma^0}$ is also bounded by $\frac{2}{3}r$. We are going to show that the first vertex of $\mathrm{Sd}_{1,\mathfrak{U}}^n(\sigma)$ converges to $\mathbf{b}_{\sigma^0}$. Therefore we define:

Definition 5.1.12. *For any affine 2-simplex* $\sigma = (v_0, v_1, v_2)$ *the* $(1,1)$-*barycentric diameter* $d_{(1,1)}(\sigma)$ *is the diameter of the simplex* $(v_0, \mathbf{b}_1(\sigma))$.

Making use of this notion we examine the change of the $(1,1)$-mesh after 1-$\mathfrak{U}$-relative subdivision.

Proposition 5.1.13. *For any affine simplex τ in $A_2(1,r)$ the $(1,1)$-diameter of $\mathbf{b}_\tau \tau^0$ is bounded by $\frac{1}{3} d_{1,1}(\tau)$.*

Proof. Let $\tau = (v_0, v_1, v_2)$ be an affine simplex in $A_2(1,r)$. The distance of $\mathbf{b}_\tau$ to the barycenter of the 1-end τ^0 computes to

$$\|\mathbf{b}_\tau - \mathbf{b}_{\tau^0}\| = \|\frac{1}{3}(v_0 + v_1 + v_2) - \frac{1}{2}(v_1 + v_2)\|$$

$$= \|\frac{1}{3}v_0 - \frac{1}{6}(v_1 + v_2)\|$$

$$\leq \frac{1}{6}\|v_0 - v_1\| + \frac{1}{6}\|v_0 - v_2\| \leq \frac{2}{6} d_{1,1}(\tau) = \frac{1}{3} d_{1,1}(\tau).$$

$\square$

Corollary 5.1.14. *For any affine 2-simplex τ in $A_2(1,r)$ the $(1,1)$-barycentric diameter of $\mathrm{Sd}_{1,\mathfrak{U}}^n(\tau)$ converges to 0.*

Restricting ourselves to open coverings $\mathfrak{U}$ of a simplex we obtain the a very useful property.

Theorem 5.1.15. *For any affine 2-simplex $\tau \in A_2(1,r)$ in Δ_q and open covering $\mathfrak{U}$ of τ there exists a subdivision $\mathrm{Sd}_{1,\mathfrak{U}}^n(\tau)$ of τ consisting only of $\mathfrak{U}$-small simplices.*

Proof. Let $\tau = (v_0, v_1, v_2)$ be an affine simplex in $A_2(1,r)$. Since the 1-end τ^0 is $\mathfrak{U}$-small it is contained in an open set $U \in \mathfrak{U}$. Because the image $\tau^0(\Delta_1)$ is compact there exists an ϵ-neighbourhood $U_\epsilon(\tau)$ of $\tau^0(\Delta_1)$ that is contained in U. Since the $(1,1)$-barycentric diameter of $\mathrm{Sd}_{1,\mathfrak{U}}^n(\tau)$ converges to 0 there exists an $n \in \mathbb{N}$ such that the $(1,1)$-barycentric diameter of $\mathrm{Sd}_{1,\mathfrak{U}}^n(\tau)$ is less than ϵ. But then the simplex $\mathrm{Sd}_{1,\mathfrak{U}}^n(\tau)$ is either $\mathfrak{U}$-small or lies completely in the convex open set $U_\epsilon(\tau^0)$. In both cases it is $\mathfrak{U}$-small. $\square$

Corollary 5.1.16. *For any affine 2-simplex τ in Δ_q and open covering $\mathfrak{U}$ of τ the $\mathfrak{U}$-mesh of $\mathrm{Sd}_{\mathfrak{U}}^n(\tau)$ converges to 0.*

Proof. Let $\tau = (v_0, v_1, v_2)$ be an affine simplex of diameter r in Δ_q and $\mathfrak{U}$ an open covering of τ. The $\mathfrak{U}$-subdivision of τ consists of a chain c in $A_2(1,r)$ and a chain c' in $L_{2,\mathfrak{U}}(\Delta_q, \frac{2}{3}r)$. By Theorem 5.1.15 there exists a 1-$\mathfrak{U}$-relative subdivision $\mathrm{Sd}_{1,\mathfrak{U}}^n(c)$ of c that lies in $L_2(\Delta_q, \mathfrak{U})$. Because the subdivisions $\mathrm{Sd}_{\mathfrak{U}}^n(c)$ and $\mathrm{Sd}_{1,\mathfrak{U}}^n(c)$ differ only by simplices of $\mathfrak{U}$-diameter less than $\frac{2}{3}r$ the chain $\mathrm{Sd}_{\mathfrak{U}}^n(c)$ is an element of $L_{2,\mathfrak{U}}(\Delta_q, \frac{2}{3}r)$. Therefore $\mathrm{Sd}_{\mathfrak{U}}^{p+1}(\tau)$ lies in $L_{2,\mathfrak{U}}(\Delta_q, \frac{2}{3}r)$, i.e. $\mathrm{mesh}_{\mathfrak{U}}(\mathrm{Sd}_{\mathfrak{U}}^{p+1}(\tau)) \leq \frac{2}{3}\mathrm{diam}(\tau)$. Thus by repeated $\mathfrak{U}$-relative subdivision the mesh of a chain can be made arbitrary small. $\square$

With this helpful convergence property of the $\mathfrak{U}$-mesh we obtain the existence of a $\mathfrak{U}$-small repeated $\mathfrak{U}$-relative subdivision for open coverings $\mathfrak{U}$.

Theorem 5.1.17. *For any affine 2-simplex τ and open covering $\mathfrak{U}$ of τ there exists a smallest natural number $s(\tau)$ such that the chain $\mathrm{Sd}_{\mathfrak{U}}^{s(\tau)}(\tau)$ consists only of $\mathfrak{U}$-small simplices.*

Proof. Let $\tau = (v_0, v_1, v_2)$ be an affine simplex in Δ_q and $\mathfrak{U}$ an open covering of τ. Since the image $\tau(\Delta_2)$ is compact there exists a Lebesgue number ϵ to the covering $\mathfrak{U}$. By Theorem 5.1.15 there exists a $n \in \mathbb{N}$ such that the $\mathfrak{U}$-mesh of $\mathrm{Sd}_{\mathfrak{U}}^n(\tau)$ is smaller than ϵ. Thus any simplex in $\mathrm{Sd}_{\mathfrak{U}}^n(\tau)$ is $\mathfrak{U}$-small. $\square$

Corollary 5.1.18. *For any open covering $\mathfrak{U}$ of Δ_q the $\mathfrak{U}$-relative subdivision $\mathrm{Sd}_{\mathfrak{U}}$ induces nilpotent endomorphisms on the quotient groups $A_2(1,r)/L_{2,\mathfrak{U}}(\Delta_q, \frac{2}{3}r)$, $L_{2,\mathfrak{U}}(\Delta_q, r)/L_{2,\mathfrak{U}}(\Delta_q, 0)$ and $L_2(\Delta_q)/L_{2,\mathfrak{U}}(\Delta_q, 0)$.*

5.2 The General Case

In this section we prove the existence of a $\mathfrak{U}$-small subdivision $\mathrm{Sd}^n_{\mathfrak{U}}(\sigma)$ for affine simplices σ of arbitrary dimension (provided $\mathfrak{U}$ is an open covering). This is done by induction on the dimension of the affine simplices. At first we generalise the notion of the groups $A_2(1, r)$.

Definition 5.2.1. *The group $\tilde{A}_p(l, r)$ is the subgroup of $L_{p,\mathfrak{U}}(\Delta_q, r)$ generated by all affine p-simplices with $\mathfrak{U}$-small $(p-l)$-end whose l-diameter is at most $\frac{p}{p+1}r$.*

Notice that the group $\tilde{A}_p(0, r)$ is the group $L_p(\Delta_q, \mathfrak{U})$ generated by all $\mathfrak{U}$-small affine simplices and $\tilde{A}_p(p, r)$ is the group generated by all simplices of $\mathfrak{U}$-diameter at most $\frac{p}{p+1}r$. Furthermore any of the groups $\tilde{A}_p(l, r)$ contains the group of all $\mathfrak{U}$-small simplices.

Definition 5.2.2. *The group $A_p(l, r)$ is the subgroup of $L_p(\Delta_q, r)$ generated by all subgroups $\tilde{A}_p(l', r)$ for $l' \geq l$ in $L_p(\Delta_q, r)$:*

$$A_p(l, r) := \left\langle \bigcup_{l' \geq l} \tilde{A}_p(l', r) \right\rangle_{L_p(\Delta_q, r)}$$

So $A_p(l, r)$ is the group generated by all affine p-simplices $\sigma = (v_0, \ldots, v_p)$ for which there is an $l' \geq l$ such that the $(p-l')$-end $(v_l, \ldots, v_p)$ of σ is $\mathfrak{U}$-small and the l'-diameter of σ is less than $\frac{p}{p+1}r$. Therefore all simplices in $A_p(l, r)$ are especially of l-diameter less than $\frac{p}{p+1}r$. Furthermore the group $\tilde{A}_p(p, r) = A_p(p, r)$ is the group $L_{p,\mathfrak{U}}(\Delta_q, r)$ generated by all affine p-simplices of $\mathfrak{U}$-diameter less than $\frac{p}{p+1}r$. So

$$L_{p,\mathfrak{U}}(\Delta_q, r) \geq A_p(1, r) \geq \cdots \geq A_p(p, r) = L_{p,\mathfrak{U}}\left(\Delta_q, \frac{\mu}{p+1}r\right)$$

is a decreasing filtration of $L_{p,\mathfrak{U}}(\Delta_q, r)$. For $p = 2$ this is a 2-step filtration:

$$L_{2,\mathfrak{U}}(\Delta_q, r) \geq A_2(1, r) \geq L_{2,\mathfrak{U}}\left(\Delta_q, \frac{2}{3}r\right)$$

In this case Theorem 5.1.15 and Theorem 5.1.17 translate as

Theorem 5.2.3. *For any affine 2-simplex $\sigma \in \tilde{A}_2(l, r)$ and open covering $\mathfrak{U}$ of σ there exists a $\mathfrak{U}$-relative subdivision $\mathrm{Sd}^n_{\mathfrak{U}}(\sigma)$ consisting of only $\mathfrak{U}$-small simplices.*

Corollary 5.2.4. *For any open covering $\mathfrak{U}$ of Δ_q the relative subdivision induces a pointwise nilpotent homomorphism $\tilde{A}_2(l, r)/L_2(\Delta_q, \mathfrak{U}) :\to \tilde{A}_2(l, r)/L_2(\Delta_q, \mathfrak{U})$.*

Theorem 5.2.5. *For any open covering $\mathfrak{U}$ of Δ_q the relative subdivision induces a pointwise nilpotent homomorphism of the associated graded group of $A_2(*, r)$.*

We are going to show that the relative subdivision induces in general a nilpotent homomorphism of the associated graded group of $A_p(*, r)$ if $\mathfrak{U}$ is an open covering of Δ_q. This will be done in a similar fashion to the case $p = 2$. So we define:

Definition 5.2.6. *Let $\mathfrak{U}$ be a family in Δ_q. For any natural number $1 \leq l \leq p+1$ the l-$\mathfrak{U}$-subdivision on p-simplices is the unique endomorphism of the group $L_p(\Delta_q)$ that is the identity on simplices of dimension less than l and inductively defined on the formula*

$$\mathrm{Sd}_{l,\mathfrak{U}}(\sigma) := b_\sigma \left(\sum_{i<l} (-1)^i \, \mathrm{Sd}_{l-1,\mathfrak{U}}(\sigma^i) \right)$$

for any simplices σ of dimension higher than l.

Note that the 0-$\mathfrak{U}$-relative subdivision is the usual $\mathfrak{U}$-relative subdivision on p-simplices. Obviously the subdivision $\mathrm{Sd}_{l,\mathfrak{U}}$ does not change the l-ends of simplices, i.e. for any simplex σ and any simplex τ in the chain $\mathrm{Sd}_{l,\mathfrak{U}}$ the l-ends of σ and τ are the same. Next we examine the behaviour of the filtration $A_p(*,r)$ under relative subdivision. Therefore we have a look at the faces of an affine simplex σ.

Lemma 5.2.7. *For any affine simplex σ in $\tilde{A}_p(l,r)$ and $l \leq i \leq p$ the i-th face σ^i of σ is an element of $\tilde{A}_{p-1}(l,r)$.*

Proof. Let $\sigma = (v_0,\ldots,v_p)$ be an affine p-simplex in $A_p(l,r)$. Since the $(p\text{-}l)$-end $(v_l,\ldots,v_p)$ of σ is $\mathfrak{U}$-small so is the $(p\text{-}1\text{-}l)$-end $(v_l,\ldots,\hat{v}_i,\ldots,v_p)$ of σ^i. Furthermore all the distances $\|v_j - v_{j'}\|$ for $j < l$ and $j' \neq i$ are smaller than $\frac{p}{p+1}r$ because the simplex σ was assumed to be an element of $\tilde{A}_p(l,r)$. Thus σ^i lies in $\tilde{A}_{p-1}(l,r)$. $\square$

With the above Lemma we are now able to prove the invariance of the filtration $A_p(*,r)$ under relative subdivision.

Proposition 5.2.8. *The filtration $A_p(*,r)$ of $L_{p,\mathfrak{U}}(\Delta_q,r)$ is $\mathrm{Sd}_{k,\mathfrak{U}}$-invariant for any $0 < k \leq p$.*

Proof. This is proved by induction on p. For $p = 1$ the filtration of $L_1(\Delta_q)$ consists only of the group $A_1(1,r)$. This is simply the group $L_{1,\mathfrak{U}}(\Delta_q, \frac{1}{2}r)$ which is invariant under any relative subdivision. So the hypothesis is true in the case $p = 1$.

Assume the hypothesis to be true for all $p' < p$ and let σ be an affine p-simplex in $\tilde{A}_p(l,r)$. By definition the $\mathfrak{U}$-relative subdivision of σ given by

$$\mathrm{Sd}_{k,\mathfrak{U}}(\sigma) = \sum_{0 \leq i < k} b_\sigma\, \mathrm{Sd}_{k-1,\mathfrak{U}}(\sigma^i).$$

So there are to cases to be distinguished:

If $i < l$ then the face σ^i lies in $A_{p-1}(l\text{-}1,r)$. By the inductive assumption the k-$\mathfrak{U}$-relative subdivision $\mathrm{Sd}_{\mathfrak{U}}(\sigma^i)$ of σ^i is also an element of $A_{p-1}(l\text{-}1,r)$. Hence the chain $b_\sigma\, \mathrm{Sd}_{k,\mathfrak{U}}(\sigma^i)$ is an element of $A_p(l,r)$.

If $l \leq i$ then the face σ^i is an element of $A_{p-1}(l,r)$ by Lemma 5.2.7. So the k-$\mathfrak{U}$-relative subdivision of σ^i also lies in $A_{p-1}(l,r)$ by the inductive assumption. By Lemma 5.1.4 the distance of b_σ to any vertex of σ is bounded by $\frac{p}{p+1}r$. So the same is true for the distance of b_σ to any vertex of a simplex in $\mathrm{Sd}_{k,\mathfrak{U}}(\sigma^i)$. Therefore the chain $b_\sigma\, \mathrm{Sd}_{k,\mathfrak{U}}(\sigma^i)$ is an element of $A_p(l+1,r) \leq A_p(l,r)$.

Because all the chains $b_\sigma\, \mathrm{Sd}_{k,\mathfrak{U}}(\sigma^i)$ lie in $A_p(l,r)$ the k-$\mathfrak{U}$-relative subdivision leaves the group $A_p(l,r)$ invariant. $\square$

From the second part of this proof we obtain another property of relative subdivisions:

Corollary 5.2.9. *For $1 \leq k \leq l \leq p$ and any affine simplex σ in $A_p(l,r)$ the chain $b_\sigma\, \mathrm{Sd}_{k,\mathfrak{U}}(\sigma^i)$ is an element of $A_p(l+1,r)$.*

Making use of this we now examine the induced maps on the quotient groups $A_p(l,r)/A_p(l+1,r)$.

Proposition 5.2.10. *For $1 \leq l < l' \leq p+1$ the homomorphism $(\mathrm{Sd}_{l',\mathfrak{U}} - \mathrm{Sd}_{l,\mathfrak{U}})$ maps $A_p(l,r)$ into $A_p(l+1,r)$.*

Proof. We prove this by induction on p. For $p = 2$ and $l = 1$ the hypothesis is true by Lemma 5.1.4. Assume the hypothesis to be true for all $p' < p$ and let $\sigma \in \tilde{A}_p(l,r)$ be an affine p-simplex and $1 \leq l < l' \leq p+1$. The difference of the l-$\mathfrak{U}$-relative and the l'-$\mathfrak{U}$-relative subdivision of σ is given by

$$\left(\mathrm{Sd}_{l',\mathfrak{u}} - \mathrm{Sd}_{l,\mathfrak{u}}\right)(\sigma) = \sum_{i<l}(-1)^i \mathbf{b}_\sigma \,\mathrm{Sd}_{l'-1,\mathfrak{u}}(\sigma^i) - \sum_{i<l'}(-1)^i \mathbf{b}_\sigma \,\mathrm{Sd}_{l-1,\mathfrak{u}}(\sigma^i)$$

$$= \sum_{i<l}(-1)^i \mathbf{b}_\sigma \left(\mathrm{Sd}_{l-1,\mathfrak{u}} - \mathrm{Sd}_{l'-1,\mathfrak{u}}\right)(\sigma^i)$$

$$+ \sum_{l\leq i<l'}(-1)^i \mathbf{b}_\sigma \,\mathrm{Sd}_{l'-1,\mathfrak{u}}(\sigma^i).$$

For any $i > l$ the face σ^i of σ is an element of $A_{p-1}(l,r)$. Because the l'-$\mathfrak{u}$-relative subdivision leaves the filtration $A_p(*,r)$ invariant the chain $\mathrm{Sd}_{l'-1,\mathfrak{u}}(\sigma^i)$ lies also in $A_{p-1}(l,r)$. So the chain $\mathbf{b}_\sigma \,\mathrm{Sd}_{l'-1,\mathfrak{u}}(\sigma^i)$ lies in $A_p(l+1,r)$. Therefore the difference between the two relative subdivisions computes to

$$\left(\mathrm{Sd}_{l',\mathfrak{u}} - \mathrm{Sd}_{l,\mathfrak{u}}\right)(\sigma) = \sum_{i<l}(-1)^i \left(\mathrm{Sd}_{l'-1,\mathfrak{u}} - \mathrm{Sd}_{l-1,\mathfrak{u}}\right)(\sigma^i) \bmod A_p(l+1,r).$$

But for $i < l$ the faces σ^i lie in $A_{p-1}(l-1,r)$. So the chains $(\mathrm{Sd}_{l'-1,\mathfrak{u}} - \mathrm{Sd}_{l-1,\mathfrak{u}})(\sigma^i)$ are contained in $A_{p-1}(l,r)$ by the inductive assumption. Therefore all the chains $\mathbf{b}_\sigma(\mathrm{Sd}_{l'-1,\mathfrak{u}} - \mathrm{Sd}_{l-1,\mathfrak{u}})(\sigma^i)$ are elements of $A_p(l+1,r)$. Thus $\mathrm{Sd}_{l',\mathfrak{u}} - \mathrm{Sd}_{l,\mathfrak{u}}$ maps $A_p(l,r)$ into $A_p(l+1,r)$. $\qquad\square$

Corollary 5.2.11. *The relative subdivisions of* $\mathrm{Sd}_\mathfrak{u}$ *and* $\mathrm{Sd}_{l,\mathfrak{u}}$ *induce the same endomorphism of the quotient group* $A_p(l,r)/A_p(l+1,r)$.

We will now prove the nilpotency of the endomorphism induced by $\mathrm{Sd}_\mathfrak{u}$ on the quotient groups $A_p(l,r)/A_p(l+1,r)$, provided that $\mathfrak{u}$ is an open covering of Δ_q. This implies especially the nilpotency of the induced map on the quotient group $L_{p,\mathfrak{u}}(\Delta_q,r)/L_{p,\mathfrak{u}}(\Delta_q,\frac{p}{p+1}r)$, which in turn guarantees that the $\mathfrak{u}$-mesh of $\mathrm{Sd}_\mathfrak{u}^n(\sigma)$ converges to 0. Then the existence of a $\mathfrak{u}$-small relative subdivision follows easily.

Corollary 5.2.11 assures us, that to prove the nilpotency of the induced map on the quotient group $A_p(l,r)/A_p(l+1,r)$ we only have to consider the l-relative subdivision. So we proceed to prove the nilpotency of the by $\mathrm{Sd}_{l,\mathfrak{u}}$ induced map on $A_p(l,r)/A_p(l+1,r)$. This will be done in the next section.

5.3 Subsequent Filtrations

In order to prove the nilpotency of the induced endomorphism we define a decreasing filtration of subgroups $B_p(l,\left(\frac{p}{p+1}\right)^n,r'')$ of $A_p(l,r'')$ (also containing $L_p(\Delta_q,\mathfrak{u})$) and restrict ourselves to the case $B_p(l,r',r'')/B_p(l,\frac{p}{p+1}r',r'')$. The nilpotency of $\mathrm{Sd}_{l,\mathfrak{u}}$ is then proved by induction on l, using a further filtration $C_p(*,r,l,r',r'')$ of the group $B_p(l,r',r'')$. This filtration is constructed analogously to the filtration $A_p(*,r'')$ of $L_p(\Delta_q,r'')$. For this we need another notion of the 'obstruction size' of affine simplices.

Definition 5.3.1. *Let* $\sigma = (v_0,\dots,v_p)$ *be an affine p-simplex in* Δ_q. *The (k,l)-diameter is is the diameter of the k-simplex* $(v_0,\dots,v_{k-1},b_l(\sigma))$ *if σ is not $\mathfrak{u}$-small and 0 otherwise:*

$$d_{k,l}(\sigma) := \begin{cases} 0 & \text{if } \sigma \text{ is } \mathfrak{u}-small \\ \mathrm{diam}_\mathfrak{u}\,(v_0,\dots,v_{k-1},\mathbf{b}_l(\sigma)) & \text{if } \sigma \text{ is not } \mathfrak{u}-small \end{cases}$$

The $(0,l)$-diameter of any simplex is defined to be 0.

Definition 5.3.2. *For $k,l \leq p$ the (k,l)-mesh of any chain c of p-simplices is the maximal (k,l)-diameter of a simplex in c:*

$$\mathrm{mesh}_{(k,l)}(c) := \max\{d_{k,l}(\sigma) \mid c(\sigma) \neq 0\}$$

For $k' \leq k$ the (k', l) diameter of any simplex σ is smaller than the (k, l)-diameter of σ. Especially the (l, l)-diameter is smaller than the l-diameter, so $d_{k,l}(\sigma) \leq d_l(\sigma)$ for $k \leq l$ and any affine simplex σ. Observe further, that any simplex σ in $A_p(l, r)$ has (l, l)-diameter $\leq \frac{p}{p+1}$. This motivates the following definition:

Definition 5.3.3. *For $l \leq p$ and $r', r'' \in \mathbb{R}$ the group $B_p(l, r', r'')$ is the subgroup of $\tilde{A}_p(l, r'')$ generated by all simplices of (l, l)-mesh at most r':*

$$B_p(l, r', r'') := \mathrm{mesh}_{(l,l)}^{-1}(r') \cap \tilde{A}_p(l, r'')$$

These groups $B_p(l, r', r'')$ for different r' form a decreasing chain of subgroups of $A_p(l, r'')$. We will show later that that the group $B_p(l, r', r'')$ is mapped into the subgroup $B_p(l, \frac{p}{p+1} r', r'')$ under repeated l-$\mathfrak{U}$-subdivision. To prove this we use a filtration of $B_p(l, r, r'')$. We proceed as in the construction of the filtration $A_p(*, r)$.

Definition 5.3.4. *For $1 \leq k, l \leq p$ and $r, r', r'' \in \mathbb{R}$ the group $\tilde{C}_p(k, r, l, r', r'')$ is the subgroup of $B_p(l, r', r'')$ generated by all affine simplices with $\mathfrak{U}$-small $(p-k)$-end whose (k, l)-barycentric diameter is at most r:*

$$\tilde{C}_p(k, r, l, r', r'') := \mathrm{mesh}_{(k,l)}^{-1}(r) \cap B_p(l, r', r'')$$

Notice that the group $\tilde{C}_p(0, r, l, r', r'')$ is the group $L_p(\Delta_q), \mathfrak{U}$ of $\mathfrak{U}$-small simplices and the group $\tilde{C}_p(l, r, l, r', r'')$ is the group $B_p(l, \hat{r}, r'')$ where $\hat{r} = \min\{r, r'\}$.

Definition 5.3.5. *The group $C_p(k, r, l, r', r'')$ is the subgroup of $L_p(\Delta_q)$ generated by all subgroups $\tilde{C}_p(k', r, l, r', r'')$ for $k \leq k' \leq l$:*

$$C_p(k, r, l, r', r) := \left\langle \bigcup_{k \leq k' \leq l} \tilde{C}_p(k', r, l, r', r'') \right\rangle_{L_p(\Delta_q)}$$

So for $r \leq r'$ and $1 \leq k \leq l$ the groups $C_p(k, r, l, r', r'')$ form a decreasing filtration of $B_p(l, r', r'')$:

$$B_p(l, r', r'') \geq C_p(1, r, l, r', r'') \geq \cdots \geq C_p(l, r, l, r', r'') = B_p(l, r, r'')$$

Note that any of these groups contains the group $L_p(\Delta_q, \mathfrak{U})$ of all $\mathfrak{U}$-small simplices.

As in the case of $A_p(l, r)$ the groups $B_p(l, *, r'')$ and the filtration $C_p(*, r, l, r', r'')$ of $B_p(l, r', r'')$ turn out to be invariant under l-$\mathfrak{U}$-relative subdivision. For a proof hereof we examine the change of the (k, l)-mesh of a chain under l-$\mathfrak{U}$-relative subdivision. Since the new vertices occurring here are barycenters of faces, we first have a look at the distance of different barycenters.

Lemma 5.3.6. *For any affine p-simplex σ and any natural numbers $0 \leq k \leq l \leq p$ the distance of the barycenters $\mathbf{b}_k(\sigma)$ and $\mathbf{b}_l(\sigma)$ is bounded by $\frac{p}{p+1} d_{(l,l)}(\sigma)$.*

Proof. Let $\sigma = (v_0, \ldots, v_p)$ be an affine p-simplex. The distance between $\mathbf{b}_k(\sigma)$ and $\mathbf{b}_l(\sigma)$ is given by

$$\|\mathbf{b}_k(\sigma) - \mathbf{b}_l(\sigma)\| = \left\| \frac{1}{n-k+1} \sum_{k \le i \le p} v_i - \frac{1}{n-l+1} \sum_{l \le j \le p} v_j \right\|$$

$$= \left\| \frac{1}{n-k+1} \sum_{k \le i < l} v_i - \left(\frac{1}{n-l+1} - \frac{1}{n-k+1} \right) \sum_{l \le j \le p} v_j \right\|$$

$$= \frac{1}{n-k+1} \cdot \left\| \sum_{k \le i < l} v_i - \frac{l-k}{n-l+1} \sum_{l \le j \le p} v_j \right\|$$

$$= \frac{1}{n-k+1} \left\| \sum_{l \le i < k} v_i - (l-k)\mathbf{b}_k(\sigma) \right\|$$

$$\le \frac{1}{n-k+1} \left\| \sum_{k \le i < l} (v_i - \mathbf{b}_l(\sigma)) \right\|$$

$$\le \frac{l-k}{n-k+1} d_{(l,l)}(\sigma), \le \frac{p}{p+1} d_{(l,l)}(\sigma),$$

which was to be proved. $\qquad\qquad\qquad\qquad\qquad\qquad\qquad\qquad\qquad\qquad\Box$

This implies especially that the distance of $\mathbf{b}_\sigma = \mathbf{b}_0(\sigma)$ to $\mathbf{b}_l(\sigma)$ is bounded by $\frac{p}{p+1} d_{l,l}(\sigma)$. So we conclude:

Proposition 5.3.7. *For any affine simplex σ in the subgroup $\tilde{C}_p(k, \frac{p}{p+1}r', l, r', r'')$ of $L_p(\Delta_q)$ the $(k+1, l+1)$-diameter of $\mathbf{b}_\sigma\sigma$ is bounded by $\frac{p}{p+1} d_{l,l}(\sigma)$.*

Proof. Let $\sigma = (v_0, \ldots, v_p) \in \tilde{C}_p(k, \frac{p}{p+1}r', l, r', r'')$ be an affine p-simplex. The distance of the vertex $\mathbf{b}_\sigma$ of $\mathbf{b}_\sigma\sigma$ to any of the vertices v_j, $j < k$, is given by

$$\|\mathbf{b}_\sigma - v_j\| = \left\| \frac{p}{p+1} \left(\sum_{j' < l} v'_j + (l+1) \cdot \mathbf{b}_l(\sigma) \right) - v'_j \right\|$$

$$= \left\| \frac{p}{p+1} \left(\sum_{j' < l} v'_j + (l+1) \cdot \mathbf{b}_l(\sigma) \right) - \frac{n-l+l+1}{p+1} v'_j \right\|$$

$$\le \frac{p}{p+1} \sum_{j' < l} \|v'_j - v_j\| + \frac{l+1}{p+1} \|\mathbf{b}_l(\sigma) - v'_j\|$$

$$\le \frac{l-1}{p+1} d_{(k,l)}(\sigma) + \frac{l+1}{p+1} d_{(k,l)}(\sigma) = \frac{p}{p+1} d_{(k,l)}(\sigma) \le \frac{p}{p+1} d_{l,l}(\sigma)$$

By Lemma 5.3.6 the distance of $\mathbf{b}_\sigma$ to $\mathbf{b}_l(\sigma)$ is also bounded by $\frac{p}{p+1} d_{l,l}(\sigma)$. Since the $\mathfrak{U}$-diameter of the simplex $(v_0, \ldots, v_{k-1}, \mathbf{b}_l\sigma)$ is exactly the (k,l)-diameter $d_{k,l}(\sigma)$ and $d_{k,l}(\sigma) \le \frac{p}{p+1} d_{l,l}(\sigma)$ the $\mathfrak{U}$-diameter of $(\mathbf{b}_\sigma, v_0, \ldots, v_{k-1}, \mathbf{b}_l(\sigma))$ is also bounded by $\frac{p}{p+1} d_{l,l}(\sigma)$. $\qquad\qquad\qquad\Box$

Corollary 5.3.8. *For any affine simplex $\sigma \in \tilde{C}_p(k, \frac{p}{p+1}r', l, r', r'')$ the simplex $\mathbf{b}_\sigma\sigma$ is contained in $\tilde{C}_{p+1}(k+1, \frac{p}{p+1}r', l+1, r', r'')$.*

Corollary 5.3.9. *For any affine simplex $\sigma \in \tilde{C}_p(k, \frac{p}{p+1}r', l, r', r'')$ and $i < k$ the simplex $\mathbf{b}_\sigma\sigma^i$ is contained in $\tilde{C}_p(k, \frac{p}{p+1}r', l, r', r'')$.*

Corollary 5.3.10. *For any affine simplex $\sigma \in \tilde{C}_p(k, \frac{p}{p+1}r', l, r', r'')$ and $k \le i < l$ the simplex $\mathbf{b}_\sigma \sigma^i$ lies in $\tilde{C}_p(k+1, \frac{p}{p+1}r', l, r', r'')$.*

Proposition 5.3.11. *The filtration $C_p(*, \frac{p}{p+1}r', l, r', r'')$ of $B_p(l, r', r'')$ is $\mathrm{Sd}_{l',\mathfrak{u}}$ invariant for any $0 \le l' \le l$.*

Proof. For $p = 0$ all relative subdivisions are the identity so the hypothesis is true in this case. If $l = 0$ then the filtration is trivial and consists only of the group $\tilde{C}_p(0, r, l, r', r'')$ which is the group $L_p(\Delta_q, \mathfrak{u})$ of all $\mathfrak{u}$-small affine p simplices which is invariant under any relative subdivision. Therefore the hypothesis is true for $p = l = 0$ and for $l = 0$ and arbitrary p. Assume the hypothesis to be true for all $p' < p$ and and $\sigma \in \tilde{C}_p(k, r, l, r', r'')$ be an affine p-simplex ($l' \le l \le p$). The $\mathfrak{u}$-relative subdivision of σ is given by

$$\mathrm{Sd}_{l',\mathfrak{u}}(\sigma) = \sum_{0 \le i < k} b_\sigma \, \mathrm{Sd}_{l'-1,\mathfrak{u}}(\sigma^i).$$

So there are to cases to be distinguished.

If $i < k$ then the face σ^i is lies in $\tilde{C}_{p-1}(k\text{-}1, \frac{p}{p+1}r, l\text{-}1, r, r'')$. By the inductive assumption the $(l'\text{-}1)$-$\mathfrak{u}$-relative subdivision $\mathrm{Sd}_{l-1,\mathfrak{u}}(\sigma^i)$ of σ^i is also an element of $\tilde{C}_{p-1}(k-1, \frac{p}{p+1}r, l-1, r, r'')$. By Corollary 5.3.9 the chain $\mathbf{b}_\sigma \sigma^i$ lies in $\tilde{C}_p(k, \frac{p}{p+1}r, l, r, r'')$. Because the images of all simplices in the chain $\mathrm{Sd}_{l-1,\mathfrak{u}}(\sigma^i)$ are contained in the image of σ the same is true for $\mathbf{b}_\sigma \, \mathrm{Sd}_{l-1,\mathfrak{u}}(\sigma^i)$.

If $k \le i \le l'$ then the face σ^i is an element of $\tilde{C}_{p-1}(k, \frac{p}{p+1}r, l - 1, r, r'')$ and the chain $\mathrm{Sd}_{l'-1,\mathfrak{u}}(\sigma^i)$ also lies in $\tilde{C}_{p-1}(k, \frac{p}{p+1}r, l - 1, r, r'')$ by the inductive assumption. By Corollary 5.3.10 the simplex $\mathbf{b}_\sigma \sigma^i$ lies in $\tilde{C}_{p-1}(k, \frac{p}{p+1}r, l, r, r'')$ and so does $\mathbf{b}_\sigma \, \mathrm{Sd}_{l'-1,\mathfrak{u}}(\sigma^i)$. So any chain $\mathbf{b}_\sigma \, \mathrm{Sd}_{l'-1,\mathfrak{u}}(\sigma^i)$ lies in $A_n(k, \frac{p}{p+1}r, l, r, r'')$ for $i < l'$. This completes the inductive step. □

From the proof of the previous Proposition we infer:

Lemma 5.3.12. *Any l-$\mathfrak{u}$-relative subdivision maps the group $\tilde{C}_p(l, r, l, r, r')$ into the group $\tilde{C}_p(1, \frac{p}{p+1}r, l, r, r')$.*

Lemma 5.3.13. *The groups $B_p(l, r', r'')$ are invariant under l-$\mathfrak{u}$-relative subdivision.*

Proof. Since $B_p(l, r', r'') = C_p(l, r', l, r', r'')$ and $C_p(1, \frac{p}{p+1}, l, r', r'')$ is a subgroup of $B_p(l, r', r'')$ this is an immediate consequence of Lemma 5.3.12. □

We now prove that any affine p-simplex $\sigma \in A_p(l, r)$ is mapped into $L_p(\Delta_q)$ under repeated l-$\mathfrak{u}$-relative subdivision (if $\mathfrak{u}$ is an open covering of σ). This is essentially the statement of the follwing Theorem for the case $p = 0, 1$ and 2:

So we do assume from now on that Theorem??? and corollary ??? hold for all $p' < p$ and $\mathfrak{u}$ is an open covering of σ. To start the induction we first prove the hypothesis for the case $l = 1$.

Lemma 5.3.14. *For any affine p-simplex σ in Δ_q the $(1,1)$-mesh of $\mathrm{Sd}_{1,\mathfrak{u}}^k(\sigma)$ converges to 0.*

Proof. Let σ be an affine p-simplex σ in Δ_q and $\mathfrak{u}$ be a family of subsets of Δ_q. From Lemma 5.3.6 we know that the $(1,1)$-mesh of $\mathrm{Sd}_{1,\mathfrak{u}}(c)$ is bounded by $\frac{p}{p+1}d_{1,1}(c)$ for any chain c. Thus $\mathrm{Sd}_{1,\mathfrak{u}}^k(\sigma) \le (\frac{p}{p+1})^k \mathrm{diam}_{\mathfrak{u}}(\sigma)$ and the $(1,1)$-mesh of $\mathrm{Sd}_{1,\mathfrak{u}}^k(\sigma)$ converges to 0. □

In analogy to the case $p = 2$ we observe that this Lemma implies the nilpotency of the endomorphism of $A_p(1, r'')/L_p(\Delta_q)$ induced by $\mathrm{Sd}_{1,\mathfrak{U}}$.

Lemma 5.3.15. *If $\sigma \in A_p(1, r'')$ is an affine p-simplex in Δ_q and $\mathfrak{U}$ an open covering of σ such that the $(1, 1)$-mesh of $\mathrm{Sd}_{1,\mathfrak{U}}(\sigma)$ converges to 0, then there exists a subdivision $\mathrm{Sd}_{1,\mathfrak{U}}^k(\sigma)$ consisting only of $\mathfrak{U}$-small simplices.*

Proof. Let $\sigma = (v_0, \ldots, v_p)$ be an affine simplex in $A_p(1, r'')$. Since the $p-1$-end σ^0 is $\mathfrak{U}$-small it is contained in an open set $U \in \mathfrak{U}$. Because the image $\sigma^0(\Delta_1)$ is compact there exists an ϵ-neighbourhood $U_\epsilon(\tau)$ of $\sigma^0(\Delta_1)$ that is contained in U. Since the $(1, 1)$-barycentric diameter of $\mathrm{Sd}_{1,\mathfrak{U}}^n(\sigma)$ converges to 0 there exists an $n \in \mathbb{N}$ such that the $(1, 1)$-barycentric diameter of $\mathrm{Sd}_{1,\mathfrak{U}}^n(\sigma)$ is less than ϵ. But then the simplex $\mathrm{Sd}_{1,\mathfrak{U}}^n(\sigma)$ is either $\mathfrak{U}$-small or lies completely in the convex open set $U_\epsilon(\sigma^0)$. In both cases it is $\mathfrak{U}$-small. $\square$

We now prove the analogous result that the (l, l)-mesh of repeated l-$\mathfrak{U}$-relative subdivisions $\mathrm{Sd}_{l,\mathfrak{U}}^n(c)$ of a chain $c \in A_p(l, r'')$ converges to zero. This especially implies that the chains $\mathrm{Sd}_{l,\mathfrak{U}}^n(c)$ are $\mathfrak{U}$-small for large n. Since this has already been shown for $l = 1$ we can use as the induction hypothesis that the statement is true for all $l' \leq l$. For the inductive step we utilise the filtration $C_p(*, \frac{p}{p+1}r', l, r', r'')$ to show that for any simplex $\sigma \in A_p(l, r)$ the (l, l)-mesh of $\mathrm{Sd}_{l,\mathfrak{U}}^{n'}(\sigma)$ converges to 0. Here we use the induction hypothesis on l to show that the (k, l)-mesh of $\mathrm{Sd}_{k,\mathfrak{U}}^n(\sigma)$ converges to 0 if σ is contained in $C_p(k, r, l, r', r'')$.

Lemma 5.3.16. *For any affine p-simplex $\sigma \in \tilde{C}_p(k, r, l, r', r'')$, $1 \leq k < l \leq p$ and open covering $\mathfrak{U}$ of σ the (k, l)-mesh of $\mathrm{Sd}_{k,\mathfrak{U}}^n(\sigma)$ converges to 0. In addition there exists a subdivision $\mathrm{Sd}_{k,\mathfrak{U}}^{n'}(\sigma)$ consisting only of $\mathfrak{U}$-small simplices.*

Proof. Let $\sigma \in \tilde{C}_p(k, r, l, r', r'')$ be an affine p-simplex and $\mathfrak{U}$ an open covering of σ. If $\hat{r}$ denotes the maximum of $\{r, r', r''\}$ then the simplex σ lies in $A_p(k, \hat{r})$. By the inductive assumptions on l there exists a k-$\mathfrak{U}$-relative subdivision $\mathrm{Sd}_{k,\mathfrak{U}}^n(\sigma)$ of σ consisting only of $\mathfrak{U}$-small simplices. Thus $\mathrm{mesh}_{k,l}\,\mathrm{Sd}_{k,\mathfrak{U}}^n(\sigma) = 0$. $\square$

Lemma 5.3.17. *For any affine p-simplex $\sigma \in \tilde{C}_p(k, r, l, r', r'')$, $1 \leq k < l \leq p$ and open covering $\mathfrak{U}$ of σ there exists a k-$\mathfrak{U}$-relative subdivision $\mathrm{Sd}_{k,\mathfrak{U}}^n(\sigma)$ of σ lying in $C_p(k+1, r, l, r', r'')$.*

Proof. Let $\sigma = (v_0, \ldots, v_p)$ be an affine simplex in Δ_q and $\mathfrak{U}$ an open covering of σ. Since the image $\sigma(\Delta_q)$ is compact there exists a Lebesgue number ϵ to the covering $\mathfrak{U}$. By Lemma 5.3.16 there exists an $n \in \mathbb{N}$ such that the $\mathfrak{U}$-mesh of $\mathrm{Sd}_{k,\mathfrak{U}}^n(\sigma)$ is smaller than ϵ. Thus any simplex in $\mathrm{Sd}_{k,\mathfrak{U}}^n(\sigma)$ is $\mathfrak{U}$-small. $\square$

Lemma 5.3.18. *For any affine p-simplex $\sigma \in \tilde{C}_p(k, r, l, r', r'')$, $1 \leq k < l \leq p$ and open covering $\mathfrak{U}$ of σ there exists a l-$\mathfrak{U}$-relative subdivision $\mathrm{Sd}_{l,\mathfrak{U}}^n(\sigma)$ of σ lying in $C_p(k+1, r, l, r', r'')$.*

Proof. Since the k-$\mathfrak{U}$-relative and the l-$\mathfrak{U}$-relative subdivision differ only by chains in $C_p(k+1, r, l, r', r'')$ this is a consequence of the above Lemma. $\square$

Lemma 5.3.19. *Any affine p-simplex $\sigma \in C_p(1, \frac{p}{p+1}r', l, r', r'')$ is mapped into $C_p(l, \frac{p}{p+1}r', l, r', r'')$ under repeated l-$\mathfrak{U}$-subdivision.*

Proof. This is an repeated application of Lemma 5.3.18. $\square$

Lemma 5.3.20. *For any affine p-simplex $\sigma \in \tilde{C}_p(l, r, l, r, r'')$ there exists a l-$\mathfrak{U}$-relative subdivision $\mathrm{Sd}_{l,\mathfrak{U}}^n(\sigma)$ of σ lying in $C_p(l, \frac{p}{p+1}r, l, r, r'')$.*

Proof. Let $\sigma \in \tilde{C}_p(l, r, l, r, r'')$ be an affine p-simplex and $\mathfrak{U}$ an open covering of σ. By Lemma 5.3.12 the l-$\mathfrak{U}$-relative subdivision maps σ into $C_p(1, \frac{p}{p+1} r, l, r, r'')$. By Lemma 5.3.19 the chain $\mathrm{Sd}_{l,\mathfrak{U}}(\sigma)$ is mapped into $C_p(l, \frac{p}{p+1} r, l, r, r'')$ under repeated l-$\mathfrak{U}$-subdivision. $\qquad\square$

Recall that the numbers r resp. $\frac{p}{p+1} r$ denote the (l, l)-mesh, so this completes the inductive step. Based on the induction hypothesis we have shown that the (l, l)-mesh of any simplex converges $\sigma \in A_p(l, r)$ converges to zero:

Proposition 5.3.21. *For any affine p-simplex $\sigma \in A_p(l, r)$ and open covering $\mathfrak{U}$ of σ the (l, l)-mesh of $\mathrm{Sd}_{l,\mathfrak{U}}^n(\sigma)$ converges to 0.*

Corollary 5.3.22. *For any affine p-simplex $\sigma \in A_p(l, r)$ and open covering $\mathfrak{U}$ of σ there exists a l-$\mathfrak{U}$-relative subdivision $\mathrm{Sd}_{l,\mathfrak{U}}^n(\sigma)$ consisting only of $\mathfrak{U}$-small simplices.*

Proof. Let $\sigma = (v_0, \ldots, v_p)$ be an affine simplex in $A_p(l, r'')$. Since the $p - l$-end $\mathbf{p} - \mathbf{l}(\sigma)$ of σ is $\mathfrak{U}$-small it is contained in an open set $U \in \mathfrak{U}$. Because the image $K = \mathbf{b}_{p-l}(\sigma)(\Delta_{p-l})$ is compact there exists an ϵ-neighbourhood $U_\epsilon(K)$ of K that is contained in U. Since the (l, l)-barycentric diameter of $\mathrm{Sd}_{l,\mathfrak{U}}^n(\sigma)$ converges to 0 there exists an $n \in \mathbb{N}$ such that the $(l, 1l)$-barycentric diameter of $\mathrm{Sd}_{l,\mathfrak{U}}^n(\sigma)$ is less than ϵ. But then the simplex $\mathrm{Sd}_{l,\mathfrak{U}}^n(\sigma)$ is either $\mathfrak{U}$-small or lies completely in the convex open set $U_\epsilon(K)$. In both cases it is $\mathfrak{U}$-small. $\qquad\square$

Lemma 5.3.23. *For any affine p-simplex $\sigma \in A_p(l, r)$ and open covering $\mathfrak{U}$ of σ there exists a $\mathfrak{U}$-relative subdivision $\mathrm{Sd}_{\mathfrak{U}}^n(\sigma)$ lying in $A_p(l + 1, r)$.*

Proof. This folows from the fact that the subdivisions $\mathrm{Sd}_{l,\mathfrak{U}}$ and $\mathrm{Sd}_{\mathfrak{U}}$ induce the same endomorphism of the quotient group $A_p(l, r)/A_p(l + 1, r)$ by Lemma 5.2.11. $\qquad\square$

Theorem 5.3.24. *For any affine p-simplex σ in Δ_q and open covering $\mathfrak{U}$ of σ there exists a natural number k such that*

$$\mathrm{mesh}_{\mathfrak{U}}\left(\mathrm{Sd}_{\mathfrak{U}}^k(\sigma)\right) \leq \frac{p}{p+1} \mathrm{diam}_{\mathfrak{U}}(\sigma).$$

Corollary 5.3.25. *For any affine p-simplex σ in Δ_q and open covering $\mathfrak{U}$ of σ the $\mathfrak{U}$-mesh of $\mathrm{Sd}_{\mathfrak{U}}^n(\sigma)$ converges to 0.*

So we finally have proved that, after a finite number of steps, the repeated relative subdivision leads to a $\mathfrak{U}$-small subdivision.

Theorem 5.3.26. *For any affine p-simplex σ in Δ_q and any open covering $\mathfrak{U}$ of σ there exists a smallest natural number $s(\sigma)$ such that all simplices in $\mathrm{Sd}_{\mathfrak{U}}^{s(\sigma)}(\sigma)$ are $\mathfrak{U}$-small.*

Definition 5.3.27. *Let σ be an affine p-simplex in euclidean m-space and $\mathfrak{U}$ an open covering of σ. The $\mathfrak{U}$-subdivision number $s_{\mathfrak{U}}(\sigma)$ is the smallest natural number such that all simplices in $\mathrm{Sd}_{\mathfrak{U}}^{s(\sigma)}(\sigma)$ are $\mathfrak{U}$-small:*

$$s_{\mathfrak{U}}(\sigma) := \min\{k \in \mathbb{N} \mid \mathrm{Sd}_{\mathfrak{U}}^k(\sigma) \in S_p(\Delta_q, \mathfrak{U})\}$$

If the covering $\mathfrak{U}$ is known from the context, we drop the subscript $\mathfrak{U}$ and write $s(\sigma)$ for the subdivision number.

Theorem 5.3.28. *For p-simplex σ in a topological apce X and any open covering $\mathfrak{U}$ of σ there exists a smallest natural number $s(\sigma)$ such that all simplices in $\mathrm{Sd}_{\mathfrak{U}}^{s(\sigma)}(\sigma)$ are $\mathfrak{U}$-small.*

Definition 5.3.29. *Let σ be a p-simplex in the topological space X and $\mathfrak{U}$ an open covering of σ. The $\mathfrak{U}$-subdivision number $s_{\mathfrak{U}}(\sigma)$ is the smallest natural number such that all simplices in $\mathrm{Sd}_{\mathfrak{U}}^{s(\sigma)}(\sigma)$ are $\mathfrak{U}$-small:*

$$s_{\mathfrak{U}}(\sigma) := \min\{k \in \mathbb{N} \mid \mathrm{Sd}_{\mathfrak{U}}^k(\sigma) \in S_p(\Delta_q, \mathfrak{U})\}$$

5.4 The Chain Map Csd$_\mathfrak{U}$

In this section we establish the relation between the complexes $S_*(X)$ and $S_*(X,\mathfrak{U})$, where we assume that $\mathfrak{U}$ is an open covering of the space X. The main result is that the $\mathfrak{U}$-relative subdivision provides us with a chain homotopy inverse Csd$_\mathfrak{U}$ for the inclusion $\iota : S_*(X,\mathfrak{U}) \hookrightarrow S_*(X)$. We prove this by explicitly constructing the map Csd$_\mathfrak{U}$.

Definition 5.4.1. *The complete $\mathfrak{U}$-subdivision* Csd$_\mathfrak{U}$ *on p-simplices is the unique homomorphism $S_p(X) \to S_p(X,\mathfrak{U})$ that is given by*

$$\mathrm{Csd}_\mathfrak{U}(\sigma) := \mathrm{Sd}_\mathfrak{U}^{s(\sigma)}(\sigma)$$

for any affine p-simplex σ.

Lemma 5.4.2. *The complete $\mathfrak{U}$-relative subdivisions* Csd$_\mathfrak{U}$ *form a chain endomorphism $L_*(\Delta_q) \to L_*(\Delta_q,\mathfrak{U}) \le L_*(\Delta_q)$.*

Proof. We prove $\partial \,\mathrm{Csd}_\mathfrak{U}(\sigma) = \mathrm{Csd}_\mathfrak{U}(\partial\sigma)$ for simplices of arbitrary dimension by induction on the $\mathfrak{U}$-subdivision number of σ. Since the complete subdivision is the identity on all $\mathfrak{U}$-small simplices the operations Csd$_\mathfrak{U}$ and ∂ commute here. If $\partial\,\mathrm{Csd}_\mathfrak{U}(\sigma) = \mathrm{Csd}_\mathfrak{U}(\partial\sigma)$ holds for all simplices of subdivision number $s' < s$ and σ is an simplex with subdivision number $s(\sigma) = s$ we obtain

$$
\begin{aligned}
\partial\,\mathrm{Csd}_\mathfrak{U}(\sigma) &= \partial\,\mathrm{Sd}_\mathfrak{U}^{s(\sigma)}(\sigma) \\
&= \partial\,\mathrm{Sd}_\mathfrak{U}^{s(\sigma)-1}[\mathbf{b}_\sigma\,\mathrm{Sd}_\mathfrak{U}(\partial\sigma)] \\
&= \mathrm{Sd}_\mathfrak{U}^{s(\sigma)-1}\,\partial[\mathbf{b}_\sigma\,\mathrm{Sd}_\mathfrak{U}(\partial\sigma)] \\
&= \mathrm{Sd}_\mathfrak{U}^{s(\sigma)-1}\,\mathrm{Sd}_\mathfrak{U}(\partial\sigma)] \\
&= \mathrm{Sd}_\mathfrak{U}^{s(\sigma)}(\partial\sigma) \\
&= \mathrm{Csd}_\mathfrak{U}(\partial\sigma)
\end{aligned}
$$

This completes the inductive step. $\qquad\square$

Theorem 5.4.3. *Any complete relative subdivision on $L_*(\Delta_q)$ is chain homotopic to the identity.*

Proof. The proof is analogous to that of Theorem 4.2.7. At first we define for every standard q-simplex Δ_q and family $\mathfrak{U}$ in Δ_q a chain homotopy $\widetilde{ch}_\mathfrak{U}$ from Csd$_\mathfrak{U}$ to the identity. Let $\widetilde{ch}_{\mathfrak{U},0} : L_0(\Delta_q) \to L_1(\Delta_q)$ be the constant function 0 and $\widetilde{ch}_{\mathfrak{U},p} : L_p(\Delta_q) \to L_{p+1}(\Delta_q)$ be defined by induction on the formula

$$\widetilde{ch}_{\mathfrak{U},p}(\sigma) = \mathbf{b}_\sigma\left(\mathrm{Csd}_\mathfrak{U}(\sigma) - \sigma - \widetilde{ch}_{\mathfrak{U},p-1}(\partial\sigma)\right),$$

for affine simplices σ in Δ_q. We will prove $\partial\circ\widetilde{ch}_\mathfrak{U} + \widetilde{ch}_\mathfrak{U}\circ\partial = \mathrm{Csd}_\mathfrak{U} -\mathrm{id}$ by induction on p. For any 0-simplex σ we get $\partial\widetilde{ch}_{\mathfrak{U},0}(\sigma) + \widetilde{ch}_{\mathfrak{U},-1}\partial(\sigma) = 0 = (\mathrm{Csd}_\mathfrak{U} -\mathrm{id})(\sigma)$. Assuming the formula to be true for chains of degree $< p$ we observe

$$\partial\widetilde{ch}_{\mathfrak{U},p-1}\partial = (\mathrm{Csd}_\mathfrak{U} -\mathrm{id} - \widetilde{ch}_{\mathfrak{U},p-2}\circ\partial)\circ\partial = (\mathrm{Csd}_\mathfrak{U} -\mathrm{id})\circ\partial,$$

by the inductive assumption. So that we can compute

$$\partial \widetilde{ch}_{\mathfrak{U},p}(\sigma) = \partial \left(\mathbf{b}_\sigma(\mathrm{Csd}_{\mathfrak{U}}(\sigma) - \sigma - \widetilde{ch}_{\mathfrak{U},p-1}(\partial\sigma)) \right)$$

$$= \left(\mathrm{Csd}_{\mathfrak{U}}(\sigma) - \sigma - \widetilde{ch}_{\mathfrak{U},p-1}(\partial\sigma) \right) - \mathbf{b}_\sigma \left(\partial \mathrm{Csd}_{\mathfrak{U}}(\sigma) - \partial\sigma - \partial\widetilde{ch}_{\mathfrak{U},p-1}\partial(\sigma) \right)$$

$$= \left(\mathrm{Csd}_{\mathfrak{U}}(\sigma) - \sigma - \widetilde{ch}_{\mathfrak{U},p-1}(\partial\sigma) \right) - \mathbf{b}_\sigma \left(\mathrm{Csd}_{\mathfrak{U}}(\partial\sigma) - \partial\sigma - \partial\widetilde{ch}_{\mathfrak{U},p-1}\partial(\sigma) \right)$$

$$= (\mathrm{Csd}_{\mathfrak{U}} - \mathrm{id})(\sigma) - \widetilde{ch}_{\mathfrak{U},p-1}(\partial\sigma) - 0(\partial\sigma),$$

for any p-simplex σ. Hence $\partial \circ \widetilde{ch}_{\mathfrak{U}} + \widetilde{ch}_{\mathfrak{U}} \circ \partial = \mathrm{Csd}_{\mathfrak{U}} - \mathrm{id}$. $\square$

Corollary 5.4.4. *Any complete relative subdivision on $L_*(\Delta_q)$ induces isomorphisms in the homology of $L_*(\Delta_q)$.*

Lemma 5.4.5. *The complete $\mathfrak{U}$-relative subdivisions $\mathrm{Csd}_{\mathfrak{U}}$ form a chain endomorphism $S_*(X) \to S_*(X, \mathfrak{U}) \le S_*(X)$.*

Proof. This follows from the fact that the $\mathfrak{U}$-relative subdivision is a chain map.

To prove an analogue of Theorem 5.4.3 for the chain $S_*(X)$ of singular simplices, one needs to generalise Lemma 4.2.9 and corollary 4.2.10 for complete relative subdivision.

Proposition 5.4.6. *For any affine simplex $f : \Delta_q \to \Delta_{q'}$ and open covering $\mathfrak{U}$ of $\Delta_{q'}$ the following square commutes.*

$$\begin{array}{ccc}
L_p(\Delta_q) & \xrightarrow{\ L_p(f)\ } & L_p(\Delta_{q'}) \\
{\scriptstyle \widetilde{ch}_{f^{-1}(\mathfrak{U}),p}}\Big\downarrow & & \Big\downarrow{\scriptstyle \widetilde{ch}_{\mathfrak{U},p}} \\
L_{p+1}(\Delta_q) & \xrightarrow[\ L_{p+1}(f)\ ]{} & L_{p+1}(\Delta_{q'})
\end{array} \qquad (5.1)$$

Proof. The proof is analogous to the proof of the proof of Proposition 4.2. We prove $L_{p+1}(f) \circ \widetilde{ch}_{f^{-1}(\mathfrak{U}),0} = \widetilde{ch}_{\mathfrak{U},0} \circ L_p(f)$ by induction on p. For $p = 0$ the homomorphisms $\widetilde{ch}_{\mathfrak{U},0}$ and $\widetilde{ch}_{f^{-1}(\mathfrak{U}),0}$ vanish so equality holds in this case. Assume the hypothesis to be true for all $p' < p$ and let σ be an affine p-simplex in Δ_q. Using the inductive definition of the homotopies $\tilde{h}$ we obtain

$$L_{p+1}(f)\widetilde{ch}_{f^{-1}(\mathfrak{U})}(\sigma) = L_{p+1}(f) \left[\mathbf{b}_\sigma \left(\mathrm{Sd}_{f^{-1}(\mathfrak{U})}(\sigma) - \sigma - \widetilde{ch}_{f^{-1}(\mathfrak{U})}(\partial\sigma) \right) \right]$$

$$= \mathbf{b}_{f\sigma} \left[L_p(f) \left(\mathrm{Csd}_{f^{-1}(\mathfrak{U})}(\sigma) - \sigma - \widetilde{ch}_{f^{-1}(\mathfrak{U})}(\partial\sigma) \right) \right]$$

$$= \mathbf{b}_{f\sigma} \left(\mathrm{Csd}_{\mathfrak{U}}(L_p(f)(\sigma) - L_p(f)(\sigma) - \widetilde{ch}_{\mathfrak{U},p-1}(\partial L_p(f)\sigma) \right)$$

$$= \widetilde{ch}_{\mathfrak{U},p}(f\sigma) = \widetilde{ch}_{\mathfrak{U},p} \circ L_p(f)(\sigma)$$

for any affine p-simplex σ in Δ_q. This completes the inductive step. $\square$

Corollary 5.4.7. *For any affine p-simplex σ and open covering $\mathfrak{U}$ of Δ_q the following equation holds:*

$$\widetilde{ch}_{\sigma^{-1}(\mathfrak{U}),p-1}(\mathrm{id}^i_{\Delta_p}) = L_p(\mathrm{id}^i_{\Delta_p}) \circ \widetilde{ch}_{(\sigma^i)^{-1}(\mathfrak{U}),p-1}(\mathrm{id}_{\Delta_{p-1}})$$

Proof. Let σ be an affine p-simplex in Δ_q and $\mathfrak{U}$ an open covering of Δ_q. The i-th face of σ is given by $\sigma^i = \sigma \circ \mathrm{id}^i_{\Delta_p}$. This implies $(\sigma^i)^{-1}(\mathfrak{U}) = (\mathrm{id}^i_{\Delta_p})^{-1}\sigma^{-1}(\mathfrak{U})$. We apply Proposition 5.4.6 to the affine simplex $\mathrm{id}^i_{\Delta_p}$ and the open covering $\sigma^{-1}(\mathfrak{U})$ of Δ_p to obtain

$$\widetilde{ch}_{\sigma^{-1}(\mathfrak{U}),p-1} \circ L_p(\mathrm{id}^i_{\Delta_p}) = L_p(\mathrm{id}^i_{\Delta_p}) \circ \widetilde{ch}_{(\mathrm{id}^i_{\Delta_p})^{-1}\sigma^{-1}(\mathfrak{U}),p} = L_p(\mathrm{id}^i_{\Delta_p}) \circ \widetilde{ch}_{(\sigma^i)^{-1}(\mathfrak{U}),p}.$$

Applying the operators on both sides to the affine simplex $\mathrm{id}_{\Delta_{p-1}}$ leads to the desired equality:

$$\widetilde{ch}_{\sigma^{-1}(\mathfrak{U}),p-1}(\mathrm{id}^i_{\Delta_p}) = \widetilde{ch}_{\sigma^{-1}(\mathfrak{U}),p-1} \circ L_{p-1}(\mathrm{id}^i_{\Delta_p})(\mathrm{id}_{\Delta_{p-1}})$$
$$= L_p(\mathrm{id}^i_{\Delta_p}) \circ \widetilde{ch}_{(\sigma^i)^{-1}(\mathfrak{U}),p}(\mathrm{id}_{\Delta_{p-1}})$$

$\square$

Theorem 5.4.8. *Any complete relative subdivision on $S_*(X)$ is chain homotopic to the identity.*

Proof. We utilise the chain homotopies $\widetilde{ch}_\mathfrak{U}$ for open coverings $\mathfrak{U}$ of the standard simplices to define homomorphisms $ch_p : S_p(X) \to S_{p+1}(X)$. Let ch_p be the unique homomorphism that is given by

$$ch_p(\sigma) = S_{p+1}(\sigma)\left(\widetilde{ch}_{\sigma^{-1}(\mathfrak{U}),p}(\mathrm{id}_{\Delta_p})\right)$$

for any singular p-simplex σ in X. We will prove that ch is a chain homotopy from Csd$_\mathfrak{U}$ to the identity. This is done by induction on p. For any 0-simplex σ we get

$$\partial ch_0(\sigma) + ch_{-1}\partial(\sigma) = \partial S_1(\sigma)\left(\widetilde{ch}_{\sigma^{-1}(\mathfrak{U}),0}(\mathrm{id}_{\Delta_p})\right) + S_0(\sigma)\widetilde{ch}_{\sigma^{-1}(\mathfrak{U}),-1}\partial(\mathrm{id}_{\Delta_p})$$
$$= 0 + 0 = (\mathrm{Csd}_\mathfrak{U} -\mathrm{id})(\sigma).$$

Assuming the formula to be true for chains of degree $< p$ we obtain

$$(\partial ch_p + ch_{p-1}\partial)(\sigma) = \partial S_{p+1}(\sigma)\widetilde{ch}_{\sigma^{-1}(\mathfrak{U}),p}(\mathrm{id}_{\Delta_p})$$
$$+ \sum_{i=0}^{p}(-1)^i S_p(\sigma^i)\widetilde{ch}_{(\sigma^i)^{-1}(\mathfrak{U}),p-1}(\mathrm{id}_{\Delta_{p-1}})$$

But

$$S_p(\sigma^i)\widetilde{ch}_{(\sigma^i)^{-1}(\mathfrak{U}),p-1}(\mathrm{id}_{\Delta_{p-1}}) = S_p(\sigma) \circ S_p(\mathrm{id}^i_{\Delta_p}) \circ \widetilde{ch}_{(\sigma^i)^{-1}(\mathfrak{U}),p-1}(\mathrm{id}_{\Delta_{p-1}})$$
$$= S_p(\sigma)\widetilde{ch}_{\sigma^{-1}(\mathfrak{U}),p-1}(\mathrm{id}^i_{\Delta_p}) \quad \text{By Corollary 5.4.7}$$

So all in all we get

$$(\partial ch_p + ch_{p-1}\partial)(\sigma) = S_p(\sigma)\partial\widetilde{ch}_{\sigma^{-1}(\mathfrak{U}),p}(\mathrm{id}_{\Delta_p}) + \sum_{i=0}^{p}(-1)^i S_p(\sigma)\widetilde{ch}_{\sigma^{-1}(\mathfrak{U}),p-1}(\mathrm{id}^i_{\Delta_p})$$
$$= S_p(\sigma)\left(\partial\widetilde{ch}_{\sigma^{-1}(\mathfrak{U}),p}(\mathrm{id}_{\Delta_p}) + \sum_{i=0}^{p}(-1)^i\widetilde{ch}_{\sigma^{-1}(\mathfrak{U}),p-1}(\mathrm{id}^i_{\Delta_p})\right)$$
$$= S_p(\sigma)\left(\partial\widetilde{ch}_{\sigma^{-1}(\mathfrak{U}),p}(\mathrm{id}_{\Delta_p}) + \widetilde{ch}_{\sigma^{-1}(\mathfrak{U}),p-1}(\partial\mathrm{id}_{\Delta_p})\right)$$
$$= S_p(\sigma)\left(\mathrm{Csd}_{\sigma^{-1}(\mathfrak{U}),p}(\mathrm{id}_{\Delta_p}) - \mathrm{id}_{\Delta_p}\right)$$
$$= \mathrm{Csd}_\mathfrak{U}(\sigma) - \sigma$$

for any affine p-simplex σ. Thus ch is a chain homotopy from Csd$_\mathfrak{U}$ to id. $\square$

Corollary 5.4.9. *The complete subdivision induces isomorphisms in singular homology.*

Corollary 5.4.10. *The complete subdivision induces isomorphisms in singular co-homology*

The advantage of using relative subdivision lies in the fact that $\mathfrak{U}$-small faces of simplices will be unchanged by Csd, so exactly those faces which are not $\mathfrak{U}$-small are subdivided. This will be of fundamental importance in investigating the extendibility of local cocycles in Chapter 15.

Lemma 5.4.11. *For any family $\mathfrak{B}$ in X the subcomplex $S_*(X, \mathfrak{B})$ of $S_*(X)$ is invariant under complete relative subdivision.*

Proof. Let $\sigma \in S_*(X, \mathfrak{B})$ $\mathfrak{U}$ and $\mathfrak{B}$ be two families in X. We prove $\mathrm{Csd}_{\mathfrak{U}}(\sigma) \in S_*(X, \mathfrak{B})$ for $\mathfrak{B}$-small simplices of arbitrary dimension by induction on the subdivision number. If $s(\sigma) = 0$, i.e. σ is $\mathfrak{U}$-small, the complete relative subdivision $\mathrm{Csd}_{\mathfrak{U}}$ maps σ to σ. So all $\mathfrak{U}$-small simplices in $S_*(X, \mathfrak{B})$ are mapped into $S_*(X, \mathfrak{B})$. Assume the the hypothesis $\mathrm{Csd}_{\mathfrak{U}}(\tau) \in S_*(X, \mathfrak{B})$ holds for all $\mathfrak{B}$-small simplices τ with subdivision number $s'(\tau) < s(\sigma)$. The complete relative subdivision of σ is given by

$$
\begin{aligned}
\mathrm{Csd}_{\mathfrak{U}}(\sigma) &= \mathrm{Sd}_{\mathfrak{U}}^{s(\sigma)}(\sigma) \\
&= \mathrm{Sd}_{\mathfrak{U}}^{s(\sigma)-1}\left(\mathrm{Sd}_{\mathfrak{U}}(\sigma)\right) \\
&= \mathrm{Csd}_{\mathfrak{U}}\left(\mathrm{Sd}_{\mathfrak{U}}(\sigma)\right).
\end{aligned}
$$

The subcomplex $S_*(X, \mathfrak{B})$ is invariant under relative subdivision and the simplices in the chain $\mathrm{Sd}_{\mathfrak{U}}(\sigma)$ all have subdivision number $< S(\sigma)$. So we can apply the induction hypothesis to obtain $\mathrm{Csd}_{\mathfrak{U}}(\sigma) \in S_*(X, \mathfrak{B})$. □

Like for the relative subdivision we obtain that for any two families $\mathfrak{U}$ and $\mathfrak{B}$ in X the complete $\mathfrak{U}$-relative subdivision restricts to an endomorphism of $S_*(X, \mathfrak{B})$. This endomorphism is also chain homotopic to the identity.

Theorem 5.4.12. *Any restriction of a complete relative subdivision to the complex $S_*(X, \mathfrak{B})$ is chain homotopic to the identity.*

Proof. The proof is analogous to that of Theorem 5.4.8. We only have to verify that the maps $\widetilde{ch}_{\mathfrak{U},p}$ map $\mathfrak{B}$-small simplices to $\mathfrak{B}$-small simplices. Let $\mathfrak{U}$ be an open covering of the space X and $\mathfrak{B}$ a family in X. The homomorphisms ch_p are given by

$$
ch_p(\sigma) = S_{p+1}(\sigma)\left(\widetilde{ch}_{\sigma^{-1}(\mathfrak{U}),p}(\mathrm{id}_{\Delta_p})\right)
$$

for any singular p-simplex σ in X. If a p-simplex σ is $\mathfrak{B}$-small, then the affine simplex id_{Δ_p} is $\sigma^{-1}(\mathfrak{B})$-small. This implies that the chain $\widetilde{ch}_{\sigma^{-1}(\mathfrak{U}),p}(\mathrm{id}_{\Delta_p})$ lies in $L_{p+1}(\Delta_p, \sigma^{-1}(\mathfrak{B}))$. Hence the chain $S_{p+1}(\sigma)\left(\widetilde{ch}_{\sigma^{-1}(\mathfrak{U}),p}(\mathrm{id}_{\Delta_p})\right)$ lies in $S_{p+1}(X, \mathfrak{B})$. □

Corollary 5.4.13. *Any restriction of a complete relative subdivision to the complex $S_*(X, \mathfrak{B})$ induces an isomorphism in homology.*

5.5 The Vertex Transformation

There exists a natural map $H_{AS}(X; V) \to H(X; V)$ from the classical Alexander-Spanier cohomology $H_{AS}(X; V)$ of a topological space X into its singular homology $H(X; V)$. Below we verify that the construction of this map carries over to yield a morphism between topological cohomology theories $H_{AS}^t(-; V)$ and $H^t(-; V)$.

Recall that the inclusion $\mathcal{O} \hookrightarrow \mathbf{Top}$ of the simplicial category as a subcategory of discrete topological spaces is a discrete cosimplicial space. There exists an embedding $i : \mathcal{O} \hookrightarrow \Delta$ of cosimplicial topological spaces which maps an element $i \in [n]$ to the vertex $\mathbf{e}_i$ of Δ_n. This embedding induces a natural transformation

$$C(i, -) : C(\Delta, -) \to C([n], -) = \mathrm{SS}, \quad C(i, X) : C(\Delta, X) \to \mathrm{SS}(X)([n]) \cong X^{n+1},$$

between the singular simplicial space functor $C(\Delta, -)$ and the function space functor $C([n], -)$. For each topological space X the n-th component $C(i([n]), X)$ of the morphism $C(i, X)$ of simplicial spaces maps a singular n-simplex σ in X to the simplex $(\sigma(\mathbf{e}_0), \ldots, \sigma(\mathbf{e}_n))$ of its vertices in X^{n+1}.

Lemma 5.5.1. *If $\mathfrak{U}$ is a family of subsets of a topological space X, then morphism $C(i, X)$ restricts to a morphism $C(i, X, \mathfrak{U}) : C(\Delta, X, \mathfrak{U}) \to \mathrm{SS}(X, \mathfrak{U})$ of simplicial topological spaces, i.e. it maps $\mathfrak{U}$-small singular simplices to $\mathfrak{U}$-small simplices in $\mathrm{SS}(X)$.*

Proof. If $\mathfrak{U}$ is a family of subsets of X and σ is a $\mathfrak{U}$-small singular n-simplex, then it is contained in one of the sets $U \in \mathfrak{U}$. In this case the simplex $(\sigma(\mathbf{e}_0), \ldots, \sigma(\mathbf{e}_n))$ is contained in U^{n+1} and thus $\mathfrak{U}$-small as well. $\qquad\square$

Definition 5.5.2. *The natural transformation $C(i, -)$ is called the* vertex transformation *and is denoted by λ. If $\mathfrak{U}$ is a family of subsets of a topological space X, then the morphism $C(i, X, \mathfrak{U})$ is called the* vertex map *and is denoted by $\lambda_{X, \mathfrak{U}}$.*

In case a family of subsets of a topological space X contains the space X itself, we drop the subscript $\mathfrak{U}$ and simply write λ_X for the morphism $C(\Delta, X) \to \mathrm{SS}(X)$ of simplicial topological spaces.

If V is a topological group we can an apply the contravariant function space functor $C(-, V)$ to the natural transformation λ to obtain a natural transformation $C(\lambda, V) : A(-; V) \to S^t(-; V)$. Similarly, if $\mathfrak{U}$ is a family of subsets of X, then the restricted morphism $\lambda_{X, \mathfrak{U}} : C(\Delta, X, \mathfrak{U}) \to \mathrm{SS}(X, \mathfrak{U})$ of simplicial topological spaces induces a morphism $C(\lambda_{X, \mathfrak{U}}, V) : A^t(X, \mathfrak{U}; V) \to S^t(X, \mathfrak{U}; V)$ of cosimplicial topological groups. This is functorial in the pairs $(X, \mathfrak{U})$ of the category $\mathbf{TF}$ (cf. Definition 2.3.23) by construction: If $\mathfrak{V} \prec \mathfrak{U}$ is a refinement of $\mathfrak{U}$, then the inclusion $\mathrm{SS}(X, \mathfrak{V}) \hookrightarrow \mathrm{SS}(X, \mathfrak{U})$ induces restriction morphisms $S^t(X, \mathfrak{U}; V) \to S^t(X, \mathfrak{V}; V)$ and $A^t(X, \mathfrak{U}; V) \to A^t(X, \mathfrak{V}; V)$ and the diagram

$$
\begin{array}{ccc}
A^t(X, \mathfrak{U}; V) & \xrightarrow{C(\lambda_{X, \mathfrak{U}}, V)} & S^t(X, \mathfrak{U}; V) \\
\downarrow & & \downarrow \\
A^t(X, \mathfrak{V}; V) & \xrightarrow{C(\lambda_{X, \mathfrak{V}}, V)} & S^t(X, \mathfrak{V}; V)
\end{array}
$$

is commutative. These morphisms do not directly descend to the colimit groups, but if $\mathfrak{U}$ and $\mathfrak{V}$ are open coverings, then we can compose the vertex maps $\lambda_{X, \mathfrak{U}}$ and $\lambda_{X, \mathfrak{V}}$ with the complete relative subdivision morphisms $\mathrm{Csd}_{\mathfrak{U}}$ and $\mathrm{Csd}_{\mathfrak{V}}$ respectively to extend the commutative diagram on the right hand side:

$$
\begin{array}{ccccc}
A^t(X, \mathfrak{U}; V) & \xrightarrow{C(\lambda_{X, \mathfrak{U}}, V)} & S^t(X, \mathfrak{U}; V) & \xrightarrow{\mathrm{Csd}_{\mathfrak{U}}} & S^t(X; V) \\
\downarrow & & \downarrow & & \downarrow{\scriptstyle \mathrm{Csd}_{\mathfrak{V}}} \\
A^t(X, \mathfrak{V}; V) & \xrightarrow{C(\lambda_{X, \mathfrak{V}}, V)} & S^t(X, \mathfrak{V}; V) & \xrightarrow{\mathrm{Csd}_{\mathfrak{V}}} & S^t(X; V)
\end{array}
$$

Restricting ourselves to Alexander-Spanier cocycles, we obtain form each $n \in \mathbb{N}$ the commutative diagram

$$Z^{tn}(X,\mathfrak{U};V) \xrightarrow{C(\lambda_{X,\mathfrak{U},n},V)} Z^{tn}(X,\mathfrak{U};V) \xrightarrow{\mathrm{Csd}_{\mathfrak{U}}^{n}} Z^{tn}(X;V)$$

$$\downarrow \qquad\qquad\qquad \downarrow \qquad\qquad\qquad \downarrow \mathrm{Csd}_{\mathfrak{V}}$$

$$Z^{tn}(X,\mathfrak{V};V) \xrightarrow{C(\lambda_{X,\mathfrak{V},n},V)} Z^{tn}(X,\mathfrak{V};V) \xrightarrow{\mathrm{Csd}_{\mathfrak{V}}^{n}} Z^{tn}(X;V)$$

Any complete relative subdivision induces the identity on the topologised singular cohomology ${}^{t}H(X;V)$. Thus, descending to the factor spaces on the right hand side we obtain the commutative diagram

$$Z^{tn}(X,\mathfrak{U};V) \xrightarrow{C(\lambda_{X,\mathfrak{U},n},V)} H^{tn}(X,\mathfrak{U};V) \xrightarrow[\cong]{\mathrm{Csd}_{\mathfrak{U}}^{n}} H^{tn}(X;V)$$

$$\downarrow \qquad\qquad\qquad \downarrow{\cong} \qquad\qquad\qquad \|$$

$$Z^{tn}(X,\mathfrak{V};V) \xrightarrow{C(\lambda_{X,\mathfrak{V},n},V)} H^{tn}(X,\mathfrak{V};V) \xrightarrow[\cong]{\mathrm{Csd}_{\mathfrak{V}}^{n}} H^{tn}(X;V)$$

So the homomorphisms $\mathrm{Csd}_{\mathfrak{U}}^{n} C(\lambda_{X,\mathfrak{U},n},V) : Z^{tn}(X,\mathfrak{U};V) \to H^{tn}(X;V)$ for different open coverings $\mathfrak{U}$ form a cocone under the groups $Z^{zn}(X;V)$. As a consequence they all factor through a unique homomorphism $\lambda_{AS}^{n} : Z_{AS}^{tn}(X;V) \to H^{tn}(X;V)$. Since this homomorphism is trivial on Alexander-Spanier coboundaries by construction, it induces a homomorphism $\lambda_{AS}^{n} : H_{AS}^{n}(X;V) \to H^{tn}(X;V)$. The homomorphisms λ_{AS}^{n} connect the singular topological cohomology and the Alexander-Spanier topological cohomology. Moreover this homomorphism is part of a natural transformation $\lambda_{AS}^{*} : H_{AS}^{t}(-;-) \to H^{t}(-;-)$.

Remark 5.5.3. The homomorphism $H_{AS}(X;V) \to H(X;V)$ of the underlying (abstract) groups will also be denoted by λ_{AS}^{n}.

Likewise one can show that – under mild assumptions on the space X – there exists a transformation between the topological cohomology theories $H_{AS}^{t}(-;-)$ and $H^{t}(-;-)$ which is a right inverse to $H(\lambda_{AS})$. This transformation will be constructed in Chapter 3.

Relative Subdivision In Transformation Semi-Groups

In this chapter we examine how the relative subdivision can be applied to transformation semi-groups (G, X). This in particular includes the equivariance of the subdivision chain map $\mathrm{Sd}_{\mathfrak{U}} : S_*(X) \to S_*(X)$ and the complete relative subdivision chain map $\mathrm{Csd}_{\mathfrak{U}} : S_*(X) \to S_*(X, \mathfrak{U})$ for G-invariant open coverings $\mathfrak{U}$. It is especially fruitful in the case of highly connected spaces. Here the relative subdivision can be used to extend equivariant Alexander-Spanier cocycles to global equivariant cocycles (see section 6.2 below), a case which includes the extension of germs of (homogeneous) group cocycles.

6.1 Transformation Semi-Groups

Let (G, X) be a transformation semi-group and $\mathfrak{U}$ be a family of subsets of X. The action of G on X induces an action of G on the chain complex $S_*(X)$ of singular simplices. We first observe that the $\mathfrak{U}$-relative subdivision operators are equivariant whenever the family $\mathfrak{U}$ is G-invariant.

Lemma 6.1.1. *For any G-invariant family $\mathfrak{U}$ in the G-space X the $\mathfrak{U}$-relative subdivisions $\mathrm{Sd}_{\mathfrak{U}} : S_n(X) \to S_n(X)$ are $\mathbb{Z}[G]$-module homomorphisms.*

Proof. Let $\mathfrak{U}$ be an G-invariant family in X, i.e. $g.U \in \mathfrak{U}$ for all $g \in G$ and $U \in \mathfrak{U}$. For any $g \in G$ and any n-simplex σ in X the $\mathfrak{U}$-relative subdivision of $g.\sigma$ is given by

$$\mathrm{Sd}_{\mathfrak{U}}(g.\sigma) = S_n(g.\sigma)\mathrm{LSd}_{(g.\sigma)^{-1}(\mathfrak{U})}(\Delta_n)$$

Since the family $\mathfrak{U}$ is G-invariant, we have $(g.\sigma)^{-1}(\mathfrak{U}) = \sigma^{-1}(\mathfrak{U})$. Furthermore the simplices $S_n(g.\sigma)(\tau)$ and $g.(S_n(\sigma)(\tau))$ are equal for arbitrary n-simplices τ in Δ_n. So we obtain

$$\mathrm{Sd}_{\mathfrak{U}}(g.\sigma) = g.S_n(\sigma)\mathrm{LSd}_{\sigma^{-1}(\mathfrak{U})}(\Delta_n) = g.\,\mathrm{Sd}_{\mathfrak{U}}(\sigma),$$

which was to be proved. $\square$

As a consequence the $\mathfrak{U}$-relative subdivision is a morphism $\mathrm{Sd}_{\mathfrak{U}} : S_*(X) \to S_*(X)$ of chain complexes in $\mathbb{Z}[\mathbf{G}]\mathbf{mod}$. Next we want to assure that the subdivision is actually chain homotopic to the identity in the category $\mathbb{Z}[\mathbf{G}]\mathbf{mod}$.

Theorem 6.1.2. *For any G-invariant family $\mathfrak{U}$ in the G-space X the $\mathfrak{U}$-relative subdivision of the complex $S_*(X)$ is chain homotopic to the identity in $\mathbb{Z}[\mathbf{G}]\mathbf{mod}$.*

Proof. The proof is analogous to that of Theorem 4.2.11. We only have to verify that the homotopy h defined in this proof consists actually of $\mathbb{Z}[G]$-module homomorphisms. Let $\mathfrak{U}$ be a G-invariant family in X. Please recall from the proof of Theorem 4.2.11 that the homomorphisms $h_n : S_n(X) \to S_{n+1}(X)$ are defined via

$$h_n(\sigma) := S_{n+1}(\sigma)\left(\tilde{h}_{\sigma^{-1}(\mathfrak{U}),n}(\mathrm{id}_{\Delta_n})\right)$$

for n-simplices σ in X. The map $\tilde{h}_{\sigma^{-1}(\mathfrak{U}),n}$ used here depends on the family $\sigma^{-1}(\mathfrak{U})$ in X. Since the family $\mathfrak{U}$ in X is G-invariant, the families $\sigma^{-1}(\mathfrak{U})$ and $(g.\sigma)^{-1}(\mathfrak{U})$ are equal for any $g \in G$. Thus we obtain

$$\begin{aligned}
h_n(g.\sigma) &= S_{n+1}(g.\sigma)\left(\tilde{h}_{(g.\sigma)^{-1}(\mathfrak{U}),n}(\mathrm{id}_{\Delta_n})\right)\\
&= S_{n+1}(g.\sigma)\left(\tilde{h}_{(\sigma^{-1}(\mathfrak{U}),n}(\mathrm{id}_{\Delta_n})\right)\\
&= g.\left[S_{n+1}(\sigma)\left(\tilde{h}_{(g.\sigma)^{-1}(\mathfrak{U}),n}(\mathrm{id}_{\Delta_n})\right)\right]\\
&= g.h_n(\sigma).
\end{aligned}$$

Therefore the maps h_n are $\mathbb{Z}[G]$-module homomorphisms and $\mathrm{Sd}_{\mathfrak{U}}$ is chain homotopic to the identity in $\mathbb{Z}[\mathbf{G}]\mathbf{mod}$. □

Corollary 6.1.3. *For any G-invariant open covering $\mathfrak{U}$ of the G-space X the $\mathfrak{U}$-relative subdivision induces the identity $\mathbb{Z}[G]$-module isomorphism of $H_*(X)$ on singular homology.*

Moreover, if G is a group and the family $\mathfrak{U}$ is G-invariant, then a singular n-simplex σ in X is $\mathfrak{U}$-small whenever any of its translates $g.\sigma$ is $\mathfrak{U}$-small. So we note:

Lemma 6.1.4. *For any G-invariant open covering $\mathfrak{U}$ of the space X of a transformation group (G, X) the $\mathfrak{U}$-subdivision numbers of singular simplices in X are unchanged by the action of G, i.e. the morphism $s_{\mathfrak{U}} : S(X) \to \mathbb{R}$ into the constant simplicial trivial G-module R is a morphism of simplicial zg-modules.*

Lemma 6.1.5. *For any G-invariant open covering $\mathfrak{U}$ of the G-space X the complete $\mathfrak{U}$-relative subdivision $\mathrm{Csd}_{\mathfrak{U}} : S_*(X) \to S_*(X)$ of the complex $S_*(X)$ is a $\mathbb{Z}[G]$-module homomorphism.*

Proof. The proof is analogous to that of Lemma 6.1.1. □

Theorem 6.1.6. *For any G-invariant open covering $\mathfrak{U}$ of the homogeneous space X the complete $\mathfrak{U}$-relative subdivision of the complex $S_*(X)$ is chain homotopic to the identity in $\mathbb{Z}[\mathbf{G}]\mathbf{mod}$.*

Proof. Let $\mathfrak{U}$ be a G-invariant open covering of the G-space X. The complete $\mathfrak{U}$-relative subdivision $\mathrm{Csd}_{\mathfrak{U}}$ is chain homotopic to the identity by Theorem 5.4.8. We only need to verify, that the homomorphisms ch defined in the proof of Theorem 5.4.8 are actually $\mathbb{Z}[G]$-module homomorphisms. These homomorphisms are given by

$$ch_p(\sigma) = S_{p+1}(\sigma)\left(\widetilde{ch}_{\sigma^{-1}(\mathfrak{U}),p}(\mathrm{id}_{\Delta_p})\right)$$

for any singular p-simplex σ in X. Since the covering $\mathfrak{U}$ is G-invariant, the coverings $\sigma^{-1}(\mathfrak{U})$ and $(g.\sigma)^{-1}(\mathfrak{U})$ coincide for any $g \in G$. Therefore we obtain

$$\begin{aligned}
ch_p(g.\sigma) &= S_{p+1}(g.\sigma)\left(\widetilde{ch}_{(g.\sigma)^{-1}(\mathfrak{U}),p}(\mathrm{id}_{\Delta_p})\right)\\
&= S_{p+1}(g.\sigma)\left(\widetilde{ch}_{(\sigma)^{-1}(\mathfrak{U}),p}(\mathrm{id}_{\Delta_p})\right)\\
&= g.\left(S_{p+1}(\sigma)\,\widetilde{ch}_{(\sigma)^{-1}(\mathfrak{U}),p}(\mathrm{id}_{\Delta_p})\right)\\
&= g.(ch_p(\sigma))
\end{aligned}$$

Thus the maps ch_p are $\mathbb{Z}[G]$-module homomorphisms. □

Corollary 6.1.7. *For any G-invariant open covering $\mathfrak{U}$ of the G-space X the complete $\mathfrak{U}$-relative subdivision induces the identity $\mathbb{Z}[G]$-module isomorphism of $H_*(X)$ on singular homology.*

Let $i_* : S_*(,\mathfrak{U}) \hookrightarrow S_*(X)$ denote the inclusion of the subcomplex $S_*(X,\mathfrak{U})$ consisting of all $\mathfrak{U}$)-small singular simplices. Since this inclusion is a right inverse to the complete $\mathfrak{U}$-relative subdivision $\mathrm{Csd}_\mathfrak{U} : S_*(X) \hookrightarrow S_*(,\mathfrak{U})$ we observe:

Corollary 6.1.8. *For any G-invariant open covering $\mathfrak{U}$ of X the morphisms i_* and $\mathrm{Csd}_\mathfrak{U}$ of chain complexes in $\mathbb{Z}[\mathbf{G}]\mathbf{mod}$ are homotopy inverses of each other.*

In the following we consider two different families $\mathfrak{U}$ and $\mathfrak{V}$ of subsets of X and observe the obvious. (Though it is obvious it has to be proven to be referred to.)

Lemma 6.1.9. *For any two families $\mathfrak{U}$ and $\mathfrak{B}$ in the homogeneous space X the subcomplex $S_*(X,\mathfrak{B})$ is $\mathrm{Sd}_\mathfrak{U}$-invariant.*

Proof. Let $\mathfrak{U}$ and $\mathfrak{B}$ be two families in X. The $\mathfrak{U}$-relative subdivision of a $\mathfrak{B}$-small n-simplex σ is given by

$$\mathrm{Sd}_\mathfrak{U}(\sigma) = S_n(\sigma)\mathrm{LSd}_{\sigma^{-1}(\mathfrak{U})}(id_{\Delta_n}).$$

So the image $\tau(\Delta_n)$ of any simplex τ in the chain $\mathrm{Sd}_\mathfrak{U}(\sigma)$ is contained in the image of σ. Thus the chain $\mathrm{Sd}_\mathfrak{U}(\sigma)$ consists of $\mathfrak{B}$-small simplices only.   □

Lemma 6.1.10. *For any G-invariant family $\mathfrak{B}$ in the G-space X the subcomplex $S_*(X,\mathfrak{B})$ is G-invariant, i.e. $S_*(X,\mathfrak{B})$ is a subcomplex of $S_*(X)$ in $\mathbb{Z}[\mathbf{G}]\mathbf{mod}$.*

Proof. Let $\mathfrak{B}$ be an G-invariant family in the G-space X. For any $\mathfrak{B}$-small n-simplex σ there exists by definition a set $B \in \mathfrak{B}$ such that $\sigma(\Delta_n) \subset B$. Because $\mathfrak{B}$ is G-invariant we have $g.\sigma(\Delta_n) \subset g.B \in \mathfrak{B}$ for all $g \in G$. So the action of G maps $\mathfrak{B}$-small simplices to $\mathfrak{B}$-small simplices.   □

Lemma 6.1.11. *For any two G-invariant families $\mathfrak{U}$ and $\mathfrak{B}$ in the G-space X the $\mathfrak{U}$-relative subdivision of $S_*(X)$ restricts to an endomorphism of the chain complex $S_*(X,\mathfrak{B})$ in $\mathbb{Z}[\mathbf{G}]\mathbf{mod}$.*

Proof. Let $\mathfrak{U}$ and $\mathfrak{B}$ be two G-invariant families in X. By Lemma 6.1.10 the complex $S_*(X,\mathfrak{B})$ is a subcomplex of $S_*(X)$ in $\mathbb{Z}[\mathbf{G}]\mathbf{mod}$. By Lemma 6.1.9 this subcomplex is invariant under the $\mathbb{Z}G$-module endomorphism $\mathrm{Sd}_\mathfrak{U}$ of $S_*(X)$. Thus the $\mathfrak{U}$-relative subdivision restricts to an endomorphism of the chain complex $S_*(X,\mathfrak{B})$ in $\mathbb{Z}[\mathbf{G}]\mathbf{mod}$.   □

Theorem 6.1.12. *For any two G-invariant families $\mathfrak{U}$ and $\mathfrak{B}$ in the G-space X the restriction of the $\mathfrak{U}$-relative subdivision to the complex $S_*(X,\mathfrak{B})$ is chain homotopic to the identity in $\mathbb{Z}[\mathbf{G}]\mathbf{mod}$.*

Proof. Let $\mathfrak{U}$ and $\mathfrak{B}$ be two G-invariant families in X. By Theorem 6.1.2 the $\mathfrak{U}$-relative subdivision $\mathrm{Sd}_\mathfrak{U}$ is chain homotopic to the identity in $\mathbb{Z}[\mathbf{G}]\mathbf{mod}$. We only need to verify, that the homotopy used there maps $\mathfrak{B}$-small simplices to chains of $\mathfrak{B}$-small simplices. Let $\sigma \in S_n(X,\mathfrak{B})$ be an n-small simplex in X. The chain $h_n(\sigma)$ is given by

$$\mathrm{Sd}_\mathfrak{U}(\sigma) = S_n(\sigma)\mathrm{LSd}_{\sigma^{-1}(\mathfrak{U})}(id_{\Delta_n}).$$

Thus all the images of simplices in the chain $h_n(\sigma)$ are contained in the image of σ, hence $\mathfrak{B}$-small.   □

Let V be a G-module. Tensoring the chain complexes in $\mathbb{Z}[\mathbf{G}]\mathbf{mod}$ with the fixed $\mathbb{Z}[G]$-module V in $\mathbb{Z}[\mathbf{G}]\mathbf{mod}$ preserves inclusions and homotopies, so we obtain:

Corollary 6.1.13. *For any two G-invariant families $\mathfrak{U}$ and $\mathfrak{B}$ in the G-space X the restriction of the $\mathfrak{U}$-relative subdivision to the complex $S_*(X, \mathfrak{B}) \otimes_{\mathbb{Z}[G]} V$ is chain homotopic to the identity in $\mathbb{Z}[\mathbf{G}]\mathbf{mod}$.*

Corollary 6.1.14. *For any two G-invariant families $\mathfrak{U}$ and $\mathfrak{B}$ in the G-space X the $\mathfrak{U}$-relative subdivision of $S_*(X, \mathfrak{B})$ induces a $\mathbb{Z}[G]$-module isomorphism of the homology $H_*(X, \mathfrak{B})$.*

Lemma 6.1.15. *For any G-invariant family $\mathfrak{B}$ in X and G-invariant open covering $\mathfrak{U}$ of the G-space X the complete $\mathfrak{U}$-relative subdivision of the complex $S_*(X, \mathfrak{B})$ is a morphism of chain complexes in $\mathbb{Z}[\mathbf{G}]\mathbf{mod}$.*

Proof. Let $\mathfrak{B}$ be an G-invariant family in X and $\mathfrak{U}$ be an open covering of X. The complete $\mathfrak{U}$-subdivision of an n-simplex $\sigma \in S_n(X, \mathfrak{B})$ is given by

$$\mathrm{Csd}_{\mathfrak{U}}(\sigma) = \mathrm{Sd}_{\mathfrak{U}}^{s(\sigma)}(\sigma).$$

Since the $\mathfrak{U}$-subdivision number is unchanged by the action of G and the $\mathfrak{U}$-relative subdivision $\mathrm{Sd}_{\mathfrak{U}}$ is a $\mathbb{Z}[G]$-module homomorphism we obtain

$$\begin{aligned} \mathrm{Csd}_{\mathfrak{U}}(g.\sigma) &= \mathrm{Sd}_{\mathfrak{U}}^{s(g.\sigma)}(g.\sigma) \\ &= \mathrm{Sd}_{\mathfrak{U}}^{s(\sigma)}(g.\sigma) \\ &= g.\,\mathrm{Sd}_{\mathfrak{U}}^{s(\sigma)}(\sigma), \end{aligned}$$

for any $g \in G$. Thus $\mathrm{Csd}_{\mathfrak{U}}$ is a $\mathbb{Z}[G]$-module homomorphism.   □

Theorem 6.1.16. *For any G-invariant family $\mathfrak{B}$ in X and G-invariant open covering $\mathfrak{U}$ of the G-space X the complete $\mathfrak{U}$-relative subdivision of the complex $S_*(X, \mathfrak{B})$ is chain homotopic to the identity in $\mathbb{Z}[\mathbf{G}]\mathbf{mod}$.*

Proof. Let $\mathfrak{B}$ be an G-invariant family in X and $\mathfrak{U}$ a G-invariant open covering of X. The restriction of the $\mathfrak{U}$-relative subdivision to $S_*(X, \mathfrak{B})$ is chain homotopic to the identity by Theorem 5.4.12. In the proof of Theorem 6.1.6 we have actually shown, that the homomorphisms ch_p of the chain homotopy are $\mathbb{Z}[G]$-module homomorphisms. Thus the complete $\mathfrak{U}$-relative subdivision of the complex $S_*(X, \mathfrak{B})$ is chain homotopic to the identity in $\mathbb{Z}[\mathbf{G}]\mathbf{mod}$.   □

Corollary 6.1.17. *For any G-invariant family $\mathfrak{B}$ in X and G-invariant open covering $\mathfrak{U}$ of the G-space X the complete $\mathfrak{U}$-relative subdivision of $S_*(X, \mathfrak{B})$ induces a $\mathbb{Z}[G]$-module isomorphism of $H_*(X, \mathfrak{B})$.*

6.2 Subdivision of Local Simplices and Local Cochains

In this section we assume that the transformation semi-group (G, X) admits an equivariant system $\hat{\sigma} : U \to \hom_{\mathbf{Top}}(\Delta, X)$ of simplices defined on a G-invariant simplicial subspace U of $\mathrm{SS}(X)$ that consists of neighbourhoods of the diagonals. Examples of such spaces are locally contractible groups G (e.g. Lie groups) or strongly equivariantly locally contractible G-spaces. Note that we do not assume the system $\hat{\sigma}$ of simplices to be continuous (so we use the notation $\hom_{\mathbf{Top}}(\Delta, X)$ to denote the simplicial set of singular simplices in X.) In the following we fix the simplicial space U and the equivariant system $\hat{\sigma} : U \to \hom(\Delta, X)$ of simplices defined on U.

Definition 6.2.1. *The points in the spaces $U([n])$ are called U-local n-simplices.*

We use the equivariant system $\hat{\sigma}$ of simplices to relate the local simplices in U and the singular simplices in X. Similar to the simplicial free abelian group $S(X)$ on the simplicial set $\hom_{\textbf{Top}}(\Delta, X)$ we consider the simplicial free abelian group on the simplicial set underlying U. To ease notation we avoid the usage of the forgetful functor $\textbf{Top} \to \textbf{Set}$ in this section and shall use the symbol U for the simplicial space and the underlying simplicial set. The particular meaning will be clear from the context. Likewise we will use the symbol $\text{SS}(X)$ for the underlying simplicial set of the simplicial topological space $\text{SS}(X)$.

Definition 6.2.2. *The simplicial free abelian group on the simplicial set of all U-local simplices in U is denoted by* $\mathbb{Z}^{(U)}$.

The n-th component of this simplicial group is the free Abelian group $\mathbb{Z}^{(U([n]))}$ on the set $U([n])$ of all U-local n-simplices in X. The morphism $\hat{\sigma} : U \to \hom_{\textbf{Top}}(\Delta, X)$ of simplicial sets extends uniquely to a morphism

$$\hat{\sigma} : \mathbb{Z}^{(U)} \to S(X) = \mathbb{Z}^{(\hom_{\textbf{Top}}(\Delta, X))}$$

of simplicial abelian groups. As indicated above we will denote this morphism by $\hat{\sigma}$ also. The morphism $\hat{\sigma}$ assigns to each point $\mathbf{x} = (x_0, \ldots, x_n)$ in $U([n])$ a singular n-simplex in X whose vertices are the points $x_0, \ldots, x_n$.

Lemma 6.2.3. *The morphism* $\hat{\sigma} : \mathbb{Z}^{(U)} \to S(X)$ *is a morphism of simplicial* $\mathbb{Z}[G]$-*modules.*

Proof. The system $\hat{\sigma} : U \to \hom_{\textbf{Top}}(\Delta, X)$ of simplices was assumed to be equivariant. As a consequence the morphism $\hat{\sigma} : \mathbb{Z}^{(U)} \to S(X)$ is equivariant as well. $\square$

The morphism $\hat{\sigma}$ in conjunction with the vertex map $\lambda : \hom(\Delta, X) \to \text{SS}(X)$ can be used to define a $\mathfrak{U}$-relative subdivision of U-local simplices for any family $\mathfrak{U}$ of subsets of X. Here fore we consider the chain complexes $\mathbb{Z}_*^{(U)}$ and $S_*(X)$ of G-modules associated to the simplicial $\mathbb{Z}[G]$-modules $\mathbb{Z}^{(U)}$ and $S(X)$. On the level of chain complexes one obtains the equivariant morphisms

$$\mathbb{Z}_*^{(U)} \xrightarrow{\hat{\sigma}_*} S_*(X) \xrightarrow{\text{Sd}_{\mathfrak{U}} *} S_*(X) \xrightarrow{\lambda_*} \mathbb{Z}_*^{(\text{SS}(X))}$$

of chain complexes. Note that the property that $\hat{\sigma}$ is a morphism of simplicial sets guarantees that any subdivision $(\lambda \, \text{Sd}_{\mathfrak{U}} \, \hat{\sigma})_n(\mathbf{x})$ of a U-local n-simplex is again U-local. Therefore the composition of all the morphisms in the above chain of morphisms can be corestricted to the complex $\mathbb{Z}_*^{(U)}$. In this fashion one obtains an endomorphism of the chain complex $\mathbb{Z}_*^{(U)}$. The latter endomorphism is equivariant by construction.

Definition 6.2.4. *The endomorphism* $\lambda_* \, \text{Sd}_{\mathfrak{U}*} \, \hat{\sigma}_*$ *of the chain complex* $\mathbb{Z}_*^{(U)}$ *is called the* $\mathfrak{U}$-*relative subdivision on* $\mathbb{Z}_*^{(U)}$ *It is denoted by* $\text{Sd}_{\mathfrak{U}*}$.

Proposition 6.2.5. *If* $\mathfrak{U}$ *is* G-*invariant then the* $\mathfrak{U}$-*relative subdivision on* $\mathbb{Z}_*^{(U)}$ *is chain homotopic to the identity in* $\mathbb{Z}[G]$-*mod.*

Proof. This follows from the fact that the $\mathfrak{U}$-relative subdivision on $S_*(X)$ is chain homotopic to the identity in $\mathbb{Z}[\textbf{G}]\textbf{mod}$ if $\mathfrak{U}$ s G-invariant (Lemma 6.1.2). $\square$

Corollary 6.2.6. *For any* G-*invariant family* $\mathfrak{U}$ *in the* G-*space* X *the* $\mathfrak{U}$-*relative subdivision of* $\mathbb{Z}_*^{(U)}$ *induces an isomorphism in homology.*

In this way one can use the subdivision of singular simplices to subdivide simplices $(x_0, \ldots, x_n)$ into "smaller" simplices in U. If $\mathfrak{U}$ is an open covering of X one can also use the complete $\mathfrak{U}$ relative subdivision. In this case one obtains the chain

$$\mathbb{Z}_*^{(U)} \xrightarrow{\hat{\sigma}_*} S_*(X) \xrightarrow{\operatorname{Csd}_{\mathfrak{U}*}} S_*(X, \mathfrak{U}) \xrightarrow{\lambda_*} \mathbb{Z}_*^{(U \cap \operatorname{SS}(X, \mathfrak{U}))} \leq \mathbb{Z}_*^{(U)}$$

of morphisms of chain complexes. This morphism is also equivariant. Furthermore it has image in the sub chain complex $\mathbb{Z}_*^{(U \cap \operatorname{SS}(X, \mathfrak{U}))}$ of U-local $\mathfrak{U}$-small chains.

Definition 6.2.7. *The endomorphism $\lambda_* \operatorname{Csd}_{\mathfrak{U}*} \hat{\sigma}_*$ of $\mathbb{Z}_*^{(U)}$ is called the* complete $\mathfrak{U}$-relative subdivision *on $\mathbb{Z}_*^{(U)}$ It is denoted by $\operatorname{Csd}_{\mathfrak{U}*}$.*

Proposition 6.2.8. *If $\mathfrak{U}$ is G-invariant then the complete $\mathfrak{U}$-relative subdivision on $\mathbb{Z}_*^{(U)}$ is chain homotopic to the identity in $\mathbb{Z}[G]$-mod.*

Proof. This follows from the fact that the complete $\mathfrak{U}$-relative subdivision on $S_*(X)$ is chain homotopic to the identity in $\mathbb{Z}[\mathbf{G}]\mathbf{mod}$ if the family $\mathfrak{U}$ is G-invariant (Lemma 6.1.6). $\qquad\qquad\square$

Corollary 6.2.9. *For any G-invariant family $\mathfrak{U}$ in the G-space X the complete $\mathfrak{U}$-relative subdivision of $\mathbb{Z}_*^{(U)}$ induces an $\mathbb{Z}[G]$-module isomorphism in the homology of $\mathbb{Z}_*^{(U)}$.*

The $\mathfrak{U}$-relative subdivision on $\mathbb{Z}_*^{(U)}$ can now be used to subdivide local cochains. Here the effect on the cochains is the opposite. Rather than being defined on a "smaller" set the subdivided cochain have a larger domain in general. This is due to the nature of the the contravariant hom-set functor $\hom_{\mathbf{Top}}(-, V)$ (, where V denotes any abelian group of coefficients).

Definition 6.2.10. *A continuous function $f : U' \to V$ defined on a subset U' of X^{n+1} is called a U'-local n-cochain. If U' is a simplicial subspace of $\operatorname{SS}(X)$ then cosimplicial abelian group $A(X, U'; V) := C(U'; V)$ is called the* cosimplicial group of U'-local cochains.

In order to obtain an action of G on these cochains we require G to be a group. We naturally identify the cosimplicial group $A(X, U; V)$ of U-local cochains with the cosimplicial group $\hom_{\mathbf{Ab}}(\mathbb{Z}^{(U)}, V)$. Applying the functor $\hom_{\mathbf{Top}}(-, V)$ to the endomorphism $\operatorname{Sd}_{\mathfrak{U}*}$ of $\mathbb{Z}_*^{(U)}$ yields an endomorphism

$$\hom_{\mathbf{Top}}(\operatorname{Sd}_{\mathfrak{U}*}, V) : A^*(X, U; V) \to A^*(X, U; V)$$

of cochain complexes, which restricts to an endomorphism of the cochain complex of all U-local cochains.

Definition 6.2.11. *The morphism $\hom_{\mathbf{Top}}(\operatorname{Sd}_{\mathfrak{U}*}, V)$ is called the $\mathfrak{U}$-relative subdivision on $A^*(X, U; V)$. It is denoted by $\operatorname{Sd}_{\mathfrak{U}}^*$.*

Proposition 6.2.12. *If $\mathfrak{U}$ is G-invariant then the $\mathfrak{U}$-relative subdivision on the chain complex $A^*(X, U; V)$ is chain homotopic to the identity in $\mathbb{Z}[\mathbf{G}]\mathbf{mod}$.*

Proof. This follows from the fact that the $\mathfrak{U}$-relative subdivision on $\mathbb{Z}^{(U)}$ is chain homotopic to the identity in $\mathbb{Z}[\mathbf{G}]\mathbf{mod}$ if $\mathfrak{U}$ is G-invariant (Proposition 6.2.5). $\quad\square$

Corollary 6.2.13. *For any G-invariant family $\mathfrak{U}$ in the G-space X the $\mathfrak{U}$-relative subdivision of $A^*(X, U; V)$ induces an isomorphism in homology.*

Here as well one can use the complete $\mathfrak{U}$-relative subdivision, provided that $\mathfrak{U}$ is an open covering of X. In this case one obtains a morphism

$$\hom_{\mathbf{Top}}(\mathrm{Csd}_{\mathfrak{U}*}, V) : A^*(X, U \cap \mathrm{SS}(X, \mathfrak{U}); V) \to A^*(X, U; V)$$

of cochain complexes, which restricts to an endomorphism of the cochain complex of all U-local cochains. Note that the subdivision has "enlarged" the domain of the functions: The $U \cap \mathrm{SS}(X, \mathfrak{U})$-local functions are extended to U-local functions.

Definition 6.2.14. *For an open covering $\mathfrak{U}$ of X the morphism $\hom_{\mathbf{Top}}(\mathrm{Csd}_{\mathfrak{U}*}, V)$ is called the* complete $\mathfrak{U}$-relative subdivision on $A^*(X, U; V)$. *It is denoted by* $\mathrm{Csd}_{\mathfrak{U}}^*$.

The so 'subdivided' cochains can be restricted to their original domain. The restriction morphism $R^* : A^*(X, U; V) \to A^*(X, U \cap \mathrm{SS}(X, \mathfrak{U}); V)$ is equivariant and the composition $\mathrm{Csd}_{\mathfrak{U}}^* R^*$ is an equivariant endomorphism of the cochain complex $A^*(X, U; V)$. Similar to the subdivision of local chains we observe:

Proposition 6.2.15. *If $\mathfrak{U}$ is G-invariant then $\mathrm{Csd}_{\mathfrak{U}}^* R^*$ is chain homotopic to the identity in $\mathbb{Z}[\mathbf{G}]\mathbf{mod}$.*

Proof. This follows from the fact that the complete $\mathfrak{U}$-relative subdivision on $\mathbb{Z}^{(U)}$ is chain homotopic to the identity in $\mathbb{Z}[\mathbf{G}]\mathbf{mod}$ if $\mathfrak{U}$ is G-invariant (Proposition 6.2.8). $\qquad\square$

Corollary 6.2.16. *For any G-invariant family $\mathfrak{U}$ in the G-space X the $\mathfrak{U}$-relative subdivision of $A^*(X, U; V)$ induces an isomorphism in homology.*

In case the first $n+1$ components $U_0, \ldots, U_n$ of the simplicial subspace U of $\mathrm{SS}(X)$ are the spaces $X, \ldots, X^{n+1}$ itself, the complete $\mathfrak{U}$-relative subdivision $\mathrm{Csd}_{\mathfrak{U}}^*$ on $A^*(X, U; V)$ in these dimensions are morphisms

$$\hom_{\mathbf{Top}}(\mathrm{Csd}_{\mathfrak{U}n}, V) : A^n(X, \mathrm{SS}(X, \mathfrak{U}); V) \to A^n(X; V),$$

i.e. every $\mathrm{SS}(X, \mathfrak{U})$-local n-cochain is being extended to a global cochain. Furthermore, since $\mathrm{Csd}_{\mathfrak{U}}^*$ is a cochain map, $\mathrm{SS}(X, \mathfrak{U})([n])$-local n-cocycles are extended to global n-cocycles. In addition all the morphisms and chain maps used are equivariant, so we summarise:

Theorem 6.2.17. *If the first $(n+1)$ components $U_0, \ldots, U_n$ of the simplicial subspace U of $\mathrm{SS}(X)$ are the spaces $X, \ldots, X^{n+1}$ itself, then the complete $\mathfrak{U}$-relative subdivision $\mathrm{Csd}_{\mathfrak{U}}^*$ extends any $\mathrm{SS}(\mathfrak{U})([n])$-local equivariant n-cocycle to a global equivariant cocycle.*

Because the restriction and the subdivision morphisms are equivariant, they can be restricted to morphisms

$$R^* : A_{eq}^*(X, U; V) \to A_{eq}^*(X, U \cap \mathrm{SS}(X, \mathfrak{U}); V)$$
$$\mathrm{Csd}_{\mathfrak{U}}^* : A_{eq}^*(X, U \cap \mathrm{SS}(X, \mathfrak{U}); V) \to A_{eq}^*(X, U; V)$$

of the subcomplexes of equivariant cochains.

Lemma 6.2.18. *The restriction of the endomorphism $\mathrm{Csd}_{\mathfrak{U}}^* R^*$ to the subcomplex $A_{eq}^*(X, U; V)$ is chain homotopic to the identity.*

Proof. This follows from the fact that $\mathrm{Csd}_{\mathfrak{U}}^* R^*$ is cochain homotopic to the identity in $\mathbb{Z}[\mathbf{G}]\mathbf{mod}$. Cochain homotopies between $\mathrm{Csd}_{\mathfrak{U}}^* R^*$ and the identity in $\mathbb{Z}[\mathbf{G}]\mathbf{mod}$-mod are equivariant and thus restrict to a cochain homotopies on the subcomplex of equivariant cochains. $\qquad\square$

As a consequence the morphism $\mathrm{Csd}_{\mathfrak{U}}^* R^*$ induces an isomorphism in cohomology. So the cohomology class of an equivariant n-cochain f in $A_{eq}^n(X, U; V)$ in the cohomology of the cochain complex $A_{eq}^*(X, U; V)$ is uniquely determined by the restriction of f to the subspace $U([n]) \cap \mathrm{SS}(X, \mathfrak{U})([n])$ of $U([n])$.

Lemma 6.2.19. *If the first $(n+1)$ components $U_0, \ldots, U_n$ of the simplicial subspace U of $\mathrm{SS}(X)$ are the spaces $X, \ldots, X^{n+1}$ itself, then the complete subdivision homomorphism $\mathrm{Csd}_{\mathfrak{U}}^n : A^n(X, \mathrm{SS}(\mathfrak{U}); V) \to A^n(X; V)$ descends to an isomorphism $\mathrm{Csd}_{\mathfrak{U}}^n : H_{AS,eq}^n(X; V) \to H_{eq}^n(X; V)$ in cohomology.*

Proof. Assume that the first $(n+1)$ components $U_0, \ldots, U_n$ of the simplicial subspace U of $\mathrm{SS}(X)$ are the spaces $X, \ldots, X^{n+1}$ itself. Then the U-local equivariant n-cocycles $Z_{eq}^n(X, U; V)$ are the global equivariant n-cocycles $Z_{eq}^n(X; V)$ and the same is true for the cohomology groups $H_{eq}^n(X; V)$ and $H_{eq}^n(X, U; V)$. The cohomology class $[F_{eq}] \in H_{eq}^n(X; V)$ of an equivariant n cocycle F_{eq} is uniquely determined by its restriction $[R^n F_{eq}] \in H_{eq}^n(X, \mathfrak{U}; V)$ But the covering $\mathfrak{U}$ can be chosen arbitrarily fine, so the class only depends on the cohomology class of the germ of F_{eq} at the diagonal in $H_{AS,eq}(X; V)$. $\qquad\square$

7

Abelian Extensions of Topological Groups

Given a topological group G, an abelian divisible topological group A and a local group cocycle f, under which circumstances is there an extension to a global cocycle F on G? For $n = 2$ this is equivalent to the existence of an extension

$$0 \to A \to A \times_F G \to G \to 1$$

of G by A which is locally given by f. Especially when G and A are Lie groups and f is known to be a local integral of a Lie algebra cocycle $\omega \in Z^2(\mathfrak{g}, \mathfrak{a})$ (which always exists if A is a quotient of a sequentially complete TVS by a discrete subgroup) then this equivalent to an integration of the extension

$$0 \to \mathfrak{a} \to \mathfrak{a} \oplus_\omega \mathfrak{g} \to \mathfrak{g} \to 0$$

of Lie algebras to an extension of Lie groups. Here the constructed extension would be a locally trivial fibration with base space G. For this case the question has been answered by [Nee04] (, where A is assumed to be a quotient of a sequentially complete topological vector space by a discrete subgroup). In this chapter we consider this problem for connected locally contractible topological groups and establish a necessary and sufficient condition for the germ of a local group 2-cocycle f to be extendible to a global group cocycle F by deriving an exact sequence

$$Z^2(G; V) \to Z^2_{AS,eq}(G; V)^G \to H^2_{AS}(G; V) \oplus \hom(\pi_1(M), H^1(G; V)).$$

This is the main result of this chapter. It is a generalisation of the result for locally smooth Lie group cocycles in [Nee04] to arbitrary topological groups G which are locally contractible. The proof relies on the existence of an equivariant system of singular 2-simplices derived in Section 3, the relative subdivision introduced in Chapter 4, the complete relative subdivision proven to exist in Chapter 5 and the observations concerning relative subdivision in transformation groups made in Chapter 6.

The following considerations are of general nature for topological groups G and do not rely on the differential structure provided by Lie groups. The constructions made do rely on the existence of a universal simply connected covering group $\tilde{G}$. So the group G is assumed to be connected and semi-locally simply connected, which implies the existence of a universal covering group $\tilde{G}$. (These conditions are always satisfied by connected groups that are CW-complexes or locally contractible, e.g. Lie groups, loop groups thereof, Kac-Moody groups etc.).

7.1 Liftings to the Universal Covering Group

Let G be a topological group with simply connected covering group $\tilde{G}$ and A be a G-module. Recall that the complexes $C^*(G; A)$ of group cochains and $A^*(G; A)^G$ of equivariant (or 'homogeneous')cochains are isomorphic (cf. [,]). The same is true for the limit complexes of germs of group cochains and the equivariant Alexander-Spanier cochains. The equivariant cochain associated with a local group cochain f will be denoted by f_{eq}. We start with considering local group 2-cocycles which are defined on an identity neighbourhood $U \subset G$. At first we assume the local cocycle f to be extendible to a global cocycle and deduce some necessary conditions therefore. To distinguish between group cochains and equivariant cochains we shall use the the symbol $Z_{grp}^n(G; A)$ for the abelian group of group n-cocycles and $Z_{grp,lc}^n(G; A)$ for group n-cocycles, which are continuous on an identity neighbourhood in G.

Lemma 7.1.1. *If the germ of a local group cocycle $f \in Z_{grp}^n(U; A)$ is extendible to a global cocycle, then the cohomology class $[f_{eq}] \in H_{AS}^n(G; A)$ is trivial.*

Proof. If there exist an identity neighbourhood V in G such that the restriction $f_{|V}$ is the restriction of a global cocycle F to V, then the cohomology classes $[f_{eq}]$ and $[F_{eq}]$ in $H_{AS}^n(G; A)$ coincide. Since the complex $A^*(G; A)$ is exact, the equivariant cochain F_{eq} is a boundary in $A^*(G; A)$. Hence the germ of F_{eq} on the diagonal is a boundary as well. Thus the cohomology classes $[f_{eq}]$ and $[F_{eq}]$ in $H_{AS}^n(G; A)$ are trivial.   □

In the following we assume the continuous local group cocycle $f \in Z_{grp,c}^2(U; A)$ to be the restriction of a global cocycle F to the identity neighbourhood U. The global cocycle F determines a locally trivial (cf. Definition 1.4.50) extension

$$0 \longrightarrow A \longrightarrow A \times_F G \longrightarrow G \longrightarrow 1$$

of G by A, which is locally given by f. Furthermore F can be lifted to a cocycle $\tilde{F}$ on the universal covering group $\tilde{G}$ of G:

$$\tilde{F} = q_G^*(F) = F \circ (q_G \times q_G)$$

Here q_G denotes the quotient map $q_G : \tilde{G} \to G$. This cocycle gives rise to a locally trivial extension of $\tilde{G}$ by A, which can be mapped onto the original extension by the quotient maps $\mathrm{id} \times q_G$ and q_G:

$$\begin{array}{ccccccccc}
0 & \longrightarrow & A & \longrightarrow & A \times_{\tilde{F}} \tilde{G} & \longrightarrow & \tilde{G} & \longrightarrow & 1 \\
& & \downarrow{\scriptstyle \mathrm{id}} & & \downarrow{\scriptstyle \mathrm{id} \times q_G} & & \downarrow{\scriptstyle q_G} & & \\
0 & \longrightarrow & A & \longrightarrow & A \times_F G & \longrightarrow & G & \longrightarrow & 1 .
\end{array}$$

The kernel of this map is the normal subsequence $0 \to 0 \to 0 \times_{\tilde{F}} \pi_1(G) \to \pi_1(G) \to 1$ of the extension of $\tilde{G}$, so that the commutative diagram

$$\begin{array}{ccccccccc}
& & 0 & & 0 & & & & \\
& & \downarrow & & \downarrow & & & & \\
0 & \longrightarrow & 0 \times_{\tilde{F}} \pi_1(G) & \longrightarrow & \pi_1(G) & \longrightarrow & 1 & & \\
& & \downarrow & & \downarrow & & & & \\
0 \longrightarrow A & \longrightarrow & A \times_{\tilde{F}} \tilde{G} & \longrightarrow & \tilde{G} & \longrightarrow & 1 & & \\
& & \downarrow & & \downarrow & & & & \\
0 \longrightarrow A & \longrightarrow & A \times_F G & \longrightarrow & G & \longrightarrow & 1 & & \\
& & \downarrow & & \downarrow & & \downarrow & & \\
& & 0 & & 1 & & 1 & &
\end{array}$$

has exact columns and rows. In addition, the surjection $\mathrm{id} \times q_G : A \times_{\tilde{F}} \tilde{G} \to A \times_F G$ is a covering with discrete covering group $0 \times_{\tilde{F}} \pi_1(G)$. So we observe:

Lemma 7.1.2. *If a 2-cocycle $\tilde{F} \in Z_{grp,lc}^2(\tilde{G}; A)$ is induced by a cocycle 2-cocycle F in $Z_{grp,lc}^2(G; A)$ via inflation, then $A \times_{\tilde{F}} \pi_1(G)$ is abelian and the extension*

$$0 \longrightarrow A \longrightarrow A \times_{\tilde{F}} \pi_1(G) \longrightarrow \pi_1(G) \longrightarrow 1$$

splits by a homomorphism $s : \pi_1(G) \to Z(A \times_{\tilde{F}} \tilde{G}) \cap (A \times_{\tilde{F}} \pi_1(G))$.

Proof. If a cocycle $\tilde{F} \in Z^2_{grp,lc}(\tilde{G}; A)$ is induced by a cocycle $F \in Z^2_{grp,lc}(G; A)$ via inflation, then the extension $A \to A \times_{\tilde{F}} \pi_1(G) \to \pi_1(G)$ of $\pi_1(G)$ by A is the pullback of the extension $A \to A \times_F G \to G$ of G by A along the trivial homomorphism $q_G \circ i : \pi_1(G) \to G$, where $i : \pi_1(G) \to \tilde{G}$ denotes the inclusion of $\pi_1(G)$ into $\tilde{G}$ as a central subgroup. Therefore the former extension of $\pi_1(G)$ by A is trivial and the map $s : \pi_1(G) \to A \times_{\tilde{F}} \tilde{G}, g \mapsto (0, g)$ is the desired splitting homomorphism.   □

Proposition 7.1.3. *If a cocycle* $h \in Z^2_{grp,lc}(\tilde{G}; A)$ *is cohomologous (in* $C^*_{grp,lc}(\tilde{G}; A)$*) to an inflated cocycle* $\tilde{F}$*, then* $A \times_{\tilde{h}} \pi_1(G)$ *is abelian and the extension*

$$0 \longrightarrow A \longrightarrow A \times_h \pi_1(G) \longrightarrow \pi_1(G) \longrightarrow 1$$

splits by a homomorphism $s : \pi_1(G) \to Z(A \times_h \tilde{G}) \cap (A \times_{\tilde{F}} \pi_1(G))$.

Proof. This follows from Lemma 7.1.2 because the extensions of topological groups determined by h and $\tilde{F}$ are isomorphic.   □

Please recall that the surjective homomorphism $\pi : \tilde{G} \to G$ induces a pullback morphism on locally trivial extensions of G by A. This pullback morphism maps any locally trivial extension H of G by A to a locally trivial extension $\tilde{H}$ of $\tilde{G}$ by A which is a covering of the extension H

$$
\begin{array}{ccccccccc}
0 & \longrightarrow & A & \longrightarrow & \tilde{H} & \longrightarrow & \tilde{G} & \longrightarrow & 1 \\
& & \downarrow & & \downarrow & & \downarrow & & \\
0 & \longrightarrow & A & \longrightarrow & H & \longrightarrow & G & \longrightarrow & 1
\end{array},
$$

where the group A is identically mapped onto itself. Furthermore any such covering extension $\tilde{H}$ of $\tilde{G}$ by A where A is mapped identically onto itself is a pullback of an extension of G by A along the quotient homomorphism $q_G : \tilde{G} \to G$.

Theorem 7.1.4. *An extension* $A \to A \times_h \tilde{G} \to \tilde{G}$ *of* $\tilde{G}$ *by* A *is a pullback extension if and only if the subextension* $0 \to A \to A \times_h \pi_1(G) \to \pi_1(G) \to 1$ *splits by a homomorphism* $s : \pi_1(G) \to Z(A \times_h \tilde{G}) \cap (A \times_{\tilde{F}} \pi_1(G))$.

Proof. Only the backward implication requires proof. Let $h \in C^2_{grp,lc}(\tilde{G}; A)$ be a group cocycle. Suppose the extension $0 \to A \to A \times_h \pi_1(G) \to \pi_1(G) \to 1$ splits by a homomorphism $s : \pi_1(G) \to Z(A \times_h \tilde{G}) \cap (A \times_{\tilde{F}} \pi_1(G))$. In this case the discrete subgroup $s(\pi_1(G))$ is central, hence normal, in $A \times_h \tilde{G}$, so the exact sequence

$$0 \to 0 \to s(\pi_1(G)) \to \pi_1(G) \to 1$$

of discrete groups is normal in the exact sequence

$$0 \to A \to A \times_h \tilde{G} \to \tilde{G} \to 1.$$

Furthermore the intersection $A \cap s(\pi_1(G)) = 1$ is trivial. Hence we obtain a commutative diagram

$$
\begin{array}{ccccccccc}
 & & 0 & & 0 & & 0 & & \\
 & & \downarrow & & \downarrow & & \downarrow & & \\
0 & \longrightarrow & 0 & \longrightarrow & s(\pi_1(G)) & \longrightarrow & \pi_1(G) & \longrightarrow & 1 \\
 & & \downarrow & & \downarrow & & \downarrow & & \\
0 & \longrightarrow & A & \longrightarrow & A \times_h \tilde{G} & \longrightarrow & \tilde{G} & \longrightarrow & 1 \\
 & & \downarrow & & \downarrow & & \downarrow & & \\
0 & \longrightarrow & A & \longrightarrow & H & \longrightarrow & G & \longrightarrow & 1 \\
 & & \downarrow & & \downarrow & & \downarrow & & \\
 & & 0 & & 1 & & 1 & &
\end{array}
$$

with exact rows and columns, where the surjection $A \times_h \tilde{G} \to H$ is a covering map. Therefore the extension $0 \to A \to A \times_h \pi_1(G) \to \pi_1(G) \to 1$ of $\tilde{G}$ by A is a pullback of an extension of G by A along q_G. $\square$

Remark 7.1.5. The property of an extension $0 \to A \to A \times_h \pi_1(G) \to \pi_1(G) \to 1$ to be isomorphic to a pullback extension along q_G is equivalent to the cocycle $h \in Z^2_{grp,lc}(\tilde{G}; A)$ to be cohomologous (in $C^*_{grp,lc}(\tilde{G}; A)$) to an inflated cocycle.

The quotient map $A \times_h \tilde{G} \to \tilde{G}$ will subsequently be denoted by q and the induced action of $\tilde{G}$ on A will be denoted by $\tilde{\varrho}$.

Corollary 7.1.6. *If a cocycle $h \in Z^2_{grp}(\tilde{G}; A)$ is cohomologous (in $C^*_{lc}(\tilde{G}; A)$) to an inflated cocycle $\tilde{F}$, then $\pi_1(G)$ is contained in the image $q(Z(A \times_h \tilde{G}))$.*

For a better understanding of this corollary we have a closer look at the central elements in $A \times_F \tilde{G}$. It turns out that being central can be expressed by a condition on an antisymmetrised cochain associated with F.

Definition 7.1.7. *For any group G we define a map $\alpha : C^2(G; A) \to C^2(G; A)$ by*

$$\alpha : C^2(G; A) \to C^2(G; A), \quad \alpha(F)(g_1, g_2) := F(g_1, g_2) - F(g_2, g_1),$$

which we call the antisymmetrising map.

The mapping groups groups $C^2(G; A)$ and $C^1(G, C^1(G; A))$ are isomorphic, via the "exponential map" which assigns to a function $f \in C^2(G; A) = C^1(G \times G, A)$ its adjoint $\hat{f} \in C^1(G, C^1(G; A))$, which is given by $\hat{f}(g)(g') = f(g, g')$ for all elements $g, g' \in G$. This allows us to view an antisymmetrised cocycle $\alpha(F)$ as an element of $C^1(G, C^1(G; A))$ by identifying it with the adjoint cochain $\hat{\alpha}(F) \in C^1(G, C^1(G; A))$. Doing this and considering the universal covering group $\tilde{G}$, we find (an interpretation is given below):

Lemma 7.1.8. *For any group cocycle $\tilde{F} \in Z^2_{grp}(\tilde{G}; A)$ the adjoint $\hat{\alpha}(\tilde{F})$ of the antisymmetrised cochain $\alpha(\tilde{F})$ maps the subgroup $Z(\tilde{G}) \cap \ker \tilde{\varrho}$ into $Z^1_{grp}(\tilde{G}; A)$.*

Proof. Let $\tilde{F}$ be a cocycle in $C^2(\tilde{G}; A)$ and h be in $Z(\tilde{G}) \cap \ker \tilde{\varrho}$. The coboundary of $\hat{\alpha}(\tilde{F})(h)$ is given by

$$
\begin{aligned}
\delta\hat{\alpha}(h)(g_1, g_2) &= g_1.[\hat{\alpha}(h)(g_2)] - \hat{\alpha}(h)(g_1 g_2) + \hat{\alpha}(h)(g_1) \\
&= g_1.\tilde{F}(h, g_2) - g_1.\tilde{F}(g_2, h) - \tilde{F}(h, g_1 g_2) + \tilde{F}(g_1 g_2, h) \\
&\quad + \tilde{F}(h, g_1) - \tilde{F}(g_1, h) \\
&= g_1.\tilde{F}(h, g_2) - [\tilde{F}(g_1 h, g_2) - \tilde{F}(g_1 h, g_2)] + [\tilde{F}(g_1, h g_2) - \tilde{F}(g_1, h g_2)] \\
&\quad - \tilde{F}(g_1, h) - g_1.\tilde{F}(g_2, h) - \tilde{F}(h, g_1 g_2) + \tilde{F}(g_1 g_2, h) + \tilde{F}(h, g_1) \\
&= g_1.\tilde{F}(h, g_2) - \tilde{F}(g_1 h, g_2) + \tilde{F}(g_1, h g_2) - \tilde{F}(g_1, h) \\
&\quad + \tilde{F}(g_1 h, g_2) - \tilde{F}(g_1, h g_2) - g_1.\tilde{F}(g_2, h) \\
&\quad - \tilde{F}(h, g_1 g_2) + \tilde{F}(g_1 g_2, h) + \tilde{F}(h, g_1) \\
&= \delta\tilde{F}(g_1, h, g_2) - g_1.\tilde{F}(g_2, h) + \tilde{F}(g_1 g_2, h) - \tilde{F}(g_1, h g_2) \\
&\quad - \tilde{F}(h, g_1 g_2) + \tilde{F}(g_1 h, g_2) + \tilde{F}(h, g_1).
\end{aligned}
$$

The summand $\delta\tilde{F}(g_1, h, g_2)$ vanishes because $\tilde{F}$ was assumed to be a cocycle. Furthermore h commutes with all elements of $\tilde{G}$, so we have $\tilde{F}(g_1, h g_2) = \tilde{F}(g_1, g_2 h)$ etc. Making use of this we obtain

$$\delta[\hat{\alpha}(\tilde{F})(h)](g_1, g_2) = -g_1.\tilde{F}(g_2, h) + \tilde{F}(g_1 g_2, h) - \tilde{F}(g_1, g_2 h) + \tilde{F}(g_1, g_2)$$
$$- \tilde{F}(g_1, g_2) + \tilde{F}(g_1 h, g_2) - \tilde{F}(h, g_1 g_2) + \tilde{F}(h, g_1)$$
$$= -\delta \tilde{F}(g_1, g_2, h) - \tilde{F}(g_1, g_2) + \tilde{F}(hg_1, g_2) - \tilde{F}(h, g_1 g_2) + \tilde{F}(h, g_1)$$
$$= 0 + \delta \tilde{F}(h, g_1, g_2) = 0,$$

where the last but one equality follows from the fact that h is contained in $\ker \tilde{\varrho}$ and thus acts trivially on the element $\tilde{F}(hg_1, g_2)$. So the coboundary of $\hat{\alpha}(\tilde{F})(h)$ vanishes i.e. $\hat{\alpha}(\tilde{F})(h)$ is a 1-cocycle. $\qquad \square$

This can be interpreted as follows: The restriction of the group 2-cocycle $\tilde{F}$ on $\tilde{G}$ to the central subgroup $Z(\tilde{G}) \cap \ker \tilde{\varrho}$ of $\tilde{G}$ describes a central extension

$$0 \to A \to A \times_{\tilde{F}} (Z(\tilde{G}) \cap \ker \tilde{\varrho}) \to Z(\tilde{G}) \cap \ker \tilde{\varrho} \to 0$$

of the abelian group $Z(\tilde{G}) \cap \ker \tilde{\varrho}$ by the abelian group A. Therefore this extension is 2-step nilpotent. The antisymmetrised cocycle $\alpha(\tilde{F})$ is the biadditive commutator map of the group $A \times_{\tilde{F}} (Z(\tilde{G}) \cap \ker \tilde{\varrho})$ which takes values in the subgroup A only because $(Z(\tilde{G}) \cap \ker \tilde{\varrho})$ is abelian.

Lemma 7.1.9. *An element (a, g) in $A \times_{\tilde{F}} (Z(\tilde{G}) \cap \ker \tilde{\varrho})$ is central in $A \times_{\tilde{F}} \tilde{G}$ if and only if $\hat{\alpha}(\tilde{F})(g)$ is the coboundary $\delta(a)$.*

Proof. Let (a, g) be an element of $A \times_{\tilde{F}} Z(\tilde{G}) \cap \ker \tilde{\varrho}$. Since g lies in the kernel of ϱ, the element (a, g) commutes with every element $(a, 1) \in A \times_{\tilde{F}} \{1\}$. So (a, g) is central if and only if it commutes with every element $(0, g')$ of $A \times_{\tilde{F}} \tilde{G}$, i.e. if

$$(1, g')(a, g) = (g'.a + \tilde{F}(g', g), g'g) = (a, g)(1, g') = (a + \tilde{F}(g, g'), gg')$$

for all $g' \in \tilde{G}$. Because the element g was assumed to be central, the equality $g'g = gg'$ holds for all g' in G. So (a, g) is central iff

$$g'.a - a = \tilde{F}(g, g') - \tilde{F}(g', g) = \hat{\alpha}(\tilde{F})(g)(g')$$

holds for all $g' \in \tilde{G}$. This is exactly the condition for $\hat{\alpha}(\tilde{F})(g)$ to be the coboundary $\delta(a)$. $\qquad \square$

Proposition 7.1.10. *The subgroup $\pi_1(G)$ of $\tilde{G}$ is contained in the image of the centre $Z(A \times_{\tilde{F}} \tilde{G})$ under the quotient map q if and only if the image $[\hat{\alpha}(\tilde{F})(\pi_1(G))]$ is trivial in $H^1(\tilde{G}; A)$.*

Proof. Let g be in $\pi_1(G)$. If g is contained in $q(Z(A \times_{\tilde{F}} \tilde{G}))$, then there exists an element $a \in A$ such that (a, g) is central. Therefore the map $\hat{\alpha}(\tilde{F})(g)$ is a coboundary. So if $\pi_1(G)$ lies in the image of $Z(A \times_{\tilde{F}} \tilde{G})$ then the image $[\hat{\alpha}(\tilde{F})(\pi_1(G))]$ is trivial in $H^1(\tilde{G}; A)$. Conversely assume $[\hat{\alpha}(\tilde{F})(\pi_1(G))] = 0 \in H^1(\tilde{G}; A)$. Then for any given $g \in \pi_1(G)$ the map $\hat{\alpha}(\tilde{F})(g)$ is a coboundary, i.e there exists an $a \in A$ such that $\hat{\alpha}(\tilde{F})(g) = \delta(a)$. So the element (a, g) is central in $A \times_{\tilde{F}} G$ and $g \in q(Z(A \times_{\tilde{F}} \tilde{G}))$. $\qquad \square$

Remark 7.1.11. For Lie groups G and topological vector spaces A the cohomology groups $H^1_{grp,ls}(\tilde{G}; A)$ and $H^1(\mathfrak{g}; \mathfrak{a})$ are naturally isomorphic. This suggests that the assignment $\tilde{F} \mapsto \hat{\alpha}(\tilde{F})_{|\pi_1(G)}$ is a kind of "topological flux homomorphism". General topological flux homomorphisms like this will be introduced in Chapter 13.

Definition 7.1.12. *For every group 2-cocycle $\tilde{F} \in Z^2_{grp}(\tilde{G}; A)$ the homomorphism $\hat{\alpha}(\tilde{F}) : \pi_1(G) \to H^1_{grp}(\tilde{G}; A)$ is called the* flux homomorphism *associated with $\tilde{F}$.*

Lemma 7.1.13. *Exact 1-cocycles in $Z^1_{grp}(\tilde{G}; A)$ are inflated coboundaries, i.e. if the cohomology class $[h] \in H^1(\tilde{G}; A)$ of a cocycle $h \in Z^1_{grp}(\tilde{G}; A)$ is trivial then the cocycle h is an inflated coboundary.*

Proof. Let $h \in Z^1_{grp}(\tilde{G}; A)$ be a group 1-cocycle with vanishing cohomology class $[h] = 0 \in H^1(\tilde{G}; A)$. In this case the cocycle h is a coboundary, i.e., there exists an element $a \in A$ such that $h = \delta a$. The fundamental group $\pi_1(G)$ acts trivially on A. As a consequence we have

$$h(\gamma g) = (\gamma g).a - a = \gamma.(g.a) - a = g.a - a$$

for all $\gamma \in \pi_1(G)$ and all $g \in \tilde{G}$. Thus the cocycle h is the inflation of the 1-coboundary $\delta a \in Z^1(G; A)$ on G. $\qquad\square$

Corollary 7.1.14. *The subgroup $\pi_1(G)$ is contained in the image of $Z(A \times_{\tilde{F}} \tilde{G})$ under q if and only if $\hat{\alpha}(\tilde{F})$ factors through a map $Z(\tilde{G}) \cap \ker \tilde{\varrho} \to B^1(G; A)$.*

Proof. This a consequence of the preceding proposition taking into account that any coboundary in $C^1(G; A)$ is automatically continuous. $\qquad\square$

So far for any group 2-cocycle h the condition $[\hat{\alpha}(h)(\pi_1(G)] = 0 \in H^1(\tilde{G}; A)$ (resp. in $H^1(G; A)$) is necessary for h to be cohomologous to an inflated cocycle. In the following we show that this condition is actually sufficient if the abelian group A is divisible.

Lemma 7.1.15. *If the image $[\hat{\alpha}(\tilde{F})(\pi_1(G))]$ is trivial in $H^1(\tilde{G}; A)$ then the restriction of $\tilde{F}$ onto $\pi_1(G)$ is symmetric and the group $A \times_{\tilde{F}} \pi_1(G)$ is abelian.*

Proof. If the image $[\hat{\alpha}(\tilde{F})(\pi_1(G))]$ is trivial in $H^1(\tilde{G}; A)$ then cohomology class of the the restriction $\hat{\alpha}(\tilde{F})(\pi_1(G)) \in H^1(\pi_1(G); A)$ to the subgroup $\pi_1(G)$ of $\tilde{G}$ is also trivial. Since the fundamental group $\pi_1(G)$ acts trivially on A this implies $\hat{\alpha}(\tilde{F})(\pi_1(G)) = 0 \in Z^1(\pi_1(G); A)$. Therefore the antisymmetrised map $\alpha(\tilde{F})$ vanishes on $\pi_1(G) \times \pi_1(G)$. This is equivalent to $\tilde{F}$ being symmetric on $\pi_1(G) \times \pi_1(G)$. Because A and $\pi_1(G)$ are abelian groups this implies the commutativity of the group $A \times_{\tilde{F}} \pi_1(G)$. $\qquad\square$

Proposition 7.1.16. *If A is divisible then the extension $A \times_{\tilde{F}} \pi_1(G)$ of $\pi_1(G)$ by A splits by a homomorphism $s : \pi_1(G) \to Z(A \times_{\tilde{F}} \tilde{G}) \cap q^{-1}(\pi_1(G))$ if and only if $\pi_1(G) \subset q(Z(A \times_{\tilde{F}} \tilde{G}))$.*

Proof. Let the fundamental group $\pi_1(G)$ be contained in the image $q(Z(A \times_{\tilde{F}} \tilde{G}))$ and let $H = q^{-1}(\pi_1(\tilde{G})) \cap Z(A \times_{\tilde{F}} \tilde{G})$ denote the inverse image of the fundamental group in $Z(A \times_{\tilde{F}} \tilde{G})$. Since the groups A, H and $\pi_1(G)$ are abelian and A is divisible the extension

$$0 \longrightarrow A \cap H \longrightarrow H \longrightarrow \pi_1(G) \longrightarrow 0$$

splits by a homomorphism $s : \pi_1(G) \to H$. Because H is a subgroup of $A \times_{\tilde{F}} \pi_1(G)$ the extension

$$0 \longrightarrow A \longrightarrow A \times_{\tilde{F}} \pi_1(G) \longrightarrow \pi_1(G) \longrightarrow 0$$

splits as well by s. Conversely, assume such a splitting homomorphism s exists. Then s has codomain in the group $H \leq Z(A \times_{\tilde{F}} \tilde{G})$. Thus the image of $Z(A \times_{\tilde{F}} \tilde{G})$ under q contains the subgroup $\pi_1(G)$. $\qquad\square$

Theorem 7.1.17. *For locally split extension $0 \to A \to A \times_{\tilde{F}} \tilde{G} \to \tilde{G} \to 1$ of $\tilde{G}$ by a divisible group A the following are equivalent:*

1. The cocycle $\tilde{F}$ is cohomologous (in $C^*_{lc}(\tilde{G}; A)$) to an inflated cocycle.
2. The restricted extension $0 \to A \to A \times_{\tilde{F}} \pi_1(G) \to \pi_1(G)$ splits by a homomorphism $s : \pi_1(G) \to Z(A \times_{\tilde{F}} \tilde{G})$
3. $\pi_1(G) \subset q(Z(A \times_{\tilde{F}} \tilde{G}))$
4. The flux vanishes, i.e. $[\hat{\alpha}(\tilde{F})(\pi_1(G))] = 0 \in H^1(\tilde{G}; A)$.
5. The map $\hat{\alpha}(\tilde{F}) : \pi_1(G) \to Z^1(\tilde{G}; A)$ is induce via inflation by a map into $Z^1_{lc}(G; A)$ with trivial cohomology class in $H^1_{lc}(G; A)$.

Proof. The assertions 1 and 2 are equivalent by Lemma 7.1.4. The implication "2 $\Rightarrow$ 3" is due to Corollary 7.1.6 and the converse implication follows from Proposition 7.1.16. The equivalence of the assertions 3 and 4 is the statement of Proposition 7.1.10 whereas the equivalence of 4 and 5 follows from Corollary 7.1.14.   □

7.2 Extensions of Local Group Cocycles

We are now going to construct an extension of the germ of a given local group cocycle f on $U \subset G$ using the existence of an extension of the local cocycle $\tilde{f} = f \circ (q_G \times q_G)$. Let U be an elementary identity neighbourhood in G and $f : U \times U \to A$ be a local group cocycle. The identity component of the inverse image $q_G^{-1}(U)$ will be denoted by $\tilde{U}$ and the inflated local cocycle on $\tilde{U}$ will be denoted by $\tilde{f}$.

Proposition 7.2.1. *For any equivariant open covering $\mathfrak{U}$ of G and every equivariant $\mathfrak{U}$-local cocycle $f_{eq} \in Z^2(G, \mathfrak{U}; A)^G$ with trivial cohomology class $[f_{eq}]$ in $H^2_{AS}(G; A)$, there exists a global equivariant cocycle $\tilde{F}_{eq} \in Z^2(\tilde{G}; A)^{\tilde{G}}$ which restricts to the inflated local cocycle $\tilde{f}_{eq} = q_G^*(f_{eq})$ on a neighbourhood of the diagonal $D_3\tilde{G}$ in $\tilde{G} \times \tilde{G} \times \tilde{G}$.*

Proof. Let $\tilde{\mathfrak{U}}$ denote the inverse image of the open covering $\mathfrak{U}$ under the quotient homomorphism. Since the universal covering group $\tilde{G}$ is locally contractible, there exists a locally continuous system $\hat{\sigma}_2$ of simplices defined on $\tilde{G} \times \tilde{G} \times \tilde{G}$. By Theorem 6.2.17 the complete $\tilde{\mathfrak{U}}$-relative subdivision of $\tilde{\mathfrak{U}}$-local cochains maps $Z^2(\tilde{G}, \tilde{\mathfrak{U}}; A)$ into $Z^2(\tilde{G}, \tilde{\mathfrak{U}}_{V_2}; A) = Z^2(\tilde{G}; A)$. Because this homomorphism is equivariant, it maps $\tilde{f}_{eq} \in Z^2(\tilde{G}, \tilde{\mathfrak{U}}; A)$ to a global equivariant cocycle $\tilde{F}_{eq} \in Z^2(G; A)$. Since the $\tilde{\mathfrak{U}}$-relative subdivision is the identity on all $\tilde{\mathfrak{U}}$-small simplices, the cocycle $\tilde{F}_{eq}$ restricts to $\tilde{f}_{eq}$ on a neighbourhood of the diagonal in $\tilde{G}^3$.   □

Corollary 7.2.2. *For every local group cocycle $f \in Z^2_{grp}(U; A)$ with trivial cohomology class $[f_{eq}]$ in $H^2_{AS}(G; A)$, there exists a global group cocycle $\tilde{F} \in C^2(\tilde{G}; A)$ which restricts to $\tilde{f} = f \circ (\pi \times \pi)$ on an identity neighbourhood of $\tilde{G}$.*

Proof. Let $f \in Z^2_{grp}(U; A)$ be local group cocycle with vanishing cohomology class $[f_{eq}]$ in $H^2_{AS}(G; A)$. By the preceding proposition there exists a global equivariant cocycle $\tilde{F} \in Z^2(\tilde{G}; A)^{\tilde{G}}$ which restricts to the inflated equivariant cocycle $\tilde{f}_{eq}$ on a neighbourhood of the diagonal. The corresponding group cocycle $\tilde{F}$ then restricts to the local group cocycle $\tilde{f}$ on an identity neighbourhood in $\tilde{G}$.   □

As we have shown, the vanishing of the cohomology class $[f_{eq}] \in H^2_{AS}(G; A)$ associated with a local group cocycle f is sufficient for the existence of a global group cocycle $\tilde{F}$ on $\tilde{G}$ that restricts to $\tilde{f}$ on an identity neighbourhood in $\tilde{G}$. On the other hand, the triviality of this cohomology class is necessary for f to be extendible to a global cocycle on G. Henceforth we shall assume the cohomology class $[f_{eq}] \in H^2_{AS}(G; A)$ to be trivial and the group cocycle $\tilde{F}$ on $\tilde{G}$ to be given by $\tilde{F} = \mathrm{Csd}_{\tilde{\mathfrak{U}}}(\tilde{f})$. As we shall see this particular choice of extension of $\tilde{f}$ is not a restriction.

Lemma 7.2.3. *The cohomology class* $[h] \in H^2_{grp,lc}(\tilde{G}; A)$ *of a lifting of a local group cocycle* $f \in Z^2_{grp,c}(U; A)$ *to a global group cocycle* $\tilde{F}$ *on* $\tilde{G}$ *is uniquely determined by the germ of* f *at* 1.

Proof. Let $f \in C^2(U; A)$ be a local group cocycle and let f_{eq} denote the corresponding equivariant cocycle. The complex $C^*(\tilde{G}; A)$ of group cochains is isomorphic to the complex $A^*(\tilde{G}; A)^{\tilde{G}}$ of equivariant global cochains. Furthermore the complex $C^*(\tilde{U}; A)$ of $\tilde{U}$-local group cochains is isomorphic to the complex $(A^*(\tilde{G}, \mathfrak{U}_{\tilde{U}}; A)^{\tilde{G}}$ of equivariant $\mathfrak{U}_{\tilde{U}}$-local cochains. These isomorphisms intertwine the restriction homomorphisms from global to local cochains:

$$
\begin{array}{ccc}
C^*(\tilde{G}; A) & \xrightarrow{\;\;\mathrm{Eq}\;\;} & A^*(\tilde{G}; A)^{\tilde{G}} \\
{\scriptstyle\mathrm{Res}}\Big\downarrow & & \Big\downarrow{\scriptstyle\mathrm{Res}} \\
C^*(\tilde{U}; A) & \xrightarrow{\;\;\mathrm{Eq}\;\;} & A^*(\tilde{G}, \mathfrak{U}_{\tilde{U}}; A)^{\tilde{G}}
\end{array}
$$

By Lemma 6.2.19 the cohomology class $[h_{eq}] \in H^2_{AS,eq}(\tilde{G}; A)$ of an extension h_{eq} of the $\mathfrak{U}_{\tilde{U}}$-local equivariant cocycle $\tilde{f}_{eq}$ is uniquely determined by the germ of $\tilde{f}_{eq}$ at the diagonal. This germ is in turn uniquely determined by the germ of f_{eq} at the diagonal which is uniquely determined by the germ of the local group cocycle f at the identity. Thus the assertion follows. $\qquad\square$

Corollary 7.2.4. *The isomorphy class of an extension* $0 \to A \to A \times_{\tilde{F}} \tilde{G} \to \tilde{G} \to 1$ *induced by a local cocycle* $f \in C^2_{grp,c}(U; A)$ *is uniquely determined by the germ of* f *at the identity.*

So we know that $\tilde{F} = \mathrm{Csd}_{\tilde{\mathfrak{U}}}(\tilde{f})$ is an extension of the germ of $\tilde{f}$ on $\tilde{G}$. In order for $\tilde{F}$ to be induced by inflation, the group $A \times_{\tilde{F}} \pi_1(G)$ has to be abelian and the extension

$$
0 \longrightarrow A \longrightarrow A \times_{\tilde{F}} \pi_1(G) \longrightarrow \pi_1(G) \longrightarrow 1
$$

has to split by a homomorphism $s : \pi_1(G) \to Z(A \times_{\tilde{F}} \tilde{G})$. As we will see, the first condition is automatically satisfied if $[f_{eq}] = 0$ in $H^2_{AS}(G; A)$. This requires some geometrical considerations which we provide in the following.

Let X be a topological space and M be a compact manifold of dimension 2 and $\hat{\sigma}_2 : G^3 \to C(\Delta_2, \tilde{G})$ and $\hat{\sigma}_1 : G^2 \to C(\Delta_1, \tilde{G})$ be compatible equivariant systems of simplices on $\tilde{G}$ (cf. Definition 3.0.15).

Definition 7.2.5. *A triangulation of a map* $\varphi : M \to X$ *is a singular chain of the form* $\sum \varphi \circ \tau_i$, *where* $\{\tau_i\}$ *is a triangulation of the manifold* M.

Definition 7.2.6. *A singular chain* $c \in S_2(X)$ *is called a* square *if it is a triangulation of a continuous map* $I^2 \to G$. *Likewise the chain is called a* cylinder *resp. a* torus *if it is a triangulation of a continuous map* $\mathbb{S}^1 \times I \to X$ *resp.* $\mathbb{S}^1 \times \mathbb{S}^1 \to X$.

Theorem 7.2.7. *For any central elements* $g_1 \in Z(\tilde{G})$ *and arbitrary element* $g_2 \in \tilde{G}$ *the map* $\psi(g_1, g_2) : I^2 \to \tilde{G}$ *defined by*

$$
\psi(g_1, g_2)(x, y) = \begin{cases} \hat{\sigma}_2(1, g_1, g_1 g_2)(1 - x, x - y, y) & \text{if } x \geq y \\ \hat{\sigma}_2(1, g_2, g_1 g_2)(1 - y, y - x, x) & \text{if } x < y \end{cases}
$$

is continuous and the singular chain $\hat{\sigma}_2(1, g_1, g_1 g_2) - \hat{\sigma}_2(1, g_2, g_1 g_2)$ *is a triangulation of* $\psi(g_1, g_2)$ *with boundary* $g_1.\hat{\sigma}_1(1, g_2) - g_2.\hat{\sigma}_1(1, g_1) + \hat{\sigma}_1(1, g_1) - \hat{\sigma}_1(1, g_2)$.

Proof. Consider two elements $g_1, g_2 \in \tilde{G}$ where $g_1 \in Z(\tilde{G})$ is contained in the centre of $\tilde{G}$. The first face $\hat{\sigma}_1(1, g_1g_2)$ of $\hat{\sigma}_2(1, g_1, g_1g_2)$ coincides with the first face $\hat{\sigma}_1(1, g_1g_2)$ of $\hat{\sigma}_2(1, g_2, g_1g_2)$.

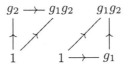

As a consequence, the chain $\hat{\sigma}_2(1, g_1, g_1g_2) - \hat{\sigma}_2(1, g_2, g_1g_2)$ is a triangulation of the square $\psi(g_1, g_2)$ obtained by gluing the simplices $\hat{\sigma}_2(1, g_1, g_1g_2)$ and $\hat{\sigma}_2(1, g_2, g_1g_2)$ along their common first faces.

$$\psi(g_1, g_2)(x, y) = \begin{cases} \hat{\sigma}_2(1, g_1, g_1g_2)(1-x, x-y, y) & \text{if } x \geq y \\ \hat{\sigma}_2(1, g_2, g_1g_2)(1-y, y-x, x) & \text{if } x \leq y \end{cases}$$

$$\tag{7.1}$$

Let $Q(g_1, g_2)$ denote the chain $\hat{\sigma}_2(1, g_1, g_1g_2) - \hat{\sigma}_2(1, g_2, g_1g_2)$. The boundary of $Q(g_1, g_2)$ is given by

$$\begin{aligned} \partial Q(g_1, g_2) &= \partial \hat{\sigma}_2(1, g_1, g_1g_2) - \partial \hat{\sigma}_2(1, g_2, g_1g_2) \\ &= \hat{\sigma}_1(g_1, g_1g_2) - \hat{\sigma}_1(1, g_1g_2) + \hat{\sigma}_1(1, g_1) \\ &\quad - \hat{\sigma}_1(g_2, g_1g_2) + \hat{\sigma}_1(1, g_1g_2) - \hat{\sigma}_1(1, g_2). \end{aligned}$$

Since the maps $\hat{\sigma}_1$ are equivariant and the first faces cancel out this simplifies to the chain

$$\partial Q(g_1, g_2) = g_1.\hat{\sigma}_1(1, g_2) - g_2.\hat{\sigma}_1(1, g_1) + \hat{\sigma}_1(1, g_1) - \hat{\sigma}_1(1, g_2).$$

This is the coboundary of the square $Q(g_1, g_2)$ indicated above in Picture 7.1. $\quad\square$

Note that the equivariant system $\hat{\sigma}_2$ of 2-simplices is defined on the whole space G^3. Therefore the functions $\psi(g_1, g_2)$ are defined for all $g_1 \in Z(\tilde{G})$ and $g_2 \in \tilde{G}$.

Corollary 7.2.8. *For any two elements $g_1 \in Z(\tilde{G})$, $g_2 \in \tilde{G}$ and any open covering $\mathfrak{U}$ of $\tilde{G}$ the chain $\mathrm{Csd}_{\mathfrak{U}} Q(g_1, g_2)$ is a square in $\tilde{G}$ and its boundary is the chain $\mathrm{Csd}_{\mathfrak{U}}(\sigma(1, g_1)) - \mathrm{Csd}_{\mathfrak{U}}(\sigma(1, g_2)) + g_1.\mathrm{Csd}_{\mathfrak{U}}(\sigma(1, g_2)) - g_2.\mathrm{Csd}_{\mathfrak{U}}(\sigma(1, g_1)).$*

Lemma 7.2.9. *For any two elements $g_1 \in \pi_1(G)$ and $g_2 \in \tilde{G}$ the square $Q(g_1, g_2)$ projects to a cylinder in G.*

Proof. If the vertex g_1 of $Q(g_1, g_2)$ is contained in the fundamental group $\pi_1(G)$, then the zeroth face $\hat{\sigma}_1(g_1, g_1g_2) = g_1.\hat{\sigma}_1(1, g_2)$ of $\hat{\sigma}_2(1, g_1, g_1g_2)$ projects to the same simplex as the second face $\hat{\sigma}_1(1, g_2)$ of $\hat{\sigma}_2(1, g_2, g_1g_2)$. Thus the continuous map

$$(x, y) \mapsto \begin{cases} \hat{\sigma}_2(1, g_1, g_1g_2)(1-x, x-y, y) & \text{if } x \geq y \\ \hat{\sigma}_2(1, g_2, g_1g_2)(1-y, y-x, x) & \text{if } x \leq y \end{cases}$$

factors through a map $\mathbb{S}^1 \times I \to G$. $\quad\square$

Corollary 7.2.10. *For any two elements $g_1 \in \pi_1(\tilde{G})$, $g_2 \in \tilde{G}$ and any left-invariant open covering $\mathfrak{U}$ of $\tilde{G}$ the square $\mathrm{Csd}_{\mathfrak{U}} Q(g_1, g_2)$ projects to a cylinder in G.*

In an analogous fashion one notices the properties of the squares $Q(g_1, g_2)$ with all vertices lying in the fundamental group $\pi_1(G)$:

Lemma 7.2.11. *For any two elements $g_1, g_2 \in \pi_1(G)$ the square $Q(g_1, g_2)$ projects to a torus in G.*

Corollary 7.2.12. *For any two elements $g_1, g_2 \in \pi_1(G)$ and any left-invariant open covering $\mathfrak{U}$ of $\tilde{G}$ the square $\mathrm{Csd}_{\mathfrak{U}}\, Q(g_1, g_2)$ projects to a torus in G.*

We remind the reader that the global cocycle $\tilde{F} \in Z^2(\tilde{G}; A)$ can be antisymmetrised and in this way gives rise to a cochain $\hat{a}(\tilde{F}) \in C^1(\tilde{G}; C^1(\tilde{G}; A))$. The cochain $\hat{a}(\tilde{F})$ can be expressed with the help of the squares $Q(g_1, g_2)$.

Lemma 7.2.13. *The antisymmetrised cochain $\alpha(\tilde{F})$ is related to the equivariant cocycle $\tilde{F}_{eq}$ via $\hat{a}(\tilde{F})(g_1, g_2) = \tilde{F}_{eq}(\lambda_* Q(g_1, g_2))$.*

Proof. Inserting the isomorphism $C^*(\tilde{G}; A) \cong A^*(G; A)^G$ into the definition of $\alpha(\tilde{F})$ one obtains

$$
\begin{aligned}
\alpha(\tilde{F})(g_1, g_2) &= \tilde{F}(g_1, g_2) - \tilde{F}(g_2, g_1) \\
&= \tilde{F}_{eq}(1, g_1, g_1 g_2) - \tilde{F}_{eq}(1, g_2, g_2 g_1) \\
&= \tilde{F}_{eq}(\lambda_* Q(g_1, g_2)).
\end{aligned}
$$

Thus the cochain $\alpha(\tilde{F})$ is given by evaluation of the equivariant cocycle $\tilde{F}_{eq}$ on the abstract simplices corresponding to the squares $Q(g_1, g_2)$.   □

Lemma 7.2.14. *The cochain $\alpha(\tilde{F})$ is given by evaluation on the $\tilde{\mathfrak{U}}$-relative subdivision of the squares $Q(g_1, g_2)$ via $\alpha(\tilde{F})(g_1, g_2) = f(\lambda_* \pi_* \,\mathrm{Csd}_{\tilde{\mathfrak{U}}}\, Q(g_1, g_2))$.*

Proposition 7.2.15. *If the cohomology class $[f_{eq}] \in H^2_{AS}(G; A)$ associated with a local group cocycle $f \in C^2(U; A)$ is trivial, then the induced cocycle $\tilde{F}$ is symmetric on $\pi_1(G) \times \pi_1(G)$, i.e. $\alpha(\tilde{F})(\pi_1(G) \times \pi_1(G)) = 0$.*

Proof. Let $f \in C^2(U; A)$ be a local group cocycle and f_{eq} be the corresponding equivariant cocycle. We assume the cohomology class $[f_{eq}] \in H^2_{AS}(G; A)$ to be trivial. Please recall that $\pi_1(G)$ is a central subgroup of $\tilde{G}$. By Lemma 7.2.11 any square $Q(g_1, g_2)$ with vertices $g_1, g_2 \in \pi_1(G)$ projects to a torus in G as does the subdivision $\mathrm{Csd}_{\tilde{\mathfrak{U}}}\, Q(g_1, g_2)$. Because the cochain $\alpha(\tilde{F})$ is given by evaluation of f_{eq} on the square $\mathrm{Sd}_{\tilde{\mathfrak{U}}}\, Q(g_1, g_2))$ and the evaluation of Alexander-Spanier cochains vanishes on cycles the cochain $\alpha(\tilde{F})$ is trivial on $\pi_1(G) \times \pi_1(G)$. This forces the cocycle $\tilde{F}$ to be symmetric on $\pi_1(G) \times \pi_1(G)$.   □

Corollary 7.2.16. *If the cohomology class $[f_{eq}] \in H^2_{AS}(G; A)$ associated with a local group cocycle $f \in C^2(V; A)$ vanishes, then the extension*

$$
0 \longrightarrow A \longrightarrow A \times_{\tilde{F}} \pi_1(G) \longrightarrow \pi_1(G) \longrightarrow 1
$$

is abelian.

Proposition 7.2.17. *If an extension $\tilde{F} \in C^2(\tilde{G}; A)$ of $\tilde{f}$ to a global group cocycle is cohomologous to an inflated cocycle then the germ of the the local group cocycle f is cohomologous to an extendible germ.*

Proof. Let $\tilde{F} \in C^2(\tilde{G}; A)$ be an extension of $\tilde{f}$ to a global group cocycle and suppose that $\tilde{F}$ is is cohomologous to an inflated group 2 cocycle $q_G^* h$. Then there exists a group 1-cochain $c \in C^1(\tilde{G}; A)$ such that $\tilde{F} = q_G^* h + \delta c$. Since the restriction $\pi_{|\tilde{U}}$ is a homeomorphism this implies

$$
f = \pi_{\tilde{U}}^{-1*} \tilde{f} = \pi_{\tilde{U}}^{-1*} q_G^* h_{|\tilde{U} \times \tilde{U}} + \pi_{\tilde{U}}^{-1*} (\delta c_{|\tilde{U} \times \tilde{U}}).
$$

Restricting ourselves to a smaller identity neighbourhood V satisfying $V \cdot V \subset U$ we obtain

$$f_{|V \times V} = h_{|V \times V} + \delta(\pi_{\tilde{V}}^{-1*} c_{|\tilde{V}}),$$

where $\tilde{V}$ denotes the identity component of the inverse image $q_G^{-1}(V)$. So the local group cocycle $f_{|V \times V}$ is cohomologous to the local cocycle $h_{|V \times V}$. $\square$

Corollary 7.2.18. *If an extension $\tilde{F} \in C^2(\tilde{G}; A)$ of $\tilde{f}$ to a global group cocycle is cohomologous to an inflated cocycle then germ of the the local group cocycle f is extendible to a global group cocycle.*

Proof. Since all germs of 2-boundaries are extendible to global boundaries this is a consequence of the preceding lemma. $\square$

Theorem 7.2.19. *If the group A is divisible then short sequence*

$$Z^2(G; V) \to Z^2_{AS,eq}(G; V)^G \to H^2_{AS}(G; V) \oplus \hom(\pi_1(M), H^1(G; V)).$$

is exact, i.e. the germ of a local group cocycle $f \in C^2(U; A)$ is extendible to a global cocycle if and only if $[f_{cq}] = 0$ in $H^2_{AS}(G; A)$ and $[\hat{\alpha}(\tilde{F})(\pi_1(G))] = 0$ in $H^1(G; A)$.

Proof. The conditions $[f_{eq}] = 0 \in H^2_{AS}(G; A)$ and $[\hat{\alpha}(\tilde{F})(\pi_1(G))] = 0 \in H^1(G; A)$ are necessary by Theorem 7.1.17. On the other hand, if the cohomology class $[f_{eq}]$ is trivial then the germ of the inflated local group cocycle $\tilde{f}$ is extendible to a global group cocycle $\tilde{F}$. The condition $[\hat{\alpha}(\tilde{F})(\pi_1(G))] = 0$ in $H^1(G; A)$ then implies that the subgroup $\pi_1(G) \leq \tilde{G}$ is contained in the image $q(Z(A \times_{\tilde{F}} \tilde{G}))$ (Lemma 7.1.10). By Corollary 7.2.18 this is equivalent for f to be extendible to a global group cocycle F on G. $\square$

8

The Rational Singular Homology of Arc-Wise Connected H-Spaces

In this chapter we prove a general version of the "Cartan-Serre Theorem" and the "Milnor-Moore Theorem" on the rational singular homology $H(X;\mathbb{Q})$ of an arc-wise connected H-space X without any additional restrictions on the topological space X. The classical versions of these Theorems assures that for any H-space of finite type the rational Hurewicz homomorphism extends to an isomorphism

$$\text{hur} : \pi_*(X) \otimes \mathbb{Q} \xrightarrow{\cong} P_*(X;\mathbb{Q})$$

of Lie algebras onto the space of primitives in the singular Hopf algebra $H(X;\mathbb{Q})$ and this isomorphism extends to an isomorphism

$$U(\pi_*(X) \otimes Q) \xrightarrow{\cong} H(X;\mathbb{Q})$$

of Hopf algebras, where $U(\pi_*(X)\otimes\mathbb{Q})$ denotes the universal enveloping algebra of the graded rational Lie algebra $\pi_*(X)\otimes\mathbb{Q}$. Below we prove an analogous theorem without the assumption that $H(X;\mathbb{Q})$ is of finite type. First we recall the construction of the coalgebra and algebra structure of the rational singular homology $H(X;\mathbb{Q})$ and the Hopf algebra structure of the rational singular homology $H(\Omega X;\mathbb{Q})$ of the loop space ΩX of a simply connected topological space X. This has very useful consequence for the structure of the rational singular homology of loop spaces of simply connected H-spaces (Theorem 8.3.9) which in particular includes loop groups of simply connected topological groups e.g. simply connected Lie groups. This is then used to derive a "Cartan-Serre Theorem" for arc-wise connected topological groups (Theorem 8.4.10) and and then derive an even more general structure theorem on the rational singular homology of arc-wise connected H-spaces:

Theorem (The Structure of the Rational Singular Homology of Arc-wise Connected H-spaces). *The rational Hurewicz homomorphism for an arc-wise connected H-space X satisfying the separation axion T_1 is a Lie algebra isomorphism of $\pi(X)\otimes\mathbb{Q}$ onto the graded Lie algebra $P(X;\mathbb{Q})$ of primitive elements in the Hopf algebra $H(X;\mathbb{Q})$ and extends to an isomorphism of graded Hopf algebras from the universal enveloping algebra $U(\pi(X)\otimes\mathbb{Q})$ of the rational graded Lie algebra $\pi(X)\otimes\mathbb{Q}$ onto the singular graded Hopf algebra $H(X;\mathbb{Q})$:*

$$
\begin{array}{ccc}
\pi(G) \otimes \mathbb{Q} & \xrightarrow[\cong]{hur} & P(G;\mathbb{Q}) \\
\downarrow & & \downarrow \\
U(\pi(G) \otimes \mathbb{Q}) & \xrightarrow{\cong} & H(G;\mathbb{Q})
\end{array}
$$

In particular the rational singular homology $H(X;\mathbb{Q})$ of an arc-wise connected H-space X is a primitively generated Hopf algebra. Its subspace $P(X;\mathbb{Q})$ of primitive *elements is the image $hur(\pi(X) \otimes \mathbb{Q})$ and the multiplication in $H(X;\mathbb{Q})$ is induced by the multiplication in X.*

This structure theorem is the main result of this chapter. It extends the classical "Cartan-Serre Theorem" to full generality without relying on such very restrictive requirements as the H-space X to be of finite type. It enables us to test the triviality of differential forms on an infinite dimensional Lie group G by integration over products of smooth spheres in G. This will particularly useful in characterising the triviality of flux homomorphisms in terms of period homomorphisms, both of which will be introduced in Chapter 13. The triviality of these homomorphisms will later turn out to be essential for the characterisation of the integrability of equivariant closed forms to smooth equivariant global global cocycles or the extensibility of equivariant Alexander-Spanier cocycles to global equivariant cocycles.

8.1 The Homology Coalgebra

In this section we recall the classical differential graded coalgebra structure for the (discrete) singular differential graded Abelian group $\bigoplus S_n(X)$ of a topological space X. More generally we consider the (discrete) singular simplicial group $S \otimes R(X)$ with coefficients in a commutative ring R and the corresponding differential graded R-module $\bigoplus S_n(X) \otimes R$ of singular simplices in a topological space X. We start with coefficients $\mathbb{Z}$ and later generalise the result to arbitrary rings R of coefficients.

Let X and Y be topological spaces and consider the singular simplicial set $\hom_{\mathbf{Top}}(\Delta, X \times Y)$ of the product space $X \times Y$. Each set $\hom_{\mathbf{Top}}(\Delta_n, X \times Y)$ of singular n-simplices in the product space $X \times Y$ is naturally isomorphic to the product of the sets $\hom_{\mathbf{Top}}(\Delta_n, X)$ and $\hom_{\mathbf{Top}}(\Delta_n, Y)$ of singular n-simplices in X resp. Y. The n-th singular Abelian group $S(X \times Y)([n])$ of the product space $X \times Y$ is by definition the free Abelian group on the set $\hom_{\mathbf{Top}}(\Delta_n, X \times Y)$. It is therefore naturally isomorphic to the free Abelian group on the product of the sets $\hom_{\mathbf{Top}}(\Delta_n, X)$ and $\hom_{\mathbf{Top}}(\Delta_n, Y)$:

$$S(X \times Y)([n]) \cong \mathrm{F}_{\mathbf{Ab}}\left(\hom_{\mathbf{Top}}(\Delta_n, X) \times \hom_{\mathbf{Top}}(\Delta_n, Y)\right)$$

The free Abelian group on a binary product set is naturally isomorphic to the tensor product (over $\mathbb{Z}$) of the free Abelian groups on the factors. In the above case these are the n-th singular group $S(X)([n])$ of X and the n-th singular group $S(Y)([n])$ of Y respectively. Therefore the n-th singular group $S(X \times Y)([n])$ of the product space $X \times Y$ is naturally isomorphic to the tensor product of the free Abelian groups $S(X)([n])$ and $S(Y)([n])$:

$$S(X \times Y)([n]) \cong S(X)([n]) \otimes S(Y)([n])$$

Moreover the boundary and degeneracy morphisms of the singular simplicial Abelian group $S(X \times Y)$ correspond to the tensor product of the corresponding boundary and degeneracy morphisms under these isomorphisms, i.e. the above isomorphisms assemble to an isomorphism

$$\varphi_{X,Y} : S(X \times Y) \xrightarrow{\cong} S(X) \otimes S(Y)$$

of simplicial Abelian groups. The simplicial Abelian group on the right hand side can also be described as the diagonal of a bisimplicial Abelian group: The tensor products $S(X)([n]) \otimes S(X)([m])$ together with the boundary and degeneracy morphisms for both factors form a bisimplicial Abelian group. This bisimplicial Abelian group is the functor

$$\otimes \circ \langle S(X), S(Y) \rangle : \mathcal{O}^{op} \times \mathcal{O}^{op} \to \mathbf{RMod}.$$

Collecting all the natural isomorphisms above we can identify the singular simplicial group of $X \times Y$ as the diagonal of the bisimplicial Abelian group described above:

$$\varphi_{X,Y} : S(X \times Y) \xrightarrow{\cong} S(X) \otimes S(Y) = \mathrm{diag}(\otimes \circ \langle S(X), S(Y) \rangle)$$

To avoid confusion we recall some subtleties on the notion of the objects involved. The chain complex of Abelian groups associated to the simplicial group $S(X)$ is denoted by $S_*(X)$ and the morphism of chain complexes induced by the morphism φ of simplicial Abelian groups is denoted by φ_*. Passing to the chain complexes associated to the simplicial Abelian groups above yields an isomorphism

$$\varphi_{X,Y*} : S_*(X \times Y) \xrightarrow{\cong} S_*(X) \otimes S_*(Y) = (\mathrm{diag}(\otimes \circ \langle S(X), S(Y) \rangle))_*$$

between the singular chain complex $S_*(X \times Y)$ and the chain complex associated to the simplicial Abelian group $\mathrm{diag}(\otimes \circ \langle S(X), S(Y) \rangle)$. The latter chain complex is the domain of the Alexander-Whitney homomorphism (cf. F.8.6)

$$AW_* : (\mathrm{diag}(\otimes \circ \langle S(X), S(Y) \rangle))_* \to \mathrm{Tot}(\otimes \circ \langle S(X), S(Y) \rangle)_{*,*}$$

whose codomain is the total complex of the double complex $(\otimes \circ \langle S(X), S(Y) \rangle)_{*,*}$ associated to the bisimplicial Abelian group $\otimes \circ \langle S(X), S(Y) \rangle$. The composition of the natural isomorphism φ_* with the Alexander-Whitney morphism yields a morphism

$$AW_* \circ \varphi_{X,Y*} : S_*(X \times Y) \to \mathrm{Tot}(\otimes \langle S(X), S(Y) \rangle)_{*,*}$$

of chain complexes. We pass further to the differential graded Abelian groups associated to these chain complexes. The codomain complex is the total complex of the double complex with Abelian group $S_p(X) \otimes S_q(X)$ at (p,q). The differential graded Abelian group associated to this chain complex is the tensor product of the differential graded Abelian groups $\bigoplus_p S_p(X)$ and $\bigoplus_q S_q(Y)$. We thus have obtained a morphism

$$AW := \bigoplus_{n \in \mathbb{N}} (AW_n \circ \varphi_{X,Y,n}) : \bigoplus_n S_n(X \times Y) \to \left(\bigoplus_p S_p(X) \right) \otimes \left(\bigoplus_q S_q(X) \right)$$

of differential graded Abelian groups, which we simply will denote by AW for the rest of this chapter. By the Eilenberg-Zilber Theorem (cf. Theorem F.8.12), the morphism AW of graded Abelian groups induces an isomorphism in homology. It can be used to define a comultiplication on the singular graded Abelian group $\bigoplus S_n(X)$ of a topological space X as follows: Consider the diagonal map $D_2 : X \to X \times X$. This map induces a morphism

$$\bigoplus_{n \in \mathbb{N}} S_n(D_2) : \bigoplus_{n \in \mathbb{N}} S_n(X) \to \bigoplus_{n \in \mathbb{N}} S_n(X \times X)$$

of differential graded Abelian groups. Composing this morphism with the morphism AW of differential graded Abelian groups results in the morphism

$$D := AW \circ \bigoplus_{n \in \mathbb{N}} S_n(D_2) : \bigoplus_{n \in \mathbb{N}} S_n(X) \to \left(\bigoplus_p S_p(X) \right) \otimes \left(\bigoplus_q S_q(X) \right) \tag{8.1}$$

of differential graded Abelian groups. Since the Alexander-Whitney morphism is associative (cf. Lemma F.8.16), the morphism D is a comultiplication on the differential graded Abelian group $\bigoplus_{n \in \mathbb{N}} S_n(X)$. Moreover can show:

Proposition 8.1.1. *The morphism D turns the graded Abelian group $\bigoplus_{n\in\mathbb{N}} S_n(X)$ into a differential graded coalgebra with counit $\epsilon : \bigoplus_{n\in\mathbb{N}} S_n(X) \to \mathbb{Z}$ induced by the continuous function $X \to \{*\}$.*

Proof. A proof can be found in [FHT01a, Chap. 4 (b)]. □

Corollary 8.1.2. *The morphism D of differential graded Abelian groups induces a morphism*

$$H(D) : H(X) = H\left(\bigoplus_{n\in\mathbb{N}} S_n(X)\right) \to H\left(\left(\bigoplus_{p\in\mathbb{N}} S_p(X)\right) \otimes \left(\bigoplus_{q\in\mathbb{N}} S_q(X)\right)\right)$$

of graded Abelian homology groups.

The homology of the singular Abelian graded group $\bigoplus_n S_n(X)$ on the left hand side in Equation 8.1 resp. Corollary 8.1.2 is the singular homology $H(X)$ of the topological space X (understood as a graded Abelian group). The homology of the differential graded Abelian group on the right hand side can be computed using the Kuenneth Theorem, which states that the natural sequence

$$0 \to \bigoplus_{p+q=n} H_p(X) \otimes H_q(Y) \xrightarrow{\psi} H_n(\mathrm{Tot} \otimes \circ \langle S_*(X), S_*(Y)\rangle) \to$$

$$\to \bigoplus_{p+q=n-1} \mathrm{Tor}_1^{\mathbb{Z}}(H_p(X), H_q(Y)) \to 0$$

of Abelian groups is split exact for each $n \in \mathbb{N}$ (cf. [Wei94, Theorem 3.6.3]). If the singular homology $H_*(X)$ of X is torsion free then the first homomorphism in the Kuenneth exact sequence is an isomorphism of Abelian groups. So we observe:

Lemma 8.1.3. *If the singular homology $H(X)$ is torsion free, then the coalgebra structure of $\bigoplus_n S_n(X)$ descends to a coalgebra structure in singular homology.*

Proof. Let the singular homology $H(X)$ of the topological space X be torsion free. The Kuenneth Formula ensures that the natural homomorphism

$$\psi : \left(\bigoplus_p H_p(X)\right) \otimes \left(\bigoplus_q H_q(X)\right) \to \bigoplus_n H_n\left(\left(\bigoplus_p S_p(X)\right) \otimes \left(\bigoplus_q S_q(X)\right)\right)$$

of graded Abelian groups is an isomorphism. So $\psi^{-1} \circ H(D)$ is a morphism of graded Abelian groups. This morphism of graded Abelian groups turns the singular homology $H(X)$ into a graded coalgebra. (A more detailed derivation can be found in [Whi78, chapter III.7].) □

A similar result can be obtained by using rings R of coefficients which turn the singular homology $H(X;R)$ with coefficients R into a flat R-module (e. g. $\mathbb{Q}$ or $\mathbb{R}$). The construction of the coalgebra structure is analogous to the above observations. Let R be a ring of coefficients. The tensor product without subscript will denote the tensor product in the category $\mathbb{Z}\mathbf{mod} = \mathbf{Ab}$ of Abelian groups. The tensor product in the category $\mathbf{RMod}$ of R-modules will be denoted by $-\otimes_R -$. Recall that the singular simplicial group with coefficients R of a topological space X is obtained by tensoring the singular simplicial Abelian group $S(X)$ with the ring R of coefficients. It is a simplicial R-module, which is denoted by $S(X;R)$ for the rest of this chapter. The functor

$$S(-;R) : \mathbf{Top} \to \mathbf{RMod}^{\mathcal{O}^{op}}, \quad X \mapsto S(X;R) := S(X) \otimes R, \quad f \mapsto S(f) \otimes \mathrm{id}_R$$

is also called the singular simplicial R-module functor.

As before we consider topological spaces X and Y. The n-th singular R-module $S(X \times Y; R)([n])$ of the product space $X \times Y$ is the free R-module on the set $\hom_{\mathbf{Top}}(\Delta_n, X \times Y)$. It is therefore naturally isomorphic to the free R-module on the product of the sets $\hom_{\mathbf{Top}}(\Delta_n, X)$ and $\hom_{\mathbf{Top}}(\Delta_n, Y)$:

$$S(X \times Y; R)([n]) \cong F_{\mathbf{RMod}}\left(\hom_{\mathbf{Top}}(\Delta_n, X) \times \hom_{\mathbf{Top}}(\Delta_n, Y)\right)$$

The free R-module on a binary product set is naturally isomorphic to the tensor product (over R) of the free R-modules on the factors. In the above case these are the n-th singular R-module $S(X; R)([n])$ of X and the n-th singular R-module $S(Y; R)([n])$ of Y respectively. Therefore the n-th singular R-module $S(X \times Y; R)([n])$ of the product space $X \times Y$ is naturally isomorphic to the tensor product of the R-modules $S(X; R)([n])$ and $S(Y; R)([n])$:

$$S(X \times Y; R)([n]) \cong S(X; R)([n]) \otimes_R S(Y; R)([n])$$

Moreover the boundary and degeneracy morphisms of the singular simplicial R-module $S(X \times Y; R)$ correspond to the tensor product of the corresponding boundary and degeneracy morphisms under these isomorphisms, i.e. the above isomorphisms assemble to an isomorphism

$$\varphi_{X,Y} : S(X \times Y; R) \xrightarrow{\cong} S(X; R) \otimes_R S(Y; R)$$

of simplicial R-modules. The simplicial R-module on the right hand side can also be described as the diagonal of a bisimplicial R-module: The tensor products $S(X; R)([n]) \otimes_R S(X; R)([m])$ together with the boundary and degeneracy morphisms for both factors form a bisimplicial R-module. This bisimplicial R-module is the functor

$$\otimes_R \circ \langle S(X; R), S(Y; R) \rangle : \mathcal{O}^{op} \times \mathcal{O}^{op} \to \mathbf{RMod}.$$

Collecting all the natural isomorphisms above we can identify the singular simplicial R-module of $X \times Y$ as the diagonal of the bisimplicial R-module described above:

$$\varphi_{X,Y} : S(X \times Y; R) \xrightarrow{\cong} S(X; R) \otimes_R S(Y; R) = \mathrm{diag}(\otimes_R \circ \langle S(X; R), S(Y; R) \rangle)$$

To avoid confusion we recall some subtleties on the notion of the objects involved. The chain complex of R-modules associated to the simplicial R-module $S(X; R)$ is denoted by $S_*(X; R)$ and the morphism of chain complexes induced by the morphism φ of simplicial R-modules is denoted by φ_*. Passing to the chain complexes associated to the simplicial R-modules above yields an isomorphism

$$\varphi_{X,Y*} : S_*(X \times Y; R) \xrightarrow{\cong} S_*(X; R) \otimes_R S_*(Y; R)) = (\mathrm{diag}(\otimes_R \circ \langle S(X; R), S(Y; R) \rangle))_*$$

between the singular chain complex $S_*(X \times Y; R)$ and the chain complex associated to the simplicial R-module $\mathrm{diag} \otimes_R \circ \langle S(X; R), S(Y; R) \rangle$. The latter chain complex is the domain of the Alexander-Whitney homomorphism (cf. F.8.6)

$$AW_* : (\mathrm{diag}(\otimes_R \circ \langle S(X; R), S(Y; R) \rangle))_* \to \mathrm{Tot}(\otimes_R \circ \langle S(X; R), S(Y; R) \rangle)_{*,*}$$

whose codomain in the total complex of the double complex associated to the bisimplicial R-module $\otimes_R \circ \langle S(X; R), S(Y; R) \rangle$. The composition of the natural isomorphism φ_* with the Alexander-Whitney morphism yields a morphism

$$AW_* \circ \varphi_{X,Y*} : S_*(X \times Y; R) \to \mathrm{Tot}(\otimes_R \langle S(X; R) \times S(Y; R) \rangle)_{*,*}$$

of chain complexes. We pass further to the differential graded R-modules associated to these chain complexes. The codomain complex is the total complex of the double complex with R-module $S_p(X; R) \otimes_R S_q(X; R)$ at (p, q). The differential graded R-module associated to this chain complex is the tensor product of the differential graded R-modules $\bigoplus_p S_p(X; R)$ and $\bigoplus_q S_q(Y; R)$. Thus we have obtained a morphism

$$\bigoplus_{n \in \mathbb{N}} (AW_n \circ \varphi_{X,Y,n}) : \bigoplus_n S_n(X \times Y; R) \to \left(\bigoplus_p S_p(X; R)\right) \otimes_R \left(\bigoplus_q S_q(X; R)\right)$$

of differential graded R-modules, which we simply will denote by AW for the rest of this chapter. By the Eilenberg-Zilber Theorem (cf. Theorem F.8.12), the morphism AW of graded R-modules induces an isomorphism in homology. Similar to the case of integer coefficients it can be used to define a comultiplication on the singular graded R-module $\bigoplus S_n(X; R)$ of a topological space X. Let $D_2 : X \to X \times X$ denote the diagonal map. This map induces a morphism

$$\bigoplus_{n \in \mathbb{N}} S_n(D_2; R) : \bigoplus_{n \in \mathbb{N}} S_n(X; R) \to \bigoplus_{n \in \mathbb{N}} S_n(X \times X; R)$$

of differential graded R-modules. Composing this morphism with the morphism AW of differential graded R-modules results in the morphism

$$D := AW \circ \bigoplus_{n \in \mathbb{N}} S_n(D_2; R) : \bigoplus_{n \in \mathbb{N}} S_n(X; R) \to \left(\bigoplus_p S_p(X; R)\right) \otimes_R \left(\bigoplus_q S_q(X; R)\right)$$

(8.2)

of differential graded R-modules. Since the Alexander-Whitney morphism is associative (cf. Lemma F.8.16), the morphism D is a comultiplication on the differential graded R-module $\bigoplus_{n \in \mathbb{N}} S_n(X; R)$. Moreover can show:

Proposition 8.1.4. *The morphism D turns the graded R-module $\bigoplus_{n \in \mathbb{N}} S_n(X; R)$ into a differential graded coalgebra with counit $\epsilon : \bigoplus_{n \in \mathbb{N}} S_n(X; R) \to R$ induced by the continuous function $X \to \{*\}$.*

Proof. A proof can be found in [FHT01a, Chap. 4 (b)].   □

Corollary 8.1.5. *The morphism D of differential graded R-modules induces a morphism*

$$H(D) : H\left(\bigoplus_{n \in \mathbb{N}} S_n(X; R)\right) \to H\left(\left(\bigoplus_{p \in \mathbb{N}} S_p(X; R)\right) \otimes_R \left(\bigoplus_{q \in \mathbb{N}} S_q(X; R)\right)\right)$$

of graded R-modules.

The homology of the R-module on the left hand side is the topological singular homology $H(X; R)$ with coefficients R of the topological space X (understood as a graded R-module). If the ring R is a principal ideal domain then the homology of the differential graded R-module on the right hand side can be computed using the Kuenneth Theorem, which states that the sequence

$$0 \to \bigoplus_{p+q=n} H_p(X; R) \otimes H_q(Y; R) \to H_n(\mathrm{Tot}(S_*(X; R) \otimes S_*(Y; R)) \to$$

$$\to \bigoplus_{p+q=n-1} \mathrm{Tor}_1^R(H_p(X; R), H_q(Y; R)) \to 0$$

of graded R-modules is split exact in this case.

Lemma 8.1.6. *If R is a principal ideal domain and the singular homology $H(X;R)$ with coefficients R is torsion free, then the coalgebra structure of $\bigoplus_n S_n(X;R)$ descends to a coalgebra structure in singular homology.*

Proof. Let R be a principal ideal domain and the singular homology $H(X;R)$ with coefficients R of the topological space X be torsion free. Then all the the modules $\mathrm{Tor}_1^R(H_p(X;R), H_q(X;R))$ are trivial and the Kuenneth Formula ensures that the natural homomorphism

$$\psi : H(X;R) \otimes_R H(X;R) \to \bigoplus_n H_n\left(\left(\bigoplus_p S_p(X;R)\right) \otimes \left(\bigoplus_q S_q(X;R)\right)\right)$$

of R-modules is an isomorphism. So the morphism $\psi^{-1} \circ H(D)$ of graded R-modules turns the singular homology $H(X;R)$ with coefficients R into a graded coalgebra. (A more detailed derivation can be found in [Whi78, chapter III.7].) □

Corollary 8.1.7. *The real and rational singular homologies $H(X;\mathbb{R})$ resp. $H(X;\mathbb{Q})$ of a topological space X are graded coalgebras over $\mathbb{R}$ resp $\mathbb{Q}$.*

8.2 Products Induced by Continuous Functions

In this section we recall products in singular homology which are induced by continuous functions of topological spaces (e.g. the multiplication $\mu : G \times G \to G$ of a topological group G). The main tool here fore is the Eilenberg-Zilber transformation applied to the singular differential graded groups of topological spaces. Recall that these morphisms are associative, i.e. for any three topological spaces X, Y, and Z the diagram

$$\left(\bigoplus_p S_p(X)\right) \otimes \left(\bigoplus_q S_q(Y)\right) \otimes \left(\bigoplus_r S_r(Z)\right) \xrightarrow{\mathrm{id} \otimes EZ} \left(\bigoplus_p S_p(X)\right) \otimes \left(\bigoplus_q S_q(Y \times Z)\right)$$

$$\downarrow{EZ \otimes \mathrm{id}} \qquad\qquad\qquad\qquad\qquad \downarrow{EZ}$$

$$\left(\bigoplus_p S_p(X \times Y)\right) \otimes \left(\bigoplus_r S_r(Z)\right) \xrightarrow{\quad EZ \quad} \bigoplus_p S_p(X \times Y \times Z)$$

commutes (cf. Lemma F.8.18). Furthermore, the Eilenberg-Zilber morphisms are natural in the singular differential graded groups involved (cf. Section F.8). Therefore, if $f : X \times Y \to Z$ is a continuous function, then the Eilenberg-Zilber morphism intertwines the morphisms $\mathrm{id} \otimes \bigoplus_q S_q(f)$ and $\bigoplus_q S_q(\mathrm{id}_X \times f)$ of differential graded groups, i.e. the following diagram is a commutative:

$$\left(\bigoplus_p S_p(X)\right) \otimes \left(\bigoplus_q S_q(X \times Y)\right) \xrightarrow{\mathrm{id} \otimes \left(\bigoplus_q S_q(f)\right)} \left(\bigoplus_p S_p(X)\right) \otimes \left(\bigoplus_q S_q(Z)\right)$$

$$\downarrow{EZ} \qquad\qquad\qquad\qquad\qquad\qquad \downarrow{EZ}$$

$$\bigoplus_n S_n(X \times X \times Y) \xrightarrow{\bigoplus_n S_n(\mathrm{id}_X \times f)} \bigoplus_n S_n(X \times Z)$$

$$(8.3)$$

Let X, Y and Z be topological spaces and $f : X \times Y \to Z$ be a continuous function. The function f induces a morphism

$$\mathrm{hom}_{\mathbf{Top}}(\Delta, f) : \mathrm{hom}_{\mathbf{Top}}(\Delta, X \times Y) \to \mathrm{hom}_{\mathbf{Top}}(\Delta, Z)$$

of singular simplicial sets. Passing to the free Abelian groups on these sets results in a morphism

$$S(f) : S(X \times Y) \to S(Z)$$

of singular simplicial groups. The morphism $S(f)$ in turn induces a morphism $S_*(f) : S_*(X \times Y) \to S_*(Z)$ between the singular chain complexes and a morphism $\bigoplus_n S_n(f)$ between the singular differential graded Abelian groups $\bigoplus_n S_n(X \times Y)$ and $\bigoplus_n S_n(Z)$ associated to the singular simplicial groups $S(X \times Y)$ and $S(Z)$. The latter homomorphism $\bigoplus_n S_n(f)$ of differential graded Abelian groups can be composed with the Eilenberg-Zilber morphism EZ to yield a chain of morphisms of differential graded Abelian groups

$$\left(\bigoplus_p S_p(X)\right) \otimes \left(\bigoplus_q S_q(Y)\right) \xrightarrow{EZ} \bigoplus_{n \in \mathbb{N}} S_n(X \times Y) \xrightarrow{\bigoplus_n S_n(f)} \bigoplus S_n(Z),$$

the composition of which is denoted by f_{alg}. In the case $X = Y = Z$ the map f_{alg} is a product on the singular differential graded topological group $\bigoplus_n S_n(X)$ of X, turning the latter into differential graded magma.

Proposition 8.2.1. *If X is a topological semi-group with multiplication μ then the product $\mu_{alg} = \bigoplus_n S_n(\mu) \circ EZ$ on the differential graded group $\bigoplus_n S_n(X)$ is associative.*

Proof. It is to prove that the associativity of $\mu : X \times X \to X$ implies the associativity of the morphism μ_{alg} of differential graded Abelian groups. This is a consequence of the associativity of the Eilenberg-Zilber morphism EZ (cf. Proposition F.8.17). Let σ_p, σ_q and σ_r be singular simplices of dimension p resp. q resp. r in X. Recall that the Eilenberg-Zilber morphism intertwines the morphisms $\mathrm{id} \otimes \bigoplus_q S_q(\mu)$ and $\bigoplus_q S_q(\mathrm{id}_X \times \mu)$ of differential graded groups (cf. diagram 8.3). In conjunction with the associativity of $S(\mu)$ this implies the equations

$$\mu_{alg}(\sigma_p \otimes (\mu_{alg}(\sigma_q \otimes \sigma_r))) =$$

$$\left(\bigoplus_n S_n(\mu)\right) \circ EZ(\sigma_p \otimes \mu_{alg}(\sigma_q \otimes \sigma_r))$$

$$= \left(\bigoplus_n S_n(\mu)\right) \circ EZ \circ \left(\mathrm{id} \otimes \left[\left(\bigoplus_n S_n(\mu)\right) \circ EZ\right]\right)(\sigma_p \otimes (\sigma_q \otimes \sigma_r))$$

$$= \left(\bigoplus_n S_n(\mu)\right) \circ EZ \circ \left[\mathrm{id} \otimes \left(\bigoplus_n S_n(\mu)\right)\right](\sigma_p \otimes (EZ(\sigma_q \otimes \sigma_r)))$$

$$= \left(\bigoplus_n S_n(\mu)\right) \circ \left(\bigoplus_n S_n(\mathrm{id}_X \times \mu)\right) \circ EZ(\sigma_p \otimes (EZ(\sigma_q \otimes \sigma_r)))$$

$$= \left(\bigoplus_n S_n(\mu \circ (\mathrm{id}_X \times \mu))\right) \circ EZ(\sigma_p \otimes (EZ(\sigma_q \otimes \sigma_r)))$$

$$= \left(\bigoplus_n S_n(\mu \circ (\mu \times \mathrm{id}_X))\right) \circ EZ(\sigma_p \otimes (EZ(\sigma_q \otimes \sigma_r)))$$

Using the associativity of the Eilenberg-Zilber morphism EZ the last expression transforms to

$$\mu_{alg}(\sigma_p \otimes (\mu_{alg}(\sigma_q \otimes \sigma_r))) =$$

$$\left(\bigoplus_n S_n(\mu \circ (\mathrm{id}_X \times \mu)) \right) \circ EZ(\sigma_p \otimes (EZ(\sigma_q \otimes \sigma_r)))$$

$$= \left(\bigoplus_n S_n(\mu \circ (\mu \times \mathrm{id}_X)) \right) \circ EZ(\sigma_p \otimes (EZ(\sigma_q \otimes \sigma_r)))$$

$$= \left(\bigoplus_n S_n(\mu \circ (\mu \times \mathrm{id}_X)) \right) \circ EZ((EZ(\sigma_p \otimes \sigma_q)) \otimes \sigma_r)$$

$$= \left(\bigoplus_n S_n(\mu) \right) \circ \left(\bigoplus_n S_n(\mu \times \mathrm{id}_X) \right) \circ EZ((EZ(\sigma_p \otimes \sigma_q)) \otimes \sigma_r)$$

$$= \left(\bigoplus_n S_n(\mu) \right) \circ EZ \circ \left[\left(\bigoplus_n S_n(\mu) \right) \otimes \mathrm{id} \right] ((EZ(\sigma_p \otimes \sigma_q)) \otimes \sigma_r)$$

$$= \left(\bigoplus_n S_n(\mu) \right) \circ EZ \circ \left(\left[\left(\bigoplus_n S_n(\mu) \right) EZ \right] \otimes \mathrm{id} \right) ((\sigma_p \otimes \sigma_q) \otimes \sigma_r)$$

$$= \mu_{alg} \left(\mu_{alg}(\sigma_p \otimes \sigma_q) \otimes \sigma_r \right),$$

i.e. the product μ_{alg} on the singular differential graded Abelian group $\bigoplus_n S_n(X)$ is associative. □

Lemma 8.2.2. *If X is a topological monoid with multiplication μ then the product $\mu_{alg} = \bigoplus_n S_n(\mu) \circ EZ$ turns $\bigoplus_n S_n(X)$ into a differential graded algebra.*

Proof. Since the product μ_{alg} on the singular differential graded group $\bigoplus_n S_n(G)$ is associative by Proposition 8.2.1 it only remains to verify the existence of a unit for the multiplication μ_{alg}. Let $e \in G$ denote the neutral element in the monoid G. The constant 0-simplex $e \in S_0(G)$ is a unit for the multiplication μ_{alg}. □

Lemma 8.2.3. *If (G, X) is a transformation monoid with action $\varrho : G \times X \to X$ then $\varrho_{alg} = \bigoplus_n S_n(\varrho) \circ EZ$ is an action of the differential graded algebra $\bigoplus_n S_n(G)$ on the differential graded group $\bigoplus_n S_n(X)$.*

Proof. It only is to verify that the morphism $\varrho_{alg} = \bigoplus_n S_n(\varrho) \circ EZ$ of differential graded groups is associative in G, i.e. that for all singular simplices σ_p, σ_q of dimension p, and q in G and singular simplices σ_r of dimension r in X the equation

$$\varrho_{alg}(\sigma_p \otimes \varrho_{alg}(\sigma_q \otimes \sigma_r)) = \varrho_{alg}(\varrho_{alg}(\sigma_p \otimes \sigma_q) \otimes \sigma_r)$$

is satisfied. The proof thereof is analogous to that of Lemma 8.2.1. □

Example 8.2.4. Every differential graded singular group $\bigoplus_n S_n(G)$ of a topological group G is a differential graded module over itself.

Example 8.2.5. If (E, p, B) is a principal bundle with structure group G then the singular differential graded algebra $\bigoplus_n S_n(G)$ acts on the singular differential graded group $\bigoplus_n S_n(E)$ of the total space E of the bundle, i.e. the latter is a (differential graded) module over the first one.

Let X and Y be topological spaces. If the ring R of coefficients is a principal ideal domain and either the singular homology $H(X; R)$ with coefficients R of the topological space X or the singular homology $H(Y; R)$ is a torsion free R-module, then the Kuenneth Theorem ensures that the morphism

$$\psi : H(X;R) \otimes_R H(Y;R) \to H\left(\left(\bigoplus_{p \in \mathbb{N}} S_p(X;R)\right) \otimes_R \left(\bigoplus_{q \in \mathbb{N}} S_q(X;R)\right)\right)$$

of graded R-modules is an isomorphism. In particular for the case of a topological monoid G and $Y = X = G$ this implies that the algebra structure of the singular differential graded R-module $\bigoplus_n S_n(G;R)$ of a topological monoid G descends to singular homology. Moreover the associativity of the multiplication $\mu : G \times G \to G$ need only be homotopy associative:

Lemma 8.2.6. *If X is a connected H-space with multiplication $\mu : X \times X \to X$ then the singular homology $H(X;R)$ with coefficients in a principal ideal domain R is a graded algebra with multiplication $H(\mu_{alg})\psi$ and unit the cohomology class of a singular 0-simplex e in X.*

Proof. See [Whi78, Theorem III.7.12] for a proof. $\square$

The algebra structure of the graded group $H(X;R)$ of a H-space X is compatible with the coalgebra structure introduced in the last section:

Proposition 8.2.7. *If X is a connected H-space with multiplication $\mu : X \times X \to X$ with torsion free singular homology $H(X;R)$ with coefficients in a principal ideal domain R then $H(X;R)$ is a graded cocommutative Hopf algebra with multiplication $H(\mu_{alg})\psi$, unit the cohomology class of a singular 0-simplex e, comultiplication $H(D)\psi$ and counit $H(\epsilon)$.*

Proof. A proof can be found in [Whi78, Theorem II.7.15]. $\square$

Corollary 8.2.8. *The singular homology $H(X;\mathbb{K})$ of a connected H-space with coefficients a field $\mathbb{K}$ is a cocommutative Hopf-algebra.*

8.3 The Loop Space Homology Algebra

Let $(X,*)$ be a simply connected pointed space. The loop space ΩX of a simply connected pointed topological space $(X,*)$ is an arc-wise connected H-space. Therefore the homology algebras of such spaces have a rich structure.

Theorem 8.3.1. *The homology $H(\Omega X;\mathbb{K})$ of the loop space of a simply connected space X with coefficients a field $\mathbb{K}$ is a graded Hopf algebra with multiplication induced by the concatenation of paths in X.*

Proof. This is a consequence of Corollary 8.2.8. $\square$

Since the total space PX of the path-loop fibration is contractible, the connecting homomorphism ∂_* of this fibration is an isomorphism $\partial_* : \pi_*(X) \cong \pi_{*-1}(\Omega X)$. This enables us to transfer the Whitehead product $[\cdot,\cdot]_W$ on $\pi_*(X)$ to a bracket $[\cdot,\cdot]$ in $\pi_*(\Omega X)$ by setting

$$[\alpha,\beta] := (-1)^{\deg\alpha+1}\partial_*[\partial_*^{-1}\alpha, \partial_*^{-1}\beta]_W \quad \text{for } \alpha,\beta \in \pi_*(\Omega X).$$

The product $[\cdot,\cdot]$ is called the *Samelson product*. It is remarkable that this bracket makes the homotopy $\pi_*(\Omega X)$ into a graded Lie algebra:

Theorem 8.3.2. *The homotopy $\pi_*(\Omega X)$ endowed with the bracket $[\cdot,\cdot]$ is a graded Lie algebra over the ring $\mathbb{Z}$ of integers.*

Proof. This an application of Theorem C.3.1 to the 0-connected H-space ΩX. $\square$

Corollary 8.3.3. *The rational homology $\pi_*(\Omega X)\otimes\mathbb{Q}$ endowed with the bracket $[\cdot,\cdot]$ is a graded Lie algebra over the field of rational numbers.*

Definition 8.3.4. *The graded Lie-algebra $\pi_*(\Omega X)$ with bracket $[\cdot,\cdot]$ as above is denoted by L_X. The graded Lie-algebra $\pi_*(\Omega X)\otimes\mathbb{Q}$ with bracket $[\cdot,\cdot]$ as above is denoted by $L_X\otimes\mathbb{Q}$.*

The rational singular graded Hopf algebra $H(\Omega X;\mathbb{Q})$ of the loop space ΩX is an associative algebra and thus a graded Lie algebra in the usual way. In particular the space $P(\Omega X;\mathbb{Q})$ of primitive elements in $H(\Omega X;\mathbb{Q})$ is a graded Lie subalgebra of $H(\Omega X;\mathbb{Q})$ (cf. [FHT01a, Chapter 21 (b)]). It turns out that the Lie algebra $L_X\otimes\mathbb{Q}$ is sufficient for a complete description of the rational homology Hopf algebra $H(\Omega X;\mathbb{Q})$ and its Lie subalgebra $P(\Omega X;\mathbb{Q})$.

Theorem 8.3.5. *The Hurewicz homomorphism for ΩX is a Lie algebra isomorphism of $L_X\otimes\mathbb{Q}$ onto the graded Lie algebra $P(\Omega X;\mathbb{Q})$ of primitive elements in the cocommutative Hopf algebra $H_*(\Omega X;\mathbb{Q})$ and extends to an isomorphism of graded Hopf algebras from the universal enveloping algebra $U(L_X\otimes\mathbb{Q})$ of the rational graded homotopy lie algebra $L_X\otimes\mathbb{Q}$ onto the singular graded Hopf algebra $H_*(\Omega X;\mathbb{Q})$:*

$$
\begin{array}{ccc}
L_X\otimes\mathbb{Q} & \xrightarrow[\cong]{hur} & P_*(\Omega X;\mathbb{Q})\\
\downarrow & & \downarrow\\
U(L_X\otimes\mathbb{Q}) & \xrightarrow{\cong} & H_*(\Omega X;\mathbb{Q})
\end{array}
$$

Proof. A proof can be found in [FHT01a, Theorem 21.5]. □

A special case are the loop spaces ωX with trivial Samelson product. If the Samelson product for the graded Lie algebra $\pi(\Omega X)$ are trivial, then the latter graded Lie algebra is Abelian. Similarly, if the rational Samelson product for the graded rational Lie algebra $\pi(\Omega X)\otimes\mathbb{Q}$ are trivial, then the rational graded Lie algebra $\pi(\Omega X)\otimes\mathbb{Q}$ is Abelian.

Proposition 8.3.6. *The rational singular homology $H(\Omega X;\mathbb{Q})$ of a loop space ΩX with trivial Samelson product is a commutative and cocommutative Hopf algebra which, as a commutative graded algebra, is the free graded commutative algebra on the graded Abelian lie algebra $P(\Omega X;\mathbb{Q})\cong\pi(\Omega X)\otimes\mathbb{Q}$ of primitives in $H(\Omega X;\mathbb{Q})$, i.e. it is the tensor product*

$$H(\Omega X;\mathbb{Q})\cong\Lambda(L_X\otimes\mathbb{Q})_{odd}\otimes_\mathbb{Q} S(L_X\otimes\mathbb{Q})_{even}$$

of the exterior algebra $\Lambda(L_X\otimes\mathbb{Q})_{odd}$ on the odd dimensional part of $L_X\otimes\mathbb{Q}$ with the Symmetric algebra $S(L_X\otimes\mathbb{Q})_{even}$ on the even dimensional part of $L_X\otimes\mathbb{Q}$.

Proof. If the rational graded Lie algebra $\pi(\Omega X)\otimes\mathbb{Q}$ is Abelian, then the universal enveloping algebra $U(L_X\otimes\mathbb{Q})$ is a free commutative graded algebra (cf. [FHT01a, Chapter 21 (a)]). This free graded commutative algebra is the tensor product of the exterior algebra $\Lambda(L_X\otimes\mathbb{Q})_{odd}$ on the odd dimensional part of $L_X\otimes\mathbb{Q}$ with the Symmetric algebra $S(L_X\otimes\mathbb{Q})_{even}$ on the even dimensional part of $L_X\otimes\mathbb{Q}$:

$$U(L_X\otimes\mathbb{Q})\cong\Lambda(L_X\otimes\mathbb{Q})_{odd}\otimes_\mathbb{Q} S(L_X\otimes\mathbb{Q})_{even}$$

The algebra $U(L_X\otimes\mathbb{Q})$ is naturally a cocommutative Hopf algebra (cf. [FHT01a, Chapter 21 (a)]). Since the Hopf algebras $U(L_X\otimes\mathbb{Q})$ and $H(\Omega X;\mathbb{Q})$ are isomorphic, the rational singular homology of ΩX also is a commutative and cocommutative Hopf algebra, which, as an algebra, is the free graded commutative algebra on the space $P(\Omega X;\mathbb{Q})\cong L_X\otimes\mathbb{Q}$ of primitives. □

This situation for example arises if the the simply connected space X is a H-space itself. In this case then all the Whitehead products for X vanish:

Proposition 8.3.7 (Arkowitz). *The Whitehead product $[\cdot,\cdot]_W$ for H-spaces is trivial.*

Proof. The proof of [Ark71, Proposition 4.6] works for all H-spaces.   □

This in particular implies that the Samelson products on the loop space ΩX of any connected H-space are trivial. As a consequence we observe:

Corollary 8.3.8. *The graded homotopy Lie algebra L_X and its rational version $L_X \otimes \mathbb{Q}$ of loop spaces of simply connected H-spaces are Abelian.*

This ultimately leads to the rational algebra $H(\Omega X; \mathbb{Q})$ being a free graded commutative algebra:

Theorem 8.3.9. *The rational homology $H(\Omega X; \mathbb{Q})$ of the loop space ΩX of any simply connected H-space X is a commutative and cocommutative Hopf algebra which, as a commutative graded algebra, is the free graded commuatative algebra on the graded Abelian lie algebra $P(\Omega X; \mathbb{Q}) \cong \pi(\Omega X) \otimes \mathbb{Q}$ of primitives in $H(\Omega X; \mathbb{Q})$, i.e. it is the tensor product*

$$H(\Omega X; \mathbb{Q}) \cong \Lambda(L_X \otimes \mathbb{Q})_{odd} \otimes_{\mathbb{Q}} S(L_X \otimes \mathbb{Q})_{even}$$

of the exterior algebra $\Lambda(L_X \otimes \mathbb{Q})_{odd}$ on the odd dimensional part of $L_X \otimes \mathbb{Q}$ with the Symmetric algebra $S(L_X \otimes \mathbb{Q})_{even}$ on the even dimensional part of $L_X \otimes \mathbb{Q}$.

Proof. If X is a simply connected H-space then all the Samelson products on $\pi(\Omega X)$ and $\pi(\Omega X) \otimes \mathbb{Q}$ are trivial and the graded Lie algebra $\pi(\Omega X) \otimes \mathbb{Q}$ is Abelian by Lemma 8.3.8. Thus an application of Proposition 8.3.6 yields the desired result.   □

8.4 The Rational Homology of H-spaces

The structure Theorem 8.3.5 on the rational singular homology of connected loop-spaces can be used to derive an analogous result for the rational homology of homotopy associative H-spaces. This will be done using classifying spaces of topological monoids resp. H-spaces. We start with considering the case of topological groups and later generalise the procedure to arbitrary homotopy associative H-spaces. Let G be a topological group. The Milnor construction provides a locally trivial principal bundle

$$G \longrightarrow EG \longrightarrow BG$$

with fibre G and contractible total space EG. We recall some relations of the homotopy of X and the homotopy the loop space ΩBG of the Milnor classifying space BG of G.

Lemma 8.4.1. *The classifying space of any connected topological group G is simply connected.*

Proof. The universal X-bundle $EG \to BG$ is a Serre fibration with contractible total space EG. The exactness of the long exact homotopy sequence

$$\cdots \longrightarrow \pi_1(EG) \longrightarrow \pi_1(BG) \longrightarrow \pi_0(G) \longrightarrow \pi_0(EG) \longrightarrow \cdots$$

and the triviality of the homotopy groups $\pi_n(EG)$ leads to the isomorphism $\pi_*(BX) \cong \pi_{*-1}(X)$. Since the fibre G is connected, the classifying space BG is simply connected.   □

The fact that the total space of the universal bundle EG is contractible provides a map $\varphi : G \to \Omega BG$ (cf. [Sam54]), whose construction we recall now. Let (E, p, B) be a bundle of pointed spaces with contractible (pointed) total space and fibre F. By assumption there exists a contraction $\Phi : E \times I \to E$ onto the base point $*$ of E. The adjoint map

$$\hat{\Phi} : E \to C(I, E)$$

of E assigns to each point e in E a path $\hat{\Phi}(e)$ which starts at e and ends at the base point $* \in E$. For each point $f \in F$ the starting point of the path $\hat{\phi}(f)$ is contained in F, so the path $\hat{\phi}(f)$ in E projects to a loop in B. In this way one obtains a continuous function

$$\varphi := C(I, p) \circ \hat{\Phi}_{|F} : F \to \Omega B \tag{8.4}$$

from the fibre F into the loop space on the base space B.

Theorem 8.4.2. *If (EG, p, BG) is a universal bundle for the topological group G then $\varphi : G \to \Omega BG$ is an H-space homomorphism which is a weak homotopy equivalence, i.e. the following diagram in* **Toph** *is commutative:*

$$
\begin{array}{ccc}
G \times G & \xrightarrow{\;[\varphi \times \varphi]\;} & \Omega B_G \times \Omega B_G \\
{\scriptstyle [\mu]}\Big\downarrow & & \Big\downarrow \\
G & \xrightarrow[\cong]{\;[\varphi]\;} & \Omega B_G
\end{array}
\tag{8.5}
$$

Proof. See [Sam54] for a proof.

Lemma 8.4.3. *The map $\varphi : G \to \Omega BG$ transfers the Samelson product on $\pi_*(G)$ to the Samelson product on $\pi_*(\Omega BG)$ and the Pontrjagin product in $H(G; R)$ to the Pontrjagin product in $H(\Omega BG; R)$.*

Proof. This is an immediate consequence of the commutativity of Diagram 8.5. $\square$

Since the Lie bracket on the graded Lie algebras $\pi(G)$ and $\pi(\Omega BG)$ is given by the Samelson product (cf. Theorem C.3.1), the above Lemma further implies:

Lemma 8.4.4. *The map φ induces isomorphisms $\pi(\varphi) : \pi(G) \xrightarrow{\cong} \pi(\Omega BG)$ and $\pi(\varphi) \otimes \mathbb{Q} : \pi(G) \otimes \mathbb{Q} \xrightarrow{\cong} \pi(\Omega BG) \otimes \mathbb{Q}$ of graded Lie algebras.*

Theorem 8.4.5. *The weak homotopy equivalence φ induces an isomorphism between the homology groups $H_*(G; \mathbb{Q})$ and $H_*(\Omega BG; \mathbb{Q})$.*

Proof. Since $\pi_*(\varphi)$ is an isomorphism, the Whitehead-Serre Theorem implies that the induced map $H_*(\varphi; \mathbb{Q})$ is an isomorphism as well. $\square$

Theorem 8.4.6. *The map φ induces an isomorphism between the graded homology algebras $H(G; \mathbb{Q})$ and $H(\Omega BG; \mathbb{Q})$.*

Proof. The map $H(\varphi; \mathbb{Q})$ is an isomorphism of graded vector spaces (Theorem 8.4.5). By Lemma 8.4.3 it interchanges with products hence is an isomorphism of graded algebras. $\square$

Proposition 8.4.7. *The map $H(\varphi; \mathbb{Q})$ is an isomorphism of graded coalgebras.*

Proof. The comultiplication in $\bigoplus_n S_n(G; \mathbb{Q})$ and $\bigoplus_n S_n(\Omega BG; \mathbb{Q})$ is given by the morphism $D = AW \circ \bigoplus_n S_n(D_2)$ presented in Section 8.1. Since the Alexander-Whitney morphism AW and the diagonal function D_2 are natural in the two spaces involved, the commutativity of Diagram 8.5 implies that $H(\varphi; \mathbb{Q})$ is an isomorphism of graded coalgebras. $\square$

Theorem 8.4.8. *The weak homotopy equivalence φ induces an isomorphism between the graded Hopf algebras $H(G;\mathbb{Q})$ and $H(\Omega B_G;\mathbb{Q})$.*

Proof. Since $H(\varphi;\mathbb{Q})$ is an isomorphism of graded algebras as well as an isomorphism of graded coalgebras, it is an isomorphism of graded Hopf algebras.   □

Lemma 8.4.9. *The isomorphism $H(\varphi;\mathbb{Q})$ maps the image $hur_G(\pi(G)\otimes\mathbb{Q})$ onto the subspace $P(\Omega B_G;\mathbb{Q})$ of primitive elements in $H(\Omega B_G;\mathbb{Q})$.*

Proof. The map φ is a weak homotopy equivalence, i.e. an isomorphism in homotopy. Because the Hurewicz homomorphism is natural transformation from π_* to H_*, this implies

$$H_*(\varphi;\mathbb{Q})\circ hur_G(\pi_*(G)) = hur_{\Omega B_G}\circ\pi_*(\varphi)(\pi_*(G)) = hur_{\Omega B_G}\pi_*(\Omega B_G),$$

By Theorem 8.3.5 the subspace $P(\Omega B_G;\mathbb{Q})$ of primitive elements in $H(\Omega B_G;\mathbb{Q})$ is the image $hur_{\Omega B_G}(\pi(\Omega B_G)\otimes\mathbb{Q})$.   □

Theorem 8.4.10 (The Structure of the Rational Singular Homology of Arc-wise Connected Topological Groups). *The rational Hurewicz homomorphism for an arc-wise connected topological group G is a Lie algebra isomorphism of $\pi(G)\otimes\mathbb{Q}$ onto the graded Lie algebra $P(G;\mathbb{Q})$ of primitive elements in the Hopf algebra $H(G;\mathbb{Q})$ and extends to an isomorphism of graded Hopf algebras from the universal enveloping algebra $U(\pi(G)\otimes\mathbb{Q})$ of the rational graded Lie algebra $\pi(G)\otimes\mathbb{Q}$ onto the singular graded Hopf algebra $H(G;\mathbb{Q})$:*

$$
\begin{array}{ccc}
\pi(G)\otimes\mathbb{Q} & \xrightarrow[\cong]{hur} & P(G;\mathbb{Q}) \\
\downarrow & & \downarrow \\
U(\pi(G)\otimes\mathbb{Q}) & \xrightarrow{\cong} & H(G;\mathbb{Q})
\end{array}
$$

In particular the rational singular homology $H(G;\mathbb{Q})$ of a connected topological group G is a primitively generated Hopf algebra. Its subspace $P(G;\mathbb{Q})$ of primitive elements is the image $hur(\pi(G)\otimes\mathbb{Q})$ and the multiplication in $H(G;\mathbb{Q})$ is induced by the multiplication in G.

Proof. Let G be a connected topological group, Φ be a contraction of the total space EG of the universal bundle (EG,p,BG) and $\varphi = C(I,p)\circ\hat{\Phi}$ the morphism of H spaces defined in Equation 8.4. By Lemma 8.4.4 the induced morphism of graded Lie algebras $\pi(\varphi)\otimes\mathbb{Q}:\pi(G)\otimes\mathbb{Q}\to\pi(\Omega BG)\otimes\mathbb{Q}$ is an isomorphism. It therefore extends to an isomorphism $U(\pi(G)\otimes\mathbb{Q})\to U(\pi(\Omega BG)\otimes\mathbb{Q})$ of universal enveloping algebras. In addition the induced morphism $H(\varphi;\mathbb{Q})$ in rational singular homology an isomorphism of graded Hopf algebras by Theorem 8.4.8. We thus have obtained a commutative diagram

$$
\begin{array}{ccc}
U(\pi(G)\otimes\mathbb{Q}) & \xrightarrow{\cong} & U(\pi(\Omega BG)\otimes\mathbb{Q}) \\
\downarrow & & \downarrow{\scriptstyle\cong} \\
H(G;\mathbb{Q}) & \xrightarrow[H(\varphi;\mathbb{Q})]{\cong} & H(\Omega BG;\mathbb{Q})
\end{array}
$$

in which the left vertical morphism is the unique extension of the rational Hurewicz morphism. All but this left vertical morphism are known to be isomorphisms of graded Hopf algebras. Therefore the rational Hurewicz morphism for G extends to an isomorphism $U(\pi(G)\otimes\mathbb{Q})\cong H(G;\mathbb{Q})$. The Hopf algebra $H(\Omega BG;\mathbb{Q})$ is primitively generated, here the subspace $P(\Omega BG; \mathbb{Q})$ of primitive elements is the image $hur(\pi(\Omega BG) \otimes \mathbb{Q})$ (Theorem 8.3.5). By Lemma 8.4.9 this image corresponds to the image $\mathrm{hur}(\pi(G) \otimes \mathbb{Q})$ under the isomorphism $H(\varphi; \mathbb{Q})$. Therefore the Hopf algebra $H(G; \mathbb{Q})$ also is primitively generated and the space of primitives in $H(G; \mathbb{Q})$ is the image $hur_G(\pi(G) \otimes \mathbb{Q})$. $\qquad\square$

Corollary 8.4.11. *The rational homology $H(G; \mathbb{Q})$ of an Abelian topological group G is the free commutative graded algebra on the rational Lie algebra $\pi(G) \otimes \mathbb{Q}$, i.e. it is the tensor product*

$$H(G; \mathbb{Q}) \cong \Lambda(\pi(G) \otimes \mathbb{Q})_{odd} \otimes_{\mathbb{Q}} S(\pi(G) \otimes \mathbb{Q})_{even}$$

of the exterior algebra $\Lambda(\pi(G) \otimes \mathbb{Q})_{odd}$ on the odd dimensional part of $\pi(G) \otimes \mathbb{Q}$ with the Symmetric algebra $S(\pi(G) \otimes \mathbb{Q})_{even}$ on the even dimensional part of $\pi(G) \otimes \mathbb{Q}$.

Proof. If the group G is Abelian then the Samelson products on $\pi(G) \otimes \mathbb{Q}$ are trivial, i.e. the rational Lie algebra $\pi(G) \otimes \mathbb{Q}$ is Abelian. Thus the Hopf algebra $H(G; \mathbb{Q})$ is the universal enveloping algebra of the Abelian graded Lie algebra $\pi(G) \otimes \mathbb{Q}$, hence the free graded commutative algebra on $\pi(G) \otimes \mathbb{Q}$ (cf. Proposition 8.3.6). $\qquad\square$

A similar conclusion can be drawn for loop groups of topological groups. Any topological group G is a H-space and so is its arc-component of the identity. Thus Theorem 8.3.9 applies, which shows that the rational singular homology $H(\Omega G; \mathbb{Q})$ of the loop group ΩG of any topological group is a free graded commmutative algebra.

All the above results for topological groups can be generalised to H-spaces. Here one needs an analogue to the universal bundle for topological groups. Fortunately such an analogue exists, although it need not be a Serre fibration but only a more general version thereof:

Definition 8.4.12. *A continuous surjection $p : E \to B$ is called a* quasi-fibration *if p induces isomorphisms $\pi_*(E, p^{-1}(b), e) \cong \pi_*(B, b)$ for all points $b \in B$ and $e \in p^{-1}(b)$.*

One of the main properties of quasi-fibrations is that one still has a long exact sequence of homotopy groups

$$\cdots \to \pi_{n+1}(B, b) \xrightarrow{\partial} \pi_n(p^{-1}(b), e) \to \pi_n(E, e) \to \pi_n(B, b) \xrightarrow{\partial} \cdots$$

for all points $b \in B$ and e in the fibre $p^{-1}(x)$ (cf. [DT58, 1.4]).

Lemma 8.4.13 (Dold-Lashof). *Every H-space X is the fibre of a quasi fibration $EX \to BX$ such that X is contractible to a point within EX.*

Proof. See [DL59, Corollary 2.4] for a proof. $\qquad\square$

Theorem 8.4.14 (Dold-Lashof). *Every H-space X satisfying the separation axiom T_1 is the fibre of a quasi fibration $EX \to BX$ with acyclic total space EX such that X is contractible to a point within EX.*

Proof. Start with the fibration $EX \to BX$ from Lemma 8.4.13 and use the construction [DL59, 3.4]. $\qquad\square$

Using the quasi-fibration $p : EX \to BX$ with fibre X and acyclic total space EX guaranteed to exist for every H-space X by Theorem 8.4.14 the construction of a weak homotopy equivalence $\varphi : X \to \Omega BX$ can be made as before. Let $p : E \to B$ be a quasi-fibration with fibre F and $* \in F \subset E$ be a base point. Assume that there exists a contraction $\Phi : X \times I \to EX$ of X to $*$ in the total space EX leaving $*$ fixed. The continuous function $\varphi := C(I, p) \circ \hat{\Phi} : X \to \Omega BX$ then has the following property:

Proposition 8.4.15. *If the fibre F of a quasi-fibration $p : E \to B$ is contractible to a point $* \in F$ within E (leaving $*$ fixed) then $\varphi : X \to \Omega B$ induces split monomorphisms $\pi_n(\varphi)$ in homotopy.*

Proof. This is a generalisation of Theorem 8.4.2 which has been proven in [Sug55, Proposition] for Serre fibrations. The proof also works for quasi-fibrations.   □

If X is an H-space satisfying the separation axiom T_1 and $EX \to BX$ the bundle with acyclic total space EX from Theorem 8.4.14 then the split monomorphisms $\pi_n(\varphi)$ are isomorphisms, so we observe:

Lemma 8.4.16. *The map $\varphi : X \to \Omega BX$ is a weak homotopy equivalence.*

Proposition 8.4.17. *The map $\varphi : X \to \Omega BX$ is a morphism of H-spaces.*

Proof. The proof of [Sug55, Theorem 1] carries over to quasi-fibrations.   □

The preceding two Propositions allow us to generalise the results on the rational singular homology of arc-wise connected topological groups to arc-wise connected H-spaces satisfying the separation axiom T_1:

Theorem 8.4.18 (The Structure of the Rational Singular Homology of Arc-wise Connected H-spaces). *The rational Hurewicz homomorphism for an arc-wise connected H-space X satisfying the separation axion T_1 is a Lie algebra isomorphism of $\pi(X) \otimes \mathbb{Q}$ onto the graded Lie algebra $P(X; \mathbb{Q})$ of primitive elements in the Hopf algebra $H(X; \mathbb{Q})$ and extends to an isomorphism of graded Hopf algebras from the universal enveloping algebra $U(\pi(X) \otimes \mathbb{Q})$ of the rational graded Lie algebra $\pi(X) \otimes \mathbb{Q}$ onto the singular graded Hopf algebra $H(X; \mathbb{Q})$:*

$$
\begin{array}{ccc}
\pi(G) \otimes \mathbb{Q} & \xrightarrow[\cong]{hur} & P(G; \mathbb{Q}) \\
\downarrow & & \downarrow \\
U(\pi(G) \otimes \mathbb{Q}) & \xrightarrow[\cong]{} & H(G; \mathbb{Q})
\end{array}
$$

In particular the rational singular homology $H(X; \mathbb{Q})$ of an arc-wise connected H-space X is a primitively generated Hopf algebra. Its subspace $P(X; \mathbb{Q})$ of primitive elements is the image $\mathrm{hur}(\pi(X) \otimes \mathbb{Q})$ and the multiplication in $H(X; \mathbb{Q})$ is induced by the multiplication in X.

Proof. The proof is analogous to that of Theorem 8.4.10.   □

9

Geometric Realisation

In this chapter some results concerning the classical geometric realisation functor $|-|$ as a functor $\mathbf{Set}^{\mathcal{O}^{op}} \to \mathbf{kTop}$ or $\mathbf{kTop}^{\mathcal{O}^{op}} \to \mathbf{kTop}$ (as defined for example in Mays book [May92] or the work [GZ67] of Gabriel and Zisman) are generalised to a functor $\mathbf{Top}^{\mathcal{O}^{op}} \to \mathbf{Top}$ (see Appendix F for a definition of the simplicial category $\mathcal{O}$ and simplicial objects). This includes in particular the preservation of certain finite limits and the existence of a right adjoint. These generalisations will be needed in the characterisation of the rational homology of Eilenberg-Mac Lane spaces and for the construction of Eilenberg-Mac Lane spaces which are abelian topological groups in Chapter 11. The results obtained in this chapter are among the most important in this work. They lay the foundations for the results on classifying bundles and spaces in Chapter 11, the construction of Eilenberg-MacLane spaces which are abelian topological groups in Chapter 11 and the derivation of the structure Theorem on the rational singular homology algebra of Eilenberg-MacLane spaces in Chapter 11. These result furthermore allow us to construct higher analogues of simply connected coverings and enables the construction of Whitehead and Postnikov towers consisting of principal bundles with abelian structure group in Chapter 12.

In the first section we introduce the general geometric realisation functor $|-|$. Then the preservation of special kinds of morphisms by the functor $|-|$ is studied, where we in particular prove the preservation of *injectivity, surjectivity, denseness of images, proclusions, identifications* and *closedness of morphisms into Hausdorff (simplicial) spaces* by the general geometric realisation functor $|-| : \mathbf{Top}^{\mathcal{O}^{op}} \to \mathbf{Top}$. Afterwards we consider the restriction of the geometric realisation functor $|-|$ to important subcategories of $\mathbf{Top}$ such as the category $\mathbf{CGTop}^{\mathcal{O}^{op}}$ of compactly generated simplicial topological spaces and the category $\mathbf{kTop}^{\mathcal{O}^{op}}$ of compactly Hausdorff generated simplicial topological spaces. Here we observe that these restrictions corestrict to functors

$$\mathbf{CGTop}^{\mathcal{O}^{op}} \to \mathbf{CGTop} \quad \text{and} \quad \mathbf{kTop}^{\mathcal{O}^{op}} \to \mathbf{kTop}$$

respectively. Thereafter we show that the geometric realisation $|X|$ of a simplicial topological space is the colimit of the geometric realisations $|X_k|$ of its k-skeleta. All of these observations are generalisations from the classical geometric realisation of simplicial sets (regarded as discrete simplicial spaces) to the general setting.

In the second section we prove the cocontinuouity of the general geometric realisation functor $|-| : \mathbf{Top}^{\mathcal{O}^{op}} \to \mathbf{Top}$ by explicitly constructing a right adjoint. This right adjoint is the simplicial singular space functor

$$C(\Delta, -) : \mathbf{Top} \to \mathbf{Top}$$

where the function spaces are equipped with the compact open topology. This the first main result in this chapter:

Theorem. *The singular simplicial space functor $C(\Delta, -)$ is right adjoint to the geometric realisation functor $|-|$.*

This observation generalises the important well known fact, that the geometric realisation of simplicial sets regarded as a functor $\mathbf{Set}^{\mathcal{O}^{op}} \to \mathbf{kTop}$ is cocontinuous. Our result is far more general and is new even for discrete simplicial spaces, because we do not corestrict the geometric realisation functor to the category $\mathbf{kTop}$ of k-spaces. Last but not least the counit $\epsilon : |C(\Delta, -)| \to \mathrm{id}_{\mathbf{Top}}$ of this adjunction is shown to be a left inverse to the inclusion $i : \mathrm{id}_{\mathbf{Top}} \hookrightarrow |C(\Delta, -)|$ of topological spaces into the geometric realisation of their singular simplicial spaces:

Lemma. *The natural transformation i is a right inverse of ϵ, i.e. $\epsilon i = \mathrm{id}_{\mathbf{Top}}$ and the transformations fit into the commutative diagram*

$$\mathrm{id}_{\mathbf{Top}} \xrightarrow{\;\;i\;\;} |C(\Delta, -)| \xrightarrow{\;\;\epsilon\;\;} \mathrm{id}_{\mathbf{Top}} \;.$$
$$\mathrm{id}$$

It will be shown later in Chapter 10 that the counit ϵ actually consists of homotopy equivalences.

In Section 3 the preservation of finite limits is examined. After showing that the general geometric realisation functor $|-| : \mathbf{Top}^{\mathcal{O}^{op}} \to \mathbf{Top}$ preserves equalisers we prove that it also preserves certain finite products. The classical theorem concerning the question of the preservation of finite products is [Mil57, Theorem 2] which was proven by Milnor. It states that the restriction $|-| : \mathbf{Set}^{\mathcal{O}^{op}} \to \mathbf{kHaus}$ preserves finite products of at most countable simplicial sets. In addition, if X and Y are simplicial sets and either CW-complex $|X|$ or $|Y|$ is locally finite, then the product $X \times Y$ is also preserved. In Section 2 below we show that the finiteness condition there can be weakened to a local countability condition:

Proposition. *If X and Y are discrete simplicial spaces and either X or Y is locally countable, then the natural map $|X \times Y| \to |X| \times |Y|$ is a homeomorphism.*

We then proceed to show that the geometric realisation of a k_ω-space is a k_ω-space, a fact that will be useful for the classifying space constructions considered in Chapter 11. Thereafter we generalise the above Proposition even more and observe that the geometric realisation functor

$$|-| : \mathbf{Set}^{\mathcal{O}^{op}} \to \mathbf{kHaus}$$

preserves finite limits. All this culminates in the following theorem:

Theorem. *Finite products of polytopes are polytopes.*

These results enable us to show that the geometric realisation of discrete simplicial algebras of arbitrary type are topological algebras of the same type (Theorem 9.3.52). In particular the geometric realisation of an (abelian) discrete simplicial group is an (Abelian) topological group. Concerning the geometric realisation of discrete simplicial algebras we then observe:

Theorem. *The corestricted functor $|-| : \mathbf{Alg}_{(\Omega, E)}^{\mathcal{O}^{op}} \to \mathbf{TopAlg}_{(\Omega, E)}$ preserves finite limits. If the category $\mathbf{Alg}_{(\Omega, E)}$ contains a zero object, then the restriction $|-| : \mathbf{Alg}_{(\Omega, E)}^{\mathcal{O}^{op}} \to \mathbf{TopAlg}_{(\Omega, E)}$ of the geometric realisation functor preserves kernels.*

We then return to the question whether or when the geometric realisation preserves products and prove that it does not preserve finite products in general but binary products with locally finite discrete simplicial spaces are always preserved. Last but not least we generalise a classical result concerning the geometric realisation of simplicial sets by proving:

Theorem. *The geometric realisation functor* $|-| : \mathbf{Top}^{\mathcal{O}^{op}} \to \mathbf{Top}$ *preserves homotopy, i.e. if f and g are homotopic maps of simplicial spaces then the induced maps $|f|$ and $|g|$ are homotopic as well.*

All these results are new in the more general context of geometric realisation of simplicial topological spaces.

9.1 The Geometric Realisation Functor

In this section we start with recalling the definition of the geometric realisation functor. Let $X \in \mathbf{Top}^{\mathcal{O}^{op}}$ be a simplicial space. We consider the spaces $X([n]) \times \Delta_n$ and glue them together along restricted boundary and degeneracy maps. For this purpose we define an equivalence relation $\sim$ on the disjoint union

$$\coprod_{n \in \mathbb{N}} X([n]) \times \Delta_n$$

via the requirement that two points $(x, s) \in X([n]) \times \Delta_n$ and $(y, t) \in X([m]) \times \Delta_m$ be equivalent if there exists a morphism $\alpha : [m] \to [n]$ in the category $\mathcal{O}$ such that $X(\alpha)(y) = x$ and $s = \Delta(\alpha)(t)$. The equivalence class of a point (x, s) in the disjoint union is denoted by $[(x, s)]$.

Definition 9.1.1. *The identification space $\coprod X([n]) \times \Delta_n / \sim$ is called the* geometric realisation *of X. It is denoted by $|X|$. The quotient map*

$$\coprod X([n]) \times \Delta_n \twoheadrightarrow |X| = \coprod X([n]) \times \Delta_n / \sim$$

onto $|X|$ is denoted by q_X.

Every morphism $f : X \to Y$ between simplicial spaces X and Y induces a continuous function $|f| : |X| \to |Y|$ between the identification spaces. These assignments constitute a functor from the category $\mathbf{Top}^{\mathcal{O}^{op}}$ of simplicial topological spaces to the category $\mathbf{Top}$ of topological spaces. This functor $|-|$ is called the *geometric realisation functor*. It has an alternate description by coends: If X is a simplicial space, then the geometric realisation $|X|$ of X is the coequaliser of the maps

$$\coprod_{\alpha:[n]\to[m]} X([m]) \times \Delta_n \xrightarrow[\coprod \mathrm{id}_{X([m])} \times \Delta(\alpha)]{\coprod X(\alpha) \times \mathrm{id}_{\Delta_n}} \coprod_{n} X([n]) \times \Delta_n,$$

i.e. it is a coend of the functor $X \times \Delta : \mathcal{O}^{op} \times \mathcal{O} \to \mathbf{Top}$. (This has been observed for the geometric realisation of simplicial sets in [ML70], but is equally true for the general geometric realisation functor.) Note, that the category $\mathbf{Top}$ is cocomplete, so every functor

$$\mathcal{O}^{op} \times \mathcal{O} \to \mathbf{Top}$$

has a coend in $\mathbf{Top}$ (cf. [ML98, Chapter IX.5, Corollary 2]). Therefore the description as a coend can also serve as a definition that implies the functoriality $|-|$. As usual the coend of a functor $T : \mathcal{O}^{op} \times \mathcal{O} \to \mathbf{Top}$ is denoted by

$$\int^{[n]} T([n],[n]).$$

If $S : \mathcal{O}^{op} \to \textbf{Top}$ and $S' : \mathcal{O} \to \textbf{Top}$ are two functors we also write $S\square_{\mathcal{O}}S'$ for the coend of the functor $\Pi \circ (S \times S')$. This is sometimes called a 'tensor product' of the functors S and S'. Let

$$T = \prod \circ (\mathrm{Ev} \times \Delta) : \left(\textbf{Top}^{\mathcal{O}^{op}} \times \mathcal{O}^{op}\right) \times \mathcal{O} \to \textbf{Top}$$

be the functor which associates to each triple $(X, [n], [m])$ of objects the product $X([n]) \times \Delta_m$ of the evaluation $X([n])$ with the standard simplex Δ_m.

Definition 9.1.2. *The geometric realisation functor* $|-| : \textbf{Top}^{\mathcal{O}^{op}} \to \textbf{Top}$ *is the unique functor* $\textbf{Top}^{\mathcal{O}^{op}} \to \textbf{Top}$ *with object function*

$$|X| = \int^{[n]} X([n]) \times \Delta_n = \int^{[n]} T(X, [n], [n])$$

such that the components of the extranatural transformation $T \to \int^{[n]} T(X, [n], [n])$ *form a natural transformation* $T(-, [n], [n]) \to |-|$ *for each* $[n]$.

The existence of the geometric realisation functor $|-|$ in the above definition is guaranteed by a general existence theorem (see [ML98, Theorem IX.7.2]) In many cases (see for example the books [GZ67], [May92]) one works in a subcategory of **Top**, e.g. the category **CGTop** of compactly generated spaces, the category **CGHaus** of compactly generated Hausdorff spaces, the category **kTop** of k-spaces or the category of weak Hausdorff k-spaces. These categories fit into the general scheme of categories $\textbf{Top}_{\mathbf{C}}$ of **C**-generated spaces (see B.6), where **C** is a subcategory of **Top**. The categories $\textbf{Top}_{\mathbf{C}}$ have the advantage that they are coreflective hulls of the category **C** of generating spaces, so there exists a functor

$$\mathrm{C} : \textbf{Top} \to \textbf{Top}_{\mathbf{C}}$$

called the *coreflector*, which is right adjoint to the inclusion $\textbf{Top}_{\mathbf{C}} \hookrightarrow \textbf{Top}$. This functor equips every topological space with a finer topology. Furhermore, being a right adjoint, it preserves all limits. When working in such a coreflective subcategory $\textbf{Top}_{\mathbf{C}}$, taking products with standard simplices in **Top** could yield spaces that are not contained in the subcategory $\textbf{Top}_{\mathbf{C}}$. So the above construction may lead outside the category $\textbf{Top}_{\mathbf{C}}$. Because any category $\textbf{Top}_{\mathbf{C}}$ of **C**-generated spaces is cocomplete (B.6.23), this can be corrected by applying the coreflector $\mathrm{C} : \textbf{Top} \to \textbf{Top}_{\mathbf{C}}$ to the space

$$\coprod_{n \in \mathbb{N}} X([n]) \times \Delta_n$$

before taking the quotient. In this way one obtains a similar functor $|-|_{\mathrm{C}}$ that takes values in the subcategory $\textbf{Top}_{\mathbf{C}}$. The abstract definition is analogous to the previous one:

Definition 9.1.3. *The C-geometric realisation functor* $|-|_{\mathrm{C}} : \textbf{Top}^{\mathcal{O}^{op}} \to \textbf{Top}_{\mathbf{C}}$ *is the unique functor* $\textbf{Top}^{\mathcal{O}^{op}} \to \textbf{Top}_{\mathbf{C}}$ *with object function*

$$|X|_{\mathrm{C}} = \int^{[n]} \mathrm{C}(X([n]) \times \Delta_n) = \int^{[n]} \mathrm{C}T(X, [n], [n])$$

such that the components of the dinatural transformation $\mathrm{C}T \to \int^{[n]} \mathrm{C}T(X, [n], [n])$ *form a natural transformation* $\mathrm{C}T(-, [n], [n]) \to |-|_{\mathrm{C}}$ *for each* $[n]$.

Here the C-geometric realisation $|X|_C$ of a simplicial topological space X is an object in the coreflective subcategory $\mathbf{Top}_C$. This functor $|-|_C$ then restricts to an functor $\mathbf{Top}_C^{\mathcal{O}^{op}} \to \mathbf{Top}_C$. The C-geometric realisation functor may have the disadvantage that it is not the restriction of the geometric realisation functor $|-|$ to the subcategory $\mathbf{Top}_C$. We will see, however, that both functors agree on useful subcategories of $\mathbf{Top}$ (e.g. the category $\mathbf{CGTop}$ of compactly generated spaces and the category $\mathbf{kTop}$ of k-spaces).

The restriction of the geometric realisation functor $|-|$ to the sub-category $\mathbf{Set}$ of discrete topological spaces takes values in the sub-category of CW-complexes ([Mil57, Theorem 1]) which are k-spaces. This restriction is known to preserve many universal categorical constructions when also corestricted to the sub-category $\mathbf{kTop}$ of k-spaces in $\mathbf{Top}$ (see [GZ67]). The goal of this section is to show that the general geometric realisation functor preserves monomorphisms, epimorphisms, proclusions and closed inclusions. We start with the preservation of injectivity and surjectivity:

Lemma 9.1.4. *The geometric realisation functor* $|-| : \mathbf{Top}^{\mathcal{O}^{op}} \to \mathbf{Top}$ *preserves injectivity and surjectivity of maps.*

Proof. Let $f : X \to Y$ be a morphism of simplicial spaces. Consider the commutative diagram

$$\coprod X([n]) \times \Delta_n \xrightarrow{\coprod f_n \times \mathrm{id}_{\Delta_n}} \coprod Y([n]) \times \Delta_n , \tag{9.1}$$

with vertical maps q_X and q_Y down to $|X| \xrightarrow{|f|} |Y|$

where $\coprod f_n \times \mathrm{id}$ is injective if and only if f is injective and $\coprod f_n \times \mathrm{id}$ is surjective if and only f is surjective. If f is injective, then any two points in the image of $\coprod f_n \times \mathrm{id}_{\Delta_n}$ are identified by q_Y if and only they are identified by q_X, hence the map $|f|$ is injective. If the map f is surjective, then $q_X \circ f$ is surjective, hence $|f|$ is surjective as well. $\square$

Any morphism $f : X \to Y$ of simplicial spaces is a monomorphism if and only each map $f([n])$ is a monomorphism; it is an epimorphism if and only each map $f([n])$ is an epimorphism (cf. F.4.7 and F.4.8). That is to say that the injective maps are the monomorphisms in the category $\mathbf{Top}^{\mathcal{O}^{op}}$ of simplicial spaces and the surjective maps are the epimorphisms in $\mathbf{Top}^{\mathcal{O}^{op}}$. Therefore one concludes:

Lemma 9.1.5. *The geometric realisation functor preserves monomorphisms and epimorphisms.*

Closely related to epimorphisms of topological spaces (and topological algebras) algebras are morphisms with dense image. These are the epimorphisms in $\mathbf{Haus}^{\mathcal{O}^{op}}$. Concerning this kind of morphisms one observes:

Lemma 9.1.6. *If* $f : X \to Y$ *is a morphism of simplicial spaces with dense image then* $|f| : |X| \to |Y|$ *has dense image as well.*

Proof. Let $f : X \to Y$ be a morphism of simplicial spaces with dense image. Any non-empty open set $U \subset |Y|$ has an open inverse image $q_Y^{-1}(U)$ in $\coprod Y([n]) \times \Delta_n$. Because U was assumed to be non-empty, the inverse image $q_Y^-(U)$ is nonempty as well. Therefore at least one of the open sets $q_Y^{-1}(U) \cap Y([n]) \times \Delta_n$, $n \in \mathbb{N}$ is not empty. So there exists a basic open set $V \times W \subset q_Y^{-1}(U) \cap Y([n]) \times \Delta_n$ for some $n \in \mathbb{N}$. Since the morphism f has dense image by assumption, the image $f_n(X([n])$ meets the open set $V \subset Y([n])$, hence there exists a point $x \in X([n])$ whose image $f_n(x)$ is contained in V. Choose an arbitrary point $t \in W \subset \Delta_n$. Then

$$|f|([(x,t)]) = (|f| \circ q_X)(x,t) = (q_Y \circ f_n)(x,t) \subset q_Y(V \times W) \subset q_Y(q_Y^{-1}(U) = U,$$

i.e. the image of $|f|$ meets the open set U. Because this is true for all non-empty open subsets U of Y the continuous map $|f|$ has dense image. $\square$

Next one observes that proclusions and quotient mappings are also preserved by the geometric realisation functor:

Proposition 9.1.7. *The geometric realisation functor $|-|$ preserves proclusions.*

Proof. Let $f : X \to Y$ be a proclusion of simplicial spaces. Since all the functions $f([n]) : X([n]) \to Y([n])$ are proclusions, the resulting function

$$\coprod f([n]) \times \mathrm{id}_{\Delta_n} : \coprod X([n]) \times \Delta_n \to \coprod Y([n]) \times \Delta_n$$

is a proclusion as well. So the function $|f|$ between the identification spaces $|X|$ and $|Y|$ in Diagram 9.1 is induced by a proclusion. Let U be a subset of $|Y|$. It is open in $|Y|$ if and only if the inverse image $q_Y^{-1}(U)$ is open in $\coprod Y([n]) \times \Delta_n$. Because the function $\coprod f([n]) \times \mathrm{id}_{\Delta_n}$ is a proclusion, this inverse image is open if and only if the inverse image $(q_Y \circ (\coprod f([n]) \times \mathrm{id}_{\Delta_n}))^{-1}(U)$ is open in $\coprod X([n]) \times \Delta_n$. On the other hand, the inverse image $|f|^{-1}(U)$ is open in $|X|$ exactly if the inverse image $q_X^{-1}|f|^{-1}(U)$ is open. Because the maps $|f| \circ q_X$ and $q_Y \circ \coprod f([n]) \times \mathrm{id}_{\Delta_n}$ coincide, a subset U of $|Y|$ is open if and only if its inverse image $|f|^{-1}(U)$ is open in $|X|$. Thus $|f|$ is a proclusion. $\square$

Corollary 9.1.8. *The geometric realisation functor preserves identifications.*

Proof. Identifications are surjective proclusions. Because the geometric realisation functor preserves epimorphisms and proclusions it also preserves identifications. $\square$

Example 9.1.9. If $f : G \twoheadrightarrow G'$ is a quotient morphism of simplicial topological groups then the function $|f| : |G| \twoheadrightarrow |G'|$ is a quotient map.

For further study of the geometric realisation functor we note that the disjoint union $F(X) := \coprod X([n]) \times \Delta_n$ used in the definition of the geometric realisation $|X|$ of a simplicial space X is the object part of a functor $F : \mathbf{Top}^{\mathcal{O}^{op}} \to \mathbf{Top}$. On morphisms $f : X \to Y$ of simplicial spaces X and Y, this functor is given by

$$F(f) := \coprod f_n \times \mathrm{id}_{\Delta_n} : F(X) \to F(Y).$$

As is apparent from the definition, the functor F preserves injectivity, surjectivity openness, and closedness of simplicial maps. In addition, it preserves all colimits:

Lemma 9.1.10. *The functor F is cocontinuous.*

Proof. the standard simplicial spaces Δ_n are compact. The products funcors $- \times \Delta_n$ are left adjoint to the function space functors $C(\Delta_n, -)$, hence cocontinuous. The disjoint union functor also preserves colimits, so F is cocontinuous. $\square$

The inclusion of the subspace $X([n]) \times \Delta_n$ into the space $F(X)$ will be denoted by i_n. It is a natural transformation from the evaluation at $[n]$ to F. Furthermore the quotient maps $q_X : F(X) \twoheadrightarrow |X|$ form a natural transformation $q : F \to |-|$.

Let $\mathbf{C}$ be a subcategory of $\mathbf{Top}$ that is either productive and generates all the standard simplices Δ_n or contains all the standard simplices, e.g. the category $\mathbf{CGTop}$ of compactly generated spaces, the category $\mathbf{CGHaus}$ of compactly generated Hausdorff spaces or the category $\mathbf{kTop}$ of k-spaces (all of which in fact satisfy both conditions). Please recall that the product of a **C**-generated space with an exponential generating space in **Top** coincides with the product in **Top**$_\text{C}$ (see B.6.21). The same is true for the product of a **C**-generated space with an exponential **C**-generated space if **C** is productive (see Lemma B.6.31). Thus, if X is a simplicial **C**-generated space, the spaces

$$X([n]) \times \Delta_n$$

also are **C**-generated. Therefore the product functors $- \times \Delta_n$ leave the sub-category **Top**$_\text{C}$ of **Top** invariant. Hence the functor F restricts and corestricts to a functor

$$F : \mathbf{Top_C}^{\mathcal{O}^{op}} \to \mathbf{Top_C}.$$

An application to the categories **CGTop**, **CGHaus** and **kTop** shows that the functor F especially restricts and corestricts to functors

$$\begin{aligned}
F_{|\mathbf{Haus}} \quad &: \mathbf{Haus}^{\mathcal{O}^{op}} \to \mathbf{Haus}, \qquad F_{|\mathbf{CGTop}} : \mathbf{CGTop}^{\mathcal{O}^{op}} \to \mathbf{CGTop} \\
F_{|\mathbf{CGHaus}} &: \mathbf{CGHaus}^{\mathcal{O}^{op}} \to \mathbf{CGHaus}, \; F_{|\mathbf{kTop}} \quad : \mathbf{kTop}^{\mathcal{O}^{op}} \to \mathbf{kTop}
\end{aligned}.$$

Because any identification space of a **C**-generated space is again a **C**-generated space (B.6.22) we have proved:

Lemma 9.1.11. *If a subcategory* **C** *of* **Top** *contains all standard simplices then the geometric realisation functor restricts to a functor* $|-| : \mathbf{Top_C}^{\mathcal{O}^{op}} \to \mathbf{Top_C}$. *The same conclusion holds if* **C** *is productive and all standard simplices are* **C**-generated.

Lemma 9.1.12. *The geometric realisation* $|X|$ *of a simplicial k-space* X *also is a k-space, i.e. the geometric realisation restricts to a functor* $\mathbf{kTop}^{\mathcal{O}^{op}} \to \mathbf{kTop}$.

Example 9.1.13. Any discrete simplicial space is compactly Hausdorff generated. Thus the geometric realisations of such simplicial spaces are k-spaces.

Example 9.1.14. Any metrisable simplicial space is a k-space. Therefore the geometric realisations of metrisable simplicial spaces are k-spaces. This includes the large classes of metrisable simplicial manifolds, e.g. simplicial classical finite dimensional Lie-groups, simplicial Banach Lie-groups and more generally simplicial connected Lie-groups satisfying the first axiom of countability.

Lemma 9.1.15. *The geometric realisation* $|X|$ *of a simplicial compactly generated space* X *also is compactly generated, i.e. the geometric realisation restricts to a functor* $\mathbf{CGTop}^{\mathcal{O}^{op}} \to \mathbf{CGTop}$.

With these results one can now compare the functors $|-|_\text{C}$ and $|-|$ on categories **C** which satisfy either of the above condiftions.

Proposition 9.1.16. *If the category* **C** *contains the standard simplicies or is productive and the standard simplicies are* **C**-generated, *then the functors* $|-|_\text{C}$ *and* $|-| \circ \text{C}$ *on* **Top** *coincide.*

Proof. Let X be a simplicial space. If **C** satisfies either of the conditions then the product of the **C**-generated space $CX([n])$ with the standard simplex Δ_n in **Top** is **C**-generated. Thus the spaces

$$F(CX) = \coprod CX([n]) \times \Delta_n = \coprod C\,[X([n]) \times \Delta_n] = C\left(\coprod CX([n]) \times \Delta_n\right)$$

and $CF(X)$ coincide and so do their quotient spaces $|X|_\text{C}$ and $|CX|$. □

Corollary 9.1.17. *If the category* **C** *contains the standard simplicies or is productive and the standard simplicies are* **C**-generated, *then the functors* $|-|_\text{C}$ *and* $|-|$ *on coincide on* **Top**$_\text{C}$.

Corollary 9.1.18. *The functors* $|-|_k$ *and* $|-|$ *on coincide on* **kTop**, *and the functors* $|-|_{CG}$ *and* $|-|$ *on coincide on* **CGTop**.

So descending to the quotient $|X|$ of $F(X)$ preserves the properties of X being **C**-generated, if **C** contains all the standard simplices. In the following we examine which further properties of the functor F are preserved by descending to the geometric realisation $|-|$ via this natural transformation q. At first we recall some classical set theoretic properties of the above constructions $F(X)$ and the quotient space $|X|$. These will be needed to prove that the space $X([0])$ can be considered to be a closed subspace of the geometric realisation $|X|$ under certain conditions.

Definition 9.1.19. *A point* $\mathbf{t} \in \Delta_n$ *is called* degenerate *if there exists a monomorphism* $\epsilon : [m] \to [n]$ *in the category* $\mathcal{O}$ *such that the image* $\Delta(\epsilon)(\Delta_m)$ *contains* $\mathbf{t}$. *Otherwise the point* $\mathbf{t}$ *is called* non-degenerate.

Lemma 9.1.20. *For every* $\mathbf{t} \in \Delta_n$ *there exists an* $m \leq n$, *a unique non-degenerate point* $\mathbf{t}' \in \Delta_m$ *and a unique monomorphism* $\epsilon_{\mathbf{t}} : [m] \hookrightarrow [n]$ *such that* $\Delta(\epsilon_{\mathbf{t}})(\mathbf{t}') = \mathbf{t}$.

Proof. See the proof of [EZ50, 8.3] with reversed roles of epimorphisms and monomorphisms. $\square$

The monomorphism ϵ_t occurring in the above lemma can be uniquely expressed by a composition of face maps $\epsilon = \epsilon_{i_1} \cdots \epsilon_{i_{m-n}}$ with $0 \leq i_1 \leq \cdots \leq i_{m-n} \leq m$ (cf. F.1.6). The epimorphism $\eta = \eta_{i_{m-n}} \cdots \eta_{i_1}$ is a left inverse of ϵ. We denote this left inverse by η_t. The non-degenerate element $\mathbf{t}'$ such that $\Delta(\epsilon)(\mathbf{t}') = \mathbf{t}$ can then also be written as

$$\mathbf{t}' = \Delta(\eta_{\mathbf{t}})\Delta(\epsilon)(\mathbf{t}') = \Delta(\eta_{\mathbf{t}})(\mathbf{t}).$$

Similar considerations can be made for points $x \in X([n])$. Here we note that the functor X reverses the roles of monomorphisms and epimorphisms.

Definition 9.1.21. *A point* $x \in X([n])$ *is called* degenerate *if there exists a epimorphism* $\eta : [n] \to [m]$ *in the category* $\mathcal{O}$ *such that the image* $X(\eta)(X([m]))$ *contains* x. *Otherwise the point* x *is called* non-degenerate.

Lemma 9.1.22. *For every element* $x \in X([n])$ *there exists a unique non-degenerate point* $x' \in X([m])$ *for some* m *and a unique epimorphism* $\eta_x : [n] \to [m]$ *such that* $X(\eta)(x') = x$.

Proof. See the proof of [EZ50, 8.3]. $\square$

Example 9.1.23. If X is a constant simplicial space then every element $x \in X([n])$ is the image of $X(\eta)(x)$ for any morphism $\eta : [n] \to [0]$ in $\mathcal{O}$. Therefore every point $x \in X([n])$ is its own unique non-degenerate representative in $X([0])$.

The epimorphism η_x occuring in the previous lemma has a unique representation as a composition of degeneracy maps $\eta_{i_1} \cdots \eta_{i_{m-n}}$ with $0 \leq i_1 < \cdots \leq i_{m-n} < m$ (cf. F.1.6). The epimorphism $\epsilon = \epsilon_{i_{m-n}} \cdots \epsilon_{i_1}$ is a right inverse of η. The right inverse is a monomorphism which we denote by ϵ_x. The non-degenerate element x' such that $X(\eta_x)(x') = x$ can then also be written as

$$x' = X(\epsilon_x)X(\eta_x)(x') = X(\epsilon_x)(x).$$

Definition 9.1.24. *A point* $(x, \mathbf{t}) \in F(X)$ *is called* non-degenerate *if* x *and* $\mathbf{t}$ *are non-degenerate.*

Proposition 9.1.25. *Every equivalence class in $F(X)$ has a unique non-degenerate representative. The unique non-degenerate representative $(x', \mathbf{t}')$ of a class $[(x, \mathbf{t})]$, $(x, \mathbf{t}) \in F(X)$ is given by*

$$x' = [X(\epsilon_{\Delta(\eta_x)(\mathbf{t})})X(\epsilon_x)](x), \quad \mathbf{t}' = [\Delta(\eta_{\Delta(\eta_x)(\mathbf{t})})\Delta(\eta_x)](\mathbf{t}).$$

Proof. Since the considerations made here are purely set theoretic, the proof of [May92, Lemma 14.2] carries over word by word. □

Corollary 9.1.26. *The equivalence class of a non-degenerate element $(x', \mathbf{t}')$ in $X([l]) \times \Delta_l$ is given by the disjoint union*

$$[(x', \mathbf{t}')] = \coprod_{n \in \mathbb{N}} \left(\bigcup_{\eta : [n] \twoheadrightarrow [m]} \bigcup_{\epsilon : [l] \hookrightarrow [m]} (X(\eta)[X(\epsilon)^{-1}(x')]) \times (\Delta(\eta)^{-1}[\Delta(\epsilon)(\mathbf{t}')]) \right),$$

where morphisms ϵ and η are understood to be monomorphisms and epimorphisms respectively as indicated by the arrows.

Using the non-degenerate representatives of the equivalence classes in $F(X)$ as above enables us to show that the maps $q_X \circ i_0$ are natural injections:

Lemma 9.1.27. *The natural inclusion $X([0]) \cong X([0]) \times \Delta_0 \hookrightarrow F(X)$ induces a natural injective function $q_X \circ i_0 : X([0]) \cong X([0]) \times \Delta_0 \to |X|$.*

Proof. Let $p = (x, 1)$ and $p' = (x', 1)$ be two points in $X([0]) \times \Delta_0$ and assume their images under q_X to be equal, i.e. $q_X \circ i_0(p) = q_X \circ i_0(p')$. Then the points p and p' are contained in the same equivalence class. The uniqueness of the non-degenerate representative of this equivalence class forces the points p and p' to be equal. Thus the map $q_X \circ i_0$ is injective. Its naturality follows from the naturality of q_X and that of i_0. □

Example 9.1.28. If X is a constant simplicial space, then every point $p \in |X|$ has a unique non-degenerate representative $(x, 1)$ in $X([0]) \times \Delta_0$. Thus the natural transformation $X([0]) \cong X([0]) \times \Delta_0 \to |X|$ is surjective, hence a bijection.

For any subcategory $\mathbf{C}$ of $\mathbf{Top}$ one can apply the coreflector $\mathrm{C} : \mathbf{Top} \to \mathbf{Top_C}$ to the inclusion $X([0]) \hookrightarrow F(X)$. In this way one obtains a natural injective function of $\mathbf{C}$-generated spaces:

Lemma 9.1.29. *The natural inclusion $\mathrm{C}X([0]) \cong \mathrm{C}X([0]) \times \Delta_0 \hookrightarrow \mathrm{C}F(X)$ induces a natural injective function $\mathrm{C}X([0]) \cong \mathrm{C}X([0]) \times \Delta_0 \to |X|_{\mathrm{C}}$.*

We go on to show that these injections are closed embeddings under additional conditions on the simplicial space X. First, note that all standard simplices Δ_n are compact Hausdorff spaces. Therefore for every morphism α in $\mathcal{O}$ the function $\Delta(\alpha)$ is closed. A similar requirement on the maps $X(\eta)$ ensures that the injection $q_X \circ i_0$ is a closed embedding.

Proposition 9.1.30. *If the functions $X(\eta)$ are closed for every epimorphism η in the category $\mathcal{O}$, then the inclusion $X([0]) \times \Delta_0 \hookrightarrow F(X)$ induces a natural closed embedding $q_X \circ i_0 : X([0]) \times \Delta_0 \hookrightarrow |X|$.*

Proof. Since the continuous function $q_X \circ i_0$ is injective, it suffices to show that in addition it is closed. Let $A \subset X([0])$ be a closed subset of $X([0])$. The subspace $q_X^{-1} q_X(A \times \Delta_0)$ of $F(X)$ is the union of all equivalence classes whose non-degenerate representative is contained in $X([0]) \times \Delta_0$:

$$q_X^{-1}q_X(A \times \Delta_0) = \coprod_{n\in\mathbb{N}} \left(\bigcup_{\eta:[n]\twoheadrightarrow[m]} \bigcup_{\epsilon:[0]\hookrightarrow[m]} X(\eta)(X(\epsilon)^{-1}(A)) \times \Delta(\eta)^{-1}(\Delta(\epsilon)(\Delta_0)) \right)$$

Because the maps $\Delta(\epsilon)$ are closed and all maps $X(\eta)$ were assumed to be closed as well, the sets $X(\eta)(X(\epsilon)^{-1}(A))$ and $\Delta(\eta)^{-1}(\Delta(\epsilon)(\Delta_0))$ are closed and so are their Cartesian products in $X([n]) \times \Delta_n$. Furthermore the union in the brackets on the right is finite, so the set $q_X^{-1}q_X(A \times \Delta_0) \cap X([n]) \times \Delta_n$ is the finite union of closed sets, hence closed. Thus the image $q_X(A \times \Delta_0) = q_X(q_X^{-1}q_X(A \times \Delta_0))$ is closed.   □

By the same line of reasoning on can ontain an analogous result for the geometric realisation $|CX|$ of the image of a simplicial space under the coreflector C:

Proposition 9.1.31. *If the functions* $CX(\eta)$ *are closed for every epimorphism* η *in the category* $\mathcal{O}$, *then the inclusion* $CX([0]) \times \Delta_0 \hookrightarrow F(CX)$ *induces a natural closed embedding* $q_{CX} \circ C(i_0) : CX([0]) \times \Delta_0 \hookrightarrow |CX|$.

Proof. Since the continuous function $q_{CX} \circ C(i_0)$ is injective, it suffices to show that in addition it is closed. Let $A \subset CX([0])$ be a closed subset of $CX([0])$. The subspace $q_{CX}^{-1}q_{CX}(A \times \Delta_0)$ of $F(CX)$ is the union of all equivalence classes whose non-degenerate representative is contained in $CX([0]) \times \Delta_0$. The union of these equivalence classes is a disjoint union

$$\tilde{q}^{-1}\tilde{q}(A \times \Delta_0) = \coprod_{n\in\mathbb{N}} A_n$$

of subspaces A_n of the spaces $CX([n]) \times \Delta_n$. Each of these subspaces is given by a finite union

$$A_n = \bigcup_{\eta:[n]\twoheadrightarrow[m]} \bigcup_{\epsilon:[0]\hookrightarrow[m]} [(CX)(\eta)]([CX(\epsilon)]^{-1}(A)) \times \Delta(\eta)^{-1}(\Delta(\epsilon)(\Delta_0)),$$

so in order to prove that the set $q_{CX}^{-1}q_{CX}(A \times \Delta_0)$ is closed it suffices to show that each of the subspaces $C[X(\eta)](C[X(\epsilon)]^{-1}(A)) \times \Delta(\eta)^{-1}(\Delta(\epsilon)(\Delta_0))$ occuring in the finite union defining A_n is closed in $CX([n]) \times \Delta_n$. Because the maps $\Delta(\epsilon)$ are closed and all maps $CX(\eta)$ were assumed to be closed as well, the sets $C[X(\eta)](C[X(\epsilon)]^{-1}(A))$ and $\Delta(\eta)^{-1}(\Delta(\epsilon)(\Delta_0))$ are closed and so are their Cartesian products in $CX([n]) \times \Delta_n$. Furthermore the union in the brackets on the right is finite, so the set $q_{CX}^{-1}q_{CX}(A \times \Delta_0) \cap CX([n]) \times \Delta_n$ is the finite union of closed sets, hence closed. Thus the image $q_{CX}(A \times \Delta_0) = q_{CX}(q_{CX}^{-1}q_{CX}(A \times \Delta_0))$ is closed.   □

Corollary 9.1.32. *If the functions* $CX(\eta)$ *are closed for every epimorphism* η *in the category* $\mathcal{O}$, *then the inclusion* $CX([0]) \times \Delta_0 \hookrightarrow CF(X)$ *induces a natural closed embedding* $CX([0]) \times \Delta_0 \hookrightarrow |X|_C$.

Proof. The composition of the injective function $CX([0]) \times \Delta_0 \to |X|_C$ with the natural continuous bijective map $|X|_C \to |CX|$ is closed by Proposition 9.1.31. Therfore the map $CX([0]) \times \Delta_0 \to |X|_C$ is closed as well.   □

So if every map $X([\eta])$ is closed, the space $X([0])$ can be considered to be a closed subspace of the geometric realisation $|X|$ via the natural homeomorphisms

$$X([0]) \cong X([0]) \times \Delta_0 \cong (q_X \circ i_0)(X \times \Delta_0) \subset |X|.$$

Example 9.1.33. Let Y be a topological space and X be the constant simplicial space Y. For every morphism α in $\mathcal{O}$ the function $X(\alpha)$ is the identity of Y, hence closed. So the continuous bijection $Y = X([0]) \to |X|$ is also closed, which forces it to be a homeomorphism. Therefore every topological space Y is homeomorphic to the geometric realisation $|X|$ of the constant simplicial space Y.

The requirement that the degeneracy maps are closed for all epimorphisms η in $\mathcal{O}$ turns out to be always satisfied for Hausdorff simplicial spaces. This is due to the following facts:

Lemma 9.1.34. *If a topological space Y is dominated by a topological space Z via*

$$Y \xrightarrow{s} Z \xrightarrow{p} Y, \quad ps = \mathrm{id}_Y,$$

then the section $s : Y \to Z$ is an embedding.

Proof. Let Y and Z be topological spaces and $p : Z \to Y$ and $s : Y \to Z$ continuous functions satisfying $ps = \mathrm{id}_Y$. The section s is a continuous bijection onto its image with inverse the restriction $p_{|s(Y)}$. Thus it is a homeomorphism onto its image, i.e. an embedding. $\qquad\square$

Proposition 9.1.35. *If a topological space Y is dominated by a Hausdorff topological space Z via*

$$Y \xrightarrow{s} Z \xrightarrow{p} Y, \quad ps = \mathrm{id}_Y,$$

then the embedding $s : Y \to Z$ is closed.

Proof. Let Y and Z be topological spaces and assume that Z is Hausdorff. If $p : Z \to Y$ and $s : Y \to Z$ are continuous functions satisfying $ps = \mathrm{id}_Y$, then the image $s(Y)$ is the subspace

$$s(Y) = \{z \mid z = sp(z)\} \leq Z$$

of Z. That is to say it is the largest subspace of Z on which the continuous functions id_Z and sp coincide. Since Z is Hausdorff, this subspace is closed. Therefore s is an embedding onto a closed subspace of Z, hence a closed map. $\qquad\square$

Corollary 9.1.36. *The degeneracy maps of Hausdorff simplicial spaces are closed.*

Proof. Let X be a Hausdorff simplicial space and $X(\eta_i) : X([n]) \to X([n+1])$ be a degeneracy map. Then $X([n])$ is dominated by $X([n+1])$ via

$$X([n]) \xrightarrow{X(\eta_i)} X([n+1]) \xrightarrow{X(\epsilon_i)} X([n]), \quad X(\epsilon_i)X(\eta_i) = \mathrm{id}_{X([n])}.$$

Because X was assumed to be Hausdorff, the function $X(\eta_i)$ has to be closed. Since for every epimorphism η in $\mathcal{O}$ the function $X(\eta)$ is a composition of degeneracy maps, all these functions $X(\eta)$ are closed. $\qquad\square$

Corollary 9.1.37. *The injections $X([0]) \times \Delta_0 \hookrightarrow |X|$ and $CX([0]) \times \Delta_0 \hookrightarrow |X|_C$ are closed embeddings for all Hausdorff simplicial spaces X.*

Example 9.1.38. If X is a Hausdorff topological space then the diagonal function $X \to X \times X$ is closed. So every degeneracy map $SS(\eta_i) : X^n \to X^{n+1}$ of the simplicial space $SS(X)$ is closed as well. Thus the space X can be naturally identified with the closed subspace $q_{SS(X)} \circ i_0(X \times \Delta_0)$ of the geometric realisation $|SS(X)|$.

The crucial fact needed to derive the result in the preceding corollary is the closedness of the diagonal map $X \to X \times X$ for Hausdorff spaces X. When working in a subcategory $\mathbf{Top}_C$ of $\mathbf{Top}$ this is the proper generalisation of Hausdorffness. By the same line of reasoning as before we obtain:

Corollary 9.1.39. *If X is a $\mathbf{C}$-generated simplicial space and the diagonal map $X \hookrightarrow X \times_C X$ is closed, then the degeneracy maps of X are closed.*

Corollary 9.1.40. *The injection $X([0]) \times \Delta_0 \hookrightarrow |X|_C$ is a closed embeddings for all $\mathbf{C}$-generated simplicial spaces with closed diagonal in $\mathbf{Top}_C$.*

Example 9.1.41. If X is a weak Hausdorff k-space then its diagonal function in **kTop** is closed. Thus the space X can be naturally identified with the closed subspace $q_{kSS(X)} \circ i_0(X \times \Delta_0)$ of the geometric realisation $|kSS(X)|$.

We proceed to derive a similar result for closed simplicial subspaces in place of the subspace $X([0])$.

Lemma 9.1.42. *For any simplicial subspace X of of a simplicial space X' the non-degenerate points of $F(X)$ are non-degenerate in $F(X')$.*

Proof. Let X be a simplicial subspace of the simplicial space X' and $(x, \mathbf{t})$ be a point in $X([n]) \times \Delta_n$ which is non-degenerate in $F(X)$. This implies especially that $\mathbf{t}$ is non-degenerate. Assume $(x, \mathbf{t})$ to be degenerate in $F(X')$. Because $\mathbf{t}$ is non-degenerate this forces the point x to be degenerate in X'. Therefore there exists an epimorphism $\eta_x : [n] \to [m]$ in the category $\mathcal{O}$ and a point $x' \in X'([m])$ such that $X'(\eta_x)(x') = x$. The epimorphism η_x has a right inverse ϵ_x in the category $\mathcal{O}$. Thus

$$x' = X'(\epsilon)X'(\eta)(x') = X'(\epsilon)(x) = X(\epsilon)(x),$$

i.e. x' is contained in $X([m])$ in contradiction to the non-degeneracy of $(x, \mathbf{t})$. So the assumption that $(x, \mathbf{t})$ is degenerate in $F(X')$ leads to a contradiction, hence $(x, \mathbf{t})$ is non-degenerate in $F(X')$ as well. □

Lemma 9.1.43. *For any simplicial subspace X of X' the underlying set of the simplicial subspace $q_{X'}^{-1} q_{X'}(F(X))$ is given by the disjoint union*

$$q_{X'}^{-1} q_{X'}(F(X)) = \coprod_{n \in \mathbb{N}} \left(\bigcup_{\eta:[n] \twoheadrightarrow [m]} \bigcup_{\epsilon:[l] \hookrightarrow [m]} [X(\eta)X(\epsilon)^{-1}X([l])] \times [\Delta(\eta)^{-1}\Delta(\epsilon)(\Delta_l)] \right),$$

where for each $n \in \mathbb{N}$ union occurring to the right is finite.

Proof. This is another application of Corollary 9.1.26. Again, the finiteness of the union follows from the finiteness of the hom-sets in $\mathcal{O}$ and the fact that for each $n \in \mathbb{N}$ there exist only finitely many numbers m such that $l \leq m \leq n$. □

Corollary 9.1.44. *If X is a closed simplicial subspace of X' and the functions $X'(\eta)$ are closed for every epimorphism η in the category $\mathcal{O}$, then $q_{X'}^{-1} q_{X'}(F(X))$ is closed in $F(X')$.*

Corollary 9.1.45. *If X is a closed simplicial subspace of a Hausdorff simplicial space X', then $q_{X'}^{-1} q_{X'}(F(X))$ is closed in $F(X')$.*

Thus if X is a closed simplicial subspace of the simplicial space X' and the functions $X'(\eta)$ are closed for every epimorphism η in $\mathcal{O}$, the inclusion $X \hookrightarrow X'$ induces a injective map $|X| \to |X'|$ onto a closed subspace of $|X'|$. Moreover the conditions on the maps $X'(\eta)$ are always satisfied if the simplicial space X' is Hausdorff. We now show that the condition that all maps $X'(\eta)$ be closed suffices for a closed simplicial map $f : X \to X'$ to induce a closed map $|f| : |X| \to |X'|$. Here fore an examination of the $q_{X'}$-saturation of a simplicial subspace of X' is necessary. The description of the the $q_{X'}$-saturation $q_{X'}^{-1} q_{X'}(X)$ of a simplicial subspace X of X' can be better understood by studying the functions on the power-sets involved. So motivated we introduce for every monomorphism $\epsilon : [m] \hookrightarrow [n]$ in $\mathcal{O}$ a function

$$F_\epsilon : \mathcal{P}(X([m]) \times \Delta_m) \to \mathcal{P}(X([n]) \times \Delta_n)$$

which describes the first mapping of power-sets occurring in Lemma 9.1.43. Guided by the formula in Lemma 9.1.43 we define the value of F_ϵ at a point S in the set $\mathcal{P}(X([m]) \times \Delta_m)$ to be

$$F_\epsilon(S) := \{(x,t) \mid \exists t' \in \Delta_m : \Delta(\epsilon)(t') = t \wedge (X(\epsilon)(x),t') \in S\}.$$

This definition can be simplified using the left inverse η to the monomorphism ϵ. The image $F_\epsilon(S)$ of a set $S \in \mathcal{P}(X([m]) \times \Delta_m)$ can then be written as the intersection

$$F_\epsilon(S) = [X(\epsilon) \times \Delta(\eta)]^{-1}(S) \cap [X([n]) \times \Delta(\epsilon)(\Delta_m)],$$

where the subspace $\Delta(\epsilon)(\Delta_m)$ is closed, because the map $\Delta(\alpha)$ is closed for every morphism α in $\mathcal{O}$. The so defined maps F_ϵ turn out to always map closed sets to closed sets.

Lemma 9.1.46. *The functions F_ϵ map closed sets to closed sets.*

Proof. Let $\epsilon : [m] \hookrightarrow [n]$ be a monomorphism in $\mathcal{O}$ and $A \subset X([m]) \times \Delta_m$ be closed. Because the subspace $X([n]) \times \Delta(\epsilon)(\Delta_m)$ is closed, the set

$$F_\epsilon(A) = [X(\epsilon) \times \Delta(\eta)]^{-1}(A) \cap [X([n]) \times \Delta(\epsilon)(\Delta_m)]$$

is an intersection of closed subspaces, hence closed. $\qquad\square$

In an analogy to the maps F_ϵ we define for every epimorphism $\eta : [n] \twoheadrightarrow [m]$ in $\mathcal{O}$ a map

$$F_\eta : \mathcal{P}(X([m]) \times \Delta_m) \to \mathcal{P}(X([n]) \times \Delta_n),$$

which describes the second map of power-sets occuring in lemma 9.1.43. Guided by the formula in Lemma 9.1.43 we define these maps via

$$F_\eta(S) := \{(x,t) \mid \exists x' \in X([m]) : X(\eta)(x')(x) = x \wedge (x', \Delta(\eta)(t)) \in S\}.$$

This definition can be simplified as well. Here fore we use the right inverse ϵ to the epimorphism η. The image $F_\eta(S)$ of a set $S \in \mathcal{P}(X([m]) \times \Delta_m)$ can then be written

$$F_\eta(S) = [X(\epsilon) \times \Delta(\eta)]^{-1}(S) \cap [X(\eta)(X([m])) \times \Delta_n].$$

That these functions map closed sets to closed sets can be insured by a condition on the maps $X(\eta)$:

Lemma 9.1.47. *If $\eta : [n] \twoheadrightarrow [m]$ is an epimorphism such that $X(\eta)$ is closed then F_η maps closed sets to closed sets.*

Proof. Let $\eta : [n] \to [m]$ be a monomorphism in $\mathcal{O}$ such that $X(\eta)$ is closed and A be closed subset of $X([m]) \times \Delta_m$. Since $X(\eta)$ is closed the subspace $X(\eta)(A)$ of $X([n])$ is also closed. Therefore the set

$$F_\epsilon(A) = [X(\epsilon) \times \Delta(\eta)]^{-1}(A) \cap [X(\eta)(X([n])) \times \Delta_n]$$

is an intersection of closed subspaces, hence closed. $\qquad\square$

Collecting the above results for the case that all the maps $X'(\eta)$ are closed we are ready to prove the final theorem of this section.

Theorem 9.1.48. *If $f : X \to X'$ is a closed morphism of simplicial spaces and the degeneracy maps of X' are closed, then the induced function $|f| : |X| \to |X'|$ is closed.*

Proof. Let $f : X \to X'$ is a closed morphism of simplicial spaces and assume the functions $X'(\eta)$ to be closed for every epimorphism η in the category $\mathcal{O}$. For every closed subset $A \subset |X|$ of $|X|$, the inverse image $q_X^{-1}(A)$ is a closed subset of $F(X)$. Therefore it is the disjoin union

$$A = \coprod_{n \in \mathbb{N}} A_n$$

of closed subsets $A_n \subset X([n]) \times \Delta_n$. The inverse image $A' = q_{X'}^{-1}(|f|(A))$ of $|f|(A)$ under $q_{X'}$ is the disjoint union

$$A' = \coprod_{n \in \mathbb{N}} A'_n$$

of subsets $A_n \subset X'([n]) \times \Delta_n$. The image $|f|(A)$ is closed if and only if A' is closed. The subset A' in turn is closed if and only if each set A'_n is a closed subset of $X([n]) \times \Delta_n$. The sets A'_n an A_n are related by the equation

$$A'_n = \bigcup_{\eta: [n] \twoheadrightarrow [m]} \bigcup_{\epsilon: [l] \hookrightarrow [m]} F_\eta F_\epsilon(A_l).$$

The maps F_ϵ map closed sets to closed sets. That the maps F_η have this property as well is guaranteed by the assumption that all the functions $X([\eta])$ are closed. Therefore each set A'_n is a finite union of closed sets, hence closed.   □

Corollary 9.1.49. *The geometric realisation preserves the closedness of morphisms into Hausdorff spaces.*

Corollary 9.1.50. *The restriction of the geometric realisation functor to the subcategory* $\mathbf{Haus}^{\mathcal{O}^{op}}$ *of simplicial Hausdorff spaces preserves closedness of morphisms.*

Corollary 9.1.51. *The geometric realisation preserves the closedness of morphisms into weak Hausdorff k-spaces.*

Proof. This follows from the fact that for k-spaces the property of being weak Hausdorff is equivalent to the diagonal map being closed in $\mathbf{kTop}$.

Corollary 9.1.52. *If* $X \hookrightarrow X'$ *is a closed embedding of simplicial spaces and all degeneracy maps of* X' *are closed, then the induced map* $|X| \hookrightarrow |X'|$ *is a closed embedding.*

Example 9.1.53. If X is a closed subspace of the Hausdorff topological space X', then the geometric realisation $|SS(X)|$ embeds as a closed topological subspace in the geometric realisation $|SS(X')|$.

Example 9.1.54. If X is a closed k-subspace of the weak Hausdorff k-space X' then the geometric realisation $|kSS(X)|$ embeds as a closed topological subspace in the geometric realisation $|kSS(X')|$.

We now turn our attention to the topology of the geometric realistaion $|X|$ of a simplicial space X. It can be described by use of a filtration of the simplicial space X, which we now define:

Definition 9.1.55. *The k-skeleton* X_k *of a simplicial space* X *is the simplicial subspace*

$$X_k([n]) = \bigcup_{\substack{\eta: [n] \twoheadrightarrow [m] \\ m \leq k}} X(\eta)(X([m]))$$

of X consisting of all points whose unique non degenerate representative is contained in one of the topological spaces $X([0]), \ldots, X([k])$.

Note that taking the k-skeleton of a simplicial space provides a functor

$$-_k : \mathbf{Top}^{\mathcal{O}^{op}} \to \mathbf{Top}^{\mathcal{O}^{op}}, \quad X \mapsto X_k,$$

which assigns to every simplicial space its k-skeleton.

The k-skeleta X_k of a simplicial space X form an increasing sequence of sub-spaces. Moreover, observe that the first k spaces $X_k([1]), \ldots, X_k([k])$ of the k-skeleton coincide with the first k spaces $X([0]), \ldots, X([k])$ of X: In the case $n \leq k$ the n-th space $X_k([n])$ of the k-skeleton coinatains the subspace $X(\mathrm{id}_{[n]})X([n]) = X([n])$ by defnition; thus it is the space $X([n])$ itself. As a consequence each space $X([n])$ is the (finite) union of the subspaces $X_k(n])$, $0 \leq k \leq n$, so the k-skeleta X_k form a filtration of the simplicial space X. As this is true for every simplicial space we note the result for the k-skeleton functor:

Lemma 9.1.56. *The identiy on* $\mathbf{Top}^{\mathcal{O}^{op}}$ *is the colimit of the k-skelta: colim$-_k =$* $\mathrm{id}_{\mathbf{Top}^{\mathcal{O}^{op}}}$

The embedding $X_k \hookrightarrow X$ of the k-skeleton X_k of X into X induces an embedding $F(X_k) \hookrightarrow F(X)$. These subspaces are also called skeleta:

Definition 9.1.57. *The the space $F(X_k)$ is called the k-skelton of $F(X)$.*

Since the k-skeleta X_k of X form a filtration of X, the k-skeleta $F(X_k)$ of $F(X)$ provide a filtration of $F(X)$. The topology of $F(X)$ turns out to be the limit topology w.r.t. this filtration:

Lemma 9.1.58. *The functor F is the colimit of the subfunctors $F \circ -_k$, i.e. each space $F(X)$ is the colimit of its subspaces $F(X_k)$.*

Proof. This follows from the cocontinuuity of the functor F (c.f. 9.1.10). $\square$

Because the equivalence relation on each k-skeleton $F(X_k)$ is the restriction of the equivalence relation on $F(X)$ the embedding $F(X_k) \to F(X)$ descends to continuous injection $|X_k| \to |X|$.

Proposition 9.1.59. *The geometric realisation functor $|-|$ is the colimut of the functors $|-_k|$, i.e. for every simplicial topological space X the geometric realisaation $|X|$ is the colimit of the spaces $|X_k|$.*

Proof. To prove that $|X|$ is the colimit of the spaces $|X_k|$ it has to be verified that a subset $U \subset |X|$ is open if and only if the inverse images in $|X_k|$ are open for all $k \in \mathbb{N}$. The set U is open in $|X|$ exactly if the inverse image q_X^{-1} of U under the quotient map q_X is open in $F(X)$. This inverse image $q_X^{-1}(U)$ in turn is open in $F(X)$ if and only if its intersection $q_X^{-1}(U) \cap F(X_k)$ with the k-skeletons of $F(X)$ are open. These intersections at last are open if and only the sets $q_{X_k}(q_X^{-1}(U) \cap F(X_k))$ are open. Therefore the subset $U \subset |X|$ is open if and onloy if its inverse images in the spaces $|X_k|$ are open. $\square$

As in the case of simplicial morphisms one important class of simplicial spaces are those simplicial spaces X with closed degeneracy maps $X(\eta)$. Here an analogous result to that of Proposition 9.1.30 can be derived:

Lemma 9.1.60. *If the morphisms $X(\eta)$ are closed for all epimorphisms η in $\mathcal{O}$ then the k-skeleton X_k of X is a closed simplicial subspace of X.*

Proof. Let the morphisms $X(\eta)$ be closed for all epimorphisms η in $\mathcal{O}$. A space $X_k([n])$ of the k-skeleton is by definition the space

$$X_k([n]) = \bigcup_{\substack{\eta:[n]\twoheadrightarrow[m] \\ m\leq k}} X(\eta)(X([m])).$$

The union on the right hand side is a union of closed subspaces by assumption. Since this union is finite, the space $X_k([n])$ is a closed subspace of the space $X([n])$.   □

Corollary 9.1.61. *If the morphisms $X(\eta)$ are closed for all epimorphisms η in $\mathcal{O}$ then the k-skeleton $F(X_k)$ of $F(X)$ is a closed subspace of $F(X)$.*

Example 9.1.62. If X is the standard simplicial space $\mathrm{SS}(Y)$ of a Hausdorff space Y then the maps $X(\eta)$ are closed for all morphisms η in $\mathcal{O}$. So the k-skeletons X_k provide a filtration by closed simplical subspaces.

Proposition 9.1.63. *If the morphisms $X(\eta)$ are closed for all epimorphisms η in $\mathcal{O}$ then embeddings $X_k \hookrightarrow X$ of the k-skeleta induce closed embeddings $|X_k| \hookrightarrow |X|$.*

Proof. Let the morphisms $X(\eta)$ be closed for all epimorphisms η in $\mathcal{O}$. By Lemma 9.1.60 each k-skeleton X_k is a closed simplicial subspace of X. By Theorem 9.1.48 the injections are closed. Being closed injections they are closed embeddings.   □

Corollary 9.1.64. *If the morphisms $X(\eta)$ are closed for all epimorphisms η in $\mathcal{O}$ then the geometric realisation $|X|$ is the colimit of its closed subspaces $|X_k|$.*

Example 9.1.65. If X is the standard simplicial space $\mathrm{SS}(Y)$ of a Hausdorff space Y then the maps $X(\eta)$ are closed for all morphisms η in $\mathcal{O}$. So the geometric realisations $|X_k|$ of the k-skeleta X_k provide a filtration of $|X|$ by closed subspaces.

9.2 Cocontinuity of the Geometric Realisation Functor

Recall that the category **Set** can be considered to be a full sub-category of the category **Top** of topological spaces by equipping every set with the discrete topology. Thus every simplicial set can be regarded as a simplicial topological space. The restriction of the geometric realisation functor $|-|$ to the sub-category $\mathbf{Set}^{\mathcal{O}^{op}}$ has been extensively studied. For example it is known that it has a right adjoint (see [MLM94, Theorem I.5.2]). We now generalise this result to the category **Top** by proving the existence of a right adjoint $R : \mathbf{Top} \to \mathbf{Top}^{\mathcal{O}^{op}}$ to the geometric realisation functor $|-| : \mathbf{Top}^{\mathcal{O}^{op}} \to \mathbf{Top}$. In analogy to the discrete case we start by defining

$$R : \mathbf{Top} \to \mathbf{Top}^{\mathcal{O}^{op}}, \quad R = C(\Delta, -),$$

where Δ is the standard cosimplicial space. This functor assigns the singular simplicial space $R(Y) = C(\Delta, Y)$ to a topological space Y. In the following it is shown that this functor R is a right adjoint to the geometric realisation functor, i.e. that there exists an isomorphism

$$\hom(|X|, Y) \cong \hom(X, R(Y))$$

which is natural in X and Y. We start with studying the set $\hom(|X|, Y)$ for a simplicial space X and a topological space Y. Recall that the continuous maps $|X| \to Y$ are in one to one correspondence to continuous maps $F(X) \to Y$ that are single valued on the fibres of the quotient map $q_X : F(X) \to |X|$. Such maps can be restricted to the subspaces $X([n]) \times \Delta_n$ of $F(X)$ to obtain a family $\{f_n\}_{n\in\mathbb{N}}$ of maps $f_n : X([n]) \times \Delta_n \to Y$ that satisfy a compatibility condition. This motivates the next definitions.

Definition 9.2.1. *A family $\{f_n\}_{n\in\mathbb{N}}$ of continuous functions $f_n : X([n]) \times \Delta_n \to Y$ is called* compatible *if the equalities*

$$f_m(X(\alpha)(x), t) = f_n(x, \Delta(\alpha)(t)) \tag{9.2}$$

are satisfied for all $[m]$, $[n]$ and morphisms $\alpha : [m] \to [n]$ in $\mathcal{O}$.

For X a simplicial space and Y a topological space we consider families of continuous functions $f_n : X([n]) \times \Delta_n \to Y$ to be points in the product space $\prod C(X([n]) \times \Delta_n, Y)$. The closed subspace consisting of all compatible families of such functions $\{f_n\}$ will be used throughout this section. Therefore we give it a name.

Definition 9.2.2. *For each simplicial space X and topological space Y the space of all compatible families $f_n : X([n]) \times \Delta_n \to Y$ is denoted by $\mathrm{Cp}(X, Y)$*

The families $\{f_n\}_{n\in\mathbb{N}}$ of continuous functions $f_n : X([n]) \times \Delta_n \to Y$ correspond bijectively to the continuous functions $f : \coprod X([n]) \times \Delta_n \to Y$ via the natural isomorphism

$$\psi_{X,Y} : \prod C(X([n]) \times \Delta_n, Y) \xrightarrow{\cong} C\left(\coprod X([n]) \times \Delta_n, Y\right) = C(F(X), Y)$$

The inverse $\psi_{X,Y}^{-1}$ of this isomorphism assigns to each continuous map $f : F(X) \to Y$ the family $\{f \circ i_n\}_{n\in\mathbb{N}}$ of restrictions to the open and closed subspaces $X([n]) \times \Delta_n$. This isomorphism allows us to identify the closed subspace $\mathrm{Cp}(X, Y)$ of all functions satisfying the Equations 9.2 with a closed subspace of $C(F(X), Y)$. The identification

$$\mathrm{Cp}(X, Y) \cong \psi_{X,Y}(\mathrm{Cp}(X, Y)) \subset C(F(X), Y), \quad \{f\}_{n\in\mathbb{N}} \mapsto \coprod f_n$$

is the restriction of the natural isomorphism ψ to a closed subspace, hence natural. The introductory remarks at the beginning of this section can now be made precise using the concept of compatibility of families of maps. The compatibility of a family of maps expresses the fact that it induces a continuous function $|X| \to Y$ and vice versa (see below).

Lemma 9.2.3. *For any continuous map $f : |X| \to Y$ from the geometric realisation of a simplicial space X into a topological space Y the family of maps $\{f \circ q_X \circ i_n\}_{n\in\mathbb{N}}$ is compatible.*

Proof. Let X be a simplicial space and Y be a topological space. For any continuous function $f : |X| \to Y$ the map $f \circ q_X$ is single valued on all q_X-saturated subsets of $\coprod X([n]) \times \Delta_n$, i.e. it satisfies the equality $(f \circ q_X)(X(\alpha)(x), t) = (f \circ q_x)(x, \Delta(\alpha)(t))$ for all morphisms $\alpha : [m] \to [n]$ in $\mathcal{O}$, all $x \in X([n])$ and all $t \in \Delta_m$. This implies the equality

$$f \circ q_X \circ i_m(X(\alpha)(x), t) = (f \circ q_X)(X(\alpha)(x), t)$$
$$= (f \circ q_X)(x, \Delta(\alpha)(t))$$
$$= f \circ q_X \circ i_n(x, \Delta(\alpha)(t))$$

for all morphisms $\alpha : [m] \to [n]$ in $\mathcal{O}$, all $x \in X([n])$ and all $t \in \Delta_m$. Thus the family of maps $\{f \circ q_X \circ i_n\}_{n\in\mathbb{N}}$ is compatible. $\square$

Thus the continuous function $\psi_{X,Y}^{-1} \circ C(q_X, Y)$ has range in the closed subspace $\mathrm{Cp}(X, Y)$ of $\prod C(X([n]) \times \Delta_n, Y)$. Since both ψ and q are natural transformations, the functions $\psi_{X,Y}^{-1} \circ C(q_X, Y)$ form a natural transformation as well. Furthermore, one deduces from the definition of the spaces $\mathrm{Cp}(X, Y)$ of compatible families that these spaces are object parts of a bifunctor $\mathrm{Cp} : \mathbf{Top}^{\mathcal{O}^{op}} \times \mathbf{Top} \to \mathbf{Top}$. (For the curious reader this is carried out at the end of this section.)

Definition 9.2.4. *The corestriction of the natural transformation $\psi^{-1} \circ C(q,-)$ to* $\mathrm{Cp}(-,-)$ *is denoted by* τ.

It may be instructive to explicitly work out how the maps $\tau_{X,Y}$ interchange with simplicial maps $f: X \to X'$ and $g: Y \to Y'$. This is done at the end of this section (see Lemmata 9.2.39 and 9.2.40).

Lemma 9.2.5. *All maps $\tau_{X,Y}$ are continuous bijections.*

Proof. Let X be a simplicial space and Y be a topological space. Because the quotient map q_X is surjective, the function $C(q_X, Y) : C(|X|, Y) \to C(F(X), Y)$ is injective. So the map $\tau_{X,Y} = \psi_{X,Y}^{-1} \circ C(q_X, Y)$ is injective as well. It remains to observe that every compatible family of maps $\{f \circ q_X \circ i_n\}_{n \in \mathbb{N}}$ assembles to a continuous function $\coprod X([n]) \times \Delta_n \to Y$ that is single valued on all q_X-saturated subsets of $\coprod X([n]) \times \Delta_n$. This follows from the definition of compatibility. Thus the map $\tau_{X,Y}$ is in addition surjective, hence bijective. $\qquad\square$

So for every continuous function $f : |X| \to Y$ there exists a unique compatible family $\tau_{X,Y}(f)$ of maps such that $\coprod \tau(f)(n)$ descends to f and every compatible family is induced by a unique continuous function $f : |X| \to Y$. These observations can be summarised in the following commutative diagram

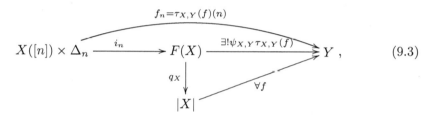

$$\tag{9.3}$$

where the maps q_X, i_n, ψ and $\tau_{X,Y}$ are natural.

We now turn our attention to some special simplicial spaces and compatible families of functions. Let Y be a topological space and consider the simplicial space $C(\Delta, Y)$ and the families of maps $f_n : C(\Delta_n, Y) \times \Delta_n \to Y$. A particularly useful such family of maps is the family of evaluation maps $\mathrm{ev}_{n,Y} : C(\Delta_n, Y) \times \Delta_n \to Y$. So this family of maps is a natural candidate for a compatible family. As it turns out, it is not only compatible but also has other useful properties.

Lemma 9.2.6. *For any topological space Y, the family $\{\mathrm{ev}_{n,Y}\}_{n \in \mathbb{N}}$ of evaluation maps $\mathrm{ev}_{n,Y} : C(\Delta_n, Y) \times \Delta_n \to Y$ is compatible.*

Proof. Let Y be a topological space and $\alpha : [m] \to [n]$ be a morphism on $\mathcal{O}$. The evaluation of a map $f \in C(\Delta_n, Y)$ at a point $\Delta(\alpha)(t)$, $t \in \Delta_m$ is given by

$$\mathrm{ev}_{n,Y}(f, \Delta(\alpha)(t)) = f(\Delta(\alpha)(t)) = [C(\Delta(\alpha), Y)(f)](t) = \mathrm{ev}_{m,Y}(C(\Delta(\alpha), Y)(f), t),$$

hence the family $\{\mathrm{ev}_{n,Y}\}_{n \in \mathbb{N}}$ of evaluation maps is compatible. $\qquad\square$

Corollary 9.2.7. *For any space Y, the map $\coprod \mathrm{ev}_{n,Y} : \coprod C(\Delta_n, Y) \times \Delta_n \to Y$ descends to a map $\epsilon_Y : |C(\Delta_n, Y)| \to Y$ making the following diagram commutative:*

$$
\begin{array}{ccc}
\coprod C(\Delta_n, Y) \times \Delta_n & & \\
\big\downarrow {\scriptstyle q_{C(\Delta, Y)}} & \searrow {\scriptstyle \coprod \mathrm{ev}_{n,Y}} & \\
|C(\Delta, Y)| & \xrightarrow[\epsilon_Y]{} & Y
\end{array}
$$

We proceed to show that the map ϵ_Y is a universal arrow in the comma category $|-| \downarrow Y$ of objects $|-|$-over Y. This universal arrow will then be the counit of an adjunction between $|-|$ and $R = C(\Delta, -)$. It is known that evaluation is a natural transformation $\mathrm{ev}_n : C(\Delta_n, -) \times \Delta_n \to \mathrm{id}_{\mathbf{Top}}$ for each $n \in \mathbb{N}$ (cf. Lemma B.5.5). A quick calculation shows that the same is true for their coproduct $\coprod \mathrm{ev}_n$:

Lemma 9.2.8. *The coproduct $\coprod \mathrm{ev}_n$ of the natural transformations ev_n is a natural transformation $F \circ C(\Delta, -) \to \mathrm{id}_{\mathbf{Top}}$.*

Proof. Let $f : Y \to Z$ be any continuous function between the topological spaces Y and Z. Every point in $\coprod C(\Delta_n, Y) \times \Delta_n$ is contained in one of the subspaces $C(\Delta_n, Y) \times \Delta_n$. Let (σ, t) be a point in $\coprod C(\Delta_n, Y) \times \Delta_n$ and $C(\Delta_k, Y) \times \Delta_k$ be the subspace containing (σ, t). The image of $(\coprod \mathrm{ev}_n)(\sigma, t)$ under f is given by

$$\left(f \circ \coprod \mathrm{ev}_{n,Y} \right)(\sigma, t) = f \circ \mathrm{ev}_{k,Y}(\sigma, t)$$
$$= f(\sigma(t))$$
$$= (f \circ \sigma)(t)$$
$$= [C(\Delta_k, f)(\sigma)](t)$$
$$= [\mathrm{ev}_{k,Z} \circ (C(\Delta_k, f) \times \mathrm{id}_{\Delta_k})](\sigma, t)$$
$$= \left[\coprod \mathrm{ev}_{n,Z} \right] \circ \left[\coprod C(\Delta_n, f) \times \mathrm{id}_{\Delta_n} \right](\sigma, t).$$

Therefore for each continuous function $f : X \to Y$ the diagram

$$
\begin{array}{ccc}
F(C(\Delta, Y)) & \xrightarrow{\coprod \mathrm{ev}_{n,Y}} & Y \\
{\scriptstyle F(C(\Delta, f))}\downarrow & & \downarrow{\scriptstyle f} \\
F(C(\Delta, Z)) & \xrightarrow[\coprod \mathrm{ev}_{n,Z}]{} & Z
\end{array}
$$

commutes, hence $\coprod \mathrm{ev}_n$ is a natural transformation $F \circ C(\Delta_n, -) \to \mathrm{id}_{\mathbf{Top}}$. $\qquad\square$

Because the evaluation maps $\mathrm{ev}_{n,Y}$ are compatible their naturality in Y carries over to the induced functions ϵ_Y. This is to say that the induced maps ϵ_Y also form a natural transformation:

Lemma 9.2.9. *The collection of morphisms ϵ_Y constitutes a natural transformation $\epsilon : |C(\Delta, -)| \to \mathrm{id}_{\mathbf{Top}}$.*

Proof. Let $f : Y \to Z$ be a continuous function between the topological spaces Y and Z. For any point $p \in |C(\Delta, Y)|$ there exists a natural number $n \in \mathbb{N}$ and a point $(\sigma, t) \in C(\Delta_n, Y) \times \Delta_n$ which is mapped to p under the quotient map $q_{C(\Delta, Y)}$. From the previous observations one now infers the equalities

$$(f \circ \epsilon_Y)(p) = \left[f \circ \coprod \mathrm{ev}_{n,Y} \right](\sigma, t) \qquad\qquad \text{by Corollary 9.2.7}$$
$$= \left[\coprod \mathrm{ev}_{n,Z} \right] \circ \left[\coprod (C(\Delta, f) \times \mathrm{id}_{\Delta_n}) \right](\sigma, t) \qquad \text{by Lemma 9.2.8}$$
$$= \epsilon_Z \circ q_{C(\Delta, Z)} \circ \left[\coprod C(\Delta_n, f) \times \mathrm{id}_{\Delta_n} \right](\sigma, t) \qquad \text{by Corollary 9.2.7}$$
$$= \epsilon_Z \circ |C(\Delta, f)| \circ q_{C(\Delta, Y)}(\sigma, t) \qquad\qquad \text{since } q_{C(\Delta, -)} \text{ is natural}$$
$$= \epsilon_Z \circ |C(\Delta, f)|(p).$$

The naturality of the morphism $\coprod \mathrm{ev}_n$ then implies the naturality of ϵ. $\qquad\square$

Recall that for any three topological spaces X, Y and Z there exists a continuous map $\varphi_{X,Y,Z} : C(X \times Y, Z) \to C(X, C(Y, Z))$ which assigns to each contiunous function $f \in C(X \times Y, Z)$ its *adjoint* function $\hat{f}$ that is given by

$$[\hat{f}(x)](y) = f(x, y)$$

for all $x \in X$ and $y \in Y$. These maps $\varphi_{X,Y,Z}$ are a natural transformation of trifunctors $\mathbf{Top} \times \mathbf{Top} \times \mathbf{Top} \to \mathbf{Top}$ (cf. Lemma B.5.11). If the spaces X and Y are locally compact or X is Hausdorff and Y is locally compact, the maps $\varphi_{X,Y,Z}$ are homeomorphisms (cf. B.5.18). So especially the product maps

$$\prod \varphi_{X([n]), \Delta_n, Y} : \prod C(X([n]) \times \Delta_n, Y) \xrightarrow{\cong} \prod C(X([n]), C(\Delta_n, Y))$$

are natural in X and Z. Our special interest lies in the restriction of the product map $\prod \varphi_{X([n]), \Delta_n, Y}$ to the subspace of compatible families $f_n : X([n]) \times \Delta_n \to Y$. We assign this restriction its its own symbol:

Definition 9.2.10. *The restriction of the natural transformation $\prod \varphi_{-([n]), \Delta_n, -}$ to the subfunctor $\mathrm{Cp}(-, -)$ of compatible families is denoted by $\varphi_{-,-}$.*

Since $\varphi_{X,Y}$ is a restriction of a natural transformation to a sub-functor, it is still natural in X and Y. Its restrictions to $\prod_{n \in \mathbb{N}} C(-([n]) \times \Delta_n, -)_{|\mathbf{Lc}^{\mathcal{O}^{op}} \times \mathbf{Top}}$ and $\prod_{n \in \mathbb{N}} C(-([n]) \times \Delta_n, -)_{|\mathbf{Haus}^{\mathcal{O}^{op}} \times \mathbf{Top}}$ are natural isomorphisms by B.5.18. In the following we examine the effect of the natural transformation $\prod \varphi_{X([n]), \Delta_n, Y}$ on compatible families of maps.

Lemma 9.2.11. *For every family $\{f_n\}_{n \in \mathbb{N}}$ of maps $f_n : X([n]) \times \Delta_n \to Y$ the family $\{\hat{f}_n\}_{n \in \mathbb{N}}$ is a morphism of simplicial spaces if and only if $\{f_n\}_{n \in \mathbb{N}}$ is compatible.*

Proof. Let $\{f_n\}_{n \in \mathbb{N}}$ be a family of maps $f_n : X([n]) \times \Delta_n \to Y$. The family $\{\hat{f}_n\}_{n \in \mathbb{N}}$ is a morphism of simplicial spaces if and only if the equality

$$\hat{f}_m(X(\alpha)(x)) = C(\Delta(\alpha), Y) \circ \hat{f}_n(x)$$

is satisfied for all morphisms $\alpha : [m] \to [n]$ in $\mathcal{O}$. We reformulate the condition for the family $\{\hat{f}_n\}_{n \in \mathbb{N}}$ to be a morphism of simplicial spaces. Let $\alpha : [m] \to [n]$ be a morphism in $\mathcal{O}$ and x and t be points of $X([n])$ and Δ_m respectively. The condition

$$\hat{f}_m(X(\alpha)(x))(t) = [C(\Delta(\alpha), Y) \circ \hat{f}_n(x)](t)$$

is by definition equivalent to the condition

$$f_m(X(\alpha)(x), t) = f_n(x, \Delta(\alpha)(t)).$$

Therefore the family $\{\hat{f}_n\}_{n \in \mathbb{N}}$ is a morphism of simplicial spaces if and only if the family $\{f_n\}_{n \in \mathbb{N}}$ is compatible. $\square$

Corollary 9.2.12. *For any compatible family of functions $\{f_n\}_{n \in \mathbb{N}} \in \mathrm{Cp}(X, Y)$, the family of maps $\coprod \hat{f}_n \times \mathrm{id}_{\Delta_n}$ descends to a function $|\hat{f}| : |X| \to |C(\Delta, Y)|$ making the following diagram commutative:*

$$
\begin{array}{ccc}
\coprod X([n]) \times \Delta_n & \xrightarrow{\coprod \hat{f}_n \times \mathrm{id}_{\Delta_n}} & \coprod C(\Delta_n, Y) \times \Delta_n \\
{\scriptstyle q_X} \downarrow & & \downarrow {\scriptstyle q_{C(\Delta, Y)}} \\
|X| & \xrightarrow[|\hat{f}|]{} & |C(\Delta, Y)|
\end{array}
$$

The preceding Lemma shows that the set of simplicial maps $X \to C(\Delta, Y)$ is a closed subspace of the product space $\prod C(X([n]), C(\Delta_n, Y))$. In general, for simplicial spaces X and X', we denote the closed subspace of $\prod C(X(n)), X'([n]))$ consisting of simplicial maps by $C(X, X')$. The underlying set of this space is the set $\hom(X, X')$ of morphisms from X to X'. With this notation we obtain:

Corollary 9.2.13. *The natural transformation* $\varphi : \mathrm{Cp}(-, -) \to C(-, C(\Delta, -))$ *consists of continuous bijections.*

Corollary 9.2.14. *The restrictions of* φ *to the functors* $\mathrm{Cp}(-, -)_{|\mathbf{Lc}^{\mathcal{O}^{op}} \times \mathbf{Top}}$ *and* $\mathrm{Cp}(-, -)_{|\mathbf{Haus}^{\mathcal{O}^{op}} \times \mathbf{Top}}$ *are natural isomorphisms.*

We now turn our attention to the family of evaluation maps ev_n and their role in the above transformations.

Lemma 9.2.15. *For every continuous function* $f : |X| \to Y$ *the induced compatible family* $\{f_n\}_{n \in \mathbb{N}} = \tau_{X,Y}(f)$ *satisfies the equation* $\coprod \mathrm{ev}_n \circ \left(\coprod \hat{f}_n \times \mathrm{id}_{\Delta_n} \right) = f \circ q_X$, *i.e. the following diagram is commutative:*

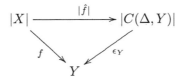

Proof. Let $f : |X| \to Y$ be a continuous function and let $\{f_n\}_{n \in \mathbb{N}} = \tau_{X,Y}(f)$ be its associated compatible family. Let (x, t) be a point in $\coprod X([n]) \times \Delta_n$ and $X([k]) \times \Delta_k$ be the subspace containing (x, t). The image of (x, t) under $\coprod f_n$ is given by

$$\left[\coprod f_n \right] (x, t) = f_k(x, t) = \hat{f}_k(x)(t) = \mathrm{ev}_k(\hat{f}_k(x), t)$$
$$= \left[\coprod \mathrm{ev}_n \circ \coprod (\hat{f}_n \times \mathrm{id}_{\Delta_n}) \right] (x, t).$$

This equation holds for arbitrary points (x, t) in $F(X)$, thus the stated equality follows. $\qquad\square$

Lemma 9.2.16. *For any function* $f : |X| \to Y$, *the function* $|\hat{f}|$ *satisfies the equation* $f = \epsilon_Y \circ |\hat{f}|$, *i.e. the following diagram is commutative:*

$$
\begin{array}{ccc}
|X| & \xrightarrow{\ |\hat{f}|\ } & |C(\Delta, Y)| \\
& f \searrow \quad \swarrow \epsilon_Y & \\
& Y &
\end{array}
$$

Proof. Let $f : X \to Y$ be a continuous function and denote the functions $\tau_{X,Y}(n)$ by f_n. By composing the commutative Diagrams 9.3, 9.2.7, 9.2.12 and 9.2.15, one obtains the commutative diagram

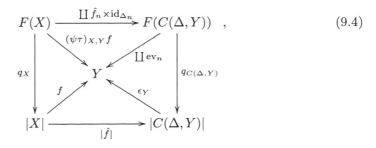

$$(9.4)$$

where the outer square is Diagram 9.2.12, the upper triangle is Diagram 9.2.15, the left triangle is Diagram 9.3 and the right triangle is Diagram 9.2.7. The bottom triangle now yields the desired equality $\epsilon_Y \circ |\hat{f}| = f$.   □

Summarising the previous considerations there exists for every continuous function $f : |X| \to Y$ a compatible family $\tau(f)$ denoted by $\{f_n\}$ inducing a function $\hat{f}$ such that the diagram in Lemma 9.2.16 commutes. Conversely, for each continuous function $g : |X| \to |C(\Delta, Y)|$, there exists a continuous function $f = \epsilon_Y \circ g$ such that $g = |\hat{f}|$ and the above diagram in Lemma 9.2.16 is commutative.

Proposition 9.2.17. *For any topological space Y, the morphism ϵ_Y is a terminal object in the comma category $|-| \downarrow Y$ of objects $|-|$-over Y.*

Proof. By Lemma 9.2.16 every continuous function $f : |X| \to Y$ from the geometric realisation of a simplicial space X to Y factors through $|C(\Delta, Y)|$ via $f = \epsilon_Y \circ |\hat{f}|$. It remains to show that $\hat{f}$ is the unique morphism of simplicial spaces such that $f = \epsilon_Y \circ |\hat{f}|$. Let $g : X \to C(\Delta, Y)$ be another morphism of simplicial spaces and assume $\epsilon_Y \circ |\hat{g}| = f$. A diagram chase in the commutative diagram

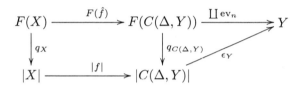

reveals that this implies the equalities

$$\left(\coprod \mathrm{ev}_n\right) \circ F(g) = \epsilon_Y \circ |g| \circ q_X = \epsilon_Y \circ |\hat{f}| \circ q_X = \left(\coprod \mathrm{ev}_n\right) \circ F(\hat{f}).$$

So for every point $(x, t) \in X([n]) \times \Delta_n$ the evaluations of $g_n(x)$ and $\hat{f}_n(x)$ at t coincide, forcing the functions $g_n(x)$ and $\hat{f}_n(x)$ to be equal. Therefore the morphisms $\hat{f}$ and g are equal. Thus the arrow $|\hat{f}| : f \to \epsilon_Y$ in the comma category $|-| \downarrow Y$ is unique.   □

Theorem 9.2.18. *The functor $C(\Delta, -)$ is right adjoint to the geometric realisation functor $|-|$. The counit of this adjunction is the natural transformation ϵ.*

Proof. The natural transformation ϵ is a terminal object in the comma category $|-| \downarrow Y$ and thus the counit of an adjunction from $\mathbf{Top}^{\mathcal{O}^{op}}$ to $\mathbf{Top}$, where the geometric realisation functor $|-|$ is left adjoint to the singular cosimplicial space functor $C(\Delta, -)$ (cf. [ML98, Ch. IV.1, Theorem 2]).   □

Corollary 9.2.19. *The singular simplicial space functor $C(\Delta, -)$ is continuous, i.e. it preserves limits.*

Corollary 9.2.20. *The geometric realisation functor $|-|$ is cocontinuous, i.e. it preserves colimits.*

Example 9.2.21. Consider the discrete simplicial space $\mathrm{hom}(-, [n])$ and a topological space Y. The natural transformation $\varphi_{-,-} \circ C(q_-, -)$ provides a continuous bijection $C(\Delta_n, Y) \to C(\mathrm{hom}(-, [n]), C(\Delta, Y))$.

Example 9.2.22. If G is a simplicial topological group which is a colimit of the simplicial groups G_i, then $|G| = \mathrm{colim}|G_i|$.

Actually, the long way to the proof of Theorem 9.2.18 provides us with a slightly more general result: The natural isomorphism of the adjunction between the geoemtric realisation functor $|-|$ and its right adjoint $C(\Delta, -)$ is the underlying transformation of the composition

$$C(|-|,-) \xrightarrow{C(q_-,-)} \mathrm{Cp}(-,-) \xrightarrow{\varphi_-,-} C(-,C(\Delta,-))$$

of natural transformations beween functors into **Top**. Thus the natural transformation $\varphi_{-,-} \circ C(q_-,-) : C(|-|,-) \to C(-,C(\Delta,-))$ consists of continuous bijections. A more careful examination reveals that this carries over to n-ads. For better readability we abbreviate an n-ad $(X_1,\ldots,X_n)$ of simplicial topological spaces by **X** and an n-ad $(Y_1,\ldots,Y_n)$ be an n-ad of topological spaces by **Y**. The functor

$$F : \mathbf{Top}^{\mathcal{O}^{op}} \to \mathbf{Top}, \quad F(X) = \coprod X([n]) \times \Delta_n$$

extends to a functor of n-ads, assigning to each n-ad **X** of simplicial topological spaces the n-ad $(F(X_1),\ldots,F(X_n))$ of topological spaces. The geometric realisation functor assigns to the n-ad **X** of simplicial spaces the n-tuple $(|X_1|,\ldots,|X_n|)$ of topological spaces. This might not be an n-ad, but if $j_2,\ldots,j_n$ denote the inclusions of the simplicial subspaces $X_2,\ldots,X_n$ into X_1 then $(|X_1|,|j_2|(|X_2|),\ldots,|j_n|(|X_n|))$ is an n-ad of topological spaces. If the inclusions $j_2,\ldots,j_n$ are closed, then the induced functions $|j_2|,\ldots,|j_n|$ are closed embeddings. In this case we omit the inclusions in the notation and write $(|X_1|,|X_2|,\ldots,|X_n|)$ for the resulting n-ad.

Example 9.2.23. The i-th face map $\epsilon_i : [n] \hookrightarrow [n+1]$ induces an inclusion $\hom(-,\epsilon_i)$ of the simplicial space $\hom(-,[n])$ into the simplicial space $\hom(-,[n+1])$. The geometric realisation of this inclusion embeds Δ_n as the i-th face of Δ_{n+1}.

Example 9.2.24. Consider the discrete simplicial space $\hom(-,[n+1])$ and its simplicial subspace $\mathrm{Bd}\,\hom(-,[n+1])$ generated by the subspaces $\hom(-,\epsilon_i)(\hom(-,[n]))$, $i = 0,\ldots,n+1$. The geometric realisation of the inclusion of $\mathrm{Bd}\,\hom(-,[n+1])$ into $\hom(-,[n+1])$ embeds $|\mathrm{Bd}\,\hom(-,[n])|$ as the topological boundary $\partial\Delta_{n+1}$ of the geometric realisiation Δ_{n+1} of $\hom(-,[n+1])$.

The several types of functions spaces occuring in the construction of the adjunction of $|-|$ and $C(\Delta,-)$ can be generalised to n-ads:

Definition 9.2.25. *For every n-ad **X** of simplicial spaces and each n-ad **Y** of topological spaces we denote by $C((|X_1|,|j_2|(|X_2|),\ldots,|j_n|(|X_n|)),\mathbf{Y})$ the subspace of $C(|X_1|,Y_1)$ consisting of all functions that map the subspace $j_i(|X_i|)$ of $|X_1|$ into Y_i for all $1 \leq i \leq n$.*

Definition 9.2.26. *For every n-ad **X** of simplicial spaces and each n-ad **Y** of topological spaces we denote by $\mathrm{Cp}(\mathbf{X},\mathbf{Y})$ the subspace of $\mathrm{Cp}(X_1,Y_1)$ consisting of the compatible families that map $F(X_i)$ into Y_i for all $1 \leq i \leq n$.*

We wish to compare these function spaces and examine how the natural transformations $C(q_-,-)$ and $\varphi_{-,-}$ can be adapted to them. At first we examine the restriction of $C(q_{X_1},Y_1)$ to functions of n-ads.

Lemma 9.2.27. *For each n-ad **X** of simplicial spaces and each n-ad **Y** of topological spaces the natural map $C(q_{X_1},Y_1) : C(|X_1|,Y_1) \to \mathrm{Cp}(X_1,Y_1)$ restricts to a continuous bijection $C((|X_1|,|j_2|(|X_2|),\ldots,|j_n|(|X_n|)),\mathbf{Y}) \to \mathrm{Cp}(\mathbf{X},\mathbf{Y})$.*

The natural transformation $\varphi_{-,-} : \mathrm{Cp}(-,-) \to C(-,C(\Delta,-))$ also restricts to the subspaces of mappings of n-ads:

Lemma 9.2.28. *The natural map* φ_{X_1,Y_1} *restricts to a continous bijection from* $Cp(\mathbf{X},\mathbf{Y})$ *onto* $C(\mathbf{X}, C(\Delta, \mathbf{Y}))$.

All in all one can summarise these observations in the follwing conclusion:

Theorem 9.2.29. *The natural transformation* $\varphi_{-,-} \circ C(q_-,-)$ *generalises to a natural transformation of n-ads. This generalisation also consists of continous bijections.*

Example 9.2.30. Consider the discrete simplicial topological space $\hom(-,[n+1])$ and its simplicial subspace $\mathrm{Bd}\,\hom(-,[n+1])$ generated by all the simplicial subspaces $\hom(-,\epsilon_i)(\hom(-,[n]))$, $i = 0, \dots, n+1$. If $(Y, *)$ is a based topological space, then the natural transformation $\varphi_{-,-} \circ C(q_-,-)$ provides a continous bijection

$$C((\Delta_{n+1}, \partial\Delta_{n+1}), (Y, *)) \rightarrow C((\hom(-,[n]), \mathrm{Bd}\,\hom(-,[n])), (C(\Delta, Y), *))$$

between the function spaces in $\mathbf{Top}(2)$ and $\mathbf{Top}^{\mathcal{O}^{op}}(2)$ respectively.

It has been shown so far, that the general geometric realisation functor $|-|$ on $\mathbf{Top}^{\mathcal{O}^{op}}$ is left adjoint to the singular simplicial space functor $C(\Delta, -)$ on $\mathbf{Top}$. Furthermore the counit $\epsilon : |C(\Delta, -)| \rightarrow \mathrm{id}_{\mathbf{Top}}$ of this adjunction is induced by evaluation and is explicitely given by

$$\epsilon_Y : |C(\Delta, Y)| \rightarrow Y, \quad [(\sigma, t)] \mapsto \sigma(t)$$

for each topological space Y. In addition to the counit ϵ there exists a natural transformation $i : \mathrm{id}_{\mathbf{Top}} \rightarrow |C(\Delta, -)|$. It is constructed as follows: The natural isomorphism $Y \cong C(\Delta_0, Y)$ assigns to each point $y \in Y$ the constant 0-simplex y in Y. It induces an embedding $Y \hookrightarrow F(C(\Delta, Y))$. Composition with the natural quotient map $q_{C(\Delta, y)} : F(C(\Delta, Y)) \rightarrow |-|$ yields a natural map $Y \rightarrow |C(\Delta, Y)|$ which we denote by i_Y.

Lemma 9.2.31. *The maps* $i_Y : Y \rightarrow |C(\Delta, Y)|$, $y \mapsto [(y, 1)]$ *form a natural transformation* $\mathrm{id}_{\mathbf{Top}} \rightarrow |C(\Delta, -)|$. *The map* i_Y *assigns to a point* y *of a topological space* Y *the equivalence class* $[(y, 1)] \in |C(\Delta, Y)|$ *of the constant 0-simplex* y *in* Y.

Proof. By construction, the maps i_Y are given by $i_Y(y) = q((y, 1)) = [(y, 1)]$, where the y in $(y, 1)$ denotes the constant 0-simplex y. Furthermore i is the composition of the natural isomorphisms $\mathrm{id}_{\mathbf{Top}} \cong C(\Delta_0, -)$, $i_0 : C(\Delta, -) \hookrightarrow F(C(\Delta, -))$ and $q : F(C(\Delta, -)) \rightarrow |C(\Delta, -))|$, so it is a natural transformation as well. $\square$

Lemma 9.2.32. *The natural transformation* i *is a right inverse of* ϵ, *i.e.* $\epsilon i = \mathrm{id}_{\mathbf{Top}}$ *and the transformations fit into the commutative diagram*

$$\mathrm{id}_{\mathbf{Top}} \xrightarrow{\quad i \quad} |C(\Delta, -)| \xrightarrow{\quad \epsilon \quad} \mathrm{id}_{\mathbf{Top}} \,.$$
$$\mathrm{id}$$

It will be shown in Chapter 10 that the counit ϵ actually consits of homotopy equivalences.

In the remaining part of this section this we show that $Cp(-, -)$ is the object part of a bifunctor $\mathbf{Top}^{\mathcal{O}^{op}} \times \mathbf{Top} \rightarrow \mathbf{Top}$. This is done using the fact that $Cp(X, Y)$ is a closed subspace of $\prod C(X([n]) \times \Delta_n, Y)$ for simplicial spaces X and topological spaces Y. Since the assignment

$$(X, Y) \mapsto \prod C(X([n]) \times \Delta_n, Y), \quad (f, g) \mapsto \prod C(f([n]) \times \mathrm{id}_{\Delta_n}, g)$$

is a bifunctor $\mathbf{Top}^{op} \times \mathbf{Top} \to \mathbf{Top}$, it suffices to show that $\mathrm{Cp}(-,-)$ is the object part of a subfunctor (that is to say a subobject in the functor category). To prove this, it has to be verified that the functions $\prod C(f([n]) \times \mathrm{id}_{\Delta_n}, g)$ above map the subspaces of compatible families into each other. We begin with studying the object part $\mathrm{Cp}(X, Y)$ for a fixed simplicial space X.

Lemma 9.2.33. *For every compatible family* $\{f_n\}_{n \in \mathbb{N}} \in \mathrm{Cp}(X, Y)$ *and continuous map* $g : Y \to Y'$, *the family of maps* $g \circ f_n : X([n]) \times \Delta_n \to Y'$ *is compatible.*

Proof. Let $f_n : X([n]) \times \Delta_n \to Y$ be an compatible family of maps and $g : Y \to Y'$ be continuous. For each morphism $\alpha : [m] \to [n]$ in $\mathcal{O}$ and points $x \in X([n])$ and $t \in \Delta_m$, the equality

$$(g \circ f_m)(X(\alpha)(x), t) = g(f_m(X(\alpha)(x), t)) = g(f_n(x, \Delta(\alpha)(t))) = (g \circ f_n)(x, \Delta(\alpha)(t))$$

is a consequence of the definition of compatibility of the family $\{f_n\}_{n \in \mathbb{N}}$. Thus the family $\{g \circ f_n\}_{n \in \mathbb{N}}$ is compatible. $\square$

So for any simplicial space X and any map $g : Y \to Y'$ between topological spaces Y and Y', the continuous function $\prod C(X([n]) \times \Delta_n, g)$ maps the subspace $\mathrm{Cp}(X, Y')$ into the subspace $\mathrm{Cp}(X, Y')$. The restriction and corestriction of this continuous function to the subspaces $\mathrm{Cp}(X, Y)$ and $\mathrm{Cp}(X, Y')$ is also continuous. It is part of the definition of a functor $\mathrm{Cp}(X, -)$.

Definition 9.2.34. *For every continuous function* $g : Y \to Y'$ *the continuous function* $\mathrm{Cp}(X, Y) \to \mathrm{Cp}(X, Y')$, $\{f_n\}_{n \in \mathbb{N}} \mapsto \{g \circ f_n\}_{n \in \mathbb{N}}$ *is denoted by* $\mathrm{Cp}(X, g)$.

Lemma 9.2.35. *For every simplicial space* X *the assignment* $Y \mapsto \mathrm{Cp}(X, Y)$ *and* $g \mapsto \mathrm{Cp}(X, g)$ *is a covariant functor* $\mathrm{Cp} : \mathbf{Top} \to \mathbf{Top}$.

Proof. Let X be a simplicial space and Y be a topological space. Consider the identity id_Y of Y. The map $\mathrm{Cp}(X, \mathrm{id}_Y)$ maps a compatible family $\{f_n\}_{n \in \mathbb{N}} \in \mathrm{Cp}(X, Y)$ to the compatible family

$$\mathrm{Cp}(X, \mathrm{id}_Y)(\{f_n\}_{n \in \mathbb{N}}) = \{\mathrm{id}_Y \circ f_n\}_{n \in \mathbb{N}} = \{f_n\}_{n \in \mathbb{N}}.$$

Thus the assignment $g \mapsto \mathrm{Cp}(X, g)$ maps the identity id_Y to the identity $\mathrm{id}_{\mathrm{Cp}(X, Y)}$. Furthermore, if $g : Y \to Y'$ and $h : Y' \to Y''$ are continuous functions, the evaluation of $\mathrm{Cp}(X, g \circ h)$ on a compatible family $\{f_n\}_{n \in \mathbb{N}}$ is given by

$$\begin{aligned}
\mathrm{Cp}(X, g \circ h)(\{f_n\}_{n \in \mathbb{N}}) &= \{(g \circ h) \circ f_n\}_{n \in \mathbb{N}} \\
&= \{g \circ (h \circ f_n)\}_{n \in \mathbb{N}} \\
&= [\mathrm{Cp}(X, g) \circ \mathrm{Cp}(X, h)](\{f_n\}_{n \in \mathbb{N}}).
\end{aligned}$$

So the assignment $g \mapsto \mathrm{Cp}(X, g)$ preserves concatenation of morphisms, hence $\mathrm{Cp}(X, -)$ is a covariant functor $\mathbf{Top} \to \mathbf{Top}$. $\square$

So far it has been proved that $\mathrm{Cp}(X, -)$ is a (covariant) functor $\mathbf{Top} \to \mathbf{Top}$. It remains to show that for each topological space the assignment $X \mapsto C(X, Y)$ is the object part of a contravariant functor $\mathbf{Top}^{\mathcal{O}^{op}} \to \mathbf{Top}$. This is done analogously to the first case.

Lemma 9.2.36. *For every compatible family* $f_n : X([n]) \times \Delta_n \to Y$ *and morphism* $h : X' \to X$ *of simplicial spaces the family* $f_n \circ (h_n \times \mathrm{id}_{\Delta_n}) : X'([n]) \times \Delta_n \to Y$ *is compatible.*

Proof. Let X be a simplicial space, Y a topological space, $\{f_n\}_{n\in\mathbb{N}} \in \mathrm{Cp}(X,Y)$ be a compatible family of maps and $h : X' \to X$ be a morphism of simplicial spaces. For every morphism $\alpha : [m] \to [n]$ in $\mathcal{O}$ and every pair $(X,t) \in X'([n]) \times \Delta(m)$ the evaluation of $f_m \circ (h_m \times \mathrm{id}_{\Delta_m})$ at $(X'(\alpha)(x),t)$ is given by

$$
\begin{aligned}
f_m \circ (h_m \times \mathrm{id}_{\Delta_m})(X'(\alpha)(x),t) &= f_m(h_m(X'(\alpha)(x)),t) \\
&= f_m(X(\alpha)(h_n(x)),t) \\
&= f_n(h_n(x),\Delta(\alpha)(t)) \\
&= f_n \circ (h_n \times \mathrm{id}_{\Delta_n})(x,\Delta(\alpha)(t)).
\end{aligned}
$$

Therefore the family of functions $\{f_n \circ (h_n \times \mathrm{id}_{\Delta_n})\}_{n\in\mathbb{N}}$ is compatible as well. □

So for any map $f : X \to X'$ between simplicial spaces X and X' and any topological space Y the continuous function $\prod C(f([n]) \times \Delta_n, \mathrm{id}_Y)$ maps the subspace $\mathrm{Cp}(X',Y)$ into the subspace $\mathrm{Cp}(X,Y)$. The restriction and corestriction of this continuous function to the subspaces $\mathrm{Cp}(X',Y)$ and $\mathrm{Cp}(X,Y)$ is also continuous and is part of the definition of a functor $\mathrm{Cp}(-,Y)$.

Definition 9.2.37. *For every morphism $h : X' \to X$ of simplicial spaces the continuous map $\mathrm{Cp}(X,Y) \to \mathrm{Cp}(X',Y)$, $\{f_n\}_{n\in\mathbb{N}} \mapsto \{f_n \circ (h_n \times \mathrm{id}_{\Delta_n})\}_{n\in\mathbb{N}}$ is denoted by $\mathrm{Cp}(h,Y)$.*

Proposition 9.2.38. *The assignment $(X,Y) \mapsto \mathrm{Cp}(X,Y)$, $(f,g) \mapsto \mathrm{Cp}(f,g)$ is a bifunctor $\mathbf{Top}^{\mathcal{O}^{op}} \times \mathbf{Top} \to \mathbf{Top}$, which is contravariant in the first and covariant in the second argument.*

Proof. It has already been observed that $\mathrm{Cp}(X,-) : \mathbf{Top} \to \mathbf{Top}$ is a covariant functor. It remains to show that the assignments $X \mapsto \mathrm{Cp}(X,Y)$, $f \mapsto \mathrm{Cp}(f,Y)$ constitute a contravariant functor for any topological space Y. Let X be a simplicial space, Y be a topological space and id_X the identity of X. The evaluation of $\mathrm{Cp}(f,Y)$ at a compatible family $\{f_n\}_{n\in\mathbb{N}} \in \mathrm{Cp}(X,Y)$ is given by

$$
\mathrm{Cp}(f,Y)(\{f_n\}_{n\in\mathbb{N}}) = \{f_n \circ (\mathrm{id}_{X([n])} \times \mathrm{id}_{\Delta_n})\}_{n\in\mathbb{N}} = \{f_n\}_{n\in\mathbb{N}}.
$$

Therefore $\mathrm{Cp}(\mathrm{id}_X,Y)$ is the identity map of $\mathrm{Cp}(X,Y)$. Let $g : X'' \to X'$ and $h : X' \to X$ be maps of simplicial spaces. The evaluation of $\mathrm{Cp}(g \circ h,Y)$ at at a compatible family $\{f_n\}_{n\in\mathbb{N}} \in \mathrm{Cp}(X,Y)$ is given by

$$
\begin{aligned}
\mathrm{Cp}(g \circ h,Y)(\{f_n\}_{n\in\mathbb{N}}) &= \{f_n \circ ((g \circ h)_n \times \mathrm{id}_{\Delta_n})\}_{n\in\mathbb{N}} \\
&= \{f_n \circ ((g_n \circ h_n) \times \mathrm{id}_{\Delta_n})\}_{n\in\mathbb{N}} \\
&= \{f_n \circ (g_n \times \mathrm{id}_{\Delta_n}) \circ (h_n \times \mathrm{id}_{\Delta_n})\}_{n\in\mathbb{N}} \\
&= [\mathrm{Cp}(h,Y) \circ \mathrm{Cp}(g,Y)](\{f_n\}_{n\in\mathbb{N}}).
\end{aligned}
$$

Thus Cp preserves the concatenation of morphisms reversing their order and therefore is a contravariant functor. □

Now we explicitly show that τ is a natural transformation by explicitly computing how τ interchanges with simplicial maps $f : X \to X'$ and functions $g : Y \to Y'$, as advertised before.

Lemma 9.2.39. *For every topological space Y the collection of maps $\tau_{-,Y}$ is a natural transformation $\tau : C(|-|,Y) \to \mathrm{Cp}(-,Y)$.*

Proof. Let $h : X' \to X$ be a morphism of simplicial spaces and $\{f_n\}_{n\in\mathbb{N}} \in \mathrm{Cp}(X,Y)$ be a compatible family of maps. This compatible family corresponds to a unique map $f : |X| \to Y$ such that $f_n = f \circ q_X \circ i_n$ for all $n \in \mathbb{N}$. From the commutativity of the diagram

$$
\begin{array}{ccccccc}
X'([n]) \times \Delta_n & \xrightarrow{\ i_n\ } & \coprod X'([n]) \times \Delta_n & \xrightarrow{\ q_{X'}\ } & |X'| & & \\
\Big\downarrow{\scriptstyle h_n \times \mathrm{id}_{\Delta_n}} & & \Big\downarrow{\scriptstyle \coprod h_n \times \mathrm{id}_{\Delta_n}} & & \Big\downarrow{\scriptstyle |h|} & \searrow^{C(|h|,Y)(f)} & \\
X([n]) \times \Delta_n & \xrightarrow{\ i_n\ } & \coprod X([n]) \times \Delta_n & \xrightarrow{\ q_X\ } & |X| & \xrightarrow{\ f\ } & Y
\end{array}
$$

$$ f_n $$

we infer the equality $C(|h|,Y)(f) \circ q_{X'} \circ i_n = f \circ q_X \circ i_n \circ (h_n \times \mathrm{id}_{\Delta_n})$. Since this is true for all $n \in \mathbb{N}$ we conclude

$$
\begin{aligned}
\tau_{X',Y}(C(|h|,Y))(f) &= \{f \circ q_X \circ i_n \circ (h_n \times \mathrm{id}_{\Delta_n})\}_{n \in \mathbb{N}} \\
&= \{f \circ q_X \circ i_n\}_{n \in \mathbb{N}} \\
&= \mathrm{Cp}(h,Y)(\tau_{X,Y}(f)).
\end{aligned}
$$

Therefore the following diagram commutes:

$$
\begin{array}{ccc}
C(|X|,Y) & \xrightarrow{\ \tau_{X,Y}\ } & \mathrm{Cp}(X,Y) \\
\Big\downarrow{\scriptstyle C(|h|,Y)} & & \Big\downarrow{\scriptstyle \mathrm{Cp}(h,Y)} \\
C(|X'|,Y) & \xrightarrow{\ \tau_{X',Y}\ } & \mathrm{Cp}(X',Y)
\end{array}
$$

Thus the collection $\tau_{-,Y}$ is a natural transformation $C(|-|,Y) \to \mathrm{Cp}(-,Y)$. $\square$

Lemma 9.2.40. *For every simplicial space X the collection of maps $\tau_{X,-}$ is a natural transformation $\tau : C(|X|, \ ) \to \mathrm{Cp}(X, \ )$.*

Proof. Let X be a simplicial space, Y be a topological space and $f : |X| \to Y$ be a continuous function. The function f is mapped to a compatible family $\{f_n\}_{n \in \mathbb{N}}$ under $\tau_{X,Y}$. Applying the definition of $\mathrm{Cp}(X,g)$ the compatible family $\tau_{X,Y} \circ C(X,g)(f)$ computes to

$$
\begin{aligned}
[\tau_{X,Y'} \circ C(X,g)](f) &= \tau_{X,Y}(g \circ f) \\
&= \{g \circ f \circ q_X \circ i_n\}_{n \in \mathbb{N}} \\
&= \{g \circ (f \circ q_X \circ i_n)\}_{n \in \mathbb{N}} \\
&= \{g \circ f_n\}_{n \in \mathbb{N}} \\
&= \mathrm{Cp}(X,g)(\{f_n\}_{n \in \mathbb{N}}) = \mathrm{Cp}(X,g)(\tau_{X,Y}(f))
\end{aligned}
$$

Thus $\tau_{X,Y'} \circ C(X,g) = \mathrm{Cp}(X,g) \circ \tau_{X,Y}$, i.e. $\tau_{X,\cdot}$ is a natural transformation $C(|X|,-) \to \mathrm{Cp}(X,-)$. $\square$

9.3 Preservation of Finite Limits

In this section we examine the behaviour of the general geometric realisation functor $|-| : \mathbf{Top}^{\mathcal{O}^{op}} \to \mathbf{Top}$ applied to finite limits. As will be shown below, geometric realisation does preserve arbitrary equalisers. Concerning products, it turns out (in Theorem 9.3.91), not to preserve products in general, but when restricted to certain categories of simplicial spaces it does. So the restriction of the geometric realisation to these categories preserves finite limits. (The proof uses the fact that finite limits can canonically be constructed by means of finite equalisers and finite products. Therefore the preservation of finite equalisers and finite products implies the preservation of all finite limits.)

Proposition 9.3.1. *The geometric realisation functor preserves equalisers.*

Proof. Let J be a family $f_i : X \to Y$ of morphisms of simplicial spaces. Since all limits in the category of simplicial spaces are are obtained by computing the limits point-wise (cf. [ML98, Chapter V.3, Theorem 1]), the equaliser $\mathrm{Eq}(J)$ of this family is the sub-simplicial space

$$\mathrm{Eq}(J)([n]) = \mathrm{Eq}(J) \subset X([n])$$

of X with all morphisms the morphisms of X restricted and corestricted to the subspaces $\mathrm{Eq}(f_i([n])) \subset X([n])$. The inclusion i of the sub-simplicial space $\mathrm{Eq}(J)$ of X in X induces an injective map $|i| : |\mathrm{Eq}(J)| \to |X|$ which fits with each induced map $|f_i| : |X| \to |Y|$ into a commutative diagram:

$$
\begin{array}{ccccc}
F(\mathrm{Eq}(J)) & \xrightarrow{F(i)} & F(X) & \xrightarrow{F(f_i)} & F(Y) \\
\downarrow{\scriptstyle q_{\mathrm{Eq}(J)}} & & \downarrow{\scriptstyle q_X} & & \downarrow{\scriptstyle q_Y} \\
|\mathrm{Eq}(J)| & \xrightarrow{|i|} & |X| & \xrightarrow{|f_i|} & |Y|
\end{array}
$$

Therefore the geometric realisation $|\mathrm{Eq}(J)|$ maps into the subspace $\mathrm{Eq}(|J|)$ of X. It remains to show that every point in the equaliser $\mathrm{Eq}(|f_i|)$ is contained in the image of $|i|$. This is a purely set theoretic argument which already has been carried in the case of discrete simplicial spaces (see [GZ67, Chapter III.3.3]). This proof is also valid in the general case, thus the geometric realisation functor preserves equalisers.   □

The remaining part of this section is devoted to the proof of the preservation of finite products. Let X and Y be simplicial spaces. The projections pr_1 and pr_2 of $X \times Y$ onto the factors X and Y induce functions $|\mathrm{pr}_1| : |X \times Y| \to |X|$ and $|\mathrm{pr}_2| : |X \times Y| \to |Y|$ which assemble to a natural continuous function

$$\Phi_{X,Y} = (|\mathrm{pr}_1| \times |\mathrm{pr}_2|) \circ D : |X \times Y| \to |X| \times |Y|,$$

where $D : |X \times Y| \to |X \times Y| \times |X \times Y|$ is the diagonal map. It is known that this map is a homeomorphism if the spaces X and Y are discrete and countable (see [May92, Chapter III, Theorem 14.3]). Since all arguments -except for the continuity- in the proof thereof are purely set theoretical, the proof that $\Phi_{X,Y}$ is a bijection carries over word by word. So in order to prove that the geometric realisation functor preserves finite products, it suffices to show that the inverse map $\Phi_{X,Y}^{-1}$ is continuous.

To illustrate the nature of this inverse map $\Phi_{X,Y}^{-1}$ we first examine the case of standard simplicial spaces. Let X and Y be topological spaces and consider the continuous bijection $\Phi : |\mathrm{SS}(X) \times \mathrm{SS}(Y)| \to |\mathrm{SS}(X)| \times |\mathrm{SS}(Y)|$ induced by the projections. A point $[(x, t)]$ of the geometric realisation $|X|$ can be identified with the step function

$$\gamma_{[(\mathbf{x},t)]} : [0,1) \to X, \quad \gamma_{[(x,t)]}(s) := \begin{cases} x_0 & \text{if} & s < t_0 \\ x_1 & \text{if} & t_0 \leq s < t_0 + t_1 \\ \vdots & \vdots & \vdots \\ x_n & \text{if} & t_0 + \cdots + t_{n-1} \leq s < 1 \end{cases}$$

(cf. [BM78, Remark p. 217]). Note that the step function $\gamma_{[(\mathbf{x},t)]}$ is independent of the representative of $(\mathbf{x}, t)$ chosen. Analogously the points of $|Y|$ can be identified with step functions $[0,1) \to Y$. The function Φ^{-1} maps $([(\mathbf{x}, t)], [(\mathbf{y}, t')])$ to the point that corresponds to the step-function $(\gamma_{[(\mathbf{x},t)]}, \gamma_{[(\mathbf{y},t')]}) : [0,1) \to X \times Y$.

Example 9.3.2. Let $(\mathbf{x}, \mathbf{t}) \in X^2 \times \Delta_1$ and $(\mathbf{y}, \mathbf{t}') \in Y^2 \times \Delta_1$ be two points in $F(\mathrm{SS}(X))$ and $F(\mathrm{SS}(Y))$ respectively. We distinguish the different cases $t_0 > t_0'$, $t_0 < t_0'$ and $t_0 = t_0'$. In the first case the step-function corresponding to $\Phi^{-1}([(\mathbf{x}, \mathbf{t})], [(\mathbf{y}, \mathbf{t}')])$ is given by

$$\gamma_{\Phi^{-1}([(\mathbf{x},\mathbf{t})],[(\mathbf{y},\mathbf{t})])}(s) = \begin{cases} (x_0, y_0) & \text{if} \quad\quad s < t_0' \\ (x_0, y_1) & \text{if} \quad t_0' \leq s < t_0 \\ (x_1, y_1) & \text{if} \quad t_0 \leq s < 1 \end{cases}.$$

The image of $([(\mathbf{x}, \mathbf{t})], [(\mathbf{y}, \mathbf{t}')])$ under the map Φ^{-1} is the point in $|\mathrm{SS}(X) \times \mathrm{SS}(Y)|$ which corresponds to this step-function, i.e.

$$\Phi^{-1}([(\mathbf{x}, \mathbf{t})], [(\mathbf{y}, \mathbf{t}')]) = [((x_0, y_0), (x_0, y_1), (x_1, y_1)), (t_0', t_0 - t_0', t_1)].$$

In the second case the roles of $\mathbf{t}$ and $\mathbf{t}'$ are interchanged, so the step function is given by

$$\gamma_{\Phi^{-1}([(\mathbf{x},\mathbf{t})],[(\mathbf{y},\mathbf{t}')])}(s) = \begin{cases} (x_0, y_0) & \text{if} \quad\quad s < t_0 \\ (x_1, y_0) & \text{if} \quad t_0 \leq s < t_0' \\ (x_1, y_1) & \text{if} \quad t_0' \leq s < 1 \end{cases}.$$

Here the the point in $|\mathrm{SS}(X) \times \mathrm{SS}(Y)|$ that corresponds to this step function is the point

$$\Phi^{-1}([(\mathbf{x}, \mathbf{t})], [(\mathbf{y}, \mathbf{t}')]) = [((x_0, y_0), (x_1, y_0), (x_1, y_1)), (t_0, t_0' - t_0, t_1')].$$

In the last case the point $\mathbf{t}$ and $\mathbf{t}'$ are equal, so the step function is the function

$$\gamma_{\Phi^{-1}([(\mathbf{x},\mathbf{t})],[(\mathbf{y},\mathbf{t}')])}(s) = \begin{cases} (x_0, y_0) & \text{if} \quad\quad s < t_0 \\ (x_1, y_1) & \text{if} \quad t_0 \leq s < 1 \end{cases}.$$

In this case the image of $([(\mathbf{x}, \mathbf{t})], [(\mathbf{y}, \mathbf{t}')])$ under Φ^{-1} is the point

$$\Phi^{-1}([(\mathbf{x}, \mathbf{t})], [(\mathbf{y}, \mathbf{t}')]) = [((x_0, y_0), (x_1, y_1)), (t_0, t_1')].$$

Note that this point can also be written as $[((x_0, y_0), (x_0, y_1), (x_1, y_1)), (t_0, 0, t_1')]$ or $[((x_0, y_0), (x_1, y_0), (x_1, y_1)), (t_0, 0, t_1')]$. The distinction of these cases corresponds to a triangulation of the space $\Delta_1 \times \Delta_1$ by affine simplices as depicted below:

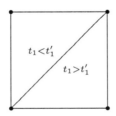

The the upper left triangle is the image of an affine simplex τ_0 and the lower right triangle is the image of an affine simplex τ_1 which are defined to be

$$\tau_0 = < (\mathbf{e}_0, \mathbf{e}_0), (\mathbf{e}_0, \mathbf{e}_1), (\mathbf{e}_1, \mathbf{e}_1) >: \Delta_2 \to \Delta_1 \times \Delta_1$$
$$\tau_0(\mathbf{t}) = (t_0 + t_1)\mathbf{e}_0 + t_2\mathbf{e}_1, t_0\mathbf{e}_0 + (t_1 + t_2)\mathbf{e}_1$$
$$\tau_1 = < (\mathbf{e}_0, \mathbf{e}_0), (\mathbf{e}_1, \mathbf{e}_0), (\mathbf{e}_1, \mathbf{e}_1) >: \Delta_2 \to \Delta_1 \times \Delta_1$$
$$\tau_1(\mathbf{t}) = t_0\mathbf{e}_0 + (t_1 + t_2)\mathbf{e}_1, (t_0 + t_1)\mathbf{e}_0 + t_2\mathbf{e}_1.$$

The step-function in the last of the three cases is the point wise limit from either side of the diagonal. Note, also, that a change of the index of the vertices $\mathbf{e}_i, \mathbf{e}_j$ corresponds to a change of the index of the x_i's and y_j's in the above defined step-functions. So the inverse function Φ^{-1} can be reconstructed on each triangle using the affine simplices τ_0 and τ_1. Observe that these affine simplices can also be written as

$$\tau_0 = [\Delta(\eta_0) \times \Delta(\eta_1)] \circ D : \Delta_2 \to \Delta_1 \times \Delta_1 \quad \text{and}$$
$$\tau_1 = [\Delta(\eta_1) \times \Delta(\eta_0)] \circ D : \Delta_2 \to \Delta_1 \times \Delta_1,$$

where $D : \Delta_2 \to \Delta_2 \times \Delta_2$ denotes the diagonal map.

Now let X and Y be arbitrary simplicial spaces. As the above example suggests, the inclusions $X([p]) \times Y([q]) \times \Delta_p \times \Delta_q \cong X([p]) \times \Delta_p \times Y([q]) \times \Delta_q \hookrightarrow F(X) \times F(Y)$ followed by the functions $(q_X \times q_Y)$ and $\Phi_{X,Y}^{-1}$

$$
\begin{array}{ccc}
(X([p]) \times Y([q])) \times (\Delta_p \times \Delta_q) & \longrightarrow & F(X) \times F(Y) \\
& & \downarrow{\scriptstyle q_X \times q_Y} \\
& & |X| \times |Y| \xrightarrow[\Phi_{X,Y}^{-1}]{} |X \times Y|
\end{array}
\qquad ,
$$

can be explicitly constructed in a purely combinatorial way. This is carried out below. Here we use the fact that there exists a canonical triangulation of the space $\Delta_p \times \Delta_q$ by affine $(p+q)$-simplices

$$< (\mathbf{e}_{i_0}, \mathbf{e}_{j_0}), (\mathbf{e}_{i_1}, \mathbf{e}_{j_1}), \ldots, (\mathbf{e}_{i_{p+q}}, \mathbf{e}_{j_{p+q}}) >,$$

where either $i_{k+1} = i_k$ and $j_{k+1} = j_k + 1$ or $i_{k+1} = i_k + 1$ and $j_{k+1} = j_k$ for all $0 \leq k < p+q$ (see [EML53, p. 63]). Any such affine simplex can be described by the subset $I = \{k < n \mid i_k = i_{k+1}\}$. Using this index set, the above affine simplex, now denoted by by τ_I, can be expressed in terms of face and degeneracy maps:

$$\tau_I = [\Delta(\eta_{i_0}, \cdots \eta_{i_{q-1}}) \times \Delta(\eta_{j_0}, \cdots \eta_{j_{p-1}})] \circ D,$$

where $J = \{j_0 < \cdots < j_{p-1}\}$ is the complement of I in $[n-1]$ and $D : \Delta_n \to \Delta_n \times \Delta_n$ is the diagonal map. The triangulation of $\Delta_p \times \Delta_q$ by the affine simplices τ_I is used below to prove that the inverse map

$$\Phi_{X,Y}^{-1} : |X| \times |Y| \to |X \times Y|$$

is continuous for special classes of simplicial spaces X and Y. This is done by constructing continuous functions $\varphi_{n,I}$ on the spaces

$$X([p]) \times Y([q]) \times \tau_I(\Delta_n) \subset X([p]) \times Y([q]) \times \Delta_p \times \Delta_q \cong X([p]) \times \Delta_p \times Y([q]) \times \Delta_y$$

that have codomain in $F(X \times Y)$. This induces functions $q_{X \times Y} \circ \varphi_{n,I}$ into the geometric realisation $|X \times Y|$. The functions $q_{X \times Y} \circ \varphi_{n,I}$ will then be shown to coincide on the intersections of their domains, thus giving rise to continuous functions

$$\varphi_{p,q} : X([p] \times \Delta_p \times Y([q]) \times \Delta_q \to |X \times Y|$$

whose coproduct is then a continuous function $\coprod \varphi_{p,q} : F(X) \times F(Y) \to |X \times Y|$. This coproduct will then be shown to factor through the product $|X| \times |Y|$, where the induced function $\varphi : |X| \times |Y|$ is then identified as the inverse $\Phi_{X,Y}^{-1}$ of $\Phi_{X,Y}$.

Motivated by the above example and the canonical triangulation of $\Delta_p \times \Delta_q$ we begin by defining:

Definition 9.3.3. *For every subset* $I = \{i_0 < \cdots < i_{q-1}\}$ *of* $[n-1]$, *the* I-*degeneracy on* $[n]$ *is the epimorphism*

$$\eta_{n,I} : [n] \to [n-q], \quad \eta_I = \eta_{i_1} \cdots \eta_{i_{q-1}}.$$

Definition 9.3.4. *For every subset I of $[n-1]$, the affine n-simplex of type I is the simplex*

$$\tau_I := [\Delta(\eta_I) \times \Delta(\eta_{[n-1]\setminus I})] \circ D : \Delta_n \to \Delta_p \times \Delta_q,$$

where $D : \Delta_n \to \Delta_n \times \Delta_n$ is the diagonal map.

Example 9.3.5. The two affine simplices in example 9.3.2 are the simplices $\tau_{2,\{0\}}$ and $\tau_{2,\{1\}}$, whose image is the upper left triangle and the lower right triangle respectively.

The finite triangulation of the product space $\Delta_p \times \Delta_q$ induces a finite decomposition of the space $X([p]) \times Y([q]) \times \Delta_p \times \Delta_q$ into the closed subspaces $X([p]) \times Y([q]) \times \tau_I(\Delta_n)$ where I ranges over all q-element subsets of $[p+q-1]$. We are going to define continuous functions

$$\varphi_{n,I} : X([p]) \times Y([q]) \times \tau_I(\Delta_n) \to |X \times Y|$$

on these subspaces that will then assemble to a continuous map on the whole space $X([p]) \times Y([q]) \times \Delta_p \times \Delta_q$. Permuting the factors in this product space is a homeomorphism. So we will obtain continuous functions

$$\varphi_{p,q} : X([p]) \times \Delta_p \times Y([q]) \times \Delta_q \cong X([p]) \times Y([q]) \times \Delta_p \times \Delta_q \to |X \times Y|$$

The coproduct of these functions is a continuous map $F(X) \times F(Y) \to |X \times Y|$ which will be shown to factor through the geometric realisations $|X| \times |Y|$ to give the map $\Phi_{X,Y}^{-1}$.

Definition 9.3.6. *For $p+q = n$ and $I = \{i_0 < \ldots < i_{q-1}\} \subset [n-1]$ the I-function on the space $X([p]) \times Y([q]) \times \tau_I(\Delta_n)$ is the continuous function*

$$\varphi_{n,I} := X(\eta_I) \times Y(\eta_{[n-1]\setminus I}) \times \tau_I^{-1} : X([p]) \times Y([q]) \times \tau_{n,I}(\Delta_n) \to X([n]) \times Y([n]) \times \Delta_n.$$

Example 9.3.7. The simplices $\tau_{2,\{0\}}$ and $\tau_{2,\{1\}}$ in example 9.3.2 which triangulate the space $\Delta_1 \times \Delta_1$ have the inverse maps

$$\tau_{2,\{0\}}^{-1}(\mathbf{t}, \mathbf{t}') = (t_0', t_0 - t_0', t_1)$$
$$\tau_{2,\{1\}}^{-1}(\mathbf{t}, \mathbf{t}') = (t_0, t_0' - t_0, t_1').$$

The functions $\varphi_{2,\{0\}}$ and $\varphi_{2,\{1\}}$ for this example are therefore given by

$$\varphi_{2,\{0\}}(\mathbf{x}, \mathbf{y}, \mathbf{t}, \mathbf{t}') = ((x_0, y_0), (x_0, y_1), (x_1, y_1), (t_0', t_0 - t_0', t_1)) \quad \text{and}$$
$$\varphi_{2,\{1\}}(\mathbf{x}, \mathbf{y}, \mathbf{t}, \mathbf{t}') = ((x_0, y_0), (x_1, y_0), (x_1, y_1), (t_0, t_0' - t_0, t_1')).$$

These are the points previously obtained by describing the map Φ^{-1} via step functions.

In order to prove that the functions $q_{X \times Y} \varphi_{n,I}$ into $|X \times Y|$ assemble to a continuous function, one needs to ensure that these functions coincide on the intersections of their domains. The following lemma is needed for this.

Lemma 9.3.8. *For two $p+q = n$ and affine simplices τ_I and $\tau_{I'}$ in $\Delta_p \times \Delta_q$, the intersection of their images is the image of a common face, i.e. there exist monomorphisms $\epsilon, \epsilon' : [m] \hookrightarrow [n]$ such that $\tau_I \Delta(\epsilon) = \tau_{I'} \Delta(\epsilon')$ and $\tau_I(\Delta_n) \cap \tau_{I'}(\Delta_n) = \tau_I \Delta(\epsilon)(\Delta_m) = \tau_{I'} \Delta(\epsilon')(\Delta_m)$.*

Proof. Let $p, q \in \mathbb{N}$ be natural numbers, $n = p+q$ and I and I' two q-element subsets of $[n-1]$. Because the affine simplices τ_I and $\tau_{I'}$ both have zeroth vertex $(\mathbf{e}_0, \mathbf{e}_0)$, the intersection of their images is never empty. The affine simplices τ_J, where J ranges over all q-element subsets of $[n-1]$, form a triangulation of $\Delta_p \times \Delta_q$. therefore the intersection $\tau_I(\Delta_n) \cap \tau_{I'}(\Delta_n)$ is the image of a common face of τ_I and $\tau_{I'}$. That is to say, there exists $m \leq n$ and monomorphisms $\epsilon, \epsilon' : [m] \hookrightarrow [n]$ such that

$$\tau_I \Delta(\epsilon) = \tau_{I'} \Delta(\epsilon') \quad \text{and} \quad \tau_I(\Delta_n) \cap \tau_{I'}(\Delta_n) = \tau_I \Delta(\epsilon)(\Delta_m) = \tau_{I'} \Delta(\epsilon')(\Delta_m).$$

This is what was to be proved. $\square$

Proposition 9.3.9. *The functions* $q_{X \times Y} \circ \varphi_{n,I} : X([p]) \times Y([q]) \times \Delta_n \to |X \times Y|$ *coincide on the intersections of their domains.*

Proof. Let X and Y be simplicial spaces, $p, q \in \mathbb{N}$ be given natural numbers and $n = p + q$. For any two q-element subsets I and I' of $[n-1]$ there exists $m \leq n$ and monomorphisms $\epsilon, \epsilon' : [m] \hookrightarrow [n]$ satisfying $\tau_I \Delta(\epsilon) = \tau_{I'} \Delta(\epsilon')$ and

$$\tau_I(\Delta_n) \cap \tau_{I'}(\Delta_n) = \tau_I \Delta(\epsilon)(\Delta_m) = \tau_{I'} \Delta(\epsilon')(\Delta_m).$$

In order to prove that the functions $q_{X \times Y} \circ \varphi_{n,I}$ and $q_{X \times Y} \circ \varphi_{n,I'}$ coincide on the intersection of their domains, it suffices to show that

$$\varphi_{p,q,I}(x, y, \mathbf{t}, \mathbf{t}') \sim \varphi_{p,q,I'}(x, y, \mathbf{t}, \mathbf{t}')$$

for all $x \in X([p])$, $y \in Y([q])$ and all pairs $(\mathbf{t}, \mathbf{t}')$ in $\tau_I(\Delta_n) \cap \tau_{I'}(\Delta_n)$. Let $(\mathbf{t}, \mathbf{t}')$ be such a pair. Then there exists $\hat{\mathbf{t}} \in \Delta_m$ such that $\tau_I \Delta(\epsilon)(\hat{\mathbf{t}}) = (\mathbf{t}, \mathbf{t}') = \tau_{I'} \Delta(\epsilon')(\hat{\mathbf{t}})$ The image of a point $(x, y, \mathbf{t}, \mathbf{t}')$ in $F(X \times Y)$ under $\varphi_{p,q,I}$ is by definition given by

$$\begin{aligned}
\varphi_{p,q,I}(x, y, \mathbf{t}, \mathbf{t}') &= (X(\eta_I)(x), Y(\eta_{[n-1]\setminus I})(y), \tau_I^{-1}(\mathbf{t}, \mathbf{t}')) \\
&- (X(\eta_I)(x), Y(\eta_{[n-1]\setminus I})(y), \Delta(\epsilon)(\hat{\mathbf{t}})) \\
&\sim (X(\eta_I \epsilon)(x), Y(\eta_{[n-1]\setminus I} \epsilon)(y), \hat{\mathbf{t}}) \\
&= (X(\eta_{I'} \epsilon')(x), Y(\eta_{[n-1]\setminus I'} \epsilon')(y), \hat{\mathbf{t}}) \\
&\sim (X(\eta_{I'})(x), Y(\eta_{[n-1]\setminus I'})(y), \Delta(\epsilon')\hat{\mathbf{t}}) \\
&= (X(\eta_{I'})(x), Y(\eta_{[n-1]\setminus I'})(y), \tau_{I'}^{-1}(\mathbf{t}, \mathbf{t}')) \\
&= \varphi_{p,q,I'}(x, y, \mathbf{t}, \mathbf{t}')
\end{aligned}$$

Therefore the images of the point $(x, y, \mathbf{t}, \mathbf{t}')$ under $\varphi_{n,I}$ and $\varphi_{n,I'}$ are equivalent in $F(X \times Y)$, hence the images of $(x, y, \mathbf{t}, \mathbf{t}')$ under $q_{X \times Y} \circ \varphi_{n,I}$ and $q_{X \times Y} \circ \varphi_{n,I'}$ coincide. This is true for all q-element subsets I and I' and all points $(x, y, \mathbf{t}, \mathbf{t}')$ in the intersection of the domains of $\varphi_{n,I}$ and $\varphi_{n,I'}$. Thus all the functions $q_{X \times Y} \circ \varphi_{n,I}$ for different q-element subsets I of $[n-1]$ coincide on the intersections of their domains. $\square$

Corollary 9.3.10. *The I-functions on* $X([p]) \times Y([q]) \times \Delta_p \times \Delta_q$ *assemble to a continuous map into* $|X \times Y|$:

$$X([p]) \times \Delta_p \times Y([q]) \times \Delta_q \xrightarrow{\cong} X([p]) \times Y([q]) \times \Delta_p \times \Delta_q$$

with diagonal arrow to $|X \times Y|$ and vertical arrow $\cup \varphi_{n,I}$ to $|X \times Y|$.

Definition 9.3.11. *The unique function on* $X([p]) \times \Lambda_p \times Y([q]) \times \Delta_q$ *that restricts to* $\varphi_{n,I} \circ i_{p,q}^{-1}$ *for* $n = p + q$ *and each subset* $I = \{i_0 < \cdots < i_{q-1}\}$ *of* $[n-1]$ *is denoted by* $\varphi_{p,q}$.

The continuous function $\coprod \varphi_{p,q}$ factors through the product $|X| \times |Y|$ exactly if the the families of functions $\varphi_{p,q}(-,(\mathbf{y},\mathbf{t}'))$ and $\varphi_{p,q}((\mathbf{x},\mathbf{t}),-)$ are compatible for fixed $(\mathbf{x},\mathbf{t}) \in F(X)$ and $(\mathbf{y},\mathbf{t}') \in F(Y)$. The verification of this property consists of a few technical computations. In order to make these computations better readable, the numbers p, q and $n = p+q$ as well as the q-element subset $I = \{i_0 < \cdots < i_{q-1}\}$ of $[n-1]$ are fixed in the following. They will be used without further reference.

Lemma 9.3.12. *For every face map $\epsilon_k : [p] \hookrightarrow [p+1]$ there exists an $0 \leq l \leq q-1$ such that $i_{l-1} < k+l$ if $l \geq 1$, $k+l \leq i_l$ and the equation*

$$\epsilon_k \eta_I = \eta_{i_0} \cdots \eta_{l-1} \eta_{i_l+1} \cdots \eta_{i_{q-1}+1}\epsilon_{k'} = \eta_{I'}\epsilon_{k'}$$

is satisfied for $k' = k+l$ and $I' = \{i_0 < \cdots < i_{l-1} < i_l + 1 < \cdots < i_{q-1} + 1\}$.

Proof. Let $\epsilon_k : [p] \hookrightarrow [p+1]$ be the k-th face map. The expression $\epsilon_k \eta_I$ can be rewritten using the simplicial identities F.1

$$\eta_j \epsilon_k = \epsilon_k \eta_{j-1} \quad \text{if } k < j.$$

If $k < i_0$ then these identities imply that the face and degeneracy maps can be exchanged by at the same time raising the index of the degeneracy maps. So one obtains the equality

$$
\begin{aligned}
\epsilon_k \eta_I &= \epsilon_k \eta_{i_0} \cdots \eta_{i_{q-1}} \\
&= \eta_{i_0+1}\epsilon_k \eta_{i_1} \cdots \eta_{i_{q-1}} \\
&= \eta_{i_0+1} \cdots \eta_{i_{q-1}+1}\epsilon_k \\
&= \eta_{I'}\epsilon_k,
\end{aligned}
$$

where $I' = \{i_0 + 1, \ldots, i_{q-1} + 1\}$. In this case $l = 0$ has the required properties. If $k \geq i_0$ the above simplicial identities imply that the expression $\epsilon_k \eta_I$ can be rewritten as

$$\epsilon_k \eta_I = \epsilon_k \eta_{i_0} \cdots \eta_{q-1} = \eta_{i_0}\epsilon_{k+1}\eta_{i_1} \cdots \eta_{q-1},$$

where the index of the monomorphism has been raised by one. Note that the assumption $i_0 \leq k$ especially implies the inequality $i_0 < k+1$. After repeating this process a maximum number of times one has shown the equality

$$\epsilon_k \eta_I = \eta_{i_0} \cdots \eta_{i_{l-1}}\epsilon_{k'}\eta_{i_l} \cdots \eta_{i_{q-1}}$$

with $k' = k+l$ and $i_{l-1} < k' < i_l$. Again using the simplicial identities above, one can exchange the last epimorphism η_l etc. with $\epsilon_{k'}$ by at the same time raising the index of the epimorphisms η_i to finally arrive at the equality

$$\epsilon_k \eta_I = \eta_{i_0} \cdots \eta_{i_{l-1}} \eta_{i_l+1} \cdots \eta_{i_{q-1}+1}\epsilon_{k'} = \eta_{I'}\epsilon_{k'}$$

where I' is the ordered subset $\{i_0 < \cdots < i_{l-1} < i_l + 1 < \cdots < i_{q-1} + 1\}$ of $[n]$. The condition $i_{l-1} < k' < i_l + 1$ is here satisfied by construction. $\square$

Lemma 9.3.13. *For every face map $\epsilon_k : [p] \hookrightarrow [p+1]$ there exists a q-element subset I' of $[n]$ and a face map $\epsilon_{k'} : [n] \to [n+1]$ such that $\eta_k \eta_I = \eta_{I'}\epsilon_{k'}$ and $\eta_{[n-1]\setminus I} = \eta_{[n]\setminus I'}\epsilon_{k'}$, i.e. the following diagram commutes*

$$
\begin{array}{ccc}
[n] & \xrightarrow{(\eta_I \times \eta_{[n-1]\setminus I})D} & [p] \times [q] \\
{\scriptstyle \epsilon_{k'}}\downarrow & & \downarrow{\scriptstyle \epsilon_k \times \mathrm{id}_{[q]}} \\
[n+1] & \xrightarrow{(\eta_{I'} \times \eta_{[n]\setminus I'})D} & [p+1] \times [q]
\end{array}
$$

,

where D denotes the diagonal maps.

Proof. Let $\epsilon_k : [p] \hookrightarrow [p+1]$ be a face map. By the preceding lemma there exists $0 \leq l \leq q-1$ such that $i_{l-1} < k' \leq i_l$ and

$$\epsilon_k \eta_I = \eta_{i_0} \eta_{i_{l-1}} \eta_{i_l+1} \cdots \eta_{i_{q-1}+1} \epsilon_{k'} = \eta_{I'} \epsilon_{k'}$$

for $k' = k+l$ and the ordered subset $I' = \{i_0 < \cdots i_{l-1} < i_l+1 < \cdots < i_{q-1}+1\}$ of $[n]$. Denote the complement if I in $[n-1]$ by J and let $J' = [n] \setminus I'$ be the complement of I' in $[n]$. Observe that the condition $i_{l-1} < k' < i_l$ implies that k' is contained in J'. The elements of J' can be expressed by the elements $j_0 < \cdots < j_{p-1}$ of J via

$$\begin{aligned}
J' = [n] \setminus I' &= [n] \setminus (\{i_0, \ldots, i_{l-1}\} \cup \{i_l+1, \ldots, i_{q-1}+1\}) \\
&= \{j_0, \ldots, j_{m-1}, k', j_m+1, \ldots, i_{p-1}+1\},
\end{aligned}$$

where m is minimal such that $j_m > k'$. Using this identity the composite $\eta_{[n] \setminus I'} \epsilon_{k'}$ can be rewritten as

$$\begin{aligned}
\eta_{[n] \setminus I'} \epsilon_{k'} &= \eta_{j_0} \cdots \eta_{j_{m-1}} \eta_{k'} \eta_{j_m+1} \cdots \eta_{j_{p-1}+1} \epsilon_{k'} \\
&= \eta_{j_0} \cdots \eta_{j_{m-1}} (\eta_{k'} \epsilon_{k'}) \eta_{j_m} \cdots \eta_{j_{p-1}} \\
&= \eta_{j_0} \cdots \eta_{j_{m-1}} \eta_{j_m} \cdots \eta_{j_{p-1}} = \eta_{[n-1] \setminus I}.
\end{aligned}$$

Therefore the degeneracy maps η_I and $\eta_{I'}$ and the face map $\epsilon_{k'}$ satisfy the equations $\epsilon_k \eta_I = \eta_{i'} \epsilon_{k'}$ and $\eta_J = \eta_{J'} \epsilon_{k'}$, which are the required properties. $\square$

Proposition 9.3.14. *For every monomorphism $\epsilon : [p] \hookrightarrow [p']$ there exists a q-element subset I' of $[n']$ for $n' = p' + q$ and a monomorphism $\epsilon' : [n] \hookrightarrow [n']$ such that $\epsilon \eta_I = \eta_{I'} \epsilon'$ and $\eta_{[n-1] \setminus I} = \eta_{[n'] \setminus I'} \epsilon'$, i.e. the following diagram commutes*

$$
\begin{array}{ccc}
[n] & \xrightarrow{(\eta_I \times \eta_{[n-1] \setminus I})D} & [p] \times [q] \\
{\scriptstyle \epsilon'} \downarrow & & \downarrow {\scriptstyle \epsilon \times \mathrm{id}_{[q]}} \\
[n'] & \xrightarrow{(\eta_{I'} \times \eta_{[n'-1] \setminus I'})D} & [p'] \times [q]
\end{array}
\quad ,
$$

where the D's denote the diagonal maps.

Proof. Every monomorphism $\epsilon : [p] \hookrightarrow [p']$ is a composition of coface maps. Thus the statement follows from Lemma 9.3.13. $\square$

Lemma 9.3.15. *For every degeneracy map $\eta_k : [p] \twoheadrightarrow [p-1]$ there exists a natural number l such that $0 \leq l \leq n-1$, $i_{l-1} < k+l$ if $l \geq 1$, $k+l \leq i_l$ and the equation*

$$\eta_k \eta_I = \eta_{i_0} \cdots \eta_{i_{l-1}} \eta_{i_l-1} \cdots \eta_{i_{q-1}-1} \eta_{k'} = \eta_{I'} \eta_{k'}$$

is satisfied for $k' = k+l$ and $I' = \{i_0 < \cdots < i_{l-1} < i_l+1 < i_{q-1}+1\}$.

Proof. Let $\eta_k : [p] \twoheadrightarrow [p-1]$ the k-th degeneracy map. The expression $\eta_k \eta_I$ can be rewritten using the simplicial identities F.1

$$\eta_k \eta_j = \eta_j \eta_{k+1} \quad \text{for } k \geq j.$$

If $k < i_0$ then these identities imply that for $0 \leq m \leq q-1$ the degeneracy maps η_k and η_{i_m} can be exchanged by at the same time raising the index of the degeneracy maps η_{i_m}. So one obtains the equality

$$\begin{aligned}
\eta_k \eta_I &= \eta_{i_0+1} \eta_k \eta_{i_1} \cdots \eta_{i_{q-1}} \\
&= \eta_{i_0+1} \cdots \eta_{i_{q-1}+1} \eta_k \\
&= \eta_{I'} \eta_k,
\end{aligned}$$

where $I' = \{i_0 + 1, \ldots, i_{q-1} + 1\}$. In this case $l = 0$ has the required properties. If $k \geq i_0$ the above expression $\eta_k \eta_I$ can be rewritten as

$$\eta_k \eta_I = \eta_k \eta_{i_0} \cdots \eta_{q-1} = \eta_{i_0} \eta_{k+1} \eta_{i_1} \cdots \eta_{q-1},$$

where the index of the epimorphism η_k has been raised by one. Note that the assumption $i_0 \leq k$ especially implies $i_0 < k + 1$. After repeating this process a maximum number of times one has shown the equality

$$\eta_k \eta_I = \eta_{i_0} \cdots \eta_{i_{l-1}} \eta_{k'} \eta_{i_l} \cdots \eta_{i_{q-1}}$$

with $k' = k + l$ and $i_{l-1} < k' < i_l$. Again using the simplicial identities above, one can exchange the last degeneracy maps η_{i_m} etc. with $\eta_{k'}$ by at the same time lowering the index of the epimorphisms η_{i_m} to finally arrive at the equality

$$\eta_k \eta_I = \eta_{i_0} \cdots \eta_{i_{l-1}} \eta_{i_l - 1} \cdots \eta_{i_{q-1}-1} \eta_{k'} = \eta_{I'} \eta_{k'}$$

where I' is the ordered subset $\{i_0 < \cdots < i_{l-1} < i_l - 1 < \cdots < i_{q-1} - 1\}$ of $[n]$. The condition $i_{l-1} < k' \leq i_l$ is satisfied by construction. □

Lemma 9.3.16. *For every degeneracy map $\eta_k : [p] \twoheadrightarrow [p-1]$ there exists a q-element subset I' of $[n-2]$ and an face map $\eta_{k'} : [n] \to [n-1]$ such that $\eta_k \eta_I = \eta_{I'} \eta_{k'}$ and $\eta_{[n-1]\backslash I} = \eta_{[n-2]\backslash I'} \eta_{k'}$, i.e. the following diagram commutes*

$$
\begin{array}{ccc}
[n] & \xrightarrow{\ (\eta_I \times \eta_{[n-1]\backslash I})D\ } & [p] \times [q] \\
{\scriptstyle \eta_{k'}} \downarrow & & \downarrow {\scriptstyle \eta_k \times \mathrm{id}_{[q]}} \\
[n-1] & \xrightarrow{\ (\eta_{I'} \times \eta_{[n-2]\backslash I'})D\ } & [p-1] \times [q]
\end{array}
\qquad,
$$

where D denotes the diagonal maps.

Proof. Let $\epsilon_k : [p] \twoheadrightarrow [p-1]$ be an epimorphism. By the preceding lemma there exists $0 \leq l \leq q - 1$ such that $i_{l-1} < k + l \leq i_l$ and

$$\eta_k \eta_I = \eta_{i_0} \cdots \eta_{l-1} \eta_{i_l - 1} \cdots \eta_{i_{q-1}-1} \epsilon_{k'} = \eta_{I'} \eta'_k$$

for $k' = k + l$ and the ordered subset $I' = \{i_0 < \cdots i_{l-1} < i_l + 1 < \cdots < i_{q-1} + 1\}$ of $[n]$. I' and let $J = [n] \backslash I'$ be the complement of I' in $[n]$. Denote the complement of I in $[n-1]$ by J and let $J' = [n-2] \backslash I'$ be the complement of I' in $[n-2]$. The condition $i_{l-1} < k' < i_l$ implies that k' is contained in $J' = [n-2] \backslash I'$. The elements of J' can be expressed by the elements $j_0 < \cdots < j_{p-1}$ of J via

$$
\begin{aligned}
J' = [n] \backslash I' &= [n] \backslash (\{i_0, \ldots, i_{l-1}\} \cup \{i_l + 1, \ldots, i_{q-1} + 1\}) \\
&= \{j_0, \ldots, j_{m-1}, k', j_m + 1, \ldots, i_{p-1} + 1\},
\end{aligned}
$$

where m is minimal such that $j_m > k'$. Using this equality the composite $\eta_{J'} \epsilon_{k'}$ can be rewritten as

$$
\begin{aligned}
\eta_{J'} \epsilon_{k'} &= \eta_{j_0} \cdots \eta_{j_{m-1}} \eta_{k'} \eta_{j_m - 1} \cdots \eta_{j_{p-1}-1} \epsilon_{k'} \\
&= \eta_{j_0} \cdots \eta_{j_{m-1}} (\eta_{k'} \epsilon_{k'}) \eta_{j_m} \cdots \eta_{j_{p-1}} \\
&= \eta_{j_0} \cdots \eta_{j_{m-1}} \eta_{j_m} \cdots \eta_{j_{p-1}} = \eta_{[n-1]\backslash I}.
\end{aligned}
$$

Therefore the I' degeneracy $\eta_{I'}$ and the degeneracy map $\eta_{k'}$ satisfy the equations $\eta_k \eta_I = \eta_{I'} \eta_{k'}$ and $\eta_J = \eta_{J'} \eta_{k'}$, which are the required properties. □

Proposition 9.3.17. *For every epimorphism* $\eta : [p] \twoheadrightarrow [p']$, *there exists a q-element subset* I' *of* $[n']$ *for* $n' = p' + q$ *and an epimorphism* $\eta' : [n] \twoheadrightarrow [n']$ *such that* $\eta\eta_I = \eta_{I'}\eta'$ *and* $\eta_{[n-1]\setminus I} = \eta_{[n-2]\setminus I'}\eta'$, *i.e. the following diagram commutes*

$$
\begin{array}{ccc}
[n] & \xrightarrow{(\eta_I \times \eta_{[n-1]\setminus I})D} & [p'] \times [q] \\
\downarrow{\scriptstyle\eta'} & & \downarrow{\scriptstyle\eta \times \mathrm{id}_{[q]}} \\
[n'] & \xrightarrow{(\eta_{I'} \times \eta_{[n'-1]\setminus I'})D} & [p'] \times [q]
\end{array} \quad ,
$$

where the diagonal maps are denoted by D.

Proof. Every epimorphism is a composition of degeneracy maps. Thus the statement follows from Lemma 9.3.16. □

Proposition 9.3.18. *For all morphisms* $\alpha : [p] \to [p']$, $\beta : [q] \to [q']$ *there exists a morphism* $\gamma : [n] \to [p'+q']$ *and a unique q'-element subset* I' *of* $[p'+q'-1]$ *such that* $\alpha\eta_I = \eta_{I'}\gamma$ *and* $\beta\eta_{[n-1]\setminus I} = \eta_{[n-2]\setminus I'}\gamma$, *i.e. the following diagram is commutative*

$$
\begin{array}{ccc}
[n] & \xrightarrow{(\eta_I \times \eta_{[n-1]\setminus I})D} & [p] \times [q] \\
\downarrow{\scriptstyle\gamma} & & \downarrow{\scriptstyle\alpha \times \beta} \\
[p' + q'] & \xrightarrow{(\eta_{I'} \times \eta_{[p'+q'-1]\setminus I'})D} & [p'] \times [q']
\end{array} \quad ,
$$

where D denotes the diagonal maps.

Proof. We first consider the special cases $\beta = \mathrm{id}_{[q]}$ and $\alpha = \mathrm{id}_{[p]}$. Let $\alpha : [p] \to [p']$ be a morphism in $\mathcal{O}$. This morphism has a unique epi-monic factorisation

$$ \alpha = \epsilon\eta, $$

where $\eta : [p] \twoheadrightarrow [\hat{p}]$ is an epimorphism and $\epsilon : [\hat{p}] \hookrightarrow [p']$ is a monomorphism. By Proposition 9.3.17 there exists an epimorphism $\eta' : [n] \twoheadrightarrow [\hat{p} + q]$ and a subset $\hat{I}$ of $[\hat{p} + q - 1]$ such that

$$ \eta\eta_I = \eta'\eta_{\hat{I}} \quad \text{and} \quad \eta_{[n-1]\setminus I} = \eta_{\hat{I}}\eta. $$

By Proposition 9.3.14 there exists a monomorphism $\epsilon' : [\hat{p} + q] \hookrightarrow [p' + q]$ and a subset I of $[p' + q - 1]$ such that

$$ \eta\eta_I = \eta'\eta_{I'} \quad \text{and} \quad \eta_{[n-1]\setminus I} = \eta_{I'}\eta. $$

Composing these identities shows that $\gamma = \epsilon'\eta'$ and I' are the required morphism and Index set in the case $\beta = \mathrm{id}_{[q]}$. The existence of the required morphism γ and index set I' in the case $\alpha = \mathrm{id}_{[p]}$ follows from symmetry. The general existence is now a consequence of the existence in both special cases. □

Corollary 9.3.19. *For all morphisms* $\alpha : [p] \to [p']$ *and* $\beta : [q] \to [q']$, *there exists a morphism* $\gamma : [n] \to [p' + q']$ *and a q'-element subset* I' *of* $[p' + q' - 1]$ *such that* $\tau'_I \Delta(\gamma) = [\Delta(\alpha) \times \Delta(\beta)]\tau_I$, *i.e. the following diagram is commutative:*

$$
\begin{array}{ccc}
\Delta_{p+q} & \xrightarrow{\tau_I} & \Delta_p \times \Delta_q \\
\downarrow{\scriptstyle\Delta(\gamma)} & & \downarrow{\scriptstyle\Delta(\alpha) \times \Delta(\beta)} \\
\Delta_{p'+q'} & \xrightarrow{\tau_{I'}} & \Delta_{p'} \times \Delta_{q'}
\end{array}
$$

Proposition 9.3.20. *For all points* $(x, \mathbf{t}) \in F(X)$ *and* $(y, \mathbf{t}') \in F(Y)$, *the families of functions* $\varphi_{p,q}(-, (y, \mathbf{t}'))$ *and* $\varphi_{p,q}(x, \mathbf{t}), -)$ *are compatible.*

Proof. Let $(y, \mathbf{t}') \in Y([q]) \times \Delta_q$ be a point of $F(Y)$. It suffices to show that the family $\varphi_{p,q}(-, (y, \mathbf{t}'))$ is compatible. The compatibility of the family $\varphi_{p,q}((x, \mathbf{t}), -)$ for points $(x, \mathbf{t})$ in $F(X)$ then follows from symmetry. To show that the maps $\varphi_{p,q}(-, (y, \mathbf{t}'))$ are compatible one needs to verify, that the images of equivalent points in $F(X)$ coincide. Let $(\mathbf{x}, \mathbf{t}) \in X([p]) \times \Delta_p$ and $(\hat{\mathbf{x}}, \hat{\mathbf{t}}) \in X([p']) \times \Delta_{p'}$ be points of $F(X)$ and let $\alpha : [p] \to [p']$ be a morphism in $\mathcal{O}$ such that

$$X(\alpha)(\hat{x}) = x \quad \text{and} \quad \Delta(\alpha)(\mathbf{t}) = \hat{\mathbf{t}},$$

i.e. the points $(x, \mathbf{t})$ and $(\hat{x}, \hat{\mathbf{t}})$ are equivalent. Denote the sum $p + q$ by n and the sum $p' + q$ by n'. The point $(\mathbf{t}, \mathbf{t}') \in \Delta_p \times \Delta_q$ is contained in the image of an affine n simplex τ_I. Let $\mathbf{t}''$ denote the inverse image of $(\mathbf{t}, \mathbf{t}')$ under τ_I. By Proposition 9.3.18 there exists a morphism $\gamma : [n] \to [n']$ and a subset I' if $[n']$ such that

$$\alpha \eta_I = \eta_{I'} \gamma \quad \text{and} \quad \eta_{[n-1] \backslash I} = \eta_{[n'-1] \backslash I'} \gamma.$$

Inserting these equations in the definition of the map $\varphi_{p,q}$ yields

$$
\begin{aligned}
\varphi_{p,q}(x, y, \mathbf{t}, \mathbf{t}') &= [X(\eta_I)(x), Y(\eta_{[n-1] \backslash I})(y), \tau_I^{-1}(\mathbf{t}, \mathbf{t}')] \\
&= [X(\eta_I)(x), Y(\eta_{[n-1] \backslash I})(y), \mathbf{t}''] \\
&= [X(\eta_I) X(\alpha)(\hat{x}), Y(\eta_{[n-1] \backslash I})(y), \mathbf{t}''] \\
&= [X(\alpha \eta_I)(\hat{x}), Y(\eta_{[n-1] \backslash I})(y), \mathbf{t}''] \\
&= [X(\eta_{I'} \gamma)(\hat{x}), Y(\eta_{[n'-1] \backslash I'} \gamma)(y), \mathbf{t}''] \\
&= [X(\gamma) X(\eta_{I'})(\hat{x}), Y(\gamma) Y(\eta_{[n'-1] \backslash I'})(y), \mathbf{t}''] \\
&= [X(\eta_{I'})(\hat{x}), Y(\eta_{[n'-1] \backslash I'})(y), \Delta(\gamma) \mathbf{t}''] \\
&= [X(\eta_{I'})(\hat{x}), Y(\eta_{[n'-1] \backslash I'})(y), \tau_{I'}^{-1}(\Delta(\alpha) \mathbf{t}, \mathbf{t}')] \\
&= \varphi_{p',q}(\hat{x}, y, \hat{\mathbf{t}}, \mathbf{t}'),
\end{aligned}
$$

where the last equality follows from the identity $\tau_{I'} \Delta(\gamma)(\mathbf{t}'') = (\Delta(\alpha)(\mathbf{t}), \mathbf{t}')$. Thus the image of equivalent points under $\coprod_p \varphi_{p,q}(-, (y, \mathbf{t}')$ is the same, hence the family $\varphi_{p,q}(-, (y, \mathbf{t}')$ is compatible. By symmetry, the families $\varphi_{p,q}((x, \mathbf{t}), -)$ are also compatible for all $(x, \mathbf{t})$ in $F(X)$. $\square$

Corollary 9.3.21. *If the product* $q_X \times q_Y$ *of the quotient maps* $q_X : F(X) \to |X|$ *and* $q_Y : F(Y) \to |Y|$ *is a quotient, then the map* $\coprod \varphi_{p,q}$ *factors uniquely through a map* $\varphi : |X| \times |Y| \to |X \times Y|$.

Proof. Let X and Y be simplicial spaces with geometric realisations $|X|$ and $|Y|$ respectively. The map $\coprod \varphi_{p,q} : F(X) \times F(Y) \to |X \times Y|$ factors uniquely through the quotient space $F(X) \times F(Y)/K(\coprod \varphi_{p,q})$ of $F(X) \times F(Y)$ defined by the equivalence relation $K(\coprod \varphi_{p,q})$. Since the continuous map $\Phi : |X \times Y| \to |X| \times |Y|$ is a bijection, the relation $K(\coprod \varphi_{p,q})$ is the the product of the relations on $F(X)$ and $F(Y)$ respectively. So if the product $q_X \times q_Y$ is a quotient map, the spaces $F(X) \times F(Y)/K(\coprod \varphi_{p,q})$ and $|X| \times |Y|$ are homeomorphic. In that case there exists a unique map $\varphi : |X| \times |Y| \to |X \times Y|$ such that $\coprod \varphi_{p,q} = \varphi \circ (q_X \times q_Y)$. $\square$

Proposition 9.3.22. *If the map* $\coprod \varphi_{p,q} : F(X) \times F(Y) \to |X \times Y|$ *factors through a map* $\varphi : |X| \times |Y| \to |X \times Y|$ *then* $\Phi_{X,Y}$ *is a left inverse of* φ.

Proof. Let $\coprod \varphi_{p,q}$ factor through a map $\varphi : |X| \times |Y| \to |x \times Y|$, $p = ([\mathbf{x}, \mathbf{t}], [\mathbf{y}, \mathbf{t}'])$ be a point of $|X \times Y|$ with $(\mathbf{x}, \mathbf{t}) \in X([p]) \times \Delta_p$ and $(\mathbf{y}, \mathbf{t}') \in Y([q]) \times \Delta_q$. Denote the sum $p + q$ by n and let I be a q-element subset of $[n-1]$ such that the pair $(\mathbf{t}, \mathbf{t}') \in \Delta_p \times \Delta_q$ is contained in the image of τ_I. Because τ_I is a homeomorphism onto its image there exists a unique point $\hat{\mathbf{t}}$ such that $\tau_I(\hat{\mathbf{t}}) = (\mathbf{t}, \mathbf{t}')$. Observe, that by the definition of τ_I the equations

$$\Delta(\eta_I)(\hat{\mathbf{t}}) = \mathbf{t} \quad \text{and} \quad \Delta(\eta_{[n-1]\setminus I})(\hat{\mathbf{t}}) = \mathbf{t}'$$

are automatically satisfied. Using this fact, the image of p under φ can be expressed as

$$\varphi(p) = [X(\eta_I)(x), Y(\eta_{[n-1]\subset I})(y), \tau_I^{-1}((\mathbf{t}, \mathbf{t}'))] = [X(\eta_I)(x), Y(\eta_{[n-1]\subset I})(y), \hat{\mathbf{t}}].$$

Applying the map $\Phi_{X,Y}$ on both ides of this equation yields

$$\begin{aligned}
(\Phi_{X,Y} \circ \varphi)(p) &= ([X(\eta_I)(\mathbf{x}), \hat{\mathbf{t}}], [Y(\eta_{[n-1]\setminus I})(\mathbf{y}), \hat{\mathbf{t}}]) \\
&= ([\mathbf{x}, \Delta(\eta_I)(\hat{\mathbf{t}})], [\mathbf{y}, \Delta(\eta_{[n-1]\setminus I})(\hat{\mathbf{t}})]) \\
&= ([\mathbf{x}, \mathbf{t}], [\mathbf{y}, \mathbf{t}']) = p.
\end{aligned}$$

Since this is true for all points $p \in |X \times Y|$ the map $\Phi_{X,Y}$ is a left inverse to φ. $\square$

If the map $\coprod \varphi_{p,q}$ does not factor through $|X| \times |Y|$ it still induces a map $\tilde{\varphi}$ on the underlying sets, which is the inverse of the underlying map $U\Phi_{X,Y}$ of sets.

Proposition 9.3.23. *The function $\Phi_{X,Y} : |X \times Y| \to |X| \times |Y|$ is an isomorphism if and only if the map $\coprod \varphi_{p,q}$ factors through a map $\varphi : |X| \times |Y| \to |X \times Y|$.*

Proof. If the map $\coprod \varphi_{p,q} : F(X) \times F(Y) \to |X \times Y|$ factors through $|X| \times |Y|$ then the induced continuous function $\varphi : |X| \times |Y| \to |X \times Y|$ is an inverse of the continuous bijection $\Phi_{X,Y}$ by Proposition 9.3.22. Conversely, if $\Phi_{X,Y}$ is a homeomorphism then the inverse $\tilde{\varphi}$ of the map $U\Phi_{X,Y}$ of sets stems from a continuous map $\varphi : |X| \times |Y| \to |X \times Y|$. Since the map $U \coprod \varphi_{p,q}$ of sets factors through $\tilde{\varphi}$ this implies that $\coprod \varphi_{p,q}$ factors through $\varphi : |X| \times |Y| \to |X \times Y|$. $\square$

The preservation of products will be examined in various cases (of increasing difficulty). One way to ensure that $\Phi_{X,Y}$ is an isomorphism is to show that the product of the quotient maps q_X and q_Y is again a quotient map. At first we observe the behaviour for discrete simplicial spaces X and Y. Here we make use of the following facts:

Theorem 9.3.24. *The geometric realisation $|X|$ of a discrete simplicial space X is a CW-complex having one n-cell for each non-degenerate point in $X([n])$.*

Proof. See [Mil57, Theorem 1] for a proof.

Lemma 9.3.25. *Geometric realisation restricts to a functor $\mathbf{Set}^{\mathcal{O}^{op}} \to \mathbf{kHaus}$, i.e. the geometric realisation $|X|$ of a discrete simplicial space X is compactly generated and Hausdorff.*

Proof. The geometric realisation of discrete simplicial spaces is Hausdorff by [GZ67, Proposition III.1.8]. In addition any discrete simplicial space is compactly generated. The geometric realisation of compactly generated simplicial spaces is also compactly generated by Lemma 9.1.12. Thus the geometric realisation of a discrete simplicial space is compactly generated and Hausdorff. $\square$

The restriction $|\ \ | : \mathbf{Sct}^{\mathcal{O}^{op}} \to \mathbf{kHaus}$ is known to preserve finite products of at most countable simplicial sets ([Mil57, Theorem 2]). In addition, if X and Y are simplicial sets and either CW-complex $|X|$ or $|Y|$ is locally finite, the product $X \times Y$ is also preserved ([Mil57, Theorem 2]). The finiteness condition there can be weakened to a countability condition.

Definition 9.3.26. *A discrete simplicial space X is called* locally countable *if every point $p \in |X|$ in the geometric realisation is contained in at most countably many simplices (of the form $q_X(\{x\} \times \Delta_n))$.*

The condition that a simplicial set X is locally countable implies that each point in the geometric realisation $|X|$ has a neighbourhood, which is the limit of an ascending sequence of compacta, hence it is a locally k_ω-space and thus a k-space. Since (locally) k_ω spaces are (locally) countably compact, we observe:

Lemma 9.3.27. *If X and Y are discrete simplicial spaces and either X or Y is locally countable, then the product $|X| \times |Y|$ of the geometric realisations $|X|$ and $|Y|$ is a k-space.*

Proof. We assume w.l.o.g. that X is locally countable, so the geometric realisation $|X|$ of X is a locally countably compact k-space. The geometric realisation $|Y|$ of Y is a quotient of a sequential space, hence sequential as well. Every product of a locally countably compact k-space with a sequential space is k-space (cf. [Tan05, Theorem 1.3]), so the product $|X| \times |Y|$ is a k-space. $\qquad\square$

Lemma 9.3.28. *If X and Y are discrete simplicial spaces and either X or Y is locally countable, then the product $q_X \times q_Y$ of the quotient maps q_X and q_Y is a quotient map in* **Top***.*

Proof. Let X and Y be discrete simplicial spaces geometric realisations $|X|$ and $|Y|$ respectively. By Lemma B.6.29 the product map

$$q_X \times_k q_Y : F(X) \times F(Y) \to \mathrm{k}(|X| \times |Y|)$$

in **kTop** is a quotient map. Because the the product space $|X| \times |Y|$ is a k-space by Lemma 9.3.27, the product $\mathrm{k}(|X| \times |Y|)$ in **kTop** coincides with the product $|X| \times |Y|$ in **Top**. Thus the product $q_X \times q_Y$ of the quotient maps is a quotient map onto $|X| \times |Y|$ in **Top**. $\qquad\square$

Lemma 9.3.29. *If X and Y are discrete simplicial spaces and either X or Y is locally countable, then the map $\coprod \varphi_{p,q} : F(X) \times F(Y) \to |X \times Y|$ factors uniquely through a map $\varphi : |X| \times |Y| \to |X \times Y|$:*

$$
\begin{array}{ccc}
F(X) \times F(Y) & & \\
\Big\downarrow{\scriptstyle q_X \times q_Y} & \searrow^{\coprod \varphi_{p,q}} & \\
|X| \times |Y| & \xrightarrow{\ \ \varphi\ \ } & |X \times Y|
\end{array}
$$

Proof. Let X and Y be discrete simplicial spaces with geometric realisations $|X|$ and $|Y|$ respectively. If either X or Y is locally countable, the product $q_X \times q_Y$ of the quotient maps q_X and q_Y is itself a quotient by Lemma 9.3.28. Thus the map $\coprod \varphi_{p,q}$ factors uniquely through the quotient space $|X| \times |Y|$ of $F(X) \times F(Y)$ by Corollary 9.3.21. $\qquad\square$

Proposition 9.3.30. *If X and Y are discrete simplicial spaces and either X or Y is locally countable, then $\Phi_{X,Y} : |X \times Y| \to |X| \times |Y|$ is a homeomorphism.*

Proof. It suffices to identify the inverse $\Phi_{X,Y}^{-1}$ as the continuous function φ. An application of the (not necessarily continuous) map $\Phi_{X,Y}^{-1}$ on the equality of maps $\Phi_{X,Y} \circ \varphi = \mathrm{id}_{|X \times Y|}$ yields

$$\Phi_{X,Y}^{-1} = \Phi_{X,Y}^{-1} \circ \mathrm{id}_{|X| \times |Y|} = \Phi_{X,Y}^{-1} \circ \Phi_{X,Y} \circ \varphi = \mathrm{id}_{|X \times Y|} \circ \varphi = \varphi,$$

where the function φ is continuous. Thus $\Phi_{X,Y}$ is a homeomorphism. $\qquad\square$

The line of reasoning above, using that the product space $|X| \times |Y|$ is a k-space, can also be used for simplicial k_ω-spaces. Recall that finite products of k_ω-spaces are k-spaces (cf. Lemma B.10.7) and that quotients and countable disjoint unions of k_ω-spaces are k_ω-spaces, (cf. Lemma B.10.3 resp. Lemma B.10.4).

Example 9.3.31. Every connected locally compact topological group is a k_ω-space. Thus every simplicial connected locally compact topological group is a simplicial k_ω-space.

Proposition 9.3.32. *If X is a simplicial k_ω-space, then $F(X)$ and $|X|$ are k_ω-spaces as well, in particular the geometric realisation functor restricts and corestricts to a functor $| - | : \mathbf{k_\omega Top}^{\mathcal{O}^{op}} \to \mathbf{k_\omega Top}$.*

Proof. If X is a simplicial k_ω-space then the products $X([n]) \times \Delta_n$ of the k_ω-spaces $X([n])$ with the compact standard simplices are k_ω-spaces. Therefore the disjoint union $F(X) = \coprod X([n]) \times \Delta_n$ also is a k_ω-space and so is its quotient space $|X|$. $\square$

Corollary 9.3.33. *If X and Y are simplicial k_ω-spaces with Hausdorff geometric realisations $|X|$ and $|Y|$, then the product $|X| \times |Y|$ is a k_ω-space.*

Proof. This follows from the fact that finite products of Hausdorff k_ω-spaces are k_ω-spaces (cf. Lemma B.10.7). $\square$

Lemma 9.3.34. *If X and Y are simplicial k_ω-spaces with Hausdorff geometric realisations $|X|$ and $|Y|$, then the product $q_X \times q_Y$ of the quotient maps q_X and q_Y is a quotient map in $\mathbf{Top}$.*

Proof. Let X and Y be simplicial k_ω-spaces with Hausdorff geometric realisations $|X|$ and $|Y|$ respectively. By Lemma B.6.29 the product map

$$q_X \times_k q_Y : F(X) \times F(Y) \to \mathrm{k}(|X| \times |Y|)$$

in $\mathbf{kTop}$ is a quotient map. Because the product $|X| \times |Y|$ is a k-space by Corollary 9.3.33, the product $\mathrm{k}(|X| \times |Y|)$ in $\mathbf{kTop}$ coincides with the product $|X| \times |Y|$ in $\mathbf{Top}$. Thus the product $q_X \times q_Y$ of the quotient maps is a quotient map onto the product $|X| \times |Y|$ in $\mathbf{Top}$. $\square$

Proposition 9.3.35. *If X and Y are simplicial k_ω-spaces with Hausdorff geometric realisations $|X|$ resp. $|Y|$, then the map $\coprod \varphi_{p,q} : F(X) \times F(Y) \to |X \times Y|$ factors uniquely through a map $\varphi : |X| \times |Y| \to |X \times Y|$:*

$$
\begin{array}{ccc}
F(X) \times F(Y) & & \\
\Big\downarrow{\scriptstyle q_X \times q_Y} & \searrow {\scriptstyle \coprod \varphi_{p,q}} & \\
|X| \times |Y| & \xrightarrow{\ \varphi\ } & |X \times Y|
\end{array}
$$

Proof. Let X and Y be simplicial k_ω-spaces with Hausdorff geometric realisations $|X|$ and $|Y|$ respectively. The product $q_X \times q_Y$ of the quotient maps q_X and q_Y is itself a quotient by Lemma 9.3.34. Thus the map $\coprod \varphi_{p,q}$ factors uniquely through the quotient space $|X| \times |Y|$ of $F(X) \times F(Y)$ by Corollary 9.3.21. $\square$

Lemma 9.3.36. *If X and Y are simplicial k_ω-spaces with Hausdorff geometric realisations $|X|$ resp. $|Y|$, then $\Phi_{X,Y}$ is a homeomorphism.*

Proof. It suffices to identify the inverse $\Phi_{X,Y}^{-1}$ as the continuous function φ. An application of the (not necessarily continuous) map $\Phi_{X,Y}^{-1}$ on the equality of maps $\Phi_{X,Y} \circ \varphi = \mathrm{id}_{|X \times Y|}$ yields

$$\Phi_{X,Y}^{-1} = \Phi_{X,Y}^{-1} \circ \mathrm{id}_{|X| \times |Y|} = \Phi_{X,Y}^{-1} \circ \Phi_{X,Y} \circ \varphi = \mathrm{id}_{|X \times Y|} \circ \varphi = \varphi, \qquad (9.5)$$

where the function φ is continuous. Thus $\Phi_{X,Y}$ is a homeomorphism. $\square$

We turn back to the geometric realisation of discrete simplicial spaces. The special nature of these simplicial spaces and their geometric realisations allows us to generalise Proposition 9.3.30 even further, dropping the assumption of local countability. The main reason here fore is the fact that for each discrete simplicial space X the space $F(X)$ is a disjoint union of standard simplices, i.e. a disjoint union of compact uniform Hausdorff spaces. This implies that taking products with the locally compact space $F(X)$ is a cocontinuous endofunctor $F(X) \times - : \mathbf{Top} \to \mathbf{Top}$ which in particular preserves quotients (cf. Proposition B.5.12). We note an immediate consequence:

Lemma 9.3.37. *For any discrete simplicial space X and any simplicial space Y the map $\coprod \varphi_{p,q} : F(X) \times F(Y) \to |X \times Y|$ factors uniquely through $\mathrm{id}_X \times q_Y$, i.e. there exists a unique continuous function ψ making the following diagram commutative:*

$$
\begin{array}{ccc}
F(X) \times F(Y) & & \\
\mathrm{id}_x \times q_Y \Big\downarrow & \searrow^{\coprod \varphi_{p,q}} & \\
F(X) \times |Y| & \xrightarrow{\;\;\psi\;\;} & |X \times Y|
\end{array}
$$

Proof. Since $F(X)$ is a locally compact space, the product $\mathrm{id}_X \times q_Y$ of id_X with the quotient q_Y is a quotient map. Because all point that are identified by $\mathrm{id}_X \times q_Y$ also are identified by $\coprod \varphi_{p,q}$ the latter function factors uniquely through $\mathrm{id}_X \times q_Y$. □

In the following we assume X and Y to be discrete simplicial spaces and ψ to be the unique function satisfying $\psi(\mathrm{id}_X \times q_Y) = \coprod \varphi_{p,q}$. The codomain $|X \times Y|$ of ψ is polytope, hence a completely regular Hausdorff space. The first and the second factor of the domain $F(X) \times |Y|$ of ψ also are completely regular, hence so is their product $F(X) \times |Y|$. Because every continuous function between completely regular spaces is uniformly continuous w.r.t. the maximal uniform structures the function ψ is uniformly continuous w.r.t. the maximal uniform structure on $F(X) \times |Y|$. The first factor $F(X)$ of the domain of ψ is a disjoint union of standard simplices, which are completely regular compact Hausdorff spaces. We first observe some generalities on compact uniform spaces and then prove a generalisation of Proposition 9.3.30.

Proposition 9.3.38. *If K and Z are completely regular topological spaces endowed with the maximal uniform structure and K is compact, then the product uniform structure on $K \times Z$ coincides with the maximal uniform structure on the completely regular space $K \times Z$.*

Proof. The maximal uniform structure on a completely regular space is induced by all continuous pseudometrics. Therefore it suffices to show that for every continuous pseudometric ϱ on $K \times Z$ there exists continuous pseudometrics ϱ_K and ϱ_Z on K resp. Z satisfying

$$\varrho \le \varrho_K \mathrm{pr}_K + \varrho_Z \mathrm{pr}_Z$$

where pr_K and pr_Z denote the projections onto K resp. Z. (This inequality ensures that the entourages of $\varrho_K \mathrm{pr}_K + \varrho_Z \mathrm{pr}_Z$ are contained in those of ϱ.) Because the space K is compact, the function

$$\varrho_Z : Z \times Z \to \mathbb{R}, \quad \varrho_z(z, z') = \max_{k,k' \in K} \varrho((k, z), (k, z'))$$

is continuous. Furthermore it is positive definite by definition and satisfies the triangle inequality because ϱ does. Thus for an arbitrary pseudometric ϱ_K on K the inequality $\varrho \le \varrho_K \mathrm{pr}_K + \varrho_Z \mathrm{pr}_Z$ is satisfied. □

As observed before, the space $F(X)$ is a disjoint union $\coprod K_i$ of completely regular compact Hausdorff spaces K_i. We denote the restriction of ψ to the subspace $K_i \times |Y|$ by ψ. The preceding proposition shows that the maximal uniform structure on the completely regular space $K_i \times |Y|$ is the product uniform structure when K_i and $|Y|$ are equipped with the maximal uniform structures separately.

Corollary 9.3.39. *The function ψ_i is uniformly continuous w.r.t. the product of the maximal uniform structures on $K_i \times |Y|$ and any compatible uniform structure on $|X \times Y|$.*

This enables us to use the topology of uniform convergence on function spaces to show that the restriction $|-| : \mathbf{Set}^{\mathcal{O}^{op}} \to \mathbf{kHaus}$ of the geometric realisation functor to the category $\mathbf{Set}$ of discrete simplicial spaces preserves finite products. Recall that for any two uniform spaces Z and Z' the set of functions $Z \to Z'$ can be equipped with a uniformity whose whose entourages are defined to be the sets

$$U(V) := \{(g, g') \mid \forall z \in Z : (g(z), g'(z)) \in V\},$$

where V is an entourage in $Z' \times Z'$ (cf. [Isb64, Chapter III]). The resulting uniform space is denoted by $U(Z, Z')$ and its topology is the topology of uniform convergence on Z. For three arbitrary uniform spaces K, Z and Z', any uniformly continuous function $f : K \times Z \to Z'$ (w.r.t. the product uniform structure on $K \times Z$) and each point $k \in K$ the function $\hat{f}(k) := f(k, -)$ is uniformly continuous. The main feature of the uniform structure on $U(Z, Z')$ we want to use is the following property:

Proposition 9.3.40. *If K, Z and Z' are uniform spaces and K is compact then the correspondence $f \mapsto \hat{f}$ is an isomorphism $U(K \times Z, Z') \cong U(K, U(Z, Z'))$.*

Proof. See [Isb64, Theorem III.24] and [Isb64, Theorem III.26] for a proof.   □

Recall that there exists a decomposition $F(X) = \coprod_{i \in I} K_i$ of the space $F(X)$ into completely regular compact Hausdorff spaces K_i and the restriction of ψ to the subspace $K_i \times |Y|$ is denoted by ψ_i, so $\psi = \coprod_{i \in I} \psi_i$ is the corresponding decomposition of ψ. An application of Proposition 9.3.40 to the functions ψ_i shows:

Proposition 9.3.41. *For each $i \in I$ the adjoint $\hat{\psi}_i$ of ψ_i is a uniformly continuous function $\psi_i : K_i \to U(|Y|, |X \times Y|)$ when K_i and $|Y|$ are endowed with the maximal uniform structure.*

Proof. By Corollary 9.3.39 each function $\psi_i : K_i \times |Y| \to |X \times Y|$ is uniformly continuous w.r.t. the product uniformity on $K_i \times |Y|$, if K_i and $|Y|$ are endowed with the maximal uniform structures. Thus Proposition 9.3.40 implies that its adjoint $\hat{\psi}_i : K_i \to U(|Y|, |X \times Y|)$ is uniformly continuous.   □

Corollary 9.3.42. *The disjoint union $\hat{\psi} = \coprod \hat{\psi}_i : F(X) \to U(|Y|, Z)$ of the uniformly continuous functions $\hat{\psi}_i$ is uniformly continuous.*

Lemma 9.3.43. *The map $\hat{\psi} : F(X) \to U(|Y|, |X \times Y|)$ factors uniquely through q_X, i.e. there exists a continuous map $\hat{\varphi} : |X| \to U(|Y|, |X \times Y|)$ making the following diagram commutative:*

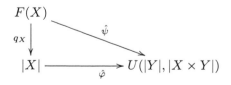

Proof. The continuous function $\hat{\psi}$ identifies all points that are identified by the quotient map q_X, hence $\hat{\psi}$ factors uniquely through the quotient map q_X.   □

the continuous function $\hat{\varphi}$ is the adjoint of the not necessarily continuous map $\varphi : |X| \times |Y| \to |X \times Y|$ whose continuity we wish to show. This then completes the last step in the proof that the restriction $|-| : \mathbf{Set}^{\mathcal{O}^{op}} \to \mathbf{kHaus}$ of the geometric realisation functor preserves finite products. It uses a backward direction of Proposition 9.3.40, where the resulting map is continuous but need not be uniformly continuous:

Lemma 9.3.44. *If Z, Z' and Z'' are uniform spaces and $\hat{f} : Z \to U(Z', Z'')$ a uniformly continuous function then its adjoint $f : Z \times Z' \to Z''$ is continuous.*

Proof. See [Isb64, Theorem III.22] and [Isb64, Theorem III.26] for a proof.   □

Proposition 9.3.45. *If X and Y are discrete simplicial spaces, then the continuous function $\coprod \varphi_{p,q} : F(X) \times F(Y) \to |X \times Y|$ factors uniquely through $q_X \times q_Y$, i.e. there exists a continuous function $\varphi : |X| \times |Y| \to |X \times Y|$ making the following diagram commutative:*

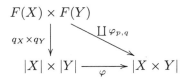

Proof. Let X and Y be discrete simplicial spaces with geometric realisations $|X|$ and $|Y|$ respectively. By Lemma 9.3.37 the map $\coprod \varphi_{p,q}$ factors through a function $\psi : F(X) \times |Y| \to |X \times Y|$. By Corollary 9.3.42 the adjoint $\hat{\psi}$ is a uniformly continuous map $F(X) \to U(|Y|, |X \times Y|)$. By Lemma 9.3.43 this continuous map factors uniquely through q_X, i.e. there exists a unique continuous function $\hat{\varphi}$ satisfying $\hat{\psi} = \hat{\varphi} q_X$. The continuous function $\hat{\varphi}$ is uniformly continuous w.r.t. the maximal uniform structure on the completely regular space $|X|$, so an application of Lemma 9.3.44 shows the map φ itself is continuous, i.e. $\coprod \varphi_{p,q}$ factors through $q_X \times q_Y$ via $\varphi(q_X \times q_Y) = \coprod \varphi_{p,q}$.   □

Corollary 9.3.46. *If X and Y are discrete simplicial spaces then the continuous function $\Phi_{X,Y} : |X \times Y| \to |X| \times |X|$ is a homeomorphism.*

Proof. If X and Y are discrete simplicial spaces then the map $\coprod \varphi_{p,q}$ factors uniquely through a map $\varphi : |X| \times |Y| \to |X \times Y|$ by Proposition 9.3.45 and so $\Phi_{X,Y}$ is a homeomorphism by Proposition 9.3.22.   □

Proposition 9.3.47. *The restriction of the natural transformation $\Phi_{X,Y}$ to the functor $|-|_{|\mathbf{Set}^{\mathcal{O}^{op}} \times \mathbf{Set}^{\mathcal{O}^{op}}}$ is a natural isomorphism.*

Proof. For discrete simplicial spaces X and Y the function $\Phi_{X,Y}$ is an isomorphism by Corollary 9.3.46.   □

Corollary 9.3.48. *The restriction $|-|_{|\mathbf{Set}^{\mathcal{O}^{op}}}$ of the geometric realisation to discrete simplicial spaces preserves finite products.*

Theorem 9.3.49. *The restriction $|-| : \mathbf{Set}^{\mathcal{O}^{op}} \to \mathbf{kHaus}$ of the geometric realisation functor preserves finite limits.*

Proof. The restriction $|-|_{|\mathbf{Set}^{\mathcal{O}^{op}}}$ preserves finite products by Corollary 9.3.48 and arbitrary equalisers by Proposition 9.3.1. Thus it preserves finite limits.   □

This generalises the result of Milnor concerning the product of polytopes with countable vertex set or locally finite polytopes ([Mil57, Theorem 2]) to arbitrary polytopes:

Theorem 9.3.50. *Finite products of polytopes are polytopes.*

Proof. Let $X_1, \ldots, X_n$ be polytopes. These polytopes X_i are by definition homeomorphic to geometric realisations of discrete simplicial spaces K_i respectively. Therefore the finite product $X_1 \times \cdots \times X_n$ is naturally homeomorphic to the geometric realisation $|K_1 \times \cdots K_n|$ of the simplicial space $K_1 \times \cdots \times K_n$.   □

The non-existence of a proof for the preservation of finite limits under the restriction $|-| : \mathbf{Set}^{\mathcal{O}^{op}} \to \mathbf{kHaus}$ was the obstruction to realise topological algebras (especially topological groups) as polytopes with uncountable vertex set. With the help of Theorem 9.3.49 this issue can now be resolved:

If ω is a binary operation on the simplicial discrete topological space X then the continuous function

$$|\omega| \circ \Phi_{X,Y}^{-1} : |X| \times |X| \to |X|$$

is a binary operation on the geometric realisation $|X|$. Similarly n-ary operations on X induce n-ary operations on the geometric realisation $|X|$. Note that a 0-ary operation $\omega : X^0 \to X$ in $\mathbf{Top}^{\mathcal{O}^{op}}$ is a morphism $* \to X$ from the constant simplicial space $*$ into X. This morphism induces a continuous function $* \cong |*| \to |X|$ from the singleton space $*$ into $|X|$, i.e. the geometric realisation of a 0-ary operation is a 0-ary operation. The Theorem below summarises these observations.

Theorem 9.3.51. *The geometric realisation of a discrete simplicial topological structure A of type Ω is a polytope that is a topological structure A of type Ω.*

If E is a set of equations satisfied by the operations $\omega \in \Omega$ of a discrete simplicial Ω-structure A then the geometric realisations $|\omega|$ of these operations also satisfy the equations in E. So we find:

Theorem 9.3.52. *The geometric realisation $|A|$ of a discrete simplicial topological (Ω, E)-algebra is a polytope that is a topological (Ω, E)-algebra A, i.e. the geometric restriction functor restricts to a functor $|-| : \mathbf{Alg}_{(\Omega, E)}^{\mathcal{O}^{op}} \to \mathbf{TopAlg}_{(\Omega, E)}$.*

Example 9.3.53. If G is a discrete simplicial topological monoid (or group) then the polytope $|G|$ is a topological monoid (group). If G is commutative then $|G|$ is commutative as well.

Example 9.3.54. If R is a discrete simplicial topological ring (unital ring) then the polytope $|R|$ is a topological ring (unital ring). If R is commutative then $|R|$ is commutative as well.

The preservations of finite limits under the restricted geometric realisation functor $|-| : \mathbf{Set}^{\mathcal{O}^{op}} \to \mathbf{Top}$ leads to the preservation of finite limits under the restricted and corestricted functor $|-| : \mathbf{Alg}_{(\Omega, E)}^{\mathcal{O}^{op}} \to \mathbf{TopAlg}_{(\Omega, E)}$. This is proved in two steps:

Lemma 9.3.55. *The restricted functor $|-| : \mathbf{Alg}_{(\Omega, E)}^{\mathcal{O}^{op}} \to \mathbf{Top}$ preserves finite limits.*

Proof. The embedding $U : \mathbf{Alg}_{(\Omega, E)} \to \mathbf{Top}$ of the category of (Ω, E)-algebras into $\mathbf{Top}$ as discrete spaces has a left adjoint, which assigns to each space the free (Ω, E)-algebra on its set of connected components. Thus limits in $\mathbf{Alg}_{(\Omega, E)}$ coincide with limits in $\mathbf{Top}$, when $\mathbf{Alg}_{(\Omega, E)}$ is regarded as a subcategory of discrete spaces in $\mathbf{Top}$. Since all limits in the categories $\mathbf{Alg}_{(\Omega, E)}^{\mathcal{O}^{op}}$ and $\mathbf{Top}^{\mathcal{O}^{op}}$ are are obtained by computing the limits point-wise (cf. [ML98, Chapter V.3, Theorem 1]), the limits in $\mathbf{Alg}_{(\Omega,E)}^{\mathcal{O}^{op}}$ and $\mathbf{Top}^{\mathcal{O}^{op}}$ also coincide. Let $J : \mathbf{D} \to \mathbf{Alg}_{(\Omega,E)}$ be a functor from a finite category $\mathbf{D}$ into $\mathbf{Alg}_{(\Omega,E)}^{\mathcal{O}^{op}}$. Identifying the category $\mathbf{Alg}_{(\Omega,E)}$ with its image under U and the category $\mathbf{Alg}_{(\Omega,E)}^{\mathcal{O}^{op}}$ with its image in $\mathbf{Top}^{\mathcal{O}^{op}}$ one obtains the equality

$$|\lim J| = \lim |J|$$

of limits. (Here the preservation of finite limits by the general geometric realisation functor has been used.) Thus the restricted functor $|-| : \mathbf{Alg}_{(\Omega,E)}^{\mathcal{O}^{op}} \to \mathbf{TopAlg}_{(\Omega,E)}$ preserves finite limits.   □

Theorem 9.3.56. *The corestricted functor* $|-| : \mathbf{Alg}_{(\Omega,E)}^{\mathcal{O}^{op}} \to \mathbf{TopAlg}_{(\Omega,E)}$ *preserves finite limits.*

Proof. This is a consequence of the fact that the category $\mathbf{TopAlg}_{(\Omega,E)}$ is a reflective subcategory of $\mathbf{Top}$ (the inclusion functor being right adjoint to the free topological (Ω, E)-algebra functor). Let $J : \mathbf{D} \to \mathbf{Alg}_{(\Omega,E)}$ be a functor from a finite category $\mathbf{D}$ into $\mathbf{Alg}_{(\Omega,E)}^{\mathcal{O}^{op}}$, and consider the composition $|J|^{|\mathbf{TopAlg}_{(\Omega,E)}} = |-|^{\mathbf{TopAlg}_{(\Omega,E)}} \circ J$. Because the inclusion $U : \mathbf{TopAlg}_{(\Omega,E)} \to \mathbf{Top}$ is continuous, the underlying space of the limit of $|J|^{\mathbf{TopAlg}_{(\Omega,E)}}$ in $\mathbf{TopAlg}_{(\Omega,E)}$ is the limit of $|J|$ in $\mathbf{Top}$. Together with the preceding Theorem one obtains the chain of equalities

$$U|\lim J|^{|\mathbf{TopAlg}_{(\Omega,E)}} = |\lim J| = \lim |J|$$
$$= \lim U|J|^{|\mathbf{TopAlg}_{(\Omega,E)}} = U \lim |J|^{|\mathbf{TopAlg}_{(\Omega,E)}}.$$

Since the embedding $\mathbf{TopAlg}_{(\Omega,E)} \hookrightarrow \mathbf{Top}$, creates limits, this implies that the restriction $|-| : \mathbf{Alg}_{(\Omega,E)}^{\mathcal{O}^{op}} \to \mathbf{TopAlg}_{(\Omega,E)}$ preserves finite limits.   □

Lemma 9.3.57. *The restriction* $|-| : \mathbf{Alg}_{(\Omega,E)}^{\mathcal{O}^{op}} \to \mathbf{TopAlg}_{(\Omega,E)}$ *of the geometric realisation functor preserves monomorphisms.*

Proof. A monomorphism $f : A \to A'$ in $\mathbf{Alg}_{(\Omega,E)}^{\mathcal{O}^{op}}$ is a morphism of simplicial algebras, whose components $f([n]) : A([n]) \to A'([n])$ are monomorphisms in $\mathbf{Alg}_{(\Omega,E)}$. The monomorphisms in $\mathbf{Alg}_{(\Omega,E)}$ are the injective homomorphisms, so any monomorphism $f : A \to A'$ of simplicial algebras is injective. By Lemma 9.1.4 the induced homomorphism $|f| : |A| \to |A'|$ in $\mathbf{TopAlg}_{(\Omega,E)}$ is injective as well, hence is a monomorphism in $\mathbf{TopAlg}_{(\Omega,E)}$.   □

Corollary 9.3.58. *The restriction* $|-| : \mathbf{Alg}_{(\Omega,E)}^{\mathcal{O}^{op}} \to \mathbf{kHausAlg}_{(\Omega,E)}$ *of the geometric realisation functor preserves monomorphisms.*

Corollary 9.3.59. *The restriction* $|-| : \mathbf{Grp}^{\mathcal{O}^{op}} \to \mathbf{kHausGrp}$ *of the geometric realisation functor preserves monomorphisms.*

Corollary 9.3.60. *The restriction* $|-| : \mathbf{Ab}^{\mathcal{O}^{op}} \to \mathbf{kHausAb}$ *of the geometric realisation functor preserves monomorphisms.*

Corollary 9.3.61. *The restriction* $|-| : \mathbf{Ring}^{\mathcal{O}^{op}} \to \mathbf{kHausRing}$ *of the geometric realisation functor preserves monomorphisms.*

Corollary 9.3.62. *The restriction* $|-| : \mathbf{RMod}^{\mathcal{O}^{op}} \to \mathbf{kHausRMod}$ *of the geometric realisation functor preserves monomorphisms.*

Since the epimorphisms in general varieties are not necessarily surjective, the analogue of the preceding Theorem might not be true. However, if all epimorphisms in a variety $\mathbf{Alg}_{(\Omega,E)}$ are surjective, the analogue holds.

Proposition 9.3.63. *If all epimorphisms in* $\mathbf{Alg}_{(\Omega,\mathrm{E})}$ *are surjective, then restrictions* $|-| : \mathbf{Alg}_{(\Omega,\mathrm{E})}^{\mathcal{O}^{op}} \to \mathbf{kHausAlg}_{(\Omega,\mathrm{E})}$ *and* $|-| : \mathbf{Alg}_{(\Omega,\mathrm{E})}^{\mathcal{O}^{op}} \to \mathbf{TopAlg}_{(\Omega,\mathrm{E})}$ *of the geometric realisation functor preserves epimorphisms.*

Proof. Suppose that all epimorphisms in $\mathbf{Alg}_{(\Omega,\mathrm{E})}$ are surjective. An epimorphism $\varphi : A \to A'$ in $\mathbf{Alg}_{(\Omega,\mathrm{E})}^{\mathcal{O}^{op}}$ is a morphism of simplicial algebras, whose components $\varphi([n]) : A([n]) \to A'([n])$ are epimorphisms in $\mathbf{Alg}_{(\Omega,\mathrm{E})}$. The epimorphisms in $\mathbf{Alg}_{(\Omega,\mathrm{E})}$ are surjective by assumption, so φ is surjective. Because the geometric realisation functor preserves surjectivity, the geometric realisation $|\varphi|$ of an epimorphism φ in $\mathbf{Alg}_{(\Omega,\mathrm{E})}^{\mathcal{O}^{op}}$ is a surjective homomorphism in $\mathbf{kHausAlg}_{(\Omega,\mathrm{E})}$ or $\mathbf{TopAlg}_{(\Omega,\mathrm{E})}$, hence an epimorphism in both categories. □

Corollary 9.3.64. *The restriction* $|-| : \mathbf{Grp}^{\mathcal{O}^{op}} \to \mathbf{kHausGrp}$ *of the geometric realisation functor preserves epimorphisms.*

Proof. This is due to the fact that epimorphisms in $\mathbf{Grp}$ are exactly the surjective homomorphisms. (See Lemma E.1.5.) □

Moreover any set based category containing a null object and in which subobjects are kernels (e.g. set based Abelian categories) has the property that epimorphisms are surjective. So we additionally obtain:

Corollary 9.3.65. *The restriction* $|-| : \mathbf{Ab}^{\mathcal{O}^{op}} \to \mathbf{kHausAb}$ *of the geometric realisation functor preserves epimorphisms.*

Corollary 9.3.66. *The restriction* $|-| : \mathbf{RMod}^{\mathcal{O}^{op}} \to \mathbf{kHausRMod}$ *of the geometric realisation functor preserves epimorphisms.*

Consider a category $\mathbf{Alg}_{(\Omega,\mathrm{E})}$ of algebras of a fixed type. The singleton set equipped with the unique algebra structure is a terminal object in $\mathbf{Alg}_{(\Omega,\mathrm{E})}$. If the category $\mathbf{Alg}_{(\Omega,\mathrm{E})}$ contains a null object, then this is an algebra with underlying singleton set. The null object in $\mathbf{Alg}_{(\Omega,\mathrm{E})}^{\mathcal{O}^{op}}$ then is the constant simplicial algebra $\{*\}$. Its geometric realisation is again the discrete algebra $\{*\}$ in $\mathbf{TopAlg}_{(\Omega,\mathrm{E})}$, which is a null object in $\mathbf{TopAlg}_{(\Omega,\mathrm{E})}$.

The kernel of a homomorphism $\varphi : A \to A'$ in $\mathbf{TopGrp}$ is the equaliser of φ and the zero homomorphism $0 : A \to A'$ in $\mathbf{Alg}_{(\Omega,\mathrm{E})}$. Because $\mathbf{Alg}_{(\Omega,\mathrm{E})}$ is a reflective subcategory of $\mathbf{Top}$ (the inclusion as discrete spaces being right adjoint to the free algebra on the connected components functor, the underlying discrete spaces of limits in $\mathbf{Alg}_{(\Omega,\mathrm{E})}$ are the limits of the underlying discrete spaces in $\mathbf{Top}$. So the underlying discrete space of the kernel of φ in $\mathbf{Alg}_{(\Omega,\mathrm{E})}$ is the equaliser of the maps of discrete spaces underlying φ and 0. This allows us to conclude:

Lemma 9.3.67. *If the category* $\mathbf{Alg}_{(\Omega,\mathrm{E})}$ *contains a zero object, then the restriction* $|-| : \mathbf{Alg}_{(\Omega,\mathrm{E})}^{\mathcal{O}^{op}} \to \mathbf{TopAlg}_{(\Omega,\mathrm{E})}$ *of the geometric realisation functor preserves kernels.*

Proof. Let $\varphi : A \to A'$ be a homomorphism in $\mathbf{Alg}_{(\Omega,\mathrm{E})}^{\mathcal{O}^{op}}$ and $\ker(\varphi) : N \to A$ be its kernel. The geometric realisation functor $|-|$ maps the zero morphism $0 : A \to A'$ to the zero homomorphism $|0| : |A| \to |A'|$. Furthermore it preserves finite limits by Theorem 9.3.49. Thus the geometric realisation $|\ker(\varphi)|$ is the equaliser of $|\varphi|$ and $0 : |A| \to |A'|$ in $\mathbf{Top}$. Because the embedding $\mathbf{TopAlg}_{(\Omega,\mathrm{E})} \hookrightarrow \mathbf{Top}$ creates limits, $|\mathrm{coker}(\varphi)|$ is also the equaliser in $\mathbf{TopAlg}_{(\Omega,\mathrm{E})}$, i.e. it is the kernel of $|\varphi|$. □

Lemma 9.3.68. *If the category* $\mathbf{Alg}_{(\Omega,\mathrm{E})}$ *contains a zero object, then the restriction* $|-| : \mathbf{Alg}_{(\Omega,\mathrm{E})}^{\mathcal{O}^{op}} \to \mathbf{kHausAlg}_{(\Omega,\mathrm{E})}$ *of the geometric realisation functor preserves kernels.*

Proof. Assume that $\mathbf{Alg}_{(\Omega,\mathrm{E})}$ contains a null object and let $\varphi : A \to A'$ a morphism in $\mathbf{Alg}_{(\Omega,\mathrm{E})}^{\mathcal{O}^{op}}$. The geometric realisation $|\ker(\varphi)|$ of the kernel of φ is the kernel of $|\varphi|$ in $\mathbf{TopAlg}_{(\Omega,\mathrm{E})}$ by Lemma 9.3.67. So it remains to show, that this kernel also is the kernel of $|\varphi|$ in the subcategory $\mathbf{kHausAlg}_{(\Omega,\mathrm{E})}$. Because the geometric realisation $|A'|$ is Hausdorff, the singleton image of the zero homomorphism $|0|$ is closed, hence the equaliser of $|\varphi|$ and $|0|$ is the embedding $N \hookrightarrow A$ of a closed subalgebra in $\mathbf{Top}$. Since closed subspaces of compactly generated spaces are compactly generated, the algebra N is also compactly generated. It is Hausdorff as a subspace of a Hausdorff space, hence contained in $\mathbf{kHausAlg}_{(\Omega,\mathrm{E})}$. Therefore $|\ker(\varphi)|$ is also the equaliser of $|\varphi|$ and $|0|$ in $\mathbf{kHausAlg}_{(\Omega,\mathrm{E})}$, i.e. the kernel of $|\varphi|$ in $\mathbf{kHausAlg}_{(\Omega,\mathrm{E})}$. $\qquad\square$

Corollary 9.3.69. *The restriction* $|-| : \mathbf{Grp}^{\mathcal{O}^{op}} \to \mathbf{kHausGrp}$ *of the geometric realisation functor preserves kernels.*

Corollary 9.3.70. *The restriction* $|-| : \mathbf{Ab}^{\mathcal{O}^{op}} \to \mathbf{kHausAb}$ *of the geometric realisation functor preserves kernels.*

Corollary 9.3.71. *The restriction* $|-| : \mathbf{RMod}^{\mathcal{O}^{op}} \to \mathbf{kHausRMod}$ *of the geometric realisation functor preserves kernels.*

An analogous results holds for cokernels in categories $\mathbf{Alg}_{(\Omega,\mathrm{E})}$ with null object. A cokernel of a homomorphism $\varphi : A \to A'$ in $\mathbf{Alg}_{(\Omega,\mathrm{E})}$ is the coequaliser of φ and the zero homomorphism $0 : A \to A'$. The preservation of arbitrary colimits by the geometric realisation functor leads to the preservation of cokernels.

Lemma 9.3.72. *If the category* $\mathbf{Alg}_{(\Omega,\mathrm{E})}$ *contains a null object, then the restriction* $|-| : \mathbf{Alg}_{(\Omega,\mathrm{E})}^{\mathcal{O}^{op}} \to \mathbf{TopAlg}_{(\Omega,\mathrm{E})}$ *of the geometric realisation functor preserves cokernels.*

Proof. Let $\varphi : A \to A'$ be a morphism in $\mathbf{Alg}_{(\Omega,\mathrm{E})}^{\mathcal{O}^{op}}$ and $\mathrm{coker}(\varphi) : A' \to B$ be its cokernel. The cokernel $\mathrm{coker}(\varphi) : A' \to B$ in the category $\mathbf{Alg}_{(\Omega,\mathrm{E})}^{\mathcal{O}^{op}}$ is the coequaliser of φ and the zero morphism $0 : A \to A'$. Because the geometric realisation preserves colimits the homomorphism $|\mathrm{coker}(\varphi)| : |A'| \to |B|$ is the coequaliser of $|\varphi|$ and the zero homomorphism $0 : |A| \to |A'|$ in $\mathbf{Top}$. Because the forgetful functor $U : \mathbf{TopAlg}_{(\Omega,\mathrm{E})} \to \mathbf{Top}$ of the embedding creates colimits, the homomorphism $|\mathrm{coker}(\varphi)| : |A| \to |B|$ is the coequaliser of $|\varphi|$ and the zero homomorphism $0 : |A| \to |A'|$ in $\mathbf{TopAlg}_{(\Omega,\mathrm{E})}$. Thus the corestricted functor $|-| : \mathbf{Alg}_{(\Omega,\mathrm{E})}^{\mathcal{O}^{op}} \to \mathbf{TopAlg}_{(\Omega,\mathrm{E})}$ preserves cokernels. $\qquad\square$

Corollary 9.3.73. *If the category* $\mathbf{Alg}_{(\Omega,\mathrm{E})}$ *contains a null object, then the restriction* $|-| : \mathbf{Alg}_{(\Omega,\mathrm{E})}^{\mathcal{O}^{op}} \to \mathbf{kHausAlg}_{(\Omega,\mathrm{E})}$ *of the geometric realisation functor preserves cokernels.*

Proof. Let $\varphi : A \to A'$ be a morphism in $\mathbf{Alg}_{(\Omega,\mathrm{E})}^{\mathcal{O}^{op}}$ and $\mathrm{coker}(\varphi) : A' \to B$ be its cokernel. The geometric realisation $|B|$ of B is a compactly generated Hausdorff algebra, so the homomorphism $|\mathrm{coker}(\varphi)|$ is a morphism in $\mathbf{kHausAlg}_{(\Omega,\mathrm{E})}$. Being the coequaliser of 0 and $|\varphi|$ in $\mathbf{TopAlg}_{(\Omega,\mathrm{E})}$ it also is the coequaliser in $\mathbf{kHausAlg}_{(\Omega,\mathrm{E})}$. Thus the restriction $|-| : \mathbf{Alg}_{(\Omega,\mathrm{E})}^{\mathcal{O}^{op}} \to \mathbf{kHausAlg}_{(\Omega,\mathrm{E})}$ of the geometric realisation functor preserves cokernels. $\qquad\square$

Corollary 9.3.74. *The restriction* $|-| : \mathbf{Grp}^{\mathcal{O}^{op}} \to \mathbf{kHausGrp}$ *of the geometric realisation functor preserves cokernels.*

Corollary 9.3.75. *The restriction* $|-| : \mathbf{Rng}^{\mathcal{O}^{op}} \to \mathbf{kHausRng}$ *of the geometric realisation functor preserves cokernels.*

Corollary 9.3.76. *The restriction* $|-|: \mathbf{RMod}^{\mathcal{O}^{op}} \to \mathbf{kHausRMod}$ *of the geometric realisation functor preserves cokernels.*

Lemma 9.3.77. *The restriction* $|-|: \mathbf{Ab}^{\mathcal{O}^{op}} \to \mathbf{kHausAb}$ *of the geometric realisation functor is additive.*

Proof. Let $f, h : G \to G'$ be two morphisms of simplicial Abelian monoids and $[(g, t)] \in |G|$ be a point of $|G|$. We write $+$ for the addition in G and G' as well as for the additions in the geometric realisations $|G|$ and $|G'|$. The evaluation of the induced morphism $|f + h|$ at the point $[(g, t)]$ is given by

$$|f + h|([(g,t)]) = [((f + h)(g,t)] = [(f(g) + h(g), t)].$$

It remains to show the equality $[(f(g) + h(g), t)] = [(f(g), t)] + [(h(g), t)]$. For this observe that the image of the point $[(((f(g), h(g)), t)]$ under $\Phi_{G,G}$ is the pair

$$\Phi_{G,G}([((f(g), h(g)), t)]) = ((f(g), t), (h(g), t)) \in |G| \times |G|,$$

so the addition of the points $[(f(g), t)]$ and $[(h(g), t)]$ in G' is given by

$$\begin{aligned}
[(f(g), t)] + [(h(g), t)] &= |+| \Phi^{-1}([(f(g), t)], [(h(g), t)]) = |+| [(((f(g), h(g)), t)] \\
&= |+| q_G((f(g), h(g)), t) = q_{G'} F(+)((f(g), h(g)), t) \\
&= q_{G'}(f(g) + h(g), t) = [(f(g) + h(g), t)].
\end{aligned}$$

Therefore restriction $|-| : \mathbf{Ab}^{\mathcal{O}^{op}} \to \mathbf{kHausAb}$ of the geometric realisation functor is additive. $\qquad\square$

Moreover, the theorem on the preservation of finite limits under $|-|$ can not only be applied to algebraic structures of arbitrary operational type, but also to any kind of structure which can be considered to be a (possibly continuous) functor from a category $\mathbf{D}$ to $\mathbf{Set}$:

Lemma 9.3.78. *For any simplicial object X in the functor category $\mathbf{Set}^{\mathbf{D}}$ for a category $\mathbf{D}$ the composition $|-| \circ X$ is an element in $\mathbf{kHaus}^{\mathbf{D}}$. If X is continuous then $|-| \circ X$ is continuous.*

Examples of such structures are simplicial discrete transformation groups, G-modules for a group G and groupoids. Thus one observes:

Theorem 9.3.79. *The geometric realisation of a simplicial discrete transformation group, G-module or groupoid is a transformation group, G-module or topological groupoid respectively.*

Analogously, if M is a simplicial module over a simplicial ring R, then the geometric realisation $|M|$ of M is a topological module over the topological ring $|R|$. In case of a constant simplicial topological ring R we obtain:

Lemma 9.3.80. *The geometric realisation functor $|-|$ restricts to functors $|-|:$ $\mathbf{RMod}^{\mathcal{O}^{op}} \to \mathbf{TopRMod}$ and $|-| : \mathbf{ModR}^{\mathcal{O}^{op}} \to \mathbf{TopModR}$*

This especially implies that the geometric realisation of a simplicial vector space over a field $\mathbf{K}$ is a topological vector space over $\mathbf{K}$.

In the following we examine the preservation of products for arbitrary simplicial spaces X and Y. The first case we consider is that of a constant simplicial space X and an arbitrary simplicial space Y. Recall that the geometric realisation $|X|$ of the constant simplicial space X is naturally isomorphic to the space $X([0])$ (cf.

Example 9.1.33). If $\Phi_{X,Y}$ is an isomorphism this suggests that the product map $\mathrm{id}_{X([0])} \times q_Y$ also is a quotient mapping. We verify this conception below.

Henceforth we shall always assume X to be a constant simplicial space. This assumption leads to the equalities

$$F(X \times Y) = \coprod_{n \in \mathbb{N}} X([n]) \times Y([n]) \times \Delta_n = \coprod_{n \in \mathbb{N}} X([0]) \times Y([n]) \times \Delta_n$$

$$= X([0]) \times \left(\coprod_{n \in \mathbb{N}} Y([n]) \times \Delta_n \right) = X([0]) \times F(Y).$$

The product of the natural embedding $i_0 : X([0]) \hookrightarrow F(X)$ with the identity of $F(Y)$ yields a natural embedding $X([0]) \times F(Y) \to F(X) \times F(Y)$. All together these maps fit into a commutative diagram:

$$
\begin{array}{ccc}
F(X \times Y) \cong X([0]) \times F(Y) & \xrightarrow{i_0 \times \mathrm{id} F(Y)} & F(X) \times F(Y) \\
{\scriptstyle q_{X \times Y}} \downarrow & & \downarrow {\scriptstyle q_X \times q_Y} \\
|X \times Y| & \xrightarrow{\quad \Phi \quad} & |X| \times |Y| \cong X([0]) \times |Y|
\end{array}
$$

Taking the identification $|X| \cong X([0])$ into account one concludes that the composition $\Phi \circ (q_{X \times Y})$ equals the product map $\mathrm{id}_{X([0])} \times q_Y$ on $X([0]) \times F(Y)$.

Lemma 9.3.81. *If X, Y are simplicial spaces, X is a constant simplicial space and $\Phi : |X \times Y| \to |X| \times |Y|$ an isomorphism, then the product map $\mathrm{id}_{X([0])} \times q_Y$ is a quotient map.*

Proof. Let X, Y be simplicial spaces and assume X to be a constant simplicial space. If Φ is an isomorphism then the composition

$$\mathrm{id}_{X([0])} \times q_Y = \Phi \circ q_{X \times Y} : X([0]) \times F(Y) \to X([0]) \times |Y|$$

of the isomorphism Φ with the quotient map $q_{X \times Y}$ is a quotient. $\square$

The property of a quotient map that its product with arbitrary identities is again a quotient has been studied in [DK70]. It can be characterised in the following way:

Theorem 9.3.82 (Day,Kelly). *Binary products of a continuous surjection f from Y onto Y' with arbitrary identity maps are quotients if and only if f satisfies the following condition:*

Property 9.3.83. Given any point $y' \in Y'$, a neighbourhood V of y' and an open covering $\{U_\alpha\}_{i \in A}$ of $f^{-1}(V)$ there exists a finite set $\alpha_1, \ldots, \alpha_n$ such that the set

$$\bigcap_{W \text{ is nbh. of } f(\cup U_{\alpha_i})} W$$

is a neighbourhood of y'.

Proof. See [DK70, Theorem 2] for a proof.

The fact that the space $F(Y)$ for simplicial space Y is an infinite disjoint union suggests that the spaces $F(Y)$ do in general not have the above Property 9.3.83. In the following we will use the fact that $F(Y)$ is an infinite disjoint union to show that the general geometric restriction functor $\mathbf{Top}^{\mathcal{O}^{op}} \to \mathbf{Top}$ does not preserve products. At first we observe that Property 9.3.83 is necessary for products to be preserved:

Proposition 9.3.84. *If the product functors* $- \times Y$ *and* $- \times |Y|$ *intertwine the geometric realisation functor then the quotient map* q_Y *has Property 9.3.83.*

Proof. If the product functors $- \times Y$ and $- \times |Y|$ intertwine the geometric realisation functor then Lemma 9.3.81 implies that $\mathrm{id}_X \times q_Y$ is a quotient map for every topological space X. By Theorem 9.3.82 the quotient map q_Y has Property 9.3.83.   □

In case the quotient space $|Y|$ is Hausdorff the condition 9.3.83 can be reformulated using the notion of bi-quotients:

Definition 9.3.85. *A function* $f : X \to Y$ *between topological spaces X and Y is called a* bi-quotient, *if for every point $y \in Y$ and any open cover $\{U_i\}_{i \in I}$ of $f^{-1}(y)$ finitely many images $f(U_i)$ cover a neighbourhood of y.*

The important Properties of biquotients are summarised in the following two observations:

Proposition 9.3.86 (Day,Kelly). *If Y' is Hausdorff and $f : Y \to Y'$ satisfies condition 9.3.83 then f is a biquotient.*

Proof. See [DK70, Proposition 4] for a proof.

Theorem 9.3.87 (Michael). *A map $f : Y \to Y'$ onto a Hausdorff space Y' is a biquotient if and only if its product with any identity map is a quotient map.*

Proof. A proof can be found in [Mic68, theorem 1.3].

Together with the preceding considerations these Properties of biquotients combine to give the following Theorem:

Theorem 9.3.88. *If $|Y|$ is Hausdorff then the product functor $- \times Y$ intertwines the geometric realisation functor if and only if $q_Y : F(Y) \to |Y|$ is a biquotient.*

Proof. If $|Y|$ is Hausdorff and $- \times Y$ intertwines the geometric realisation functor then q_Y satisfies condition 9.3.83 and thus is a biquotient by Proposition 9.3.86. Conversely, if q_Y is a biquotient then its product with any identity map is again a quotient map by Theorem 9.3.87.   □

Although the geometric restriction functor behaves nicely when restricted to certain subcategories we shall see that it does not preserve products in general.

Proposition 9.3.89. *If Y is a Hausdorff space containing more that one point and $O = \coprod O_n \subset \mathrm{SS}(Y)$ a non-empty compatible open subset of $F(Y)$ then there exists a sequence $\{p_{k+n}\}_{n \in \mathbb{N}}$ of non-degenerate points in O_{k+n} respectively.*

Proof. Let Y be a non singleton Hausdorff space and $O = \coprod O_n \subset \mathrm{SS}(Y)$ be a non-empty compatible open subset of $F(Y)$. Let k be the smallest number such that O_k is not empty. (This number exists because O is non-empty.) Since $\coprod O_n$ is compatible, all points in O_k are non-degenerate. Let p_k be such a non-degenerate point in O_k. We show by induction that there exists a non-degenerate point $p_{k+n+1} \in O_{k+n}$ for each non-degenerate point $p_{k+n} \in O_{k+n}$. Let $p_{k+n} \in O_{k+n}$ be a non-degenerate point in O_{n+k}. This point is of the form $(\mathbf{y}, \mathbf{t})$ with $\mathbf{y} \in Y^{n+k+1}$ and $\mathbf{t} \in \Delta_{k+n}$ non-degenerate. Since Y contains more than one point there exists a point $y_{k+n+1} \in Y$ such that $y_{k+n} \neq y_{k+n+1}$, i.e. $\mathbf{y}' = (y_0, \dots, y_{k+n+1})$ is non-degenerate. The points $(\mathbf{y}, \mathbf{t})$ and $(\mathbf{y}', \hat{\mathbf{t}}) = (\mathbf{y}', (t_0, \dots, t_{k+n}, 0))$ are equivalent. Because O is compatible this implies that $(\mathbf{y}', \hat{\mathbf{t}})$ is contained in O_{k+n+1}. Since O_{k+n+1} is open, there exists a basic open neighbourhood $V \times W$ of $(\mathbf{y}', \hat{\mathbf{t}})$ in $O_{k+n+1} \subset Y([k+n+1]) \times \Delta_{k+n+1}$. The open set $V \subset \Delta_{k+n+1}$ meets the interior of Δ_{k+n+1}. Thus there exists a non-degenerate $\mathbf{t}'$ in V and $(\mathbf{y}', \mathbf{t}')$ is a non-degenerate point in O_{n+k+1}.   □

Lemma 9.3.90. *If Y is a Hausdorff space containing more that one point then the image $q_Y(O)$ of any finite union $O = \coprod_{n=1}^{m} O_n$ of open subsets $O_n \in Y([n]) \times \Delta_n$ has empty interior.*

Proof. Let Y be Hausdorff and $O = \coprod_{n=1}^{m} O_n$ be a finite union of open subsets $O_n \in Y([n]) \times \Delta_n$. If the interior U of $q_Y(O)$ were non-empty, then its inverse image $V = q_Y^{-1}(U)$ were a compatible open subset of $F(Y)$. By Proposition 9.3.89 there would exist a sequence $\{p_{k+n}\}_{n\in\mathbb{N}}$ of non-degenerate points in V_{k+n} respectively. The images of these points were contained in U by construction, hence each point p_{k+n} would be equivalent to some point in O. Since all the points p_{k+n} were non-degenerate, this is impossible for $k + n > m$. Thus the image $q_Y(O)$ has empty interior. $\square$

Theorem 9.3.91. *The geometric realisation functor $|-| : \mathbf{Top}^{\mathcal{O}^{op}} \to \mathbf{Top}$ does not preserve products.*

Proof. Let Y be a Hausdorff space containing more than one point and let p be a point in $\mathrm{SS}(Y)$. Then $\{\mathrm{SS}(Y)([n]) \times \Delta_n\}_{n\in\mathbb{N}}$ is an open covering of $q_Y^{-1}(p)$. By Lemma 9.3.90 no neighbourhood of p is covered by finitely many images $q_Y(Y([n]) \times \Delta_n)$. Therefore q_Y is not a biquotient. By Theorem 9.3.87 the product functor $-\times\mathrm{SS}(Y)$ does not intertwine the geometric realisation functor, hence the geometric realisation functor does not preserve products in general. $\square$

Though the natural quotient morphisms $q_Y : F(Y) \to Y$ are not biquotients in general, there exists an important class of spaces where the natural quotient maps q_Y are always biquotients.

Definition 9.3.92. *A discrete simplicial space is called* locally finite, *if each point $p \in X([n])$ is contained in the boundary of only finitely many non-degenerate points.*

Example 9.3.93. The discrete simplicial spaces $\hom(-, [n])$ are all locally finite.

We denote category of locally finite discrete simplicial spaces and maps between them by $\mathbf{Set}_{lf}$. It is a very convenient category because the locally finiteness of a discrete simplicial space X has strong influences on the topology of its geometric realisation:

Theorem 9.3.94. *For a discrete simplicial space X the following are equivalent:*

1. *The space X is locally finite.*
2. *The geometric realisation $|X|$ is locally compact.*
3. *The geometric realisation $|X|$ is metrisable.*
4. *The geometric realisation $|X|$ is first countable.*

Proof. See [tD91, Satz 2.4] for a proof.

Example 9.3.95. The geometric realisation of the discrete simplicial topological space $\hom(-, [n])$ is the standard simplex Δ_n. It is a compact metric space.

Definition 9.3.96. *A function $F : Y \to Y'$ is called* compact covering *if every compact subset of Y' is the image of some compact subset of Y.*

Lemma 9.3.97. *The natural quotient map $q_X : F(X) \to |X|$ of a discrete simplicial space is compact covering.*

Proof. Let X be a discrete simplicial space and $K \subset |X|$ be compact. Since K is compact, it meets only finitely many cells of the CW-complex $|X|$ (cf. [tD91, Satz 2.3]). Each such cell is the image of a standard simplex $\{s\} \times \Delta_n \subset F(X)$ with $s \in X([n])$. Let S denote the finite set of simplices, such that the image $q_X(\{s\} \times \Delta_n)$ meets K. Then K is contained in the image of the set $\bigcup_{s \in S} \{s\} \times \Delta_n$ and this set is a finite union of compact sets, hence compact. The inverse image $q_Y^{-1}(K)$ is a closed subset of this compact set, so K is the image of the compact subset $\bigcup_{s \in S} \{s\} \times \Delta_n \cap q_Y^{-1}(K)$. □

Lemma 9.3.98. *If X is a locally finite discrete simplicial space, then the natural map q_X is a bi-quotient.*

Proof. Let X be a locally finite discrete simplicial space. The geometric realisation $|X|$ of X is Hausdorff and locally compact by Theorem 9.3.94. In addition, the quotient map q_X is compact covering, so [Mic68, Proposition 3.3] guarantees that q_X is a bi-quotient. □

Proposition 9.3.99. *If X and Y are simplicial spaces and Y is discrete and locally finite, then the product $q_X \times q_Y$ of the quotient maps is a quotient map.*

Proof. Let X and Y be simplicial spaces and assume Y to be discrete and locally finite. The quotient map q_Y is a biquotient by Lemma 9.3.98, so the product map $\mathrm{id}_{F(X)} \times q_Y$ is a quotient map. Because the space $|Y|$ is locally compact, the product functor $- \times |Y|$ is cocontinuous; hence it preserves colimits and therefore $q_X \times \mathrm{id}_{|Y|}$ is a quotient map as well. Since the composition of quotient maps is again a quotient map, the product $q_X \times q_Y = (q_X \times \mathrm{id}_{|Y|}) \circ (\mathrm{id}_{F(X)} \times q_Y)$ is a quotient map as well. □

Corollary 9.3.100. *If Y is a locally finite discrete simplicial space, then the map $\coprod \varphi_{p,q} : F(X) \times F(Y) \to |X \times Y|$ factors through a map $\varphi : |X| \times |Y| \to |X \times Y|$.*

Theorem 9.3.101. *The restriction of Φ to $|-| \circ \prod : \mathbf{Top}^{\mathcal{O}^{op}} \times \mathbf{Set}_{lf}^{\mathcal{O}^{op}} \to \mathbf{Top}$ is a natural isomorphism.*

Example 9.3.102. The functions $\Phi : |X \times \hom(-, [n])| \to |X| \times \Delta_n$ are natural isomorphisms for all simplicial spaces X.

We now turn our attention to the preservation of homotopy. Recall (cf. Definition F.6.1) that a homotopy from a morphism $f : X \to Y$ of simplicial spaces to a morphism $g : X \to Y$ consists of continuous functions $h_{n,0}, \ldots, h_{n,n}$ such that $X(\eta_0)h_{n,0} = f([n])$, $X(\eta_{n+1})h_{n,n} = g([n])$ and

$$Y(\epsilon_i)h_{n,j} = \begin{cases} h_{n-1,j-1}X(\epsilon)_i & \text{if } i < j \\ Y(\epsilon_i)h_{n,i-1} & \text{if } i = j \neq 0 \ , \\ h_{n-1,j}X(\epsilon_{i-1}) & \text{if } i > j+1 \end{cases}$$

$$Y(\eta_i)h_j = \begin{cases} h_{n+1,j+1}X(\eta_i) & \text{if } i \leq j \\ h_{n+1,j}X(\eta_{i-1}) & \text{if } i > j \end{cases} \ .$$

As in the case of Abelian categories or the category of sets the notion of simplicial homotopy in $\mathbf{Top}^{\mathcal{O}^{op}}$ can be expressed by use of the simplicial set $\hom(-, [1])$ regarded as a discrete simplicial space. Recall that the simplicial space $\hom(-, [0])$ is a singleton simplicial space. The face maps $\epsilon_0 : [0] \hookrightarrow [1]$ and $\epsilon_1 : [0] \hookrightarrow [1]$ in $\mathcal{O}$ induce embeddings

$$\hom(-, \epsilon_0) : * = \hom(-, [0]) \hookrightarrow \hom(-, [1])$$
$$\hom(-, \epsilon_1) : * = \hom(-, [0]) \hookrightarrow \hom(-, [1])$$

of simplicial spaces. Identifying the simplicial space X with the simplicial product space $X \times \hom(-, [0])$ one obtains embeddings $j_{X,0} = \mathrm{id}_X \times \hom(-, \epsilon_0)$ and $j_{X,1} = \mathrm{id}_X \times \hom(-, \epsilon_1)$ of X into the product space $X \times \hom(-, [1])$. By the use of these embeddings the simplicial homotopies in $\mathbf{Top}^{\mathcal{O}^{op}}$ can be characterised in the following way:

Theorem 9.3.103. *There exists a one-to-one correspondence between simplicial homotopies $f \simeq g : X \to X'$ in $\mathbf{Top}^{\mathcal{O}^{op}}$ and morphisms $h : X \times \hom(-, [1]) \to X'$ of simplicial topological spaces making the following diagram commutative:*

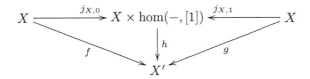

Proof. The proof of [May92, Proposition 6.2] uses maps that are continuous in our context. It thus carries over word by word. □

The morphism $h : X \times \hom(-, [1]) \to X'$ of simplicial spaces in the above characterisation is also called a *homotopy* from f to g.

Theorem 9.3.104. *The geometric realisation functor $| - | : \mathbf{Top}^{\mathcal{O}^{op}} \to \mathbf{Top}$ preserves homotopy, i.e. if f and g are homotopic maps of simplicial spaces then the induced maps $|f|$ and $|g|$ are homotopic as well.*

Proof. Let f and g be homotopic maps $X \to X'$ of simplicial spaces. We identify the space X with the simplicial space $X \times \hom(-, [0])$. By Theorem 9.3.103, there exists a map $h : X \times \hom(-, [1]) \to Y$ of simplicial spaces such that $h \circ \hom(-, \epsilon_0) = f$ and $h \circ \hom(-, \epsilon_1) = g$. The geometric realisation of $\hom(-, [1])$ is the standard simplex Δ_1. So the composition

$$|X| \times \Delta_1 \xrightarrow{\Phi^{-1}} |X \times \hom(-, [1])| \xrightarrow{|h|} Y$$

is a homotopy from $|f|$ to $|g|$. □

For further reference we note that the geometric realisation functor restricted to varius categories of topological groups, preserves kernels and cokernels, when considered as a functor into $\mathbf{Top}$.

Lemma 9.3.105. *The restriction $| - | : \mathbf{TopGrp}^{\mathcal{O}^{op}} \to \mathbf{Top}$ of the geometric realisation functor preserves monomorphisms.*

Proof. A monomorphism $f : G \to G'$ in $\mathbf{TopGrp}^{\mathcal{O}^{op}}$ is a morphism of simplicial topological groups, whose components $f([n]) : G([n]) \to G'([n])$ are monomorphisms in $\mathbf{TopGrp}$. The monomorphisms in $\mathbf{TopGrp}$ are the injective homomorphisms, so any monomorphism $f : G \to G'$ of simplicial topological groups is injective. By Lemma 9.1.4 the induced homomorphism $|f| : |G| \to |G'|$ in $\mathbf{Top}$ is injective as well, hence is a monomorphism in $\mathbf{Top}$. □

Lemma 9.3.106. *The restriction $| - | : \mathbf{TopGrp}^{\mathcal{O}^{op}} \to \mathbf{Top}$ of the geometric realisation functor preserves epimorphisms.*

Proof. An epimorphism $f : G \to G'$ in $\mathbf{TopGrp}^{\mathcal{O}^{op}}$ is a morphism of simplicial topological groups, whose components $f([n]) : G([n]) \to G'([n])$ are epimorphisms in $\mathbf{TopGrp}$. The epimorphisms in $\mathbf{TopGrp}$ are known to be exactly the surjective homomorphisms. (See Lemma E.1.5.) Thus the geometric realisation $|\varphi|$ of an epimorphism φ in $\mathbf{TopGrp}^{\mathcal{O}^{op}}$ is a surjective homomorphism in $\mathbf{Top}$, hence an epimorphism. □

This proof applies to all subcategories of **TopGrp** in which the epimorphisms are the surjective homomorphisms. So we find:

Lemma 9.3.107. *The restriction of the geometric realisation functor to the categories* **TopAb**$^{\mathcal{O}^{op}}$ *and* **CompHausGrp**$^{\mathcal{O}^{op}}$ *preserve epimorphisms.*

Proof. The epimorphisms in **TopAb** and **CompHausGrp** are exactly the surjective homomorphisms. (See Lemma E.1.4 and [Pog70] for proofs thereof.)   □

10

Preservation of Homotopy

10.1 Preservation of Homotopy

In this section we show that the simplicial space functor $C(\Delta, -) : \mathbf{Top} \to \mathbf{Top}^{\mathcal{O}^{op}}$ preserve homotopies. In addition it is shown that the counit $\epsilon : |C(\Delta, -)| \to \mathrm{id}_{\mathbf{Top}}$ of the adjunction between the geometric realisation functor functor $|-|$ and the singular simplicial space functor $C(\Delta, -)$ provides a homotopy equivalence $|C(\Delta, -)| \simeq \mathrm{id}_{\mathbf{Top}}$. Therefore every topological space X is naturally equivalent to the geometric realisation $|C(\Delta, X)|$ of its singular simplicial space $C(\Delta, X)$, i.e. it is up to homotopy recovered by the geometric realisation $|C(\Delta, X)|$ of its singular simplicial space $C(\Delta, X)$. In particular application of the singular simplicial space functor $C(\Delta, -)$ does not loose any information on the homotopy type of topological spaces. Summarising we observe that all morphisms in the commutative diagram

$$\mathrm{id}_{\mathbf{Top}} \xrightarrow{\quad i \quad} |C(\Delta, -)| \xrightarrow{\quad \epsilon \quad} \mathrm{id}_{\mathbf{Top}} ,$$
$$\mathrm{id}$$

are homotopy equivalences. This generalises the classical Theorem concerning the geometric realisation of simplicial sets $\mathbf{Set}^{\mathcal{O}^{op}} \to \mathbf{kHaus}$ which states that the geometric realisation $|\hom_{\mathbf{Top}}(\Delta, X)|$ of the discrete singular simplicial set $\hom(\Delta, X)$ of a CW-complex X is homotopy equivalent to X itself. This classical theorem is proved by abstract nonsense using the Whitehead-Serre Theorem, which is not available for arbitrary topological spaces. We prove the generalisation by explicitly constructing the homotopy equivalence and showing that it is natural in the topological space X.

We start with the counit $\epsilon : |C(\Delta, -)| \to \mathrm{id}_{\mathbf{Top}}$. Recall that the counit ϵ fits into the commutative diagram

$$\mathrm{id}_{\mathbf{Top}} \xrightarrow{\quad i \quad} |C(\Delta, -)| \xrightarrow{\quad \epsilon \quad} \mathrm{id}_{\mathbf{Top}} ,$$
$$\mathrm{id}$$

where for each topological space Y the function $i_Y : Y \to |C(\Delta, Y)|$ is given by assigning to a point $y \in Y$ the equivalence class $[(y, 1)] \in |C(\Delta, Y)|$ of the constant 0-simplex y in Y. The counit ϵ_Y is induced by evaluation and given by the assignment

$$\epsilon_Y : |C(\Delta, Y)| \to Y, \quad [(\sigma, t)] \mapsto \sigma(t).$$

The homotopy equivalence of $|C(\Delta, -)|$ and $\mathrm{id}_{\mathbf{Top}}$ will be shown by explicitly constructing a functor

$$H : |C(\Delta, -)| \times I \to |C(\Delta, -)|$$

such that $H(-,1) = \mathrm{id}_{|C(\Delta,-)|}$ and $H(-,0) = i\epsilon$. This then implies that H_Y is a homotopy $\mathrm{id}_{|C(\Delta,Y)|} \simeq i_Y\epsilon_Y$ for each topological space Y. In addition the homotopy H_Y should be a loop homotopy if Y is a (partial) topological algebra. Anticipating the result in Theorem 10.1.15 the homotopy H_Y will be given by the map

$$|C(\Delta,Y)| \times I \longrightarrow |C(\Delta,Y)|, \quad ([(\sigma,t)],s) \mapsto [(t' \mapsto \sigma(st + (1-s)t'),t)]$$

which yet is not known to be well-defined. In the following we show that the above map is indeed well defined and a homotopy $\mathrm{id}_{|C(\Delta,Y)|} \simeq i_Y\epsilon_Y$. We begin with the description of the composition of arguments s, t and t' in the above suggestion for the homotopy H_Y:

Definition 10.1.1. *For each natural number $n \in \mathbb{N}$ we denote with h_n the function*

$$I \times \Delta_n \times \Delta_n \to \Delta_n, \quad (s,t,t') \mapsto st + (1-s)t'.$$

The maps h_n turn out to commute with boundary and degeneracy maps, if these maps are applied simultaneously to all factors Δ_n. For the precise description we will need the diagonal functor $\mathcal{O} \to \mathcal{O} \times \mathcal{O}$ which we denote by $D_{\mathcal{O}}$. The functor

$$\mathcal{O} \times \mathcal{O} \to \mathbf{Top}, \quad ([n],[m]) \mapsto I \times \Delta_n \times \Delta_m, \quad (\alpha,\beta) \mapsto \mathrm{id}_I \times \Delta(\alpha) \times \Delta(\beta),$$

will be denoted by $I \times \Delta \times \Delta$. It is a bisimplicial object in **Top**. Composition of this bifunctor with the diagonal functor $D_{\mathcal{O}}$ yields a simplicial space $(I \times \Delta \times \Delta)D_{\mathcal{O}}$.

Lemma 10.1.2. *The maps h_n form a natural transformation $h : (I \times \Delta \times \Delta)D_{\mathcal{O}} \to \Delta$.*

Proof. Let $\alpha : [m] \to [n]$ be a morphism in $\mathcal{O}$, $t, t' \in \Delta_m$ be points in Δ_m and $s \in I$ be a point in I. The evaluation of $h_n(I \times \Delta \times \Delta)D_{\mathcal{O}}(\alpha)$ at (s,t,t') is by definition given by

$$
\begin{aligned}
[h_n(I \times \Delta \times \Delta)D_{\mathcal{O}}(\alpha)](s,t,t') &= h_n[I \times \Delta(\alpha) \times \Delta(\alpha)](s,t,t') \\
&= h_n(s, \Delta(\alpha)(t), \Delta(\alpha)(t')) \\
&= s[\Delta(\alpha)(t)] + (1-s)[\Delta(\alpha)(t')] \\
&= \Delta(\alpha)[st + (1-s)t'] \\
&= \Delta(\alpha)(h_m(s,t,t')).
\end{aligned}
$$

So the functions h_n intertwine the maps $(I \times \Delta \times \Delta)D_{\mathcal{O}}(\alpha)$ and $\Delta(\alpha)$ and thus form a natural transformation $(I \times \Delta \times \Delta)D_{\mathcal{O}} \to \Delta$. $\square$

Note, that the functions h_n pick the first argument t and the second argument t' for the boundary values $s = 1$ and $s = 0$ respectively:

Lemma 10.1.3. *The restrictions $h(1,-,-)$ and $h(0,-,-)$ of the natural transformation h are the projections onto the second and onto the last factor of $I \times \Delta \times \Delta$ respectively:*

$$h(1,-,-) = pr_2 : (I \times \Delta \times \Delta)D_{\mathcal{O}} \to \Delta, \quad h(0,-,-) = pr_3 : (I \times \Delta \times \Delta)D_{\mathcal{O}} \to \Delta$$

The natural transformation h can be inserted in the bifunctor $C(-,-)$ to obtain a natural transformation

$$C(h,-) : C(\Delta,-) \to C((I \times \Delta \times \Delta)D_{\mathcal{O}},-)$$

of bifunctors. Recall that the exponential maps of function spaces form a natural transformation $\varphi : C(- \times -,-) \to C(-,C(-,-))$ of trifunctors (cf. B.5.11). So the exponential maps $\varphi_{I \times \Delta_n, \Delta_n, Y}$ form natural transformations

$$\varphi_{I \times \Delta_n, \Delta_n, -} : C(I \times \Delta_n \times \Delta_n, -) \to C(I \times \Delta_n, C(\Delta_n, -)), \quad f \mapsto \hat{f}$$

of functors for each $n \in \mathbb{N}$. Furthermore the standard simplices Δ_n are locally compact, so the transformations $\varphi_{I \times \Delta_n, \Delta_n, -}$ are natural isomorphisms (cf. B.5.18). The composition of these natural isomorphisms with the functor $C(I \times \Delta \times \Delta, -)D_{\mathcal{O}}$ turns out also to be natural in $[n]$:

Lemma 10.1.4. *The exponential functions $\varphi_{I \times \Delta_n, \Delta_n, -}$ form a natural isomorphism $\varphi : C((I \times \Delta \times \Delta)D_{\mathcal{O}}, -) \to C(I \times \Delta, C(\Delta, -)) \circ (D_{\mathcal{O}} \times \mathrm{id}_{\mathbf{Top}})$ of bifunctors.*

Proof. The naturality follows from the fact that $\varphi : C(- \times -, -) \to C(-, C(-, -))$ is a natural transformation of trifunctors $\mathbf{Top}^{op} \times \mathbf{Top}^{op} \times \mathbf{Top} \to \mathbf{Top}$. That each of the maps $\varphi_{I \times \Delta_n, \Delta_n, Y}$ is a homeomorphism follows from the local compactness of the standard simplices (cf. B.5.18). $\square$

The natural transformations φ and $C(h, -)$ can be composed to obtain a natural transformation $\varphi : C(h, -) : C(\Delta, -) \to C(I \times \Delta, C(\Delta, -)) \circ (D_{\mathcal{O}} \times \mathrm{id}_{\mathbf{Top}})$.

Lemma 10.1.5. *All the evaluation maps*

$$\mathrm{ev}_{I \times \Delta_n} : C(I \times \Delta_n, C(\Delta_n, Y)) \times I \times \Delta_n \to C(\Delta_n, Y), \quad (f, s, t) \mapsto f(s, t)$$

form a natural transformation $\mathrm{ev}_n : C(I \times \Delta_n, C(\Delta_n, -)) \times I \times \Delta_n \to C(\Delta_n, -)$.

Proof. Because the standard simplices are locally compact, each of the evaluations $\mathrm{ev}_{I \times \Delta_n} : C(I \times \Delta_n, -) \times I \times \Delta_n \to \mathrm{id}_{\mathbf{Top}}$ is a natural transformation (by Lemma B.5.5). Composing the natural transformation $\mathrm{ev}_{I \times \Delta_n}$ with the functor $C(\Delta_n, -)$ yields a natural transformation $\mathrm{ev}_n : C(I \times \Delta_n, C(\Delta_n, -)) \times I \times \Delta_n \to C(\Delta_n, -)$. $\square$

These natural transformations can be composed with the natural transformations $\varphi \times \mathrm{id}_{I \times \Delta}$ and $C(h, -) \times I \times \Delta$. First, the composition of the latter natural transformations can be written as

$$(\varphi \times \mathrm{id}_{I \times \Delta}) \circ (C(h, -) \times I \times \Delta) = (\varphi_{I \times \Delta_n, \Delta_n, -} C(h_n, -)) \times \mathrm{id}_{I \times \Delta_n}$$

Then, for each $n \in \mathbb{N}$ the composition with the natural transformation $\mathrm{ev}_{I \times \Delta_n}$ yields a family of natural transformations:

$$\mathrm{ev}_{I \times \Delta_n} \circ [(\varphi_{I \times \Delta_n, \Delta_n, -} C(h_n, -)) \times \mathrm{id}_{I \times \Delta_n}] : C(\Delta_n, -) \times I \times \Delta_n \to C(\Delta_n, -)$$

Definition 10.1.6. *The transformation $\mathrm{ev}_{I \times \Delta_n}[(\varphi_{I \times \Delta_n, \Delta_n, -} C(h_n, -)) \times \mathrm{id}_{I \times \Delta_n}]$ is denoted by $\tilde{h}_n$.*

Lemma 10.1.7. *For each topological space Y the map $\tilde{h}_{n,Y}$ is explicitly given by*

$$\tilde{h}_{n,Y} : C(\Delta_n, Y) \times I \times \Delta_n \to C(\Delta_n, Y), \quad \tilde{h}_{n,Y}(\sigma, s, t) = (t' \mapsto \sigma(st + (1 - s)t')).$$

Proof. Let Y be a topological space, $\sigma \in C(\Delta_n, Y)$ be an n-simplex in Y, $t \in \Delta_n$ be a point in the n-dimensional standard simplex and $s \in I$ a point in the unit interval. The image of the point $(\sigma, s, t) \in C(\Delta_n, Y) \times I \times \Delta_n$ under the map $\tilde{h}_{n,Y}$ is given by

$$\begin{aligned}
\tilde{h}_{n,Y}(\sigma, s, t) &= [\mathrm{ev}_n(\varphi_{I \times \Delta_n, \Delta_n, Y} \times \mathrm{id}_{I \times \Delta_n})][C(h_n, Y) \times \mathrm{id}_{I \times \Delta_n}](\sigma, s, t) \\
&= [\mathrm{ev}_n(\varphi_{I \times \Delta_n, \Delta_n, Y} \times \mathrm{id}_{I \times \Delta_n})](C(h_n, Y)(\sigma), s, t) \\
&= [\mathrm{ev}_n(\varphi_{I \times \Delta_n, \Delta_n, Y} \times \mathrm{id}_{I \times \Delta_n})](\sigma \circ h_n, s, t) \\
&= \mathrm{ev}_n(\varphi_{I \times \Delta_n, \Delta_n, Y}(\sigma \circ h_n), s, t) \\
&= [(\varphi_{I \times \Delta_n, \Delta_n, Y})(\sigma \circ h_n)](s, t) \\
&= \widehat{\sigma \circ h_n}(s, t),
\end{aligned}$$

where $\widehat{\sigma \circ h_n} : \Delta_n \times I \to C(\Delta_n, Y)$ is the adjoint map to $\sigma \circ h_n : I \times \Delta_n \times \Delta_n \to Y$. Thus the evaluation of the map $\tilde{h}_{n,Y}(\sigma, s, t)$ at a point $t' \in \Delta_n$ is given by

$$\tilde{h}_{n,Y}(\sigma, s, t)(t') = \widehat{\sigma \circ h_n}(s, t)(t') = \sigma \circ h_n(s, t, t') = \sigma(st + (1 - s)t').$$

This is the equality that was to be proved.   □

The image of a point $(\sigma, t) \in C(\Delta_n, Y) \times \Delta_n$ under the evaluation map is a point $\sigma(t) \in Y$. Using the natural isomorphism $Y \cong C(\Delta_0, Y)$ this point can be regarded a constant 0-simplex. Alternatively it can be regarded as a constant n-simplex. Formally this is described by applying the map $C(\Delta(\eta_{[n]\setminus[n-1]}), Y)$ to the constant 0-simplex $\sigma(t)$. (Recall that $\eta_{[n]\setminus[n-1]}$ is the unique morphism $[n] \twoheadrightarrow [0]$.) Interpreting the evaluation map in this sense one obtains:

Lemma 10.1.8. *The restrictions $\tilde{h}_n(-, 1, -)$ and $\tilde{h}_n(-, 0, -)$ of the natural transformations $\tilde{h}_n$ are the transformation $C(\Delta(\eta_{[n]\setminus[n-1]}), Y)\mathrm{ev}_n$ and the projection pr_1 onto the first factor of $C(\Delta_n, -) \times \Delta_n$ respectively.*

Proof. Let Y be a topological space and $\sigma \in C(\Delta_n, Y)$ be a singular n-simplex in Y. From the preceding Lemma we infer the equalities

$$\tilde{h}_{n,Y}(\sigma, 1, t)(t') = \sigma(t + 0 \cdot t') = \sigma(t) \quad \text{and} \quad \tilde{h}_{n,Y}(\sigma, 0, t)(t') = \sigma(0 \cdot t + t') = \sigma(t')$$

for all points t, t' in the standard n-simplex Δ_n. So the map $\tilde{h}_{n,Y}(\sigma, 1, t)$ is the constant n-simplex $\sigma(t)$ in Y and the map $\tilde{h}_{n,Y}(\sigma, 0, t)$ is the n-simplex σ in Y. Therefore the maps $\tilde{h}_{n,Y}(-, 1, -)$ and $C(\Delta(\eta_{[n]\setminus[n-1]}), Y)\mathrm{ev}_n$ coincide and the map $\tilde{h}_{n,Y}(-, 0, -)$ is the projection pr_1 onto the first factor of $C(\Delta_n, Y) \times I \times \Delta_n$.   □

Lemma 10.1.9. *The natural transformations $\tilde{h}_n$ satisfy the equations*

$$\tilde{h}_m \circ (C(\Delta(\alpha), -) \times \mathrm{id}_{I \times \Delta_m}) = C(\Delta(\alpha), -) \circ \tilde{h}_n \circ (\mathrm{id}_{C(\Delta_n, -) \times I} \times \Delta(\alpha))$$

for all morphisms $\alpha : [m] \to [n]$ in $\mathcal{O}$.

Proof. Let Y be a topological space, $\alpha : [m] \to [n]$ be a morphism in $\mathcal{O}$. For each singular n-simplex $\sigma \in C(\Delta_n, Y)$ in Y and each point $(s, t) \in I \times \Delta_m$ the evaluation of $\tilde{h}_{m,Y}$ at $(C(\Delta(\alpha), Y)(\sigma), s, t)$ in $C(\Delta_m, Y) \times I \times \Delta_m$ is given by

$$\begin{aligned}
\tilde{h}_{m,Y}(C(\Delta(\alpha), Y)(\sigma), s, t) &= \tilde{h}_{m,Y}(\sigma \circ \Delta(\alpha), s, t) \\
&= (t' \mapsto [\sigma \circ \Delta(\alpha)](st + (1 - s)t')) \\
&= (t' \mapsto \sigma(s[\Delta(\alpha)(t)] + (1 - s)[\Delta(\alpha)(t')])) \\
&= \left[\tilde{h}_{n,Y}(\sigma, s, \Delta(\alpha)(t)) \right] (\Delta(\alpha)(t')) \\
&= \left[C(\Delta(\alpha), Y) \left(\tilde{h}_{n,Y}(\sigma, s, \Delta(\alpha)(t)) \right) \right] (t') \\
&= \left[C(\Delta(\alpha), Y)\tilde{h}_{n,Y}(\mathrm{id}_{C(\Delta_n, -) \times I} \times \Delta(\alpha))(\sigma, s, t) \right] (t'),
\end{aligned}$$

which proves the stated equality.   □

Note that the projection $\mathrm{pr}_3 : C(\Delta, -) \times I \times \Delta \to \Delta$ onto the last factor is a natural transformation of functors $\mathcal{O} \times \mathbf{Top} \times \mathcal{O} \to \mathbf{Top}$. Therefore the composition $(\tilde{h}_n, \mathrm{pr}_3) : C(\Delta_n, -) \times I \times \Delta_n \to C(\Delta_n, -) \times \Delta_n$ is a natural transformation for each $n \in \mathbb{N}$. The coproduct of these natural transformations is a natural transformation

$$\coprod (\tilde{h}_n, \mathrm{pr}_3) : \coprod_{n \in \mathbb{N}} C(\Delta_n, -) \times I \times \Delta_n \to \coprod_{n \in \mathbb{N}} C(\Delta_n, -) \times \Delta_n.$$

The domain of this natural transformation is the functor $F(C(\Delta, -) \times I)$, where F is the functor which was used in the definition of the geometric realisation functor (cf. the discussion preceding Lemma 9.1.10). The natural transformation $\coprod(\tilde{h}_n, \mathrm{pr}_3)$ can be composed with the natural transformation $q : F(C(\Delta, -)) \to |C(\Delta, -)|$ to obtain another natural transformation

$$q \circ \coprod(\tilde{h}_n, \mathrm{pr}_3) : F(C(\Delta, -) \times I) \to |C(\Delta, -)|.$$

The restriction of this natural transformation onto the subfunctors $C(\Delta_n, -) \times I \times \Delta_n$ of $F(C(\Delta, -) \times I)$ is the natural transformation $q \circ (h_n, \mathrm{pr}_3)$ It will be given a name:

Definition 10.1.10. *The natural transformation $q \circ (h_n, \mathrm{pr}_3)$ is denoted by $\tilde{H}_n$.*

Proposition 10.1.11. *For every topological space Y, the family $\tilde{H}_{n,Y}$ is compatible.*

Proof. Let Y be a topological space and let $(\sigma, s, t) \in C(\Delta_n, Y) \times I \times \Delta_n$ and $(\sigma', s, t') \in C(\Delta_m, Y) \times I \times \Delta_m$ be two points in $F(C(\Delta, Y) \times I)$. Suppose that there exists a morphism $\alpha : [m] \to [n]$ in $\mathcal{O}$ such that $[C(\Delta(\alpha), Y) \times \mathrm{id}_I](\sigma, s) = (\sigma', s)$ and $\Delta(\alpha)(t') = t$. Then the image of (σ', s, t') under $\tilde{H}_{m,Y}$ is given by

$$
\begin{aligned}
\tilde{H}_{m,Y}(\sigma', s, t') &= [(\tilde{h}_{m,Y}(\sigma', s, t'), t')] \\
&= [(\tilde{h}_{m,Y}(C(\Delta(\alpha), Y)(\sigma, s, t')), t')] \\
&= [(\tilde{h}_{m,Y}(\sigma, s, \Delta(t')) \circ \Delta(\alpha), t')] \\
&= [(\tilde{h}_{m,Y}(\sigma, s, \Delta(t')), \Delta(\alpha)(t'))] \\
&= [(\tilde{h}_{m,Y}(\sigma, s, t), t)] \\
&- \tilde{H}_{n,Y}(\sigma, s, t).
\end{aligned}
$$

Thus the image of equivalent points in $F(C(\Delta, Y \times I)$ under $\coprod \tilde{H}_{n,Y}$ coincides, i.e. the family $\tilde{H}_{n,Y}$ is compatible. $\qquad\square$

Corollary 10.1.12. *For every topological space Y, the map $\coprod \tilde{H}_{n,Y}$ descends to a continuous function $\tilde{H}_Y : |C(\Delta, Y) \times I| \to |C(\Delta, Y)|$ making the following diagram commutative:*

$$
\begin{array}{ccc}
F(C(\Delta, Y) \times I) & & \\
\Big\downarrow{\scriptstyle q_{C(\Delta, Y) \times I}} & \searrow{\scriptstyle \coprod \tilde{H}_{n,Y}} & \\
|C(\Delta, Y) \times I| & \xrightarrow[\tilde{H}_Y]{} & |C(\Delta, Y)|
\end{array}
$$

Corollary 10.1.13. *The natural transformation $\coprod \tilde{H}_n$ descends to a natural transformation $\tilde{H} : |C(\Delta, -) \times I| \to |C(\Delta, -)|$.*

Recall that the product functor $- \times I : \mathbf{Top} \to \mathbf{Top}$ is cocontinuous, i.e. it commutes with colimits. The geometric realisation of a simplicial topological space is a coend which is a special colimit. Thus the functions

$$\psi_Y : |C(\Delta, Y)| \times I \cong |C(\Delta, Y) \times I|, \quad ([(\sigma, t)], s) \mapsto [((\sigma, s), t)]$$

are homeomorphisms for all topological spaces Y. These homeomorphisms form a natural isomorphism $\psi : |C(\Delta, -)| \times I \to |C(\Delta, -) \times I|$. This natural isomorphism ψ can be composed with the natural transformation $\tilde{H}$ to obtain a natural transformation $|C(\Delta, Y)| \times I \to |C(\Delta, Y)|$.

Definition 10.1.14. *The natural transformation $\tilde{H}\psi$ is denoted by H.*

Note that for each topological space Y the function H_Y is given by

$$H_Y([(\sigma,t)],s) = [(t' \mapsto \sigma(st + (1-s)t'),t)]$$

as suggested at the beginning of this section. Furthermore these maps are natural in Y by construction: For every continuous map $f : Y \to Y'$ between topological spaces Y and Y' the diagram

$$
\begin{array}{ccc}
|C(\Delta,Y)| & \xrightarrow{\ \epsilon_Y\ } & Y \\
{\scriptstyle |C(\Delta,f)|}\downarrow & & \downarrow{\scriptstyle f} \\
|C(\Delta,Y')| & \xrightarrow{\ \epsilon_{Y'}\ } & Y'
\end{array}
$$

is commutative. Each of the maps H_Y turns out to be a homotopy $\mathrm{id}_{|C(\Delta,Y)|} \simeq i_Y \epsilon_Y$:

Theorem 10.1.15. *The natural transformation H is a homotopy from $\mathrm{id}_{|C(\Delta,-)|}$ to $i\epsilon$, i.e. for each space Y the map H_Y is a homotopy $\mathrm{id}_{|C(\Delta,Y)|} \simeq i_Y \epsilon_Y$.*

Proof. Let Y be a topological space and $[(\sigma,t)]$ be a point in $|C(\Delta,Y)|$. The image of this point under the map $H_Y(-,1)$ is given by

$$H_Y([(\sigma,t)],1) = [(\sigma(t),t)],$$

where $\sigma(t)$ denotes the constant n-simplex $\sigma(t) \in Y$. The constant n-simplex is the image of the constant 0-simplex $\sigma(t)$ under the map $C(\Delta(\eta_{[n]\setminus[n-1]}),Y)$. This implies that the points

$$(\sigma(t),t) = (C(\Delta(\eta_{[n]\setminus[n-1]}),Y)(\sigma(t)),t) \quad \text{and} \quad (\sigma(t),\Delta(\eta_{[n]\setminus[n-1]})(t) = (\sigma(t),1)$$

in $F(C(\Delta,Y))$ are equivalent. So we obtain the equalities

$$
\begin{aligned}
H_Y([(\sigma,t)],1) = [(\sigma(t),t)] &= [(C(\Delta(\eta_{[n]\setminus[n-1]}),Y)(\sigma)(t),t)] \\
&= [(\sigma(t),\Delta(\eta_{[n]\setminus[n-1]}),Y)(t)] \\
&= [(\sigma(t),1)] = i_Y(\sigma(t)) = i_Y \epsilon_Y[(\sigma,t)],
\end{aligned}
$$

which are valid for all points $[(\sigma,t)]$ of $|C(\Delta,Y) \times I|$. So the maps $H_Y(-,1)$ and $i_Y \epsilon_Y$ coincide. The image of the point $[(\sigma,t)]$ under the map $H_Y(-,0)$ is given by

$$H_Y([(\sigma,t)],0) = [(t' \mapsto \sigma(0 \cdot t + 1t'),t)] = [(\sigma,t)],$$

i.e. $H_Y(-,0)$ is the identity on $|C(\Delta,Y)|$. Therefore H_Y is a homotopy from $\mathrm{id}_{|C(\Delta,Y)|}$ to $i_Y \epsilon_Y$. Since this is true for every topological space Y and the natural transformation H is built up from the homotopies H_Y it is itself a homotopy from $\mathrm{id}_{|C(\Delta,-)|}$ to $i\epsilon$. $\square$

Theorem 10.1.16. *The counit $\epsilon : |C(\Delta,-)| \to \mathrm{id}_{\mathbf{Top}}$ is a homotopy equivalence.*

Proof. The composition ϵi is the identity $\mathrm{id}_{\mathbf{Top}}$ by Lemma 9.2.32. The composition $i\epsilon$ is homotopic to the identity $\mathrm{id}_{|C(\Delta,-)|}$ by Theorem 10.1.15. Thus the counit ϵ is a homotopy equivalence. $\square$

Corollary 10.1.17. *Every topological space Y is naturally homotopy equivalent to the geometric realisation $|C(\Delta,Y)|$ of its singular simplicial space $C(\Delta,Y)$.*

So all in all we have shown that the maps in the commutative diagram

$$\mathrm{id}_{\textbf{Top}} \xrightarrow{\ \ i\ \ } |C(\Delta,-)| \xrightarrow{\ \ \epsilon\ \ } \mathrm{id}_{\textbf{Top}}$$
$$\mathrm{id}$$

are homotopy equivalences.

In the remaining part of this section we prove that the singular simplicial space functor $C(\Delta,-) : \textbf{Top} \to \textbf{Top}^{\mathcal{O}^{op}}$ preserves loop homotopies and loop homotopy equivalences. To show this we need some preparatory observations. Let Y be a topological space and consider the functions

$$\psi_{n,Y} : C(\Delta_n, Y) \to C(\Delta_n \times I, Y \times I), \quad \sigma \mapsto (\sigma \times \mathrm{id}_I).$$

Proposition 10.1.18. *For every topological space Y, the maps $\psi_{n,Y}$ form an embedding $C(\Delta, Y) \hookrightarrow C(\Delta \times I, Y \times I)$ of simplicial spaces.*

Proof. Consider the cosimplicial space $\Delta \times I$ and let pr_1 and pr_2 denote the projections onto the first and second factor respectively (We regard the unit interval I as a constant cosimplicial space here.) The projection pr_1 induces an embedding

$$C(\mathrm{pr}_1, Y) : C(\Delta, Y) \to C(\Delta \times I, Y)$$

of simplicial spaces. The codomain $C(\Delta \times I, Y)$ of this embedding can be embedded into the product $C(\Delta \times I, Y) \times C(\Delta \times I, I)$ as the slice trough pr_2:

$$C(\Delta \times I, Y) \hookrightarrow C(\Delta \times I, Y) \times C(\Delta \times I, I), \quad f \mapsto (f, \mathrm{id}_I).$$

The latter product space in turn is naturally isomorphic to the simplicial space $C(\Delta \times I, Y \times I)$, thus establishing an embedding $\psi_Y : C(\Delta, Y) \hookrightarrow C(\Delta \times I, Y \times I)$ of simplicial spaces. This embedding is explicitly given by

$$\psi_{n,Y} : C(\Delta_n, Y) \to C(\Delta \times I, Y \times I), \quad \sigma \mapsto (\sigma \times \mathrm{id}_I).$$

which recovers the definition of the maps $\psi_{n,Y}$. So these maps form an embedding of simplicial spaces. $\qquad\square$

By observing that the previous constructions all are natural in the topological space Y one obtains the following generalisation:

Proposition 10.1.19. *The embeddings ψ_Y of simplicial spaces form an embedding $C(\Delta, -) \hookrightarrow C(\Delta \times I, - \times I)$ of bifunctors.*

Proof. Consider the cosimplicial space $\Delta \times I$ and let pr_1 and pr_2 denote the projections onto the first and second factor respectively (We again regard I as a constant cosimplicial space here.) The projection pr_1 induces an embedding

$$C(\mathrm{pr}_1, -) : C(\Delta, -) \to C(\Delta \times I, -)$$

of functors. The codomain $C(\Delta \times I, -)$ of this embedding can be embedded into the product $C(\Delta \times I, -) \times C(\Delta \times I, I)$ as the slice trough pr_2:

$$C(\Delta \times I, -) \hookrightarrow C(\Delta \times I, -) \times C(\Delta \times I, I),$$

The latter product of bifunctors in turn is naturally isomorphic to the bifunctor $C(\Delta \times I, - \times I)$, thus establishing an embedding $\psi : C(\Delta, -) \hookrightarrow C(\Delta \times I, - \times I)$ of bifunctors. Evaluation at a topological space Y recovers the embedding ψ_Y of simplicial spaces. $\qquad\square$

Corollary 10.1.20. *For all topological space Y and Y' and every continuous function $H : Y \times I \to Y'$ the composition*

$$\tilde{H} = C(\Delta \times I, H) \circ \psi_Y : C(\Delta, Y) \to C(\Delta \times I, Y'), \quad \tilde{H}(\sigma) = ((t, s) \mapsto H(\sigma(t), s))$$

is a morphism of simplicial spaces.

Each of the spaces $\Delta_n \times I$ is canonically triangulated by the affine $n+1$-simplices

$$\tau_{n+1,i} = [\Delta(\eta_i) \times \Delta(\eta_0 \cdots \hat{\eta}_i \cdots \eta_n)]D_{\Delta_{n+1}} : \Delta_n \to \Delta_n \times I$$

as described in Section 9.3. These affine simplices $\tau_{n+1,i}$ induce continuous maps

$$C(\tau_{n+1,i}, Y') : C(\Delta_n \times I, Y') \to C(\Delta_{n+1}, Y')$$

for each topological space Y'. Combining these maps with the morphism $\tilde{H}$ of simplicial spaces induced by a homotopy $H : Y \times I \to Y'$ yields continuous functions

$$H_{n,i} := C(\tau_{n+1,i}, Y') \circ \tilde{H} : C(\Delta_n, Y) \to C(\Delta_{n+1}, Y').$$

These functions will be shown to form a simplicial homotopy $C(\Delta, H(-, 1)) \simeq C(\Delta, H(-, 0))$. We start by observing the behaviour of the maps $\tau_{n+1,i}$ under the face and degeneracy maps of the singular simplicial space $C(\Delta, \Delta_n \times I)$. Thereafter we will transfer the results to the functions $h_{n,i}$. The face of the affine $n + 1$-simplex $\tau_{n+1,i} : \Delta_{n+1} \to \Delta_n \times I$ is an affine n-simplex in $\Delta_n \times I$. There are two extreme possibilities which occur:

Lemma 10.1.21. *For each $n \in \mathbb{N}$ the functions $\tau_{n+1,0}$ and $\tau_{n+1,n}$ satisfy the equations $\tau_{n+1,0}\Delta(\epsilon_0) = s_1$ and $\tau_{n+1,n}\Delta(\epsilon_{n+1}) = s_0$.*

Proof. Let $n \in \mathbb{N}$ be any natural number. The composition of $\tau_{n+1,0}$ and the coface operator $\Delta(\epsilon_0)$ on Δ_n is given by

$$
\begin{aligned}
\tau_{n+1,0}\Delta(\epsilon_0) &= [\Delta(\eta_0) \times \Delta(\eta_1 \cdots \eta_n)]D_{\Delta_{n+1}}\Delta(\epsilon_0) \\
&= [\Delta(\eta_0) \times \Delta(\eta_1 \cdots \eta_n)][\Delta(\epsilon_0) \times \Delta(\epsilon_0)]D_{\Delta_n} \\
&= [\Delta(\eta_0)\Delta(\epsilon_0) \times \Delta(\eta_1 \cdots \eta_n)\Delta(\epsilon_0)]D_{\Delta_n} \\
&= [\Delta(\eta_0\epsilon_0) \times \Delta(\eta_1 \cdots \eta_n\epsilon_0)]D_{\Delta_n} \\
&= [\mathrm{id}_{\Delta_n} \times (0, 1)]D_{\Delta_n} = s_1
\end{aligned}
$$

Thus the composition $\tau_{n+1,0}\Delta(\epsilon_0)$ is the embedding $s_1 : \Delta_n \to \Delta_n \times I$ as the slice through 1 in I. A similar computation shows that the composition of $\tau_{n+1,n}$ and the coface operator $\Delta(\epsilon_{n+1})$ on Δ_n is given by

$$
\begin{aligned}
\tau_{n+1,n}\Delta(\epsilon_{n+1}) &= [\Delta(\eta_n) \times \Delta(\eta_0 \cdots \eta_{n-1})]D_{\Delta_{n+1}}\Delta(\epsilon_{n+1}) \\
&= [\Delta(\eta_n) \times \Delta(\eta_0 \cdots \eta_{n-1})][\Delta(\epsilon_{n+1}) \times \Delta(\epsilon_{n+1})]D_{\Delta_n} \\
&= [\Delta(\eta_n)\Delta(\epsilon_{n+1}) \times \Delta(\eta_0 \cdots \eta_{n-1})\Delta(\epsilon_{n+1})]D_{\Delta_n} \\
&= [\Delta(\eta_{n+1}\epsilon_{n+1}) \times \Delta(\eta_0 \cdots \eta_{n-1}\epsilon_{n+1})]D_{\Delta_n} \\
&= [\mathrm{id}_{\Delta_n} \times (1, 0)]D_{\Delta_n} = s_0,
\end{aligned}
$$

i.e. the composition $\tau_{n+1,0}\Delta(\epsilon_0)$ is the embedding $s_0 : \Delta_n \to \Delta_n \times I$ as the slice through 0 in I.   □

Lemma 10.1.22. *For each natural number $n \in \mathbb{N}$ the maps $h_{n,0}$ and $h_{n,n}$ satisfy the equations $\partial_0 h_{n,0} = C(\Delta_n, H(-, 1))$ and $\partial_{n+1} h_{n,n} = C(\Delta_n, H(-, 0))$.*

Proof. Let $n \in \mathbb{N}$ be any natural number. We utilise the result from the preceding Lemma. The composition of $h_{n,0}$ and the face operator $\partial_0 = C(\Delta(\epsilon_0), Y')$ on $C(\Delta_n, Y')$ is given by

$$
\begin{aligned}
\partial_0 h_{n,0} &= C(\Delta(\epsilon_0), Y') \circ C(\tau_{n+1,0}, Y') \circ \tilde{H} \\
&= C(\tau_{n+1,0}\Delta(\epsilon_0), Y') \circ \tilde{H} \\
&= C(s_1, Y') \circ \tilde{H} \\
&= (t' \mapsto H(\sigma(t'), 1)) \\
&= C(\Delta_n, H(-, 1)),
\end{aligned}
$$

which proves the first equality. A similar computation shows that the composition of $h_{n,n}$ and the face operator $C(\Delta(\epsilon_{n+1}, Y'))$ on $C(\Delta_{n+1}, Y')$ is given by

$$
\begin{aligned}
\partial_{n+1} h_{n,n} &= C(\Delta(\epsilon_{n+1}), Y') \circ C(\tau_{n+1,n}, Y') \circ \tilde{H} \\
&= C(\tau_{n+1,n}\Delta(\epsilon_{n+1}), Y') \circ \tilde{H} \\
&= C(s_0, Y') \circ \tilde{H} \\
&= (t' \mapsto H(\sigma(t'), 0)) \\
&= C(\Delta_n, H(-, 0))
\end{aligned}
$$

which establishes the second equality. $\qquad\square$

Lemma 10.1.23. *The maps $\tau_{n,j}$ satisfy the equations $\tau_{n,j}\Delta(\epsilon_i) = [\Delta(\epsilon_i) \times \mathrm{id}_I]\tau_{j-1}$ for all $n \in \mathbb{N}$ and $i < j$.*

Proof. Fix a natural number $n \in \mathbb{N}$ and let $0 \le i < j \le n$ be arbitrary. The affine simplex $\tau_{n,j}\Delta(\epsilon_i)$ is by definition given via

$$
\begin{aligned}
\tau_{n,j}\Delta(\epsilon_i) &= [\Delta(\eta_j) \times \Delta(\eta_0 \cdots \hat{\eta}_j \cdots \eta_n)]D_{\Delta_{n+1}}\Delta(\epsilon_i) \\
&= [\Delta(\eta_j) \times \Delta(\eta_0 \cdots \hat{\eta}_j \cdots \eta_n)][\Delta(\epsilon_i) \times \Delta(\epsilon_i)]D_{\Delta_n} \\
&= [\Delta(\eta_j)\Delta(\epsilon_i) \times \Delta(\eta_0 \cdots \hat{\eta}_j \cdots \eta_n)\Delta(\epsilon_i)]D_{\Delta_n} \\
&= [\Delta(\eta_j \epsilon_i) \times \Delta(\eta_0 \cdots \hat{\eta}_j \cdots \eta_n \epsilon_i)]D_{\Delta_n}.
\end{aligned}
$$

Using the simplicial identities $\eta_j \epsilon_i = \epsilon_i \eta_{j-1}$ for $i < j$ (cf. F.1) one observes $\Delta(\eta_j \epsilon_i) = \Delta(\epsilon_i \eta_{j-1})$. Repeated application of this simplicial identity yields

$$
\begin{aligned}
\eta_0 \cdots \hat{\eta}_j \cdots \eta_n \epsilon_i &= \eta_0 \cdots \hat{\eta}_j \cdots \eta_{n-1}\epsilon_i\eta_{n-1} \\
&= \eta_0 \cdots \hat{\eta}_j \epsilon_i \eta_j \cdots \eta_{n-1} \\
&= \eta_0 \cdots \eta_i \epsilon_i \eta_i \cdots \hat{\eta}_{j-1}\eta_j \cdots \eta_{n-1} \\
&= \eta_0 \cdots \hat{\eta}_{j-1}\eta_j \cdots \eta_{n-1}.
\end{aligned}
$$

Inserting these results in the expression for $\tau_{n,j}\Delta(\epsilon_i)$ one obtains the equality

$$
\tau_{n,j}\Delta(\epsilon_i) = [\Delta(\epsilon_i \eta_{j-1}) \times \Delta(\eta_0 \cdots \hat{\eta}_{j-1} \cdots \eta_{n-1}\epsilon_i)]D_{\Delta_n} = [\Delta(\epsilon_i) \times \mathrm{id}_I]\tau_{n,j-1}.
$$

This the equation that was to be proved. $\qquad\square$

Lemma 10.1.24. *The maps $h_{n,j}$ satisfy the equations $\partial_i h_{n,j} = h_{n,j-1}\partial_i$ for all $n \in \mathbb{N}$ and all $i < j$.*

Proof. Let $n \in \mathbb{N}$ be any natural number. We utilise the result from the preceding Lemma. The composition of $h_{n,j}$ and the face operator $\partial_i = C(\Delta(\epsilon_i), Y')$ on $C(\Delta_n, Y')$ is given by

$$\begin{aligned}
\partial_i h_{n,j} &= \partial_i [C(\tau_{n,j}, Y')\tilde{H}] \\
&= C(\Delta(\epsilon_i), Y')C(\tau_{n,j}, Y')\tilde{H} \\
&= C(\tau_{n,j}\Delta(\epsilon_i), Y')\tilde{H} \\
&= C([\Delta(\epsilon_i) \times \mathrm{id}_I]\tau_{n,j-1}, Y')\tilde{H} \\
&= C(\tau_{n,j-1}, Y')C(\Delta(\epsilon_i) \times \mathrm{id}_I, Y')\tilde{H} \\
&= C(\tau_{n,j}, Y')\tilde{H}\partial_i \\
&= h_{n,j-1}\partial_i.
\end{aligned}$$

This establishes the equations to be proved.   □

Lemma 10.1.25. *The maps $\tau_{n,j}$ satisfy the equations $\tau_{n,i}\Delta(\epsilon_i) = \tau_{n,i-1}\Delta(\epsilon_i)$ for all $n \in \mathbb{N}$ and $i \neq 0$.*

Proof. Fix a natural number $n \in \mathbb{N}$ and let $0 \leq i \leq n$ be arbitrary. The affine simplex $\tau_{n,i}\Delta(\epsilon_i)$ is by definition given via

$$\begin{aligned}
\tau_{n,i}\Delta(\epsilon_i) &= [\Delta(\eta_i) \times \Delta(\eta_0 \cdots \hat{\eta}_i \cdots \eta_n)]D_{\Delta_{n+1}}\Delta(\epsilon_i) \\
&= [\Delta(\eta_i) \times \Delta(\eta_0 \cdots \hat{\eta}_i \cdots \eta_n)][\Delta(\epsilon_i) \times \Delta(\epsilon_i)]D_{\Delta_n} \\
&= [\Delta(\eta_i)\Delta(\epsilon_i) \times \Delta(\eta_0 \cdots \hat{\eta}_i \cdots \eta_n)\Delta(\epsilon_i)]D_{\Delta_n} \\
&= [\Delta(\eta_i\epsilon_i) \times \Delta(\eta_0 \cdots \hat{\eta}_i \cdots \eta_n\epsilon_i)]D_{\Delta_n}.
\end{aligned}$$

Using the simplicial identities $\eta_j\epsilon_i = \mathrm{id}$ for $i = j$ and $i = j+1$ (cf. F.1) one observes the equality $\Delta(\eta_j\epsilon_i) = \mathrm{id}_{\Delta_n} = \Delta(\eta_{i-1}\epsilon_i)$. Repeated application of the simplicial identity $\eta_j\epsilon_i = \epsilon_i\eta_{j-1}$ for $i < j$ yields

$$\begin{aligned}
\eta_0 \cdots \hat{\eta}_i \cdots \eta_n\epsilon_i &= \eta_0 \cdots \hat{\eta}_i \cdots \eta_{n-1}\epsilon_i\eta_{n-1} \\
&= \eta_0 \cdots \hat{\eta}_i\epsilon_i\eta_i \cdots \eta_{n-1} \\
&= \eta_0 \cdots (\eta_{i-1}\epsilon_i)\eta_i \cdots \eta_{n-1} \\
&= \eta_0 \cdots \eta_{i-2} \cdot \mathrm{id} \cdot \eta_i \cdots \eta_{n-1} \\
&= \eta_0 \cdots \eta_{i-2}(\eta_i\epsilon_i)\eta_i \cdots \eta_{n-1} \\
&= \eta_0 \cdots \eta_{i-2}\eta_i\eta_{i+1}\epsilon_i\eta_{i+1} \cdots \eta_{n-1} \\
&= \eta_0 \cdots \hat{\eta}_{i-1} \cdots \eta_n\epsilon_i.
\end{aligned}$$

Collecting the above results one obtains the equality

$$\tau_{n,i}\Delta(\epsilon_i) = [\Delta(\eta_{i-1}\epsilon_i) \times \Delta(\eta_0 \cdots \hat{\eta}_{i-1} \cdots \eta_n\epsilon_i)]D_{\Delta_n} = \tau_{n,i-1}\Delta(\epsilon_i).$$

This is the equality that was to be established.   □

Lemma 10.1.26. *The maps $h_{n,j}$ satisfy the equations $\partial_i h_{n,i} = \partial_i h_{n,i-1}$ for all $n \in \mathbb{N}$ and $i \neq 0$.*

Proof. Let $n \in \mathbb{N}$ be any natural number and $0 \leq i \leq n$ be arbitrary. We utilise the result from the preceding Lemma. The composition of $h_{n,i}$ and the face operator $\partial_i = C(\Delta(\epsilon_i), Y')$ on $C(\Delta_n, Y')$ is given by

$$\begin{aligned}
\partial_i h_{n,i} &= \partial_i [C(\tau_{n,i}, Y')\tilde{H}] \\
&= C(\Delta(\epsilon_i), Y')C(\tau_{n,i}, Y')\tilde{H} \\
&= C(\tau_{n,i}\Delta(\epsilon_i), Y')\tilde{H} \\
&= C(\tau_{n,i-1}\Delta(\epsilon_i), Y')\tilde{H} \\
&= \partial_i C(\tau_{n,i-1}, Y')\tilde{H} \\
&= \partial_i h_{n,i-1}.
\end{aligned}$$

So all in all we have proved the equality $\partial_i h_{n,i} = \partial_i h_{n,i-1}$ for all $0 \leq i \leq n$.   □

Lemma 10.1.27. *The maps $\tau_{n,j}$ satisfy the equations $\tau_{n,j}\Delta(\epsilon_i) = \tau_{n,j}\Delta(\epsilon_{i-1})$ for all $n \in \mathbb{N}$ and $i > j+1$.*

Proof. Fix a natural number $n \in \mathbb{N}$ and let $0 \le j < i \le n$ be arbitrary. The affine simplex $\tau_{n,j}\Delta(\epsilon_i)$ is by definition given via

$$
\begin{aligned}
\tau_{n,j}\Delta(\epsilon_i) &= [\Delta(\eta_j) \times \Delta(\eta_0 \cdots \hat{\eta}_j \cdots \eta_n)]D_{\Delta_{n+1}}\Delta(\epsilon_i) \\
&= [\Delta(\eta_j) \times \Delta(\eta_0 \cdots \hat{\eta}_j \cdots \eta_n)][\Delta(\epsilon_i) \times \Delta(\epsilon_i)]D_{\Delta_n} \\
&= [\Delta(\eta_j)\Delta(\epsilon_i) \times \Delta(\eta_0 \cdots \hat{\eta}_j \cdots \eta_n)\Delta(\epsilon_i)]D_{\Delta_n} \\
&= [\Delta(\eta_j\epsilon_i) \times \Delta(\eta_0 \cdots \hat{\eta}_j \cdots \eta_n\epsilon_i)]D_{\Delta_n}.
\end{aligned}
$$

Using the simplicial identities $\eta_j\epsilon_i = \epsilon_{i-1}\eta_j$ for $i > j+1$ (cf. F.1) one observes $\Delta(\eta_j\epsilon_i) = \Delta(\epsilon_{i-1}\eta_j)$. Repeated application of the simplicial identity $\eta_j\epsilon_i = \epsilon_i\eta_{j-1}$ for $i < j$ (cf. F.1) yields

$$
\begin{aligned}
\eta_0 \cdots \hat{\eta}_j \cdots \eta_n\epsilon_i &= \eta_0 \cdots \hat{\eta}_j \cdots \eta_{n-1}\epsilon_i\eta_{n-1} \\
&= \eta_0 \cdots \hat{\eta}_j \cdots (\eta_{i+1}\epsilon_i)\eta_{i+1} \cdots \eta_{n-1} \\
&= \eta_0 \cdots \hat{\eta}_j \cdots \eta_i \cdot \text{id} \cdot \eta_{i+1} \cdots \eta_{n-1} \\
&= \eta_0 \cdots \hat{\eta}_j \cdots \eta_{n-1}.
\end{aligned}
$$

Inserting these results in the expression for $\tau_{n,j}\Delta(\epsilon_i)$ yields the equality

$$
\tau_{n,j}\Delta(\epsilon_i) = [\Delta(\epsilon_{i-1}\eta_j) \times \Delta(\eta_0 \cdots \hat{\eta}_j \cdots \eta_{n-1})]D_{\Delta_n} = [\Delta(\epsilon_{i-1}) \times \text{id}_I]\tau_{n-1,j}.
$$

So the equality $\tau_{n,j}\Delta(\epsilon_i) = \tau_{n,j}\Delta(\epsilon_{i-1})$ holds for all $n \in \mathbb{N}$ and all $n \ge i > j+1 > 0$. $\square$

Lemma 10.1.28. *The maps $h_{n,j}$ satisfy the equations $\partial_i h_{n,j} = h_{n,j}\partial_{i-1}$ for all $n \in \mathbb{N}$ and $i > j+1$.*

Proof. Let $n \in \mathbb{N}$ be any natural number and $0 < j+1 < i \le n$ be arbitrary. We utilise the result from the preceding Lemma. The composition of $h_{n,j}$ and the face operator $\partial_i = C(\Delta(\epsilon_i), Y')$ on $C(\Delta_n, Y')$ is given by

$$
\begin{aligned}
\partial_i h_{n,j} &= \partial_i[C(\tau_{n,j}, Y')\tilde{H}] \\
&= C(\Delta(\epsilon_i), Y')C(\tau_{n,j}, Y')\tilde{H} \\
&= C(\tau_{n,j}\Delta(\epsilon_i), Y')\tilde{H} \\
&= C([\Delta(\epsilon_{i-1}) \times \text{id}_I]\tau_{n-1,j}, Y')\tilde{H} \\
&= C(\tau_{n-1,j}, Y')C(\Delta(\epsilon_{i-1}) \times \text{id}_I, Y')\tilde{H} \\
&= C(\tau_{n,j}, Y')\tilde{H}\partial_{i-1} \\
&= h_{n,j-1}\partial_i,
\end{aligned}
$$

which establishes the equation to be proved. $\square$

We summarise the results of Lemmata 10.1.24, 10.1.26 and 10.1.28 in the following Proposition:

Proposition 10.1.29. *The maps $h_{n,i}$ satisfy the equations*

$$
\partial_i h_{n,j} = \begin{cases} h_{n-1,j-1}\partial_i & \text{if } i < j \\ \partial_i h_{n,i-1} & \text{if } i = j \neq 0 \\ h_{n-1,j}\partial_{i-1} & \text{if } i > j+1 \end{cases}.
$$

We now turn our attention to the composition of the maps $h_{n,j}$ with the degeneracy maps $C(\Delta(\eta_i), Y')$ on $C(\Delta_n, Y')$.

Lemma 10.1.30. *The maps $\tau_{n,j}$ satisfy the equations $\tau_{n,j}\Delta(\eta_i) = \Delta(\eta_i)\tau_{n,j+1}$ for all $n \in \mathbb{N}$ and $i \leq j$.*

Proof. Fix a natural number $n \in \mathbb{N}$ and let $0 \leq i \leq jn$ be arbitrary. The affine simplex $\tau_{n,j}\Delta(\eta_i)$ is by definition given by

$$
\begin{aligned}
\tau_{n,j}\Delta(\eta_i) &= [\Delta(\eta_j) \times \Delta(\eta_0 \cdots \hat{\eta}_j \cdots \eta_n)]D_{\Delta_{n+1}}\Delta(\eta_i) \\
&= [\Delta(\eta_j) \times \Delta(\eta_0 \cdots \hat{\eta}_j \cdots \eta_n)][\Delta(\eta_i) \times \Delta(\eta_i)]D_{\Delta_n} \\
&= [\Delta(\eta_j\eta_i) \times \Delta(\eta_0 \cdots \hat{\eta}_j \cdots \eta_n\eta_i)]D_{\Delta_n}.
\end{aligned}
$$

Using the simplicial identities $\eta_j\eta_i = \epsilon_i\eta_{j+1}$ for $i \leq j$ (cf. F.1) one observes $\Delta(\eta_j\eta_i) = \Delta(\eta_i\eta_{j+1})$. Repeated application of this simplicial identity yields

$$
\begin{aligned}
\eta_0 \cdots \hat{\eta}_j \cdots \eta_n\eta_i &= \eta_0 \cdots \hat{\eta}_j \cdots \eta_{n-1}\eta_i\eta_{n+1} \\
&= \eta_0 \cdots \eta_{i-1}\eta_i\eta_{i+1} \cdots \hat{\eta}_{j+1} \cdots \eta_{n+1} \\
&= \eta_0 \cdots \hat{\eta}_{j+1} \cdots \eta_{n+1}.
\end{aligned}
$$

Collecting the results one obtains the equality

$$
\tau_{n,j}\Delta(\eta_i) = [\Delta(\eta_i\eta_{j+1}) \times \Delta(\eta_0 \cdots \hat{\eta}_{j+1} \cdots \eta_{n+1})]D_{\Delta_n} = [\Delta(\eta_i) \times \mathrm{id}_I]\tau_{n+1,j+1}
$$

for all $n \in \mathbb{N}$ And all $0 \leq i \leq j \leq n$. $\square$

Lemma 10.1.31. *For all $n \in \mathbb{N}$ and $0 \leq i \leq j \leq n$ the maps $h_{n,j}$ satisfy the equations $C(\Delta(\eta_i), Y')h_{n,j} = h_{n+1,j+1}C(\Delta(\eta_i), Y)$.*

Proof. Let $n \in \mathbb{N}$ be any natural number and $0 \leq i \leq j \leq n$ be arbitrary. We utilise the result from the preceding Lemma. The composition of $h_{n,j}$ and the degeneracy operator $C(\Delta(\eta_i), Y')$ on $C(\Delta_n, Y')$ is given by

$$
\begin{aligned}
C(\Delta(\eta_i), Y')h_{n,j} &= C(\Delta(\eta_i), Y')[C(\tau_{n,j}, Y')\tilde{H}] \\
&= C(\tau_{n,j}\Delta(\eta_i), Y')\tilde{H} \\
&= C([\Delta(\eta_i) \times \mathrm{id}_I]\tau_{n+1,j+1}, Y')\tilde{H} \\
&= C(\tau_{n+1,j+1}, Y')C(\Delta(\eta_i) \times \mathrm{id}_I, Y')\tilde{H} \\
&= C(\tau_{n+1,j+1}, Y')\tilde{H}C(\Delta(\eta_i), Y) \\
&= h_{n+1,j+1}C(\Delta(\eta_i), Y').
\end{aligned}
$$

This is the equation we wanted to prove. $\square$

Lemma 10.1.32. *The maps $\tau_{n,j}$ satisfy the equations $\tau_{n,j}\Delta(\eta_i) = \Delta(\eta_{i-1})\tau_{n,j}$ for all $n \in \mathbb{N}$ and $i > j$.*

Proof. Fix a natural number $n \in \mathbb{N}$ and let $0 \leq j < i \leq n$ be arbitrary. The affine simplex $\tau_{n,j}\Delta(\eta_i)$ is by definition given by

$$
\begin{aligned}
\tau_{n,j}\Delta(\eta_i) &= [\Delta(\eta_j) \times \Delta(\eta_0 \cdots \hat{\eta}_j \cdots \eta_n)]D_{\Delta_{n+1}}\Delta(\eta_i) \\
&= [\Delta(\eta_j) \times \Delta(\eta_0 \cdots \hat{\eta}_j \cdots \eta_n)][\Delta(\eta_i) \times \Delta(\eta_i)]D_{\Delta_n} \\
&= [\Delta(\eta_j\eta_i) \times \Delta(\eta_0 \cdots \hat{\eta}_j \cdots \eta_n\eta_i)]D_{\Delta_n}
\end{aligned}
$$

Using the simplicial identities $\eta_j\eta_i = \epsilon_{i-1}\eta_j$ for $i > j + 1$ (cf. F.1) one observes $\Delta(\eta_j\eta_i) = \Delta(\eta_{i-1}\eta_j)$. Repeated application of this simplicial identity $\eta_j\eta_i = \epsilon_i\eta_{j+1}$ for $i \leq j$ yields

$$\eta_0 \cdots \hat{\eta}_j \cdots \eta_n \eta_i = \eta_0 \cdots \hat{\eta}_j \cdots \eta_{n-1} \eta_i \eta_{n+1}$$
$$= \eta_0 \cdots \hat{\eta}_j \cdots \eta_i \eta_i \eta_{i+2} \cdots \eta_{n+1}$$
$$= \eta_0 \cdots \hat{\eta}_j \cdots \eta_i \eta_{i+1} \cdots \eta_{n+1}.$$

Inserting the above result in the expression for $\tau_{n,j}\Delta(\eta_i)$ on finally obtains the equality

$$\tau_{n,j}\Delta(\eta_i) = [\Delta(\eta_{i-1}\eta_j) \times \Delta(\eta_0 \cdots \hat{\eta}_j \cdots \eta_{n+1})]D_{\Delta_n} = [\Delta(\eta_{i-1}) \times \mathrm{id}_I]\tau_{n+1,j},$$

which is the equality that was to be proved.

Lemma 10.1.33. *For all $n \in \mathbb{N}$ and $0 \le j < i \le n$ the maps $h_{n,j}$ satisfy the equations $C(\Delta(\eta_i), Y')h_{n,j} = h_{n+1,j}C(\Delta(\eta_{i-1}), Y)$.*

Proof. Let $n \in \mathbb{N}$ be any natural number and $0 \le j < i \le n$ be arbitrary. We utilise the result from the preceding Lemma. The composition of $h_{n,j}$ and the degeneracy operator $C(\Delta(\eta_i), Y')$ on $C(\Delta_n, Y')$ is given by

$$C(\Delta(\eta_i), Y')h_{n,j} = C(\Delta(\eta_i), Y')[C(\tau_{n,j}, Y')\tilde{H}]$$
$$= C(\tau_{n,j}\Delta(\eta_i), Y')\tilde{H}$$
$$= C([\Delta(\eta_{i-1}) \times \mathrm{id}_I]\tau_{n+1,j}, Y')\tilde{H}$$
$$= C(\tau_{n+1,j}, Y')C(\Delta(\eta_{i-1}) \times \mathrm{id}_I, Y')\tilde{H}$$
$$= C(\tau_{n+1,j}, Y')\tilde{H}C(\Delta(\eta_{i-1}), Y)$$
$$= h_{n+1,j}C(\Delta(\eta_{i-1}), Y').$$

So we have proved the equality $C(\Delta(\eta_i), Y')h_{n,j} = h_{n+1,j}C(\Delta(\eta_{i-1}), Y)$ for all $n \in \mathbb{N}$ and all $0 \le j < i \le n$. $\square$

Theorem 10.1.34. *The maps $h_{n,i}$ form a simplicial homotopy from $C(\Delta, H(-,1))$ to $C(\Delta, H(-,0))$.*

Proof. By Lemma 10.1.22 the maps $h_{n,i}$ fulfil the conditions $\partial_0 h_{n,0} = C(\Delta, H(-,1))$ and $\partial_{n+1} h_{n,n} = C(\Delta, H(-,0))$. By Proposition 10.1.29 they satisfy the equalities

$$\partial_i h_{n,j} = \begin{cases} h_{n-1,j-1}\partial_i & \text{if } i < j \\ \partial_i h_{n,i-1} & \text{if } i = j \neq 0 \\ h_{n-1,j}\partial_{i-1} & \text{if } i > j \end{cases}$$

By collecting the results in Lemmmata 10.1.31 and 10.1.33 one verifies that in addition they satisfy the conditions

$$C(\Delta(\eta_i), Y')h_j = \begin{cases} h_{n+1,j+1}C(\Delta(\eta_i), Y') & \text{if } i \le j \\ h_{n+1,j}C(\Delta(\eta_{i-1}), Y') & \text{if } i > j \end{cases}.$$

Altogether these are the equalities which define simplicial homotopies. Thus the maps $h_{n,i}$ form a simplicial homotopy from $C(\Delta, H(-,1))$ to $C(\Delta, H(-,0))$. $\square$

Abelian Hausdorff Eilenberg-Mac Lane Groups

In this chapter we exploit the new results on the geometric realisation of simplicial topological spaces obtained in the previous chapters. This is prepared in Section 1, where we derive preservation properties of the standard simplicial space functor $SS : \mathbf{Top} \to \mathbf{Top}^{\mathcal{O}^{op}}$. These will be needed to prove the analogous properties of the enveloping space functor E to be constructed in Section 2. At first we observe that the standard simplicial space functor SS preserves Hausdorffness and restricts and corestricts to a functor

$$\mathbf{k}_\omega \mathbf{Top} \to \mathbf{k}_\omega \mathbf{Top}^{\mathcal{O}^{op}} \quad \text{and} \quad \mathbf{k}_\omega \mathbf{Haus} \to \mathbf{k}_\omega \mathbf{Haus}^{\mathcal{O}^{op}}.$$

The most important preservation properties derived are the preservation of *monomorphisms*, *epimorphisms*, *open proclusions*, *open identifications*, *(closed) embeddings* and the preservation of *limits*. The preservation of limits in particular implies that for each category $\mathbf{TopAlg}_{(\Omega,E)}$ of topological algebras of type (Ω, E) the standard simplicial space functor SS restricts and corestricts to functors

$$SS : \mathbf{TopAlg}_{(\Omega,E)} \to \mathbf{TopAlg}_{(\Omega,E)}^{\mathcal{O}^{op}} \qquad SS : \mathbf{k}_\omega \mathbf{TopAlg}_{(\Omega,E)} \to \mathbf{k}_\omega \mathbf{TopAlg}_{(\Omega,E)}^{\mathcal{O}^{op}}$$

$$SS : \mathbf{k}_\omega \mathbf{HausAlg}_{(\Omega,E)} \to \mathbf{k}_\omega \mathbf{HausAlg}_{(\Omega,E)}^{\mathcal{O}^{op}}.$$

A kind of topological algebras we are particularly interested in are topological groups. concerning these we prove that the restriction and corestriction

$$SS : \mathbf{TopGrp} \to \mathbf{TopGrp}^{\mathcal{O}^{op}}$$

of the standard simplicial space functor preserves *monomorphisms*, *epimorphisms*, *proclusions*, *kernels*, *cokernels*, exact sequences and extensions. In addition the standard simplicial space of an Abelian topological group is Abelian. Moreover the application of the standard simplicial space functor is compatible with (semi-)group actions, i.e. one obtains functor

$$SS : \mathbf{TrGrp} \to \mathbf{TrGrp}^{\mathcal{O}^{op}} \quad \text{and} \quad SS : \mathbf{TrSemGrp} \to \mathbf{TrSemGrp}^{\mathcal{O}^{op}}.$$

In particular the standard simplicial space $SS(X)$ of the G-space X of a transformation (semi-)group (G, X) is a simplicial G-space. Where possible, these results are also be shown for the standard simplicial object functor in subcategories $\mathbf{Top}_C$ of $\mathbf{Top}$. Here we are primarily interested in the category $\mathbf{CGTop}$ of compactly generated spaces and the category $\mathbf{kTop}$ of k-spaces as well as several 'Hausdorff' versions of them.

In the second section we use the preparations made in Section 1 to construct an enveloping space functor $E : \mathbf{Top} \to \mathbf{Top}$ which assigns to each topological space a contractible enveloping space EX. The enveloping space EX is the geometric realisation $|SS(X)|$ of the standard simplicial space $SS(X)$ of X. This construction has already been used in the category $\mathbf{kTop}$ of k-spaces (cf. [Seg68]). Here we verify that the result generalises to the category $\mathbf{Top}$ of topological spaces and that the contraction of EX is natural in X. Furthermore we show that the enveloping space functor $E : \mathbf{Top} \to \mathbf{Top}$ preserves *monomorphisms, epimorphisms, equalisers, denseness of images, open proclusions, Hausdorffness* and *closed embeddings into Hausdorff spaces*. In addition we prove that the enveloping space functor E restricts and corestricts to an endofunctor

$$E : \mathbf{k}_\omega\mathbf{Haus} \to \mathbf{k}_\omega\mathbf{Haus}$$

of the category $\mathbf{k}_\omega\mathbf{Haus}$ of k_ω Hausdorff spaces and this endofunctor also preserves *equalisers* and *finite limits*. In particular, for every type (Ω, E) of algebras the enveloping space functor restricts and corestricts to an endofunctor

$$E : \mathbf{k}_\omega\mathbf{HausAlg}_{(\Omega,\mathrm{E})} \to \mathbf{k}_\omega\mathbf{HausAlg}_{(\Omega,\mathrm{E})},$$

which includes the category $\mathbf{k}_\omega\mathbf{HausGrp}$ of k_ω Hausdorff groups. This category contains real and complex finite dimensional Lie groups and real and complex Kac-Moody groups (cf. [GGH06]) and is therefore of interest in itself. The endofunctor $E : \mathbf{k}_\omega\mathbf{HausGrp} \to \mathbf{k}_\omega\mathbf{HausGrp}$ so obtained is then shown to preserve *kernels, cokernels* and to be *exact*.

In Section 3 In this section we introduce a construction of universal bundles and classifying spaces for (G, A)-bundles, where G and A are Hausdorff k_ω topological groups. This construction relies on the enveloping space functor E introduced in the previous section. In addition to the functor E we introduce another endofunctor B of $\mathbf{Top}$ resp. $\mathbf{k}_\omega\mathbf{Haus}$ and a natural transformation $E \to B$ which consists of quotient maps. The main results of this section are the following theorem and its consequences:

Theorem. *If A is a locally contractible Hausdorff k_ω-group, then the principal bundle $EA \to BA$ is numerable, hence a universal A-bundle and BA is a classifying space for A.*

Thus for each locally contractible k_ω Hausdorff group A there exists a universal bundle $EA \to BA$ whose total space EA is a k_ω Hausdorff group containing the fibre A as a subgroup. This in particular is shown to hold for the direct limit groups $\mathbf{SO} = \lim SO_n\mathbf{SO} = \lim SO_n$, $\mathbf{U} = \lim U_n$, $\mathbf{SU} = \lim SU_n$, $\mathbf{PU} = \lim PU_n$ and real and complex Kac-Moody groups. Moreover, if the group A is Abelian, we prove that the classifying space BA is an Abelian k_ω Hasdorff group as well and the sequence

$$0 \hookrightarrow A \to EA \twoheadrightarrow BA \to 0$$

is exact and locally trivial. This allows the construction to be iterated as in [Seg70], which we do not carry out here.

In the fourth section we consider Eilenberg-Mac Lane spaces. Given a countable discrete Abelian group π and $n \in \mathbb{N}$ it is known that there exist Eilenberg Mac Lane spaces $K(\pi, n)$ which are Abelian topological groups. This has been proved by Ivanov in [Iva85]. He uses free Abelian topological groups constructed in [DT58]. This construction leads to the existence of the Eilenberg Mac Lane spaces $K(\pi, n)$ which are Abelian topological groups. Unfortunately his proof relies on the countability of the group π. Different constructions using classifying spaces have been found by [Mil57, Theorem 3] and [McC69]. These constructions also relied on the countability of the group π. In this chapter we give a general natural procedure to construct compactly generated locally contractible Hausdorff Eilenberg-Mac Lane spaces which are Abelian topological groups or even CW-complexes. This extends the result concerning countable Abelian groups to the full class of Abelian groups. The new result is then used to prove a generalised "Cartan-Serre" Theorem on the rational (co)homology of Eilenberg Mac Lane spaces. This theorem does not rely on the finite dimensionality of the rational homotopy or homology groups, which is the strong restriction of the classical result of Cartan and Serre. We derive an even more general result by presenting a functorial construction of Abelian topological groups with prescribed homotopy groups. The construction establishes a correspondence between the category **AbGrLieAlg** of Abelian graded lie algebras and the category CW-**TopAb** of Abelian topological groups of the homotopy type of CW-complexes. It requires no prerequisites and only relies on the preservation of finite products under the geometric realisation functor $| - | : \mathbf{Set}^{\mathcal{O}^{op}} \to \mathbf{Top}$ (Theorem 9.3.49). This in particular includes a general procedure to construct Abelian Eilenberg Mac Lane topological groups $K(\pi, n)$.

In the last section we use the results obtained in Section 4 to derive the following structure theorem on the rational singular homology of Eilenberg-Mac Lane spaces:

Theorem. *The rational homology $H(X; \mathbb{Q})$ of any Eilenberg-Mac Lane space $K(\pi, n)$ is the exterior algebra over $\pi \otimes \mathbb{Q}$ if n is odd and the polynomial algebra on $\pi \otimes \mathbb{Q}$ if n is even.*

11.1 Standard Simplicial Spaces

In this section we derive preservation properties of the standard simplicial space functor SS : **Top** $\to$ **Top**$^{\mathcal{O}^{op}}$. These will be needed to prove the analogous properties of the enveloping space functor E to be constructed in the next section. We are interested in properties analogous to that of the geometric realisation functor $| - |$. These are the preservation of monomorphisms, epimorphisms, open proclusions, open identifications, (closed) embeddings and the preservation of limits. Where possible, the results will also be shown for the standard simplicial object functor in subcategories **Top**$_C$ of **Top**. Here we are primarily interested in the category **CGTop** of compactly generated spaces and the category **kTop** of k-spaces as well as several 'Hausdorff' versions of them. We start with collecting properties of the restriction of SS to various subcategories of **Top**. First, any product of Hausdorff spaces is a Hausdorff space. Thus the standard simplicial space functor SS restricts to a functor

$$SS_{|\mathbf{Haus}} : \mathbf{Haus} \to \mathbf{Haus}^{\mathcal{O}^{op}}.$$

Also, please recall the definition of the category **Top**$_C$ of **C**-generated spaces, where **C** is a subcategory of **Top** and that of the coreflector C : **Top** $\to$ **Top**$_C$. Straight from the definitions one derives:

Lemma 11.1.1. *For any subcategory* **C** *of* **Top** *the functor* CSS *restricts to an endofunctor* CSS : **Top**$_C$ $\to$ **Top**$_C^{\mathcal{O}^{op}}$.

Prominent examples are the category **CGTop** of compactly generated spaces, the category **kTop** of k-spaces and the category **CGHaus** of compactly generated Hausdorff spaces (which is the category of Hausdorff k-spaces by Lemma B.9.9). So especially the functors CGSS : **Top** $\to$ **CGTop** and kSS : **Top** $\to$ **kTop** restrict to endofunctors

$$CGSS : \mathbf{CGTop} \to \mathbf{CGTop}^{\mathcal{O}^{op}}, \qquad kSS : \mathbf{kTop} \to \mathbf{kTop}^{\mathcal{O}^{op}}$$

$$\text{and} \quad CGSS, kSS : \mathbf{CGHaus} = \mathbf{kHaus} \to \mathbf{CGHaus}^{\mathcal{O}^{op}} = \mathbf{kHaus}^{\mathcal{O}^{op}},$$

where the former two functors coincide on the category **CGHaus** = **kHaus**. the appropriate notion analogous to Hausdorffness in the category **Top**$_C$ of **C**-generated spaces is the closedness of the diagonal map. This property is alway preserved by the standard simplicial space functor.

Proposition 11.1.2. *The functor* $\mathrm{CSS} : \mathbf{Top_C} \to \mathbf{Top_C}^{\mathcal{O}^{op}}$ *preserves the closedness of the diagonal map.*

Proof. Let X be $\mathbf{C}$-generated space with closed diagonal map $D : X \to \mathrm{C}(X \times X)$. The n-th component of $\mathrm{CSS}(D)([n]) = D \times_\mathbf{C} \cdots \times_\mathbf{C} D$ is a finite product of closed injections in $\mathbf{Top_C}$. Finite products of closed injections in $\mathbf{Top}$ are closed by Lemma B.1.2, so the product map $D \times \cdots \times D : (X \times_\mathbf{C} X) \to (X \times_\mathbf{C} X)^n$ in $\mathbf{Top}$ is closed. Since the topology on $\mathrm{C}(X \times_\mathbf{C} X)^n$ is finer than the topology of the product $(X \times_\mathbf{C} X)^n$ in $\mathbf{Top}$ the product map $D \times_\mathbf{C} \cdots \times_\mathbf{C} D$ in $\mathbf{Top_C}$ is closed as well. $\square$

Corollary 11.1.3. *The standard simplicial k-space functor* kSS *restricts to a functor* $\mathrm{kSS} : \mathbf{whkTop} \to \mathbf{whkTop}^{\mathcal{O}^{op}}$.

Moreover finite products of Hausdorff k_ω-spaces in $\mathbf{Top}$ are already Hausdorff k_ω-spaces (cf. B.10.7). So the standard simplicial space of a Hausdorff k_ω-space is a simplicial Hausdorff k_ω-space. We note the consequence:

Lemma 11.1.4. *The standard simplicial space functor restricts and corestricts to a functor* $\mathrm{SS} : \mathbf{k_\omega Haus} \to \mathbf{k_\omega Haus}^{\mathcal{O}^{op}}$.

The reader is reminded that products in $\mathbf{Top_C}$ are obtained by applying the coreflector $\mathrm{C} : \mathbf{Top} \to \mathbf{Top_C}$ to the usual product in $\mathbf{Top}$ (see B.6.25). So the standard simplicial $\mathbf{C}$-generated space $\mathrm{CSS}(X)$ of a $\mathbf{C}$-generated space X is given by

$$\mathrm{CSS}(X)([n]) = X \times_\mathbf{C} X \cdots \times_\mathbf{C} X = \mathrm{C}(X^{n+1}),$$

which shows that the functor CSS restricts to the standard simplicial object functor on $\mathbf{Top_C}$, i.e. its construction is analogous to that of SS, just with products in $\mathbf{Top_C}$ instead of $\mathbf{Top}$. For 'nice' categories $\mathbf{C}$ the standard simplicial $\mathbf{C}$-generated space functor is cocontinuous:

Proposition 11.1.5. *If the category* $\mathbf{C}$ *is productive, then* $\mathrm{CSS} : \mathbf{Top_C} \to \mathbf{Top_C}^{\mathcal{O}^{op}}$ *is cocontinuous.*

Proof. If the subcategory $\mathbf{C}$ of $\mathbf{Top}$ is productive, its coreflective hull $\mathbf{Top_C}$ is cartesian closed (see B.6.27). Thus $- \times_\mathbf{C} - : \mathbf{Top_C}^2 \to \mathbf{Top_C}$ has a right adjoint. Therefore all binary, hence all finite products $\mathbf{Top_C}^n \to \mathbf{Top_C}$ are cocontinuous (cf. Corollary B.6.28). $\square$

Corollary 11.1.6. *If the category* $\mathbf{C}$ *is productive, then* $\mathrm{CSS} : \mathbf{Top_C} \to \mathbf{Top}^{\mathcal{O}^{op}}$ *is cocontinuous.*

Proof. The inclusion $\mathbf{Top_C} \hookrightarrow \mathbf{Top}$ is cocontinuous (see Corollary B.6.12). So if $\mathbf{C}$ is productive, the functor CSS is the composition of cocontinuous functors, hence cocontinuous as well. $\square$

Corollary 11.1.7. *The restricted and corestricted functors* $\mathrm{kSS} : \mathbf{kTop} \to \mathbf{kTop}^{\mathcal{O}^{op}}$ *and* $\mathrm{CGSS} : \mathbf{CGTop} \to \mathbf{CGTop}^{\mathcal{O}^{op}}$ *preserve all colimits.*

Proof. This follows from the fact that the categories $\mathbf{Comp}$ and $\mathbf{CompHaus}$ are productive (Lemmata B.9.5 and B.9.5). $\square$

Example 11.1.8. If $X = \mathrm{colim} X_i$ is the colimit of the simplicial k-spaces X_i in $\mathbf{kTop}$ then $\mathrm{kSS}(X)$ is the colimit of the simplicial k-spaces $\mathrm{kSS}(X_i)$ in $\mathbf{kTop}$. Because colimits in $\mathbf{kTop}$ coincide with those in $\mathbf{Top}$ the simplicial space $\mathrm{kSS}(X)$ also is the colimit of the simplicial spaces $\mathrm{kSS}(X_i)$ in $\mathbf{Top}$.

Note that products of injective or surjective maps are injective or surjective respectively. Thus we observe:

Lemma 11.1.9. *The standard simplicial* **C**-*generated space functor* CSS *preserves injectivity and surjectivity.*

More generally, for any subcategory **C** of **Top** the monomorphisms in **Top**$_\mathbf{C}$ are the injective functions in **Top**$_\mathbf{C}$ (cf. B.6.9) and the epimorphisms in **Top**$_\mathbf{C}$ are the surjective functions in **Top**$_\mathbf{C}$ (cf. B.6.16). So one can conclude:

Lemma 11.1.10. *The standard simplicial* **C**-*generated space functor* CSS *on* **Top** *preserves monomorphisms and epimorphisms.*

Since finite products of open maps are open, the standard simplicial space functor SS preserves the openness of maps. Moreover, finite products of inclusions also are inclusions, so the functor SS preserves inclusions of subspaces as well (and thus preserves open embeddings which are open inclusions). While the analogue is not true for proclusions in general, it still holds for open proclusions:

Proposition 11.1.11. *The standard simplicial space functor* SS : **Top** $\to$ **Top**$^{\mathcal{O}^{op}}$ *preserves open proclusions and open identifications.*

Proof. Let $p : X \to Y$ be an open proclusion. Finite products of open proclusions are open proclusions (by Theorem B.1.6). Therefore all the components SS$(p)([n])$ of the simplicial map SS(p) : SS$(X) \to$ SS(Y) are open proclusions. This is the definition of the simplicial map SS(p) being an open proclusion. The preservation of open identifications follows analogously. Here one uses the fact that finite products of open identifications are identifications (Corollary B.1.7). $\square$

Corollary 11.1.12. *The restriction* SS : **TopGrp** $\to$ **Top** *of the standard simplicial space functor* SS *to the category of topological groups preserves identifications.*

Proof. An identification $q : G \to G'$ in **TopGrp** is a quotient homomorphism. Quotient maps in **TopGrp** are open. Because the standard simplicial space functor SS preserves surjectivity and open proclusions the morphism SS(q) is an open identification as well. $\square$

Corollary 11.1.13. *The restriction of the standard simplicial space functor* SS *to the category* **TopGrp** *of topological groups preserves proclusions.*

Proof. Any proclusion $f : G \to G'$ in **TopGrp** is an identification onto the open image $f(G) \subset G'$, hence proclusions in **TopGrp** are automatically open. Therefore the functor SS : **TopGrp** $\to$ **TopGrp**$^{\mathcal{O}^{op}}$ preserves proclusions. $\square$

Proposition 11.1.14. *If* **C** *is productive then the standard simplicial* **C**-*generated space functor* CSS *on* **Top**$_\mathbf{C}$ *preserves proclusions and identifications.*

Proof. Let $p : X \to Y$ be a proclusion in **Top**$_\mathbf{C}$. If **C** is productive then finite products (in **Top**$_\mathbf{C}$) of proclusions in are proclusions (by Proposition B.6.30). Therefore all the components CSS$(p)([n])$ of the simplicial map CSS(p) : CSS$(X) \to$ CSS(Y) are proclusions. This is the definition of the simplicial map CSS(p) being a proclusion. The preservation of identifications follows analogously. Here one uses the fact that – under the premise that **C** is productive – finite products of identifications are identifications (cf. Corollary B.6.29). $\square$

Corollary 11.1.15. *If* **C** *is productive then the restriction* CSS : **CGrp** $\to$ **CGrp**$^{\mathcal{O}^{op}}$ *preserves quotients.*

Corollary 11.1.16. *If* **C** *is productive then the restriction* CSS : **CGrp** $\to$ **CGrp**$^{\mathcal{O}^{op}}$ *preserves proclusions.*

Furthermore, like the geometric realisation functor, the standard simplicial space functor SS preserves the closedness of injective maps.

Proposition 11.1.17. *The standard simplicial space functor* SS : **Top** $\to$ **Top**$^{\mathcal{O}^{op}}$ *preserves closed embeddings.*

Proof. Finite products of closed injections are closed injections by Lemma B.1.2. Therfore the standard simplicial space functor SS : **Top** $\to$ **Top**$^{\mathcal{O}^{op}}$ preserves closed injections. $\square$

This preservation of closedness carries over to the standard simplicial object functors CGSS : **CGTop** $\to$ **CGTop**$^{\mathcal{O}^{op}}$ and kSS : **kTop** $\to$ **kTop**$^{\mathcal{O}^{op}}$:

Lemma 11.1.18. *A closed embedding* $X \hookrightarrow Y$ *of* X *in* Y *induces closed embeddings* CGSS$(X) \to$ CGSS(Y) *and* kSS$(X) \to k$SS(Y) *of simplicial spaces.*

Proof. In view of the preceding corollary it suffices to observe that the coreflectors CG and k preserve closed embeddings. (See Lemma B.7.4 and Lemma B.9.4.) $\square$

In contrast to the geometric realisation functor $|-|$ the standard simplicial space functor SS does not only preserve equalisers but preserves all limits, as shall be proved below. We will derive the general result concerning the simplicial **C**-generated space functor CSS for an arbitrary subcategory **C** of **Top**.

Lemma 11.1.19. *The standard simplicial space functors* SS : **Top** $\to$ **Top**$^{\mathcal{O}^{op}}$ *on* **Top** *and* CSS : **Top**$_\mathbf{C}$ $\to$ **Top**$_\mathbf{C}$$^{\mathcal{O}^{op}}$ *on* **Top**$_\mathbf{C}$ *preserve products.*

Proof. This is a consequence of the fact that any permutation of the index set of a product induces a natural isomorphism in **Top** or **Top**$_\mathbf{C}$ respectively. $\square$

The continuity of the standard simplicial space functor SS : **Top** $\to$ **Top**$^{\mathcal{O}^{op}}$ now follows from the preservation of arbitrary equalisers:

Theorem 11.1.20. *The standard simplicial space functor* SS : **Top** $\to$ **Top**$^{\mathcal{O}^{op}}$ *is continuous, i.e. it preserves all limits.*

Proof. Since the standard simplicial space functor SS preserves products, it only remains to show that it also preserves equalisers. Because all limits in the category **Top**$^{\mathcal{O}^{op}}$ of simplicial topological spaces are computed point wise (cf. [ML98, Chapter V.3,Theorem 1]), it is sufficient to prove that each functor SS$(-)([n])$ preserves all equalisers. Let J be a family of continuous functions $f_i : X \to Y$ indexed by the set I. The equaliser of the family SS$(J)([n])$ is given by

Eq(SS$(J)([n])) =$
$\quad \{\mathbf{x} \in X^{n+1} \mid \forall i, i' \in I : \text{SS}([n])(f_i)(\mathbf{x}) = \text{SS}([n])(f_{i'})(\mathbf{x})\}$
$\quad = \{\mathbf{x} \in X^{n+1} \mid \forall i, i' \in I : (f_i(x_0) = f_{i'}(x_0) \wedge \cdots \wedge f_i(x_n) = f_{i'}(x_n))\}$
$\quad = \{\mathbf{x} \in X^{n+1} \mid (\forall i, i' \in I : f_i(x_0) = f_{i'}(x_0)) \wedge \cdots \wedge (\forall i, i' \in I : f_i(x_n) = f_{i'}(x_n))\}$
$\quad = \text{SS}(\text{Eq}(\text{SS}(J))([n]).$

Therefore the standard simplicial space functor SS preserves all equalisers and so also preserves all limits. $\square$

Corollary 11.1.21. *The standard* **C***-generated space functor* CSS : **Top** $\to$ **Top**$_\mathbf{C}$$^{\mathcal{O}^{op}}$ *on* **Top** *is continuous for any subcategory* **C** *of* **Top**.

Proof. The standard **C**-generated space functor CSS is a composition of continuous functors, hence continuous as well.   □

Lemma 11.1.22. *The standard* **C**-*generated space functor* CSS : $\mathbf{Top_C} \to \mathbf{Top_C}^{\mathcal{O}^{op}}$ *on* $\mathbf{Top_C}$ *is continuous for any subcategory* **C** *of* **Top**.

Proof. Let D be a diagram in $\mathbf{Top_C}$ and let $i : \mathbf{Top_C} \hookrightarrow \mathbf{Top}$ denote the inclusion functor. The limit $\lim_{\mathbf{Top_C}} D$ in $\mathbf{Top_C}$ is the image of the limit of the corresponding diagram $i \circ D$ in **Top** under the coreflector C : $\mathbf{Top} \to \mathbf{Top_C}$ (cf. B.6.15):

$$\lim_{\mathbf{Top_C}} D = \mathrm{C} \lim_{\mathbf{Top}} (i \circ D)$$

The coreflector C : $\mathbf{Top} \to \mathbf{Top_C}$ intertwines the functors SS on **Top** and CSS on $\mathbf{Top_C}$. Using this fact and the continuity of the coreflector C one can apply the functor CSS to the above equality to obtain:

$$\mathrm{CSS} \lim_{\mathbf{Top_C}} D = \mathrm{CSSC} \lim_{\mathbf{Top}}(i \circ D) = \mathrm{CCSS} \lim_{\mathbf{Top}}(i \circ D) = \mathrm{CSS} \lim_{\mathbf{Top}}(i \circ D)$$

$$= \mathrm{C} \lim_{\mathbf{Top}} \mathrm{SS}(i \circ D) = \lim_{\mathbf{Top_C}} \mathrm{CSS}(i \circ D) = \lim_{\mathbf{Top_C}} \mathrm{CSS}D$$

Thus the standard **C**-generated space functor on $\mathbf{Top_C}$ is continuous.   □

Note that the limit of the empty diagram is the singleton space $\{*\}$, which is the terminal object in **Top**. The standard simplicial space SS$(*)$ of the singleton is isomorphic to the constant simplicial space. So a 0-ary operation $* \to X$ on a topological space X induces a 0-ary operation. $* \cong \mathrm{SS}(*) \to \mathrm{SS}(X)$ on the simplicial space SS(X). If ω is an n-ary operation on the topological space X then the functions

$$\widetilde{\mathrm{SS}}(\omega)([m]) : \mathrm{SS}(X)([m])^n = (X^{m+1})^n \cong (X^n)^{m+1} \to X^{m+1} = \mathrm{SS}(X)([m])$$

$$\widetilde{\mathrm{SS}}(\omega)([m])(\mathbf{x}_1,\ldots,\mathbf{x}_n) = (\omega(x_{10},\ldots x_{n0}),\ldots,\omega(x_{1m},\ldots x_{nm}))$$

form an n-ary operation on the simplicial space SS(X). Analogously every n-ary operation in $\mathbf{Top_C}$ on a **C**-generated space X induces an n-ary operation on the standard simplicial **C**-generated space CSS(X). As a consequence the standard simplicial **C**-generated space functor SS preserves the structure of (total) topological algebras. So for operational structures of type Ω in $\mathbf{Top_C}$ one obtains:

Theorem 11.1.23. *The standard simplicial* **C**-*generated space* CSS(A) *of an operational structure* A *of type* Ω *in* $\mathbf{Top_C}$ *is a simplicial operational structure of type* Ω *in* $\mathbf{Top_C}$.

If E is a set of equations satisfied by the operations $\omega \in \Omega$ of a simplicial Ω-structure A then the operations $\widetilde{\mathrm{SS}}(\omega)$ also satisfy the equations in E. So we find:

Theorem 11.1.24. *The standard simplicial* **C**-*generated space* CSS(A) *of a model* A *for* (Ω, E) *in* $\mathbf{Top_C}$ *is a simplicial model for* (Ω, E) *in* $\mathbf{Top_C}^{\mathcal{O}^{op}}$.

Corollary 11.1.25. *The standard simplicial* **C**-*generated space functor* CSS *restricts to functors* CSS : $\mathbf{CGrp} \to \mathbf{CGrp}^{\mathcal{O}^{op}}$ *and* CSS : $\mathbf{Top_C Ab} \to \mathbf{Top_C Ab}^{\mathcal{O}^{op}}$.

Corollary 11.1.26. *The standard simplicial* **C**-*generated space functor* CSS *restricts to functors* $\mathbf{Top_C Rng} \to \mathbf{Top_C Rng}^{\mathcal{O}^{op}}$ *and* $\mathbf{Top_C Ring} \to \mathbf{Top_C Ring}^{\mathcal{O}^{op}}$

This includes the special case $\mathbf{C} = \mathbf{Top}$, so the standard simplicial space functor SS on $\mathbf{Top}$ restricts to functors

$$\text{SS} : \mathbf{TopAlg}_{(\Omega,\mathrm{E})} \to \mathbf{TopAlg}_{(\Omega,\mathrm{E})}^{\mathcal{O}^{op}}$$

in general and to functors

$$\text{SS} : \mathbf{TopGrp} \to \mathbf{TopGrp}^{\mathcal{O}^{op}} \qquad \text{SS} : \mathbf{TopAb} \to \mathbf{TopAb}^{\mathcal{O}^{op}}$$

$$\text{SS} : \mathbf{TopRng} \to \mathbf{TopRng}^{\mathcal{O}^{op}} \qquad \text{SS} : \mathbf{TopRing} \to \mathbf{TopRing}^{\mathcal{O}^{op}}$$

in particular. Because the standard simplicial space functor SS restricts to a functor $\text{SS} : \mathbf{k}_\omega\mathbf{Haus} \to \mathbf{k}_\omega\mathbf{Haus}^{\mathcal{O}^{op}}$ by Lemma 11.1.4, this implies that SS restricts and corestricts to functors

$$\text{SS} : \mathbf{k}_\omega\mathbf{HausAlg}_{(\Omega,\mathrm{E})} \to \mathbf{k}_\omega\mathbf{HausAlg}_{(\Omega,\mathrm{E})}^{\mathcal{O}^{op}}$$

in general and to functors

$$\text{SS} : \mathbf{k}_\omega\mathbf{HausGrp} \to \mathbf{k}_\omega\mathbf{HausGrp}^{\mathcal{O}^{op}} \quad \text{SS} : \mathbf{k}_\omega\mathbf{HausAb} \to \mathbf{k}_\omega\mathbf{HausAb}^{\mathcal{O}^{op}}$$

$$\text{SS} : \mathbf{k}_\omega\mathbf{HausRng} \to \mathbf{k}_\omega\mathbf{HausRng}^{\mathcal{O}^{op}} \quad \text{SS} : \mathbf{k}_\omega\mathbf{HausRing} \to \mathbf{k}_\omega\mathbf{HausRing}^{\mathcal{O}^{op}}$$

in particular.

Lemma 11.1.27. *The restriction of the standard simplicial space functor SS to the category* $\mathbf{TopGrp}$ *preserves monomorphisms and epimorphisms.*

Proof. Monomorphisms in $\mathbf{TopGrp}$ are injective homomorphisms. The injectivity of maps is preserved by the standard simplicial space functor SS, so SS maps monomorphisms to injective homomorphisms in $\mathbf{TopGrp}^{\mathcal{O}^{op}}$. These injective homomorphisms in turn are monomorphisms.

The preservation of epimorphisms follows analogously: Epimorphisms in $\mathbf{TopGrp}$ are surjective (by Lemma E.1.5). The surjectivity of maps is preserved by the standard simplicial space functor SS. Thus SS maps epimorphisms to surjective homomorphisms in $\mathbf{TopGrp}^{\mathcal{O}^{op}}$. These surjective homomorphisms in turn are epimorphisms. □

Lemma 11.1.28. *The restriction of the standard simplicial $\mathbf{C}$-generated space functor CSS to the category* $\mathbf{TopGrp}$ *preserves monomorphisms and epimorphisms.*

Proof. This follows from the observation that the injective and the surjective homomorphisms in $\mathbf{TopGrp}$ are the monomorphisms and the epimorphisms in $\mathbf{TopGrp}$ respectively.

Note that the simplicial space functor SS on $\mathbf{TopGrp}$ preserves the terminal object, which is the trivial group. Because the kernel of a homomorphism $\varphi : G \to H$ in $\mathbf{TopGrp}$ is the equaliser of φ and the trivial homomorphism $1 : G \to H$, we conclude:

Lemma 11.1.29. *The restriction of the standard simplicial space functor SS to the category* $\mathbf{TopGrp}$ *preserves kernels.*

Analogously, a cokernel of a homomorphism $\varphi : G \to H$ in $\mathbf{TopGrp}$ is the cocqualiser of the homomorphism φ and the trivial homomorphism $1 : G \to H$. So we obtain:

Lemma 11.1.30. *The restriction of the standard simplicial space functor SS to the category* $\mathbf{TopGrp}$ *preserves cokernels.*

Proof. A cokernel in the category **TopGrp** is a quotient homomorphism. Quotient homomorphisms in **TopGrp** are homomorphisms which are open identifications. Open identifications are preserved under the standard simplicial space functor SS by Proposition 11.1.11. Thus the image of a cokernel $q : G \to G'$ in **TopGrp** is a surjective morphism $SS(q) : SS(G) \to G'$ in **TopGrp**$^{\mathcal{O}^{op}}$ which is an open identification, i.e. a cokernel.   □

The trivial group $\{1\}$ is always **C**-generated. Because the coreflector C is continuous, it preserves equalisers. So we infer:

Lemma 11.1.31. *The restriction of the standard simplicial **C**-generated space functor CSS to the category **TopGrp** preserves kernels.*

Lemma 11.1.32. *The restriction SS : **TopGrp** → **TopGrp**$^{\mathcal{O}^{op}}$ preserves exact sequences.*

Proof. The restriction of SS to the category **TopGrp** preserves kernels and cokernels. Therefore it preserves exact sequences.   □

Lemma 11.1.33. *The restriction $SS_{|\mathbf{TopGrp}}$: **TopGrp** → **TopGrp**$^{\mathcal{O}^{op}}$ preserves extensions.*

Proof. The restriction of CSS to the category **CGrp** preserves exact sequences and cokernels (i.e. quotients). Thus it preserves extensions.   □

The **C**-generated standard simplicial space functor CSS can always be corestricted to the category **Top**$_C$ of **C**-generated spaces. This corestriction has more convienient preservation properties in general.

Lemma 11.1.34. *The restriction CSS : **CGrp** → **CGrp**$^{\mathcal{O}^{op}}$ preserves monomorphisms.*

Proof. This follows from the fact that monomorphisms in **CGrp** are injective (cf. B.6.9).   □

Lemma 11.1.35. *If **C** is productive then the restriction CSS : **CGrp** → **CGrp**$^{\mathcal{O}^{op}}$ preserves epimorphisms.*

Proof. If the category **C** is productive, then epimorphisms in **CGrp** are surjective by Lemma B.6.16. The surjectivity of maps is preserved by the functor CSS, so CSS maps epimorphisms to surjective homomorphisms in **CGrp**$^{\mathcal{O}^{op}}$. These surjective morphisms in turn are epimorphisms.   □

Note that the **C**-generated space functor on **CGrp** also preserves the terminal object $\{1\}$. Because the kernel of a homomorphism $\varphi : G \to H$ in **CGrp** is the equaliser of φ and the trivial homomorphism $1 : G \to H$, we conclude:

Lemma 11.1.36. *The restriction CSS : **CGrp** → **CGrp**$^{\mathcal{O}^{op}}$ preserves kernels.*

Proof. Let $\varphi : G \to G'$ be a homomorphism in **CGrp** and $\ker(\varphi) : N \to G$ be its kernel. The standard simplicial **C**-generated space functor CSS maps the trivial morphism $1 :\to G$ to the trivial morphism $CSS(1) : 1 \cong CSS(1) \to G$. Furthermore it is continuous by Corollary 11.1.22. Thus the morphism $CSS(\ker(\varphi))$ is the equaliser of $1 : SS(1) \to CSS(G)$ and $CSS(\varphi)$, i.e. it is its kernel.   □

Lemma 11.1.37. *If **C** is productive then the restriction CSS : **CGrp** → **CGrp**$^{\mathcal{O}^{op}}$ preserves cokernels.*

Proof. A cokernel in the category **CGrp** is a quotient homomorphism. If **C** is productive then identifications are preserved under the standard simplicial **C**-generated space functor CSS by Proposition 11.1.14. Thus the standard simplicial **C**-generated space functor CSS : **CGrp** → **CGrp**$^{\mathcal{O}^{op}}$ preserves cokernels.   □

Lemma 11.1.38. *If* **C** *is productive then the restriction* CSS : **CGrp** → **CGrp**$^{\mathcal{O}^{op}}$ *preserves exact sequences.*

Proof. If **C** is productive then the restriction of CSS to the category **CGrp** preserves kernels and cokernels. Therefore it preserves exact sequences.   □

Lemma 11.1.39. *If* **C** *is productive then the restriction* CSS$_{|\mathbf{CGrp}}$ *preserves extensions.*

Proof. The restriction of CSS to the category **CGrp** preserves exact sequences and cokernels (i.e. quotients). Thus it preserves extensions.   □

Restricting ourselves to the category **CGTop** (**CGHaus**) of compactly generated (Hausdorff) spaces, or the category **kTop** (**kHaus** = **CGHaus**) of (Hausdorff) k-spaces yields the classical result:

Corollary 11.1.40. *The standard simplicial c.g.-space functor* CGSS *restricts to functors* **CGTopGrp** → **CGTopGrp**$^{\mathcal{O}^{op}}$ *and* **CGTopAb** → **CGTopAb**$^{\mathcal{O}^{op}}$.

Corollary 11.1.41. *The standard simplicial k-space functor* kSS *restricts to functors* kSS : **kTopGrp** → **kTopGrp**$^{\mathcal{O}^{op}}$ *and* kSS : **kTopAb** → **kTopAb**$^{\mathcal{O}^{op}}$.

Corollary 11.1.42. *The standard simplicial c.g.-space functor* CGSS *and the standard simplicial k-space functor* kSS *restrict to functors*

$$\text{CGSS}, \text{kSS} : \mathbf{CGHausGrp} = \mathbf{kHausGrp} \to \mathbf{CGHausGrp}^{\mathcal{O}^{op}} = \mathbf{kHausGrp}^{\mathcal{O}^{op}}$$

and these restrictions coincide.

Example 11.1.43. If $G = \text{colim}_{\mathbf{kTop}} G_i$ is the colimit of the simplicial k-groups G_i in **kTop** then kSS(X) is the colimit of the simplicial k-groups kSS(X_i) in **kTop**. Because colimits in **kTop** coincide with those in **Top** the simplicial k-group kSS(X) also is the colimit of the simplicial k-groups kSS(X_i) in **Top**.

Lemma 11.1.44. *The restriction* SS : **kHausGrp** → **kHausGrp**$^{\mathcal{O}^{op}}$ *preserves kernels.*

Proof. The kernel ker φ of a morphism $\varphi : G \to G'$ in **kHausGrp** is the equaliser of φ and the trivial morphism $G \to \{1\} \leq G$ in **kHausGrp**. It is the inclusion of the inverse image $N = \varphi^{-1}(1)$ which is a closed subgroup of a compactly generated group and therefore compactly generated (cf. Theorem B.7.3). This also is the the kernel of φ in the category **TopGrp**. By Lemma 11.1.29 the standard simplicial space functor preserves kernels, so SS(ker φ) : SS(N) $\hookrightarrow$ SS(G) is the kernel of SS(φ) in the category **TopGrp**$^{\mathcal{O}^{op}}$. Consequently the morphism SS(ker φ) is the kernel of SS(φ) in the category **kHaus**$^{\mathcal{O}^{op}}$.   □

Lemma 11.1.45. *The restriction* SS : **kHausGrp** → **kHausGrp**$^{\mathcal{O}op}$ *preserves cokernels.*

Proof. The cokernel coker φ of a morphism $\varphi : G \to G'$ in **kHausGrp** is the coequaliser of φ and the trivial morphism $G \to \{1\} \leq G$ in **kHausGrp**. It is the quotient $q : G \to G/N$ by the closure $N = \overline{\varphi(G)}$ of the image of φ. The standard simplicial space functor preserves cokernels by Lemma 11.1.30, so the morphism SS(q) : SS(G') $\hookrightarrow$ SS(G'/N) is the cokernel of the inclusion of SS(N) into SS(G) in the category **TopGrp**$^{\mathcal{O}^{op}}$. Because SS(G) is dense in SS(N) this also is the cokernel of SS(φ) in the category **kHaus**$^{\mathcal{O}^{op}}$.   □

Proposition 11.1.46. *The restriction* SS : **kHausGrp** → **kHausGrp**$^{\mathcal{O}^{op}}$ *is exact.*

Proof. This is a consequence of the preceding two lemmata.   □

This can be generalised to structures that can be expressed by a (continuous) functor from a category **D** into **Top**:

Proposition 11.1.47. *For any functor* $F : \mathbf{D} \to \mathbf{Top}$, *the composition* SS ∘ F *is a simplicial object in the category* **Top**$^{\mathbf{D}}$. *If the functor* F *is continuous then* SS ∘ F *is continuous.*

Corollary 11.1.48. *The standard simplicial* **C**-*generated space* CSS(G) *of a* **C**-*groupoid* G *is a simplicial* **C**-*groupoid.*

Corollary 11.1.49. *The standard simplicial space functor* SS *restricts to a functor* SS : **HTop** → **HTop**$^{\mathcal{O}^{op}}$. *It restricts further to simplicial objects in the categories of homotopy associative* H-*spaces and homotopy Abelian* H-*spaces.*

For a topological module M over a topological ring R let ϱ denote the action $\varrho : R \times M \to M$ of R on M. Then SS(M) is a simplicial Abelian topological group and $\widetilde{SS}(\varrho) : SS(R) \times SS(M) \to SS(M)$ is an action of SS(R) on SS(M). Thus SS(M) is a module over the simplicial ring SS(R) ∈ **TopRng**$^{\mathcal{O}^{op}}$. We can embed the ring R as the constant simplicial space into SS(R) via the diagonal embedding. In this way one obtains an action of the ring R on the simplicial Abelian group SS(M), hence we notice:

Lemma 11.1.50. *For any topological ring* R *the standard simplicial space functor* SS *restrict to functors* SS : **RMod** → **RMod**$^{\mathcal{O}^{op}}$ *and* SS : **ModR** → **ModR**$^{\mathcal{O}^{op}}$.

For the case that the ring R is actually a topological field we conclude:

Lemma 11.1.51. *For any topological field* $\mathbb{K}$ *the standard simplicial space functor* SS *restricts to a functor* SS : $\mathbb{K}$**Vec** → $\mathbb{K}$**Vec**$^{\mathcal{O}^{op}}$.

Applying Proposition 11.1.47 to the category **TrGrp** of transformation groups and the category **TrSemGrp** of transformation semi-groups yields a functor that maps each transformation (semi-)group (G, X) to the simplicial transformation (semi-)group (SS(G), SS(X)):

Corollary 11.1.52. *The standard simplicial space functor* SS *restricts to functors* SS : **TrGrp** → **TrGrp**$^{\mathcal{O}^{op}}$ *and* SS : **TrSemGrp** → **TrSemGrp**$^{\mathcal{O}^{op}}$

Let (G, X) be a transformation (semi-)group. The topological (semi-)group G can be regarded as a constant simplicial topological semi-group. This constant simplicial topological group G can be embedded in in the simplicial topological (semi-)group SS(G) via the diagonal embedding. In this way one obtains an action of G on the simplicial topological space SS(X). So we observe:

Corollary 11.1.53. *For any topological group* G *the standard simplicial space functor* SS *restricts to a functor* SS : **GTop** → **GTop**$^{\mathcal{O}^{op}}$.

Corollary 11.1.54. *For any topological semi-group* G *the standard simplicial space functor* SS *restricts to a functor* SS : **GTop** → **GTop**$^{\mathcal{O}^{op}}$.

Again, restricting ourselves to the category **CGTop** of compactly generated spaces or the category **kTop** of k-spaces yields analogous results for these categories:

Corollary 11.1.55. *The standard simplicial compactly generated space functor* CGSS *restricts to a functor* CGSS : **CGTrGrp** → **CGTrGrp**$^{\mathcal{O}^{op}}$.

Corollary 11.1.56. *The standard simplicial k-space functor* kSS *restricts to a functor* kSS : **kTrGrp** → **kTrGrp**$^{\mathcal{O}^{op}}$.

11.2 Contractible Enveloping Spaces

In this section we construct an endofunctor E : **Top** → **Top** which assigns to every topological space X a contractible enveloping space EX. Then we prove preservation properties of the functor E. In analogy to the classical results obtained in the categories **CGTop** and **kTop** (whose product topologies differ from the one in **Top**), we are able to prove that the functor E preserves Hausdorffness in general as well as finite limits of Hausdorff k_ω-spaces in **Top**. In addition the enveloping space EX of a compactly generated Hausdorff space is again compactly generated and Hausdorff. As a consequence we show that for any locally contractible Hausdorff k_ω-topological group G there exists a classifying bundle

$$G \to EG \to BG$$

whose total space EG is a Hausdorff k_ω-topological group. This in particular includes all real and complex finite dimensional Lie groups, all real and complex Kac-Moody groups and all direct limits of Hausdorff k_ω-groups (cf. [GGH06]). The contractions on the enveloping spaces are shown to be natural, i.e. there exists a natural transformation $E \times I \to E$ that restricts to the identity on $E \times \{1\}$ and to a constant map on $E \times \{0\}$ The construction presented here is a generalisation of the classical results and contains them as special cases. The relation of E to the classical constructions will be explained.

Definition 11.2.1. *The functor* $|\cdot| \circ \mathrm{SS} :$ **Top** → **Top** *is be denoted by* E. *It is called the* enveloping space functor.

In a similar way one also obtains functors $|-|_\mathrm{C} \circ \mathrm{SS}$ and $|-| \circ \mathrm{CSS}$ that corestrict to the category **Top**$_\mathrm{C}$ of **C**-generated spaces. Their connection to the functor CE is given by the following observations.

Lemma 11.2.2. *For every topological space X the map $|\mathrm{SS}(X)|_\mathrm{C} \to EX$ factors uniquely through CEX:*

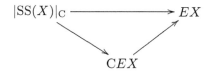

Proof. Being a quotient space of the **C**-generated space $CF(\mathrm{SS}(X))$, the space $|\mathrm{SS}(X)|_\mathrm{C}$ is **C**-generated. Thus the continuous bijection $|\mathrm{SS}(X)|_\mathrm{C} \to EX$ factors uniquely through CEX (cf. Lemma B.6.11). □

The following shows that the functors $|-| \circ \mathrm{CSS}$ and $|-|_\mathrm{C} \circ \mathrm{SS}$ coincide if the category **C** satisfies one of the usual conditions, which is e.g. satisfied for **C** = **Comp** or **C** = **CompHaus**.

Lemma 11.2.3. *For any subcategory **C** of **Top** that either contains the standard simplices or is productive and such that the standard simplices are **C**-generated the functors $|-| \circ \mathrm{CSS}$ and $|-|_\mathrm{C} \circ \mathrm{SS}$ coincide.*

Proof. In both cases the product functors $- \times \Delta_n$ on **Top** restrict to endofunctors of **Top**$_\mathrm{C}$. Therefore the spaces $CF(\mathrm{SS}(X))$ and $F(\mathrm{CSS}(X))$ coincide and so do their quotients $|\mathrm{SS}(X)|_\mathrm{C}$ and $|\mathrm{CSS}(X)|$. ⊓

We now turn our attention to the preservation of special morphisms under E. As it turns out, most of the preservation properties of the standard simplicial space functor SS carry over to E.

Lemma 11.2.4. *The functor E preserves monomorphisms and epimorphisms.*

Proof. This follows from the fact that both the standard simplicial space functor SS and the geometric realisation functor $|-|$ preserve monomorphisms and epimorphisms by Lemma 11.1.10 and Lemma 9.1.5 respectively.   □

Lemma 11.2.5. *The enveloping space functor E preserves the denseness of images.*

Proof. The standard simplicial space functor preserves denseness of subspaces. The geometric realisation functor $|-|$ preserves denseness of images by Lemma 9.1.6. Thus, if $f : X \to Y$ is a continuous function with dense image $f(X)$ in Y then the morphism $SS(f) : SS(X) \to SS(Y)$ of simplicial topological spaces has dense image as well and so has the function $Ef : EX \to EY$.   □

Lemma 11.2.6. *The functor E preserves open proclusions.*

Proof. The standard simplicial space functor SS preserves open proclusions by Proposition 11.1.11. The geometric realisation functor preserves proclusions in general by Proposition 9.1.7. Thus the composition $E = |-| \circ SS$ also preserves open proclusions   □

As a consequence the functor E preserves proclusions on all subcategories of **Top** in which proclusions are open in general:

Corollary 11.2.7. *The restriction of E to the category **TopGrp** of topological groups preserves proclusions.*

Theorem 11.2.8. *The enveloping space functor $E : **Top** \to **Top**$ preserves homotopy equivalence of maps.*

Proof. The standard simplicial space functor $SS : **Top** \to **Top**$ preserves homotopy equivalence by Theorem 2.5.6. The geometric realisation functor $|-|$ on $**Top**^{\mathcal{O}^{op}}$ preserves homotopy by Theorem 9.3.104. Therefore the composition $E = |-| \circ SS$ preserves homotopy as well.   □

Theorem 11.2.9. *The enveloping space functor E preserves Hausdorffness.*

Proof. Let X be a topological space. There exists a continuous bijection $EX \to X'$ onto a space X', which is Hausdorff if X is Hausdorff. (See [BM78, p. 217].) Thus the enveloping space EX is Hausdorff whenever X is Hausdorff.   □

Corollary 11.2.10. *The enveloping space functor E restricts and corestricts to a functor $E : \mathbf{k}_\omega\mathbf{Haus} \to \mathbf{k}_\omega\mathbf{Haus}^{\mathcal{O}^{op}}$.*

Proof. The standard simplicial space $SS(X)$ of a Hausdorff k_ω-space X is a simplicial Hausdorff k_ω-space by Lemma 11.1.4. So the enveloping space EX is a k_ω-space by Proposition 9.3.32. Because the enveloping space functor E in addition preserves Hausdorffness (Theorem 11.2.9), the enveloping space EX of a Hausdorff k_ω-space X also is a Hausdorff k_ω-space.   □

For any topological space X the space $SS(X)([0]) = X^1$ is naturally homeomorphic to the space X itself. Recall, that the inclusion $i_0 : X \times \Delta_0 \hookrightarrow F(SS(X))$ induces a natural injection $q \circ i_0 \circ (- \times \Delta_0) : X \to EX = |SS(X)|$ by Lemma 9.1.27. By Proposition 9.1.30 these injections are closed embeddings if the maps $X(\eta)$ are closed for all epimorphisms η in $\mathcal{O}$. This condition is satisfied exactly if the space X is Hausdorff. More generally one observes:

Proposition 11.2.11. *The maps $CSS(X)(\eta)$ are closed for every epimorphism η in $\mathcal{O}$ if and only if the diagonal map $CX \to C(X \times X)$ is closed.*

Proof. Let Y be a simplicial space. Every epimorphism η in $\mathcal{O}$ is a composition of degeneracy maps. Therefore it suffices to prove that the functions $\mathrm{CSS}(X)(\eta)$ are closed for all degeneracy maps η iff the diagonal map $CX \to C(X \times X)$ is closed. The function $\mathrm{CSS}(X)(\eta_i)$ for a degeneracy map $\eta_i : [n+1] \to [n]$ is given by

$$\mathrm{CSS}(X)(\eta_i)(x_0,\ldots,x_n) = (x_0,\ldots,x_i,x_i,\ldots,x_n),$$

i.e. it is a product of identity maps and the diagonal map $CX \to C(X \times X)$. Thus the maps $\mathrm{SS}(X)(\eta)$ are closed for every epimorphism η in $\mathcal{O}$ if and only if the diagonal map $CX \to C(X \times X)$ is closed. $\square$

Corollary 11.2.12. *The maps $\mathrm{SS}(X)(\eta)$ are closed for every epimorphism η in $\mathcal{O}$ if and only if X is Hausdorff.*

Corollary 11.2.13. *If X is a weak Hausdorff k-space then the maps $k\mathrm{SS}(X)(\eta)$ are closed for every epimorphism η in $\mathcal{O}$.*

Proof. This is due to the fact that k-spaces are weak Hausdorff if and only if their diagonal map in **kTop** is closed (cf. B.9.11). $\square$

Another example are compactly generated spaces with the property that every compact subspace is Hausdorff. These spaces are called $LM - T_2$-spaces. The compactly generated $LM - T_2$-spaces are exactly those with closed diagonal map (cf. B.7.8). So we infer:

Corollary 11.2.14. *If X is a compactly generated $LM - T_2$-space then the maps $\mathrm{CGSS}(X)(\eta)$ are closed for every epimorphism η in $\mathcal{O}$.*

These criteria for the closedness of the maps $X(\eta)$ imply that the natural maps $q_{\mathrm{SS}(X)} \circ i_0$ are closed embeddings in each of the above cases.

Proposition 11.2.15. *If the diagonal map $CX \to C(X \times X)$ is closed, then the natural injection $q_{\mathrm{CSS}(X)} \circ C(i_0) \circ (- \times \Delta_0) : CX \to |\mathrm{CSS}(X)|$ is a closed embedding.*

Proof. Assume the diagonal map $CX \to C(X \times X)$ to be closed. By Lemma 11.2.11 the functions $\mathrm{CSS}(X)(\eta)$ are closed for all epimorphisms η in $\mathcal{O}$. Thus the natural injection $q_{\mathrm{CSS}(X)} \circ C(i_0) : CX \times \Delta_0 \to |\mathrm{CSS}(X)|$ is a closed embedding by Proposition 9.1.31 and so is $q_{\mathrm{CSS}(X)} \circ C(i_0) \circ (- \times \Delta_0)$. $\square$

Corollary 11.2.16. *On the category **Haus** of Hausdorff spaces the natural transformation $q \circ i_0 \circ (- \times \Delta_0) : \mathrm{id}_{\mathbf{Top}} \to E$ consists of closed embeddings.*

Corollary 11.2.17. *The restriction of $q \circ i_0 \circ (- \times \Delta_0) : \mathrm{id}_{\mathbf{Top}} \to |k\mathrm{SS}(-)|$ to the category of weak Hausdorff k-spaces consists of closed embeddings.*

Corollary 11.2.18. *The restriction of $q \circ i_0 \circ (- \times \Delta_0) : \mathrm{id}_{\mathbf{Top}} \to |\mathrm{CGSS}(-)|$ to the category of $LM - T_2$ compactly generated spaces consists of closed embeddings.*

A similar conclusion can be drawn for the functor $|-|_{\mathrm{C}} \circ \mathrm{SS}$:

Lemma 11.2.19. *If the diagonal map $CX \to C(X \times X)$ is closed, then the natural injection $CX \to |\mathrm{SS}(X)|_{\mathrm{C}}$ is a closed embedding.*

Proof. The composition of the injective function $CX \times \Delta_0 \hookrightarrow |\mathrm{SS}(X)|_{\mathrm{C}}$ with the natural continuous bijective map $|\mathrm{SS}(X)|_{\mathrm{C}} \to |\mathrm{CSS}(X)|$ is closed by Proposition 9.1.31. Therefore the map $CX \times \Delta_0 \to |\mathrm{SS}(X)|_{\mathrm{C}}$ is closed as well. $\square$

Recall that any closed embedding $X \hookrightarrow Y$ of topological spaces induces a closed embedding $\mathrm{SS}(X) \hookrightarrow \mathrm{SS}(Y)$ of simplicial spaces. By Corollary 9.1.52 this closed embedding descends to a closed embedding $EX \hookrightarrow EY$ if the morphisms $Y(\eta)$ are closed for all epimorphisms η in $\mathcal{O}$. The previous considerations show that this condition is satisfied in many cases.

Lemma 11.2.20. *A closed embedding $X \hookrightarrow Y$ of X in a Hausdorff space Y induces a closed embedding of geometric realisations $EX \hookrightarrow EY$.*

Proof. Let $i : X \hookrightarrow Y$ be a closed embedding of X in the Hausdorff space Y. By Corollary 11.2.12 the simplicial morphism $\mathrm{SS}(i)$ also is a closed embedding. Because the space Y is Hausdorff, the morphisms $X(\eta)$ are closed for all epimorphisms η in $\mathcal{O}$. Thus (by Corollary 9.1.52) the induced mapping $EX \hookrightarrow EY$ is a closed embedding. $\qquad\square$

Because the coreflectors $\mathrm{CG} : \mathbf{Top} \to \mathbf{CGTop}$ and $k : \mathbf{Top} \to \mathbf{kTop}$ preserve closed embeddings, this result carries over to the functors CGSS and kSS.

Lemma 11.2.21. *A closed embedding $X \hookrightarrow Y$ of X into a Hausdorff space Y induces a closed embedding of geometric realisations $|\mathrm{CGSS}(X)| \hookrightarrow |\mathrm{CGSS}(Y)|$ and $|\mathrm{kSS}(X)| \hookrightarrow |\mathrm{kSS}(Y)|$*

In an similar fashion one observes that the above observations still have analogues for compactly or compactly Hausdorff generated spaces.

Lemma 11.2.22. *A closed embedding $X \hookrightarrow Y$ of X into a weak Hausdorff k-space Y induces a closed embedding of geometric realisations $|\mathrm{kSS}(X)| \hookrightarrow |\mathrm{kSS}(Y)|$.*

Lemma 11.2.23. *A closed embedding $X \hookrightarrow Y$ of X into a compactly generated $LM - T_2$-space Y induces a closed embedding $|\mathrm{CGSS}(X)| \hookrightarrow |\mathrm{CGSS}(Y)|$.*

It is known that any topological group G satisfies the separation axiom T_3, so if the points are closed it automatically is Hausdorff and the diagonal map is closed.

Lemma 11.2.24. *A closed embedding $X \hookrightarrow G$ in a T_1-topological group induces a closed embedding $EX \to EG$.*

A similar result can be proved for **C**-groups, i.e. group objects in the category $\mathbf{Top}_\mathbf{C}$ of **C**-generated spaces. We start with noting:

Lemma 11.2.25. *The diagonal map $D : X \to \mathrm{C}(X \times X)$ of a **C**-generated space X is closed if and only if its image $D(X)$ is closed in $\mathrm{C}(X \times X)$.*

Proof. If the diagonal map $DX \to \mathrm{C}(X \times X)$ of a **C**-generated space X is closed, then its image $D(X)$ is closed in $\mathrm{C}(X \times X)$. Conversely, assume the image $D(X)$ to be closed in $\mathrm{C}(X \times X)$ and let $A \subset X$ be a closed subset of A. The inverse image $pr_1^{-1}(A)$ of A under the projection pr_1 onto the first factor of $\mathrm{C}(X \times X)$ is closed. This implies that the intersection

$$pr_1^{-1}(A) \cap D(X) = D(A)$$

is closed in $\mathrm{C}(X \times X)$. Thus the image of A under the diagonal map D is closed in $\mathrm{C}(X \times X)$. Because this holds for every closed subset A of X the diagonal map $D : X \to \mathrm{C}(X \times X)$ is closed. $\qquad\square$

Proposition 11.2.26. *The diagonal map $G \to \mathrm{C}(G \times G)$ of any **C**-group G that satisfies the separation axiom T_1 is closed.*

Proof. Let G be a **C**-group that satisfies the separation axiom T_1. The image $D(G) \subset C(G \times G)$ of the diagonal map D is the inverse image

$$D(G) = [\mu \circ (\mathrm{id}_G \times_{\mathbf{C}} i)]^{-1}(\{1\})$$

where i denotes the inversion in G and 1 is the identity in G. Since G was assumed to satisfy the separation axiom T_1 the point $1 \in G$ is closed and so is his inverse image $D(X)$ under the continuous map $\mu \circ (\mathrm{id}_G \times_{\mathbf{C}} i)$. Thus the diagonal map D in $\mathbf{Top_C}$ is closed. □

Corollary 11.2.27. *The natural injection* $q_{\mathrm{CSS}(G)} \circ i_0 : G \to |\mathrm{CSS}(G)|$ *is a closed embedding for any* **C**-*group* G *satisfying the separation axiom* T_1.

Corollary 11.2.28. *A closed embedding* $X \hookrightarrow G$ *into a* **C**-*group* G *satisfying the separation axiom* T_1 *induces a closed embedding of the geometric realisations* $|\mathrm{CSS}(X)| \hookrightarrow |\mathrm{CSS}(G)|$.

In the above observations we have explored the preservation properties of E concerning special classes of morphisms and the nature of the transformation $\mathrm{id}_{\mathbf{Top}} \to E$. We now turn ourselves to the preservation of limits. collecting all the information gathered in the previous section and chapter we obtain:

Lemma 11.2.29. *The enveloping space functor* $E : \mathbf{Top} \to \mathbf{Top}$ *preserves equalisers.*

Proof. The standard simplicial space functor SS on **Top** is continuous by Theorem 11.1.20. The geometric realisation functor $|-|$ preserves equalisers by Proposition 9.3.1. Thus the composition $E = |-| \circ \mathrm{SS}$ preserves equalisers. □

Lemma 11.2.30. *The restriction* $E : k_\omega \mathbf{Haus} \to k_\omega \mathbf{Haus}$ *preserves equalisers.*

Proof. This follows from the facts that equalisers of Hausdorff spaces are closed Hausdorff subspaces and closed subspaces of k_ω-spaces are k_ω-spaces. □

Theorem 11.2.31. *The enveloping space functor* $E : \mathbf{Top} \to \mathbf{Top}$ *preserves finite limits of Hausdorff* k_ω-*spaces.*

Proof. Since the enveloping space functor preserves equalisers by Lemma 11.2.30 it only is to show that E preserves binary (hence finite) products of Hausdorff k_ω-spaces. The standard simplicial space functor SS on **Top** is continuous by Theorem 11.1.20, so it preserves finite products. By Lemma 11.1.4 it maps binary products $X \times Y$ of Hausdorff spaces X and Y to binary products $\mathrm{SS}(X) \times \mathrm{SS}(Y)$ of standard simplicial k_ω-spaces. Since the geometric realisations EX and EY thereof are Hausdorff by Theorem 11.2.9, the spaces $EX \times EY = |\mathrm{SS}(X)| \times |\mathrm{SS}(Y)|$ and $|\mathrm{SS}(X \times Y)|$ are naturally isomorphic by Lemma 9.3.36. □

Corollary 11.2.32. *The restriction* $E : k_\omega \mathbf{Haus} \to k_\omega \mathbf{Haus}$ *preserves finite limits.*

Proof. This follows from the fact that the finite limits in $k_\omega \mathbf{Haus}$ and **Top** coincide. □

This has immediate applications for (total) Hausdorff k_ω-topological algebras: The geometric realisation $E*$ of the standard simplicial space $\mathrm{SS}(*)$ of the singleton $*$ is a singleton space. Therefore a 0-ary operation $* \to X$ on a Hausdorff k_ω-space X in $k_\omega \mathbf{Haus}$ induces a 0-ary operation. $* \cong E* \to EX$ on the enveloping space EX. If ω is an n-ary operation on the Hausdorff k_ω-space X then the function

$$\tilde{E}(\omega) : (EX)^n \cong E(X^n) \to EX$$

$$\tilde{E}(\omega)[(\mathbf{x}_1, \ldots, \mathbf{x}_n), t)] = [(\omega(x_{1\,0}, \ldots x_{n\,0}), \ldots, \omega(x_{1\,m}, \ldots x_{n\,m}))]$$

is an n-ary operation on the space EX. So we observe:

Theorem 11.2.33. *The enveloping space functor E restricts and corestricts to an endofunctor of the category $\mathbf{k}_\omega\mathbf{HausAlg}_\Omega$ of Hausdorff k_ω-algebras of operational type Ω.*

If E is any set of equations satisfied by the operations $\omega \in \Omega$ of a Hausdorff k_ω-topological algebra A of operational type Ω, then the induced operations on EA also satisfy these equations. So we obtain:

Theorem 11.2.34. *The functor E restricts and corestricts to an endofunctor of the category $\mathbf{k}_\omega\mathbf{HausAlg}_{(\Omega,\mathrm{E})}$ of Hausdorff k_ω-algebras of operational type (Ω, E).*

Corollary 11.2.35. *The enveloping space functor functor E restricts to endofunctors of the categories $\mathbf{k}_\omega\mathbf{HausSemGrp}$, $\mathbf{k}_\omega\mathbf{HausGrp}$, and $\mathbf{k}_\omega\mathbf{HausAb}$*

Corollary 11.2.36. *The enveloping space functor E restricts to endofunctors of the categories $\mathbf{k}_\omega\mathbf{HausMon}$ of Hausdorff k_ω-monoids and the category $\mathbf{k}_\omega\mathbf{HausMonAb}$ of Abelian Hausdorff k_ω-monoids.*

Corollary 11.2.37. *The enveloping space functor E restricts to endofunctors of the categories $\mathbf{k}_\omega\mathbf{HausGrp}$ and $\mathbf{k}_\omega\mathbf{HausAb}$.*

Example 11.2.38. Connected finite dimensional real or complex Lie groups G are Hausdorff k_ω-spaces. Thus the enveloping space EG of a finite dimensional real or complex Lie group G is a Hausdorff k_ω-group.

Example 11.2.39. Real or complex Kac-Moody groups are Hausdorff k_ω-groups (cf. [GGH06]). Thus the enveloping space EG of a real or complex Kac-Moody group G is a Hausdorff k_ω-group.

Corollary 11.2.40. *The enveloping functor functor E restricts to an endofunctor of the category $\mathbf{k}_\omega\mathbf{HausRng}$ of Hausdorff k_ω-Rings.*

Note that the standard simplicial space functor SS on $\mathbf{k}_\omega\mathbf{HausGrp}$ preserves the terminal object, which is the trivial group $\{1\}$. Because the kernel of a homomorphism $\varphi : G \to H$ in $\mathbf{k}_\omega\mathbf{HausGrp}$ is the equaliser of φ and the trivial homomorphism 1, we conclude:

Lemma 11.2.41. *The restriction $E : \mathbf{k}_\omega\mathbf{HausGrp} \to \mathbf{k}_\omega\mathbf{HausGrp}$ preserves kernels.*

Analogously, a cokernel of a homomorphism $\varphi : G \to H$ in $\mathbf{k}_\omega\mathbf{HausGrp}$ is the coequaliser of the homomorphism φ and the trivial homomorphism 1. So we obtain:

Lemma 11.2.42. *The restriction $E : \mathbf{k}_\omega\mathbf{HausGrp} \to \mathbf{k}_\omega\mathbf{HausGrp}$ preserves cokernels.*

Proof. The cokernel of a morphism $\varphi : G \to G'$ of Hausdorff k_ω-groups in $\mathbf{k}_\omega\mathbf{HausGrp}$ is the quotient $q : G' \to G'/N$ by the closure $N = \overline{\varphi(G)}$ of the image of φ. Because the enveloping space functor preserves open proclusions and epimorphisms, the homomorphism $Eq : EG' \to E(G'/N)$ is a quotient homomorphism. Since the restriction of E to $\mathbf{kHausGrp}$ preserves kernels, the kernel of this quotient morphism is the inclusion $EN \hookrightarrow EG'$ (which also is a closed embedding by Lemma 11.2.20). Because the enveloping space functor preserves denseness of images, the closure of $E\varphi(EG)$ in EG' is the closed subgroup EN of EG. Therefore Eq is the cokernel of $E\varphi$ in $\mathbf{k}_\omega\mathbf{HausGrp}$. $\square$

Proposition 11.2.43. *The restriction $E : \mathbf{k}_\omega\mathbf{HausGrp} \to \mathbf{k}_\omega\mathbf{HausGrp}$ is exact.*

Proof. This follows from the preceding two lemmata.   □

Lemma 11.2.44. *If* **C** *is productive, then the restriction* E : **CGrp** → **CGrp** *preserves epimorphisms.*

Proof. If **C** is productive, then the epimorphisms in **CGrp** are surjective by Lemma E.2.16. The surjectivity of maps is preserved by the functor E by Lemma 11.2.4, so E maps epimorphisms to surjective homomorphisms in **CGrp** which are epimorphisms.   □

Like the result for the geometric realisation functor, Theorem 11.2.34 can be generalised to structures that can be considered to be a functor from a category **D** into $\mathbf{k}_\omega\mathbf{Haus}$:

Theorem 11.2.45. *For any functor* $F : \mathbf{D} \to \mathbf{k}_\omega\mathbf{Haus}$ *the composition* $E \circ F$ *is a simplicial object in the category* $\mathbf{k}_\omega\mathbf{Haus}^{\mathbf{D}}$. *If the functor* F *preserves finite limits then* $E \circ F$ *preserves finite limits.*

Corollary 11.2.46. *The functor* E *restricts to an endofunctor of the category of Hausdorff* k_ω-*topological groupoids.*

Corollary 11.2.47. *The enveloping space functor* E *restricts to an endofunctor* $E : \mathbf{Hk}_\omega\mathbf{Haus} \to \mathbf{Hk}_\omega\mathbf{Haus}$. *It restricts further to an endofunctor of the category of homotopy associative compactly Hausdorff* k_ω-*Hopf-spaces and to one of the category of homotopy Abelian Hausdorff* k_ω-*Hopf-spaces.*

For Hausdorff k_ω-topological module M over a Hausdorff k_ω-topological ring R let ϱ denote the action $\varrho : R \times M \to M$ of R on M. Then EM is a simplicial Abelian Hausdorff k_ω-topological group and $\tilde{E}(\varrho) : ER \times EM \to EM$ is an action of ER on EM. Thus EM is a module over the Hausdorff k_ω topological ring ER. The ring R is a subring of ER via the natural injection $q_{\mathrm{SS}(R)} \circ i_0 : R \to ER$. So EM also is a module over the original ring R, and we obtain:

Lemma 11.2.48. *For any Hausdorff* k_ω-*ring* R *the functor* E *restricts to functors* $E : \mathbf{k}_\omega\mathbf{HausRMod} \to \mathbf{k}_\omega\mathbf{HausRMod}$ *and* $E : \mathbf{k}_\omega\mathbf{HausModR} \to \mathbf{k}_\omega\mathbf{HausModR}$.

For the special case of topological vector spaces over a topological field $\mathbb{K}$ this specialises to:

Lemma 11.2.49. *For any Hausdorff* k_ω-*topological field* $\mathbb{K}$ *the functor* E *restricts to a functor* $E : \mathbf{k}_\omega\mathbf{Haus}\mathbb{K}\mathbf{Vec} \to \mathbf{k}_\omega\mathbf{Haus}\mathbb{K}\mathbf{Vec}$.

Example 11.2.50. If V is a Hausdorff k_ω vector space over a finite discrete field $\mathbb{K}$ then EV also is a vector space over $\mathbb{K}$.

Example 11.2.51. If V is a real or complex Hausdorff k_ω-topological vector space, then EV is a real resp. complex topological vector space.

We now show that these results transfer to the functor $|-| \circ \mathrm{SS}$ on $\mathbf{Top}_{\mathbf{C}}$ if the category **C** is productive. This recovers the classical result for compactly generated Hausdorff groups in [ML70] as a special case.

Theorem 11.2.52. *If the category* **C** *is productive then the restriction* $|-| \circ \mathrm{CSS}$ *to* $\mathbf{Top}_{\mathbf{C}}$ *is cocontinuous.*

Proof. If the category **C** is productive then the standard simplicial space functor CSS on $\mathbf{Top}_{\mathbf{C}}$ is cocontinuous. In this case the functor $|-| \circ (\mathrm{CSS})$ is the composition of two cocontinuous functors, hence cocontinuous as well.   □

Corollary 11.2.53. *The endofunctors* $|-| \circ \text{CGSS}$ *on* **CGTop** *and* $|-| \circ \text{kSS}$ *on* **kTop** *are cocontinuous.*

It has been shown by Mac Lane in [ML70, Sec. 9] that for a compactly generated Hausdorff group G the geometric realisation $|\text{CGSS}(G)|$ is contractible in **CGTop**. The construction given there relies on the representation of $|\text{CGSS}(G)|$ as a coend of the functor $\text{CGSS}(G) \times \Delta : \mathcal{O}^{op} \times \mathcal{O} \to \textbf{CGTop}$. The cocontinuity of the geometric realisation in **Top** (Corollary 9.2.20) and the fact that the construction of Mac Lane is categorical makes it possible to generalise it to arbitrary topological spaces. The key in constructing the contraction is an endofunctor of the simplicial category $\mathcal{O}$, whose definition we here recall:

Definition 11.2.54. *The* right shift $T : \mathcal{O} \to \mathcal{O}$ *is defined via*

$$T[n] = [n+1], \quad T(\alpha)(k) = \begin{cases} 0 & \textit{if } k = 0 \\ 1 + \alpha(k-1) & \textit{if } k > 0 \end{cases} \quad \textit{for all } \alpha : [m] \to [n].$$

This functor was already used by Mac Lane in his article [ML70]. (Be aware that this article contains a misprint in the definition of T.) With the help of this translation functor one can construct a natural transformation $h : \Delta \times I \to \Delta T$ by defining

$$h_n : \Delta([n]) \times I \to \Delta T([n]), \quad h_n(s, \mathbf{t}) = (s, (1-s)t_0, \ldots, (1-s)t_n).$$

(cf. [ML70, Sec. 9]). The functions h_n are homotopies between the the inclusion $h_n(0, -) = \Delta(\epsilon_0)$ of Δ_n into Δ_{n+1} as the zeroth face of Δ_{n+1} and the constant maps $h_n(1, -) = \mathbf{e}_0$. These homotopies form a natural transformation $h : \Delta \times I \to \Delta T$ (See [ML70, Sec. 9]).

A similar transformation can be constructed for the standard simplicial complex of a based space X (with basepoint $*$). Here we define a natural transformation $h'_X : \text{SS}(X) \to \text{SS}(X)T$ via

$$h'_X([n])(\mathbf{x}) = (*, x_0, \ldots, x_n), \quad \text{and} \quad h'_X([n])(\alpha)(\mathbf{x})_i = x_{T(\alpha)(i)}$$

for all $0 \leq i \leq n+1$. This natural transformation has the following important property: Let $\eta_{[n+1] \setminus \{0\}} : [n+1] \to [0]$ be the $[n+1] \setminus \{0\}$-degeneracy and $\epsilon_0 : [n] \hookrightarrow [n+1]$ be the zeroth face map. Then the natural functions $h'_X([n])$ satisfy the equations

$$\text{SS}(X)(\epsilon_0)h'_X([n]) = \text{id}_{X^{n+1}} \quad \text{and} \quad \text{SS}(X)(\eta_{[n+1] \setminus \{0\}})h'_X([n]) = *,$$

where $*$ denotes the constant map $*$ into $X = \text{SS}(X)([0])$. This property will be needed to show that the enveloping space EX of any topological space X is contractible.

Lemma 11.2.55. *The maps h'_X form a natural transformation* $\text{SS} \to \text{SS}(-) \circ T$ *between the functors* $\text{SS}, \text{SS}(-) \circ T : \textbf{Top} \to \textbf{Top}^{\mathcal{O}^{op}}$.

For any topological space X the two natural transformations h'_X and h can be combined to obtain a natural function between the coends of the functors $\text{SS}(X) \times (\Delta \times I)$ and $\text{SS}(X)T \times \Delta T$:

Lemma 11.2.56. *The natural transformation $h'_X \times h$ induces a natural transformation $h'_X \times_{\mathcal{O}} h : \text{SS}(X) \times_{\mathcal{O}} (\Delta \times I) \to \text{SS}(X)T \times_{\mathcal{O}} \Delta T$.*

Proof. This follows from the universal property of coends. (Taking coends here is a functor $\textbf{Top}^{\mathcal{O}^{op} \times \mathcal{O}} \to \textbf{Top}$). $\qquad\qquad\square$

There also exists a natural map $SS(X)T \times_{\mathcal{O}} \Delta T \to EX$. It can be constructed as follows: Consider the universal wedge $\psi_X : SS(X) \times \Delta \to EX$. The maps $\psi_{X,m,n}$ of the dinatural transformation ψ_X can be composed with the identities

$$\tau_{X,m,n} : SS(X)T([m]) \times \Delta T([n]) \xrightarrow{=} SS(X)([m+1]) \times \Delta([n+1]).$$

to obtain maps $\psi_{X,m+1,n+1} \circ \tau_{X,m,n} : SS(X)T([m]) \times \Delta T([n]) \to EX$. We note that the maps $\tau_{X,m,n}$ do not form a dinatural transformation but are natural in X for fixed objects $[m], [n]$ in $\mathcal{O}$. The composition $\psi_X \circ \tau_{X,-,-}$ however turns out to be a dinatural transformation:

Proposition 11.2.57. *The maps $\psi_{X,m+1,n+1}\tau_{X,m,n}$ form a dinatural transformation $SS(X)T \times \Delta T \to EX$.*

Proof. It is to verify that for each morphism $\alpha : [m] \to [n]$ in $\mathcal{O}$ the following diagram is commutative:

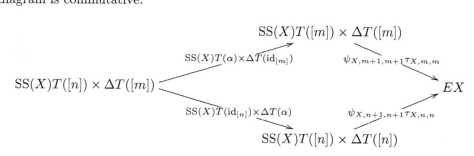

Let $\alpha : [m] \to [n]$ be a morphism in $\mathcal{O}$ and $(\mathbf{x}, t) \in X^{n+2} \times \Delta_{m+1}$ be a point of $SS(X)T([n]) \times \Delta([m])$. The image of this point $(\mathbf{x}, t)$ under the composition $\psi_{X,m+1,m+1}\tau_{X,m,m}(SS(X)T(\alpha) \times \Delta(\mathrm{id}_{[m]}))$ is given by

$$\psi_{X,m+1,m+1}\tau_{X,m,m}(SS(X)T(\alpha) \times \Delta(\mathrm{id}_{[m]}))(\mathbf{x}, t) =$$
$$\psi_{X,m+1,m+1}\tau_{X,m,m}(SS(X)T(\alpha)(x), t)$$
$$= [(SS(X)T(\alpha)(\mathbf{x}), t)]$$
$$= [(SS(X)(T(\alpha))(\mathbf{x}), t)]$$
$$= [(SS(X)(\mathbf{x}), \Delta(T(\alpha))(t)]$$
$$= [(SS(X)(\mathbf{x}), \Delta T(\alpha)(t)]$$
$$= \psi_{X,n+1,n+1}\tau_{X,n,n}(\mathbf{x}, \Delta T(\alpha)(t))$$
$$= \psi_{X,n+1,n+1}\tau_{X,n,n}(\mathrm{id}_{SS(X)T([n])} \times \Delta T(\alpha))(\mathbf{x}, t).$$

Thus the above diagram is commutative and so the maps $\psi_{X,m+1,n+1} \circ \tau_{X,m,n}$ form a wedge $SS(X)T \times \Delta T \to EX$. ☐

The dinatural transformation formed by the maps $\psi_{X,m+1,n+1}\tau_{X,m,n}$ will be denoted by $(\psi\tau)_X$.

Corollary 11.2.58. *The dinatural transformation $(\psi\tau)_X$ factors through a map $\int^n (\psi\tau)_X : SS(X)T \times_{\mathcal{O}} \Delta T \longrightarrow EX$.*

Proof. This follows from the fact that wedge $SS(X)T \times \Delta T \to SS(X)T \times_{\mathcal{O}} \Delta T$ is universal amongst all wedges $SS(X)T \times \Delta T \to Y$ to constant functors Y. ☐

Lemma 11.2.59. *The maps $\int^n (\psi\tau)_X$ form a natural transformation $\int^n (\psi\tau)$ from $SS(-)T \times_{\mathcal{O}} \Delta T$ to the enveloping space functor E.*

Proof. Let $f : X \to Y$ be a continuous function between the topological spaces X and Y. It is to verify that the diagram

$$
\begin{array}{ccc}
\mathrm{SS}(X)T \times_\mathcal{O} \Delta T & \xrightarrow{\mathrm{SS}(f)T \times_\mathcal{O} \Delta T(\mathrm{id})} & \mathrm{SS}(Y)T \times_\mathcal{O} \Delta T \\
{\scriptstyle \int^n \psi_X \tau_X} \downarrow & & \downarrow {\scriptstyle \int^n \psi_Y \tau_Y} \\
EX & \xrightarrow{Ef} & EY
\end{array}
$$

is commutative. Let $\omega_X : \mathrm{SS}(X)T \times \Delta T \to \mathrm{SS}(X)T \times_\mathcal{O} \Delta T$ denote the universal wedge from the functor $\mathrm{SS}(X)T \times \Delta T$ to its coend $\mathrm{SS}(X)T \times_\mathcal{O} \Delta T$. For every point $p \in \mathrm{SS}(X)T \times_\mathcal{O} \Delta T$ there exists an $n \in \mathbb{N}$ and a point $(\mathbf{x}, t) \in \mathrm{SS}(X)T([n]) \times \Delta T([n])$ such that p is the image of $(\mathbf{x}, t)$ under the natural map

$$
\omega_{X,[n]} : \mathrm{SS}(X)T([n]) \times \Delta T([n]) \to \mathrm{SS}(X)T \times_\mathcal{O} \Delta T
$$

Using the naturality of $\omega_{-,[n]}$ and $\tau_{-,n,n}$ for fixed $[n]$ in $\mathcal{O}$ one obtains the following diagram

where all but the front square are known to be commutative. A diagram chase reveals the following equalities:

$$
\left(\int^n \psi_Y \tau_Y \right) (\mathrm{SS}(f)T \times_\mathcal{O} \Delta T(\mathrm{id}))(p) =
$$

$$
\left(\int^n \psi_Y \tau_Y \right) (\mathrm{SS}(f)T \times_\mathcal{O} \Delta T(\mathrm{id}))\omega_{X,[n]}(\mathbf{x}, t)
$$

$$
= \left(\int^n \psi_Y \tau_Y \right) \omega_{Y,[n]}(\mathrm{SS}(f)T([n]) \times \Delta T(\mathrm{id}_{[n]}))(\mathbf{x}, t)
$$

$$
= \psi_{Y,[n+1],[n+1]} \tau_{Y,[n],[n]}(\mathrm{SS}(f)T([n]) \times \Delta T(\mathrm{id}_{[n]}))(\mathbf{x}, t)
$$

$$
= \psi_{Y,[n+1],[n+1]}(\mathrm{SS}(f)([n+1]) \times \Delta(\mathrm{id}_{[n+1]}))\tau_{Y,[n],[n]}(\mathbf{x}, t)
$$

$$
= Ef\psi_{X,[n+1],[n+1]}\tau_{X,[n],[n]}(\mathbf{x}, t)
$$

$$
= Ef \left(\int^n \psi_X \tau_X \right) \omega_{X,[n]}(\mathbf{x}, t)
$$

$$
= Ef \left(\int^n \psi_Y \tau_Y \right) (p)
$$

The definition of the geometric realisation via coends makes it possible to construct a natural contraction of the space EX with the help of the natural transformations $h \times_\mathcal{O} h'$ and $\int^n \psi_X \tau_X$.

Theorem 11.2.60. *There exists a natural transformation $H : E \times I \to E$ such that $H \circ \{0\} = \mathrm{id}_{\mathbf{Top}_*}$ and $H \circ \{1\} = * \subseteq E$ in $\mathbf{Top}_*$.*

Proof. Let $*$ denote the basepoint of X. Combining the natural transformations $h' \times_\mathcal{O} h$ and $\int^n (\psi\tau)_X$ yields a natural map

$$\mathrm{SS}(X) \times_\mathcal{O} (\Delta \times I) \xrightarrow{h' \times_\mathcal{O} h} \mathrm{SS}(X)T \times_\mathcal{O} \Delta T \xrightarrow{\int^n \psi_X \tau_X} \mathrm{SS}(X) \times_\mathcal{O} \Delta = EX.$$

Because the unit interval I is compact the product functor $- \times I$ is cocontinuous. Therfore it preserves all coequalisers and thus commutes with the coend functor. So we obtain a natural homeomorphism

$$EX \times I \cong \int^n X([n]) \times \Delta([n]) \times I$$

Combining this natural homeomorphism with the natural transformation $(h \times_\mathcal{O} h) \circ \int^n (\psi\tau)_X$ yields a natural transformation

$$H : EX \times I \longrightarrow EX$$

This natural transformation H is the desired one. $\square$

Corollary 11.2.61. *The enveloping space EX of any based topological space X is contractible and these contractions are natural in X.*

11.3 Universal Bundles and Classifying Spaces

In this section we introduce a construction of universal bundles and classifying spaces for (G, A)-bundles, where G and A are Hausdorff k_ω topological groups. We start by considering bi-transformation groups. All G-spaces are left G-spaces unless specified otherwise.

Definition 11.3.1. *A simplicial (G, A)-space is an object in* $\mathbf{GTopA}^{\mathcal{O}^{op}}$.

Thus simplicial (G, A)-spaces are simplicial spaces equipped with commuting left and right actions of G resp. A. Furthermore these actions have to intertwine the boundary and degeneracy maps. Any simplicial (G, A)-space can be considered as a simplicial bi-transformation group by regarding the groups G and A as constant simplicial topological groups.

Example 11.3.2. For every (G, A)-space X the standard simplicial space $\mathrm{SS}(X)$ comes naturally with commuting left and right actions of G resp. A. These are the diagonal actions, which are given by

$$G \times X^{n+1} \to X^{n+1}, \quad (g, (x_0, \ldots, x_n)) \mapsto (g.x_0, \ldots, g.x_n)$$

and

$$\mathrm{C}X^{n+1} \times A \to X^{n+1}, \quad ((x_0, \ldots, x_n), a) \mapsto (x_0.a, \ldots, x_n.a).$$

These actions intertwine the boundary and degeneracy maps and thus $\mathrm{SS}(X)$ is a simplicial (G, A)-space. By applying the corefelctor C one obtains a simplicial $(\mathrm{C}G, \mathrm{C}A)$-space $\mathrm{CSS}(X)$ in $\mathbf{Top}_\mathrm{C}$ for any class $\mathbf{C}$ of topological spaces.

The simplicial (G, A) spaces $\mathrm{SS}(X)$ can be identified with the simplicial bi-transformation group $(G, \mathrm{SS}(X), A)$, where G and A are constant simplicial groups. The action of the constant simplicial group G on $\mathrm{SS}(X)$ is here given via the diagonal inclusion $G \hookrightarrow \mathrm{SS}(G)$. The action of the constant simplicial group A on $\mathrm{SS}(X)$ is given analogously.

Given a simplicial (G, A)-space X one can form the simplicial orbit spaces $G\backslash X$ and X/A. Since the actions of G and A on the simplicial space X commute, there is the possibility for the action of G on X to descend to an action of G on the latter orbit space X/A.

Lemma 11.3.3. *The action of G on X descends to an action of G on X/A and the action of A on X descends to an action of A on $G\backslash X$.*

Proof. Let $\varrho : G \times X \to X$ denote the action of G on X and let $q : X \to X/A$ denote the orbit map of the action of A. The quotient map $q : X \to X/A$ is open (cf. [tD87, Proposition 3.1 (iv)]). Therefore the product $\mathrm{id}_G \times q$ is a finite product of open identifications, hence an identification (by Corollary B.1.6). As a consequence the composition $q\varrho$ factors uniquely through the quotient map $\mathrm{id}_G \times q$:

$$
\begin{array}{ccc}
G \times X & \xrightarrow{\;\varrho\;} & X \\
{\scriptstyle \mathrm{id}_G \times q}\downarrow & & \downarrow{\scriptstyle q} \\
G \times (X/A) & \xrightarrow[\varrho/A]{} & X/A
\end{array}
$$

The unique function $\varrho/A : G \times X/A \to X/A$ satisfying $q\varrho = (\varrho/A)(\mathrm{id}_G \times q)$ then is a continuous action of G on X/A. The proof that the action of A on X induces a continuous action of A on $G\backslash X$ is completely analogous. $\qquad\square$

So the simplicial orbit spaces $G\backslash X$ and X/A are simplicial right A-spaces and simplicial left G-spaces respectively. Furthermore, as shall be proved below, all the group actions on the simplicial spaces X, $G\backslash X$ and X/A descend to the geometric realisations $|X|$, $|G\backslash X|$ and $|X/A|$ of these simplicial spaces. As usual we let the groups G and A act trivially on the standard simplices Δ_n, so the action of G and A on X induces an action of G and A on the disjoint union $F(X) = \coprod X([n]) \times \Delta_n$.

Lemma 11.3.4. *The equivalence relation on $F(X)$ defining the identification map $q_X : F(X) \twoheadrightarrow |X|$ is invariant under the actions of G and A.*

Proof. The equivalence relation $\sim$ on $F(X)$ is the transitive hull of the defining relations

$$(\exists \alpha : [m] \to [n]) : X(\alpha) = x' \wedge \Delta(\alpha)(t') = t \Rightarrow (x,t) \sim (x',t')$$

for all $(x,t) \in X([n]) \times \Delta_n$ and $(x',t') \in X([m]) \times \Delta_m$. So it suffices to show the invariance of the set of defining relations under the groups A and G. Since the space X is a simplicial (G, A)-space, the actions of G and A intertwine the the maps $X(\alpha)$ for all morphisms α in $\mathcal{O}$. Thus the set of defining relations is invariant under the actions of G and A. $\qquad\square$

Lemma 11.3.5. *If G is locally compact then the action of G on $F(X)$ descends to an action of G on $|X|$. Similarly, if A is locally compact then the action of A on $F(X)$ descends to an action of A on $|X|$.*

Proof. Let $\varrho : G \times F(X) \to F(X)$ denote the induced action of G on $F(X)$ and let $q_X : F(X) \to |X|$ denote the quotient map. Assume G to be locally compact. Because taking the product with the locally compact space G is a cocontinuous functor $G \times - : \mathbf{Top} \to \mathbf{Top}$, the product map $\mathrm{id}_G \times q$ is a quotient map. As a consequence the composition $q\varrho$ factors uniquely through the quotient map $\mathrm{id}_G \times q$:

$$
\begin{array}{ccc}
G \times F(X) & \xrightarrow{\;\varrho\;} & F(X) \\
{\scriptstyle \mathrm{id}_G \times q_X}\downarrow & & \downarrow{\scriptstyle q_X} \\
G \times |X| & \xrightarrow[\varrho']{} & |X|
\end{array}
$$

The unique function $\varrho' : G \times |X| \to |X|$ satisfying $q\varrho = (\varrho/A)(\mathrm{id}_G \times q)$ is a continuous action of G on $|X|$. The proof for locally compact groups A is analogous. $\qquad\square$

Lemma 11.3.6. *If G is a Hausdorff k_ω-group and X is a simplicial Hausdorff k_ω-space with Hausdorff geometric realisation $|X|$, then the action of G on X descends to an action of G on $|X|$.*

Proof. Let $\varrho : G \times F(X) \to F(X)$ denote the induced action of G on X and let $q_X : X \to |X|$ denote the quotient map onto the geometric realisation $|X|$ of X. Assume the spaces G and X to be Hausdorff and of type k_ω. Then the space $F(X)$ is as well as its quotient space $|X|$ are k_ω-spaces by Proposition 9.3.32. Since products of quotient maps in **kTop** are quotient maps, the action ϱ descends to an action $\varrho' : G \times_k |X| \to |X|$ in **kTop**. If the spaces $|X|$ Hausdorff in addition, then the product $G \times |X|$ of the Hausdorff k_ω-spaces is a Hausdorff k_ω-space by Lemma B.10.7, and in particular a k-space. In this case the product $G \times_k |X|$ in **kTop** coincides with the product in **Top** and the action ϱ' is an action in **Top**. □

Lemma 11.3.7. *If A is a Hausdorff k_ω-group, X is a simplicial Hausdorff k_ω-space with Hausdorff geometric realisation $|X|$, then the action of A on X descends to an action of A on $|X|$.*

Proof. The proof is analogous to that of Lemma 11.3.6. □

Lemma 11.3.8. *If G and A are Hausdorff k_ω-groups and X is a simplicial Hausdorff k_ω-space with Hausdorff geometric realisation $|X|$, then the actions of G and A on X descend to actions of on $|X|$, i.e. $|X|$ is a (G, A)-space.*

Descending to orbit spaces is a functor **TrGrp** $\to$ **Top** from the category of transformation groups to the category of topological spaces. So from every simplicial (G, A)-space X one obtains the simplicial G-space X/A, the simplicial right A-space $G\backslash X$ and the simplicial space $G\backslash X/A$. From the construction of the quotient space $|X|$ from $F(X)$ one infers:

Proposition 11.3.9. *The orbit spaces $G\backslash|X|$, and $|X|/A$ and $G\backslash|X|/A$ are naturally isomorphic to the geometric realisations $|G\backslash X|$, $|X/A|$ and $|G\backslash X/A|$ of the simplicial spaces $G\backslash X$, X/A and $G\backslash X/A$ respectively.*

Remark 11.3.10. We do not require group actions to be continuous for taking orbit spaces, since the latter are independent of the topology of the groups.

Corollary 11.3.11. *The orbit spaces $G\backslash EX$ and EX/A and $G\backslash EX/A$ are the geometric realisations of the simplicial spaces $G\backslash SS(X)$, $SS(X)/A$ and $G\backslash SS(X)/A$ respectively.*

This can now be applied to standard simplicial spaces. If X is a (G, A)-space then its standard simplicial space $SS(X)$ is a simplicial (G, A)-space. The geometric realisation of this simplicial space is the enveloping space EX. The action of G and A on EX might not be continuous. In any case we note:

Corollary 11.3.12. *The orbit spaces $G\backslash EX$ and EX/A and $G\backslash EX/A$ are the geometric realisations of the simplicial spaces $G\backslash SS(X)$, $SS(X)/A$ and $G\backslash SS(X)/A$ respectively.*

These constructions can especially be applied to simplicial (G, A)-spaces of the form $X = SS(A)$, where G acts on the space A and this action commutes with right translations. Here the group A acts via the diagonal (right) action on the standard simplicial space $SS(A)$. So we obtain a simplicial fibre structure

$$A \text{------} SS(A) \text{------} SS(A)/A$$

with fibre A where the projection onto the base $SS(A)/A$ is the orbit map. If A is abelian then the projection onto the base $SS(A)/A$ is a group homomorphism with kernel the diagonal subgroup A of $SS(A)$.

Lemma 11.3.13. *The simplicial space* $SS(A)/A$ *is a simplicial G-space.*

Proof. This is a special case of Lemma 11.3.3.   □

The simplicial space $SS(A)/A$ and the above bundle (without G-action) is known as the 'bar-construction'. This fibre structure leads to a fibre structure of the geometric realisations as follows: Because the standard simplices Δ_n are compact, each orbit map $SS(A)([n]) \to SS(A)([n])/A$ for a component $SS(A)([n])$ of $SS(A)$ induces an orbit map $SS(A)([n]) \times \Delta_n \to (SS(A)([n]) \times \Delta_n)/A = (SS(A)/A)([n]) \times \Delta_n$. The disjoint union of these orbit maps is again an orbit map. All in all one obtains the following commutative diagram

$$
\begin{array}{ccccccc}
A & \longrightarrow & F(SS(A)) & \longrightarrow & F(SS(A)/A) & \xrightarrow{\ \cong\ } & F(SS(A)/A) \\
\downarrow{\scriptstyle q_A} & & \downarrow{\scriptstyle q_{SS(A)}} & & \downarrow{\scriptstyle q_{SS(A)}/A} & & \downarrow{\scriptstyle q_{SS(A)/A}} \\
A & \longrightarrow & EA & \longrightarrow & EA/A & \xrightarrow{\ \cong\ } & |SS(A)/A|
\end{array}
\quad ,
$$

in which the subspace A of EA is the fibre of the orbit map $EA \to EA/A$. We give the base space a name:

Definition 11.3.14. *The geometric realisation of* $SS(A)/A$ *is denoted by* BA.

So far we have obtained a contractible enveloping A-space EA of every topological group A. Moreover the quotient onto the orbits of the action of A is a fibre structure

$$A \text{\textemdash\textemdash} EA \text{\textemdash\textemdash} BA$$

with fibre A itself. If this fibre structure is a numerable bundle, then this bundle is a classifying bundle of A.

Proposition 11.3.15. *If A is a locally contractible group then EA and BA are locally contractible as well and $EA \to BA$ is a locally trivial bundle with fibre A.*

Proof. See [Seg70, Proposition A.1, Proposition A.2] for a proof.   □

Corollary 11.3.16. *If A is a locally contractible group then $EA \to BA$ is a principal A-bundle.*

Theorem 11.3.17. *If A is a locally contractible Hausdorff k_ω-group, then the principal bundle $EA \to BA$ is numerable, hence a universal A-bundle and BA is a classifying space for A.*

Proof. Let A be a locally contractible Hausdorff k_ω-group. The bundle $EA \to BA$ is a locally trivial principal A-bundle by Corollary 11.3.16. The base space BA of this bundle is a Hausdorff k_ω-space and therefore paracompact (cf. [Mor56, Lemma 5]). Thus the principal bundle $EA \to BA$ is numerable, hence a universal A bundle and BA is a classifying space for A.   □

Example 11.3.18. If G is a connected finite dimensional Lie group, then EG is a Hausdorff k_ω-group and the bundle $EG \to BG$ is universal.

More generally every countable direct limit of Hausdorffff k_ω groups is a Hausdorff k_ω-group (cf. [GGH06, Proposition 5.4]). This leads to the following important examples:

Example 11.3.19. The diretc limit group **SO** $= \lim SO_n$ is a Hausdorff k_ω-group. Therefore the **SO**-bundle $E\mathbf{SO} \to B\mathbf{SO}$ is universal and the total space $E\mathbf{SO}$ of this bundle is a Hausdorff k_ω-group.

Example 11.3.20. The diretc limit group $\mathbf{U} = \lim U_n$ is a Hausdorff k_ω-group. Therefore the $\mathbf{U}$-bundle $E\mathbf{U} \to B\mathbf{U}$ is universal and the total space $E\mathbf{U}$ of this bundle is a Hausdorff k_ω-group.

Example 11.3.21. The diretc limit group $\mathbf{SU} = \lim SU_n$ is a Hausdorff k_ω-group. Therefore the $\mathbf{SU}$-bundle $E\mathbf{SU} \to B\mathbf{SU}$ is universal and the total space $E\mathbf{SU}$ of this bundle is a Hausdorff k_ω-group.

Example 11.3.22. The diretc limit group $\mathbf{PU} = \lim PU_n$ is a Hausdorff k_ω-group. Therefore the $\mathbf{PU}$-bundle $E\mathbf{PU} \to B\mathbf{PU}$ is universal and the total space $E\mathbf{PU}$ of this bundle is a Hausdorff k_ω-group.

Example 11.3.23. Real and complex Kac-Moody groups are Hausdorff k_ω-groups (cf. [GGH06]). Thus the universal enveloping space EG of a real or complex Kac-Moody group G is a Hausdorff k_ω group and $EG \to BG$ is a universal G-bundle.

Lemma 11.3.24. *If A is Hausdorff, then the fibre A is closed in EA and BA is a T_1-space.* ☐

Proof. If A is Hausdorff, then the simplicial space $SS(A)$ is Hausdorff and the simplicial subspace $A \subseteq SS(A)$ is closed. By Lemma 11.2.20 the inclusion $A = |A| \hookrightarrow |EA|$ of the fibre A is a closed embedding as well. In this case the base space BA is a T_1-space. ☐

If the group A is a Hausdorff k_ω-group, then the enveloping space EA is a Hausdorff k_ω-group as well by Corollary 11.2.35. It then contains the group A as a closed subspace. The inclusion turns out to be a group homomorphism, more generally we observe:

Lemma 11.3.25. *If A is a Hausdorff k_ω-algebra of type (Ω, E), then EA is a Hausdorff k_ω algebra of type (Ω, E) and the inclusion $A \hookrightarrow EA$ is a morphism in $\mathbf{k_\omega HausAlg}_{(\Omega,E)}$.*

Proof. Let A be a Hausdorff k_ω-algebra of type (Ω, E). The diagonal inclusion $A \hookrightarrow SS(A)$ is a morphism of simplicial algebras of type (Ω, E). As a consequence the induced inclusion $A \cong |A| \hookrightarrow |SS(A)| = EA$ is a morphism of Hausdorff k_ω-algebras. ☐

If A is an abelian group in $\mathbf{k_\omega Haus}$, then the projection onto the base space BA is the quotient of the Abelian topological group EA by the normal subgroup A of EA. In this case the base space BA is an abelian topological group as well. Moreover it also is a Hausdorff k_ω-group:

Lemma 11.3.26. *If A is an abelian Hausdorff k_ω-group, then the base space BA is an abelian compactly generated Hausdorff group as well.*

Proof. If A is a k_ω-group then BA is a quotient of the k_ω-space EA, hence a k_ω-space. If in addition A is Hausdorff, then subgroup $A \leq EA$ is closed, hence the quotient group BA is Hausdorff. ☐

Theorem 11.3.27. *For every abelian Hausdorff k_ω-group A the sequence of abelian Hausdorff k_ω-groups*

$$0 \hookrightarrow A \to EA \twoheadrightarrow BA \to 0$$

is exact and locally trivial.

This makes it possible to transfer the results concerning the continuous cohomology of Hausdorff k-groups obtained in [Seg70] to Hausdorff k_ω-groups without working in the category $\mathbf{kTop}$ of k-spaces.

11.4 Abelian Topological Groups with Prescribed Homotopy Groups

In this section we present a functorial construction of abelian topological groups with prescribed homotopy groups. The construction establishes a correspondence between the category **AbGrLieAlg** of abelian graded lie algebras and the category CW-**TopAb** of abelian topological groups of the homotopy type of CW-complexes. It requires no prerequisites and only relies on the preservation of finite products under the geometric realisation functor $|-| : \mathbf{Set}^{\mathcal{O}^{op}} \to \mathbf{Top}$ (Theorem 9.3.49). This in particular includes a general procedure to construct Abelian Eilenberg Mac Lane topological groups $K(\pi, n)$.

In this section all graded Lie algebras will be $\mathbb{N}$-graded and the category of abelian $\mathbb{N}$-graded Lie algebras will be denoted by **AbGrLieAlg**. Consider an $\mathbb{N}$-graded abelian Lie-algebra $A = \bigoplus_{n=0}^{\infty} A_n$. There exists a natural and trivial way to assign to each such graded abelian Lie-algebra A in **AbGrLieAlg** a simplicial abelian group A with homotopy $\pi_n(A) = A_n$. It is given by the positive chain complex of abelian groups whose homology is exactly A. It is given by the functor

$$F : \mathbf{AbGrLieAlg} \to \mathbf{Ch}_{\geq 0}(\mathbf{Ab}), \quad F(A) = (A_*, 0)$$

which assigns to every graded abelian Lie-algebra A the chain complex

$$\cdots A_{n+1} \xrightarrow{0} A_n \xrightarrow{0} A_{n+1} \cdots$$

with zero differentials. Recall the Dold-Kan correspondence (cf. Theorem F.7.3), which establishes an equivalence of the category $\mathbf{Ab}^{\mathcal{O}^{op}}$ of simplicial Abelian groups and the category $\mathbf{Ch}_{\geq 0}(\mathbf{Ab})$ of positive chain complexes of Abelian groups. It consists of a pair of adjoint functors

$$(N-)_* : \mathbf{Ab}^{\mathcal{O}^{op}} \to \mathbf{Ch}_{\geq 0}(\mathbf{Ab}), \quad \text{and} \quad K : \mathbf{Ch}_{\geq 0}(\mathbf{Ab}) \to \mathbf{Ab}^{\mathcal{O}^{op}}$$

the first of which is the normalised chain complex functor. We consider the homotopy $\pi(A)$ of a simplicial Abelian group A as an Abelian graded Lie algebra. In this way one obtains a functor $\pi : \mathbf{Ab}^{\mathcal{O}^{op}} \to \mathbf{AbGrLieAlg}$ from the category $\mathbf{Ab}^{\mathcal{O}^{op}}$ back into the category **AbGrLieAlg** of Abelian graded Lie algebras. The composition of this functor π with the functors K and $F : \mathbf{AbGrLieAlg} \to \mathbf{Ch}_{\geq 0}(\mathbf{Ab})$ has the following important property:

Proposition 11.4.1. *The endofunctors* $\pi K F : \mathbf{Ab}^{\mathcal{O}^{op}} \to \mathbf{AbGrLieAlg}$ *and* $\mathrm{id}_{\mathbf{AbGrLieAlg}}$ *are naturally isomorphic.*

Proof. Let A be an Abelian graded Lie algebra. The homology $H(FA)$ of the chain complex FA is the graded Abelian group A itself. Furthermore the Dold-Kan correspondence ensures the existence of natural isomorphisms $\pi_n K(FA) \cong H_n(FA)$, so all in all we obtain a natural isomorphism $\pi K F \cong \mathrm{id}_{\mathbf{Ab}}\mathbf{GrLieAlg}$. □

This ensures that the homotopy $\pi(KFA)$ of the simplicial Abelian group KFA associated to a graded Abelian Lie algebra is exactly the graded Abelian group A. So we observe:

Corollary 11.4.2. *For every graded Abelian Lie algebra A there exists a simplicial Abelian group KFA with homotopy $\pi(KFA) \cong A$.*

We endow all Abelian groups with the discrete topology and in this way consider the category **Ab** of Abelian groups as a subcategory of **Top**. This turns all simplicial Abelian groups into discrete simplicial spaces. Making further use of the fact that the geometric realisation of discrete simplicial spaces preserves homotopy, we can show:

Lemma 11.4.3. *The endofunctors $\pi_n| - |KFA : \mathbf{AbGrLieAlg} \to \mathbf{AbGrLieAlg}$ and $H_n : A \mapsto A_n$ are naturally isomorphic.*

Proof. All simplicial Abelian groups are fibrant simplicial sets (cf. Lemma [Wei94, 8.2.8]) with Abelian homotopy groups. The composition $\pi_n|-|$ of the n-th homotopy group functor $\pi_n : \mathbf{Top} \to \mathbf{Grp}$ with the geometric realisation functor $| - |$ on fibrant simplicial sets is naturally isomorphic to the homotopy group functor π_n on fibrant simplicial sets (cf. [May92, Theorem 16.1]). Therefore the composition $\pi_n|-|KFA$ is naturally isomorphic to $\pi_n KFA$, which in turn is naturally isomorphic to $H_n : A \mapsto A_n$ by Corollary 11.4.2. $\qquad\square$

By Theorem 9.3.52 the geometric realisation of a discrete Abelian group KFA is a CW-complex which is a topological Abelian group. This especially implies that the functor $| - |KFA$ has domain in the category $\mathbf{kHausAb}$ of compactly generated Hausdorff Abelian groups. The homotopy Lie algebra of an Abelian topological group is Abelian because the Samelson products of homotopy Abelian H-spaces are trivial, hence the homotopy lie algebra functor on $\mathbf{kHaus}$ has codomain $\mathbf{AbGrLieAlg}$. Therefore we can compose the functors π and $| - |KF$ to obtain an endofunctor $\pi| - |KF$ of the category $\mathbf{AbGrLieAlg}$ of Abelian graded lie algebras. We collect the above observations:

Theorem 11.4.4. *The endofunctors $\pi| - |KF : \mathbf{LieAlg} \to \mathbf{AbGrLieAlg}$ and $\mathrm{id}_{\mathbf{AbGrLieAlg}}$ are naturally isomorphic.*

This is to say that the functor $| - |KF : \mathbf{AbGrLieAlg} \to \mathbf{kHausAb}$ assigns to each Abelian graded Lie algebra $A = \bigoplus_n A_n$ an Abelian topological group $|KFA|$ with homotopy groups $\pi_n(|KFA|) = A_n$. In particular we observe:

Corollary 11.4.5. *For every sequence A_n of Abelian groups there exists an Abelian topological group G which has homotopy groups $\pi_n(G) = A_n$.*

Moreover this construction is functorial in the homotopy groups A_n prescribed, i.e. if $f : A \to B$ is a morphism of Abelian graded Lie algebras then the geometric realisation $|kf(f)| : |KFA| \to |\pi KFB|$ of the morphism $Kf(f)$ of simplicial Abelian groups is a homomorphism of Abelian topological groups, and the homomorphisms $\pi_n(KFf)$ in the homotopy of these topological groups correspond to the morphisms $f_n : A_n \to B_n$.

For each $n \in \mathbb{N}$ and Abelian group π one can consider the Abelian graded Lie algebra A whose only nontrivial component $A_n = \pi$ is in dimension n. The homology of the chain complex associated to this graded Abelian Lie algebra is trivial except in dimension n, where it is exactly the group π. So we observe:

Theorem 11.4.6. *For every discrete Abelian group π and natural number $n \in \mathbb{N}$ there exists a CW-complex of type $K(\pi, n)$ which is an Abelian topological group.*

11.5 The Rational Homology of Eilenberg-MacLane Spaces

The universal construction of Eilenberg-MacLane spaces presented in the last section allows us to prove a version of the "Cartan-Serre Theorem" for all Eilenberg-Mac Lane spaces regardless of the dimension of the rational homotopy groups or homology groups, i.e. we do not have to assume that the topological spaces in question are of finite type. We continue to consider a discrete Abelian group π. For $n \geq 1$ the Eilenberg-MacLane spaces $K(\pi, n)$ are connected. Furthermore, by Theorem 11.4.6 there exists an Eilenberg-MacLane space $K(\pi, n)$ which is a CW-complexe and an abelian topological group.

Proposition 11.5.1. *The rational homology $H(X;\mathbb{Q})$ of an Eilenberg MacLane-space $K(\pi,n)$ which is an abelian topological group is the exterior algebra over $\pi\otimes\mathbb{Q}$ if n is odd and the polynomial algebra on $\pi\otimes\mathbb{Q}$ if n is even.*

Proof. This is an application of Theorem 8.4.11. $\square$

Corollary 11.5.2. *The rational homotopy of any $K(\pi,n)$ which is a CW-complex is the exterior algebra over $\pi\otimes\mathbb{Q}$ if n is odd and the polynomial algebra on $\pi\otimes\mathbb{Q}$ if n is even.*

Proof. Since any CW-complexes which are Eilenberg-MacLane spaces of type (π,n) are weakly homotopy equivalent (cf. [Whi78, Theorem V.7.2]), their rational homology algebras are isomorphic by the Whitehead-Serre Theorem, so the result in Proposition 11.5.1 carries over to arbitrary CW-complexes of type (π,n). $\square$

Theorem 11.5.3. *The rational homology $H(X;\mathbb{Q})$ of any Eilenberg-MacLane space $K(\pi,n)$ is the exterior algebra over $\pi\otimes\mathbb{Q}$ if n is odd and the polynomial algebra on $\pi\otimes\mathbb{Q}$ if n is even.*

Proof. Consider an Eilenberg-MacLane space X. There exists a CW-complex Y which is weakly homotopy equivalent to X. By the Whitehead-Serre theorem the rational homology algebras $H(X;\mathbb{Q})$ and $H(Y;\mathbb{Q})$ are isomorphic. By the preceding corollary the algebra $H(Y;\mathbb{Q})$ is the exterior algebra over $\pi\otimes\mathbb{Q}$ if n is odd and the polynomial algebra on $\pi\otimes\mathbb{Q}$ if n is even. Thus the stated result follows. $\square$

Towers of Principal Bundles

In this chapter we construct special versions of Whitehead and Postnikov towers. Each stage of these towers will consist of principal bundles with abelian structure group. The special version of the Whitehead tower generalises the notion of universal coverings: Recall that to each semi-locally simply connected space X there exists a principal bundle

$$\pi_1(X) \longrightarrow \tilde{X} \longrightarrow X$$

whose total space $\tilde{X}$ is the universal covering space of X and whose fibre is the discrete fundamental group $\pi_1(X)$ of X. Moreover every arc-wise connected covering space Y of X with fundamental group $\pi_1(Y)$ is the orbit space of the action of a subgroup $N \cong \pi_1(Y)$ of $\pi_1(X)$ on $\tilde{X}$. Simple connectivity of topological spaces is a very useful property that makes many constructions possible. (The existence of universal covering groups for example ensures the existence of a left adjoint to the Lie algebra functor for classical Lie groups.) This makes covering spaces a very useful tool in general topology. Higher connectivity is even more desirable, so it would be very convenient to have a construction of principal bundles killing higher homotopy groups analogous to that of universal covering spaces. Unfortunately no such generalisation (using principal bundles) to higher dimensions is known. The Whitehead tower to be constructed generalises the notion of universal coverings and resolves this issue.

In contrast the special Postnikov tower to be constructed eases the computation of homology groups. It finally enables us to make fundamental observations concerning the rational homology and cohomology of CW-complexes and topological spaces with abelian fundamental group.

12.1 Connecting Homomorphisms and Čech-Cohomology

In this section we study principal bundles (E, G, X) whose structure group G is an Eilenberg-MacLane space. Our primary interest lies in the homotopy and homology groups of these spaces. As a final result we establish for n-connected (but not $n+1$-connected) base spaces X and $\pi_{n+1}(X) \cong \pi_n(G)$ a one-to-one correspondence

$$\check{H}^1(X, \mathcal{G}) \xrightarrow{\cong} \hom(\pi_{n+1}(X), \pi_n(G))$$

between the first Čech cohomology of the base space with coefficients the sheaf of continuous G-valued functions and the candidates for connecting homomorphisms $\pi_{n+1}(X) \to \pi_n(G)$ in the long exact sequence of homotopy groups associated to the bundle (E, G, X). This especially implies that any group homomorphism from $\pi_{n+1}(X)$ to $\pi_n(G)$ is the connecting homomorphism of a G-principal bundle over

X. This allows the construction of Whitehead and Postnikov towers in the later sections of this chapter.

The case $n = 0$ corresponds to principal bundles with discrete structure group, i.e. to covering spaces. The situation here has been well studied and is completely understood. In contrast, for $n > 0$ little is known. (The main reason therefore being the previous unavailability of general Eilenberg-MacLane spaces $K(A, n)$ which are CW-complexes and abelian topological groups.) Since all higher homotopy groups are abelian, we only consider abelian homotopy groups. Let G be an $(n-1)$-connected topological group (which is assumed to be discrete in the case $n = 0$) with first non-trivial homotopy group $\pi_n(G) = A$ and

$$G \quad\text{------}\quad EG \quad\text{------}\quad BG$$

be a classifying bundle for G. The total space EG of the universal bundle is contractible and the only nontrivial part of the long exact sequence of homotopy groups are the homotopy groups of G and of BG and natural connecting homomorphisms

$$\cdots \to 0 \to \pi_{k+1}(BG) \xrightarrow[\cong]{\partial} \pi_k(G) \to 0 \to \cdots$$

which are natural isomorphism. We denote this connecting isomorphisms by φ_{k+1}. The classifying space BG is n-connected and its first non-trivial homotopy group $\pi_{n+1}(BG)$ is isomorphic to $\pi_n(G) = A$ via φ_{n+1}. We will identify the abelian groups A and $\pi_{n+1}(BG)$ without further notice. The numerable G-principal bundles over any topological space X are classified by the set $[X, BG]$ of homotopy classes of maps from X to BG. Any G-principal bundle (E, p, X) over X is uniquely described by a classifying map $f : B \to BG$. This map identifies the bundle (E, p, X) as a pullback of the universal bundle (EG, q, BG) by f:

$$
\begin{array}{ccc}
G & & G \\
| & & | \\
E & \xrightarrow{\tilde{f}} & EG \\
{\scriptstyle p}\big| & & \big|{\scriptstyle q} \\
X & \xrightarrow{f} & BG
\end{array}
$$

Here $\tilde{f}$ is a lifting of f, so that $(\tilde{f}, f)$ is a morphism of G-principal bundles. In order to use the long exact sequence of homotopy groups we have to pick a base point $*$ in X, a base point in the fibre $p^{-1}(\{*\})$, a base point $*$ in BG and a base point in the fibre $q_E^{-1}(\{*\})$. We do this in a consistent way such that the functions f and $\tilde{f}$ are morphism of based spaces. Thus in the following we will assume all classifying maps f and their liftings $\tilde{f}$ to be morphism of based spaces. The restriction and corestriction of the G-map $\tilde{f}$ to the fibres $q_E^{-1}(\{*\})$ and G over the base points $* \in X$ and $* \in BG$ then is an isomorphism of G-spaces. This allows us to identify the fibres $q_E^{-1}(\{*\})$ and G over the base points $* \in X$ and $* \in BG$ respectively. So the above pullback diagram can be rewritten as

$$
\begin{array}{ccc}
G & = & G \\
| & & | \\
E & \xrightarrow{\tilde{f}} & EG \\
{\scriptstyle p}\big| & & \big|{\scriptstyle q} \\
X & \xrightarrow{f} & BG
\end{array}
$$

The bundle morphism $(\tilde{f}, f)$ induces a homomorphism between the long exact sequences of homotopy groups associated to both bundles. This allows a description of the connecting homomorphism $\pi_{n+1}(X) \to \pi_n(G)$ by the classifying map:

Lemma 12.1.1. *If f is a based classifying map of the G-principal bundle (E, p, X) then the connecting homomorphism $\partial : \pi_{n+1}(X) \to \pi_n(G)$ coincides with the homomorphism $\varphi_{n+1} \circ \pi_{n+1}(f)$.*

Proof. Let f be a based classifying map of the G-principal bundle (E, p, X) and let $\tilde{f} : E \to EG$ be a based lifting of f. Consider the morphism of long exact sequences of homotopy groups induced by the morphism $(\tilde{f}, f)$ of G-principal bundles:

$$\begin{array}{ccccccccc}
\cdots \longrightarrow & \pi_{n+1}(E) & \longrightarrow & \pi_{n+1}(X) & \overset{\partial}{\longrightarrow} & \pi_n(G) & \longrightarrow & \pi_n(E) & \longrightarrow \cdots \\
& \downarrow{\pi_{n+1}(\tilde{f})} & & \downarrow{\pi_{n+1}(f)} & & \| & & \downarrow{\pi_n(\tilde{f})} & \\
\cdots \longrightarrow & 0 & \longrightarrow & \pi_{n+1}(BG) & \overset{\varphi_{n+1}}{\underset{\cong}{\longrightarrow}} & \pi_n(G) & \longrightarrow & 0 & \longrightarrow \cdots
\end{array}$$

Here we used the identification of the fibres over the base points, so the map $\tilde{f}$ restricts to the identity on these fibres, hence $\pi_n(\tilde{f}_{|G}) = \pi_n(\mathrm{id}_G) = \mathrm{id}_{\pi_n(G)}$. The middle commutative square results in the equality $\partial = \varphi_{n+1} \circ \pi_{n+1}(f)$. $\square$

Because the induced homomorphism $\pi_{n+1}(f)$ between the homotopy groups $\pi_{n+1}(X)$ and $\pi_{n+1}(BG)$ only depends on the based homotopy class of f we thus have established a map

$$\psi_{n+1,X} : [X, BG]^0 \to \hom(\pi_{n+1}(X), \pi_n(G)), \quad \psi_{n+1,X}(f) = \varphi_{n+1} \circ \pi_{n+1}(f)$$

which associates to every homotopy equivalence class of maps $X \to BG$ of based spaces the connecting homomorphism $\partial : \pi_{n+1}(X) \to \pi_n(G)$ of the induced G-principal bundle over X. Since a different choice of base points changes the map ψ_{n+1} by an automorphism of $\pi_{n+1}(G)$, one has to restrict oneself to based maps $X \to BG$. This is not necessary if G is an abelian compactly generated topological group (e.g. an Eilenberg-MacLane space which is a CW-complex). In this case one can use the classifying bundle $EG \to BG$ constructed in Section 11.3 whose classifying space is an abelian k-group. For this classifying space construction the neutral element $0 \in BG$ is a natural choice of a base point. In addition all maps $X \to BG$ are homotopic to based ones, because BG is arc-wise connected. Thus we observe:

Lemma 12.1.2. *If G compactly generated and abelian, then we have established a map*

$$\psi_{n+1,X} : [X, BG] \to \hom(\pi_{n+1}(X), \pi_n(G)), \quad \psi_{n+1,X}(f) = \varphi_{n+1} \circ \pi_{n+1}(f)$$

which associates to every homotopy equivalence class of maps $X \to BG$ the connecting homomorphism $\partial : \pi_{n+1}(X) \to \pi_n(G)$ of the induced G-principal bundle over X.

Moreover the k-group structure of BG then induces the structure of an abelian group on $[X, BG]$ which is given by the point-wise multiplication in BG. Endowing the set $[X, BG]$ with this group structure we note that the morphism ψ_{n+1} preserves this group structure:

Lemma 12.1.3. *If G is compactly generated and abelian, then the map $\psi_{n+1,X}$ is a group homomorphism.*

Proof. The neutral element in in $[X, BG]$ is represented by the constant zero map $0 : X \rightarrow \{0\} \subset BG$. The pullback bundle corresponding to the constant classifying map 0 is the trivial G-principal bundle $(X \times G, G, X)$ over X. Its connecting homomorphism is the zero homomorphism

$$0 : \pi_{n+1}(X) \rightarrow \pi_n(G),$$

which is the neutral element in $\hom(\pi_{n+1}(X); \pi_n(G))$. If f and g are continuous maps $X \rightarrow BG$ then the product $[f] \cdot [g]$ in the group $[X; BG]$ is represented by the point-wise product $f \cdot g$ of the maps f and g. The addition in the homotopy groups of connected topological groups is induced by the group multiplication (see [Bre97b, Theorem VII.4.2] for a proof). This implies the equalities

$$\pi_{n+1}(f \cdot g)([s]) = [(f \cdot g)s] = [(fs) \cdot (gs)] = \pi_{n+1}(f)([s]) + \pi_{n+1}(g)([s])$$

for every based map $s : \mathbb{S}^{n+1} \rightarrow X$ representing an element of $\pi_{n+1}(X)$. The last term in the above equation is the sum of the classes $\pi_{n+1}(f)([s])$ and $\pi_{n+1}(g)([s])$ in $\pi_{n+1}(BG)$. Consequently the image of the product $[f] \cdot [g]$ under $\psi_{n+1,X}$ is given by

$$\begin{aligned}
\psi_{n+1,X}([f] \cdot [g]) &= \varphi_{n+1}\pi_{n+1}(f \cdot g) \\
&= \varphi_{n+1}[\pi_{n+1}(f) \cdot \pi_{n+1}(g)] \\
&= [\varphi_{n+1}\pi_{n+1}(f)] \cdot [\varphi_{n+1}\pi_{n+1}(g)] \\
&= \psi_{n+1,X}(f) + \psi_{n+1,X}(g),
\end{aligned}$$

so the map $\psi_{n+1,X}$ also preserves products. Preserving in addition neutral elements, it is a group homomorphism. $\qquad \square$

In the following we restrict ourselves to abelian topological groups G which are CW-complexes and Eilenberg-MacLane spaces of type (A, n). This in particular implies that the classifying space BG is an Eilenberg-MacLane space of type $(n+1, A)$ and of the homotopy type of a CW-complex. We would like to identify all homomorphisms in the group $\hom(\pi_{n+1}(X), \pi_n(G))$ as connecting homomorphisms of a G-principal bundle over X. This is the case if and only if the homomorphism $\psi_{n+1,X}$ is surjective, which motivates our further investigation of this homomorphism.

For simplicity we first assume that the base space X of the bundle (E, p, X) is a CW-complex. The general case will be treated at the end of this section. The singular cohomology on the full subcategory of **Top** with objects all CW-complexes is represented by any CW-Eilenberg-MacLane spectrum, i.e. there exists a natural isomorphism $[-, K(A, n+1)] \cong H^{n+1}(-; A)$. Each isomorphism

$$[X, K(A, n+1)] \cong H^{n+1}(X; A), \quad f \mapsto H^{n+1}(f)(i),$$

assigns to the homotopy class of a continuous function $f : X \rightarrow K(A, n+1)$ the pullback of the characteristic class $i \in H^{n+1}(K(A, n+1); A)$ (cf. [Bre97b, Theorem VII.12.1]). Since the base space BG of the universal G-principal bundle is a CW-complex and an Eilenberg MacLane space of type $(A, n+1)$, this implies that the natural map

$$\alpha_{n+1} : [X, BG] \cong H^{n+1}(BG; A), \quad \alpha_{n+1}(f) = H^{n+1}(f)(i),$$

which assigns to every homotopy class of a classifying map $f : X \rightarrow BG$ the pullback of the characteristic class $i \in H^{n+1}(BG; A)$ is an isomorphism. Thus the singular cohomology group $H^{n+1}(X; A)$ classifies all G-principal bundles over X. The map α_{n+1} can be described in more detail as follows: Because the classifying space BG is

n-connected, the n-th homology group $H_n(X)$ is trivial. Therefore the short exact sequence

$$0 \to \mathrm{Ext}^1(H_n(BG); A) \to H^{n+1}(BG; A) \xrightarrow{\beta_{n+1}} \hom(H_{n+1}(BG); A)) \to 0$$

from the universal coefficient theorem for cohomology provides an isomorphism β_{n+1} between the groups $H^{n+1}(BG; A)$ and $\hom(H_{n+1}(BG); A))$. In addition the Hurewicz-homomorphism

$$\mathrm{hur}_{n+1} : \pi_{n+1}(BG) \to H_{n+1}(BG)$$

in dimension $n + 1$ is an isomorphism by the Hurewicz Theorem (cf. [Bre97b, Corollary VII.10.8]). Applying the contravariant functor $\hom(-, A)$ to this isomorphism yields an isomorphism between the abelian groups $\hom(H_{n+1}(BG); A)$ and $\hom(\pi_{n+1}(BG); A)$; the latter one being the endomorphism group $\hom(A; A)$ of A. We thus have obtained a chain

$$H^{n+1}(BG; A) \xrightarrow[\cong]{\beta_{n+1}} \hom(H_{n+1}(BG); A) \xrightarrow[\cong]{\hom(\mathrm{hur}_{n+1}:A)} \hom(\pi_{n+1}(BG); A)$$

of group isomorphisms. The characteristic class $i \in H^{n+1}(BG; A)$ is the image of the identity id_A under the isomorphism $\beta_{n+1}^{-1} \hom(\mathrm{hur}_{n+1}; A)^{-1}$. Recall that all homomorphisms in the short exact sequence of the universal coefficient theorem are natural in the spaces involved. So for every classifying map $f : X \to BG$ we obtain a commutative diagram

$$
\begin{array}{ccc}
H^{n+1}(BG; A) & \xrightarrow[\cong]{\beta_{n+1}} & \hom(H_{n+1}(BG); A)) \\
{\scriptstyle H^{n+1}(f;A)}\downarrow & & \downarrow{\scriptstyle \hom(H_{n+1}(f);A)} \\
H^{n+1}(X; A) & \xrightarrow{\beta_{n+1}} & \hom(H_{n+1}(X); A)
\end{array}
\quad ,
$$

where the homomorphisms β_{n+1} are natural and the upper horizontal arrow is an isomorphism. This diagram can be enlarged using the Hurewicz-homomorphisms; since these form a natural transformation $\mathrm{hur}_* : \pi_* \to H_*$, we obtain another commutative diagram:

$$
\begin{array}{ccc}
\pi_{n+1}(X) & \xrightarrow{\mathrm{hur}_{n+1}} & H_{n+1}(X) \\
{\scriptstyle \pi_{n+1}(f)}\downarrow & & \downarrow{\scriptstyle H_{n+1}(f)} \\
\pi_{n+1}(BG) & \xrightarrow[\cong]{\mathrm{hur}_{n+1}} & H_{n+1}(BG)
\end{array}
\quad ,
$$

where the lower horizontal arrow is an isomorphism. Applying the contravariant functor $\hom(-, A)$ to the latter diagram yields a diagram that can be spliced to the one derived from the universal coefficient theorem, where we abbreviate the homomorphisms $\hom(\mathrm{hur}_{n+1}; A)$ by γ_{n+1}:

$$
\begin{array}{ccccccc}
[BG, BG] & \xrightarrow{\alpha_{n+1}} & H^{n+1}(BG; A) & \xrightarrow[\cong]{\beta_{n+1}} & \hom(H_{n+1}(BG); A)) & \xrightarrow[\cong]{\gamma_{n+1}} & \hom(\pi_{n+1}(BG); A) \\
{\scriptstyle [f,BG]}\downarrow & & {\scriptstyle H^{n+1}(f)}\downarrow & & {\scriptstyle \hom(H_{n+1}(f);A)}\downarrow & & {\scriptstyle \hom(\pi_{n+1}(f);A)}\downarrow \\
[X, BG] & \xrightarrow{\alpha_{n+1}} & H^{n+1}(X; A) & \xrightarrow{\beta_{n+1}} & \hom(H_{n+1}(X); A) & \xrightarrow{\gamma_{n+1}} & \hom(\pi_{n+1}(X); A)
\end{array}
$$

Here we have additionally appended a commutative square expressing the naturality of the transformation α_{n+1} on the left.

Proposition 12.1.4. *If $f : X \to BG$ is a continuous function, then the image of $\mathrm{id}_A \in \hom(\pi_{n+1}(BG), A)$ under the homomorphism $\hom(\pi_{n+1}(f), A)$ coincides with the image of f under the homomorphism $\gamma_{n+1} \circ \beta_{n+1} \circ \alpha_{n+1}$.*

Proof. The characteristic class $i \in H^{n+1}(BG; A)$ is the image of the identity map id_A under the isomorphism $\beta_{n+1}^{-1} \circ \hom(\mathrm{hur}_{n+1}; A)$. The class $\alpha_{n+1}(f)$ is the pullback $H^{n+1}(f)(i)$ of the characteristic class i by f. A diagram chase in the above diagram reveals the equalities

$$
\begin{aligned}
\hom(\pi_{n+1}(f), A)(\mathrm{id}_A) &= [\hom(\pi_{n+1}(f), A) \circ \gamma_{n+1} \circ \beta_{n+1}](i) \\
&= [\gamma_{n+1} \circ \hom(H_{n+1}(f); A) \circ \beta_{n+1}](i) \\
&= [\gamma_{n+1} \circ \beta_{n+1} \circ H^{n+1}(f)](i) \\
&= [\gamma_{n+1} \circ \beta_{n+1} \circ \alpha_{n+1}](f).
\end{aligned}
$$

In the last step we made use of the definition $\alpha_{n+1}(f) = H^{n+1}(f)(i)$ of the homomorphism α_{n+1}. $\qquad\square$

The commutative diagram can even be more enlarged by applying the natural transformation $\hom(-, \varphi_{n+1}) : \hom(-, A) \to \hom(-, \pi_n(G))$ on the right hand side. In this way we obtain a chain $\hom(-, \varphi_{n+1})\gamma_{n+1}\beta_{n+1}\alpha_{n+1}$ of natural transformations.

Theorem 12.1.5. *The group homomorphism $\psi_{n+1,X}$ coincides with the homomorphism $\hom(\pi_{n+1}(X); \varphi_{n+1})\gamma_{n+1} \circ \beta_{n+1} \circ \alpha_{n+1}$*

Proof. Let $f : X \to BG$ be a continuous function. The image of the homotopy equivalence class $[f]$ in $[X, BG]$ under the homomorphism $\psi_{n+1,X}$ is the element $\varphi_{n+1} \circ \pi_{n+1}(f)$ of $\hom(\pi_{n+1}(X), \pi_n(G))$. The homomorphism $\pi_{n+1}(f)$ is the image of $\mathrm{id}_A \in \hom(\pi_{n+1}(BG); A)$ under the homomorphism $\hom(\pi_{n+1}(f); A)$, so all in all we obtain

$$
\psi_{n+1,X}([f]) = \varphi_{n+1} \circ \pi_{n+1}(f) = [\hom(\pi_{n+1}(X), \varphi_{n+1}) \circ \hom(\pi_{n+1}(f), A)](\mathrm{id}_A).
$$

Using Proposition 12.1.4 these equations can be rewritten as

$$
\begin{aligned}
\psi_{n+1,X}([f]) &= \hom(\pi_{n+1}(X); \varphi)\hom(\pi_{n+1}(f); A)(\mathrm{id}_A) \\
&= [\hom(\pi_{n+1}(X); \varphi)\gamma_{n+1} \circ \beta_{n+1} \circ \alpha_{n+1}](f).
\end{aligned}
$$

Since this is true for all possible classifying maps $f : X \to BG$, the homomorphisms $\psi_{n+1,X}$ and $\hom(\pi_{n+1}(X); \varphi_{n+1})\gamma_{n+1} \circ \beta_{n+1} \circ \alpha_{n+1}$ coincide. $\qquad\square$

Corollary 12.1.6. *The group homomorphisms $\psi_{n+1,X}$ form a natural transformation $\psi_{n+1,X} : [-, BG] \to \hom(\pi_{n+1}(-), \pi_n(G))$.*

Proof. By the preceding Theorem each function $\psi_{n+1,X}$ is a composition of homomorphisms, which all are natural in X. Thus the maps $\psi_{n+1,X}$ form a natural transformation. $\qquad\square$

Theorem 12.1.7. *If the CW-complex X is n-connected then the group homomorphism $\psi_{n+1,X} : [X, BG] \to \hom(\pi_{n+1}(X); \pi_n(G))$ is an isomorphism.*

Proof. If the CW-complex X is n-connected, then its n-th homology group $H_n(X)$ is trivial. In this case the the natural homomorphism β_{n+1} in the short exact sequence

$$
0 \to \mathrm{Ext}^1(H_n(X); A) \to H^{n+1}(X; A) \xrightarrow{\beta_{n+1}} \hom(H_{n+1}(X); A)) \to 0
$$

from the universal coefficient theorem is an isomorphism. In addition the Hurewicz-homomorphism $\mathrm{hur}_{n+1} : \pi_{n+1}(X) \to H_{n+1}(X)$ in dimension $n+1$ is an isomorphism, causing the map $\gamma_{n+1} : \mathrm{hom}(H_{n+1}(X); A) \to \mathrm{hom}(\pi_{n+1}(X), A)$ to be an isomorphism as well. By Theorem 12.1.5 the homomorphism $\psi_{n+1,X}$ then is a composition of the isomorphisms $\mathrm{hom}(\pi_{n+1}(X); \varphi_{n+1})$, γ_{n+1}, β_{n+1} and α_{n+1}, hence an isomorphism as well. $\square$

Corollary 12.1.8. *If the CW-complex X is n-connected, then every G-principal bundle is –up to equivalence– uniquely determined by its connecting homomorphism $\partial : \pi_{n+1}(X) \to \pi_n(G)$.*

Corollary 12.1.9. *If X is an n-connected CW-complex then every homomorphism in $\mathrm{hom}(\pi_{n+1}(X), \pi_n(G))$ can be realised as the connecting homomorphism of a G-principal bundle over X.*

This characterisation of G-principal bundles via their connecting homomorphism can be used to construct bundles that kill the first non-trivial homotopy groups. We recall some notation:

Definition 12.1.10. *A continuous function $f : X \to Y$ is called n-connective, if X is n-connected and $\pi_k(f)$ is an isomorphism for all $k > n$.*

The main feature of a bundle (E, p, X) over the topological space X with $(n+1)$-connective bundle projection p is that the bundle kills all the homotopy groups $\pi_k(X)$, $k \leq n$ of X. The preceding discussion of connecting homomorphisms ensures the existence of such principal bundles which kill the first non-trivial homotopy group:

Theorem 12.1.11. *For every n-connected CW-complex X there exists a principal bundle over X with polyhedral abelian structure group and $(n+1)$-connective bundle projection killing the $(n+1)$-th homotopy group of X.*

Proof. If X is a connected CW-complex with first non-trivial homotopy group $\pi_{n+1}(X) = A$, let G be an Eilenberg-MacLane space of type (A, n) that is an abelian topological group (which is guaranteed to exist by Theorem 11.4.6). The homotopy groups $\pi_{n+1}(X)$ and $\pi_n(G)$ are isomorphic and by Corollary 12.1.9 there exists a principal G-bundle over X such that the connecting homomorphism $\partial : \pi_{n+1}(X) \to \pi_n(G)$ is an isomorphism. Since all homotopy group $\pi_k(G)$ of G for $k \neq n$ are trivial, the exactness of the long homotopy sequence

$$\cdots \to 0 \to \pi_{n+1}(E) \to \pi_{n+1}(X) \overset{\partial}{\underset{\cong}{\to}} \pi_n(G) \to \pi_n(E) \to \pi_n(X) = 0 \to \cdots$$

implies that the $n+1$-th homotopy group $\pi_{n+1}(E)$ of the total space of this bundle is trivial and that the bundle projection p induces isomorphisms $\pi_k(p)$ for all $k \neq n+1$. Thus E is $(n+1)$-connected and the the bundle projection is $(n+1)$-connective. $\square$

It is known that every semi-locally simply connected topological space has a universal (i.e. simply connected) covering space. If the fundamental group of a CW-complex X is abelian, we can prove the converse implication:

Lemma 12.1.12. *If the CW-complex X is arc-wise connected and the fundamental group $\pi_1(X)$ is abelian, then there exists a universal covering space $\tilde{X}$ of X and X is semi-locally simply connected.*

Proof. Let X be an arc-wise connected topological space with abelian fundamental group $\pi_1(X)$. By Theorem 12.1.11 there exists a $K(\pi_1(X), 0)$-principal bundle $(\tilde{X}, p, X)$ over X with simply connected total space $\tilde{X}$. The total space $\tilde{X}$ of this bundle then is a universal covering space of X. The topological space X is therefore semi-locally simply connected. $\square$

If X is an n-connected CW-complex with first non-trivial homotopy group A in dimension $n+1$ and G a polyhedral Eilenberg-MacLane abelian topological group of type (A, n) then the $(n+1)$-connective bundle over X constructed above is a bundle corresponding to the isomorphism id : $\pi_{n+1}(X) = A \to \pi_n(G) = A$. The homotopy class of a classifying map $f : X \to BG$ of this bundle is given by $\psi_{n+1,X}^{-1}(\mathrm{id}_A)$.

Theorem 12.1.13. *If X is an n-connected CW-complex and G a polyhedral Eilenberg-MacLane abelian topological group of type $(\pi_{n+1}(X), n)$ then any G-principal bundle with $(n + 1)$-connective bundle projection over X is universal among all G-principal bundles over X.*

Proof. Let (E', G, X) be a G-principal bundle over X induce by the classifying map $f' : B \to BG$ and let $f : B \to BG$ be the classifying map of a bundle (E, p, X) with $(n+1)$-connective bundle projection. The homomorphism $\pi_{n+1}(f')$ factors through the isomorphism $\pi_{n+1}(f)$:

$$\pi_{n+1}(X) \xrightarrow[\cong]{\pi_{n+1}(f)} \pi_{n+1}(BG)$$

with maps $\pi_{n+1}(f')$ and $\gamma = \pi_{n+1}(f')\pi_{n+1}(f)^{-1}$ to $\pi_{n+1}(BG)$.

Because the Eilenberg-MacLane space BG is of the homotopy type of a CW-complex (cf. [Mil67, Section 2]) there exists a continuous map $h : BG \to BG$ inducing the homomorphism γ, i.e. $\gamma = \pi_{n+1}(h)$. The connecting homomorphism of the bundle induced by the continuous map $h \circ f$ is given by

$$\varphi_{n+1} \circ \pi_{n+1}(hf) = \varphi_{n+1} \circ \pi_{n+1}(h)\pi_{n+1}(f) = \varphi_{n+1}\pi_{n+1}(f'),$$

so it coincides with the connecting homomorphism of the bundle (E', G, X). Because any G-principal bundle over X is –up to equivalence– uniquely determined by its connecting homomorphism (Corollary 12.1.8), the bundles (E', G, X) and (h^*f^*E) are equivalent. Thus the the pullback bundle $(E', G, X) = f'^*(EG) - h^*f^*(EG)$ of (EG, G, BG) by f' also is equivalent to the pullback bundle of $f^*(EG)$ by h.   □

Definition 12.1.14. *A principal bundle (E, p, X) over an n-connected CW-complex X with polyhedral structure group G of type (π_{n+1}, n) and $(n+1)$-connective bundle projection is called an* universal $(n + 1)$-connective G-bundle *over X.*

Corollary 12.1.15. *IF X is an n-connected CW-complex and G the polyhedral abelian Eilenberg-MacLane group of type $(\pi_{n+1}(X), n)$ then all universal $(n + 1)$-connective G-principal bundles are equivalent.*

There exists a procedure to kill the first non-trivial homotopy group $\pi_{n+1}(X)$ of an arc-wise connected topological space X, which is commonly used in the construction of Postnikov towers (see the construction in [Bre97b, Chapter VII.13]). It uses the path-loop fibration $PK(\pi_{n+1}(X), n + 1) \to K(\pi_{n+1}(X), n + 1)$ with fibre $\Omega K(\pi_{n+1}(X), n + 1)$ over an Eilenberg-MacLane space of type $(\pi_{n+1}, n + 1)$. Here the principal fibration over X killing the homotopy group $\pi_{n+1}(X)$ is induced by a classifying map $f : X \to K(\pi_{n+1}(X), n + 1)$. Using the polyhedral abelian Eilenberg-MacLane groups constructed in Section 11.4, this construction can be modified to generalse Theorem 12.1.11 to arbitrary topological spaces:

Theorem 12.1.16. *For every n-connected topological space X there exists a principal bundle over X with abelian polyhedral structure group of type $(\pi_{n+1}(X), n)$ and $(n + 1)$-connective bundle projection.*

Proof. Let X be an n-connected topological space and G denote the polyhedral Eilenberg-MacLane space of type (π_{n+1}, n) which is an abelian topological group constructed in Section 11.4. The base space BG of the universal G bundle is an Eilenberg-MacLane space of type $(\pi_{n+1}(X), n+1)$ and of the homotopy type of a CW-complex (cf. [Mil67, Section 2]). Applying the standard construction to the principal G-bundle $EG \to BG$ instead to an arbitrary fibration of the form $PK(\pi_{n+1}(X), n+1) \to K(\pi_{n+1}(X), n+1)$ then the resulting principal fibration over X is a G-principal bundle $\qquad\qquad\qquad\qquad\qquad\qquad\qquad$ □

12.2 Whitehead and Postnikov Towers

The principal bundles constructed in the previous section can be used to obtain special forms of two important constructions in algebraic topology: Whitehead and Postnikov towers. Recall that a Whitehead tower over an arc-wise connected space X is a series of Serre fibrations

$$\vdots$$

$$
\begin{array}{ccc}
K(\pi_2(X), 1) & \rule[0.5ex]{3em}{0.4pt} & X_2 \\
& & \big| \\
K(\pi_1(X), 0) & \rule[0.5ex]{3em}{0.4pt} & X_1 \\
& & \big| \\
& & X_0 = X
\end{array}
$$

such that each fibration $X_{n+1} \to X_n$ is $(n+1)$-connective. (The fibre of the fibration $X_{n+1} \to X_n$ then is necessarily an Eilenberg MacLane space of type $(\pi_{n+1}(X), n)$.)

Since it has been shown that one can kill abelian homotopy groups by principal bundles with fibre a polyhedral abelian topological group we conclude:

Theorem 12.2.1. *Over every arc-wise connected space X with abelian fundamental group $\pi_1(X)$ there exists a Whitehead tower such that all fibrations $X_{n+1} \to X_n$ are principal bundles with polyhedral abelian structure group.*

If the fundamental group $\pi_1(X)$ of a topological space X is not abelian but the space X is semi-locally simply connected, then there exists a universal covering space $\tilde{X}$ of X. In this case one can choose X_1 to be the covering space $\tilde{X}$ and the bundle $X_1 \to X$ to be the projection of the covering space $\tilde{X}$ onto X. All other principal bundles $X_{n+1} \to X_n$ for $n > 0$ can be constructed as before. Thus we observe:

Theorem 12.2.2. *Over every arc-wise connected semi-locally simply connected space X there exists a Whitehead tower such that all fibrations $X_{n+1} \to X_n$ are principal bundles with polyhedral structure group. These structure groups are abelian for all $n \geq 1$.*

A similar result can be obtained concerning Postnikov towers. Before we start deriving these results, we recall some definitions. The mapping cylinder of a continuous function f is denoted by I_f.

Definition 12.2.3. *A continuous function $f : X \to Y$ is called n-connected, if the pair (I_f, X) is n-connected.*

The mapping cylinder I_f of a continuous function $f : X \to Y$ is homotopy equivalent to the space Y, so its homotopy groups fit into the long exact sequence

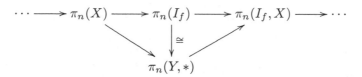

From this long exact sequence of homotopy groups one infers that a continuous function $f : X \to Y$ between arc-wise connected spaces X and Y is n-connected if and only if $\pi_k(f)$ is an isomorphism for all $k < n$ and an epimorphism for $k = n$. The dual notion to n-connectedness is called n-anticonnectedness:

Definition 12.2.4. *A topological space X is called n-anticonnected, if the homotopy groups $\pi_k(X)$, $k \geq n$ are trivial.*

Definition 12.2.5. *A continuous function $f : X \to Y$ is called n-anticonnected, if its mapping fibre is n-anticonnected.*

Dually to the notion of n-connectedness one obtains a characterisation of n-anticonnected maps between arc-wise connected spaces X and Y in terms of the homomorphisms $\pi_k(f)$: A continuous function $f : X \to Y$ is n-anticonnected if and only if $\pi_k(f)$ is an isomorphism for all $k > n$ and a monomorphism for $k = n$.

Recall that a Postnikov factorisation of a map $f : X \to B$ between arc-wise connected spaces X and Y is natural transformation $\{f_n\}$ from the constant functor X to an inverse system of fibrations $X_{n+1} \to X_n$ such that each space X_n is n-anticonnected, $f_0 = f$ the maps f_n are n-equivalences and the fibres of the projections $X_{n+1} \to X_n$ are an Eilenberg MacLane space of type $(\pi_{n+1}(X), n+1)$ respectively. A Postnikov factorisation of a map $f : X \to B$ is usually depicted as follows:

$$\vdots$$

$$X_2 \longrightarrow K(\pi_2(X), 2)$$

$$f_1 \quad X_1 \longrightarrow K(\pi_1(X), 1)$$

$$X \dashrightarrow X_0 = B$$

As remarked before Theorem 12.1.16 the standard construction of Postnikov towers uses principal fibrations which are obtained by pulling back path-loop fibrations over Eilenberg-MacLane spaces. It has been shown in the proof of Theorem 12.1.16 that the principal fibrations in each step of the Postnikov tower can be replaced by principal bundles with abelian polyhedral structure group. So we summarise:

Theorem 12.2.6. *For every simple map $f : X \to Y$ between arc-wise connected spaces there exists a Postnikov factorisation satisfying the following conditions:*

1. *All fibrations $X_{n+1} \to X_n$ are principal bundles with polyhedral structure group.*
2. *The structure groups of the principal bundles $X_{n+1} \to X_n$ are abelian for $n > 0$. If the fundamental group $\pi_1(X)$ is abelian, then the structure group of the bundle $X_1 \to X_0$ is abelian as well.*

As special case arises for continuous functions $f : X \to \{*\}$ into a singleton space. These lead to Postnikov towers.

Definition 12.2.7. *A Postnikov tower of an arc-wise connected space X is a Postnikov factorisation of the map $X \twoheadrightarrow \{*\}$.*

As a consequence to Theorem 12.2.6 one obtains Postnikov towers which consist of principal fibre bundles:

Corollary 12.2.8. *For every arc-wise connected topological space X there exists a Postnikov tower of X satisfying the following conditions:*

1. *All fibrations $X_{n+1} \to X_n$ are principal bundles with polyhedral structure group.*
2. *The structure groups of the principal bundles $X_{n+1} \to X_n$ are abelian for $n > 0$. If the fundamental group $\pi_1(X)$ is abelian, then the structure group of the bundle $X_1 \to X_0$ is abelian as well.*

13

The Generalised Flux Homomorphism

The subject of this chapter is the de Rham complex for infinite dimensional flat smooth fibre bundles. We start with a reminder on the differential calculus of infinite dimensional manifolds proposed by BERTRAM, GLÖCKNER and NEEB in their exposition [BGN04], where the maps of class C^0 are the continuous ones. We assume that the reader is familiar with the notions of differential calculus, manifolds smooth bundles and Lie groups developed in this article and with the topological of spaces of C^k-functions with the C^k-compact open topology introduced in the follow up [Glö04]. At first we observe some topological properties of manifolds in the above sense and then we recall some fundamental theorems on function spaces, which we later generalise to a larger class of manifolds. To be precise we generalise the exponential law

$$\hom_{\mathbf{Mf}}(M \times N, V) \cong \hom_{\mathbf{Mf}}(M, C^\infty(N, V))$$

to the case of infinite dimensional manifolds M and N which have the property that arbitrary finite products $W^p \times W'^q$ of the model spaces W of M and W' of N are k-spaces. This extends the recent result [Glö04, Proposition 12.6 (b)] of Glöckner to a more general setting.

Following these general consideration on differential calculus we derive a decomposition of the topological de Rham spaces ${}^t\Omega^n(E; V)$ of the total space E of a flat smooth bundle. This is a straightforward generalisation from the classical finite dimensional context to the topological de Rham cohomology of infinite dimensional bundles, where we use the notion of infinite dimensional calculus introduced above. This decomposition is then used to construct a double complex $\Omega^{*,*}(E; V)$ consisting of spaces of differential forms. This double complex in turn is the necessary tool used in deriving a Leray-Serre spectral sequence

$$\Omega^p(B, {}^t\Omega^q(F; V)) \Rightarrow H_{dR}^{p+q}(E; V)$$

for flat infinite dimensional bundles with base manifold B and fibre F.

The double complex $\Omega^{*,*}(E; V)$ will also be necessary to construct a generalisation of the flux homomorphism of symplectic topology to infinite dimensional bundles. Here the diffeomorphism group of a (compact) manifold is replaced by an infinite dimensional Lie group G acting on a G-bundle of infinite dimensional manifolds. After constructing the flux homomorphisms we also define general period maps. These are necessary tools to be used to characterise the integrability of equivariant differential forms in Chapter 16 and the extensibility of equivariant Alexander-Spanier cocycles in Chapter 15.

13.1 Differential Calculus on Infinite Dimensional Manifolds

We use the differential calculus carefully introduced by BERTRAM, GLÖCKNER and NEEB in their exposition [BGN04], where the maps of class C^0 are the continuous ones. We assume that the reader is familiar with the notions of differential calculus, manifolds smooth bundles and Lie groups developed in this article and with the topological of spaces of C^k-functions with the C^k-compact open topology introduced in the follow up [Glö04]. At first we observe some topological properties of manifolds in the above sense and then we recall some fundamental theorems on function spaces, which we later generalise to a larger class of manifolds. All topological vector spaces and manifolds considered in this Section are vector spaces over a fixed topological field $\mathbb{K}$ (e.g. $\mathbb{K} = \mathbb{C}$ or $\mathbb{K} = \mathbb{R}$).

Lemma 13.1.1. *Every regular C^k-manifold $(0 \leq k \leq \infty)$ is completely regular.*

Proof. Let M be a regular C^k-manifold, $p \in M$ a point and A be a closed subset of M, not containing the point p. It is to show that there exists a continuous function $f : M \to I$, which takes the value 1 at p and whose support is disjoint to A. Let V be the modelling space of M and $\varphi : U \to V$ be a chart defined on an open neighbourhood of p. We assume w.l.o.g. that U is disjoint to A. Because M is regular, there exists an open neighbourhood W of p, whose closure $\overline{W}$ is contained in the open set U. The chart φ is a homeomorphism onto an open subset U' of the modelling space V. It maps the closed neighbourhood $\overline{W}$ of p to a closed neighbourhood $\overline{W'}$ of $\varphi(p)$. Because topological vector spaces are completely regular, there exists a continuous function $f' : V \to I$ whose support is contained in $\overline{W'}$ and which takes the value 1 at $\varphi(p)$. Then the function

$$f(m) := \begin{cases} f\varphi(m) & \text{for } m \in U \\ 0 & \text{for } m \in M \setminus \varphi^{-1}(\operatorname{supp} f') \end{cases}$$

has support in $\overline{W}$ and is continuous on both open sets U and $M \setminus \varphi^{-1}(\operatorname{supp} f')$, hence it is continuous on M. Furthermore the value of f at p is 1 and the support of f is contained in U, therefore disjoint to A. $\square$

Recall that for all topological spaces X, Y and Z there exists an injective continuous function $\varphi_{X,Y,Z} : C(X \times Y, Z) \to C(X, C(Y, Z))$ which stems from a natural transformation $\varphi : C(- \times -, -) \to C(-, C(-, -))$ (cf. Section B.5). In case the topological spaces X and Y are regular manifolds, the special nature of manifolds forces these functions to be homeomorphisms, provided the product $X \times Y$ is a k-space:

Lemma 13.1.2. *If M and N are regular C^k-manifolds whose product $M \times N$ is a k-space, then the natural transformation $\varphi_{M,N,-} : C(M \times N, -) \to C(M, C(N, -))$ is a natural isomorphism.*

Proof. Since the C^k-manifolds are completely regular by Lemma 13.1.1, the compact subspaces of M and N are completely regular as well. Thus Lemma B.5.17 applies, which shows that $\varphi_{M,N,-}$ is a natural isomorphism. $\square$

Example 13.1.3. Topological vector spaces are completely regular. Therefore, if V_1 and V_2 are topological vector spaces, the natural transformation $\varphi_{V_1,V_2,-}$ is a natural isomorphism.

Example 13.1.4. Open subspaces of topological vector spaces are completely regular. So for open subspaces U_1 and U_2 of topological vector spaces V_1 and V_2 respectively the the natural transformation $\varphi_{U_1,U_2,-}$ is a natural isomorphism.

The crucial argument in the proof of the preceding lemma is the regularity of the compact subspaces of the C^k-manifolds M and N, which is ensured by the complete regularity of these manifolds. For these kind of manifolds we observe:

Lemma 13.1.5. *Regular Manifolds which satisfy the separation axiom T_1 are T_3-spaces.*

Proof. Let M be a regular C^k-manifold ($0 \leq k \leq \infty$) and V be the modelling space of M. If M is a T_1-space, i.e. if the points of M are closed, then the complete regularity of M implies that M is Hausdorff and thus T_3.   □

Conversely, if a C^k-manifold M is Hausdorff, then so are all its compact subspaces. In this case the compact subspaces of M are T_4-spaces and thus regular. This enables us to replace the regularity in Lemma 13.1.2 by the assumption that the manifolds considered are Hausdorff:

Lemma 13.1.6. *If M and N are Hausdorff C^k-manifolds whose product $M \times N$ is a k-space, then the natural transformation $\varphi_{M,N,-} : C(M \times N, -) \to C(M, C(N, -))$ is a natural isomorphism.*

Proof. Since the C^k-manifolds M and N are Hausdorff, their compact subspaces are regular. Thus Lemma B.5.17 applies, which shows that the natural transformation $\varphi_{M,N,-}$ is a natural isomorphism.   □

Proposition 13.1.7. *If M is a locally compact C^k-manifold $0 \leq k \leq \infty$ and W a Hausdorff $\mathbb{K}$-vector space, then the evaluation map $\mathrm{ev} : C^k(M,W) \times M \to W$ is of class C^k.*

Proof. The proof of [Glö04, Lemma 11.1] only relies on the local compactness of N (but not on the preceding observations) and thus carries over to our slightly more general case.   □

Lemma 13.1.8. *If M and N are C^{k+r}-manifolds ($0 \leq k,r \leq \infty$) and W is a Hausdorff topological $\mathbb{K}$-vector space, then for each map $f : M \times N \to W$ of class C^{k+r} the adjoint map $\hat{f} : M \to C^r(N,W)$ is of class C^k. Furthermore the linear function $\varphi_{M,N,W} : C^{k+r}(M \times N, W) \to C^k(M, C^r(N,W))$ is continuous.*

Proof. A proof appears in [Glö04, Lemma 12.1].   □

This in particular implies that the adjoint map $\hat{f} : M \to C^\infty(N,W)$ of a smooth function $f : M \times N \to W$ is itself smooth. If the manifold N is locally compact, then the converse implication also holds. For locally compact Hausdorff topological fields $\mathbb{K}$ and vector spaces W this is proved in [Glö04, Proposition 12.2]. Since we do not assume the field $\mathbb{K}$ to be locally compact, we have to reprove this statement. For further reference and a more detailed view we split the proof into smaller pieces.

Lemma 13.1.9. *If M is a C^k-manifold $0 \leq k \leq \infty$, N is a locally compact C^r-manifold, $0 \leq r \leq \infty$, W is a Hausdorff $\mathbb{K}$-vector space and $\hat{f} : M \to C^r(N,W)$ is a function of class C^k, then its adjoint map $f : M \times N \to W$ is of class $C^{\min\{k,r\}}$.*

Proof. Let M be a C^k-manifold, N be a locally compact C^r-manifold, $0 \leq k, r \leq \infty$, W be a Hausdorff $\mathbb{K}$-vector space and $\hat{f} : M \to C^r(N,W)$ be a k-times continuously differentiable function. The adjoint map $f : M \times N \to W$ of $\hat{f}$ can be written as the composition $f = \mathrm{ev}(\hat{f} \times \mathrm{id}_N)$

$$M \times N \xrightarrow{\hat{f} \times \mathrm{id}_N} C^r(N,W) \times N \xrightarrow{\mathrm{ev}} W$$

of the C^k-function $\hat{f} \times \mathrm{id}_N$ and the evaluation $C^r(N,W) \times N \xrightarrow{\mathrm{ev}} W$, which is of class C^r by Proposition 13.1.7. Therefore the composition $f = \mathrm{ev}(\hat{f} \times \mathrm{id}_N)$ is at least $\min\{k,r\}$-times continuously differentiable.   □

This in particular implies that for smooth manifolds M and N and a Hausdorff $\mathbb{K}$-vector-space W where N is locally compact the adjoint map $f : M \times N \to W$ of a smooth function $\hat{f} : M \to C^\infty(N, W)$ is smooth itself:

Lemma 13.1.10. *If M and N are smooth manifolds, N is locally compact and W is a Hausdorff $\mathbb{K}$-vector space, then $\varphi_{M,N,W} : C^\infty(M \times N, W) \to C^\infty(M, C^\infty(N, W))$ is a continuous bijection.*

Proof. If M and N are smooth manifolds and W is a Hausdorff $\mathbb{K}$-vector space, then the map $\varphi_{M,N,W} : C^\infty(M \times N, W) \to C^\infty(M, C^\infty(N, W)))$ is a continuous by Lemma 13.1.8. Since it is always an injection, it remains to prove the surjectivity. Each smooth function $\hat{f} : M \to C^\infty(N, W)$ is of class C^k for all $k \in \mathbb{N}$. Therefore, if N is locally compact, the adjoint function $f : M \times N \to W$ of such a smooth function $\hat{f} : M \to C^\infty(N, W)$ is of class C^k for all $k \in \mathbb{N}$ by Lemma 13.1.9, i.e. it is smooth. Thus the function $\varphi_{M,N,W} : C^\infty(M \times N, W) \to C^\infty(M, C^\infty(N, W))$ is surjective. $\square$

Proposition 13.1.11. *If M and N are C^k-manifolds, N is locally compact and W is a Hausdorff $\mathbb{K}$-vector space, then $\varphi_{M,N,W} : C^\infty(M \times N, W) \to C^\infty(M, C^\infty(N, W))$ is an isomorphism of topological $\mathbb{K}$-vector spaces.*

Proof. The proof is a variation of the proof of Proposition 12.2 b) in [Glö04]. Using Proposition 13.1.7 in place of [Glö04, Proposition 11.1], the proof of [Glö04, Proposition 12.2 b)] carries over to our slightly more general setting. $\square$

We are going to generalise this result to the case where M and N are manifolds with model spaces V_1 and V_2 such that all finite products $V_1^p \times V_2^q \times \mathbb{K}$ are k-spaces. Here fore we adapt the notation of the article [BGN04]. Thus, if U is an (open) subset of a topological vector space V over the topological field $\mathbb{K}$, we denote by $U^{[1]}$ the (open) subset $U^{[1]} := \{(u, v, t) \in U \times V \times \mathbb{K} \mid u + tv \in U\}$. Recall that a continuous function $f : U \to W$ into another topological $\mathbb{K}$-vector space W is called *continuously differentiable* or *of class C^1* if the continuous difference quotient function

$$U^{[1]} \cap (U \times V \times \mathbb{K}^*) \to W, \quad (u, v, t) \mapsto \frac{f(u + tv) - f(u)}{t}$$

is extendible to a continuous function $f^{[1]} : U^{[1]} \to W$. If the codomain space W is Hausdorff, then limits in W are unique and so is the extension $f^{[1]}$ of the difference quotient function. For Hausdorff codomains this is generalised inductively: The (open) subset $U^{[2]}$ is defined to be the (open) set $U^{[1]\,[1]}$ and generally, the set $U^{[k+1]}$ is defined to be the set $U^{[k]\,[1]}$ for all $k \in \mathbb{N}$. A continuously differentiable function $f : U \to W$ is said to be of *class C^2*, if $f^{[1]} : U^{[1]} \to W$ is of class C^1. More generally f is said to of *class C^{k+1}*, if f is of class C^k and $f^{[k]} : U^{[k]} \to W$ is of class C^1, in which case we define $f^{[k+1]} := f^{[k]\,[1]}$. This is equivalent to $f^{[1]}$ being of class C^k (cf. [BGN04, Remark 4.2]).

Lemma 13.1.12 (Glöckner). *For any open subset U of a $\mathbb{K}$-vector space and Hausdorff $\mathbb{K}$-vector space W, the linear maps $C^{k+1}(U, W) \to C(U, W) \times C^k(U_1^{[1]}, W)$, $f \mapsto (f, f^{[1]})$ are closed embeddings for all $0 \le k \le \infty$. In particular they are smooth.*

Proof. A proof can be found in [Glö04, Article 3, Lemma 4.3]. $\square$

Lemma 13.1.13 (Glöckner). *If U and V are open subsets of $\mathbb{K}$-vector spaces and $g : U \to W$ is a function of class C^k and W a Hausdorff $\mathbb{K}$-vector space, then the pullback map $C^k(g, W) : C^k(V, W) \to C^k(U, W)$, $f \mapsto fg$ is a continuous $\mathbb{K}$-linear map. In particular it is smooth.*

Proof. See [Glö04, Article 3, Lemma 4.4] for a proof.   □

Let V_1, V_2 and W be topological $\mathbb{K}$-vector spaces and $f : U \to W$ be a function of class C^1 defined on an open subset U of $V_1 \times V_2$. Then the functions $f(u_1, -)$ and $f(-, u_2)$ are of class C^1 as well (wherever they are defined). Let U_1 and U_2 denote the open subsets

$$U_1 := \{(u_1, u_2, v_1, t) \in U \times V_1 \times \mathbb{K} \mid (u_1 + tv_1, u_2) \in U\}$$
$$\text{and}\quad U_2 := \{(u_1, u_2, v_2, t) \in U \times V_2 \times \mathbb{K} \mid (u_1, u_2 + tv_2) \in U\}$$

of $U \times V_1 \times \mathbb{K}$ and $U \times V_2 \times \mathbb{K}$ respectively. The continuous difference quotient function $f^{[1]} : U^{[1]} \to V$ may be restricted to continuous functions

$$\frac{\Delta f}{\Delta_1} : U_1 \to W, \quad \frac{\Delta f}{\Delta_1}(u_1, u_2, v_1, t) = f^{[1]}(u_1, u_2, v_1, 0, t)$$
$$\text{and}\quad \frac{\Delta f}{\Delta_2} : U_2 \to W, \quad \frac{\Delta f}{\Delta_2}(u_1, u_2, v_2, t) = f^{[1]}(u_1, u_2, 0, v_2, t).$$

The continuous functions $\frac{\Delta f}{\Delta_1}$ on U_1 and $\frac{\Delta f}{\Delta_1}$ on U_2 are continuous extensions of the partial difference quotient functions

$$U_1 \cap (U \times V_1 \times \mathbb{K}^*) \to W, \quad (u_1, u_2, v_1, t) \mapsto \frac{f(u_1 + tv_1, u_2) - f(u_1, u_2)}{t}$$
$$U_2 \cap (U \times V_2 \times \mathbb{K}^*) \to W, \quad (u_1, u_2, v_2, t) \mapsto \frac{f(u_1, u_2 + tv_2) - f(u_1, u_2)}{t}$$

to U_1 and U_2 respectively. The restriction to $t = 0$ of these continuous functions defines continuous functions on $U \times V_1$ and $U \times V_2$ respectively, which are called the *partial derivatives* $d_1 f$ and $d_2 f$ of f. The derivative $df : U \times (V_1 \times V_2) \to W$ then decomposes into these partial derivatives:

$$df(u_1, u_2)(v_1, v_2) = d_1 f(u_1, u_2)(v_1) + d_2 f(u_1, u_2)(v_2)$$

Conversely, using the same notation, if the difference quotient functions exist and are continuously extendible to U_1 and U_2 respectively, then f is of class C^1:

Lemma 13.1.14 (Bertram, Glöckner, Neeb). *If V_1, V_2 and W are topological $\mathbb{K}$-vector spaces and $f : U \to W$ is a continuous function defined on an open subset U of $V_1 \times V_2$ whose partial difference quotient functions*

$$U_1 \cap (U \times V_1 \times \mathbb{K}^*) \to W, \quad (u_1, u_2, v_1, t) \mapsto \frac{f(u_1 + tv_1, u_2) - f(u_1, u_2)}{t}$$
$$\text{and}\quad U_2 \cap (U \times V_2 \times \mathbb{K}^*) \to W, \quad (u_1, u_2, v_2, t) \mapsto \frac{f(u_1+, u_2 + t_2) - f(u_1, u_2)}{t}$$

are extendible to continuous functions on the open subsets U_1 and U_2 respectively, then f is of class C^1 and its derivative df is given by the sum of the partial derivatives: $df(u_1, u_2)(v_1, v_2) = d_1 f(u_1, u_2)(v_1) + d_2 f(u_1, u_2)(v_2)$.

Proof. See [BGN04, Lemma 3.9] for a proof.   □

Since (continuous) differentiability only depends on the local behaviour of the functions, we note the consequence for functions between C^1-manifolds:

Lemma 13.1.15. *If M_1, M_2 and N are C^1-manifolds with model spaces V_1, V_2 and W respectively and $f : M_1 \times M_2 \to N$ is a continuous function, then f is continuously differentiable if and only if for all charts $\varphi_1 : O_1' \to O_1 \subset V_1$ on M_1,*

$\varphi_2 : O_2' \to O_2 \subset V_2$ *on* M_2 *and* $\psi : O_3' \to O_3 \subset W$ *on* N *the difference quotient functions*

$$(u_1, u_2, v_1, t) \mapsto \frac{\psi f(\varphi_1 \times \varphi_2)^{-1}(u_1 + tv_1, u_2) - \psi f(\varphi_1 \times \varphi_2)^{-1}(u_1, u_2)}{t}$$

$$\text{and} \quad (u_1, u_2, v_1, t) \mapsto \frac{\psi f(\varphi_1 \times \varphi_2)^{-1}(u_1+, u_2 + t_2) - \psi f(\varphi_1 \times \varphi_2)^{-1}(u_1, u_2)}{t}$$

on the subsets $[(O_1 \times O_2) \cap (\varphi_1 \times \varphi_2)^{-1} f^{-1}(O_3')] \times V_1 \times \mathbb{K}^*$ *of* $O_1 \times O_2 \times V_1 \times \mathbb{K}$ *and* $[(O_1 \times O_2) \cap (\varphi_1 \times \varphi_2)^{-1} f^{-1}(O_3')] \times V_2 \times \mathbb{K}^*$ *of* $O_1 \times O_2 \times V_2 \times \mathbb{K}$ *respectively are extendible to continuous functions* $\frac{\Delta \psi f(\varphi_1 \times \varphi_2)^{-1}}{\Delta_1}$ *and* $\frac{\Delta \psi f(\varphi_1 \times \varphi_2)^{-1}}{\Delta_2}$ *on the open subsets* $U_1 := \{(u_1, u_2, v_1, t) \in U \times V_1 \times \mathbb{K} \mid (u_1 + tv_1, u_2) \in U \cap f^{-1}(U_3)\}$ *and* $U_2 := \{(u_1, u_2, v_2, t) \in U \times V_1 \times \mathbb{K} \mid (u_1, u_2 + tv_2) \in U \cap f^{-1}(U_3)\}$ *respectively.*

Lemma 13.1.16. *If* U_1 *and* U_2 *are open subsets of topological* $\mathbb{K}$*-vector spaces* V_1 *and* V_2 *respectively, all finite products* $V_1^p \times V_2^q \times \mathbb{K}$ *are* k*-spaces,* W *is a Hausdorff topological* $\mathbb{K}$*-vector space and* $\hat{f} : U_1 \to C^1(U_2, W)$ *a continuously differentiable function, then its adjoint function* $f : U_1 \times U_2 \to V$ *is continuously differentiable.*

Proof. Let U_1 and U_2 be open subsets of topological $\mathbb{K}$-vector spaces such that all finite products $V_1^p \times V_2^q \times \mathbb{K}$ are k-spaces, W be a Hausdorff $\mathbb{K}$-vector space and $\hat{f} : U_1 \to C^1(U_2, W)$ be a continuously differentiable function. Since the injection $C^1(U_2, W) \to C(U_2, W)$ is continuous by Lemma 13.1.12, the adjoint map f of $\hat{f}$ is continuous by Lemma 13.1.2. For points $(u_1, u_2, v_1, v_2, t) \in (U_1 \times U_2)^{[1]}$ the difference quotient of the function f can be rewritten as

$$\frac{f(u_1 + tv_1, u_2 + t) - f(u_1, u_2)}{t} = \frac{f(u_1 + tv_1, u_2 + t) - f(u_1, u_2 + tv_2)}{t}$$
$$+ \frac{f(u_1, u_2 + tv_2) - f(u_1, u_2)}{t}$$
$$= \hat{f}^{[1]}(u_1, v_1, t)(u_2 + tv_2) + \hat{f}(u_1)^{[1]}(u_2, v_2, t).$$

The set $U_1^{[1]} \times U_2$ is an open subset of the k-space $V_1 \times V_1 \times \mathbb{K} \times V_2$, hence a k-space as well. The adjoint map of the continuous function $\hat{f}^{[1]} : U_1^{[1]} \to C^1(U_2, W)$ is continuous by Lemma 13.1.2. Therefore the first summand in the above decomposition of the difference quotient is the evaluation of the continuous function

$$(U_1 \times U_2)^{[1]} \to W, \quad (u_1, u_2, v_1, v_2, t) \mapsto \hat{f}^{[1]}(u_1, v_1, t)(u_2 + tv_2).$$

at the point (u_1, u_2, v_1, v_2, t). Similarly, the set $U_1 \times U_2^{[1]}$ is an open subset of the k-space $V_1 \times V_2 \times V_2 \times \mathbb{K}$, hence a k-space. The injection $C^1(U_2, W) \to C(U_2^{[1]}, W)$ is continuous. Therefore the function $U_1 \to C(U_2^{[1]}, W), u_1 \mapsto \hat{f}(u_1)^{[1]}$ is continuous and so is its adjoint function $U_1 \times U_2^{[1]} \to W, (u_1, u_2, v_2, t) \mapsto \hat{f}(u_1)^{[1]}(u_2, v_2, t)$ by Lemma 13.1.2. The latter function can be regarded as a continuous function on the open set $(U_1 \times U_2)^{[1]}$ by composing it with the embedding $U_1 \times U_2^{[1]} \hookrightarrow (U_1 \times U_2)^{[1]}$, $(u_1, u_2, v_2, t) \mapsto (u_1, u_2, 0, v_2, t)$. Thus, the difference quotient above is extendible to a continuous function on $(U_1 \times U_2)^{[1]}$, i.e. f is continuously differentiable. $\square$

Remark 13.1.17. As the proof of the preceding Lemma shows, for any continuously differentiable function $f : U_1 \times U_2 \to W$ the function $f^{[1]}$ can expressed in terms of the continuous functions $\hat{f}^{[1]} : U_1^{[1]} \to C^1(U_2, W)$ and $\hat{f}(-)^{[1]} : U_1 \to C(U_2^{[1]}, W)$ via $f(u_1, u_2, v_1, v_2, t) = \hat{f}^{[1]}(u_1, v_1, t)(u_2 + tv_2) + \hat{f}(u_1)^{[1]}(u_2, v_2, t)$.

Proposition 13.1.18. *If M_1 and M_2 are C^k-manifolds, $1 \leq k \leq \infty$, with model spaces V_1 and V_2 respectively, all finite products $V_1^p \times V_2^q \times \mathbb{K}$ are k-spaces, W is a Hausdorff $\mathbb{K}$-vector space and $\hat{f} : M_1 \to C^1(M_2, W)$ is a C^1-function, then its adjoint map $f : M_1 \times M_2 \to W$ is of class C^1.*

Proof. The function f is of class C^1 if and only if it is locally of class C^1. This is guaranteed by Lemma 13.1.16. $\square$

Lemma 13.1.19. *If U_1 and U_2 are open subsets of topological $\mathbb{K}$-vector spaces V_1 and V_2 respectively, all finite products $V_1^p \times V_2^q \times \mathbb{K}$ are k-spaces, W is a Hausdorff topological $\mathbb{K}$-vector space and $\hat{f} : U_1 \to C^k(U_2, W)$ a function of class C^k, then its adjoint function $f : U_1 \times U_2 \to V$ is of class C^k as well.*

Proof. The proof is by induction on k. The cases $k = 0$ and $k = 1$ are proved in Lemmata 13.1.2 and 13.1.16. We assume the assertion to be true for all $k' \leq k$ and prove it for $k + 1$. Let U_1 and U_2 be open subsets of topological $\mathbb{K}$-vector spaces V_1 and V_2 respectively, all finite products $V_1^p \times V_2^q \times \mathbb{K}$ be k-spaces, W be a Hausdorff $\mathbb{K}$-vector space and $\hat{f} : U_1 \to C^{k+1}(U_2, W)$ a function of class C^{k+1}. The injection $C^{k+1}(U_2, W) \to C^k(U_2, W)$ is continuous, so the inductive hypothesis implies that the adjoint function $f : U_1 \times U_2 \to V$ of $\hat{f}$ is of class C^k. It remains to prove that the function $f^{[1]} : (U_1 \times U_2)^{[1]} \to W$ is of class C^k as well. The evaluation of the continuous function $f^{[1]}$ at a point $(u_1, u_2, v_1, v_2, t) \in (U_1 \times U_2)^{[1]}$ is given by

$$f^{[1]}(u_1, u_2, v_1, v_2, t) = \hat{f}^{[1]}(u_1, v_1, t)(u_2 + tv_2) + \hat{f}(u_1)^{[1]}(u_2, v_2, t) \tag{13.1}$$

(cf. Remark 13.1.17). The map $\hat{f} : U_1 \to C^{k+1}(U_2, W)$ is of class C^{k+1} by assumption. Therefore the function $f^{[1]} : U_1^{[1]} \to C^{k+1}(U_2, W)$ is of class C^k. Its adjoint map $U_1^{[1]} \times U_2 \to W$ is of class C^k by the inductive assumption and the function $V_2^2 \times \mathbb{K}, (u_2, v_2, t) \to u_2 + tv_2$ is a composition of smooth maps, hence smooth as well. So the map

$$(U_1 \times U_2)^{[1]} \to W, \quad (u_1, u_2, v_1, v_2, t) \mapsto \hat{f}^{[1]}(u_1, v_1, t)(u_2 + tv_2)$$

is a composition of functions of class C^k, hence of class C^k as well. The linear map $C^{k+1}(U_2, W) \to C^k(U_2^{[1]}, W), g \mapsto g^{[1]}$ is continuous by Lemma 13.1.12, hence smooth. Therefore the function $U_1 \times U_2^{[1]} \to W, (u_1, u_2, v_1, t) \mapsto \hat{f}(u_1)(u_2, v_2, t)$ is of class C^k by the inductive assumption. So the map

$$(U_1 \times U_2)^{[1]} \to W, \quad (u_1, u_2, v_1, v_2, t) \mapsto \hat{f}(u_1)(u_2, v_2, t)$$

is a composition of maps of class C^k, hence of class C^k itself. All in all the decomposition of $f^{[1]}$ in Equation 13.1 identifies $f^{[1]}$ as a sum of maps of class C^k; therefore $f^{[1]}$ is of class C^k as well. As a consequence the function f is of class C^{k+1}. This completes the inductive step. $\square$

Corollary 13.1.20. *If U_1 and U_2 are open subsets of topological $\mathbb{K}$-vector spaces V_1 and V_2 respectively, all finite products $V_1^p \times V_2^q \times \mathbb{K}$ are k-spaces, W is a Hausdorff topological $\mathbb{K}$-vector space and $\hat{f} : U_1 \to C^\infty(U_2, W)$ a smooth function, then its adjoint function $f : U_1 \times U_2 \to V$ is smooth as well.*

Corollary 13.1.21. *If U_1 and U_2 are open subsets of topological $\mathbb{K}$-vector spaces V_1 and V_2 respectively, all finite products $V_1^p \times V_2^q \times \mathbb{K}$ are k-spaces and W is a Hausdorff topological $\mathbb{K}$-vector space, then $\varphi_{U_1, U_2, W} : C^\infty(U_1 \times U_2, W) \to C^\infty(U_1, C^\infty(U_2, W))$ is a continuous bijection.*

Proposition 13.1.22. *If M_1 and M_2 are C^k-manifolds, $0 \leq k \leq \infty$, with model spaces V_1 and V_2 respectively, all finite products $V_1^p \times V_2^q \times \mathbb{K}$ are k-spaces, W is a Hausdorff $\mathbb{K}$-vector space and $\hat{f} : M_1 \to C^k(M_2, W)$ is a C^k-function, then its adjoint map $f : M_1 \times M_2 \to W$ is of class C^k.*

Proof. Let M_1 and M_2 be C^k-manifolds, $0 \leq k \leq \infty$, with model spaces V_1 and V_2 respectively, all finite products $V_1^p \times V_2^q \times \mathbb{K}$ be k-spaces, W be a Hausdorff $\mathbb{K}$-vector space and $\hat{f} : M_1 \to C^k(M_2, W)$ be a function of class C^k. The adjoint function $f : M_1 \times M_2 \to W$ of f is of class C^k if and only if it is locally of class C^k, which we are going to prove now. Let $\varphi_1 : U_1' \to U_1 \subseteq V_1$ and $\varphi_1 : U_1' \to U_1 \subseteq V_1$ be charts of M_1 and M_2 onto open subsets U_1 of V_1 and U_2 of V_2 respectively. It is to prove that the function

$$f_{|U_1 \times U_2}(\varphi_1 \times \varphi_2)^{-1} : U_1 \times U_2 \to W$$

is of class C^k. The function $\hat{f}_{|U_1}\varphi_1^{-1} : U_1 \to C^k(M_2, W)$ is of class C^k by assumption. Furthermore the restriction map $(- \, |_{U_2'}) : C^k(M_2, W) \to C^k(U_2', W)$ is of class C^k by Lemma 13.1.13. Therefore the composition $(- \, |_{U_2'})\hat{f} \circ \varphi^{-1} : U_1 \to C^k(U_2', W)$ is of class C^k as well. Since $\varphi_2 : U_2 \to U_2'$ is a diffeomorphism, this implies that the function $C^k(\varphi_2^{-1}, W)\hat{f}\varphi_1^{-1} : U_1 \to C^k(U_2, W)$ is of class C^k. The adjoint map of this function is the function $f_{|U_1 \times U_2}(\varphi_1 \times \varphi_2)^{-1}$ which we have to show to be of class C^k. By Lemma 13.1.19 this adjoint map is of class C^k. So f is locally of class C^k and therefore of class C^k. $\qquad \square$

Corollary 13.1.23. *If M_1 and M_2 are smooth manifolds with model spaces V_1 and V_2 respectively, all finite products $V_1^p \times V_2^q \times \mathbb{K}$ are k-spaces, W is a Hausdorff $\mathbb{K}$-vector space, then $\varphi_{M_1, M_2, W} : C^\infty(M_1 \times M_2, W) \to C^\infty(M_1, C^\infty(M_2, W))$ is a continuous bijection.*

13.2 Fibre Bundles with Connection

Let $F \hookrightarrow E \xrightarrow{p} B$ be a locally trivial bundle of smooth manifolds. Since the exterior algebra bundle cannot be used for arbitrary infinite dimensional manifolds, we describe an alternate approach to describe differential forms. For this purpose, we use direct sums of tangent bundles.

Definition 13.2.1. *The* vertical bundle VE *of the bundle $E \xrightarrow{p} B$ is the Kernel of the map $Tp : TE \to TB$ of vector bundles:*

$$VE := Ker \, Tp$$

The vertical bundle VE is a closed vector subbundle of TE. Since the fibre bundle $E \xrightarrow{p} B$ is locally trivial, the vertical bundle VE is a locally splitting vector subbundle of TE. The inclusion of VE in TE will be denoted by i_{ver}.

Definition 13.2.2. *A* connection *on the bundle $E \xrightarrow{p} B$ is a bundle valued 1-form $\Phi \in \Omega^1(TE; VE)$ such that $Im\, \Phi = VE$ and $\Phi \circ \Phi = \Phi$.*

This is equivalent to Φ being a projection $\Phi : TE \to VE$ of vector bundles.

Example 13.2.3. Every locally trivial bundle $E \to B$ over a smoothly paracompact base manifold B admits a connection form Φ.

Example 13.2.4. Every flat bundle $E \to B$ admits a connection form Φ.

Definition 13.2.5. *The* horizontal bundle HE *of the bundle $E \xrightarrow{p} B$ is the Kernel of the connection $\Phi : TE \to TE$:*

$$HE := Ker\ \Phi$$

The horizontal bundle HE is a closed vector subbundle of TE. Its inclusion map into TE will be denoted by i_{hor}. Since the kernel HE of the projection Φ is the image of the projection $\mathrm{id}_{TE} - \Phi$, the bundle map $\mathrm{id}_{TE} - \Phi$ projects TE onto HE. This leads to the following observation:

Lemma 13.2.6. *The tangent bundle TE of E splits into the direct sum $HE \oplus VE$.*

This decomposition of the tangent bundle TE induces a decomposition of the n-fold sum of the vector bundle TE over E.

Definition 13.2.7. *For a vector bundle $E' \to B'$ over a topological space B' we denote the the n-fold sum of this vector bundle over B' by $E'^{\oplus n}$. For a smooth manifold M we denote by*

$$T^{\oplus n}M := \bigoplus_{i=1}^{n} TM$$

the n-fold sum of the tangent bundle of M. Elements $(X_1, \ldots, X_n)$ of $T^{\oplus n}M$ will be abbreviated by $\mathbf{X}$. Smooth sections of $T^{\oplus n}M$ will be called n-vector fields.

Note that we do not require the tangent vectors $X_1, \cdots, X_n$ to be linearly independent in the above definition. The n-fold sum of vector bundles is a functor from the category $\mathbf{VB}$ of vector bundles into itself. As a consequence $T^{\oplus n}$ is a functor from the category $\mathbf{Mf}$ of manifolds to the category $\mathbf{VB}$.

Definition 13.2.8. *For a smooth manifold M and $p \le n$ we denote by $i_{\ell,p}$ inclusion*

$$T^{\oplus p}M \hookrightarrow T^{\oplus n}M, \quad (X_1, \ldots, X_p) \mapsto (X_1, \ldots, X_p, 0, \ldots, 0)$$

which identifies $T^{\oplus p}M$ with the first p summands of $T^{\oplus n}M$ and for $q \le n$ by $i_{r,q}$ the inclusion

$$T^{\oplus q}M \hookrightarrow T^{\oplus n}M, \quad (X_1, \ldots, X_q) \mapsto (0, \ldots, 0, X_1, \ldots, X_q)$$

which identifies $T^{\oplus q}M$ with the last q summands of $T^{\oplus n}M$.

As before we consider a bundle $F \hookrightarrow E \xrightarrow{p} B$ of smooth manifolds with connection form Φ, vertical bundle VE and horizontal bundle HE.

Definition 13.2.9. *The sum of the vector bundles $HE^{\oplus p}$ and $VE^{\oplus q}$ will be denoted by $T^{p,q}E$:*
$$T^{p,q}E := \left(HE^{\oplus p} \right) \oplus \left(VE^{\oplus q} \right)$$

For $p + q = n$ we regard the bundles $T^{p,q}E$ as closed vector subbundles of $T^{\oplus n}E$. The canonical inclusion $i_{\ell,p} \circ i_{hor} + i_{r,q} \circ i_{ver}$ of $T^{p,q}E$ in $T^{\oplus n}E$ will be denoted by $i_{p,q}$. The projection of $T^{\oplus n}E$ onto the closed subbundle $T^{p,q}E$ will be denoted by $\pi_{p,q}$. In analogy to vector fields on TE one can consider smooth sections of the bundles $T^{p,q}E$. These will be used throughout this chapter.

Definition 13.2.10. *A n-vector field of type (p,q) is a smooth section of the vector bundle $T^{p,q}E$. The n-vector fields of type $(n,0)$ are called* horizontal n-vector fields *and n-vector fields of type $(0,n)$ are called* vertical n-vector fields. *We use the same notation as for elements $\mathbf{X} \in T^{\oplus n}E$.*

Let $F' \hookrightarrow E' \to B'$ be another smooth bundle and $f : E \to E'$ be a smooth map. The restriction of $(T^{\oplus n}f) : T^{\oplus n}E \to T^{\oplus n}E'$ to the subbundles $T^{p,q}E$ leads to a decomposition of the map $(T^{\oplus n}f) : T^{\oplus n}E \to T^{\oplus n}E'$.

Definition 13.2.11. *The composition $(T^{\oplus n}f) \circ i_{p,q}$ for $p + q = n$ will be denoted by $T^{p,q}f$:*

$$
\begin{array}{ccc}
T^{p,q}E & \xrightarrow{\;i_{p,q}\;} & T^{\oplus n}E \\
 & \searrow {\scriptstyle T^{p,q}f} & \downarrow {\scriptstyle T^{\oplus n}f} \\
 & & T^{\oplus n}E'
\end{array}
$$

The maps $T^{p,q}f$ are smooth, they can be inserted in the contravariant functor $C^k(-, V)$, $0 \le k \le \infty$ for any smooth manifold V:

$$
C^k(T^{\oplus n}E', V) \xrightarrow{\;C^k(T^{p,q}f, V)\;} C^k(T^{p,q}E, V)
$$

For any manifold E' and Hausdorff topological vector space V the groups $\Omega^n(E', V)$ of the de Rham complex are closed subspaces of the spaces $C^\infty(T^{\oplus n}E', V)$. Therefore it is natural to examine the effect of $T^{p,q}f$ on these subspaces. This will be carried out in the following sections.

13.3 Differential Forms

In this section we derive a topological version of the decomposition of differential forms on fibre bundles $E \to B$ with connection form. We only consider differential forms with values in a Hausdorff topological vector space V. The spaces $\Omega^n(M; V)$ of differential forms on a smooth manifold M with values in V will always be given the C^∞-compact open topology. The The decomposition of the vector bundles $T^{\oplus n}E$ presented in the last section leads to a decomposition of differential forms on E in the following way: A differential n-form ω on E can be regarded as a function $\Lambda^n TE \to V$ from the exterior bundle $\Lambda^n TE$ on E into V which is linear on each fibre $\Lambda^n TE_e$ over a point e of E. This bundle ha a purely algebraic (i.e. non-topological) decomposition

$$
\Lambda^n TE = \Lambda^n(HE \oplus VE) \cong \bigoplus_{p+q=n} \Lambda^p HE \vee \Lambda^q VE
$$

which leads to the usual algebraic decomposition $\Omega^n(M, V) \cong \bigoplus_{p+q=n} \Omega^{p,q}(M, V)$ of the (abstract) vector space $\Omega^n(E; V)$ into the direct sum of subspaces $\Omega^{p,q}(E, V)$ of $\Omega^n(E, V)$. In this decomposition a differential n-form ω is mapped to a sum $\sum \omega_{p,q}$ in which the form $\omega_{p,q}$ corresponds to the restriction of ω to $\Lambda^p HE \vee \Lambda^q VE$. Below we verify that this purely algebraic decomposition in fact leads to an isomorphism

$$
\Omega^n(M, V) \cong \bigoplus_{p+q=n} \Omega^{p,q}(M, V)
$$

of topological vector spaces. To construct the subspaces for this decomposition in the topological context we use the action of the symmetric group S_n on $T^{\oplus n}E$. The group S_n acts by vector bundle isomorphisms on $T^{\oplus n}E$ via permutations of the components of elements $\mathbf{X} = (X_1, \ldots, X_n) \in T^{\oplus n}E$:

$$
S_n \times T^{\oplus n}E \to T^{\oplus n}E, \quad (\sigma, (X_1, \ldots X_n)) \mapsto (X_{\sigma(1)}, \ldots, X_{\sigma(n)})
$$

This action induces an action of S_n on the function space $C^\infty(T^{\oplus n}E, V)$ which contains the de Rham space $\Omega^n(E; V)$ as a subspace. The latter subspace $\Omega^n(M; V)$

is a closed subspace if the vector space V is Hausdorff. Since a permutation of the arguments of a differential form can only change the sign, the spaces $\Omega^n(E;V)$ are invariant under the action of S_n on $C^\infty(T^{\oplus n}E, V)$. With the help of the action of S_n we can now define an antisymmetrising operator on these function spaces.

Definition 13.3.1. *The antisymmetrising operator α_n on $C^\infty(T^{\oplus n}E, V)$ is given by*

$$\alpha_n(f)(X_1, \ldots X_n) := \frac{1}{n!} \sum_{\sigma \in S_n} \operatorname{sgn}(\sigma) f(X_{\sigma(1)}, \ldots X_{\sigma(n)}).$$

Because antisymmetric maps are invariant under α_n we obtain the following Lemma:

Lemma 13.3.2. *The operators α_n and $\frac{n!}{p!q!}\alpha_n \circ C^\infty(\pi_{p,q}, V)$ on $C^\infty(T^{\oplus n}E; V)$ are idempotent.*

Since differential forms on E are antisymmetric, we observe that α_n restricts to the identity on $\Omega^n(E;V)$.

Lemma 13.3.3. *For any function $f \in C^\infty(T^{p,q}E, V)$ that is multilinear in $HE^{\oplus p}$ and $VE^{\oplus q}$ respectively the map $\alpha_n(f \circ \pi_{p,q})$ is a differential n-form on E.*

Proof. Let $f \in C^\infty(T^{p,q}E, V)$ be a function that is multilinear in $HE^{\oplus p}$ and $VE^{\oplus q}$ respectively. The projection $\pi_{p,q} : TE \to T^{p,q}E$ is a vector bundle homomorphism. Furthermore the map $(X_1, \ldots X_n) \mapsto (X_{\sigma(1)}, \ldots X_{\sigma(n)})$ induced by a permutation $\sigma \in S_n$ is a vector bundle isomorphism. Therefore the map $f \circ \pi_{p,q}$ is multilinear in $i_{\ell,p}(T^{\oplus p}E)$ and $i_{r,q}(T^{\oplus q}E)$ respectively, hence multilinear in $T^{\oplus n}E$. Thus the antisymmetrised map $\alpha_n(f \circ \pi_{p,q})$ is a smooth multilinear antisymmetric function on $T^{\oplus n}E$, i.e. a differential n-form on E. $\square$

Definition 13.3.4. *An n-form of type (p,q) on E is an n-form $\omega_{p,q} \in \Omega^n(E;V)$ that satisfies the identity $\omega = \frac{n!}{p!q!}\alpha_n(\omega \circ \pi_{p,q})$. The closed subspace of $\Omega^n(E;V)$ consisting of n-forms of type (p,q) is denoted by $\Omega^{p,q}(E;V)$:*

$$\Omega^{p,q}(E;V) := \Omega^n(E;V)^{\frac{n!}{p!q!}\alpha_n \circ C^\infty(\pi_{p,q}, V)}$$

If the bundle E is trivial with base space B, fibre F and the connection form Φ is the projection $TE \cong TB \times TF \to TF$ we also denote the de Rham space $\Omega^{p,q}(E;V)$ by $\Omega^{p,q}(B, F; V)$.

The n-forms of type (p,q) are the forms $\omega_{p,q}$ which are trivial on $(X_1, \ldots, X_n)$ unless exactly p of the tangent vectors X_i are horizontal and q of the tangent vectors X_i are vertical. The spaces $\Omega^{p,q}(E;V)$ are closed subspaces of the de Rham spaces $\Omega^n(E;V)$. Applying Lemmata 13.3.2 and 13.3.3 to differential n-forms we obtain:

Lemma 13.3.5. *For any n-form $\omega \in \Omega^n(E;V)$ the map $\frac{n!}{p!q!}\alpha_n(\omega \circ \pi_{p,q})$ is a differential n-form of type (p,q) on E.*

These forms will play a central role in the sequel. They will be used to decompose de Rham cohomology and to derive a version of the Leray-Serre spectral sequence for bundles with non paracompact base manifolds.

Definition 13.3.6. *For any n-form $\omega \in \Omega^n(E;V)$ the associated n-forms of type (p,q) are the forms*

$$\omega_{p,q} := \frac{n!}{p!q!}\alpha_n(\omega \circ \pi_{p,q}).$$

Proposition 13.3.7. *The topological vector space $\Omega^n(E;V)$ of n-forms on E decomposes into the direct sum $\bigoplus_{p+q=n} \Omega^{p,q}(E;V)$ of its closed subspaces $\Omega^{p,q}(E,V)$ and the projections onto these subspaces $\Omega^{p,q}(E;V)$ are given by $\frac{n!}{p!q!}\alpha_n \circ C^\infty(\pi_{p,q},V)$, i.e. every n-form ω decomposes into its projections $\omega_{p,q}$ under the isomorphism*

$$\Omega^n(E;V) \xrightarrow{\cong} \bigoplus_{p+q=n} \Omega^{p,q}(E;V), \quad \omega \mapsto \sum_{p+q=n} \omega_{p,q}.$$

Proof. The projections onto the closed subspaces $\Omega^{p,q}(E,V)$ are continuous. Since the decomposition is an algebraic isomorphism of vector spaces which is a sum of continuous linear functions, it is an isomorphism of topological vector spaces.   □

Since an n-form $\omega_{p,q}$ of type (p,q) associated to an n-form ω on E corresponds to the restriction to the subbundle $\Lambda^p HE \wedge \Lambda^q VE$ in the the above decomposition of the exterior bundle, the n-form $\omega_{p,q}$ of type (p,q) can be regarded as a multilinear function

$$HE^{\oplus p} \oplus VE^{\oplus q} = T^{p,q}E \to V,$$

which is alternating in $HE^{\oplus p}$ and $VE^{\oplus q}$ separately. The vector space of all such forms is a closed subspace $A^{p,q}$ of the function space $C^\infty(T^{p,q}E,V)$. The smooth inclusion $T^{p,q}E \hookrightarrow T^{\oplus n}E$ induces a smooth function $C^\infty(T^{\oplus n}E,V) \to C^\infty(T^{p,q},V)$, which restricts to a continuous linear bijection $\Omega^{p,q}(M;V) \to A^{p,q}$, i.e. the topological vector subspace $A^{p,q}$ of $C^\infty(T^{p,q},V)$ is isomorphic to $\Omega^{p,q}(M,V)$. Henceforth we shall identify the closed subspaces $\Omega^{p,q}(E;V)$ of $\Omega^n(E;V)$ with these closed subspaces $A^{p,q}$ of the vector spaces $C^\infty(T^{p,q}E,V)$.

The projections $\frac{n!}{p!q!}\alpha_n \circ C^\infty(\pi_{p,q},V)$ onto the subspaces $\Omega^{p,q}(E;V)$ of $\Omega^n(E;V)$ (for $p+q=n$) will be denoted by $\pi^{p,q}$. In many cases the de Rham spaces $\Omega^{p,q}(E;V)$ form a first quadrant double complex, where the differentials in the horizontal and vertical direction are motivated by the exterior derivatives on B and F respectively. To make this construction better readable we introduce some notation.

Definition 13.3.8. *For any p-vector field $\mathbf{X}$ on E and differential n-form ω we denote with $i_{\mathbf{X}}\omega$ the differential $(n-p)$-form*

$$i_{\mathbf{X}}\omega := \omega(\mathbf{X},-) = i_{X_p}\cdots i_{X_1}\omega : T^{\oplus(n-p)}E \to V.$$

Likewise, for any element $\mathbf{X} \in T^{\oplus p}E$ at $e \in E$ we denote with $i_{\mathbf{X}}\omega_e$ the function

$$i_{\mathbf{X}}\omega_e := \omega_e(\mathbf{X},-) = i_{X_p}\cdots i_{X_1}\omega_e : T^{\oplus(n-p)}E_e \to V.$$

Analogously, for any q-vector field $\mathbf{Y}$ on E, we denote by $r_{\mathbf{Y}}\omega$ the differential $(n-q)$-form

$$r_{\mathbf{Y}}\omega := \omega(-,\mathbf{Y}) = (-1)^{pq} i_{Y_p}\cdots i_{Y_1}\omega : T^{\oplus(n-q)}E \to V.$$

and for any element $\mathbf{Y} \in T^{\oplus q}E_e$ at $e \in E$, we denote by $r_{\mathbf{Y}}\omega_e$ the function

$$r_{\mathbf{Y}}\omega_e := \omega_e(-,\mathbf{Y}) = (-1)^{pq} i_{Y_p}\cdots i_{Y_1}\omega_e : T^{\oplus(n-q)}E_e \to V.$$

We use the same notation for differential forms $\omega_{p,q}$ of type (p,q), horizontal p-vector fields $\mathbf{X}$ and vertical q-vector fields $\mathbf{Y}$.

If the connection Φ is flat, then the horizontal bundle HE is involutive, i.e. for any two horizontal vector fields X_1 and X_2 the vector field $[X_1, X_2]$ is also horizontal. This allows the definition of horizontal and vertical differentials on $\Omega^{*,*}(E;V)$.

Definition 13.3.9. *Let the connection Φ on the bundle $p : E \to B$ be flat. The horizontal differential $d_h^{p,q} : \Omega^{p,q}(E;V) \to \Omega^{p+1,q}(E;V)$ is given by*

$$d_h^{p,q}\omega_{p,q}(X_0,\ldots,X_p,\mathbf{Y}) = (dr_\mathbf{Y}\omega_{p,q})(\mathbf{X})$$
$$= \sum_{i=0}^{p}(-1)^i X_i.\omega_{p,q}(X_0,\ldots,\hat{X}_i,\ldots,X_p,\mathbf{Y})$$
$$+ \sum_{i<j}(-1)^{i+j}\omega_{p,q}([X_i,X_j],X_0,\ldots,\hat{X}_i,\ldots,\hat{X}_j,\ldots,X_p,\mathbf{Y})$$

for any n-form of type (p,q) and any $(n+1)$-vector field $(\mathbf{X},\mathbf{Y}) = (X_0,\ldots,X_p,\mathbf{Y})$ of type $(p+1,q)$ on E. The vertical differential $d_v^{p,q} : \Omega^{p,q}(E;V) \to \Omega^{p,q+1}(E;V)$ *is given by*

$$d_v^{p,q}\omega_{p,q}(\mathbf{X},Y_0,\ldots,Y_q) = (-1)^p(di_\mathbf{X}\omega_{p,q})(\mathbf{Y})$$
$$= \sum_{i=0}^{q-1}(-1)^{p+i}Y_i.\left(\omega_{p,q}(\mathbf{X},Y_0,\ldots,\hat{Y}_i,\ldots,Y_q)\right)$$
$$+ \sum_{i<j}(-1)^{p+i+j}\omega_{p,q}(\mathbf{X},[Y_i,Y_j],Y_0,\ldots,\hat{Y}_i,\ldots,\hat{Y}_j,\ldots,Y_{q-1})$$

for any n-form of type (p,q), horizontal p-vector field $\mathbf{X} = (X_1,\ldots,X_p)$ and $(n+1)$-vector field $(\mathbf{X},\mathbf{Y}) = (\mathbf{X},Y_0,\ldots,Y_q)$ of type $(p,q+1)$ on E.

In cases where the domain of the differentials $d_h^{p,q}$ or $d_v^{p,q}$ is obvious we will omit the superscript (p,q).

Proposition 13.3.10. *If connection Φ is flat, then the horizontal and vertical differentials d_h and d_v anticommute, i.e. they satisfy the equation $d_h d_v + d_v d_h = 0$.*

Proof. Let $\omega_{p,q}$ be a differential n-form of type (p,q) and $(X_0,\ldots,X_p,Y_0,\ldots,Y_q)$ be an element of $T^{p+1,q+1}E$. We extend this element to a local $(n+2)$-vector field of type $(p+1,q+1)$. The evaluation of $d_h d_v \omega_{p,q}$ at $(\mathbf{X},\mathbf{Y})$ is given by

$$d_h d_v \omega_{p,q}(\mathbf{X},\mathbf{Y}) =$$
$$\sum_{i}(-1)^i X_i.(d_v\omega_{p,q})(X_0,\ldots,\hat{X}_i,\ldots,X_p,\mathbf{Y})$$
$$+ \sum_{i<j}(-1)^{i+j}(d_v\omega_{p,q})([X_i,X_j],X_0,\ldots,\hat{X}_i,\ldots,\hat{X}_j,\ldots,X_p,\mathbf{Y})$$
$$= \sum_{i,k}(-1)^{p+i+k}X_i.(Y_k.\omega_{p,q}(X_0,\ldots,\hat{X}_i,\ldots,X_p,Y_0,\ldots,\hat{Y}_k,\ldots,X_q))$$
$$+ \sum_{\substack{i \\ k<l}}(-1)^{p+i+k+l}X_i.\omega_{p,q}(X_0,\ldots,\hat{X}_i,\ldots,X_p,[Y_k,Y_l],Y_0,\ldots,\hat{Y}_k,\ldots,\hat{Y}_l,\ldots,X_q))$$
$$+ \sum_{\substack{i<j \\ k}}(-1)^{p+i+j+k}Y_k.\omega_{p,q}([X_i,X_j],X_0,\ldots,\hat{X}_i,\ldots,\hat{X}_j,\ldots,X_p,\mathbf{Y})$$
$$+ \sum_{\substack{i<j \\ k<l}}(-1)^{p+i+j+k+l}\omega_{p,q}([X_i,X_j],X_0,\ldots,\hat{X}_i,\ldots,\hat{X}_j,\ldots,X_p,[Y_k,Y_l],Y_0,\ldots,\hat{Y}_k,\ldots,\hat{Y}_l,\ldots,X_q).$$

Likewise, the evaluation of $d_v d_h \omega_{p,q}$ at $(\mathbf{X},\mathbf{Y})$ is given by

$$d_v d_h \omega_{p,q}(\mathbf{X}, \mathbf{Y}) =$$
$$\sum_k (-1)^{p+1+k} Y_k.(d_h\omega_{p,q}(\mathbf{X}, Y_0, \ldots, \hat{Y}_k, \ldots, Y_q))$$
$$+ \sum_{k<l} (-1)^{p+1+k+l}(d_h\omega_{p,q}(\mathbf{X}, [Y_k, Y_l], Y_0, \ldots, \hat{Y}_k, \ldots, \hat{Y}_l, \ldots, Y_q)$$
$$= \sum_{i,k} (-1)^{p+1+i+k} X_i.(Y_k.\omega_{p,q}(X_0, \ldots, \hat{X}_i, \ldots, X_p, Y_0, \ldots, \hat{Y}_k, \ldots, X_q))$$
$$+ \sum_{\substack{k \\ i<j}} (-1)^{p+1+i+j+k} Y_k.\omega_{p,q}([X_i, X_j], X_0, \ldots, \hat{X}_i, \ldots, \hat{X}_j, \ldots, X_p, \mathbf{Y})$$
$$+ \sum_{\substack{k<l \\ i}} (-1)^{p+1+i+k+l} X_i.\omega_{p,q}(X_0, \ldots, \hat{X}_i, \ldots, X_p, [Y_k, Y_l], Y_0, \ldots, \hat{Y}_k, \ldots, \hat{Y}_l, \ldots, X_q))$$
$$+ \sum_{\substack{k<l \\ i<j}} (-1)^{p+1+i+j+k+l} \omega_{p,q}([X_i, X_j], X_0, \ldots, \hat{X}_i, \ldots, \hat{X}_j, \ldots, X_p, [Y_k, Y_l], Y_0, \ldots, \hat{Y}_k, \ldots, \hat{Y}_l, \ldots, X_q$$

These two expressions differ only in sign, hence $d_h d_v + d_v d_h = 0$.   □

Corollary 13.3.11. *If the connection Φ is flat, then the spaces $\Omega^{*,*}(E; V)$ form a double complex with horizontal differentials d_h and vertical differentials d_v.*

The differential D of the total complex $\text{Tot}(\Omega^{*,*}(E; V))$ is then given by the sum

$$D = \sum_{p+q=n} d_h^{p,q} + d_v^{p,q} : \bigoplus_{p+q=n} \Omega^{p,q}(E; V) \longrightarrow \bigoplus_{p+q=n+1} \Omega^{p,q}(E; V).$$

Definition 13.3.12. *A bundle $p : E \to B$ has the good field extension property (GFEP) if every element $(\mathbf{X}, \mathbf{Y}) \in T^{p,q}E$ can locally be extended to a n-vector field of type (p, q) such that $[X_i, Y_j] = 0$ for all $1 \leq i \leq p$ and $1 \leq j \leq q$.*

Important examples of bundles with the GFEP are flat bundles. Our main purpose for introducing the GFEP is the following observation:

Proposition 13.3.13. *If the bundle $p : E \to B$ has the GFEP, then the differentials on $\Omega^{*,*}(E; V)$ and $\Omega^*(E; V)$ are related via*

$$d_h\omega_{p,q} + d_v\omega_{p+1,q-1} = (d_E\omega)_{p+1,q}$$

for any n-form ω on E (where $p + q = n$), i.e. the maps $\sum_{p+q=n} \pi_{p,q}$ assemble to an isomorphism $\Omega^*(E; V) \to \text{Tot}\,\Omega^{*,*}(E; V)$ of chain complexes.

Proof. Suppose ω is an n-form on E and $p + q = n$. Let $e \in E$ be a point in E and elements $\mathbf{X} = (X_0, \ldots, X_p) \in HE^{\oplus(p+1)}$ $\mathbf{Y} = (Y_0, \ldots, Y_{q-1}) \in VE^{\oplus(q+1)}$ be given. We extend these elements to local horizontal and vertical $(p + 1)$- resp. q-vector fields respectively such that the vector fields X_i and Y_j centralise each other. The evaluation of the sum $d_h\omega_{p,q} + d_v\omega_{p+1,q-1}$ at $(\mathbf{X}, \mathbf{Y})$ is given by

$$(d_h\omega_{p,q} + d_v\omega_{p+1,q-1})(\mathbf{X}, \mathbf{Y}) =$$
$$\sum_{i=0}^p (-1)^i X_i.\omega_{p,q}(X_0, \ldots, \hat{X}_i, \ldots, X_p, \mathbf{Y})$$
$$+ \sum_{i<j} (-1)^{i+j} \omega_{p,q}([X_i, X_j], X_0, \ldots, \hat{X}_i, \ldots, \hat{X}_j, \ldots, X_p, \mathbf{Y})$$
$$+ \sum_{i=0}^{q-1} (-1)^{p+1+i} Y_i.\left[\omega_{p+1,q-1}(\mathbf{X}, Y_0, \ldots, \hat{Y}_i, \ldots, Y_{q-1})\right]$$
$$+ \sum_{i<j} (-1)^{p+1+i+j} \omega_{p+1,q-1}(\mathbf{X}, [Y_i, Y_j], Y_0, \ldots, \hat{Y}_i, \ldots, \hat{Y}_j, \ldots, Y_{q-1}).$$

In view of $\omega_{p,q}(\mathbf{X}, \mathbf{Y}) = \omega(\mathbf{X}, \mathbf{Y})$ for $(\mathbf{X}, \mathbf{Y}) \in T^{p,q}E$, the above expression can be reformulated:

$$(d_h\omega_{p,q} + d_v\omega_{p+1,q-1})(\mathbf{X}, \mathbf{Y}) =$$

$$\sum_{i=0}^{p}(-1)^i X_i.\omega(X_0, \ldots, X_i, \ldots, X_p, \mathbf{Y})$$

$$+\sum_{i=0}^{q-1}(-1)^{i+p+1}Y_i.\omega(\mathbf{X}, Y_0, \ldots, \widehat{Y}_i, \ldots, Y_{q-1})$$

$$+\sum_{i<j}(-1)^{i+j}\omega([X_i, X_j], X_0, \ldots, \widehat{X}_i, \ldots, \widehat{X}_j, \ldots, X_p, \mathbf{Y})$$

$$+\sum_{i<j}(-1)^{i+j}\omega([Y_i, Y_j], \mathbf{X}, Y_0, \ldots, \widehat{Y}_i, \ldots, \widehat{Y}_j, \ldots, Y_{q-1})$$

Due to the special choice of extensions of $\mathbf{X}$ and $\mathbf{Y}$ to horizontal $(p+1)$ resp. vertical q-vector fields the horizontal and vertical vector fields commute, i.e. $[X_i, Y_j] = 0$ for all i, j. Thus the sum

$$\sum_{\substack{0\leq i\leq p\\0\leq j\leq q}}(-1)^{i+(j+p+1)}\omega([X_i, Y_j], X_0, \ldots, \widehat{X}_i, \ldots, X_p, Y_1, \ldots, \widehat{Y}_j, \ldots, Y_{q-1})$$

vanishes. Adding zero to the right hand side of the expression for $d_h\omega_{p,q}+d_v\omega_{p+1,q-1}$ yields the equalities

$$(d_h\omega_{p,q} + d_v\omega_{p+1,q-1})(\mathbf{X}, \mathbf{Y}) =$$

$$\sum_{i=0}^{p}(-1)^i X_i.\omega(X_0, \ldots, \widehat{X}_i, \ldots, X_p, \mathbf{Y})$$

$$+\sum_{i=0}^{q-1}(-1)^{p+i+1}Y_i.\omega(\mathbf{X}, Y_0, \ldots, \widehat{Y}_i, \ldots, Y_{q-1})$$

$$+\sum_{i<j}(-1)^{i+j}\omega([X_i, X_j]), X_0, \ldots, \widehat{X}_i, \ldots, \widehat{X}_j, \ldots, X_p, \mathbf{Y})$$

$$+\sum_{i<j}(-1)^{i+j}\omega([Y_i, Y_j], \mathbf{X}, Y_0, \ldots, \widehat{Y}_i, \ldots, \widehat{Y}_j, \ldots, Y_{q-1})$$

$$+\sum_{\substack{0\leq i\leq p\\1\leq j\leq q}}(-1)^{i+(j+p+1)}\omega([X_i, Y_j], X_0, \ldots, \widehat{X}_i, \ldots, X_p, Y_0, \ldots, \widehat{Y}_j, \ldots, Y_{q-1})$$

$$= d_E\omega(\mathbf{X}, \mathbf{Y}) = (d\omega)_{p+1,q}(\mathbf{X}, \mathbf{Y}).$$

Since the n-form ω was arbitrary, the claimed equality follows.   □

Corollary 13.3.14. *If the bundle $p : E \to B$ is flat, then $\sum_{p+q=*}\pi^{p,q}$ is an isomorphism of chain complexes.*

Proof. By Proposition 13.3.7 the maps $\sum_{p+q=n}\pi^{p,q}$ are isomorphisms of topological vector spaces. Therefore we only need to prove that $\sum_{p+q=*}\pi^{p,q}$ is a chain map. Let ω be a differential n-form on E. The differential $D\sum_{p+q=n}\pi^{p,q}$ computes to

$$D\sum_{p+q=n}\pi^{p,q}(\omega) = D\sum_{p+q=n}\omega_{p,q}$$

$$= \sum_{p+q=n}(d_h + d_v)\omega_{p,q}$$

$$= \sum_{p+q=n+1}(d_E\omega)_{p,q} = \sum_{p+q=n+1}\pi^{p,q}(\omega)$$

Thus the map $\sum_{p+q=*} \pi^{p,q}$ is a morphism of chain complexes and therefore an isomorphism of chain complexes.   □

To the double complex $\Omega^{*,*}(E;V)$ one can associate two spectral sequences $_IE_r^{*,*}$ and $_{II}E_r^{*,*}$ which are induced by the column-wise and the row-wise filtrations of the total complex $\text{Tot}\,\Omega^{*,*}(E;V)$ respectively (cf. [McC01, Theorem 2.15]). Concerning these spectral sequences we observe:

Theorem 13.3.15 (The Leray-Serre spectral sequence). *If the spaces F and B are smoothly paracompact, then the spectral sequence $_IE_r^{*,*}$ of the double complex $\Omega^{p,q}(E;V)$ converges to the de Rham cohomology $H_{dR}(E;V)$ of E. If the bundle $p : E \to B$ has the GFEP then both spectral sequences $_IE_r^{*,*}$ and $_IIE_r^{*,*}$ of the double complex $\Omega^{*,*}(E;V)$ converge to the de Rham cohomology $H_{dR}(E;V)$ of E.*

Proof. In case the spaces F and B are smoothly paracompact, the convergence of the spectral sequence $_IE_r^{*,*}$ is a classical result. If the bundle $p : E \to B$ has the GFEP, then the complexes $\text{Tot}(\Omega^{*,*}(E;V))$ and $\Omega^*(E;V)$ are isomorphic. Since both spectral sequences $_IE_r^{*,*}$ and $_IIE_r^{*,*}$ converge to the cohomology of the total complex $\text{Tot}\,\Omega^{*,*}(E;V)$ (cf. [McC01, Theorem 2.15]) the result follows.   □

For finite dimensional manifolds B and F these are classical results. In case of base manifolds B that are not smoothly paracompact this is new as it is for fibres that are not finite dimensional.

13.4 Induced Maps on Tangent Bundles

Let $F \hookrightarrow E \xrightarrow{p} B$ be a bundle of smooth manifolds with flat connection form Φ. We assume further that this bundle has the GFEP and consider a smooth map $f : E \to M$ into a smooth manifold M. In this section we examine the pullback of differential forms by the smooth map f.

Lemma 13.4.1. *The image of a V-valued differential $(p + q)$-form on M under $C^\infty(T^{p,q}f, V)$ is linear and alternating on $HE^{\oplus p}$ and $VE^{\oplus q}$ respectively.*

Proof. Let ω be a $(p+q)$-form on M, $X_1, \ldots, X_p$ be horizontal tangent vectors and $Y_1, \ldots, Y_q$ be vertical tangent vectors at $e \in E$. We denote the corresponding elements in $HE^{\oplus p}$ and $VE^{\oplus q}$ by $\mathbf{X}$ and $\mathbf{Y}$ respectively. The function $C^\infty(T^{p,q}f, V)(\omega)$ is given by

$$C^\infty(T^{p,q}f, V)(\omega)\,(\mathbf{X}, \mathbf{Y})) = \omega\,(T^{p,q}f\,(\mathbf{X}, \mathbf{Y})))$$
$$= \omega\,(Tf(X_1), \ldots, Tf(X_q), Tf(Y_1), \ldots, Tf(Y_q))$$

Therefore the map $C^\infty(T^{p,q}f, V)(\omega)$ is linear and alternating in the arguments $X_1, \ldots, X_p$ and $Y_1, \ldots, Y_q$ respectively.   □

Lemma 13.4.2. *The image of a differential n-form $\omega \in \Omega^n(M;V)$ on M under $\alpha_n \circ C^\infty(\pi_{p,q} \circ T^{p,q}f, V)$ is the n-form $(f^*\omega)_{p,q}$ of type (p,q) on E (for $n = p + q$).*

Definition 13.4.3. *The corestriction of $\alpha_{p+q} \circ C^\infty(\pi_{p,q} \circ T^{p,q}f, V)$ to the subspace $\Omega^{p,q}(E;V)$ will be denoted by $f^{p,q}$.*

The map $f^{p,q}$ maps each n-form ω on M to the associated n-form $(f^*\omega)_{p,q}$ of type (p,q) of $f^*\omega$. The generality is motivated by the following examples.

Example 13.4.4. Let G be a Lie group, M be a manifold and $\varrho : G \times M \to M$ be a smooth action of G on M. Then one cannot use the usual spectral sequences to connect the cohomologies of G and M for which one uses smooth paracompactness and the existence of a good cover of M. On the other hand one wants at least to generalise the flux homomorphism of symplectic topology, where $G = \mathrm{Symp}(M, \omega)_0$, relating the cohomology of M with paths in G.

The maps $f^{p,q} : \Omega^n(M; V) \to \Omega^{p,q}(E; V)$ for $p + q = n$ can be added to obtain a linear map $\sum_{p+q=n} f^{p,q} : \Omega^n(M; V) \to \mathrm{Tot}(\Omega^{*,*}(E; V))^n$. These linear maps in turn form a homomorphism between the complexes $\Omega^*(M; V)$ and $\mathrm{Tot}(\Omega^{*,*}(E; V))$:

Lemma 13.4.5. *The map $\sum_{p+q=*} f^{p,q}$ is a morphism of chain complexes.*

Proof. The map $\sum_{p+q=*} f^{p,q}$ is a composition of the morphisms $\sum_{p+q=*} (\pi^{p,q})$ and f^* of chain complexes and thus a morphism of chain complexes as well. □

13.5 Flat Bundles

In this section we utilise an exponential map for the de Rham spaces $\Omega^{p,q}(E; V)$ of flat bundles. This is done using a double complex $\Omega^*(B, \Omega^*(F; V))$, where $\Omega^*(F; V)$ denotes the de Rham complex of the fibre. The spectral sequence associated to this double complex will later be used to define a generalised flux homomorphism. We first restrict ourselves to trivial bundles an then generalise the results to flat bundles.

Let $p : E = B \times F \to B$ be a trivial bundle. Because the tangent functor preserves products, the product structure of E further transfers to the tangent bundle $TE{:}TE \cong TB \times TF$. Identifying the manifold B with its image under the zero section $B \to TB$, we regard B as a closed submanifold of TB etc.. The vertical bundle VE then is the submanifold $B \times TF$ of E and the horizontal bundle is the submanifold $TB \times F$ of E. So far, we have obtained the following decomposition of the tangent bundle TE:

$$TE \cong TB \times TF \cong (TB \times F) \oplus (B \times TF)$$

There is a canonical connection form Φ on E which is the projection onto the second summand in the above decomposition. A similar decomposition can be obtained for the spaces $T^{p,q}E$.

Lemma 13.5.1. *If $B \times F \to B$ is a trivial bundle then the spaces $T^{p,q}E$ decompose into the Cartesian products $(T^{\oplus p}B) \times (T^{\oplus q}F)$.*

Proof. This is an immediate consequence of the above decomposition of the tangent bundle TE:

$$\begin{aligned}
T^{p,q}E &= HE^{\oplus p} \oplus VE^{\oplus q} \\
&= (TB \times F)^{\oplus p} \oplus (B \times TF)^{\oplus q} \\
&\cong (T^{\oplus p}B \times F) \oplus (B^{\oplus q} \times TF) \\
&\cong (T^{\oplus p}B) \times (T^{\oplus q}F)
\end{aligned}$$

□

As a consequence one can use one half of the "exponential law" for the function spaces $C^\infty(T^{p,q}E, V)$. We denote the function which assigns to each smooth function $g \in C^\infty(T^{p,q}E, V)$ its adjoint $\hat{g} \in C^\infty(T^pB, C^\infty(T^qF, V))$ by $\varphi^{p,q}$:

$$\varphi^{p,q} : C^\infty(T^{p,q}E, V) = C^\infty(T^{\oplus p}B \times T^{\oplus q}F, V) \to C^\infty(T^{\oplus p}B, C^\infty(T^{\oplus q}F, V))$$

Because the de Rham spaces $\Omega^{p,q}(E;V)$ are closed subspaces of these function spaces, one has an exponential map $\varphi^{p,q} : \Omega^{p,q}(E;V) \to C^\infty(T^{\oplus p}B, C^\infty(T^{\oplus q}F, V))$. As one expects, the image of the de Rham spaces under the maps $\varphi^{p,q}$ are actually contained in subspaces of differential forms:

Lemma 13.5.2. *The image of the de Rham space $\Omega^{p,q}(E;V)$ under the exponential map $\varphi^{p,q}$ is contained in $\Omega^p(B; \Omega^q(F;V))$.*

Proof. Let $\omega_{p,q}$ be a differential form of type (p,q) on E, i.e. a smooth multilinear function $\omega_{p,q} : (T^{\oplus p}B) \times (T^{\oplus q}F) \to V$ that is alternating in $T^{\oplus p}B$ and $T^{\oplus q}F$ respectively. For any element $\mathbf{X} \in HE^{\oplus p}$ the function $\varphi^{p,q}(\omega_{p,q})(\mathbf{X}) = i_\mathbf{X}\omega_{p,q}$ is multilinear and alternating in $T^{\oplus q}F$. Therefore $\varphi^{p,q}(\omega)$ takes values in $\Omega^q(F;V)$. Furthermore $\varphi^{p,q}(\omega_{p,q})$ is multilinear and alternating in $T^{\oplus p}B$ since $\omega_{p,q}$ is. Hence $\varphi^{p,q}(\omega_{p,q})$ is a differential p-form on B with values in $\Omega^q(F;V)$. $\square$

Definition 13.5.3. *The de Rham spaces $\Omega^p(\Omega^q(F;V))$ are denoted by $\hat{\Omega}^{p,q}(E;V)$ and the symbol for a differential form $\hat{\omega}_{p,q}$ in $\Omega^p(\Omega^q(F;V))$ is always written with a check on it.*

Using this notation, the function $\varphi^{p,q}$ assigns to a differential n-form of type (p,q) its adjoint differential p-form $\hat{\omega}_{p,q}$ on B with values in $\Omega^q(F;V)$. The exterior derivative d on $\Omega^*(B; \Omega^q(F;V))$ is a horizontal differential d_h on $\hat{\Omega}^{*,q}(E;V)$ for all q. Likewise, the exterior derivative on $\Omega^q(F;V)$ induces a vertical differential d_v on $\hat{\Omega}^{p,q}(E;V)$ via

$$(d_v\hat{\omega}_{p,q})(\mathbf{X}) := (-1)^p d(\hat{\omega}_{p,q}(\mathbf{X})), \quad \text{for all } \mathbf{X} \in T^{\oplus p}B.$$

As in the case of the double complex $\Omega^{*,*}(E;V)$ these differentials anticommute.

Proposition 13.5.4. *The horizontal and vertical differentials on $\hat{\Omega}^{*,*}(E;V)$ anticommute, i.e. $d_h d_v + d_v d_h = 0$.*

Proof. The proof is essentially the same as the proof of proposition 13.3.10. One only has to make some slight changes. Let $\hat{\omega}_{p,q}$ be a differential p-form in $\hat{\Omega}^{p,q}(E;V)$, $\mathbf{X} = (X_0, \ldots, X_p) \in T^{\oplus p+1}B \times F$ be horizontal at $e \in E$ and $\mathbf{Y} = (Y_0, \ldots, Y_q) \in B \times T^{\oplus q+1}F$ vertical. We extend $\mathbf{X}$ and $\mathbf{Y}$ to local horizontal $(p+1)$- and vertical $(q+1)$-vector fields respectively. The evaluation of $(d_h d_v \hat{\omega}_{p,q})(\mathbf{X})$ at $\mathbf{Y}$ is given by

$$(d_h d_v \omega_{p,q})(\mathbf{X})(\mathbf{Y}) =$$
$$\sum_i (-1)^i X_i.(d_v\hat{\omega}_{p,q})(X_0, \ldots, \hat{X}_i, \ldots, X_p)(\mathbf{Y})$$
$$+ \sum_{i<j} (-1)^{i+j}(d_v\hat{\omega}_{p,q})([X_i,X_j], X_0, \ldots, \hat{X}_i, \ldots, \hat{X}_j, \ldots, X_p)(\mathbf{Y})$$
$$= \sum_{i,k} (-1)^{p+i+k} X_i.(Y_k.\hat{\omega}_{p,q}(X_0, \ldots, \hat{X}_i, \ldots, X_p, Y_0, \ldots, \hat{Y}_k, \ldots, X_q))$$
$$+ \sum_{\substack{i \\ k<l}} (-1)^{p+i+k+l} X_i.\hat{\omega}_{p,q}(X_0, \ldots, \hat{X}_i, \ldots, X_p)([Y_k,Y_l], Y_0, \ldots, \hat{Y}_k, \ldots, \hat{Y}_l, \ldots, X_q))$$
$$+ \sum_{\substack{i<j \\ k}} (-1)^{p+i+j+k} Y_k.\hat{\omega}_{p,q}([X_i,X_j], X_0, \ldots, \hat{X}_i, \ldots, \hat{X}_j, \ldots, X_p)(\mathbf{Y})$$
$$+ \sum_{\substack{i<j \\ k<l}} (-1)^{p+i+j+k+l} \hat{\omega}_{p,q}([X_i,X_j], X_0, \ldots, \hat{X}_i, \ldots, \hat{X}_j, \ldots, X_p)([Y_k,Y_l], Y_0, \ldots, \hat{Y}_k, \ldots, \hat{Y}_l, \ldots, X_q).$$

Likewise, the evaluation of $d_v d_h \hat{\omega}_{p,q}(\mathbf{X})$ at $(\mathbf{Y})$ is given by

$$(d_v d_h \hat\omega_{p,q})(\mathbf{X})(\mathbf{Y}) =$$

$$\sum_k (-1)^{p+1+k} Y_k.(d_h\hat\omega_{p,q}(\mathbf{X})(Y_0,\ldots,\hat{Y}_k,\ldots,Y_q))$$

$$+ \sum_{k<l} (-1)^{p+1+k+l}(d_h\hat\omega_{p,q}(\mathbf{X})([Y_k,Y_l],Y_0,\ldots,\hat{Y}_k,\ldots,\hat{Y}_l,\ldots,Y_q)$$

$$= \sum_{i,k} (-1)^{p+1+i+k} X_i.(Y_k.\hat\omega_{p,q}(X_0,\ldots,\hat{X}_i,\ldots,X_p)(Y_0,\ldots,\hat{Y}_k,\ldots,X_q))$$

$$+ \sum_{\substack{k \\ i<j}} (-1)^{p+1+i+j+k} Y_k.\hat\omega_{p,q}([X_i,X_j],X_0,\ldots,\hat{X}_i,\ldots,\hat{X}_j,\ldots,X_p)(\mathbf{Y})$$

$$+ \sum_{\substack{k<l \\ i}} (-1)^{p+i+k+l} X_i.\hat\omega_{p,q}(X_0,\ldots,\hat{X}_i,\ldots,X_p)([Y_k,Y_l],Y_0,\ldots,\hat{Y}_k,\ldots,\hat{Y}_l,\ldots,X_q))$$

$$+ \sum_{\substack{k<l \\ i<j}} (-1)^{p+i+j+k+l} \hat\omega_{p,q}([X_i,X_j],X_0,\ldots,\hat{X}_i,\ldots,\hat{X}_j,\ldots,X_p)([Y_k,Y_l],Y_0,\ldots,\hat{Y}_k,\ldots,\hat{Y}_l,\ldots,X_q).$$

These two expressions differ only in sign, hence $d_h d_v + d_v d_h = 0$.   □

Corollary 13.5.5. *The spaces $\hat\Omega^{*,*}(E;V)$ form a double complex with horizontal differentials d_h and vertical differentials d_v.*

The differential D of the total complex $\mathrm{Tot}\,\hat\Omega^{*,*}(E;V)$ is therefore given by the sum

$$D = \sum_{p+q=n} d_h^{p,q} + d_v^{p,q} : \bigoplus_{p+q=n} \Omega^{p,q}(E;V) \longrightarrow \bigoplus_{p+q=n+1} \Omega^{p,q}(E;V).$$

The spectral sequences induced by the column-wise and the row-wise filtrations of the total complex $\mathrm{Tot}\,\hat\Omega^{*,*}(E;V)$ will be denoted by $_I\hat E_r^{*,*}$ and $_{II}\hat E_r^{*,*}$ respectively.

Proposition 13.5.6. *The functions $\varphi^{p,q} : \Omega^{p,q}(E,V) \to \Omega^p(B,\Omega^q(F,V))$ form a morphism of double complexes $\varphi^{*,*} : \Omega^{*,*} \to \hat\Omega^{*,*}$.*

Proof. The maps $\varphi^{p,q}$ intertwine the vertical and horizontal differentials by definition.   □

Corollary 13.5.7. *The map $\mathrm{Tot}(\varphi^{*,*}) : \mathrm{Tot}\,\Omega^{*,*}(E;V) \to \mathrm{Tot}\,\hat\Omega^{*,*}(E;V)$ between the total complexes of $\Omega^{*,*}(E;V)$ and $\hat\Omega^{*,*}(E:V)$ is a morphism of chain complexes.*

Corollary 13.5.8. *The morphism $\mathrm{Tot}(\varphi^{*,*})$ induces a homomorphism of spectral sequences $_I E_r^{*,*} \to_I \hat E_r^{*,*}$.*

If the exponential law holds for the function $\varphi^{p,q} : \Omega^{p,q}(E;V) \to \Omega^p(B;\Omega^q(F;V))$ then the latter is an continuous linear bijection of topological vector spaces. This also happens if F is locally compact (cf. Proposition 13.1.11) or if all finite products of the modelling spaces V_1 of B and V_2 of F and the ground field $\mathbb{K}$ are k-spaces (cf. Corollary 13.1.23). These examples include the large class of metrisable manifolds, e.g. the spaces $C^\infty(M,N)$ of smooth mappings between finite dimensional manifolds M and N. In these cases one can work with the double complexes $\Omega^{*,*}(E;V)$ and $\Omega^*(B;\Omega^*(F;V))$ interchangeably:

Proposition 13.5.9. *If F is locally compact or all finite products of the modelling spaces V_1 of B and V_2 of F and the ground field are k-spaces then the function $\varphi^{*,*}$ is an isomorphism of topological double complexes and $\mathrm{Tot}(\varphi^{*,*})$ is an isomorphism of topological chain complexes.*

Let $f : E \to M$ be a smooth map. If ω is an n-form on M then the pullback form $f^*\omega$ can be decomposed into the sum $\sum_{p+q=n} f^{p,q}\omega$. The function $\varphi^{p,q}$ maps a form $f^{p,q}\omega$ to a p-form $\widehat{f^{p,q}\omega}$ on on B. In this fashion the total map $\mathrm{Tot}(\varphi^{*,*})$ can be applied after the complex homomorphism $\sum_{p+q=*} f^{p,q}$ to obtain a morphism $\Omega^*(M;V) \to \mathrm{Tot}(\Omega^*(B;\Omega^*(F;V)))$ of chain complexes. This morphism of chain complexes will later enable us to define the generalised flux homomorphism.

Lemma 13.5.10. *If the sequence $B_{dR}^*(F;V) \hookrightarrow \Omega^*(F;V) \to H_{dR}^*(F;V)$ splits topologically, then the first stage terms of the spectral sequence $_I\hat{E}_r^{*,*}$ are given by $_I\hat{E}_1^{p,q} = \Omega^p(B, H_{dR}^q(F;V))$ and the second stage terms are given by $\hat{E}_2^{p,q} = H_{dR}^p(B, H_{dR}^q(F;V))$.*

This can be generalised to flat bundles $E \to B$ with fibre F provided that the short exact sequences $B_{dR}^n(F;V) \hookrightarrow Z_{dR}^n(F;V) \twoheadrightarrow H_{dR}^n(F;V)$ split topologically.

Consider a smooth locally trivial bundle $p : E \to B$ with fibre F and two bundle charts $\varphi_i : p^{-1}(U_i) \to U_i \times F$ and $\varphi_j : p^{-1}(U_j) \to U_j \times F$. The change of coordinates is given by

$$\varphi_i \circ \varphi_j^{-1} : U_{ij} \times F \to U_{ij} \times F, \quad (b,f) \mapsto (b, \varphi_{ij}(b)(f)),$$

where φ_{ij} is a transition function $\varphi_{ij} : U_{ij} \to \mathrm{Diff}(F)$.

Definition 13.5.11. *A (locally trivial) fibre bundle is called* flat, *if the transition functions $\varphi_{ij} : U_{ij} \to \mathrm{Diff}(F)$ are constant.*

The advantage of flat bundles $E \to B$ is the fact that the fundamental group $\pi_1(B)$ of the base manifold acts on the fibre via a homomorphism $\pi_1(B) \to \mathrm{Diff}(F)$ and the deRham cohomology spaces $H_{dR}(F_b;V)$ of the fibres F_b over the points $b \in B$ become a smooth system of local coefficients. Recall that such a system of local coefficients is called *simple*, if the action of $\pi_1(B)$ on $H_{dR}(F_b;V)$ is trivial. We note the consequence:

Lemma 13.5.12. *If the bundle $E \to B$ is flat and the system $H_{dR}(F_b;V)$ of local coefficients is simple, then the change of coordinates induces an isomorphism in $\Omega^p(U_{ij}, \Omega^*(F;V))$ which descents to the identity on $\Omega^p(U_{ij}, H_{dR}^*(F;V))$.*

Example 13.5.13. If $E \to B$ is a flat bundle over a simply connected base manifold B, then the the system $H_{dR}(F_b;V)$ of local coefficients is simple.

Example 13.5.14. Any diffeomorphism in the identity component $\mathrm{Diff}(F)_0$ induces the identity in the de Rham cohomology $H_{dR}^*(F;V)$ of F. Thus, if the transition functions φ_{ij} take values in $\mathrm{Diff}(F)_0$ only, the action of $\pi_1(B)$ on the spaces $H_{dR}(F_b;V)$ is trivial.

Example 13.5.15. If $E \to B$ is a locally trivial smooth principal bundle whose fibre is a connected Lie group G and the transition functions φ_{ij} are constant, then the action of $\pi_1(B)$ on the spaces $H_{dR}(F_b;V)$ is trivial.

If $E \to B$ is a flat bundle with fibre F and the smooth system $H_{dR}(F_b;V)$ of local coefficients is trivial, then the p-forms $\hat{\omega}_{p,q} \in \hat{\Omega}^p(U_i; H_{dR}^q(F;V))$ on the open sets U_i with values in the de Rham cohomology $H_{dR}^q(F;V)$ of the fibre can be glued together to obtain a p-form in $\Omega^p(B; H_{dR}^q(F;V))$. Provided that the sequences

$$B_{dR}^n(F;V) \hookrightarrow Z_{dR}^n(F;V) \twoheadrightarrow H_{dR}^n(F;V) \tag{13.2}$$

split topologically, this implies that the spectral sequence $_I\hat{E}_r^{*,*}$ for the bundle $E \to B$ exists from the first stage on:

Lemma 13.5.16. *If the bundle $E \to B$ with fibre F is flat, the system $H_{dR}(F_b; V)$ of coefficients is trivial and the exact sequences $B^n(F; V) \hookrightarrow \Omega^n(F; V) \to H^n_{dR}(F; V)$ split topologically, then the spectral sequence $_I\hat{E}^{*,*}_r$ exists from the first stage on and the second stage terms of $\hat{E}^{*,*}_r$ are given by $E^{p,q}_2 = H^p_{dR}(B, H^q_{dR}(F; V))$.*

Proof. If the short exact the sequences $B^n(F; V) \hookrightarrow \Omega^n(F; V) \to H^n_{dR}(F; V)$ split topologically, the spectral sequence $_I\hat{E}^{*,*}_r$ for the bundles $p^{-1}(U_i) \to U_i$ exist and have first stage term $_I\hat{E}^{p,q}_1 = \Omega^p(U_i, H^q_{dR}(F; V))$. If the bundle $E \to B$ is flat and the system $H_{dR}(F_b; V)$ of local coefficients is trivial, then these sequences can be glued from the first stage on to obtain the spectral sequence $_I\hat{E}^{*,*}_r$, $r \geq 1$ for the bundle $E \to B$. $\qquad\square$

The morphism $\sum_{p+q=*} \hat{f}^{p,q}$ of chain complexes will allow us to define a generalised version of the flux homomorphism of symplectic topology. This will be done in the next section.

13.6 The Flux Homomorphism

In this section we examine the effect of the homomorphism $\sum_{p+q=*} f^{p,q}$ in cohomology. We recall that for any manifold E and Mackey complete topological vector space V there exists a pairing $k : S_{\infty,p}(E) \times \Omega^p(E; V) \to V$, which is given by integration of p-forms over smooth p-simplices:

$$k(\sigma, \omega) = \int_\sigma \omega \qquad \text{for } \omega \in \Omega_{dR}(E; V), \ \sigma \in S_{\infty,p}(E)$$

This pairing induces the de Rham homomorphisms $k^* : \Omega^*(E; V) \to S^*_\infty(E; V)$. Combined with the chain map $\sum_{p+q=*} f^{p,q}$ these homomorphisms will enable us to define a general notion of the flux homomorphism. We first restrict ourselves to trivial bundles and then generalise the results to flat bundles.

Let $F \hookrightarrow E = B \times F \to B$ be a trivial bundle. Composing the de Rham homomorphism for $\Omega^*(B; \Omega^q(F; V))$ with the functions $\varphi^{*,q}$ yields homomorphisms

$$k^p \circ \varphi^{p,q} : \Omega^{p,q}(B \times F; V) \to S^p_\infty(B; \Omega^q(F; V))$$

and a chain map $\Omega^{*,*}(E; V) \to S^*_\infty(B; \Omega^*(F; V))$. Let $f : E \to M$ be a smooth map. The morphism $\sum_{p+q=*} f^{p,q} : \Omega^*(M; V) \to \text{Tot}\Omega^{*,*}(E; V)$ of chain complexes can be composed with the morphism $\text{Tot}(k^*\varphi^{*,*})$:

Definition 13.6.1. *The morphism $\mathcal{F}_f : \Omega^*(M; V) \to \text{Tot}(S^*_\infty(B; \Omega^*(F; V)))$ of chain complexes given by $\mathcal{F}_f := k^* \circ \text{Tot}(\varphi^{*,*}) \circ \sum_{p+q=*} f^{p,q}$ is called the* generalised flux homomorphism *for f.*

The (p, q)-component of $\mathcal{F}_f$ will be denoted by $\mathcal{F}^{p,q}_f$. The name flux homomorphism is justified by the following observations applied to the special case of the action of $\text{Symp}(M)_0$ on a compact manifold M. For a fixed differential n-form ω on M the maps $\mathcal{F}^{p,q}_f(\omega)$ are homomorphisms between the singular groups $S_p(B)$ and the de Rham spaces $\Omega^q(F; V)$. These maps almost commute with the differentials on both complexes.

Lemma 13.6.2. *For any closed n-form ω on M the map $\mathcal{F}^{p,q}_f(\omega)$ satisfies*

$$d_h \mathcal{F}^{p,q}_f(\omega) = -d_v \left(\mathcal{F}^{p+1,n-p+1}_f(\omega) \right),$$

i.e. the maps $\mathcal{F}^{p,q}(\omega) : S_{\infty,p}(B) \to \Omega^{n-p}(F; V)$ commute with differentials up to sign.

Proof. Let ω be a closed n-form on M. Because ω is closed and $\mathcal{F}_f$ is a chain map we have $D\mathcal{F}_f(\omega) = \mathcal{F}_f(d\omega) = 0$. Therefore the $(p+1, n-p)$ component of $D\mathcal{F}_f(\omega)$ which is given by

$$D\mathcal{F}_f(\omega)^{p+1,n-p} = d_h \mathcal{F}_f^{p,n-p}(\omega) + d_v \mathcal{F}^{p+1,n-p-1}(\omega)$$

vanishes (cf. Proposition 13.3.13). Hence the stated equality follows.   □

The preceding lemma implies that the maps $\mathcal{F}_f^{p,n-p}(\omega)$ map cycles to cocycles and boundaries to boundaries. Hence they induce a homomorphism from the singular homology of the base B to the de Rham cohomology of the fibre F.

Corollary 13.6.3. *For any closed n-form ω on M the maps $\mathcal{F}_f^{p,q}(\omega)$ induce homomorphisms $\mathcal{F}_f^{p,q} : H_p(B) \to H_{dR}^{n-p}(F;V)$ from the homology of B into the de Rham-cohomology of the fibre F.*

An analogous result is obtained by fixing a singular cycle $z \in Z_p(B)$ and varying the differential form ω on M:

Lemma 13.6.4. *For any singular p-cycle $z \in S_{\infty,p}(B)$ in B the evaluation map $\mathrm{ev}_z \circ \mathcal{F}_f^{p,*} : \Omega^*(M;V) \to \Omega^{*-p}(F;V)$ is a chain map up to sign.*

Proof. Let $z \in S_{\infty,p}(B)$ be a cycle. Because $\mathcal{F}_f^{*,*}$ is chain map we know that

$$\left[d_h \mathcal{F}_f^{p-1,n-p+1}(\omega) + d_v \mathcal{F}_f^{p,n-p}(\omega) \right](z) = \left[\mathcal{F}_f^{p,n+1-p}(d\omega) \right](z)$$

for any n-form ω on M. Since z is a cycle the first summand on the left hand side vanishes. Hence the desired equality

$$\mathcal{F}^{p,n+1-p}(d\omega)(z) = d_v \mathcal{F}^{p,n+1-p}(\omega)(z) = \pm d_v \mathcal{F}^{p,n-p}(\omega)(z)$$

follows.   □

Lemma 13.6.5. *The evaluation $\mathrm{ev} \circ \mathcal{F}_f^{p,n-p} : S_{\infty,p}(B) \times \Omega^n(M;V) \to \Omega^{n-p}(F;V)$ is given by*

$$[\mathcal{F}_f^{p,n-p}(\omega)(\sigma)](\mathbf{Y}) = \int_{\Delta^p} \sigma^*\left[f^*\omega(-, \mathbf{Y}) \right]$$

for any p-simplex $\sigma \in S_{\infty,p}(B)$ and vertical tangent $(n-p)$-vector $\mathbf{Y} \in T^{\oplus n-p}F$.

Proof. Let $\sigma \in S_{\infty,p}(B)$ be a p-simplex in B and $\mathbf{Y}$ be an element in $T^{\oplus(n-p)}F$. The evaluation of $\mathcal{F}_f^{p,q}(\omega)(\sigma)$ at $\mathbf{Y}$ is given by

$$\left[\mathcal{F}_f^{p,q}(\omega)(\sigma) \right](\mathbf{Y}) = \left[\int_{\Delta^p} \sigma^* f^{p,q}(\omega) \right](\mathbf{Y}) = \int_{\Delta^p} [\sigma^* f^{p,q}(\omega)(\mathbf{Y})]$$

$$= \int_{\Delta^p} \sigma^* [(f^*\omega)(-, \mathbf{Y})]$$

This is the claimed equality.   □

Corollary 13.6.6. *If (G, M) is a smooth transformation group, $E = G \times M$ and $f = \varrho : G \times M \to M$ the smooth Lie group action on the manifold M then the map $\mathrm{ev} \circ \mathcal{F}_f^{p,n-p} : S_{\infty,p}(G) \times \Omega^n(M;V) \to \Omega^{n-p}(M;V)$ is given by*

$$[\mathrm{ev} \circ \mathcal{F}_f^{p,n-p}](\omega)(\sigma)(\mathbf{Y}) = \int_{\Delta^p} \sigma^* [\varrho^*\omega(-, \mathbf{Y})]$$

Corollary 13.6.7. *The integration of the $(n{-}p)$-form $\mathcal{F}_f^{p,n-p}(\omega)(\sigma)$ over an $(n{-}p)$-simplex τ is given by*

$$\int_\tau \mathcal{F}_f^{p,n-p}(\omega)(\sigma) = \int_{f\circ(\sigma\times\tau)} \omega. \qquad (13.3)$$

In the special case that M is a compact symplectic manifold with symplectic form ω and $G = \mathrm{Symp}(M,\omega)_0$ this reduces to the flux homomorphism of symplectic topology. Here the homomorphism $\mathrm{ev} \circ \mathcal{F}_f^{1,1}(\omega) : \pi_1(G) = H_1(G) \to H_{dR}^1(M;V)$ is given by

$$\left[\mathrm{ev} \circ \mathcal{F}_f^{1,1}(\omega)\right]([\gamma])([\tau]) = \int_{\gamma\cdot\tau} \omega$$

for representatives $\gamma \in \pi_1(G)$, $\tau \in \pi_1(M)$. Note however, that the above definition of the Flux does neither rely on topological properties of M nor on the non-degeneracy of the form ω. Furthermore the definition given here results in homomorphisms on the (co)chain level, whereas the usual definition of the flux only gives a homomorphism on the (co)homology level.

Example 13.6.8. Let G be a Lie group. Then G acts on itself by left translation and the action $\varrho : G \times G \to G$ is actually the group multiplication and a trivial fibre bundle. In this case formula 13.3 simplifies to

$$\int_\tau \mathcal{F}_\varrho^{p,n-p}(\omega)(\sigma) = \int_{\sigma\cdot\tau} \omega$$

for any n-form ω on G, any smooth simplex $\sigma \in S_{p,\infty}(G)$ and any smooth simplex $\tau \in S_{n-p}(G)$. The domain of integration $\sigma \cdot \tau$ is the point-wise product of σ and τ in G.

13.7 Equivariance of $\sum_{p+q=*} f^{p,q}$ and the Flux

In this section we consider manifolds which are acted upon by a Lie group G. Let E be a smooth G-bundle admitting a G-equivariant connection form Φ. The equivariance of Φ implies the invariance of the subbundles $T^{p,q}E$ under G. In this section we examine the effect of the equivariance of maps $f : E \to M$ into smooth G-manifolds M.

Let $F \hookrightarrow E \to B$ be a smooth G-bundle with G-equivariant connection form Φ. All the bundles and projections introduced in the previous sections are equivariant. This is due to the fact that the image of an G-equivariant map $f : X \to Y$ between G-spaces X and Y is a G-invariant subset of Y.

Lemma 13.7.1. *The vertical and horizontal bundles VE and HE are G-invariant subbundles of TE the projections onto which are equivariant.*

Proof. The projection onto VE is the connection form Φ which was assumed to be equivariant. Thus the image VE of Φ is a G-invariant subbundle. Likewise the projection onto HE is given by the equivariant map $\mathrm{id}_{TE} - \Phi$. Hence the subbundle HE is equivariant as well. $\square$

Because the group G acts fibre-wise on the bundles $T^{\oplus n}E$ the functions $T^{\oplus n}f$ are equivariant if the function f is equivariant. The same can be concluded for almost all maps introduced in the previous sections.

Lemma 13.7.2. *The subbundles $T^{p,q}E$ of $T^{\oplus n}E$ are G-invariant and the projections $\pi_{p,q} : T^{\oplus n}E \to T^{p,q}E$ are equivariant.*

Proof. Since the projections $\pi_{p,q}$ are onto it suffices to prove the equivariance of these. We denote by $\chi = \mathrm{id}_{TE} - \Phi$ the projection onto the horizontal subbundle HE. Recalling the definitions of $T^{p,q}E$ and $\pi_{p,q}$ one sees that $\pi_{p,q}$ is given by

$$\pi_{p,q} : T^{\oplus n}E = T^{\oplus p}E \oplus T^{\oplus q}E \to HE^{\oplus p} \oplus VE^{\oplus q} = T^{p,q}E$$
$$(X_1, \ldots, X_n) \mapsto (\chi(X_1), \ldots, \chi(X_p), \Phi(X_{p+1}), \ldots, \Phi(X_n)).$$

Because the group G acts fibre-wise on TE and the projections Φ and χ are equivariant, the projections $\pi_{p,q}$ are equivariant as well. $\square$

Please recall that the action of G on TE induces an action of G on the G spaces $T^{p,q}E$ and on $T^{\oplus n}E$. The last space is also acted upon by the permutation group S_n. We consider the permutation group S_n to be a discrete trivial G-space.

Lemma 13.7.3. *The action $S_n \times (T^{\oplus n}E) \to T^{\oplus n}E$ is equivariant.*

Proof. Let $\mathbf{X} \in T^{\oplus n}E$ be given and $s \in S_n$ a permutation. Because the group G acts fibre-wise on TE and trivial on S_n we have

$$(g.s).(g.\mathbf{X}) = s.(g.\mathbf{X}) = s.(g.X_1, \ldots, g.X_n) = (g.X_{s(1)}, \ldots, g.X_{s(n)}) = g.(s.\mathbf{X})$$

for all $g \in G$. Thus the action of S_n on $T^{\oplus n}E$ is equivariant. $\square$

The equivariance of a map $f : M \to N$ between G-manifolds M and N implies the equivariance of the map $C^\infty(Tf, V)$ for any G vector space V. As a consequence we note the next lemma:

Lemma 13.7.4. *The induced action $S_n \times C^\infty(T^{\oplus n}E, V) \to C^\infty(T^{\oplus n}E, V)$ is equivariant.*

This especially implies that the antisymmetrising maps α_n are equivariant. Let M be a G manifold and $f : E \to M$ be a smooth equivariant map. The equivariance of f carries over to the various maps induced on the tangent spaces.

Lemma 13.7.5. *If $f : E \to M$ is equivariant, then $T^{p,q}f$ is equivariant.*

Proof. The maps $i_{p,q}$ and Tf are equivariant. Since S_n acts fibre-wise on the bundle $T^{\oplus n}E$ the map $T^{\oplus n}f$ is equivariant. Therefore the maps $T^{p,q}f = T^{\oplus n}f \circ i_{p,q}$ are equivariant as well. $\square$

The implications go even further:

Proposition 13.7.6. *If $f : E \to M$ is an equivariant smooth function, then the morphism $\sum_{p+q=*} f^{p,q} : \Omega^*(M; V) \to \mathrm{Tot}\,\Omega^{*,*}(E; V)$ of chain complexes is equivariant.*

Proof. The maps $f^{p,q}$ are the compositions of the equivariant maps α_n, $C^\infty(T^{p,q}f, V)$ and $C^\infty(\pi_{p,q}, V)$ and thus are equivariant. $\square$

Example 13.7.7. Let $\varrho : G \times M \to M$ be a Lie group action of the Lie group G on the smooth manifold M. If we consider the adjoint action of G on itself, the the action ϱ is equivariant:

$$\varrho(g.(g', m)) = \varrho(gg'g^{-1}, g.m) = gg'g^{-1}g.m = gg'.m = g.\varrho(g', n)$$

Example 13.7.8. Let G be a Lie group and $\mu : G \times G \to G$ be the group multiplication, i.e. the action by left translation. If we consider the adjoint action of G on itself for the first factor and the action by left translation for the second factor G then the multiplication μ is equivariant.

Lemma 13.7.9. *If $\omega \in \Omega^n(M; V)$ is an equivariant n-form, then the forms $f^{p,n-p}(\omega)$ are equivariant as well.*

Proof. Let $\omega \in \Omega^n(M; V)$ be an equivariant n-form. For any $g \in G$ the equivariance of $f^{p,n-p}$ results in

$$g.(f^{p,n-p}(\omega)) = f^{p,n-p}(g.\omega) = f^{p,n-p}(\omega).$$

Thus the forms $f^{p,n-p}(\omega)$ are equivariant. □

Lemma 13.7.10. *The morphism $\varphi^{*,*} : \Omega^{*,*}(B \times F; V) \to \Omega^*(B, \Omega^*(F; V))$ of double complexes is equivariant.*

Proof. Let $\omega \in \Omega^{p,q}(B \times F; V)$ be an n-form of type (p,q). For a horizontal tangent p-vector $\mathbf{X} = (X_1, \ldots, X_p) \in T^{\oplus p}B$ and a vertical tangent q-vector $\mathbf{Y} = (Y_1, \ldots, Y_q) \in T^{\oplus q}F$ we denote with $(\mathbf{X}, \mathbf{Y})$ the element $i_{\ell,n}(\mathbf{X}) + i_{r,n}(\mathbf{Y})$ in $T^{p,q}E$. If $g \in G$ is an arbitrary group element we have

$$(g.\hat{\omega}_{p,q})\,(\mathbf{X}) = g.\left[\hat{\omega}_{p,q}(g^{-1}\mathbf{X})\right]$$

By the definition of the action of G on $\Omega^q(M; V)$ we have

$$\begin{aligned}
g.\left[\hat{\omega}_{p,q}(g^{-1}\mathbf{X})\right](\mathbf{Y}) &= g.\left[\hat{\omega}_{p,q}(g^{-1}\mathbf{X})(g^{-1}\mathbf{Y})\right] \\
&= g.\left[\omega_{p,q}(g^{-1}\mathbf{X}, g^{-1}\mathbf{Y})\right] \\
&= (g.\omega)_{p,q}\,(\mathbf{X}, \mathbf{Y}).
\end{aligned}$$

Thus the map $\varphi^{p,q} : \Omega^{p,q}(E, V) \to \hat{\Omega}^{p,q}(E, V)$ is equivariant. □

Corollary 13.7.11. *If $f : E \to M$ is equivariant, then the flux $\mathcal{F}_f$ for f is equivariant.*

We now consider the special case $E = M \times G$, where (G, M) is a smooth transformation group with action $\varrho : G \times M \to M$. Let $t : M \times G \to G \times M$ be the coordinate flip $(m, g) \mapsto (g, m)$ and consider the smooth map

$$f = \varrho t : M \times G \to M, \quad (m, g) \mapsto g.m.$$

If we take the first factor M of the product to be a trivial G-space and equip the second factor G with the action by left translation, then the map $f = \varrho t$ is equivariant. In this case, the restriction of the flux to the spaces $\Omega^*(M; V)^G$ of equivariant forms on M can be corestricted to a G-invariant subcomplex of $\hat{\Omega}^{*,*}(E; V)$.

Lemma 13.7.12. *The morphism $\mathrm{Tot}(\varphi^{*,*}) \circ \sum_{p+q=*}(\varrho t)^{p,q}$ of chain complexes restricts to a morphism $\Omega^*(M; V)^G \to \mathrm{Tot}(\Omega^*(M; \Omega^*(G; V)^G))$, where $\Omega^*(G; V)^G$ denotes the complex of forms which are equivariant w.r.t. the action of G by left translation.*

Proof. Let $\omega \in \Omega^n(M; G)^G$ be an equivariant n-form on M, $p + q = n$ and $\mathbf{X} \in T^{\oplus p}M$ be given. The evaluation of $\varphi^{p,q} \circ (\varrho t)^{p,q}(\omega)$ at a point $\mathbf{Y} \in T^{\oplus q}G$ is given by

$$\begin{aligned}
\varphi^{p,q} \circ (\varrho t)^{p,q}(\omega)(\mathbf{X})(g.\mathbf{Y}) &= f^{p,q}\omega(\mathbf{X}, g.\mathbf{Y}) \\
&= f^{p,q}\omega(g.\mathbf{X}, g.\mathbf{Y}) \quad \text{since } M \text{ is a trivial } G\text{-space} \\
&= g.[f^{p,q}\omega(\mathbf{X}, \mathbf{Y})].
\end{aligned}$$

Thus the form $\varphi^{p,q} \circ f^{p,q}(\omega)$ takes values in the subspace $\Omega^q(G; V)^G \cong C^q(\mathfrak{g}; V)$ of left-equivariant q-forms on G. □

Thus for each equivariant differential n-form ω on M the p-forms $\varphi^{p,q}(\varrho t)^{p,q}(\omega)$ on M can be corestricted to the subspaces $\Omega^q(G;V)^G$ of $\Omega^q(G;V)$ which are isomorphic to the spaces $C^q(\mathfrak{g};V)$ of Lie algebra cochains. As a consequence the flux components $\mathcal{F}_{\varrho t}^{p,q}(\omega): S_p(M) \to \Omega^q(G;V)$ may be corestricted to $\Omega^q(G;V)^G$ also, so the flux $\mathcal{F}_{\varrho t}(\omega)$ can be regarded as an element of the total complex of $S^*(M;\Omega^*(G;V)^G)$.

Lemma 13.7.13. *For any equivariant closed n-form ω on M the corestricted map $\mathcal{F}_{\varrho t}^{p,q}(\omega)$ satisfies*

$$d_h \mathcal{F}_{\varrho t}^{p,q}(\omega) = -d_v\left(\mathcal{F}_{\varrho t}^{p+1,n-p+1}(\omega)\right),$$

i.e. the maps $\mathcal{F}_{\varrho t}^{p,q}(\omega): S_{\infty,p}(M) \to \Omega^{n-p}(G;V)^G$ commute with differentials up to sign and induces a homomorphism $H_p(M) \to H_{dR,eq}^q(G;V) \cong H_c(\mathfrak{g};V)$.

If we consider the action via ϱ on the first factor of $M \times G$ and the adjoint action on the second factor, then the map $f = \varrho t$ is also equivariant. Hence the forms $\varphi^{p,q} \circ f^{p,q}(\omega)$ are equivariant if ω is equivariant. All in all this results in

Theorem 13.7.14. *The morphism $\mathrm{Tot}(\varphi^{*,*}) \circ \sum_{p+q=*} f^{p,q}$ of chain complexes restricts to a morphism $\Omega^*(M;V)^G \to \mathrm{Tot}\Omega^*(M;\Omega^*(G;V)^G)^G$ where $\Omega^*(G;V)^G$ is the space of left-equivariant q-forms on G, G acts on the first factor by left translation and on the second factor by conjugation.*

Proof. Let G act on itself by conjugation and on M via ϱ. The flux of a G-invariant form is also G-invariant, so the homomorphism $\mathrm{Tot}(\varphi^{*,*}) \circ \sum_{p+q=*} f^{*,*}$ corestricts to a homomorphism

$$\Omega^*(M;V)^G \to \Omega^*(M;\Omega^*(G;V))^G.$$

If ω is a G-invariant n-form on M and $p+q = n$ then the maps $\varphi^{p,q} \circ f^{p,q}(\omega)$ are independent of the action of G on M chosen. So these maps actually take values in the subcomplex $\Omega^*(M;G)^G$ of left equivariant differential forms on G. Thus the result follows. $\square$

Consider the even more special case $M = G$ and $\varrho = \mu: G \times G \to G$. With the preceding theorem we obtained a homomorphism

$$\Omega^n(G,V)^G \to \Omega^p(G,\Omega^{n-p}(G,V)^G)^G,$$

where the action of G on $\Omega^{n-p}(G,V)^G$ is induced by the adjoint action on G and the action of G on V. Since the equivariant forms with respect to left translation can be identified with continuous forms on the Lie algebra $\mathfrak{g}$ of G we have obtained homomorphisms

$$C^n(\mathfrak{g},V) \to C^p(\mathfrak{g},C^{n-p}(\mathfrak{g},V)),$$

where $\mathfrak{g}$ acts on itself by the adjoint action. These are (up to a factor $(-1)^{p(n-p)}$) the 'transfer' maps of [Nee06].

13.8 Period Homomorphisms

We consider a trivial bundle $E = B \times F \to B$ of smooth manifolds and a smooth function $f: E \to M$ into another smooth manifold M. For a closed n-form ω on M the components $\mathcal{F}_f^{p,q}(\omega)$ of the flux induce homomorphisms $H_p(B) \to H_{dR}^q(F;V)$ for all $p+q = n$ (by Lemma 13.6.3). These homomorphisms can be composed with the Hurewicz homomorphisms $\mathrm{hur}_p: \pi_p(B) \to H_p(B)$.

Definition 13.8.1. *The composition* $\mathcal{F}_f^{p,q}(\omega)\mathrm{hur}_p : \pi_p(B) \to H_{dR}^q(F;V)$ *is called the* p-*th period map of the flux* $\mathcal{F}_f^{p,q}(\omega)$.

If the manifold F is connected, then the zeroth de Rham cohomology group $H_{dR}^0(F;V)$ can be identified with the space V of coefficients. In this case the component $\mathcal{F}_f^{n,0}(\omega)$ of bidegree $(n,0)$ of the flux $\mathcal{F}_f^{n,0}(\omega)$ may be regarded as function $H_p(B) \to V$. This function is then given by

$$\mathcal{F}_f^{n,0}(\omega)(z) = \int_{f \circ (z \times \{f_0\})} \omega = \int_z [f(-,f_0)]^* \omega$$

for any point $f_0 \in F$. (Since the function $f : E \to M$ is smooth, different choices for f_0 lead to cohomologous n-forms; therefore the above integral over cycles z does not depend on this choice.)

Example 13.8.2. Let (G,M) be a smooth transformation group, $E = G \times M$ and $f = \varrho : G \times M \to M$ the action of G on M. If the manifold M is connected, then the de Rham cohomology group $H_{dR}^0(M;V)$ can be identified with the space V of coefficients. In this case the flux $\mathcal{F}_\varrho^{n,0}(\omega) : H_n(G) \to V$ is then given by

$$\mathcal{F}_\varrho^{n,0}(\omega)(z) = \int_{\varrho \circ (z \times \{m_0\})} \omega = \int_z [\varrho(-,m_0)]^* \omega$$

for any point $x_0 \in X$.

Let (G,M) be a smooth transformation group, $E = G \times M$, denote the action $\varrho : G \times M \to M$ of G on M by ϱ and let $t : M \times G \to G \times M$ the coordinate flip $(m,g) \mapsto (g,m)$. We consider the smooth map $f = \varrho t : M \times G \to M$. As has been shown in Lemma 13.7.12, the flux $\mathcal{F}_{\varrho t}(\omega)$ of an equivariant n-form ω on M can be corestricted to $S^*(M;\Omega^*(G;V)^G)$. Doing so, we obtain different period maps, which are now given by

$$\mathrm{per}_{\varrho t}(\omega)_p := \mathcal{F}_{\varrho t}^{p,q}(\omega)\mathrm{hur}_p : \pi_p(M) \to H_{dR,eq}^q(G;V) \cong H_c^q(\mathfrak{g};V).$$

Example 13.8.3. If the Lie group G is connected, then the de Rham cohomology group $H_{dR,eq}^0(G;V)$ can still be identified with the space V of coefficients. In this case the flux $\mathcal{F}_{\varrho t}^{n,0}(\omega) : H_n(M) \to V$ is then given by

$$\mathcal{F}_{\varrho t}^{n,0}(\omega)(z) = \int_{\varrho t \circ (z \times \{g_0\})} \omega = \int_{\varrho \circ (\{g_0\} \times z)} \omega = \int_{g_0 . z} \omega$$

for any point $g_0 \in G$. The period map $\mathrm{per}_{\varrho \tau}(\omega)_n$ is then given by the integration of the n-form ω over the spheres in M:

$$\mathrm{per}_{\varrho \tau}(g)_n : \pi_n(M) \to V, \quad [s] \mapsto \int_s \omega.$$

13.9 The Flux Homomorphism for Singular (Co)Homology

In this section we define a version of the flux homomorphism for singular homology and cohomology. This has the advantage, that it can be applied to arbitrary transformation groups, not relying on a smooth structure.

Let G be a topological group and $X \times Y \to X$ be a trivial bundle of G-spaces and $f : X \times Y \to Z$ be a continuous function into another G-space. Consider the singular simplicial Abelian groups $S(X)$, $S(Y)$ and $S(Z)$. The continuous function

f induces a morphism $S(f) : (X \times Y) \to S(Z)$ of simplicial Abelian groups, which in turn induces a morphism $\bigoplus_n S_n(f) : \bigoplus_n S_n(X \times Y) \to S_n(Z)$ of differential graded Abelian groups. This morphism may be composed with the Eilenberg-Zilber morphism. In this fashion we obtain a chain of morphisms of differential graded Abelian groups

$$\left(\bigoplus_p S_p(X) \right) \otimes \left(\bigoplus_q S_q(Y) \right) \xrightarrow{EZ} \bigoplus_{n \in \mathbb{N}} S_n(X \times Y) \xrightarrow{\bigoplus_n S_n(f)} \bigoplus S_n(Z),$$

(cf. Section 8.2) the composition of which is denoted by f_{alg}. The components $S_p(X) \otimes S_q(Y) \to S_{p+q}(Z)$ of this morphism will be denoted by $f^{p,q}$.

Lemma 13.9.1. *If f is equivariant, then f_{alg} and all the morphisms $f^{p,q}$ are equivariant.*

Proof. If f is equivariant, then the homomorphism $\bigoplus_n S_n(f)$ is equivariant. The Eilenberg-Zilber morphism is a morphism of dga's in $\mathbb{Z}[\mathbf{G}]\mathbf{mod}$, hence the composition f_{alg} is equivariant. □

Applying the hom-functor $\hom_{\mathbf{Ab}}(-; V)$ for an Abelian group V of coefficients yields a morphism

$$\hom \left(\bigoplus S_n(Z), V \right) \xrightarrow{\hom(f_{alg}, V)} \hom \left(\left(\bigoplus_p S_p(X) \right) \otimes \left(\bigoplus_q S_q(Y) \right), V \right)$$

of differential graded coalgebras. the codomain can be rewritten using the natural isomorphism $\varphi_{A,B} : \hom(A \otimes B, V) \cong \hom(A, \hom(B, V))$ in $\mathbf{Ab}$. So finally obtain a morphism

$$\hom \left(\bigoplus S_n(Z), V \right) \xrightarrow{\bigoplus_{p,q} \alpha_{S_p(X), S_q(Y)} \hom(f_{alg}, V)} \hom \left(\bigoplus_p S_p(X), \hom \left(\bigoplus_q S_q(Y), V \right) \right)$$

of differential groups, where the codomain is the differential graded group associated to the total complex of the double complex $\hom(S_*(X), \hom(S_*(Y), V)) \cong S^*(X; S^*(Y; V))$ (and the differential increases the dimension).

Definition 13.9.2. *The morphism $\bigoplus_{p,q} \alpha_{S_p(X), S_q(Y)} \hom(f_{alg}, V)$ is called the (singular) flux homomorphism for f and denoted by $\mathcal{F}_f$.*

Remark 13.9.3. We use the same notation as in the smooth context. The two versions of the flux will be applied in different cases, where it is clear from the context which one is to be used.

In the following we assume the group V of coefficients to be a $\mathbb{Z}[G]$-module.

Lemma 13.9.4. *If f is equivariant, then the flux $\mathcal{F}_f$ is equivariant.*

Proof. If f is equivariant, then the morphism f_{alg} is equivariant, hence $\hom(f_{alg}, V)$ is equivariant. Since the morphisms $\alpha_{S_p(X), S_q(Y)}$ are equivariant, the assertion follows. □

If $g \in S_n(Z; V)$ is a singular n-cochain, then the component of bidegree (p, q) of the flux $\mathcal{F}_f(g)$ in $S^p(X; S^q(Y; V))$ is denoted by $\mathcal{F}_f(g)^{p,q}$, so the flux $\mathcal{F}_f(y)$ splits into the sum $\sum_{p+q=n} \mathcal{F}_f(g)^{p,q}$. Alternatively to a morphism of differential graded groups, the flux homomorphism can be regarded as a morphism $S^*(X; V) \to \mathrm{Tot} S^*(X; S^*(Y; V))$ of cochain complexes.

Lemma 13.9.5. *For any singular n-cocycle g in Z the map $\mathcal{F}_f^{p,q}(g)$ satisfies*

$$d_h \mathcal{F}_f^{p,q}(z) = -d_v \left(\mathcal{F}_f^{p+1,n-p+1}(g) \right),$$

i.e. the maps $\mathcal{F}_f^{p,q}(g) : S_p(X) \to S^{n-p}(Y; V)$ commute with differentials up to sign.

Proof. Let g be a singular n-cocycle in Z. Because g is a cycle and $\mathcal{F}_f$ is a cochain map we have $D\mathcal{F}_f(g) = \mathcal{F}_f(dg) = 0$. Therefore the $(p+1, n-p)$ component of $D\mathcal{F}_f(g)$ which is given by

$$D\mathcal{F}_f(g)^{p+1,n-p} = d_h \mathcal{F}_f^{p,n-p}(g) + d_v \mathcal{F}^{p+1,n-p-1}(g)$$

vanishes. Hence the stated equality follows. □

The preceding lemma implies that the maps $\mathcal{F}_f^{p,n-p}(g)$ map cycles to cocycles and boundaries to boundaries. Hence they induce homomorphisms from the singular homology of the base X to the singular cohomology of the fibre Y.

Corollary 13.9.6. *For any singular n-cocycle g on X the maps $\mathcal{F}_f^{p,q}(g)$ induce homomorphisms $\mathcal{F}_f^{p,q} : Z_p(X) \to Z^{n-p}(Y; V)$ and $\mathcal{F}_f^{p,q} : H_p(X) \to H^{n-p}(Y; V)$ from the homology of X into the singular cohomology of the fibre Y.*

Lemma 13.9.7. *If g is a singular n-cocycle in Z, $p + q = n$ and the space X is contractible, then the flux $\mathcal{F}_f(g) : Z_p(X) \to Z^q(Y; V)$ induces the zero morphism $Z_p(X) \to H^q(Y; V)$*

Proof. The flux $\mathcal{F}_f(g) : Z_p(X) \to Z^q(Y; V)$ assigns to each singular p-cycle z in $Z_p(X)$ a singular q-cocycle $z' \in Z^q(Y; V)$. If X is contractible, then every singular p-cycle $z \in Z_p(X)$ is a boundary $z = \partial c$. Because $\mathcal{F}_f(g)$ commutes with differentials up to sign, this implies $\mathcal{F}_f(g)(z) = \mathcal{F}_f(g)(\partial c) = \pm\partial \mathcal{F}_f(g)(c)$. The cohomology class of the coboundary $\pm\partial \mathcal{F}_f(g)(c)$ is the zero class. □

We now consider the special case $E = X \times G$, where (G, M) is a transformation group with action $\varrho : G \times X \to X$. Let $t : X \times G \to G \times X$ be the coordinate flip $(x, g) \mapsto (g, x)$ and consider the continuous function

$$f = \varrho t : X \times G \to X, \quad (x, g) \mapsto g.x.$$

If we take the first factor X of the product to be a trivial G-space and equip the second factor G with the action by left translation, then the map $f = \varrho t$ is equivariant. In this case, the restriction of the flux to the groups $S^*(X; V)^G$ of equivariant singular cochains on X can be corestricted to a G-invariant subcomplex of $S^*(X, S^*(G; V))$.

Lemma 13.9.8. *The morphism $\sum_{p+q=*} \mathcal{F}_f^{p,q}$ of chain complexes restricts to a morphism $S^*(M; V)^G \to \mathrm{Tot}(S^*(X; \Omega^*(G; V)^G))$, where $S^*(G; V)^G$ denotes the complex of singular cochains which are equivariant w.r.t. the action of G.*

Proof. Let $c \in S^n(M; G)^G$ be an equivariant singular cochain on X, $p + q = n$ and $\sigma \in S_p(X)$ be given. The evaluation of $[\mathcal{F}_f^{p,q}(c)](\sigma)$ at a singular q-simplex τ in G is given by

$$[\mathcal{F}_f^{p,q}(c)](\sigma)(g.\tau) = c((g.\tau).\sigma) = c(g.(\tau.\sigma)) = g.[c(\tau.\sigma)]$$

since the Alexander-Whitney morphism is associative. Thus the singular q-cochain form $[\mathcal{F}_f^{p,q}(c)](\sigma)$ takes values in the subgroup $S^q(G; V)^G$ of G-equivariant singular q-cochains on G. □

Lemma 13.9.9. *For any equivariant singular n-cochain c on X the corestricted map $\mathcal{F}_{\varrho t}^{p,q}(c)$ satisfies the identities*

$$d_h \mathcal{F}_{\varrho t}^{p,q}(c) = -d_v \left(\mathcal{F}_{\varrho t}^{p+1,n-p+1}(c) \right),$$

i.e. the maps $\mathcal{F}_{\varrho t}^{p,q}(c) : S_p(M) \to S^{n-p}(G;V)^G$ commute with differentials up to sign and induces a homomorphism $H_p(M) \to H_{sing,eq}^q(G;V)$ into the cohomology of the complex of equivariant singular cochains.

Proof. This follows from Lemma 13.9.5. □

13.10 Period Homomorphisms Revisited

We consider a trivial bundle $E = X \times Y \to X$ of topological spaces and a continuous function $f : E \to Z$ into another topological space. For a singular n-cocycle g on X the components $\mathcal{F}_f^{p,q}(g)$ of the flux induce homomorphisms $H_p(X) \to H^q(Y;V)$ for all $p + q = n$ (by Lemma 13.9.6. These homomorphisms can be composed with the Hurewicz homomorphisms $\text{hur}_p : \pi_p(X) \to H_p(X)$.

Definition 13.10.1. *The composition $\mathcal{F}_f^{p,q}(g)\text{hur}_p : \pi_p(X) \to H^q(Y;V)$ is called the p-th period map of the flux $\mathcal{F}_f^{p,q}(g)$.*

If the topological space Y is connected, then the zeroth cohomology group $H^0(Y;V) = Z^0(Y;V)$ can be identified with the space V of coefficients. In this case the component $\mathcal{F}_f^{n,0}(g)$ of bidegree $(n,0)$ of the flux $\mathcal{F}_f^{n,0}(g)$ may be regarded as function $Z_p(X) \to V$. This function is then given by

$$\mathcal{F}_f^{n,0}(g)(z) = g(f^{n,0}(z,y_0))$$

for any point $y_0 \in Y$. (Since the function $f : E \to X$ is continuous, different choices for y_0 lead to cohomologous n-cocycles; therefore the above expression does not depend on this choice.)

Example 13.10.2. Let (G,X) be a transformation group, $E = G \times X$ and $f = \varrho : G \times X \to X$ the action of G on X. If the space X is connected, then the singular cohomology group $Z^0(X;V) = H^0(M;V)$ can be identified with the space V of coefficients. In this case the flux $\mathcal{F}_\varrho^{n,0}(g) : Z_n(G) \to V$ is then given by

$$\mathcal{F}_\varrho^{n,0}(g)(z) = g(z.x_0)$$

for any point $x_0 \in X$.

Let (G,X) be a transformation group, $E = G \times X$, denote the action $\varrho : G \times X \to X$ of G on X by ϱ and let $t : X \times G \to G \times X$ the coordinate flip $(x,g) \mapsto (g,x)$. We consider the continuous function $f = \varrho t : X \times G \to X$. Similar to Lemma 13.7.12, the flux $\mathcal{F}_{\varrho t}(g)$ of an equivariant singular n-cocycle g on X can be corestricted to $S^*(X; S^*(Y;V)^G)$. Doing so, we obtain different period maps, which are now given by

$$\text{per}_{\varrho t}(\omega)_p := \mathcal{F}_{\varrho t}^{p,q}(g)\text{hur}_p : \pi_p(X) \to H_{eq}^q(G;V)$$

Example 13.10.3. If the topological group G is connected, then the singular cohomology group $Z^0(G;V) = H_{eq}^0(G;V)$ can still be identified with the space V of coefficients. In this case the flux $\mathcal{F}_{\varrho t}^{n,0}(g) : H_n(G) \to V$ is then given by

$$\mathcal{F}_{\varrho t}^{n,0}(g)(z) = g(z.x_0)$$

for any point $x_0 \in G$. The period map $\text{per}_{\varrho \tau}(\omega)_n$ is then given by the integration of the n-form ω over the spheres in M:

$$\text{per}_{\varrho \tau}(\omega)_n : \pi_n(M) \to V, \quad [s] \mapsto g(s)$$

Spectral Sequences for Transformation Groups

Let G be a topological group, X be a G-space and V be a G-module. To compute the equivariant $\mathfrak{U}$-local cohomology $H^*_{eq}(X, \mathfrak{U}; V)$ for some open G-invariant cover of X we define a suitable spectral sequence. The first approach to this problem was described in two articles [vE62a], [vE62b] of W.T. van Est. He considered a transformation group (G, X) and a G-vector space V and imposed the restriction of local contractibility on the group G and finite dimensionality of the cohomology modules $H^*_{AS}(G; V)$. With these restrictions he constructed a spectral sequence

$$E_2^{p,q} = H_c^p(G; H_{AS}^q(G; V)) \Rightarrow H_{AS,eq}^{p+q}(G; V),$$

where $H_c^p(G, H_{AS}^q(G; V))$ denotes the group cohomology with continuous cochains with values in the vector space $H_{AS}^q(G; V)$. Another prominent example of a similar spectral sequence connects the smooth group cohomology and the Lie algebra cohomology of a finite dimensional Lie group G:

$$E_2^{p,q} = H_s^p(G; H_{dR}^q(G; V)) \Rightarrow H_c^{p+q}(\mathfrak{g}; V)$$

This spectral sequence straightforwardly generalises to infinite dimensional Lie groups whose de Rham complex splits when given the C^∞-compact-open topology. A common example of Lie groups of this type are diffeomorphism groups of compact manifolds (cf. [Beg87]). There also exists a version for properly discontinuous actions of a group G on a finite dimensional manifold M:

$$E_2^{p,q} = H^p(G; H_{dR}(M; V)) \Rightarrow H_{dR,eq}^{p+q}(M; V),$$

We will construct a slightly more general spectral sequence for arbitrary transformation groups (G, X), G-modules V and a version for smooth actions on (possibly infinite dimensional) manifolds. Instead of considering Vietoris cohomology $H^*(\Gamma_u; V)$ as can Est did, we work with the equivariant $\mathfrak{U}$-local cohomology here. Using an elementary approach similar to van Est, we prove that the above restrictions on G and the dimension of the modules $H_{dR}^q(M; V)$ are unnecessary, i.e. we neither have to restrict ourselves to finite dimensional Lie groups G, manifolds M and vector spaces V nor do we have to assume the manifolds G, M or V to be (smoothly) paracompact.

We begin with a smooth transformation group (G, M), a topological vector space V that is a G-module and construct a spectral sequence

$$E_0^{p,q} = A_{s,dR}^{p,q}(M, M; V)^G \Rightarrow H_{dR,eq}^{p+q}(M; V)$$

linking the smooth equivariant cohomology $H_{s,eq}^*(M; V)$ of M to the equivariant de Rham cohomology $H_{dR,eq}^*(M; V)$. The general setting goes back to van Est. In addition to the spectral sequence described there, we explicitly construct a row contraction h, which gives further information on the edge maps. The result we obtain is known only for finite dimensional Lie groups acting on themselves by left translation, but not in the generality obtained below. The spectral sequence so obtained enables us to use classical spectral sequence arguments for infinite dimensional smooth transformation groups (G, M), which has not been possible until now.

The smoothness of an equivariant cocycle is a major restriction. In general smooth cocycles are not the appropriate concept to describe fibre bundles or Lie group extensions. This is so because the global smoothness of cocycles amounts to the existence of global smooth sections. A globally smooth group 2-cocycle $f \in C^2(G; A)$ of a Lie group G for example describes an isomorphy class of group extensions

$$0 \to A \to A \times_f G \to G \to 1$$

that admit a global smooth section, i.e. $A \times_f G = A \times G$ as manifolds. Since we are interested in the more general class of locally trivial bundles (resp. group extensions that are principal fibre bundles) and not necessarily smooth extensions of local cocycles, we introduce a new double complex, that is appropriate for describing these objects. This double complex then gives rise to a spectral sequence

$$H^p_{eq}(X; H^q(X, \mathfrak{U}; V)) \Rightarrow H^{p+1}_{eq}(X, \mathfrak{U}; V)$$

linking the cohomology $H^*_{eq}(X; V)$ of global equivariant cochains on the G-space of a transformation group (G, X) to the equivariant $\mathfrak{U}$-local cohomology $H_{eq}(X, \mathfrak{U}; V)$ of X. This spectral sequence generalises the one for smooth transformation groups. It enables us to consider arbitrary equivariant $\mathfrak{U}$-local cocycles and examine their extensibility to global equivariant cocycles. This in particular includes the extensibility of (germs of) local group cocycles to global group cocycles.

The spectral sequences constructed in this chapter will be used to characterise the integrability of equivariant differential forms in Chapter 16 resp. the extensibility of equivariant $\mathfrak{U}$-local cocycles to global equivariant cocycles in Chapter 15.

14.1 Smooth Cochains and Differential Forms

In this section we discuss a topologised version $^tA_{AS,s}(M; V)$ of the Alexander-Spanier cosimplicial group for smooth manifolds M and verify that the classical derivation morphism $\tau^* : A^*_{AS,s}(M; V) \to \Omega^*(M; V)$ from the smooth Alexander-Spanier complex into the de Rham complex stems from a morphism $^t\tau^* : {}^tA^*_{AS,s}(M; V) \to {}^t\Omega^*(M; V)$ of cochain complexes of topological vector spaces.

We consider smooth manifolds M and Hausdorff topological vector spaces V of coefficients, where we consider vector spaces and manifolds over a fixed topological ground field $\mathbb{K}$. The category of topological $\mathbb{K}$-vector spaces is denoted by $\mathbf{TopKVec}$ and the category of abstract vector spaces over the field underlying $\mathbb{K}$ is denoted by $\mathbb{K}\mathbf{Vec}$.

A (trivial) topologised cohomology of a manifold M with coefficient group V can be obtained with the help of the function space functor $C^\infty(-, V)$. All these function spaces are understood to be equipped with the C^∞-compact-open topology. Applying the contravariant function space functor $C^\infty(-, V)$ to the standard simplicial manifold $SS(M)$ yields the cosimplicial topological vector space $C^\infty(SS(M), V)$.

Definition 14.1.1. *The cosimplicial topological vector space $C^\infty(SS(M), V)$ is denoted by $^tA_s(M; V)$. It is called the smooth standard cosimplicial topological vector space with coefficients V of M. The bifunctor $A(-; -) = C^\infty(SS(-), -)$:*

Mf × **Top**$\mathbb{K}$**Vec** → **Top**$\mathbb{K}$**Vec**$^{\mathcal{O}}$ *is called the* smooth standard cosimplicial topological vector space functor. *The elements of* $A(M;V)([n])$ *are called smooth* n-*cochains.*

We note that the restriction of $^tA(-;V)$ to the category of open subsets of a smooth manifold M (whose morphisms are the inclusions) is a presheaf of cosimplicial topological vector spaces on M by construction.

Lemma 14.1.2. *The smooth standard cosimplicial topological vector space* $^tA(M;V)$ *of a smooth manifold* M *is Hausdorff and completely regular.*

Proof. Let M be a smooth manifold. Each of the function spaces $C^\infty(\mathrm{SS}(X)([n]),V)$ is a topological group and therefore completely regular. Furthermore each function space $C^\infty(\mathrm{SS}(X)([n]),V)$ is automatically Hausdorff, because the coefficient space V is Hausdorff. $\qquad\square$

Passing to the cochain complex associated to the cosimplicial topological vector space $^tA(M;V)$ yields another functor $^tA^*(-;V) : \mathbf{Mf} \to \mathbf{coCh}(\mathbf{Top}\mathbb{K}\mathbf{Vec})$ into the category $\mathbf{coCh}(\mathbf{Top}\mathbb{K}\mathbf{Vec})$ of cochain complexes of topological vector spaces. By Theorem 1.1.41 these cochain complexes are proper, i.e. the differentials are quotient maps onto their image.

Definition 14.1.3. *The functor* $^tA^*(-;V) : \mathbf{Mf} \to \mathbf{coCh}(\mathbf{Top}\mathbb{K}\mathbf{Vec})$ *is called the* smooth standard topological cochain complex functor with coefficients V *and the cochain complex* $^tA^*(M;V)$ *associated to a smooth manifold* M *is called the* smooth standard topological cochain complex (with coefficients V) *associated to the manifold* M. *The bifunctor* $^tA^*(-;-) : \mathbf{Mf} \times \mathbf{Top}\mathbb{K}\mathbf{Vec} \to \mathbf{coCh}(\mathbf{Top}\mathbb{K}\mathbf{Vec})$ *is called the* smooth standard topological cochain complex functor.

Recall that the category **Top** of topological spaces contains the subcategory **Top**$_*$ of based spaces (cf. Example A.4.4). Analogously to the discrete versions of the standard topological cochain complexes we observe:

Lemma 14.1.4. *Every smooth standard cochain complex* $^tA^*(M;V)$ *in the category* **coCh(Top$\mathbb{K}$Vec)** *of cochain complexes of topological $\mathbb{K}$-vector spaces associated to a smooth manifold* M *is split exact and this splitting is functorial on the subcategory* **Mf**$_*$ × **Top$\mathbb{K}$Vec** *of* **Mf** × **Top$\mathbb{K}$Vec**.

Proof. The proof is analogous to that of Lemma 2.5.10. $\qquad\square$

Analogously to the Alexander-Spanier cosimplicial group of a topological space one can restrict oneself to subsets of simplices $(x_0,\ldots,x_n)$ in the manifolds $\mathrm{SS}(M)([n])$, which 'fit' in families of open subsets of M:

Definition 14.1.5. *Let* $\mathfrak{U}$ *be a family of open subsets of a smooth manifold* M. *An* n-*simplex* $(m_0,\ldots,m_n)$ *in* $\mathrm{SS}(M)([n])$ *is called* $\mathfrak{U}$-*small if there exists a set* $U \in \mathfrak{U}$ *such that the product space* U^{n+1} *contains* $(m_0,\ldots,m_n)$.

Definition 14.1.6. *The submanifold of* $\mathrm{SS}(M)([n])$ *consisting only of* $\mathfrak{U}$-*small simplices is denoted by* $\mathrm{SS}(M;\mathfrak{U})([n])$.

In case there is no restriction on the n-simplices (e.g. if the family $\mathfrak{U}$ contains the manifold M,) we simply write $\mathrm{SS}(M)([n])$ for the open submanifold of $\mathfrak{U}$-small n-simplices. Since boundaries and degeneracies of $\mathfrak{U}$-small simplices are $\mathfrak{U}$-small as well, the submanifolds $\mathrm{SS}(M,\mathfrak{U})([n])$ form a simplicial submanifold of the standard simplicial manifold $\mathrm{SS}(M)$ of the manifold M.

Definition 14.1.7. *The simplicial manifold* $\mathrm{SS}(X,\mathfrak{U})$ *is called the* $\mathfrak{U}$-*small standard simplicial manifold of* M.

Definition 14.1.8. *The cosimplicial topological vector space $C^\infty(\mathrm{SS}(M,\mathfrak{U});V)$ is denoted by $^tA_s(M,\mathfrak{U};V)$. It is called the $\mathfrak{U}$-small smooth standard cosimplicial topological vector space of M.*

Lemma 14.1.9. *The $\mathfrak{U}$-small smooth standard cosimplicial topological vector space $^tA(M,\mathfrak{U};V)$ of a smooth manifold M is Hausdorff and completely regular.*

Proof. Let M be a smooth manifold. Each of the function spaces $C^\infty(\mathrm{SS}(M,\mathfrak{U})([n]),V)$ is a topological group and therefore completely regular. Furthermore each function space $C^\infty(\mathrm{SS}(M,\mathfrak{U})([n]),V)$ is Hausdorff, because the coefficient space V is Hausdorff. ∎

Let $\mathfrak{U}$ and $\mathfrak{V}$ be families of open subsets of a smooth manifold M. We write $\mathfrak{V} \prec \mathfrak{U}$ if $\mathfrak{V}$ is a refinement of $\mathfrak{U}$. In this case every $\mathfrak{V}$-small simplex also is $\mathfrak{U}$-small, so the simplicial manifold $\mathrm{SS}(M,\mathfrak{V})$ is a simplicial submanifold of the simplicial manifold $\mathrm{SS}(M,\mathfrak{U})$. The inclusion as a simplicial submanifold induces a restriction morphism $i_{\mathfrak{V},\mathfrak{U}} :\, ^tA(M,\mathfrak{U};V) \to\, ^tA(M,\mathfrak{V};V)$ of cosimplicial topological vector spaces. Thus the $\mathfrak{U}$-small smooth standard cosimplicial topological vector space $^tA(M,\mathfrak{U};V)$ for different families $\mathfrak{U}$ of open subsets of M (with the relation "is refined by") form a directed set of cosimplicial topological vector spaces.

Definition 14.1.10. *The colimit topological vector space $\mathrm{colim}_{\mathfrak{U}}^t A(M,\mathfrak{U};V)$ where $\mathfrak{U}$ ranges over all open coverings of M is denoted by $^tA_{AS,s}(M;V)$. It is called the smooth Alexander-Spanier cosimplicial topological vector space with coefficients V of M. The elements of $A_{AS,s}^t(M;V)([n])$ are called smooth Alexander-Spanier n-cochains.*

This is the cosimplicial topological vector space of germs of smooth V-valued functions defined on a neighbourhood of the diagonal. In case the smooth manifold M is smoothly paracompact, the smooth Alexander-Spanier cosimplicial topological vector space $^tA_{AS,s}(M;V)$ of M can also be obtained from the smooth standard cosimplicial topological vector space $^tA(M;V)$ by factoring out the cosimplicial subspace formed by the subspaces

$$^tA_0(M;V)([n]) := \{f \in A(M;V)([n]) \mid \exists \text{ open covering } \mathfrak{U} \text{ of } M : f_{|\mathrm{SS}(M,\mathfrak{U})([n])} = 0\}$$

of functions that vanish on a neighbourhood of the diagonal. The morphisms forming the colimit cocone under the directed set of cosimplicial topological vector spaces $^tA_S(M,\mathfrak{U};V)$ will be denoted by $\varrho_{\mathfrak{U}} : {}^tA_s(M,\mathfrak{U};V) \to {}^tA_{AS,s}(M;V)$.

Let M and N be smooth manifolds and $f : X \to Y$ be a smooth function. For each open covering $\mathfrak{U}$ of N the pullback family $f^{-1}(\mathfrak{U})$ is an open covering of M. Because the push forward family $ff^{-1}(\mathfrak{U})$ refines $\mathfrak{U}$, the morphism $\mathrm{SS}(f)$ of simplicial manifolds restricts and corestricts to a morphism

$$\mathrm{SS}(f)\big|_{\mathrm{SS}(M,f^{-1}(\mathfrak{U}))}^{\mathrm{SS}(N,\mathfrak{U})} : \mathrm{SS}(X,f^{-1}(\mathfrak{U})) \to \mathrm{SS}(N,\mathfrak{U})$$

of simplicial manifolds. An application of the contravariant functor $C^\infty(-,V)$ yields a morphism $m_{f,\mathfrak{U}} :\, ^tA_s(N,\mathfrak{U};V) \to\, ^tA_s(M,f^{-1}(\mathfrak{U});V)$ of cosimplicial topological vector spaces. The morphisms $m_{f,\mathfrak{U}}$ for different open coverings $\mathfrak{U}$ of Y can be composed with the colimit morphisms $\varrho_{f^{-1}(\mathfrak{U})} : A_s(M,f^{-1}(\mathfrak{U});V) \to\, ^tA_{AS,s}(M;V)$. In this way one obtains a morphism $\varrho_{f^{-1}(\mathfrak{U})}m_{f,\mathfrak{U}} : A_S(N,\mathfrak{U};V) \to\, ^tA_{AS,s}(M;V)$ for every open covering $\mathfrak{U}$ of N. We assert that those morphisms of cosimplicial topological vector spaces are compatible with the morphisms $i_{\mathfrak{U},\mathfrak{V}} : A_s(N,\mathfrak{V};V) \to A_s(N,\mathfrak{U};V)$ for open coverings $\mathfrak{U} \prec \mathfrak{V}$ of Y.

Proposition 14.1.11. *For every smooth function $f : M \to N$ the collection of morphisms $\varrho_{f^{-1}(\mathfrak{U})} m_{f,\mathfrak{U}}$ for open coverings $\mathfrak{U}$ of N form a cocone under the diagram of topological vector spaces ${}^t A(N, \mathfrak{U}; V)$, i.e. the equations $\varrho_{f^{-1}(\mathfrak{U})} m_{f,\mathfrak{U}} = \varrho_{f^{-1}(\mathfrak{V})} m_{f,\mathfrak{V}} i_{\mathfrak{V},\mathfrak{U}}$ are satisfied for all open coverings $\mathfrak{V} \prec \mathfrak{U}$ of N.*

Proof. The proof is analogous to that of Proposition 2.5.17.   □

Corollary 14.1.12. *A smooth function $f : M \to N$ between smooth manifolds M and N induces a unique morphism ${}^t A_{AS,s}(f; V) : {}^t A_{AS,s}(N; V) \to {}^t A_{AS,s}(M; V)$ of cosimplicial topological vector spaces satisfying $\varrho_{f^{-1}(\mathfrak{U})} m_{f,\mathfrak{U}} = {}^t A_{AS,s}(f; V) \varrho_{\mathfrak{U}}$ for all open coverings $\mathfrak{U}$ of N.*

The identity function $\mathrm{id}_M : M \to M$ of any smooth manifold M induces the identity morphism ${}^t A_{AS,s}(\mathrm{id}_M; V) = \mathrm{id}_{A^t_{AS,s}(M;V)}$. In addition the assignment $f \mapsto A^t_{AS,s}(f; V)$ preserves composition of morphisms by construction, but reverses their order. So we note:

Lemma 14.1.13. *The assignments $M \mapsto {}^t A_{AS,s}(M; V)$ and $f \mapsto {}^t A^t_{AS,s}(f; V)$ form a contravariant functor ${}^t A_{AS,s}(-; V) : \mathbf{Mf} \to \mathbf{Top}\mathbb{K}\mathbf{Vec}^{\mathcal{O}}$.*

Moreover all the constructions made are natural in the coefficient space V. Thus the contravariant functors ${}^t A_{AS,s}(-; V)$ for all topological vector spaces V are part of a bifunctor ${}^t A_{AS,s}(-; -) : \mathbf{Mf} \times \mathbf{Top}\mathbb{K}\mathbf{Vec} \to \mathbf{Top}\mathbb{K}\mathbf{Vec}^{\mathcal{O}}$.

Definition 14.1.14. *The bifunctor ${}^t A_{AS,s}(-; -) : \mathbf{Mf} \times \mathbf{Top}\mathbb{K}\mathbf{Vec} \to \mathbf{Top}\mathbb{K}\mathbf{Vec}^{\mathcal{O}}$ is called the* smooth Alexander-Spanier cosimplicial topological vector space functor. *For each coefficient space V the functor ${}^t A_{AS,s}(-; V) : \mathbf{Mf} \to \mathbf{Top}\mathbb{K}\mathbf{Vec}^{\mathcal{O}}$ is called the* smooth Alexander-Spanier cosimplicial topological vector space functor with coefficients V.

Regarding homotopies in the category $\mathbf{Mf}$ of manifolds (i.e. homotopies which are smooth functions) we observe that the smooth Alexander-Spanier cosimplicial topological vector space functor ${}^t A_{AS,s}(-; -) : \mathbf{Mf} \times \mathbf{Top}\mathbb{K}\mathbf{Vec} \to \mathbf{Top}\mathbb{K}\mathbf{Vec}^{\mathcal{O}}$ behaves as expected:

Theorem 14.1.15. *The smooth Alexander-Spanier cosimplicial topological vector space functor ${}^t A_{AS,s}(-; -)$ preserves homotopy.*

Proof. The proof is analogous to that of Theorem 2.5.22.   □

Considering the cochain complexes the associated to the smooth Alexander-Spanier cosimplicial topological vector spaces ${}^t A_{AS,s}(M; V)$ of manifolds M yields another functor ${}^t A^*_{AS,s}(-; -) := (-)^* \circ {}^t A_{AS,s}(-, -) : \mathbf{Mf} \times \mathbf{Top}\mathbb{K}\mathbf{Vec} \to \mathbf{coCh}(\mathbf{Top}\mathbb{K}\mathbf{Vec})$ into the category $\mathbf{coCh}(\mathbf{Top}\mathbb{K}\mathbf{Vec})$ of cochain complexes of topological vector spaces. Theorem 1.1.41 guarantees that the cochain complexes so obtained are proper.

Definition 14.1.16. *The bifunctor ${}^t A^*_{AS,s}(-, -) : \mathbf{Mf} \times \mathbf{Top}\mathbb{K}\mathbf{Vec} \to \mathbf{coCh}(\mathbf{Top}\mathbb{K}\mathbf{Vec})$ is called the* smooth Alexander-Spanier topological cochain complex functor. *For each smooth manifold M and coefficient space V the cochain complex ${}^t A^*_{AS,s}(M; V)$ is called the* smooth Alexander-Spanier topological cochain complex with coefficients V of the manifold M.

Lemma 14.1.17. *The bifunctor ${}^t A^*_{AS}(-; -)$ preserves homotopy.*

Proof. This follows from the fact that cosimplicially homotopic morphisms of cosimplicial topological vector spaces induce cochain homotopic morphisms between the associated cochain complexes (cf. Lemma F.6.6).   □

Taking the cohomology of the topological Alexander-Spanier cochain complexes yields a functor $\mathbf{Top} \times \mathbf{TopAb} \to \mathbf{TopAb}$.

Definition 14.1.18. *The bifunctor ${}^t\!H_{AS,s}(-;-) := H \circ {}^t\!A^*_{AS,s}(-;-)$ is called the* smooth Alexander-Spanier topological cohomology functor. *For each coefficient space V the functor ${}^t\!H_{AS,s}(-;V)$ is called the* smooth Alexander-Spanier topological cohomology functor with coefficients V. *The graded topological vector space ${}^t\!H_{AS,s}(M;V)$ associated to a manifold M is called its* topological smooth Alexander-Spanier cohomology *of M with coefficients V.*

Lemma 14.1.19. *The topological smooth Alexander-Spanier cohomology functor is homotopy invariant.*

Proof. This is a direct consequence of Lemma 14.1.17.   □

If one applies the forgetful functor $U : \mathbf{Top}\mathbb{K}\mathbf{Vec} \to \mathbb{K}\mathbf{Vec}$ to ${}^t\!A_{AS,s}(X;V)$ one obtains the underlying cosimplicial vector space defining ordinary smooth Alexander-Spanier cohomology with coefficients UV. As a consequence one obtains:

Lemma 14.1.20. *The composition of the forgetful functor $U : \mathbf{Top}\mathbb{K}\mathbf{Vec} \to \mathbb{K}\mathbf{Vec}$ with the topological cohomology functor ${}^t\!H_{AS,s}(-;V)$ (with coefficients V) is the classical smooth Alexander-Spanier cohomology functor $H_{AS,s}(-;UV)$ with coefficients UV.*

Proof. The forgetful functor $U : \mathbf{Top}\mathbb{K}\mathbf{Vec} \to \mathbb{K}\mathbf{Vec}$ preserves kernels and cokernels.   □

Recall the derivation morphism $A^*_{AS,s}(M;V) \to \Omega^*(M;V)$ which is equivariant w.r.t. any group action on M and V. Below we verify that the homomorphisms $\tau^n : A^n_{AS,s}(M;V) \to \Omega^n(M;V)$ define continuous linear maps ${}^t\tau^n$ between the topological vector spaces ${}^t\!A^n_{AS,s}(M;V)$ and ${}^t\Omega^n(M;V)$ and thus define a morphism ${}^t\tau^* : {}^t\!A^*_{AS,s}(M;V) \to {}^t\Omega^*(M;V)$ of cochain complexes of topological vector spaces. We start the proof thereof by making some preparatory observations concerning function spaces and (partial) derivatives. Recall that for any open subset U of a topological vector space any any Hausdorff topological vector space V the spaces $C^k(U,V)$ are given the initial topology w.r.t. all the linear mappings

$$C^k(U,V) \to C(U^{[j]},V), \quad f \mapsto f^{[j]}$$

for $j \le k$ (cf. the notation in Section 13.1 and in [BGN04]). This in particular implies the continuity of the linear maps $C^{k+1}(U,V) \to C^k(U^{[1]},V)$.

Lemma 14.1.21. *For any open subset U of a topological vector space W, any Hausdorff topological vector space V the linear map $C^{k+1}(U,V) \to C^k(U \times W, V)$, $f \mapsto df$ is continuous.*

Proof. Let U be an open subset of a topological vector space W and V be a Hausdorff topological vector space. The linear map $C^{k+1}(U,V) \to C^k(U^{[1]},V)$, $f \mapsto f^{[1]}$ is continuous. The product space $U \times W$ can be identified with the subspace of the domain $U \times W \times \{0\}$ of

$$U^{[1]} := \{(u,v,t) \in U \times V \times \mathbb{K} \mid u + tv \in U\}.$$

of $f^{[1]}$. The inclusion $i : U \times W \to U^{[1]}$, $(u,w) \mapsto (u,w,0)$ is a continuous linear map and induces a continuous restriction map $C^k(U^{[1]},V) \to C^k(U \times W, V)$ (cf. [Glö04, Lemma 4.4]). The composition of the continuous linear map $f \mapsto f^{[1]}$ with this restriction map is exactly the linear map $f \mapsto df$, which is therefore continuous.   □

Lemma 14.1.22 (Bertram, Glöckner, Neeb). *For any C^{k+1}-manifold M, any Hausdorff C^{k+1}-manifold N and any C^{k+1}-function $f : M \to N$, the tangent map $Tf : TM \to TN$ is of class C^k.*

Proof. See [BGN04, Section 8] for a proof. □

Lemma 14.1.23. *(Glöckner) For any $0 \le k \le l \le \infty$, C^l-manifold M, atlas $\mathcal{A}$ of charts $\varphi : U_\varphi \to W_\varphi$ defined on open subsets U_φ of M and any Hausdorff topological vector space V the topology on $C^k(M, V)$ is initial w.r.t. the pullback maps $C^k(\varphi^{-1}, V) : C^k(M, V) \to C^k(W_\varphi, V)$.*

Proof. See [Glö04, Lemma 4.9, Article 3] for a proof. □

Proposition 14.1.24. *For any C^{k+1}-manifold M and any Hausdorff topological vector space V the linear map $C^{k+1}(M, V) \to C^k(TM, V)$, $f \mapsto df$ is continuous.*

Proof. Let M be a C^{k+1}-manifold and V be a Hausdorff topological vector space. For any atlas $\mathcal{A}$ on M the topology on the vector space $C^{k+1}(M, V)$ is the initial topology w.r.t. all the pullback maps $C^{k+1}(\varphi^{-1}, V)$, $\phi \in \mathcal{A}$ by Lemma 14.1.23. Each chart $\varphi \in \mathcal{A}$ is a function from an open subset U_φ of M onto an open subset W_φ of the model space of M. The tangent space TU_φ is considered as an open subset of the tangent manifold TM. The collection of open subsets $\{TU_\varphi \mid \varphi \in \mathcal{A}\}$ is an open covering of TM and $\{T\varphi \mid \varphi \in \mathcal{A}\}$ is an atlas of the tangent manifold TM. The topology on $C^k(TM, V)$ is therefore initial w.r.t. the pullback maps $C^k((T\varphi)^{-1}, TV)$, $\phi \in \mathcal{A}$. As a consequence the linear map $C^{k+1}(M, V) \to C^k(TM, V)$, $f \mapsto df$ is continuous if and only if the maps $C^{k+1}(M, V) \to C^k(TW_\varphi, V)$, $f \mapsto df \circ (T\varphi)^{-1}$ are continuous for all charts $\varphi \in \mathcal{A}$. For each such chart $\varphi \in \mathcal{A}$ the commutativity of the diagram

$$
\begin{array}{ccc}
C^{k+1}(M, V) & \xrightarrow{\;f \mapsto df\;} & C^k(TM, V) \\
{\scriptstyle C^{k+1}(\varphi^{-1}, V)} \downarrow & & \downarrow {\scriptstyle C^k((T\varphi)^{-1}, V)} \\
C^{k+1}(W_\varphi, V) & \xrightarrow[\;f \mapsto df\;]{} & C^k(TW_\varphi, V)
\end{array}
$$

shows that the linear map $C^{k+1}(M, V) \to C^k(TU, V)$, $f \mapsto df \circ (T\varphi)^{-1}$ is the composition of the continuous linear pullback map $C^{k+1}(\varphi^{-1}, V)$ with the linear map $C^{k+1}(W_\varphi, V) \to C^k(TW_\varphi, V)$, which is continuous by Lemma 14.1.21. Therefore the composition $C^{k+1}(M, V) \to C^k(TM, V)$, $f \mapsto df$ is continuous. □

Corollary 14.1.25. *For any smooth manifold M and any Hausdorff topological vector space V the linear map $C^\infty(M, V) \to C^\infty(TM, V)$, $f \mapsto df$ is continuous.*

Lemma 14.1.26. *For smooth manifolds M_0, M_1, M_2 and a Hausdorff topological vector space V the partial derivative map*

$$d_1 : C^\infty(M_0 \times M_1 \times M_2, V) \to C^\infty(M_0 \times TM_1 \times M_2, V)$$

is continuous.

Proof. Let M_0, M_1 and M_2 be smooth manifolds and V be a Hausdorff topological vector space. The partial derivative map d_1 is the composition of the continuous linear map

$$C^\infty(M_0 \times M_1 \times M_2, V) \to C^\infty(T(M_0 \times M_1 \times M_2), V) \cong C^\infty(TM_0 \times TM_1 \times TM_2, V),$$

$$f \mapsto df$$

with the continuous map $C^\infty(TM_0 \times TM_1 \times TM_2, V) \to C^\infty(M_0 \times TM_1 \times M_2, V)$ induced by the inclusion $M_0 \times TM_1 \times M_2 \hookrightarrow TM_0 \times TM_1 \times TM_2$. Therefore the partial derivative map d_1 is continuous. □

This can readily be applied to finite products M^{n+1} of a smooth manifold M, where we denote points of M^{n+1} by $\mathbf{m} = (m_0, \ldots, m_n)$, i.e. the first index is zero. In particular the partial derivative of a smooth function $f : M^{n+1} \to V$ into a Hausdorff topological vector space V w.r.t. the last component of the product will be denoted by d_n etc.

Lemma 14.1.27. *For a smooth manifold M, a Hausdorff topological vector space V and $p \leq n$ the linear map $d_{n-p} \cdots d_n : C^\infty(M^{n+1}, V) \to C^\infty(M^p \times TM^{n+1-p}, V)$ is continuous.*

Proof. This follows from a repeated application of Lemma 14.1.26.   □

The composition $d_{n-p} \cdots d_n : C^\infty(M^{n+1}, V) \to C^\infty(M^p \times TM^{n+1-p}, V)$ of partial derivatives can now be used to describe the derivation homomorphisms τ^n. For each $n \in \mathbb{N}$ and smooth function $f : M^{n+1} \to V$ the differential n-form $\tau^n(f)$ on M is given by the formula

$$\tau^n(f)(X_1, \ldots, X_n) = \sum_{s \in S_n} \operatorname{sgn}(s)[d_1 \ldots d_n f]\big(\pi_{TM}(X_1), X_{s(1)}, \ldots, X_{s(n)}\big)$$

where $X_1, \ldots, X_n$ are tangent vectors at a common point in M (cf. appendix G.2). Since antisymmetrisation is a continuous operation on $C^\infty(TM^{\oplus n}M, V)$ the derivation homomorphism τ^n is defines a homomorphism ${}^t\tau^n :{}^t A_s^n(M;V) \to {}^t\Omega^n(M;V)$ of topological vector spaces. The fact that the underlying maps τ^n of the homomorphisms ${}^t\tau^n$ form a cochain map ensures that ${}^t\tau^* : A_s^*(M;V) \to {}^t\Omega^*(M;V)$ is a cochain map in the category $\mathbf{coCh}(\mathbf{Top}\mathbb{K}\mathbf{Vec})$ of cochain complexes of topological $\mathbb{K}$-vector spaces.

Definition 14.1.28. *The morphism ${}^t\tau^* : A_s^*(M;V) \to {}^t\Omega^*(M;V)$ of cochain complexes of topological vector spaces is called the derivation morphism. The homomorphisms ${}^t\tau^n : A_s^n(M;V) \to {}^t\Omega^n(M;V)$ are called derivation homomorphisms.*

Definition 14.1.29. *A differential n-form on a smooth manifold M is called integrable to a smooth n-cochain if it is contained in the image of the derivation homomorphism ${}^t\tau^t A_{AS,s}^*(M;V) \to {}^t\Omega^n(M;V)$. A closed differential n-form ω is called integrable to a smooth n-cocycle if there exists an n-cocycle $f \in A_s^n(M;V)$ with image ${}^t\tau^n(f) = \omega$.*

Now suppose that G is a group acting via diffeomorphisms on a manifold M. The action of G on M induces an action of G on TM. If the Hausdorff topological vector space V is a representation space of V (i.e. if G acts by continuous linear transformations on V), then the derivative map $d : C^\infty(M, V) \to C^\infty(TM, V)$ is equivariant. The same is true for a repeated application of partial derivative maps. As a consequence we observe:

Lemma 14.1.30. *If a group G acts on M by diffeomorphisms and on V by continuous linear transformations, then the morphism ${}^t\tau^*$ of cochain complexes is equivariant.*

Remark 14.1.31. The actions of G on M and V is not required to be smooth or continuous here.

So far the derivation homomorphisms ${}^t\tau^n$ are defined on global smooth cochains only. It remains to observe that they generalise to homomorphisms defined on germs of smooth functions. To this end we note the following generalisation of Lemma 14.1.23:

Lemma 14.1.32. *For $0 \leq k \leq l \leq \infty$, a C^l-manifold M with open cover $\mathfrak{U}$ and any Hausdorff topological vector space V the restriction maps $C^k(M, V) \to C^k(U, V)$ for all $U \in \mathfrak{U}$ induce an embedding*

$$C^k(M, V) \to \prod_{U \in \mathfrak{U}} C^k(U, V), \quad f \mapsto \prod_{U \in \mathfrak{U}} f_{|U}$$

of $C^k(M; V)$ as a closed subspace of the product space $\prod_{U \in \mathfrak{U}} C^k(U, V)$.

Proof. The proof of [Glö04, Article 3, Lemma 4.12] carries over to our slightly more general setting. □

Corollary 14.1.33. *For each C^l-manifold M, open cover $\mathfrak{U}$ of M, Hausdorff topological vector space V and every $k \leq l$, the topology on $C^k(M, V)$ is initial w.r.t. all the restriction maps $C^k(M, V) \to C^k(U, V)$, $U \in \mathfrak{U}$.*

So we observe that for each open cover $\mathfrak{U}$ of M the neighbourhood $\mathrm{SS}(M, \mathfrak{U})([n])$ of the diagonal in M^{n+1} is covered by the open sets U^{n+1}, $U \in \mathfrak{U}$. Therefore the C^∞-compact open topology on $A_s^n(M, \mathfrak{U}; V)$ is the initial topology w.r.t. the pullback maps $A_s^n(M, \mathfrak{U}; V) \to C^\infty(U^{n+1}, V)$ for all $U \in \mathfrak{U}$. Furthermore the open cover $\mathfrak{U}$ of M gives rise to an open cover $T^{\oplus n}U$ of the n-fold direct sum $T^{\oplus n}M$ of the tangent bundle TM of M. Consequently the space $C^\infty(T^{\oplus n}M, V)$ of smooth V-valued functions on $T^{\oplus n}M$ carries the initial topology w.r.t. all pullback maps $C^\infty(T^{\oplus n}M, V) \to C^\infty(T^{\oplus n}U, V)$. Since the topological vector space ${}^t\Omega^n(M, V)$ of differential n-forms on M is a closed subspace of $C^\infty(M, V)$, it carries the initial topology w.r.t. all the pullback maps ${}^t\Omega^n(M, V) \to {}^t\Omega^n(U_\varphi, V)$.

Lemma 14.1.34. *For every open cover $\mathfrak{U}$ of a smooth manifold M, the derivation homomorphism ${}^t\tau^n : {}^tA_s^n(M, \mathfrak{U}; V) \to {}^t\Omega^n(M, V)$ is continuous.*

Proof. Let M be a smooth manifold, $\mathfrak{U}$ be an open cover of M and V be a Hausdorff topological vector space. Since the space ${}^t\Omega^n(M, V)$ carries the initial topology w.r.t. all the restriction maps ${}^t\Omega^n(M, V) \to {}^t\Omega^n(U, V)$, $U \in \mathfrak{U}$, the derivation homomorphism ${}^t\tau^n$ is continuous if and only if the linear maps $A_s^n(M, \mathfrak{U}; V) \to {}^t\Omega^n(U, V)$, $f \mapsto {}^t\tau^n(f)_{|U}$ are continuous for all $U \in \mathfrak{U}$. For each such open set $U \in \mathfrak{U}$ the commutativity of the diagram

$$
\begin{array}{ccc}
{}^tA_s^n(M, \mathfrak{U}; V) & \xrightarrow{\quad {}^t\tau^n \quad} & {}^t\Omega^n(M, V) \\
{\scriptstyle f \mapsto f_{|U^{n+1}}} \downarrow & & \downarrow {\scriptstyle \omega \mapsto \omega_{|U}} \\
{}^tA_s^n(U; V) = C^\infty(U^{n+1}, V) & \xrightarrow[\quad {}^t\tau^n \quad]{} & {}^t\Omega^n(U, V)
\end{array}
$$

shows that the linear map ${}^tA_s^n(M, \mathfrak{U}; V) \to {}^t\Omega^n(U, V)$, $f \mapsto {}^t\tau^n(f)_{|U}$ is the composition of the continuous linear restriction map ${}^tA_s^n(M, \mathfrak{U}; V) \to {}^tA_s^n(U; V)$, $f \mapsto f_{|U^{n+1}}$ with the continuous linear derivation homomorphism ${}^tA_s^n(M; V) \to {}^t\Omega^n(U; V)$. Therefore the composition ${}^tA_s^n(M, \mathfrak{U}; V) \to {}^t\Omega^n(U; V)$, $f \mapsto {}^t\tau^n(f)_{|U}$ is continuous. □

For every open refinement $\mathfrak{V}$ of an open cover $\mathfrak{U}$ of M the continuous linear restriction maps $A_s^n(M, \mathfrak{U}; V) \to A_s^n(M, \mathfrak{V}; V)$ intertwine the derivation homomorphisms ${}^tA_s^n(M, \mathfrak{U}; V) \to {}^t\Omega^n(M; V)$ and ${}^tA_s^n(M, \mathfrak{V}; V) \to {}^t\Omega^n(M; V)$. As a consequence there exists a unique continuous linear map ${}^t\tau^n : A_{AS,s}^n(M; V) \to {}^t\Omega^n(M; V)$ whose underlying map of (non topological) vector spaces is the classical derivation homomorphism τ^n. The fact, that the underlying maps τ^n form a morphism of cochain complexes ensures that the same is true for the continuous linear maps ${}^t\tau^n$.

Definition 14.1.35. *The morphism $^t\tau^* : A^*_{AS,s}(M;V) \to {}^t\Omega^*(M;V)$ of cochain complexes of topological vector spaces is called the derivation morphism.*

Now suppose that G is a group acting via diffeomorphisms on a manifold M and on V by continuous linear transformations. For a fixed group element g and open coverings $\mathfrak{V} \prec \mathfrak{U}$ of M the linear maps $^tA^n_s(M;\mathfrak{U};V) \to {}^tA^n_s(M,g.\mathfrak{U};V)$, $f \mapsto g.f$ for open coverings $\mathfrak{U}$ of M intertwine the restriction maps $^tA^n_s(M,\mathfrak{U};V) \to {}^tA^n_s(M,\mathfrak{V};V)$ and thus induce a continuous linear map $^tA^n_{AS,s}(M;V) \to {}^tA^n_{AS,s}(M;V)$, $[f] \mapsto [g.f]$. Hence the group G acts by continuous linear transformations on the topological vector space $^tA^n_{AS,s}(;V)$ of smooth Alexander-Spanier n-cochains on M. Moreover this action is compatible with the differential on the cochain complex $^tA^*_{AS,s}(M;V)$. In addition the action of G on M induces an action of G on TM, hence on $^t\Omega^*(M;V)$. The derivative map $d : C^\infty(M,V) \to C^\infty(TM,V)$ is equivariant. The same is true for a repeated application of partial derivative maps. As a consequence we observe:

Lemma 14.1.36. *If a group G acts on M by diffeomorphisms and on V by continuous linear transformations, then the morphism $^t\tau^* : A^*_{AS,s,eq}(M;V) \to {}^t\Omega^*(M;V)$ of cochain complexes is equivariant.*

As a consequence we note that the derivation homomorphisms $^t\tau^n$ restrict and corestrict to homomorphisms $^tA^n_{AS,s}(M;V)^G \to {}^t\Omega^n(M;V)^G$ from the spaces of equivariant smooth Alexander-Spanier cochains $A^n_{AS,s}(M;V)^G$ into the spaces $^t\Omega^n(M;V)^G$ of equivariant differential forms on M.

14.2 A Spectral Sequence of van Est Type for Smooth Cochains

In this section we present a spectral sequence relating the equivariant smooth global cochains of a G-manifold M to the equivariant differential forms on M. Let (G,M) be a smooth transformation group and V be a Hausdorff $\mathbb{K}[G]$-module of coefficients (where we consider vector spaces and manifolds over the topological ground field $\mathbb{K}$.) The spectral sequence to be constructed below arises a spectral sequence obtained by filtrating a double complex. It is a generalisation of the spectral sequence

$$H^p_s(G; H^q_{dR}(G;V)) \Rightarrow H^{p+q}(\mathfrak{g};V)$$

for finitely dimensional Lie groups G with Lie algebra $\mathfrak{g}$ and G-module V which was (implicitly) presented by van Est in [vE53] and [vE55b]. In contrast to van Est we will neither restrict ourselves to finite dimensional Lie groups G, manifolds M and vector spaces V nor will we assume the manifolds G, M or V to be (smoothly) paracompact.

At first we introduce a double complex whose adjoint complex is the double complex $A^*_s(M;\Omega^*(M;V))^G$ of smooth global equivariant cochains on M with values in the de Rham complex $\Omega^*(M;V)$ of M. The total complex of this new double complex will then compute the equivariant de Rham cohomology of M. All function spaces occurring in this Section are spaces of smooth functions. They are equipped with C^∞-compact open topology. As usual we abbreviate points $(m_0,\ldots,m_p)$ in M^{p+1} by $\mathbf{m}$ and elements $(X_1,\ldots,X_q)$ in $T^{\oplus q}M$ by $\mathbf{X}$.

Definition 14.2.1. *For $p,q \geq 0$ the subspace of $C^\infty(M^{p+1} \times T^{\oplus q}M;V)$ consisting of all smooth functions f whose adjoint $\hat{f} : M^{p+1} \to C^\infty(T^{\oplus q}M,V)$ has codomain in the space $\Omega^q(M,V)$ of differential q-forms will be denoted by $^tA^{p,q}_{s,dR}(M;V)$.*

Because the subspace ${}^t A_{s,dR}^{p,q}(M;V)$ is the fixed-point set of a continuous linear self map, it is a closed linear subspace of $C^\infty(M^{p+1} \times T^{\oplus q}M;V)$. We endow it with the subspace topology (which also is the C^∞-compact open topology). The action of G on M induces a smooth action of G on the tangent space TM and on the Whitney sums $T^{\oplus n}M$. This in turn induces a not necessarily smooth action of G on the topological vector spaces ${}^t A_{s,DR}^{*,*}(M;V)$ which is given by

$$g.f(\mathbf{m}, \mathbf{X}) = g.(g^{-1}\mathbf{m}, g^{-1}\mathbf{X}).$$

The spaces M^{p+1} are the components of the standard simplicial G-manifold $\mathrm{SS}(M)$ of M. The face and degeneracy morphisms of this simplicial G-manifold turn each ${}^t A^{*,q}(M;V)$ into a cosimplicial topological G-vector vector space. In particular the coface maps induce a coboundary operator:

Definition 14.2.2. *The horizontal differential* $d_h^{p,q} : {}^t A_{s,dR}^{p,q}(M;V) \to {}^t A_{s,dR}^{p+1,q}(M;V)$ *is the coboundary induced by the cosimplicial topological G-vector space structure on* ${}^t A^{*,q}(M;V)$. *It is given by*

$$d_h^{p,q}(f)(m_0, \ldots, m_{p+1}, \mathbf{X}) := \sum_{i=0}^{p+1}(-1)^i f(m_0, \ldots, \hat{m}_i, \ldots, m_{p+1}, \mathbf{X})$$

The differential of the de Rham complex $\Omega^*(M;V)$ induces a vertical differential on the G-spaces $A^{p,*}(M;V)$. It is defined via exterior differentiation taking the column-wise sign change into account.

Definition 14.2.3. *The vertical differential on* $d_v^{p,q} : {}^t A_{s,dR}^{p,q}(M;V) \to {}^t A_{s,dR}^{p,q+1}(M;V)$ *is given by*

$$d_v^{p,q}(f)(\mathbf{m}, X_0, \ldots, X_q) := (-1)^p d[\hat{f}(\mathbf{m})](\mathbf{X})$$
$$= \sum_{i=0}^{q}(-1)^{p+i} X_i.f(\mathbf{m}, X_0, \ldots, \hat{X}_i, \ldots, X_q)$$
$$+ \sum_{i<j}(-1)^{p+i+j} f(\mathbf{m}, [X_i, X_j], X_0, \ldots, \hat{X}_i, \ldots, \hat{X}_j, \ldots, X_q)$$

Arising from two different independent structures, the differentials defined on the cochains and those defined on forms commute; since the vertical and horizontal coboundaries on the spaces ${}^t A_{s,dR}^{p,q}(M;V)^G$ are defined with a column-wise sign change, we note:

Proposition 14.2.4. *The horizontal and vertical differential anticommute, i.e.* $d_h d_v + d_v d_h = 0$.

Proof. Let $f_{p,q} \in {}^t A_{s,dR}(M;V)$ be a cochain of bidegree (p,q). The evaluation of $d_h d_v f^{p,q}$ at a point $(\mathbf{m}, \mathbf{X}) = (m_0, \ldots, m_{p+1}, X_0, \ldots, X_q)$ is given by

$$d_h d_v f^{p,q}(\mathbf{m}, \mathbf{X}) =$$
$$\sum_i (-1)^i (d_v f^{p,q})(m_0, \ldots, \hat{m}_i, \ldots, m_{p+1}, \mathbf{X})$$
$$= \sum_{i,k}(-1)^{p+i+k} X_k.f^{p,q}(m_0, \ldots, \hat{m}_i, \ldots, m_{p+1}, X_0, \ldots, \hat{X}_k, \ldots, X_q)$$
$$+ \sum_{\substack{i \\ k<l}}(-1)^{p+i+k+l} f^{p,q}(m_0, \ldots, \hat{m}_i, \ldots, m_{p+1}, [X_k, X_l], X_0, \ldots, \hat{X}_k, \ldots, \hat{X}_l, \ldots, X_q).$$

Likewise, the evaluation of $d_v d_h f^{p,q}$ at $(\mathbf{m}, \mathbf{X})$ is given by

$$d_v d_h f^{p,q}(\mathbf{m}, \mathbf{X}) =$$

$$\sum_k (-1)^{p+1+k} X_k.(d_h f^{p,q}(\mathbf{m}, X_0, \ldots, \hat{X}_k, \ldots, X_q))$$

$$+ \sum_{k<l} (-1)^{p+1+k+l}(d_h f^{p,q}(\mathbf{m}, [X_k, X_l], X_0, \ldots, \hat{X}_k, \ldots, \hat{X}_l, \ldots, X_q)$$

$$= \sum_{i,k} (-1)^{p+1+i+k} X_k.f^{p,q}(m_0, \ldots, \hat{m}_i, \ldots, m_{p+1}, X_0, \ldots, \hat{X}_k, \ldots, X_q))$$

$$+ \sum_{\substack{k<l \\ i}} (-1)^{p+1+i+k+l} f^{p,q}(m_0, \ldots, \hat{m}_i, \ldots, m_{p+1}, [X_k, X_l], X_0, \ldots, \hat{X}_k, \ldots, \hat{X}_l, \ldots, X_q)).$$

These two expressions differ only in sign, hence $d_h d_v + d_v d_h = 0$. ☐

Corollary 14.2.5. *The modules $^t A^{p,q}_{s,dR}(M; V)$ form a double complex in* **GTopAb**.

We restrict ourselves to the sub-double complex $^t A^{*,*}_{s,dR}(M; V)^G$ of G-equivariant functions and consider the underlying double complex $A^{*,*}_{s,dR}(M; V)^G$ of (abstract) vector spaces. The double complex $A^{*,*}_{s,dR}(M; V)^G$ can be filtrated column-wise and row-wise. This leads to two spectral sequences $_I E^{*,*}_*$ and $_{II} E^{*,*}_*$ respectively (cf. [McC01, Theorem 2.15]). Since the double complex $A^{*,*}_{s,dR}(M; V)^G$ is a first quadrant double complex, the spectral sequences $_I E^{*,*}_*$ and $_{II} E^{*,*}_*$ converge to the cohomology of the total complex $\mathrm{Tot} A^{*,*}_{s,dR}(M; V)^G$:

Proposition 14.2.6. *The spectral sequences $_I E^{*,*}_*$ and $_{II} E^{*,*}_*$ converge to the cohomology of the total complex $\mathrm{Tot} A^{*,*}_{s,dR}(M; V)^G$:*

$$_I E^{*,*}_* \Rightarrow H^*(\mathrm{Tot} A^{*,*}_{s,dR}(M; V)^G), \quad _{II} E^{*,*}_* \Rightarrow H^*(\mathrm{Tot} A^{*,*}_{s,dR}(M; V)^G),$$

Proof. See [McC01, Theorem 2.15] for a proof. ☐

In the following the spectral sequence $_{II} E^{*,*}_*$ is studied further. We start with an augmentation both of rows and columns. Please note that for each $q \geq 0$ the kernel of the horizontal differential $d_h : A^{0,q}_{s,dR}(M; V)^G \to A^{1,q}_{s,dR}(M; V)^G$ consists of those equivariant functions $f : M \times T^{\oplus q} M \to V$ in $A^{0,q}_{s,dR}(M; V)^G$ that are independent of the first variable. Hence this kernel is isomorphic to the de Rham space $\Omega^q(M; V)^G$ of equivariant q-forms on M. One can therefore augment each row via

$$0 \longrightarrow \Omega^q(M; V)^G \xrightarrow{i^q} A^{*,q}_{s,dR}(M; V)^G, \quad i^q(\omega)(m_0, \mathbf{X}) = \omega(\mathbf{X}),$$

where the augmentation i^q is an embedding. Furthermore the augmentations i^q intertwine the differentials on the complexes $\Omega^*(M; V)^G$ and $A^{*,q}_{s,dR}(M; V)^G$. Analogously one can augment the columns $A^{p,*}_{s,dR}(M; V)^G$ with the kernel of the vertical differential $d_v : A^{p,0}_{s,dR}(M; V)^G \to A^{p,1}_{s,dR}(M; V)^G$. Here the kernel is isomorphic to the vector space $A^p_s(M; V)^G$ of smooth global equivariant p-cochains

$$0 \longrightarrow A^p_s(M; V)^G \xrightarrow{j^p} A^{p,*}_{s,dR}(M; V)^G,$$

and the augmentations j^p intertwine the differential on the complexes $A^*_s(M; V)^G$ and $A^{p,*}_{s,dR}(M; V)^G$. These augmentations may be arranged as an augmented double complex:

$$\begin{array}{cccccc}
\vdots & \vdots & \vdots & \vdots & \\
\Omega^3(M;V)^G \hookrightarrow & A^{0,3}_{s,dR}(M;V)^G & A^{1,3}_{s,dR}(M;V)^G & A^{2,3}_{s,dR}(M;V)^G & \cdots \\
\Omega^3(M;V)^G \hookrightarrow & A^{0,2}_{s,dR}(M;V)^G & A^{1,2}_{s,dR}(M;V)^G & A^{2,2}_{s,dR}(M;V)^G & \cdots \\
\Omega^3(M;V)^G \hookrightarrow & A^{0,1}_{s,dR}(M;V)^G & A^{1,1}_{s,dR}(M;V)^G & A^{2,1}_{s,dR}(M;V)^G & \cdots \\
\\
& A^0_s(M;V)^G & A^1_s(M;V)^G & A^2_s(M;V)^G & \cdots
\end{array}$$

We now construct a contracting homotopy operator on the augmented row-complexes $0 \to \Omega^q(M;V) \hookrightarrow A^{*,q}_{s,dR}(M;V)$ of this double complex. We begin with maps on the components of the simplicial G-manifold $\mathrm{SS}(M) \times T^{\oplus q}M$. Here we define functions h_p via

$$h_p : M^p \times T^{\oplus q}M \to M^{p+1} \times T^{\oplus q}M, \quad h(m_0, \ldots m_p, \mathbf{X}) := (m_0, \ldots m_p, \pi_{TM}(X_0), \mathbf{X})$$

The functions h_p are smooth and equivariant by definition. They induce homomorphisms h^* on the rows $A^{*,q}_{s,dR}(M;V)^G$ of the double complex $A^{*,*}_{s,dR}(M;V)^G$ via $h^p(f)(\mathbf{m}, \mathbf{X}) = (-1)^p f(h_p(\mathbf{m}, \mathbf{X}))$.

Proposition 14.2.7. *The homomorphisms h^p form a row contraction for each augmented row $0 \hookrightarrow \Omega^q(M;V)^G \to A^{*,q}(M;V)^G$.*

Proof. Let $f \in A^{p,q}_{s,dR}(M;V)$ be a (p,q)-cochain. The evaluation of the cochain $(d_h \circ h^p + h^{p+1} \circ d_h)(f)$ at a point $(\mathbf{m}, \mathbf{X}) \in M^{p+2} \times T^{\oplus q+1}M$ is given by

$$[(d_h \circ h^p + h^{p+1} \circ d_h)(f)](\mathbf{m}, \mathbf{X}) = [d_h \circ h^p(f)](\mathbf{m}, \mathbf{X}) + [h^{p+1} \circ d_h(f)](\mathbf{m}, \mathbf{X})$$

$$= \sum_{i=0}^{p} (-1)^i h^p(f)(m_0, \ldots, \hat{m}_i, \ldots, m_p, \mathbf{X})$$

$$+ d_h(f)(m_0, \ldots, m_p, \pi_{TM}(X_0), \mathbf{X})$$

$$= \sum_{i=0}^{p} (-1)^{i+p} f(m_0, \ldots, \hat{m}_i, \ldots, m_p, \pi_{TM}(X_0), \mathbf{X})$$

$$+ \sum_{i=0}^{p} (-1)^{i+(p+1)} f(m_0, \ldots, \hat{m}_i, \ldots, m_p, \pi_{TM}(\mathbf{X}), \mathbf{X})$$

$$+ (-1)^{(p+1)+(p+1)} f(m_0, \ldots, m_p, \mathbf{X})$$

$$= 0 + (-1)^{2(p+1)} f(\mathbf{m}, \mathbf{X}) = f(\mathbf{m}, \mathbf{X}).$$

Thus the linear maps h^p satisfy the equations $\mathrm{id} = d_h \circ h^p + h^{p+1} \circ d_h$, hence h^* is a row contraction. $\square$

Corollary 14.2.8. *The rows of the double complex $A^{*,*}_{s,dR}(M;V)^G$ are split exact except at $A^{0,*}(M;V)^G$.*

Corollary 14.2.9. *The spectral sequence $_{II}E^{*,*}_*(M;V)$ collapses at the first stage.*

The rows $A^{*,q}_{s,dR}(M;V)^G$ of the double complex $A^{*,*}_{s,dR}(M;V)^G$ are exact except at degree 0, where the cohomology $H^0(A^{*,q}_{s,dR}(M;V)^G)$ is non-trivial. Therefore the only non-trivial first stage entries are those in the zeroth row:

Lemma 14.2.10. *The first stage modules* $_{II}E^{p,q}_{s,dR}(M;V)$ *are given by*

$$_{II}E^{p,q}_1 \cong \begin{cases} \Omega^q(M;V)^G & \text{for } p = 0 \\ 0 & \text{for } p \neq 0 \end{cases}$$

Proof. The kernel of the horizontal differential $d_h : A^{0,q}_{s,dR}(M;V)^G \to A^{1,q}_{s,dR}(M;V)^G$ consists of those equivariant functions which are independent of the first variable. Thus the assertion $_{II}E^{0,q}_1 = \ker d_h \cong \Omega^q(M;V)^G$ follows. $\square$

Theorem 14.2.11. *The spectral sequences* $_{II}E^{*,*}_*(M;V)$ *and* $_{II}E^{*,*}_*(M;V)$ *converge to the equivariant de Rham cohomology* $H^*_{dR,eq}(M;V)$.

Proof. Since both spectral sequences converge to the cohomology of the total complex $\mathrm{Tot}\,A^{*,*}_{s,dR}(M;V)^G$ it is sufficient to show that $_{II}E^{*,*}_*(M;V)$ converges to the equivariant de Rham cohomology $H^*_{dR,eq}(M;V)$. By Corollary 14.2.9 the spectral sequence collapses at the first stage and by Lemma 14.2.10 this spectral the only non-trivial modules are those of the zeroth column

$$_{II}E^{0,q}_1 = \Omega^q(M;V)^G.$$

Therefore the final stage modules are equal to the second stage modules. These compute to

$$_{II}E^{p,q}_\infty =_{II} E^{p,q}_2 \cong \begin{cases} H^q_{dR,eq}(M;V) & \text{for } p = 0 \\ 0 & \text{for } p > 0, \end{cases}$$

hence the spectral sequence $_{II}E^{*,*}_r$ converges to the equivariant de Rham cohomology of M. By Proposition 14.2.6 the spectral sequence $_IE^{*,*}_r$ then also converges to the equivariant de Rham cohomology. $\square$

We now turn our attention to the spectral sequence $_IE^{*,*}_*$ obtained by filtrating the double complex $A^{*,*}_{s,dR}(M;V)^G$ column-wise. The elements of the function spaces $A^{p,q}_{s,dR}(M;V)$ are smooth maps $f : M^{p+1} \times T^{\oplus q}M \to V$. In many cases the "exponential law" holds, i.e. the functions

$$\varphi_{M^{p+1},T^{\oplus q}M} : A^{p,q}_{s,dR}(M;V) \to A^p_s(M;{}^t\Omega^q(M;V)), \quad \varphi_{M^{p+1},T^{\oplus q}M}(f) = \hat{f}$$

which assign to each map $f \in A^{p,q}_s(M;V)$ its adjoint map $\hat{f}$ in $A^p_s(M;{}^t\Omega^q(M;V))$ are bijections of function spaces. To shorten the notation we abbreviate the homomorphisms $\varphi_{M^{p+1},T^{\oplus q}M}$ by $\varphi^{p,q}$. Since these functions $\varphi^{p,q}$ are G-equivariant by construction, they then assemble to an isomorphism of double complexes:

Lemma 14.2.12. *If the manifold M is locally compact, then $\varphi^{*,*}$ is an isomorphism of double complexes in* $\mathbb{K}[G]\mathbf{mod}$.

Proof. Since The homomorphisms $\varphi^{p,q} : A^{p,q}_{s,dR}(M;V) \to A^p_s(M;{}^t\Omega^q(M;V))$ are G-equivariant, it only is to prove that they are bijections. If M is locally compact then the maps $\varphi_{p,q}$ also are isomorphisms by Corollary 13.1.11. Therefore each of the homomorphisms $\varphi_{M^{p+1},T^{\oplus q}M} : A^{p,q}_{s,dR}(M;V) \to A^p_s(M;{}^t\Omega^q(M;V))$ is an isomorphism of $\mathbb{K}[G]$-modules in this case. $\square$

This can be readily generalised to manifolds, whose model space W and all finite products $W^n \times \mathbb{K}^m$ are k-spaces:

Proposition 14.2.13. *If all finite products $W^n \times \mathbb{K}^m$ of the model space W of M and the ground field $\mathbb{K}$ are k-spaces, then the homomorphisms*

$$\varphi_{M^{p+1},T^{\oplus q}M,V} : A^{p,q}_{s,dR}(M;V) \to A^p_s(M;{}^t\Omega^q(N;V))$$

are isomorphisms.

Proof. The homomorphisms $\varphi_{M^{p+1},T^{\oplus q}M,V} : A^{p,q}_{s,DR}(M;V) \to A^p_s(M;{}^t\Omega^q(M;V))$ of vector spaces are always injective. Under the assumptions on the modelling space W of M Corollary 13.1.23 implies that they are bijections. $\qquad\qquad\Box$

Corollary 14.2.14. *If all finite products $W^n \times \mathbb{K}^m$ of the model space W of the manifold M and the ground field $\mathbb{K}$ are k-spaces, then $\varphi^{*,*}$ is an isomorphism of double complexes in $\mathbb{K}[G]mod$.*

Example 14.2.15. Metrisable spaces are k-spaces. Since (higher) tangent bundles of metrisable manifolds are metrisable, the double complexes $A^{*,*}_{s,dR}(M;V)^G$ and $A^*_s(M;{}^t\Omega^*(M;V))^G$ are isomorphic if the manifold M is metrisable.

Example 14.2.16. Finite products of Hausdorff k_ω-spaces are k_ω-spaces (cf. Lemma B.10.7). Thus for each Hausdorff k_ω-manifold M the double complexes $A^{*,*}_{s,dR}(M;V)^G$ and $A^*_s(M;{}^t\Omega^*(M;V))^G$ are isomorphic.

Example 14.2.17. All countable colimits of k_ω-spaces are k_ω-spaces (cf. Proposition B.10.5). Thus for every Hausdorff manifold M which is a countable colimit of k_ω-spaces the double complexes $A^{*,*}_{s,dR}(M;V)^G$ and $A^*_s(M;{}^t\Omega^*(M;V))^G$ are isomorphic.

Let (G,M) be a smooth transformation group for which the double complexes $A^{*,*}_{s,dR}(M;V)^G$ and $A^*_s(M;{}^t\Omega^*(M;V))^G$ are naturally isomorphic via the map $f \mapsto \hat{f}$ and consider the spectral sequence ${}_IE^{*,*}_*$ associated to the column-wise filtration of the double complex $A^{*,*}_{s,dR}(M;V)^G$. Since the double complexes $A_{s,dR}(M;V)^G$ and $A^*_s(M;{}^t\Omega^*(M;V))^G$ are isomorphic, we can replace the first with the second one. The differential on the zeroth page ${}_IE^{*,*}_0$ of the spectral sequence ${}_IE^{*,*}_*$ is the vertical differential d_v, which is (up to a column-wise sign change) induced by the equivariant differential $d^t\Omega$ of the de Rham complex ${}^t\Omega^*(M;V)$ via $d^{p,q}_h = (-1)^p A^p_s(M;d^t\Omega)^G$. If the short sequences

$$0 \to {}^tB^q_{dR}(M;V) \hookrightarrow {}^tZ^q_{dR}(M;V) \twoheadrightarrow {}^tH_{dR}(M;V) \to 0$$

in **GTop$\mathbb{K}$Vec** are split exact, then the first stage modules of the spectral sequence ${}_IE^{*,*}_*$ can be expressed by equivariant smooth functions with values in the the the de Rham cohomology $H^*_{dR}(M;V)$ modules of M:

Proposition 14.2.18. *If all finite products $W^n \times \mathbb{K}^m$ of the model space W of the manifold M and the ground field $\mathbb{K}$ are k-spaces and the short exact sequences $0 \to {}^tB^q_{dR}(M;V) \hookrightarrow {}^tZ^q_{dR}(M;V) \twoheadrightarrow {}^tH_{dR}(M;V) \to 0$ in **GTop$\mathbb{K}$Vec** split, then the first and second stage modules of ${}_IE^{*,*}_*$ are given by*

$$ {}_IE^{p,q}_1 = A^p_s(M;{}^tH^q_{dR}(M;V))^G \quad and \quad {}_IE^{p,q}_2 = H^p_{s,eq}(M;{}^tH^q_{dR}(M;V)) $$

Proof. Under the assumption that all finite products $W^n \times \mathbb{K}^m$ of the model space W of the manifold M and the ground field $\mathbb{K}$ are k-spaces the spectral sequences ${}_IW^{*,*}_*$ and ${}_I\hat{E}^{*,*}_*$ are isomorphic (by Corollary 14.2.14). If in addition the sequences $0 \to {}^tB^q_{dR}(M;V) \hookrightarrow {}^tZ^q_{dR}(M;V) \twoheadrightarrow {}^tH_{dR}(M;V) \to 0$ in **GTop$\mathbb{K}$Vec** are split exact, then the application of each function space functor $A^p_s(M;-)^G$ yields a split short exact sequence

$$0 \to A^p_s(M;{}^tB^q_{dR}(M;V))^G \hookrightarrow A^p_s(M;{}^tZ^q_{dR}(M;V))^G \twoheadrightarrow A^p_s(M;{}^tH_{dR}(M;V))^G \to 0$$

of vector spaces, in which the first space $A^p_s(M;{}^tZ^q_{dR}(M;V))^G$ is the image of the vertical differential $d_v : A^p_s(M;{}^t\Omega^{q-1}(M;V))^G \to A^p_s(M;{}^t\Omega^q(M;V))^G$ and the middle space $A^p_s(M;{}^tZ^q_{dR}(M;V))^G$ is the kernel of the vertical differential. As a consequence the morphism $A^p_s(M;{}^tZ^q_{dR}(M;V))^G \twoheadrightarrow A^p_s(M;{}^tH_{dR}(M;V))^G$ induced by the quotient map ${}^t Z^q_{dR}(M;V) \twoheadrightarrow {}^t H_{dR}(M;V)$ is a quotient map itself. Since the first stage modules ${}_I \hat{E}^{p,q}_1$ are the quotients

$$ {}_I \hat{E}^{p,q}_1 = \ker d_v / \operatorname{Im} d_v = A^p_s(M; {}^t Z^q_{dR}(M;V))^G / A^p_s(M; {}^t B^q_{dR}(M;V))^G $$

they are isomorphic to the modules $A^p_s(M; {}^t H^q_{dR}(M;V))^G$. This proves the assertion on the first stage modules. The second assertion is an immediate consequence.   □

The spectral sequence ${}_I E^{*,*}_*$ relates subcomplex $C^*_s(M;V)^G$ of smooth global equivariant cochains of the total complex $\operatorname{Tot} A^{*,*}_{s,dR}(M;V)^G$ to the subcomplex of equivariant de Rham cochains: Each smooth global equivariant n-cocycle F on M can be identified with the $(n,0)$-cocycle $j^n(F)$ in $\operatorname{Tot} A^{*,*}_{s,dR}(M;V)^G$; this $(n,0)$-cocycle is cohomologous to the image $i^n(\omega)$ of an equivariant closed n-form ω on M. This equivalence is given by a homomorphism $D^* : A^*_s(M;V)^G \to \Omega^*(M;V)^G$ of chain complexes, which we are going to describe now. We first consider the double complex $A^{*,*}_{s,dR}(M;V)$ and later on restrict ourselves to equivariant cochains.

Let F be a smooth global n-cocycle on M an consider the cocycle $j^n(F)$ of bidegree $(0,n)$ in $\operatorname{Tot} A^{*,*}_{s,dR}(M;V)$. Since the row $A^{*,0}_{s,dR}(M;V)$ is exact and h^* provides a row contraction, i.e. $d_h h^* + h^* d_h = \operatorname{id}$, this cocycle $j^n(F)$ in $\operatorname{Tot} A^{*,*}_{s,dR}(M;V)$ is the horizontal coboundary of the cochain $h^n(F^{n,0})$ of bidegree $(n-1,0)$. The negative of vertical boundary $-d_v(h^n(j^n(F))$ of this $(n-1,0)$-cochain is cohomologous to the cocycle $j^n(F)$ in the total complex:

Lemma 14.2.19. *If G is a topological group, $F^{p,q}$ is a cocycle of bidegree (p,q) in the total complex of a double complex $A^{*,*}$ in $\mathbf{GTopAb}$, the cocycle $F^{p,q}$ is the horizontal coboundary $d_h \psi^{p-1,q}$ of a cochain of bidegree $(p-1,q)$, then the vertical coboundary $d_v \psi^{p-1,q}$ is a cocycle in $\operatorname{Tot} A^{*,*}$ which is cohomologous to $-F^{p,q}$. If $F^{p,q}$ is a cocycle of bidegree (p,q) in the total complex of a double complex $A^{*,*}$ in $\mathbf{GTopAb}$, the cocycle $F^{p,q}$ is the vertical coboundary $d_v \psi^{p,q-1}$ of a cochain of bidegree $(p,q-1)$, then the horizontal coboundary $d_h \psi^{p,q-1}$ is a cocycle in $\operatorname{Tot} A^{*,*}$ which is cohomologous to $-F^{p,q}$.*

Proof. Let G be a topological group, $A^{*,*}$ be a double complex in $\mathbf{GTopAb}$ and $F^{p,q}$ be a cocycle of bidegree (p,q) in the total complex $\operatorname{Tot} A^{*,*}$, which is the horizontal coboundary $d_h \psi^{p-1,q}$ of a cochain of bidegree $(p-1,q)$. The anticommutativity of the horizontal and vertical differentials implies the equations

$$ d_h d_v \psi^{p-1,q} = -d_h d_v \psi^{p-1,q} = d_v F^{p,q} = 0, $$

i.e. the horizontal coboundary of $d_v \psi^{p-1,q}$ is trivial. Since the vertical coboundary $d_v d_v \psi^{p-1,q}$ is also trivial, the cochain $d_v \psi^{p-1,q}$ is a cocycle in the total complex $\operatorname{Tot} A^{*,*}$. Furthermore the difference $d_v \psi^{p-1,q} - (-F^{p,q})$ is exactly the boundary $D \psi^{p-1,q} = d_v \psi^{p-1,q} + d_h \psi^{p-1,q}$, so $-F^{p,q}$ is cohomologous to $d_v \psi^{p-1,q}$. The other case is proved analogously.   □

We denote the cocycle $j^n(F)$ in the total complex $\operatorname{Tot} A^{*,*}_{s,dR}(M;V)$ by $F^{n,0}$, the cochain $d_h(F^{n,0})$ by $\psi^{n-1,0}$ and the cocycle $d_v \psi^{n-1,1}$ by $F^{n-1,1}$. All in all, we have found a cocycle $F^{n-1,-1}$ of bidegree $(n-1,1)$, which is cohomologous to $-F^{n,0}$ in the total complex $\operatorname{Tot} A^{*,*}_{s,dR}(M;V)$, where the situation can be depicted as follows:

$$
\begin{array}{ccc}
0 & & \\
\uparrow{\scriptstyle d_v} & & \\
F^{n-1,1} \xrightarrow{\ d_h\ } & 0 & \\
\uparrow{\scriptstyle d_v} & \uparrow{\scriptstyle d_v} & \\
\psi^{n-1,0} = h^n(F^{n,0}) \xrightarrow{\ d_h\ } F^{n,0} \xrightarrow{\ d_h\ } 0
\end{array}
$$

Recall that the cocycle F is a smooth function on M^{n+1} and we write points in M^{n+1} as $\mathbf{m} = (m_0, \ldots, m_n)$, beginning with index 0. The cocycle $F^{n-1,1}$ is a smooth function $M^n \times TM \to V$. It can be explicitly described using the partial derivative of F in the last (i.e. the n-th) coordinate:

Lemma 14.2.20. *The evaluation of the cocycle $F^{n-1,1} : M^n \times TM \to V$ at a point $(\mathbf{m}, X) \in M^n \times TM$ is given by $F^{n-1,1}(\mathbf{m}, X_1) = -[d_n F](\mathbf{m}, X_1)$.*

Proof. By the definition of the row contraction h, the cochain $\psi^{n-1,0}$ is given by $\psi^{n-1,0}(\mathbf{m}, m') = (-1)^n F^{n,0}(\mathbf{m}, m')$. The evaluation of the vertical coboundary $F^{n-1,1}$ of this cochain at a point $(\mathbf{m}, X) \in M^n \times TM$ is by definition given by

$$F^{n-1,1}(\mathbf{m}, X_1) = (d_v \psi^{n-1,0})(\mathbf{m}, X_1) = (-1)^{n-1} d[\widehat{\psi^{n-1,0}(\mathbf{m})}](X_1)$$
$$= (-1)^{n-1} d[(-1)^n F(m_0, \ldots, m_{n-1}, -)](X_1)$$
$$= -[d_n F](\mathbf{m}, X_1).$$

This proves the assertion. $\square$

The cocycle $F^{n-1,1}$ in the total complex $\mathrm{Tot} A^{*,*}_{s,dR}(M;V)$ is itself a horizontal boundary $d_h \psi^{n-2,1}$ of the cochain $\psi^{n-2,1} := h^{n-1}(F^{n-1,1})$ of bidegree $(n-2,1)$, because the row $A^{*,1}_{s,dR}(M;V)$ is exact and h^* provides a row contraction. So the above process can be repeated to obtain a cocycle $F^{n-2,2} := d_v \psi^{n-2,1}$ in the total complex of $A^{*,*}_{s,DR}(M;V)$ that is cohomologous to $F^{n,0}$. Proceeding inductively we obtain cochains $\psi^{n-1,0}, \ldots, \psi^{0,n-1}$ of total degree $n-1$ and cocycles $F^{n,0}, \ldots, F^{n,0}$ satisfying the equations $d_v \psi^{p-1,q} = F^{p,q}$ and $d_v \psi^{p-1,q} = F^{p-1,q+1}$ for all p, q such that $p + q = n$:

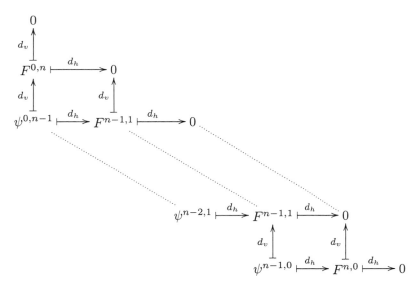

This may be expressed by cochains ψ, $F^{0,n}$ and $F^{n,0}$ in the total complex of double complex $A^{*,*}_{s,dR}(M;V)$: The alternating sum $\psi := \sum_{p'+q'=n-1} (-1)^{q'} \psi^{p',q'}$ is a cochain of total degree $n-1$ in the total complex $\mathrm{Tot} A^{*,*}_{s,dR}(M;V)$. Its coboundary in the total complex is the cocycle

$$D\psi = \sum_{p'+q'=n-1} (-1)^{q'} D\psi^{p',q'} = \sum_{p'+q'=n-1} (-1)^{q'} (d_h + d_v)\psi^{p',q'}$$
$$= \sum_{p'+q'=n-1} (-1)^{q'} (F^{p'+1,q'} + F^{p',q'+1}) = F^{n,0} + (-1)^{n-1} F^{0,n}.$$

This shows that the cocycles $F^{n,0}$ and $(-1)^n F^{0,n}$ in the total complex of the double complex $A_{s,dR}^{*,*}(M;V)$ are cohomologous. The cocycle $F^{0,n}$ is the image of a differential n-form ω on M under the augmentation $i^n : \Omega^n(M;V) \hookrightarrow A_{s,dR}^{*,n}(M;V)$ of the row $A_{s,dR}^{*,n}(M;V)$:

Lemma 14.2.21. *Every cochain $F^{0,n}$ of bidegree $(0,n)$ in the total complex of $A_{s,dR}^{*,*}(M;V)$ whose horizontal coboundary is trivial is the image $i^n(\omega)$ of a differential n-form ω on M under the augmentation $i^n : \Omega^n(M;V) \hookrightarrow A_{s,dR}^{*,n}(M;V)$.*

Proof. Let $F^{n,0}$ be a cochain $F^{0,n}$ of bidegree $(0,n)$ in the total complex of the double complex $A_{s,dR}^{*,*}(M;V)$. The evaluation of the horizontal coboundary $d_h F^{0,n}$ at a point $(m_0, m_1, \mathbf{X}) \in M^2 \times T^{\oplus n}M$ is given by

$$d_h F^{0,n}(m_0, m_1, \mathbf{X}) = F^{0,n}(m_1, \mathbf{X}) - F^{0,n}(m_0, \mathbf{X}).$$

Thus, if the horizontal coboundary $d_h F^{0,n}$ of $F^{0,n}$ is trivial, then the function $F^{0,n}$ is independent of the first variable $m_0 \in M$. In this case $\omega(\mathbf{X}) := F(m_0, \mathbf{X})$ defines an n-form on M, which is independent of the point m_0 chosen. The image $i^n(\omega)$ of this n-form is exactly the cochain $F^{0,n}$. □

Lemma 14.2.22. *Every cochain $F^{n,0}$ of bidegree $(n,0)$ in the total complex of $A_{s,dR}^{*,*}(M;V)$ whose vertical coboundary is trivial is the image $j^n(F)$ of a smooth equivariant global cocycle F under the augmentation $j^n : A_s^n(M;V) \hookrightarrow A_{s,dR}^{n,*}(M;V)$.*

Proof. The proof is similar to that of the previous lemma. □

Since the horizontal and vertical differentials as well as the row contractions all are G-equivariant, this procedure restricts to the sub-double complex $A_{s,dR}^{*,*}(M;V)^G$, i.e. if the n-cocycle F on M is equivariant, then so are all the cochains $\psi^{p',q'}$ and $F^{p,q}$. In particular the process for obtaining the cochains $\psi^{n-1,0}, \ldots, \psi^{0,n-1}$ and the cocycles $F^{0,n}, \ldots, F^{n,0}$ described above is $\mathbb{K}[G]$-linear in the starting cocycle F on M. This suggests that the row contraction h^* can be used to construct a morphism

$$p^* : \mathrm{Tot} A_{s,dR}^{*,*}(M;V) \to \Omega^*(M;V)$$

of chain complexes in $\mathbb{K}[G]\mathbf{mod}$ which is a left inverse for the inclusion $\Omega^*(M;V) \hookrightarrow \mathrm{Tot} A_{s,dR}^{*,*}(M;V)$ induced by the augmentation $i^* : \Omega^*(M;V) \hookrightarrow A_{s,dR}^{0,*}(M;V)$ and which is a homotopy equivalence of complexes. This is a general observation concerning double complexes with exact rows. The suggested morphism p^* has been constructed in the general case (see the book [BT82]):

Proposition 14.2.23. *There exists a morphism $p^* : \mathrm{Tot} A_{s,dR}^{*,*}(M;V) \to \Omega^*(M;V)$ of chain complexes in $\mathbb{K}[G]\mathbf{mod}$ which is explicitly given by the formula*

$$p^n(f^{p,q}) = (-d_v \circ h)^p(f^{p,q}) - \left[h \circ (-d_v \circ h)^{p-1} d_v \right](f^{p,q}) - \left[h \circ (-d_v \circ h)^p d_h \right](f^{p,q})$$

for cochains $f^{p,q} \in A_{s,dR}^{p,q}(M;V)$ of bidegree (p,q) $(p+q=n)$. It is a left inverse for the inclusion $\Omega^*(M;V) \hookrightarrow \mathrm{Tot} A_{s,dR}^{*,*}(M;V)$ and a homotopy equivalence.

Proof. The morphism p^* is constructed in [BT82, Proposition 9.5], without considering an action of a group G (i.e. for a trivial group action). Since the vertical and horizontal differentials as well as the row contraction h are equivariant, the morphism p^* given by the above formula is equivariant as well. □

If the cochain $F^{p,q}$ is a cocycle (i.e. both the vertical and horizontal boundaries of $f^{p,q}$ are trivial), then this formula reduces to the much simpler expression

$$p^{p+q}(F^{p,q}) = (-1)^p (d_v \circ h)^p (F^{p,q}),$$

which recovers our procedure introduced above. The morphism p^* may be restricted to the subcomplex $j^*(A_s^*(M;V))$ of the total complex $\mathrm{Tot}A_{s,dR}^{*,*}(M;V)$. This yields a morphism $p^* \circ j^* : A_s^*(M;V) \to \Omega^*(M;V)$ from the complex $A_s^*(M;V)$ of global smooth cochains on M to the de Rham complex $\Omega^*(M;V)$ of M. Since the vertical coboundary of any cochain $j^n(f)$ is trivial, we observe:

Lemma 14.2.24. *The morphism $p^* j^* : A_s^*(M;V) \to \Omega^*(M;V)$ is explicitly given by*

$$p^n j^n(f) = (-d_v \circ h)^p j^n(f) - [h \circ (-d_v \circ h)^p d_h] j^n(f)$$

for cochains $f^n \in A_s^n(M;V)$ of degree n.

The calculations below will reveal that the restriction of the homomorphism $p^* j^*$ to cocycles is the derivation homomorphism $\tau^* : Z_s^*(M;V) \to Z_{dR}^*(M;V)$.

Lemma 14.2.25. *The homomorphisms d_v and h satisfy the equations*

$$[(d_v \circ h + h \circ d_v)f](\mathbf{m}, \mathbf{X}) = \sum_{i=0}^{q+1} (-1)^i [d_p f](\mathbf{m}, X_i, X_0, \ldots, \hat{X}_i, \ldots, X_{q+1})$$

for all cochains $f \in A_{s,dR}^{p,q}(M;V)$ and all points $(\mathbf{m}, \mathbf{X})$ in $M^p \times T^{\oplus q+1}M$.

Proof. By the definition of the row contraction h, the cochain $h^p(f)$ is given by $h^p(f)(\mathbf{m}, \mathbf{X}) = (-1)^p f(\mathbf{m}, \pi_{TM}(X_0), \mathbf{X})$. Consider a point $(\mathbf{m}, \mathbf{X}) \in M^p \times T^{\oplus q+1}M$, let $m' = \pi_{TM}(X_1)$ the projection of the tangent vectors onto M and extend the tangent vectors $X_1, \ldots, X_{q+1}$ locally to smooth vector fields. The evaluation of the vertical coboundary $d_v h^p(f)$ at the point $(\mathbf{m}, \mathbf{X}) \in M^p \times T^{\oplus q+1}M$ is by definition given by

$$
\begin{aligned}
[d_v h^p(f)](\mathbf{m}, \mathbf{X}) &= \sum_{i=1}^{q+1} (-1)^{p-1+i+1} X_i . [h^p(f)(\mathbf{m}, X_1, \ldots, \hat{X}_i, \ldots, X_{q+1}) \\
&\quad + \sum_{i,j} (-1)^{p-1+i+j} h^p(f)(\mathbf{m}, [X_i, X_j], X_1, \ldots, \hat{X}_i, \ldots, \hat{X}_j, \ldots, X_{q+1}) \\
&= \sum_{i=1}^{q+1} (-1)^i X_i . [f(\mathbf{m}, \pi_{TM}(X_i), X_1, \ldots, \hat{X}_i, \ldots, X_{q+1}) \\
&\quad + \sum_{i,j} (-1)^{i+j-1} f(\mathbf{m}, m', [X_i, X_j], X_1, \ldots, \hat{X}_i, \ldots, \hat{X}_j, \ldots, X_{q+1}) \\
&= \sum_{i=1}^{q+1} (-1)^i X_i . [f(\mathbf{m}, m', X_1, \ldots, \hat{X}_i, \ldots, X_{q+1})] \\
&\quad + \sum_{i=1}^{q+1} (-1)^i [d_p f](\mathbf{m}, X_i, X_1, \ldots, \hat{X}_i, \ldots, X_{q+1}) \\
&\quad + \sum_{i,j} (-1)^{i+j-1} f(\mathbf{m}, m', [X_i, X_j], X_1, \ldots, \hat{X}_i, \ldots, \hat{X}_j, \ldots, X_{q+1})
\end{aligned}
$$

The evaluation of the cochain $h^p d_v(f)$ at the point $(\mathbf{m}, \mathbf{X}) \in M^p \times T^{\oplus q+1}M$ is by definition given by

$$[h^p d_v f](\mathbf{m}, \mathbf{X}) = \sum (-1)^p (d_v f)(m_0, \ldots, m_{p-1}, m', \mathbf{X})$$

$$= \sum_i (-1)^{p+p+i+1} X_i.[f(\mathbf{m}, m', X_1, \ldots, X_{q+1})]$$

$$+ \sum_{i,j} (-1)^{p+p+i+j} f(\mathbf{m}, m', [X_i, X_j], X_1, \ldots, \hat{X}_i, \ldots, \hat{X}_j, \ldots, X_{q+1})$$

$$= \sum_i (-1)^{i+1} X_i.[f(\mathbf{m}, m', X_1, \ldots, X_{q+1})]$$

$$+ \sum_{i,j} (-1)^{i+j} f(\mathbf{m}, m', [X_i, X_j], X_1, \ldots, \hat{X}_i, \ldots, \hat{X}_j, \ldots, X_{q+1})$$

The two expression for $h^p d_v f$ and $d_v h^p f$ differ in sign and in the additional term $\sum_{i=1}^{q+1} (-1)^i [d_p f](\mathbf{m}, X_i, X_1, \ldots, \hat{X}_i, \ldots, X_{q+1})$. This proves the assertion. □

If the vertical coboundary of a cochain $f^{p,q}$ of bidegree (p, q) is trivial, the preceding Lemma implies that the image $d_v h^p f^{p,q}$ of $f^{p,q}$ under $d_v \circ h^p$ is given by the formula

$$d_v \circ h(f)(\mathbf{m}, \mathbf{X}) = \sum_{i=1}^{q+1} (-1)^i [d_p f](\mathbf{m}, X_i, X_1, \ldots, \hat{X}_i, \ldots, X_{q+1}).$$

As is apparent form the construction, the vertical coboundary $d_v(d_v h(f)) = d_v^2 h(f)$ of the new cochain $d_v h(f)$ is also trivial, so this process may be iterated. (cf. the construction of the cocycles $F^{n,0}, \ldots, F^{0,n}$). Doing so, we make the following observation:

Proposition 14.2.26. *For any cochain $f \in A_{s,dR}^{p,q}(M; V)$ with trivial vertical coboundary and any $k \leq p$ the image of f under $(d_v \circ h)^k$ is given by*

$$[(d_v \circ h)^k f](\mathbf{m}, \mathbf{X}) = \frac{1}{q!} \sum_{\sigma \in S_{k+q}} \operatorname{sign}(\sigma) [d_{p-k+1} \cdots d_p f](\mathbf{m}, X_{\sigma(1)}, \ldots, X_{\sigma(k+q)})$$

for all tangent $(q + k)$-vectors $\mathbf{X} \in T^{\oplus q+k} M$.

Proof. This is proved by induction on k. For $k = 1$ the assertion has been proved in Lemma 14.2.25. Let $f \in A_{s,dR}^{p,q}(M; V)$ be a cochain with trivial vertical coboundary and assume the assertion to be true for $0 \leq k$. Consider the cochain $[(d_v \circ h)^{k+1} f]$. Let $(\mathbf{m}, \mathbf{X})$ be a point in $M^{p-k} \times T^{\oplus k+q+1}$. By Lemma 14.2.25 the evaluation of the cochain $[(d_v \circ h)^{k+1} f]$ at the point $(\mathbf{m}, \mathbf{X})$ is given by

$$[(d_v \circ h)^{k+1} f](\mathbf{m}, \mathbf{X}) = [(d_v h)(d_v \circ h)^k f](\mathbf{m}, \mathbf{X})$$

$$= \sum (-1)^i [d_{p-k}(d_v \circ h)^{k+1} f](\mathbf{m}, X_i, X_1, \ldots, \hat{X}_i, \ldots, X_q)$$

$$= \frac{1}{q!} \sum_{\substack{\sigma \in S_{q+k} \\ 0 \leq i \leq k}} (-1)^i \operatorname{sgn}(\sigma) [d_{p-k} d_{p-k+1} \cdots d_p f](\mathbf{m}, X_i, X_{\sigma(1)}, \ldots, \hat{X}_{\sigma(i)}, \ldots, X_{\sigma(k+q)})$$

The bijection $(1, \ldots, q+k+1) \mapsto (i, 1, \ldots, i-1, i+1, \ldots, q+k+1)$ is a cyclic permutation of the first i entries. It has sign $(-1)^i$. The composition of this permutation with a permutation σ of the last $q + k$ entries is a permutation of $(1, \ldots, q+k+1)$ with sign $(-1)^i \operatorname{sign}(\sigma)$. The compositions of the cyclic permutation with all the permutations of the last $q + k$ entries are all the permutations which map 1 to i. All in all the sum runs over all permutations of $(1, \ldots, q+k+1)$ where each derivative is weighted by the sign of the permutation σ applied. This proves the assertion. □

Corollary 14.2.27. *The restriction $(d_v h)^* j^* : Z_s^*(M; V) \rightarrow Z_{dR}^*(M; V)$ is the derivation homomorphism $\tau^* : Z_s^*(M; V) \rightarrow Z_{dR}^*(M; V)$.*

As a consequence we observe that a closed differential n-form on M is integrable to a global smooth cocycle if and only if it is in the image of the homomorphism $(d_v h)^n j^n : Z_s^n(M;V) \to Z_{dR}^n(M;V)$. Similarly, an equivariant closed n-form ω on M is integrable to an equivariant smooth global n-cocycle if and only if there exists an equivariant smooth global n-cocycle F on M with image $\tau^n(F) = \omega$. In this case the cochain $j^n(F)$ is a cocycle in the total complex of $A_{s,dR}^{*,*}(M;V)^G$ and the same is true for the cochains $F^{p,q} = (d_v h)^q j^n(F)$, where h denotes the the row contractions of $A_{s,dR}^{*,*}(M;V)^G$. In addition the cochains $\psi^{p',q'} = (d_V h)^{q'} h j^n(F)$ satisfy the equations $D\psi^{p',q'} = F^{p',q'+1} + F^{p'+q,q'}$:

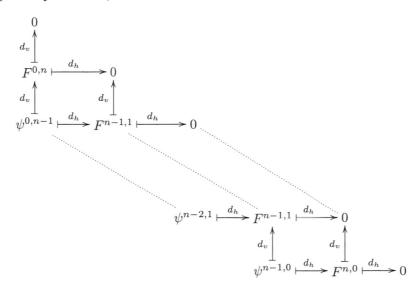

Theorem 14.2.28. *The second stage modules on the horizontal axis of the spectral sequence $_I E_*^{*,*}$ are given by $_I E_2^{n,0} \cong H_{s,eq}^n(M;V)$. The edge homomorphisms $_I E_2^{n,0} \cong H_{s,eq}^n(M;V) \to {}_I E_\infty^{n,0} \le H_{dR,eq}^n(X;V)$ and $H_{dR,eq}^n(X,A) \to {}_I E_\infty^{0,n}$ are induced by the derivation homomorphism $\tau^* : A_s^*(M;V)^G \to \Omega^*(M;V)^G$ and the inclusion $i^* : \Omega^*(M;V)^G \to {}_I E_*^{0,*}(M;V)$ respectively.*

Proof. Only the first assertions require proof. The kernel of the vertical differential $d_v : A_{s,dR}^{n,0}(M;V)^G \to A_{s,dR}^{n,1}(M;V)^G$ is the image of the homomorphism $j^n : A_s^n(M;V)^G \hookrightarrow A_{s,dR}^{n,0}(M;V)^G$. Since the double complex $A_{s,dR}^{*,*}(M;V)^G$ is a first quadrant double complex, the image of the vertical differential in $A_{s,dR}^{n,0}(M;V)^G$ trivial. Therefore the first stage modules $_I E_1^{n,0}$ are the submodules of $A_{s,dR}^{n,0}(M;V)^G$ consisting of the equivariant cochains with trivial vertical coboundary, i.e. the the augmentation j^n provides an isomorphism

$$j^n : A_s^n(M;V)^G \xrightarrow{\cong} {}_I E_1^{n,0}. \tag{14.1}$$

The augmentation $j^n : A_s^n(M;V)^G \hookrightarrow A_{s,dR}^{n,0}(M;V)^G$ intertwines the differential on $A_s^*(M;V)^G$ and the horizontal differential on $A_{s,dR}^{*,0}(M;V)^G$. Therefore the cohomology of the complexes $A_s^*(M;V)^G$ and $(_I E_1^{*,0}, d_h)$ coincide, i.e. the second stage modules on the horizontal axis are determined by the isomorphism

$$H^n(j^*) : H_{s,eq}^n(M;V) \xrightarrow{\cong} {}_I E_2^{n,0}.$$

The homotopy equivalence $p : \mathrm{Tot} A^{*,*}(M;V)^G \to \Omega^*(M;V)^G$ maps each cocycle $j^n(F)$ in $A_{s,dR}^{n,0}$ to the cocycle $D^n(F)$ in $\Omega^n(M;V)$, hence $H^n(p^*)$ maps each class $[j^n(F)]$ in $_I E_2^{n,0}$ to the class $[D^n(F)]$ in $H_{dR,eq}(M;V)$. $\square$

14.3 Non-smooth Cochains

The smoothness of an equivariant cocycle is a major restriction. In general smooth cocycles are not the appropriate concept to describe fibre bundles or Lie group extensions. This is so because the global smoothness of cocycles amounts to the existence of global smooth sections. A globally smooth 2 cocycle $f \in C^2(G; A)$ for example describes an isomorphy class of group extensions

$$0 \to A \to A \times_f G \to G \to 1$$

that admit a global smooth section, i.e. $A \times_f G = A \times G$ as manifolds. Since we are interested in the more general class of locally trivial bundles (resp. group extensions that are principal fibre bundles) and not necessarily smooth extensions of local cocycles, we introduce a new double complex, that is appropriate for describing these objects.

Let (G, X) be a transformation group, V be a G-module and $\mathfrak{U}$ be on open covering of X. We denote the cosimplicial G-module of $\mathfrak{U}$-local cochains $\mathrm{SS}(X, \mathfrak{U}) \to V$ by $A(X, \mathfrak{U}; V)$ and the cosimplicial G-module of global cochains on X with values in V by $A(X; V)$. (These are the non-topological versions to the cosimplicial topological G-modules $A^t(X, \mathfrak{U}; V)$ and $A^t(X; V)$ introduced in Section 2.5.) The cohomology of the complex $A^*(X, \mathfrak{U}; V)$ associated to V is called the $\mathfrak{U}$-*local cohomology*. If $\mathfrak{U}$ is a G-invariant open covering, then the cohomology of the complex $A^*(X, \mathfrak{U}; V)^G$ is called the *equivariant $\mathfrak{U}$-local cohomology*. Combining both constructions $A(X; V)$ and $A(X, \mathfrak{U}; V)$ yields a bisimplicial G-module

$$A(X; A(X, \mathfrak{U}; V)),$$

which in turn gives rise to a double complex $A^*(X; A^*(X, \mathfrak{U}; V))$. The elements of the modules $A^p(X; A^q(X, \mathfrak{U}; V))$ are the functions $X^{p+1} \to A^q(X, \mathfrak{U}; V)$, i.e. the global p-cochains with values in the $\mathbb{Z}[G]$-module $A^q(X, \mathfrak{U}; V)$ of $\mathfrak{U}$-local q-cochains. We consider the $\mathbb{Z}[G]$-modules of the adjoin functions $X^{p+1} \times \mathrm{SS}(X, \mathfrak{U})([q]) \to V$. The also form a bisimplicial $\mathbb{Z}[G]$-module. We call it the *bisimplicial $\mathbb{Z}[G]$-module adjoint* to $A(X; A(X, \mathfrak{U}; V))$.

Definition 14.3.1. *The double complex associated to the bisimplicial $\mathbb{Z}[G]$-module adjoint to $A(X; A(X, \mathfrak{U}; V))$ is denoted by $A^{*,*}(X, \mathfrak{U}; V)$.*

This double complex is a first quadrant double complex, which is isomorphic to the first quadrant double complex $A^*(X; A^*(X, \mathfrak{U}; V))$. The horizontal and vertical differentials on $A^{*,*}(X, \mathfrak{U}; V)$ are induced by the usual differentials on the complexes $A^*(X; A^q(X, \mathfrak{U}; V))$ and $A^*(X, \mathfrak{U}; V)$ respectively. Therefore the horizontal differential $d_h^{p,q}$ on $A^{p,q}(X, \mathfrak{U};; V))$ is given by

$$(d_h^{p,q} f)(x_0, \ldots, x_{p+1}, \mathbf{y}) := \sum_{0 \le i \le n+1} (-1)^i f(x_0, \ldots, \hat{x}_i, \ldots, x_{p+1}, \mathbf{y}),$$

and the vertical differential $d_v^{p,q}$ on $A^{p,q}(X, \mathfrak{U}; V))$ is given by

$$(d_v^{p,q} f)(\mathbf{x}, y_0, \ldots, y_{q+1}) := \sum_{0 \le j \le n+1} (-1)^{p+j} f(\mathbf{x}, y_0, \ldots, \hat{y}_j, \ldots, y_{q+1}).$$

Analogously to the procedure in the last section we restrict ourselves to the sub-complex $A^{*,*}(X, \mathfrak{U}; V)^G$ of equivariant cochains, where we assume the open covering $\mathfrak{U}$ to be equivariant. The associated spectral sequences to the double complex $A^{*,*}(X, \mathfrak{U}; V)^G$ will be denoted by $_I E_*^{*,*}$ and $_{II} E_*^{*,*}$. Since the double complex $A^{*,*}(X, \mathfrak{U}; V)^G$ is a first quadrant double complex its associated spectral sequences converge to the cohomology of the total complex:

Proposition 14.3.2. *The spectral sequences $_IE_*^{*,*}$ and $_{II}E_*^{*,*}$ converge to the cohomology of the total complex $\mathrm{Tot}(A^{*,*}(X,\mathfrak{U};V))$.*

$$_IE_*^{*,*} \Rightarrow H^*(\mathit{Tot}(A^{*,*}(X,\mathfrak{U};V))), \quad _{II}E_*^{*,*} \Rightarrow H^*(\mathit{Tot}(A^{*,*}(X,\mathfrak{U};V))),$$

Proof. See [McC01, Theorem 2.15] for a proof. □

Fortunately all the constructions introduced in the previous sections go through for this more general double complex as well. At first we construct a contracting homotopy h^p on each row $A^{*,q}(X,\mathfrak{U};V)$ by defining

$$h^p : A^{p,q}(X,\mathfrak{U};V) \to A^{p-1,q}(X,\mathfrak{U};V)$$
$$h^p(f)(x_0,\ldots,x_{p-1},\mathbf{y}) = (-1)^p f(x_0,\ldots,x_{p-1},y_0,\mathbf{y}).$$

The homomorphisms h^p are equivariant by definition, hence they restrict to functions $A^{p,q}(X,\mathfrak{U};V)^G \to A^{p-1,q}(X,\mathfrak{U};V)^G$.

Proposition 14.3.3. *The maps h^p form contracting homotopy on the augmented row-complexes $0 \to A^q(X,\mathfrak{U};V)^G \to A^{*,q}(X,\mathfrak{U};V)^G$.*

Proof. Let $f \in A^{p,q}(X,\mathfrak{U};V)$ be given. The evaluation of $[d_h h^p + h^{p+1} d_h](f)$ at a point $(\mathbf{x},\mathbf{y}) \in X^{p+1} \times \mathrm{SS}(X,\mathfrak{U}([n])$ is given by

$$
\begin{aligned}
[d_h h^p + h^{p+1} d_h](f)(\mathbf{x})(\mathbf{y}) &= \sum_{i=0}^{p}(-1)^i (h^p f)(x_0,\ldots,\hat{x}_i,\ldots,x_p)(\mathbf{y}) \\
&\quad + (-1)^{p+1}(d_h f)(x_0,\ldots,x_p,y_0)(\mathbf{y}) \\
&= \sum_{i=0}^{p}(-1)^{p+i} f(x_0,\ldots,\hat{x}_i,\ldots,x_p,y_0)(\mathbf{y}) \\
&\quad + \sum_{i=0}^{p}(-1)^{p+i+1} f(x_0,\ldots,\hat{x}_i,\ldots,x_p,y_0)(\mathbf{y}) \\
&\quad + (-1)^{p+1}(-1)^{(p+1)} f(\mathbf{x})(\mathbf{y}) \\
&= f(\mathbf{x})(\mathbf{y}),
\end{aligned}
$$

which proves the equality $d_h h^p + h^{p+1} d_h = \mathrm{id}$. Thus h^* is a row contraction. □

Corollary 14.3.4. *The rows of the double complex $A^{*,*}(X,\mathfrak{U};V)$ are split exact, except at the modules $A^{0,q}(X,\mathfrak{U};V)$.*

Corollary 14.3.5. *The spectral sequence $_{II}E_*^{*,*}$ collapses at the first stage.*

Lemma 14.3.6. *The first stage modules $_{II}E^{p,q}(X,\mathfrak{U};V)$ are given by*

$$_{II}E_1^{p,q} = \begin{cases} A^q(X,\mathfrak{U};V)^G & \text{for } p = 0 \\ 0 & \text{for } p \neq 0 \end{cases}$$

Proof. The kernel of the horizontal differential $d_h : A^{0,q}(M,\mathfrak{U};V) \to A^{1,q}(M,\mathfrak{U};V)$ consists of those functions which are independent of the first variable. Thus the assertion $_{II}E_1^{0,q} = \ker d_h \cong A^q(M,\mathfrak{U};V)^G$ follows. □

Theorem 14.3.7. *The spectral sequence $_{II}E_*^{*,*}(X,\mathfrak{U};V)$ converges to the equivariant $\mathfrak{U}$-local cohomology $H_{eq}^*(X,\mathfrak{U};V)$.*

Proof. By the preceding Lemma the stage 2 modules $_{II}E_2^{0,q} =_{II} E_\infty^{0,q}$ are the modules $H_{eq}^q(X,\mathfrak{U};V)$ and all other modules are trivial. Thus the spectral sequence converges to the equivariant $\mathfrak{U}$-local cohomology of X. □

Since the spectral sequences $_IE_r^{*,*}$ and $_{II}E_r^{*,*}$ both converge to the cohomology of the total complex $\text{Tot}A^{*,*}(X,\mathfrak{U};V)^G$, the latter cohomology has to be isomorphic to the equivariant $\mathfrak{U}$-local cohomology $H_{eq}(X,\mathfrak{U};V)$ on X. The complex of $\mathfrak{U}$-local equivariant cochains can be embedded into the total complex $\text{Tot}A^{*,*}(X,\mathfrak{U};V)^G$ via

$$i^q : A^q(X,\mathfrak{U};V)^G \hookrightarrow A^{0,q}(X,\mathfrak{U};V)^G \le (\text{Tot}A^{*,*}(X,\mathfrak{U};V))^q, \quad (i^q f)(x_0,\mathbf{y}) = f(\mathbf{y})$$

Moreover, because the rows of the double complex $A^{*,*}(X,\mathfrak{U};V)^G$ are exact (by Corollary 14.3.4) each cocycle in the total complex $\text{Tot}A^{*,*}(X,\mathfrak{U};V)^G$ is cohomologous to a cocycle in the image of i^* and the morphism i^* induces an isomorphism in cohomology. (See [BT82, Chapter 2,§8] for an explicit explanation.) The complex $A^*(X;V)^G$ of global equivariant cochains can also be embedded into the total complex. The embedding is given by

$$j^p : A^p(X;V)^G \hookrightarrow A^{p,0}(X;V)^G \le (\text{Tot}A^{*,*}(X,\mathfrak{U};V))^p, \quad (j^p f)(\mathbf{x},y_0) = f(\mathbf{x}).$$

In this way the total complex contains both the equivariant global cochains and the equivariant $\mathfrak{U}$-local cochains. We assert that every global equivariant n-cocycle in the total complex is cohomologous to its restriction $f_{|SS(X,\mathfrak{U})([n])}$, which is a $\mathfrak{U}$-local n-cocycle:

Proposition 14.3.8. *The image $j^n(f)$ of a global equivariant n-cocycle f on X is cohomologous to the image $i^n(f_{|SS(X,\mathfrak{U})([n])})$ of the $\mathfrak{U}$-local cocycle $f_{|SS(X,\mathfrak{U})([n])}$.*

Proof. Let f be a global equivariant cocycle on X and define cochains $\psi^{p,q}$ of bidegree (p,q), $p+q = n-1$ via $\psi^{p,q}(\mathbf{x},\mathbf{y}) = (-1)^p f(\mathbf{x},\mathbf{y})$ for points $\mathbf{x} \in X^{p+1}$ and $\mathbf{y} \in SS(X,\mathfrak{U})([q])$. The vertical coboundary of the cochain $\psi^{p,q}$ is given by

$$[d_v\psi^{p,q}](\mathbf{x},y_0,\dots,y_{q+1}) = (-1)^p(-1)^p \sum(-1)^i f(\mathbf{x},y_0,\dots,\hat{y}_i,\dots,y_q)$$
$$= -\sum(-1)^{p+1+i} f(x_0,\dots,\hat{x}_i,\dots,x_p,\mathbf{y})$$
$$= d_h\psi^{p-1,q+1}$$

The anticommutativity of the horizontal and the vertical differential ensures that the coboundary of the cochain $\sum_{p+q=n-1}(-1)^p\psi^{p,q}$ is the cochain $j^n(f) - i^n(f_{|SS(X,\mathfrak{U})([p])})$ in the total complex. Thus the cocycles $j^n(f)$ and $i^n(f_{|SS(X,\mathfrak{U})([n])})_{|SS(X,\mathfrak{U})([n])}$ are cohomologous. $\square$

Remark 14.3.9. This proof also shows that an equivariant $\mathfrak{U}$-local n-cocycle f in $Z^n(X,\mathfrak{U};V)^G$, is extendible to a global equivariant cocycle if and only if it can successively be "shifted" to cohomologous cocycles $g^{1,n-1}$, $g^{2,n-2}$, ... $g^{n,0}$ of bidegree $(1,n-1)$, $(2,n-2)$ etc. in the total complex $\text{Tot}A^{*,*}(X,\mathfrak{U};V)$.

This allows us to derive the following generalisation of van Ests result on the extensibility of local group cocycles:

Theorem 14.3.10. *If the space X is n-connected, then to every equivariant $\mathfrak{U}$-local n-cocycle f there exists a refinement $\mathfrak{V}$ of $\mathfrak{U}$ and a global equivariant n-cocycle F, whose restriction to $SS(X,\mathfrak{V})([n])$ coincides with the restriction of f to $SS(X,\mathfrak{V})([n])$.*

Proof. Let $f \in Z^n(X,\mathfrak{U};V)^G$ be an equivariant $\mathfrak{U}$-local cocycle and consider the image $i^n(f) \in A^0(X;A^n(X,\mathfrak{U};V))^G$ in the total complex $\text{Tot}A^{*,*}(X,\mathfrak{U};V)^G$. Because the space X is n-connected, the n-th Alexander-Spanier cohomology $H_{AS}^n(X;V)$ is trivial. Therefore there exists an open refinement $\mathfrak{U}_1$ of $\mathfrak{U}$ such that the restriction $f_{|SS(X,\mathfrak{U}_1)([n])}$ is d_v-exact, i.e. an image of a cochain $g^{0,n-1} \in A^{0,n-1}(X,\mathfrak{U};V)^G$

under the vertical differential. The image $f^{1,n-1} = -d_h g^{0,n-1}$ of this cochain under the horizontal differential is cohomologous to the restriction $i^n(f)_{|SS(X,\mathfrak{U}_1)([n])}$ in the total complex $\mathrm{Tot}A^{*,*}(X,\mathfrak{U}_1;V)^G$. The anticommutativity of the horizontal and vertical differential implies the equalities

$$d_v f^{1,n-1} = d_v d_h g^{0,n-1} = -d_h d_v g^{0,n-1} = -d_h i^n(f_{|SS(x,\mathfrak{U}_1)([n])}) = 0,$$

i.e. the cochain $f^{1,n-1}$ is a cocycle. Repeated application of this procedure yields a cocycle $f^{n,0} \in A^{n,0}(X,\mathfrak{V};V)$ for some refinement $\mathfrak{V}$ of $\mathfrak{U}$, which is cohomologous to the restriction $f_{|SS(X,\mathfrak{V})([n])}$ of $i^n(f)$ in the total complex $\mathrm{Tot}A^{*,*}(X,\mathfrak{V};V)^G$. Since the vertical differential of this cocycle is trivial, it has to be independent of the variable y_0, hence is the image of a cocycle $F \in Z^n(X,\mathfrak{U})$ under j^n. This is a global equivariant cocycle whose restriction to $SS(X,\mathfrak{V})([n])$ has the same cohomology class in $H_{eq}(X,\mathfrak{V};V)$, hence F can be altered by a coboundary to obtain a cocycle whose restriction to $SS(X,\mathfrak{V})([n])$ coincides with that of f. $\qquad\square$

As has just been demonstrated, the spectral sequence $_IE_*^{*,*}$ relates subcomplex $A_s^*(M;V)^G$ of global equivariant cochains of the total complex $\mathrm{Tot}A^{*,*}(X,\mathfrak{U};V)^G$ to the subcomplex of equivariant $\mathfrak{U}$-local cochains: Each global equivariant n-cocycle f on X can be identified with the $(n,0)$-cocycle $j^n(f)$ in $\mathrm{Tot}A^{*,*}(M,\mathfrak{U};V)^G$; this $(n,0)$-cocycle is cohomologous to the image $i^n(f_{|SS(X,\mathfrak{U})([n])})$ of the equivariant $\mathfrak{U}$-local cocycle $f_{|SS(X,\mathfrak{U})([n])}$. This correspondence can be described in terms of the row contraction h:

Proposition 14.3.11. *There morphism $p^* : \mathrm{Tot}A^{*,*}(X,\mathfrak{U};V)^G \to A^*(X,\mathfrak{U};V)^G$ of chain complexes in $\mathbb{R}[G]\mathbf{mod}$ given by the formula*

$$p^n(f^{p,q}) = (-d_v \circ h)^p(f^{p,q}) - \left[h \circ (-d_v \circ h)^{p-1} d_v \right] (f^{p,q}) - \left[h \circ (-d_v \circ h)^p d_h \right] (f^{p,q})$$

for cochains $f^{p,q} \in A^{p,q}(X;V)^G$ of bidegree (p,q) $(p+q=n)$ is a left inverse for the inclusion $A^*(X,\mathfrak{U};V)^G \hookrightarrow \mathrm{Tot}A^{*,*}(X,\mathfrak{U};V)^G$ and a homotopy equivalence.

Proof. The morphism p^* is constructed in [BT82, Proposition 9.5], without considering an action of a group G (i.e. for a trivial group action). Since the vertical and horizontal differentials as well as the row contraction h are equivariant, the morphism p^* given by the above formula is equivariant as well. $\qquad\square$

If the cochain $F^{p,q}$ is a cocycle (i.e. both the vertical and horizontal boundaries of $f^{p,q}$ are trivial), then this formula reduces to the much simpler expression

$$p^{p+q}(f^{p,q}) = (-1)^p(d_v \circ h)^p(f^{p,q}).$$

The morphism p^* may be restricted to the subcomplex $j^*(A^*(X;V)^G)$ of the total complex $\mathrm{Tot}A^{*,*}(X,\mathfrak{U};V)^G$. This yields a morphism $A^*(X;V)^G \to A^*(X,\mathfrak{U};V)^G$ from the complex $A^*(X;V)^G$ of global equivariant cochains on X to the complex $A^*(X,\mathfrak{U};V)^G$ of $\mathfrak{U}$-local equivariant cochains on X. Since the vertical coboundary of any cochain $j^n(f)$ is trivial, we observe:

Lemma 14.3.12. *The morphism $p^* j^* : A^*(M;V)^G \to A^*(M,\mathfrak{U};V)^G$ is explicitly given by*

$$p^n j^n(f) = (-d_v \circ h)^p j^n(f) - \left[h \circ (-d_v \circ h)^p d_h \right] j^n(f)$$

for n-cochains f in $A^n(M;V)^G$.

As in the last section we see that any cocycle $f^{0,n}$ of bidegree $(0,n)$ in the total complex $\mathrm{Tot}A^{*,*}(X,\mathfrak{U};V)^G$ is essentially a $\mathfrak{U}$-local equivariant n-cocycle on X:

Lemma 14.3.13. *Every cochain $f^{0,n}$ of bidegree $(0, n)$ in the total complex of $A^{*,*}(X, \mathfrak{U}; V)^G$ whose horizontal coboundary is trivial is the image $i^n(f)$ of an equivariant $\mathfrak{U}$-local n-cochain on X under the augmentation $i^n : A^n(M, \mathfrak{U}; V)^G \hookrightarrow A^{*,n}(M, \mathfrak{U}; V)^G$.*

Proof. Let $f^{n,0}$ be a cochain $f^{0,n}$ of bidegree $(0, n)$ in the total complex of the double complex $A^{*,*}(X, \mathfrak{U}; V)^G$. The evaluation of the horizontal coboundary $d_h f^{0,n}$ at a point $(x_0, x_1, \mathbf{y}) \in X^2 \times \mathrm{SS}(X, \mathfrak{U})([n])$ is given by

$$d_h f^{0,n}(x_0, x_1, \mathbf{y}) = f^{0,n}(x_1, \mathbf{y}) - F^{0,n}(x_0, \mathbf{y}).$$

Thus, if the horizontal coboundary $d_h f^{0,n}$ of $f^{0,n}$ is trivial, then the function $f^{0,n}$ is independent of the first variable $x_0 \in X$. In this case $f(\mathbf{y}) := f^{0,n}(x_0, \mathbf{y})$ defines an equivariant n-cochain on M, which is independent of the point x_0 chosen. The image $i^n(f)$ of this n-form is exactly the cochain $F^{0,n}$. □

Lemma 14.3.14. *Every cochain $f^{n,0}$ of bidegree $(n, 0)$ in the total complex of $A^{*,*}(X, \mathfrak{U}; V)^G$ whose vertical coboundary is trivial is the image $j^n(f)$ of an equivariant global cocycle f under the augmentation $j^n : A^n(M; V)^G \hookrightarrow A^{n,*}(X, \mathfrak{U}; V)^G$.*

Proof. The proof is similar to that of the previous lemma. □

Recall from the proof of Proposition 14.3.8 that for each global equivariant n-cocycle f on X there exist cochains $\psi^{n-1,0}, \ldots, \psi^{0,n-1}$ of bidegree $(n-1, 0), \ldots, (0, n-1)$ respectively such that $d_v \psi^{p,q} = d_h \psi^{p-1,q+1}$ and the coboundary $D\psi$ of the cochain $\sum_{p+q=n-1} (-1)^p \psi^{p,q}$ is the cocycle $j^n(f) - (-1)^n i^n(f_{|\mathrm{SS}(X,\mathfrak{U})([n])})$. Defining cochains $f^{p,q}$ of bidegree (p, q) via $f^{p,q} = d_v \psi^{p+1,q-1}$ or $f^{p,q} = d_h \psi^{p-1,q+1}$ one can depict this situation as follows:

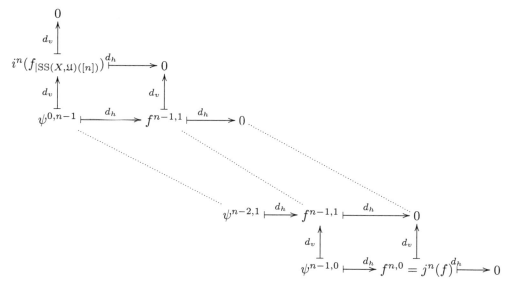

Thus if f is an equivariant $\mathfrak{U}$-local cocycle which is the restriction of a global equivariant cocycle, then cochains $\psi^{p,q}$ and $f^{p,q}$ as depicted above have to exist.

Equivariant Extension of Local Cocycles

The purpose of this chapter is to relate the cohomology $H_{eq}(M; V)$ of equivariant global cochains on the G-space X of a transformation group (G, X) of special type to the equivariant $\mathfrak{U}$-local cohomology $H_{eq}(X, \mathfrak{U}; V)$ for an open cover $\mathfrak{U}$ of X. This includes the relation of the group cohomology $H(G; V)$ of a topological group G and the cohomology of local group cochains as a special case. In contrast to the integrability of differential forms considered in the next chapter there are no difficulties arising from the non-existence of classical theorems in the infinite dimensional context. Also, we do not have any restrictions on the coefficient space V, which here is just an abstract Abelian group.

The cohomologies $H_{eq}(X; V)$ and $H_{eq}(X, \mathfrak{u}; V)$ of a transformation group (G, X) of special type are related by characterising the extendibility of equivariant $\mathfrak{U}$-local cocycles on X to global equivariant cocycles; we call an equivariant $\mathfrak{U}$-local cocycle on X *extendible* to a global equivariant cocycle if is the restriction of a global equivariant cocycle on X. Under different assumptions on the space X the necessary and sufficient condition for a G-equivariant $\mathfrak{U}$-local cocycle $[f] \in Z_{eq}^n(M, \mathfrak{U}; V)^G$ ($n > 0$) to be extendible to a global equivariant cocycle will be shown to be the triviality of its flux.

This characterisation of extensibility uses the flux and the period maps introduced in Section 13.6. The main results of this chapter are the characterisation of the extensibility of equivariant $\mathfrak{U}$-local cocycles to equivariant global cocycles under varying connectedness-conditions on the space X:

1. Concerning equivariant Alexander-Spanier 1-cocycles we observe the exactness of the sequence

$$Z^1(X; V)^G \to Z_{AS}^1(X; V)^G \xrightarrow{\ \mathrm{per}_\varrho(-)_1\ } \hom(\pi_1(X), V),$$

which describes the extensibility of an equivariant Alexander-Spanier 1-cocycle $[f]$ on X in terms of the period map $\pi_1(X) \to V$, $[s] \mapsto \lambda^1([f])(s)$.
2. The extendibility of (germs of) $\mathfrak{U}$-local n-cocycles on $(n-1)$-connected spaces X is described by an exact sequence

$$Z^n(X; V)^G \to \mathrm{colim}_\mathfrak{u} Z^n(X, \mathfrak{U}; V)^G \xrightarrow{\ \mathrm{per}_{\varrho t}(-)_n\ } \hom(\pi_n(X); V^G)$$

If the space X is not $(n-1)$-connected but only $(n-2)$-connected, we add another term and a flux homomorphism on the right to obtain an exact sequence

$$Z^n(X; V)^G \to \mathrm{colim}_\mathfrak{u} Z^n(X, \mathfrak{U}; V)^G \xrightarrow{\ \mathrm{per}_{\varrho t}(-)_n\ } \hom(Z_n(X), V^G) \oplus \hom(Z_{n-1}(X), Z^1(G; V)^G)$$

Both result generalise the characterisation of the integrability of continuous Lie algebra cocycles to locally smooth Lie group cocycles in [Nee02a] to general transformation groups. In contrast to the procedure there we do neither require any of the spaces X or V to be Hausdorff. Moreover we consider equivariant $\mathfrak{U}$-local cocycles in any dimension:

In any of the above cases (the germ of) an equivariant $\mathfrak{U}$-local n-cocycle is extendible to a global equivariant cocycle if and only if all relevant flux homomorphisms are trivial. If the coefficients V are uniquely divisible (i.e. if they are $\mathbb{Q}$-modules), then one can w.l.o.g. switch to the rational version of the flux homomorphisms.

It should be remarked that the above characterisations of the extensibility of equivariant $\mathfrak{U}$-local cocycles do not require any condition on the G-space to be satisfied. They hold in full generality.

The proof of the above statements on the extensibility of equivariant Alexander-Spanier cocycles relies on the spectral sequence $_IE^{p,q} \Rightarrow H^{p+q}_{eq}(X, \mathfrak{U}; V)$ introduced in Section 14.3 and on the flux and period homomorphisms introduced in Chapter 13.

15.1 Extensions and Flux

In this section we consider transformation groups (G, X) and derive necessary conditions for an equivariant local cocycle on X to be extendible to a global equivariant cocycle. These necessary conditions are expressible via the flux homomorphism introduced in Section 13.6. After considering local equivariant cocycles we turn our attention to equivariant differential forms on smooth transformation groups and drive a similar condition for these forms to be integrable to locally smooth global equivariant cocycles.

Throughout this section V will denote a real vector space of coefficients, i.e. we work over the ground field $\mathbb{K} = \mathbb{R}$ of real numbers. We use the enveloping space functor $E : \mathbf{Top} \to \mathbf{Top}$ of [BM78]. The main tool in this section is the natural transformation $i : \mathrm{id}_{\mathbf{Top}} \to E$ which embeds every topological space X in a space EX.

Lemma 15.1.1. *The enveloping space EX is naturally contractible.*

Proof. See [BM78] for a proof. □

Recall (cf. [BM78]), that the Elements of the topological space $E(X)$ are step functions in X, i.e. functions $\gamma : [0, 1) \to E(X)$ which are constant of the intervals $[t_{i-1}, t_i)$ of a partition $0 = t_0 < t_1 < \cdots t_n = 1$ of the half open interval $[0, 1)$.

Remark 15.1.2. The underlying set of this enveloping space functor is the same as that of the one constructed in Section 11.2, but the topology differs from that used there in general.

Lemma 15.1.3. *The enveloping space functor $E : \mathbf{Top} \to \mathbf{Top}$ preserves products, i.e. for any two spaces X and Y the spaces $E(X \times Y)$ and $E(X) \times E(Y)$ are naturally isomorphic. The isomorphism $\tau : E \circ \times \to \times \circ (E \times E)$ is given by*

$$\tau_{X,Y} : E(X \times Y) \to E(X) \times E(Y), \quad \tau(\gamma) = (pr_X \circ \gamma, pr_Y \circ \gamma).$$

Proof. A proof can be found in [BM78]. □

Lemma 15.1.4. *For any transformation group (G, X) the pair $(E(G), E(X))$ is a transformation group.*

Proof. Let $\varrho : G \times X \to X$ denote the action of G on X. The action of $E(G)$ on $E(X)$ can now be defined as the composition of $\tau_{G,X}^{-1}$ and $E(\varrho)$:

$$
\begin{array}{ccc}
E(G) \times E(X) & & \\
\tau_{G,X}^{-1} \downarrow & \searrow & \\
E(G \times X) & \xrightarrow[E(\varrho)]{} & E(X)
\end{array}
$$

The naturality of the homomorphisms $\tau_{G,X}$ ensure that $E(\varrho) \circ \tau_{G,X}^{-1}$ is an action of $E(G)$ on $E(X)$. $\qquad\square$

The action $E(\varrho) \circ \tau_{G,X}^{-1}$ of EG on EX can be directly expressed in terms of the action of G on X. For elements $g \in E(G)$ and $x \in E(X)$ the step function $g.x$ can be computed via

$$g.x(t) = [E(\varrho)(\tau_{G,X}^{-1}(g,x))](t) = \varrho[\tau^{-1}(g,x)(t)] = g(t).x(t),$$

for any $t \in [0,1)$.

The group G can be embedded into $E(G)$ as the subgroup of constant functions. We denote this embedding by i_G. Identifying G with its image $i_g(G)$ we can restrict the action of $E(G)$ to an action of G on $E(X)$:

$$
\begin{array}{ccc}
G \times E(X) & \xrightarrow{i_G \times \mathrm{id}_{E(X)}} & E(G) \times E(X) \\
& \searrow & \downarrow E(\varrho) \circ \tau_{G,X}^{-1} \\
& & E(X)
\end{array}
$$

So we observe:

Lemma 15.1.5. *The group G acts an the space $E(X)$ via $E(\varrho) \circ \tau_{G,X}^{-1} \circ (i_G \times \mathrm{id}_{E(X)})$.*

Lemma 15.1.6. *The inclusion $(G,X) \hookrightarrow (E(G), E(X))$ is an inclusion of transformation groups, i.e. the following diagram commutes*

$$
\begin{array}{ccc}
G \times X & \xrightarrow{\varrho} & X \\
i_G \times i_X \downarrow & & \downarrow i_X \\
E(G) \times E(X) & \xrightarrow[E(\varrho) \circ \tau_{G,X}^{-1}]{} & E(X)
\end{array},
$$

and $i_X(X)$ is a G-invariant subspace of $E(X)$.

This implies that any action $\varrho : G \times X \to X$ of a topological group G on a space X is the restriction of the action $E(\varrho) \circ \tau_{G,X}^{-1}$ of $E(G)$ on $E(X)$. Therefore we shall also denote this action by ϱ.

We now turn our attention to the question whether a local equivariant cocycle in X is extendible to a global equivariant cocycle, where we first observe the situation for equivariant Alexander-Spanier cocycles on universal enveloping spaces. Let (G,X) be a transformation group. Because the space EX is contractible, the flux morphism

$$\mathcal{F}_\varrho : A_{AS}^*(EX;V) \to \mathrm{Tot}S^*(G;S^*(EX;V))$$

of the action of $E(G)$ on $E(X)$ has trivial image in $\mathrm{Tot}S^*(G;H^*(EX;V))$ except for degree 0. In the following we prove that the flux $\mathcal{F}_\varrho(f)$ of any equivariant Alexander-Spanier n-cocycle $f \in Z_{AS}^n(EX;V)$, $n > 0$ that is extendible to a global equivariant cocycle is also trivial. This will be done by 'pulling back the condition on EX to the space X along the embedding $i_X : E \hookrightarrow EX$. To accomplish this, we construct a natural transformation

$$s : A(X; V) \to A(EX; V).$$

To start with a simple case, we first assume that the coefficient group V is a real vector space. This allows us to define the following maps:

We now consider the inclusion $i_X : X \hookrightarrow E(X)$ of G-spaces. Recall that this inclusion induces cosimplicial maps $A(i_X; V) : A(E(X); V) \to A(X; V)$ and $A_{AS}(i_X; V) : A_{AS}(E(X); V) \to A_{AS}(X; V)$ of cosimplicial G-modules. The cosimplicial morphism i intertwines the germ map so the following diagram commutes:

$$
\begin{array}{ccc}
A(E(X); V) & \xrightarrow{\;A(i_X;V)\;} & A(X; V) \\
\downarrow & & \downarrow \\
A_{AS}(E(X); V) & \xrightarrow{\;A_{AS}(i_X;V)\;} & A_{AS}(X; V)
\end{array}
$$

Since $E(X)$ is contractible, the morphism $A(E(X); V)^G \to A_{AS}(E(X); V)^G$ of cosimplicial groups an epimorphism from the equivariant cohomology onto the equivariant Alexander-Spanier cohomology of $E(X)$ by Theorem 14.3.10. The corresponding diagram in cohomology now reads:

$$
\begin{array}{ccc}
H_{eq}^*(E(X); V) & \xrightarrow{\;H_{eq}^*(i_X;V)\;} & H_{eq}^*(X; V) \\
\downarrow & & \downarrow \\
H_{AS,eq}^*(E(X); V) & \xrightarrow{\;H_{AS,eq}^*(i_X;V)\;} & H_{AS,eq}^*(X; V)
\end{array}
$$

In case the coefficient group V is a real topological vector space, we now construct an equivariant right inverse $s : A(X; V) \to A(E(X); V)$ to the morphism $A(i_X; V)$ induced by the inclusion i_X. To avoid confusion of elements of X with those of $E(X)$ and for better readability the universal space $E(X)$ is denoted by Y. Elements of Y are denoted by y here. Recall that elements of Y can be identified with step functions $[0, 1) \to X$ that are constant on a partition $[0 = t_0, t_1), \ldots, [t_{n-1}, t_n)$ of the half open interval $[0, 1)$.

Definition 15.1.7. *A partition $\mathcal{P}$ of $[0, 1)$ subordinate to a path $\gamma \in Y = E(X)$ is a partition of $[0, 1)$ such that γ is constant on each set $P \in \mathcal{P}$. Likewise a partition subordinate to $\mathbf{y} \in Y^{n+1}$ is a partition of $[0, 1)$ that is subordinate to every y_i, $0 \leq i \leq n$.*

The set of partitions subordinate to a given $\gamma \in Y$ is a directed set, with join the common refinement of partitions. Note that for each $\gamma \in Y$ there exists a least partition of $[0, 1)$ subordinate to γ. This is a finite partition by half open intervals. It can be described in terms of the length of the half-open intervals:

Definition 15.1.8. *An element $\mathbf{t} \in \Delta_n$ subordinate to $\gamma \in Y = E(X)$ is a point $\mathbf{t} \in \Delta_n$ such that $[0, t_0), [t_0, t_0 + t_1), \ldots, [\sum_{i<n} t_i, 1)$ is a partition of $[0, 1)$ subordinate to γ. Likewise a point $\mathbf{t} \in \Delta_n$ subordinate to $\mathbf{y} \in Y^{n+1}$ is a point that is subordinate to every y_i, $0 \leq i \leq n$.*

The use of the smallest partition subordinate to a given $\gamma \in Y$ allows us to define maps $s^n : A^n(X; V) \to A^n(EX; V)$:

Definition 15.1.9. *For any global n-cochain $F \in A^n(X;V)$ on X we define a global n-cochain $s(F) \in A^n(E(X);V)$ whose value on points $\mathbf{y} \in Y^{n+1}$ is given by*

$$s(F)(y_0, \ldots, y_n) := \sum_{j=0}^{k-1} |t_{j+1} - t_j| \cdot F(y_0(t_j), \ldots, y_n(t_j)),$$

where $[0 = t_0, t_1), \ldots, [t_{k-1}, t_k = 1)$ is the smallest partition of $[0,1)$ subordinate to $\mathbf{y}$.

It should be noticed that the partition used in the definition of the cochain $s(F)$ above need not be the smallest one subordinate to $\mathbf{y}$:

Lemma 15.1.10. *For any cochain $F \in A^n(X;V)$, any point $\mathbf{y} \in Y^{n+1}$ and any partition $0 = t_0 < t_1 < \cdots < t_k = 1$ subordinate to $\mathbf{y}$ the evaluation of $s(F)$ at $\mathbf{y}$ is given by*

$$s(F)(y_0, \ldots, y_n) := \sum_{j=0}^{k-1} |t_{j+1} - t_j| \cdot F(y_0(t_j), \ldots, y_n(t_j)),$$

i.e. the partition used in the definition of $s(F)$ can be replaced by any finer partition of $[0,1)$.

As the evaluation homomorphism $\mathrm{ev}_t : E(X) \to X$ at a given time $t \in [0,1)$ is G-equivariant and the homomorphisms s^n are defined via summation of such evaluations, the homomorphisms s^n turn out to be G-equivariant as well.

Lemma 15.1.11. *The maps s^n are G-equivariant.*

Proof. Let $F \in A^n(X;V)$ be an n-cochain. For any $g \in G$ the evaluation of $g.s(F)$ at a point $\mathbf{y} \in Y^{n+1}$ is given by

$$[g.s(F)](y_0, \ldots, y_n)] = g.[s(F)(g^{-1}.y_0, \ldots, g^{-1}.y_n)]$$

$$= g.\left[\sum_{j=0}^{k-1} |t_{j+1} - t_j| \cdot F(g^{-1}.y_0(t_j), \ldots, g^{-1}.y_n(t_j)) \right]$$

$$= \sum_{j=0}^{k-1} |t_{j+1} - t_j| \cdot g.[F(g^{-1}.y_0(t_j), \ldots, g^{-1}.y_n(t_j))]$$

$$= s(g.F)(y_0, \ldots, y_n),$$

where $0 = t_0 < t_1 < \cdots < t_k = 1$ is any partition of $[0,1)$ subordinate to $\mathbf{y}$. Thus the maps s^n are G-equivariant. $\qquad\square$

Another straightforward consequence of the definition of s is the intertwining with boundary maps.

Lemma 15.1.12. *The map $s : A(X;V) \to A(Y;V)$ is a cosemisimplicial map of G-modules.*

Proof. Let $\mathbf{y} \in Y^{n+2}$ be given and $0 = t_0 < t_1 < \cdot < t_k = 1$ be a partition of $[0,1)$ subordinate to $\mathbf{y}$. By the definition of $s(F)$ the value of i-th face of $s(F)$ on a point $\mathbf{y} \in Y^{n+2}$ is given by

$$(\partial_i s(F))(y_0, \ldots, y_n) = s(F)(y_0, \ldots, \hat{y}_i, \ldots, y_n)$$

$$= \sum_{j=0}^{k-1} |t_{j+1} - t_j| \cdot F(y_0(t_j), \ldots, \hat{y}_i(t_j), \ldots, y_n(t_i))$$

$$= \sum_{j=0}^{k-1} |t_{j+1} - t_j| \cdot (\partial_i F)(y_0(t_j), \ldots, y_n(t_i)) = s(\partial_i F)(\mathbf{y}),$$

Therefore s intertwines the face operators, hence is a cosemisimplicial map. $\qquad\square$

Fortunately the cosemisimplicial map s also intertwines the degeneracy operators. So all in all one obtains:

Proposition 15.1.13. *The map s is a morphism of cosimplicial G-modules.*

Proof. Since the homomorphisms s^n are equivariant and intertwine the face operators, it only is to prove that they also intertwine the degeneracy operators. This also follows straight from the definition of s. Let $\mathbf{y} \in Y^n$ be given and let $0 = t_0 < t_1 < \cdots < t_k = 1$ be a partition of $[0,1)$ subordinate to $\mathbf{y}$. By the definition of $s(F)$ the value of i-th degeneracy of $s(F)$ at $\mathbf{y} \in Y^n$ is given by

$$(\delta_i s(F))(y_0, \ldots, y_n) = s(F)(y_0, \ldots, y_i, y_i, \ldots, y_n)$$

$$= \sum_{j=0}^{k-1} |t_{j+1} - t_j| \cdot F(y_0(t_j), \ldots, y_i(t_j), y_i(t_j), \ldots, y_n(t_i))$$

$$= \sum_{j=0}^{k-1} |t_{j+1} - t_j| \cdot (\delta_i F)(y_0(t_j), \ldots, y_n(t_i)) = s(\delta_i F)(\mathbf{y}).$$

Thus s intertwines the face as well as the degeneracy operators, hence is a cosimplicial map. $\square$

Theorem 15.1.14. *The morphism $s : A(X;V) \to A(E(X);V)$ is a right inverse of $A(i_X;V) : A(E(X);V) \to A(X;V)$ in $\mathbb{Z}[\mathbf{G}]\mathbf{mod}^{\mathcal{O}}$*

Proof. Let F be a cochain in $A^n(X;V)$. For any point $\mathbf{x} \in X^{n+1}$ we denote the constant paths $(x_0, \ldots, x_n)$ also by $\mathbf{x}$. These constant paths are constant on the partition $0 = t_0 < t_1 = 1$ of $[0,1)$. Using this partition of $[0,1)$ the value of $i^n s^n(F)$ at the point $\mathbf{x} \in Y^{n+1}$ is given by

$$i^n s^n(F)(\mathbf{x}) = s^n(F)(\mathbf{x}) = 1 \cdot F(x_0(0), \ldots, x_n(0)) = F(\mathbf{x}). \qquad (15.1)$$

Thus the morphism s is a right inverse of i. $\square$

Summarising the above observations we see that the following diagram in the category $\mathbb{Z}[\mathbf{G}]\mathbf{mod}^{\mathcal{O}}$ of cosimplicial G-modules commutes:

$$
\begin{array}{ccc}
A(E(X);V) & \underset{s}{\overset{A(i_X;V)}{\rightleftarrows}} & A(X;V) \\
\downarrow & & \downarrow \\
A_{AS}(E(X);V) & \xrightarrow{A_{AS}(i_X;V)} & A_{AS}(X;V)
\end{array}
$$

Lemma 15.1.15. *If an equivariant cocycle $f \in Z_{AS}^n(X;V)^G$ is extendible to a global cocycle $F \in Z^n(X;V)^G$ then it can be extended to an equivariant Alexander-Spanier cocycle on $E(X)$.*

Proof. If a cocycle $f \in Z_{AS}^n(X;V)^G$ is extendible to a global cocycle $F \in Z^n(X;V)^G$ then the commutativity of the above diagram implies that f is the restriction $A_{AS}^n(i_X;)([s(F)])$ of the Alexander-Spanier cocycle $[s(F)] \in Z_{AS}^n(E(X);V)^G$. $\square$

Now follows an examination of the implications of the above lemma for the extensibility of Alexander-Spanier cocycles on X. We first work out the consequences for the flux of the action $\varrho : G \times X \to X$ and then we study the flux of $\varrho t : X \times G \to X$.

Consider the functor $S(G; S(-;V))$ on the category **GTop** of G-spaces, which assigns to each G-space X the bicosimplicial $\mathbb{Z}[G]$-module $S(G; S(X;V))$. The inclusion $i_X : X \hookrightarrow E(X)$ of G-spaces induces a morphism

$$i^{*,*} = S^*(G; S^*(i_X; V)) : S^*(G, S^*(E(X); V)) \to S^*(G, S^*(X; V))$$

of double complexes in $\mathbb{Z}[\mathbf{G}]\mathbf{mod}$, which in turn induces a morphism $\mathrm{Tot}(i^{*,*})$ of total complexes in $\mathbb{Z}[\mathbf{G}]\mathbf{mod}$. The morphism $i^{*,*}$ restricts and corestricts to a morphism $i^{*,*} : S^*(G, S^*(EX; V))^G \to S^*(G, S^*(X; V))^G$ in the category $\mathbf{Ab}$ of Abelian groups and the morphism $\mathrm{Tot}(i^{*,*})$ of total complexes restricts and corestricts to a morphism $\mathrm{Tot}(i^{*,*}) : \mathrm{Tot}S^*(G, S^*(EX; V))^G \to \mathrm{Tot}S^*(G, S^*(X; V))^G$ of total complexes. The property of this morphism we want to use is the following:

Lemma 15.1.16. *The morphisms* $\mathrm{Tot}(i^{*,*})$ *and* $S(i_X; V)$ *intertwine the flux homomorphisms, i.e. the following diagram in* $\mathbb{Z}[\mathbf{G}]\mathbf{mod}$ *commutes:*

$$
\begin{array}{ccc}
S^*(E(X); V) & \xrightarrow{\quad \mathcal{F}_\varrho \quad} & \mathrm{Tot}S^*(G, S^*(E(X); V)) \\
{\scriptstyle S^*(i_X; V)} \downarrow & & \downarrow {\scriptstyle \mathrm{Tot}(i_X^{*,*})} \\
S^*(X; V) & \xrightarrow{\quad \mathcal{F}_\varrho \quad} & \mathrm{Tot}S^*(G, S^*(X; V))
\end{array}
$$

Proof. This is a consequence of the fact that the flux homomorphism $\mathcal{F}_\varrho$ is natural in the action ϱ. $\qquad\square$

Recall the vertex maps $C(\lambda_{X,\mathfrak{U}}, V) : A(X, \mathfrak{U}; V) \to S(X, \mathfrak{U}; V)$ of cosimplicial G-modules for every open cover $\mathfrak{U}$ of X which assign to each $\mathfrak{U}$-local n-cochain f in $A^n(X, \mathfrak{U}; V)$ the singular n-cochain $\sigma \mapsto f(\sigma(\mathbf{e}_0), \dots, \sigma(\mathbf{e}_n))$. Composition of this morphism with the complete $\mathfrak{U}$-relative subdivision $\mathrm{Csd}_\mathfrak{U}$ yields a morphism $\mathrm{Csd}_\mathfrak{U} C(\lambda_{X,\mathfrak{U}}, V) : A(X, \mathfrak{U}; V) \to S(X; V)$ which we simply denote by $\tilde{\lambda}_\mathfrak{U}$.

Theorem 15.1.17. *If a $\mathfrak{U}$-local n-cocycle $f \in A^n(X, \mathfrak{U}; V)$ is extendible to a global n-cocycle $F \in A^n(X; V)$ then the restrictions $\mathcal{F}_\varrho(\tilde{\lambda}_\mathfrak{U}^n f)^{p,q}_{|Z_p(G)} : Z_p(G) \to H^q(X; V)$ of the flux $\mathcal{F}_\varrho(\tilde{\lambda}_\mathfrak{U}^n f)^{p,q}$ to singular p-cycles are trivial for all $0 \le p \le n$.*

Proof. If a local n-cocycle $f \in Z^n(X, \mathfrak{U}; V)$, is extendible to a global n-cocycle $F \in A^n(X; V)$, then its Alexander-Spanier cohomology class $[f] \in H^n_{AS}(X; V)$ is trivial and so is the singular cohomology class $[\tilde{\lambda}_\mathfrak{U}^n f] \in H^n(X; V)$. Therefore the evaluation of $\tilde{\lambda}_\mathfrak{U}^n f$ at singular n-cycles is trivial, which implies the triviality of the flux component $\mathcal{F}_\varrho(\tilde{\lambda}_\mathfrak{U}^n f)^{n,0}$. Moreover, for $n > 0$, the global cocycle F is the coboundary dH of a global $(n-1)$-cochain $H \in A^{n-1}(X; V)$. The restriction of H to the neighbourhood $\mathrm{SS}(X, \mathfrak{U})([n-1])$ then is a $\mathfrak{U}$-local cochain with coboundary f. The singular cochain $\lambda_\mathfrak{U}^{n-1} h \in S^{n-1}(X; V)$ can be extended to a singular cochain $H' \in S^{n-1}(E(X); V)$. The coboundary dH' of this singular cochain is a singular n-cocycle on $E(X)$ which extends the singular n-cocycle $\lambda_\mathfrak{U}^n f$ on X. The commutativity of the above diagram implies the equalities

$$\mathcal{F}_\varrho(\tilde{\lambda}_\mathfrak{U}^n(f)) = \mathcal{F}_\varrho(S^n(i_X; V)(dH')) = \mathrm{Tot}(i^{*,*})\mathcal{F}_\varrho(dH')$$

i.e. the image of the flux $\mathcal{F}_\varrho(\tilde{\lambda}_\mathfrak{U}^n f)$ of the singular cocycle $\tilde{\lambda}_\mathfrak{U}^n f$ in $\mathrm{Tot}S^*(G; H^*(X; V))$ is the image of the class of $\mathcal{F}_\varrho(dH')$ in $\mathrm{Tot}S^*(G; S^*(E(X); V))$ under the morphism induced by $\mathrm{Tot}(i^{*,*})$. Therefore the class of $\mathcal{F}_\varrho(\tilde{\lambda}_\mathfrak{U}^n f)$ in $\mathrm{Tot}S^*(G; H^*(X; V))$ is contained in the image of $\mathrm{Tot}S^*(\mathrm{id}_G, H^*(i_X; V))$. For each $p < n$, i.e $q > 0$ the singular cohomology $H^q(E(X); V)$ is trivial, because $E(X)$ is contractible. As a consequence the maps $\mathcal{F}_\varrho(\tilde{\lambda}_\mathfrak{U}^n f)^{p,q}_{|Z_p(G)} : Z_p(G) \to H^q(X; V)$ are also trivial for all $0 \le p < n$. $\quad\square$

Corollary 15.1.18. *If an equivariant $\mathfrak{U}$-local n-cocycle $f \in A^n(X, \mathfrak{U}; V)^G$ is extendible to an equivariant global n-cocycle $F \in A^n(X; V)^G$ then the restrictions $\mathcal{F}_\varrho(\tilde{\lambda}_\mathfrak{U}^n f)^{p,q}_{|Z_p(G)} : Z_p(G) \to H^q(X; V)$ of the flux $\mathcal{F}_\varrho(\tilde{\lambda}_\mathfrak{U}^n f)^{p,q}$ to singular p-cycles are trivial for all $0 \le p \le n$.*

A sharper observation can be made for the flux of the map $\varrho t : X \times G \to X$, $(x, g) \mapsto g.x$. For this flux we obtain the commutative diagram

$$
\begin{array}{ccc}
S^*(E(X); V) & \xrightarrow{\;\mathcal{F}_{\varrho t}\;} & \mathrm{Tot}S^*(E(X), S^*(G; V)) \\
{\scriptstyle S^*(i_X;V)}\Big\downarrow & & \Big\downarrow{\scriptstyle \mathrm{Tot}(i_X^{*,*})} \\
S^*(X; V) & \xrightarrow{\;\mathcal{F}_{\varrho t}\;} & \mathrm{Tot}S^*(X, S^*(G; V))
\end{array}
$$

in $\mathbb{Z}[\mathbf{G}]\mathbf{mod}$. By Lemma 13.9.9 the restriction of the flux morphism $\mathcal{F}_{\varrho t}$ to equivariant singular cochains has codomain in the subcomplex $\mathrm{Tot}S^*(G, S^*(X; V)^G)$ of $\mathrm{Tot}S^*(G, S^*(X; V)^G)$. This shows that the diagram

$$
\begin{array}{ccc}
S^*(E(X); V)^G & \xrightarrow{\;\mathcal{F}_{\varrho t}\;} & \mathrm{Tot}S^*(E(X), S^*(G; V)^G) \\
{\scriptstyle S^*(i_X;V)}\Big\downarrow & & \Big\downarrow{\scriptstyle \mathrm{Tot}(i_X^{*,*})} \\
S^*(X; V)^G & \xrightarrow{\;\mathcal{F}_{\varrho t}\;} & \mathrm{Tot}S^*(X, S^*(G; V)^G)
\end{array}
$$

is commutative as well. In a similar fashion to before we observe:

Theorem 15.1.19. *If an equivariant $\mathfrak{U}$-local n-cocycle $f \in A^n(X, \mathfrak{U}; V)^G$ is extendible to a global equivariant n-cocycle $F \in A^n(X; V)^G$ then the flux components $\mathcal{F}_{\varrho t}(\tilde{\lambda}_{\mathfrak{U}}^n f)^{p,q}$ induce the zero homomorphisms $H_p(X) \to H_{sing,eq}^q(G; V)$ for $p > 0$.*

Proof. If an equivariant $\mathfrak{U}$-local local n-cocycle $f \in Z^n(X, \mathfrak{U}; V)^G$ is extendible to a global equivariant n-cocycle $F \in Z^n(X; V)^G$, then the singular cohomology classes of $\tilde{\lambda}_{\mathfrak{U}}^n f = f \circ \lambda_{X,\mathfrak{U}} \mathrm{Csd}_{\mathfrak{U}}$ and $\tilde{\lambda}_X^n F = F \circ \lambda_X$ in the equivariant singular cohomology $H_{sing,eq}^n(X; V)$ coincide because the complete relative subdivision $\mathrm{Csd}_{\mathfrak{U}}$ is chain homotopic to the identity in $\mathbb{Z}[\mathbf{G}]\mathbf{mod}$. As a consequence the induced maps $H_p(X) \to H_{sing,eq}^q(G; V)$ coincide as well. The equivariant singular n-cocycle $\tilde{\lambda}_X^n F$ is the restriction of the equivariant singular cocycle $\tilde{\lambda}_{E(X)}^n s^n(F)$. The commutativity of the above diagram shows that the homomorphism $H_p(X) \to H_{sing,eq}^q(G; V)$ induced by the singular n-cocycle $\tilde{\lambda}_X^n F$ is the composition

$$
H_p(X) \xrightarrow{\;H_p(i_X)\;} H_p(EX) \to H_{sing,eq}^q(G; V),
$$

where the second homomorphism is induced by the flux component $\mathcal{F}_{\varrho}(\tilde{\lambda}_{E(X)}^n s^n(F))$. Since the singular homology $H_p(EX)$ is trivial for $p > 0$, this shows that the homomorphism $H_p(X) \to H_{sing,eq}^q(G; V)$ induced by the singular n-cocycle $\tilde{\lambda}_X^n F$ is the zero homomorphism for $p > 0$. $\qquad\square$

These are a necessary conditions for an equivariant $\mathfrak{U}$-local n-cocycle f to be extendible to a global equivariant cocycle. For smooth transformation groups (G, X) a similar result can be derived regarding the flux of differential forms. For this purpose we assume the manifold M to be smoothly paracompact. This is especially helpful for studying the integrability of Lie algebra cocycles to locally smooth group cocycles.

Theorem 15.1.20. *If a closed differential n-form on a smoothly paracompact G-manifold M can be integrated to a locally smooth global n-cocycle on M, then the flux components $\mathcal{F}_{\varrho}(\omega)^{p,q}$ induce the zero homomorphisms $H_{p,\infty}(G) \to H_{dR}^q(M; V)$.*

Proof. Let (G, M) be a smooth transformation group M be smoothly paracompact and ω a closed differential n-form on M. The smooth paracompactness implies that the de Rham cohomology $H_{dR}(M;V)$, the smooth Alexander-Spanier cohomology $H_{AS,s}(M;V)$ and the smooth singular cohomology $H_\infty(M;V)$ are isomorphic under the standard homomorphisms. If ω is integrable to a locally smooth global n-cocycle on M, then these isomorphisms imply that the singular n-cocycle $k^n(\omega)$, which is obtained by integrating ω over smooth singular n-simplices has trivial cohomology class, i.e. it is a coboundary of a smooth singular $(n-1)$-cochain $h \in S^{n-1}(M;V)$. This especially implies that the integration of ω over smooth singular n-cycles is trivial, which proves the triviality of the induced homomorphism

$$H_{n,\infty}(G) \to H_{dR}^0(M;V) \cong V .$$

In the case $n > 0$ the smooth singular $(n-1)$-cochain h on M can be extended to a singular $(n-1)$-cochain H on $E(M)$. The coboundary $F = DH$ of this singular $(n-1)$-cochain is a singular n-cocycle on $E(M)$ which restricts to a singular n-cocycle f on M, which in turn restricts to the smooth singular cocycle $k^n(\omega)$. As before we observe that this implies the equalities

$$\mathcal{F}_\varrho(f) = \mathcal{F}_\varrho(S^n(i_M;V)(F)) = \mathrm{Tot}(i^{*,*})\mathcal{F}_\varrho(F)$$

i.e. the image of the flux $\mathcal{F}_\varrho(f)$ of f in $\mathrm{Tot}S^*(G;H^*(M;V))$ is the image of the class of $\mathcal{F}_\varrho(f)$ in $\mathrm{Tot}S^*(G;S^*((E(M);V))$ under the morphism induced by $\mathrm{Tot}(i^{*,*})$. Therefore the class of $\mathcal{F}_\varrho(f)$ in $\mathrm{Tot}S^*(G;H^*(M;V))$ is contained in the image of $\mathrm{Tot}S^*(\mathrm{id}_G,H^*(i_M;V))$. For each $p < n$, i.e $q > 0$ the singular cohomology $H^q(E(M);V)$ is trivial, because $E(M)$ is contractible. As a consequence the homomorphisms $Z_p(G) \to H^q(M;V)$ induced by the flux component $\mathcal{F}_\varrho(f)^{p,q}$ are trivial for all $p < n$. The value of the flux component $\mathcal{F}_\varrho(f)^{p,q}$ at a smooth singular p-simplex τ is a singular q-cochain, whose value at a smooth singular q-cycle z is by definition given by

$$\mathcal{F}_\varrho(f)^{p,q}(\tau)(z) = f(\tau.z) = k^n(\omega)(\tau.z) = \int_{\tau.z} \omega = \int_z \mathcal{F}_\varrho(\omega)^{p,q}(\tau).$$

Since these values are zero for all smooth singular p-simplices τ and smooth singular q-cycles z, the differential forms $\mathcal{F}_\varrho(\omega)^{p,q}(\tau)$ are coboundaries for all smooth singular p-simplices τ, so the induced homomorphisms $H_{p,\infty}(G) \to H_{dR}^q(M;V)$ also are trivial for all $0 \le p < n$.   □

Corollary 15.1.21. *If a closed equivariant differential n-form on a smoothly paracompact G-manifold M can be integrated to a locally smooth equivariant global n-cocycle on M, then the restrictions $\mathcal{F}_\varrho(\omega)^{p,q}_{|Z_p(G)} : Z_{p,\infty}(G) \to H_{dR}^q(M;V)$ of the flux $\mathcal{F}_\varrho(\omega)^{p,q}$ to singular p-cycles are trivial for $p < n$.*

Since the coefficients V are assumed to be real vector spaces, any homomorphism $Z_p(G) \to Z^q(X;V)$ factors uniquely through the group $Z_p(G;\mathbb{Q})$ of rational singular p-cycles in G. Likewise, any homomorphism $Z_p(X) \to Z^q(G;V)$ factors uniquely through the group $Z_p(X;\mathbb{Q})$ of rational singular p-cycles in X. Consequently, for any $\mathfrak{U}$-local equivariant n-cocycle f on X the components

$$\mathcal{F}_{\varrho t}(\lambda_{\mathfrak{U}}^n f)^{p,q} : S_p(X) \to S^q(G;V)^G$$

$$\text{and} \qquad \mathcal{F}_{\varrho t}(\lambda_{\mathfrak{U}}^n f)^{p,q} : Z_p(X) \to Z^q(G;V)^G$$

of the flux $\mathcal{F}(\lambda_{\mathfrak{U}}^n f)$ factor uniquely through the groups $S_p(X;\mathbb{Q}) = S_p(X) \otimes \mathbb{Q}$ and $Z_p(X;\mathbb{Q}) = Z_p(X) \otimes \mathbb{Q}$. This can be subsumed in the commutative diagram

$$
\begin{array}{ccc}
Z_p(X) & \xrightarrow{\mathcal{F}(\lambda_{\mathfrak{U}}^n f)^{p,q}} & Z^q(G;V)^G \\
{\scriptstyle z \mapsto z \otimes 1}\downarrow & & \| \\
Z_p(X) \otimes \mathbb{Q} & \xrightarrow{\mathcal{F}(\lambda_{\mathfrak{U}}^n f)^{p,q} \otimes \mathrm{id}_{\mathbb{Q}}} & Z^q(G;V)^G \otimes \mathbb{Q} \cong Z^q(G;V)^G
\end{array}
\qquad (15.2)
$$

In particular, as is apparent from the above diagram, the flux component $\mathcal{F}(\lambda_{\amalg}^n f)^{p,q}$ is trivial if and only if its rational version $\mathcal{F}(\lambda_{\amalg}^n f)^{p,q} \otimes \mathrm{id}_{\mathbb{Q}}$ is trivial. The above commutative diagram induces a diagram in (co)homology which may be extended on left by the (rational) Hurewicz-homomorphism to obtain the following commutative diagram:

$$
\begin{array}{ccccc}
& \xrightarrow{\mathrm{per}_{\varrho t}(\lambda_{\amalg}^n f)_n} & & & \\
\pi_n(X) & \xrightarrow{\mathrm{hur}_n} & H_p(X) & \xrightarrow{\mathcal{F}(\lambda_{\amalg}^n f)^{p,q}} & H^q_{sing,eq}(G;V) \\
{\scriptstyle s \mapsto s \otimes 1}\Big\downarrow & & {\scriptstyle z \mapsto z \otimes 1}\Big\downarrow & & \Big\| \\
\pi_n(X) \otimes \mathbb{Q} & \xrightarrow{\mathrm{hur}_n \otimes \mathrm{id}_{\mathbb{Q}}} & H_p(X;\mathbb{Q}) & \xrightarrow{\mathcal{F}(\lambda_{\amalg}^n f)^{p,q} \otimes \mathrm{id}_{\mathbb{Q}}} & H^q_{sing,eq}(G;V), \\
& \xrightarrow{\mathrm{per}_{\varrho t}(\lambda_{\amalg}^n f)_n \otimes \mathrm{id}_{\mathbb{Q}}} & & &
\end{array}
\tag{15.3}
$$

where the lowest horizontal arrow is the rational version $\mathrm{per}_{\varrho t}(\lambda_{\amalg}^n f)_n \otimes \mathrm{id}_{\mathbb{Q}}$ of the period homomorphism $\mathrm{per}_{\varrho t}(\omega)_n$. We write $H_{sing,eq}(X;V)$ to denote the cohomology of equivariant singular cochains in order to distinguish it from the cohomology $H_{eq}(X;V)$ of global equivariant cochains on X. The rational period homomorphism is trivial if and only if the the period homomorphism $\mathrm{per}_{\varrho t}(\omega)_n$ itself is trivial. We record the result:

Lemma 15.1.22. *The flux and period homomorphisms are trivial if and only if their rational versions are trivial.*

In the case $G = X$ of an arc-wise connected topological group acting on itself by left translations one can show that the triviality of all the period homomorphisms $\mathrm{per}_{\varrho t}(\tilde{\lambda}_{\amalg}^n f)_p$ implies the triviality of all the induced homomorphisms $H_p(G) \to H^q_{sing,eq}(X;V)$ as will be shown below. For this purpose we use the characterisation of the rational singular homology of H-spaces in Theorem 8.4.18, which ensures that the cohomology class of a V-valued singular n-cocycle h on a connected H-space X is trivial if and only if the evaluations of h on all products of spheres

$$
s_1 \cdot s_2 \cdots s_k
$$

in X whose dimension add up to n are trivial.

Lemma 15.1.23. *If $G = X$ is an arc-wise connected topological group acting on itself by left translations, h an equivariant singular n-cocycle on X and all period homomorphisms $\mathrm{per}_{\varrho t}(h)_p : \pi_p(X) \to H^q_{sing,eq}(X;V)$ are trivial, then h is a singular coboundary.*

Proof. Let X be an arc-wise connected topological group acting on itself by left translations, h be an equivariant singular n-cocycle on X and assume all the period homomorphisms $\mathrm{per}_{\varrho t}(h)_p$ to be trivial. By Theorem 8.4.18 it suffices to show that the evaluation of h on any product $s_1 \cdot s_2 \cdots s_k$ of spheres in X whose dimension add up to n is trivial. Let $s_1 \cdot s_2 \cdots s_k$ be such a product, where the last sphere s_k has dimension p. The evaluation of h at the singular n-cycle $s_1 \cdot s_2 \cdots s_k$ is given by

$$
\begin{aligned}
h(s_1 \cdot s_2 \cdots s_k) &= h([s_1 \cdot s_2 \cdots s_{k-1}] \cdot s_k) \\
&= \mathcal{F}_{\varrho t}(h)(s_k)(s_1 \cdot s_2 \cdots s_{k-1}) \\
&= \mathrm{per}_{\varrho t}(h)_p(s_k)(s_1 \cdot s_2 \cdots s_{k-1}),
\end{aligned}
$$

which is trivial by assumption. Thus the singular n-cocycle h is a coboundary. $\quad\square$

In a similar way we obtain the desired result on flux and period homomorphisms for the singular n-cocycle h:

Proposition 15.1.24. *If $G = X$ is a connected topological group acting on itself by left translations, h an equivariant singular n-cocycle on X and all the period homomorphisms $\mathrm{per}_{\varrho t}(h)_p : \pi_p(X) \to H^q_{sing,eq}(X; V)$ are trivial, then the induced homomorphisms $H_p(X) \to H^q_{sing,eq}(G; V)$ of the flux homomorphisms are trivial as well.*

Proof. Let X be a connected topological group acting on itself by left translations, h a singular n-cocycle on X and assume all the period homomorphisms $\mathrm{per}_{\varrho t}(h)_p$ to be trivial. If z_p is a singular p-cocycle then there exists a product $s_1 \cdot s_2 \cdots s_k$ of spheres of total dimension p and a singular $(p + 1)$-cochain c in X such that the equality $dc = z - s_1 \cdot s_2 \cdots s_k$ holds rationally. Let p' be the dimension of the last sphere s_k in the product $s_1 \cdot s_2 \cdots s_k$. The evaluation of the flux homomorphism $\mathcal{F}_{\varrho t}(h)^{p,q}(z)$ at a singular $q = n - p$-cocycle z_q is given by

$$\begin{aligned}
\mathcal{F}_{\varrho t}(h)^{p,q}(z_p)(z_q) &= \mathcal{F}_{\varrho t}(h)^{p,q}(s_1 \cdot s_2 \cdots s_k)(z_q) \\
&= h(z_q \cdot s_1 \cdot s_2 \cdots s_k) = \mathrm{per}_{\varrho t}(h)_{p'}(s_k)(z_q \cdot s_1 \cdot s_2 \cdots s_{k-1}) = 0,
\end{aligned}$$

where p' is the dimension of the sphere s_k in X. Therefore all the homomorphisms $H_p(X) \to H^q_{sing,eq}(G; V)$ induced by the flux components $\mathcal{F}_{\varrho t}(h)^{p,q}$ are trivial. $\quad\square$

These result can now be applied to the singular n-cocycle $\tilde{\lambda}^n_{\mathfrak{U}} f$ associated to an equivariant $\mathfrak{U}$-local n-cocycle on X:

Corollary 15.1.25. *If $G = X$ is a connected topological group acting on itself by left translations, $f \in A^n(X, \mathfrak{U}; V)^G$ a $\mathfrak{U}$-local equivariant n-cocycle on X and all the period homomorphisms $\mathrm{per}_{\varrho t}(\tilde{\lambda}^n_{\mathfrak{U}} f)_p : \pi_p(X) \to H^q_{sing,eq}(X; V)$ are trivial, then the homomorphisms $H_p(X) \to H^q_{sing,eq}(G; V)$ induced by the flux homomorphisms $\mathcal{F}_{\varrho t}(\tilde{\lambda}^n_{\mathfrak{U}} f)^{p,q}$ are trivial as well.*

Similar results can be derived for the flux of equivariant differential forms on Lie groups acting on themselves by left translation. Here we observe:

Lemma 15.1.26. *If an equivariant differential n-form ω on a G-manifold M is exact, then the restrictions $\mathcal{F}_{\varrho t}(\omega)^{p,q}|_{Z_{p,\infty}(M)} : Z_{p,\infty}(M) \to Z^q_{dR}(M; V)^G$ have codomain in the subgroup $B^q_{dR}(M; V)^G$ of equivariant exact forms.*

Proof. Let ω be an equivariant differential n-form on a G-manifold M and η be a differential $(n - 1)$-form satisfying $d\eta = \omega$. The equalities

$$\mathcal{F}_{\varrho t}(\omega)^{p,q} = \mathcal{F}_{\varrho t}(d\eta)^{p,q} = d_h \mathcal{F}_{\varrho t}(\eta)^{p-1,q} + d_v \mathcal{F}_{\varrho t}(\eta)^{p,q-1}$$

imply that the restriction of the flux component $\mathcal{F}_{\varrho t}(\omega)^{p,q}$ to smooth singular p-cycles coincides with the restriction of $d_v \mathcal{F}_{\varrho t}(\eta)^{p,q-1}$ to smooth singular p-cycles. This proves the assertion. $\quad\square$

Proposition 15.1.27. *If G is a connected Lie group acting on itself by left translations, ω a closed equivariant differential n-form on G and all the period homomorphisms $\mathrm{per}_{\varrho t}(\omega)_p : \pi_p(G) \to H^q_{dR,eq}(G; V) \cong H^q_c(\mathfrak{g}; V)$ are trivial, then the homomorphisms $H_p(G) \to H^q_{dR,eq}(G; V) \cong H^q_c(\mathfrak{g}; V)$ induced by the flux homomorphisms are trivial as well.*

Proof. Let G be a connected Lie group acting on itself by left translations, ω a closed differential n-form on G and assume all the period homomorphisms $\mathrm{per}_{\varrho t}(\omega)_p$ to be trivial. If z_p is a smooth singular p-cocycle then there exists a product $s_1 \cdot s_2 \cdots s_k$ of smooth spheres of total dimension p and a smooth singular $(p + 1)$-cochain c in G such that the equality $dc = z - s_1 \cdot s_2 \cdots s_k$ holds rationally. Let p' be the dimension of the last sphere s_k in the product $s_1 \cdot s_2 \cdots s_k$. The evaluation of the flux component $\mathcal{F}_{\varrho t}(\omega)^{p,q}$ at z_p is given by

$$\mathcal{F}_{\varrho t}(\omega)^{p,q}(z_p) = \mathcal{F}_{\varrho t}(\omega)^{p,q}(s_1 \cdot s_2 \cdots s_k)$$
$$= \mathcal{F}_{\varrho t}\left(\mathcal{F}_{\varrho t}(\omega)^{p',n-p'}(s_k)\right)^{p-p'}(s_1 \cdot s_2 \cdots s_{k-1})$$

where p' is the dimension of the sphere s_k in X. Since the equivariant differential $n - p'$-form $\mathcal{F}_{\varrho t}(\omega)^{p',n-p'}(s_k)$ is exact by assumption, the differential q-form $\mathcal{F}_{\varrho t}(\omega)^{p,q}(z_p)$ is trivial as well. $\qquad\square$

In the next sections we examine, whether the triviality of the flux as shown above is sufficient to ensure the extendibility of equivariant local cocycles. Equivariant differential forms are studied in the next chapter.

15.2 1-Cocycles

As before we consider a transformation group (G, X) but we restrict ourselves to transformation groups with free action. Recall that the the cosimplicial groups $A(X, \mathfrak{U}; V)$ of $\mathfrak{U}$-local cochains for open coverings $\mathfrak{U}$ together with the restrictions morphisms $\varrho_{\mathfrak{B}, \mathfrak{U}} : A(X, \mathfrak{U}; V) \to A(X, \mathfrak{B}; V)$ for $\mathfrak{B} \prec \mathfrak{U}$ form a directed system of cosimplicial Abelian groups whose colimit is the Alexander-Spanier cosimplicial group $A_{AS}(X; V)$. The morphisms $A(X, \mathfrak{U}; V) \to A_{AS}(X; V)$ which assign to each local cochain its germ at the diagonal are denoted by ϱ. We assume the space X to be arc-wise connected. In this case the relative subdivision provides an equivariant section $B^1_{AS}(X; V) \to Z^1(X; V)$ to the germ map $\varrho : Z(X; V) \to B_{AS}(X; V)$ as shall be shown below. In the following we will identify an $\mathfrak{U}$-local n-cochains f with the unique homomorphism $\mathbb{Z}^{(\mathrm{SS}(X,\mathfrak{U})([n]))} \to V$ of Abelian groups induced by f. The topological space X is arc-wise connected, so by Lemma 3.0.5 there exists a equivariant map $\hat{\sigma}_1 : X \times X \to \hom_{\mathbf{Top}}(\Delta_1, X)$ such that $\hat{\sigma}_1(x_0, x_1)$ is a path joining x_0 to x_1. The complete $\mathfrak{U}$-relative subdivision $\mathrm{Csd}_{\mathfrak{U}}$ of these paths provides a function

$$\mathrm{Csd}^1_{\mathfrak{U}} : A^1(X, \mathfrak{U}; V) \to A^1(X; V), \quad \mathrm{Csd}^1_{\mathfrak{U}}(f) := f \circ \lambda_1 \circ \mathrm{Csd}_{\mathfrak{U}} \circ \hat{\sigma}_1$$

which assigns to each $\mathfrak{U}$-local 1-cochain a global 1-cochain on X and is called the *complete $\mathfrak{U}$-relative subdivision of 1-cochains* (cf. Section 6.2).

Lemma 15.2.1. *For any G-invariant open cover $\mathfrak{U}$ of X the homomorphism $\mathrm{Csd}^1_{\mathfrak{U}}$ is a $\mathbb{Z}[G]$-module homomorphism.*

Proof. Let $\mathfrak{U}$ be a G-invariant open covering of X. In this case the complete $\mathfrak{U}$-relative subdivision of singular simplices is an equivariant homomorphism $\mathrm{Csd}_{\mathfrak{U}} : S^1(X) \to S^1(X, \mathfrak{U})$ by Lemma 6.1.1. Therefore map $\lambda_1 \circ \mathrm{Csd}_{\mathfrak{U}} \circ \hat{\sigma}_1$ is equivariant as a composition of equivariant maps. Hence the homomorphism $\mathrm{Csd}^1_{\mathfrak{U}}$ is equivariant as well. $\qquad\square$

Lemma 15.2.2. *For any open cover $\mathfrak{U}$ of X the extension $\mathrm{Csd}^1_{\mathfrak{U}}(f)$ of a cochain $f \in A^1(X, \mathfrak{U}; V)$ is a cocycle if and only if f is a coboundary.*

Proof. Let $f \in A^1(X, \mathfrak{U}; V)$ be a cochain whose extension $\mathrm{Csd}^1_{\mathfrak{U}}(f)$ is a cocycle. Since the complex $A^*(X; V)$ is exact, the extension $\mathrm{Csd}^1_{\mathfrak{U}}(f)$ is the coboundary dh of a cochain $h \in A^0(X; V)$. The restriction $\varrho_{\mathfrak{U}, X}(h)$ then is a 0-cochain in $A^0(X, \mathfrak{U}; V)$ with coboundary f. Conversely, if $f \in A^1(X, \mathfrak{U}; V)$ is the coboundary of a 1-cochain $h \in A^0(X, \mathfrak{U}; V)$ then the coboundary of any extension $\tilde{h} \in A^0(X; V)$ of h is a 2-cocycle which restricts to f. $\qquad\square$

We are only interested in local and Alexander-Spanier cocycles which can be extended to global cocycles. These are necessarily coboundaries and the preceding Lemma allows ensures that the restriction of $\mathrm{Csd}_{\mathfrak{U}}^1$ to the submodule $B^1(X, \mathfrak{U}; V)$ of 1-boundaries has codomain in the group of global 1-cocycles. The fact that the complete $\mathfrak{U}$-relative subdivision of local 1-cocycles is equivariant furthermore implies that $\mathrm{Csd}_{\mathfrak{U}}$ maps equivariant 1-coboundaries to equivariant global cocycles. Since the path $\hat{\sigma}_1(x_0, x_1)$ for a $\mathfrak{U}$-small 1-simplex (x_0, x_1) need not be $\mathfrak{U}$-small in general, the $\mathfrak{U}$-relative subdivision $\mathrm{Csd}_{\mathfrak{U}} f$ of a $\mathfrak{U}$-local cochain $f \in A^1(X, \mathfrak{U}; V)$ does not necessarily restrict to f on the neighbourhood $\mathrm{SS}(X, \mathfrak{U})([1])$ of the diagonal in $X \times X$. We will now show that the restriction of $\mathrm{Csd}_{\mathfrak{U}}$ to the group $B^1(X, \mathfrak{U}; V)$ of coboundaries does have this property.

Lemma 15.2.3. *For any two open covers $\mathfrak{B} \prec \mathfrak{U}$ of the topological space X the complete $\mathfrak{B}$-relative subdivision $\mathrm{Csd}_{\mathfrak{B}}^1 : A^1(X, \mathfrak{U}; V) \to A^1(X, \mathfrak{U}; V)$ of $\mathfrak{U}$-local 1-cochains restricts to the identity on $B^1(X, \mathfrak{U}; V)$.*

Proof. Let $\mathfrak{B} \prec \mathfrak{U}$ be open covers of the topological space X. The image of a $\mathfrak{U}$-local cochain $f \in A^1(X, \mathfrak{U}; V)$ under the homomorphism $\mathrm{Csd}_{\mathfrak{B}}^1$ is by definition given by

$$\mathrm{Csd}_{\mathfrak{B}}(f)(x_0, x_1) = (f \lambda_1 \, \mathrm{Csd}_{\mathfrak{B}} \, \hat{\sigma}_1)(x_0, x_1) = f(\lambda_1 \, \mathrm{Csd}_{\mathfrak{B}} \, \hat{\sigma}_1(x_0, x_1)),$$

where the subdivision $\mathrm{Csd}_{\mathfrak{B}} \, \hat{\sigma}_1(x_0, x_1)$ divides the path $\hat{\sigma}_1(x_0, x_1)$ from x_0 to x_1 into smaller segments. The boundary $\partial \, \mathrm{Csd}_{\mathfrak{B}} \, \hat{\sigma}_1(x_0, x_1)$ of the singular 1-chain $\mathrm{Csd}_{\mathfrak{B}} \, \hat{\sigma}_1(x_0, x_1)$ is the singular chain $x_1 - x_0$. If the cochain f is a coboundary dh then this especially implies the equalities

$$\begin{aligned}
[\mathrm{Csd}_{\mathfrak{B}} \, f - f](x_0, x_1) &= [f \lambda_1 \, \mathrm{Csd}_{\mathfrak{B}} \, \hat{\sigma}_1 - f](x_0, x_1) \\
&= dh(\lambda_1 \, \mathrm{Csd}_{\mathfrak{B}} \, \hat{\sigma}_1(x_0, x_1)) - dh(x_0, x_1) \\
&= h(\lambda_1 \partial \, \mathrm{Csd}_{\mathfrak{B}} \, \hat{\sigma}_1(x_0, x_1)) - [h(x_1) - h(x_0)] \\
&= [h(x_1) - h(x_0)] - [h(x_1) - h(x_0)] = 0 \,.
\end{aligned}$$

Thus the homomorphism $\mathrm{Csd}_{\mathfrak{B}}^1$ restricts to the identity on the group $B^1(X; V)$ of $\mathfrak{U}$-local 1-coboundaries. $\square$

In the case $\mathfrak{B} = \mathfrak{U}$ this implies that for each $\mathfrak{U}$-local 1-coboundary f the restriction of the global 1-cocycle $\mathrm{Csd}_{\mathfrak{U}}^1 f$ to the neighbourhood $\mathrm{SS}(X, \mathfrak{U})([1])$ of the diagonal in $X \times X$ coincides with f, i.e. $\mathrm{Csd}_{\mathfrak{U}}^1 f$ is an extension of f to a global 1-cocycle.

Lemma 15.2.4. *For any open cover $\mathfrak{U}$ of the topological space X the restriction $\mathrm{Csd}_{\mathfrak{U}|B^1(X,\mathfrak{U};V)}^1 : B^1(X, \mathfrak{U}; V) \to Z^1(X; V)$ is a right inverse to the restriction homomorphism $Z^1(X; V) \to B^1(X, \mathfrak{U}; V)$.*

Proof. Let $\mathfrak{U}$ be an open cover of X and $f \in B^1(X, \mathfrak{U}; V)$ a $\mathfrak{U}$-local 1-coboundary. Then there exists a 0-chain (i.e. a V-valued function) h on X whose coboundary dh restricts to f. From the previous Lemma we now infer the equalities

$$\begin{aligned}
\mathrm{Csd}_{\mathfrak{U}}^1(f)(x_0, x_1) &= f \lambda_1 \, \mathrm{Csd}_{\mathfrak{U}}(x_0, x_1) = (dh) \lambda_1 \, \mathrm{Csd}_{\mathfrak{U}}(x_0, x_1) \\
&= (dh)(x_0, x_1) = f(x_0, x_1)
\end{aligned}$$

for all points (x_0, x_1) in the neighbourhood $\mathrm{SS}(X, \mathfrak{U})([1])$ of the diagonal. This proves the assertion. $\square$

Proposition 15.2.5. *For any two open covers $\mathfrak{B} \prec \mathfrak{U}$ of X the homomorphisms $\mathrm{Csd}_{\mathfrak{U}|B^1(\mathfrak{U};V)}^1$ and $\mathrm{Csd}_{\mathfrak{B}|B^1(\mathfrak{B};V)}^1$ intertwine the restriction $\varrho_{\mathfrak{B},\mathfrak{U}}$.*

Proof. Let $\mathcal{B} \prec \mathfrak{U}$ be open covers of X. For every point (x_0, x_1) in X^2 the singular 1-cochains $\mathrm{Csd}_{\mathcal{B}}\,\hat{\sigma}_1(x_0, x_1)$ and $\mathrm{Csd}_{\mathfrak{U}}\,\hat{\sigma}_1(x_0, x_1)$ have boundary $x_1 - x_0$, so the singular 1-cochain $\mathrm{Csd}_{\mathcal{B}}\,\hat{\sigma}_1(x_0, x_1) - \mathrm{Csd}_{\mathfrak{U}}\,\hat{\sigma}_1(x_0, x_1)$ is a singular 1-cocycle. Each element $f \in B^1(X, \mathfrak{U}; V)$ is a coboundary $f = dh$ of a 0-cochain h. The evaluation of the cochain $\mathrm{Csd}^1_{\mathcal{B}}\,\varrho_{\mathcal{B},\mathfrak{U}}(f) - \mathrm{Csd}^1_{\mathfrak{U}}(f)$ at a point (x_0, x_1) is given by

$$
\begin{aligned}
[\mathrm{Csd}^1_{\mathcal{B}}\,\varrho_{\mathcal{B},\mathfrak{U}}(f) - \mathrm{Csd}^1_{\mathfrak{U}}(f)](x_0, x_1) &= f(\lambda_1\,\mathrm{Csd}_{\mathcal{B}}\,\hat{\sigma}_1(x_0, x_1) - \lambda_1\,\mathrm{Csd}_{\mathfrak{U}}\,\hat{\sigma}_1(x_0, x_1)) \\
&= h(\partial[\lambda_1\,\mathrm{Csd}_{\mathcal{B}}\,\hat{\sigma}_1(x_0, x_1) - \lambda_1\,\mathrm{Csd}_{\mathfrak{U}}\,\hat{\sigma}_1(x_0, x_1)]) \\
&= h(0) = 0
\end{aligned}
$$

for all points $(x_0, x_1) \in X^2$. Thus the homomorphisms $\mathrm{Csd}^1_{\mathfrak{U}|B^1(\mathfrak{U};V)}$ and $\mathrm{Csd}^1_{\mathcal{B}|B^1(\mathcal{B};V)}$ intertwine the restriction $\varrho_{\mathcal{B},\mathfrak{U}}$. $\qquad\square$

Because the sections $\mathrm{Csd}^1_{\mathfrak{U}}$ for open covers $\mathfrak{U}$ of X intertwine the restriction homomorphisms they form a cocone

$$
\cdots \longrightarrow B^1(\mathfrak{U};V) \xrightarrow{\varrho_{\mathcal{B},\mathfrak{U}}} B^1(\mathcal{B};V) \xrightarrow{\varrho_{\mathfrak{C},\mathcal{B}}} B^1(\mathfrak{C};V) \longrightarrow \cdots
$$

$$
\mathrm{Csd}^1_{\mathfrak{U}} \searrow \quad \Big\downarrow \mathrm{Csd}^1_{\mathcal{B}} \quad \swarrow \mathrm{Csd}^1_{\mathfrak{C}}
$$

$$
B^1(X;V)
$$

in the category $\mathbb{Z}[\mathbf{G}]\mathbf{mod}$. Therefore they pass through the colimit $B^1_{AS}(X;V)$ and induce a homomorphism $s^1 : B^1_{AS}(X;V) \to Z^1(X;V)$ which is a right inverse to the germ map ϱ. So we obtain:

Corollary 15.2.6. *There exists an equivariant right inverse s^1 to the germ map $\varrho : Z^1(X;V) \to B^1_{AS}(X;V)$ in $\mathbb{Z}[\mathbf{G}]\mathbf{mod}$.*

Lemma 15.2.7. *If X is arc-wise connected and $h \in A^0(X;V)$ satisfies $dh = 0$ on a neighbourhood $\mathrm{SS}(X, \mathfrak{U})([1])$ of the diagonal, then h is constant.*

Proof. Let X be arc-wise connected, $h \in A^0(X;V)$ and let the restriction of the boundary dh to the neighbourhood $\mathrm{SS}(X, \mathfrak{U})([1])$ of the diagonal vanish. For any two points x and y in X there is a path γ joining x to y. The complete $\mathfrak{U}$-relative subdivision $\mathrm{Csd}_{\mathfrak{U}}\,\gamma$ of γ consists of paths whose images are contained in $\mathrm{SS}(X, \mathfrak{U})([1])$. As a consequence we have

$$
h(x) - h(y) = dh(x, y) = dh(\lambda_1(\gamma)) = dh(\lambda_1(\mathrm{Csd}_{\mathfrak{U}}\,\gamma)) = 0.
$$

Since this is true for arbitrary points x and y in X the function h is constant. $\quad\square$

Proposition 15.2.8. *If X is arc-wise connected then the extension $s^1(f)$ of a cocycle $f \in B^1_{AS}(X;V)$ to a global cocycle is unique.*

Proof. Let $f \in Z^1_{AS}(X;V)$ be an Alexander-Spanier cocycle. Suppose there exist extensions $h_1 \in Z^1(X;V)$ and $h_2 \in Z^1(X;V)$ of f to global cocycles. Because the complex $A^*(X;V)$ is exact, these cocycles are boundaries of 0-chains g_1 and g_2 respectively: $h_1 = dg_1$ and $h_2 = dg_2$. Then the difference $d(g_1 - g_2)$ vanishes on a neighbourhood W of the diagonal:

$$
d(g_1 - g_2)_{|W} = (dg_1)_{|W} - (dg_2)_{|W} = f - f = 0
$$

By lemma 15.2.7 the cochain $g_1 - g_2$ is a constant function and thus the difference $h_1 - h_2 = d(g_1 - g_2)$ vanishes. $\qquad\square$

Corollary 15.2.9. *For any arc-wise connected space X the restriction homomorphism $Z^1(X;V) \to Z^1_{AS}(X;V)$ is injective.*

Corollary 15.2.10. *For any arc-wise connected space X and open covering $\mathfrak{U}$ the restriction homomorphism $Z^1(X;V) \to Z^1(X,\mathfrak{U};V)$ is injective.*

Theorem 15.2.11. *For any arc-wise connected topological G-space X the sequence of $\mathbb{Z}G$-modules*

$$0 \to Z^1(X;V) \xrightarrow{\varrho} Z^1_{AS}(X;V) \to H^1_{AS}(G;V) \to 0$$

is split exact in $\mathbb{Z}[\mathbf{G}]\mathbf{mod}$ and the right inverse $s^1 : B^1_{AS}(X;V) \to Z^1(X;V)$ to the restriction homomorphism is unique. Furthermore the sequences

$$0 \to Z^1(X;V) \xrightarrow{\varrho_{\mathfrak{U},X}} Z^1(X,\mathfrak{U};V) \to H^1(X,\mathfrak{U};V) \to 0$$

$$\text{and} \qquad 0 \to Z^1(X;V) \xrightarrow{\varrho_{\mathfrak{U},X}} Z^1(X,\mathfrak{U};V) \to H^1_{AS}(G;V)$$

of $\mathbb{Z}[G]$-modules are exact for any open cover $\mathfrak{U}$ of X and the restriction of a right inverse $B^1(X,\mathfrak{U},V) \to Z^1(X;V)$ to the restriction homomorphism $\varrho_{\mathfrak{U},X}$ is unique.

Corollary 15.2.12. *If X is arc-wise connected then an equivariant local 1-cocycle f_{eq} is extendible to a global equivariant cocycle if and only if its Alexander-Spanier cohomology class $[f_{eq}] \in H^1_{AS}(X;V)$ is trivial. In this case the extension to a global equivariant 1-cocycle is unique.*

15.3 Observations for the General Case

In this section we examine the general extensibility of equivariant local and Alexander Spanier cocycles on the space X of a transformation group (G,X) to global equivariant cocycles on X. We use the spectral sequence of van Est type

$$H^p_{eq}(X; H^q(X,\mathfrak{U};V)) \Rightarrow H^{p+q}_{eq}(X,\mathfrak{U};V)$$

which was introduced in Chapter 14 to prove the existence of extensions in certain cases. Here we give necessary and sufficient conditions on the cocycles f for such extensions to exist. These conditions are the triviality of a flux in cohomology. In the special case of connected topological groups groups acting on themselves by left translation this can expressed by "integration" of the flux over spheres, i.e. by period maps

$$\pi_p(G) \to H^{n-p}_{sing,eq}(G;V).$$

We restrict ourselves to connected transformation groups (G,X) which admit a continuous equivariant map $p : X \to G$. These are exactly the transformation group (G,X) whose orbit bundle $X \to G\backslash X$ is a trivial G-principal bundle. We fix the transformation group (G,X) and an equivariant map $p : X \to G$.

Example 15.3.1. Every total space E of a trivial G-principal bundle $E \twoheadrightarrow B = E/G$ is of this form.

Example 15.3.2. Every topological group G acts freely on itself by left translation. The identity map $p = \mathrm{id}_G$ is an equivariant continuous map $G \to G$. The complex of global equivariant cochains here is isomorphic to the complex of group cochains. This in particular includes all Lie groups.

As before we consider a transformation group (G,X). An immediate requirement for a n-cocycle f to be extendible to a global cocycle is the following:

Lemma 15.3.3. *If an Alexander-Spanier cocycle f on X is extendible to a global n-cocycle F on X, then f is an Alexander-Spanier coboundary. Likewise, if $\mathfrak{U}$ is a G-invariant open cover of X and $f \in A^n(X, \mathfrak{U}; V)$ an $\mathfrak{U}$-local cocycle on X which is extendible to a global cocycle F on X, then f is a coboundary in $A^*(X, \mathfrak{U}; V)$.*

Proof. Let f be an Alexander-Spanier n-cocycle which is extendible to a global n-cocycle on X. Since the standard complex $A^*(X; V)$ is exact, there exists a global $(n-1)$-cochain F' in $A_s^{n-1}(X; V)$ whose coboundary is F. If f' denotes the germ of F' at the diagonal, then the cocycle f is the coboundary of f'. The proof for local cocycles is analogous. $\square$

Lemma 15.3.4. *If $\mathfrak{U}$ is a G-invariant open covering of X and the germ of an equivariant $\mathfrak{U}$-local cochain $f \in A^n(X, \mathfrak{U}; V)$ at the diagonal in X^{n+1} is extendible to an equivariant global cochain F on X, then there exists a G-invariant open refinement $\mathfrak{B}$ of $\mathfrak{U}$ such that the restriction $f_{|\mathrm{SS}(X, \mathfrak{B}; V)([n])}$ coincides with the restriction $F_{|\mathrm{SS}(X, \mathfrak{B}; V)([n])}$ of the cochain F.*

Proof. Let $\mathfrak{U}$ be a G-invariant open covering of X and the germ of an equivariant $\mathfrak{U}$-local cocycle $f \in A^n(X, \mathfrak{U}; V)$ on X be extendible to an equivariant global cocycle F on X. Then there exists an open cover $\mathfrak{B}' \prec \mathfrak{U}$ such that the restrictions $f_{|\mathrm{SS}(X, \mathfrak{B}'; V)([n])}$ and $F_{|\mathrm{SS}(X, \mathfrak{B}'; V)([n])}$ coincide. Since the cocycles f and F both are equivariant, they also coincide on the G-invariant open covering $\mathfrak{B} = G.\mathfrak{B}'$ of X, which still is a refinement of $\mathfrak{U}$. $\square$

Example 15.3.5. If $X = G$ is a topological group acting on itself by left translations and f is a local group n-cocycle whose germ at the identity is extendible to a global group cocycle F, then there exists a neighbourhood V of the identity such that the restrictions $f_{|V^{n+1}}$ and $F_{|V^{n+1}}$ coincide.

Recall that we have fixed an equivariant map $p : X \to G$. We expect the the extensibility of equivariant $\mathfrak{U}$-local n-cocycles f only to depend on their cohomology classes $[f] \in H_{eq}^n(X, \mathfrak{U}; V)$. To prove this we first observe:

Lemma 15.3.6. *If $\mathfrak{U}$ is a G-invariant open covering of X and $F' \in A^n(X; V)$ is an extension of an equivariant $\mathfrak{U}$-local n-cochain f on X, then the cochain*

$$F : X^{n+1} \to V, \quad F(\mathbf{x}) = p(x_0). \left[F'(p(x_0)^{-1}.\mathbf{x}) \right]$$

is an equivariant extension of f on X. In particular every equivariant $\mathfrak{U}$-local cochain is extendible to a global equivariant cochain.

Proof. Let $\mathfrak{U}$ be a G-invariant open covering of X, F' be an extension of an equivariant $\mathfrak{U}$-local n-cochain f on X and F be defined as above. The cochain F is equivariant by definition. The restriction map $\varrho_{\mathfrak{U}, X} : A^n(X; V) \to A^n(X, \mathfrak{U}; V)$ is equivariant. This implies the equalities

$$\begin{aligned}
\varrho_{\mathfrak{U}, X}(F)(\mathbf{x}) &= p(x_0). \left[\varrho_{\mathfrak{U}, X}(F) \left(p(x_0)^{-1}.\mathbf{x} \right) \right] \\
&= p(x_0). \left[\varrho_{\mathfrak{U}, X}(F') \left(p(x_0)^{-1}.\mathbf{x} \right) \right] \\
&= p(x_0). \left[f \left(p(x_0)^{-1}.\mathbf{x} \right) \right] \\
&= f(\mathbf{x}),
\end{aligned}$$

for all $\mathfrak{U}$-local simplices $\mathbf{x} = (x_0, \ldots, x_n)$ in X^{n+1}. So F is an equivariant extension of f. This proves the assertion. $\square$

Lemma 15.3.7. *If $\mathfrak{U}$ is a G-invariant open covering of X then every equivariant $\mathfrak{U}$-local n-cochain can be extended to a global equivariant n-cochain.*

Proof. Let $\mathfrak{U}$ be a G-invariant open covering of X and f be an equivariant $\mathfrak{U}$-local n-cochain on X. Extend f arbitrarily to a global n-cochain F' and define a global equivariant n-cochain F on X via

$$F : X^{n+1} \to V, \quad F(\mathbf{x}) = p(x_0).\left[f(p(x_0)^{-1}.\mathbf{x})\right] .$$

By Lemma 15.3.6 this cochain F is an equivariant extension of f.   □

The same construction allows us to extend coboundaries of equivariant local cochains to global equivariant cocycles:

Lemma 15.3.8. *If $\mathfrak{U}$ is a G-invariant open covering of X then every coboundary of a $\mathfrak{U}$-local equivariant cochain on X can be extended to a global equivariant cocycle.*

Proof. Let $\mathfrak{U}$ is a G-invariant open covering of X and $f \in A^n(X,\mathfrak{U};V)$ be the coboundary of a $\mathfrak{U}$-local equivariant cochain $h \in A^{n-1}(X,\mathfrak{U};V)$. The cochain h can be extended to a global equivariant cochain H whose coboundary H is an n-cocycle and an equivariant extension of f.   □

Proposition 15.3.9. *If $\mathfrak{U}$ is a G-invariant open covering of X then the extendibility of an $\mathfrak{U}$-local equivariant n-cocycle f on X to a an equivariant global n-cocycle F on X only depends on its cohomology class $[f] \in H^n_{eq}(X,\mathfrak{U};V)$.*

Proof. Let $\mathfrak{U}$ be a G-invariant open covering of X and f be a $\mathfrak{U}$-local equivariant n-cocycle which is extendible to a global equivariant n-cocycle F on X. It is to show that an $\mathfrak{U}$-local equivariant n-cocycle f' with the same cohomology class $[f'] = [f]$ in $H^n_{eq}(X,\mathfrak{U};V)$ is also extendible to a global equivariant n-cocycle. If an $\mathfrak{U}$-local n-cocycle f' is cohomologous to f in $A^*(X,\mathfrak{U};V)^G$, then there exists an equivariant $\mathfrak{U}$-local $(n-1)$-chain h with coboundary $dh = f - f'$. The coboundary dh is extendible to a global equivariant cocycle H by Lemma 15.3.8. The equivariant n-cocycle $F-H$ satisfies the equality $(F-H)_{|SS(X,\mathfrak{U})([n])} = f - dh = f'$, i.e. the equivariant n-cocycle f' is also extendible to a global equivariant cocycle.   □

Lemma 15.3.10. *If $\mathfrak{U}$ is a G-invariant open covering of X, $\mathfrak{B}$ a refinement of $\mathfrak{U}$ and f an equivariant $\mathfrak{U}$-local n-cochain on X whose restriction to $SS(X,\mathfrak{B})([n])$ is extendible to a global n-cochain, then there exists a common refinement $\mathfrak{V}$ of $\mathfrak{U}$ and $\mathfrak{B}$ such that $f_{|SS(X,G.\mathfrak{V})([n])}$ is extendible to a global equivariant cochain.*

Proof. Let $\mathfrak{U}$ be a G-invariant open covering of X, $\mathfrak{B}$ a refinement of $\mathfrak{U}$ and f an equivariant $\mathfrak{U}$-local n-cochain on X whose restriction to $SS(X,\mathfrak{B})([n])$ is extendible to a global n-cochain F' on X. We assert that there exists a G-invariant refinement $\mathfrak{V}$ of $\mathfrak{U}$ and $\mathfrak{B}$ such that the equivariant cochain

$$F : X^{n+1} \to V, \quad F(\mathbf{x}) = p(x_0).\left[F'(p(x_0)^{-1}.\mathbf{x})\right]$$

is an equivariant extension of $f_{|sss(X,\mathfrak{V})([n])}$. For this purpose we observe that the continuous map

$$h : X^{n+1} \to X^{n+1}, \quad h(\mathbf{x}) = p(x_0)^{-1}.\mathbf{x}$$

pulls the open neighbourhood $SS(X,\mathfrak{B})([n])$ of the diagonal in X^{n+1} back to an open neighbourhood W of the diagonal in X^{n+1}. There exists an open covering $\mathfrak{V}$ of X such that the neighbourhood $SS(X,\mathfrak{V})([n])$ of the diagonal in X^{n+1} is contained in W and $SS(X,\mathfrak{B})([n])$. The evaluation of F at a point $\mathbf{x} \in SS(X,\mathfrak{V})([n])$ is given by

$$\begin{aligned} F(\mathbf{x}) &= p(x_0).\left[\varrho_{\mathfrak{U},X}(F')\left(p(x_0)^{-1}.\mathbf{x}\right)\right] \\ &= p(x_0).\left[f\left(p(x_0)^{-1}.\mathbf{x}\right)\right] \\ &= f(\mathbf{x}), \end{aligned}$$

i.e. the restriction of F to $\mathrm{SS}(X, \mathfrak{V})([n])$ coincides with the restriction of f to $\mathrm{SS}(X, \mathfrak{V})([n])$. Since both F and f are equivariant, they also coincide on the larger open G-invariant neighbourhood $G.\mathrm{SS}(X, \mathfrak{V})([n]) = \mathrm{SS}(X, G.\mathfrak{V})([n])$ of the diagonal in X^{n+1}. $\qquad\square$

Having made these preparatory observations, we now utilise the spectral sequences $H_{eq}^p(X; H^q(X, \mathfrak{U}; V)) \Rightarrow H_{eq}^{p+q}(X, \mathfrak{U}; V)$ for different G-invariant open coverings $\mathfrak{U}$ to describe the extensibility of equivariant local and equivariant Alexander-Spanier cocycles on X to global cocycles. Recall that these spectral sequence were constructed using the double complexes $A^{*,*}(X, \mathfrak{U}; V)^G$ introduced in Section 14.3. The horizontal and the vertical differential in this double complex are denoted by d_h and d_v respectively. The differential of the total complex will be denoted by D. The total complex $\mathrm{Tot}A^{*,*}(X, \mathfrak{U}; V)^G$ may be filtrated column-wise and row-wise. By Theorem 14.3.7 the spectral sequences $_IE^{*,*}$ and $_{II}E^{*,*}$ obtained by these filtrations both converge to the equivariant $\mathfrak{U}$-local cohomology $H_{eq}^*(X, \mathfrak{U}; V)$ of X. Moreover, as has been shown in Section 14.3, the the augmentations

$$0 \longrightarrow A^p(X; V)^G \xrightarrow{j^p} A^{p,*}(X, \mathfrak{U}; V)^G$$

of the columns $A^{p,*}(X, \mathfrak{U}; V)^G$ of the double complex $A^{*,*}(X, \mathfrak{U}; V)^G$ induce an embedding $A^*(X; V)^G \hookrightarrow \mathrm{Tot}A^{*,*}(X, \mathfrak{U}; V)^G$ of the complex $A^*(X; V)^G$ of equivariant global cochains into the total complex $\mathrm{Tot}A^{*,*}(X, \mathfrak{U}; V)^G$. Similarly, the complex $A^*(X, \mathfrak{U}; V)^G$ of equivariant $\mathfrak{U}$-local cochains on X can be embedded into the total complex $\mathrm{Tot}A^{*,*}(X, \mathfrak{U}; V)^G$ via the augmentations

$$i^q : A^q(X, \mathfrak{U}; V)^G \hookrightarrow A^{0,q}(X, \mathfrak{U}; V)^G, \quad i^q(f)(x_y, \mathbf{y}) = f(\mathbf{y}).$$

of the rows of the double complex. The augmentation i^q embeds an $\mathfrak{U}$-local equivariant q-cocycle f on X as a cocycle of bidegree $(0, q)$ in the total complex $\mathrm{Tot}A^{*,*}(X, \mathfrak{U}; V)^G$. The restriction morphism $\varrho_{\mathfrak{U}, X} : A^*(X; V)^G \to A^*(X, \mathfrak{U}; V)^G$ maps an equivariant global n-cochain F on X to an equivariant $\mathfrak{U}$-local n-cochain f on X. This n-cochain f is a cocycle whenever F is a cocycle. If this is the case, then the cocycle $j^n(F)$ in $\mathrm{Tot}A^{*,*}(X; V)^G$ is cohomologous to the cocycle $i^n(f)$. (See Section 14.3 for a discussion of this topic). This is the case if and only if there exists a cochain ψ of total degree $n-1$ with coboundary $D\psi = i^n(F) - j^n(F)$. The latter condition is is equivalent to the existence of cocycles $i^n(f) = F^{0,n}, \ldots, F^{n,0} = (-1)^{n-1}j^n(F)$ of bidegree $(0, n), (1, n-1), \ldots, (n, 0)$ and cochains $\psi^{0,n-1}, \ldots, \psi^{n-1,0}$ of bidegree $(0, n-1), (1, n-2), \ldots, (n-1, 0)$ respectively such that $D\psi^{p-1,q} = F^{p,q} + F^{p-1,q+1}$:

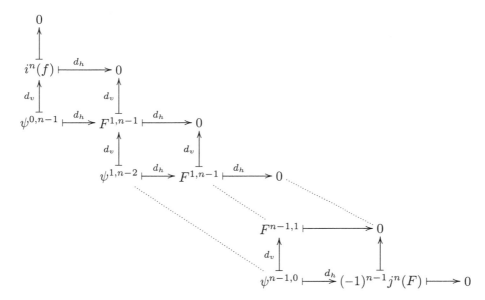

Here the cochain ψ is the alternating sum $\psi = \sum_{p'+q'=n-1}(-1)^{q'}\psi^{p',q'}$. Conversely, an $\mathfrak{U}$-local equivariant n-cocycle f on X is the image $\varrho^n_{\mathfrak{U},X}(F)$ of a global equivariant cocycle F on X if and only if the cocycle $i^n(f)$ of bidegree $(0,n)$ in the complex $\mathrm{Tot}A^{*,*}(X;V)^G$ is cohomologous to a cocycle $j^n(F)$ of bidegree $(n,0)$:

Lemma 15.3.11. *An equivariant $\mathfrak{U}$-local n-cocycle f on X is extendible to an equivariant global cocycle if and only if $i^n(f)$ is cohomologous to a cocycle $j^n(F)$ of bidegree $(n,0)$.*

Proof. The "only if" implication has been proved in Section 14.3. If the cocycle $i^n(f)$ is cohomologous to a cocycle $j^n(F)$ of bidegree $(n,0)$, then there exists a cochain ψ of total degree $n-1$ with coboundary $D\psi = i^n(f) - j^n(F)$. Using the homotopy equivalence $p^* : \mathrm{Tot}A^{*,*}(X,\mathfrak{U};V)^G \to A^*(X,\mathfrak{U};V)^G$ we see that this implies that the cocycles $p^n(j^n(F)) = \varrho_{\mathfrak{U},X}(F)$ and $p^n(i^n(f)) = f$ are cohomologous in $A^*(X,\mathfrak{U};V)^G$. By Proposition 15.3.9 the extensibility of the $\mathfrak{U}$-local cocycle f only depends on the cohomology class $[f] \in H^n_{eq}(X,\mathfrak{U};V)$, so f is extendible to a global equivariant cocycle on X. This proves the backward implication. $\qquad\square$

Moreover, if the $\mathfrak{U}$-local equivariant n-cocycle f on X is the restriction $\varrho^n_{\mathfrak{U},X}(F)$ of an equivariant global n-cocycle, the the cochains $F^{p,q}$ in the above diagram can be chosen to be given by the formula $F^{p,q} = (d_v h)^q j^n(F)$, where h denotes the row contractions of $A^{*,*}(X,\mathfrak{U};V)^G$ (see Section 14.3). We will show that this implies that each of the cocycles $F^{p,q}$ is "extendible" to a global cocycle of bidegree (p,q) in the total complex of the complex $A^*(X, A^*(X,V))^G$. To shorten notation we denote the double complex $A^*(X, A^*(X,V))$ by $A^{*,*}(X,V)$ and its sub double complex $A^*(X, A^*(X,V))^G$ by $A^{*,*}(X,V)^G$. By definition the horizontal and vertical differentials on the double complex $A^{*,*}(X,V)$ are given by

$$d_h : A^{p,q}(X;V) \to A^{p+1,q}(X;V)$$

$$(d_h f)(x_0,\ldots,x_{p+1},\mathbf{y}) = \sum_{i=0}^{p+1}(-1)^i f(x_0,\ldots,\hat{x}_i,\ldots,x_{p+1},\mathbf{y})$$

and by

$$d_v : A^{p,q}(X;V) \to A^{p,q+1}(X;V),$$

$$(d_h f)(\mathbf{x}, y_0,\ldots,y_{q+1}) = (-1)^p \sum_{i=0}^{q+1}(-1)^i f(\mathbf{x}, y_0,\ldots,\hat{y}_i,\ldots,y_{q+1})$$

respectively. These horizontal and vertical differentials are equivariant by construction. Similar to the double complex $A^{*,*}(X, \mathfrak{U}; V)$ we note that for each $q \geq 0$ the kernel of the horizontal differential $d_h : A^{0,q}(X; V)^G \to A^{1,q}(X; V)^G$ consists of those equivariant functions $f : X \times X^{q+1} \to V$ in $A^{0,q}(X; V)^G$ that are independent of the first variable. Hence this kernel is isomorphic to the space $A^q(X; V)^G$ of equivariant functions. One can therefore augment each row via

$$0 \longrightarrow A^q(X; V)^G \xrightarrow{i^q} A^{*,q}(X; V)^G, \quad i^q(f)(x_0, \mathbf{y}) = f(\mathbf{y}),$$

where the augmentation i^q is an embedding. Furthermore the augmentations i^q intertwine the differentials on the complexes $A^*(X; V)^G$ and $A^{*,q}(X; V)^G$. Analogously one can augment the columns $A^{p,*}(X; V)^G$ with the kernel of the vertical differential $d_v : A^{p,0}(X; V)^G \to A^{p,1}(X; V)^G$. Here the kernel is isomorphic to the group $A^p(X; V)^G$ of global equivariant p-cochains

$$0 \longrightarrow A^p(X; V)^G \xrightarrow{j^p} A^{p,*}(X; V)^G,$$

and the augmentations j^p intertwine the differential on the complexes $A^*(X; V)^G$ and $A^{p,*}(X; V)^G$. Also, similar to the double complex $A^{*,*}(X, \mathfrak{U}; V)$ one can construct row contractions h^p for each augmented row of the double complex $A^{*,*}(X; V)$ by defining

$$h^p : A^{p,q}(X, \mathfrak{U}; V) \to A^{p-1,q}(X, \mathfrak{U}; V)$$
$$h^p(f)(x_0, \ldots, x_{p-1}, \mathbf{y}) = (-1)^p f(x_0, \ldots, x_{p-1}, y_0, \mathbf{y}).$$

The homomorphisms h^p are equivariant by definition, hence they restrict to functions $A^{p,q}(X; V)^G \to A^{p-1,q}(X; V)^G$.

Proposition 15.3.12. *The maps h^p form an equivariant contracting homotopy on the augmented row-complexes $0 \to A^q(X; V)^G \to A^{*,q}(X; V)^G$.*

Proof. Let f be a cochain of bidegree (p, q) in $A^{p,q}(X; V)$. The evaluation of the cochain $[d_h h^p + h^{p+1} d_h](f)$ at a point $(\mathbf{x}, \mathbf{y}) \in X^{p+1} \times X^{q+1}$ is given by

$$[d_h h^p + h^{p+1} d_h](f)(\mathbf{x}, \mathbf{y}) = \sum_{i=0}^{p} (-1)^i (h^p f)(x_0, \ldots, \hat{x}_i, \ldots, x_p, \mathbf{y})$$
$$+ (-1)^{p+1} (d_h f)(x_0, \ldots, x_p, y_0, \mathbf{y})$$
$$= \sum_{i=0}^{p} (-1)^{p+i} f(x_0, \ldots, \hat{x}_i, \ldots, x_p, y_0, \mathbf{y})$$
$$+ \sum_{i=0}^{p} (-1)^{p+i+1} f(x_0, \ldots, \hat{x}_i, \ldots, x_p, y_0, \mathbf{y})$$
$$+ (-1)^{p+1} (-1)^{p+1} f(\mathbf{x}, \mathbf{y})$$
$$= f(\mathbf{x}, \mathbf{y}),$$

which proves the equality $d_h h^p + h^{p+1} d_h = \mathrm{id}$. Thus h^* is a row contraction. $\square$

Corollary 15.3.13. *The rows of the double complex $A^{*,*}(X; V)^G$ are split exact, except at the modules $A^{0,q}(X; V)^G$.*

In contrast to the double complex $A^{*,*}(X, \mathfrak{U}; V)$ we can apply the same procedure to the columns of the double complex $A^{*,*}(X; V)$ to show:

Proposition 15.3.14. *The homomorphisms $h'^q : A^{p,q}(X; V) \to A_{s,s}^{p-1,q}(X; V)$ defined by $h'^q(f)(\mathbf{x}, \mathbf{y}) = f(\mathbf{x}, x_0, \mathbf{y})$ form an equivariant contracting homotopy of the augmented columns $0 \to A^q(X; V) \to A^{*,q}(X; V)$.*

Proof. The proof is analogous to that of Proposition 15.3.12.   □

Observe that the Abelian group $A^{p,0}(X, \mathfrak{U}; V)^G$ is the group of equivariant functions $f : X^{p+1} \times X \to V$ and the same is true for the group $A^{p,0}(X; V)^G$, so the Abelian groups $A^{p,0}(X, \mathfrak{U}; V)^G$ and $A^{p,0}(X; V)^G$ are identical. In addition the augmentations $A^p(X, \mathfrak{U}; V)^G \hookrightarrow A^{p,*}(X, \mathfrak{U}; V)^G$ and $A^p(X; V)^G \hookrightarrow A^{p,*}(X; V)^G$ coincide. Furthermore the restriction homomorphisms

$$\varrho_{\mathfrak{U},X}^{p,q} : A^{p,q}(X; V)^G \to A^{p,q}(M, \mathfrak{U}; V)^G, \quad \varrho_{\mathfrak{U},X}^{p,q} f(\mathbf{x}, \mathbf{y}) = f(\mathbf{x}, \mathbf{y})$$

which restricts a global cochain f of bidegree (p, q) in the last $(q + 1)$ elements to the subspace $\mathrm{SS}(X, \mathfrak{U})([q])$ intertwine the horizontal and vertical differentials. Therefore the maps $\varrho_{\mathfrak{U},X}^{p,q}$ for a morphism $\varrho_{\mathfrak{U},X}^{*,*} : A^{*,*}(X; V)^G \to A^{*,*}(M, \mathfrak{U}; V)^G$ of double complexes. This morphism of double complexes can be used to examine the cocycles $F^{p,q} = (d_v h)^q j^n(F)$ associated to an equivariant global n-cocycle F on X.

Definition 15.3.15. *The morphism $\varrho_{\mathfrak{U},X}^{*,*} : A^{*,*}(X; V)^G \to A^{*,*}(X, \mathfrak{U}; V)^G$ of double complexes is called the* restriction morphism.

Definition 15.3.16. *A cochain $F^{p,q}$ of bidegree (p, q) in the total complex of the double complex $A^{*,*}(X, \mathfrak{U}; V)^G$ is called* extendible to an equivariant global cochain *of bidegree (p, q) if it is in the image of the restriction morphism $\varrho_{\mathfrak{U},X}^{p,q}$. A cocycle $F^{p,q}$ of bidegree (p, q) in the total complex of the double complex $A^{*,*}(X, \mathfrak{U}; V)^G$ is called* extendible to an equivariant global cocycle of bidegree (p, q) *if there exists a cocycle $f^{p,q} \in A^{p,q}(X; V)^G$ with image $\varrho_{\mathfrak{U},X}^{p,q}(f^{p,q}) = F^{p,q}$.*

Lemma 15.3.17. *The restriction homomorphisms $\varrho_{\mathfrak{U},X}^{p,q}$ intertwine the row contractions on $A^{*,*}(X; V)^G$ and $A^{*,*}(X, \mathfrak{U}; V)^G$.*

Proof. This follows from the fact that the row contractions on $A^{*,*}(X; V)^G$ and $A^{*,*}(X, \mathfrak{U}; V)^G$ are defined by the same formula.   □

Recall that the cocycles $F^{p,q}$ associated to an equivariant global n-cocycle F on X are given by $F^{p,q} = (d_v h)^q j^n(F)$. Using the above observations we can finally note:

Lemma 15.3.18. *The cocycles $F^{p,q}$ are extendible to equivariant cocycles of bidegree (p, q) respectively. In particular each cocycle $F^{p,q}$ is the image $\varrho_{\mathfrak{U},X}^{p,q}((d_v h)^q j^n(F))$, where $j^n(F)$ is considered as a cocycle in the total complex of the double complex $A^{*,*}(X; V)^G$.*

Proof. The restriction homomorphisms $\varrho_{\mathfrak{U},X}^{p,q}$ intertwine the vertical differentials by construction. Since they also intertwine the row contractions on the double complexes $A^{*,*}(X; V)^G$ and $A^{*,*}(X, \mathfrak{U}; V)^G$ we observe the equalities

$$F^{p,q} = (d_v h)^q j^n(F) = F^{p,q} = (d_v h)^q \varrho_{\mathfrak{U},X}^{n,0} j^n(F) = \varrho_{\mathfrak{U},X}^{p,q}(d_v h)^q j^n(F),$$

which proves the assertion.   □

A similar conclusion can be drawn for the cochains $\psi^{p',q'} = ((d_v h)^{q'} h j^n(F))$ in the total complex of the double complex $A^{*,*}(M, \mathfrak{U}; V)^G$:

Lemma 15.3.19. *The cochains $\psi^{p',q'}$ are extendible to global equivariant cochains of bidegree (p', q') respectively. In particular each cochain $\psi^{p',q'}$ is the image $\varrho_{\mathfrak{U},X}^{p',q'}((d_v h)^{q'} h j^n(F))$, where $j^n(F)$ is considered as a cocycle in the total complex of the double complex $A^{*,*}(X; V)^G$.*

Proof. The proof is analogous to that of the previous Lemma.   □

Each of the cocycles $F^{p,q}$ is cohomologous to $F^{n,0}$ and extendible to a global cocycle. Below we observe that an extendible cocycle of bidegree (p,q) in $\text{Tot}A^{*,*}(X,\mathfrak{U};V)^G$ always is cohomologous to a cocycle of bidegree $(p+q,0)$.

Lemma 15.3.20. *If a cocycle $F^{p,q}$, $q > 0$ of bidegree (p,q) in $A^{p,q}(X,\mathfrak{U};V)^G$ is extendible to a global equivariant cocycle in $A^{p,q}(X;V)^G$, then it is cohomologous to a cocycle $F^{p+1,q-1}$ of bidegree $(p+1,q-1)$ in $A^{p,q}(X,\mathfrak{U};V)^G$ which is extendible to a equivariant global cocycle in $A^{p+1,q-1}(X;V)^G$.*

Proof. If a cocycle $F^{p,q}$ of bidegree (p,q) is extendible to a equivariant global cocycle $f^{p,q}$, then the application of the column contraction h' of $A^{p,*}(X;V)^G$ to $f^{p,q}$ yields a cochain $h'f^{p,q}$ of bidegree $(p,q-1)$ whose vertical differential is $f^{p,q}$. The equation

$$D(h'f^{p,q}) = d_h(h'f^{p,q}) + d_v(h'f^{p,q})$$

shows that the cocycle $f^{p,q}$ is cohomologous to the cocycle $d_h(h'f^{p,q})$ in the total complex $\text{Tot}A^{*,*}(X;V)^G$. An application of the morphism $\text{Tot}\varrho_{\mathfrak{U},X}^{*,*}$ now shows that the cocycle $F^{p,q} = \tau^{p,q}(f^{p,q})$ in the total complex $\text{Tot}A^{*,*}(X,\mathfrak{U};V)^G$ is cohomologous to the cocycle $\varrho_{\mathfrak{U},X}^{p+1,q-1}(d_hh'f^{p,q})$. □

Lemma 15.3.21. *If a cocycle $F^{p,q}$ of bidegree (p,q) in $\text{Tot}A^{*,*}(X,\mathfrak{U};V)^G$ is extendible to an equivariant global cocycle of bidegree (p,q) then it is cohomologous to a cocycle $F^{p+q,0}$ of bidegree $(p+q,0)$ in $\text{Tot}A^{*,*}(X,\mathfrak{U};V)^G$.*

Proof. This follows from a repeated application of Lemma 15.3.20. □

Thus in order to show that a cocycle $F^{p,q}$ of bidegree (p,q) in the total complex of $A^{*,*}(X,\mathfrak{U};V)^G$ is cohomologous to a cocycle of bidegree $(p+q,0)$ it suffices to verify that it is extendible to an equivariant global cocycle of bidegree (p,q) in the total complex of $A^{*,*}(X;V)^G$.

In some rare cases the columns of the double complex $A^{*,*}(X,\mathfrak{U};V)^G$ are exact. In this case every cocycle $F^{p,q}$ of bidegree (p,q) is the vertical coboundary of a cochain $\psi^{p,q-1}$ of bidegree $(p,q-1)$. Observe that for every cochain $F^{p,q}$ of bidegree (p,q) with trivial vertical coboundary $d_vF^{p,q}$ in the total complex of $A^{*,*}(X,\mathfrak{U};V)^G$ the adjoint function $\hat{F}^{p,q} : X^{p+q} \to A^q(X,\mathfrak{U};V)$ takes values in the $\mathfrak{U}$-local q-cocycles and thus factors to a function $\hat{F}^{p,q} : X^{p+q} \to H^q(X,\mathfrak{U};V)$.

Lemma 15.3.22. *If $F^{p,q}$ is a cochain $F^{p,q}$ of bidegree (p,q) in $A^{p,q}(X,\mathfrak{U};V)^G$ with trivial vertical coboundary and the induced function $F^{p,q} : X^{q+1} \to H^q(X,\mathfrak{U};V)$ is trivial, then $F^{p,q}$ is the vertical coboundary $d_v\psi^{p,q-1}$ of a cochain $\psi^{p,q-1}$ of bidegree $(p,q-1)$ in $A^{p,q-1}(X,\mathfrak{U};V)^G$.*

Proof. Let $F^{p,q} \in A^{p,q}(M,\mathfrak{U};V)^G$ be a cochain of bidegree (p,q) with trivial vertical coboundary $d_vF^{p,q}$ and assume the induced function $\hat{F}^{p,q} : X^{p+1} \to H^q(X,\mathfrak{U};V)$ to be trivial. Then the adjoint function $\hat{F}^{p,q} : X^{p+1} \to A^q(X,\mathfrak{U};V)$ takes values in the $\mathfrak{U}$-local q-coboundaries $B^q(X,\mathfrak{U};V)$. Let $s : B^q(X,\mathfrak{U};V) \to A^{q-1}(X,\mathfrak{U};V)$ be a right inverse to the differential $d : A^{q-1}(X,\mathfrak{U};V) \to B^q(X,\mathfrak{U};V)$. The equivariant function $\psi^{p,q-1} : X^{p+1} \times \text{SS}(X,\mathfrak{U})([q-1]) \to V$ defined via

$$\psi^{p,q-1}(\mathbf{x},\mathbf{y}) = p(x_0). \left[s\hat{F}^{p,q}\left(p(x_0)^{-1}.\mathbf{x} \right)\left(p(x_0)^{-1}.\mathbf{y} \right) \right]$$

is an equivariant cochain of bidegree $(p,q-1)$ in $A^{p,q-1}(X,\mathfrak{U};V)^G$. We assert that the vertical coboundary of $\psi^{p,q-1}$ is the cochain $F^{p,q}$. For all points $\mathbf{x}$ in X^{p+1} and $\mathbf{y}$ in $\text{SS}(X,\mathfrak{U})([q])$ the value of $d_v\psi^{p,q-1}$ at $(\mathbf{x},\mathbf{y})$ computes to

$$d_v\psi^{p,q-1}(\mathbf{x},\mathbf{y}) = p(x_0).\left[d_v\psi^{p,q-1}\left(p(x_0)^{-1}.\mathbf{x}\right)\left(p(x_0)^{-1}.\mathbf{y}\right)\right]$$
$$= p(x_0).\left[d_v s\hat{F}^{p,q}\left(p(x_0)^{-1}.\mathbf{x}\right)\left(p(x_0)^{-1}.\mathbf{y}\right)\right]$$
$$= p(x_0).\left[\hat{F}^{p,q}\left(p(x_0)^{-1}.\mathbf{x})\right)\left(p(x_0)^{-1}.\mathbf{y}\right)\right]$$
$$= F^{p,q}(\mathbf{x},\mathbf{y}),$$

since the cochains $\psi^{p,q-1}$ and $F^{p,q}$ are equivariant. This proves that $F^{p,q}$ is the vertical coboundary of $\psi^{p,q-1}$. $\qquad\square$

Lemma 15.3.23. *If the $\mathfrak{U}$-local cohomology $H^q(X,\mathfrak{U};V)$ of X is trivial, then every cochain $F^{p,q}$ of bidegree (p,q) in $A^{p,q}(X,\mathfrak{U};V)^G$ with trivial vertical coboundary $d_vF^{p,q}$ is the vertical coboundary $d_v\psi^{p,q-1}$ of a cochain $\psi^{p,q-1}$ of bidegree $(p,q-1)$ in $A^{p,q-1}(X,\mathfrak{U};V)^G$.*

Proof. Let $F^{p,q} \in A^{p,q}(M,\mathfrak{U};V)^G$ be a cochain of bidegree (p,q) with trivial vertical coboundary $d_vF^{p,q}$ and assume the cohomology $H^q(X,\mathfrak{U};V)$ to be trivial. Then the induced function $\hat{F}^{p,q}: X^{p+1} \to H^q(X,\mathfrak{U};V)$ is trivial. The existence of the desired cochain $\psi^{p,q-1}$ now follows from Lemma 15.3.22. $\qquad\square$

Corollary 15.3.24. *If the the cohomology $\mathrm{colim}\,H(A^*(X,\mathfrak{U};V))$, where $\mathfrak{U}$ ranges over left X-invariant open coverings embeds into the singular cohomology and the q-th singular cohomology $H^q(X;V)$ is trivial, there exists a G-invariant open covering $\mathfrak{B} \prec \mathfrak{U}$ of X such that the restriction $\varrho_{\mathfrak{B},\mathfrak{U}}f$ is a coboundary.*

Corollary 15.3.25. *If the the cohomology $H(\mathrm{colim}\,A^*(X,\mathfrak{U};V))$, where $\mathfrak{U}$ ranges over left X-invariant open coverings embeds into the singular cohomology and X is n-connected, then for every G-invariant open covering $\mathfrak{U}$ and cochain $F^{p,q}$ of bidegree (p,q), $q < n$ with trivial vertical coboundary in the total complex of $A^{*,*}(X,\mathfrak{U};V)^G$ there exists a finer G-invariant open covering $\mathfrak{B}$ of X such that the restriction $\varrho_{\mathfrak{B},\mathfrak{U}}^{p,q}F^{p,q}$ of $F^{p,q}$ in $A^{p,q}(X,\mathfrak{B};V)^G$ is the vertical coboundary of a cochain $\psi^{p,q-1}$ in $A^{p,q}(X,\mathfrak{B};V)^G$.*

Recall that for locally contractible topological groups G acting on themselves by left translation the natural homomorphism $H(\mathrm{colim}_\mathfrak{U} A^*(G,\mathfrak{U};V)) \to H(G;V)$, where $\mathfrak{U}$ ranges over all G-invariant open coverings, is injective (by Corollary 3.0.85). Therefore the triviality of the cohomology class $[f] \in H^n(\mathrm{colim}_\mathfrak{U} A^*(G,\mathfrak{U};V))$ of a $\mathfrak{B}$-local cocycle $f \in A^n(G,\mathfrak{B};V)$ can be tested by evaluating the corresponding singular n-cochain at singular n-cycles. In particular the triviality of the q-th singular cohomology $H^q(X;V)$ implies the triviality of $H^q(\mathrm{colim}_\mathfrak{U} A^*(G,\mathfrak{U};V))$. So we note:

Corollary 15.3.26. *If X is a connected locally contractible topological group with trivial q-th singular cohomology, $\mathfrak{U}$ a left X-invariant open covering of X and f a $\mathfrak{U}$-local q-cocycle then there exists a left X-invariant open covering $\mathfrak{B} \prec \mathfrak{U}$ of X such that the restriction $\varrho_{\mathfrak{B},\mathfrak{U}}f$ is a coboundary.*

Corollary 15.3.27. *If $G = X$ is an n-connected locally contractible topological group acting on itself by left translation then for every G-invariant open covering $\mathfrak{U}$ and cochain $F^{p,q}$ of bidegree (p,q), $q < n$ with trivial vertical coboundary in the total complex of $A^{*,*}(X,\mathfrak{U};V)^G$ there exists a finer G-invariant open covering $\mathfrak{B}$ of X such that the restriction $\varrho_{\mathfrak{B},\mathfrak{U}}^{p,q}F^{p,q}$ of $F^{p,q}$ in $A^{p,q}(X,\mathfrak{B};V)^G$ is the vertical coboundary of a cochain $\psi^{p,q-1}$ in $A^{p,q}(X,\mathfrak{B};V)^G$.*

Recall that, by Lemma 14.2.19, the horizontal coboundary $d_h\psi^{p,q-1}$ of a cochain $\psi^{p,q-1}$ is a cocycle whenever the vertical coboundary $d_v\psi^{p,q-1}$ is a cocycle. Thus Lemma 15.3.23 can be applied repeatedly:

Corollary 15.3.28. *If $\mathfrak{U}$ is a G-invariant open cover of X, $F^{p,q}$ is a cocycle of bidegree (p,q) and total degree n in the total complex of $A^{*,*}(X,\mathfrak{U};V)$ and the $\mathfrak{U}$-local cohomologies $H^l(X,\mathfrak{U};V)$ are trivial for $l \leq q$, then there exist cocycles $F^{p+1,q-1},\ldots,F^{n,0}$ of bidegree $(p+1,q-1),\ldots,(n,0)$ and cochains $\psi^{p,q-1},\ldots,\psi^{n-1,0}$ of bidegree $(p,q-2),\ldots,(n-1,0)$ satisfying $D\psi^{p',q'} = F^{p'+1,q'} + F^{p',q'+1}$. In particular $F^{p,q}$ is cohomologous to $(-1)^q F^{n,0}$:*

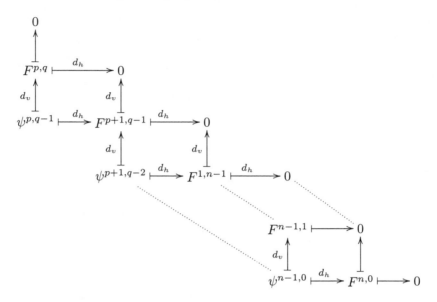

Similarly, by repeated application of Corollary 15.3.27 we obtain:

Corollary 15.3.29. *If $X = G$ is an q-connected locally contractible topological group acting on itself via left translations, $\mathfrak{U}$ a G-invariant open covering of X and $F^{p,q}$ is a cocycle of bidegree (p,q) and total degree n in the total complex of $A^{*,*}(X,\mathfrak{U};V)$, then there exists an open G-invariant cover $\mathfrak{B} \prec \mathfrak{U}$ of X, cocycles $F^{p+1,q-1},\ldots,F^{n,0}$ of bidegree $(p+1,q-1),\ldots,(n,0)$ in the total complex of $A^{*,*}(X,\mathfrak{B};V)$ and cochains $\psi^{p,q-1},\ldots,\psi^{n-1,0}$ of bidegree $(p,q-2),\ldots,(n-1,0)$ satisfying $D\psi^{p',q'} = F^{p'+1,q'} + F^{p',q'+1}$. In particular the restriction $\varrho_{\mathfrak{B},\mathfrak{U}}^{p,q} F^{p,q}$ is cohomologous to $(-1)^q F^{n,0}$.*

Corollary 15.3.30. *If $X = G$ is an n-connected locally contractible topological group acting on itself by left translations, $\mathfrak{U}$ an open G-invariant cover of X and f an equivariant $\mathfrak{U}$-local n-cocycle, then there exists an open G-invariant cover $\mathfrak{B} \prec \mathfrak{U}$ of X such that the restriction $\varrho_{\mathfrak{B},\mathfrak{U}}f$ of f is extendible to a global equivariant cocycle.*

Proof. If $X = G$ is an n-connected locally contractible topological group acting on itself by left translations, then Corollary 15.3.28 implies the existence of a finer G-invariant open cover $\mathfrak{B}$ of X such that the restriction $\varrho_{\mathfrak{B},\mathfrak{U}}f$ is cohomologous to a cocycle $F^{n,0}$ of bidegree $(n,0)$ in the total complex of $A^{*,*}(X,\mathfrak{B};V)^G$. By Lemma 15.3.11 this restriction is extendible to a global equivariant n-cocycle. □

Example 15.3.31. If $X = G$ is an n-connected Lie group acting on itself via left translation, then the germ of every equivariant local n-cocycle on G is extendible to a global equivariant cocycle.

Example 15.3.32. If $X - G$ is a simply connected finite dimensional Lie group acting on itself via left translation, then the germ of every equivariant local 1-cocycle on G is extendible to a global equivariant cocycle. Since finite dimensional Lie groups have trivial second homotopy group, the same is true for local equivariant 2-cocycles.

Example 15.3.33. If $X = \Omega G$ is the loop group of a simply connected finite dimensional Lie group G, then X is simply connected as well. If the group G acts on $X = \Omega G$ via the adjoint action on G, then the germ of every equivariant local 1-cocycle on $X = \Omega G$ is extendible to a global equivariant 1-cocycle on ΩG. If one lets ΩG act on itself via left translation, then the germ of every equivariant local 1-cocycle on X is extendible to a global equivariant cocycle.

As we have seen before, a $\mathfrak{U}$-local equivariant n-cocycle f on X is extendible to a global equivariant n-cocycle if and only if there exist cocycles $i^n(f) = F^{0,n}, \dots, F^{n,0}$ of bidegree $(0,n), (1,n-1), \dots, (n,0)$ in the total complex of $A^{*,*}(X, \mathfrak{U}; V)^G$ and cochains $\psi^{0,n-1}, \dots, \psi^{n-1,0}$ of bidegree $(0,n-1), (1,n-2), \dots, (n-1,0)$ respectively such that $D\psi^{p-1,q} = F^{p,q} + F^{p-1,q+1}$:

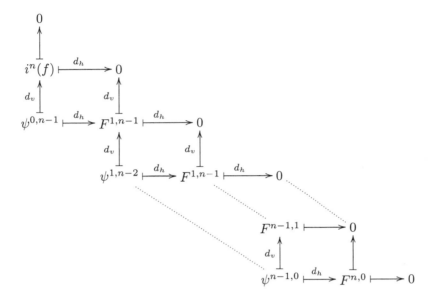

Since the columns of the double complex $A^{*,*}(X, \mathfrak{U}; V)^G$ are not exact in general, we have to find conditions on f and X which ensure the existence of the cochains $\psi^{p',q'}$ and $F^{p,q}$. This will be done step by step, i.e. we first describe conditions for $\psi^{0,n-1}$ to exist and then proceed to examine the cocycle $F^{1,n-1} = d_v\psi^{0,n-1}$ in order to find conditions for $\psi^{1,n-2}$ to exist etc.. In each of these steps we possibly restrict ourselves to local cochains defined on a smaller neighbourhood of the diagonal. This does not affect the extensibility of the germ of the local cocycle we start with.

Recall that being a coboundary is a necessary condition on equivariant local or equivariant Alexander-Spanier n-cocycles to be extendible to a global equivariant cocycle (cf. Lemmata 15.3.3 and 15.3.4). This condition will turn out to be necessary and sufficient for the existence of the cochain $\psi^{0,n-1}$, i.e. for the first step in the above described process. Let $\mathfrak{U}$ be a G-invariant open covering of X and f be an equivariant $\mathfrak{U}$-local n-cocycle on X. Throughout the rest of this chapter we make the assumption that the homomorphism $\mathrm{colim}_{\mathfrak{U}} H(X, \mathfrak{U}; V) \to H(X; V)$ from the colimit cohomology $\mathrm{colim}\, H(X, \mathfrak{U}; V)$ of X where $\mathfrak{U}$ ranges over open coverings into the singular cohomology is injective. This can be accomplished by requiring the space $X = G$ to be a connected topological group acting on itself via left translations. The additional requirement that the coefficients are divisible then enables us to test the triviality of the germ $[f] \in \mathrm{colim}\, H(X, \mathfrak{U}; V)$ by evaluating the singular n-cocycle $\tilde{\lambda}_{\mathfrak{U}}^n f$ on singular n-cycles. This evaluation on singular n-cycles can be expressed in terms of the flux morphism $\mathcal{F}_{\varrho t}(\tilde{\lambda}_{\mathfrak{U}}^n f)$ associated to the equivariant function

$$\varrho t : X \times G \to X, \quad (x,g) \mapsto g.x,$$

where $t : X \times G \to G \times X$ denotes the coordinate flip and we let the group G act on itself via the adjoint action. The function ϱt is the bundle projection of a left G-principal bundle. The bundle projection ϱt induces a flux homomorphism from the equivariant singular cochain complex $S^*(X; V)^G$ into the total complex of the double complex $S^*(X; S^*(G; V)^G)$ (see Section 13.9 for an introduction of flux morphism for singular cochains):

$$\mathcal{F}_{\varrho t} : S^*(X; V)^G \to \mathrm{Tot} S^*(X; S^*(G; V)^G)$$

where we have corestricted the flux to the coefficient spaces $S^q(G; V)^G$ of left equivariant singular cochains on G (cf. Lemma 13.7.12). In the following we will only use this corestricted version of the flux $\mathcal{F}_{\varrho t}(\tilde{\lambda}^n_{\mathfrak{U}} f)$.

The image $\mathcal{F}_{\varrho t}(\tilde{\lambda}^n_{\mathfrak{U}} f)$ of the equivariant singular n-cocycle $\tilde{\lambda}^n_{\mathfrak{U}} f$ on X under the flux homomorphism $\mathcal{F}_{\varrho t}$ is a cocycle in the total complex $\mathrm{Tot} S^*(X; S^*(G; V)^G)$. We denote the component of bidegree (p, q) of the flux $\mathcal{F}_{\varrho t}(\tilde{\lambda}^n_{\mathfrak{U}} f)$ by $\mathcal{F}_{\varrho t}(\tilde{\lambda}^n_{\mathfrak{U}} f)^{p,q}$. This situation can be depicted as follows:

We use the component $\mathcal{F}_{\varrho t}(\tilde{\lambda}^n_{\mathfrak{U}} f)^{n,0}$ of bidegree $(n, 0)$ of the flux $\mathcal{F}_{\varrho t}(\tilde{\lambda}^n_{\mathfrak{U}} f)$ to describe a necessary and sufficient condition for the germ $[f] \in \mathrm{colim}\, A^n(X, \mathfrak{U}; V)$ of f to be a coboundary. This component $\mathcal{F}_{\varrho t}(\tilde{\lambda}^n_{\mathfrak{U}} f)^{n,0}$ is an equivariant homomorphism

$$\mathcal{F}_{\varrho t}(\tilde{\lambda}^n_{\mathfrak{U}} f)^{n,0} : S_n(X) \to S^0(G; V)^G$$
$$\mathcal{F}_{\varrho t}(\tilde{\lambda}^n_{\mathfrak{U}} f)^{0,n} \left(\sum a_i \tau_i \right) (g) = \sum a_i \tilde{\lambda}^n_{\mathfrak{U}} f(g.\tau_i) = g. \sum a_i \tilde{\lambda}^n_{\mathfrak{U}} f(\tau_i)$$

As explained in Lemma 13.9.5 and Corollary 13.9.6 the homomorphisms $\mathcal{F}_{\varrho t}(\tilde{\lambda}^n_{\mathfrak{U}} f)^{p,q}$ map singular cycles to singular cocycles and restrict to homomorphisms

$$\mathcal{F}_{\varrho t}(\tilde{\lambda}^n_{\mathfrak{U}} f)^{p,q} : Z_p(X) \to Z^q(G; V)^G$$

which descend to homomorphisms

$$\mathcal{F}_{\varrho t}(\tilde{\lambda}^n_{\mathfrak{U}} f)^{p,q} : H_p(X) \to H^q_{sing,eq}(G; V).$$

In the special case $p = n$, $q = 0$ the codomain $Z^0(X; V)^G$ of the restricted flux component $\mathcal{F}_{\varrho t}(\tilde{\lambda}^n_{\mathfrak{U}} f)^{n,0}_{|Z_n(X)}$ of bidegree $(n, 0)$ of the flux $\mathcal{F}_{\varrho t}(\tilde{\lambda}^n_{\mathfrak{U}} f)$ can be identified with the G-invariants V^G in V. So the homomorphism $\mathcal{F}_{\varrho t}(\tilde{\lambda}^n_{\mathfrak{U}} f)^{n,0}$ is given by

$$\mathcal{F}_{\varrho t}(\tilde{\lambda}^n_{\mathfrak{U}} f)^{n,0} : Z_n(X) \to V^G, \quad z \mapsto \tilde{\lambda}^n_{\mathfrak{U}} f(z), \tag{15.4}$$

i.e. by evaluation of the singular n-cocycle $\tilde{\lambda}_{\mathfrak{U}}^{n} f$ on the singular n-cycles in X. The cohomology group $H_{eq}^{0}(G; V)$ is isomorphic to the group $Z^{0}(G; V)^{G} \cong V^{G}$, because the 0-boundaries are trivial. Moreover, since the n-chain $\tilde{\lambda}_{\mathfrak{U}}^{n} f$ is a cocycle, the evaluation in formula 15.4 only depends on the homotopy classes of the cycles z; as consequence we see directly that the flux descends to a homomorphism $\mathcal{F}_{\varrho t}(\tilde{\lambda}_{\mathfrak{U}}^{n} f)^{n,0} : H_{n}(X) \to V^{G}$ and the latter is trivial if and only if the restriction $\mathcal{F}_{\varrho t}(\tilde{\lambda}_{\mathfrak{U}}^{n} f)^{n,0} : Z_{n}(X) \to V^{G}$ is trivial. So we observe:

Lemma 15.3.34. *The germ $[f] \in \operatorname{colim} Z^{n}(X, \mathfrak{U}; V)$ of an equivariant $\mathfrak{U}$-local n-cocycle f on X is a coboundary if and only if the component $\mathcal{F}_{\varrho t}(\tilde{\lambda}_{\mathfrak{U}}^{n} f)^{n,0} : Z_{n}(X) \to V^{G}$ of the flux is trivial if and only if the induced map $\mathcal{F}_{\varrho t}(\tilde{\lambda}_{\mathfrak{U}}^{n} f)^{n,0} : H_{n}(X) \to H_{sing,eq}^{0}(G; V) \cong V^{G}$ is trivial.*

Proof. Because the homomorphism $\operatorname{colim} H^{n}(X, \mathfrak{U}; V) \to H^{n}(X; V)$ was assumed to be injective, the cocycle $[f]$ is a coboundary if and only if the evaluation of the singular cocycle $\tilde{\lambda}_{\mathfrak{U}}^{n} f \in Z^{n}(X; V)$ on singular n-cycles is trivial. Thus formula 15.4 shows that this cocycle is trivial if and only if the flux component $\mathcal{F}_{\varrho t}(\tilde{\lambda}_{\mathfrak{U}}^{n} f)^{n,0}$ is trivial. $\square$

If the coefficients V are a rational vector space, then this fact may also be stated in terms of the rational singular homology of X: In this case the singular groups $S^{*}(G; V)^{G}$ and $H_{eq}(X; V)$ also are rational vector spaces. This allows us to identify the groups $S^{*}(G; V)^{G}$ and $S^{*}(G; V)^{G} \otimes_{\mathbb{Z}} \mathbb{Q}$ and the groups $H_{eq}(X; V)$ and $H_{eq}(X; V) \otimes_{\mathbb{Z}} \mathbb{Q}$, which we will do in the following, provided V is a rational vector space. (As usual we will omit the subscript '$\mathbb{Z}$' when dealing with tensor products over the ring $\mathbb{Z}$ of integers.) Therefore any homomorphism from an Abelian group A into one of these groups annihilates torsion and factors uniquely through the rationalisation $A \otimes \mathbb{Q}$. Consequently, for any $\mathfrak{U}$-local equivariant n-cocycle f on X the components

$$\mathcal{F}_{\varrho t}(\tilde{\lambda}_{\mathfrak{U}}^{n} f)^{p,q} : S_{p}(X) \to S^{q}(G; V)^{G}$$
$$\text{and} \quad \mathcal{F}_{\varrho t}(\tilde{\lambda}_{\mathfrak{U}}^{n} f)^{p,q} : Z_{p}(X) \to Z^{q}(G; V)^{G}$$

of the flux $\mathcal{F}(\tilde{\lambda}_{\mathfrak{U}}^{n} f)$ factor uniquely through the rational singular groups $S_{p}(X; \mathbb{Q}) = S_{p}(X) \otimes \mathbb{Q}$ and $Z_{p}(X; \mathbb{Q}) = Z_{p}(X) \otimes \mathbb{Q}$. This can be subsumed in the commutative diagram

$$
\begin{array}{ccc}
Z_{p}(X) & \xrightarrow{\quad \mathcal{F}(\tilde{\lambda}_{\mathfrak{U}}^{n} f)^{p,q} \quad} & Z^{q}(G; V)^{G} \\
{\scriptstyle z \mapsto z \otimes 1} \downarrow & & \Big\| \\
Z_{p}(X) \otimes \mathbb{Q} & \xrightarrow{\mathcal{F}(\tilde{\lambda}_{\mathfrak{U}}^{n} f)^{p,q} \otimes \operatorname{id}_{\mathbb{Q}}} & Z^{q}(G; V)^{G} \otimes \mathbb{Q} \cong Z^{q}(G; V)^{G} .
\end{array}
\tag{15.5}
$$

In particular, as is apparent from the above diagram, the flux component $\mathcal{F}(\tilde{\lambda}_{\mathfrak{U}}^{n} f)^{p,q}$ is trivial if and only if its rational version $\mathcal{F}(\tilde{\lambda}_{\mathfrak{U}}^{n} f)^{p,q} \otimes \operatorname{id}_{\mathbb{Q}}$ is trivial. So we note:

Lemma 15.3.35. *If V is a real vector space then the germ $[f] \in \operatorname{colim} Z^{n}(X, \mathfrak{U}; V)$ of an equivariant $\mathfrak{U}$-local n-cocycle f on X is a coboundary if and only if the component $\mathcal{F}_{\varrho t}(\tilde{\lambda}_{\mathfrak{U}}^{n} f)^{n,0} \otimes \operatorname{id}_{\mathbb{Q}} : Z_{n}(X; \mathbb{Q}) \to V^{G}$ of the flux is trivial if and only if the induced map $\mathcal{F}_{\varrho t}(\tilde{\lambda}_{\mathfrak{U}}^{n} f)^{n,0} \otimes \operatorname{id}_{\mathbb{Q}} : H_{n}(X; \mathbb{Q}) \to H_{sing,eq}^{0}(G; V) \cong V^{G}$ is trivial.*

Since the condition that the germ $[f] \in \operatorname{colim} Z^{n}(X, \mathfrak{U}; V)$ is a coboundary is necessary for the germ of f at the diagonal to be extendible to a global equivariant n-cocycle F on X, we see again that the vanishing of the $(0, n)$-flux in (co)homology is a necessary condition for this to be the case. In the following we show that $[f]$ being a coboundary (or equivalently, the triviality of the $(n,0)$-flux $\mathcal{F}_{\varrho t}(\tilde{\lambda}_{\mathfrak{U}}^n f)^{n,0} : H_0(G) \to H^n_{sing,eq}(X;V))$ is a sufficient condition for the existence of a G-invariant refinement $\mathfrak{B}$ of $\mathfrak{U}$ and a cochain $\psi^{n-1,0}$ in $A^{0,n-1}(X,\mathfrak{B};V)^G$ with vertical coboundary $d_v\psi^{n-1,0} = i^n(f_{|SS(X,\mathfrak{U})([n])})$.

Lemma 15.3.36. *The germ $[f] \in \operatorname{colim} Z^n(X,\mathfrak{U};V)$ of an equivariant $\mathfrak{U}$-local n-cocycle on X is a coboundary if and only if there exists an open G-invariant refinement $\mathfrak{B}$ of $\mathfrak{U}$ and a cochain $\psi^{0,n-1}$ of bidegree $(0,n-1)$ in the total complex $\operatorname{Tot} A^{*,*}(X,\mathfrak{B};V)^G$ with vertical coboundary $d_v\psi^{0,n-1} = i^n(f_{|SS(X,\mathfrak{U})([n])})$.*

Proof. Assume that the germ $[f] \in \operatorname{colim} Z^n(X,\mathfrak{U};V)$ of f is a coboundary. Then there exists an open G-invariant refinement $\mathfrak{B}$ of $\mathfrak{U}$ and a global $(n-1)$-cochain H on X whose coboundary dH restricts to $f_{|SS(X,\mathfrak{B})([n])}$ on $SS(X,\mathfrak{B})([n])$. Then the function $\psi^{0,n-1} : X \times X^n \to V$ defined via

$$\psi^{0,n-1}(x_0, \mathbf{y}) := [p(x_0).H](\mathbf{y}) = p(x_0).\left[H\left(p(x_0)^{-1}.\mathbf{y}\right)\right]$$

is equivariant and thus defines an cochain of bidegree $(0,n-1)$ in $A^{0,n-1}(X,\mathfrak{B};V)^G$. Furthermore the evaluation of the vertical differential of the cochain $\psi^{0,n-1}$ at a point $(x_0,\mathbf{y})$ in $X \times SS(X,\mathfrak{B})([n])$ computes to

$$(d_v\psi^{0,n-1})(x_0,\mathbf{y}) = d[p(x_0).H](\mathbf{y}) = [p(x_0).dH](\mathbf{y})$$
$$= [p(x_0).f](\mathbf{y}) = f(\mathbf{y}) = i^n(f)(x_0,\mathbf{y}),$$

since the cocycle f is equivariant. So for every equivariant $\mathfrak{U}$-local n-cocycle f on X whose germ $[f] \in \operatorname{colim} Z^n(X,\mathfrak{U};V)$ is a coboundary there exists a G-invariant refinement $\mathfrak{B}$ of $\mathfrak{U}$ and a cochain $\psi^{0,n-1} \in A^{0,n-1}(X,\mathfrak{B};V)^G$ with vertical coboundary $i^n(f_{|SS(X,\mathfrak{U})([n])})$. Conversely, assume that there exists an open G-invariant refinement $\mathfrak{B}$ of $\mathfrak{U}$ and a cochain $\psi^{0,n-1} \in A^{0,n-1}(X,\mathfrak{B};V)^G$ of bidegree $(0,n-1)$ with vertical differential $d_v\psi^{0,n-1} = i^n(f_{|SS(X,\mathfrak{U})([n])})$. Fix a point $x \in X$ and let the $(n-1)$-cochain h on X be defined by $h(\mathbf{y}) = \psi^{0,n-1}(x,\mathbf{y})$. The coboundary of the $(n-1)$-chain h on X is given by

$$dh(\mathbf{y}) = d_h\psi^{0,n-1}(x,\mathbf{y}) = i^n(f)(x,\mathbf{y}) = f(\mathbf{y}),$$

i.e. it is exactly the n-cocycle f on $SS(X,\mathfrak{B})([n])$. Thus the $\mathfrak{B}$-local cocycle $f_{|SS(X,\mathfrak{U})([n])}$ is a coboundary. $\square$

If the topological space X is $(n-1)$-connected, then the n-th Hurewicz homomorphism $\operatorname{hur}_n : \pi_n(X) \to H_n(X)$ is an epimorphism. In this case the flux homomorphism $\mathcal{F}_{\varrho t}(\tilde{\lambda}_{\mathfrak{U}}^n f)^{n,0} : H_n(X) \to V^G$ is trivial if and only if its composition with the Hurewicz homomorphism hur_n is trivial. Recall that this composition is the period homomorphism

$$\operatorname{per}_{\varrho t}(\tilde{\lambda}_{\mathfrak{U}}^n f)_n = \operatorname{hur}_n \mathcal{F}_{\varrho t}(\tilde{\lambda}_{\mathfrak{U}}^n f)^{n,0} : H_n(X) \to V^G, \quad [s] \mapsto \tilde{\lambda}_{\mathfrak{U}}^n f \operatorname{hur}_n([s])$$

by definition. As a consequence we note:

Lemma 15.3.37. *If the topological space X is $(n-1)$-connected, then the germ $[f]$ in $\operatorname{colim} Z^n(X,\mathfrak{U};V)$ of an $\mathfrak{U}$-local equivariant n-cocycle f is a coboundary if and only if the period homomorphism $\operatorname{per}_{\varrho t}(\tilde{\lambda}_{\mathfrak{U}}^n f)_n$ is trivial.*

If the coefficient group is a rational vector space, then the codomain $Z^q(G;V)^G$ of any period homomorphism $\operatorname{per}_{\varrho t}(\tilde{\lambda}_{\mathfrak{U}}^n f)_p$ is a rational vector space, we also can w.l.o.g. switch to the rational version: The commutative diagram 15.5 descends to (co)homology and may there be extended on left by the (rational) Hurewicz-homomorphism to obtain the following commutative diagram:

$$
\begin{array}{ccccc}
& & \xrightarrow{\quad\operatorname{per}_{\varrho t}(\tilde\lambda_{\mathfrak{U}}^n f)_n\quad} & & \\
\pi_n(X) & \xrightarrow{\quad\mathrm{hur}_n\quad} & H_p(X) & \xrightarrow{\mathcal{F}(\tilde\lambda_{\mathfrak{U}}^n f)^{p,q}} & H^q_{sing,eq}(G;V) \qquad (15.6)\\[2pt]
{\scriptstyle s\mapsto s\otimes1}\Big\downarrow & & {\scriptstyle z\mapsto z\otimes1}\Big\downarrow & & \Big\| \\[2pt]
\pi_n(X)\otimes\mathbb{Q} & \xrightarrow{\mathrm{hur}_n\otimes\mathrm{id}_{\mathbb{Q}}} & H_p(X;\mathbb{Q}) & \xrightarrow{\mathcal{F}(\tilde\lambda_{\mathfrak{U}}^n f)^{p,q}\otimes\mathrm{id}_{\mathbb{Q}}} & H^q_{sing,eq}(G;V)^G, \\
& & \xrightarrow{\quad\operatorname{per}_{\varrho t}(\tilde\lambda_{\mathfrak{U}}^n f)_n\otimes\mathrm{id}_{\mathbb{Q}}\quad} & &
\end{array}
$$

where the lowest horizontal arrow is the rational version $\operatorname{per}_{\varrho t}(\tilde\lambda_{\mathfrak{U}}^n f)_n\otimes\mathrm{id}_{\mathbb{Q}}$ of the period homomorphism $\operatorname{per}_{\varrho t}(\tilde\lambda_{\mathfrak{U}}^n f)_n$. This rational period homomorphism is trivial if and only if the the period homomorphism $\operatorname{per}_{\varrho t}(\tilde\lambda_{\mathfrak{U}}^n f)_n$ itself is trivial. We record the result:

Lemma 15.3.38. *If the space X is $(n-1)$-connected and V is a rational vector space, then the germ $[f]\in\operatorname{colim} Z^n(X,\mathfrak{U};V)$ of an $\mathfrak{U}$-local equivariant n-cocycle f is a coboundary if and only if the rational period homomorphism $\operatorname{per}_{\varrho t}(\tilde\lambda_{\mathfrak{U}}^n f)_n\otimes\mathrm{id}_{\mathbb{Q}}$ is trivial.*

Recall that the rational vector spaces $Z^n(X,\mathfrak{U};V)^G$ and $Z^n(X;V)^G$ can be identified with the tensor products $Z^n(X,\mathfrak{U};V)^G\otimes\mathbb{Q}$ and $Z^n(X;V)^G\otimes\mathbb{Q}$ respectively. Using these identifications we observe:

Proposition 15.3.39. *If the space X is $(n-1)$-connected then the sequence*

$$
Z^n(X;V)^G \to \operatorname{colim} Z^n(X,\mathfrak{U};V)^G \xrightarrow{\operatorname{per}_{\varrho t}(-)_n} \hom(\pi_n(X);V^G)
$$

is exact, i.e. the germ $[f]\in\operatorname{colim} Z^n(X,\mathfrak{U};V)^G$ of an equivariant local n-cocycle f on X is integrable to a global equivariant cocycle if and only if its period homomorphism $\operatorname{per}_{\varrho t}(\tilde\lambda_{\mathfrak{U}}^n f)_n$ is trivial. If in addition V a rational vector space, then the sequence

$$
Z^n(X;V)^G \to \operatorname{colim} Z^n(X,\mathfrak{U};V)^G \xrightarrow{\operatorname{per}_{\varrho t}(-)_n\otimes\mathrm{id}_{\mathbb{Q}}} \hom(\pi_n(X)\otimes\mathbb{Q};V^G)
$$

is exact as well, i.e. the germ $[f]\in\operatorname{colim} Z^n(X,\mathfrak{U};V)^G$ of an equivariant local n-cocycle f on X is integrable to a global equivariant cocycle if and only if its rational period homomorphism $\operatorname{per}_{\varrho t}(\tilde\lambda_{\mathfrak{U}}^n f)_n\otimes\mathrm{id}_{\mathbb{Q}}$ is trivial.

Proof. Let X be $(n-1)$-connected. If the germ $[f]\in\operatorname{colim} Z^n(X,\mathfrak{U};V)^G$ of an equivariant local n-cocycle f on X is extendible to a global equivariant cocycle, then its is a coboundary by Lemma 15.3.3, hence the period homomorphism $\operatorname{per}_{\varrho t}(\tilde\lambda_{\mathfrak{U}}^n f)_n$ and its rational version $\operatorname{per}_{\varrho t}(\tilde\lambda_{\mathfrak{U}}^n f)_n\otimes\mathrm{id}_{\mathbb{Q}}$ are trivial by Lemmata 15.3.37 and 15.3.38. Conversely, if these period homomorphisms are trivial, then the germ $[f]\in\operatorname{colim} Z^n(X,\mathfrak{U};V)^G$ is a coboundary and there exists a G-invariant open refinement $\mathfrak{B}$ of $\mathfrak{U}$ on X such that $i^n(f_{|\mathrm{SS}(X,\mathfrak{U})([n])})$ is cohomologous to a cocycle $-F^{1,n-1}=-d_h\psi^{0,n-1}$ in the total complex of $A^{*,*}(X,\mathfrak{B};V)^G$. The latter cocycle is then cohomologous to a cocycle of bidegree $(n,0)$ by repeated application of Corollary 15.3.25. This implies that the germ $[f]\in\operatorname{colim} Z^n(X,\mathfrak{U};V)^G$ of f is extendible to a global equivariant cocycle. $\qquad\square$

Example 15.3.40. If $X=G$ is a connected locally contractible group acting on itself via left translation, then a germ $[f]\in\operatorname{colim} Z^1(X,\mathfrak{U};V)^G$ of an equivariant local 1-cocycle f on X is extendible to a global equivariant 1-cocycle, if and only if the period homomorphism $\operatorname{per}_{\varrho t}(\tilde\lambda_{\mathfrak{U}}^n f)_1:\pi_1(G)\to V^G$ is trivial. If V is a rational vector space then this is equivalent to the rational period homomorphism $\operatorname{per}_{\varrho t}(\tilde\lambda_{\mathfrak{U}}^n f)_1\otimes\mathrm{id}_{\mathbb{Q}}:\pi_1(G)\otimes\mathbb{Q}\to V^G$ being trivial.

Example 15.3.41. If $X = G$ is a simply connected locally contractible group acting on itself via left translation, then a germ $[f] \in \operatorname{colim} Z^2(X, \mathfrak{U}; V)^G$ of an equivariant local 2-cocycle f on X is extendible to a global equivariant 2-cocycle, if and only if the period homomorphism $\operatorname{per}_{\varrho t}(\tilde{\lambda}^n_{\mathfrak{U}} f)_2 : \pi_2(G) \to V^G$ is trivial. If V is a rational vector space then this is equivalent to the rational period homomorphism $\operatorname{per}_{\varrho t}(\tilde{\lambda}^n_{\mathfrak{U}} f)_1 \otimes \operatorname{id}_{\mathbb{Q}} : \pi_1(G) \otimes \mathbb{Q} \to V^G$ being trivial. If G is a simply connected finite dimensional Lie group then every germ $[f] \in \operatorname{colim} Z^2(X, \mathfrak{U}; V)^G$ of an equivariant local 2-cocycle f on X is extendible to a global equivariant 1-cocycle and a germ $[f] \in \operatorname{colim} Z^3(X, \mathfrak{U}; V)^G$ of an equivariant local 3-cocycle f on X is extendible to a global equivariant 3-cocycle, if and only if the period homomorphism $\operatorname{per}_{\varrho t}(\tilde{\lambda}^n_{\mathfrak{U}} f)_3 : \pi_3(G) \to V^G$ is trivial. If V is a rational vector space then this is equivalent to the rational period homomorphism $\operatorname{per}_{\varrho t}(\tilde{\lambda}^n_{\mathfrak{U}} f)_3 \otimes \operatorname{id}_{\mathbb{Q}} : \pi_3(G) \otimes \mathbb{Q} \to V^G$ being trivial.

Example 15.3.42. If $X = f_s G$ is the loop group of a simply connected locally compact Lie group G, then M is simply connected as well. If the group G acts on $M = f_s G$ via the adjoint action on G, then every germ $[f] \in \operatorname{colim} Z^2(X, \mathfrak{U}; V)^G$ of an equivariant local 2-cocycle f on X is extendible to a global equivariant 2-cocycle, if and only if the period homomorphism $\operatorname{per}_{\varrho t}(\tilde{\lambda}^n_{\mathfrak{U}} f)_2 : \pi_2(G) \to V^G$ is trivial. If V is a rational vector space then this is equivalent to the rational period homomorphism $\operatorname{per}_{\varrho t}(\tilde{\lambda}^n_{\mathfrak{U}} f)_2 \otimes \operatorname{id}_{\mathbb{Q}} : \pi_2(G) \otimes \mathbb{Q} \to V^G$ being trivial. If one lets $f_s G$ act on itself via left translation, then every germ $[f] \in \operatorname{colim} Z^2(X, \mathfrak{U}; V)^G$ of an equivariant local 2-cocycle f on X is extendible to a global equivariant 2-cocycle, if and only if the period homomorphism $\operatorname{per}_{\varrho t}(\tilde{\lambda}^n_{\mathfrak{U}} f)_2 : \pi_2(G) \to V^G$ is trivial. If V is a rational vector space then this is equivalent to the rational period homomorphism $\operatorname{per}_{\varrho t}(\tilde{\lambda}^n_{\mathfrak{U}} f)_2 \otimes \operatorname{id}_{\mathbb{Q}} : \pi_2(G) \otimes \mathbb{Q} \to V^G$ being trivial.

In the remaining part of this section we assume that the $(0, n)$-flux $\mathcal{F}_{\varrho t}(\tilde{\lambda}^n_{\mathfrak{U}} f)^{0,n}$ is trivial such that there exists a G-invariant refinement $\mathfrak{U}_0$ of $\mathfrak{U}$ and a cochain $\psi^{0,n-1}$ in $A^{0,n-1}(X, \mathfrak{U}_0; V)^G$ with vertical coboundary $d_v \psi^{0,n-1} = i^n(f_{|SS(X,\mathfrak{U}_0)([n])})$. In particular we will understand $\psi^{0,n-1}$ to be the cochain constructed as in Lemma 16.6.35. We denote the horizontal coboundary $d_h \psi^{0,n-1}$ of $\psi^{0,n-1}$ by $F^{1,n-1}$ will work with the cochains $\psi^{0,n-1}$ and $F^{1,n-1}$ without further reference. Lemma 14.2.19 ensures that the horizontal coboundary of the cochain $F^{1,n-1} := d_h \psi^{0,n-1}$ is indeed a cocycle in the total complex $\operatorname{Tot} A^{*,*}(X, \mathfrak{U}_0; V)^G$. The cocycle $F^{1,n-1}$ is then expected to represent the local n-cocycle f in some way:

Lemma 15.3.43. *The cocycle $i^n(f_{|SS(X,\mathfrak{U}_0)([n])})$ in the total complex of $A^{*,*}(X, \mathfrak{U}_0; V)^G$ is cohomologous to the cocycle $-F^{1,n-1} \in Z^{1,n-1}(X, \mathfrak{U}_0; V)^G$.*

Proof. The difference cochain $i^n(f_{|SS(X,\mathfrak{U}_0)([n])}) - (-F^{1,n-1})$ is the coboundary $D\psi^{0,n-1} = d_h \psi^{0,n-1} + d_v \psi^{0,n-1}$ in the total complex $\operatorname{Tot} A^{*,*}(X, \mathfrak{U}_0; V)^G$, so the cocycles $i^n(f_{|SS(X,\mathfrak{U}_0)([n])})$ and $-F^1$ in the total complex $\operatorname{Tot} A^{*,*}(X, \mathfrak{U}_0; V)^G$ are cohomologous. $\square$

Lemma 15.3.44. *If the cocycle $i^n(f_{|SS(X,\mathfrak{U}_0)([n])})$ is cohomologous to a cocycle $-\tilde{F}^{1,n-1}$ of bidegree $(1, n-1)$ in $\operatorname{Tot} A^{*,*}(X, \mathfrak{U}_0; V)^G$, then there exists a cochain $c^{0,n-1}$ in $A^{0,n-1}(X, \mathfrak{U}_0; V)^G$ such that $d_v c^{0,n-1} = i^n(f_{|SS(X,\mathfrak{U}_0)([n])})$ and $d_h c^{0,n-1}$ is cohomologous to $\tilde{F}^{1,n-1}$.*

Proof. Assume that there exists a cocycle $\tilde{F}^{1,n-1}$ of bidegree $(1, n-1)$ which is cohomologous to $-i^n(f_{|SS(X,\mathfrak{U}_0)([n])})$ in the total complex $\operatorname{Tot} A^{*,*}(X, \mathfrak{U}_0; V)^G$. Because the cocycles $i^n(f_{|SS(X,\mathfrak{U}_0)([n])})$ and $-\tilde{F}^{1,n-1}$ are cohomologous, there exists a cochain

$c = \sum_{p'+q'=n-1} c^{p',q'}$ with coboundary $Dc = i^n(f_{|SS(X,\mathfrak{U}_0)([n])}) + \tilde{F}^{1,n-1}$ in the total complex $\mathrm{Tot}A^{*,*}(X,\mathfrak{U}_0;V)^G$. In particular the vertical coboundary $d_v c^{0,n-1}$ is exactly the cocycle $i^n(f_{|SS(X,\mathfrak{U}_0)([n])})$. The equalities

$$D \sum_{p'+q'=n-1,p'>0} c^{p',q'} = i^n(f_{|SS(X,\mathfrak{U}_0)([n])}) + \tilde{F}^{1,n-1} - Dc^{0,n-1}$$

$$= i^n(f_{|SS(X,\mathfrak{U}_0)([n])}) + \tilde{F}^{1,n-1} - d_h c^{0,n-1} - d_v c^{0,n-1}$$

$$= i^n(f_{|SS(X,\mathfrak{U}_0)([n])}) + \tilde{F}^{1,n-1} - d_h c^{0,n-1} - i^n(f_{|SS(X,\mathfrak{U}_0)([n])})$$

$$= (-d_h c^{0,n-1}) + \tilde{F}^{1,n-1}$$

then show that the cocycle $d_h c^{0,1}$ is cohomologous to $\tilde{F}^{1,n-1}$. $\qquad\square$

Lemma 15.3.45. *If there exists a G-invariant open covering $\mathfrak{U}_0 \prec \mathfrak{U}$ such that the cocycle $i^n(f_{|SS(X,\mathfrak{U}_0)([n])})$ is cohomologous to a cocycle $-\tilde{F}^{1,n-1}$ of bidegree $(1,n-1)$ in $\mathrm{Tot}A^{*,*}(X,\mathfrak{U}_0;V)^G$, then the germ $[f] \in \mathrm{colim}\, Z^n(X,\mathfrak{U}_0;V)$ is a coboundary and $\tilde{F}^{1,n-1}$ is cohomologous to $F^{1,n-1}$.*

Proof. Assume that there exists a G-invariant open covering $\mathfrak{U}_0 \prec \mathfrak{U}$ of X such that the cocycle $i^n(f_{|SS(X,\mathfrak{U}_0)([n])})$ is cohomologous to a cocycle $-\tilde{F}^{1,n-1}$ of bidegree $(1,n-1)$ in $\mathrm{Tot}A^{*,*}(X,\mathfrak{U}_0;V)^G$. By the preceding Lemma there exists a cochain $c^{0,n-1}$ of bidegree $(0,n-1)$ satisfying $d_v c^{0,n-1} = i^n(f_{|SS(X,\mathfrak{U}_0)([n])})$ and such that $d_h c^{0,n-1}$ is cohomologous to $\tilde{F}^{1,n-1}$. Fix a point x in X and define an $(n-1)$-chain $h \in A^{n-1}(M,\mathfrak{U}_0;V)$ via

$$h(\mathbf{y}) = c^{0,n-1}(x,\mathbf{y}).$$

The coboundary dh of h is given by the vertical coboundary of the cochain $c^{0,n-1}$:

$$dh(\mathbf{y}) = (d[c^{0,n-1}(x,-)])(\mathbf{y}) = [d_v c^{0,n-1}](x,\mathbf{y}) = i^n(f_{|SS(X,\mathfrak{U}_0)([n])})(x,\mathbf{y}) = f(\mathbf{y})$$

Thus the n-cocycle $f_{|SS(X,\mathfrak{U}_0)([n])}$ is a coboundary. Since both cocycles $-\tilde{F}^{1,n-1}$ and $-F^{1,n-1}$ are cohomologous to $i^n(f_{|SS(X,\mathfrak{U}_0)([n])})$, the cocycles $\tilde{F}^{1,n-1}$ and $F^{1,n-1}$ have to be cohomologous. $\qquad\square$

Summarising the three preceding Lemmata we observe:

Lemma 15.3.46. *The germ $[f] \in \mathrm{colim}\, Z^n(X,\mathfrak{U};V)$ of a $\mathfrak{U}$-local n-cocycle is a coboundary if and only if there exists an open G-invariant refinement $\mathfrak{U}_0$ of $\mathfrak{U}$ such that $i^n(f_{|SS(X,\mathfrak{U}_0)([n])})$ is cohomologous to a cocycle of bidegree $(1,n-1)$ in the total complex of $A^{*,*}(X,\mathfrak{U}_0;V)^G$.*

Proposition 15.3.47. *The following conditions on the $\mathfrak{U}$-local n-cocycle f on X are equivalent:*

1. *The the germ $[f] \in \mathrm{colim}\, Z^n(X,\mathfrak{U};V)$ of f is a coboundary.*
2. *The flux component $\mathcal{F}_{\varrho t}(\tilde{\lambda}_{\mathfrak{U}}^n f)^{n,0} : Z_n(X) \to V^G$ of f is trivial.*
3. *The induced map $\mathcal{F}_{\varrho t}(\tilde{\lambda}_{\mathfrak{U}}^n f)^{n,0} : H_n(X) \to V^G$ in (co)homology is trivial.*
4. *There exists an open G-invariant refinement $\mathfrak{U}_0$ of $\mathfrak{U}$ and a cochain $\psi^{0,n-1}$ of bidegree $(0,n-1)$ in the total complex $\mathrm{Tot}A^{*,*}(X,\mathfrak{U}_0;V)^G$ with vertical differential $d_v \psi^{0,n-1} = i^n(f_{|SS(X,\mathfrak{U}_0)([n])})$.*
5. *There exists an open G-invariant refinement $\mathfrak{U}_0$ of $\mathfrak{U}$ such that the cocycle $i^n(f_{|SS(X,\mathfrak{U}_0)([n])})$ is cohomologous to a cocycle $-F^{1,n-1}$ of bidegree $(1,n-1)$ in the total complex $\mathrm{Tot}A^{*,*}(X,\mathfrak{U}_0;V)^G$.*

Proof. It has already been shown in Lemma 15.3.34 that the germ $[f] \in \operatorname{colim} Z^n(X, \mathfrak{U}; V)$ of an equivariant $\mathfrak{U}$-local n-cocycle f on X is a coboundary if and only if the $(0, n)$-flux $\mathcal{F}_{\varrho t}(\tilde{\lambda}_{\mathfrak{U}}^n f)^{n,0} : Z_n(X) \to V^G$ is trivial if and only if the induced map in (co)homology is trivial, which proves the equivalence of 1, 2 and 3. The equivalence of 1 and 4 is shown in Lemma 15.3.36. Finally, the equivalence of 1 and 5 has been shown in Lemma 15.3.46.   □

If the topological space X is $(n-1)$-connected then the conditions in Proposition 16.6.46 are equivalent to the triviality of the period map $\operatorname{per}_{\varrho t}(\tilde{\lambda}_{\mathfrak{U}}^n f)_n : \pi_n(X) \to V^G$. In this case the the germ $[f] \in \operatorname{colim} Z^n(X, \mathfrak{U}; V)$ of a $\mathfrak{U}$-local cocycle f is extendible to a global equivariant cocycle if and only if the period map $\operatorname{per}_{\varrho t}(\tilde{\lambda}_{\mathfrak{U}}^n f)_n$ is trivial (Proposition 15.3.39).

If the topological space X is $(n-2)$-connected and $n \geq 3$, then the Hurewicz homomorphism $\operatorname{hur}_n : \pi_n(X) \to H_n(X)$ is an epimorphism. In this case one can test whether $[f] \in \operatorname{colim} Z^n(X, \mathfrak{U}; V)$ is a coboundary by evaluating $\tilde{\lambda}_{\mathfrak{U}}^n f$ on n-spheres $\mathbb{S}^n \to M$ only. This makes it possible to reformulate the question whether the germ $[f] \in \operatorname{colim} Z^n(X, \mathfrak{U}; V)$ is a coboundary using the period map. Recall that the period homomorphism for the flux $\mathcal{F}_{\varrho t}^{n,0}(\tilde{\lambda}_{\mathfrak{U}}^n f) : H_n(X) \to H_{eq}^0(G; V) = V^G$ is given by

$$\operatorname{per}_{\varrho t}(\tilde{\lambda}_{\mathfrak{U}}^n f)_n = [\mathcal{F}_{\varrho t}^{n,0}(\tilde{\lambda}_{\mathfrak{U}}^n f) \circ \operatorname{hur}_n](s) = [\tilde{\lambda}_{\mathfrak{U}}^n f]([s])$$

i.e. by "integration" of $\tilde{\lambda}_{\mathfrak{U}}^n f$ over n-spheres in X.

Lemma 15.3.48. *If the period homomorphism* $\operatorname{per}_{\varrho t}(\tilde{\lambda}_{\mathfrak{U}}^n f)_n : \pi_n(X) \to V^G$ *is trivial, $n \geq 3$ and the space X is $(n-2)$-connected, then the conditions of Proposition 15.3.47 are satisfied, i.e. $[f] \in \operatorname{colim} Z^n(X, \mathfrak{U}; V)$ is a coboundary and there exists an open G-invariant refinement $\mathfrak{U}_0$ of $\mathfrak{U}$ such that $i^n(f_{|\operatorname{SS}(X, \mathfrak{U}_0)([n])})$ is cohomologous to a cocycle $-F^{1, n-1}$ in $\operatorname{Tot}A^{*,*}(X, \mathfrak{U}_0; V)^G$.*

Proof. If the space X is $(n-2)$-connected and $n \geq 3$, then the Hurewicz homomorphism $\operatorname{hur}_n : \pi_n(X) \to H_n(X)$ is an epimorphism. Therefore the assumption that the period homomorphism $\operatorname{per}_{\varrho t}(\tilde{\lambda}_{\mathfrak{U}}^n f)_n : \pi_n(X) \to V^G$ is trivial implies that the germ $[f] \in \operatorname{colim} Z^n(X, \mathfrak{U}; V)$ is a coboundary thus all the conditions of Proposition 15.3.47 are satisfied.   □

Example 15.3.49. If X is simply connected and f is an equivariant $\mathfrak{U}$-local 3-cocycle on X, then there exists an open G-invariant refinement $\mathfrak{U}_0$ of $\mathfrak{U}$ such that $i^n(f_{|\operatorname{SS}(X, \mathfrak{U}_0)([n])})$ is cohomologous to a cocycle $-F^{1,2}$ of bidegree $(1, 2)$ in $\operatorname{Tot}A^{*,*}(X, \mathfrak{U}_0; V)^G$ if and only if the period map $\operatorname{per}_{\varrho t}(\tilde{\lambda}_{\mathfrak{U}}^n f) : \pi_3(X) \to V^G$ is trivial.

Recall that the germ $[f] \in \operatorname{colim} Z^n(X, \mathfrak{U}; V)$ has to be a coboundary in order to be extendible to a global equivariant cocycle. Moreover, if the germ $[f] \in \operatorname{colim} Z^n(X, \mathfrak{U}; V)$ is a coboundary, then the extensibility of $[f]$ can be decided by examination of the cocycle $F^{1, n-1} = d_h \psi^{0, n-1}$:

Lemma 15.3.50. *If the germ $[f] \in \operatorname{colim} Z^n(X, \mathfrak{U}; V)$ is a coboundary, then it is extendible to a global equivariant n-cocycle if and only if there exists a G-invariant open cover $\mathfrak{B} \prec \mathfrak{U}_0$ of X such that the restriction $\varrho_{\mathfrak{B}, \mathfrak{U}_0}^{1, n-1} F^{1, n-1}$ is cohomologous to a cocycle of bidegree $(0, n)$ in $\operatorname{Tot}A^{*,*}(X, \mathfrak{B}; V)^G$.*

Proof. Lemma 15.3.11 shows that the germ $[f] \in \operatorname{colim} Z^n(X, \mathfrak{U}; V)$ is extendible to a global equivariant cocycle if and only if there exists an open G-invariant covering $\mathfrak{B} \prec \mathfrak{U}$ such that the cocycle $i^n(f_{|\operatorname{SS}(X, \mathfrak{B})([n])})$ is cohomologous to a cocycle of bidegree $(0, n)$ in $\operatorname{Tot}A^{*,*}(X, \mathfrak{B}; V)^G$. If the restriction $f_{|\operatorname{SS}(X, \mathfrak{U}_0)([n])}$

is a coboundary, then the cocycles $i^n(f_{|SS(X,\mathfrak{U}_0)([n])})$ and $-F^{1,n-1}$ are cohomologous. In this case $i^n(f_{|SS(X,\mathfrak{B})([n])})$ is cohomologous to a cocycle of bidegree $(0,n)$ in $\mathrm{Tot}A^{*,*}(X,\mathfrak{B};V)^G$ if and only if the restriction $\varrho_{\mathfrak{B},\mathfrak{U}_0}(-F^{1,n-1})$ and thus the restriction $\varrho_{\mathfrak{B},\mathfrak{U}_0}F^{1,n-1}$ is cohomologous to a cocycle of bidegree $(0,n)$ in $\mathrm{Tot}A^{*,*}(X,\mathfrak{B};V)^G$. $\qquad\square$

This observation also motivates the extension of the definition of extensibility to arbitrary cochains in the double complex $A^{*,*}(X,\mathfrak{U};V)$, which was introduced in the last section. Using this notion we observe:

Lemma 15.3.51. *If the restriction $f_{|SS(X,\mathfrak{U}_0)([n])}$ is a coboundary and the cocycle $F^{1,n-1}$ is extendible to an equivariant global cocycle of bidegree $(1,n-1)$ in $A^{1,n-1}(X,\mathfrak{U}_0;V)^G$, then $f_{|SS(X,\mathfrak{U}_0)([n])}$ is extendible to an equivariant global n-cocycle.*

Proof. If the restriction $f_{|SS(X,\mathfrak{U}_0)([n])}$ is a coboundary and the cocycle $F^{1,n-1}$ is extendible to an equivariant global cocycle, then the latter is cohomologous to a cocycle $F^{n,0}$ of bidegree $(n,0)$ in the total complex of $A^{*,*}(X,\mathfrak{U}_0;V)$ by Lemma 15.3.21. Since the cocycle $i^n(f_{|SS(X,\mathfrak{U}_0)([n])})$ is cohomologous to $-F^{1,n-1}$ it also is cohomologous to $F^{n,0}$ in this case. By Lemma 15.3.11 the restriction $f_{|SS(X,\mathfrak{U}_0)([n])}$ is then extendible to a global equivariant n-cocycle. $\qquad\square$

15.4 $(n-2)$-Connected Spaces

In this section we continue to consider a transformation group (G,M) which admits an equivariant continuous map $p:X\to G$. Like in the last section we require the homomorphisms $\mathrm{colim}_{\mathfrak{U}}H^n(X,\mathfrak{U};V)\to H^n(X;V)$ where $\mathfrak{U}$ ranges over all G-invariant open coverings of X to be injective.

Example 15.4.1. If G is a locally SULC topological group acting on itself by left translations, then the identity $p=\mathrm{id}_G$ is continuous and equivariant. Furthermore the homomorphisms $\mathrm{colim}\,H^n(X,\mathfrak{U};V)\to H^n(X;V)$ is injective by Corollary 3.0.76. This in particular includes all real and complex Lie groups (cf. Lemma 3.0.51).

We fix an equivariant map $X\to G$ for later use. Let $\mathfrak{U}$ be a G-invariant open cover of X and f be a $\mathfrak{U}$-local n-cocycle on X. We already know that the existence of an open G-invariant cover $\mathfrak{U}_0\prec\mathfrak{U}$ such that $f_{|SS(X,\mathfrak{U}_0)([n])}$ is a coboundary is necessary for the germ $[f]\in\mathrm{colim}_{\mathfrak{V}}A^n(X,\mathfrak{V};V)$ to be extendible to a global equivariant cocycle. Therefore we assume the cover $\mathfrak{U}_0$ and an $\mathfrak{U}_0$-local $(n-1)$-cochain h with coboundary $dh=f_{|SS(X,\mathfrak{U}_0)([n])}$ to be given and let the cochain $\psi^{0,n-1}$ in $A^{0,n-1}(X,\mathfrak{U}_0;V)$ be defined via

$$\psi^{0,n-1}(x_0,\mathbf{y}) := [p(x_0).h](\mathbf{y})$$

so the vertical coboundary of $\psi^{0,n-1}$ in the double complex $A^{*,*}(X,\mathfrak{U}_0;V)^G$ is the cocycle $i^n(f_{|SS(X,\mathfrak{U}_0)([n])})$. We denote the horizontal coboundary $d_v\psi^{0,n-1}$ of $\psi^{0,n-1}$ by $F^{1,n-1}$. The cochain $F^{1,n-1}$ is a cocycle in the total complex $\mathrm{Tot}A^{*,*}(X,\mathfrak{U}_0;V)^G$ by Lemma 14.2.19 and the cocycles $i^n(f_{|SS(X,\mathfrak{U}_0)([n])})$ and $-F^{1,n-1}$ are cohomologous in the total complex of $A^{*,*}(X,\mathfrak{U}_0;V)^G$ by Lemma 16.6.42. The situation in this double complex can be depicted as follows:

$$
\begin{array}{ccc}
0 & & \\
\uparrow{\scriptstyle d_v} & & \\
i^n(f_{|\mathrm{SS}(X,\mathfrak{U}_0)([n])}) & \longmapsto & 0 \\
\uparrow{\scriptstyle d_v} & & \uparrow{\scriptstyle d_v} \\
\psi^{0,n-1} \overset{d_h}{\longmapsto} & F^{1,n-1} & \overset{d_h}{\longrightarrow} 0
\end{array}
$$

We observe that the cocycle $F^{1,n-1}$ in the total complexes of $A^{*,*}(X,\mathfrak{U}_0;V)^G$ can be explicitly written down in terms of the local $(n-1)$-cochain h

$$
F^{1,n-1}(x_0,x_1,\mathbf{y}) = \psi^{0,n-1}(x_1,\mathbf{y}) - \psi^{0,n-1}(x_0,\mathbf{y}) = [p(x_1).h](\mathbf{y}) - [p(x_0).h](\mathbf{y})
$$

We now examine the cocycle $F^{1,n-1}$ of bidegree $(1,n-1)$ and search for conditions on f for the cocycle $F^{1,n-1}$ to be the vertical coboundary $d_v\psi^{1,n-2}$ of a cochain $\psi^{1,n-2}$ of bidegree $(1,n-2)$, where we might need to restrict $F^{1,n-1}$ to a smaller neighbourhood of the diagonal. As one may have expected, the necessary and sufficient condition for this to happen is a condition on the flux component $\mathcal{F}_{\varrho t}(\lambda_{\mathfrak{U}}^n f)^{n-1,1}$.

First assume that there exists an open cover $\mathfrak{U}_1 \prec \mathfrak{U}_0$ and a cochain $\psi^{1,n-2}$ of bidegree $(1,n-2)$ in $A^{1,n-2}(X,\mathfrak{U}_1;V)^G$ whose vertical coboundary $d_v\psi^{1,n-2}$ is the restriction $\varrho_{\mathfrak{U}_1,\mathfrak{U}_0}^{1,n-1}F^{1,n-1}$ of the cocycle $F^{1,n-1}$. In this case the horizontal coboundary $F^{2,n-2} := d_h\psi^{1,n-2}$ of the cochain $\psi^{1,n-2}$ is a cocycle by Lemma 14.2.19 and this cocycle $F^{2,n-2}$ is cohomologous to the cocycle $-\varrho_{\mathfrak{U}_1,\mathfrak{U}_0}^{1,n-1}F^{1,n-1}F^{1,n-1}$ by Lemma 14.2.19:

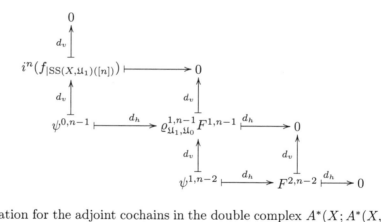

The situation for the adjoint cochains in the double complex $A^*(X; A^*(X,\mathfrak{U}_1;V))^G$ is the same: The adjoint cocycle $\varrho_{\mathfrak{U}_1,\mathfrak{U}_0}^{1,n-1}\hat{F}^{1,n-1}$ in $A^1(X; A^{n-1}(X,\mathfrak{U}_1;V))^G$ of $\varrho_{\mathfrak{U}_1,\mathfrak{U}_0}^{1,n-1}F^{1,n-1}$ is the vertical coboundary of the adjoint cochain $\hat{\psi}^{1,n-2} \in A^1(X; A^{n-2}(X,\mathfrak{U}_1;V))^G$ of $\psi^{1,n-2}$ in the double complex $A^*(X; A^*(X,\mathfrak{U}_1;V))^G$. So all the the cochains $\varrho_{\mathfrak{U}_1,\mathfrak{U}_0}^{1,n-1}\hat{F}^{1,n-1}(x_0,x_1)$ for points $(m_0,m_1) \in X^2$ are coboundaries, i.e. the the adjoint cocycle $\hat{F}^{1,n-1} = d_v\hat{\psi}^{1,n-1} : X^2 \to A^{n-1}(X,\mathfrak{U}_1;V)$ takes values in the group $B^{n-1}(X,\mathfrak{U}_1;V)$ of $\mathfrak{U}_1$-local coboundaries. So we note:

Lemma 15.4.2. *If a restriction $\varrho_{\mathfrak{U}_1,\mathfrak{U}_0}^{1,n-1}F^{1,n-1}$ of the cocycle $F^{1,n-1}$ is the vertical coboundary $d_v\psi^{1,n-2}$ of a cochain of bidegree $(1,n-2)$, then it can be corestricted to $B^{n-1}(X,\mathfrak{U}_1;V)$ and its image in the vector space $Z^1(X; H^{n-1}(X,\mathfrak{U}_1;V))^G$ is trivial.*

This shows that the triviality of the cohomology classes $[\varrho_{\mathfrak{U}_1,\mathfrak{U}_0}^{1,n-1}\hat{F}^{1,n-1}(m_0,m_1)]$ in the $\mathfrak{U}_1$-local cohomology $H^{n-1}(X,\mathfrak{U}_1;V)$ for some G-invariant open cover $\mathfrak{U}_1 \prec \mathfrak{U}_0$

for all points $(x_0, x_1) \in X^2$ is a necessary condition for some restriction $\varrho_{\mathfrak{U}_1,\mathfrak{U}_0}^{1,n-1} F^{1,n-1}$ to be the vertical coboundary $d_v \psi^{1,n-2}$ of a cochain $\psi^{1,n-2}$ of bidegree $(1, n-2)$. The converse implication is also true:

Lemma 15.4.3. *If there exists an G-invariant open cover $\mathfrak{U}_1 \prec \mathfrak{U}_0$ such that the image of the restriction $\varrho_{\mathfrak{U}_1,\mathfrak{U}_0}^{1,n-1} \hat{F}^{1,n-1}$ in the group $A^1(X; H^{n-1}(X, \mathfrak{U}_1; V))^G$ is trivial, then the cocycle $\varrho_{\mathfrak{U}_1,\mathfrak{U}_0}^{1,n-1} F^{1,n-1}$ is the vertical coboundary of a cochain $\psi^{1,n-2}$ in $A^1(X; A^{n-2}(X, \mathfrak{U}_1; V))^G$.*

Proof. Let $\mathfrak{U}_1 \prec \mathfrak{U}_0$ be a G-invariant open cover of X such that the image of the restriction $\varrho_{\mathfrak{U}_1,\mathfrak{U}_0}^{1,n-1} \hat{F}^{1,n-1}$ in the group $A^1(X; H^{n-1}(X, \mathfrak{U}_1; V))^G$ is trivial. Let $s : B^{n-1}(X, \mathfrak{U}_1; V) \to A^{n-2}(X, \mathfrak{U}_1; V)$ a not necessarily equivariant right inverse to the differential $d : A^{n-2}(X, \mathfrak{U}_1; V) \to A^{n-1}(X, \mathfrak{U}_1; V)$. The equivariant function $\psi^{1,n-2}$ in $A^{1,n-2}(X, \mathfrak{U}_1; V)^G$ defined via

$$\hat{\psi}^{1,n-2}(\mathbf{x}, \mathbf{y}) = p(x_0). \left[s\hat{F}^{1,n-1} \left(p(x_0)^{-1}.\mathbf{x}) \right) \left(p(x_0)^{-1}.\mathbf{y} \right) \right]$$

is an equivariant cochain of bidegree $(1, n-2)$ in $A^{1,n-2}(X, \mathfrak{U}_1; V)^G$ whose vertical coboundary is the restriction $\varrho_{\mathfrak{U}_1,\mathfrak{U}_0}^{1,n-1} F^{1,n-1}$ of the cocycle $F^{1,n-1}$. $\qquad \square$

Similar to the coboundary condition on f for the existence of $\psi^{0,n-1}$ the last but one Lemma implies a necessary condition for some restriction $\varrho_{\mathfrak{U}_1,\mathfrak{U}_0}^{1,n-1} F^{1,n-1}$ to be the vertical coboundary of a cochain $\psi^{1,n-2}$ of bidegree $(1, n-2)$ in the double complex $A^{*,*}(X, \mathfrak{U}_1; V)^G$. The previous Lemma then shows that the triviality of the cohomology classes $[\varrho_{\mathfrak{U}_1,\mathfrak{U}_0}^{1,n-1} \hat{F}^{1,n-1}(x_0, x_1)]$ in the $\mathfrak{U}_1$-local cohomology $H^{n-1}(X, \mathfrak{U}_1; V)$ for some open G-invariant open covering $\mathfrak{U}_1 \prec \mathfrak{U}_0$ of X also is sufficient for $\varrho_{\mathfrak{U}_1,\mathfrak{U}_0}^{1,n-1} F^{1,n-1}$ to be the vertical coboundary $d_v \psi^{1,n-2}$ of a cochain $\psi^{1,n-2}$ of bidegree $(1, n-2)$. Similar to the extensibility conditions in the last section, this condition can be expressed in terms of the flux $\mathcal{F}_{\varrho t}(\lambda_{\mathfrak{U}}^n)$. As a first step to prove this we observe:

Lemma 15.4.4. *If the action of G on the coefficients V is trivial, then for all points $(x_0, x_1) \in X^2$ the evaluation of the singular $(n-1)$-cocycle $\lambda_{\mathfrak{U}_0}^{n-1} \hat{F}^{1,n-1}(x_0, x_1)$ on singular $(n-1)$-cycles $z \in Z_{n-1}(X)$ is given by*

$$\lambda_{\mathfrak{U}_0}^{n-1} \hat{F}^{1,n-1}(x_0, x_1) = \mathcal{F}_{\varrho t}(\lambda_{\mathfrak{U}_0}^n f)^{n-1,1}(z)(\gamma)$$

for any path γ from $p(x_0)$ to $p(x_1)$ in G.

Proof. Let (x_0, x_1) be a point in X^2 and $z \in Z_{n-1}(X)$ be a singular $(n-1)$-cycle in X. The evaluation of the singular cocycle $\lambda_{\mathfrak{U}_0}^{n-1} \hat{F}^{1,n-1}(x_0, x_1)$ at the singular cycle $z \in S_{n-1}(X)$ is by definition given by

$$\lambda_{\mathfrak{U}_0}^{n-1} \hat{F}^{1,n-1}(x_0, x_1)(z) = \left(\lambda_{\mathfrak{U}_0}^{n-1}[p(x_1).h] - \lambda_{\mathfrak{U}_0}^{n-1}[p(x_0).h] \right)(z).$$

If the action of G on the coefficients V is trivial and $g \in G$ any group element, then the $(n-1)$-cochain $g.h$ coincides with the $(n-1)$-cochain $h \circ \varrho(g^{-1}, -)$. Therefore the formula above simplifies to

$$\lambda_{\mathfrak{U}_0} \hat{F}^{1,n-1}(m_0, m_1)(z) = h \left(\lambda_{n-1} \operatorname{Csd}_{\mathfrak{U}_0}[p(x_1)^{-1}.z] - \lambda_{n-1} \operatorname{Csd}_{\mathfrak{U}_0}[p(x_0)^{-1}.z] \right).$$

Since the singular $(n-1)$-chain z is a cycle, the chain $[p(x_1)^{-1}.z] - [p(x_0)^{-1}.z]$ is the coboundary of $\gamma^{-1}.z$ for any path γ from $p(x_0)$ to $p(x_1)$ in G. Therefore we observe:

$$\begin{aligned} \lambda_{\mathfrak{U}_0} \hat{F}^{1,n-1}(x_0, x_1)(z) &= h(\lambda_{n-1} \operatorname{Csd}_{\mathfrak{U}_0}(\partial \gamma^{-1}).z) = h(\lambda_{n-1} \operatorname{Csd}_{\mathfrak{U}_0} \partial(\gamma^{-1}.z)) \\ &= h(\partial \lambda_{n-1} \operatorname{Csd}_{\mathfrak{U}_0}(\gamma^{-1}.z)) = \lambda_{\mathfrak{U}_0}^n f(\gamma^{-1}.z) \\ &= \mathcal{F}_{\varrho t}(\lambda_{\mathfrak{U}_0} f)^{n-1,1}(z)(\gamma) \end{aligned}$$

This proves the assertion. $\qquad \square$

This observation enables us to express the triviality all of the cohomology classes $[\hat{F}^{1,n-1}(x_0, x_1)]$ in $H^{n-1}(X, \mathfrak{U}_1; V)$ in terms of the flux component $\mathcal{F}_{\varrho t}(\lambda^n_{\mathfrak{U}_0} f)^{1,n-1}$:

Lemma 15.4.5. *If the cocycle $F^{1,n-1}$ is the vertical coboundary of a cochain $\psi^{1,n-2}$ of bidegree $(1, n-2)$ and the action of G on the coefficients V is trivial, then the flux $\mathcal{F}^{n-1,1}_{\varrho t}(\lambda^n_{\mathfrak{U}_0} f) : Z_{n-1}(X) \to Z^1(G; V)^G$ is trivial.*

Proof. If the cocycle $F^{1,n-1}$ is the vertical coboundary of a cochain $\psi^{1,n-2}$ of bidegree $(1, n-2)$, then the adjoint cocycle $\hat{F}^{1,n-1}$ is the vertical coboundary of the adjoint cochain $\hat{\psi}^{1,n-2}$ of $\psi^{1,n-2}$. In this case the $(n-1)$-cocycle $\hat{F}^{1,n-1}(m_0, m_1)$ is a coboundary for every point $(x_0, x_1) \in X^2$. This implies that all the evaluations

$$\mathcal{F}_{\varrho t}(\lambda^n_{\mathfrak{U}_0} f)^{n-1,1}(z)(\gamma)$$

are trivial for all paths γ in G and all singular $(n-1)$-cycles z in X. This in turn implies that for every singular $(n-1)$-cycle z in X the equivariant 1-cocycle $\mathcal{F}^{n-1,1}_{\varrho t}(\lambda^n_{\mathfrak{U}_0} f)(z)$ on G is trivial. Therefore the flux component $\mathcal{F}_{\varrho t}(\lambda^n_{\mathfrak{U}_0} f)^{n-1,1} : Z_{n-1}(X) \to Z^1(G; V)^G$ is trivial. $\quad\square$

Lemma 15.4.6. *If the the flux $\mathcal{F}^{n-1,1}_{\varrho t}(\lambda^n_{\mathfrak{U}_0} f) : Z_{n-1}(M) \to Z^1(G; V)^G$ is trivial and the action of G on the coefficients V is trivial, then there exists a G-invariant open covering $\mathfrak{U}_1 \prec \mathfrak{U}_0$ such that the cohomology classes $[\hat{F}^{1,n-1}(x_0, x_1)]$ in $H^{n-1}(X, \mathfrak{U}_1; V))^G$ are trivial for all points $(x_0, x_1) \in X^2$.*

Proof. Let the flux $\mathcal{F}^{n-1,1}_{\varrho t}(\lambda^n_{\mathfrak{U}_0} f) : Z_{n-1}(X) \to Z^1(X; V)^G$ and the action of G on the coefficients V be trivial. Let $(x_0, x_1) \in X^2$ be given. By Lemma 15.4.4 the evaluation of $\hat{F}^{1,n-1}(x_0, x_1)$ at a singular $(n-1)$-cycle z in X is given by

$$\hat{F}^{1,n-1}(x_0, x_1)(z) = \mathcal{F}_{\varrho t}(\lambda^n_{\mathfrak{U}_0} f)^{n-1,1}(z)(\gamma)$$

for any path γ from $p_G \varrho^{-1}_{|G \times N}(m_0)$ to $p_G \varrho^{-1}_{|G \times N}(x_1)$ in G. These evaluations vanish by assumption. Since the homomorphism $\mathrm{colim}\, H^{n-1}(X, \mathfrak{U}; V) \to H^{n-1}(X; V)$ is injective, this implies that there exists a G-invariant open covering $\mathfrak{U}_1 \prec \mathfrak{U}_0$ such that the cohomology class $[\hat{F}^{1,n-1}(x_0, x_1)]$ in $H^{n-1}(X, \mathfrak{U}_2; V))^G$ is trivial. $\quad\square$

Proposition 15.4.7. *If V is a trivial G-module and the function, then some restriction $\varrho^{1,n-1}_{\mathfrak{U}_1, \mathfrak{U}_0} F^{1,n-1}$ of the cocycle $F^{1,n-1}$ is the vertical coboundary $d_v \psi^{1,n-2}$ of a cochain $\psi^{1,n-2}$ if and only if the flux $\mathcal{F}^{n-1,1}_{\varrho t}(\lambda^n_{\mathfrak{U}_0} f) : Z_{n-1}(X) \to Z^1(G; V)^G$ is trivial.*

Proof. If some some restriction $\varrho^{1,n-1}_{\mathfrak{U}_1, \mathfrak{U}_0} F^{1,n-1}$ of the cocycle $F^{1,n-1}$ is the vertical coboundary $d_v \psi^{1,n-2}$ of a cochain $\psi^{1,n-2}$, then the flux component $\mathcal{F}^{n-1,1}_{\varrho t}(\lambda^n_{\mathfrak{U}_0} f) : Z_{n-1}(X) \to Z^1(G; V)^G$ is trivial by Lemma 15.4.5. Conversely, if this flux component is trivial, then the adjoint cocycle $\hat{F}^{1,n-1}$ has trivial image in $H^{n-1}(X; V)$ by Lemma 15.4.6. In this case the cocycle $\varrho^{1,n-1}_{\mathfrak{U}_1, \mathfrak{U}_0} F^{1,n-1}$ is the vertical coboundary $d_v \psi^{1,n-2}$ of a cochain $\psi^{1,n-2}$ of bidegree $(1, n-2)$ by Lemma 15.4.3. $\quad\square$

Proposition 15.4.8. *If V is a trivial G-module, f an $\mathfrak{U}$-local equivariant 2-cocycle on X, the flux homomorphisms $\mathcal{F}_{\varrho t}(\lambda^n_{\mathfrak{U}} f)^{2,0} : Z_2(X) \to Z^0(G; V)^G \cong V^G$ and $\mathcal{F}_{\varrho t}(\lambda^n_{\mathfrak{U}} f)^{1,1} : Z_1(X) \to Z^1(G; V)^G$ are trivial then there exists an open G-invariant covering $\mathfrak{U}_1 \prec \mathfrak{U}$ of X such that there exist cocycles $F^{0,2} = i^n(f_{|SS(X, \mathfrak{U}_1)([2])}), F^{1,1}, F^{2,n-2}$ of bidegree $(0, 2), (1, 1)$ and $(2, 0)$ in the total complex of $A^{*,*}(X, \mathfrak{U}_1; V)$ respectively and cochains $\psi^{0,1}, \psi^{1,0}$ of bidegree $(0, 1)$ and $(1, 0)$ respectively such that $D\psi^{p',q'} = F^{p'+1} + F^{p',q'+1}$ for all $p' + q' = 1$:*

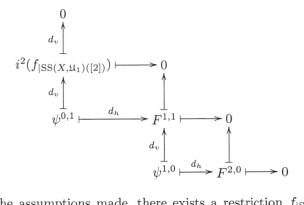

Proof. Under the assumptions made, there exists a restriction $f_{|\mathrm{SS}(X,\mathfrak{U}_0)([2])}$ of f which is a coboundary. Therefore there exists a $\mathfrak{U}_0$-local 1-chain h on X whose coboundary dh is the restriction $f_{|\mathrm{SS}(X,\mathfrak{U}_0)([2])}$ of f. The cocycle $i^2(f_{|\mathrm{SS}(X,\mathfrak{U}_0)([2])}$ in $\mathrm{Tot}A^{*,*}(X,\mathfrak{U}_0;V)^G$ is then the vertical coboundary of the cochain $\psi^{0,1}$ defined via

$$\psi^{0,1}(x_0,\mathbf{y}) := (\mathrm{pr}_G \varrho^{-1}_{|G\times N}(x_0).h)(\mathbf{y}).$$

By Proposition 15.4.7 the cocycle $F^{1,1} := d_h\psi^{0,1}$ is the vertical coboundary of a cochain $\psi^{1,0}$, whose horizontal boundary $F^{2,0}$ then is a cocycle. $\square$

Theorem 15.4.9. *If V is a trivial G-module and f is a $\mathfrak{U}$-local equivariant 2-cocycle with trivial flux homomorphisms*

$$\mathcal{F}^{2,0}_{\varrho t}(\lambda^n_{\mathfrak{U}}f) : Z_2(X) \to Z^0(G;V)^G \cong V^G$$

and $\quad \mathcal{F}^{1,1}_{\varrho t}(\lambda^n_{\mathfrak{U}}f) : Z_1(X) \to Z^1(G;V)^G$

then some restriction $f_{|\mathrm{SS}(X,\mathfrak{U}_1)([2])}$ of f is extendible to a global equivariant 2-cocycle on X.

Proof. Under the assumptions made, the cocycle $i^2(f_{|\mathrm{SS}(X,\mathfrak{U}_2)([2])})$ for a closed equivariant 2-form ω on M with trivial flux homomorphisms is cohomologous to the cocycle $F^{2,0}$ from the preceding proposition. Thus $f_{|\mathrm{SS}(X,\mathfrak{U}_2)([2])}$ is extendible to a global equivariant cocycle on X.

Example 15.4.10. If G is a locally contractible topological group and f a $\mathfrak{U}$-local equivariant cocycle with values in a trivial G-module V some restriction of the local cocycle f is extendible to a global equivariant cocycle which describes a central extension if for some open G-invariant covering $\mathfrak{U}_1$ of G both period homomorphisms

$$\mathcal{F}^{2,0}_{\varrho t}(\lambda^n_{\mathfrak{U}_1}f) : Z_2(X) \to V^G \qquad \text{and} \qquad \mathcal{F}^{1,1}_{\varrho t}(\lambda^n_{\mathfrak{U}_1}f) : Z_1(X) \to Z^1(G;V)^G$$

are trivial.

Theorem 15.4.11. *If V is a trivial G-module, f an $\mathfrak{U}$-local equivariant n-cocycle on an $(n-2)$-connected space X, $n \geq 2$, the flux homomorphisms $\mathcal{F}^{n,0}_{\varrho t}(\lambda^n_{\mathfrak{U}}f) : Z_n(X) \to Z^0(G;V)^G \cong V^G$ and $\mathcal{F}^{n-1,1}_{\varrho t}(\lambda^n_{\mathfrak{U}}f) : Z_{n-1}(X) \to Z^1(G;V)^G$ are trivial, then there exists a G-invariant open covering $\mathfrak{U}_1 \prec \mathfrak{U}$ of X and cocycles $F^{0,n} = i^n(f_{|\mathrm{SS}(X,\mathfrak{U}_1)([n])})$, $F^{1,n-1}$, $F^{2,n-2}$ of bidegree $(0,n),(1,n-1)$ and $(2,n-2)$ in the total complex of $A^{*,*}(X,\mathfrak{U}_1;V)^G$ respectively and cochains $\psi^{0,n-1}$, $\psi^{1,n-2}$ of bidegree $(0,n-1)$ and $(1,n-2)$ respectively such that $D\psi^{p',q'} = F^{p'+1} + F^{p',q'+1}$ for all $p'+q'=n-1$:*

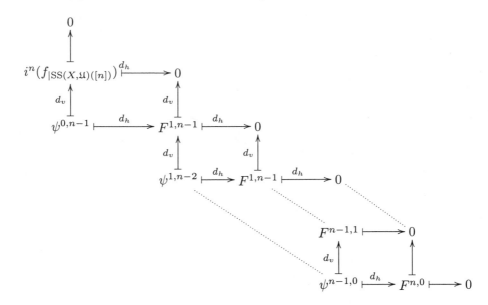

Proof. Under the assumptions made, the cochains $\psi^{0,n-1}$, $F^{1,n-1}$, $\psi^{1,n-2}$ and $F^{2,n-2}$ exist by the last Proposition. Because X is $(n-2)$-connected, the cocycle $F^{2,n-2}$ is extendible to a global equivariant cocycle of bidegree $(2, n-2)$, which implies the assertion.   □

Corollary 15.4.12. *If V is a trivial G-module, X is $(n-2)$-connected, $\mathfrak{U}$ an open G-invariant covering of X and f a $\mathfrak{U}$-local equivariant n-cocycle on X, $n \geq 2$, with trivial flux homomorphisms*

$$\mathcal{F}^{n,0}_{\varrho t}(\lambda^n_{\mathfrak{U}} f) \; : Z_n(X) \to Z^0(G; V)^G \cong V^G$$

and $\quad \mathcal{F}^{n-1,1}_{\varrho t}(\lambda^n_{\mathfrak{U}} f) : Z_{n-1}(X) \to Z^1(G; V)^G$

then some restriction of f is extendible to a global equivariant n-cocycle on X.

Example 15.4.13. If V is a trivial G-module and X a simply connected G-space, then some restriction of a $\mathfrak{U}$-local equivariant 3-cocycle f on X is extendible to a global equivariant 3-cocycle if both period homomorphisms

$$\mathcal{F}^{3,0}_{\varrho t}(\lambda^n_{\mathfrak{U}} f) : Z_3(X) \to V^G \qquad \text{and} \qquad \mathcal{F}^{1,2}_{\varrho t}(\lambda^n_{\mathfrak{U}} f) : Z_2(X) \to Z^1(G; V)^G$$

are trivial.

Integrability of Differential Forms

The purpose of this chapter is to relate the cohomology $H_{ls,eq}(M;V)$ of locally smooth global equivariant cochains and $H_{s,eq}(M;V)$ smooth equivariant global cochains on the G-manifold M of a smooth transformation group (G,M) of special type to the equivariant de Rham cohomology $H_{dR,eq}(M;V)$ of M. This includes the relation of the locally smooth and smooth group cohomologies $H_{ls}(G;V)$ and $H_s(G;V)$ of a Lie group G and the continuous Lie algebra cohomology $H_c(\mathfrak{g};V)$ of its Lie algebra $\mathfrak{g}$ as a special case. A great difficulty arising in the general context of infinite dimensional manifolds M and Lie groups G is the the fact that neither of them needs to be smoothly paracompact, so de Rahm's Theorem is not available. Even Abelian Lie-groups that are Banach spaces (e.g. $C([0,1],\mathbb{R})$ or $l^1(\mathbb{N},\mathbb{R})$) need not be smoothly paracompact. Another difficulty arising in infinite dimensional analysis is the absence of the "exponential law" for spaces of smooth functions. Here we can rely on the observations concerning k-spaces (e.g. metrisable vector spaces) in Section 13.1 or ensure the exponential law of certain function spaces in different way. The coefficient space V is required to be Mackey complete to ensure the existence of integrals of V-valued differential forms over smooth singular simplices.

The cohomologies $H_{ls,eq}(M;V)$ resp. $H_{s,eq}(M;V)$ and $H_{dR,eq}(M;V)$ of a smooth transformation group (G,M) of special type are related by characterising the integrability of closed equivariant forms on M to smooth global equivariant cocycles; we call an equivariant close n-form ω on M *integrable* to a locally smooth equivariant global cocycle if it is in the image of the derivation homomorphism

$$Z_{ls}^n(M;V)^G \xrightarrow{\tau^n} Z_{dR}^n(M;V)^G.$$

Under additional assumptions on the manifold M, the necessary and sufficient condition for a G-equivariant closed n-form $\omega \in Z_{dR}^n(M;V)^G$ ($n > 0$) to be integrable to a (locally) smooth global cocycle will be shown to be the triviality of its flux. All (topological) vector spaces and manifolds considered in this chapter will be vector spaces or manifolds over the real or complex numbers.

The characterisation of integrability is done in in terms of the flux and the period maps introduced in Section 13.6. The main results of this chapter are the characterisation of the integrability of closed equivariant forms to locally smooth and smooth equivariant global cocycles under varying connectedness-conditions on the manifold M:

1. Concerning equivariant closed 1-forms we observe the exactness of the sequence

$$Z_s^1(M;V)^G \to Z_{dR}^1(M;V)^G \xrightarrow{\mathrm{per}_{\varrho t}(-)_1} \hom(\pi_1(M), V^G),$$

   which describes the integrability of a closed equivariant 1-form ω on M in terms of the period map $\pi_1(M) \to V^G$, $[s] \otimes 1 \mapsto \int_s \omega$.

2. For closed equivariant n-forms on $(n-1)$-connected manifolds M with additional structure (such as Lie groups) a similar sequence

$$Z_{ls}^n(M;V)^G \rightarrow Z_{dR}^n(M;V)^G \xrightarrow{\mathrm{per}_{\varrho t}(-)_n} \mathrm{hom}(\pi_n(M),V)$$

turns out to be exact. If the manifold M is not $(n-1)$-connected but $(n-2)$-connected, we obtain an exact sequence

$$Z_{ls}^n(M;V)^G \rightarrow Z_{dR}^2(M;V)^G \rightarrow \mathrm{hom}(Z_n(M),V^G) \oplus \mathrm{hom}(Z_{n-1}(M),Z_c^1(\mathfrak{g};V))$$

Both result generalise the corresponding observations concerning smooth Lie group cocycles and continuous Lie algebra cocycles in [Nee02a] to more general transformation groups. In contrast to the procedure there we do neither require any of the spaces $\mathfrak{g}$ or V to be locally convex nor do we require M or G to be Hausdorff. Moreover we extend these results by characterising the integrability of equivariant closed n-forms on $\lfloor \frac{n-1}{2} \rfloor$-connected manifolds(with additional structure). Here a sequence of the form

3.

$$Z_{ls}^n(M;V)^G \rightarrow Z_{dR}^n(M;V)^G \rightarrow \bigoplus_{p \leq \lfloor \frac{n-1}{2} \rfloor + 1}^n \mathrm{hom}(Z_p(M),Z_c^{n-p}(\mathfrak{g};V))$$

is exact. In all cases the homomorphisms $Z_{dR}^n(M;V)^G \rightarrow \mathrm{hom}(Z_p(M),Z_c^{n-p}(\mathfrak{g};V))$ occurring in the above sequences are flux homomorphisms. Thus in any of the above cases an equivariant closed n-form is integrable to a locally smooth global equivariant n-cocycle if and only if all relevant flux homomorphisms are trivial. The last result (3) is completely new and was not even known for finite dimensional Lie groups.

Under the additional assumptions that the exponential law holds for the natural linear continuous maps $C^\infty(M^p \times T^{\oplus q}M,V) \rightarrow C^\infty(M^p,C^\infty(T^{\oplus q}M,V))$ and that the topological de Rham complex ${}^t\Omega^*(M;V)$ of M splits we can also prove the analogon of (2) for globally smooth cocycles, i.e. if the manifold M is $(n-2)$-connected, then the sequence

$$Z_{ls}^n(M;V)^G \rightarrow Z_{dR}^2(M;V)^G \rightarrow \mathrm{hom}(Z_n(M),V^G) \oplus \mathrm{hom}(Z_{n-1}(M),Z_c^1(\mathfrak{g};V))$$

is exact.

It should be mentioned that Hilbert manifolds satisfy any of the above conditions on the manifold M and its de Rham complex. This in particular includes all finite dimensional manifolds. Moreover, any diffeomorphism group $\mathrm{Diff}(M)$ of a compact manifold M has split de Rham complex (cf. [Beg87, Theorem 7.5]) and the group $\Omega_s(M)$ of smooth loops in a separable Hilbert manifold M has split de Rham complex. This is also true for the space $C^\infty(M,N)$ of smooth functions from a compact manifold M into a separable metrisable Hilbert manifold N (cf. [Beg87, Theorem 7.6]).

The proof of the above statements on the integrability of closed equivariant forms relies on the spectral sequences introduced in Section 14.2 and on the flux and period homomorphisms introduced in chapter 13.

16.1 $(n-1)$-Connected Manifolds

We begin pour considerations by defining integrability as indicated in the introduction and then examine smooth transformation groups with smooth equivariant local systems of singular simplices, e.g. Lie groups. Let (G, M) be a smooth transformation group and V be a Hausdorff G-vector space of coefficients. Global n-cochains $F \in A^n(M; V)$ which are smooth on a neighbourhood of the diagonal are called *locally smooth*. The cosimplicial subgroup of the cosimplicial group $A(M; V)$ of global cochains on M consisting of locally smooth cochains is denoted by $A_{ls}(M; V)$ and its cohomology is denoted by $H_{ls}(M; V)$. Recall the equivariant derivation morphism

$$\tau^* : A^*_{AS,s}(M; V) \to \Omega^*(M; V),$$

which assigns to each smooth Alexander-Spanier n-cochain F on M a differential n-form $\tau^n(F)$ on M (cf. Section 14.1 and Appendix G.2). In particular the n-form $\tau^n(F)$ is closed if F is a cocycle and equivariant if F is equivariant. Composing the natural complex morphism $A^*_{ls}(M; V) \to A^*_{AS,s}(M; V)$ with the derivation morphism $\tau^* : A^*_{AS,s}(M; V) \to \Omega^*$ yields a morphism

$$A^*_{ls}(M; V) \to \Omega^*(M; V), \quad F \mapsto \tau^*([F])$$

which we also denote by τ^* and called *derivation morphism*

Definition 16.1.1. *A differential n-form ω on M is said to be* integrable to a locally smooth global cochain *if it is in the image of the equivariant derivation homomorphism $\tau^n : A^n_{ls}(M; V) \to \Omega^n(M; V)$. It is called* integrable to a locally smooth global cocycle *if there exists a locally smooth global n-cocycle $F \in Z^n_{ls}(M; V)$ on M with image $\tau^n(F) = \omega$.*

The integrability of equivariant differential forms to equivariant smooth local cocycles is defined analogously:

Definition 16.1.2. *An equivariant differential n-form is called* integrable to an equivariant locally smooth global cochain *if it is the image $\tau^n(F)$ of an equivariant locally smooth global n-cochain $F \in A^n_{ls}(M; V)^G$ on M. It is called* integrable to a locally smooth equivariant global cocycle *if it is the image $\tau^n(F)$ of an equivariant locally smooth global n-cocycle $F \in Z^n_{ls}(M; V)^G$ on M.*

Example 16.1.3. If $G = M$ is a Lie group (with Lie algebra $\mathfrak{g}$) acting on itself via left translation, then the complex $\Omega^*(G; V)^G$ of equivariant differential forms can be identified with the complex $C^*(\mathfrak{g}; V)$ of Lie algebra cochains. Similarly the complex $A^*_{ls}(G; V)^G$ of locally smooth equivariant global cochains can be identified with the complex $C^*_{ls}(G; V)$ of locally smooth group cochains. A Lie algebra cocycle then is integrable to a locally smooth group n-cocycle if and only if its corresponding equivariant closed differential n-form is integrable to a locally smooth equivariant global n-cocycle.

To ensure the existence of integrals of differential forms over smooth singular simplices we assume the coefficients V to be Mackey complete and Hausdorff. The action of G on M is denoted by ϱ. If there exists an open covering $\mathfrak{U}$ of M and a smooth local system of singular n-simplices $\hat{\sigma}_n : \mathrm{SS}(M, \mathfrak{U})([n]) \to C^\infty(\Delta_n, M)$ on the neighbourhood $\mathrm{SS}(M, \mathfrak{U})$ of the diagonal in M^{n+1}, then the latter can be used to construct smooth local cochains from n-forms. This construction is given by the homomorphism

$$s^n : \Omega^n(M; V) \to A^n_s(M, \mathfrak{U}; V), \quad s^n(\omega)(\mathbf{m}) = \int_{\hat{\sigma}_n(\mathbf{m})} \omega.$$

The linear function s^n is a composition of smooth functions and therefore smooth as well. For Lie groups this has already been observed in [Fuc03].

Example 16.1.4. If G is a Lie group acting on itself via left translation, then there exists an equivariant smooth local system of n-simplices on a neighbourhood $\mathrm{SS}(G, \mathfrak{U})([n])$ of the diagonal in G^{n+1} (cf. Lemma 3.0.94). Thus the above procedure is applicable and provides a homomorphism $s^n : \Omega^n(M; V) \to A_s^n(M, \mathfrak{U}; V)$.

Example 16.1.5. If (G, M) is a smooth transformation group and M is a complete Riemannian manifold, then there exists a smooth local system of n-simplices on a neighbourhood $\mathrm{SS}(M, \mathfrak{U})([n])$ of the diagonal in M^{n+1}. Thus the above procedure is applicable and provides a homomorphism $s^n : \Omega^n(M; V) \to A_s^n(M, \mathfrak{U}; V)$.

Lemma 16.1.6. *The homomorphism $s^n : \Omega^n(M; V) \to A_s^n(M, \mathfrak{U}; V)$ is equivariant whenever $\mathfrak{U}$ is G-invariant and $\hat{\sigma}_n$ is equivariant.*

Proof. Let $\mathfrak{U}$ be a G-invariant open covering of M, $\hat{\sigma}_n : \mathrm{SS}(M, \mathfrak{U})([n]) \to C^\infty(\Delta_n, M)$ be an equivariant smooth system of simplices and ω be an n-form on M and $g \in G$ be a group element. The evaluation of the cochain $s^n(g.\omega)$ on a point $\mathbf{m} \in M^{n+1}$ is given by

$$s^n(g.\omega)(\mathbf{m}) = \int_{\hat{\sigma}_n(\mathbf{m})} g.\omega = g. \left[\int_{\hat{\sigma}_n(\mathbf{m})} \lambda_{g^{-1}}^* \omega \right] = g. \left[\int_{g^{-1}.\hat{\sigma}_n(\mathbf{m})} \omega \right]$$

$$= g. \left[\int_{\hat{\sigma}_n(g^{-1}.\mathbf{m})} \omega \right] = g.[s^n(\omega)(g^{-1}.\mathbf{m})] = [g.s^n(\omega)](\mathbf{m}),$$

hence the homomorphism s^n intertwines the action of G, i.e. it is G-equivariant.   □

As a consequence the homomorphism $s^n : \Omega^n(M; V) \to A_s^n(M, \mathfrak{U}; V)$ maps equivariant n-forms to $\mathfrak{U}$-local equivariant smooth functions. For special systems $\hat{\sigma}_n$ of n-simplices, the differential forms obtained from these local smooth cochains coincide with the differential forms one has started with:

Lemma 16.1.7. *If $\hat{\sigma}_n : \mathrm{SS}(M, \mathfrak{U})([n]) \to C^\infty(\Delta, M)$ is an integrating smooth local system of simplices, then the homomorphism*

$$s^n : \Omega^n(M; V) \to A_s^n(M, \mathfrak{U}; V), \quad s^n(\omega)(\mathbf{m}) = \int_{\hat{\sigma}_n(\mathbf{m})} \omega$$

is a right inverse to the derivation homomorphism $\tau^n : A_s^n(M, \mathfrak{U}; V) \to \Omega^n(M; V)$, i.e. the latter homomorphism is a split epimorphism in **Ab**. *If, in addition, $\mathfrak{U}$ is G-invariant and $\hat{\sigma}_n$ is equivariant, then s^n also is a right inverse in* $\mathbb{Z}[\mathbf{G}]\mathbf{mod}$.

Proof. If $\hat{\sigma}_n$ is an integrating smooth local system of n-simplices, then s^n is a right inverse of the derivation homomorphism by Definition. If the system $\hat{\sigma}_n$ is equivariant in addition, then s^n is equivariant by Lemma 16.1.6, hence a homomorphism in $\mathbb{Z}[\mathbf{G}]\mathbf{mod}$ and thus a right inverse to τ^n in the latter category.   □

Example 16.1.8. If (G, M) is a smooth transformation group and M is a complete Riemannian manifold with G-invariant Riemannian structure, then there exists an integrating equivariant smooth local system of n-simplices. Thus the above procedure is applicable and provides a homomorphism $s^n : \Omega^n(M; V) \to A_s^n(M, \mathfrak{U}; V)$ in $\mathbb{Z}[\mathbf{G}]\mathbf{mod}$.

Example 16.1.9. If G is a Lie group acting on itself via left translation, then there exists an integrating equivariant smooth local system of n-simplices on on a neighbourhood $\mathrm{SS}(G, \mathfrak{U})([n])$ of the diagonal in G^{n+1}. Thus the above procedure is applicable and provides a homomorphism $s^n : \Omega^n(M; V) \to A_s^n(M, \mathfrak{U}; V)$ in $\mathbb{Z}[\mathbf{G}]\mathbf{mod}$.

The homomorphism in the previous example restricts to a homomorphism between the subgroups of G-invariants. The complex of left equivariant differential forms on a Lie group G can be identified with the complex of Lie algebra cochains. Thus for every Lie group there exists an integrating equivariant smooth local system of n-simplices on on a neighbourhood $\mathrm{SS}(G,\mathfrak{U})([n])$ of the diagonal in G^{n+1} and a homomorphism $C_c^n(\mathfrak{g};V) \cong \Omega^n(M;V)^G \to A_s^n(M,\mathfrak{U};V)$ in **Ab**. So we observe:

Lemma 16.1.10. *Every Lie algebra cochain with values in a Mackey complete topological vector space can be integrated to a local group cochain.*

Remark 16.1.11. This result is not new. It has already been observed in [Fuc03].

We are interested in necessary and sufficient conditions for the $\mathfrak{U}$-local smooth functions $s^n(\omega)$ to be n-cocycles. For compatible systems $\hat{\sigma}_n$ of singular n-simplices and closed n-forms this can be tested by integrating the n-forms over n-spheres: Let ω be a closed n-form and $t : M \times G \to G \times M$ be the coordinate flip $(m,g) \mapsto (g,m)$. Recall that the composition of the flux $\mathcal{F}_{\varrho t}(\omega)^{n,0} : H_n(M) \to H^0_{dR,eq}(G;V) = V^G$ with the Hurewicz homomorphism $\mathrm{hur}_n : \pi_n(M) \to H_n(M)$ is the n-th period map

$$\mathrm{per}_{\varrho t}(\omega)_n : \pi_n(M) \to V^G, \quad [s] \mapsto \int_s \omega$$

(cf. Example 13.8.3). As we will see, the triviality of this homomorphism amounts to the cochain $s^n(\omega)$ being a cocycle:

Lemma 16.1.12. *If $\hat{\sigma}_n : \mathrm{SS}(M,\mathfrak{U})([n]) \to C^\infty(\Delta_n, M)$ is a compatible smooth local system of singular n-simplices and the period map $\mathrm{per}_{\varrho t}(\omega)_n : \pi_n(M) \to V^G$ of a closed n-form ω on M is trivial, then the cochain $s^n(\omega)$ is a $\mathfrak{U}$-local cocycle.*

Proof. Let ω be a closed n-form on M and assume the period map $\mathrm{per}_{\varrho t}(\omega)_n$ to be trivial. The evaluation of $ds^n(\omega)$ at a point $\mathbf{m} = (m_0, \ldots, m_{n+1}) \in \mathrm{SS}(M,\mathfrak{U})([n+1])$ is given by

$$d[s^n(\omega)](\mathbf{m}) = \sum (-1)^i s^n(\omega)(m_0, \ldots, \hat{m}_i, \ldots, m_{n+1})$$
$$= \sum (-1)^i \int_{\hat{\sigma}_n(m_0, \ldots, \hat{m}_i, \ldots, m_{n+1})} \omega.$$

The singular chain $\sum (-1)^i \hat{\sigma}_n(m_0, \ldots, \hat{m}_i, \ldots, m_{n+1})$ is the triangulation of an n-sphere in M. By assumption on the period morphism all integrals of ω over n-spheres are trivial. Therefore $s^n(\omega)$ is a cocycle. □

This condition on the period homomorphism is always satisfied if the equivariant closed differential n-form ω is exact. So we note:

Lemma 16.1.13. *If $\hat{\sigma}_n : \mathrm{SS}(M,\mathfrak{U})([n]) \to C^\infty(\Delta_n, M)$ is a compatible smooth local system of singular n-simplices and ω an exact differential n-form on M, then the cochain $s^n(\omega)$ is a $\mathfrak{U}$-local cocycle.*

Theorem 16.1.14. *If $\mathfrak{U}$ is an open cover of M and the manifold M admits an integrating smooth local system $\hat{\sigma}_n : \mathrm{SS}(M,\mathfrak{U})([n]) \to C^\infty(\Delta, M)$ of n-simplices, then there exists a right inverse $s^n : \Omega^n(M;V) \to A_s^n(M,\mathfrak{U};V)$ to the derivation homomorphism $\tau^n : A_s^n(M,\mathfrak{U};V) \to \Omega^n(M;V)$. This right inverse s^n is equivariant whenever $\mathfrak{U}$ is G-invariant and $\hat{\sigma}_n$ is equivariant. If the period homomorphism of a closed n-form ω is trivial, then the smooth $\mathfrak{U}$-local cochain $s^n(\omega)$ is a cocycle.*

Proof. Since $\hat{\sigma}_n$ is assumed to be integrating, the homomorphism s^n is a right inverse to the derivation homomorphism τ^n. If $\hat{\sigma}_n$ is equivariant, then s^n is equivariant by Lemma 16.1.6. The restriction of s^n to the kernel of the period homomorphism has codomain the space $Z^n(M, \mathfrak{U}; V)$ of $\mathfrak{U}$-local n-cocycles by Lemma 16.1.12. □

Corollary 16.1.15. *If ω is a left equivariant differential form on a Lie group G, then there exists am open neighbourhood $\mathrm{SS}(G, \mathfrak{U})([n])$ of the diagonal and a smooth $\mathfrak{U}$-local cocycle $s^n(\omega) \in A_s(M, \mathfrak{U}; V)$ whose image $\tau^n(s^n(\omega))$ under derivation homomorphism τ^n is exactly the form ω. If ω is closed and the period map $\mathrm{per}_{\varrho t}(\omega)_n$ is trivial, then $s^n(\omega)$ is a cocycle.*

This may be reformulated in terms of Lie algebra cochains and local Lie group cochains:

Corollary 16.1.16. *Every Lie group G has an neighbourhood U of the identity such that any continuous Lie algebra cochain can be integrated to a U-local group cochain and the latter is a cocycle if the period homomorphism of the Lie algebra cochain is trivial.*

Proof. This follows from the fact that every Lie group admits an integrating compatible smooth equivariant local system of simplices (Lemma 3.0.94).   □

The above theorem in particular implies that any closed n-form ω can be integrated to a $\mathfrak{U}$-local smooth n-cocycle if its period homomorphism $\mathrm{per}_{\varrho t}(\omega)_n$ is trivial. Likewise, if the integrating smooth system $\hat{\sigma}_n$ of n-simplices is equivariant, then every closed equivariant n-form ω can be integrated to a $\mathfrak{U}$-local smooth n-cocycle if its period homomorphism $\mathrm{per}_{\varrho t}(\omega)_n$ is trivial. For the special open covering $\mathfrak{U} = \{M\}$ of the manifold M the $\mathfrak{U}$-local cochains are the global ones, so we note:

Lemma 16.1.17. *If the manifold M admits a smooth integrating global system of n-simplices, $\hat{\sigma}_n : M^{n+1} \to C^\infty(\Delta, M)$, then $s^n : \Omega^n(M; V) \to A_s^n(M; V)$ is a right inverse to the derivation homomorphism $\tau^n : A_s^n(M; V) \to \Omega^n(M; V)$ and the sequence*

$$Z_s^n(M; V) \to Z_{dR}^n(M; V) \xrightarrow{\mathrm{per}_{\varrho t}(-)_n} \mathrm{hom}(\pi_n(M), V^G)$$

of Abelian groups is split exact at $Z_{dR}^n(M; V)$. If $\hat{\sigma}_n$ is equivariant, then it also is split exact in in the category $\mathbb{Z}[\mathbf{G}]\mathbf{mod}$ of G-modules.

Proof. Since the system $\hat{\sigma}_n$ of simplices is assumed to be integrating, the homomorphism s^n is a right inverse to the derivation homomorphism τ^n. Its restriction to the kernel of the period homomorphism has codomain $Z^n(M, \mathfrak{U}; V)$ by Lemma 16.1.12. Thus the kernel of the period homomorphism is contained in the image of the derivation homomorphism τ^n. Conversely, if ω the image $\tau^n(F)$ of a smooth global n-cocycle F on M, then the cocycle F is the coboundary df' of a smooth global $(n-1)$-cochain F' on M. Since the derivation morphism τ^* is a morphism of chain complexes, this forces the form ω to be the coboundary $d\tau^{n-1}(F')$ of the $(n-1)$-form $\tau^{n-1}(F')$ on M. In this case the form ω is exact, so all its periods are trivial. Therefore the image of the derivation homomorphism τ^n is contained in the kernel of the period map and the sequence

$$Z_s^n(M, \mathfrak{U}; V) \xrightarrow{\tau^n} Z_{dR}^n(M; V) \xrightarrow{\mathrm{per}_{\varrho t}(-)_n} \mathrm{hom}(\pi_n(M), V^G)$$

of Abelian groups is split exact at $Z_{dR}^n(M; V)$. If $\hat{\sigma}_n$ is equivariant then so is s^n and the sequence splits in $\mathbb{Z}[\mathbf{G}]\mathbf{mod}$.   □

The requirement of the existence of an integrating compatible smooth equivariant system $\hat{\sigma}_n : M^{n+1} \to C^\infty(\Delta, M)$ of n-simplices makes it possible to integrate differential forms to global smooth cochains and closed forms to global smooth cocycles, but it is a major restriction. Moreover, to integrate differential forms to locally smooth global cochains and closed forms to locally smooth cocycles, the equivariant system $\hat{\sigma}_n$ only needs to be smooth on a neighbourhood of the diagonal, as shall be shown below.

Now suppose that the action of G on the manifold M is free and M admits a smooth compatible equivariant local system $\hat{\sigma}_n : \mathrm{SS}(M,\mathfrak{U})([n]) \to C^\infty(\Delta_n, M)$ of singular n-simplices and is $(n-1)$-connected. In this case the underlying system $\hat{\sigma}_n : M^{n+1} \to \hom_{\mathbf{Top}}(\Delta_n, M)$ of singular simplices can be extended to a global equivariant system

$$\hat{\sigma}_n : M^{n+1} \to \hom_{\mathbf{Top}}(\Delta_n, M)$$

of singular simplices by Theorem 3.0.82. The restriction of this map to the neighbourhood $\mathrm{SS}(M,\mathfrak{U})([n])$ of the diagonal in M^{n+1} then is the map underlying the smooth system we have started with.

Example 16.1.18. If $M = G$ is an $(n-1)$-connected Lie group then there exists a system $\hat{\sigma}_n : M^n \to \hom_{\mathbf{Top}}(\Delta_n, M)$ of left invariant singular n-simplices simplices, whose restriction to a neighbourhood $\mathrm{SS}(M,\mathfrak{U})([n])$ of the diagonal in M^{n+1} is an equivariant compatible smooth local system of simplices.

This extension of the underlying map can be used to construct global cycles from $\mathfrak{U}$-local ones. For this purpose we use the complete $\mathfrak{U}$-relative subdivision of singular simplices. Recall that this operator is an equivariant chain map

$$\mathrm{Csd}_{\mathfrak{U},*} : S_*(M) \to S_*(M,\mathfrak{U})$$

which subdivides every singular n-simplex into $\mathfrak{U}$-small simplices. The composition of the equivariant vertex map $\lambda_n : C(\Delta_n, M) \to M^{n+1}$ with the complete $\mathfrak{U}$-relative subdivision of singular n-simplices and the equivariant system $\hat{\sigma}_n$ of n-simplices provides an equivariant homomorphism

$$\mathrm{Csd}_{\mathfrak{U}}^n : A^n(M,\mathfrak{U};V) \to A^n(M;V), \quad \mathrm{Csd}_{\mathfrak{U}}^n(f) := f \circ \lambda_n \circ \mathrm{Csd}_{\mathfrak{U}} \circ \hat{\sigma}_n$$

which assigns to each $\mathfrak{U}$-local 1-cochain a global n-cochain on M and is called the *complete $\mathfrak{U}$-relative subdivision of n-cochains* (cf. Section 6.2). Since the singular n-simplices $\hat{\sigma}_n(\mathbf{m})$ for a $\mathfrak{U}$-small n-simplex $\mathbf{m}$ need not be $\mathfrak{U}$-small in general, the $\mathfrak{U}$-relative subdivision $\mathrm{Csd}_{\mathfrak{U}}^n f$ of a $\mathfrak{U}$-local cochain $f \in A^n(M,\mathfrak{U};V)$ does not necessarily restrict to f on the neighbourhood $\mathrm{SS}(M,\mathfrak{U})([n])$ of the diagonal in M^{n+1}. However, the continuity of the smooth map $\hat{\sigma}_n : \mathrm{SS}(M,\mathfrak{U})([n]) \to C^\infty(\Delta_n, M)$ ensures that a restriction to a possibly smaller neighbourhood $\mathrm{SS}(M,\mathfrak{B})([n])$ of the diagonal has this property:

Lemma 16.1.19. *If $\hat{\sigma}_n : \mathrm{SS}(M,\mathfrak{U})([n]) \to C(\Delta_n, M)$ is a continuous system of n-simplices, then there exists an open refinement $\mathfrak{B}$ of $\mathfrak{U}$ such that $\hat{\sigma}_n$ restricts to a continuous map $\mathrm{SS}(M,\mathfrak{B})([n]) \to C(\Delta_n, \mathfrak{U}, M)$, i.e. for every $\mathfrak{B}$-small simplex $\mathbf{m} \in \mathrm{SS}(M,\mathfrak{B})([n])$ the singular simplex $\hat{\sigma}_n(\mathbf{m})$ is $\mathfrak{U}$-small. If $\mathfrak{U}$ is G-invariant and $\hat{\sigma}_n$ is equivariant, then $\mathfrak{B}$ can be chosen to be G-invariant.*

Proof. Let $\hat{\sigma}_n : \mathrm{SS}(M,\mathfrak{U})([n]) \to C(\Delta_n, M)$ be a continuous system of n-simplices. Since $\hat{\sigma}_n$ is continuous, the inverse image $\hat{\sigma}_n^{-1}(\mathrm{SS}(M,\mathfrak{U})([n]))$ is an open neighbourhood of the diagonal. There exists an open covering $\mathfrak{B}$ of M such that the neighbourhood $\mathrm{SS}(M,\mathfrak{B})([n])$ of the diagonal in M^{n+1} is contained in the inverse image $\hat{\sigma}_n^{-1}(\mathrm{SS}(M,\mathfrak{U})([n]))$. If $\mathfrak{U}$ is G-invariant and $\hat{\sigma}_n$ is equivariant, then the open neighbourhood $\hat{\sigma}_n^{-1}(\mathrm{SS}(M,\mathfrak{U})([n]))$ of the diagonal in M^{n+1} is G-invariant as well. In this case $\mathfrak{B}$ may be replaced by the G-invariant open cover $G.\mathfrak{B}$ of M. $\qquad\square$

This in particular implies that the complete $\mathfrak{U}$-relative subdivision $\mathrm{Csd}_{\mathfrak{U}}^n f$ of a smooth $\mathfrak{U}$-local n-cochain f is locally smooth as well.

Example 16.1.20. If $G = M$ is an $(n-1)$-connected Lie group acting on itself via left translation and $\hat{\sigma}_n : \mathrm{SS}(M,\mathfrak{U})([n])) \to C^\infty(\Delta_n)$ a smooth equivariant system of simplices then there exists an equivariant open cover $\mathfrak{B}$ of G such that the $\mathfrak{U}$-relative subdivision of $\mathfrak{U}$-local cochains does not change the values of n-cochains at $\mathfrak{B}$-small simplices $\mathbf{m} \in \mathrm{SS}(M,\mathfrak{B})([n]))$.

Since the complete $\mathfrak{U}$-relative subdivision $\mathrm{Csd}_{\mathfrak{U}}^n$ of $\mathfrak{U}$-local cochains is equivariant, it maps $\mathfrak{U}$-local equivariant cochains to global equivariant cochains. For the case of Lie groups the above example may be reformulated as:

Corollary 16.1.21. *Every Lie group G has neighbourhoods $W \subset U$ of the identity such that there exists a homomorphism*

$$C^n(U, V) \to C^n(G, V)$$

which extends the restriction $f_{|W^n}$ of a U-local group cochain f to a global group cochain.

Lemma 16.1.22. *If the system $\hat{\sigma}_n$ of singular n-simplices is integrating then the composition $\mathrm{Csd}_{\mathfrak{U}}^n s^n : \Omega^n(M; V) \to A_{ls}^n(M; V)$ is a right inverse to the derivation homomorphism $\tau^n : A_{ls}(M; V) \to \Omega^n(M; V)$. If $\mathfrak{U}$ is G-invariant and $\hat{\sigma}_n$ is equivariant, then $\mathrm{Csd}_{\mathfrak{U}}^n s^n$ is a right inverse in $\mathbb{Z}[\mathbf{G}]\mathbf{mod}$.*

Proof. If $\mathfrak{U}$ is G-invariant and the system $\hat{\sigma}_n$ of singular n-simplices is equivariant, then the composition $\mathrm{Csd}_{\mathfrak{U}}^n s^n$ is equivariant as a composition of equivariant maps. Let ω be a differential n-form on M. By the preceeding Lemma the germ of $\mathrm{Csd}_{\mathfrak{U}}^n s^n(\omega)$ at the diagonal coincides with the germ of $s^n(\omega)$ at the diagonal. If $\hat{\sigma}_n$ is integrating, then s^n is a right inverse to the derivation homomorphism $\tau^n : A_s(M, \mathfrak{U}; V) \to \Omega^n(M; V)$. This implies the equalities

$$\tau^n \, \mathrm{Csd}_{\mathfrak{U}}^n s^n(\omega) = \tau^n s^n(\omega) = \omega,$$

so the composition $\mathrm{Csd}_{\mathfrak{U}}^n s^n$ is a right inverse to the derivation homomorphism. □

Example 16.1.23. If $M = G$ is an $(n-1)$-connected Lie group acting on itself via left translations, then there exists an equivariant system $\hat{\sigma}_n : M^n \to \hom_{\mathbf{Top}}(\Delta_n, M)$ of smooth singular n-simplices simplices, whose restriction to a neighbourhood $\mathrm{SS}(M, \mathfrak{U})([n])$ of the diagonal in M^{n+1} is an equivariant compatible smooth local system of simplices. The homomorphism $\mathrm{Csd}_{\mathfrak{U}}^n s^n$ then integrates differential forms to locally smooth global cochains and equivariant differential n-forms to equivariant global cochains. In particular every Lie algebra cocycle is integrable to a locally smooth Lie group cocycle.

The composition $\mathrm{Csd}_{\mathfrak{U}}^n s^n : \Omega^n(M; V) \to A^n(M; V)$ is a homomorphism which assigns to each differential n-form ω on M a locally smooth n-cochain. We are interested in necessary and sufficient conditions for the locally smooth functions $\mathrm{Csd}_{\mathfrak{U}}^n s^n(\omega)$ to be n-cocycles. As before, this can be tested by integrating the n-forms over n-spheres:

Lemma 16.1.24. *If the period map $\mathrm{per}_{\varrho t}(\omega)_n : \pi_n(M) \to V^G$ of a closed n-form ω on M is trivial, then the cochain $\mathrm{Csd}_{\mathfrak{U}}^n s^n(\omega)$ is a global n-cocycle.*

Proof. Let ω be a closed n-form on M. The evaluation of the coboundary of the global n-cochain $\mathrm{Csd}_{\mathfrak{U}}^n s^n(\omega)$ at a point $\mathbf{m} \in M^{n+2}$ is given by

$$d[\mathrm{Csd}_{\mathfrak{U}}^n s^n(\omega)](\mathbf{m}) = \sum_{i=0}^{n+1} (-1)^i [\mathrm{Csd}_{\mathfrak{U}}^n s^n(\omega)](m_0, \ldots, \hat{m}_i, \ldots, m_{n+1})$$

$$= \sum_{i=0}^{n+1} (-1)^i [s^n(\omega)] \lambda_n \, \mathrm{Csd}_{\mathfrak{U},n} \, \hat{\sigma}_n(m_0, \ldots, \hat{m}_i, \ldots, m_{n+1}).$$

The singular chain $\sum (-1)^i \mathrm{Csd}_{\mathfrak{U},n} \, \hat{\sigma}_n(m_0, \ldots, \hat{m}_i, \ldots, m_{n+1})$ is the triangulation of an n-sphere in M. By assumption on the period morphism all integrals of ω over n-spheres are trivial. Therefore $\mathrm{Csd}_{\mathfrak{U}}^n s^n(\omega)$ is a cocycle. □

Proposition 16.1.25. *If an $(n-1)$-connected manifold M admits an integrating smooth local system $\hat{\sigma}_n : \mathrm{SS}(M, \mathfrak{U})([n]) \to C^\infty(\Delta, M)$ of singular n-simplices, then there exists an extension of $\hat{\sigma}_n$ to a (not necessarily continuous) system of singular n-simplices on M and a right inverse $\mathrm{Csd}^n_{\mathfrak{U}} s^n : \Omega^n(M; V) \to A^n_{ls}(M; V)$ to the derivation homomorphism $\tau^n : A^n_{ls}(M; V) \to \Omega^n(M; V)$. This right inverse s^n is equivariant whenever $\mathfrak{U}$ is G-invariant and $\hat{\sigma}_n$ is equivariant. If the period homomorphism of a closed n-form ω is trivial, then the smooth $\mathfrak{U}$-local cochain $s^n(\omega)$ is a cocycle.*

Proof. Since $\hat{\sigma}_n$ is assumed to be integrating, the homomorphism $\mathrm{Csd}^n_{\mathfrak{U}} s^n$ is a right inverse to the derivation homomorphism τ^n by Lemma 16.1.22. If $\hat{\sigma}_n$ is equivariant, then $\mathrm{Csd}^n_{\mathfrak{U}} s^n$ is equivariant (also by Lemma 16.1.22). The restriction of $\mathrm{Csd}^n_{\mathfrak{U}} s^n$ to the kernel of the period homomorphism has codomain the space $Z^n(M; V)$ of global n-cocycles by Lemma 16.1.24. $\qquad\square$

Corollary 16.1.26. *If G is an $(n-1)$-connected Lie group acting on itself via left translations, then there exists an equivariant right inverse to the derivation homomorphism $A^n_{als}(M; V) \to \Omega^n(M; V)$. This right inverse integrates closed differential n-forms ω with trivial period map $\mathrm{per}_{\varrho t}(\omega)_n$ to to locally smooth global cocycles. In particular every Lie algebra n-cocycle with trivial period map is integrable to a locally smooth Lie group n-cocycle.*

Proof. If G is an $(n-1)$-connected Lie group acting on itself via left translations, then there exists a compatible equivariant system $\hat{\sigma}_n : M^n \to \mathrm{hom}_{\mathbf{Top}}(\Delta_n, M)$ of singular n-simplices simplices, whose restriction to a neighbourhood $\mathrm{SS}(M, \mathfrak{U})([n])$ of the diagonal in M^{n+1} is an equivariant compatible smooth local system of singular n-simplices. The equivariant homomorphism $\mathrm{Csd}^n_{\mathfrak{U}} s^n$ then integrates closed differential n-forms ω with trivial period map $\mathrm{per}_{\varrho t}(\omega)_n$ to to locally smooth global cocycles. In particular every Lie algebra n-cocycle with trivial period map is integrable to a locally smooth Lie group n-cocycle. $\qquad\square$

In the next sections we will see how the requirement of $(n-1)$-connectedness and the use of global systems of n-simplices can be avoided by the use of the spectral sequences introduced in the previous chapter.

16.2 Differential n-Forms on $(n-2)$-Connected Manifolds

In this section we examine the integrability of equivariant differential n-forms on G-manifolds M which admit an compatible equivariant smooth local system of singular n-simplices. The main examples are Lie groups and complete Riemannian G-manifolds with G-invariant metric. In addition we require the action ϱ of G on the manifold M to be free, so that there exists an equivariant not necessarily continuous map $p : M \to G$, which we fix for later use.

Example 16.2.1. Every Lie group acts freely on itself by left translation and admits an integrating compatible smooth locally system of singular simplices (Theorem 3.0.95). The complex $A^*_{ls}(G; V)^G$ of locally smooth global equivariant cochains here is isomorphic to the complex $C^*_{ls}(G; V)$ of locally smooth group cochains.

The assumption that there exists a compatible integrating smooth equivariant local system $\hat{\sigma}_n : \mathrm{SS}(M, \mathfrak{U})([n]) \to C^\infty(\Delta_n, M)$ of singular n-simplices (where $\mathfrak{U}$ is a G-invariant open cover of M) implies the existence of compatible smooth equivariant local systems $\hat{\sigma}_k : \mathrm{SS}(M, \mathfrak{U})([k]) \to C^\infty(\Delta_k, M)$ of singular k-simplices for $k \leq n$.

We let V be a Hausdorff and Mackey complete topological vector space, so that V-valued differential forms can be integrated over singular chains. As a consequence there exist equivariant homomorphisms

$$s^k : \Omega^k(M;V) \to A_s^k(M,\mathfrak{U};V), \quad s^k(\omega)(\mathbf{m}) = \int_{\sigma_k(\mathbf{m})} \omega.$$

for $k \leq n$ intertwine the coboundary operators on the cochain complexes $\Omega^*(M;V)$ and $A_s^*(M,\mathfrak{U};V)$. Furthermore the requirement that the smooth equivariant local system $\hat{\sigma}_n$ of singular n-simplices is integrable ensures that the homomorphism $s^n : \Omega^n(M;V) \to A_s^n(M,\mathfrak{U};V)$ integrates every equivariant differential n-form ω on M to a smooth equivariant $\mathfrak{U}$-local cochain $s^n(\omega)$ in $A_s^n(M,\mathfrak{U};V)$. It has been shown in Lemma 16.1.12 that the triviality of the period map $\mathrm{per}_{\varrho t}(\omega)_n : \pi_n(M) \to V^G$ for closed forms ω implies that the smooth $\mathfrak{U}$-local cochain $s^n(\omega)$ is a cocycle. We search for necessary and sufficient conditions on the equivariant differential n-form ω for the equivariant $\mathfrak{U}$-local cocycle $s^n(\omega)$ to be extendible to a locally smooth equivariant global n-cocycle on M. A first sufficient condition is the following:

Lemma 16.2.2. *If the equivariant n-form ω is exact in the complex $\Omega^*(M;V)^G$ of equivariant differential forms on M, then $s^n(\omega)$ is extendible to a global equivariant cocycle.*

Proof. Assume ω is exact in the complex $\Omega^*(M;V)^G$. Then there exists an equivariant $(n-1)$-form η on M with exterior derivative $d\eta = \omega$. This $(n-1)$-form can be integrated to an equivariant $\mathfrak{U}$-local $(n-1)$-cochain $s^{n-1}(\eta)$. The coboundary of this equivariant u-local cochain $s^{n-1}(\eta)$ is the equivariant $\mathfrak{U}$-local cochain $s^n(\omega)$. Extend $s^{n-1}(\eta)$ to an equivariant global $(n-1)$-cochain H on M. Then the coboundary dH is an equivariant n-cocycle which extends the $\mathfrak{U}$-local cochain $s^n(\omega)$.   □

Proposition 16.2.3. *The extendibility of the $\mathfrak{U}$-local cochain $s^n(\omega)$ of an equivariant closed n-form ω on M to a global equivariant n-cocycle only depends on its cohomology class $[\omega] \in H_{dR,eq}^n(M;V)$.*

Proof. Let ω be a closed equivariant n-form whose local integral $s^n(\omega)$ is extendible to a global equivariant n-cocycle F on M. It is to show that the $\mathfrak{U}$-local integral $s^n(\omega')$ of a closed equivariant n-form ω' with the same cohomology class $[\omega'] = [\omega]$ in $H_{dR,eq}^n(M;V)$ is extendible to a global equivariant n-cocycle as well. If an n-form ω' is cohomologous to ω in $\Omega^*(M;V)^G$, then there exists an equivariant $(n-1)$-form η with coboundary $d\eta = \omega - \omega'$. The $\mathfrak{U}$-local integral $s^n(d\eta)$ of the coboundary $d\eta$ is extendible to a global equivariant n-cocycle H by Lemma 16.2.2. The equivariant n-cocycle $F - H$ is an extension of $s^n(\omega)$ to a global equivariant cocycle, hence the $\mathfrak{U}$-local n-cochain $s^n(\omega')$ is also extendible to a global equivariant n-cocycle.   □

As suggested by this result and those in Chapter 7 the triviality of the homomorphisms

$$H_{p,\infty(M)} \to H_{dR,eq}^q(G;V) \cong H_c^q(\mathfrak{g};V)$$

induced by the flux components $\mathcal{F}_{\varrho t}(\omega)^{p,q}$ might be a necessary and sufficient condition for this to be the case. We would like to show that this condition is always sufficient for $(n-1)$-connected manifolds, if the differential n-form ω is exact or the de Rham homomorphism $k^n : H_{dR}^n(M;V) \to H^n(M;V)$ is injective. For many manifolds this is a necessary condition for ω to be integrable to a locally smooth equivariant global cocycle.

Lemma 16.2.4. *If the manifold M is smoothly paracompact and the closed differential n-form ω on M is integrable to a locally smooth global n-cocycle then ω is exact.*

Proof. If the manifold M is smoothly paracompact then the classical cohomologies $H_{AS}(M;V)$ and $H_{dR}(M;V)$ are naturally isomorphic. If ω is integrable to a locally smooth global n-cocycle F then the cohomology class $[\omega] \in H_{dR}(M;V)$ corresponds to the cohomology class $[F] \in H_{AS}^n(M;V)$ under this isomorphism. The latter cohomology class is trivial. $\qquad\square$

If the latter homomorphism k^n is injective, then the exactness of a closed differential n-form ω on M can be tested by integrating ω over smooth singular n-cycles. This in turn can be expressed in terms of the flux component $\mathcal{F}_{\varrho t}(\omega)^{n,0}$. Recall that the image $\mathcal{F}_{\varrho t}(\omega)$ of an equivariant differential n-form ω on M under the flux homomorphism $\mathcal{F}_{\varrho t} : \Omega^*(M;V)^G \to \mathrm{Tot} S^*\big(M;\Omega(G;V)^G\big)$ is a cochain in the total complex of the double complex $S^*(M;\Omega^*(G;V)^G)$. (See Section 13.6 for a discussion of flux homomorphisms). We denote the component of bidegree (p,q) of the flux $\mathcal{F}_{\varrho t}(\omega)$ by $\mathcal{F}_{\varrho t}(\omega)^{p,q}$ (cf. Section 13.6). If the equivariant differential n-form ω on M is closed, then the cochain $\mathcal{F}_{\varrho t}(\omega) = \sum_{p+q=n} \mathcal{F}_{\varrho t}(\omega)^{p,q}$ in the total complex $\mathrm{Tot} S^*(M;\Omega^*(G;V)^G)$ is a cocycle:

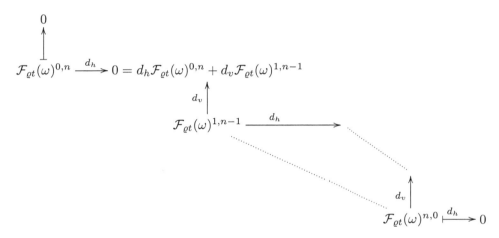

Let $\omega \in \Omega^n(M;V)^G$ be a closed equivariant differential n-form on M. We use the component $\mathcal{F}_{\varrho t}(\omega)^{n,0}$ of bidegree $(n,0)$ of the flux $\mathcal{F}_{\varrho t}(\omega)$ to describe a sufficient condition for the germ of the $\mathfrak{U}$-local cocycle $s^n(\omega)$ to be exact. The flux component $\mathcal{F}_{\varrho t}(\omega)^{n,0}$ is an equivariant homomorphism

$$\mathcal{F}_{\varrho t}(\omega)^{n,0} : S_{n,\infty}(M) \to \Omega^0(G;V)^G = C^\infty(G;V)^G$$

$$\mathcal{F}_{\varrho t}(\omega)^{0,n} \left(\sum a_i \tau_i \right)(g) = \sum a_i \int_{g.\tau_i} \omega = g.\left(\sum a_i \int_{\tau_i} \omega \right)$$

Each flux component $\mathcal{F}_{\varrho t}(\omega)^{p,q}$ has codomain the complex $\Omega^*(G;V)^G$ of left equivariant forms on G. This complex $\Omega^*(G;V)^G$ can be identified with the complex $C^*(\mathfrak{g};V)$ of Lie algebra cochains of the Lie algebra $\mathfrak{g}$ of G. As explained in Lemma 13.6.2 and Corollary 13.6.3 the homomorphisms $\mathcal{F}_{\varrho t}(\omega)^{p,q}$ map smooth singular p-cycles to de Rham q-cocycles and restrict to homomorphisms

$$\mathcal{F}_{\varrho t}(\omega)^{p,q}_{|Z_{p,\infty}(M)} : Z_{p,\infty}(M) \to Z_{dR}^q(G;V)^G \cong Z_c^q(\mathfrak{g};V)$$

which descend to homomorphisms

$$H_{p,\infty}(M) \to H_{dR,eq}^q(G;V) \cong H_c^q(\mathfrak{g};V).$$

In the special case $p = n$, $q = 0$ the codomain $Z_{dR}^0(M;V)^G \cong Z_c^0(\mathfrak{g};V) \cong V^G$ of the component $\mathcal{F}_{\varrho t}(\omega)^{n,0}$ of bidegree $(n,0)$ of the flux $\mathcal{F}_{\varrho t}(\omega)$ can be identified with the G-invariants V^G in V. So the homomorphism $\mathcal{F}_{\varrho t}(\omega)^{n,0}$ is given by

$$\mathcal{F}_{\varrho t}(\omega)^{n,0} : Z_{n,\infty}(M) \to V^G, \quad z \mapsto \int_z \omega, \tag{16.1}$$

i.e. by integration of the n-form ω over the singular n-cycles. The cohomology group $H^0_{dR,eq}(G;V) \cong H^0_c(\mathfrak{g};V)$ is isomorphic to the group $Z^0_{dR}(G;V)^G \cong V^G$, because the 0-boundaries are trivial. Moreover, since the form ω is closed, the integrals in formula 16.3 only depend on the homotopy classes of the cycles z; as consequence we see directly that the flux descends to a homomorphism $\mathcal{F}_{\varrho t}^{n,0}(\omega) : H_n(M) \to V^G$ and the latter is trivial if and only if $\mathcal{F}_{\varrho t}^{n,0}(\omega) : Z_n(M) \to V^G$ is trivial.

Lemma 16.2.5. *If ω is a closed equivariant differential n-form on M, the de Rham homomorphism $k^n : H^n_{dR}(M;V) \to H^n(M;V)$ is injective and the restricted flux component $\mathcal{F}_{\varrho t}(\omega)^{n,0} : Z_{n,\infty}(M) \to V^G$ is trivial, then ω is exact.*

Proof. Let ω be a closed equivariant n-form on M. The restricted flux component $\mathcal{F}_{\varrho t}(\omega)^{n,0} : Z_{n,\infty}(M) \to V^G$ is trivial if and only if the singular n-cocycle $k^n(\omega)$ is a coboundary. If the de Rham homomorphism $k^n : H^n_{dR}(M;V) \to H^n(M;V)$ is injective then this is equivalent to the exactness of the differential form ω.   □

Example 16.2.6. If M is smoothly paracompact, then the de Rham homomorphism $k^n : H^n_{dR}(M;V) \to H^n(M;V)$ is an isomorphism. In this case the triviality of the restriction of the flux component $\mathcal{F}_{\varrho t}(\omega)^{n,0}$ to singular n-cycles implies the exactness of ω. This includes all paracompact Hilbert- and especially all finite dimensional manifolds.

Lemma 16.2.7. *If the closed equivariant n-form ω is exact, then the smooth $\mathfrak{U}$-local cochain $s^n(\omega)$ is a coboundary in the complex $A^*_s(M,\mathfrak{U};V)$ of smooth $\mathfrak{U}$-local cochains.*

Proof. Let η be an $(n-1)$-form η on M with exterior derivative $d\eta = \omega$. Then Stokes Theorem implies that the smooth $\mathfrak{U}$-local $(n-1)$-cochain

$$s^{n-1}(\eta) \in A^{n-1}(M,\mathfrak{U};V), \quad s^{n-1}(\mathbf{m}) = \int_{\hat{\sigma}_{n-1}(\mathbf{m})} \eta$$

has coboundary $s^n(\omega)$. This proves the assertion.   □

If the manifold M is $(n-2)$-connected and $n \geq 3$, then the Hurewicz homomorphism $\mathrm{hur}_n : \pi_n(M) \to H_n(M)$ is an epimorphism. In this case the exactness of the differential n-form ω on M can also be tested by integrating over (smooth) n-spheres $\mathbb{S}^n \to M$ only. This makes the exactness of the closed differential n-form ω expressible in terms of a period map. Recall that the n-period homomorphism for the flux component $\mathcal{F}_{\varrho t}^{n,0}(\omega) : H_n(M) \to H^0_{dR,eq}(G;V) = V^G$ is given by

$$\mathrm{per}_{\varrho t}(\omega)_n = [\mathcal{F}_{\varrho t}^{n,0}(\omega) \circ \mathrm{hur}_n]([s]) = \int_s \omega$$

i.e. by integration of ω over n-spheres in M. So because the Hurewicz homomorphism hur_n for M is an epimorphism we observe:

Lemma 16.2.8. *If the de Rham homomorphism $k^n : H^n_{dR}(M;V) \to H^n(M;V)$ is injective, the period homomorphism $\mathrm{per}_{\varrho t}(\omega)_n : \pi_n(M) \to V^G$ is trivial, $n \geq 3$ and the manifold M is $(n-2)$-connected, then differential n-form ω is exact.*

Example 16.2.9. If M is a simply connected smoothly paracompact manifold and ω a closed equivariant 3-form on M with trivial period homomorphism $\mathrm{per}_{\varrho t}(\omega)_n$, then ω is exact. This especially includes all finite dimensional simply connected manifolds.

This fact may also be stated in terms of the rational singular homology of the manifold M: The space of coefficients V is a real and a rational vector space and so are all the Abelian groups $\Omega^q(G;V)^G$ and $H^q_{dR,eq}(M;V)$. This allows us to identify the groups $\Omega^q(G;V)^G$ with the groups $\Omega^q(G;V)^G \otimes_{\mathbb{Z}} \mathbb{Q}$ and the groups $H^q_{dR,eq}(M;V)$ with the groups $H^q_{dR,eq}(M;V) \otimes_{\mathbb{Z}} \mathbb{Q}$, which we will do in the following. (As usual we will omit the subscript '$\mathbb{Z}$' when dealing with tensor products over the ring $\mathbb{Z}$ of integers.) Therefore any homomorphism from an Abelian group A into one of these groups annihilates torsion and factors uniquely through the rationalisation $A \otimes \mathbb{Q}$ of A. Consequently, for any closed equivariant n-form ω on M the components

$$\mathcal{F}_{\varrho t}(\omega)^{p,q} : S_p(M) \to \Omega^q(G;V)^G \cong C^q_c(\mathfrak{g};V)$$

$$\text{and} \qquad \mathcal{F}_{\varrho t}(\omega)^{p,q} : Z_p(M) \to Z^q_{dR}(G;V)^G \cong Z^q_c(\mathfrak{g};V)$$

of the flux $\mathcal{F}(\omega)$ factor uniquely through the singular groups $S_p(M;\mathbb{Q}) = S_p(M) \otimes \mathbb{Q}$ and $Z_p(M;\mathbb{Q}) = Z_p(M) \otimes \mathbb{Q}$. This can be subsumed in the commutative diagram

$$
\begin{array}{ccc}
Z_p(M) & \xrightarrow{\ \mathcal{F}_{\varrho t}(\omega)^{p,q}\ } & Z^q_{dR}(G;V)^G \cong Z^q_c(\mathfrak{g};V) \\
{\scriptstyle z \mapsto z \otimes 1}\downarrow & & \| \\
Z_p(M) \otimes \mathbb{Q} & \xrightarrow{\ \mathcal{F}_{\varrho t}(\omega)^{p,q} \otimes \mathrm{id}_{\mathbb{Q}}\ } & Z^q_{dR}(G;V)^G \otimes \mathbb{Q} \cong Z^q_{dR}(G;V)^G \cong Z^q_c(\mathfrak{g};V).
\end{array}
$$

$$(16.2)$$

In particular, as is apparent from the above diagram, the flux component $\mathcal{F}_{\varrho t}(\omega)^{p,q}$ is trivial if and only if its rational version $\mathcal{F}(\omega)^{p,q} \otimes \mathrm{id}_{\mathbb{Q}}$ is trivial. So we note:

Lemma 16.2.10. *If the de Rham homomorphism $k^n : H^n_{dR}(M;V) \to H^n(M;V)$ is injective, then an equivariant n-form ω on M is exact if and only if the component $\mathcal{F}_{\varrho t}(\omega)^{n,0} \otimes \mathrm{id}_{\mathbb{Q}} : Z_n(M;\mathbb{Q}) \to V^G$ of the flux is trivial if and only if the induced map $\mathcal{F}_{\varrho t}(\omega)^{n,0} \otimes \mathrm{id}_{\mathbb{Q}} : H_n(M;\mathbb{Q}) \to H^0_{dR,eq}(G;V) \cong H^0_c(\mathfrak{g};V) \cong V^G$ is trivial.*

A special case are manifolds whose rational homology $H_n(M;\mathbb{Q})$ is a graded algebra generated by the images of the rational Hurewicz-homomorphisms:

Lemma 16.2.11. *If the rational homology $H(X;\mathbb{Q})$ of an n-connected topological space X is a graded algebra generated by the images $\mathrm{Im}(\mathrm{hur}_n \otimes \mathrm{id}_{\mathbb{Q}})$, then the rational Hurewicz-homomorphisms $\mathrm{hur}_k \otimes \mathrm{id}_{\mathbb{Q}} : \pi_k(X) \otimes \mathbb{Q} \to H_k(X;\mathbb{Q})$ are surjective for all $n \le k \le 2n$.*

Proof. Let the topological space X be n-connected and assume that the rational homology $H(X;\mathbb{Q})$ of X is a graded algebra generated by the images $\mathrm{Im}(\mathrm{hur}_n \otimes \mathrm{id}_{\mathbb{Q}})$ of the rational Hurewicz-homomorphisms. then every rational singular k-cycle z in $Z_k(X;\mathbb{Q})$ is a product $s_1 \cdots s_l$ of images $s_1, \ldots, s_l$ of rational spheres under the Hurewicz-homomorphisms, whose dimension add up to n. For $k \le n$ this implies that all but one of the spheres are 0-spheres and one of them is a k-sphere. Thus z is in the image of the rational Hurewicz-homomorphism in this case. $\qquad\square$

Lemma 16.2.12. *If G is an n-connected topological group, then the rational Hurewicz-homomorphisms $\pi_k(X) \otimes \mathbb{Q} \to H_k(X;\mathbb{Q})$ are surjective for all $n \le k \le 2n$.*

Proof. If G is an n-connected H-topological group, then the rational homology algebra $H(G;\mathbb{Q})$ is a primitively generated Hopf algebra by Theorem 8.4.10. Its subspace $P(G;\mathbb{Q})$ of primitive elements is the image $hur(\pi(G) \otimes \mathbb{Q})$ and the multiplication in $H(G;\mathbb{Q})$ is induced by the multiplication in G (Theorem 8.4.10). Therefore the assumption that G is n-connected implies that the rational Hurewicz-homomorphisms $\mathrm{hur}_k \otimes \mathrm{id}_{\mathbb{Q}} : \pi_k(X) \otimes \mathbb{Q} \to H_k(X;\mathbb{Q})$ are surjective for all $n \le k \le 2n$. $\qquad\square$

Example 16.2.13. If G is an n-connected Lie group, then the rational Hurewicz-homomorphisms $\mathrm{hur}_k \otimes \mathrm{id}_{\mathbb{Q}} : \pi_k(X) \otimes \mathbb{Q} \to H_k(X; \mathbb{Q})$ are surjective for all $n \leq k \leq 2n$.

Moreover we can show:

Proposition 16.2.14. *If G is a Lie group acting on itself by left translations and the de Rham homomorphism $k^n : H_{dR}^n(G; V) \to H^n(G; V)$ is injective, then the triviality of all period homomorphisms $\mathrm{per}_{\varrho t}(\omega)_p : \pi_p(G) \to H_{dR,eq}^{n-p}(G; V)$ implies the exactness of ω.*

Proof. If all the period homomorphisms $\mathrm{per}_{\varrho t}(\omega)_p$ are trivial, then the homomorphisms $H_p(G) \to H_{dR,eq}^q(G; V) \cong H_c^q(\mathfrak{g}; V)$ induced by the flux components $\mathcal{F}_{\varrho t}(\omega)^{p,q}$ are trivial by Proposition 15.1.27. The triviality of the the homomorphism $H_n(G) \to V^G$ and the injectivity of the de Rham homomorphism k^n imply the exactness of ω. $\qquad\square$

From now on we assume the equivariant differential n-form ω on M to be exact and fix a differential $(n-1)$-form η on M with exterior derivative $d\eta = \omega$. Since the action $\varrho : G \times M \to M$ of G on M is free, there exists a not necessarily continuous equivariant map $p : M \to G$. We fix such an equivariant map p and use it to define a cochain $\psi^{0,n-1}$ of bidegree $(0, n-1)$ in the double complex $A^{*,*}(M, \mathfrak{U}; V)^G$:

$$\psi^{0,n-1} : M \times \mathrm{SS}(M, \mathfrak{U})([n-1]) \to V,$$

$$\psi^{0,n-1}(m_0, \mathbf{m}') = [p(m_0).s^{n-1}(\eta)](\mathbf{m}') = \int_{\hat{\sigma}_{n-1}(\mathbf{m}')} p(m_0).\eta$$

Lemma 16.2.15. *The vertical coboundary $d_v \psi^{0,n-1}$ of $\psi^{0,n-1}$ in the double complex $A^{*,*}(M, \mathfrak{U}; V)^G$ is the cocycle $i^n(s^n(\omega))$.*

Proof. The vertical coboundary of the cochain $\psi^{0,n-1}$ is obtained by fixing the first argument m_0 and applying the coboundary operator of the complex $A^*(M, \mathfrak{U}; V)$ to the $(n-1)$-cochain $\psi^{0,n-1}(m_0, -)$:

$$d_v \psi^{0,n-1}(m_0, \mathbf{m}') = d[p(m_0).s^{n-1}(\eta)](\mathbf{m}') = p(m_0).[ds^{n-1}(\eta)](\mathbf{m}')$$
$$= [p(m_0).s^n(\omega)](\mathbf{m}') = s^n(\omega)(\mathbf{m}') = [i^n s^n(\omega)](m_0, \mathbf{m}')$$

Thus the vertical coboundary of $\psi^{0,n-1}$ is the cocycle $i^n s^n(\omega)$. $\qquad\square$

We denote the horizontal coboundary of $\psi^{0,n-1}$ in the double complex $A^{*,*}(M, \mathfrak{U}; V)^G$ by $F^{1,n-1}$. It is a cochain of bidegree $(1, n-1)$ in the total complex of the double complex $A^{*,*}(M, \mathfrak{U}; V)^G$, which is a cocycle by Lemma 14.2.19. The situation in the double complex $A^{*,*}(M, \mathfrak{U}; V)^G$ can be depicted as follows:

$$
\begin{array}{ccccc}
& 0 & & & \\
& \uparrow {\scriptstyle d_v} & & & \\
i^n s^n(\omega) & \overset{d_h}{\longmapsto} & 0 & & \\
& \uparrow {\scriptstyle d_v} & & \uparrow {\scriptstyle d_v} & \\
\psi^{0,n-1} & \overset{d_h}{\longmapsto} & F^{1,n-1} & \overset{d_h}{\longmapsto} & 0
\end{array}
$$

The cocycle $-F^{1,n-1}$ is cohomologous to the cocycle $i^n s^n(\omega)$ in $\mathrm{Tot} A^{*,*}(M, \mathfrak{U}; V)^G$ by Lemma 14.2.19. So we observe:

Lemma 16.2.16. *The equivariant $\mathfrak{U}$-local cocycle $s^n(\omega)$ is extendible to a global equivariant cocycle if and only if $F^{1,n-1}$ is cohomologous to a cocycle $F^{n,0}$ of bidegree $(n,0)$ in the total complex $\mathrm{Tot}A^{*,*}(M,\mathfrak{U};V)^G$.*

Proof. The equivariant $\mathfrak{U}$-local cocycle $s^n(\omega)$ is extendible to a global equivariant cocycle if and only if the cocycle $i^n s^n(\omega)$ in $\mathrm{Tot}A^{*,*}(M,\mathfrak{U};V)^G$ is cohomologous to a cocycle $F^{n,0}$ of bidegree $(n,0)$ in $\mathrm{Tot}A^{*,*}(M,\mathfrak{U};V)^G$. Since the cocycles $i^n s^n(\omega)$ and $-F^{1,n-1}$ are cohomologous in the total complex $\mathrm{Tot}A^{*,*}(M,\mathfrak{U};V)^G$ this is equivalent to $F^{1,n-1}$ being cohomologous to a cocycle $F^{n,0}$ of bidegree $(n,0)$ in the total complex $\mathrm{Tot}A^{*,*}(M,\mathfrak{U};V)^G$. $\qquad\square$

We now examine under which conditions on ω the cocycle $F^{1,n-1}$ is the vertical coboundary of a cochain $\psi^{1,n-2}$ or at least cohomologous to a cocycle $F^{2,n-2}$ of bidegree $(2,n-2)$. At first we derive a necessary condition for the latter to be the case. Thereafter we show that this condition is also sufficient. Recall that the groups $A^{p,q}(M,\mathfrak{U};V)^G$ of the double complex $A^{*,*}(M,\mathfrak{U};V)^G$ were defined to be the groups of equivariant functions $M^{p+1} \times \mathrm{SS}(M,\mathfrak{U})[q]) \to V$. Their adjoint functions $M^{p+1} \to A^q(M;\mathfrak{U};V)$ are global equivariant p-cochains with values in the group $A^q(M,\mathfrak{U};V)$ of $\mathfrak{U}$-local q-cochains. The adjoint cocycle $\hat{F}^{1,n-1}$ of the cocycle $F^{1,n-1}$ has trivial horizontal and vertical coboundaries and thus is an element of the group $Z^1(M;Z^{n-1}(M,\mathfrak{U}))^G$. Its evaluation at a point $\mathbf{m} \in M^2$ defines a cohomology class $[q \circ F^{1,n-1}(\mathbf{m})] \in Z^{n-1}(M,\mathfrak{U})$, where $q : Z^{n-1}(M,\mathfrak{U}) \to H^{n-1}(M,\mathfrak{U})$ denotes the quotient homomorphism. If the cocycle $F^{1,n-1}$ is a vertical coboundary then these cohomology classes are trivial. Moreover the 1-cocycle $q\hat{F}^{1,n-1}$ is equivariant because $\hat{F}^{1,n-1}$ is equivariant. Therefore this 1-cocycle defines a cohomology class in $H^1_{eq}(M;H^{n-1}(M,\mathfrak{U}))$ which we denote by $[q\hat{F}^{1,n-1}]$

Lemma 16.2.17. *If the cocycle $\hat{F}^{1,n-1}$ is cohomologous to a cocycle of bidegree $(n,0)$ in the total complex of $A^*(M;A^*(M,\mathfrak{U};V))^G$, then the cohomology class $[qF^{1,n-1}] \in H^1_{eq}(M;H^{n-1}(M,\mathfrak{U}))$ is trivial.*

Proof. Let the cocycle $\hat{F}^{1,n-1}$ be cohomologous to a cocycle $\hat{F}^{n,0}$ of bidegree $(n,0)$ in the total complex of $A^*(M;A^*(M,\mathfrak{U};V))^G$. Then there exists a cochain c of total degree $(n-1)$ in $A^*(M;A^*(M,\mathfrak{U};V))^G$ satisfying $Dc = \hat{F}^{1,n-1} - \hat{F}^{n,0}$. The cochain c is a sum $\sum_{p'+q'=n-1} c^{p',q'}$ of cochains of bidegree (p',q'). In particular the vertical coboundary $d_v c^{0,n-1}$ is zero, so $c^{0,n-1}$ is an element in $Z^0(M;Z^{n-1}(M,\mathfrak{U}))^G$. As a consequence cohomology class $[qd_h c^{0,n-1}] \in H^1_{eq}(M;H^{n-1}(M,\mathfrak{U}))$ of the horizontal coboundary $d_h c^{0,n-1}$ is trivial. Therefore the cocycle $d_h c^{0,n-1}$ may be subtracted from $\hat{F}^{1,n-1}$ without changing the cohomology class in $H^1_{eq}(M;H^{n-1}(M,\mathfrak{U}))$. The equalities

$$D(c - c^{0,n-1}) = \hat{F}^{1,n-1} - Dc^{0,n-1} = \hat{F}^{1,n-1} - d_h c^{0,n-1}$$

show that the cohomology class $[q\hat{F}^{1,n-1} - qd_h c^{0,n-1}]$ and thus the cohomology class $[q\hat{F}^{1,n-1}]$ in $H^1_{eq}(M;H^{n-1}(M,\mathfrak{U}))$ is trivial. $\qquad\square$

The cocycle $\hat{F}^{1,n-1}$ is cohomologous to a cocycle of bidegree $(n,0)$ in the total complex of $A^*(M;A^*(M,\mathfrak{U};V))^G$ if and only if its adjoint cocycle $F^{1,n-1}$ is cohomologous to a cocycle of bidegree $(n,0)$ in the total complex of $A^{*,*}(M,\mathfrak{U};V)^G$. So the preceding two Lemmata show:

Proposition 16.2.18. *If the $\mathfrak{U}$-local cocycle $s^n(\omega)$ is extendible to a global equivariant cocycle then the cohomology class $[q\hat{F}^{1,n-1}]$ in $H^1_{eq}(M;H^{n-1}(M,\mathfrak{U}))$ is trivial.*

If the 1-cocycle $q\hat{F}^{1,n-1} : M^2 \to H^{n-1}(M,\mathfrak{U};V)$ is the zero function, then $F^{1,n-1}$ is the vertical coboundary of a cochain $\psi^{1,n-2}$ of bidegree $(1,n-1)$ in the double complex $A^{*,*}(M, \mathfrak{U}; V)^G$ by Lemma 15.3.22. In this case the horizontal coboundary $d_h \psi^{1,n-2}$ is a cocycle in the total complex $\mathrm{Tot} A^{*,*}(M, \mathfrak{U}; V)^G$ and the cocycles $F^{1,n-1}$ and $F^{2,n-2}$ are cohomologous in this total complex (Lemma 14.2.19). Moreover we can show:

Lemma 16.2.19. *If the cohomology class* $[qF^{1,n-1}] \in H^1(M; H_{eq}^{n-1}(M, \mathfrak{U}; V))^G$ *is trivial, then* $F^{1,n-1}$ *is cohomologous to a cocycle* $F^{2,n-2}$ *in the total complex of the double complex* $A^{*,*}(M, \mathfrak{U}; V)^G$.

Proof. If the cohomology class $[qF^{1,n-1}] \in H^1(M; H_{eq}^{n-1}(M, \mathfrak{U}; V))^G$ be trivial, then there exists an equivariant 0-cochain $\hat{c} \in A^1(M; H^{n-1}(M, \mathfrak{U}; V))^G$ with coboundary $dc = qF^{1,n-1}$. Let $s : H^{n-1}(M, \mathfrak{U}; V) \to Z^{n-1}(M, \mathfrak{U}; V)$ be a not necessarily linear section and define a cochain $\hat{c}'$ in $A^1(M, Z^{n-1}(M, \mathfrak{U}; V))$ via

$$\hat{c}(m_0)(\mathbf{m}') = p(m_0). \left[s\hat{c}(p(m_0)^{-1}.m_0) \right] ((p(m_0)^{-1}.\mathbf{m}')) .$$

The equivariance of the 0-cochain $\hat{c}$ implies the equalities

$$\begin{aligned}
\hat{c}'(g.m_0) &= p(g.m_0). \left[s\hat{c}(p(g.m_0)^{-1}.[g.m_0])(p(g.m_0)^{-1}.\mathbf{m}') \right] \\
&= [g \cdot p(m_0)]. \left[s\hat{c}([g \cdot p(m_0)]^{-1}[g.m_0])([g \cdot p(m_0)]^{-1}.\mathbf{m}') \right] \\
&= g. \left[p(m_0). \left(s\hat{c}(p(m_0)^{-1}.m_0)(p(m_0)^{-1}.[g^{-1}.\mathbf{m}']) \right) \right] \\
&= g. \left[\hat{c}'(m_0)(g^{-1}.\mathbf{m}') \right] \\
&= [g. (\hat{c}'(m_0))] (\mathbf{m}') ,
\end{aligned}$$

so the 0-cochain $\hat{c}'$ is equivariant and an element of $A^1(M; Z^{n-1}(M, \mathfrak{U}; V))^G$. The equivariant 0-cochain $\hat{c}'$ factors to $\hat{c} = q\hat{c}'$. Its vertical coboundary is trivial by construction. Therefore the horizontal coboundary $d_h\hat{c}'$ in the double complex $A^*(M, A^*(M, \mathfrak{U}; V))^G$ coincides with the coboundary $D\hat{c}' = [d_h + d_v]\hat{c}'$ in the total complex of the double complex $A^*(M, A^*(M, \mathfrak{U}; V))^G$. As a consequence the horizontal coboundary $d_h\hat{c}'$ may be subtracted from $\hat{F}^{1,n-1}$ without changing the cohomology class $[\hat{F}^{1,n-1}]$ in the total complex $\mathrm{Tot} A^*(M, A^*(M, \mathfrak{U}; V))^G$. The cohomology class $q[\hat{F}^{1,n-1} - d_h\hat{c}'] = [q\hat{F}^{1,n-1} - d_h\hat{c}]$ is trivial by construction. Therefore the cocycle $\hat{F}^{1,n-1} - d_h\hat{c}'$ is cohomologous to a cocycle $F^{2,n-2}$ of bidegree $(2, n-2)$ in the total complex $\mathrm{Tot} A^*(M, A^*(M, \mathfrak{U}; V))^G$. Since the cohomology classes $[q\hat{F}^{1,n-1}]$ and $q[\hat{F}^{1,n-1} - d_h\hat{c}']$ coincide, this shows that the cocycle $\hat{F}^{1,n-1}$ is also cohomologous to $F^{2,n-2}$. $\square$

We would like to show that the cohomology class $[q\hat{F}^{1,n-1}]$ is trivial provided the flux component $\mathcal{F}_{\varrho t}(\omega)^{1,n-1} : S_{n-1}(M)\mathfrak{U}; V) \to \Omega^{n-1}(G; V)^G$ induces the zero homomorphism $H_1(M) \to H_{dR,eq}^{n-2}(G; V)$. For this purpose we recall that the natural homomorphism $H^{n-1}(M, \mathfrak{U}; V) \to H^{n-1}(M; V)$ into the $(n-1)$-singular cohomology group is injective. Therefore each cohomology class $\hat{F}^{1,n-1}(m_0, m_1)$ is trivial if and only if the corresponding singular cohomology class $[\tilde{\lambda}_{\mathfrak{U}}^{n-1}\hat{F}^{1,n-1}(m_0, m_1)]$ in $H^{n-1}(M; V)$ is trivial. If the Lie group G is connected, then two points $p(m_0)$ and $p(m_1)$ can always be joined by a smooth path γ in G. In this case the singular cochain $\tilde{\lambda}_{\mathfrak{U}}^{n-1}\hat{F}^{1,n-1}(m_0, m_1)$ in $Z^{n-1}(M; V)$ can nicely be expressed in terms of the form ω. For this purpose we recall the Eilenberg-Zilber morphism $S_p(G) \otimes S_q(M) \to S_{p+q}(M)$ for the action $\varrho : G \times M \to M$ (cf. Section 8.2) which maps a product of a singular p-simplex σ and a singular q-simplex τ in M to a singular $p + q$-chain in M, which we simply denote by $\sigma.\tau$. In this way one can integrate a differential n-form on M over the singular chain $\sigma.\tau$ provided $p + q = n$.

Lemma 16.2.20. *If the Lie group G is connected and V is a trivial $\mathbb{Z}[G]$-module, then for all points $(m_0, m_1) \in M^2$ the evaluation of the singular $(n-1)$-cocycle* $\tilde{\lambda}_{\mathfrak{U}}^{n-1}\hat{F}^{1,n-1}(m_0, m_1)$ *at a $\mathfrak{U}$-small singular $(n-1)$-cocycle* $z \in Z_n(M, \mathfrak{U})$ *is given by*

$$\hat{F}^{1,n-1}(m_0,m_1)(z) = \int_{\gamma^{-1}.(\lambda_{n-1}(z))} \omega = \int_{\gamma^{-1}} \mathcal{F}_{\varrho t}(\omega)^{n-1,1}(\sigma_{n-1}(\lambda_{n-1}(z)))$$

for any smooth path γ from $p(m_0)$ to $p(m_1)$ in G.

Proof. Let (m_0,m_1) be a point in M^2 and $z \in Z_n(M,\mathfrak{U})$ be a $\mathfrak{U}$-small singular $(n-1)$-cycle in M. The evaluation of the singular $(n-1)$-cocycle $\tilde{\lambda}_{\mathfrak{U}}^{n-1}\hat{F}^{1,n-1}(m_0,m_1)$ at the $\mathfrak{U}$-small singular $(n-1)$-cycle $z \in S_{n-1,\infty}(M)$ is by definition given by

$$\hat{F}^{1,n-1}(m_0,m_1)(z) = \int_{p(m_1)^{-1}.(\sigma_{n-1}\lambda_{n-1}(z))} \omega$$
$$- \int_{p(m_0)^{-1}.(\sigma_{n-1}\lambda_{n-1}(z))} \omega$$
$$= \int_{(p(m_1)^{-1}-p(m_1)^{-1}).\sigma_{n-1}(\lambda_{n-1}(z))} \omega$$

Because the singular n-chain z is a cycle, the singular n-chains $\sigma_{n-1}(\lambda_{n-1}(z))$ and $p(m_0)^{-1}.\sigma_{n-1}(\lambda_{n-1}(z)) - p(m_0)^{-1}.\sigma_{n-1}(\lambda_{n-1}(z))$ are cycles as well. The latter singular chain is the boundary of $\gamma^{-1}.\sigma_{n-1}(\lambda_{n-1}(z))$ for any path γ from $p(m_0)$ to $p(m_1)$. Therefore Stokes Theorem implies the equalities

$$\hat{F}^{1,n-1}(m_0,m_1)(z) = \int_{(\partial\gamma^{-1}).(\sigma_{n-1}\lambda_{n-1}(z))} \omega = \int_{\gamma^{-1}.\sigma_{n-1}(\lambda_{n-1}(z))} d\omega + 0$$
$$= \int_{\gamma^{-1}.\sigma_{n-1}(\lambda_{n-1}(z))} \omega = \int_{\gamma^{-1}} \mathcal{F}_{\varrho t}(\omega)^{n-1,1}(\sigma_{n-1}(\lambda_{n-1}(z)))$$

for any smooth path γ from $p(m_0)$ to $p(m_1)$. This proves the assertion. $\square$

Lemma 16.2.21. *If the homomorphism $H_{n-1}(M) \to H_{dR}^1(G;V)^G \cong H_c^1(\mathfrak{g};V)$ induced by $\mathcal{F}_{\varrho t}(\omega)^{n-1,1}$ is trivial, the Lie group G is connected and V is a trivial $\mathbb{Z}[G]$-module, then the cocycle $q\hat{F}^{1,n-1}(m_0,m_1) : M^2 \to H^{n-1}(M,\mathfrak{U};V)$ is the zero function.*

Proof. Let V be a trivial $\mathbb{Z}[G]$-module. The triviality of the action of G on the coefficients V ensures that the 1-coboundaries in $Z_{dR,eq}^1(G;V) \cong Z^1(\mathfrak{g};V)$ are the zero functions, so the group $Z_{dR,eq}^1(G;V)$ of closed equivariant 1-forms on G coincides with its factor group $H_{dR,eq}^1(G;V)$. If the homomorphism $H_{n-1}(M) \to H_{dR}^1(G;V)^G$ induced by the flux component $\mathcal{F}_{\varrho t}(\omega)^{n-1,1}$ is trivial, then all the evaluations

$$\hat{F}^{1,n-1}(m_0,m_1)(z) = \int_{\gamma^{-1}} \mathcal{F}_{\varrho t}(\omega)^{n-1,1}(\sigma_{n-1}\lambda_{n-1}(z))$$

of the singular $(n-1)$-cocycles $\tilde{\lambda}_{\mathfrak{U}}^{n-1}\hat{F}^{1,n-1}(m_0,m_1)$ at smooth singular $(n-1)$-cycles z in M are trivial as well. By Proposition 3.0.21 the natural homomorphism $H^{n-1}(M,\mathfrak{U};V) \to H^n(M;V)$ into the singular cohomology $H^n(M;V)$ is injective. This implies that the cohomology class $[\hat{F}^{1,n-1}(m_0,m_1)]$ in $H^{n-1}(M,\mathfrak{U};V)^G$ is trivial. $\square$

Proposition 16.2.22. *If ω is an exact equivariant differential n-form on M, $n \geq 2$, the Lie group G is connected, V a trivial $\mathbb{Z}[G]$-module and the homomorphism $H_{n-1}(M) \to H_{dR,eq}^1(G;V)^G \cong H_c^1(\mathfrak{g};V)$ induced by the flux component $\mathcal{F}_{\varrho t}(\omega)^{n-1,1}$ is trivial, then the cocycle $i^n s^n(\omega)$ is cohomologous to a cocycle $F^{2,n-2}$ of bidegree $(2, n-2)$ in the total complex $\text{Tot}A^{*,*}(M,\mathfrak{U};V)$.*

Proof. Under the assumptions made the the the cohomology class $[q\hat{F}^{1,n-1}]$ in $H^1_{eq}(M; H^{n-1}(M,\mathfrak{U}; V))$ is trivial by Lemma 16.2.21 By Lemma 16.2.19 this implies that the cocycle $F^{1,n-1} := d_h\psi^{0,n-1}$ cohomologous to a cocycle $F^{2,n-2}$ of bidegree $(2, n-2)$ in the total complex $\mathrm{Tot}A^{*,*}(M,\mathfrak{U}; V)$. $\square$

Theorem 16.2.23. *If ω is an exact equivariant differential n-form on an $(n-2)$-connected G-manifold M, $n \geq 2$, the Lie group G is connected V a trivial $\mathbb{Z}[G]$-module and the homomorphism $H_{n-1}(M) \to H^1_{dR,eq}(G; V)^G \cong H^1_c(\mathfrak{g}; V)$ induced by the flux component $\mathcal{F}_{\varrho t}(\omega)^{n-1,1}$ is trivial, then the $\mathfrak{U}$-local cocycle $i^n s^n(\omega)$ is extendible to a global equivariant cocycle.*

Proof. The cocycle $i^n s^n(\omega)$ is cohomologous to the cocycle $F^{1,n-1}$ which in turn is cohomologous to a cocycle$F^{2,n-2}$ of bidegree $(2, n-2)$ in the total complex $\mathrm{Tot}A^{*,*}(M,\mathfrak{U}; V)$ by Proposition 16.2.22. If the manifold M is $(n-2)$-connected, then the injectivity of the natural homomorphisms $H^p(M,\mathfrak{U}; V) \to H^p(M; V)$ imply that the $\mathfrak{U}$-local cohomology groups $H^p(M,\mathfrak{U}; V)$ are trivial. In this case the cocycle $F^{2,n-2}$ is cohomologous to a cocycle $F^{n,0}$ of bidegree $(n,0)$ in the total complex $\mathrm{Tot}A^{*,*}(M,\mathfrak{U}; V)$ by Corollary 15.3.28. Therefore the cocycle $i^n s^n(\omega)$ is cohomologous to a cocycle $F^{n,0}$ of bidegree $(n,0)$ in the total complex $\mathrm{Tot}A^{*,*}(M,\mathfrak{U}; V)$, hence the $\mathfrak{U}$-local cocycle $s^n(\mathfrak{U})$ is extendible to a global equivariant cocycle. $\square$

Corollary 16.2.24. *If ω is an exact equivariant differential n-form on an $(n-2)$-connected G-manifold M, $n \geq 2$, the Lie group G is connected V a trivial $\mathbb{Z}[G]$-module and the homomorphism $H_{n-1}(M) \to H^1_{dR,eq}(G; V)^G \cong H^1_c(\mathfrak{g}; V)$ induced by the flux component $\mathcal{F}_{\varrho t}(\omega)^{n-1,1}$ is trivial, then ω is integrable to a locally smooth equivariant global n-cocycle on M.*

Corollary 16.2.25. *If the Lie group G is connected, V is a trivial $\mathbb{Z}[G]$-module, M a smoothly paracompact $(n-2)$-connected G-manifold and ω an equivariant exact n-form and the homomorphism*

$$H_{n-1}(M) \to H^1_{dR}(G; V)^G \cong H^1_c(\mathfrak{g}; V)$$

induced by the flux component $\mathcal{F}_{\varrho t}^{n-1,1}$ is trivial, then ω is integrable to a locally smooth equivariant global n-cocycle on M.

Corollary 16.2.26. *If the Lie group G is connected, V is a trivial $\mathbb{Z}[G]$-module, M a smoothly paracompact $(n-2)$-connected G-manifold and ω an equivariant closed n-form and the homomorphisms*

$$H_n(M) \to H^0_{dR}(G; V)^G \cong H^0_c(\mathfrak{g}; V) \cong V^G$$
$$\text{and} \qquad H_{n-1}(M) \to H^1_{dR}(G; V)^G \cong H^1_c(\mathfrak{g}; V)$$

induced by the flux components $\mathcal{F}_{\varrho t}^{n,0}$ and $\mathcal{F}_{\varrho t}^{n-1,1}$ are trivial, then ω is integrable to a locally smooth equivariant global n-cocycle on M.

Example 16.2.27. If the Lie group G is $(n-2)$-connected, V is a trivial G-module, then an equivariant exact differential n-form ω on G is integrable to a locally smooth global equivariant n-cocycle if the homomorphism

$$H_{n-1}(M) \to Z^1_c(\mathfrak{g}; V)$$

induced by the flux component $\mathcal{F}_{\varrho t}(\omega)^{n,0}$ is trivial.

Example 16.2.28. If the smoothly paracompact Lie group G is $(n-2)$-connected and V is a trivial G-module, then a closed equivariant n-form ω on M is integrable to a locally smooth global equivariant n-cocycle if both homomorphisms

$$H_n(G) \to V^G \qquad \text{and} \qquad H_{n-1}(G) \to Z^1_c(\mathfrak{g}; V)$$

induced by the flux components $\mathcal{F}_{\varrho t}(\omega)^{n,0}$ and $\mathcal{F}_{\varrho t}(\omega)^{n-1,1}$ respectively are trivial.

Example 16.2.29. If M is a finite dimensional $(n-2)$-connected G-manifold and V is a trivial G-module, then a closed equivariant n-form ω on M is integrable to a locally smooth global equivariant n-cocycle if both homomorphisms

$$H_n(M) \to V^G \qquad \text{and} \qquad H_{n-1}(M) \to Z_c^1(\mathfrak{g}; V)$$

induced by the flux components $\mathcal{F}_{\varrho t}(\omega)^{n,0}$ and $\mathcal{F}_{\varrho t}(\omega)^{n-1,1}$ respectively are trivial.

Theorem 16.2.30. *If the Lie group G is connected, the G-manifold M is smoothly paracompact and V is a trivial $\mathbb{Z}[G]$-module then a closed equivariant ω 2-form on M is integrable to a locally smooth equivariant global n-cocycle provided the homomorphisms*

$$H_2(M) \to H_{dR}^0(G;V)^G \cong H_c^0(\mathfrak{g};V) \cong V$$
$$\text{and} \qquad H_1(M) \to H_{dR}^1(G;V)^G \cong H_c^1(\mathfrak{g};V)$$

induced by the flux components $\mathcal{F}_{\varrho t}(\omega)^{n,0}$ and $\mathcal{F}_{\varrho t}(\omega)^{n-1,1}$ respectively are trivial.

Example 16.2.31. If G is a connected smoothly paracompact Lie group with Lie algebra $\mathfrak{g}$ and $V \hookrightarrow \hat{\mathfrak{g}} \twoheadrightarrow \mathfrak{g}$ is a central extension of Lie algebras described by a 2-cocycle $\omega \in Z_c^2(\mathfrak{g};V)$, then this extension is integrable to a topologically split group extension of G by V if both period homomorphisms

$$\mathcal{F}_{\varrho t}^{2,0}(\omega) : Z_2(M) \to V^G \qquad \text{and} \qquad \mathcal{F}_{\varrho t}^{1,1}(\omega) : H_1(M) \to H_c^1(\mathfrak{g};V)$$

are trivial. For locally convex Hausdorff coefficients and Lie groups modelled on locally convex Hausdorff topological vector spaces this has been proven in [Nee02a]. Theorem 16.2.30 generalises this result to smoothly paracompact Lie groups modelled on arbitrary topological vector spaces and not necessarily locally convex coefficients V.

16.3 $(n-3)$-Connected Manifolds

In this section we continue the investigation of the integrability of closed equivariant differential forms to locally smooth global equivariant cocycles. As before we consider a smooth connected transformation group (G, M) which admits a smooth equivariant map $p : M \to G$. In addition we require the existence of compatible integrable smooth equivariant local systems $\hat{\sigma}_k : \mathrm{SS}(M, \mathfrak{U})([k]) \to C^\infty(\Delta_k, M)$, $k \leq n$ of singular k-simplices. So every differential k-form, $k \leq n$ is integrable to a locally smooth $\mathfrak{U}$-local k-cochain $s^k(\omega)$ via the equivariant homomorphism

$$s^k : \Omega^k(M;V) \to A_s^k(M, \mathfrak{U}; V)$$

and the homomorphisms s^k intertwine the coboundary operators on $\Omega^*(M;V)$ and $A_s^*(M, \mathfrak{U}; V)$. The manifold M is assumed to be connected, so the system $\hat{\sigma}_1$ is extendable to an equivariant system of smooth singular 1-simplices on M, so the equivariant homomorphism s^1 actually integrates differential 1-forms to locally smooth global 1-cochains. The coefficients space V is assumed to be a Mackey complete Hausdorff and trivial $\mathbb{Z}[G]$-module. Similar to the last section we require the Lie group G to be connected and to admit a smooth equivariant map $p : M \to G$, which we fix for the rest of this section.

Let ω be a closed equivariant n-form on M and recall that the exactness of the form ω is a necessary condition for ω to be integrable to a smooth global equivariant cocycle if the manifold M is smoothly paracompact (Lemma 16.2.4). If the de Rham homomorphism $k^n : H_{dR}^n(M;V) \to H^n(M;V)$ is injective, then the exactness of ω can be expressed in terms of the restriction

$$\mathcal{F}_{\varrho t}(\omega)^{n,0}_{|Z_{s,n}(M)} : Z_{n,\infty}(M) \to Z_{dR}^0(G;V)^G \cong Z_c^0(\mathfrak{g};V)$$

of the flux component $\mathcal{F}_{\varrho t}(\omega)^{n,0}$ which then is trivial if and only if the equivariant form ω on M is exact.

If the equivariant differential n-form ω is exact, then there exists a differential $(n-1)$-form η on M with coboundary $d\eta = \omega$. As a consequence the cocycle $i^n s^n(\omega)$ of bidegree $(0,n)$ in the double complex $A^{*,*}(M,\mathfrak{U};V)^G$ is the vertical coboundary of the cochain $\psi^{0,n-1}$ of bidegree $(0,n-1)$ defined via

$$\psi^{0,n-1}(m_0, \mathbf{m}') := s^{n-1}[p(m_0).\eta](\mathbf{m}') = \int_{\hat{\sigma}_{n-1}(\mathbf{m}')} p(m_0).\eta$$

From now on we assume the equivariant closed differential n-form ω on M to be exact and fix the form η and the cochain $\psi^{0,n-1}$ constructed therefrom. The horizontal coboundary $F^{1,n-1} := d_h \psi^{0,n-1}$ of this cochain is a cocycle of bidegree $(1,n-1)$ in the total complex of the double complex $A^{*,*}(M,\mathfrak{U};V)^G$. By Lemma 15.3.21 the extendibility of this cocycle to a cocycle in $A^{1,n-1}(M;V)^G$ implies the extendibility of the $\mathfrak{U}$-local cocycle $s^n(\omega)$ to a global equivariant cocycle and thus the integrability of the equivariant n-form ω on M to a locally smooth global equivariant n-cocycle. Rather than finding a cochain of bidegree $(1,n-2)$ with vertical coboundary $F^{1,n-1} = d_h \psi^{0,n-1}$ we will show that under an additional assumption on the flux $\mathcal{F}_{\varrho t}(\omega)$ there exists a cochain $\psi^{1,n-2}$ of bidegree $(1,n-2)$ and a cocycle $f^{1,n-1}$ of bidegree $(1,n-1)$ which is extendible to a cocycle of bidegree $(1,n-1)$ in the total complex of the double complex $A^{*,*}(M;V)^G$ and satisfies the equality $f^{1,n-1} = d_h \psi^{0,n-1} + d_h \psi^{1,n-2}$:

The fact that the cocycle $f^{1,n-1}$ is extendible to a cocycle of bidegree $(1,n-1)$ in the total complex of $A^{*,*}(M;V)^G$ then implies that it can be added to or subtracted from the cocycle $F^{1,n-1}$ without affecting the extendibility of $F^{1,n-1}$ to a cocycle in the total complex of $A^{*,*}(M;V)^G$. Therefore the cocycle $F^{1,n-1} = d_v \psi^{0,n-1}$ is extendible to a cocycle in $A^{1,n-1}(M;V)^G$ if and only if the horizontal coboundary $d_h \psi^{1,n-2}$ is extendible to a cocycle in $A^{2,n-2}(M;V)^G$. The cochain $\psi^{1,n-2}$ is given by the formula

$$\psi^{1,n-2}(m_0, m_1, \mathbf{m}') = -\mathcal{F}_{\varrho t}(\eta)^{n-2,1}(\hat{\sigma}_{n-1}(\mathbf{m}'))([p\hat{\sigma}_1(\mathbf{m})]^{-1})$$
$$= -\mathcal{F}_{\varrho}(\eta)^{1,n-2}([p\hat{\sigma}_1(\mathbf{m})]^{-1})(\hat{\sigma}_{n-1}(\mathbf{m}'))$$
$$= -\int_{[p\hat{\sigma}_1(\mathbf{m})]^{-1}.\hat{\sigma}_{n-1}(\mathbf{m}')} \eta$$

for all points $\mathbf{m} \in M^2$ and $\mathbf{m}' \in SS(M,\mathfrak{U})([n-1])$ where $\sigma_1 : M^2 \to C^\infty(\Delta_1, M)$ is an extension of the smooth local system of singular 1-simplices to a global system of smooth singular 1-simplices (which exists because M is arc-wise connected). The extendible cocycle $f^{1,n-1}$ of bidegree $(1, n-1)$ in the total complex of the double complex $A^{*,*}(M, \mathfrak{U}; V)$ will be given by

$$f^{1,n-2}(\mathbf{m}, \mathbf{m}') = \mathcal{F}_{\varrho t}(\omega)^{n-1,1}(\hat{\sigma}_{n-1}(\mathbf{m}'))([p\hat{\sigma}_1(\mathbf{m})]^{-1})$$
$$= \mathcal{F}_{\varrho}(\omega)^{1,n-1}([\hat{\sigma}_1(\mathbf{m})]^{-1})(\hat{\sigma}_{n-1}(\mathbf{m}'))$$
$$= \int_{[p\hat{\sigma}_1(\mathbf{m})]^{-1}.\hat{\sigma}_{n-1}(\mathbf{m}')} \omega$$

for all points $\mathbf{m} \in M^2$ and $\mathbf{m}' \in SS(M, \mathfrak{U})([n-1])$. Straight from these definitions we derive:

Lemma 16.3.1. *The cochains $\psi^{0,1}, \psi^{1,n-1}$ and $f^{1,n-1}$ in $A^{*,*}(M, \mathfrak{U}; V)^G$ satisfy the equation $f^{1,n-1} = d_h\psi^{0,1} + d_v\psi^{1,n-1}$.*

Proof. This follows from Stokes Theorem. The evaluation of the cochain $f^{1,n-1}$ at a point $(\mathbf{m}, \mathbf{m}') \in M^2 \times SS(M, \mathfrak{U})([n-1])$ is given by

$$f^{1,n-1}(\mathbf{m}, \mathbf{m}') = \int_{[p\hat{\sigma}_1(\mathbf{m})]^{-1}.\hat{\sigma}_{n-1}(\mathbf{m}')} \omega$$
$$= \int_{[p\hat{\sigma}_1(\mathbf{m})]^{-1}.\hat{\sigma}_{n-1}(\mathbf{m}')} d\eta$$
$$= \int_{\partial([p\hat{\sigma}_1(\mathbf{m})]^{-1}.\hat{\sigma}_{n-1}(\mathbf{m}'))} \eta$$
$$= \int_{[\partial p\hat{\sigma}_1(\mathbf{m})]^{-1}.\hat{\sigma}_{n-1}(\mathbf{m}')} \eta - \int_{[p\hat{\sigma}_1(\mathbf{m})]^{-1}.\partial\hat{\sigma}_{n-1}(\mathbf{m}')} \eta$$
$$= d_h\psi^{0,n-1}(\mathbf{m}, \mathbf{m}') + d_v\psi^{0,n-1}(\mathbf{m}, \mathbf{m}').$$

This proves the assertion. $\qquad\square$

The proof that this defines a cocycle of bidegree $(1, n-1)$ in the total complex of $A^{*,*}(M, \mathfrak{U}; V)^G$ which is extendible to a cocycle of bidegree $(1, n-1)$ in the total complex of the double complex $A^{*,*}(M; V)^G$ will occupy a large part of this section. At first we prove that the cochain $f^{1,n-1}$ is a cocycle in the total complex $\mathrm{Tot}A^{*,*}(M, \mathfrak{U}; V)$ provided some flux components (to be specified later) are trivial on singular cycles. Thereafter we show that this cocycle is extendible to a global cocycle of bidegree $(1, n-1)$ in the total complex $\mathrm{Tot}A^{*,*}(M; V)^G$. We begin with some preparatory observations:

Recall that the groups $A^{p,q}(M, \mathfrak{U}; V)^G$ of the double complex $A^{*,*}(M, \mathfrak{U}; V)^G$ were defined to be the groups of equivariant functions

$$A^{p,q}(M, \mathfrak{U}; V)^G = \{f : M^{p+1} \times SS(M, \mathfrak{U})[q]) \to V\}^G.$$

The switching of arguments yields equivariant functions $SS(M, \mathfrak{U})([q]) \times M^{p+1} \to V$. We let $\chi^{p,q}$ denote the argument switching operator, which is defined via

$$\chi^{p,q} : A^{p,q}(M, \mathfrak{U}; V) \to \{f : SS(M, \mathfrak{U})[n]) \times M^{p+1} \to V\}.$$

Let $c^{p,q}$ be a cochain of bidegree (p, q) in the double complex $A^{*,*}(M, \mathfrak{U}; V)$. The adjoint function of $\chi^{p,q}(c^{p,q})$ is an equivariant $\mathfrak{U}$-local q-cochain with values in the group $A^p(M; V)$ of global p-cochains on M. The composition of the argument switching $\chi^{p,q}$ of and taking adjoints is an isomorphism

$$\chi^{p,q} : A^{p,q}(M, \mathfrak{U}; V)^G \to A^q(M, \mathfrak{U}; A^p(M; V))^G,$$
$$\chi^{p,q}(f^{p,q})(\mathbf{m}')(\mathbf{m}) = f^{p,q}(\mathbf{m}, \mathbf{m}')$$

of Abelian groups. Modulo sign, these switchings $\chi^{p,q}$ of arguments translate the horizontal resp. vertical coboundary operators on the double complex $A^{*,*}(M,\mathfrak{U};V)^G$ into the vertical resp. horizontal coboundary operators on the double complex $A^*(M,\mathfrak{U};A^*(M;V))^G$. So we observe:

Lemma 16.3.2. *The cocycle $f^{1,n-1}$ in $A^{1,n-1}(M,\mathfrak{U};V)^G$ is extendible to a cocycle in $A^{1,n-1}(M;V)^G$ if and only if $\chi^{1,n-1}(f^{1,n-1})$ in $A^{n-1}(M;\mathfrak{U};A^1(M;V))^G$ is extendible to a cocycle in $A^{n-1}(M;A^1(M;V))^G$.*

As a consequence the extendibility of the of the cocycle $f^{1,n-1}$ to a cocycle of bidegree $(1,n-1)$ in $\mathrm{Tot}A^{*,*}(M;V)$ can be described in terms of the extendibility of an equivariant $\mathfrak{U}$-local cocycle which is the local integral of a closed equivariant differential $(n-1)$-form on M. The construction is a follows: Consider the argument switching on the differential n-forms of type (p,q) on the bundle $\mathrm{pr}_1 : G \times M \to G$, i.e. on the vector space of smooth functions

$$\omega^{p,q} : T^{\oplus p}G \times T^{\oplus q}M \to V$$

which are differential forms if either one of the two arguments is fixed (cf. Section 13.3). We also denote the argument switching on differential n-forms of type (p,q) by $\chi^{p,q}$. Here as well the composition of the argument switching $\chi^{p,q}$ of and taking adjoints is a linear map

$$\chi^{p,q} : \Omega^{p,q}(\mathrm{pr}_1 : G \times M \to G)^G \to \Omega^q(M;{}^t\Omega^p(G;V))^G,$$
$$\chi^{p,q}(\omega^{p,q})(\mathbf{X}')(\mathbf{X}) = \omega^{p,q}(\mathbf{X},\mathbf{X}')$$

of vector spaces and these switchings $\chi^{p,q}$ of arguments translate the horizontal resp. vertical coboundary operators on the double complex $\Omega^{*,*}(G \times M \to G)^G$ into the vertical resp. horizontal coboundary operators on the double complex $\Omega^*(M;{}^t\Omega^*(G;V))^G$ modulo sign. We apply this operation to the pullback $\varrho^*\omega$ of the differential n-form ω under the action ϱ. It is a differential n-form on $G \times M$ which decomposes into its components of bidegree (p,q) under the homomorphism

$$\Omega^n(G \times M;V)^G \to \bigoplus_{p+q=n} \Omega^p(G;{}^t\Omega^q(M;V))^G, \quad \omega' \mapsto \sum_{p+q=n} \omega'^{p,q}$$

(where the factor M in the product $G \times M$ is regarded as a trivial G-space). The cochain $\chi^{1,n-1}(f^{1,n-1})$ is obtained from the differential form $\chi^{1,n-1}((\varrho^*\omega)^{1,n-1})$ via integration over smooth singular simplices of the form $\hat{\sigma}_1(\mathbf{m})$ and $\hat{\sigma}_{n-1}(\mathbf{m}')$:

$$[\chi^{1,n-1}(f^{1,n-1})](\mathbf{m}')(\mathbf{m}) = f^{1,n-1}(\mathbf{m})(\mathbf{m}') = \int_{[p\sigma_1(\mathbf{m})]^{-1}.\sigma_{n-1}(\mathbf{m}')} \omega$$

$$= \int_{[p\sigma_1(\mathbf{m})]^{-1}} \left[\int_{\sigma_{n-1}(\mathbf{m}')} \chi^{1,n-1}((\varrho^*\omega)^{1,n-1})\right]$$

$$= \int_{[p\sigma_1(\mathbf{m})]^{-1}} \left[s^{n-1}\left(\chi^{1,n-1}((\varrho^*\omega)^{1,n-1})\right)(\mathbf{m}')\right],$$

where $s^{n-1} : \Omega^{n-1}(M;{}^t\Omega^1(G;V)) \to A_s^1(M,\mathfrak{U};{}^t\Omega^1(G;V)$ is the equivariant homomorphism which integrates every $(n-1)$-form on M to a smooth $\mathfrak{U}$-local cochain and every closed $(n-1)$-form to a $\mathfrak{U}$-local $(n-1)$-cocycle. If the restrictions of the flux components $\mathcal{F}_{\varrho t}(\omega)^{n-1,1}$ and $\mathcal{F}_{\varrho t}(\omega)^{n-1,1}$ to smooth singular cycles are trivial, then the cochain $f^{1,n-1}$ is a cocycle. To prove this, we first examine the effect of vanishing fluxes on the differential q-forms $\chi^{p,q}((\varrho^*\omega)^{p,q})$ on M.

Lemma 16.3.3. *If the restriction of the flux component $\mathcal{F}_{\varrho t}(\omega)^{q-1,p+1}$ to singular $(q-1)$-cycles is trivial then the vertical coboundary of $\chi^{p,q}((\varrho^*\omega)^{p,q})$ in the double complex $\Omega^*(M;{}^t\Omega^*(G;V))^G$ is trivial as well.*

Proof. The vertical coboundary of $\chi^{p,q}((\varrho^*\omega)^{p,q})$ is a q-form on M with values in $(p+1)$-forms on G. It is trivial if and only if its integration over arbitrary smooth singular q-simplices is trivial. Its integral $\int_\tau d_v\chi^{p,q}((\varrho^*\omega)^{p,q})$ over a smooth singular q-simplex τ is a differential $(p+1)$-form, which is trivial if and only if its integration over any smooth singular $(p+1)$-simplex σ is zero. The integration of the $(p+1)$-form $\int_\tau d_v\chi^{p,q}((\varrho^*\omega)^{q,p})$ over a smooth singular $(p+1)$-simplex σ computes to

$$\int_\sigma \left[\int_\tau [d_v\chi^{p,q}((\varrho^*\omega)^{p,q})]\right] = \int_\sigma d\left[\int_\tau \chi^{p,q}((\varrho^*\omega)^{p,q})\right]$$

$$= \int_{\partial\sigma}\left[\int_\tau \chi^{p,q}((\varrho^*\omega)^{p,q})\right] = \int_{(\partial\sigma).\tau}\omega$$

$$= 0 - (-1)^p\int_{\sigma.(\partial\tau)}\omega$$

$$= -(-1)^p\int_\sigma \mathcal{F}_{\varrho t}(\omega)^{q-1,p+1}(\partial\tau)$$

The boundary $\partial\tau$ is a smooth singular $(q-1)$-cycle. The assumption that the restriction of the flux component $\mathcal{F}_{\varrho t}(\omega)^{q-1,p+1}$ to singular q-cycles is trivial implies the triviality of the p-form $\mathcal{F}_{\varrho t}(\omega)^{q,p}(\partial\sigma)$ which in turn implies the triviality of the above integrals, hence the vertical coboundary of $\chi^{p,q}((\varrho^*\omega)^{p,q})$ is trivial. □

Lemma 16.3.4. *If the restriction of the flux component $\mathcal{F}_{\varrho t}(\omega)^{q,p}$ to singular q-cycles is trivial then the differential q-form $\chi^{p,q}((\varrho^*\omega)^{p,q})$ on M is closed, i.e. the horizontal coboundary of $\chi^{p,q}((\varrho^*\omega)^{q,p})$ in the double complex $\Omega^*(M;{}^t\Omega^*(G;V))^G$ is trivial.*

Proof. Let the restriction $\mathcal{F}_{\varrho t}(\omega)^{q,p}_{Z_{q,\infty}(M)}: Z_{q,\infty}(M) \to Z^p_{dR}(G;V)$ be trivial. The differential q-form $\chi^{p,q}((\varrho^*\omega)^{q,p})$ is closed if its exterior derivative is trivial, which in turn happens if and only if the integration of this derivative over arbitrary smooth singular $(q+1)$-simplices is trivial. The integration of the exterior derivative over a smooth singular $(q+1)$-simplex τ is a differential p-form on M. This p-form is trivial if and only if its integrals over all smooth singular p-simplices are trivial. The integration of the p-form $\int_\tau \chi^{p,1}((\varrho^*\omega)^{q,p})$ over a smooth singular p-simplex σ computes to

$$\int_\sigma \left[\int_\tau [d_h\chi^{p,1}((\varrho^*\omega)^{p,q})]\right] = \int_\sigma \left[\int_{\partial\tau} \chi^{p,q}((\varrho^*\omega)^{p,q})\right] = \int_{\sigma.(\partial\tau)}\omega$$

$$= \int_\sigma \mathcal{F}_{\varrho t}(\omega)^{p,q}(\partial\tau)$$

The boundary $\partial\tau$ is a smooth singular q-cycle. The assumption that the restriction of the flux component $\mathcal{F}_{\varrho t}(\omega)^{q,p}$ to singular q-cycles is trivial implies the triviality of the p-form $\mathcal{F}_{\varrho t}(\omega)^{q,p}(\partial\sigma)$ which in turn implies the triviality of the above integrals, hence the vertical coboundary of $\chi^{p,q}((\varrho^*\omega)^{p,q})$ is trivial. □

Proposition 16.3.5. *If the restrictions of the flux components $\mathcal{F}_{\varrho t}(\omega)^{q-1,p+1}$ and $\mathcal{F}_{\varrho t}(\omega)^{q,p}$ to smooth singular cycles are trivial then the q-form $\chi^{p,q}((\varrho^*\omega)^{p,q})$ on M is a cocycle of bidegree (q,p) in the double complex $\Omega^*(M;{}^t\Omega^*(G;V))^G$.*

Proof. This is a consequence of Lemma 16.3.3 and Lemma 16.3.4. □

Recall the cochain $f^{1,n-1}$ of bidegree $(1,n-1)$ in the double complex $A^{*,*}(M,\mathfrak{U};V)^G$ we have defined earlier. The important properties of the cochain $\chi^{p,q}((\varrho^*\omega)^{p,q})$ in the double complex $\Omega^*(M;{}^t\Omega^*(G;V))^G$ carry over to the cochain $f^{1,n-1}$:

Lemma 16.3.6. *If the horizontal coboundary of $\chi^{1,n-1}((\varrho^*\omega)^{1,n-1})$ is trivial, then the horizontal coboundary of $\chi^{1,n-1}(f^{1,n-1})$ is trivial as well.*

Proof. This follows from the construction of the cochain $\chi^{1,n-1}(f^{1,n-1})$ by integrating the differential $(n-1)$-form $\chi^{1,n-1}((\varrho^*\omega)^{1,n-1})$. Let the horizontal coboundary of $\chi^{1,n-1}((\varrho^*\omega)^{1,n-1})$ be trivial. The evaluation of the horizontal coboundary of the cochain $\chi^{1,n-1}(f^{1,n-1})$ at a point $(\mathbf{m}', \mathbf{m}) \in \mathrm{SS}(M, \mathfrak{U})([n]) \times M^2$ is given by

$$d_h \chi^{1,n-1}(f^{1,n-1})(\mathbf{m}', \mathbf{m}) =$$

$$\sum (-1)^i \chi^{1,n-1}(f^{1,n-1})(m'_0, \ldots, \hat{m}'_i, \ldots, m'_n, \mathbf{m})$$

$$= \sum (-1)^i \int_{[p\sigma_1(\mathbf{m})]^{-1}} \left[\int_{\sigma_{n-1}(m'_0,\ldots,\hat{m}'_i,\ldots,m'_n)} \chi^{1,n-1}((\varrho^*\omega)^{1,n-1}) \right]$$

$$= \int_{[p\sigma_1(\mathbf{m})]^{-1}} \left[\int_{\partial\sigma_n(\mathbf{m}')} \chi^{1,n-1}((\varrho^*\omega)^{1,n-1}) \right]$$

$$= \int_{[p\sigma_1(\mathbf{m})]^{-1}} \left[\int_{\sigma_n(\mathbf{m}')} d_h \chi^{1,n-1}((\varrho^*\omega)^{1,n-1}) \right] = 0$$

This proves the assertion. □

Lemma 16.3.7. *If the manifold M is simply connected and the vertical coboundary of $\chi^{1,n-1}((\varrho^*\omega)^{1,n-1})$ is trivial, then the vertical coboundary of $\chi^{1,n-1}(f^{1,n-1})$ is trivial as well.*

Proof. Let the vertical coboundary of $\chi^{1,n-1}((\varrho^*\omega)^{1,n-1})$ be trivial. The evaluation of the vertical coboundary of the cochain $\chi^{1,n-1}(f^{1,n-1})$ at a point $(\mathbf{m}', \mathbf{m})$ in $\mathrm{SS}(M, \mathfrak{U})([n-1]) \times M^3$ is given by

$$d_v \chi^{1,n-1}(f^{1,n-1})(\mathbf{m}', \mathbf{m}) =$$

$$\chi^{1,n-1}(f^{1,n-1})(\mathbf{m}', m_1, m_2) - \chi^{1,n-1}(f^{1,n-1})(\mathbf{m}', m_0, m_2)$$

$$+\chi^{1,n-1}(f^{1,n-1})(\mathbf{m}', m_1, m_2)$$

$$= \int_{[p\sigma_1(m_1,m_2)]^{-1} - [p\sigma_1(m_0,m_2)]^{-1} + [p\sigma_1(m_0,m_1)]^{-1}} \left[\int_{\sigma_{n-1}(\mathbf{m}')} \chi^{1,n-1}((\varrho^*\omega)^{1,n-1}) \right]$$

The singular cochain $\sigma_1(m_1, m_2) - \sigma_1(m_0, m_2) + \sigma_1(m_0, m_1)$ is a smooth singular 1-cycle in M. If the manifold M is simply connected, then it is the boundary of a smooth singular 2-cochain c. In this case the last expression simplifies to

$$\int_{[p\sigma_1(m_1,m_2)]^{-1} - [p\sigma_1(m_0,m_2)]^{-1} + [p\sigma_1(m_0,m_1)]^{-1}} \left[\int_{\sigma_{n-1}(\mathbf{m}')} \chi^{1,n-1}((\varrho^*\omega)^{1,n-1}) \right] =$$

$$\int_{[p\partial c]^{-1}} \left[\int_{\sigma_{n-1}(\mathbf{m}')} \chi^{1,n-1}((\varrho^*\omega)^{1,n-1}) \right]$$

$$= \int_{\partial[pc]^{-1}} \left[\int_{\sigma_{n-1}(\mathbf{m}')} \chi^{1,n-1}((\varrho^*\omega)^{1,n-1}) \right]$$

$$= \int_{[pc]^{-1}} \left[\int_{\sigma_{n-1}(\mathbf{m}')} d_v \chi^{1,n-1}((\varrho^*\omega)^{1,n-1}) \right] = 0$$

This proves the assertion. □

Proposition 16.3.8. *If the manifold M is simply connected and the restrictions of the flux components $\mathcal{F}_{\varrho t}(\omega)^{n-2,2}$ and $\mathcal{F}_{\varrho t}(\omega)^{n-1,1}$ to smooth singular cycles are trivial, then the cochain $f^{1,n-1}$ is a cocycle in the total complex $\mathrm{Tot}A^{*,*}(M, \mathfrak{U}; V)$.*

Proof. If the restrictions of the flux components $\mathcal{F}_{\varrho t}(\omega)^{n-2,2}$ and $\mathcal{F}_{\varrho t}(\omega)^{n-1,1}$ to smooth singular cycles are trivial, then the form $\chi^{1,n-1}((\varrho^*\omega)^{1,n-1})$ is a cocycle of bidegree $(n-1,1)$ in the double complex $\Omega^*(M; {}^t\Omega^*(G;V))^G$ by Proposition 16.3.5. In this case the preceding two Lemmata show that the cochain $f^{1,n-1}$ is a cocycle in the total complex of the double complex $A^{*,*}(M,\mathfrak{U};V)$.   □

So far we have shown that the triviality of the restriction of the flux components $\mathcal{F}_{\varrho t}(\omega)^{n-1,1}$ and $\mathcal{F}_{\varrho}(\omega)^{n-2,2}$ to singular cycles ensures that the cochain $f^{1,n-1}$ is a cocycle in the total complex of the double complex $A^{*,*}(M,\mathfrak{U};V))^G$. It remains to show that this cocycle is then extendable to a cocycle in $A^{1,n-1}(M,\mathfrak{U};V)^G$. For this purpose we apply our knowledge on the integrability of equivariant differential forms obtained in the last section to the differential $(n-1)$-form $\chi^{p,q}((\varrho^*\omega)^{p,q})$ on the manifold M.

Lemma 16.3.9. *If the restricted flux component $\mathcal{F}_{\varrho t}(\omega)^{q,p}_{Z_{q,\infty}(M)}$ is trivial, then the integration of the differential q-form $\chi^{p,q}((\varrho^*\omega)^{p,q})$ over smooth singular q-simplices is trivial.*

Proof. Let the restricted flux component $\mathcal{F}_{\varrho t}(\omega)^{q,p}_{Z_{q,\infty}(M)}$ be trivial. The integration of the closed q-form $\chi^{p,q}((\varrho^*\omega)^{p,q})$ over a smooth singular q-simplex $z \in Z_{q,\infty}(M)$ is a differential p-form on G. This differential p-form is trivial if and only if its integration over arbitrary smooth singular p-simplices is trivial. Let z be a smooth singular q-cycle and σ be a smooth singular p-simplex. The equalities

$$\int_\sigma \left[\int_z \chi^{p,q}((\varrho^*\omega)^{p,q})\right] = \int_{\sigma.z} \omega = \int_\sigma \mathcal{F}_{\varrho t}(\omega)^{p,q}(z)$$

show that the integration of the differential q-form $\chi^{p,q}((\varrho^*\omega)^{p,q})$ over a smooth singular q-cycle $z \in Z_{q,\infty}(M)$ is trivial.   □

If If the restrictions of the flux components $\mathcal{F}_{\varrho t}(\omega)^{q-1,p+1}$ and $\mathcal{F}_{\varrho t}(\omega)^{q,p}$ to smooth singular cycles are trivial, then the q-form $\chi^{p,q}((\varrho^*\omega)^{p,q})$ takes values in the topological vector space ${}^tZ^p_{dR}(G;V)$ of closed differential p-forms on M (by Lemma 16.3.3). Moreover we note:

Proposition 16.3.10. *If the restrictions of the flux components $\mathcal{F}_{\varrho t}(\omega)^{q-1,p+1}$ and $\mathcal{F}_{\varrho t}(\omega)^{q,p}$ to smooth singular cycles are trivial and the de Rham homomorphism $H^q_{dR}(M; {}^tZ^p_{dR}(G;V)) \to H^q(M; {}^tZ^p_{dR}(M;V))$ is injective, then the closed differential q-form $\chi^{p,q}((\varrho^*\omega)^{p,q})$ on M is exact in the complex $\Omega^*(M; {}^tZ^p_{dR}(M;V))$.*

Proof. This is a consequence of the preceding Lemma.   □

Corollary 16.3.11. *If the manifold M is smoothly paracompact and the restrictions of the flux components $\mathcal{F}_{\varrho t}(\omega)^{q-1,p+1}$ and $\mathcal{F}_{\varrho t}(\omega)^{q,p}$ to smooth singular cycles are trivial, then the closed differential q-form $\chi^{p,q}((\varrho^*\omega)^{p,q})$ on M is exact in the complex $\Omega^*(M; {}^tZ^p_{dR}(M;V))$.*

Consider the equivariant $\mathfrak{U}$-local cocycle $s^{n-1}(\chi^{1,n-1}((\varrho^*\omega)^{1,n-1}))$ which is obtained by integrating $\chi^{1,n-1}((\varrho^*\omega)^{1,n-1})$ over the smooth singular simplices $\hat{\sigma}_{n-1}(\mathbf{m}')$. If the manifold M is $(n-3)$-connected, then the extendibility of the $\mathfrak{U}$-local equivariant cocycle $s^{n-1}(\chi^{1,n-1}((\varrho^*\omega)^{1,n-1}))$ to a global equivariant cocycle can be expressed in terms of the restriction of the flux components $\mathcal{F}_{\varrho t}(\chi^{1,n-1}((\varrho^*\omega)^{1,n-1}))^{q',p'}$ of $\chi^{1,n-1}((\varrho^*\omega)^{1,n-1})$ to smooth singular cycles. The triviality of these restrictions in turn can be ensured by requiring the restrictions of the flux components $\mathcal{F}_{\varrho t}(\omega)^{q,p}$ of the equivariant differential n-form ω on M to be trivial on smooth singular cycles:

Lemma 16.3.12. *If for some $q < n$ the restriction of the flux component $\mathcal{F}_{\varrho t}(\omega)^{q,p}$ to smooth singular q-cycles is trivial, then the restriction of the flux component $\mathcal{F}_{\varrho t}(\chi^{1,n-1}((\varrho^*\omega)^{1,n-1}))^{q,p-1}$ to smooth singular q-cycles is trivial as well.*

Proof. Let the restriction of the flux component $\mathcal{F}_{\varrho t}(\omega)^{q,p}$ to smooth singular q-cycles be trivial for some $q < n$. For each smooth singular q-cycle z in M the $(p-1)$-form $\mathcal{F}_{\varrho t}(\chi^{1,n-1}((\varrho^*\omega)^{1,n-1}))^{q,p-1}(z)$ on G is trivial if and only if for each smooth singular $(p-1)$-simplex τ in G the integral

$$\int_\tau \mathcal{F}_{\varrho t}(\chi^{1,n-1}((\varrho^*\omega)^{1,n-1}))^{q,p-1}(z) = \int_{\tau.z} \chi^{1,n-1}((\varrho^*\omega)^{1,n-1})$$

is trivial. The latter integral is a differential 1-form in ${}^tZ^1_{dR}(G;V)^G$ which is trivial if and only if its integrals over arbitrary smooth singular 1-simplices are zero. Let σ be a smooth singular 1-simplex. The integration of the above differential 1-form on G over σ computes to

$$\int_\sigma \left[\int_{\tau.z} \chi^{1,n-1}((\varrho^*\omega)^{1,n-1}) \right] = \int_{\sigma.(\tau.z)} \omega = \int_{(\sigma.\tau)} \mathcal{F}_{\varrho t}(\omega)^{q,p}(z)$$

The last integrand is trivial by assumption. Therefore the restriction of the flux component $\mathcal{F}_{\varrho t}(\chi^{1,n-1}((\varrho^*\omega)^{1,n-1}))^{q,p-1}$ to smooth singular q-cycles is trivial. $\square$

Proposition 16.3.13. *If the manifold M is $(n-3)$-connected, $n > 3$, the restrictions of the flux components $\mathcal{F}_{\varrho t}(\omega)^{q,p}$ to smooth singular q-cycles are trivial for $q = n-2$ and $q = n-3$ and the de Rham homomorphism homomorphism $H^{n-1}_{dR}(M; {}^tZ^1_{dR}(M;V)) \to H^{n-1}(M; {}^tZ^1_{dR}(M;V))$ is injective, then the cocycle $f^{1,n-1}$ in the total complex $\mathrm{Tot}A^{*,*}(M,\mathfrak{U};V))^G$ is extendible to a global cocycle of bidegree $(1, n-1)$ in $\mathrm{Tot}A^{*,*}(M,\mathfrak{U};V))^G$.*

Proof. Let the manifold M be $(n-3)$-connected and assume $n > 3$. If the restrictions $\mathcal{F}_{\varrho t}(\omega)^{q,p}_{|Z_{q,\infty}(M)}$ of the flux components $\mathcal{F}_{\varrho t}(\omega)^{q,p}$ are trivial for $q = n-2$ and $q = n-1$, then the restrictions of the flux components $\mathcal{F}_{\varrho t}(\chi^{1,n-1}((\varrho^*\omega)^{1,n-1})^{1,n-1}$ and $\mathcal{F}_{\varrho t}(\chi^{1,n-1}((\varrho^*\omega)^{1,n-1})^{2,n-2}$ are trivial by the preceding Lemma. Because the de Rham homomorphism $H^{n-1}_{dR}(M; {}^tZ^1_{dR}(M;V)) \to H^{n-1}(M; {}^tZ^1_{dR}(M;V))$ is injective and the manifold M is $(n-1)-2$-connected Theorem 16.2.23 ensures that the $\mathfrak{U}$-local integral $s^{n-1}(\chi^{1,n-1}((\varrho^*\omega)^{1,n-1})$ of the exact equivariant $(n-1)$-form $\chi^{1,n-1}((\varrho^*\omega)^{1,n-1}$ can be extended to a cocycle H in $Z^{n-1}(M; Z^1_{dR}(G;V))^G$. Then the function

$$F : \mathrm{SS}(M,\mathfrak{U})([n-1]) \times M^2 \to V, \quad (\mathbf{m}',\mathbf{m}) \mapsto \int_{\sigma_1(\mathbf{m})} H(\mathbf{m}')$$

is an extension of $\chi^{1,n-1}(f^{1,n-2})$ to a global cocycle of bidegree $(n-1,1)$ in the total complex of the double complex $A^*(M; A^*(M;V))^G$. By Lemma 16.3.2 this implies the extendibility of $f^{1,n-1}$ to a cocycle of bidegree $(1, n-1)$ in the total complex of the double complex $A^{*,*}(M;V)^G$ $\square$

Corollary 16.3.14. *If the restrictions of the flux components $\mathcal{F}_{\varrho t}(\omega)^{n-2,2}$ and $\mathcal{F}_{\varrho t}(\omega)^{n-1,1}$ to smooth singular cycles are trivial, then the cocycle $F^{1,n-1} = d_v\psi^{1,n-1}$ is extendible to a cocycle in the total complex of $A^{*,*}(M;V)^G$ if and only if the cocycle $d_v\psi^{1,n-2} = F^{1,n-1} - f^{1,n-1}$ is extendible to a cocycle in $\mathrm{Tot}A^{*,*}(M;V)^G$.*

This observation allows us to replace the horizontal coboundary $F^{1,n-1}$ by the horizontal coboundary $F^{2,n-2} = d_h\psi^{1,n-2}$ of the cochain $\psi^{1,n-2}$ of bidegree $(1, n-2)$ when considering the extendibility of the $\mathfrak{U}$-local cocycle $s^n(\omega)$ to a global equivariant cocycle.

Lemma 16.3.15. *If the manifold M is simply connected and the restricted flux component $\mathcal{F}_{\varrho t}(\omega)^{n-2,2}_{|Z_{n-2,\infty}(M)}$ is trivial, then the vertical coboundary of $d_h\psi^{2,n-2}$ is trivial.*

Proof. Assume the restriction of the flux component $\mathcal{F}_{\varrho t}(\omega)^{n-2,2}$ to smooth singular $(n-2)$-cycles to be trivial. The evaluation of the vertical coboundary of $d_h\psi^{2,n-2}$ at a point $(\mathbf{m},\mathbf{m}') \in M^3 \times \mathrm{SS}(M,\mathfrak{U})([n-1])$ is given by

$$d_h d_v \psi^{1,n-2}(\mathbf{m},\mathbf{m}') = \int_{[p\sigma_1(\mathbf{m})^{-1}.\partial\sigma_{n-1}(m_1,m_2)} \eta - \int_{[p\sigma_1(m_0,m_1)]^{-1}.\partial\sigma_{n-1}(\mathbf{m}')} \eta$$
$$+ \int_{[p\sigma_1(m_0,m_1)]^{-1}.\partial\sigma_{n-1}(\mathbf{m}')} \eta\,.$$

The singular 1-chain $[p\sigma_1(m_1,m_2)]^{-1} - [p\sigma_1(m_0,m_2)]^{-1} + [p\sigma_1(m_0,m_1)]^{-1}$ is a piecewise smooth loop in G. If the manifold M is simply connected, then this singular 1-cycle is the boundary of a smooth singular 2-cochain c. The above evaluation can so be rewritten as

$$d_v d_h \psi^{1,n-2}(\mathbf{m},\mathbf{m}') = \int_{(\partial c).\partial\sigma_{n-1}(\mathbf{m}')} \eta = \int_{\partial(c.\partial\sigma_{n-1}(\mathbf{m}'))} \eta$$
$$= \int_{c.\partial\sigma_{n-1}(\mathbf{m}')} \omega = \int_c \mathcal{F}_{\varrho t}(\omega)^{n-2,2}(\partial\sigma_{n-1}(\mathbf{m}'))\,.$$

The assumption that the restriction of the flux component $\mathcal{F}_{\varrho t}(\omega)^{n-2,2}$ to smooth singular $(n-2)$-cycles is trivial implies the triviality of the above integrals, hence the vertical coboundary of $d_v\psi^{1,n-2}$ is trivial in this case. $\square$

Since the cochain $F^{2,n-2} := d_h\psi^{1,n-2}$ is itself a horizontal coboundary, the triviality of the restriction of the flux component $\mathcal{F}_{\varrho t}(\omega)^{2,n-2}$ to smooth singular $(n-2)$-cycles implies that $F^{2,n-2}$ is a cocycle in the total complex $\mathrm{Tot}A^{*,*}(M,\mathfrak{U};V)^G$.

Proposition 16.3.16. *If the manifold M is simply connected, the restrictions of the flux components $\mathcal{F}_{\varrho t}(\omega)^{n-2,2}$ and $\mathcal{F}_{\varrho t}(\omega)^{n-1,1}$ to smooth singular cycles are trivial, then the closed equivariant n-form ω on M is extendible to an equivariant smooth global n-cocycle if the cocycle $F^{2,n-2} = d_h\psi^{1,n-2}$ is extendible to a cocycle of bidegree $(2,n-2)$.*

Proof. The differential form ω is extendible to a smooth global n-cocycle if and only if the cocycle $i^n s^n(\omega)$ is cohomologous to a cocycle of bidegree $(n,0)$ in the total complex of $A^{*,*}(M;\mathfrak{U};V)^G$. This is the case if and only if the cocycle $d_h\psi^{0,n-1}$ is cohomologous to a cocycle of bidegree $(n,0)$. Since the cocycle $f^{1,n-1}$ is extendible to a global cocycle under the assumption made, it is cohomologous to a cocycle of bidegree $(n,0)$ and may be subtracted from $d_h\psi^{0,n-1}$ without changing whether the resulting cocycle is cohomologous to a cocycle of bidegree $(n,0)$ or not. Thus the cocycle $d_h\psi^{0,n-1}$ is cohomologous to a cocycle of bidegree $(n,0)$ if and only if the cocycle $d_v\psi^{1,n-2} = f^{1,n-1} - d_h\psi^{0,n-1}$ is cohomologous to a cocycle of bidegree $(n,0)$. This is equivalent to $d_h\psi^{1,n-2}$ being cohomologous to a cocycle of bidegree $(n,0)$. $\square$

Recall that we have assumed the equivariant n-form ω to be exact at the beginning of this section. Collecting all the preceding results we can now prove:

Theorem 16.3.17. *If ω is a an exact equivariant n-form on an $(n-3)$-connected G-manifold M, $n \geq 4$, the restrictions of the flux components $\mathcal{F}_{\varrho t}(\omega)^{n-2,2}$ and $\mathcal{F}_{\varrho t}(\omega)^{n-1,1}$ to smooth singular cycles are trivial and the de Rham homomorphism $H_{dR}^{n-1}(M;{}^t Z_{dR}^1(M;V)) \to H^{n-1}(M;{}^t Z_s^1(M;V))$ is injective, then ω is extendible to an equivariant smooth global cocycle.*

Proof. Under the assumptions made the equivariant closed n-form ω is extendible to a smooth equivariant global n-cocycle if the cocycle $F^{2,n-2} = d_h\psi^{1,n-2}$ of bidegree $(2, n-2)$ is cohomologous to a cocycle of bidegree $(n,0)$ in the total complex of the double complex $A^{*,*}(M, \mathfrak{U}; V)^G$. Consider the adjoint cocycle $\hat{F}^{2,n-2}$ of $F^{2,n-2}$. It induces a 2-cocycle $q\hat{F}^{2,n-2} : M^3 \rightarrow H^{n-2}(M, \mathfrak{U}; V)$, where $q : Z^{n-2}(M, \mathfrak{U}; V) \rightarrow H^{n-2}(M, \mathfrak{U}; V)$ denotes the quotient map. If the 2-cocycle $q\hat{F}^{2,n-2} : M^3 \rightarrow H^{n-2}(M, \mathfrak{U}; V)$ is trivial, then the cocycle $F^{2,n-2}$ is the vertical coboundary of a cochain $\psi^{2,n-3}$ and thus cohomologous to the cocycle $F^{3,n-3} = -d_h\psi^{2,n-3}$ of bidegree $(3, n-3)$ in the total complex $\mathrm{Tot}A^{*,*}(M, \mathfrak{U}; V)^G$ by Lemma 15.3.23. In this case the $(n-3)$-connectedness of M ensures that the cocycle $F^{3,n-3}$ is cohomologous to a cocycle $F^{n,0}$ of bidegree $(n,0)$ in the total complex $\mathrm{Tot}A^{*,*}(M, \mathfrak{U}; V)^G$ by Corollary 15.3.28. As a consequence the cocycle $i^n s^n(\omega)$ is then cohomologous to $F^{n,0}$, hence the $\mathfrak{U}$-local cocycle $s^n(\omega)$ then is extendible to a global equivariant n-cocycle on M. To prove the triviality of the 2-cocycle $q\hat{F}^{2,n-2} : M^3 \rightarrow H^{n-2}(M, \mathfrak{U}; V)$ we examine the singular $(n-2)$-cocycle $\tilde{\lambda}_\mathfrak{U}^{n-2}(\hat{F}^{2,n-2}(\mathbf{m}))$ for each point $\mathbf{m} \in M^3$. The evaluation of this singular $(n-2)$-cochain at a singular $(n-2)$-cycle $z \in Z_{n-2}(M)$ computes to

$$\int_z \hat{F}^{2,n-2}(\mathbf{m}) = \int_z [d_h\hat{\psi}^{1,n-2}(\mathbf{m})]$$

$$= \int_{[p\hat{\sigma}_1(m_1,m_2)]^{-1}\cdot[\hat{\sigma}_{n-2}\lambda_{n-2}(z)]} \eta$$

$$- \int_{[p\hat{\sigma}_1(m_0,m_2)]^{-1}\cdot[\hat{\sigma}_{n-2}\lambda_{n-2}(z)]} \eta$$

$$+ \int_{[p\hat{\sigma}_1(m_1,m_2)]^{-1}\cdot[\hat{\sigma}_{n-2}\lambda_{n-2}(z)]} \eta$$

$$= \int_{\partial[p\hat{\sigma}_2(\mathbf{m})]^{-1}\cdot[\hat{\sigma}_{n-2}\lambda_{n-2}(z)]} \eta$$

$$= \int_{\partial([p\hat{\sigma}_2(\mathbf{m})]^{-1}\cdot[\hat{\sigma}_{n-2}\lambda_{n-2}(z)])} \eta$$

$$= \int_{[p\hat{\sigma}_2(\mathbf{m})]^{-1}\cdot[\hat{\sigma}_{n-2}\lambda_{n-2}(z)]} \omega = \int_{[p\hat{\sigma}_2(\mathbf{m})]^{-1}} \mathcal{F}_{\varrho t}(\omega)^{n-2}(\hat{\sigma}_{n-2}\lambda_{n-2}(z)) = 0.$$

Because the natural homomorphism $H^{n-2}(M, \mathfrak{U}; V) \rightarrow H^{n-2}(M; V)$ into the singular cohomology group $H^n(M; V)$ is injective, this forces the 1-cocycle $q\hat{F}^{2,n-2} : M^3 \rightarrow H^{n-2}(M, \mathfrak{U}; V)$ to be trivial. Thus $s^n(\omega)$ is extendible to a global equivariant cocycle. $\qquad\square$

Corollary 16.3.18. *If the manifold M is $(n-3)$-connected, $n \geq 4$ and smoothly paracompact, then every closed equivariant differential n-form ω on M is integrable to an equivariant locally smooth global n-cocycle provided that the restricted flux homomorphisms*

$$\mathcal{F}_{\varrho t}(\omega)^{n,0}_{|Z_{n,\infty}(M)} : Z_{n,\infty}(M) \rightarrow Z^0_{dR}(G; V)^G \cong V^G$$

$$\mathcal{F}_{\varrho t}(\omega)^{n-1,1}_{|Z_{n-1,\infty}(M)} : Z_{n-1,\infty}(M) \rightarrow Z^1_{dR}(G; V)^G \cong Z^1_c(\mathfrak{g}; V)$$

$$\mathcal{F}_{\varrho t}(\omega)^{n-2,2}_{|Z_{n-2,\infty}(M)} : Z_{n-2,\infty}(M) \rightarrow Z^2_{dR}(G; V)^G \cong Z^2_c(\mathfrak{g}; V)$$

are trivial.

Example 16.3.19. If G is an $(n-3)$-connected smoothly paracompact Lie group with Lie algebra $\mathfrak{g}$, then every Lie algebra n-cocycle ω is integrable to a locally smooth group cocycle provided that the restricted flux homomorphisms

$$\mathcal{F}_{\varrho t}(\omega_{eq})^{n,0}_{|Z_{n,\infty}(M)} : Z_{n,\infty}(M) \rightarrow Z^0_{dR}(G;V)^G \cong V^G$$

$$\mathcal{F}_{\varrho t}(\omega_{eq})^{n-1,1}_{|Z_{n-1,\infty}(M)} : Z_{n-1,\infty}(M) \rightarrow Z^1_{dR}(G;V)^G \cong Z^1_c(\mathfrak{g};V)$$

$$\mathcal{F}_{\varrho t}(\omega_{eq})^{n-2,2}_{|Z_{n-2,\infty}(M)} : Z_{n-2,\infty}(M) \rightarrow Z^2_{dR}(G;V)^G \cong Z^2_c(\mathfrak{g};V)$$

of the corresponding left invariant differential n-form ω_{eq} on G are trivial.

16.4 $\lfloor \frac{n-1}{2} \rfloor$-Connected Manifolds

In this section we extend the procedure from the last section to derive a more general result on the integrability of equivariant differential n-forms for smooth transformation groups (G, M) whose manifold is $\lfloor \frac{n-1}{2} \rfloor$-connected, where

$$\lfloor r \rfloor := \max\{k \in \mathbb{Z} \mid k \leq r\}$$

is the 'floor-function'. As before we assume the existence of integrating compatible smooth local systems $\hat{\sigma}_k : \mathrm{SS}(M,\mathfrak{U})([k]) \rightarrow C^\infty(\Delta_k, M)$, $k \leq n$ of singular k-simplices for $k \leq n$ and of an equivariant smooth map $p : M \rightarrow G$. We assume the manifold M to be $\lfloor \frac{n-1}{2} \rfloor$-connected for $n \geq 1$, so the smooth local systems $\hat{\sigma}_k : \mathrm{SS}(M,\mathfrak{U})([k]) \rightarrow C^\infty(\Delta_k, M)$, of singular k-simplices can be extended to (non-smooth) systems of smooth singular simplices for $k \leq \lfloor \frac{n-1}{2} \rfloor + 1$, i.e. for each $k \leq \lfloor \frac{n-1}{2} \rfloor + 1$ and point $\mathbf{m} \in M^{k+2}$ there exists a smooth singular k-simplex $\hat{\sigma}_k(\mathbf{m})$ and the faces of these smooth singular simplices are the corresponding simplices in dimension $k-1$. This then implies that for $k \leq \lfloor \frac{n-1}{2} \rfloor + 1$ the equivariant homomorphisms $s^k : \Omega^k(M;V) \rightarrow A^k_s(M;\mathfrak{U};V)$ for extend to equivariant homomorphisms $s^k : \Omega^k(M;V) \rightarrow A^k_{ls}(M;V)$ which also intertwine the coboundary operators. In addition the coefficient space V is required to be Hausdorff and Mackey complete. We wish to prove the following theorem:

Theorem. *If the manifold M is $\lfloor \frac{n-1}{2} \rfloor$-connected, the de Rham homomorphisms $k^q : H^q_{dR}(M;V) \rightarrow H^q(M;V)$ are injective for all $\lfloor \frac{n-1}{2} \rfloor < q < n$, then the $\mathfrak{U}$-local integral $s^n(\omega)$ of an exact equivariant differential n-form ω on M is extendible to a locally smooth equivariant global n-cocycle provided that th restrictions*

$$\mathcal{F}_{\varrho t}(\omega)^{q,p}_{|Z_{q,\infty}(M)} : Z_{q,\infty}(M) \rightarrow Z^p_{dR}(G;V)^G \cong Z^p_c(\mathfrak{g};V)$$

of the flux components are trivial for all $\lfloor \frac{n-1}{2} \rfloor < q < n$.

We will prove this theorem by induction on n. For the cases $n = 1, 2$ and $n = 3$ the assertion has already been proved in theorems 16.2.23 and 16.3.17. This is the induction basis. As induction hypothesis we assume theorem 16.4.12 for all $n' \leq n$. The inductive step will occupy the rest of this section. It is a generalisation of the procedure applied in the last section.

Let ω be an equivariant differential n-form on M which is exact, i.e. there exists a differential $(n-1)$-form η on M with exterior derivative $d\eta = \omega$. We fix this form η throughout this section. With the help of the systems of smooth singular simplices we define for all $p' \leq \lfloor \frac{n-1}{2} \rfloor + 1$ cochains $\psi^{p',q'}$ bidegree (p',q') and of total degree $p' + q' = n - 1$ in $A^{*,*}(M;\mathfrak{U};V)$ by setting

$$\psi^{p',q'}(m_0, m_1, \mathbf{m}') = (-1)^{p'} \mathcal{F}_{\varrho t}(\eta)^{p',q'}(\hat{\sigma}_{q'}(\mathbf{m}'))([p\hat{\sigma}_{p'}(\mathbf{m})]^{-1})$$

$$= (-1)^{p'} \mathcal{F}_\varrho(\eta)^{p',q'}([p\hat{\sigma}_{p'}(\mathbf{m})]^{-1})(\hat{\sigma}_{q'}(\mathbf{m}'))$$

$$= (-1)^{p'} \int_{[p\hat{\sigma}_{p'}(\mathbf{m})]^{-1} \cdot \hat{\sigma}_{q'}(\mathbf{m}')} \eta$$

for all points $(\mathbf{m}, \mathbf{m}') \in M^{p'+1} \times SS(M, \mathfrak{U})([q'])$. In a similar way, but integrating the exact differential form $\omega = d\eta$ we define cochains $f^{p',q'}$ of bidegree (p, q) and of total degree $p + q = n$ in $A^{*,*}(M; \mathfrak{U}; V)$ for all $p \leq \leq \lfloor \frac{n-1}{2} \rfloor + 1$. These are defined to be given by

$$
\begin{aligned}
f^{p,q}(\mathbf{m}, \mathbf{m}') &= \mathcal{F}_{\varrho t}(\omega)^{p,q}(\hat{\sigma}_q(\mathbf{m}'))([p\hat{\sigma}_p(\mathbf{m})]^{-1}) \\
&= \mathcal{F}_{\varrho}(\omega)^{p,q}([\hat{\sigma}_p(\mathbf{m})]^{-1})(\hat{\sigma}_q(\mathbf{m}')) \\
&= \int_{[p\hat{\sigma}_p(\mathbf{m})]^{-1} \cdot \hat{\sigma}_q(\mathbf{m}')} \omega
\end{aligned}
$$

for all points $(\mathbf{m}, \mathbf{m}') \in M^{p+1} \times SS(M, \mathfrak{U})([q])$. We are going to prove that the cochains $f^{p,q}$ so defined are cocycles in the total complex of the double complex $A^{*,*}(M, \mathfrak{U}; V)^G$ provided some flux components satisfy a triviality condition. Analogously to the observation in the last section we note:

Lemma 16.4.1. *The cochains* $\psi^{p-1,q}, \psi^{p,q-1}$ *and* $f^{p,q}$, $p \leq \lfloor \frac{n-1}{2} \rfloor + 1$ *in the double complex* $A^{*,*}(M, \mathfrak{U}; V)^G$ *satisfy the equation* $f^{p,q} = d_h \psi^{p-1,q} + d_v \psi^{p,q-1}$.

Proof. This follows from Stokes Theorem. The evaluation of the cochain $f^{p,q}$ at a point $(\mathbf{m}, \mathbf{m}') \in M^{p+1} \times SS(M, \mathfrak{U})([q])$ is given by

$$
\begin{aligned}
f^{p,q}(\mathbf{m}, \mathbf{m}') &= \int_{[p\hat{\sigma}_p(\mathbf{m})]^{-1} \cdot \hat{\sigma}_q(\mathbf{m}')} \omega \\
&= \int_{[p\hat{\sigma}_p(\mathbf{m})]^{-1} \cdot \hat{\sigma}_q(\mathbf{m}')} d\eta \\
&= \int_{\partial([p\hat{\sigma}_q(\mathbf{m})]^{-1} \cdot \hat{\sigma}_q(\mathbf{m}'))} \eta \\
&= \int_{[\partial p\hat{\sigma}_p(\mathbf{m})]^{-1} \cdot \hat{\sigma}_q(\mathbf{m}')} \eta + (-1)^p \int_{[p\hat{\sigma}_p(\mathbf{m})]^{-1} \cdot \partial\hat{\sigma}_q(\mathbf{m}')} \eta \\
&= d_h \psi^{0,n-1}(\mathbf{m}, \mathbf{m}') + d_v \psi^{0,n-1}(\mathbf{m}, \mathbf{m}').
\end{aligned}
$$

This proves the assertion.   □

Recall that the composition of taking adjoints and the switching of arguments of the cochains of bidegree (p, q) in $A^{*,*}(M, \mathfrak{U}; V)^G$ is an isomorphism

$$
\begin{aligned}
\chi^{p,q} &: A^{p,q}(M, \mathfrak{U}; V)^G \to A^q(M, \mathfrak{U}; A^p(M; V))^G, \\
\chi^{p,q}(f^{p,q})(\mathbf{m}')(\mathbf{m}) &= f^{p,q}(\mathbf{m}, \mathbf{m}')
\end{aligned}
$$

of Abelian groups and that, modulo sign change, these switchings $\chi^{p,q}$ of arguments translate the horizontal resp. vertical coboundary operators on the double complex $A^{*,*}(M, \mathfrak{U}; V)^G$ into the vertical resp. horizontal coboundary operators on the double complex $A^*(M, \mathfrak{U}; A^*(M; V))^G$. Therefore we observe:

Lemma 16.4.2. *The cocycle* $f^{p,q}$, $p \leq \leq \lfloor \frac{n-1}{2} \rfloor + 1$ *in* $A^{p,q}(M, \mathfrak{U}; V)^G$ *is extendible to a cocycle in* $A^{p,q}(M; V)^G$ *if and only if* $\chi^{p,q}(f^{p,q})$ *in* $A^q(M; \mathfrak{U}; A^p(M; V))^G$ *is extendible to a cocycle in* $A^q(M; A^p(M; V))^G$.

The results on the cochain $f^{1,n-1}$ considered in the last section can now be generalised to all the cochains $f^{p,q}$, $p \leq \lfloor \frac{n-1}{2} \rfloor + 1$. For this purpose we recall the linear maps

$$
\begin{aligned}
\chi^{p,q} &: \Omega^{p,q}(\mathrm{pr}_1 : G \times M \to G)^G \to \Omega^q(M; {}^t\Omega^p(G; V))^G, \\
\chi^{p,q}(\omega^{p,q})(\mathbf{X}')(\mathbf{X}) &= \omega^{p,q}(\mathbf{X}, \mathbf{X}')
\end{aligned}
$$

of vector spaces and that these homomorphisms $\chi^{p,q}$ translate the horizontal resp. vertical coboundary operators on the double complex $\Omega^{*,*}(\mathrm{pr}_1 : G \times M \to G)^G$ into the vertical resp. horizontal coboundary operators on the double complex $\Omega^*(M; {}^t\Omega^*(G;V))^G$ modulo sign. The cochain $\chi^{p,q}(f^{p,q})$ for $p \leq \lfloor\frac{n-1}{2}\rfloor + 1$ is obtained from the differential form $\chi^{p,q}((\varrho^*\omega)^{p,q})$ via integration over smooth singular simplices of the form $\hat{\sigma}_p(\mathbf{m})$ and $\hat{\sigma}_q(\mathbf{m}')$:

$$[\chi^{p,q}(f^{p,q})](\mathbf{m}')(\mathbf{m}) = f^{p,q}(\mathbf{m})(\mathbf{m}') = \int_{[p\sigma_p(\mathbf{m})]^{-1}\cdot\sigma_q(\mathbf{m}')} \omega$$

$$= \int_{[p\sigma_p(\mathbf{m})]^{-1}} \left[\int_{\sigma_q(\mathbf{m}')} \chi^{p,q}((\varrho^*\omega)^{p,q}) \right]$$

$$= \int_{[p\sigma_p(\mathbf{m})]^{-1}} \left[s^{n-1}\left(\chi^{p,q}((\varrho^*\omega)^{p,q})\right)(\mathbf{m}') \right],$$

where $s^q : \Omega^q(M; {}^t\Omega^q(G;V)) \to A_s^q(M,\mathfrak{U}; {}^t\Omega^1(G;V)$ is the equivariant homomorphism which integrates every q-form on M to a smooth $\mathfrak{U}$-local cochain and every closed q-form to a $\mathfrak{U}$-local q-cocycle.

Lemma 16.4.3. *If the horizontal coboundary of $\chi^{p,q}((\varrho^*\omega)^{p,q})$ is trivial, then the horizontal coboundary of $\chi^{p,q}(f^{p,q})$ is trivial as well.*

Proof. This follows from the construction of the cochain $\chi^{p,q}(f^{p,q})$ by integrating the differential q-form $\chi^{p,q}((\varrho^*\omega)^{p,q})$. Let the horizontal coboundary of $\chi^{p,q}((\varrho^*\omega)^{p,q})$ be trivial. The evaluation of the horizontal coboundary of the cochain $\chi^{p,q}(f^{p,q})$ at a point $(\mathbf{m}', \mathbf{m}) \in \mathrm{SS}(M,\mathfrak{U})([q+1]) \times M^{p+1}$ is given by

$$d_h\chi^{p,q}(f^{p,q})(\mathbf{m}', \mathbf{m}) =$$

$$\sum(-1)^i\chi^{p,q}(f^{p,q})(m'_0, \ldots, \hat{m}'_i, \ldots, m'_q, \mathbf{m})$$

$$= \int_{[p\sigma_p(\mathbf{m})]^{-1}} \left[\int_{\partial\sigma_{q+1}(\mathbf{m}')} \chi^{p,q}((\varrho^*\omega)^{p,q}) \right]$$

$$= \int_{[p\sigma_p(\mathbf{m})]^{-1}} \left[\int_{\sigma_{q+1}(\mathbf{m}')} d_h\chi^{p,q}((\varrho^*\omega)^{p,q}) \right] = 0$$

This proves the assertion.   □

Lemma 16.4.4. *The vertical coboundary of $\chi^{p,q}((\varrho^*\omega)^{p,q})$, $p \leq \lfloor\frac{n-1}{2}\rfloor+1$, is trivial, then the vertical coboundary of $\chi^{p,q}(f^{p,q})$ is trivial as well.*

Proof. Let the vertical coboundary of the cochain $\chi^{p,q}((\varrho^*\omega)^{p,q})$ be trivial and assume $p \leq \lfloor\frac{n-1}{2}\rfloor$. The evaluation of the vertical coboundary of the cochain $\chi^{p,q}(f^{p,q})$ at a point $(\mathbf{m}', \mathbf{m})$ in $\mathrm{SS}(M,\mathfrak{U})([q]) \times M^{p+2}$ is given by

$$d_v\chi^{p,q}(f^{p,q})(\mathbf{m}', \mathbf{m}) = \int_{[p\partial\sigma_{q+1}(\mathbf{m})]^{-1}} \left[\int_{\sigma_q(\mathbf{m}')} \chi^{p,q}((\varrho^*\omega)^{p,q}) \right]$$

$$= \int_{\partial[p\sigma_{q+1}(\mathbf{m})]^{-1}} \left[\int_{\sigma_q(\mathbf{m}')} \chi^{p,q}((\varrho^*\omega)^{p,q}) \right]$$

$$= \int_{[p\sigma_{q+1}(\mathbf{m})]^{-1}} \left[\int_{\sigma_q(\mathbf{m}')} d_v\chi^{p,q}((\varrho^*\omega)^{p,q}) \right] = 0$$

This proves the assertion.   □

Proposition 16.4.5. *If the manifold M is simply connected and the restrictions of the flux components $\mathcal{F}_{\varrho t}(\omega)^{n-2,2}$ and $\mathcal{F}_{\varrho t}(\omega)^{n-1,1}$ to smooth singular cycles are trivial, then the cochain $f^{1,n-1}$ is a cocycle in the total complex $\mathrm{Tot} A^{*,*}(M,\mathfrak{U};V)$.*

Proof. If the restrictions of the flux components $\mathcal{F}_{\varrho t}(\omega)^{n-2,2}$ and $\mathcal{F}_{\varrho t}(\omega)^{n-1,1}$ to smooth singular cycles are trivial, then the form $\chi^{1,n-1}((\varrho^*\omega)^{1,n-1})$ is a cocycle of bidegree $(n-1,1)$ in the double complex $\Omega^*(M;{}^t\Omega^*(G;V))^G$ by Proposition 16.3.5. In this case the preceding two Lemmata show that the cochain $f^{1,n-1}$ is a cocycle in the total complex of the double complex $A^{*,*}(M,\mathfrak{U};V)$. $\qquad\square$

So far we have shown that the triviality of the restriction of the flux components $\mathcal{F}_{\varrho t}(\omega)^{q,p}$ and $\mathcal{F}_{\varrho}(\omega)^{q-1,p+1}$ to singular cycles ensures that the cochain $f^{p,q}$ is a cocycle in the total complex of the double complex $A^{*,*}(M,\mathfrak{U};V))^G$. It remains to show that this cocycle is then extendable to a cocycle in $A^{p,q}(M,\mathfrak{U};V)^G$. For this purpose we apply the induction hypothesis to the differential q-forms $\chi^{p,q}((\varrho^*\omega)^{p,q})$ on M. The the triviality of the flux of these q-forms is implied by the triviality of the flux of ω:

Lemma 16.4.6. *If for some $p > 0$ the restriction of the flux component $\mathcal{F}_{\varrho t}(\omega)^{q,p}$ to smooth singular q-cycles is trivial, then for all $k \le p$, $k+l = n$ the restriction of the flux component $\mathcal{F}_{\varrho t}(\chi^{k,l}((\varrho^*\omega)^{k,l}))^{q,p-k}$ to smooth singular q-cycles is trivial as well.*

Proof. Let the restriction of the flux component $\mathcal{F}_{\varrho t}(\omega)^{q,p}$ to smooth singular q-cycles be trivial for some $p < n$ and assume $k + l = n$, $k \le p$. For each smooth singular q-cycle z in M the $(p-k)$-form $\mathcal{F}_{\varrho t}(\chi^{k,l}((\varrho^*\omega)^{k,l}))^{q,p-1}(z)$ on G is trivial if and only if for each smooth singular $(p-k)$-simplex τ in G the integral

$$\int_\tau \mathcal{F}_{\varrho t}(\chi^{k,l}((\varrho^*\omega)^{k,l}))^{q,p-k}(z) = \int_{\tau.z} \chi^{k,l}((\varrho^*\omega)^{k,l})$$

is trivial. The latter integral is a differential k-form in ${}^tZ_{dR}^k(G;V)^G$ which is trivial if and only if its integrals over arbitrary smooth singular k-simplices are zero. Let σ be a smooth singular k-simplex. The integration of the above differential k-form on G over σ computes to

$$\int_\sigma \left[\int_{\tau.z} \chi^{k,l}((\varrho^*\omega)^{k,l}) \right] = \int_{\sigma.(\tau.z)} \omega = \int_{(\sigma.\tau)} \mathcal{F}_{\varrho t}(\omega)^{q,p}(z)$$

The last integrand is trivial by assumption. Therefore the restriction of the flux component $\mathcal{F}_{\varrho t}(\chi^{k,l}((\varrho^*\omega)^{k,l}))^{q,p-k}$ to smooth singular q-cycles is trivial. $\qquad\square$

Proposition 16.4.7. *If the restrictions of the flux components $\mathcal{F}_{\varrho t}(\omega)^{q,p}$ to smooth singular cycles are trivial for all $\lfloor \frac{n-1}{2} \rfloor < q < n$ and the de Rham homomorphism homomorphism $H_{dR}^q(M;{}^tZ_{dR}^p(M;V)) \to H^q(M;{}^tZ_{dR}^p(M;V))$ are injective for all $\lfloor \frac{n-1}{2} \rfloor < q < n$, then the cocycles $f^{p,q}$ in the total complex $\mathrm{Tot} A^{*,*}(M,\mathfrak{U};V))^G$ are extendible to a global cocycles of bidegree (p,q) in $\mathrm{Tot} A^{*,*}(M,\mathfrak{U};V))^G$.*

Proof. Let the restrictions $\mathcal{F}_{\varrho t}(\omega)^{q,p}_{|Z_{q,\infty}(M)}$ of the flux components $\mathcal{F}_{\varrho t}(\omega)^{q,p}$ be trivial for for all $\lfloor \frac{n-1}{2} \rfloor < q < n$. And consider a differential q-form $\chi^{p,q}((\varrho^*\omega)^{p,q})$ on M. By the preceding Lemma the restrictions of all the flux components $\mathcal{F}_{\varrho t}(\chi^{p,q}((\varrho^*\omega)^{p,q})^{k,q-k}$, $k \le q$ to singular cycles are trivial. Because the de Rham homomorphisms $H_{dR}^p(M;{}^tZ_{dR}^q(M;V)) \to H^p(M;{}^tZ_{dR}^q(M;V))$ are injective for all $0 < p << \lfloor \frac{n-1}{2} \rfloor$ and the manifold M is $\lfloor \frac{n-1}{2} \rfloor$-connected our induction hypothesis applies to these differential forms. Therefore the $\mathfrak{U}$-local integral $s^q(\chi^{p,q}((\varrho^*\omega)^{p,q}))$ of the exact equivariant q-form $\chi^{p,q}((\varrho^*\omega)^{p,q})$ can be extended to a cocycle H in $Z^q(M;Z_{dR}^p(G;V))^G$. Then the function

$$F : \mathrm{SS}(M,\mathfrak{U})([q]) \times M^{p+1} \to V, \quad (\mathbf{m'}, \mathbf{m}) \mapsto \int_{\sigma_p(\mathbf{m})} H(\mathbf{m'})$$

is an extension of $\chi^{p,q}(f^{p,q})$ to a global cocycle of bidegree (p,q) in the total complex of the double complex $A^*(M; A^*(M;V))^G$. By Lemma 16.4.2 this implies the extendibility of $f^{p,q}$ to a cocycle of bidegree (p,q) in the total complex of the double complex $A^{*,*}(M;V)^G$ $\qquad\square$

Lemma 16.4.8. *If the restriction of the flux component $\mathcal{F}_{\varrho t}(\omega)^{n-\lfloor \frac{n-1}{2}\rfloor-1, \lfloor \frac{n-1}{2}\rfloor+1}$ to smooth singular $(n-\lfloor \frac{n-1}{2}\rfloor-1)$-cycles is trivial, the the vertical coboundary of $d_h\psi^{\lfloor \frac{n-1}{2}\rfloor, n-\lfloor \frac{n-1}{2}\rfloor-1}$ is trivial.*

Proof. Let the restriction of the flux component $\mathcal{F}_{\varrho t}(\omega)^{n-\lfloor \frac{n-1}{2}\rfloor-1, \lfloor \frac{n-1}{2}\rfloor+1}$ to smooth singular $(n-\lfloor \frac{n-1}{2}\rfloor-1)$-cycles be trivial. The evaluation of the vertical coboundary of $d_h\psi^{\lfloor \frac{n-1}{2}\rfloor, n-\lfloor \frac{n-1}{2}\rfloor-1}$ at a point $(\mathbf{m},\mathbf{m'}) \in M^{\lfloor \frac{n-1}{2}\rfloor+2} \times \mathrm{SS}(M,\mathfrak{U})([n-\lfloor \frac{n-1}{2}\rfloor])$ is given by

$$d_v d_h \psi^{\lfloor \frac{n-1}{2}\rfloor, n-\lfloor \frac{n-1}{2}\rfloor-1}(\mathbf{m},\mathbf{m'}) =$$

$$\int_{[p\partial\sigma_{\lfloor \frac{n-1}{2}\rfloor+1}(\mathbf{m})]^{-1}.\partial\sigma_{n-\lfloor \frac{n-1}{2}\rfloor}(\mathbf{m'})} \eta$$

$$= \int_{\partial([p\sigma_{\lfloor \frac{n-1}{2}\rfloor+1}(\mathbf{m})]^{-1}.\partial\sigma_{n-\lfloor \frac{n-1}{2}\rfloor}(\mathbf{m'}))} \eta$$

$$= \int_{[p\sigma_{\lfloor \frac{n-1}{2}\rfloor+1}(\mathbf{m})]^{-1}.\partial\sigma_{n-\lfloor \frac{n-1}{2}\rfloor}(\mathbf{m'})} \omega$$

$$= \int_{[p\sigma_{\lfloor \frac{n-1}{2}\rfloor+1}(\mathbf{m})]^{-1}.\partial\sigma_{n-\lfloor \frac{n-1}{2}\rfloor}(\mathbf{m'})} \omega$$

$$= \int_{[p\sigma_{\lfloor \frac{n-1}{2}\rfloor+1}(\mathbf{m})]^{-1}} \mathcal{F}_{\varrho t}(\omega)^{n-\lfloor \frac{n-1}{2}\rfloor-1, \lfloor \frac{n-1}{2}\rfloor+1}(\partial\sigma_{n-\lfloor \frac{n-1}{2}\rfloor}(\mathbf{m'})).$$

The assumption that the restriction of the flux component $\mathcal{F}_{\varrho t}(\omega)^{n-\lfloor \frac{n-1}{2}\rfloor-1, \lfloor \frac{n-1}{2}\rfloor+1}$ to smooth singular $(n-\lfloor \frac{n-1}{2}\rfloor-1)$-cycles is trivial implies the triviality of the above integrals, hence the vertical coboundary of $d_h\psi^{\lfloor \frac{n-1}{2}\rfloor, n-\lfloor \frac{n-1}{2}\rfloor-1}$ is trivial in this case. $\qquad\square$

We denote the horizontal coboundary of $\psi^{\lfloor \frac{n-1}{2}\rfloor, n-\lfloor \frac{n-1}{2}\rfloor-1}$ by $F^{\lfloor \frac{n-1}{2}\rfloor+1, n-\lfloor \frac{n-1}{2}\rfloor-1}$. Since this cochain is itself a horizontal coboundary, the triviality of the restriction of the flux component $\mathcal{F}_{\varrho t}(\omega)^{n-\lfloor \frac{n-1}{2}\rfloor-1, \lfloor \frac{n-1}{2}\rfloor+1}$ to smooth singular $(n-\lfloor \frac{n-1}{2}\rfloor-1)$-cycles implies that it is a cocycle in the total complex $\mathrm{Tot}A^{*,*}(M,\mathfrak{U};V)^G$. Furthermore, as has been shown, the triviality of the restriction of the flux components $\mathcal{F}_{\varrho t}(\omega)^{q,p}$ to smooth singular cycles for all $p \le \lfloor \frac{n-1}{2}\rfloor+1$ implies that each of the cocycles $f^{p,q}$ is extendible to a global cocycle of bidegree (p,q) in $A^{p,q}(M;V)^G$. Therefore every one of these cocycles is cohomologous to a cocycle of boidegree $(n,0)$ in $A^{*,*}(M;V)^G$ and thus can be neglected when considering the extendibility to global cocycles:

Lemma 16.4.9. *If the restrictions of the flux components $\mathcal{F}_{\varrho t}(\omega)^{q,p}$ to smooth singular cycles are trivial for all $\lfloor \frac{n-1}{2}\rfloor < q < n$ and the de Rham homomorphism homomorphism $H^q_{dR}(M; {}^tZ^p_{dR}(M;V)) \to H^q(M; {}^tZ^p_{dR}(M;V))$ are injective for all $\lfloor \frac{n-1}{2}\rfloor < q < n$, then the cocycle $i^n s^n(\omega)$ of bidegree $(n,0)$ in the total complex of the double complex $A^{*,*}(M,\mathfrak{U};V)^G$ is cohomologous to a cocycle of bidegree $(n,0)$ if and only if the cocycle $d_h\psi^{\lfloor \frac{n-1}{2}\rfloor, n-\lfloor \frac{n-1}{2}\rfloor-1}$ is cohomologous to a cocycle of bidegree $(n,0)$.*

Proof. Let the restrictions of the flux components $\mathcal{F}_{\varrho t}(\omega)^{q,p}$ to smooth singular cycles be trivial for all $\lfloor \frac{n-1}{2} \rfloor < q < n$ and the de Rham homomorphism homomorphism $H^q_{dR}(M; {}^t Z^p_{dR}(M; V)) \to H^q(M; {}^t Z^p_{dR}(M; V))$ are injective for all $\lfloor \frac{n-1}{2} \rfloor < q < n$. Then the cochains $f^{p,q}$ are cocycles in the total complex $\mathrm{Tot} A^{*,*}(M, \mathfrak{U}; V)^G$. We define cochains ψ and f in this total complex by adding all the cochains $\psi^{p',q'}$ resp. $f^{p,q}$:

$$\psi := \sum_{\substack{p'+q'=n-1 \\ p \le \lfloor \frac{n-1}{2} \rfloor}} \psi^{p',q'}, \quad f := \sum_{\substack{p+q=n \\ 1 \le p \le \lfloor \frac{n-1}{2} \rfloor}} f^{p,q}$$

We assert that the coboundary of the cochain ψ in the total complex of the double complex $A^{*,*}(M, \mathfrak{U}; V)^G$ is the sum $i^n s^n(\omega) + f + d_h \psi^{\lfloor \frac{n-1}{2} \rfloor, n - \lfloor \frac{n-1}{2} \rfloor - 1}$. The coboundary $D\psi$ of the cochain ψ in the total complex $\mathrm{Tot} A^{*,*}(M, \mathfrak{U}; V)^G$ computes to

$$D\psi = \sum_{p'+q'=n-1, p \le \lfloor \frac{n-1}{2} \rfloor} [d_h + d_v] \psi^{p',q'}$$

$$= d_v \psi^{0,n-1} + \sum_{p+q=n, 1 \le p \le \lfloor \frac{n-1}{2} \rfloor} [d_h \psi^{p-1,q} + d_v \psi^{p,q-1}] + d_h \psi^{\lfloor \frac{n-1}{2} \rfloor, n - \lfloor \frac{n-1}{2} \rfloor - 1}$$

$$= i^n s^n(\omega) + f + d_h \psi^{\lfloor \frac{n-1}{2} \rfloor, n - \lfloor \frac{n-1}{2} \rfloor - 1}$$

The cocycle $D\psi$ is a coboundary and thus cohomologous to the zero cocycle of bidegree $(n, 0)$. Furthermore each of the cocycles $f^{p,q}$ is cohomologous to a cocycle of bidegree $(n, 0)$ in the total complex $\mathrm{Tot} A^{*,*}(M, \mathfrak{U}; V)^G$. Thefore the difference cocycle

$$D\psi - f = i^n s^n(\omega) + d_h \psi^{\lfloor \frac{n-1}{2} \rfloor, n - \lfloor \frac{n-1}{2} \rfloor - 1}$$

is also cohomologous to a cocycle of bidegree $(n, 0)$. This implies that the cocycle $i^n s^n(\omega)$ is cohomologous to a cocycle of bidegree $(n, 0)$ if and only if the cocycle $d_h \psi^{\lfloor \frac{n-1}{2} \rfloor, n - \lfloor \frac{n-1}{2} \rfloor - 1}$ is cohomologous to a cocycle of bidegree $(n, 0)$. $\square$

We are now ready to complete the infuctive step:

Proposition 16.4.10. *If the restriction of $\mathcal{F}_{\varrho t}(\omega)^{n - \lfloor \frac{n-1}{2} \rfloor - 1, \lfloor \frac{n-1}{2} \rfloor + 1}$ to smooth singular cycles is trivial, then the cocycle $F^{\lfloor \frac{n-1}{2} \rfloor + 1, n - \lfloor \frac{n-1}{2} \rfloor - 1}$ is cohomologous to a cocycle of bidegree $(n, 0)$ in the total complex of the double complex $A^{*,*}(M, \mathfrak{U}; V)^G$.*

Proof. Tho shorten the notation we define $k = \lfloor \frac{n-1}{2} \rfloor + 1$ and $l = n - \lfloor \frac{n-1}{2} \rfloor - 1$. Consider the function the adjoint cocycle $\hat{F}^{k,l}$ of $F^{k,l}$. It induces a k-cocycle $q\hat{F}^{k,l} : M^{k+1} \to H^l(M, \mathfrak{U}; V)$, where $q : Z^l(M, \mathfrak{U}; V) \to H^l(M, \mathfrak{U}; V)$ denotes the quotient map. If the k-cocycle $q\hat{F}^{k,l} : M^{k+1} \to H^l(M, \mathfrak{U}; V)$ is trivial, then the cocycle $F^{k,l}$ is the vertical coboundary of a cochain $\psi^{k,l-1}$ and thus cohomologous to the cocycle $F^{k+1,l-1} = -d_h \psi^{k,l-1}$ of bidegree $(k+1, l-1)$ in the total complex $\mathrm{Tot} A^{*,*}(M, \mathfrak{U}; V)^G$ by Lemma 15.3.23. In this case the $(l-1)$-connectedness of M ensures that the cocycle $F^{k+1,l-1}$ is cohomologous to a cocycle $F^{n,0}$ of bidegree $(n, 0)$ in the total complex $\mathrm{Tot} A^{*,*}(M, \mathfrak{U}; V)^G$ by Corollary 15.3.28. As a consequence the cocycle $i^n s^n(\omega)$ is then cohomologous to $F^{n,0}$, hence the $\mathfrak{U}$-local cocycle $s^n(\omega)$ then is extendible to a global equivariant n-cocycle on M. To prove the triviality of the k-cocycle $q\hat{F}^{k,l} : M^{k+1} \to H^l(M, \mathfrak{U}; V)$ we examine the singular k-cocycle $\tilde{\lambda}^k_{\mathfrak{U}}(\hat{F}^{k,l}(\mathbf{m}))$ for each point $\mathbf{m} \in M^{k+1}$. The evaluation of this singular k-cochain at a singular k-cycle $z \in Z_k(M)$ computes to

$$\int_z \hat{F}^{k,l}(\mathbf{m}) = \int_z [d_h \hat{\psi}^{k,l}(\mathbf{m})]$$

$$= \int_{[\partial p \hat{\sigma}_k(\mathbf{m})]^{-1}.[\hat{\sigma}_l \lambda_l(z)]} \eta$$

$$= \int_{\partial([p \hat{\sigma}_k(\mathbf{m})]^{-1}.[\hat{\sigma}_l \lambda_l(z)])} \eta$$

$$= \int_{[p \hat{\sigma}_k(\mathbf{m})]^{-1}.[\hat{\sigma}_l \lambda_l(z)]} \omega$$

$$= \int_{[p \hat{\sigma}_k(\mathbf{m})]^{-1}} \mathcal{F}_{\varrho t}(\omega)^{k,l}(\hat{\sigma}_l \lambda_l(z)) = 0 .$$

Because the natural homomorphism $H^l(M, \mathfrak{U}; V) \to H^l(M; V)$ into the singular cohomology group $H^l(M; V)$ is injective, this forces the k-cocycle $q\hat{F}^{k,l} : M^{k+1} \to H^l(M, \mathfrak{U}; V)$ to be trivial. Thus $s^n(\omega)$ is extendible to a global equivariant cocycle. □

Corollary 16.4.11. *If the restrictions of the flux components $\mathcal{F}_{\varrho t}(\omega)^{q,p}$ to smooth singular cycles are trivial for all $\lfloor \frac{n-1}{2} \rfloor < q < n$ and the de Rham homomorphism homomorphism $H^q_{dR}(M; {}^t Z^p_{dR}(M; V)) \to H^q(M; {}^t Z^p_{dR}(M; V))$ are injective for all $\lfloor \frac{n-1}{2} \rfloor < q < n$, then the $\mathfrak{U}$-local equivariant cocycle $s^n(\omega)$ is extendible to a global equivariant n-cocycle on M.*

This completes the inductive step. All in all we shown:

Theorem 16.4.12. *If the manifold M is $\lfloor \frac{n-1}{2} \rfloor$-connected, the de Rham homomorphisms $k^q : H^q_{dR}(M; V) \to H^q(M; V)$ are injective for all $\lfloor \frac{n-1}{2} \rfloor < q < n$, then the $\mathfrak{U}$-local integral $s^n(\omega)$ of an exact equivariant differential n-form ω on M is extendible to a locally smooth equivariant global n-cocycle provided that th restrictions*

$$\mathcal{F}_{\varrho t}(\omega)^{q,p}_{|Z_{q,\infty}(M)} : Z_{q,\infty}(M) \to Z^p_{dR}(G; V)^G \cong Z^p_c(\mathfrak{g}; V)$$

of the flux components are trivial for all $\lfloor \frac{n-1}{2} \rfloor < q < n$.

16.5 Integrability to Smooth Global Cocycles

In the remaining sections we examine the integrability of equivariant differential forms to smooth equivariant global cocycles.

We begin pour considerations by defining integrability to global smooth equivariant cocycles and then consider smooth transformation groups with smooth equivariant systems of singular simplices, e.g. Lie groups. Let (G, M) be a smooth transformation group and V be a Hausdorff G-vector space of coefficients. Recall the equivariant derivation morphism

$$\tau^* : A^*_{AS,s}(M; V) \to \Omega^*(M; V),$$

which assigns to each smooth Alexander-Spanier n-cochain f on M a differential n-form $\tau^n(f)$ on M (cf. Section 14.1 and Appendix G.2). In particular the n-form $\tau^n(f)$ is closed if f is a cocycle and equivariant if f is equivariant. Composing the natural complex morphism $A^*_s(M; V) \to A^*_{AS,s}(M; V)$ with the derivation morphism τ^* yields a morphism

$$A^*_s(M; V) \to \Omega^*(M; V), \quad F \mapsto \tau^*([F])$$

which we also denote by τ^*.

Definition 16.5.1. *A differential n-form ω on M is said to be* integrable to a smooth global cochain *if it is in the image of the equivariant derivation homomorphism $\tau^n : A_s^*(M;V) \to \Omega^*(M;V)$. It is called* integrable to a smooth global cocycle *if there exists a smooth global n-cocycle $s \in Z_s^n(M;V)$ on M with image $\tau^n(f) = \omega$.*

The integrability of equivariant differential forms to equivariant smooth global cocycles is defined analogously:

Definition 16.5.2. *A differential n-form is called* integrable to an equivariant smooth global cochain *if it is the image $\tau^n(f)$ of an equivariant smooth global n-cochain $f \in A_s^n(M;V)^G$ on M. It is called* integrable to a smooth equivariant global cocycle *if it is the image $\tau^n(f)$ of an equivariant smooth global n-cocycle $f \in Z_s^n(M;V)^G$ on M.*

To ensure the existence of integrals of differential forms over smooth singular simplices we assume the coefficients V to be Mackey complete. The action of the Lie group G on M is denoted by ϱ. If there exists an integrating smooth system of n-simplices $\hat{\sigma}_n : M^{n+1} \to C^\infty(\Delta_n, M)$ on M, then the latter can be used to integrate equivariant differential n-forms to smooth global equivariant cochains. The construction is given by the homomorphism

$$s^n : \Omega^n(M;V) \to A_s^n(M,\mathfrak{U};V), \quad s^n(\omega)(\mathbf{m}) = \int_{\hat{\sigma}_n(\mathbf{m})} \omega.$$

The linear function s^n is a composition of smooth functions and therefore smooth as well. For Lie groups this has already been observed in [Fuc03]. It has been observed in Lemma 16.1.12 that the resulting cochain $s^n(\omega)$ for a closed differential n-form ω is a cocycle if its period map

$$\mathrm{per}_{\varrho t}(\omega)_n : \pi_n(M) \to V^G, \quad [s] \mapsto \int_s \omega$$

is trivial. So we note:

Lemma 16.5.3. *If the manifold M is smoothly paracompact and admits a smooth integrating global system of n-simplices, $\hat{\sigma}_n : M^{n+1} \to C^\infty(\Delta, M)$ of singular n-simplices, then $s^n : \Omega^n(M;V) \to A_s^n(M;V)$ is a right inverse to the derivation homomorphism $\tau^n : A_s^n(M;V) \to \Omega^n(M;V)$ and the sequence*

$$Z_s^n(M;V) \to Z_{dR}^n(M;V) \xrightarrow{\mathrm{per}_{\varrho t}(-)_n} \hom(\pi_n(M), V^G)$$

of Abelian groups is exact and splits at $Z_{dR}^n(M;V)$. If $\hat{\sigma}_n$ is equivariant, then it also is split exact in in the category $\mathbb{Z}[\mathbf{G}]\mathbf{mod}$ of G-modules.

Proof. If the period map of a closed differential n-form ω is trivial, then $s^n(\omega)$ is a smooth global cocycle. Conversely, if the form ω is integrable to a smooth global cocycle, then the isomorphy of the Alexander-Spanier and the singular homology implies that the integration of ω over smooth singular cocycles is trivial, hence the period map $\mathrm{per}_{\varrho t}(\omega)_n$ is trivial as well. Furthermore s^n is equivariant if $\hat{\sigma}_n$ is equivariant. $\qquad\square$

The requirement of the existence of an integrating compatible smooth equivariant system $\hat{\sigma}_n : M^{n+1} \to C^\infty(\Delta, M)$ of n-simplices is a major restriction. In the following sections we will see that this restriction can be avoided by the use of the spectral sequence introduced in Chapter 14.

16.6 Observations for the General Case

In the following sections we examine the general integrability of closed equivariant n-forms on manifolds M of a smooth transformation group (G, M) to smooth global equivariant cochains on M. Without assuming the existence of a smooth global equivariant system of simplices, the forms can not directly integrated to global equivariant cochains as in the previous section. Instead we use the spectral sequence of van Est type

$$H^p_{s,eq}(M; H^q_{dR}(M; V)) \Rightarrow H^{p+q}_{dR,eq}(M; V)$$

which was introduced in Chapter 14 to prove the existence of integrals in certain cases. Here we give necessary and sufficient conditions on the form ω for such integrals to exist. These conditions are the triviality of a flux in (co)homology. In the special case of Lie groups acting on themselves by left translation this can expressed by integration of the flux over spheres, i.e. by period maps

$$\pi_p(G) \to H^{n-p}_{dR,eq}(G; V) \cong H^{n-p}_c(\mathfrak{g}; V).$$

We restrict ourselves to connected smooth transformation groups (G, M) which admit a smooth equivariant map $p : M \to G$.

Example 16.6.1. Every total space E of a trivial G-principal bundle $E \twoheadrightarrow B = E/G$ is of this form.

Example 16.6.2. Every Lie group acts on itself by left translation and the orbit bundle $G \to G\backslash G$ is trivial. The complex of smooth global equivariant cochains here is isomorphic to the complex of smooth group cochains.

An immediate requirement for a closed differential n-form ω on M to be integrable to a smooth global cocycle is the following:

Lemma 16.6.3. *If a closed n-form ω on M is the image $\tau^n(F)$ of a smooth global cocycle F on M, then ω is exact.*

Proof. Let F be a smooth global n-cocycle on M whose image $\tau^n(F)$ is the differential form ω. Since the smooth standard complex $A^*_s(M; V)$ is exact, there exists a smooth $(n-1)$-cochain F' in $A^{n-1}_s(M; V)$ whose coboundary is F. Let $\eta \in \Omega^{n-1}(M; V)$ be the differential form $\tau^{n-1}(F')$. Because the derivation morphism $\tau^* : A^*_s(M; V) \to \Omega^*(M; V)$ is a morphism of chain complexes, the coboundary

$$d\eta = d\tau^{n-1}(F') = \tau^n(dF') = \tau^n(F) = \omega$$

of the $(n-1)$-form η is exactly the n-form ω, hence ω is exact.   □

Example 16.6.4 (Beggs). If M is a compact manifold and N a separable finite dimensional manifold, then the de Rham complex $\Omega^*(C^\infty(M, N); V)$ of the manifold $C^\infty(M, N)$ of smooth functions from M into N has split de Rham complex (cf. [Beg87, Theorem 7.4]).

Example 16.6.5. The smooth path space $P_s M = C^\infty(I, M)$ and the smooth loop space $\Omega_s M = C^\infty_*(\mathbb{S}^n, M)$ of a finite dimensional separable manifold with base point M has split de Rham complex. Therefore the path-loop fibration

$$
\begin{array}{ccc}
\Omega_s M & \relbar\joinrel\relbar & P_s M \\
 & & \big\downarrow \\
 & & M
\end{array}
$$

of a finite dimensional manifold M is a fibration of smooth manifolds with split topological de Rham complex. This in particular includes all finite dimensional Lie groups.

Example 16.6.6. The diffeomorphism group $\mathrm{Diff}(M)$ of a compact manifold M has split de Rham complex (cf. [Beg87, Theorem 7.5]).

Example 16.6.7. If M is a compact manifold and N is an infinite dimensional separable metrisable Hilbert manifold, then $C^\infty(M; N)$ has split de Rham complex (cf. [Beg87, Theorem 7.6]). Therefore the smooth path space $P_s M = C_*^\infty(I, M)$ and the smooth loop space $\Omega_s M = C_*^\infty(\mathbb{S}^n, M)$ of a separable Hilbert manifold M with base point has split topological de Rham complex. As a consequence path-loop fibration

$$\Omega_s M \text{ ------ } P_s M$$
$$\downarrow$$
$$M$$

of a separable Hilbert manifold M is a fibration of smooth manifolds with split topological de Rham complexes. This in particular includes all connected Hilbert Lie groups.

We would like to show that the the integrability of a closed equivariant n-form ω to a global smooth cocycle only depends on its cohomology class $[\omega] \in H^n_{dR,eq}(M; V)$. For this purpose we first observe:

Lemma 16.6.8. *If the cochain $f \in A^n_s(M; V)$ is an integral of an equivariant n-form ω on M (i.e. $\tau^n(f) = \omega$), then the cochain*

$$F : M^{n+1} \to V, \quad F(\mathbf{m}) = p(m_0). \left[f(p(m_0)^{-1}.\mathbf{m}) \right]$$

is an equivariant integral of ω on M.

Proof. Let f be an integral of an equivariant n-form ω on M and F be defined as above. The cochain F is equivariant by definition. The derivation homomorphism $\tau^n : A^n_s(M; V) \to \Omega^n(M; V)$ and the map $p : M \to G$ are equivariant. Therefore the evaluation of $\tau^n(F)$ at a point $\mathbf{X}$ in $T^{\oplus n} M$ computes to

$$\begin{aligned}
\tau^n(F)(\mathbf{X}) &= [p\pi_{TM}(X_0)]. \left[\tau^n(F) \left([p\pi_{TM}(X_0)]^{-1}.\mathbf{X} \right) \right] \\
&= [p\pi_{TM}(X_0)]. \left[\tau^n(f) \left([p\pi_{TM}(X_0)]^{-1}.\mathbf{X} \right) \right] \\
&= [p\pi_{TM}(X_0)]. \left[\omega \left([p\pi_{TM}(X_0)]^{-1}.\mathbf{X} \right) \right] \\
&= \omega(\mathbf{X}),
\end{aligned}$$

so F is an equivariant integral of ω. This proves the assertion. $\square$

Lemma 16.6.9. *If M is smoothly paracompact then every equivariant differential n-form on M can be integrated to a global equivariant cochain.*

Proof. If ω is an equivariant n-form on M, integrate ω arbitrarily to a smooth function $f : M^{n+1} \to V$ (which is possible if M is smoothly paracompact) and define a smooth global equivariant n-cochain F on M via

$$F : M^{n+1} \to V, \quad F(\mathbf{m}) = p(m_0). \left[f(p(m_0)^{-1}.\mathbf{m}) \right].$$

By Lemma 16.6.8 this cochain F is an equivariant integral of ω. $\square$

The same construction allows us to integrate coboundaries of equivariant differential forms to smooth global equivariant cocycles:

Lemma 16.6.10. *If M is smoothly paracompact then every coboundary of an equivariant differential form on M can be integrated to a smooth global equivariant cocycle.*

Proof. If ω is an equivariant n-cocycle that is the coboundary of an $(n-1)$-form η, integrate η arbitrarily to a smooth function $f : M^n \to V$ (which is possible if M is smoothly paracompact) and define a smooth global equivariant $(n-1)$-cochain $F' : M^n \to V$ via

$$F'(\mathbf{m}) = p(m_0). \left[f(p(m_0)^{-1}.\mathbf{m}) \right]$$

This is a smooth equivariant integral of η to M^n. The coboundary of F' is a smooth global equivariant cocycle which is an integral of ω.   □

Proposition 16.6.11. *If M is smoothly paracompact then the integrability of an equivariant closed n-form ω on M to a smooth equivariant global n-cocycle F on M only depends on its cohomology class in $H^n_{dR,eq}(M;V)$.*

Proof. Let ω be a closed equivariant n-form which is integrable to a smooth global equivariant n-cocycle F on M. It is to show that a closed equivariant n-form ω' with the same cohomology class $[\omega'] = [\omega] \in H^n_{dR,eq}(M;V)$ is integrable to a smooth equivariant n-cocycle as well. If an n-form ω' is cohomologous to ω in $\Omega^*(M;V)^G$, then there exists an equivariant $(n-1)$-form η with coboundary $d\eta = \omega - \omega'$. The coboundary $d\eta$ is integrable to a smooth global equivariant cocycle H by Lemma 16.6.10. The equivariant n-cocycle $F - H$ satisfies $\tau^n(F - H) = \omega - \eta = \omega'$, i.e. the closed n-form ω' is integrable as well.   □

Having made these preparatory observations, we now utilise the spectral sequence $H^p_{s,eq}(M;H^q_{dR}(M;V)) \Rightarrow H^{p+q}_{dR,eq}(M;V)$ to describe the integrability of equivariant closed forms on M to smooth global cocycles. For this purpose we make the additional requirement that the topological de Rham complex ${}^t\Omega^*(M;V)$ of the manifold M splits and that M to is smoothly paracompact. The latter assumption ensures that the de Rham morphism $k^* : \Omega^*(M;V) \to S^*(M;V)$ and the derivation morphism $\tau^* : A^*_s(M;V) \to \Omega^*(M;V)$ induce an isomorphisms in cohomology (cf. [War83, Theorem 5.23, Corollary 5.23]). Furthermore, every differential n-form is integrable to a global smooth n-cochain F under this assumption.

Recall that this spectral sequence was constructed using the double complex $A^{*,*}_{s,dR}(M;V)^G$ introduced in Section 14.2. The horizontal and the vertical differential in this double complex are denoted by d_h and d_v respectively. The differential of the total complex will be denoted by D. The total complex $\mathrm{Tot}A^{*,*}_{s,dR}(M;V)^G$ may be filtrated column-wise and row-wise. By Theorem 14.2.11 the spectral sequences $_I E^{*,*}_{s,dR}$ and $_{II} E^{*,*}_{s,dR}$ obtained by these filtrations both converge to the equivariant de Rham cohomology $H^*_{dR,eq}(M;V)$ of M. Moreover, as has been shown in Section 14.2, the the augmentations

$$0 \longrightarrow A^p_s(M;V)^G \xrightarrow{j^p} A^{p,*}_{s,dR}(M;V)^G$$

of the columns $A^{p,*}_{s,dR}(G, M;V)^G$ of the double complex $A^{*,*}_{s,dR}(M;V)^G$ induce an embedding $A^*_s(M;V)^G \hookrightarrow \mathrm{Tot}A^{*,*}_{s,dR}(M;V)^G$ of the complex of equivariant smooth global cochains into the total complex $\mathrm{Tot}A^{*,*}_{s,dR}(M;V)^G$. Similarly, the complex $\Omega^*(M;V)^G$ of equivariant forms on M can be embedded into the total complex $\mathrm{Tot}A^{*,*}_{s,dR}(M;V)^G$ via the augmentations

$$i^q : \Omega^q(M;V)^G \hookrightarrow A^{0,q}_{s,dR}(M;V)^G, \quad i^q(\omega)(m_0, \mathbf{X}) = \omega(\mathbf{X}).$$

of the rows of the double complex. The augmentation i^q embeds a closed equivariant q-form ω on M as a cocycle of bidegree $(0,q)$ in the total complex $\mathrm{Tot}A^{*,*}_{s,dR}(M;V)^G$. The derivation morphism $\tau^* : A^*_s(M;V)^G \to \Omega^*(M;V)^G$ maps an equivariant global smooth n-cochain F on M to an equivariant n-form ω on M. This n-form ω is closed whenever F is closed. If this is the case, then the cocycle $j^n(F)$ in

$\mathrm{Tot} A^{*,*}_{s,dR}(M;V)^G$ is cohomologous to the cocycle $i^n(\omega)$. (See Section 14.2 for a discussion of this topic). This is the case if and only if there exists a cochain ψ of total degree $n - 1$ with coboundary $D\psi = i^n(F) - j^n(F)$. The latter condition is is equivalent to the existence of cocycles $i^n(\omega) = F^{0,n}, \ldots, F^{n,0} = (-1)^{n-1} j^n(F)$ of bidegree $(0,n), (1, n-1), \ldots, (n,0)$ and cochains $\psi^{0,n-1}, \ldots, \psi^{n-1,0}$ of bidegree $(0, n-1), (1, n-2), \ldots, (n-1, 0)$ respectively such that $D\psi^{p-1,q} = F^{p,q} + F^{p-1,q+1}$:

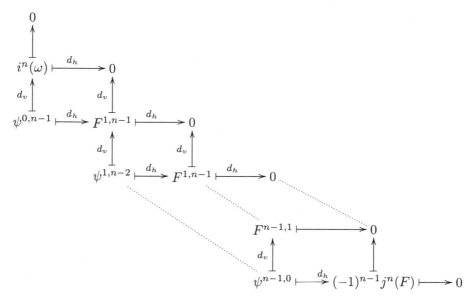

Here the cochain ψ is the alternating sum $\psi = \sum_{p'+q'=n-1} (-1)^{q'} \psi^{p',q'}$. Conversely, an equivariant closed n-form ω on M is the image $\tau^n(F)$ of a smooth global equivariant cocycle F on M if and only if the cocycle $i^n(\omega)$ of bidegree $(0,n)$ in the complex $\mathrm{Tot} A^{*,*}_{s,dR}(M;V)$ is cohomologous to a cocycle $j^n(F)$ of bidegree $(n,0)$:

Lemma 16.6.12. *An equivariant closed n-form ω on M is integrable to an equivariant smooth global cocycle if and only if $i^n(\omega)$ is cohomologous to a cocycle $j^n(F)$ of bidegree $(n,0)$.*

Proof. The "only if" implication has been proved in Section 14.2. If the cocycle $i^n(\omega)$ is cohomologous to a cocycle $j^n(F)$ of bidegree $(n,0)$, then there exists a cochain ψ of total degree $n - 1$ with coboundary $D\psi = i^n(\omega) - j^n(F)$. Using the homotopy equivalence $p^* : \mathrm{Tot} A^{*,*}_{s,dR}(M;V)^G \to \Omega^*(M;V)^G$ we see that this implies that the cocycles $p^n(j^n(F)) = \tau^n(F)$ and $p^n(i^n(\omega)) = \omega$ are cohomologous in $\Omega^*(M;V)^G$. By Proposition 16.6.11 the integrability of the form ω only depends on the cohomology class $[\omega] \in H^n_{dR,eq}(M;V)$, so ω is integrable to a smooth equivariant global cocycle on M. This proves the backward implication. $\square$

Moreover, if the closed equivariant n-form ω on M is the image $\tau^n(F)$ of an equivariant smooth global n-cocycle, the the cochains $F^{p,q}$ in the above diagram can be chosen to be given by the formula $F^{p,q} = (d_v h)^q j^n(F)$, where h denotes the row contractions of $A^{*,*}_{s,dR}(M;V)^G$ (see Section 14.2). These defining equations can be reformulated using the derivation homomorphisms ${}^t\tau^q : {}^tA_s(M;V) \to {}^t\Omega^q(M;V)$. Recall that the cocycle $j^n(F)$ in the total complex of the double complex $A^{*,*}_{s,dR}(M;V)^G$ is explicitly given by $j^n(F)(\mathbf{m}, m'_0) = F(\mathbf{m})$ and the cocycles $F^{p,q}$ are given by

$$F^{p,q}(\mathbf{m}, \mathbf{X}) = \sum_{\sigma \in S_q} \mathrm{sign}(\sigma)[d_1 \cdots d_n f](\mathbf{m}, X_1, \ldots, X_q)$$

as has been proved in Proposition 14.2.26. This cocycle is obtained by applying a "derivation homomorphism" to the last $q + 1$-arguments of the cocycle F. This motivates us to consider the following double complex: We define $A^{p,q}_{s,s}(M;V)$ to be the (abstract) vector space underlying the topological vector space $C^\infty(M^{p+1} \times M^{q+1}, V)$. The spaces $A^{p,q}_{s,s}(M;V)$ form a double complex whose horizontal and vertical differentials are given by

$$d_h : A^{p,q}_{s,s}(M;V) \to A^{p+1,q}_{s,s}(M;V)$$

$$(d_h f)(m_0, \ldots, \mathbf{m}') = \sum_{i=0}^{p+1} (-1)^i f(m_0, \ldots, \hat{m}_i, \ldots, m_{p+1}, \mathbf{m}')$$

and by

$$d_v : A^{p,q}_{s,s}(M;V) \to A^{p,q+1}_{s,s}(M;V),$$

$$(d_h f)(\mathbf{m}, m'_0, \ldots, m'_{q+1}) = (-1)^p \sum_{i=0}^{q+1} (-1)^i f(\mathbf{m}, m'_0, \ldots, \hat{m}'_i, \ldots, m'_q)$$

respectively. Since the horizontal and vertical differentials are equivariant by construction, the subspaces $A^{p,q}_{s,s}(M;V)^G$ of equivariant functions form a sub double complex of $A^{*,*}_{s,s}(M;V)$. Similar to the double complex $A^{*,*}_{s,dR}(M;V)$ we note that for each $q \geq 0$ the kernel of the horizontal differential $d_h : A^{0,q}_{s,s}(M;V)^G \to A^{1,q}_{s,s}(M;V)^G$ consists of those smooth equivariant functions $f : M \times M^{q+1} \to V$ in $A^{0,q}_{s,s}(M;V)^G$ that are independent of the first variable. Hence this kernel is isomorphic to the space $A^q_s(M;V)^G$ of smooth equivariant functions. One can therefore augment each row via

$$0 \longrightarrow A^q_s(M;V)^G \xrightarrow{i^q} A^{*,q}_{s,s}(M;V)^G, \quad i^q(f)(m_0, \mathbf{m}') = f(\mathbf{m}'),$$

where the augmentation i^q is an embedding. Furthermore the augmentations i^q intertwine the differentials on the complexes $A^*_s(M;V)^G$ and $A^{*,q}_{s,s}(M;V)^G$. Analogously one can augment the columns $A^{p,*}_{s,s}(M;V)^G$ with the kernel of the vertical differential $d_v : A^{p,0}_{s,s}(M;V)^G \to A^{p,1}_{s,s}(M;V)^G$. Here the kernel is isomorphic to the vector space $A^p_s(M;V)^G$ of smooth global equivariant p-cochains

$$0 \longrightarrow A^p_s(M;V)^G \xrightarrow{j^p} A^{p,*}_{s,s}(M;V)^G,$$

and the augmentations j^p intertwine the differential on the complexes $A^*_s(M;V)^G$ and $A^{p,*}_{s,s}(M;V)^G$. Similar to the double complex $A^{*,*}_{s,dR}(M;V)$ one can construct row contractions h^p for each augmented row of the double complex $A^{*,*}_{s,s}(M;V)$ by defining

$$h^p : A^{p,q}_{s,s}(M;V) \to A^{p-1,q}(M;V)$$
$$h^p(f)(m_0, \ldots, m_{p-1}, \mathbf{m}') = (-1)^p f(m_0, \ldots, m_{p-1}, m'_0, \mathbf{m}').$$

The homomorphisms h^p are equivariant by definition, hence they restrict to functions $A^{p,q}_{s,s}(M;V)^G \to A^{p-1,q}_{s,s}(M;V)^G$.

Proposition 16.6.13. *The maps h^p form an equivariant contracting homotopy on the augmented row-complexes $0 \to A^q_s(M;V)^G \to A^{*,q}_{s,s}(M;V)^G$.*

Proof. Let $f \in A^{p,q}_{s,s}(M;V)$ be given. The evaluation of $[d_h h^p + h^{p+1} d_h](f)$ at a point $\mathbf{m} \in M^{p+1}$ and a point $\mathbf{m}' \in M^{q+1}$ is given by

$$[d_h h^p + h^{p+1} d_h](f)(\mathbf{m}, \mathbf{m}') = \sum_{i=0}^{p} (-1)^i (h^p f)(m_0, \ldots, \hat{m}_i, \ldots, m_p, \mathbf{m}')$$

$$+ (-1)^{p+1} (d_h f)(m_0, \ldots, m_p, m'_0, \mathbf{m}')$$

$$= \sum_{i=0}^{p} (-1)^{p+i} f(m_0, \ldots, \hat{m}_i, \ldots, m_p, m'_0, \mathbf{m}')$$

$$+ \sum_{i=0}^{p} (-1)^{p+i+1} f(m_0, \ldots, \hat{m}_i, \ldots, m_p, m'_0, \mathbf{m}')$$

$$+ (-1)^{p+1} (-1)^{p+1} f(\mathbf{m}, \mathbf{m}')$$

$$= f(\mathbf{m}, \mathbf{m}'),$$

which proves the equality $d_h h^p + h^{p+1} d_h = \mathrm{id}$. Thus h^* is a row contraction. $\qquad \square$

Corollary 16.6.14. *The rows of the double complex $A_{s,s}^{*,*}(M; V)^G$ are split exact, except at the modules $A^{0,q}(M; V)^G$.*

In contrast to the double complex $A_{s,dR}^{*,*}(M; V)$ we can apply the same procedure to the columns of the double complex $A_{s,s}^{*,*}(; V)$ to show:

Proposition 16.6.15. *The homomorphisms $h'^q : A_{s,s}^{p,q}(M; V) \to A_{s,s}^{p-1,q}(M; V)$ defined by $h'^q(f)(\mathbf{m}, \mathbf{m}') = f(\mathbf{m}, m_0, \mathbf{m}')$ form an equivariant contracting homotopy of the augmented columns $0 \to A_s^q(M; V) \to A_{s,s}^{*,q}(M; V)$.*

Proof. The proof is analogous to that of Proposition 16.6.13. $\qquad \square$

Observe that the space $A_{s,dR}^{p,0}(M; V)^G$ is the vector space of equivariant smooth functions $f : M^{p+1} \times M \to V$ and the same is true for the vector space $A_{s,s}^{p,0}(M; V)^G$, so the vector spaces $A_{s,dR}^{p,0}(M; V)^G$ and $A_{s,s}^{p,0}(M; V)^G$ are identical. In addition the augmentations $A_s^p(M; V)^G \hookrightarrow A_{s,s}^{p,*}(M; V)^G$ and $A_s^p(M; V)^G \hookrightarrow A_{s,dR}^{p,*}(M; V)^G$ coincide. We will now show that some kind of "derivation homomorphism" in the last $(q + 1)$ arguments of cochains of bidegree (p, q) leads to a morphism $\tau^{*,*} : A_{s,s}^{*,*}(M; V)^G \to A_{s,dR}^{*,*}(M; V)^G$ of double complexes which can be used to further examine the cocycles $F^{p,q} = (d_v h)^q j^n(F)$ associated to an equivariant smooth n-cocycle F on M. Here fore we consider the double complex $A_s^*(M; {}^t A_s^*(M; V))^G$ whose cochains $\hat{f}^{p,q}$ of bidegree (p, q) are smooth equivariant functions

$$\hat{f}^{p,q} : M^{p+1} \to {}^t A_s^q(M; V)$$

The adjoint function $\hat{f}^{p,q}$ of a cochain $f^{p,q} \in A_{s,s}^{*,*}(M; V)^G$ is a smooth equivariant function $M^{p+1} \to {}^t A_s^q(M; V)$, hence an element of $A_s^p(M; {}^t A_s^q(M; V))^G$. The natural functions $A_{s,s}^{p,q}(M; V)^G \to A_s^p(M; {}^t A_s^q(M; V))^G$, $f^{p,q} \mapsto \hat{f}^{p,q}$ form a morphism of double complexes. The derivation morphism ${}^t \tau^* : {}^t A_s^*(M; V) \to {}^t \Omega^*(M; V)$ induces morphisms $A_s^p(M; {}^t \tau^*)^G : A_s^p(M; {}^t A_s^*(M; V))^G \to A_s^p(M; {}^t \Omega^*(M; V))^G$ of the column complexes. These morphisms intertwine the horizontal differentials, hence the morphisms $A_s^p(M; {}^t \tau^*)^G$ assemble to a morphism

$$A_s^*(M; {}^t \tau^*)^G : A_s^*(M; {}^t A_s^*(M; V))^G \to A_s^*(M; {}^t \Omega^*(M; V))^G$$

of double complexes. The homomorphisms $\tau^{p,q}$ on the vector spaces $A_{s,s}^{p,q}(M; V)^G$ corresponding to $A_s^p(M; {}^t \tau^q)^G$ are given by

$$\tau^{p,q} : A_{s,s}^{p,q}(M; V)^G \to A_{s,dR}^{p,q}(M; V)^G, \quad \tau^{p,q}(f^{p,q})(\mathbf{m}, \mathbf{X}) = \tau^q(\hat{f}(\mathbf{m}))(\mathbf{X}).$$

The fact that the homomorphisms $A_s^p(M; {}^t \tau^q)^G$ form a morphism of double complexes ensures that the homomorphisms $\tau^{p,q}$ also assemble to a morphism of double complexes.

Definition 16.6.16. *The morphism $\tau^{*,*} : A_{s,s}^{*,*}(M;V)^G \to A_{s,dR}^{*,*}(M;V)^G$ of double complexes is called the* derivation morphism.

Definition 16.6.17. *A cochain $F^{p,q}$ of bidegree (p,q) in the total complex of the double complex $A_{s,dR}^{*,*}(M;V)^G$ is called* integrable to an equivariant smooth global cochain *of bidegree (p,q) if it is in the image of the derivation morphism $\tau^{p,q}$. A cocycle $F^{p,q}$ of bidegree (p,q) in the total complex of the double complex $A_{s,dR}^{*,*}(M;V)^G$ is called* integrable to an equivariant smooth global cocycle *of bidegree (p,q) if there exists a cocycle $f^{p,q} \in A_{s,s}^{p,q}(M;V)^G$ with image $\tau^{p,q}(f^{p,q}) = F^{p,q}$.*

Lemma 16.6.18. *The derivation homomorphisms $\tau^{p,q}$ intertwine the row contractions on $A_{s,s}^{*,*}(M;V)^G$ and $A_{s,dR}^{*,*}(M;V)^G$.*

Proof. Let $f^{p,q}$ be a cochain of bidegree (p,q) in $A_{s,s}^{p,q}(M;V)^G$. The value of the cochain $(h\tau^{p,q})(f^{p,q})$ at a point $(\mathbf{m},\mathbf{X}) \in M^p \times T^{\oplus q}M$ is given by

$$
\begin{aligned}
(h\tau^{p,q})(f^{p,q})(\mathbf{m},\mathbf{X}) &= (\tau^{p,q}(f^{p,q}))(\mathbf{m},\pi_{TM}(X_1),\mathbf{X}) \\
&= \tau^q[\hat{f}^{p,q}(\mathbf{m},\pi_{TM}(X_1))](\mathbf{X}) \\
&= \sum_{\sigma \in S_q} \mathrm{sgn}(\sigma)d_1\cdots d_q[\hat{f}^{p,q}(\mathbf{m},\pi_{TM}(X_1))](\pi_{TM}(X_1),X_{\sigma(1)},\ldots,X_{\sigma(q)}) \\
&= (\tau^{p-1,q}hf^{p,q})(\mathbf{m},\mathbf{X}).
\end{aligned}
$$

This proves the assertion. $\qquad\square$

Recall that the cocycles $F^{p,q}$ associated to an equivariant smooth n-cocycle F on M are given by $F^{p,q} = (d_v h)^q j^n(F)$. Using the above observations we can finally note:

Lemma 16.6.19. *The cocycles $F^{p,q}$ are integrable to equivariant smooth cocycles of bidegree (p,q) respectively. In particular each cocycle $F^{p,q}$ is the image $\tau^{p,q}((d_v h)^q j^n(F)$, where $j^n(F)$ is considered as a cocycle in the total complex of the double complex $A_{s,s}^{*,*}(M;V)^G$.*

Proof. The derivation homomorphisms $\tau^{p,q}$ intertwine the vertical differentials by construction. Since they also intertwine the row contractions on the double complexes $A_{s,s}^{*,*}(M;V)^G$ and $A_{s,dR}^{*,*}(M;V)^G$ we observe the equalities

$$
F^{p,q} = (d_v h)^q j^n(F) = F^{p,q} = (d_v h)^q \tau^{n,0} j^n(F) = \tau^{p,q}(d_v h)^q j^n(F),
$$

which proves the assertion. $\qquad\square$

A similar conclusion can be drawn for the cochains $\psi^{p',q'} = ((d_v h)^{q'} hj^n(F))$ in the total complex of the double complex $A_{s,dR}^{*,*}(M;V)^G$:

Lemma 16.6.20. *The cochains $\psi^{p',q'}$ are integrable to equivariant smooth cochains of bidegree (p',q') respectively. In particular each cochain $\psi^{p',q'}$ is the image $\tau^{p',q'}((d_v h)^{q'} hj^n(F))$, where $j^n(F)$ is considered as a cocycle in the total complex of the double complex $A_{s,s}^{*,*}(M;V)^G$.*

Proof. The proof is analogous to that of the previous Lemma. $\qquad\square$

Each of the cocycles $F^{p,q}$ is cohomologous to $F^{n,0}$ and integrable to a smooth global cocycle. Below we observe that an integrable cocycle of bidegree (p,q) in $\mathrm{Tot}A_{s,dR}^{*,*}(M;V)^G$ always is cohomologous to a cocycle of bidegree $(p+q,0)$.

Lemma 16.6.21. *If a cocycle $F^{p,q}$, $q > 0$ of bidegree (p,q) in $A_{s,dR}^{p,q}(M;V)^G$ is integrable to a smooth equivariant global cocycle in $A_{s,s}^{p,q}(M;V)^G$, then it is cohomologous to a cocycle $F^{p+1,q-1}$ of bidegree $(p+1,q-1)$ in $A_{s,dR}^{p,q}(M;V)^G$ which is integrable to a smooth equivariant global cocycle in $A_{s,s}^{p+1,q-1}(M;V)^G$.*

Proof. If a cocycle $F^{p,q}$ of bidegree (p,q) is integrable to a smooth equivariant global cocycle $f^{p,q}$, then the application of the column contraction h' of $A_{s,s}^{p,*}(M;V)^G$ to $f^{p,q}$ yields a cochain $h'f^{p,q}$ of bidegree $(p,q-1)$ whose vertical differential is $f^{p,q}$. The equation

$$D(h'f^{p,q}) = d_h(h'f^{p,q}) + d_v(h'f^{p,q})$$

shows that the cocycle $f^{p,q}$ is cohomologous to the cocycle $d_h(h'f^{p,q})$ in the total complex $\mathrm{Tot}A_{s,s}^{*,*}(M;V)^G$. An application of the morphism $\mathrm{Tot}\tau^{*,*}$ now shows that the cocycle $F^{p,q} = \tau^{p,q}(f^{p,q})$ in the total complex $\mathrm{Tot}A_{s,dR}^{*,*}(M;V)^G$ is cohomologous to the cocycle $\tau^{p+1,q-1}(d_hh'f^{p,q})$. $\square$

Lemma 16.6.22. *If a cocycle $F^{p,q}$ of bidegree (p,q) in $\mathrm{Tot}A_{s,dR}^{*,*}(M;V)^G$ is integrable to a smooth equivariant global cocycle of bidegree (p,q) then it is cohomologous to a cocycle $F^{p+q,0}$ of bidegree $(p+q,0)$ in $\mathrm{Tot}A_{s,dR}^{*,*}(M;V)^G$.*

Proof. This follows from a repeated application of Lemma 16.6.21. $\square$

Thus in order to show that a cocycle $F^{p,q}$ of bidegree (p,q) in the total complex of $A_{s,dR}^{*,*}(M;V)^G$ is cohomologous to a cocycle of bidegree $(p+q,0)$ it suffices to verify that it is integrable to an equivariant smooth global cocycle of bidegree (p,q) in the total complex of $A_{s,s}^{*,*}(M;V)^G$.

In some rare cases the columns of the double complex $A_{s,dR}^{*,*}(M;V)^G$ are exact. In this case every cocycle $F^{p,q}$ of bidegree (p,q) is the vertical coboundary of a cochain $\psi^{p,q-1}$ of bidegree $(p,q-1)$. The considerations leading thereto use the "adjoint" double complex. Recall that the de Rham complex $\Omega^*(M;V)$ may be endowed with the C^∞-compact-open topology, which turns it into a chain complex in **TopAb**. We denote this topological chain complex by $^t\Omega^*(M;V)$ to distinguish it from the underlying chain complex $\Omega^*(M;V)$ of (abstract) vector spaces. The linear equivariant map

$$\varphi_{M^p,T\oplus^q M,V} : A_{s,dR}^{p,q}(M;V)^G \to A_s^p(M;{}^t\Omega^q(M;V))^G$$

of function spaces, which maps each smooth global equivariant cochain $F^{p,q}$ in $A_{s,dR}^{p,q}(M;V)$ to its adjoint cochain $\hat{F}^{p,q} \in A_s^p(M;{}^t\Omega^q(M;V))$ is always injective. It is surjective if M is locally compact or all finite products of the modelling space of M are k-spaces (cf. Corollary 13.1.23). If the map $\varphi_{M^p,T\oplus^q M,V}$ is a bijection, we say that the "exponential law" holds.

Example 16.6.23. If the manifold M is finite dimensional, then the exponential law holds in this case, i.e. the function $\varphi_{M^p,T\oplus^q M,V}$ is a bijection.

Example 16.6.24. If the manifold M has metrisable modelling space W, then all finite products W^n are metrisable, hence k-spaces (cf. Proposition B.9.7). Therefore the exponential law holds in this case, i.e. the function $\varphi_{M^p,T\oplus^q M,V}$ is a bijection. This in particular includes all Riemannian manifolds.

If the exponential law holds, then the double complex $A_{s,dR}^{*,*}(M;V)$ and the double complex $A_s^*(M;{}^t\Omega^*(M;V))$ in $\mathbb{Z}[\mathbf{G}]\mathbf{mod}$ are isomorphic, i.e. one can w.l.o.g. switch freely between the double complexes $A_{s,dR}^{*,*}(M;V)^G$ and $A_s^*(M;{}^t\Omega^*(M;V))^G$. In the following all cochains in the double complex $A_s^*(M;{}^t\Omega^*(M;V))^G$ will always written with "hat" on it, while symbols without will denote cochains in the double complex $A_{s,dR}^{*,*}(M;V)^G$.

Lemma 16.6.25. *If the de Rham cohomology $H^q_{dR}(M;V)$ is trivial, then every cochain $\hat{F}^{p,q}$ of bidegree (p,q) in $A^p_s(M;{}^t\Omega^q(M;V))^G$ with trivial vertical coboundary $d_v\hat{F}^{p,q}$ is the vertical coboundary $d_v\hat{\psi}^{p,q-1}$ of a cochain $\hat{\psi}^{p,q-1}$ of bidegree $(p,q-1)$ in $A^p_s(M;{}^t\Omega^{q-1}(M;V))^G$.*

Proof. Let $\hat{F}^{p,q} \in A^p_s(M;{}^t\Omega^q(M;V))^G$ be a cochain of bidegree (p,q) with trivial vertical coboundary $d_v\hat{F}^{p,q}$ and assume the cohomology $H^q_{dR}(M;V)$ to be trivial. The condition $d_v\hat{F}^{p,q} = 0$ ensures that the function $\hat{F}^{p,q} : M^{p+1} \to {}^t\Omega^q(M;V)$ takes values in the de Rham cocycles ${}^tZ^q_{dR}(M;V)$. Since the de Rham cohomology $H^q_{dR}(M;V)$ is assumed to be trivial, this implies that the cochain $\hat{F}^{p,q}$ has codomain in the de Rham q-coboundaries ${}^tB^q(M;V)$. Furthermore we consider manifolds with split topological de Rham complex ${}^t\Omega^*(M;V)$ only, so there exists a continuous linear map $s : {}^tB^q(M;V) \to {}^t\Omega^{q-1}(M;V)$ satisfying $d_\Omega s = \mathrm{id}_{{}^tB^q(M;V)}$. In this case the equivariant function $\hat{\psi}^{p,q-1} : M^p \to {}^t\Omega^{q-1}(M;V)$ defined via

$$\hat{\psi}^{p,q-1}(\mathbf{m})(\mathbf{X}) = p(m_0).\left[s\hat{F}^{p,q}\left(p(m_0)^{-1}.\mathbf{m}\right)(p(m_0)^{-1}.\mathbf{X})\right]$$

is a cochain of bidegree $(p,q-1)$ in $A^p_s(M;{}^t\Omega^{q-1}(M;V))^G$. We assert that the vertical coboundary of $\hat{\psi}^{p,q-1}$ is the cochain $\hat{F}^{p,q}$. For all points $\mathbf{m}$ in M^{p+1} and $\mathbf{X}$ in $T^{\oplus q}M$ the value of $d_v\hat{\psi}^{p,q-1}(\mathbf{m})$ at $\mathbf{X}$ computes to

$$\begin{aligned}
\hat{\psi}^{p,q-1}(\mathbf{m})(\mathbf{X}) &= p(m_0).\left[d_v\hat{\psi}^{p,q-1}(p(m_0)^{-1}.\mathbf{m})(p(m_0)^{-1}.\mathbf{X})\right]\\
&= p(m_0).\left[d_v s\hat{F}^{p,q}\left(p(m_0)^{-1}.\mathbf{m}\right)(p(m_0)^{-1}.\mathbf{X})\right]\\
&= p(m_0).\left[\hat{F}^{p,q}\left(p(m_0)^{-1}.\mathbf{m}\right)(p(m_0)^{-1}.\mathbf{X})\right]\\
&= \hat{F}^{p,q}(\mathbf{m})(\mathbf{X}),
\end{aligned}$$

since the cochains $\hat{\psi}^{p,q-1}$ and $\hat{F}^{p,q}$ are equivariant. This proves that $\hat{F}^{p,q}$ is the vertical coboundary of the cochain $\hat{\psi}^{p,q-1}$. $\square$

Proposition 16.6.26. *If the function $\varphi_{M^p,T^{\oplus(q-1)}M,V}$ is bijective and the de Rham cohomology $H^q_{dR}(M;V)$ is trivial, then every cochain $F^{p,q}$ of bidegree (p,q) with trivial vertical coboundary $d_vF^{p,q}$ is the vertical coboundary $d_v\psi^{p,q-1}$ of a cochain $\psi^{p,q-1}$ of bidegree $(p,q-1)$.*

Proof. Let $\varphi_{M^p,T^{\oplus q}M,V}$ be bijective, $F^{p,q}$ be a cochain of bidegree (p,q) with trivial vertical coboundary $d_vF^{p,q}$ and assume the de Rham cohomology $H^q_{dR}(M;V)$ be trivial. By Lemma 16.6.25 there exists a cochain $\hat{\psi}^{p,q-1}$ of bidegree $(p,q-1)$ in $A^p_s(M;{}^t\Omega^{q-1}(M;V))^G$ whose vertical coboundary is the adjoint cochain $\hat{F}^{p,q}$ of $F^{p,q}$. Because the function $\varphi_{M^p,T^{\oplus(q-1)}M,V}$ is bijective, the adjoint cochain $\psi^{p,q-1}$ of $\hat{\psi}^{p,q-1}$ is a cochain of bidegree $(p,q-1)$ with vertical coboundary $F^{p,q}$. $\square$

Example 16.6.27. If the de Rham spaces ${}^t\Omega^*(M;V)$ are finite dimensional (e.g. if M and V are finite dimensional), then the topological de Rham complex ${}^t\Omega^*(M;V)$ splits. So, if the cohomology group $H^q_{dR}(MV)$ is trivial, every cochain $F^{p,q}$ of bidegree (p,q), $q \geq 1$ with trivial vertical coboundary $d_vF^{p,q}$ is the vertical coboundary $d_v\psi^{p,q-1}$ of a cochain $\psi^{p,q-1}$ of bidegree $(p,q-1)$.

Corollary 16.6.28. *If the functions $\varphi_{M^p,T^{\oplus(q-1)}M,V}$ are bijective for fixed p and all $q \geq 0$ and the de Rham cohomology $H^*_{dR}(M;V)$ is trivial, then the column $A^{p,*}_{s,dR}(M;V)^G$ of the double complex $A^{*,*}_{s,dR}(M;V)$ is exact.*

Recall that, by Lemma 14.2.19, the horizontal coboundary $d_h\psi^{p,q-1}$ of a cochain $\psi^{p,q-1}$ is a cocycle whenever the vertical coboundary $d_v\psi^{p,q-1}$ is a cocycle. Thus Proposition 16.6.26 can be applied repeatedly, provided its conditions are satisfied:

Corollary 16.6.29. *If $F^{p,q}$ is a cocycle of bidegree (p,q) and total degree n, the functions $\varphi_{M^k,T^{\oplus l}M,V}$ are bijective for $k + l = n - 1$, $k \geq p$ and the de Rham cohomologies $H^l_{dR}(M;V)$ are trivial for $l \leq q$, then there exist cocycles $F^{p+1,q-1},\ldots,F^{n,0}$ of bidegree $(p+1,q-1),\ldots,(n,0)$ and cochains $\psi^{p,q-1},\ldots,\psi^{n-1,0}$ of bidegree $(p,q-2),\ldots,(n-1,0)$ satisfying $D\psi^{p',q'} = F^{p'+1,q'} + F^{p',q'+1}$. In particular $F^{p,q}$ is cohomologous to $(-1)^q F^{n,0}$:*

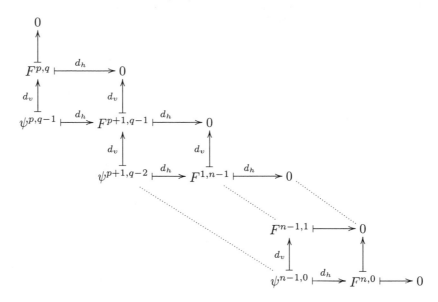

Corollary 16.6.30. *If M is n-connected and $\varphi_{M^{n-q},T^{\oplus(q-1)}M,V}$ is bijective for all $q \leq n$, then every closed n-form ω on M is integrable to a smooth equivariant global cocycle F on M.*

Proof. If M is n-connected, then the de Rham cohomology groups $H^q_{dR}(M;V)$ are trivial for all $q \leq n$. If in addition the homomorphisms $\varphi_{M^{n-q},T^{\oplus(q-1)}M,V}$ are bijective for all $q \leq n$, then repeated application of Corollary 16.6.29 shows that $i^n(\omega)$ is cohomologous to a cocycle $(-1)^n F^{n,0}$ of bidegree $(n,0)$. In this case ω is integrable to an equivariant smooth global cocycle by Lemma 16.6.12.   □

Example 16.6.31. If $M = G$ is a simply connected locally compact Lie group acting on itself via left translation, then every equivariant closed 1-form ω on G is integrable to a smooth global equivariant cocycle. Since finite dimensional Lie groups have trivial second homotopy group, the same is true for equivariant closed 2-forms.

Example 16.6.32. If $M = \Omega_s G$ is the smooth loop group of a simply connected locally compact Lie group G, then M is simply connected as well. If the group G acts on $M = \Omega_s G$ via the adjoint action on G, then every equivariant closed 1-form ω on $M = \Omega_s G$ is integrable to a smooth global equivariant cocycle on $\Omega_s G$. If one lets $\Omega_s G$ act on itself via left translation, then every equivariant closed 1-form ω on G is integrable to a smooth global equivariant cocycle.

As we have seen before, a closed equivariant n-form on M is integrable to a smooth global equivariant n-cocycle if and only if there exist cocycles $i^n(\omega) = F^{0,n},\ldots,F^{n,0}$ of bidegree $(0,n),(1,n-1),\ldots,(n,0)$ and cochains $\psi^{0,n-1},\ldots,\psi^{n-1,0}$ of bidegree $(0,n-1),(1,n-2),\ldots,(n-1,0)$ respectively such that $D\psi^{p-1,q} = F^{p,q} + F^{p-1,q+1}$:

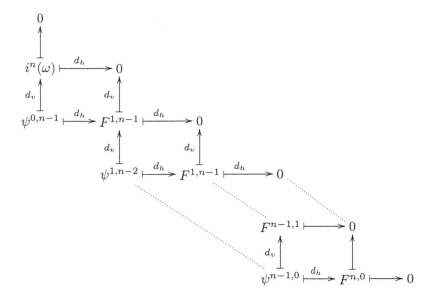

Since the columns of the double complex $A_{s,dR}^{*,*}(M;V)^G$ are not exact in general, we have to find conditions on ω and M which ensure the existence of the cochains $\psi^{p',q'}$ and $F^{p,q}$. This will be done step by step, i.e. we first describe conditions for $\psi^{0,n-1}$ to exist and then proceed to examine the cocycle $F^{1,n-1} = d_v\psi^{0,n-1}$ in order to find conditions for $\psi^{1,n-2}$ to exist etc..

Recall that the exactness of an equivariant closed n-form ω is a necessary condition for ω to be the image $\tau^n(F)$ of a smooth equivariant global n-cocycle F on M (Lemma 16.6.3). This condition will turn out to be necessary and sufficient for the existence of the cochain $\psi^{0,n-1}$, i.e. for the first step in the above described process. Since the de Rham homomorphism $H_{dR}^n(M;V) \to H^n(M;V)$ into the singular cohomology space $H^n(M;V)$ of M is an isomorphism, the exactness of the closed equivariant n-form ω on M can be tested by integrating over (smooth) singular n-cycles. This integration over singular n-cycles can be expressed in terms of the flux morphism $\mathcal{F}_{\varrho t}$ associated to the equivariant function

$$\varrho t : M \times G \to M, \quad (m,g) \mapsto g.m,$$

where $t : M \times G \to G \times M$ denotes the coordinate flip and we let the Lie group G act on itself via the adjoint action. The function ϱt is the bundle projection of a left G-principal bundle. The bundle projection ϱt induces a flux homomorphism from the equivariant de Rham complex $\Omega^*(M;V)^G$ into the total complex of the double complex $S^*(M;\Omega^*(G;V)^G)$ (see Section 13.3 for an introduction of general flux morphisms):

$$\mathcal{F}_{\varrho t} : \Omega^*(M;V)^G \to \mathrm{Tot}S^*(M;\Omega(G;V)^G) \cong \mathrm{Tot}S^*(M;C_c^*(\mathfrak{g};V)),$$

where we have corestricted the flux to the coefficient spaces $\Omega^q(G;V)^G \cong C^q(\mathfrak{g};V)$ of left equivariant forms on G (cf. Lemma 13.7.12). In the following we will only use this corestricted version of the flux $\mathcal{F}_{\varrho t}(\omega)$.

The image $\mathcal{F}_{\varrho t}(\omega)$ of an equivariant n-form ω on M under the flux homomorphism $\mathcal{F}_{\varrho t}$ is a cochain in the total complex $\mathrm{Tot}S^*(M;\Omega^*(G;V)^G)$. (See Section 13.6 for a discussion of flux homomorphisms). We denote the component of bidegree (p,q) of the flux $\mathcal{F}_{\varrho t}(\omega)$ by $\mathcal{F}_{\varrho t}(\omega)^{p,q}$ (cf. Section 13.6). If the equivariant form ω on M is closed, then the cochain $\mathcal{F}_{\varrho t}(\omega) = \sum_{p+q=n} \mathcal{F}_{\varrho t}(\omega)^{p,q}$ in the total complex $\mathrm{Tot}S^*(M;\Omega^*(G;V)^G)$ is a cocycle:

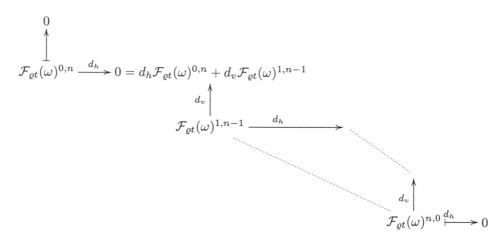

We use the component $\mathcal{F}_{\varrho t}(\omega)^{n,0}$ of bidegree $(n,0)$ of the flux $\mathcal{F}_{\varrho t}(\omega)$ to describe a necessary and sufficient condition for ω to be exact. This component $\mathcal{F}_{\varrho t}(\omega)^{n,0}$ is an equivariant homomorphism

$$\mathcal{F}_{\varrho t}(\omega)^{n,0} : S_n(M) \to \Omega^0(G;V)^G = C^\infty(G;V)^G,$$

$$\mathcal{F}_{\varrho t}(\omega)^{0,n}\left(\sum a_i \tau_i\right)(g) = \sum a_i \int_{g.\tau_i} \omega = g.\left(\sum a_i \int_{\tau_i} \omega\right)$$

The complex $\Omega^*(G;V)^G$ of left equivariant forms on G can be identified with the complex $C^*(\mathfrak{g};V)$ of Lie algebra cochains of the Lie algebra $\mathfrak{g}$ of G. As explained in Lemma 13.6.2 and Corollary 13.6.3 the homomorphisms $\mathcal{F}_{\varrho t}(\omega)^{p,q}$ map singular cycles to de Rham cocycles and restrict to homomorphisms

$$\mathcal{F}_{\varrho t}(\omega)^{p,q} : Z_p(M) \to Z^q_{dR}(G;V)^G \cong Z^q_c(\mathfrak{g};V)$$

which descend to homomorphisms

$$\mathcal{F}_{\varrho t}(\omega)^{p,q} : H_p(M) \to H^q_{dR,eq}(G;V) \cong H^q_c(\mathfrak{g};V).$$

In the special case $p = n$, $q = 0$ the codomain $Z^0_{dR}(M;V)^G \cong Z^0_c(\mathfrak{g};V)$ of the component $\mathcal{F}_{\varrho t}(\omega)^{n,0}$ of bidegree $(n,0)$ of the flux $\mathcal{F}_{\varrho t}(\omega)$ can be identified with the G-invariants V^G in V. So the homomorphism $\mathcal{F}_{\varrho t}(\omega)^{n,0}$ is given by

$$\mathcal{F}_{\varrho t}(\omega)^{n,0} : Z_n(M) \to V^G, \quad z \mapsto \int_z \omega, \tag{16.3}$$

i.e. by integration of the n-form ω over the singular n-cycles. The cohomology space $H^0_{dR,eq}(G;V)$ is isomorphic to the space $Z^0_{dR}(G;V)^G \cong V^G$, because the 0-boundaries are trivial. Moreover, since the form ω is closed, the integrals in formula 16.3 only depend on the homotopy classes of the cycles z; as consequence we see directly that the flux descends to a homomorphism $\mathcal{F}^{n,0}_{\varrho t}(\omega) : H_n(M) \to V^G$ and the latter is trivial if and only if $\mathcal{F}^{n,0}_{\varrho t}(\omega) : Z_n(M) \to V^G$ is trivial. So we observe:

Lemma 16.6.33. *An equivariant n-form ω on M is exact if and only if the component $\mathcal{F}_{\varrho t}(\omega)^{n,0} : Z_n(M) \to V^G$ of the flux is trivial if and only if the induced map $\mathcal{F}_{\varrho t}(\omega)^{n,0} : H_n(M) \to H^0_{dR,eq}(G;V) \cong H^0_c(\mathfrak{g};V) \cong V^G$ is trivial.*

Proof. Because the de Rham homomorphism $k^n : H^n_{dR}(M;V) \to H^n(M;V)$ is an isomorphism, the form ω is exact if and only if all integrals of ω over singular n-cycles are trivial. Formula 16.3 shows that this is equivalent to $\mathcal{F}_{\varrho t}(\omega)^{n,0}$ being trivial. $\square$

Recall from Section 16.2 that the triviality of the restricted flux components may also be expressed in terms of the rational flux components $\mathcal{F}(\omega)^{p,q} \otimes \mathrm{id}_{\mathbb{Q}}$, which make the diagram

$$
\begin{array}{ccc}
Z_p(M) & \xrightarrow{\ \mathcal{F}(\omega)^{p,q}\ } & Z_{dR}^q(G;V)^G \cong Z_c^q(\mathfrak{g};V) \\
{\scriptstyle z \mapsto z \otimes 1}\downarrow & & \| \\
Z_p(M) \otimes \mathbb{Q} & \xrightarrow{\ \mathcal{F}(\omega)^{p,q}\otimes \mathrm{id}_{\mathbb{Q}}\ } & Z_{dR}^q(G;V)^G \otimes \mathbb{Q} \cong Z_{dR}^q(G;V)^G \cong Z_c^q(\mathfrak{g};V).
\end{array}
$$

$$(16.4)$$

commutative. In particular, as is apparent from the above diagram, the flux component $\mathcal{F}(\omega)^{p,q}$ is trivial if and only if its rational version $\mathcal{F}(\omega)^{p,q} \otimes \mathrm{id}_{\mathbb{Q}}$ is trivial.

Lemma 16.6.34. *An equivariant n-form ω on M is exact if and only if the component $\mathcal{F}_{\varrho t}(\omega)^{n,0} \otimes \mathrm{id}_{\mathbb{Q}} : Z_n(M;\mathbb{Q}) \to V^G$ of the flux is trivial if and only if the induced map $\mathcal{F}_{\varrho t}(\omega)^{n,0} \otimes \mathrm{id}_{\mathbb{Q}} : H_n(M;\mathbb{Q}) \to H_{dR,eq}^0(G;V) \cong H_c^0(\mathfrak{g};V) \cong V^G$ is trivial.*

Proof. This has already been observed in Lemma 16.6.34. □

Since the exactness of the form ω is a necessary condition for ω to be the differential $\tau^n(F)$ of a smooth global equivariant n-cocycle F on M, we see that the vanishing of the $(0,n)$-flux in (co)homology also is a necessary condition for this to be the case. In the following we show that the exactness of ω (or equivalently, the triviality of the $(n,0)$-flux $\mathcal{F}_{\varrho t}(\omega)^{n,0} : H_0(G) \to H_{dR}^n(M;V)$) is a sufficient condition for the existence of a cochain $\psi^{n-1,0}$ with vertical coboundary $d_v\psi^{n-1,0} = i^n(\omega)$.

Lemma 16.6.35. *The equivariant n-form ω on M is exact if and only if there exists a cochain $\psi^{0,n-1}$ of bidegree $(0, n-1)$ in the total complex $\mathrm{Tot}A_{s,dR}^{*,*}(M;V)^G$ with vertical coboundary $d_v\psi^{0,n-1} = i^n(\omega)$.*

Proof. Let $p : M \to N$ be an equivariant projection onto a fundamental domain N of M so the action $\varrho : G \times M \to M$ induces a diffeomorphism $\varrho : G \times N \to M$. The function $\mathrm{pr}_G \varrho_{|G\times N}^{-1} : M \to G$ is a smooth equivariant function on M. Assume the form ω to be exact, i.e. that there exists a closed form $\eta \in \Omega^{n-1}(M;V)$ with coboundary $d\eta = \omega$. Then the function $\psi^{0,n-1} : M \times T^{\oplus(n-1)}M \to V$ defined via

$$\psi^{0,n-1}(m_0, \mathbf{X}) := [p(m_0).\eta](\mathbf{X}) = p(m_0).[\eta(p(m_0)^{-1}.\mathbf{X})]$$

is equivariant and thus defines an element in $A_{s,dR}^{0,n-1}(M;V)^G$. Furthermore the vertical differential of the cochain $\psi^{0,n-1}$ computes to

$$
\begin{aligned}
(d_v\psi^{0,n-1})(m_0), \mathbf{X}) &= d[p(m_0).\eta](\mathbf{X}) = [p(m_0).d\eta](\mathbf{X}) \\
&= [p(m_0).\omega](\mathbf{X}) = \omega(\mathbf{X}) = i^n(\omega)(m_0, \mathbf{X}),
\end{aligned}
$$

since the form ω is equivariant. So for every equivariant exact n-form ω on M there exists a cochain $\psi^{0,n-1}$ with vertical coboundary $i^n(\omega)$. Conversely, assume that there exists a cochain $\psi^{0,n-1}$ of bidegree $(0, n-1)$ with vertical differential $d_v\psi^{0,n-1} = \omega$. Fix a point $m \in M$ and let the $(n-1)$-form η on M be defined by $\eta(\mathbf{X}) = \psi^{0,n-1}(m, \mathbf{X})$. The coboundary of the $(n-1)$-form η on M is given by

$$d\eta(\mathbf{X}) = d_h\psi(m, \mathbf{X}) = i^n(\omega)(m, \mathbf{X}) = \omega(\mathbf{X}),$$

i.e. it is exactly the n-form ω on M. Thus the form ω on M is exact. □

If the manifold M is $(n-1)$-connected, then the n-th Hurewicz homomorphism $\mathrm{hur}_n : \pi_n(M) \to H_n(M)$ is an epimorphism. In this case the flux homomorphism $\mathcal{F}_{\varrho t}(\omega)^{n,0} : H_n(M) \to V^G$ is trivial if and only if its composition with the Hurewicz homomorphism hur_n is trivial. Recall that this composition is the period homomorphism

$$\mathrm{per}_{\varrho t}(\omega)_n = \mathrm{hur}_n \mathcal{F}_{\varrho t}(\omega)^{n,0} : H_n(M) \to V^G, \quad [s] \mapsto \int_s \omega$$

by definition. As a consequence we note:

Lemma 16.6.36. *If the manifold M is $(n-1)$-connected, then an equivariant closed n-form ω is exact if and only if the period homomorphism $\mathrm{per}_{\varrho t}(\omega)_n$ is trivial.*

Since the codomain $Z_{dR}^q(G; V)^G$ of any period homomorphism $\mathrm{per}_{\varrho t}(\omega)_p$ is a rational vector space, we also can w.l.o.g. switch to the rational version: The commutative diagram 16.4 gives rise to a diagram in (co)homology which may be extended on left by the (rational) Hurewicz-homomorphism to obtain the following commutative diagram:

$$
\begin{array}{c}
\overset{\mathrm{per}_{\varrho t}(\omega)_n}{\overbrace{\hspace{10cm}}} \\[-0.3em]
\begin{array}{ccccc}
\pi_n(M) & \xrightarrow{\ \mathrm{hur}_n\ } & H_p(M) & \xrightarrow{\ \mathcal{F}(\omega)^{p,q}\ } & H_{dR}^q(G; V)^G \cong H_c^q(\mathfrak{g}; V) \\[0.5em]
{\scriptstyle s \mapsto s \otimes 1}\Big\downarrow & & {\scriptstyle z \mapsto z \otimes 1}\Big\downarrow & & \Big\| \\[1em]
\pi_n(M) \otimes \mathbb{Q} & \xrightarrow{\mathrm{hur}_n \otimes \mathrm{id}_{\mathbb{Q}}} & H_p(M; \mathbb{Q}) & \xrightarrow{\mathcal{F}(\omega)^{p,q} \otimes \mathrm{id}_{\mathbb{Q}}} & H_{dR}^q(G; V)^G \cong Z_c^q(\mathfrak{g}; V),
\end{array} \\[-0.3em]
\underset{\mathrm{por}_{\varrho t}(\omega)_n \otimes \mathrm{id}_{\mathbb{Q}}}{\underbrace{\hspace{10cm}}}
\end{array}
$$

(16.5)

where the lowest horizontal arrow is the rational version $\mathrm{per}_{\varrho t}(\omega)_n \otimes \mathrm{id}_{\mathbb{Q}}$ of the period homomorphism $\mathrm{per}_{\varrho t}(\omega)_n$. This rational period homomorphism is trivial if and only if the the period homomorphism $\mathrm{per}_{\varrho t}(\omega)_n$ itself is trivial. We record the result:

Lemma 16.6.37. *If the manifold M is $(n-1)$-connected, then an equivariant closed n-form ω is exact if and only if the rational period homomorphism $\mathrm{per}_{\varrho t}(\omega)_n \otimes \mathrm{id}_{\mathbb{Q}}$ is trivial.*

Recall that the rational vector spaces $Z^n(M; V)^G$ and $Z_{dR}^n(M; V)^G$ can be identified with the tensor products $Z^n(M; V)^G \otimes \mathbb{Q}$ and $Z_{dR}^n(M; V)^G \otimes \mathbb{Q}$ respectively. Using these identifications we observe:

Proposition 16.6.38. *If the manifold M is $(n-1)$-connected and $\varphi_{M^{n-q}, T^{\oplus(q-1)}M, V}$ is bijective for all $q \leq n$, then the sequences*

$$Z^n(M; V)^G \to Z_{dR}^n(M; V)^G \xrightarrow{\ \mathrm{per}_{\varrho t}(-)_n\ } \mathrm{hom}(\pi_n(M); V^G)$$

$$Z^n(M; V)^G \to Z_{dR}^n(M; V)^G \xrightarrow{\ \mathrm{per}_{\varrho t}(-)_n \otimes \mathrm{id}_{\mathbb{Q}}\ } \mathrm{hom}(\pi_n(M) \otimes \mathbb{Q}; V^G)$$

are exact, i.e. an equivariant closed n-form ω on M is integrable to a smooth global equivariant cocycle if and only if its period homomorphism $\mathrm{per}_{\varrho t}(\omega)_n$ is trivial if and only if its rational period homomorphism $\mathrm{per}_{\varrho t}(\omega)_n \otimes \mathbb{Q}$ is trivial

Proof. Let M be $(n-1)$-connected. If the form ω is integrable to a global smooth cocycle, then its is exact by Lemma 16.6.3, hence period homomorphism $\mathrm{per}_{\varrho t}(\omega)_n$ and its rational version $\mathrm{per}_{\varrho t}(\omega)_n \otimes \mathrm{id}_{\mathbb{Q}}$ are trivial by Lemmata 16.6.36 and 16.6.37. Conversely, if these period homomorphisms are trivial, then ω is exact and $i^n(\omega)$

is cohomologous to a cocycle $-F^{1,n-1}$ of bidegree $(1, n-1)$ in $\mathrm{Tot}A^{*,*}_{s,dR}(M;V)^G$ by Lemma 16.6.35. By Corollary 16.6.29 the cocycle $-F^{1,n-1}$ is cohomologous to a cocycle $(-1)^n F^{n,0} = j^n(F)$ of bidegree $(n,0)$. Therefore $i^n(\omega)$ is cohomologous to the cocycle $j^n(F)$ of bidegree $(n,0)$, which implies that ω is integrable to a global smooth equivariant cocycle (Lemma 16.6.12). □

Corollary 16.6.39. *If G is an $(n-1)$-connected smoothly paracompact metrisable Lie group and ω is a Lie algebra n-cocycle, then ω is integrable to a smooth group cocycle if and only if the period map $\mathrm{per}_{\varrho t}(\omega_{eq})_n : \pi_n(G) \to V^G$ of its corresponding equivariant closed n-form ω_{eq} is trivial.*

Example 16.6.40. If M is a smoothly paracompact metrisable $(n-1)$-connected manifold, e.g. a smoothly paracompact $(n-1)$-connected Riemannian manifold, then the maps $\varphi_{M^{n-q}, T^{\oplus(q-1)}M, V}$ are always bijective. Thus an equivariant differential n-form ω on M is integrable to a smooth global equivariant cocycle if and only if its period map $\mathrm{per}_{\varrho t}(\omega)_n : \pi_n(M) \to V^G$ is trivial.

Example 16.6.41. If $M = G$ is a connected locally compact Lie group acting on itself via left translation, then every equivariant closed 1-form ω on G is integrable to a smooth global equivariant cocycle if and only if the period homomorphism $\mathrm{per}_{\varrho t}(\omega)_1 : \pi_1(G) \to V^G$ is trivial if and only if the rational period homomorphism $\mathrm{per}_{\varrho t}(\omega)_1 \otimes \mathrm{id}_{\mathbb{Q}} : \pi_1(G) \otimes \mathbb{Q} \to V^G$ is trivial. If G is simply connected then every equivariant closed 2-form is integrable to a smooth equivariant cocycle and an equivariant closed 3-form ω on G is integrable to a smooth global equivariant cocycle if and only if the period homomorphism $\mathrm{per}_{\varrho t}(\omega)_3 : \pi_3(G) \to V^G$ is trivial if and only if the rational period homomorphism $\mathrm{per}_{\varrho t}(\omega)_3 \otimes \mathrm{id}_{\mathbb{Q}} : \pi_3(G) \otimes \mathbb{Q} \to V^G$ is trivial.

In the remaining part of this Section we will understand $\psi^{0,n-1}$ to be the cochain constructed in the above Lemma, provided the form ω is exact. In this case we denote the horizontal coboundary $d_h \psi^{0,n-1}$ of $\psi^{0,n-1}$ by $F^{1,n-1}$. We will then work with the cochains $\psi^{0,n-1}$ and $F^{1,n-1}$ without further reference. Lemma 14.2.19 ensures that the horizontal coboundary of the cochain $F^{1,n-1} := d_h \psi^{0,n-1}$ is indeed a cocycle in the total complex $\mathrm{Tot}A^{*,*}_{s,dR}(M;V)^G$. The cocycle $F^{1,n-1}$ is then expected to represent the form ω in some way:

Lemma 16.6.42. *If the form ω is exact, then the cocycle $i^n(\omega)$ in the total complex of $A^{*,*}_{s,dR}(M;V)^G$ is cohomologous to the cocycle $-F^{1,n-1} \in Z^{1,n-1}_{s,dR}(M;V)^G$.*

Proof. Let the n-form ω be exact and consider the cocycle $F^{1,n-1}$ in the total complex $\mathrm{Tot}A^{*,*}_{s,dR}(M;V)^G$. Since the difference $i^n(\omega) - (-F^{1,n-1})$ is the coboundary $D\psi^{0,n-1} = d_h \psi^{0,n-1} + d_v \psi^{0,n-1}$ in the total complex $\mathrm{Tot}A^{*,*}_{s,dR}(M;V)^G)$, the cocycles $i^n(\omega)$ and $-F^1$ in the total complex $\mathrm{Tot}A^{*,*}_{s,dR}(M;V)^G)$ are cohomologous. □

Lemma 16.6.43. *If the cocycle $i^n(\omega)$ is cohomologous to a cocycle $-\tilde{F}^{1,n-1}$ of bidegree $(1, n-1)$ in $\mathrm{Tot}A^{*,*}_{s,dR}(M;V)^G$, then there exists a cochain $c^{0,n-1}$ in $\mathrm{Tot}A^{*,*}_{s,dR}(M;V)^G$ such that $d_v c^{0,n-1} = i^n(\omega)$ and $d_h c^{0,n-1}$ is cohomologous to $\tilde{F}^{1,n-1}$.*

Proof. Assume that there exists a cocycle $\tilde{F}^{1,n-1}$ of bidegree $(1, n-1)$ which is cohomologous to $-i^n(\omega)$ in the total complex $\mathrm{Tot}A^{*,*}_{s,dR}(M;V)^G$. Because the cocycles $i^n(\omega)$ and $-\tilde{F}^{1,n-1}$ are cohomologous, there exists a cochain $c = \sum_{p'+q'=n-1} c^{p',q'}$ with coboundary $Dc = i^n(\omega) + \tilde{F}^{1,n-1}$ in the total complex $\mathrm{Tot}A^{*,*}_{s,dR}(M;V)^G$. In particular the vertical coboundary $d_v c^{0,n-1}$ is exactly the cocycle $i^n(\omega)$. The equalities

$$D \sum_{p'+q'=n-1, p'>0} c^{p',q'} = i^n(\omega) + \tilde{F}^{1,n-1} - Dc^{0,n-1}$$

$$= i^n(\omega) + \tilde{F}^{1,n-1} - d_h c^{0,n-1} - d_v c^{0,n-1}$$
$$= i^n(\omega) + \tilde{F}^{1,n-1} - d_h c^{0,n-1} - i^n(\omega)$$
$$= (-d_h c^{0,n-1}) + \tilde{F}^{1,n-1}$$

then show that the cocycle $d_h c^{0,1}$ is cohomologous to $\tilde{F}^{1,n-1}$.   □

Lemma 16.6.44. *If the cocycle* $i^n(\omega)$ *is cohomologous to a cocycle* $-\tilde{F}^{1,n-1}$ *of bidegree* $(1, n-1)$ *in* $\mathrm{Tot}A^{*,*}_{s,dR}(M;V)^G$, *then the form* ω *is exact and* $\tilde{F}^{1,n-1}$ *is cohomologous to* $F^{1,n-1}$.

Proof. Assume that there exists a cocycle $\tilde{F}^{1,n-1}$ of bidegree $(1, n-1)$ in the total complex of $A^{*,*}_{s,dR}(M;V)^G$ such that $i^n(\omega)$ is cohomologous to $-\tilde{F}^{1,n-1}$. By the preceding Lemma there exists a cochain $c^{0,n-1}$ of bidegree $(0, n-1)$ satisfying $d_v c^{0,n-1} = i^n(\omega)$ and such that $d_h c^{0,n-1}$ is cohomologous to $\tilde{F}^{1,n-1}$. Fix a point m in M and define an $(n-1)$-form $\eta' \in \Omega^{n-1}(M;V)$ on M via

$$\eta'(\mathbf{X}) = c^{0,n-1}(m, \mathbf{X}).$$

The coboundary $d\eta'$ of η' is given by the vertical coboundary of the cochain $c^{0,n-1}$:

$$d\eta'(\mathbf{X}) = (d[c^{0,n-1}(m, -)])(\mathbf{X}) = [d_v c^{0,n-1}](m, \mathbf{X}) = i^n(\omega)(m, \mathbf{X}) = \omega(\mathbf{X})$$

Thus the n-form ω is exact. Since both cocycles $-\tilde{F}^{1,n-1}$ and $-F^{1,n-1}$ are cohomologous to $i^n(\omega)$, the cocycles $\tilde{F}^{1,n-1}$ and $F^{1,n-1}$ have to be cohomologous.   □

Summarising the three preceding Lemmata we observe:

Lemma 16.6.45. *The closed equivariant n-form ω on M is exact if and only if* $i^n(\omega)$ *is cohomologous to a cocycle of bidegree* $(1, n-1)$ *in* $A^{*,*}_{s,dR}(M;V)^G$.

Proposition 16.6.46. *The following conditions on the closed equivariant n-form ω on M are equivalent:*

1. *The form ω is exact.*
2. *The flux component $\mathcal{F}_{\varrho t}(\omega)^{n,0} : Z_n(M) \to V^G$ of ω is trivial.*
3. *The induced map $\mathcal{F}_{\varrho t}(\omega)^{n,0} : H_n(M) \to V^G$ in (co)homology is trivial.*
4. *The rational flux component $\mathcal{F}_{\varrho t}(\omega)^{n,0} \otimes \mathrm{id}_{\mathbb{Q}} : Z_n(M;\mathbb{Q}) \to V^G$ of ω is trivial.*
5. *The induced map $\mathcal{F}_{\varrho t}(\omega)^{n,0} \otimes \mathrm{id}_{\mathbb{Q}} : H_n(M;\mathbb{Q}) \to V^G$ in rational (co)homology is trivial.*
6. *There exists a cochain $\psi^{0,n-1}$ of bidegree $(0, n-1)$ in the total complex $\mathrm{Tot}A^{*,*}_{s,dR}(M;V)^G$ with vertical differential $d_v \psi^{0,n-1} = i^n(\omega)$.*
7. *The cocycle $i^n(\omega)$ is cohomologous to a cocycle $-F^{1,n-1}$ of bidegree $(1, n-1)$.*

Proof. It has already been shown in Lemma 16.6.33 that the equivariant n-form ω is exact if and only if the $(0,n)$-flux $\mathcal{F}^{0,n}_\varrho(\omega)$ of ω is trivial if and only if the induced map in (co)homology is trivial, which proves the equivalence of 1, 2 and 3. The equivalence of 1, 4 and 5 is the content of Lemma 16.6.34. The equivalence of 1 and 6 is shown in Lemma 16.6.35. Finally, the equivalence of 1 and 7 has been shown in Lemma 16.6.45.   □

If the manifold M is $(n-1)$-connected then the conditions in Proposition 16.6.46 are equivalent to the triviality of the period map $\mathrm{per}_{\varrho t}(\omega)_n : \Pi_n(M) \to V^G$. If additionally the maps $\varphi_{M^{n-q},T^{\oplus(q-1)}M,V}$ are bijective for all $q \le n$, then an equivariant closed n-form ω is integrable to a smooth global equivariant cocycle if and only if the period map $\mathrm{per}_{\varrho t}(\omega)_n$ (Proposition 16.6.38).

If the manifold M is $(n-2)$-connected and $n \geq 3$, then the Hurewicz homomorphism $\mathrm{hur}_n : \pi_n(M) \to H_n(M)$ is an epimorphism. In this case the exactness of the form ω can also be tested by integrating over n-spheres $\mathbb{S}^n \to M$ only. This makes the exactness of ω expressible in terms of a period map. Recall that the period homomorphism for the flux $\mathcal{F}_{\varrho t}^{n,0}(\omega) : H_n(M) \to H_{dR,eq}^0(G;V) = V^G$ is given by

$$\mathrm{per}_{\varrho t}(\omega)_n = [\mathcal{F}_{\varrho t}^{n,0}(\omega) \circ \mathrm{hur}_n](s) = \int_s \omega$$

i.e. by integration of ω over n-spheres in M.

Lemma 16.6.47. *If the period homomorphism* $\mathrm{per}_{\varrho t}(\omega)_n : \pi_n(M) \to V^G$ *is trivial, $n \geq 3$ and the manifold M is $(n-2)$-connected, then the conditions of Proposition 16.6.46 are satisfied, i.e. ω is exact and $i^n(\omega)$ is cohomologous to a cocycle $-F^{1,n-1}$ in* $\mathrm{Tot}A_{s,dR}^{*,*}(M;V)^G$.

Proof. If the manifold M is $(n-2)$-connected and $n \geq 3$, then the Hurewicz homomorphism $\mathrm{hur}_n : \pi_n(M) \to H_n(M)$ is an epimorphism. Therefore the assumption that the period homomorphism $\mathrm{per}_{\varrho t}(\omega)_n : \pi_n(M) \to V^G$ is trivial and the de Rham homomorphism $H_{dR}^n(M;V) \to H^n(M;V)$ is injective imply that the form ω is exact and thus all the conditions of Proposition 16.6.46 are satisfied. $\square$

Example 16.6.48. If M is simply connected and ω a closed equivariant 3-form on M, then $i^n(\omega)$ is cohomologous to a cocycle $-F^{1,2}$ of bidegree $(1,2)$ in $\mathrm{Tot}A_{s,dR}^{*,*}(M;V)^G$ if and only if the period map $\mathrm{per}_{\varrho t}(\omega) : \pi_3(M) \to V^G$ is trivial.

It has been shown that for smoothly paracompact Lie groups G acting on themselves by left translations the triviality of all period homomorphisms $\mathrm{per}_{\varrho t}(\omega)_p : \pi_p(G) \to H_{dR,eq}^{n-p}(G;V)$ implies the exactness of ω. If the group G is $\max\{k \in \mathbb{Z} \mid k \leq \frac{n}{2}\}$-connected in addition, then Lemma 16.2.11 ensures that the Hurewicz homomorphism $\mathrm{hur}_N : \pi(G) \to H_n(G)$ is surjective. So we observe:

Lemma 16.6.49. *If G is a smoothly paracompact $\max\{k \in \mathbb{Z} \mid k \leq \frac{n}{2}\}$-connected Lie group then the triviality of the period map* $\mathrm{per}_{\varrho t}(\omega)_n : \pi_n(G) \to V^G$ *implies the exactness of ω.*

Recall that the exactness of the equivariant n-form ω on M is a necessary condition for ω to be integrable to a smooth global equivariant n-cocycle on M. Moreover, if the form ω is exact, then the integrability of ω can be decided by examination of the cocycle $F^{1,n-1} = d_h \psi^{0,n-1}$ constructed in Lemma 16.6.35:

Lemma 16.6.50. *If the equivariant n-form ω on M is exact, then it is integrable to a smooth global equivariant n-cocycle if and only if $F^{1,n-1}$ is cohomologous to a cocycle of bidegree $(0,n)$ in* $\mathrm{Tot}A_{s,dR}^{*,*}(M;V)^G$.

Proof. By Lemma 16.6.12 the closed equivariant n-form ω is integrable to a smooth global equivariant n-cocycle if and only if the cocycle $i^n(\omega)$ is cohomologous to a cocycle of bidegree $(0,n)$ in $\mathrm{Tot}A_{s,dR}^{*,*}(M;V)^G$. If the form ω is exact, then the cocycles $i^n(\omega)$ and $-F^{1,n-1}$ are cohomologous. In this case $i^n(\omega)$ is cohomologous to a cocycle of bidegree $(0,n)$ in $\mathrm{Tot}A_{s,dR}^{*,*}(M;V)^G$ if and only if $-F^{1,n-1}$ and thus $F^{1,n-1}$ is cohomologous to a cocycle of bidegree $(0,n)$ in $\mathrm{Tot}A_{s,dR}^{*,*}(M;V)^G$. $\square$

This observation also motivates the extension of the definition of integrability to arbitrary cochains in the double complex $A_{s,dR}^{*,*}(M;V)$, which was introduced in the last section. Using this notion we observe:

Lemma 16.6.51. *If the equivariant n-form ω on M is exact and the cocycle $F^{1,n-1}$ is integrable to an equivariant smooth global cocycle of bidegree $(1, n-1)$ in $A^{1,n-1}_{s,s}(M;V)^G$, then ω is integrable to an equivariant smooth global n-cocycle.*

Proof. If the form ω is exact and the cocycle $F^{1,n-1}$ is integrable to an equivariant smooth global cocycle, then the latter is cohomologous to a cocycle $F^{n,0}$ of bidegree $(n,0)$ in the total complex of $A^{*,*}_{s,dR}(M;V)$ by Lemma 16.6.22. Since the cocycle $i^n(\omega)$ is cohomologous to $-F^{1,n-1}$ it also is cohomologous to $F^{n,0}$ in this case. By Lemma 16.6.12 the n-form is then integrable to a smooth global equivariant n-cocycle. $\qquad\square$

16.7 $(n-2)$-Connected Manifolds

In this section we continue to consider a smooth transformation group (G, M) with smoothly paracompact G-manifold M. Like in the last section we require the topological de Rham complex ${}^t\Omega^*(M;V)$ of M to split and assume the existence of an equivariant smooth map $p : M \to G$. We already know that the exactness of an equivariant n-form ω on M is necessary for ω to be integrable to a global equivariant cocycle. Therefore we assume the closed n-form ω on M we would like to integrate to be exact and fix an $(n-1)$-form η such that $d\eta = \omega$. Furthermore we let the cochain $\psi^{0,n-1}$ in $A^{0,n-1}_{s,dR}(M;V)$ be defined via

$$\psi^{0,n-1}(m_0, \mathbf{X}) := [p(m_0).\eta](\mathbf{X})$$

so the vertical coboundary of $\psi^{0,n-1}$ in the double complex $A^{*,*}_{s,dR}(M;V)^G$ is the cocycle $i^n(\omega)$ (cf. the proof of Lemma 16.6.35). We denote the horizontal coboundary $d_v\psi^{0,n-1}$ of $\psi^{0,n-1}$ by $F^{1,n-1}$. The cochain $F^{1,n-1}$ is a cocycle in the total complex $\mathrm{Tot}A^{*,*}_{s,dR}(M;V)^G$ by Lemma 14.2.19 and the cocycles $i^n(\omega)$ and $-F^{1,n-1}$ are cohomologous in the total complex by Lemma 16.6.42. The situation in the double complex $A_{s,dR}(M;V)^G$ can be depicted as follows:

$$
\begin{array}{ccc}
0 & & \\
\uparrow {\scriptstyle d_v} & & \\
i^n(\omega) & \longmapsto & 0 \\
\uparrow {\scriptstyle d_v} & & \uparrow {\scriptstyle d_v} \\
\psi^{0,n-1} & \overset{d_h}{\longmapsto} F^{1,n-1} & \overset{d_h}{\longrightarrow} 0
\end{array}
$$

The cocycle $F^{1,n-1}$ has trivial vertical and horizontal coboundaries, so its adjoint cocycle $\hat{F}^{1,n-1} : M^2 \to {}^t\Omega^{n-1}(M;V)$ in the total complex $\mathrm{Tot}A^*_s(M; {}^t\Omega^*(M;V))^G$ has codomain in the subspaces ${}^tZ^{n-1}_{dR}(M;V)$ of closed $(n-1)$-forms on M. We observe that the cocycles $F^{1,n-1}$ and $\hat{F}^{1,n-1}$ in the total complexes of $A^{*,*}_{s,dR}(M;V)^G$ and $A^*_s(M; {}^t\Omega^*(M;V))^G$ respectively can be explicitly written down in terms of the $(n-1)$-form η:

$$\hat{F}^{1,n-1}(m_0, m_1)(\mathbf{X}) = F^{1,n-1}(m_0, m_1, \mathbf{X}) = \psi^{0,n-1}(m_1, \mathbf{X}) - \psi^{0,n-1}(m_0, \mathbf{X})$$
$$= [p(m_1).\eta](\mathbf{X}) - [p(m_0).\eta](\mathbf{X})$$

We now examine the cocycle $F^{1,n-1}$ of bidegree $(1, n-1)$ and search for conditions on ω for the cocycle $F^{1,n-1}$ to be the vertical coboundary $d_v\psi^{1,n-2}$ of a cochain $\psi^{1,n-2}$ of bidegree $(1, n-2)$. As one may have expected, the necessary and sufficient condition for this to happen is a condition on the flux component $\mathcal{F}_{\varrho t}(\omega)^{n-1,1}$.

First assume that there exists a cochain $\psi^{1,n-2}$ of bidegree $(1, n-2)$ whose vertical coboundary $d_v\psi^{1,n-2}$ is the cocycle $F^{1,n-1}$. In this case the horizontal coboundary $F^{2,n-2} := d_h\psi^{1,n-2}$ of the cochain $\psi^{1,n-2}$ is a cocycle by Lemma 14.2.19 and this cocycle $F^{2,n-2}$ is cohomologous to the cocycle $-F^{1,n-1}$ by Lemma 14.2.19:

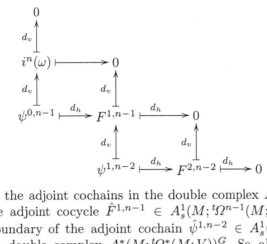

The situation for the adjoint cochains in the double complex $A_s^*(M; {}^t\Omega^*(M;V))^G$ is the same: The adjoint cocycle $\hat{F}^{1,n-1} \in A_s^1(M; {}^t\Omega^{n-1}(M;V))^G$ of $F^{1,n-1}$ is the vertical coboundary of the adjoint cochain $\hat{\psi}^{1,n-2} \in A_s^1(M; {}^t\Omega^{n-2}(M;V))^G$ of $\psi^{1,n-2}$ in the double complex $A_s^*(M; {}^t\Omega^*(M;V))^G$. So all the the cochains $\hat{F}^{n,1,n-1}(m_0, m_1)$ for points $(m_0, m_1) \in M^2$ are coboundaries, i.e. the the adjoint cocycle $\hat{F}^{1,n-1} = d_v\hat{\psi}^{1,n-1} : M^2 \to {}^t\Omega^{n-1}(M;V)$ takes values in the vector space ${}^tB_{dR}^{n-1}(M;V)$ of the Rham coboundaries in ${}^t\Omega^{n-1}(M;V)$. So we note

Lemma 16.7.1. *If the cocycle $F^{1,n-1}$ is the vertical coboundary $d_v\psi^{1,n-2}$ of a cochain of bidegree $(1, n-2)$, then it can be corestricted to ${}^tB_{dR}^{n-1}(M;V)$ and its image in the vector space $Z_s^1(M; {}^tH_{dR}^{n-1}(M;V))^G$ is trivial.*

This shows that the triviality of the cohomology classes $[\hat{F}^{1,n-1}(m_0, m_1)]$ in the topologised de Rham cohomology ${}^tH_{dR}^{n-1}(M;V)$ for all points $(m_0, m_1) \in M^2$ is a necessary condition for $F^{1,n-1}$ to be the vertical coboundary $d_v\psi^{1,n-2}$ of a cochain $\psi^{1,n-2}$ of bidegree $(1, n-2)$. If the exponential law holds for smooth V-valued functions on $M^2 \times T^{\oplus(n-1)}M$, i.e. if the homomorphism

$$\varphi_{M^2, T^{\oplus(n-1)}M, V} : A_{s,dR}^{1,n-1}(M;V)^G \to A_s^1(M; {}^t\Omega^{n-1}(M;V))^G$$

is an isomorphism, then the converse implication is also true:

Lemma 16.7.2. *If the image of $\hat{F}^{1,n-1}$ in the vector space $A_s^1(M; {}^tH_{dR}^{n-1}(M;V))^G$ is trivial, then the cocycle $\hat{F}^{1,n-1}$ is the vertical coboundary of a cochain $\hat{\psi}^{1,n-2}$ in $A_s^1(M; {}^t\Omega^{n-2}(M;V))^G$.*

Proof. Let the image of $\hat{F}^{1,n-1}$ in the vector space $A_s^1(M; {}^tH_{dR}^{n-1}(M;V))^G$ be trivial. If the topological de Rham complex ${}^t\Omega^*(M;V)$ splits, then there exists a continuous linear function $s : {}^t\Omega^{n-1}(M;V) \to {}^t\Omega^{n-2}(M;V)$ which satisfies the equation $(d_\Omega s)|_{{}^tB^{n-1}(M;V)} = \mathrm{id}_{{}^tB^{n-1}(M;V)}$. In this case the equivariant function $\hat{\psi}^{1,n-2} : M^2 \to {}^t\Omega^{q-1}(M;V)$ defined via

$$\hat{\psi}^{1,n-2}(\mathbf{m})(\mathbf{X}) = [p(m_0)^{-1}].\left[s\hat{F}^{1,n-1}\left(p(m_0)^{-1}.\mathbf{m}\right)\right)(p(m_0)^{-1}.\mathbf{X})\right]$$

is a cochain of bidegree $(1, n-2)$ in $A_s^1(M; {}^t\Omega^{n-2}(M;V))^G$ whose vertical coboundary is the cocycle $\hat{F}^{1,n-1}$. □

Similar to the exactness condition on ω for the existence of $\psi^{0,n-1}$ the above Lemma implies a necessary condition for $F^{1,n-1}$ to be the vertical coboundary of a cochain $\psi^{1,n-2}$ of bidegree $(1, n-2)$ in the double complex $A_{s,dR}^{*,*}(M;V)^G$:

Lemma 16.7.3. *If the image of $\hat{F}^{1,n-1}$ in the vector space $A_s^1(M; {}^tH_{dR}^{n-1}(M;V))^G$ is trivial and $\varphi_{M^2,T^{\oplus(n-1)}M,V}$ is an isomorphism, then $F^{1,n-1}$ is the vertical coboundary of a cochain $\psi^{1,n-2}$ in $A_{s,dR}^{1,n-2}(M;V)^G$.*

Proof. If the image of $\hat{F}^{1,n-1}$ in the vector space $A_s^1(M; {}^tH_{dR}^{n-1}(M;V))^G$ is trivial, then there exists a cochain $\hat{\psi}^{1,n-2}$ with vertical coboundary $\hat{F}^{1,n-1}$ by Lemma 16.7.2. If in addition the homomorphism $\varphi_{M^2,T^{\oplus(n-1)}M,V}$ is bijective, then the adjoint function $\psi^{1,n-2}$ of $\hat{\psi}^{1,n-1}$ is a cochain in $A_{s,dR}^{1,n-2}(M;V)^G$. The vertical coboundary of the latter cochain is the cocycle $F^{1,n-1}$. $\qquad\square$

This shows that – under the assumption that M is smoothly paracompact – the triviality of the cohomology classes $[\hat{F}^{1,n-1}(m_0,m_1)]$ in the de Rham cohomology $H_{dR}^{n-1}(M;V)$ for all points $(m_0,m_1) \in M^2$ also is sufficient for $F^{1,n-1}$ to be the vertical coboundary $d_v\psi^{1,n-2}$ of a cochain $\psi^{1,n-2}$ of bidegree $(1, n-2)$. Similar to the integrability conditions in the last section, this condition can be expressed in terms of the flux $\mathcal{F}_{\varrho t}(\omega)$. As a first step to prove this we observe:

Lemma 16.7.4. *If the action of G on the coefficients V is trivial, then for all points $(m_0,m_1) \in M^2$ the integration of the $(n-1)$-form $\hat{F}^{1,n-1}(m_0,m_1)$ over a smooth singular $(n-1)$-cycle $z \in Z_n(M)$ is given by*

$$\int_z \hat{F}^{1,n-1}(m_0,m_1) = \int_{\gamma^{-1}.z} \omega = \int_{\gamma^{-1}} \mathcal{F}_{\varrho t}(\omega)^{n-1,1}(z)$$

for any smooth path γ from $p(m_0)$ to $p(m_1)$ in G.

Proof. Let (m_0,m_1) be a point in M^2 and $z \in Z_{n-1}(M)$ be a smooth singular $(n-1)$-cycle in M. The integration of the cocycle $\hat{F}^{1,n-1}(m_0,m_1)$ over the singular chain $z \in S_{n-1}(M)$ is by definition given by

$$\int_z \hat{F}^{1,n-1}(m_0,m_1) = \int_z [p(m_1).\eta] - [p(m_0).\eta].$$

If the action of G on the coefficients V is trivial and $g \in G$ any group element, then the $(n-1)$-form $g.\eta$ coincides with the $(n-1)$-form $\eta \circ T\varrho(g^{-1},-)$. Therefore the integral above simplifies to

$$\int_z \hat{F}^{1,n-1}(m_0,m_1) = \int_{p(m_1)^{-1}.z-p(m_0)^{-1}.z} \eta.$$

If the singular chain z is a cycle then the chain $p(m_1)^{-1}.z - p(m_0)^{-1}.z$ is the coboundary of the chain $\gamma^{-1}.z$ for any path γ from $p(m_0)$ to $p(m_1)$ in G. Therefore Stokes Theorem implies the equalities

$$\int_z \hat{F}^{1,n-1}(m_0,m_1) = \int_{(\partial\gamma^{-1}).z} \eta = \int_{\gamma^{-1}.z} d\eta + 0 = \int_{\gamma^{-1}.z} \omega = \int_{\gamma^{-1}} \mathcal{F}_{\varrho t}(\omega)^{n-1,1}(z).$$

This proves the assertion. $\qquad\square$

This observation enables us to express the triviality all of the cohomology classes $[\hat{F}^{1,n-1}(m_0,m_1)]$ in $H_{dR}^{n-1}(M;V)$ in terms of the flux component $\mathcal{F}_{\varrho t}(\omega)^{1,n-1}$:

Lemma 16.7.5. *If the cocycle $F^{1,n-1}$ is the vertical coboundary of a cochain $\psi^{1,n-2}$ of bidegree $(1, n-2)$ and the action of G on the coefficients V is trivial, then the flux $\mathcal{F}_{\varrho t}^{n-1,1}(\omega) : Z_{n-1}(M) \to Z_{dR}^1(G;V)^G \cong Z_c^1(\mathfrak{g};V)$ is trivial.*

Proof. If the cocycle $F^{1,n-1}$ is the vertical coboundary of a cochain $\psi^{1,n-2}$ of bidegree $(1, n-2)$, then the adjoint cocycle $\hat{F}^{1,n-1}$ is the vertical coboundary of the adjoint cochain $\hat{\psi}^{1,n-2}$ of $\psi^{1,n-2}$. In this case the $(n-1)$-form $\hat{F}^{1,n-1}(m_0, m_1)$ is exact for every point $(m_0, m_1) \in M^2$. This implies that all the integrals

$$\int_{\gamma^{-1}.z} \omega = \int_{\gamma^{-1}} \mathcal{F}_{\varrho t}^{n-1,1}(\omega)(z)$$

are trivial for all smooth paths γ in G and all smooth singular $(n-1)$-cycles z in M. This in turn implies that for every smooth singular $(n-1)$-cycle z in M the equivariant 1-form $\mathcal{F}_{\varrho t}^{n-1,1}(\omega)(z)$ on G is trivial. Therefore the flux component $\mathcal{F}_{\varrho t}(\omega)^{n-1,1} : Z_{n-1}(M) \to Z_{dR}^1(G; V)^G \cong Z_c^1(\mathfrak{g}; V)$ is trivial. $\square$

Lemma 16.7.6. *If the the flux $\mathcal{F}_{\varrho t}^{n-1,1}(\omega) : Z_{n-1}(M) \to Z_{dR}^1(G; V)^G \cong Z_c^1(\mathfrak{g}; V)$ is trivial and the action of G on the coefficients V is trivial, then the cohomology classes $[\hat{F}^{1,n-1}(m_0, m_1)]$ in $H_{dR}^{n-1}(M; V))^G$ are trivial for all points $(m_0, m_1) \in M^2$.*

Proof. Let the flux $\mathcal{F}_{\varrho t}^{n-1,1}(\omega) : Z_{n-1}(M) \to Z_{dR}^1(G; V)^G \cong Z_c^1(\mathfrak{g}; V)$ and the action of G on the coefficients V be trivial. Let $(m_0, m_1) \in M^2$ be given. By Lemma 16.7.4 the integration of $\hat{F}^{1,n-1}(m_0, m_1)$ over a smooth singular $(n-1)$-cycle z in M is given by

$$\hat{F}^{1,n-1}(m_0, m_1)(z) = \int_z \hat{F}^{1,n-1}(m_0, m_1) = \int_{\gamma^{-1}.z} \omega$$

for any smooth path γ from $p(m_0)$ to $p(m_1)$ in G. These integrals vanish by assumption. Since the manifold M was assumed to be smoothly paracompact, the de Rham homomorphism $H_{dR}^{n-1}(M; V) \to H^{n-1}(M; V)$ is an isomorphism. This implies that the cohomology class $[\hat{F}^{1,n-1}(m_0, m_1)]$ in $H_{dR}^{n-1}(M; V))^G$ is trivial. $\square$

Proposition 16.7.7. *If V is a trivial G-module and the function $\varphi_{M^2, T^{\oplus(n-1)}M, V}$ is an isomorphism, then the cocycle $F^{1,n-1}$ is the vertical coboundary $d_v \psi^{1,n-2}$ of a cochain $\psi^{1,n-2}$ if and only if the flux $\mathcal{F}_{\varrho t}^{n-1,1}(\omega) : Z_{n-1}(M) \to Z_{dR}^1(G; V)^G$ is trivial.*

Proof. If the cocycle $F^{1,n-1}$ is the vertical coboundary $d_v \psi^{1,n-2}$ of a cochain $\psi^{1,n-2}$, then the flux component $\mathcal{F}_{\varrho t}^{n-1,1}(\omega) : Z_{n-1}(M) \to Z_{dR}^1(G; V)^G \cong Z_c^1(\mathfrak{g}; V)$ is trivial by Lemma 16.7.5. Conversely, if this flux component is trivial, then the adjoint cocycle $\hat{F}^{1,n-1}$ has trivial image in $H_{dR}^{n-1}(M; V)$ by Lemma 16.7.6. In this case the cocycle $F^{1,n-1}$ is the vertical coboundary $d_v \psi^{1,n-2}$ of a cochain $\psi^{1,n-2}$ of bidegree $(1, n-2)$ by Lemma 16.7.2. $\square$

Proposition 16.7.8. *If V is a trivial G-module, ω an equivariant closed 2-form on M, the flux homomorphisms $\mathcal{F}_{\varrho t}(\omega)^{2,0} : Z_2(M) \to Z_{dR}^0(G; V)^G \cong Z_c^n(\mathfrak{g}; V) \cong V^G$ and $\mathcal{F}_{\varrho t}(\omega)^{1,1} : Z_1(M) \to Z_{dR}^1(G; V)^G \cong Z_c^1(\mathfrak{g}; V)$ are trivial and $\varphi_{M^2, T^{\oplus(n-1)}M, V}$ is an isomorphism, then there exist cocycles $F^{0,2} = i^n(\omega), F^{1,1}, F^{2,n-2}$ of bidegree $(0, 2), (1, 1)$ and $(2, 0)$ respectively and cochains $\psi^{0,1}, \psi^{1,0}$ of bidegree $(0, 1)$ and $(1, 0)$ respectively such that $D\psi^{p',q'} = F^{p'+1} + F^{p',q'+1}$ for all $p' + q' = 1$:*

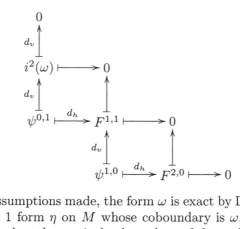

Proof. Under the assumptions made, the form ω is exact by Lemma 16.6.33. Therefore there exists an 1 form η on M whose coboundary is ω. The cocycle $i^2(\omega)$ in $\mathrm{Tot} A^{*,*}_{s,dR}(M;V)^G$ is then the vertical coboundary of the cochain $\psi^{0,1}$ defined via

$$\psi^{0,1}(m_0, \mathbf{X}) := [p(m_0).\eta](\mathbf{X}).$$

By Proposition 16.7.7 the cocycle $F^{1,1} := d_h \psi^{0,1}$ is the vertical coboundary of a cochain $\psi^{1,0}$, whose horizontal boundary $F^{2,0}$ is a cocycle by Lemma 14.2.19 $\square$

Theorem 16.7.9. *If V is a trivial G-module and $\varphi_{M^2, T^{\oplus(n-1)}M, V}$ is an isomorphism, ω is a closed equivariant 2-form with trivial flux homomorphisms*

$$\mathcal{F}^{2,0}_{\varrho t}(\omega) : Z_2(M) \to Z^0_{dR}(G;V)^G \cong Z^n_c(\mathfrak{g};V) \cong V^G$$

and $$\mathcal{F}^{1,1}_{\varrho t}(\omega) : Z_1(M) \to Z^1_{dR}(G;V)^G \cong Z^1_c(\mathfrak{g};V)$$

then ω is integrable to a smooth equivariant 2-cocycle on M.

Proof. Under the assumptions made, the cocycle $i^2(\omega)$ for a closed equivariant 2-form ω on M with trivial flux homomorphisms is cohomologous to the cocycle $F^{2,0}$ from the preceding proposition. By Lemma 14.2.22 the latter cocycle is the image $j^n(F)$ of a smooth equivariant global cocycle F on M under the augmentation $j^n : A^n_s(M;V)^G \hookrightarrow A^{n,*}_{s,dR}(M;V)$. The derivation homomorphism $\tau^2 : A^2_s(M;V)^G \to \Omega^2(M;V)^G$ maps the cocycle F to a closed equivariant 2-form ω', which is cohomologous to ω in $\Omega^*(M;V)^G$, i.e. $[\omega] = [\omega']$ in $H_{dR,eq}(M;V)$. By Proposition 16.6.11 the integrability only depends on the cohomology class $[\omega] \in H^2_{dR,eq}(M;V)$, hence ω is integrable to a smooth global equivariant cocycle. $\square$

Example 16.7.10. If V is a trivial G-module and M a locally compact G-manifold, then a closed equivariant 2-form ω on M is integrable to a smooth global equivariant cocycle if both period homomorphisms

$$\mathcal{F}^{2,0}_{\varrho t}(\omega) : Z_2(M) \to V^G \quad \text{and} \quad \mathcal{F}^{1,1}_{\varrho t}(\omega) : Z_1(M) \to Z^1_c(\mathfrak{g};V)$$

are trivial.

Example 16.7.11. If G is a Lie group with Lie algebra $\mathfrak{g}$ and $V \hookrightarrow \hat{\mathfrak{g}} \twoheadrightarrow \mathfrak{g}$ is a central extension of Lie algebras described by a 2-cocycle $\omega \in Z^2_c(\mathfrak{g};V)$, then this extension is integrable to a topologically split group extension if both period homomorphisms

$$\mathcal{F}^{2,0}_{\varrho t}(\omega) : Z_2(M) \to V^G \quad \text{and} \quad \mathcal{F}^{1,1}_{\varrho t}(\omega) : Z_1(M) \to Z^1_c(\mathfrak{g};V)$$

are trivial. For locally convex Hausdorff coefficients and Lie groups modelled on locally convex Hausdorff topological vector spaces this has been proven in [Nee02a]. Theorem 16.7.9 generalises this result to smoothly paracompact Lie groups with split de Rham complex modelled on arbitrary topological vector spaces and not necessarily locally convex Hausdorff coefficients V.

Theorem 16.7.12. *If V is a trivial G-module, ω an equivariant closed n-form on M, $n \geq 2$, the flux homomorphisms $\mathcal{F}_{\varrho t}^{n,0}(\omega) : Z_n(M) \to Z_{dR}^n(G;V)^G \cong Z_c^n(\mathfrak{g};V)$ and $\mathcal{F}_{\varrho t}^{n-1,1}(\omega) : Z_{n-1}(M) \to Z_{dR}^1(G;V)^G \cong Z_c^1(\mathfrak{g};V)$ are trivial and the topologised de Rham complex ${}^t\Omega^*(M;V)$ of M splits equivariantly and $\varphi_{M^2,T^{\oplus(n-1)}M,V}$ is an isomorphism, then there exist cocycles $F^{0,n} = i^n(\omega), F^{1,n-1}, F^{2,n-2}$ of bidegree $(0,n),(1,n-1)$ and $(2,n-2)$ respectively and cochains $\psi^{0,n-1}, \psi^{1,n-2}$ of bidegree $(0,n-1)$ and $(1,n-2)$ respectively such that $D\psi^{p',q'} = F^{p'+1} + F^{p',q'+1}$ for all $p' + q' = n-1$:*

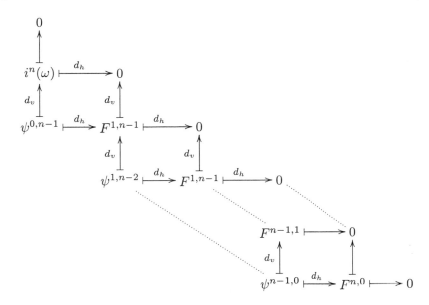

Proof. Under the assumptions made, the form ω is exact by Lemma 16.6.33. Therefore there exists an $(n-1)$ form η on M whose coboundary is ω. The cocycle $i^n(\omega)$ in $\mathrm{Tot}A_{s,dR}^{*,*}(M;V)^G$ is then the vertical coboundary of the cochain $\psi^{0,1}$ defined via

$$\psi^{0,n-1}(m_0, \mathbf{X}) := [p(m_0).\eta](\mathbf{X}).$$

By Proposition 16.7.7 the cocycle $F^{1,n-1} := d_h\psi^{0,1}$ is the vertical coboundary of a cochain $\psi^{1,n-2}$, whose horizontal boundary $F^{2,n-2}$ is a cocycle by Lemma 14.2.19. The assertion follows now from Lemma 16.6.29. $\qquad\square$

Corollary 16.7.13. *If V is a trivial G-module, M is smoothly paracompact, the topologised de Rham complex ${}^t\Omega^*(M;V)$ of M splits equivariantly, $\varphi_{M^2,T^{\oplus(n-1)}M,V}$ is an isomorphism and ω an equivariant closed n-form on M, $n \geq 2$, with trivial flux homomorphisms*

$$\mathcal{F}_{\varrho t}^{n,0}(\omega) \; : Z_n(M) \to Z_{dR}^0(G;V)^G \cong Z_c^n(\mathfrak{g};V) \cong V^G$$

and $\qquad \mathcal{F}_{\varrho t}^{n-1,1}(\omega) : Z_{n-1}(M) \to Z_{dR}^1(G;V)^G \cong Z_c^1(\mathfrak{g};V)$

then ω is integrable to a smooth equivariant n-cocycle on M.

Example 16.7.14. If V is a trivial G-module and M a simply connected locally compact G-manifold, then a closed equivariant 3-form ω on M is integrable to a smooth global equivariant 3-cocycle if both period homomorphisms

$$\mathcal{F}_{\varrho t}^{3,0}(\omega) : Z_3(M) \to V^G \qquad \text{and} \qquad \mathcal{F}_{\varrho t}^{1,2}(\omega) : Z_2(M) \to Z_c^1(\mathfrak{g};V)$$

are trivial.

A

Categories

A.1 Preliminaries on Morphisms

Lemma A.1.1. *Pullbacks of monomorphisms are monomorphisms, i.e. in any pull-back square*

$$P \xrightarrow{f'} X \quad , \tag{A.1}$$

(with vertical arrows m' from P to Y and m from X to Z, and bottom arrow $Y \xrightarrow{f} Z$)

where m is a monomorphism the morphism m' also is a monomorphism.

Proof. Let $m : X \to Z$ be a monomorphism and $f : Y \to Z$ be an arbitrary morphism with codomain Z. Assume that there exists a pullback square as in diagram A.1. We show that m' is left-cancellative. If $\alpha, \beta : P' \to P$ are morphisms satisfying the the equality $m'\alpha = m'\beta$, we can conclude

$$m'\alpha = m'\beta \Rightarrow fm'\alpha = fm'\beta \Rightarrow mf'\alpha = mf'\beta \Rightarrow f'\alpha = f'\beta,$$

where the last equation follows from the fact that m is a monomorphism and thus can be cancelled on the left. All in all we have observed that the following diagram is commutative:

$$\tag{A.2}$$

(commutative diagram with P' at top, arrow labelled $f'\alpha = f'\beta$ to X, arrow labelled $m'\alpha = m'\beta$ to Y, and inner square $P \xrightarrow{f'} X$, m' from P to Y, m from X to Z, $Y \xrightarrow{f} Z$)

Since the square in this commutative diagram is a pullback square, there exists a unique morphism $g : P' \to P$ satisfying the equations $f'g = f'\alpha = f'\beta$ and $m'g = m'\alpha = m'\beta$. Thus $g = \alpha = \beta$. $\square$

Lemma A.1.2. *Pushouts of epimorphisms are epimorphisms.*

Proof. This is the dual of the preceding Lemma. $\square$

Definition A.1.3. *The fibre product of morphisms $f_1 : X_1 \to Y$ and $f_2 : X_2 \to Y$ is the pullback of f_1 and f_2. When the morphisms f_1 and f_2 are known from the context, the colimit object is denoted $X_1 \times_Y X_2$.*

For many categories there exists a natural choice of such pullbacks. If this is the case we will always speak of *the fibre product* and understand it to be the one given by the natural choice.

Example A.1.4. In the category **Set** the fibre product of two maps $f_1 : X_1 \to Y$ and $f_2 : X_2 \to Y$ is the subset

$$X_1 \times_Y X_2 = \{(x_1, x_2) \in X_1 \times X_2 \mid f_1(x_1) = f_2(x_2)\}$$

of the product set $X_1 \times X_2$.

Example A.1.5. In the category **RMod** of modules over a ring R the fibre product of two morphisms $f_1 : M_1 \to N$ and $f_2 : M_2 \to N$ is the submodule

$$M_1 \times_N M_2 = \{(m_1, m_2) \in M_1 \times M_2 \mid f_1(m_1) = f_2(m_2)\}$$

of the product module $M_1 \times M_2$.

Example A.1.6. In the category **Top** of topological spaces the fibre product of two continuous functions $f_1 : X_1 \to Y$ and $f_2 : X_2 \to Y$ is the subspace

$$X_1 \times_Y X_2 = \{(x_1, x_2) \in X_1 \times X_2 \mid f_1(x_1) = f_2(x_2)\}$$

of the product space $X_1 \times X_2$. It is the inverse image of the diagonal in $X \times X$ under the function $f_1 \times f_2$. Thus if X is Hausdorff, then the fibre product $X_1 \times_Y X_2$ is a closed subspace of the product space $X \times X$.

Example A.1.7. In the category **TopRMod** of topological modules over a topological ring R the fibre product of two morphisms $f_1 : M_1 \to N$ and $f_2 : M_2 \to N$ is the submodule

$$M_1 \times_N M_2 = \{(m_1, m_2) \in M_1 \times M_2 \mid f_1(m_1) = f_2(m_2)\}$$

of the product module $M_1 \times M_2$. If the topological module N is Hausdorff, then the fibre product $M_1 \times_N M_2$ is a closed submodule of the product module $M_1 \times N_2$.

For any morphism $f : X \to Y$ in a category $\mathbf{C}$ the pullback of f with itself may exist. The morphisms of the corresponding cone have a special name:

Definition A.1.8. *A* kernel pair *of a morphism* $f : X \to Y$ *is the pair of morphisms* $X \times_Y X \to X$ *in the pullback cone.*

Example A.1.9. In the category **Set** of sets the kernel pair of a map $f : X \to Y$ is the pair of maps

$$X_1 \times_Y X_2 = \{(x_1, x_2) \in X_1 \times X_2 \mid f_1(x_1) = f_2(x_2)\} \to X$$

given by restricting the projections $\mathrm{pr}_1 : X \times X \to X$ and $\mathrm{pr}_2 : X \times X \to X$ onto the first and second factor respectively to the subset $X \times_Y X$ of $X \times X$.

Example A.1.10. In the category **RMod** of modules over a ring R the kernel pair of a homomorphism $f : M \to N$ is the pair of homomorphisms

$$M_1 \times_N M_2 = \{(m_1, m_2) \in M_1 \times M_2 \mid f_1(m_1) = f_2(m_2)\} \to M$$

given by restricting the projections $\mathrm{pr}_1 : M \times M \to M$ and $\mathrm{pr}_2 : M \times M \to M$ onto the first and second factor respectively to the submodule $M \times_N M$ of $M \times M$.

Example A.1.11. In the category **Top** of topological spaces the kernel pair of a continuous function $f : X \to Y$ is the pair of functions

$$X_1 \times_Y X_2 = \{(x_1, x_2) \in X_1 \times X_2 \mid f_1(x_1) = f_2(x_2)\} \to X$$

given by restricting the projections $\mathrm{pr}_1 : X \times X \to X$ and $\mathrm{pr}_2 : X \times X \to X$ onto the first and second factor respectively to the subspace $X \times_Y X$ of $X \times X$.

Example A.1.12. In the category **TopRMod** of topological modules over a topological ring R the kernel pair of a homomorphism $f : M \to N$ is the pair of homomorphisms

$$M_1 \times_N M_2 = \{(m_1, m_2) \in M_1 \times M_2 \mid f_1(m_1) = f_2(m_2)\} \to M$$

given by restricting the projections $\mathrm{pr}_1 : M \times M \to M$ and $\mathrm{pr}_2 : M \times M \to M$ onto the first and second factor respectively to the submodule $M \times_N M$ of $M \times M$.

Lemma A.1.13. *A morphism $f : X \to Y$ with kernel pair $p_1, p_2 : X \times_Y X \to X$ is a monomorphism if and only if $p_1 = p_2$.*

Proof. See [BB04, Lemma A.2.7] for a proof.   □

Let $f : X \to Y$ be a morphism with kernel pair p_1 and p_2. The morphisms p_1 and p_2 of the limit cone

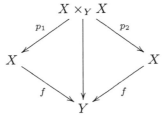

have common domain and codomain. This makes it possible to consider their coequalisers:

Definition A.1.14. *A coimage of a morphism $f : X \to Y$ in a category $\mathbf{C}$ such that the fibre product $X \times_Y X$ exists is a coequaliser*

$$\mathrm{CoIm}\, f = \mathrm{Coeq}(X \times_Y X \rightrightarrows X) : Y \to CI(f)$$

of the kernel pair of f.

Example A.1.15. In the category **Set** the coimage $\mathrm{CoIm}\, f$ of a map $f : X \to Y$ is the quotient map

$$\mathrm{CoIm}\, f : X \twoheadrightarrow X/\sim$$

onto the factor set $X/\sim$ of X by the equivalence relation $\sim$ which is generated by $x \sim y \Leftrightarrow f(x) = f(y)$. This equivalence relation $\sim$ is precisely the subset $X \times_Y X$ of $X \times X$.

Example A.1.16. In the category **RMod** of modules over a ring R the coimage of a homomorphism $f : M \to N$ is the quotient homomorphism

$$\mathrm{CoIm}\, f : M \twoheadrightarrow M/\sim$$

onto the quotient module $M/\sim$ of M by the congruence relation which is generated by $m \sim m' \Leftrightarrow f(m) = f(m')$. This congruence relation $\sim$ is given by the submodule $M \times_N M$ of $M \times M$.

Example A.1.17. In the category **Top** of topological spaces the coimage CoIm f of a continuous function $f : X \to Y$ is the quotient map

$$\text{CoIm } f : X \twoheadrightarrow X/\sim$$

onto the quotient space $X/\sim$ of X by the equivalence relation $\sim$ which is generated by $x \sim y \Leftrightarrow f(x) = f(y)$. This equivalence relation $\sim$ is given by the subspace $X \times_Y X$ of $X \times X$.

Example A.1.18. In the category **TopRMod** of topological modules over a topological ring R the coimage of a homomorphism $f : M \to N$ is the quotient homomorphism

$$\text{CoIm } f : M \twoheadrightarrow M/\sim$$

onto the quotient module $M/\sim$ of M by the congruence relation which is generated by $m \sim m' \Leftrightarrow f(m) = f(m')$. This congruence relation $\sim$ is given by the submodule $M \times_N M$ of $M \times M$.

Being a coequaliser, a coimage $\text{CoIm } f : Y \to CI(f)$ of a morphism $f : X \to Y$ is an epimorphism. Because the compositions $f\text{pr}_1$ and $f\text{pr}_2$ coincide, the morphism f factors uniquely through its coimage:

$$
\begin{array}{ccc}
X \times_Y X \underset{\text{pr}_2}{\overset{\text{pr}_1}{\rightrightarrows}} X & \xrightarrow{f} & Y \\
\text{CoIm } f \downarrow \quad \nearrow m_f & & \\
CI(f) & &
\end{array}
$$

In this chapter the codomain of a coimage $\text{CoIm } f$ of a morphism f will mostly be denoted by $CI(f)$ (as an abbreviation of the word "coimage") like in the Definition and the Diagram above.

Definition A.1.19. *The factorisation of a morphism f through a coimage* $\text{CoIm } f$ *is called* coimage factorisation *of f.*

Lemma A.1.20. *The morphism m_f in the coimage factorisation $f = m_f \text{CoIm } f$ is a monomorphism.*

Proof. A proof can be found in the proof of [BB04, Theorem A.5.3]. $\qquad\square$

A special case is the coimage of an idempotent endomorphism $f : X \to X$ of an object X. One can take powers of such morphisms without changing the coimage. In particular one observes:

Lemma A.1.21. *If $f : X \to X$ is an idempotent endomorphism of X with coimage* $\text{CoIm } f$ *then* $\text{id}_{CI(f)} = (\text{CoIm } f)m_f$, *i.e. the following diagram is commutative:*

$$
\begin{array}{ccccc}
X \times_Y X \underset{\text{pr}_1}{\overset{\text{pr}_1}{\rightrightarrows}} X & \xrightarrow{f} & X & \xrightarrow{f} & X \\
\text{CoIm } f \downarrow \quad \nearrow m_f \quad \text{CoIm } f \downarrow \quad \nearrow m_f & & & & \\
CI(f) \xrightarrow{\text{id}} CI(f) & & & &
\end{array}
$$

Proof. Let $f : X \to X$ is an idempotent endomorphism of an object X and suppose that there exists a coimage $\text{CoIm } f$ of f. From the coimage factorisation $f = m_f \text{CoIm } f$ we obtain the equalities

$$m_f \text{CoIm } f = f = f^2 = m_f(\text{CoIm } f)m_f \text{CoIm } f.$$

The coimage $\text{CoIm } f$ is right-cancellative because it is an epimorphism. The monomorphism m_f is left-cancellative. Cancelling $\text{CoIm } f$ on the right and m_f on the left we obtain the equality $\text{id}_{CI(f)} = (\text{CoIm } f)m_f$. $\qquad\square$

The concepts dual to fibre products and the coimage factorisation are fibre co-products and the image factorisation. These are obtained working in the opposite category:

Definition A.1.22. *A fibre coproduct of morphisms $f_1 : X \to Y_1$ and $f_2 : X \to Y_2$ is a pushout of f_1 and f_2. When the morphisms f_1 and f_2 are known from the context, the colimit object is denoted $Y_1 \coprod_X Y_2$.*

For many categories there exists a natural choice of such pushouts. If this is the case we will always speak of *the fibre coproduct* and understand it to be the one given by the natural choice.

Example A.1.23. In the category **Set** the fibre coproduct of two maps $f_1 : X \to Y_1$ and $f_2 : X \to Y_2$ is the factor set

$$Y_1 \coprod_X Y_2 = \left(Y_1 \coprod Y_2 \right) / \sim$$

of the disjoint union of $Y_1 \dot{\cup} Y_2$ by the equivalence relation generated by $f_1(x) \sim f_2(x)$ for all $x \in X$.

Example A.1.24. In the category **RMod** of modules over a ring R the fibre co-product of two homomorphisms $f_1 : M \to N_1$ and $f_2 : M \to N_2$ is the quotient module

$$N_1 \coprod_X N_2 = (N_1 \oplus N_2) / \sim$$

of the direct sum of $N_1 \oplus N_2$ by the congruence relation generated by $f_1(m) \sim f_2(m)$ for all $m \in M$.

Example A.1.25. In the category **Top** of topological spaces the fibre coproduct of two continuous maps $f_1 : X \to Y_1$ and $f_2 : X \to Y_2$ is the quotient space

$$Y_1 \coprod_X Y_2 = \left(Y_1 \coprod Y_2 \right) / \sim$$

of the disjoint union of $Y_1 \dot{\cup} Y_2$ by the equivalence relation generated by $f_1(x) \sim f_2(x)$ for all $x \in X$.

Example A.1.26. In the category **TopRMod** of topological modules over a topo-logical ring R the fibre coproduct of two morphisms $f_1 : M \to N_1$ and $f_2 : M \to N_2$ is the quotient module

$$N_1 \coprod_X N_2 = (N_1 \oplus N_2) / \sim$$

(carrying the quotient topology) of the direct sum of $N_1 \oplus N_2$ by the congruence relation generated by $f_1(m) \sim f_2(m)$ for all $m \in M$.

For any morphism $f : X \to Y$ in a category **C** one can consider the pushout of f with itself. This fibre coproduct is denoted by $Y \coprod_X Y$. The morphisms i_1 and i_2 of the colimit cocone

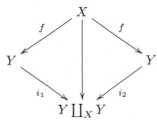

have common domain and codomain. This makes it possible to consider their equaliser (if it exists):

Definition A.1.27. *An image of a morphism $f : X \to Y$ with pushout $Y \coprod_X Y$ of f with itself is an equaliser of the two cocone morphisms $i_1 : Y \to Y \coprod_X Y$ and $i_2 : Y \to Y \coprod_X Y$.*

Note that an image of a morphism f is an equaliser and thus a monomorphism. Because the compositions $i_1 f$ and $i_2 f$ coincide, the morphism f factors uniquely through its image:

$$X \xrightarrow{\;f\;} Y \underset{i_1}{\overset{i_1}{\rightrightarrows}} Y \coprod_X Y \tag{A.3}$$

$$\begin{array}{c} e_f \searrow \quad \uparrow \text{Im} f \\ I(f) \end{array}$$

Here the domain of the image imf has been denoted by $I(f)$.

Definition A.1.28. *The factorisation $(\text{Im } f)e_f$ of a morphism f through an image $\text{Im } f$ is called* image factorisation *of f.*

Lemma A.1.29. *The morphism e_f in an image factorisation $f = (\text{Im } f)e_f$ is an epimorphism.*

Proof. This is the dual statement to Lemma A.1.20.   □

In many categories there exists a natural choice of pushouts and equalisers. If this is the case we will always speak of *the image* and understand it to be the one given by the natural choice and simply write $\text{Im } f$ for the image.

Example A.1.30. In the category **Set** a natural choice for an equaliser of two morphisms $f_1 : X \to Y$ and $f_2 : X \to Y$ is the subset

$$\text{Eq}\{f_1, f_2\} = \{x \in X \mid f_1(x) = f_2(x)\}$$

of X. If $f : X \to Y$ is a morphism in **Set** then the above choice of coequalisers makes the inclusion $f(X) \hookrightarrow Y$ a natural choice for an image of f.

Example A.1.31. In the category **RMod** of modules over a ring R a natural choice for an equaliser of two morphisms $f_1 : M \to N$ and $f_2 : M \to N$ is the submodule

$$\text{Eq}\{f_1, f_2\} = \{m \in M \mid f_1(m) = f_2(m)\}$$

of M. If $f : M \to N$ is a morphism in **RMod** then the above choice of coequalisers makes the inclusion $f(M) \hookrightarrow N$ a natural choice for an image of f.

Example A.1.32. In the category **Top** of topological spaces a natural choice for an equaliser of two morphisms $f_1 : X \to Y$ and $f_2 : X \to Y$ is the subspace

$$\text{Eq}\{f_1, f_2\} = \{x \in X \mid f_1(x) = f_2(x)\}$$

of X. If $f : X \to Y$ is a morphism in **Top** then the above choice of coequalisers makes the inclusion $f(X) \hookrightarrow Y$ of the subspace $f(X)$ a natural choice for an image of f.

Example A.1.33. In the category **TopRMod** of topological modules over a topological ring R a natural choice for an equaliser of two morphisms $f_1 : M \to N$ and $f_2 : M \to N$ is the (topological) submodule

$$\text{Eq}\{f_1, f_2\} = \{m \in M \mid f_1(m) = f_2(m)\}$$

of M. If $f : M \to N$ is a morphism in **TopRMod** then the above choice of coequalisers makes the inclusion $f(M) \hookrightarrow N$ a natural choice for an image of f.

Proposition A.1.34. *If both an image and a coimage of a morphism $f : X \to Y$ exist, then the morphism f factors uniquely through the coimage and the image making the diagram*

$$X \times_Y X \underset{\mathrm{pr}_1}{\overset{\mathrm{pr}_1}{\rightrightarrows}} X \xrightarrow{\;f\;} Y \underset{i_2}{\overset{i_1}{\rightrightarrows}} Y \coprod_X Y$$

with $\mathrm{CoIm}\,f$ down from X, $\mathrm{Im}\,f$ up into Y, and $CI(f) \longrightarrow I(f)$

commutative. The morphism f is an epimorphism exactly if the image $\mathrm{Im}\,f$ is an epimorphism if and only if the image $\mathrm{Im}\,f$ is an isomorphism.

Proof. See [KS06, Proposition 5.1.2] for a proof. The assumption that finite limits and colimits exists in the proof therein is only made to ensure the existence of the image $\mathrm{Im}\,f$ and the coimage $\mathrm{CoIm}\,f$ of f. ☐

Definition A.1.35. *The factorisation of a morphism f through a coimage $\mathrm{CoIm}\,f$ and an image $\mathrm{Im}\,f$ is called an* coimage-image factorisation *of f.*

A.2 Pointed Categories and Exact Sequences

Most of the homological algebra found in textbooks is developed in abelian categories (cf. Definition A.2.46. The categories considered in this work fail to be abelian in general, so the needed results of homological algebra have to be proven in more general categories.

Definition A.2.1. *A category $\mathbf{C}$ is called* small, *if its class $\mathrm{Obj}(\mathbf{C})$ of objects is a set.*

Definition A.2.2. *A category $\mathbf{C}$ is called an* Ab-category, *if every hom-set carries the structure of an abelian group such that composition distributes over addition.*

Example A.2.3. The category $\mathbf{Ab}$ of abelian groups is an Ab-category.

Example A.2.4. The category $\mathbf{RMod}$ of modules over a ring R is an Ab-category.

Example A.2.5. The category $\mathbf{TopAb}$ of abelian topological groups is an Ab-category.

Example A.2.6. The category $\mathbf{TopRMod}$ of topological modules over a topological ring R is an Ab-category.

Remark A.2.7. Coproducts in Ab-categories are also called *direct sums* and denoted by $\oplus$ instead of the usual symbol $\coprod$ for coproducts.

Definition A.2.8. *A* biproduct diagram *for objects C_1, C_2 in an Ab-category is a diagram*

$$C_1 \underset{i_1}{\overset{p_1}{\leftrightarrows}} C \underset{i_2}{\overset{p_2}{\rightleftarrows}} C_2 \tag{A.4}$$

whose morphisms p_1, p_2, i_1 and i_2 satisfy the identities $p_1 i_1 = \mathrm{id}_{C_1}$, $p_2 i_2 = \mathrm{id}_{C_2}$ and $i_1 p_1 + i_2 p_2 = \mathrm{id}_C$. The object C is called the biproduct *of C_1 and C_2.*

Theorem A.2.9. *In Ab-categories, two objects C_1 and C_2 have a binary product if and only they have a biproduct. In this case the object C in the biproduct diagram A.4 with projections p_1 and p_2 is a product of C_2 and C_2 while C together with the morphisms i_1 and i_2 is a coproduct of C_1 and C_2. In particular the two objects have a product if and only if they have a coproduct.*

Proof. A proof can be found in [ML98, Theorem VIII.2.2].   □

Biproducts are coproducts by definition. This special kind of coproducts can straightforwardly be generalised to more than two objects. The generalisation is called a "direct sum":

Definition A.2.10. *A* direct sum *of objects C_j, $j \in J$ in an Ab-category* **C** *is an object $\bigoplus_{j \in J} C_j$ such that there exist morphisms $i_j : C_j \to \bigoplus_{j \in J} C_j$ called* inclusions *and morphisms $p_j : \bigoplus_{j \in J} C_j \to C_j$ called* projections *satisfying the equations $p_j i_j = \mathrm{id}_{C_j}$ for all $j \in J$ and the equality $\sum_{j \in J} i_j p_j = \mathrm{id}$.*

Lemma A.2.11. *In Ab-categories, finitely many objects $C_1, \ldots, C_n$ have a direct sum if and only if they have a product. In this case the object $\bigoplus_{j=1}^n C_j$ with projections $p_1, \ldots, p_n$ is a product of $C_1, \ldots, C_n$.*

Proof. The proof is by induction on n using Theorem A.2.9.   □

Definition A.2.12. *A* terminal and initial object *is called a* zero object *or* null object. *Zero objects are denoted by 1 (or 0 in Ab-categories).*

Definition A.2.13. *A* pointed category *category is a category with a distinguished zero object.*

Example A.2.14. The category **Top**$_*$ of based spaces is a pointed category. It contains the category **Mf**$_*$ of based manifolds as a pointed subcategory.

Example A.2.15. The categories **Grp** of groups and **Ab** of abelian groups are pointed categories. More generally all varieties of algebras that have exactly one 0-ary operation are pointed categories.

Example A.2.16. The categories **LieGrp** of Lie-groups is a pointed category. More generally all classes of smooth algebras that have exactly one 0-ary operation are the objects of a pointed category.

Example A.2.17. The category **LieAlg$_R$** of Lie-algebras over a ring R is a pointed category.

Example A.2.18. The categories **TopGrp** of topological groups and **TopAb** of topological abelian groups are pointed categories. More generally all categories of topological algebras that have exactly one 0-ary operation are pointed categories.

Example A.2.19. The category **TopLieAlg$_R$** of Lie-algebras over a topological ring R is a pointed category.

Example A.2.20. The category **TrGrp**$_*$ of based transformation groups (where all spaces acted upon are based spaces) is a pointed category. It contains the category of smooth based transformation groups as a pointed subcategory.

For every two objects C and D in a pointed category there exists a unique morphism $C \to 1 \to D$ which is called the *zero morphism*. We denote this morphism by 1, except in Ab-categories, where we also write 0 for the zero morphism.

Definition A.2.21. *A* kernel *of a morphism $f : A \to B$ in a pointed category is an equaliser of f and the zero morphism $1 : A \to B$.*

Example A.2.22. If $f : G \to G'$ is a morphism of topological groups then the inclusion $f^{-1}(\{1\}) \to G$ is a kernel of f in the category **TopGrp** of topological groups. If $f : A \to A'$ is a morphism of abelian topological groups then the inclusion $f^{-1}(\{0\}) \to A$ is a kernel of f in the category **TopAb** of abelian topological groups.

Lemma A.2.23. *The pullback of a kernel of a morphism f along a morphism g is a kernel of fg, i.e. if $f : Y \to Z$ and $g : X \to Y$ are morphisms in a pointed category, $\ker f : K(f) \to Y$ is a kernel of f and the pullback*

$$
\begin{array}{ccc}
P & \xrightarrow{\;u\;} & X \\
\downarrow & & \downarrow{\scriptstyle g} \\
K(f) & \xrightarrow[\ker f]{} & Y
\end{array}
\tag{A.5}
$$

of $\ker f$ and g exists, then the morphism $u : P \to X$ in the pullback square is a kernel of fg.

Proof. Let $f : Y \to Z$ and $g : X \to Y$ be morphisms in a pointed category and assume that f has a kernel $\ker f : K(f) \to Y$. We show that the morphism u in a pullback square as Diagram A.5 has the universal property of a kernel of fg. Let such a pullback square be given and let $h : X' \to X$ me a morphism with the property $fgh = 0$. The universal property of the kernel of f implies that (gh) factors uniquely through $\ker f$ as $gh = (\ker f)h_2$, yielding the following commutative diagram:

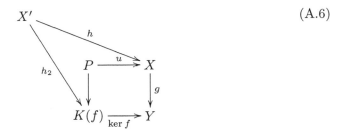

$$\tag{A.6}$$

The universal property of the pullback square now implies the existence of a morphism $h' : X' \to X$ such that $uh' = h$. The morphism u is a pullback of a monomorphism, hence a monomorphism as well (by Lemma A.1.1). So, if $h'' : X' \to X$ is another morphism with the property $uh'' = h$ one can conclude $uh' = uh''$ and thus $h' = h''$. Therefore the morphism h factors uniquely through u, which proves that u has the universal property of a kernel of fg. $\qquad\square$

Conversely, if $f : Y \to Z$ is a morphism with kernel $\ker f : K(f) \to Y$ and $g : X \to Y$ is a morphism such that a kernel $\ker fg : K(fg) \to X$ of gf exists, then the composition $fg\,\ker(fg)$ is the zero morphism. Therefore the morphism $g\,\ker(gf)$ factors uniquely through the kernel $\ker f$ of f. We assert that the resulting commutative square is a pullback square:

Lemma A.2.24. *If $f : Y \to Z$ is a morphism with kernel $\ker f : K(f) \to Y$ and $g : X \to Y$ is a morphism such that a kernel $\ker fg : K(fg) \to X$ exists, then the pullback of $\ker f$ and g also exists and is given by the morphisms $\ker(fg)$ and the unique morphism $g' : K(fg) \to K(f)$ satisfying $(\ker f)g' = g\,\ker(fg)$.*

Proof. Let $f : Y \to Z$ be a morphism with kernel $\ker f : K(f) \to Y$, $g : X \to Y$ be a morphism such that a kernel $\ker fg : K(fg) \to X$ exists, and let $g' : K(fg) \to K(f)$ denote the unique morphism satisfying $(\ker f)g' = g\,\ker(fg)$. It is to prove that the commutative diagram

$$
\begin{array}{ccc}
K(fg) & \xrightarrow{\;\ker(fg)\;} & X \\
{\scriptstyle g'}\downarrow & & \downarrow{\scriptstyle g} \\
K(f) & \xrightarrow[\ker f]{} & Y
\end{array}
$$

is a pullback square. We show that it has the required universal property. Let $\alpha : P \to X$ and $\beta : P \to Y$ be morphisms satisfying the equation $(\ker f)\beta = g\alpha$. This implies the equality $fg\alpha = f(\ker f)\beta = 0$, so α factors uniquely through the kernel $\ker(fg)$ of fg. Let $\gamma : P \to K(fg)$ denote the unique morphism satisfying $\ker(fg)\gamma = fg\alpha$. The commutativity of the above diagram implies the equalities

$$(\ker f)\beta = g\alpha = g(\ker fg)\gamma = (\ker f)g'\gamma.$$

Since the kernel $\ker f$ is a monomorphism, this implies $f\beta = g'\gamma$. So $\gamma : P \to K(fg)$ is a unique morphism satisfying $(\ker fg)\gamma = \alpha g'\gamma$ and $g'\gamma = \beta$. $\square$

Lemma A.2.25. *Pullbacks of split monomorphisms along kernels are split monomorphisms.*

Proof. Let $\ker f : K(f) \to Y$ be the kernel of a morphism f and let $m : X \to Y$ be a monomorphism with left inverse $e : Y \to X$. Assume that a pullback of $\ker f$ and g exists. By Lemma A.2.23 the pullback square has the form

$$
\begin{array}{ccc}
K(fm) & \xrightarrow{\ker(fm)} & X \\
{\scriptstyle m'}\downarrow & & \downarrow{\scriptstyle m} \\
K(f) & \xrightarrow{\ker f} & Y
\end{array}
\quad,
$$

where all morphisms are monomorphisms.

Consider the morphism $e \ker f$ from $K(f)$ to X. The composition of the morphism morphism with fm is given by

$$fm(e \ker f) = f(me)$$

Definition A.2.26. *A* cokernel *of a morphism $f : A \to B$ in a pointed category is a coequaliser of f and the zero morphism 1.*

Example A.2.27. If $f : G \to G'$ is a morphism of topological groups then the quotient map $G' \twoheadrightarrow G'/\operatorname{Im} f$ is a cokernel of f in the category **TopGrp** of topological groups. If $f : A \to A'$ is a morphism of abelian topological groups then the quotient map $A' \twoheadrightarrow A'/\operatorname{Im} f$ is a kernel of f in the category **TopAb** of abelian topological groups.

Definition A.2.28. *A sequence $1 \to A \xrightarrow{f} B \xrightarrow{g} C \to 1$ of morphisms in a pointed category is called* short exact *if f is the kernel of g and g is the cokernel of f.*

Example A.2.29. A sequence $1 \to G \xrightarrow{f} G' \xrightarrow{g} G'' \to 1$ in either category **TopGrp** or **TopAb** is short exact if and only if g is a quotient homomorphism onto G'', the homomorphism f is an embedding and $f(G) = g^{-1}(\{1\})$.

Definition A.2.30. *A homological δ-functor between pointed categories* **C** *and* **D** *is a sequence of functors $H_n : \mathbf{C} \to \mathbf{D}$ such that*

1. *For each short exact sequence $1 \to A \to B \to C \to 1$ in* **C** *there exist morphisms $\delta_n : H_n(C) \to H_{n-1}(A)$ in* **D**.
2. *For each morphism (α, β, γ) of exact sequences the morphisms δ_n satisfy the equations $\delta_n H_n(\gamma_n) = H_{n-1}(\alpha_{n-1})\delta_n$.*
3. *For each short exact sequence $1 \to A \to B \to C \to 1$ in* **C** *the sequence*

$$\cdots \xrightarrow{\delta_{n+2}} H_{n+1}(A) \to H_{n+1}(B) \to H_{n+1}(C) \xrightarrow{\delta_{n+1}} H_n(A) \to \cdots$$

is exact.

If the categories **C** *and* **D** *are additive, then the functors* H_n *are required to be additive. A homological δ-functor w.r.t. a class $\mathcal{E}$ of short exact sequences in* **C** *is a sequence of functors* $H_n : \mathbf{C} \to \mathbf{D}$ *fulfilling conditions 2 and 3 but condition 1 only for exact sequences* $1 \to A \to B \to C \to 1$ *in* $\mathcal{E}$.

Example A.2.31. The homology of chain complexes in homological (cf. A.6.11) or even abelian (cf. A.2.46) categories is a homological δ-functor.

Definition A.2.32. *A* cohomological δ-functor *between pointed categories* **C** *and* **D** *is a series of contravariant functors* $H^n : \mathbf{C} \to \mathbf{D}$ *such that*

1. *For each short exact sequence* $1 \to A \to B \to C \to 1$ *in* **C** *there exist morphisms* $\partial^n : H^n(A) \to H^{n+1}(C)$ *in* **D**.
2. *For each morphism* (α, β, γ) *of exact sequences the morphisms* ∂^n *satisfy the equations* $\partial^n H^n(\alpha_n) = H^{n+1}(\gamma_{n+1})\partial^n$.
3. *For each short exact sequence* $1 \to A \to B \to C \to 1$ *in* **C** *the sequence*

$$\cdots \xrightarrow{\partial^{n-2}} H^{n-1}(C) \to H^{n-1}(B) \to H^{n-1}(A) \xrightarrow{\partial^{n-1}} H^n(C) \to \cdots$$

is exact.

If the categories **C** *and* **D** *are additive, then the functors* H^n *are required to be additive. A cohomological δ-functor w.r.t. a class $\mathcal{E}$ of short exact sequences in* **C** *is a sequence of functors* $H^n : \mathbf{C} \to \mathbf{D}$ *fulfilling conditions 2 and 3 but condition 1 only for exact sequences* $1 \to A \to B \to C \to 1$ *in* $\mathcal{E}$.

Lemma A.2.33. *In pointed Ab-categories every coimage is a cokernel.*

Proof. Let $f : C \to C'$ be a morphism in a pointed Ab-category **C**. If a coimage $\mathrm{CoIm}\, f$ of f exists, then – by definition – the fibre product $C \times_f C$ exists and $\mathrm{CoIm}\, f$ is the coequaliser of its kernel pair $(\mathrm{pr}_1, \mathrm{pr}_2)$. Such a coequaliser is a coequaliser of the the pair $(\mathrm{pr}_1 - \mathrm{pr}_2, 0)$ which is the cokernel of the morphism $\mathrm{pr}_1 - \mathrm{pr}_2$. □

Definition A.2.34. *A morphism admitting a left inverse is called a* split monomorphism.

Definition A.2.35. *A morphism admitting a section (i.e. a right inverse) is called a* split epimorphism.

Remark A.2.36. Any section of a split epimorphism is necessarily a split monomorphism.

Proposition A.2.37. *In pointed Ab-categories, every kernel of a split epimorphism is a split monomorphism.*

Proof. Let $e : C \to C'$ be a split epimorphism with right inverse $m : C' \to C$ and kernel $\ker e : K(e) \to C$. Consider the endomorphism $(\mathrm{id}_C - me)$ of C. Staring with the equation $\mathrm{id}_{C'} = me$ one observes

$$\mathrm{id}_{C'} = em \Rightarrow e = eme \Rightarrow e - eme = 0 \Rightarrow e(\mathrm{id}_C - me) = 0.$$

Hence the endomorphism $(\mathrm{id}_C - me)$ of C factors uniquely through the kernel $\ker e$ of e. Let e' denote the morphism in the unique factorisation $(\mathrm{id}_C - me) = (\ker e)e'$. We assert that the morphism e' is a left inverse of $\ker e$. The composition of $(\mathrm{id}_C - me)$ and $\ker e$ evaluates to

$$(\mathrm{id}_C - me)\ker e = \ker e - (me)\ker e = \ker e - 0 = \ker e.$$

Substituting $(\ker e)e'$ for the endomorphism $(\mathrm{id}_C - me)$ one obtains the equality $(\ker e)e'(\ker e) = \ker e$. Since the kernel $\ker e$ is a monomorphism, this in turn implies $(\ker e)e' = \mathrm{id}_{K(e)}$. Thus the kernel $\ker e$ is a split monomorphism. □

Lemma A.2.38. *A coimage* $\text{CoIm}\, f$ *of an idempotent endomorphism* f *in a pointed Ab-category satisfies the equations* $(\text{CoIm}\, f)(\text{id} - f) = 0$ *and* $(\text{CoIm}\, f)m_{\text{id}-f} = 0$.

Proof. Let $f : X \to X$ be an idempotent endomorphism in a pointed Ab-category **C** and assume there exists a coimage $\text{CoIm}\, f : X \to CI(f)$ of f in **C**. Using the idempotence of f one obtains the equations

$$m_f \,\text{CoIm}\, f(\text{id}_X - f) = f(\text{id}_x - f) = f - f^2 = 0 = m_f 0.$$

The morphism m_f is left-cancellative, because it is a monomorphism. This implies the equality $(\text{CoIm}\, f)(\text{id}_X - f) = 0$. Here one can insert the coimage factorisation $m_{\text{id}-f} \,\text{CoIm}(\text{id}_X - f)$ for $(\text{id}_X - f)$. This results in the equality

$$(\text{CoIm}\, f)m_{\text{id}-f}\,\text{CoIm}(\text{id} - f) = (\text{CoIm}\, f)(\text{id} - f) = 0 = 0\,\text{CoIm}(\text{id} - f).$$

Being an epimorphism, the coimage $\text{CoIm}(\text{id} - f)$ is right-cancellative. Therefore the morphism $(\text{CoIm}\, f)m_{\text{id}-f}$ is the zero morphism. $\square$

Proposition A.2.39. *For any idempotent endomorphism* $f : X \to X$ *of an object* X *in an additive category such that coimages* $\text{CoIm}\, f : X \to CI(f)$ *of* f *and* $\text{CoIm}(\text{id} - f) : X \to CI(1 - f)$ *of* $\text{id} - f$ *exist, the morphism*

$$(\text{CoIm}\, f, \text{CoIm}(\text{id} - f)) : X \to CI(f) \times CI(\text{id} - f)$$

is an isomorphism with inverse morphism $\text{pr}_1 m_f + \text{pr}_1 m_{\text{id}-f}$. *Dually, if images* $\text{Im}\, f : I(f) \to X$ *and* $\text{Im}(\text{id} - f) : I(\text{id} - f) \to X$ *of* f *and* $\text{id} - f$ *and the coproduct* $I(f) \coprod I(\text{id} - f)$ *exist, then the morphism*

$$(\text{Im}\, f\,\text{pr}_1 + \text{Im}(\text{id} - f)\text{pr}_2 : I(f) \coprod I(\text{id} - f) \to X$$

is an isomorphism with inverse $(e_f, e_{\text{id}-f})$, *where* e_f *and* $e_{\text{id}-f}$ *are the epimorphisms in the image factorisations* $f = (\text{Im}\, f)e_f$ *and* $\text{id} - f = [\text{Im}(\text{id} - f)]e_{\text{id}-f}$ *respectively (cf. diagram A.3).*

Proof. Let $f : X \to X$ be an idempotent endomorphism in a pointed Ab-category **C** and assume that coimages $\text{CoIm}\, f$ and $\text{CoIm}(\text{id} - f)$ exist. Let pr_1 and pr_2 denote the projections of the product $X \times X$ onto the first and second factor respectively. Consider the morphism

$$\text{pr}_1 m_f + \text{pr}_2 m_{\text{id}-f} = (\text{pr}_1 + \text{pr}_2)(m_f \times m_{\text{id}-f}) : CI(f) \times CI(\text{id} - f) \to X.$$

The composition of this morphism with the morphism $(\text{CoIm}\, f, \text{CoIm}(\text{id} - f))$ is given by

$$(\text{pr}_1 + \text{pr}_2)(m_f \times m_{\text{id}-f})(\text{CoIm}\, f, \text{CoIm}(\text{id} - f)) =$$
$$(\text{pr}_1 + \text{pr}_2)(m_f\,\text{CoIm}\, f, m_{\text{id}-f}\,\text{CoIm}(\text{id} - f)) = (\text{pr}_1 + \text{pr}_2)(f, (\text{id} - f))$$
$$= f + \text{id} - f = \text{id}.$$

So the morphism $\text{pr}_1 m_f + \text{pr}_2 m_{\text{id}-f}$ is a left inverse of $(\text{CoIm}\, f, \text{CoIm}(\text{id} - f))$. To proceed, observe that $\text{CoIm}\, f(m_{\text{id}-f}) = 0$ and $\text{CoIm}(1 - f)m_f f = 0$ by Lemma A.2.38. The composition $(\text{CoIm}\, f, \text{CoIm}(\text{id} - f))(\text{pr}_1 + \text{pr}_2)(m_f \times m_{\text{id}-f})$ of the morphisms $(\text{CoIm}\, f, \text{CoIm}(\text{id} - f))$ and $\text{pr}_1 m_f + \text{pr}_1 m_{\text{id}-f}$ is an endomorphism

$$CI(f) \times CI(\text{id} - f) \to CI(f) \times CI(\text{id} - f)$$

of the product $CI(f) \times CI(\text{id} - f)$. Consider the composition of this endomorphism with the projection pr_1 onto the first factor:

$$\mathrm{pr}_1(\mathrm{CoIm}\,f, \mathrm{CoIm}(\mathrm{id}-f))(\mathrm{pr}_1+\mathrm{pr}_2)(m_f\times m_{\mathrm{id}-f}) =$$
$$= \mathrm{CoIm}\,f(\mathrm{pr}_1+\mathrm{pr}_2)(m_f\times m_{\mathrm{id}-f})$$
$$= \mathrm{CoIm}\,f(m_f+m_{\mathrm{id}-f}) = \mathrm{id}_{CI(f)}+0 = \mathrm{id}_{CI(f)},$$

where we have used the identity $(\mathrm{CoIm}\,f)m_f = \mathrm{id}_{CI(f)}$ (Lemma A.1.21). Analogously one considers the projection pr_2 onto the second factor and derives the equality

$$\mathrm{pr}_2(\mathrm{CoIm}\,f, \mathrm{CoIm}(\mathrm{id}-f))(\mathrm{pr}_1+\mathrm{pr}_2)(m_f\times m_{\mathrm{id}-f}) = \mathrm{id}_{CI(\mathrm{id}-f)}$$

These two equalities together imply that the endomorphism of $CI(f)\times CI(\mathrm{id}-f)$ considered is the identity. Thus the morphism $\mathrm{pr}_1 m_f + \mathrm{pr}_1 m_{\mathrm{id}-f}$ is also a right inverse of $(\mathrm{CoIm}\,f, \mathrm{CoIm}(\mathrm{id}-f))$ and the morphisms are inverse to each other. The second statement is dual to the first one. $\qquad\square$

Definition A.2.40. *A pointed Ab-category that has binary products is called an* additive category.

Example A.2.41. The category **Ab** of abelian groups is additive, while the category **Grp** of groups is not additive.

Example A.2.42. The category **RMod** of modules over a ring R is an additive category.

More generally every pointed category $\mathbf{TopAlg}_{(\Omega,\mathrm{E})}$ of topological algebras of type (Ω, E) is additive if the underlying category $\mathbf{Alg}_{(\Omega,\mathrm{E})}$ of abstract algebras is additive.

Example A.2.43. The category **TopAb** of abelian topological groups is an additive category.

Example A.2.44. The category **TopRMod** of topological modules over a topological ring R is additive.

Theorem A.2.45. *Additive categories have all finite products and all finite coproducts.*

Proof. This is an application of Theorem A.2.9. $\qquad\square$

A more common specialisation of additive categories is that of *abelian* categories. These are categories satisfying additional requirements:

Definition A.2.46. *An* abelian category *is an additive category satisfying the following conditions:*

1. *Every morphism has a kernel and a cokernel.*
2. *Every monomorphism is the kernel of its cokernel.*
3. *Every epimorphism is the cokernel of its kernel.*

Example A.2.47. The category **Ab** of abelian groups is abelian, while the category **TopAb** of abelian topological groups is not abelian.

Example A.2.48. The category **RMod** of modules over a ring R is an abelian category, while the category **TopRMod** of topological modules over a topological ring R is not abelian.

More generally every pointed category $\mathbf{TopAlg}_{(\Omega,\mathrm{E})}$ of topological algebras of type (Ω, E) fails to be abelian, even if is the underlying category $\mathbf{Alg}_{(\Omega,\mathrm{E})}$ of abstract algebras is abelian.

A.3 Regular Epimorphisms and Regular Categories

Definition A.3.1. *An epimorphism is called* regular, *if it is the coequaliser of a pair of morphisms.*

Example A.3.2. An epimorphism $e : S \to S'$ in the category **Set** of sets is regular if and only if it is a factor map onto a factor set.

Example A.3.3. An epimorphism $e : X \to X'$ in the category **Top** of topological spaces is regular if and only if it is a quotient map.

Example A.3.4. An epimorphism $e : A \to A'$ in **TopAb** is regular if and only if it is a quotient homomorphism, i.e. a cokernel. Similarly an epimorphism $e : M \to M'$ in **TopRMod** is regular if and only if it is a cokernel.

More generally, every coimage $\mathrm{CoIm}\, f$ of a morphism $f : X \to Y$ is a coequaliser by definition and thus a regular epimorphism. The same is true for cokernels of morphisms in pointed categories.

Proposition A.3.5. *Every coequaliser $f : X \to Y$ that has a kernel pair is the coequaliser of this kernel pair.*

Proof. The proof of [BB04, Proposition A.4.8] carries over.   □

Corollary A.3.6. *Every coequaliser is its own coimage, provided that it has a kernel pair. In particular every cokernel is its own coimage if it has a kernel pair.*

Proposition A.3.7. *If a morphism $f : X \to Y$ in a pointed category has a kernel, a cokernel of this kernel and a coimage, then the latter two are isomorphic.*

Proof. Let $f : X \to Y$ be a morphism in a pointed category. Assume that there exists a kernel $\ker f : K(f) \to X$ of f, a cokernel $\mathrm{coker}(\ker f) : X \to CK(\ker f)$ thereof and a coimage $\mathrm{CoIm}\, f : Y \to CI(f)$ of f. The universal property of the cokernel $\mathrm{coker}(\ker f)$ guarantees the existence of a monomorphism

By definition, the coimage $\mathrm{CoIm}\, f$ of f is a coequaliser of a kernel pair $\mathrm{pr}_1, \mathrm{pr}_2 : X \times_f X \to X$ of f.

$$\mathrm{CoIm}\, f = \mathrm{Coeq}(\mathrm{pr}_1, \mathrm{pr}_2)$$

Proposition A.3.8. *Split epimorphisms are regular.*

Proof. See [BB04, Proposition A.4.11] for a proof.

Proposition A.3.9. *Pullbacks of split epimorphisms are split epimorphisms.*

Proof. See [BB04, Proposition A.4.13] for a proof.

Definition A.3.10. *A category* **C** *is called* regular *if the following conditions are satisfied:*

1. *The category* **C** *has finite limits.*
2. *Every kernel pair has a coequaliser.*
3. *Pullbacks of regular epimorphisms along arbitrary morphisms are regular epimorphisms, i.e. if $e : Y \to Z$ is a regular epimorphism and $f : X \to Z$ is an arbitrary morphism, then the vertical morphism $f^*(e)$ in the pullback square*

$$
\begin{array}{ccc}
X \times_Z Y & \longrightarrow & Y \\
{\scriptstyle f^*(e)}\big\downarrow & & \big\downarrow{\scriptstyle e} \\
X & \xrightarrow{\ f\ } & Z
\end{array}
$$

is a regular epimorphism as well.

Example A.3.11. The category **TopAb** of abelian topological groups is a regular category.

Example A.3.12. The category **TopRMod** of topological modules over a topological ring R is a regular category.

Conditions 1 and 2 of the definition of regular categories imply the existence of fibred products and coimages. So we observe:

Theorem A.3.13. *In a regular category every morphism $f : C \to D$ factors through a coimage* $\operatorname{CoIm} f$ *as a regular epimorphism* $\operatorname{CoIm} f$ *followed by a monomorphism m_f, i.e. $f = m_f \operatorname{CoIm} f$. This factorisation is unique up to isomorphism.*

Definition A.3.14. *The factorisation $f = m_f \operatorname{CoIm} f$ of a morphism $f : C \to D$ as a regular epimorphism* $\operatorname{CoIm} f$ *followed by a monomorphism m_f is called the coimage factorisation of f.*

Beware that some authors call the monomorphism $m_f : CI(f) \to Y$ an image of f. Consequently they call the coimage factorisation an image factorisation and denote the coimage $I(f)$. To avoid confusion we always use our notation even when refering to these authors.

To each sequence $A \xrightarrow{f} B \xrightarrow{g} C$ of morphisms in a regular category one can consider the image factorisations of f and g. These fit into a commutative diagram:

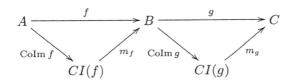

So to each sequence $A \xrightarrow{f} B \xrightarrow{g} C$ of morphism in a pointed regular category one can associate another sequence $CI(f) \to B \to CI(g)$ of morphisms. These factorisations are used to define exactness of sequences:

Definition A.3.15. *A sequence $A \xrightarrow{f} B \xrightarrow{g} C$ of morphisms in a pointed regular category is called* exact *at B if the sequence $1 \to CI(f) \to B \to CI(g) \to 1$ is a short exact sequence.*

Recall that, if a morphism $f : X \to Y$ in a pointed regular category admits an image, then the monomorphism m_f in the coimage factorisation $f = m_f \operatorname{CoIm} f$ of f factors uniquely through the image $\operatorname{Im} f$ (cf. Proposition A.3.13):

$$X \times_Y X \underset{\mathrm{pr}_1}{\overset{\mathrm{pr}_1}{\rightrightarrows}} X \xrightarrow{f} Y \underset{i_2}{\overset{i_1}{\rightrightarrows}} Y \textstyle\coprod_X Y$$

In this case the monomorphism m_f is a kernel exactly if the image $\operatorname{Im} f$ is a kernel and u is an isomorphism. Thus a sequence $A \xrightarrow{f} B \xrightarrow{g} C$ of morphisms in a pointed regular category with image $\operatorname{Im} f$ of f exact at B if the sequence

$$1 \to I(f) \xrightarrow{\operatorname{Im} f} B \to CI(g) \xrightarrow{\operatorname{CoIm} g} 1$$

is a short exact sequence.

Example A.3.16. A sequence $G \xrightarrow{f} G' \xrightarrow{g} G''$ in either category **TopGrp** or **TopAb** is exact at G' if and only if g is a quotient map $G' \to G'/\operatorname{Im} f \le G''$ onto its image and $CI(f) \cong g^{-1}(\{1\})$.

Morphisms whose image is a kernel are part of an important class of morphisms:

Definition A.3.17. *A morphism f in a pointed category is called* proper *if it has a coimage factorisation $f = m_f \operatorname{Colm} f$ and the monomorphism m_f in this coimage factorisation is a kernel.*

Recall that every morphism f in a regular category has a coimage factorisation $f = m_f \operatorname{Colm} f$ (by Theorem A.3.13), so it is proper if and only if m_f is a kernel. If a morphism $f : X \to Y$ admits an image, then the monomorphism m_f in the coimage factorisation

$$X \times_Y X \overset{\mathrm{pr}_1}{\underset{\mathrm{pr}_1}{\rightrightarrows}} X \xrightarrow{\;\;f\;\;} Y \overset{i_1}{\underset{i_2}{\rightrightarrows}} Y \coprod_X Y$$

$$\operatorname{Colm} f \downarrow \qquad m_f \qquad \uparrow \operatorname{Im} f$$

$$CI(f) \xrightarrow{\;u\;} I(f)$$

is a kernel exactly if u is an isomorphism and $\operatorname{Im} f$ is a kernel.

Example A.3.18. The image of a morphism $f : A \to A'$ of abelian groups is the kernel of its cokernel. Thus every morphism in **Ab** is proper.

In contrast to the above example, the image of a morphism $f : A \to A'$ of abelian topological groups is in general not the kernel of its cokernel. So not every morphism in **TopAb** is proper.

Example A.3.19. The image of a morphism $f : M \to M'$ of modules over a ring R is the kernel of its cokernel. Thus every morphism in **RMod** is proper.

Example A.3.20. The image of a morphism $f : M \to M'$ of topological modules over a topological ring R is in general not the kernel of its cokernel. Thus not every morphism in **TopRMod** is proper.

Definition A.3.21. *A chain complex (A_n, d_n) in a pointed regular category is called a* proper chain complex *if all the differentials d_n are proper. The category of proper chain complexes in a category $\mathbf{C}$ is denoted by* **PChC**.

Remark A.3.22. All chain complexes (A_n, d_n) in the category **Ab** of abelian groups are proper. Similarly all chain complexes (M_n, d_n) in the category **RMod** of modules over a ring R are proper.

Remark A.3.23. Not all chain complexes (A_n, d_n) in the category **TopAb** of abelian topological groups are proper. Similarly not all chain complexes (M_n, d_n) in the category **TopRMod** of topological modules over a topological ring R are proper.

A.4 The Category Pt(C) of Points of a Category C.

Definition A.4.1. *The category $\mathbf{C}_*$ of pointed objects of a category $\mathbf{C}$ with terminal object 1 is the comma category $1 \downarrow \mathbf{C}$ of objects under 1.*

Example A.4.2. A pointed object $1 \xrightarrow{f} S$ in the category **Set** of sets is also called a pointed set with base point $f(1) \in S$. A morphism $g : f \to f'$ between pointed sets $f : 1 \to S$ and $f' : 1 \to S'$ is a function $S \to S'$ which maps the base point $f(1)$ of S to the base point $f'(1)$ of S'.

Example A.4.3. Every morphism $1 \xrightarrow{f} G$ in the category **Grp** of groups maps the trivial group 1 onto the identity element of G. So a pointed object $f : 1 \to G$ in **Grp** can be identified with the group G itself. A morphism $g : f \to f'$ between pointed objects $f : 1 \to G$ and $f' : 1 \to G'$ is just a homomorphism $G \to G'$ of groups. Thus the comma category $1 \downarrow$ **Grp** groups is isomorphic to the category **Grp** itself. Similarly, the comma category $0 \downarrow$ **Ab** is isomorphic to the category **Ab** of abelian groups itself.

Example A.4.4. A pointed object $1 \xrightarrow{f} X$ in the category **Top** of topological spaces is also called a pointed (or based) topological space with base point $f(1) \in X$. A morphism $g : f \to f'$ between pointed topological spaces $f : 1 \to X$ and $f' : 1 \to X'$ is a continuous function $X \to X'$ which maps the base point $f(1)$ of X to the base point $f'(1)$ of X'. The category $1 \downarrow$ **Top** is also denoted by **Top**$_*$.

Example A.4.5. Every morphism $1 \xrightarrow{f} G$ in the category **TopGrp** of topological groups maps the trivial group 1 onto the identity element of G. So a pointed object $f : 1 \to G$ in **TopGrp** can be identified with its codomain, the group G itself. A morphism $g : f \to f'$ between pointed objects $f : 1 \to G$ and $f' : 1 \to G'$ is just a homomorphism $G \to G'$ of topological groups. Thus the comma category $1 \downarrow$ **TopGrp** is isomorphic to the category **TopGrp** itself. Similarly, the comma category $0 \downarrow$ **TopAb** is isomorphic to the category **TopAb** of abelian topological groups itself.

If **C** is an arbitrary category and I any object of **C** then the comma category **C** $\downarrow I$ of objects over I has (id_I, I) as terminal object. This allows us to consider the pointed category of this comma category:

Definition A.4.6. *The category* $\mathrm{Pt}_I(\mathbf{C})$ *of points over an object* I *in a category* **C** *is the category* $(\mathbf{C} \downarrow I)_*$ *of pointed objects of the comma category* **C** $\downarrow I$ *of objects over* I.

Proposition A.4.7. *If a category* **C** *has finite limits then the category* $\mathrm{Pt}_I(\mathbf{C})$ *of points over any object* I *in* **C** *is pointed and finitely complete.*

Proof. See [BB04, Proposition 2.1.11] for a proof.   □

Example A.4.8. The category **TopAb** of abelian topological groups has finite limits. Thus any category of points $\mathrm{Pt}_A(\mathbf{TopAb})$ over an abelian topological group A is finitely complete.

Example A.4.9. The category **TopGrp** of topological groups has finite limits. Thus any category of points $\mathrm{Pt}_G(\mathbf{TopGrp})$ over a topological group G is finitely complete.

Example A.4.10. The category **TopRMod** of topological modules over an topological ring R has finite limits. Thus any category of points $\mathrm{Pt}_M(\mathbf{TopRMod})$ over a topological R-module is finitely complete.

Example A.4.11. The category **TopAlg**$_{(\Omega, \mathrm{E})}$ topological algebras of type (Ω, E) has finite limits. Thus any category of points $\mathrm{Pt}_A(\mathbf{TopAb})$ over a topological algebra A of type (Ω, E) is finitely complete.

Definition A.4.12. *For any morphism* $f : I \to J$ *in a category* **C** *the pullback-functor* $f^* : \mathrm{Pt}_J(\mathbf{C}) \to \mathrm{Pt}_I(\mathbf{C})$ *induced by pulling back along* f *is called the* inverse image along f.

Proposition A.4.13. *If a category* **C** *has finite limits then the inverse image functors* $f^* : \mathrm{Pt}_I(\mathbf{C}) \to \mathrm{Pt}_J(\mathbf{C})$ *induced by morphisms* $f : I \to J$ *are left exact.*

Proof. See [BB04, Proposition 2.1.11] for a proof.   □

Example A.4.14. The inverse image functors along arbitrary morphisms in the categories **TopGrp**, **TopAb**, **TopRMod** or **TopAlg**$_{(\Omega,\mathrm{E})}$ are left exact.

Definition A.4.15. *The* category of points $\mathrm{Pt}(\mathbf{C})$ *of a category C is the category defined as follows:*

1. *The objects are the pairs (p, s) of a split epimorphism $p : C \to D$ with section $s : D \to C$.*
2. *If (p, s), $p : C \to D$ and (p', s'), $p' : C' \to D'$ are objects of $\mathrm{Pt}(\mathbf{C})$ then the morphisms $(p, s) \to (p', s')$ are pairs of morphisms (f, g) in $\mathbf{C}$ such that the diagram*

$$
\begin{array}{ccc}
C & \underset{s}{\overset{p}{\rightleftarrows}} & D \\
{\scriptstyle f}\downarrow & & \downarrow{\scriptstyle g} \\
C' & \underset{s'}{\overset{p'}{\rightleftarrows}} & D'
\end{array}
$$

commutes in every possible way, i.e $p' \circ f = g \circ p$ and $f \circ s = s' \circ g$.
3. *The composition of morphisms is defined by glueing the corresponding commutative diagrams together, i.e. $(f', g') \circ (f, g) = (f' \circ f, g' \circ g)$.*

Theorem A.4.16. *If a category $\mathbf{C}$ has pullbacks of split epimorphisms then the following holds:*

1. *The codomain functor $\mathrm{codom} : \mathrm{Pt}(\mathbf{C}) \to \mathbf{C}$ is a fibration.*
2. *The fibre of this fibration at an object I in $\mathbf{C}$ is the category $\mathrm{Pt}_I(\mathbf{C})$ of points over I.*
3. *the change of base functor along a morphism $f : I \to J$ is the inverse image functor $f^* : \mathrm{Pt}_I(\mathbf{C}) \to \mathrm{Pt}_J(\mathbf{C})$.*

Proof. See [BB04, Proposition 2.1.15] for a proof.   □

Definition A.4.17. *The fibration $\mathrm{codom} : \mathrm{Pt}(\mathbf{C}) \to \mathbf{C}$ for a category $\mathbf{C}$ with finite limits is called the* fibration of points of $\mathbf{C}$.

A.5 Protomodular Categories

Definition A.5.1. *A category $\mathbf{C}$ is called* protomodular *if it has pullbacks of split epimorphisms along any morphism and all inverse image functors of the fibration $\mathrm{codom} : \mathrm{Pt}(\mathbf{C}) \to \mathbf{C}$ of points reflect isomorphisms.*

Lemma A.5.2. *The category of* **Grp** *groups is pointed protomodular.*

Proof. See [BB04, Example 3.1.4] for a proof.

Lemma A.5.3. *Any abelian category is protomodular.*

Proof. See [BB04, Example 3.1.5] for a proof.

Example A.5.4. The category **Ab** of abelian groups is protomodular.

Example A.5.5. The category **RMod** modules over a ring R is protomodular.

Lemma A.5.6. *Any additive category with finite limits is protomodular.*

Proof. See [BB04, Example 3.1.13] for a proof.

Example A.5.7. The category **TopAb** of abelian topological groups is protomodular.

Example A.5.8. The category **TopRMod** topological modules over an topological ring R is protomodular.

Theorem A.5.9. *Any category* **TopAlg**$_{(\Omega,\mathrm{E})}$ *of topological algebras of type* (Ω, E) *with underlying protomodular category* **Alg**$_{(\Omega,\mathrm{E})}$ *is protomodular.*

Proof. A proof can be found in [BB04, Example 3.1.6]. □

Proposition A.5.10. *For a morphism f in a protomodular category with finite limits the following are equivalent:*

1. *The morphism f is a regular epimorphism.*
2. *The morphism f is the cokernel of its kernel: $f = \mathrm{coker}(\ker f)$.*

Proof. A proof can be found in [BB04, Proposition 3.1.23]. □

Lemma A.5.11. *In a pointed protomodular category* **C** *with finite limits a sequence*

$$1 \to A \xrightarrow{f} B \xrightarrow{g} C \to 1$$

is exact precisely if $f = \ker g$ and g is a regular epimorphism.

Proof. Let **C** be a protomodular pointed category with finite limits and assume $f = \ker g$. The morphism g is a regular epimorphism if and only if it is the cokernel of its kernel (see Proposition A.5.10). The kernel of g is the morphism f. Thus g is a regular epimorphism if and only if $g = \ker f$, i.e. if the sequence is exact. □

A.6 Homological Categories

Definition A.6.1. *A homological category is a pointed regular protomodular category.*

Example A.6.2. The category **TopAb** of abelian topological groups is a homological category.

Example A.6.3. The category **TopRMod** of topological modules over a topological ring R is a homological category.

Proposition A.6.4. *If* **C** *is a homological category then the epimorphism* $\mathrm{CoIm}\,f$ *in the image factorisation* $m_f \,\mathrm{CoIm}\,f$ *of a morphism* $f : C \to D$ *is a cokernel of f.*

Proof. The coimage factorisation exist because category **C** is regular (cf. Theorem A.3.13). By Proposition A.5.10 the epimorphism $\mathrm{CoIm}\,f$ is its own cokernel, $\mathrm{CoIm}\,f = \mathrm{coker}(\mathrm{CoIm}\,f)$. Since m_f is a monomorphism this implies $\mathrm{coker}\,f = \mathrm{coker}(m\,\mathrm{CoIm}\,f) = \mathrm{coker}(\mathrm{CoIm}\,f)$. □

Theorem A.6.5. *In a homological category the 3×3 Lemma holds, i.e. if* **C** *is a homological category,*

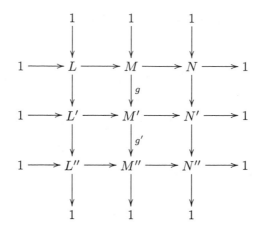

is a commutative diagram in **C** with exact rows, $g'g = 1$ and two of the columns are exact then, the remaining third column also is exact.

Proof. A proof can be found in [BB04, Theorem 4.2.7].

The Snake Lemma is not true in general, but there exists a special version in homological categories. Consider a commutative diagram

$$L \longrightarrow M \longrightarrow N \longrightarrow 1 \ ,$$
$$\downarrow f \qquad \downarrow g \qquad \downarrow h$$
$$1 \longrightarrow L' \longrightarrow M' \longrightarrow N'$$

where both horizontal sequences are exact and the vertical morphisms are proper. This commutative diagram can be enlarged by adjoining the kernels and cokernels of the homomorphisms f, g and h. All in all one obtains the following commutative diagram

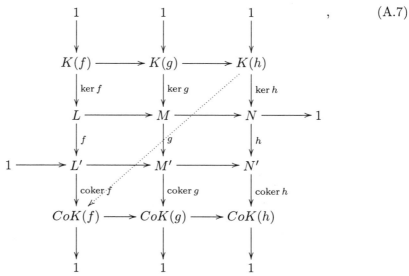

, (A.7)

in which all rows and columns are exact and in which the dotted arrow indicates a homomorphism guaranteed by the following Theorem:

Theorem A.6.6 (Snake Lemma). *Under the assumption that the morphisms f, g and h in in the above diagram A.7 are proper, there exists a unique connecting homomorphism $\delta : \ker h \to \operatorname{coker} f$ filling in as the dotted arrow such that the sequence*

$$K(f) \to K(g) \to K(h) \xrightarrow{\delta} CoK(f) \to CoK(g) \to CoK(h)$$

is exact and the so enlarged diagram is commutative.

Proof. See [BB04, Theorem 4.4.2] for a proof.

Proposition A.6.7. *The category* **Ch(C)** *of chain complexes in a homological category* **C** *is homological itself.*

If the cokernel coker f of a morphism $f : A \to B$ is a proper subobject of B we also write B/A for the cokernel.

Proof. A proof can be found in [BB04, Proposition 4.5.5].

Lemma A.6.8. *If* **C** *is a homological category and* $A \xrightarrow{f} B \xrightarrow{g} C$ *a zero sequence in* **C** *with a proper morphism* f *then the following holds:*

1. *There exists a factorisation* $j : CI(f) \to K(g)$ *which is a proper monomorphism.*
2. *The morphism* f *has a cokernel.*
3. *There exists a factorisation* $k : CoK(f) \to I(g)$ *which is a regular epimorphism.*
4. *There exist isomorphisms* $K(g)/CI(f) \cong CoK(j) \cong K(k)$.

Proof. See [BB04, Lemma 4.5.1] for a proof.   □

Definition A.6.9. *If* **C** *is a homological category and* $A \xrightarrow{f} B \xrightarrow{g} C$ *a zero sequence with a proper morphism* f *then the object* $K(g)/CI(f)$ *is called the* homology object *of the zero sequence* (f, g).

Definition A.6.10. *If* (A_*, d_n) *is a proper chain complex in a homological category then the homology object of the zero sequence* $A_{n+1} \xrightarrow{d_{n+1}} A_n \xrightarrow{d_n} A_{n-1}$ *is denoted by* $H_n(A_*)$.

Theorem A.6.11 (The long exact homology sequence). *Every short exact sequence of proper chain complexes*

$$1 \to A_* \xrightarrow{\alpha} B_* \xrightarrow{\beta} C_* \to 1$$

in a homological category **C** *induces a long exact sequence*

$$\cdots \xrightarrow{\partial} H_{n+1}(A_*) \xrightarrow{H_{n+1}(\alpha)} H_{n+1}(B_*) \xrightarrow{H_{n+1}\beta} H_{n+1}(C_*) \xrightarrow{\delta} H_n(A_*) \xrightarrow{H_n(\alpha)} \cdots$$

of homology objects.

Proof. This is a consequence of the Snake Lemma for proper morphisms of exact sequences.   □

Corollary A.6.12. *Homology is a homological δ-functor on the category of proper chain complexes of abelian topological groups.*

Corollary A.6.13. *Homology is a homological δ-functor on the category of proper chain complexes of topological modules over a topological ring R.*

B

Topological Spaces

This chapter is devoted to some subcategories of the category **Top** of topological spaces and their relation to each other.

B.1 Topological Spaces

Definition B.1.1. *The category of topological spaces and continuous functions is denoted by* **Top**.

Lemma B.1.2. *Finite products of closed injections are closed injections.*

Proof. Let $f_i : X_i \to Y_i$ be closed injective maps for $1 \leq i \leq n$ and $A \subset \prod X_i$ be a closed subset of $\prod X_i$. Because every closed subset of $\prod X_i$ is an intersection of basic closed sets, there exists an index set J and basic closed subsets $A_j \subset \prod X_i$ such that

$$A = \bigcap_{j \in J} A_j.$$

Furthermore, since the functions f_i were assumed to be injective, the product map $\prod f_i$ is injective as well and thus preserves intersections. As a consequence the image

$$\left(\prod_{i=1}^{n} f_i \right) (A) = \left(\prod_{i=1}^{n} f_i \right) \left(\prod_{i=1}^{n} f_i \right) \left(\bigcap_{j \in J} A_j \right) = \bigcap_{j \in J} \left(\prod_{i=1}^{n} f_i \right) (A_j)$$

is an intersection of the images of basic closed sets. Thus it suffices to prove that images of basic closed sets $A_j \subset \prod X_i$ under $\prod f_i$ are closed. Every such basic closed set A_j is a finite union of subbasic closed sets

$$A_j = \bigcup_{k=1}^{m} B_{j,k},$$

so it suffices to show that each image $(\prod f_i)(B_{j,k})$ of a subbasic closed set $B_{j,k}$ under $\prod f_i$ is closed in $\prod Y_i$. Every subbasic closed set $B_{j,k}$ is a product of closed sets

$$B_{j,k} = A_{j,k,0} \times \cdots \times A_{j,k,n}$$

(where all but one of the closed sets $A_{j,k,i}$ is the whole space X_i). Therefore the image of $B_{j,k}$ under $\prod f_i$ is the product of the images of the closed sets $A_{j,k,i} \subset X_i$:

$$\left(\prod f_i \right) (B_{j,k}) = \left(\prod f_i \right) \left(\prod_{i=1}^{n} A_{j,k,i} \right) = \prod_{i=1}^{n} f_i(A_{j,k,i})$$

Since the maps f_i were assumed to be closed, the product on the right hand side is closed in $\prod Y_i$. Therefore the image of a subbasic closed set $B_{j,k}$ under $\prod f_i$ is closed and thus the function $\prod f_i$ is closed as well. $\qquad\square$

Definition B.1.3. *A proclusion* is a continuous function $f : X \to Y$ such that the codomain Y has the final topology with respect to f.

The notion of proclusions slightly generalises the notion of identifications. In fact, the image of a proclusion $f : X \to Y$ is an open subspace of the codomain Y and the corestriction of f to $f(X)$ is an identification map. The complement $Y \setminus f(X)$ of its image is a discrete open subspace of Y.

Lemma B.1.4. *If $p : X \to Y$ is a proclusion and $g : Y \to Z$ any map then the final topology on Y with respect to $g \circ p$ coincides with the final topology with respect to g. In particular $g \circ p$ is a proclusion if and only g is.*

Proof. The map g is continuous if and only if $p \circ g$ is. Therefore the inverse image $g^{-1}(U)$ of a set $\subset Y$ is open on Y if and only if $(g \circ p)^{-1}(U)$ is open in X. $\qquad\square$

Theorem B.1.5. *Given proclusions $p : X \to X'$, $q : Y \to Y'$ and a function $f : X \to Y$ inducing a map $f_* : X' \to Y$ (satisfying $f_* \circ p = q \circ f$) then f_* is a proclusion if and only if f is a proclusion.*

Proof. Because p is a proclusion the map f_* is a proclusion if and only if $f_* \circ p = q \circ f$ is a proclusion. The function $q \circ f$ is a proclusion if and only if f is a proclusion. Thus f_* is a proclusion iff f is a proclusion. $\qquad\square$

Theorem B.1.6. *Finite products of open proclusions are proclusions.*

Proof. Let $p_i : X_i \to Y_i$, $i = 1, \dots, n$ be a family of open proclusions. Each space Y_i is the disjoint union of its open subspaces $p_i(X_i)$ and $Y \setminus p_i(X_i)$. Therefore the product $\prod Y_i$ is the disjoint union of open subspaces of the form

$$\prod_{i=1}^{n} Z_i,$$

where each space Z_i is either the open image $p_i(X_i)$ or its discrete open complement $Y_i \setminus p_i(X_i)$. Any subspace U of $\prod Y_i$ is open if and only if the intersection $U \cap \prod Z_i$ with each subspace of the form $\prod Z_i$ is open in $\prod Z_i$. Since all the complements $Y_i \setminus p_i(X_i)$ are discrete, this is equivalent to the intersection $U \cap \prod p_i(X_i)$ being open. The last condition in turn is equivalent to the inverse image $(\prod p_i)^{-1}(U)$ being open in $\prod X_i$. Thus a subset U of $\prod Y_i$ is open if and only if its inverse image $(\prod p_i)^-(U)$ is open in $\prod X_i$, i.e. the product map $\prod p_i$ is a proclusion. $\qquad\square$

Corollary B.1.7. *Finite products of open identifications are identifications.*

Definition B.1.8. *A space X is called* weak Hausdorff *if every image $f(K)$ of a compact Hausdorff space K under a continuous function $f : K \to X$ is closed in X.*

Every Hausdorff space is weak Hausdorff but the converse implication is not valid.

B.2 Separation Axioms

The definitions of the separation axioms in the literature are not consistent. Therefore we here provide the reader with the version that is used in this work.

Definition B.2.1. *A space X is called a T_0-space if the open sets separate the points, i.e. for two disjoint points in X there exists an open set containing exactly one of them.*

Definition B.2.2. *A space X is called a T_1-space if the points are closed.*

Definition B.2.3. *A space X is said to be* Hausdorff *or T_2 if two disjoint points always have disjoint neighbourhoods.*

Definition B.2.4. *A space X is said to be* regular, *if every point has a neighbourhood filterbase of closed sets.*

Lemma B.2.5. *Every compact subset of a regular space has a neighbourhood basis of closed sets.*

Proof. Let $K \subset X$ be a compact subset of a regular topological space X and V be a neighbourhood of X. Because X is regular, each point $k \in K$ has a neighbourhood U_x whose closure is contained in V. since K is compact, it covered by finitely many of these neighbourhoods:

$$K \subset U := \bigcup_{i=1}^{n} U_{x_i} \subset V$$

Because the union on the right is finite, the closure $\overline{U}$ of the union U is the union of the closures $\overline{U}_{x_i}$ of the sets U_{x_i}. Since all these closures are contained in V, the closure $\overline{U}$ of the neighbourhood U of K is also contained in V. $\qquad\square$

Definition B.2.6. *A topological space X is called a T_3-space, if it is Hausdorff and regular.*

Definition B.2.7. *A space X is said to be* completely regular, *if every point has a neighbourhood filterbase of supports of continuous functions $f : X \to I$. The full subcategory of* **Top** *with objects all completely regular spaces is denoted by* **Cr**

Proposition B.2.8. *Complete regularity has the following invariance properties:*

1. *Any subspace of a completely regular space is completely regular.*
2. *A product space is completely regular if and only if each of its factors is completely regular.*

Proof. See [Dug89, theorem VII.7.2] for a proof. $\qquad\square$

Corollary B.2.9. *Limits of completely regular spaces in* **Top** *are completely regular.*

Theorem B.2.10. *The category* **Cr** *of completely regular spaces is a full reflective subcategory of the category* **Top**.

Definition B.2.11. *A space X is said to be* normal, *if two disjoint closed subsets always have disjoint neighbourhoods.*

Lemma B.2.12. *For a topological space X the following properties are equivalent:*

1. *The space X is normal.*
2. *Each closed set has a neighbourhood filterbase of closed sets.*
3. *For each pair of disjoint closed sets $A, B \subset X$ there is an open set $U \subset X$ satisfying $A \subset U$ and $\overline{U} \cap B = \emptyset$.*
4. *Every two disjoint closed sets have neighbourhoods whose closure do not intersect.*

Proof. See [Dug89, Theorem VII.3.2] for a proof.   □

Lemma B.2.13.   *1. normality is invariant under continuous closed surjections.*
2. closed subspaces of normal spaces are normal.
3. Each of the factors of a normal product space is normal.

Proof. See [Dug89, Theorem VII.3.3] for a proof.   □

Definition B.2.14. *A covering $\{U_\alpha \mid \alpha \in \mathcal{A}\}$ of a space X is called* point-finite, *if every point is contained in at most finitely many sets U_α.*

Theorem B.2.15. *For a topological space X the following properties are equivalent:*

1. *The space X is normal.*
2. *For each pair $A, B \subset X$ of disjoint closed sets there exists a continuous function $f : X \to I$ satisfying $A \subset f^{-1}(0)$ and $B \subset f^{-1}(B)$.*
3. *Every continuous function $f : A \to \mathbb{R}$ on a closed subspace A of X can be extended to a continuous function F on X. Furthermore, if $f < c$ on A then the extension F can be chosen to also satisfy $F < c$.*
4. *If $\{U_\alpha \mid \alpha \in \mathcal{A}\}$ is a point-finite covering of X by open sets, then there exists a covering $\{V_\alpha \mid \alpha \in \mathcal{A}\}$ by open sets satisfying $\overline{V}_\alpha \subset U_\alpha$ for all $\alpha \in \mathcal{A}$ and $V_\alpha \neq \emptyset$ whenever $U_\alpha \neq \emptyset$.*

Proof. See [Dug89, Theorem VII.4.1], [Dug89, Theorem VII.5.1] and [Dug89, Theorem VII.6.1] for a proof.   □

Definition B.2.16. *A topological space X is called a T_4-space, if it is Hausdorff and normal.*

B.3 Covering Axioms

Definition B.3.1. *A space X is called* paracompact *if each open covering of X has an open neighbourhood-finite refinement.*

Lemma B.3.2.   *1. Paracompact Hausdorff spaces are T_4-spaces.*

Proof. See [Dug89, Theorem VIII.2.2] for a proof.   □

B.4 Compactness

Definition B.4.1. *A topological space X is called* compact *if every open covering of X has a finite subcovering.*

Definition B.4.2. *The subcategory of* **Top** *consisting of all compact spaces and continuous maps between such spaces is denoted by* **Comp**.

Note that compactness does not imply Hausdorffness. Since any finite subcovering of an open covering is a neighbourhood-finite refinement, we conclude:

Lemma B.4.3. *Compact spaces are paracompact.*

The definition of compactness has several equivalent formulations:

Lemma B.4.4. *For a topological space X the following properties are equivalent:*

1. *The space X is compact.*
2. *Every filterbase of closed sets in X has non-empty intersection.*

3. Each filterbase in X has at least one accumulation point.

4. Every maximal filterbase in X converges.

Proof. The proof in [Dug89, Theorem XI.1.3] does not rely on the Hausdorffness of X and thus carries over word by word. □

Lemma B.4.5. *Compactness has the following invariance properties:*

1. *Continuous images of compact spaces are compact.*
2. *Closed subspaces of compact spaces are compact.*
3. *A subspace of a Hausdorff space is compact if and only if it is closed.*
4. *A product $\Pi_\alpha Y_\alpha$ of topological spaces is compact if and only if each space Y_α is compact.*

Proof. See [Dug89, Theorem XI.1.4] for proofs of the various implications.

In Hausdorff spaces the compact sets behave like points and share the same separation properties:

Lemma B.4.6.   1. *Finite unions of compact subsets of a topological space are compact.*
2. *Two disjoint compact subspaces of a Hausdorff space have disjoint neighbourhoods.*
3. *Compact subsets of regular spaces have a neighbourhood filterbase of closed neighbourhoods.*
4. *If $\{U_\alpha \mid \alpha \in \mathcal{A}\}$ is a neighbourhood-finite covering of a topological space X and A is a compact subset of X, then A has a neighbourhood meeting only finitely many sets U_α.*

Lemma B.4.7. *If $f : X \times K \to Y$ is a continuous function and K is compact then the set $U' := \{x \in X \mid f(\{x\} \times K) \subset U\}$ is open in X whenever U is open in Y.*

Proof. Let $f : X \times K \to Y$ be a continuous function, K be compact and U be an open subset of Y. If x is contained in $U' := \{x \in X \mid f(\{x\} \times K) \subset U\}$ then all the points (x, k), $k \in K$ are contained in the open inverse image $f^{-1}(U)$. Therefore there exist open neighbourhoods V_k of x and W_k of k such that $V_k \times W_k$ is contained in $f^{-1}(U)$. Because K is compact, it is covered by finitely many open sets $W_{k_1}, \ldots, W_{k_n}$. The intersection $\bigcap_{i=0}^{n} V_k$ is a neighbourhood of x which is contained in U'. This shows that every point $x \in U'$ has a neighbourhood which is contained in U', hence U' is open. □

Definition B.4.8. *A space X is said to be* weak Hausdorff *if all images $f(K)$ of compact Hausdorff spaces K under continuous maps $f : K \to X$ are closed in X.*

Definition B.4.9. *A space X is called* locally compact *if every point has a neighbourhood filterbase of compact sets.*

Definition B.4.10. *The category of all locally compact spaces and all continuous functions between these spaces is denoted by* **Lc***.*

Lemma B.4.11. *A Hausdorff space is locally compact if and only if each point has a compact neighbourhood.*

Proof. See [Dug89, Theorem XI.6.2, 1⇔2] for a proof. □

Lemma B.4.12. *For a topological space X the follwing properties are equivalent:*

1. *The space X is locally compact.*
2. *Each compact subspace has a neighbourhoodbase of compact sets.*
3. *The space X has a basis of relatively compact open sets.*

Proof. The proof of the implications 1⇒2⇒3 in [Dug89, Theorem XI.6.2] does not rely on the Hausdorffness of X and thus carries over word by word. The implication 3⇒1 is trivial. □

B.5 Function Spaces

In this section we recall some basic facts concerning function spaces. The set $\hom_{\mathbf{Top}}(X,Y)$ of continuous functions between topological spaces X and Y can be equipped with various topologies. We denote the set $\hom_{\mathbf{Top}}(X,Y)$ equipped with the compact open topology by $C(X,Y)$. So we obtain a bifunctor

$$C(-,-): \mathbf{Top}^{op} \times \mathbf{Top} \to \mathbf{Top},$$

into the category $\mathbf{Top}$ of topological spaces. The composition of the functor $C(-,-)$ with the forgetful functor $U : \mathbf{Top} \to \mathbf{Set}$ returns the hom-functor on $\mathbf{Top}$. For each subset C of X and subset U of Y the subset $\{f \in C(X,Y) \mid f(C) \subset U\}$ of the function space $C(X,Y)$ is denoted by (C,U). (The sets (K,U) for compact subsets C of X and open subsets U of Y are a subbasis for the compact open topology.)

Let X, Y and Z be topological spaces. The composition of a continuous function $f \in C(X,Y)$ with a continuous function $g \in C(Y,Z)$ is a continuous function $g \circ f : X \to Y$. So composition defines a (not necessarily continuous) map

$$T : C(X,Z) \times C(X,Y) \to Z, \quad T(f,g) = g \circ f.$$

Lemma B.5.1. *The composition map T is continuous in each argument separately, i.e. for each fixed continuous function $f \in C(X,Y)$ the map*

$$T(-,f): C(Y,Z) \to Z, \quad g \mapsto g \circ f$$

is continuous and likewise, for each fixed continuous function $g \in C(Y,z)$ the map

$$T(g,-): C(X,Y) \to Z, \quad f \mapsto g \circ f$$

is continuous.

Proof. A proof can be found in [Dug89, XII.2.1].

Theorem B.5.2. *If X, Y and Z are topological spaces, Y is locally compact, then the composition $T : C(Y,Z) \times C(X,Y) \to Z$ is continuous.*

Proof. Assume Y to be locally compact and let $f \in C(X,Y)$, $g \in C(Y,Z)$ and a subbasic neighbourhood of (A,W) of gf be given. Since f and g are continuous, the inverse image $g^{-1}(W) \subset Y$ is open and the image $f(A) \subset g^{-1}(W)$ of the compact set A is compact. Since Y was assumed to be locally compact, there exists a relatively compact open set $V \subset Y$ satisfying $f(A) \subset V \subset \overline{V} \subset g^{-1}(W)$. The sets (A,V) and $(\overline{V}, g^{-1}(W))$ are subbasic neighbourhoods of f and g respectively. In addition the image $T((A,V),(\overline{V}, g^{-1}(W)))$ is contained in W by construction. $\square$

Definition B.5.3. *For any two topological spaces X and Y, the (not necessarily continuous) map $\mathrm{ev}_{X,Y} : C(X,Y) \times X \to$ defined by $\mathrm{ev}_{X,Y}(f,x) = f(x)$ is called the* evaluation map *of $C(X,Y)$.*

Theorem B.5.4. *The evaluation $\mathrm{ev}_{X,Y}(-,x) : C(X,Y) \to Y$, $F \mapsto f(x)$ at a fixed point $x \in X$ is always continuous. If X is locally compact, then the evaluation map $\mathrm{ev}_{X,Y}$ is continuous.*

Proof. See [Dug89, Theorem XII.2.4] for a proof of the first statement. For the second statement, let X be locally compact, $(f,x) \in C(X,Y) \times X$ and an open neighbourhood W of $\mathrm{ev}(f,y) = f(y)$ be given. The inverse image $f^{-1}(W)$ of W is an open neighbourhood of X. Since X was assumed to be locally compact, there exists a compact neighbourhood A of x which is contained in $f^{-1}(W)$. So (A,W) is a subbasic open neighbourhood of f in $C(X,Y)$. Furthermore, each function f' in (A,W) maps the interior $\mathring{A}$ of A into W. Therefore $(A,W) \times \mathring{A}$ is an open neighbourhood of (f,x) that is mapped into W under the evaluation map. $\square$

Lemma B.5.5. *If X is locally compact, then the evaluations $\mathrm{ev}_{X,Y}$ form a natural transformation $\mathrm{ev}_{X,-} : C(X,-) \times X \to \mathrm{id}_{\mathbf{Top}}$.*

Proof. If X is locally compact, then the evaluation map $\mathrm{ev}_{X,Y}$ is continuous for every topological space Y. Let $f : Y \to Y'$ be a continuous map. The the evaluation of $\mathrm{ev}_{X,Y'}C(X,f)$ at a point $(g,x) \in C(X,Y) \times X$ is given by

$$\mathrm{ev}_{X,Y'} \circ C(X,f)(g,x) = ev_{X,Y'}(fg,x) = fg(x) = f(\mathrm{ev}_{X,Y}(g,x)).$$

Thus the evaluation maps $\mathrm{ev}_{X,Y}$ for all topological spaces Y form a natural transformation $\mathrm{ev}_{X,-} : C(X,-) \times X \to \mathrm{id}_{\mathbf{Top}}$. □

This suggests to consider the bifunctor $C(-,-)|_{\mathbf{Lc} \times \mathbf{Top}}$ and the evaluation maps defined on its images. Let $\mathrm{pr}_1 : \mathbf{Lc} \times \mathbf{Top} \to \mathbf{Lc}$ denote the projection onto the first factor and $\mathrm{pr}_2 : \mathbf{Lc} \times \mathbf{Top} \to \mathbf{Top}$ be the projection onto the second factor. Collecting the above results we obtain:

Lemma B.5.6. *Evaluation is natural transformation $C(-,-)|_{\mathbf{Lc} \times \mathbf{Top}} \times \mathrm{pr}_1 \to \mathrm{pr}_2$.*

Proof. This follows from the fact that the underlying maps of sets form a natural transformation and all the maps involved are continuous. □

The bifunctor $C(-,-) : \mathbf{Top}^{op} \times \mathbf{Top} \to \mathbf{Top}$ can be composed with the product functor $\prod : \mathbf{Top} \times \mathbf{Top} \to \mathbf{Top}$, by inserting the latter in the first argument first. In this manner one obtains a trifunctor

$$C(-,-) \circ (\prod \times \mathrm{id}_{\mathbf{Top}}) : \mathbf{Top} \times \mathbf{Top} \times \mathbf{Top} \to \mathbf{Top},$$

which is contravariant in the first two variables and covariant in the third one.

Definition B.5.7. *The trifunctor $C(-,-) \circ (\prod \times \mathrm{id}_{\mathbf{Top}})$ is denoted by $C(- \times -,-)$.*

Given three topological spaces X, Y and Z, a function $f : X \times Y \to Z$ can be regarded as a family of maps $Y \to Z$ with X as parameter space.

Definition B.5.8. *For each function $f : X \times Y \to Z$ that is continuous in Y and fixed $x \in X$ we denote by $\hat{f}(x)$ the continuous function*

$$\hat{f}(x) : Y \to Z, \quad [\hat{f}(x)](y) = f(x,y).$$

The resulting (not necessarily continuous) map $X \to C(Y,Z)$ is denoted by $\hat{f}$. The two maps f and $\hat{f}$ are called adjoint.

A particular useful feature of the compact open topology is the following result:

Theorem B.5.9. *If $f : X \times Y \to Z$ is continuous, then its adjoint map $\hat{f}$ is also continuous. If $\hat{f} : X \to C(Y,Z)$ is continuous and either Y is locally compact or $X \times Y$ is a k-space then $f : X \times Y \to Z$ is also continuous.*

Proof. A proof of the first statement can be found in [Dug89, Theorem XII.3.1]. If Y is locally compact, then the evaluation map $\mathrm{ev}_{Y,Z} : C(Y,Z) \times Y \to Z$ is continuous by Theorem B.5.4. In this case a function $f : X \times Y \to Z$ is the composition $f = \mathrm{ev}_{Y,Z}(\hat{f} \times \mathrm{id}_Y)$:

$$X \times Y \xrightarrow{\hat{f} \times \mathrm{id}_Y} C(Y,Z) \times Y \xrightarrow{\mathrm{ev}_{Y;Z}} Z$$

Thus, if $\hat{f} : X \to C(Y,Z)$ is continuous then so is the function f itself. A proof for the case that $X \times Y$ is a k-space is given in [Dug89, Corollary XII.3.2]. □

Corollary B.5.10. *If the topological space Y is locally compact then the product functor $- \times Y : \mathbf{Top} \to \mathbf{Top}$ is left adjoint to the function space functor $C(Y, -)$.*

This can now be summarised using categorical language.

Lemma B.5.11. *The assignments $f \mapsto \hat{f}$ form a natural transformation of trifunctors $\varphi : C(- \times -, -) \to C(-, C(-, -))$.*

Proof. The underlying maps of hom-sets form a natural transformation. In view of Theorem B.5.9, it remains to show that for all topological spaces X, Y and Z the map $\varphi_{X,Y,Z} : C(X \times Y, Z) \to C(X, C(Y, Z))$, $f \mapsto \hat{f}$ is a continuous function. To prove the continuity of this map, it suffices to show that the inverse image $\varphi_{X,Y,Z}^{-1}(U)$ of subbasic open sets of the form $U = (C_1, (C_2, V))$ is open, where C_1 and C_2 are compact subsets of X and Y respectively and V is an open subset of Z. Straight from the definitions we derive the equalities

$$\hat{f} \in (C_1, (C_2, V)) \Leftrightarrow \hat{f}(C_1) \subset (C_2, V) \Leftrightarrow \hat{f}(C_1)(C_2) \subset V$$
$$\Leftrightarrow f(C_1 \times C_2) \subset V \Leftrightarrow f \in (C_1 \times C_2, V),$$

i.e. the inverse image $\varphi_{X,Y,Z}^{-1}(U)$ of subbasic open set $U = (C_1, (C_2, V))$ is the subbasic open set $(C_1 \times C_2, V)$ in $C(X \times Y, Z)$. Therefore $\varphi_{X,Y,Z}$ is continuous. $\square$

Proposition B.5.12. *If Y is a locally compact then the product functor $- \times Y$ is left adjoint to the function space functor $C(Y, -)$.*

Proof. If Y is locally compact, then Theorem B.5.9 implies that the functions $\varphi_{X,Y,Z}$ are surjective for all topological spaces X and Z. Since these functions are always injective, the natural transformation $\varphi_{-,Y,-} : C(- \times X, -) \to C(-, C(Y, -))$ consists of continuous bijections. Thus the hom-set $\hom_{\mathbf{Top}}(X \times Y, Z)$ and the hom-set $\hom_{\mathbf{Top}}(X, C(Y, Z))$ are naturally isomorphic. $\square$

Corollary B.5.13. *For each locally compact topological space X the function space functor $C(X, -)$ is continuous.*

Lemma B.5.14. *If X and Y are topological spaces and $\{U_\beta \mid \beta \in \mathcal{B}\}$ is a subbasis for the topology on Y then $\{(K, U_\beta) \mid K \text{ is compact}, \beta \in \mathcal{B}\}$ is a subbasis for the compact open topology on $C(X, Y)$. If $\mathcal{F} = \{C_\alpha \mid \alpha \in \mathcal{A}\}$ is a family of compact subsets of X such that every compact subset C of X which is contained in an open set O has a finite cover $C_{\alpha_1}, \ldots, C_{\alpha_n}$ of sets in $\mathcal{F}$ which also is contained in O, then the family $\{(C_\alpha, U) \mid, \alpha \in \mathcal{A}, U \subset Y \text{ is open}\}$ is a subbasis for the compact open topology on $C(X, Y)$.*

Proof. See [Dug89, XII.5.1] for a proof. $\square$

Lemma B.5.15. *If X, Y and Z are topological spaces and the compact subspaces of X and Y are regular, then the subsets $(C \times C', U)$ for compact subsets C, C' of X and Y respectively and open subsets U of Z form a subbasis for the compact open topology on $C(X \times Y, Z)$.*

Proof. Let the compact subspaces of X and Y be regular. We use the preceding lemma to show that the sets $(C \times C', U)$ for compact subsets C, C' of X and Y respectively and open subsets U of Z form a subbasis for the compact open topology on $C(X \times Y, Z)$. If K is a compact subset of $X \times Y$ and O an open neighbourhood of K, then every point (x, y) in K has a subbasic open neighbourhood $U_x \times U_y$ which is contained in O. Consider the projections $C := \mathrm{pr}_X(K)$ and $C' := \mathrm{pr}_Y(K)$ of K onto X resp. Y. Because the compact spaces C and C' are regular, there exist open neighbourhoods V_x of x in C and V_y of y in C' satisfying $\overline{V}_x \subset U_x \cap C$ and

$\overline{V}_y \subset U_y \cap C'$. The open neighbourhood V_x of x in C is the intersection $W_x \cap C$ of an open neighbourhood W_x of x in X with the subspace C. Likewise, the open neighbourhood V_y of y in C' is the intersection $W_y \cap C'$ of an open neighbourhood W_y of y in Y with the subspace C'. We assume w.l.o.g. $W_x \subset U_x$ and $W_y \subset U_y$. The compact set K is covered by finitely many man open sets $W_{x_1} \times W_{y_1}, \ldots, W_{x_n} \times W_{y_n}$. The sets $C_i := \overline{W}_{x_i} \cap C$ are the closures of the open sets V_{x_i} in C, hence compact. Similarly, the sets $C_i' := \overline{W}_{y_i} \cap C'$ are the closures of the open sets V_{y_i} in C' and thus compact. Furthermore the compact set K is covered by the product sets $C_i \times C_i'$, $1 \le i \le n$. The union $\bigcup_{i=1}^n C_i \times C_i'$ is contained in the open set O by construction. All in all we have shown that for every compact subset K of $X \times Y$ and open neighbourhood O of K there exist compact subsets $C_1, \ldots, C_n$ of X and $C_1', \ldots, C_n'$ of Y satisfying $K \subset \bigcup_{i=1}^n C_i \times C_i' \subset O$. By the preceding lemma the subsets $(C \times C', U)$ for compact subsets C, C' of X and Y respectively and open subsets U of Z form a subbasis for the compact open topology on $C(X \times Y, Z)$.   □

Lemma B.5.16. *If X and Y are topological spaces whose compact subspaces are regular, then the natural transformation $\varphi_{X,Y,-} : C(X \times Y, -) \to C(X, C(Y, -))$ consists of embeddings.*

Proof. Let X and Y be topological spaces whose compact subspaces are regular and Z be any topological space. By Lemma B.5.15 the sets $(C \times C', U)$ where C and C' are compact and U is open form a subbasis for the compact open topology on $C(X \times Y, Z)$. The image of such a subbasic open set $(C \times C', U)$ is the subbasic open set $(C, (C', U)) \cap \mathrm{Im}\, \varphi_{X,Y,Z}$ in the image $\mathrm{Im}\, \varphi_{X,Y,Z}$ in $C(X, C(Y, Z))$. Thus the function $\varphi_{X,Y,Z}$ is an embedding.   □

Proposition B.5.17. *If X and Y are topological spaces whose compact subspaces are regular and $X \times Y$ is a k-space, then the natural transformation $\varphi_{X,Y,-}$ is a natural isomorphism.*

Proof. By Theorem B.5.9 the natural transformation $\varphi_{X,Y,-}$ consists of continuous bijections. By the preceding lemma these bijections are embeddings, hence homeomorphisms.   □

Lemma B.5.18. *The restrictions of φ to the subfunctors $C(- \times -, -)_{|\mathbf{Lc} \times \mathbf{Lc} \times \mathbf{Top}}$, $C(- \times -, -)_{|\mathbf{Haus} \times \mathbf{Lc} \times \mathbf{Top}}$ and $C(- \times -, -)_{|\mathbf{Cr} \times \mathbf{Lc} \times \mathbf{Top}}$ are a natural isomorphisms.*

Proof. By Theorem B.5.9 the restrictions of the transformation φ to the subfunctors $C(- \times -, -)_{|\mathbf{Lc} \times \mathbf{Lc} \times \mathbf{Top}}$, $C(- \times -, -)_{|\mathbf{Haus} \times \mathbf{Lc} \times \mathbf{Top}}$ and $C(- \times -, -)_{|\mathbf{Cr} \times \mathbf{Lc} \times \mathbf{Top}}$ of $C(- \times -, -)$ are surjective. Since they are always injective, it suffices to prove that the maps $\varphi_{X,Y,Z}$ are embeddings. Using the fact that compact subspaces of the Hausdorff space X are locally compact the proof [tD91, Satz II.5.2] carries over to our cases.   □

B.6 C-generated Spaces

Let $\mathbf{C}$ be a subcategory of the category **Top** of topological spaces. We refer to the spaces in $\mathbf{C}$ as *generating spaces*.

Definition B.6.1. *A continuous function from a generating space into an arbitrary topological space X is called a* probe over X.

Definition B.6.2. *The* **C***-generated topology on a topological space X is the final topology of all probes over X. The underlying set of X equipped with the* **C***-generated topology is denoted by* $\mathbf{C}X$.

Example B.6.3. For $\mathbf{C} = \mathbf{Top}$ the category of topological spaces the $\mathbf{C}$-generated topology on a topological space X is its original topology.

Lemma B.6.4. *The $\mathbf{C}$-generated topology on a topological space X is finer than the original topology of X, i.e. the set theoretic identity map $\mathbf{C}X \to X$ is continuous.*

Corollary B.6.5. *The $\mathbf{C}$-generated topology on a discrete space is also discrete.*

Example B.6.6. The (unique) topology on the singleton space $\{*\}$ is $\mathbf{C}$-generated for any subcategory $\mathbf{C}$ of $\mathbf{Top}$.

Definition B.6.7. *A topological space is called $\mathbf{C}$-generated if $\mathbf{C}X = X$. The full subcategory of $\mathbf{Top}$ with objects all $\mathbf{C}$-generated spaces is denoted by $\mathbf{Top_C}$.*

Example B.6.8. For $\mathbf{C} = \mathbf{Top}$ the category of topological spaces the category $\mathbf{Top_C}$ is the entire category $\mathbf{Top}$.

Lemma B.6.9. *The monomorphisms in $\mathbf{Top_C}$ are the injective functions in $\mathbf{Top_C}$.*

Proof. Clearly every injective map in $\mathbf{Top_C}$ is a monomorphism. Conversely assume $f : X \to Y$ to be a monomorphism. Let $x, y \in X$ be points of X. Since the singleton space $\{*\}$ is $\mathbf{C}$-generated, the functions $i_x : * \to X, * \mapsto x$ and $i_y : * \to X, * \mapsto y$ are morphisms in $\mathbf{Top_C}$. If the images of x and y under f coincide, then the composites $f \circ i_x$ and $f \circ i_y$ are equal. Because f was assumed to be a monomorphism, the maps i_x and i_y must be equal as well, i.e. $x = y$. Thus f is injective. $\square$

Lemma B.6.10. *Every generating space is $\mathbf{C}$-generated.*

Proof. Consider the identity probe.

Any continuous function $f : X \to Y$ between topological space X and Y gives rise to a continuous function $\mathbf{C}(f) : \mathbf{C}X \to \mathbf{C}Y$, which coincides with f on the underlying set. These assignments constitute a functor

$$\mathbf{C} : \mathbf{Top} \to \mathbf{Top_C}.$$

Lemma B.6.11. *The category $\mathbf{Top_C}$ is a coreflective subcategory of $\mathbf{Top}$ with coreflector $\mathbf{C}$, i.e. the functor $\mathbf{C}$ is a right adjoint to the inclusion $\mathbf{Top_C} \to \mathbf{Top}$.*

Proof. Let X be a $\mathbf{C}$-generated space and Y be an arbitrary topological space. The continuous function $i_Y : \mathbf{C}Y \to Y$ induces an injective map

$$i_* : \hom_{\mathbf{Top_C}}(X, \mathbf{C}Y) \to \hom_{\mathbf{Top}}(X, Y)$$

which is natural in X and Y. It remains to show that i_* is surjective. Let $f : X \to Y$ be a continuous function. Then the function $\mathbf{C}(f) : X = \mathbf{C}X \to \mathbf{C}Y$ is continuous and $i_* \mathbf{C}(f) = f$. Therefore i_* is bijective. $\square$

Corollary B.6.12. *The inclusion $\mathbf{Top_C} \to \mathbf{Top}$ is cocontinuous, i.e. the colimits of $\mathbf{C}$-generated spaces in $\mathbf{Top_C}$ coincide with those in $\mathbf{Top}$.*

Corollary B.6.13. *The coreflector $\mathbf{C} : \mathbf{Top} \to \mathbf{Top_C}$ is continuous, i.e. it preserves limits.*

Example B.6.14. For $\mathbf{C} = \mathbf{Top}$ the category of topological spaces the categories $\mathbf{Top_C}$ and $\mathbf{Top}$ coincide and the coreflector $\mathbf{C} : \mathbf{Top} \to \mathbf{Top}$ is the identity on $\mathbf{Top}$.

Remark B.6.15. This especially implies that any limit in $\mathbf{Top_C}$ can be obtained by computing the limit in $\mathbf{Top}$ and then applying the coreflector $\mathbf{C} : \mathbf{Top} \to \mathbf{Top_C}$. So for example the equaliser of a family J of maps $f_i : X \to Y$ between $\mathbf{C}$-generated spaces X and Y is the image of the equaliser $\mathrm{Eq}(J)$ in $\mathbf{Top}$ under the coreflector $\mathbf{C}$.

Lemma B.6.16. *The epimorphisms in* $\mathbf{Top}_C$ *are the surjective functions in* $\mathbf{Top}_C$.

Proof. Every surjective function in $\mathbf{Top}_C$ is clearly an epimorphism. Conversely, assume $f : X \to Y$ to be an epimorphism in $\mathbf{Top}_C$. The identification map $q : Y \to Y/f(X)$ is a morphism in $\mathbf{Top}_C$ by Corollary B.6.12. Let $*$ denote the point $q(f(X))$ in $Y/f(X)$. If f is not surjective, then the quotient map $q : Y \to Y/f(X)$ and the constant map $* : Y \to *$ are distinct but the compositions $q \circ f$ and $* \circ f$ coincide. This is a contradiction to the assumption that f is an epimorphism. Thus f must be surjective. $\square$

A subspace A of a **C**-generated space Y need not be **C**-generated itself. The following shows that the subobjects in $\mathbf{Top}_C$ are the images of the subspaces in $\mathbf{Top}$ under the coreflector $C : \mathbf{Top} \to \mathbf{Top}_C$.

Lemma B.6.17. *Let A be a subspace of a **C**-generated space Y. A map $f : X \to CA$ of a **C**-generated space X into the coreflection of A is continuous if and only if $C(i) \circ f$ is continuous, where i denotes the inclusion $A \hookrightarrow Y$.*

Proof. If f is continuous then the composition $C(i) \circ f$ of continuous maps is also continuous. Conversely, assume the map $C(i) \circ f$ to be continuous. The function $C(i)$ can be rewritten as $C(i) = i \circ i_A$, where i_A denotes the continuous bijection $CA \to A$. Because i is the inclusion of a subspace the function $i_A \circ f$ is also continuous. The continuity of f follows now from Lemma B.6.11. $\square$

This especially implies that for every topological space X the probes over CX are probes over X.

Definition B.6.18. *A map $f : X \to Y$ between topological spaces X and Y is called* **C**-*continuous if the composite $f \circ p$ is continuous for every probe p over X.*

Lemma B.6.19. *Continuous functions are* **C**-*continuous.*

Proof. This follows from the fact that the **C**-generated topology on a space X is finer than the original topology.

The reverse statement can be used to characterise the **C**-generated spaces.

Lemma B.6.20. *A topological space X is* **C**-*generated if and only if the* **C**-*continuous maps on X are the continuous maps on X.*

Proof. Let X be a C-generated space and $f : X \to Y$ be a not necessarily continuous map into a topological space Z. Assume the map f to be C-continuous and let A be a closed subset of Y. By assumption the inverse image $(f \circ p)^{-1}(A)$ of $f^{-1}(A)$ under any probe p over X is closed, hence $f^{-1}(A)$ is closed in X. Because this is true for all closed subsets of Y the map f is continuous. Conversely assume that a map $f : X \to Y$ is continuous exactly if it is C-continuous. Consider the bijection of sets $f : X \to CX$. By Lemma B.6.11, this map is C-continuous. Therefore f is continuous, hence a homeomorphism, and X is C-generated. $\square$

Proposition B.6.21. *If a generating space C is exponentiable then every product $X \times C$ with a **C**-generated space X in* $\mathbf{Top}$ *is* **C**-*generated.*

Proof. Let C be an exponentiable generating space and X be a **C**-generated space. In order to prove that $X \times C$ is a **C**-generated space it suffices to show that every **C**-continuous function $f : X \times C \to Y$ into a topological space Y is continuous. Let $f : X \times C \to Y$ be such a **C**-continuous function. Denote the set $\hom_{\mathbf{Top}}(C, Y)$ equipped with the exponential topology by $C(C, Y)_{exp}$ and let p_x be the continuous probe

$$p_x : C \to X \times C, \quad c \mapsto (x, c)$$

on $X \times C$. Since f was assumed to be **C**-continuous the composite function $f \circ p_x :$ $C \to Y$ is continuous, hence an element of $C(C, Y)_{exp}$. Therefore the adjoint map $\hat{f}$ takes values in the space $C(C, Y)_{exp}$. By assumption the maps $f \circ p$ are continuous for all probes p over $X \times C$. But these functions are the compositions

$$C \xrightarrow{\mathrm{pr}_1 \circ p} X \xrightarrow{\hat{f}} C(C, Y)_{exp}.$$

Because the functions $\mathrm{pr}_1 \circ p$ range over all probes over X as p ranges over all probes over $X \times Y$, the adjoint function $\hat{f}$ is C-continuous, hence continuous. Since the space $C(C, Y)_{exp}$ is equipped with the exponential topology, this is equivalent to the continuity of f. Thus every **C**-continuous map on $X \times C$ is continuous and so $X \times C$ is a **C**-generated space. $\square$

The category $\mathbf{Top_C}$ is the coreflective hull of **C**. It has some useful properties:

Proposition B.6.22. *The category* $\mathbf{Top_C}$ *is closed under the formation of quotients and disjoint sums.*

Proof. See the proof of [ELS04, Lemma 3.2]

Corollary B.6.23. *The category* $\mathbf{Top_C}$ *is cocomplete. Its colimits coincide with those in* **Top**.

Lemma B.6.24. *Any C-generated space is the quotient of a disjoint sum of generating spaces and a topological space is C-generated if and only if it is a colimit of generating spaces in* **Top**.

Proof. See [ELS04, Lemma 3.2].

Proposition B.6.25. *The product of spaces* Y_i *in* $\mathbf{Top_C}$ *is given by* $\mathrm{C}\left(\prod Y_i\right)$, *where* $\prod Y_i$ *is the product in* **Top**.

Proof. By the use of Lemma B.6.11 one obtains for every **C**-generated space X the following chain of natural isomorphisms:

$$\hom_{\mathbf{Top_C}} \left(X, \mathrm{C} \prod Y_i \right) = \hom_{\mathbf{Top}} \left(X, \prod Y_i \right)$$
$$\cong \prod \hom_{\mathbf{Top}}(X, Y_i) \cong \prod \hom_{\mathbf{Top_C}}(X, Y_i),$$

which is what was to be proved. $\square$

Definition B.6.26. *A subcategory* **C** *of* **Top** *is called* productive *if its objects are exponentiable spaces and the product of two generating spaces is always* **C**-*generated.*

Theorem B.6.27. *If* **C** *is productive then* $\mathbf{Top_C}$ *is Cartesian closed.*

Proof. See [ELS04, Theorem 3.6 - Lemma 3.12] for a proof. $\square$

Corollary B.6.28. *If* **C** *is productive then the product functor* $\prod : \mathbf{Top_C}^n \to \mathbf{Top_C}$ *in* $\mathbf{Top_C}$ *is cocontinuous.*

Lemma B.6.29. *If* **C** *is productive then finite products (in* $\mathbf{Top_C}$) *of quotient maps are quotient maps.*

Proof. Quotient maps in $\mathbf{Top_C}$ are coequalisers. If **C** is productive then the product functor $\mathbf{Top_C}^n \to \mathbf{Top_C}$ preserves coequalisers, hence it preserves quotients. $\square$

Lemma B.6.30. *If* **C** *is productive then finite products (in* **Top$_C$***) of proclusions are proclusions.*

Proof. Let $p_i : X_i \to Y_i$, $i = 1,\ldots,n$ be a family of proclusions in **Top$_C$**. Each space Y_i is the disjoint union of its open subspaces $p_i(X_i)$ and $Y \setminus p_i(X_i)$. Therefore the product $\mathrm{C}\prod Y_i$ is the disjoint union of open subspaces of the form

$$\mathrm{C}\prod_{i=1}^{n} Z_i,$$

where each space Z_i is either the open image $p_i(X_i)$ or its discrete open complement $Y_i \setminus p_i(X_i)$. Any subspace U of $\mathrm{C}\prod Y_i$ is open if and only if the intersection $U \cap \mathrm{C}\prod Z_i$ with each subspace of the form $\mathrm{C}\prod Z_i$ is open in $\prod Z_i$. Since all the complements $Y_i \setminus p_i(X_i)$ are discrete, this is equivalent to the intersection $U \cap \mathrm{C}\prod p_i(X_i)$ being open. If **C** is productive, then the corestriction

$$\mathrm{C}\prod_{i=1}^{n} p_i : \mathrm{C}\prod_{i=1}^{n} X_i \to \mathrm{C}\prod_{i=1}^{n} p_i(X_i)$$

is a finite product of identifications, hence an identification (by Lemma B.6.29). So a subset U of $\mathrm{C}\prod Y_i$ is open if and only if its inverse image $(\mathrm{C}\prod p_i)^{-1}(U)$ is open in $\prod X_i$, i.e. the product map $\mathrm{C}\prod p_i$ is a proclusion. $\square$

Moreover for productive subcategories **C** of **Top** Proposition B.6.21 can be generalised:

Lemma B.6.31. *If* **C** *is productive, X, Y are* **C**-*generated and Y is exponential in* **Top** *then the product $X \times Y$ in* **Top** *is* **C**-*generated.*

Proof. A proof appears in [ELS04, Theorem 5.4]. $\square$

B.7 Compactly Generated Spaces

We consider the subcategory **Comp** of compact topological spaces.

Definition B.7.1. *A topological space X is called* compactly generated *if is* **Comp**-*generated. The category* **Top$_{Comp}$** *of all compactly generated spaces is abbreviated by* **CGTop**. *The coreflector* **Top** $\to$ **CGTop** *is abbreviated by* CG.

Lemma B.7.2. *The inverse images of a subset A of a topological space X under all probes $f : C \to X$ over X is open (closed) if and only if the inverse images under all injective probes are open (closed).*

Proof. If the inverse images of a subset A of a topological space X under all probes $f : C \to X$ over X is open (closed), then the inverse images under all injective probes are open (closed). Conversely, assume that the inverse images $p'^{-1}(A)$ under all injective probes are open (closed). If $p : C \to X$ is a probe over X, then p factors uniquely through the quotient space $C/K(p)$ by the equivalence relation $K(p)$ which is given by $x \sim y \Leftrightarrow f(x) = f(y)$. Let $q : C \twoheadrightarrow C/K(p)$ denote the quotient map. Since quotients of compact spaces are compact, the unique function $p' : C/K(p) \to X$ satisfying $p = p'q$ is an injective probe over X. Moreover the inverse image $p'^{-1}(A)$ of a subset A of X is closed in $C/K(p)$ if and only if the inverse image $p^{-1}(A) = q^{-1}p'^{-1}(A)$ is closed in C. Thus the inverse image $p^{-1}(A)$ of A under all probes p over X is closed. $\square$

This is to say that a topological space X is called *compactly generated* if a subset A that intersects every compact subspace C in a closed subspace of C is itself closed in X.

Theorem B.7.3. *Any closed subspace of a compactly generated space is also compactly generated.*

Proof. Let X be a compactly generated space and $A \subset X$ be a closed subspace. It is to show that $\mathrm{CG}A \to A$ is an isomorphism, i.e. every subset B of A that is closed in $\mathrm{CG}A$ is already closed in A. Let B be such a closed subset of $\mathrm{CG}A$. Since A is closed in X, the inverse image $p^{-1}(A)$ under any probe $p : C \to X$ is closed in the compact generating space C. Because all generating spaces are compact, the inverse image $p^{-1}(A)$ also is compact. The restriction $p_{|p^{-1}(A)}$ of p to the compact subspace $p^{-1}(A)$ of C is a probe on A. The inverse image $p^{-1}(B)$ is equal to the inverse image $p_{|p^{-1}(A)}^{-1}(B)$ under the probe $p_{|p^{-1}(A)}$. The latter inverse image is closed in $p^{-1}(A)$ by assumption. Being a closed subspace of a closed subspace $p^{-1}(A)$ of C it is closed in C. Thus the inverse image of B under any probe $p : C \to X$ is closed in C, i.e. B is closed in X. $\qquad\square$

Lemma B.7.4. *The coreflector* $\mathrm{CG} : \mathbf{Top} \to \mathbf{CGTop}$ *preserves closed embeddings.*

Proof. Let A be a closed subspace of a topological space X and let $i_A : A \hookrightarrow X$ denote the inclusion. To show that $\mathrm{CG}(i_A) : \mathrm{CG}A \to \mathrm{CG}X$ is closed, it suffices to prove that the image $\mathrm{CG}(i_A)(B)$ of every closed subset B of $\mathrm{CG}A$ is closed in $\mathrm{CG}X$. If $B \subseteq \mathrm{CG}A$ is such a closed subset, then every inverse image $f'^{-1}(B)$ under a probe $f' : C' \to A$ from a compact topological space C' into A is closed. If $f : C \to X$ is a probe, then the inverse image $C' := f^{-1}(A)$ is closed in C, hence a compact subspace of C. Therefore the restriction and corestriction $f_{|C'}^{|A}$ is a probe into A; The inverse image $f^{-1}(B)$ of B under the probe $f : C \to X$ coincides with the inverse image $f_{|C'}^{|A}{}^{-1}(B)$ of B under the probe $f_{|C'}^{|A}$, which is closed in C'. Because C' is a closed subspace of C, the inverse image $f^{-1}(B)$ also is closed in C. Thus the inverse image $f^{-1}(B)$ of B under any probe $f : C \to X$ is closed, hence B is closed in $\mathrm{CG}X$. $\qquad\square$

Lemma B.7.5. *The category* $\mathbf{Comp}$ *of all compact spaces is productive.*

Proof. Compact topological spaces are exponentiable in $\mathbf{Top}$. In addition products of compact spaces are compact, hence they also are $\mathbf{Comp}$-generated. Thus $\mathbf{Comp}$ is productive. $\qquad\square$

Proposition B.7.6. *The category* $\mathbf{CGHaus}$ *is a full coreflective subcategory of the category* $\mathbf{Top}$. *It is small complete, cocomplete and Cartesian closed.*

Proof. The category $\mathbf{CGHaus}$ is coreflective by construction (cf. Lemma B.6.11). It is complete, because the coreflector $\mathrm{CG} : \mathbf{Top} \to \mathbf{CGTop}$ preserves limits (by Corollary B.6.13), and cocomplete by Corollary B.6.23. Finally, since the subcategory $\mathbf{Comp}$ of compact spaces is productive (Lemma B.7.5), its coreflective hull $\mathbf{CGTop}$ is Cartesian closed by Lemma B.6.27. $\qquad\square$

Definition B.7.7. *A space is called* $LM - T_2$ *if every compact subspace is Hausdorff.*

Lemma B.7.8. *A compactly generated space* X *is* $LM - T_2$ *if and only if the diagonal map* $X \to \mathrm{CG}(X \times X)$ *in* $\mathbf{CGTop}$ *is closed.*

Proof. See [Hof79, 2.9 (8)] for a proof.

B.8 Compactly Generated Hausdorff Spaces

Definition B.8.1. *The full subcategory of* **Top** *consisting of all compactly generated Hausdorff spaces and continuous maps between such spaces is denoted by* **CGHaus**.

Lemma B.8.2. *Any closed subspace of a compactly generated Hausdorff space is also compactly generated Hausdorff.*

Proof. This is a consequence of Theorem B.7.3.   □

Proposition B.8.3. *The category* **CGHaus** *is a full coreflective subcategory of the category* **Haus**. *It is small complete, cocomplete and Cartesian closed.*

Proof. See [ML98, Proposition VII.8.1-3].

B.9 k-Spaces

Definition B.9.1. *The category* **Top$_{\mathbf{CGHaus}}$** *of all* **CGHaus**-*generated spaces is denoted by* **kTop**. *Its objects are called* k-*spaces. The coreflector* **Top** $\to$ **kTop** *is denoted by* k.

Theorem B.9.2. *Any closed subspace of a k-space is a k-space.*

Proof. Let X be a k-space and $A \subset X$ be a closed subspace. It is to show that $kA \to A$ is an isomorphism, i.e. every subset B of A that is closed in kA is already closed in A. Let B be such a closed subset of kA. Since A is closed in X, the inverse image $p^{-1}(A)$ under any probe $p : C \to X$ is closed in the compact generating space C. Because all generating spaces are compact, the inverse image $p^{-1}(A)$ also is compact. The restriction $p_{|p^{-1}(A)}$ of p to the compact subspace $p^{-1}(A)$ of C is a probe on A. The inverse image $p^{-1}(B)$ is equal to the inverse image $p_{|p^{-1}(A)}^{-1}(B)$ under the probe $p_{|p^{-1}(A)}$. The latter inverse image is closed in $p^{-1}(A)$ by assumption. Being a closed subspace of a closed subspace $p^{-1}(A)$ of C it is closed in C. Thus the inverse image of B under any probe $p : C \to X$ is closed in C, i.e. B is closed in X.   □

Lemma B.9.3. *Any open subspace of a k-space is a k-space.*

Proof. See [tD91, Satz 6.6] for a proof.   □

Lemma B.9.4. *The coreflector* k *preserves closed embeddings.*

Proof. Let A be a closed subspace of a topological space X and let $i_A : A \hookrightarrow X$ denote the inclusion. To show that $k(i_A) : kA \to kX$ is closed, it suffices to prove that the image $k(i_A)(B)$ of every closed subset B of kA is closed in kX. If $B \subseteq kA$ is such a closed subset, then every inverse image $f'^{-1}(B)$ under any probe $f' : C' \to A$ from a compact Hausdorff space C' into A is closed. If $f : C \to X$ is a probe, then the inverse image $C' := f^{-1}(A)$ is closed in C, hence a compact Hausdorff subspace of C. Therefore the restriction and corestriction $f_{|C'}^{|A}$ is a probe into A; The inverse image $f^{-1}(B)$ of B under the probe $f : C \to X$ coincides with the inverse image $f_{|C'}^{|A}{}^{-1}(B)$ of B under the probe $f_{|C'}^{|A}$, which is closed in C'. Because C' is a closed subspace of C, the inverse image $f^{-1}(B)$ also is closed in C. Thus the inverse image $f^{-1}(B)$ of B under any probe $f : C \to X$ is closed, hence B is closed in kX.   □

Lemma B.9.5. *The category* **CompHaus** *of all compact Hausdorff spaces is productive.*

Proof. The spaces in **CompHaus** are compact, so they are exponentiable in **Top**. In addition products of compact Hausdorff spaces are compact Hausdorff, hence they also are **CompHaus**-generated. Thus **CompHaus** is productive. □

Proposition B.9.6. *The category ktops is a full coreflective subcategory of the category* **Haus**. *It is small complete, cocomplete and Cartesian closed.*

Proof. The category **CGHaus** is coreflective by construction (cf. Lemma B.6.11). It is complete, because the coreflector CG : **Top** → **CGTop** preserves limits (by Corollary B.6.13), and cocomplete by Corollary B.6.23. Finally, since the subcategory **CompHaus** of compact Hausdorff spaces is productive (Lemma B.9.5), its coreflective hull **kTop** is Cartesian closed by Lemma B.6.27. □

Proposition B.9.7. *Any of the following conditions on a topological space X is sufficient for X to be a k-space:*

1. *The space X is metrisable.*
2. *the space X is first-countable, i.e. every point in X has a countable neighbourhood basis.*
3. *If the intersection $A \cap C$ is closed in C for all compact Hausdorff subspaces C of X, then A is closed in X.*

Proof. A proof can be found in [tD91, Satz 6.1]. □

Lemma B.9.8. *If X is a Hausdorff k-space, then one can restrict oneself to injective probes $p : C \to X$, i.e. a subspace A of X is closed if and only if the intersection $A \cap C$ is closed in C for all compact Hausdorff subspaces C of X.*

Proof. See [tD91, Satz 6.2] for a proof. □

Recall that compactly generated topology on a topological space X is the final topology with respect to all injective probes $p : C \to X$ over X, where C is a compact topological space. If the topological space X is Hausdorff, then every continuous map $p : C \to X$ from a compact space C into X is automatically closed. Therefore any injective probe p is a homeomorphism onto its closed image. This forces the compact space C to be Hausdorff. So the compactly generated topology on X coincides with the compactly Hausdorff generated topology and we observe:

Lemma B.9.9. *The category of Hausdorff k-spaces is the category of compactly generated Hausdorff spaces.*

Lemma B.9.10. *A k-space X is weak Hausdorff if and only if the diagonal map $X \to k(X \times X)$ in* **kTop** *is closed.*

Proof. See [McC69, Proposition 2.3] for a proof. □

Moreover one can relax the definition of weak Hausdorffness in the category **kTop** and still obtain the same subclass of spaces:

Lemma B.9.11. *A k-space X is weak Hausdorff if and only if every compact subspace is closed.*

Proof. A proof can be found in [Hof79, 1.14] (for $\mathfrak{P}$ the class of compact Hausdorff spaces and T_2' the required property).

B.10 k_ω-Spaces

There exists a variant of the category **kTop** of k-spaces, which shares many of the properties of the category **kTop**. It is formed by the class of all topological spaces whose topology is the weak topology with respect to some countable set of compact Hausdorff spaces:

Definition B.10.1. *A topological space X is called a k_ω-space if it has the weak topology w.r.t. a countable set $\{f_n(K_n)\}$ of images of compact Hausdorff spaces K_n under continuous functions $f_n : K_n \to X$. The full subcategory of* **Top** *consisting with objects all k_ω-spaces is denoted by* $\mathbf{k_\omega Top}$.

Since the images $f_n(K_n)$ in the above definition are compact subspaces of the topological space X, this implies that X has the colimit topology of the ascending sequence $\bigcup_{i=1}^{n} K_i$ of compact subspaces of X. Conversely, if X is the direct limit of an ascending sequence of compact subspaces, then it is a k_ω-space. Similar to the class of k-spaces the class of k_ω-spaces can be described as quotients of compact Hausdorff-spaces:

Lemma B.10.2. *A topological space is a k_ω-space if and only if it is the quotient of a disjoint union of at most countably many compact Hausdorff spaces.*

Proof. The proof of [Dug89, Theorem XI.9.4] generalises to k_ω-spaces.   □

Lemma B.10.3. *Quotients of k_ω-spaces are k_ω-spaces.*

Proof. Since compositions of quotient maps are quotient maps, this is a consequence of the characterisation of k_ω-spaces in Lemma B.10.2.   □

Lemma B.10.4. *Finite and countable disjoint unions of k_ω-spaces are k_ω-spaces.*

Proof. This follows from the characterisation of k_ω-spaces in Lemma B.10.2.   □

Summarising the last two lemmata we observe that the subcategory $\mathbf{k_\omega Top}$ of **Top** has the same countable colimits:

Proposition B.10.5. *Finite and countable colimits of k_ω-spaces in* **Top** *are k_ω-spaces.*

Definition B.10.6. *A topological space X is called* locally k_ω *if every point has a neighbourhood filter basis of k_ω-spaces.*

Lemma B.10.7. *Finite products of Hausdorff k_ω-spaces are Hausdorff k_ω-spaces.*

Proof. This is a special case of [GGH06, Lemma 1.1 (b)], cf. [GGH06, Proposition 4.2 (c)].   □

C

Homotopy Theory

C.1 Homotopy Theory of Topological Spaces

Definition C.1.1. *A* homotopy *from a continuous function $f : X \to Y$ to a continuous function $g : X \to Y$ is a continuous function $H : X \times I \to Y$ such that $H(-, 0) = f$ and $H(-, 1) = g$.*

The analogous definition for G-spaces (where G is a topological semi-group) is given by:

Definition C.1.2. *A* homotopy in **GTop** *or* equivariant homotopy *from a continuous equivariant function $f : X \to Y$ to a continuous equivariant function $g : X \to Y$ is a homotopy $H : X \times I \to Y$ from f to g which is equivariant when I is endowed with the trivial G-action.*

C.2 Homotopy Groups

Definition C.2.1. *A continuous function $f : X \to Y$ between based topological spaces X and Y is called n-simple, if the homomorphisms $\pi_k(f) : \pi_k(X) \to \pi_k(Y)$ are isomorphisms for all $k < n$ and the homomorphism $\pi_n(f) : \pi_n(X) \to \pi_n(Y)$ is an epimorphism.*

Theorem C.2.2 (J.H.C. Whitehead). *For a continuous function $f : X \to Y$ between arc-wise connected spaces and any $n > 0$ the following holds:*

1. *If f is n-simple, then homomorphisms $H_k(f) : H_k(X) \to H_k(Y)$ are isomorphisms for all $k < n$ and the homomorphism $H_n(f) : H_n(X) \to H_n(Y)$ is an epimorphism.*
2. *If the homomorphisms $H_k(f) : H_k(X) \to H_k(Y)$ are isomorphisms for all $k \leq n$ then the same is true for the homomorphisms $\pi_k(f)$ for all $k \leq n$.*
3. *If X and Y are simply connected, the homomorphisms $H_k(f)$ are isomorphisms for all $k < n$ and the homomorphism $H_n(f)$ is an epimorphism then f is n-simple.*

Proof. See [Bre97b, Theorem VII.11.2] for a proof. $\square$

C.3 The Homotopy Lie-algebra

Theorem C.3.1. *The Samelson product $\pi_p(X) \times \pi_q(X) \to \pi_{p+q}(X)$ of an arc-wise connected H-space turns the graded Abelian group $\pi(X)$ into a graded Lie-algebra over the integers, i.e.:*

1. *It is bilinear.*
2. *It is graded antisymmetric, i.e. it satisfies* $[x, y] = -(-1)^{\deg x \cdot \deg y}[y, x]$ *for all homogeneous elements x and y.*
3. *It satisfies the graded Jacobi-identity*
   $$(-1)^{\deg x \cdot \deg z}[x, [y, z]] + (-1)^{\deg x \cdot \deg y}[y, [z, x]](-1)^{\deg y \cdot \deg z}[z, [x, y]]$$
   for all homogeneous elements x, y and z.

Proof. See [Whi78, Theorem X.5.1,Theorem X.5.5] for proofs of the different equalities. □

Definition C.3.2. *The graded Abelian group $\pi_*(X)$ of an arc-wise connected H-space is called the* homotopy Lie-algebra of X.

Theorem C.3.3. *The Hurewicz-homomorphisms* $\mathrm{hur}_n : \pi_n(X) \rightarrow H_n(X)$ *of an arc-wise connected H-space X form a morphism of graded Lie-algebras.*

Proof. See [Whi78, Theorem X.6.3] for a proof. □

D

Bundles

D.1 Fibre Spaces

Definition D.1.1. A fibre structure *is a continuous surjection* $p : E \to B$ *between topological spaces* E *and* B.

It is common to explicitly indicate the domain E and codomain B of a fibre structure by writing (E, p, B) for a continuous surjection $p : E \to B$.

D.2 Lifting Functions

Definition D.2.1. *A fibre structure* $p : E \to B$ *is said to have the* Homotopy Lifting Property *(HLP) with respect to a topological space* X *if, given a homotopy* $H : X \times I \to B$ *and a continuous function* $h : X \to E$ *covering* $H(-, 0)$ *there exists a homotopy* $\tilde{H} : X \times I \to E$ *covering* H:

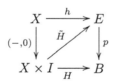

The fibre structure is said to be a fibration *for a class* $\mathcal{C}$ *of topological spaces or to have the HLP w.r.t. the class* $\mathcal{C}$ *of topological spaces if it has the HLP w.r.t. all spaces* X *in* $\mathcal{C}$. *The fibration is called* regular

Definition D.2.2. *A fibre structure* (E, p, B) *is called a* Hurewicz fibration *if it has the homotopy lifting property w.r.t. the class of all topological spaces.*

Definition D.2.3. *If* $p : E \to B$ *is a fibre structure and* $\tilde{H} : X \times I \to E$ *a homotopy covering* $H : X \times I \to B$ *then* $\tilde{H}$ *is called* stationary with H *if all paths* $\tilde{H}(x, -) : I \to E$ *are constant whenever* $H(x, -) : I \to B$ *is constant.*

Definition D.2.4. *A fibration* $p : E \to B$ *is called* regular *if for every homotopy* $H : X \times I \to B$ *there exists a covering homotopy* $\tilde{H} : X \times I \to E$ *which is stationary with* H.

Some important fibre structures in topology are path spaces. These come in different flavours. The fibre structure $p : WB \to B$ consists of the space $WB = C(I, B)$ of continuous paths and projection the evaluation at 1. Fixing a base point $*$ in B, one can also form the subbundle $p : PB \to B$, where $PB = C_*(I, B)$ is the subspace of based paths in WB.

Proposition D.2.5. *The fibre structure* $PB \xrightarrow{\text{ev}_1} B$ *over a based topological space* B *is a Hurewicz fibration.*

Proof. See [McC01, Chapter 4.3] for a proof. $\qquad\qquad\qquad\qquad\qquad\qquad$ $\square$

Let $p : E \to B$ be a fibre structure and $E' := \text{ev}_0^* E$ denote the pullback of E along the evaluation $\text{ev}_0 : WB \to B$ at 0:

$$
\begin{array}{ccc}
E' & \xrightarrow{\ pr_2\ } & E \\
{\scriptstyle pr_1}\downarrow & & \downarrow{\scriptstyle p} \\
WB & \xrightarrow[\text{ev}_0]{} & B
\end{array}
$$

The evaluation $\text{ev} : WB \times I \to B$ gives rise to a homotopy $H : E' \times I \to B$ via $H = \text{ev} \circ (pr_1 \times \text{id})$. If the projection $p : E \to B$ has the HLP with respect to E' then there exist a homotopy $\tilde{H} : E' \times I \to E$ covering H such that the following diagram is commutative:

$$
\begin{array}{ccc}
E' & \xrightarrow{\ pr_2\ } & E \\
{\scriptstyle (-,0)}\downarrow & {\scriptstyle \tilde{H}}\nearrow & \downarrow{\scriptstyle p} \\
E' \times I & \xrightarrow[H]{} & B
\end{array}
\qquad\qquad (\text{D.1})
$$

Definition D.2.6. *A lifting function* $\Lambda : E' \to PE$ *for a fibre structure* $p : E \to B$ *is the adjoint of a smooth homotopy* $\tilde{H}$ *covering the homotopy* H *in the above diagram D.1. The lifting function is called* regular *if it lifts constant paths to constant paths.*

Given the homotopy $\tilde{H}$ the corresponding lifting function is explicitly given by $\Lambda(\gamma, e)(t) := \tilde{H}((\gamma, e), t)$ This mapping satisfies the properties

$$
p \circ \Lambda(\gamma, e) = \gamma, \quad \text{and} \quad \Lambda(\gamma, e)(0) = \gamma(0) = e.
$$

Theorem D.2.7. *A fibre structure* (E, p, B) *is a Hurewicz fibration if and only if a lifting function* Λ *exists. In particular if* $f : X \times 0 \to E$ *is a continuous function and* $H : X \times I \to B$ *is a homotopy of* pf *then the homotopy* H *is lifted by* $\tilde{H}(x, t) = \Lambda(H(x, -), f(x, 0))(t)$.

Proof. See [Dug89, Theorem XX.2.2] and the proof thereof. $\qquad\qquad\qquad$ $\square$

D.3 Equivariant Lifting Functions

Let G be a topological semi-group. Restricting ourselves to the the category **GTop** of G-spaces, we define the all the notions from the previous section for the category of G-spaces.

Definition D.3.1. *A fibre structure* $p : E \to B$ *in* **GTop** *is said to have the G-Homotopy Lifting Property (GHLP) with respect to a G-space* X *if, given a G-equivariant homotopy* $H : X \times I \to B$ *and a continuous G-equivariant function* $h : X \to E$ *covering* $H(-, 0)$ *there exists a G-equivariant homotopy* $\tilde{H} : X \times I \to E$ *covering* H:

$$
\begin{array}{ccc}
X & \xrightarrow{\ h\ } & E \\
{\scriptstyle (-,0)}\downarrow & {\scriptstyle \tilde{H}}\nearrow & \downarrow{\scriptstyle p} \\
X \times I & \xrightarrow[H]{} & B
\end{array}
$$

The fibre structure is said to have the GHLP w.r.t. a class $\mathcal{C}$ *of G-spaces if it has the GHLP w.r.t. all G-spaces* X *in* $\mathcal{C}$.

Definition D.3.2. *A fibre structure* (E, p, B) *in* **GTop** *is called a* Hurewicz *G-fibration if it has the G-homotopy lifting property w.r.t. the class of all G-spaces.*

If B is a G-space then the path space $WB = C(I, B)$ also is a G-space. (The action of the semi-group G on the unit interval is assumed to be trivial here.) So the projection p of the fibre structure $p : WB \to B$ is G-equivariant. If B is a based G-space (which by definition implies that the base point is a G-fixed point), we can also form the sub G-space PB of WB.

Let $p : E \to B$ be a fibre structure in **GTop** and $E' := \mathrm{ev}_0^* E$ denote the pullback of E along the evaluation $\mathrm{ev}_0 : WB \to B$ at 0:

$$
\begin{array}{ccc}
E' & \xrightarrow{\;pr_2\;} & E \\
{\scriptstyle pr_1}\downarrow & & \downarrow{\scriptstyle p} \\
WB & \xrightarrow[\mathrm{ev}_0]{} & B
\end{array}
$$

The evaluation $\mathrm{ev} : WB \times I \to B$ gives rise to a G-equivariant homotopy $H : E' \times I \to B$ via $H = \mathrm{ev} \circ (pr_1 \times \mathrm{id})$. If the projection $p : E \to B$ has the GHLP with respect to E' then there exist a G-homotopy $\tilde{H} : E' \times I \to E$ covering H such that the following diagram is commutative:

$$
\begin{array}{ccc}
E' & \xrightarrow{\;pr_2\;} & E \\
{\scriptstyle (-,0)}\downarrow & \overset{\tilde{H}}{\nearrow} & \downarrow{\scriptstyle p} \\
E' \times I & \xrightarrow[H]{} & B
\end{array}
\qquad (D.2)
$$

Definition D.3.3. *An* equivariant lifting function $\Lambda : E' \to PE$ *for a fibre structure* $p : E \to B$ *in* **GTop** *is the adjoint of a G-equivariant homotopy* $\tilde{H}$ *covering the homotopy H in the above diagram D.2. The lifting function is called* regular *if it lifts constant paths to constant paths.*

Given the G-homotopy $\tilde{H}$ the corresponding equivariant lifting function is explicitly given by $\Lambda(\gamma, e)(t) := \tilde{H}((\gamma, e), t)$ This mapping satisfies the properties

$$
p \circ \Lambda(\gamma, e) = \gamma, \quad \text{and} \quad \Lambda(\gamma, e)(0) = \gamma(0) = e.
$$

Theorem D.3.4. *A fibre structure* (E, p, B) *is a Hurewicz G-fibration if and only if an equivariant lifting function* Λ *exists. In particular if* $f : X \times 0 \to E$ *is a continuous equivariant function and* $H : X \times I \to B$ *is a G-homotopy of* pf *then the homotopy H is lifted by* $\tilde{H}(x, t) = \Lambda(H(x, -), f(x, 0))(t)$.

Proof. The proof of [Dug89, Theorem XX.2.2] carries over to the equivariant setting. $\qquad\square$

E

Topological Algebras

In this chapter we collect some properties of the category of topological algebras and subcategories thereof. We start with recalling the definitions:

E.1 Topological Groups

Definition E.1.1. *A* topological group *is a group object in the category* **Top** *of topological spaces. The category of topological groups and continuous homomorphisms is denoted by* **TopGrp**.

Definition E.1.2. *The subcategory (of* **TopGrp***) of Abelian topological groups and continuous homomorphisms is denoted by* **TopAb**.

The forgetful functors $U : \textbf{TopGrp} \to \textbf{Top}$ and $U : \textbf{TopAb} \to \textbf{Top}$ create limits and colimits. Because the category **Top** of topological spaces is small complete and cocomplete we conclude:

Proposition E.1.3. *The category* **TopAb** *is small complete and small cocomplete.*

Lemma E.1.4. *The epimorphisms in the category* **TopAb** *are exactly the surjective homomorphisms.*

Proof. Since surjective morphisms in any set based category are epimorphisms, it remains to prove hat the epimorphisms in **TopAb** are surjective. We give a proof by contradiction. Let $\varphi : G \to H$ be a homomorphism of Abelian topological groups. The homomorphism φ followed by the quotient map $q : H \to H/\varphi(G)$ is the trivial morphism. The same is true for φ followed by the trivial morphism $0 : H \to 0 \leq H/\varphi(G)$. If φ is not surjective, then the homomorphisms q and 0 are distinct, so φ can not be an epimorphism. □

Lemma E.1.5. *The epimorphisms in the category* **TopGrp** *are exactly the surjective homomorphisms.*

Proof. Since surjective morphisms in any set based category are epimorphisms, it remains to prove hat the epimorphisms in **TopGrp** are surjective. We give a proof by contradiction, using the fact that epimorphisms in the category **Grp** of groups are surjective. Let $e : G \to G'$ be an epimorphism in **TopGrp**. Denote the forgetful functor $\textbf{TopGrp} \to \textbf{Grp}$ by U and consider the underlying group UH of the topological group H. If e were not surjective then there would exist distinct homomorphisms $\varphi, \psi : G' \to H$ of abstract groups whose composites $\varphi \circ Ue$ and $\psi \circ Ue$ would coincide. Equipping H with the indiscrete topology would yield two distinct homomorphisms $\varphi, \psi : G' \to H$ of topological groups, whose composite with e would coincide, in contradiction to the assumption that e is an epimorphism in **TopGrp**. □

Theorem E.1.6. *The forgetful functors* **TopGrp** → **Top** *and* **TopAb** → **Top** *have left adjoints* $F_{\textbf{TopGrp}}$: **Top** → **TopGrp** *and* $F_{\textbf{TopAb}}$: **Top** → **TopAb** *respectively.*

Proof. We use the forgetful functor U' : **Top** → **Set** to explicitly construct a free topological group on each topological space X. Let X be a topological space and let $F_{\textbf{Grp}}(U'X)$ be the free group on the set underlying X. The homomorphic images of $F_{\textbf{Grp}}(U'X)$ in **Grp** form a set. For each such homomorphic image G of $F_{\textbf{Grp}}(U'X)$ the topological groups in **TopGrp** which have G as underlying group do also from a set. Thus the topological groups whose underlying groups are homomorphic images of $F_{\textbf{Grp}}(U'X)$ form a set. We denote this set by S. (This is the solution set condition.) Consider the set of all homomorphisms $f : F_{\textbf{Grp}}(U'X) \to U''G$, $G \in S$ in **Grp** whose restriction to $U'X$ is the map underlying a continuous function $X \to G$. The group $F_{\textbf{Grp}}(U'X)$ equipped with the initial topology with respect to these functions is a free topological group on X. The free abelian topological group on X is constructed analogously.   □

Definition E.1.7. *The* free topological group functor **Top** → **TopGrp** *is denoted by* $F_{\textbf{TopGrp}}$. *The* free Abelian topological group functor **Top** → **TopAb** *is denoted by* $F_{\textbf{TopAb}}$.

Example E.1.8. The free abelian topological group $F_{\textbf{TopAb}}\{*\}$ on a singleton space $\{*\}$ is the discrete group $\mathbb{Z}$ of integers.

As apparent from the proof of Theorem E.1.6 the underlying group of the free topological group $F_{\textbf{TopGrp}}X$ on a space X is the free group $F_{\textbf{Grp}}U'X$ on the underlying set $U'X$ of X. Therefore the unit of the adjunction between U and $F_{\textbf{TopGrp}}$ consists of injections $X \to UF_{\textbf{TopGrp}}X$. The same is true for free abelian topological groups. Because topological groups are always completely regular but topological spaces in general are not, the natural injections $X \to UF_{\textbf{TopGrp}}X$ and $X \to UF_{\textbf{TopAb}}X$ are in general not embeddings. However, for completely regular spaces X they are embeddings:

Lemma E.1.9. *The natural injections* $X \to F_{\textbf{TopGrp}}X$ *and* $X \to F_{\textbf{TopAb}}X$ *are embeddings if and only if X is completely regular.*

Proof. See [Mor69, Theorem I.2.6] for a proof.   □

Analogously to the forgetful functors **TopGrp** → **Top** and **TopAb** → **Top** one can also consider the forgetful functors **TopGrp** → **Top**$_*$ **TopAb** → **Top**$_*$ into the category **Top**$_*$ of pointed spaces, where the identity element of a topological group serves as the base point. These functors have left adjoints as well.

Theorem E.1.10. *The forgetful functors* **TopGrp** → **Top**$_*$ *and* **TopAb** → **Top**$_*$ *have left adjoints* $F^*_{\textbf{TopGrp}}$: **Top** → **TopGrp** *and* $F^*_{\textbf{TopAb}}$: **Top** → **TopAb** *respectively.*

Proof. the proof is analogous to that of Theorem E.1.6.   □

Example E.1.11. The topological group $F^*_{\textbf{TopAb}}\{*\}$ on a singleton based space $\{*\}$ is the (discrete) trivial group $\{0\}$.

Example E.1.12. The topological group $F^*_{\textbf{TopAb}}\{x, *\}$ on the discrete based space $\{x, *\}$ is the discrete group $\mathbb{Z}$ of integers.

The choice of a base point of a topological space X turns out to be not important for the isomorphy type of the topological group $F^*_{\textbf{TopAb}}X$:

Theorem E.1.13. *Different choices p, p' of base points of a topological space X induce natural isomorphisms $\mathrm{F}^*_{\mathbf{TopAb}}(X,p) \cong \mathrm{F}^*_{\mathbf{TopAb}}(X,p')$.*

Proof. See [Mor73, Theorem 2] for a proof. □

The free groups $\mathrm{F}_{\mathbf{TopAb}}X$ and $\mathrm{F}^*_{\mathbf{TopAb}}X$ on a (based) space X differ only slightly. The relation can be described as follows: The forgetful functor $\mathbf{Top}_* \to \mathbf{Top}$ has a left adjoint $P : \mathbf{Top} \to \mathbf{Top}_*$ which assigns to each topological space X the disjoint union $PX = X \dot\cup \{*\}$ of X with an isolated base point. Composition with the functor $\mathrm{F}^*_{\mathbf{TopAb}} : \mathbf{Top}_* \to \mathbf{Top}$ yields a functor $\mathrm{F}^*_{\mathbf{TopAb}} \circ P : \mathbf{Top} \to \mathbf{TopAb}$.

Theorem E.1.14. *The free abelian topological group functor $\mathrm{F}_{\mathbf{TopAb}}$ is naturally isomorphic to $\mathrm{F}^*_{\mathbf{TopAb}} \circ P$.*

Proof. A proof can be found in [Mor73, Theorem 3]. □

Consider the forgetful functor $U : \mathbf{Top}_* \to \mathbf{Top}$. If X is a based topological space with base point p then the space PUX is the disjoint union of X with a new base point $*$. Let Y be the space $X \dot\cup \{*\}$ with base point p. The exchange of base points p and $*$ induces a natural isomorphism $\mathrm{F}^*_{\mathbf{TopAb}}PUX \to \mathrm{F}^*_{\mathbf{TopAb}}Y$ of abelian topological groups by Theorem E.1.13. The space Y is the coproduct of the based space X and the based discrete space $\{p, *\}$ in $\mathbf{Top}_*$. Because the functor $\mathrm{F}^*_{\mathbf{TopAb}}$ is a left adjoint, it preserves this coproduct. In composition with Theorem E.1.14 on obtains a chain

$$\mathrm{F}_{\mathbf{TopAb}}UX \cong \mathrm{F}^*_{\mathbf{TopAb}}PUX \cong \mathrm{F}_{\mathbf{TopAb}}Y$$
$$\cong \mathrm{F}^*_{\mathbf{TopAb}}X \oplus \mathrm{F}^*_{\mathbf{TopAb}}\{p,*\} \cong \mathrm{F}^*_{\mathbf{TopAb}}X \oplus \mathbb{Z}$$

of natural isomorphisms of abelian topological groups. So we note (cf. [Mor73, (3)]):

Corollary E.1.15. *The free abelian topological group $\mathrm{F}_{\mathbf{TopAb}}X$ on a based topological space X is naturally isomorphic to the coproduct $\mathrm{F}^*_{\mathbf{TopAb}}X \oplus \mathbb{Z}$ of the group $\mathrm{F}_{\mathbf{TopAb}}X$ with the discrete group of integers.*

The free topological group functor and the free abelian topological group functor have other useful properties. A few of them are listed below:

Lemma E.1.16. *If $f : X \to G$ is a continuous open map from the topological space X into the topological group G then the unique homomorphism $\mathrm{F}_{\mathbf{TopGrp}}X \to G$ extending f is open as well. If $f : X \to A$ is a continuous open map from the topological space X into the abelian topological group A then the unique homomorphism $\mathrm{F}_{\mathbf{TopGrp}}X \to A$ extending f is open as well.*

Proof. See [Mor69, Lemma I.2.11] for a proof. □

Lemma E.1.17. *For any topological group G the identity of G extends to a quotient homomorphism $\mathrm{F}_{\mathbf{TopGrp}}UG \to G$. Likewise for any abelian topological group A the identity of A extends to a quotient homomorphism $\mathrm{F}_{\mathbf{TopAb}}UA \to G$.*

Proof. Let G be a topological group. The identity of G extends to a surjective homomorphism $q : \mathrm{F}_{\mathbf{TopGrp}}UG \to G$. The restriction of q to G is open. By the preceding Lemma, the homomorphism is open as well, hence a quotient map. The proof for abelian topological groups is analogous. □

Lemma E.1.18. *If X is a completely regular Hausdorff space then $\mathrm{F}_{\mathbf{TopGrp}}X$ and $\mathrm{F}_{\mathbf{TopAb}}X$ are Hausdorff as well and the natural inclusions of X into $\mathrm{F}_{\mathbf{TopGrp}}X$ and $\mathrm{F}_{\mathbf{TopAb}}X$ are closed embeddings.*

Proof. See [Mor70a, Theorem I.1.1] for a proof.   □

Lemma E.1.19. *If X is a completely regular Hausdorff space then all subgroups of $F_{\mathbf{TopGrp}}X$ and $F_{\mathbf{TopAb}}X$ whose underlying group is invariant under the automorphism group of $F_{\mathbf{Grp}}U'X$ and $F_{\mathbf{TopAb}}U'X$ respectively are closed.*

Proof. See [Mor70a, Theorem I.1.5] for a proof.   □

A very special situation is an embedding $X \hookrightarrow Y$ of a subspace X of Y. For closed embedding we note:

Lemma E.1.20. *If $X \hookrightarrow Y$ is a closed embedding of completely regular Hausdorff topological spaces then the induced homomorphisms $F_{\mathbf{TopGrp}}X \to F_{\mathbf{TopGrp}}Y$ and $F_{\mathbf{TopAb}}X \to F_{\mathbf{TopAb}}Y$ have closed image.*

Proof. See [Mor70a, Theorem I.1.12] for a proof.   □

Every inclusion $i : X \hookrightarrow Y$ of a subspace induces injective homomorphisms $F_{\mathbf{TopAb}}(i) : F_{\mathbf{TopAb}}X \to F_{\mathbf{TopAb}}Y$ and $F_{\mathbf{TopGrp}}(i) : F_{\mathbf{TopGrp}}X \to F_{\mathbf{TopGrp}}Y$ of topological groups. If i is an embedding of based spaces then it induces injective homomorphisms $F^*_{\mathbf{TopAb}}(i)$ and $F^*_{\mathbf{TopGrp}}(i)$ of topological groups. These induced homomorphisms are not embeddings of topological subgroups in general. The cases in which the homomorphisms $F^*_{\mathbf{TopAb}}(i)$ and $F_{\mathbf{TopGrp}}(i)$ are inclusions of topological subgroups have been characterised:

Theorem E.1.21. *An inclusion $i : X \hookrightarrow Y$ of completely regular based spaces induces an inclusion $F^*_{\mathbf{TopAb}}(i)$ of $F_{\mathbf{TopAb}}X$ as a subgroup if and only if every bounded pseudometric on X can be extended to a pseudometric on Y.*

Proof. The proof is a corollary of [Tka83, Theorem 1].   □

Theorem E.1.22. *The inclusion $i : X \hookrightarrow Y$ of a subspace X of a completely regular T_1-space Y induces an inclusion $F_{\mathbf{TopGrp}}(i)$ of $F_{\mathbf{TopGrp}}X$ as a subgroup if and only if every bounded pseudometric on X can be extended to a pseudometric on Y.*

Proof. This general statement has been proved in [Sip00, Theorem 1]   □

The condition that pseudometrics on a subspace X of a topological space Y can be extended to the space Y has been thoroughly investigated. There are several characterisations of subspaces where this is the case:

Theorem E.1.23. *For a subspace X of a topological space Y the following are equivalent:*

1. *Any bounded continuous pseudometric on X can be extended to a bounded continuous pseudometric on Y.*
2. *Any bounded continuous pseudometric on X can be extended to a continuous pseudometric on Y.*
3. *Any continuous pseudometric on X can be extended to a continuous pseudometric on Y.*
4. *Every normal locally finite cozero-set cover of X has a refinement that is the restriction of a normal open cover of Y.*

Proof. The equivalence of these statements is proved in [Sha66, Theorem 2.1].   □

Theorem E.1.24. *For a closed subspace X of a topological space Y the following are equivalent:*

1. *Any continuous pseudometric on X can be extended to a continuous pseudometric on Y.*

2. *Every continuous map $f : X \to S$ into a metrisable subset S of a locally convex topological vector space can be extended to a function $f' : Y \to S$ if either S is complete in one of its metrics or X is the support of a continuous real valued function on Y.*

Proof. This is Theorem 2.4 of the article [Are53]. The proof presented there uses [Are52, Theorem 4.1]. □

Lemma E.1.25. *If X is a completely regular Hausdorff k_ω-space then the topological groups $\mathrm{F_{TopGrp}}X$ and $\mathrm{F_{TopAb}}X$ are k_ω-spaces.*

Proof. See [MMO73, Corollary 1] for a proof. □

Lemma E.1.26. *If a Hausdorff topological space X is a k_ω-space then the topological groups $\mathrm{F_{TopGrp}}X$ and $\mathrm{F_{TopAb}}X$ are k_ω-spaces.*

Proof. Let X be a Hausdorff k_ω-space. The free topological groups $\mathrm{F_{TopGrp}}X$ and $\mathrm{F_{TopAb}}X$ are Hausdorff by Lemma E.1.18. Therefore the images of X in $\mathrm{F_{TopGrp}}X$ and $\mathrm{F_{TopAb}}X$ are completely regular Hausdorff k_ω-spaces. Thus by Lemma E.1.25 the groups $\mathrm{F_{TopGrp}}X$ and $\mathrm{F_{TopAb}}X$ are k_ω-spaces. □

A tensor product of two abelian topological groups G and H always exists. It can be constructed analogously to the discrete case: Let G and H be abelian topological groups. Consider free abelian topological group $\mathrm{F_{TopAb}}(G \times H)$ on the product space $G \times H$ and let N be the normal subgroup of $\mathrm{F_{TopAb}}(G \times H)$ generated by all elements of the form $n(g, h) - (ng, h)$ and $n(g, h) - (g, nh)$. The biadditive map

$$\otimes : G \times H \to G \otimes H = \mathrm{F_{TopAb}}(G \times H)/N, \quad \otimes(g, h) = (g, h) + N.$$

satisfies the universal property of a tensor product. So we note: In particular, the underlying group $U(G \otimes H)$ of the tensor product coincides with the tensor product $UG \otimes UH$ in the category $\mathbf{Ab}$ of abelian groups. As usual we denote an element $(g, h) + N$ by $g \otimes h$. All in all we have observed:

Lemma E.1.27. *The map $\otimes : G \times H \to G \otimes H$ is a tensor product in the category $\mathbf{TopAb}$ of abelian topological groups. Its underlying map $U\otimes$ is the tensor product of the underlying groups UG and UH in $\mathbf{Ab}$.*

E.2 Group Objects in $\mathbf{Top_C}$

Lemma E.2.1. *The monomorphisms in $\mathbf{CGrp}$ are the injective homomorphisms.*

Proof. Clearly every injective map in $\mathbf{CGrp}$ is a monomorphism. Conversely assume $\varphi : H \to G$ to be a monomorphism. Let $x, y \in X$ be points of H. Since the trivial group $\{1\}$ is a $\mathbf{C}$-group, the homomorphisms $i_x : 1 \to X, 1 \mapsto x$ and $i_y : 1 \to X$, $1 \mapsto y$ are homomorphisms in $\mathbf{CGrp}$. If the images of x and y under f coincide, then the composites $\varphi \circ i_x$ and $\varphi \circ i_y$ are equal. Because φ was assumed to be a monomorphism, the maps i_x and i_y have to be equal as well, i.e. $x = y$. Thus φ is injective. □

Lemma E.2.2. *Quotient morphisms in $\mathbf{CGrp}$ are open.*

Proof. Let $q : G \twoheadrightarrow H$ be a quotient homomorphism in $\mathbf{Top_C}$ and let $N = \ker q$ be the kernel of q. If O is an open subset of G then the q-saturation $q^{-1}q(O) = N \cdot O$ is open as a union of open sets. Thus the image $q(O)$ is open in H. □

Let the category $\mathbf{C}$ be productive, so that $\mathbf{Top}_{\mathbf{C}}$ is Cartesian closed. The forgetful functor $\mathbf{Top}_{\mathbf{C}} \to \mathbf{Set}$ is denoted by U. For any $\mathbf{C}$-generated space X we consider the free group FUX on the underlying set UX of X. We are going to equip this group with a $\mathbf{C}$-generated topology that makes it a free $\mathbf{C}$-generated group. First we recall the construction of FUX (cf. [Rob96, Chap. 2.1]):

The set UX is joined with a set $UX^{-1} := \{x^{-1} \mid x \in X\}$ of formal 'inverses to the elements of X. Then we consider the set $\coprod_{n \in \mathbb{N}}(UX \dot\cup UX^{-1})^n$ of words in the alphabet $UX \dot\cup UX^{-1}$ and factor by an equivalence relation to obtain the free group on the set X:

$$\coprod_{n \in \mathbb{N}}(UX \dot\cup UX^{-1})^n \twoheadrightarrow FUX$$

We will verify that the analogue construction –carried out in the category $\mathbf{Top}_{\mathbf{C}}$– yields a free $\mathbf{C}$-group on the $\mathbf{C}$-space X. To accomplish this we need a space of formal 'inverses to the elements of X. We take a space X^{-1} that is homeomorphic to X via a homeomorphism $i : X \to X^{-1}$. The image of an element $x \in X$ under i is denoted by $i(x) = x^{-1}$. The homeomorphism i is then extended to an involution of the space $X \dot\cup X^{-1} \to X \dot\cup X$ by defining $i(x^{-1}) = x$. We take the $\mathbf{C}$-space $\coprod_{n \in \mathbb{N}} \mathrm{C}(X \dot\cup X^{-1})^n$ of words in the alphabet $UX \dot\cup UX^{-1}$ and factor by the same equivalence relation as above. The quotient space of this identification in $\mathbf{Top}_{\mathbf{C}}$ is denoted by $\mathrm{F}_{\mathbf{CGrp}}X$:

$$\coprod_{n \in \mathbb{N}} \mathrm{C}(X \dot\cup X^{-1})^n \twoheadrightarrow \mathrm{F}_{\mathbf{CGrp}}X$$

Note that the quotient space $\mathrm{F}_{\mathbf{CGrp}}X$ is $\mathbf{C}$-generated, because quotients of $\mathbf{C}$-generated spaces are always $\mathbf{C}$-generated. The underlying set $U\mathrm{F}_{\mathbf{CGrp}}X$ of the space $\mathrm{F}_{\mathbf{CGrp}}X$ is then the underlying set of the free group FUX by construction.

Lemma E.2.3. *The assignment $X \mapsto \mathrm{F}_{\mathbf{CGrp}}X$ is the object part of a functor $\mathrm{F}_{\mathbf{CGrp}} : \mathbf{Top}_{\mathbf{C}} \to \mathbf{CGrp}$ such that $U\mathrm{F}_{\mathbf{CGrp}} = UFU$.*

Proof. Let X and Y be two $\mathbf{C}$-generated spaces and $f : X \to Y$ be a continuous function. We define a continuous function $f' : X^{-1} \to Y^{-1}$ via $f' = i \circ f \circ i^{-1}$. The function $f \dot\cup f' : X \dot\cup X^{-1} \to Y \dot\cup Y^{-1}$ induces a continuous function

$$\coprod_{n \in \mathbb{N}} \mathrm{C}(\prod_{i=1}^{n}(f \dot\cup f')) : \coprod_{n \in \mathbb{N}} \mathrm{C}(X \dot\cup X^{-1})^n \to \mathrm{C}\coprod_{n \in \mathbb{N}} \mathrm{C}(Y \dot\cup Y^{-1})^n$$

From the construction of the free group FUX in $\mathbf{Grp}$ we know that this function is relation preserving. Thus it descends to a continuous function $\mathrm{F}_{\mathbf{CGrp}}f$ between the quotient spaces $\mathrm{F}_{\mathbf{CGrp}}X$ and $\mathrm{F}_{\mathbf{CGrp}}Y$. The equality $U\mathrm{F}_{\mathbf{CGrp}}f = UFUf$ is satisfied by construction. $\square$

For each $\mathbf{C}$-generated space X the natural injection $X \hookrightarrow \coprod \mathrm{C}(X \dot\cup X^{-1})^n$ induces a continuous function $\eta_X : X \to \mathrm{F}_{\mathbf{CGrp}}X$.

Lemma E.2.4. *The maps η_X form a natural transformation $\mathrm{id}_{\mathbf{Top}_{\mathbf{C}}} \to \mathrm{F}_{\mathbf{CGrp}}$.*

Proof. This follows from the respective property of the corresponding morphisms $UX \to FUX$ in $\mathbf{Set}$. $\square$

Lemma E.2.5. *The natural transformation $\eta : \mathrm{id}_{\mathbf{Top}_{\mathbf{C}}} \to \mathrm{F}_{\mathbf{CGrp}}X$ consists of injections.*

Proof. This follows from the fact that the natural maps $U\eta_X : UX \to U\mathrm{F}_{\mathbf{CGrp}}X$ are injections. $\square$

Lemma E.2.6. *The involution $\coprod(i \times \cdots \times i) : \coprod C(X\dot\cup X^{-1})^n \to \coprod C(X\dot\cup X^{-1})^n$ induces an involution of $\mathrm{F_{CGrp}}X$.*

Proof. It is known from the construction of FUX that the map on the underlying set induces the inversion on FUX. So the involution $\coprod(i \times \cdots \times i)$ it is relation preserving and descends to an involution on $\mathrm{F_{CGrp}}X$. □

Proposition E.2.7. *The space $\mathrm{F_{CGrp}}X$ is a **C**-monoid with multiplication μ induced by the the juxtaposition of words.*

Proof. It is known from the construction of FUX that the juxtaposition of words induces the multiplication on FUX. It remains to show the continuity of the corresponding multiplication map on $\mathrm{F_{CGrp}}X$. The juxtaposition of words

$$C(X\dot\cup X^{-1})^p \times_C C(X\dot\cup X^{-1})^p \to C(X\dot\cup X^{-1})^{p+q}$$
$$((a_1,\ldots,a_p),(b_1,\ldots,b_q)) \mapsto (a_1,\ldots,a_p,b_1,\ldots,b_q)$$

is continuous for all $p,q \in \mathbb{N}$. Because these sets form an open partition of the product space $[\coprod C(X\dot\cup X^{-1})^p] \times_C [\coprod C(X\dot\cup X^{-1})^p]$ the 'multiplication

$$\tilde\mu : \left[\coprod C(X\dot\cup X^{-1})^p\right] \times_C \left[\coprod C(X\dot\cup X^{-1})^p\right] \to \coprod (X\dot\cup X^{-1})^n$$
$$((a_1,\ldots,a_p),(b_1,\ldots,b_q)) \mapsto (a_1,\ldots,a_p,b_1,\ldots,b_q)$$

is continuous. Let $\pi : \coprod C(X\dot\cup X^{-1})^n \to \mathrm{F_{CGrp}}X$ be the quotient map onto $\mathrm{F_{CGrp}}X$ in **Top**_C. By Lemma B.6.29 the product map $\pi \times_C \pi$ is a quotient map, so the multiplication $\tilde\mu$ on $\coprod C(X\dot\cup X^{-1})^n$ induces a continuous multiplication μ on $\mathrm{F_{CGrp}}X$ in **Top**_C. □

Proposition E.2.8. *The space $\mathrm{F_{CGrp}}X$ is a **C**-group with inversion induced by the involution i and multiplication μ induced by the juxtaposition of words. Its underlying group is the free group on the underlying set UX of X.*

Proof. It is known that the underlying set $U\mathrm{F_{CGrp}}X = FUX$ is a group with multiplication Ui and multiplication $U\mu$. Thus the operations i and μ turn $\mathrm{F_{CGrp}}X$ into a **C**-group. □

Theorem E.2.9. *For every continuous map $f : X \to G$ into a **C** group G there exists a unique homomorphism $\hat f : \mathrm{F_{CGrp}}X \to G$ of **C**-groups such that $\hat f \circ \eta_X = f$.*

Proof. Let X be a **C**-generated space, G be a **C**-group and $f : X \to G$ be a continuous function. Since FUX is the free group on the set UX there exists a unique (not necessarily continuous) map $\hat f : \mathrm{F_{CGrp}}X \to G$ which is a homomorphism of the underlying groups. It remains to show that $\hat f$ is continuous. The map $\hat f$ is induced by the continuous function

$$\coprod_{n\in\mathbb{N}} C(X\dot\cup X^{-1})^n \to G, \quad (a_1,\ldots,a_n) \mapsto f(a_1)\cdots f(a_n).$$

Because $\coprod C(X\dot\cup X^{-1})^n \twoheadrightarrow \mathrm{F_{CGrp}}X$ is an identification, the induced map $\hat f$ is continuous as well. □

Corollary E.2.10. *The functor $\mathrm{F_{CGrp}} \to \mathbf{Top}_C$ is left adjoint to the grounding functor $\mathbf{CGrp} \to \mathbf{Top}_C$ and the natural transformation $\eta : \mathrm{id}_{\mathbf{Top}_C} \to \mathrm{F_{CGrp}}$ is the counit of this adjunction.*

The above observations justify the the following definition:

Definition E.2.11. *The functor* $F_{\mathbf{CGrp}} : \mathbf{Top}_C \to \mathbf{CGrp}$ *is called the* free **C**-*group functor.*

In a similar manner one obtains a free Abelian **C**-group functor:

Theorem E.2.12. *There exists a left adjoint* $F_{\mathbf{Top}_C\mathbf{Ab}}$ *to the grounding functor* $\mathbf{CGrp} \to \mathbf{Top}_C$.

Proof. The functor is constructed analogous to the functor $F_{\mathbf{CGrp}}$. Replacing the free group FUX by the free Abelian group $F_{\mathbf{TopAb}}UX$ the proofs carry over word by word.

The only difference to the construction of $F_{\mathbf{CGrp}}X$ for a **C**-generated space X is that the equivalence relation on the space $C(X\dot\cup X^{-1})^n$ is greater than that used for obtaining the **C**-group $F_{\mathbf{CGrp}}X$.

Lemma E.2.13. *If* $\mathbf{Top}_C$ *is productive then the descent to A-orbits in* $\mathbf{GTop}_C\mathbf{A}$ *is a functor* $\mathbf{GTop}_C\mathbf{A} \to \mathbf{GTop}_C$.

Proof. Let G and A be **C**-groups and X be a (G, A) space in $\mathbf{Top}_C$. We denote the action $C(G \times X) \to X$ of G on X by ϱ and the quotient map $X \to X/A$ by q. The composite $q \circ \varrho$ factors through the quotient space $C(G \times X)/A$ making the following diagram commutative:

$$
\begin{array}{ccc}
C(G \times X) & \xrightarrow{\varrho} & X \\
\downarrow & & \downarrow q \\
C(G \times X)/A & \longrightarrow & X/A
\end{array}
$$

If the spaces $C(G \times X)/A$ and $C(G \times X/A)$ are naturally homeomorphic, then the induced map $C(G \times X/A) \to X/A$ is an action of G on X/A. If $\mathbf{Top}_C$ is productive, then the product of the identity map id_G and the identification map $X \to X/A$ in $\mathbf{Top}_C$ is an identification by Lemma B.6.29. Therefore the spaces $C(G \times X)/A$ and $C(G \times X/A)$ are naturally homeomorphic. So the action ϱ factors to an action of G on X/A. $\qquad\square$

Lemma E.2.14. *If* **C** *is productive then the left translation action* $\lambda : C(G \times G)$ *of a* **C** *group descends to a left action on the right coset spaces of* G.

Lemma E.2.15. *If* **C** *is productive and* $H \leq G$ *a subgroup of* G *then the action of* G *on the coset space* (G/H) *induces an action of* G *on the free Abelian* **C**-*group* $F_{\mathbf{Top}_C\mathbf{Ab}}(G/H)$.

Proof. Consider the set $(G/H)^{-1} := \{(gH)^{-1} \mid gH \in G/H\}$ and topologise it via the identification $i : G/H \to (G/H)^{-1}, gH \mapsto (gH)^{-1}$. Using this identification the action of G on the coset space G/H carries over to an action of G on the space $G/H \dot\cup (G/H)^{-1}$ thus inducing an action

$$
G \times_C \coprod_{n\in\mathbb{N}} C(G/H\dot\cup(G/H)^{-1})^n \longrightarrow \coprod_{n\in\mathbb{N}} C(G/H\dot\cup(G/H)^{-1})^n
$$

The **C**-group $F_{\mathbf{CGrp}}(G/H)$ is the quotient of the G-space $\coprod C((G/H)\dot\cup(G/H)^{-1})^n$ by construction. Since the equivalence relation on this space is G-invariant and **C** is productive the action descends to an action of G on $F_{\mathbf{Top}_C\mathbf{Ab}}X$. $\qquad\square$

Proposition E.2.16. *If* $\mathbf{Top}_C$ *is Cartesian closed, then the epimorphisms in* $\mathbf{CGrp}$ *are exactly the surjective homomorphisms.*

Proof. Since surjective morphisms in any set based category are epimorphisms, it remains to prove hat the epimorphisms in **CGrp** are surjective. We give a proof by contradiction, where we adapt the idea of [HM98, Exercise EA3.16] for discrete groups. It suffices to show that every injective epimorphism $\varphi : H \to G$ is surjective. Let $\varphi : H \to G$ be an injective homomorphism and denote the image of φ by H'. We consider the semi-direct product

$$0 \to E \to E \rtimes G \twoheadrightarrow G \to 1$$

of **C**-groups. Assume that H' is a proper subgroup of G. If we succeed in finding a 1-cocycle $f \in Z_c^1(G, E)$ which is trivial on H' but non-trivial on G, then the two homomorphisms

$$G \to E \rtimes G, \quad g \mapsto (f(g), g) \qquad \text{and} \qquad G \to E \rtimes G, \quad g \mapsto (0, g)$$

coincide on H' but differ on G. Thus φ is not an epimorphism. Assuming that H' is a proper subgroup of G we now give an explicit construction of a G-module E and a 1-cocycle $f \in Z_c^1(G, E)$ satisfying the above assumptions. Because the colimits in **Top$_\mathrm{C}$** coincide with those in **Top**, the homogeneous space G/H' is **C**-generated. Let $\mathrm{F}_{\mathbf{Top_C Ab}}(G/H')$ be the free Abelian **C**-group over the coset space G/H' and let G act trivially on $\mathrm{F}_{\mathbf{Top_C Ab}}(G/H')$. Denote the natural injection of (G/H') into $\mathrm{F}_{\mathbf{Top_C Ab}}(G/H')$ by $\eta_{G/H'}$. The continuous function

$$p : G \to \mathrm{F}_{\mathbf{Top_C Ab}}(G/H'), \quad g \mapsto \eta_{G/H'}(gH')$$

is a point of $E = C(G, \mathrm{F}_{\mathbf{Top_C Ab}}(G/H'))_{exp}$. It is trivial on H' but not trivial on G by construction. The coboundary $f = \delta p$ is a 1-cocycle in $Z_c^1(G, E)$, which is also trivial on H'. Since H' was assumed to be a proper subgroup of G there exists an element $g \in G \setminus H'$ such that

$$(p)(g) = \eta_{G/H'}(gH') \neq 0 \in \mathrm{F}_{\mathbf{Top_C Ab}}(G/H'),$$

i.e. p is not a fixed point of G. Therefore the cocycle f is not trivial on G. $\qquad\square$

E.3 Topological Algebras

Definition E.3.1. *A* topological (Ω, E)-algebra *is an* (Ω, E)-algebra *in the category* **Top** *of topological spaces. The category of topological* (Ω, E)-algebras *and continuous homomorphisms is denoted by* **TopAlg**$_{(\Omega, \mathrm{E})}$.

Lemma E.3.2. *The category* **TopAlg**$_{(\Omega, \mathrm{E})}$ *is complete and cocomplete.*

Lemma E.3.3. *The forgetful functor* $U :$ **TopAlg**$_{(\Omega, \mathrm{E})} \to$ **Top** *creates limits and colimits.*

Lemma E.3.4. *The monomorphisms in the category* **TopAlg**$_{(\Omega, \mathrm{E})}$ *are exactly the injective homomorphisms.*

Proof. Since injective morphisms in any set based category are monomorphisms, it remains to prove hat the monomorphisms in **TopAlg**$_{(\Omega, \mathrm{E})}$ are injective. We give a proof by contradiction. If $\varphi : A \to A'$ is a homomorphism in **TopAlg**$_{(\Omega, \mathrm{E})}$, which is not injective, then there exist different points Let $a, b \in A$ with common image $\varphi(a) = \varphi(b)$. Let $F\{*\}$ be the free topological (Ω, E)-algebra on one generator and let f and g be the unique homomorphisms of (Ω, E)-algebras mapping $*$ to a and b respectively. Then the morphisms φf and φg coincide, but f and g are distinct. Thus φ can not be a monomorphism. $\qquad\square$

F

Simplicial Objects

F.1 The Simplicial Category

We now describe a category which plays a central role in topology. Let **FinOrd** denote the category of finite ordinal numbers $n = \{0, \ldots, n-1\}$ and morphisms all monotone (i.e. non-decreasing) functions. The ordinal number 0 is initial and the ordinal number 1 is terminal in this category. One can describe all morphisms in **FinOrd** in terms of so called *face-* and *degeneracy maps*.

Definition F.1.1. *The i-th face map $\epsilon_i : n \to n+1$, is the unique injective function whose image does not contain i:*

$$\epsilon_i(j) = \begin{cases} j & \text{if } j < i \\ j+1 & \text{if } j \geq i \end{cases}$$

Definition F.1.2. *The i-th face map $\eta_i : n+1 \to n$ is the unique surjective function that maps two elements to i:*

$$\eta_i(j) = \begin{cases} j & \text{if } j \leq i \\ j-1 & \text{if } j > i \end{cases}$$

There are $n+2$ different face maps $\epsilon_i : n \to n+1$ as well as $n+1$ different degeneracy maps $\eta_i : n+1 \to n$. The face and degeneracy maps satisfy the following identities ([ML98, ch. VII.5]):

$$\epsilon_j \epsilon_i = \epsilon_i \epsilon_{j-1} \quad \text{if } i < j \tag{F.1}$$

$$\eta_j \eta_i = \eta_i \eta_{j+1} \quad \text{if } i \leq j \tag{F.2}$$

$$\eta_j \epsilon_i = \begin{cases} \epsilon_i \eta_{j-1} & \text{if } i < j \\ \text{identity} & \text{if } i = j \text{ or } i = j+1 \\ \epsilon_{i-1} \eta_j & \text{if } i > j+1 \end{cases} \tag{F.3}$$

Also, every morphism $\alpha : m \to n$ in **FinOrd** has a unique epi-monic factorisation

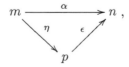

where $p+1$ is the cardinality of the image $\alpha(m)$. The morphisms ϵ and η in this factorisation can be expressed as concatenations of face and degeneracy maps respectively. If one requires the indices of the face and degeneracy maps in this representation of α to have descending and ascending order respectively, the representation is unique:

Lemma F.1.3. *Every morphism $\alpha : m \to n$ in* **FinOrd** *has a unique epi-monic factorisation $\alpha = \epsilon\eta$, where the the monomorphism ϵ is a unique composition of face maps*

$$\epsilon = \epsilon_{i_1} \cdots \epsilon_{i_s}$$

with $m > i_1 > \cdots > i_s \geq 0$, the epimorphism η is a unique composition of degeneracy maps

$$\eta = \eta_{j_1} \cdots \eta_{j_t}$$

with $0 \leq j_1 < j_2 < \cdots < j_t < m - 1$ and furthermore the equation $m + s = n + t$ is satisfied.

Proof. See [ML98, Ch. VII.5].

There exists an endofunctor of the category **FinOrd** that shifts the objects and the indices of the face and degeneracy morphisms (cf. [Wei94, 8.3.14]). It is obtained by formally adjoining an initial element $0'$ to each finite ordinal number n and then identifying the ordered set $0' < 0 < \cdots < n$ with the ordinal number $n + 1$:

Definition F.1.4. *The* right shift functor $T :$ **FinOrd** $\to$ **FinOrd** *is given by*

$$T(n) = (n + 1), \quad T(\alpha)(k) = \begin{cases} 0 & \text{if } k = 0 \\ 1 + \alpha(k-1) & \text{if } k > 0 \end{cases} \quad \text{for all } \alpha : m \to n.$$

By construction, the right shift functor T maps a face morphism $\epsilon_i : n \to n + 1$ in **FinOrd** to the face morphism $\epsilon_{i+1} : n + 1 \to n + 2$ and a degeneracy morphism $\eta_i : n+1 \to n$ to the degeneracy morphism $\eta_{i+1} : n+2 \to n+1$. Also by construction, the the face morphisms $\epsilon_0 : n \hookrightarrow n + 1$ for each $n \in \mathbb{N}$ assemble to a natural transformation $\epsilon_0 : \text{id}_{\textbf{FinOrd}} \to T$.

Definition F.1.5. *The* simplicial category $\mathcal{O}$ *is the full subcategory of* **FinOrd** *with objects all the positive ordinals. The object $\{0, \ldots, n\}$ is denoted by $[n]$.*

The category $\mathcal{O}$ has no initial object but contains a unique terminal object $[0]$.

Lemma F.1.6. *Every morphism $\alpha : [m] \to [n]$ in $\mathcal{O}$ has a unique epi-monic factorisation $\alpha = \epsilon\eta$, where the the monomorphism ϵ is a unique composition of face maps*

$$\epsilon = \epsilon_{i_1} \cdots \epsilon_{i_s}$$

with $0 \leq i_1 \leq \cdots \leq i_s \leq n$, the epimorphism η is a unique composition of degeneracy maps

$$\eta = \eta_{j_1} \cdots \eta_{j_t}$$

with $0 \leq j_1 < \cdots j_t < m$ and furthermore the equation $m + s = n + t$ is satisfied.

Proof. See [Wei94, Ch. 8.1].

The right shift functor $T :$ **FinOrd** $\to$ **FinOrd** has codomain in the full subcategory $\mathcal{O}$ of **FinOrd**. It therefore restricts to an endofunctor $T : \mathcal{O} \to \mathcal{O}$ of the simplicial category $\mathcal{O}$. As a consequence, the natural transformation $\epsilon_0 : \text{id}_{\textbf{FinOrd}} \to T$ restricts to a natural transformation between $\text{id}_{\mathcal{O}}$ and the restriction of T to $\mathcal{O}$.

F.2 Simplicial Objects

Definition F.2.1. *A* simplicial object *A in a category* A *is a functor $A : \mathcal{O}^{op} \to$ A, i.e. a contravariant functor $\mathcal{O} \to$ A. An* augmented simplicial object *A in a category* A *is a functor $A : \mathbf{FinOrd}^{op} \to$ A, i.e. a contravariant functor $\mathbf{FinOrd} \to$ A.*

Example F.2.2 (Constant (augmented) simplicial objects). Let A be a fixed object of a category **A**. The constant functor A is an augmented simplicial object in the category **A**. It restricts to a simplicial object $A : \mathcal{O}^{op} \to \mathbf{A}$.

Proposition F.2.3. *Every simplicial object A in a category* **A** *is uniquely determined by the sequence $A([n])$ and all the face and degeneracy maps $\partial_i := A(\epsilon_i)$ and $s_i := A(\eta_i)$. Conversely, given a sequence A_n, $n \in \mathbb{N}$ of objects in* **A** *together with face operators $\partial_i^{n+1} : A_{n+1} \to A_n$ and degeneracy operators $s_i^n : A_n \to A_{n+1}$, $0 \leq i \leq n$, which satisfy the 'simplicial' identities*

$$\partial_i \partial_j = \partial_{j-1} \partial_i \quad \text{if } i < j$$
$$s_i s_j = s_{j+1} s_i \quad \text{if } i \leq j$$
$$\partial_i s_j = \begin{cases} s_{j-1} \partial_i & \text{if } i < j \\ \text{identity} & \text{if } i = j \text{ or } i = j+1 \\ s_j \partial_{i-1} & \text{if } i < j \end{cases}$$

there exists a unique simplicial object A in **A** *such that $A_n = A([n])$, $\partial_i^{n+1} = A(\epsilon_i^{n+1})$ and $s_i = A(\eta_i^n)$ for all $n \in \mathbb{N}$ and all $0 \leq i \leq n$.*

Definition F.2.4. *A* cosimplicial object *A in a category* **A** *is a functor $A : \mathcal{O} \to$ A. An* augmented cosimplicial object *A in a category* **A** *is a functor $A : \mathbf{FinOrd} \to$ A.*

Example F.2.5 (Constant (augmented) cosimplicial objects). Let A be a fixed object of a category **A**. The constant functor A is an augmented cosimplicial object in the category **A**. It restricts to a simplicial object $A : \mathcal{O} \to \mathbf{A}$.

Example F.2.6. The right shift functor $T : \mathbf{FinOrd} \to \mathcal{O}$ is an augmented cosimplicial object in $\mathcal{O}$. It restricts to a cosimplicial object $T : \mathcal{O} \to \mathcal{O}$.

The category $\mathcal{O}$ has a direct geometric interpretation by affine simplices.

Definition F.2.7. *The* standard n-simplex Δ_n *is the subspace*

$$\Delta_n := \left\{ (x_0, \ldots, x_n) \in \mathbb{R}^{n+1} \mid (\forall i : 0 \leq x_i \leq 1) \wedge \sum x_i = 1 \right\}$$

of the real vector space $\mathbb{R}^{n+1}$. The vertex $(0, \ldots, 0, 1, 0, \ldots, 0)$ is denoted by $\mathbf{e}_i$.

The assignment $[n] \mapsto \Delta_n$ can be made a cosimplicial object in the category **Top** of topological spaces as follows: Let $\alpha : [m] \to [n]$ be a morphism in $\mathcal{O}$. The function $\Delta(\alpha)$ is defined to be the unique affine map which maps each vertex $\mathbf{e}_i$ to the vertex $\mathbf{e}_{\alpha(i)}$.

Definition F.2.8. *The cosimplicial object $\mathcal{O} \to$ Top constructed above is called the* standard cosimplicial space. *It is denoted by Δ.*

The objects $\Delta([n])$ are the standard simplices Δ_n by construction. For categories **C** with finite products there exists a general construction, which assigns to each object X of **C** a simplicial object in **C**.

Definition F.2.9. *The* standard simplicial object functor SS *on a category* **C** *with finite products is the functor* SS $: \mathbf{C} \to \mathbf{C}^{\mathcal{O}^{op}}$ *which is given by*

$$\mathrm{SS}(X)([n]) = \prod_{i=0}^{n} X, \quad \mathrm{SS}(X)(\alpha) := \prod_{i=0}^{m} p_{\alpha(i)} : \prod_{i=0}^{n} X \to \prod_{i=0}^{m} X,$$

for every morphism $\alpha : [m] \to [n]$ *in* $\mathcal{O}$. *Here the maps* $p_{\alpha(i)}$ *are meant to be the projections of* $\prod_{i=0}^{n} X$ *onto the* $\alpha(i)$-*th factor* X.

To each simplicial object in an abelian category **A** one can associate a chain complex $C(A)$ in the category **A**.

Definition F.2.10. *The* associated *or* unnormalised chain complex $C(A)$ *of a simplicial object* $A : \mathcal{O}^{op} \to \mathbf{A}$ *in an abelian category* **A** *is the chain complex*

$$C(A)_n(A) = A([n]), d_n = \sum_{i=0}^{n} (-1)^i A(\epsilon_i) : C(A)_n \to C(A)_{n-1}.$$

The assignment $A \to C(A)$ is the object part of a functor $C : \mathbf{A}^{\mathcal{O}^{op}} \to \mathbf{Ch}(\mathbf{A})$:

Definition F.2.11. *The functor* $C : \mathbf{A}^{\mathcal{O}^{op}} \to \mathbf{Ch}(\mathbf{A})$, $A \to C(A)$, $f \mapsto A(f)$ *is called the* associated *or* unnormalised chain complex functor.

F.3 Simplicial Sets

Definition F.3.1. *A* simplicial set *is a simplicial object in the category* **Set**, *i.e. a contravariant functor* $\mathcal{O} \to \mathbf{Set}$.

Definition F.3.2. *For any set* X *the* standard simplicial set $SS(X)$ *is the simplicial set* $SS(X) : \mathcal{O}^{op} \to \mathbf{Set}$ *which is given by*

$$SS(X)([n]) := X^{n+1}, \quad and \quad SS(X)(\alpha)(x_0, \ldots, x_n) = (x_{\alpha(0)}, \ldots, x_{\alpha(m)})$$

for any morphism $\alpha : [m] \to [n]$ *in* $\mathcal{O}$.

Definition F.3.3. *A* cosimplicial set *is a simplicial object in the category* **Set**, *i.e. a functor* $\mathcal{O} \to \mathbf{Set}$.

Example F.3.4. Let X be a topological space. Since the hom-functor $\hom(-, X)$ is contravariant, the functor $\hom(\Delta, X) := \hom(-, X) \circ \Delta : \mathcal{O}^{op} \to \mathbf{Set}$ is a simplicial set. It is called the *singular simplicial set* of X.

Definition F.3.5. *A simplicial set* x *is called* fibrant *if it satisfies the following condition for every* $n \in \mathbb{N}$:
For all $0 \le k \le n+1$ *and every* $n+1$ *elements* $x_0, \ldots, x_{k-1}, x_{k+1}, \ldots, x_{n+1} \in X([n])$ *such that* $\partial_i x_j = \partial_{j-1} x_i$ *for* $i < j$, $i, j \ne k$ *there exists an element* $x \in X([n+1])$ *such that* $\partial_i x = x_i$ *for* $i \ne k$.

Lemma F.3.6. *The singular simplicial set* $\hom(\Delta, X)$ *of a topological space* X *is fibrant.*

Proof. This is a consequence of the fact that the subspaces of a standard simplex Δ_{n+1} consisting of all but one face are retracts of Δ_{n+1}. □

Lemma F.3.7. *The underlying simplicial sets of simplicial groups are fibrant.*

Proof. See [Wei94, Lemma 8.2] for a proof. □

Definition F.3.8. *A morphism* $\pi : E \to B$ *of simplicial sets is called a* fibration *if it satisfies the following condition for all* $n \in \mathbb{N}$:
For all $b \in B([n+1])$, $0 \le k \le n+1$ *and elements* $e_0, \ldots, e_{k-1}, e_{k+1}, \ldots, e_{n+1} \in E([n])$ *such that* $\partial_i b = \pi(e_i)$ *and* $\partial_i e_j = \partial_{j-1} e_i$ *for* $i < j$, $i, j \ne k$ *there exists an element* $e \in E([n+1])$ *such that* $\pi(e) = b$ *and* $\partial_i x = x_i$ *for all* $i \ne k$.

F.4 Simplicial Spaces

Definition F.4.1. *A simplicial space is a simplicial object in the category* **Top**, *i.e. a contravariant functor* $\mathcal{O} \to$ **Top**.

Example F.4.2. Let X be a topological space. Since the functor $C(-, X)_{c.o.}$ is contravariant, the functor $C(\Delta, X)_{c.o.} := C(-, X)_{c.o.} \circ \Delta : \mathcal{O}^{op} \to$ **Top** is a simplicial space.

Definition F.4.3. *The simplicial space* $C(\Delta, X)$ *is called the* singular simplicial space of X.

The function space functor $C(-, -)$ is a bifunctor which is contravariant in the first and covariant in the second argument. The composition of this functor with the functor $\Delta \times \mathrm{id}_{\mathbf{Top}} : \mathcal{O} \times \mathbf{Top} \to \mathbf{Top}$ is a bifunctor

$$C(\Delta, -) := C(-, -) \circ (\Delta \times \mathrm{id}_{\mathbf{Top}}) : \mathcal{O}^{op} \times \mathbf{Top} \to \mathbf{Top},$$

which we regard as a functor $\mathbf{Top} \to \mathbf{Top}^{\mathcal{O}^{op}}$ using the exponential law. It assigns to every topological space X the singular simplicial space $C(\Delta, X)$ of X.

Definition F.4.4. *The functor* $C(\Delta, -)$: **Top** $\to$ **Top**$^{\mathcal{O}^{op}}$ *is called the* singular simplicial space functor.

Definition F.4.5. *For any topological space* X, *the* standard simplicial space $SS(X)$ *is the simplicial space* $SS(X) : \mathcal{O}^{op} \to$ **Top** *which is given by*

$$SS(X)([n]) := X^{n+1}, \quad and \quad SS(X)(\alpha)(x_0, \ldots, x_n) = (x_{\alpha(0)}, \ldots, x_{\alpha(m)})$$

for any morphism $\alpha : [m] \to [n]$ *in* $\mathcal{O}$.

Definition F.4.6. *A natural transformation* $\tau : F \to G$ *of functors* $F, G \in$ **Top**$^{\mathbf{D}}$ *from a category* **D** *into* **Top** *is called* open, *if the maps* $\tau_d : F(d) \to F(d)$ *are open for all objects d of* **D**. *It is called* closed *if all maps* $\tau_d : F(d) \to F(d)$ *are closed. Analogously it is called* surjective, *if the maps* $\tau_d : F(d) \to F(d)$ *are surjective for all objects d of* **D**. *It is called* injective *if all maps* $\tau_d : F(d) \to F(d)$ *are injective.*

Lemma F.4.7. *The monomorphisms in the category* **Top**$^{\mathcal{O}^{op}}$ *of simplicial spaces are the injections, i.e. a map* $F : X \to Y$ *between simplicial spaces X and Y is a monomorphism if and only if each function* $f([n]) : X([n]) \to Y([n])$ *is a monomorphism.*

Lemma F.4.8. *The epimorphisms in the category* **Top**$^{\mathcal{O}^{op}}$ *of simplicial spaces are the surjections, i.e. a map* $F : X \to Y$ *between simplicial spaces X and Y is an epimorphism if and only if each function* $f([n]) : X([n]) \to Y([n])$ *is an epimorphism.*

Proof. If a map $F : X \to Y$ between simplicial spaces X and Y is a surjection it has the required universal property, hence is an epimorphism. Conversely assume that $F : X \to Y$ is not surjective and consider the quotient map $q : Y \to Y/F(X)$ of simplicial spaces. (Here

$$q([n]) : Y([n]) \twoheadrightarrow (Y/F(X))([n]) = Y([n])/F(X)([n])$$

is the quotient obtained by collapsing the subspace $F(X)([n])$ of $Y([n])$ to a single point $* = F(X)([n]) \in Y([n])/F(X)$.) The constant map $* : Y \to F(X) \in Y/F(X)$ and the quotient map q satisfy the equation

$$q \circ F = * \circ F.$$

Because the map F was assumed not to be surjective, the maps q and $*$ are distinct and so F is not an epimorphism. □

Definition F.4.9. *A natural transformation* $\tau : F \to G$ *of functors* $F, G \in \mathbf{Top}^{\mathbf{D}}$ *from a category* $\mathbf{D}$ *into* $\mathbf{Top}$ *is a* proclusion, *if the maps* $\tau_d : F(d) \to F(d)$ *are proclusions for all objects* d *of* $\mathbf{D}$. *It is an* identification *if all maps* $\tau_d : F(d) \to F(d)$ *are identifications.*

F.4.1 Fibrant Spaces and Homotopy Extension Properties

In this section we generalise the notion of fibrant simplicial complexes to simplicial topological spaces.

Definition F.4.10. *A point* $\mathbf{x}$ *in* $X[n]^{n+1}$ *is called* k-compatible, *if it satisfies the equations* $\partial_i x_j = \partial_{j-1} x_i$ *for all* $i < j$ *and* $i \neq k \neq j$.

Example F.4.11. If X is the singleton simplicial space, then all powers of X are singleton. So all points of $X[n]$ are k-compatible for every $0 \leq k \leq n$. More generally, if X is a constant simplicial space then the face and degeneracy maps are the identity on X. Here the compatibility conditions $\partial_i x_j = \partial_{j-1} x_i$ imply $x_j = x_i$ for all $i < j$ and $i, j \neq k$. Thus the space of k-compatible points in $X[n]^{n+1}$ is the $n+1$-diagonal in X^{n+1}.

The subspace of $X[n]^{n+1}$ formed by the k-compatible points is the intersection of the equalisers $Eq(\partial_i \mathrm{pr}_j, \partial_{j-1} \mathrm{pr}_i)$ for all $i < j$ and $i \neq k \neq j$. We denote the subspace of k-compatible points in $X[n]$ by $X[n]_k$.

Example F.4.12. Consider the standard simplicial space $\mathrm{SS}(Y)$ of a topological space Y. The conditions $\partial_i \mathbf{x}_j = \partial_{j-1} \mathbf{x}_i$ translate into the equations

$$(x_{j,0}, \ldots, \hat{x}_{j,i}, \ldots, x_{j,n}) = (x_{i,0}, \ldots, \hat{x}_{i,j-1}, \ldots, x_{i,n})$$

for all $i < j$ and $i \neq k \neq j$. This especially implies $x_{i,l} = x_{j,l}$ for all $0 \leq l < i$ and $x_{i,l} = x_{j,l}$ for all $j < l \leq n+1$. These properties will be used in the proof of the Proposition below.

Definition F.4.13. *A simplicial space* X *is called* fibrant *if there exist continuous functions* $\varphi_{n,k} : X[n]_k \to X[n+1]$ *for all* $n \in \mathbb{N}$ *and all* $0 \leq k \leq n$ *such that the equations*

$$(\partial_1 \varphi_k, \ldots, \widehat{\partial_k \varphi_k}, \ldots, \partial_{n+1} \varphi_k) = \mathrm{id}_{X[n]_k} : X[n]_k \to X[n]_k$$

are satisfied for all $n \in \mathbb{N}$ *and all* $0 \leq k \leq n$.

Proposition F.4.14. *The standard simplicial space* $\mathrm{SS}(Y)$ *of any topological space* Y *is fibrant.*

Proof. Let Y be a topological space and $n \in \mathbb{N}$ be given. We define the needed continuous functions $\varphi_{n,k} : \mathrm{SS}(Y)^{n+1} \to \mathrm{SS}(Y)[n+1]$ for all $0 \leq k \leq n$ via

$$\varphi_{n,k}(\mathbf{x}_0, \ldots, \hat{\mathbf{x}}_k, \ldots, \mathbf{x}_{n+1}) = \begin{cases} (x_{n+1,0}, \ldots, x_{n+1,n}, x_{1,n}) & \text{if } k = 0 \\ (x_{n+1,0}, \ldots, x_{n+1,n}, x_{0,n}) & \text{if } 0 < k < n+1 \\ (x_{n,0}, \ldots, x_{n,n-1}, x_{0,n-1}, x_{0,n}) & \text{if } k = n+1 \end{cases}.$$

For $\mathrm{SS}(Y)$ to be fibrant the equalities $\partial_i \varphi_{n,k}(\mathbf{x}_0, \ldots, \hat{\mathbf{x}}_k, \ldots, \mathbf{x}_{n+1}) = \mathbf{x}_i$ have to be verified for all k-compatible points $(\mathbf{x}_0, \ldots, \hat{\mathbf{x}}_k, \ldots, \mathbf{x}_{n+1})$ in $\mathrm{SS}(X)^{n+1}$ and all $i \neq k$. Let $p = (\mathbf{x}_0, \ldots, \hat{\mathbf{x}}_k, \ldots, \mathbf{x}_{n+1})$ be such a k-compatible point and assume $k = 0$. The i-th boundary of the image $\varphi_{n,k}(p)$ is given by

$$\partial_i \varphi_{n,k}(\mathbf{x}_0, \ldots, \hat{\mathbf{x}}_k, \ldots, \mathbf{x}_{n+1}) = \partial_i(x_{n+1,0}, \ldots, x_{n+1,n}, x_{1,n})$$

In the case $i = n+1$ the boundary operator omits the last entry and the boundary is the point $(x_{n+1,0}, \ldots, x_{n+1,n}) = \mathbf{x}_{n+1}$. For $0 < i < n+1$ and one can apply the equality $\partial_i \mathbf{x}_{n+1} = \partial_n \mathbf{x}_i$ for $0 \neq i$ to obtain

$$\partial_i \varphi_{n,k}(\mathbf{x}_0, \ldots, \hat{\mathbf{x}}_k, \ldots, \mathbf{x}_{n+1}) = (\partial_i \mathbf{x}_{n+1}, x_{1,n}) = (\partial_n \mathbf{x}_i, x_{1,n})$$
$$= (x_{i,0}, \ldots, x_{i,n-1}, x_{1,n})$$

The compatibility condition $\partial_0 \mathbf{x}_i = \partial_{i-1} \mathbf{x}_0$ implies that the points $x_{1,n}$ and $x_{i,n}$ are equal (cf. Example F.4.12). So all in all we obtain the equations

$$\partial_i \varphi_{n,k}(\mathbf{x}_0, \ldots, \hat{\mathbf{x}}_k, \ldots, \mathbf{x}_{n+1}) = (x_{i,0}, \ldots, x_{i,n-1}, x_{1,n}) = (x_{i,0}, \ldots, x_{i,n-1}, x_{i,n}) = \mathbf{x}_i,$$

for all $0 < i < n+1$. Altogether we have proved $\partial_i \varphi_{n,k}(p) = \mathbf{x}_i$ for all $i \neq k$. This is the part of the defining equations for $SS(Y)$ being fibrant in the case $k = 0$. Now assume $0 < k < n+1$. Again, the i-th boundary of the image $\varphi_{n,k}(p)$ is given by

$$\partial_i \varphi_{n,k}(\mathbf{x}_0, \ldots, \hat{\mathbf{x}}_k, \ldots, \mathbf{x}_{n+1}) = \partial_i(x_{n+1,0}, \ldots, x_{n+1,n}, x_{0,n})$$

In the case $i = n+1$ the boundary operator omits the last entry and the boundary is the point $(x_{n+1,0}, \ldots, x_{n+1,n}) = \mathbf{x}_{n+1}$. For $i < n+1$ and $i \neq k$ one can apply the equality $\partial_i \mathbf{x}_{n+1} = \partial_n \mathbf{x}_i$ for $i \neq k$ to obtain

$$\partial_i \varphi_{n,k}(\mathbf{x}_0, \ldots, \hat{\mathbf{x}}_k, \ldots, \mathbf{x}_{n+1}) = (\partial_i \mathbf{x}_{n+1}, x_{0,n}) = (\partial_n \mathbf{x}_i, x_{0,n})$$
$$= (x_{i,0}, \ldots, x_{i,n-1}, x_{0,n}).$$

The compatibility condition $\partial_0 \mathbf{x}_i = \partial_{i-1} \mathbf{x}_0$ implies that the points $x_{0,n}$ and $x_{i,n}$ are equal (cf. Example F.4.12). So all in all we obtain the equations

$$\partial_i \varphi_{n,k}(\mathbf{x}_0, \ldots, \hat{\mathbf{x}}_k, \ldots, \mathbf{x}_{n+1}) = (x_{i,0}, \ldots, x_{i,n-1}, x_{0,n}) = (x_{i,0}, \ldots, x_{i,n-1}, x_{i,n}) = \mathbf{x}_i,$$

for all $i < n+1$, $i \neq k$. Altogether we have proved $\partial_i \varphi_{n,k}(p) = \mathbf{x}_i$ for all $i \neq k$. This is the part of the defining equations for $SS(Y)$ being vibrant in the case $0 < k < n+1$. In the case $k = n+1$ the i-th boundary of $\varphi_{n,k}(p)$ is given by

$$\partial_i \varphi_{n,k}(\mathbf{x}_0, \ldots, \mathbf{x}_n) = \partial_i(x_{n,0}, \ldots, x_{n,n-1}, x_{0,n-1}, x_{0,n})$$

For $i = n$ the boundary operator omits the last but one entry and the boundary is the point $(x_{n,0}, \ldots, x_{n,n-1}, x_{0,n}) = (x_{n,0}, \ldots, x_{n,n-1}, x_{n,n}) = \mathbf{x}_n$. In the case $i < n$ one can apply the equations $\partial_i \mathbf{x}_n = \partial_{n-1} \mathbf{x}_i$ for $i \neq k$ to obtain

$$\partial_i \varphi_{n,k}(\mathbf{x}_0, \ldots, \hat{\mathbf{x}}_k, \ldots, \mathbf{x}_{n+1}) = (\partial_i \mathbf{x}_n, x_{0,n-1}, x_{0,n})$$
$$= (\partial_{n-1} \mathbf{x}_i, x_{0,n})$$
$$= (x_{i,0}, \ldots, \hat{x}_{i,n-1}, x_{0,n-1}, x_{0,n})$$
$$= (x_{i,0}, \ldots, x_{i,n-1}, x_{i,n}) = \mathbf{x}_i,$$

where we have used the equality $x_{0,n} = x_{i,n}$ (cf. Example F.4.12). In any case the equation $\partial_i \varphi_{n,k}(p) = \mathbf{x}_i$ is satisfied for all $i \neq k$. The above results imply that the functions $\varphi_{n,k}$ satisfy the equalities F.4.13 defining vibrant simplicial spaces. Thus the standard simplicial space $SS(Y)$ of any topological space Y is vibrant.   □

Recall that the union of all but one faces of the standard simplex Δ_{n+1} is a retract of Δ_{n+1}. That is to say that for all $0 \leq k \leq n+1$ there exists a continuous function

$$r_{n,k} : \Delta_{n+1} \to \bigcup_{i \neq k} \Delta(\epsilon_i)(\Delta_n),$$

which restricts to the identity on $\cup_{i \neq k} \Delta(\epsilon_i)(\Delta_n)$.

Lemma F.4.15. *If Y is a topological space and $(\tau_0, \ldots, \hat{\tau}_k, \ldots, \tau_{n+1})$ a k-compatible point of $C(\Delta_n, Y)^{n+1}$, then the functions $\tau_i \circ \Delta(\epsilon_i)^{-1}$ coincide on the intersections of their domains.*

Proof. Let Y be a topological space and $(\tau_0, \ldots, \hat{\tau}_k, \ldots, \tau_{n+1})$ be a k-compatible point of $C(\Delta_n, Y)^{n+1}$. The domain of $\tau_i \circ \Delta(\epsilon_i)^{-1}$ is the image $\Delta(\epsilon_i)(\Delta_n)$ of the affine n-simplex $\langle \mathbf{e}_0, \ldots, \hat{\mathbf{e}}_i, \ldots, \mathbf{e}_{n+1} \rangle$ in Δ_{n+1}. The domain of another function $\tau_j \circ \Delta(\epsilon_j)^{-1}$ is the image $\Delta(\epsilon_j)(\Delta_n)$ of the affine n-simplex $\langle \mathbf{e}_0, \ldots, \hat{\mathbf{e}}_j, \ldots, \mathbf{e}_{n+1} \rangle$ in Δ_{n+1}. We can w.l.o.g. assume $i < j$. The intersection of both domains is the image of the affine $n-1$-simplex $\langle \mathbf{e}_0, \ldots, \hat{\mathbf{e}}_i, \ldots, \hat{\mathbf{e}}_j, \ldots, \mathbf{e}_{n+1} \rangle$ in Δ_{n+1}. Using the simplicial identity $\epsilon_j \epsilon_i = \epsilon_i \epsilon_{j-1}$ for $i < j$ this affine $n-1$-simplex can be either written as $\Delta(\epsilon_j)\Delta(\epsilon_i)$ or as $\Delta(\epsilon_i)\Delta(\epsilon_{j-1})$. So each point in the intersection of the domains of $\tau_i \Delta(\epsilon_i)^{-1}$ and $\tau_j \Delta(\epsilon_j)^{-1}$ is the image $\Delta(\epsilon_j)\Delta(\epsilon_i)(t)$ of a point $t \in \Delta_{n-1}$. The evaluation of $\tau_i \Delta(\epsilon_i)^{-1}$ at such a point a point is given by

$$\tau_j \Delta(\epsilon_j)^{-1} \Delta(\epsilon_j)\Delta(\epsilon_i)(t) = \tau_j \Delta(\epsilon_i)(t)$$
$$= [\partial_i \tau_j](t)$$
$$= [\partial_{j-1} \tau_i](t)$$
$$= \tau_i \Delta(\epsilon_{j-1})(t)$$
$$= \tau_i \Delta(\epsilon_i)^{-1} \Delta(\epsilon_i)\Delta(\epsilon_{j-1})(t)$$

Thus the functions $\tau_i \circ \Delta(\epsilon_i)^{-1}$ coincide on the intersections of their domains. □

Lemma F.4.15 implies that for every k-compatible point $(\tau_0, \ldots, \hat{\tau}_k, \ldots, \tau_{n+1})$ of $C(\Delta_n, Y)^{n+1}$ the functions $\tau_i \circ \Delta(\epsilon_i)^{-1}$ can be glued together to obtain a unique continuous function

$$\cup_{i \neq k} \tau_i \circ \Delta(\epsilon_i)^{-1} : \bigcup_{i \neq k} \Delta(\epsilon_i)(\Delta_n) \to Y,$$

which restricts to the function $\tau_i \Delta(\epsilon_i)^{-1}$ on the i-th face of Δ_n. Using these continuous functions and the retractions $r_{n,k}$ we are now able to prove:

Lemma F.4.16. *The singular simplicial space $C(\Delta, Y)$ of any topological space Y is fibrant.*

Proof. Let Y be a topological space and $n \in \mathbb{N}$ be given. For each $0 \leq k \leq n+1$ let $r_{n,k} : \Delta_{n+1} \to \cup_{i \neq k} \Delta(\epsilon_i)(\Delta_n)$ be a retraction of Δ_{n+1} onto the union of all but the k-th faces. We define the needed continuous functions $\varphi_{n,k}$ from $C(\Delta_n, Y)_k$ to $C(\Delta_{n+1}, Y)$ via

$$\varphi_{n,k}(\tau_0, \ldots, \hat{\tau}_k, \ldots, \tau_{n+1}) = \left(\cup_{i \neq k} \tau_i \Delta(\epsilon_i)^{-1} \right) \circ r_{n+1,k}.$$

for all k-compatible points $(\tau_0, \ldots, \hat{\tau}_k, \ldots, \tau_{n+1})$ in $C(\Delta_n, Y)^{n+1}$. Since the function $r_{n,k}$ is a retraction onto the union of all but the k-th faces, the equality $r_{n,k}\Delta(\epsilon_i) = \Delta(\epsilon_i)$ holds for each face map $\Delta_i : \Delta_n \to \Delta_{n+1}$, $i \neq k$. Therefore the composition of $\varphi_{n,k}$ and ∂_i for $i \neq k$ is given by

$$\partial_i \varphi_{n,k}(\tau_0, \ldots, \hat{\tau}_k, \ldots, \tau_{n+1}) = \partial_i \left[(\cup_{i \neq k} \tau_i) \circ r_{n+1,k} \right]$$
$$= C(\Delta(\epsilon_i), Y) \left[\left(\cup_{i \neq k} \tau_i \Delta(\epsilon_i)^{-1} \right) \circ r_{n+1,k} \right]$$
$$= \left(\cup_{i \neq k} \tau_i \Delta(\epsilon_i)^{-1} \right) \circ r_{n+1,k} \circ \Delta(\epsilon_i)$$
$$= \left(\cup_{i \neq k} \tau_i \Delta(\epsilon_i)^{-1} \right) \Delta(\epsilon_i)$$
$$= \tau_i$$

This proves we obtain the equality $(\partial_1, \ldots, \hat{\partial}_k, \ldots, \partial_{n+1})\varphi_{n,k} = \mathrm{id}_{C(\Delta_n,Y)}$ for all k-compatible points $(\tau_0, \ldots, \hat{\tau}_k, \ldots, \tau_{n+1})$ in $C(\Delta_n, Y)^{n+1}$. So the singular simplicial space $C(\Delta, Y)$ of Y is fibrant. □

F.5 Simplicial Homotopy Groups

In this section we extend the notions of simplicial homology and homotopy to functors $\mathbf{Top}^{\mathcal{O}^{op}} \to \mathbf{TopAb}$ resp. $\mathbf{Top}^{\mathcal{O}^{op}} \to \mathbf{TopGrp}$.

Definition F.5.1. *Two points* $x, x' \in X([n])$ *of a simplicial space* X *are called* homotopic *if all their faces coincide and there exists a point* $y \in X([n+1])$ *such that*

$$
X(\epsilon_i)y = \begin{cases} X(\eta_{n-1})X(\epsilon_i)(x) = X(\eta_{n-1})X(\epsilon_i)(x') & \text{if } i < n \\ x & \text{if } i = n \\ x' & \text{if } i = n+1 \end{cases}
$$

The point $y \in X([n+1])$ *is called a homotopy from* x *to* x'.

We write $x \sim x'$ if two points x and x' are homotopic. This relation is in general not an equivalence relation, but for simplicial space whose underlying simplicial set is fibrant it is:

Proposition F.5.2. *If the underlying set of the simplicial space* X *is fibrant, then* $\sim$ *is an equivalence relation.*

Proof. The proof for simplicial sets in [May92, Proposition 3.2] carries over word by word. □

Definition F.5.3. *Two points* $x, x' \in X([n])$, $n > 0$ *of a simplicial space* X *are called* homotopic relative to a subspace Y *of* X *if all but the zeroth of their faces coincide, their zeroth faces* $X(\epsilon_0)(x)$ *and* $X(\epsilon_0)(x')$ *are contained in* $Y([n-1])$, *there exists a homotopy* y *from* $X(\epsilon_0)(x)$ *to* $X(\epsilon_0)(x')$ *in* Y *and a point* $z \in X([n+1])$ *such that*

$$
X(\epsilon_i)z = \begin{cases} y & \text{if } i = 0 \\ X(\eta_{n-1})X(\epsilon_i)(x) = X(\eta_{n-1})X(\epsilon_i)(x') & \text{if } 0 < i < n \\ x & \text{if } i = n \\ x' & \text{if } i = n+1 \end{cases}
$$

The point z *is called a* homotopy relative to Y *from* x *to* x'.

We write $x \sim x'$ rel Y if two points x and x' are homotopic relative to a simplicial subspace Y. Like before, this relation is in general not an equivalence relation, but for simplicial space whose underlying simplicial sets are fibrant it is:

Proposition F.5.4. *If the underlying simplicial sets of a pair* (X, Y) *of simplicial spaces are fibrant, then* $\sim$ rel Y *is an equivalence relation.*

Proof. The proof for simplicial sets in [May92, Proposition 3.4] carries over word by word. □

Definition F.5.5. *A simplicial topological space with base point is a pair* $(X; *)$ *where* X *is a simplicial topological space and* $*$ *is a point of* $X([0])$. *the point* $*$ *is called the* base point *of* X.

Let X be a simplicial topological space with base point $* \in X([0])$. We take the points $X(\eta_0)(*), X(\eta_0\eta_0)(*), \ldots$ etc. as base points of the spaces $X([1]), X([2]), \ldots$ respectively. These base points form a simplicial subspace of X that has only one point in each dimension. This singleton simplicial subspace $*$ of X is always fibrant. Therefore the pair $(X, *)$ of simplicial spaces is fibrant if X is fibrant. The subspaces

$Z_n(X, *) \leq X([n])$ are defined to consist of exactly those points, all of whose faces are the point $*$ in $X([n-1])$, i.e.

$$Z_n(X, *) := \bigcap_{0 \leq i \leq n} X(\epsilon_i)^{-1}(*).$$

Please observe that two points $x, x' \in Z_n(X, *)$ are homotopic if and only if there exists a point $z \in X([n+1])$ such that

$$\partial_i z = \begin{cases} * & \text{if } i < n \\ x & \text{if } i = n \\ x' & \text{if } i = n+1 \end{cases}$$

Definition F.5.6. *If* $(X$ *is a simplicial space whose underlying simplicial set is fibrant and* $* \in X([0])$ *a base point, then the quotient spaces* $Z_n(X, *)/\sim$ *are denoted by* $\pi_n(X, *)$. *The space* $\pi_0(X, *)$ *is called the* space of path components *of* X.

Definition F.5.7. *A simplicial space* X *with fibrant underlying simplicial set and base point* $*$ *is called* connected, *if* $\pi_0(X, *)$ *is a singleton space. It is called* n-connected, *if* $\pi_i(X, *)$ *is a singleton space for all* $0 \leq i \leq n$.

Remark F.5.8. If X is a fibrant discrete simplicial space with base point $*$ then the discrete set $\pi_n(X, *)$ is in bijection to the underlying set of the homotopy group $\pi_n(|X|, |*|)$.

The relative version is defined analogously. Let (X, Y) be a pair of simplicial spaces and $* \in Y([0])$ be a base point of Y. We take the points $Y(\eta_0)(*)$, $Y(\eta_0 \eta_0)(*)$ etc. as base points of the spaces $Y([1]), Y([2]), \dots$ respectively. These base points form a simplicial subspace of Y that has only one point in each dimension. The singleton simplicial subspace $*$ of Y is always fibrant, so the triple $(X, Y, *)$ of simplicial spaces is fibrant if the pair (X, Y) is fibrant. We define the subspaces $Z_n(X, Y, *) \leq X([n])$ to consist of exactly those points, all but the zeroth of whose faces are the point $*$ in $Y([n-1])$, and the zeroth face is contained in $Y([n-1])$, i.e.

$$Z_n(X, Y, *) := \bigcap_{1 \leq i \leq n} X(\epsilon_i)^{-1}(*) \cap X(\epsilon_0)^{-1}(Y([n-1])).$$

One observes observes that two points $x, x' \in Z_n(X, Y, *)$ are homotopic rel Y if their zeroth faces $X(\epsilon_0)(x)$ and $X(\epsilon_0)(x')$ are contained in $Y([n-1])$, there exists a homotopy y from $X(\epsilon_0)(x)$ to $X(\epsilon_0)(x')$ in Y and a point $z \in X([n+1])$ such that

$$\partial_i z = \begin{cases} y & \text{if } i = 0 \\ * & \text{if } 0 < i < n \\ x & \text{if } i = n \\ x' & \text{if } i = n+1 \end{cases}$$

Definition F.5.9. *If* (X, Y) *is a pair of simplicial spaces with underlying fibrant simplicial sets and base point* $*$, *then the quotient spaces* $Z_n(X, Y, *)/\sim$ *rel* Y *are denoted by* $\pi_n(X, Y, *)$.

Note that the equalities $Z_n(X, *, *) = Z_n(X, *)$ imply $\pi_n(X, *, *) = \pi_n(X, *)$ for all natural numbers $n \in \mathbb{N}$.

Definition F.5.10. *A pair* (X, Y) *of simplicial spaces with underlying fibrant simplicial sets and base point* $*$ *is called* connected, *if* $\pi_0(X, Y, *)$ *is a singleton space. It is called* n-connected, *if* $\pi_i(X, , Y*)$ *is a singleton space for all* $0 \leq i \leq n$.

Remark F.5.11. If (X,Y) is a pair of fibrant discrete simplicial spaces with base point $*$ then the discrete set $\pi_n(X,Y,*)$ is in bijection to the underlying set of the homotopy group $\pi_n(|X|,|Y|,|*|)$, where $|Y|$ is a closed subspace of $|X|$.

If (X,Y) is a pair of simplicial spaces with base point $*$, then the inclusions $i : (Y,*) \hookrightarrow (X,*)$ and $j : (X,*,*) \hookrightarrow (X,Y,*)$ induces inclusions of the associated topological spaces Z_n so one obtains a chain

$$Z_n(Y,*) \hookrightarrow Z_n(X,*) \hookrightarrow Z_n(X,Y,*)$$

of embeddings of topological spaces for all $n \in \mathbb{N}$. If the underlying simplicial sets of X and Y are fibrant, these embeddings descend to continuous functions on the quotient spaces $\pi_n(Y,*)$ and $\pi_n(X,*)$ respectively. In addition the zeroth face operator $X(\epsilon_0)$ on $Z_n(X,Y,*)$ has image in the space $Z_{n-1}(Y,*)$ and preserves the equivalence relation (cf. [May92, §3]). These therefore descend to continuous functions $\pi_n(X,Y,*) \to \pi_n(Y,*)$, if the underlying sets of X and Y are fibrant. All in all one then gets an infinite chain

$$\cdots \xrightarrow{j} \pi_{n+1}(X,Y,*) \xrightarrow{\partial} \pi_n(Y,*) \xrightarrow{i} \pi_n(X,*) \xrightarrow{j} \pi_n(X,Y,*) \xrightarrow{\partial} \cdots$$

of continuous functions.

For simplicial objects in abelian categories the spaces π_n can be defined in a different way.

Definition F.5.12. *The* normalised *or* Moore chain complex NA_* *of a simplicial object A in an abelian category is the chain complex with*

$$NA_n :- \bigcap_{i=0}^{n-1} \ker[A(\epsilon_i) : A([n]) \to A([n-1])]$$

and differential $d_n = (-1)^n A(\epsilon_n)$.

The normalised chain complex NA_* is a sub-complex of the chain complex A_* associated to A. Furthermore every morphism $f : A \to B$ of simplicial objects induces a morphism $f_* : A_* \to B_*$ which restricts to a morphism $Nf_* : NA_* \to NB_*$ of normalised chain complexes. So the assignment $A \mapsto NA_*$, $f \mapsto Nf_*$ is a functor $(N-)_* : \mathbf{A}^{\mathcal{O}^{op}} \to \mathbf{Ch(A)}$:

Definition F.5.13. *The functor $(N-)_* : \mathbf{A}^{\mathcal{O}^{op}} \to \mathbf{Ch(A)}$ is called the* normalised *or* Moore chain complex *functor.*

Definition F.5.14. *The* degenerate chain sub-complex DA_* *of A_* of a simplicial object A in an abelian category $\mathbf{A}$ is the sub-complex generated by the images of the degeneracy morphisms $A(\eta_i)$:*

$$DA_n := \sum_{i=0}^{n} A(\eta_i)(A([n-1]))$$

Lemma F.5.15. *If the category $\mathbf{A}$ is abelian, then the associated chain complex functor $(-)_* : \mathbf{A}^{\mathcal{O}^{op}} \to \mathbf{Ch(A)}$ is naturally isomorphic to the direct sum of $(N-)_*$ and $(D-)_*$, i.e. the sequence of functors*

$$0 \longrightarrow (N-)_* \longrightarrow (-)_* \cong (N-)_* \oplus (D-)_* \longrightarrow (D-)_* \longrightarrow 0$$

is split exact and $(N-)_* \cong (-)_*/(D-)_*$.

Proof. See [Wei94, Lemma 8.3.7] for a proof. □

The split exactness of the above sequence implies that the induced long exact sequence

$$\cdots \longrightarrow H^{n+1} \circ D \xrightarrow{\delta} H^n \circ N \longrightarrow H^n \circ C \longrightarrow H^n \circ D \xrightarrow{\delta} H^{n-1} \circ N \longrightarrow \cdots$$

splits into short exact sequences

$$0 \longrightarrow H^n \circ N \longrightarrow H^n \circ C \longrightarrow H^n \circ D \longrightarrow 0.$$

Proposition F.5.16. *The degenerate chain complex DA_* of a simplicial object A in an abelian category $\mathbf{A}$ is acyclic.*

Proof. See [Wei94, Theorem 8.3.8] for a proof. $\qquad\square$

Corollary F.5.17. *The inclusion $(N-)_* \hookrightarrow (-)_*$ of $(N-)_*$ as a subfunctor of $(-)_*$ induces a natural isomorphism $H \circ (N-)_* \cong H \circ (-)_*$.*

F.6 Simplicial Homotopies

Definition F.6.1. *Two morphisms $f, g : A \to B$ of simplicial objects in a category $\mathbf{A}$ are called* homotopic, *if for each $n \in \mathbb{N}$ there exist morphisms $h_{n,0}, \ldots, h_{n,n}$ from $A([n])$ to $B([n+1])$ in $\mathbf{A}$ such that $B(\epsilon_0)h_{n,0} = f([n])$, $B(\epsilon_{n+1})h_{n,n} = g([n])$ and*

$$B(\epsilon_i)h_{n,j} = \begin{cases} h_{n-1,j-1}A(\epsilon_i) & \text{if } i < j \\ B(\epsilon_i)h_{n,i-1} & \text{if } i = j \neq 0 \\ h_{n-1,j}A(\epsilon_{i-1}) & \text{if } i > j+1 \end{cases}$$

$$B(\eta_i)h_{n,j} = \begin{cases} h_{n+1,j+1}A(\eta_i) & \text{if } i \leq j \\ h_{n+1,j}A(\eta_{i-1}) & \text{if } i > j \end{cases}.$$

The collection of morphisms $\{h_{n,i}\}$ is called a simplicial Homotopy *from f to g.*

We write $f \simeq g$ if to morphisms f and g of simplicial objects are homotopic. The corresponding notion for morphisms of cosimplicial objects is obtained by dualising:

Definition F.6.2. *Two morphisms $f, g : A \to B$ of cosimplicial objects in a category $\mathbf{A}$ are called* homotopic, *if the corresponding morphisms f^{op} and g^{op} of simplicial objects in the opposite category $\mathbf{A}^{op}$ are homotopic, i.e. if for each $n \in \mathbb{N}$ there exist morphisms $h^{n,0}, \ldots, h^{n,n}$ from $A([n+1])$ to $B([n])$ in $\mathbf{A}$ such that $h^{n,0}A(\epsilon_0) = f([n])$, $h^{n,n}A(\epsilon_{n+1}) = g([n])$ and*

$$h^{n,j}A(\epsilon_i) = \begin{cases} B(\epsilon_i)h^{n-1,j-1} & \text{if } i < j \\ h^{n,i-1}B(\epsilon_i) & \text{if } i = j \neq 0 \\ B(\epsilon_{i-1})h^{n-1,j} & \text{if } i > j+1 \end{cases}$$

$$h^{n,j}A(\eta_i) = \begin{cases} B(\eta_i)h^{n+1,j+1} & \text{if } i \leq j \\ B(\eta_{i-1})h^{n+1,j} & \text{if } i > j \end{cases}.$$

The collection of morphisms $\{h^{n,i}\}$ is called a simplicial Homotopy *from f to g.*

An important example of a cosimplicial homotopy involves the right shift functor $T : \mathcal{O} \to \mathcal{O}$. Consider the unique morphisms $[n] \to [0]$ for each object $[n]$ of $\mathcal{O}$. These morphisms assemble to a natural transformation $p : T \to [0]$, where 0 denotes the constant cosimplicial object $[0]$ in $\mathcal{O}$. The constant cosimplicial object $[0]$ can be embedded in T via

$$i : [0] \to T, \quad i([n])(0) = 0 \in T([n]).$$

This embedding is a right inverse to the natural transformation $p : T \to [0]$. The composition $ip : T \to T$ is an endomorphism of the cosimplicial object T in $\mathcal{O}$. For each $n\mathbb{N}$ the endomorphism $ip([n]) : T([n]) \to T([n])$ can be written as

$$ip([n]) = \epsilon_{n+1} \cdots \epsilon_1 \eta_0 \cdots \eta_n : T([n]) = [n+1] \to [n+1].$$

We assert that this endomorphism is homotopic to the identity of T:

Theorem F.6.3. *The collection* $h^{n,j} = \epsilon_j \cdots \epsilon_1 \eta_0 \cdots \eta_j : T([n+1]) \to T([n])$ *of morphisms is a cosimplicial homotopy between the identity of the cosimplicial object* T *and the endomorphism* ip *of* T.

Proof. It is to verify that the equations defining cosimplicial homotopies are satisfied. At first we observe how the morphisms $h^{n,j}$ intertwine the face morphisms. Using the simplicial identities $\eta_j \epsilon_i = \epsilon_{i-1} \eta_j$ for $i > j+1$ and $\epsilon_j \epsilon_i = \epsilon_i \epsilon_{j-1}$ for $i < j$ the composition $h^{n,n} T(\epsilon_{n+1})$ computes to

$$\begin{aligned}
h^{n,n} T(\epsilon_{n+1}) &= \epsilon_n \cdots \epsilon_1 \eta_0 \cdots \eta_n T(\epsilon_{n+1}) = \epsilon_n \cdots \epsilon_1 \eta_0 \cdots \eta_n \epsilon_{n+2} \\
&= \epsilon_n \cdots \epsilon_1 \eta_0 \cdots \eta_{n-1} \epsilon_{n+1} \eta_n = \epsilon_n \cdots \epsilon_1 \epsilon_1 \eta_0 \cdots \eta_n \\
&= \epsilon_n \cdots \epsilon_2 (\epsilon_1 \epsilon_1) \eta_0 \cdots \eta_n = \epsilon_n \cdots \epsilon_2 (\epsilon_2 \epsilon_1) \eta_0 \cdots \eta_n \\
&= \epsilon_{n+1} \epsilon_n \cdots \epsilon_1 \eta_0 \cdots \eta_n = ip([n]).
\end{aligned}$$

The composition $h^{n,0} T(\epsilon_0)$ is the endomorphism $\eta_0 T(\epsilon_0) = \eta_0 \epsilon_1 = $ id, where we have used the simplicial identity $\eta_j \epsilon_i = $ id for $j = i, i+1$. A repeated application of the simplicial identities $\eta_j \epsilon_i = \epsilon_i \eta_{j-1}$ and $\epsilon_j \epsilon_i = \epsilon_i \epsilon_{j-1}$ for $i < j$ shows that $h^{n,j} T(\epsilon_i)$ is in this case equal to the morphism

$$\begin{aligned}
h^{n,j} T(\epsilon_i) &= \epsilon_j \cdots \epsilon_1 \eta_0 \cdots \eta_j \epsilon_{i+1} = \epsilon_j \cdots \epsilon_1 \eta_0 \cdots \eta_i \epsilon_i \eta_i \cdots \eta_{j-1} \\
&= \epsilon_j \cdots \epsilon_{i+2} \epsilon_{i+1} \cdots \epsilon_1 \eta_0 \cdots \eta_{j-1} = \epsilon_j \cdots \epsilon_{i+3} \epsilon_{i+1} \epsilon_{i+1} \cdots \epsilon_1 \eta_0 \cdots \eta_{j-1} \\
&= \epsilon_{i+1} (\epsilon_{j-1} \cdots \epsilon_1 \eta_0 \cdots \eta_{j-1}) = T(\epsilon_i) h^{n-1,j-1}
\end{aligned}$$

In the case $i = j \neq 0$ the morphism $h^{n,j} T(\epsilon_i)$ the simplicial identity $\eta_j \epsilon_i = $ id for $i = j, j+1$ shows

$$\begin{aligned}
h^{n,j} T(\epsilon_i) &= h^{n,i} T(\epsilon_i) = \epsilon_i \cdots \epsilon_1 \eta_0 \cdots \eta_i \epsilon_{i+1} = \epsilon_i \cdots \epsilon_1 \eta_0 \cdots \eta_{i-1} \text{id} \\
&= \epsilon_i \cdots \epsilon_1 \eta_0 \cdots \eta_{i-1} \eta_i \epsilon_i = h^{n,i-1} T(\epsilon_{i-1}).
\end{aligned}$$

In the last case, one can use the simplicial identity $\eta_j \epsilon_i = \epsilon_{i-1} \eta_j$ for $i > j+1$ and the simplicial identity $\epsilon_j \epsilon_i = \epsilon_i \epsilon_{j-1}$ for $i < j$ to obtain the equalities

$$\begin{aligned}
h^{n,j} T(\epsilon_i) &= (\epsilon_j \cdots \epsilon_1) \eta_0 \cdots \eta_j \epsilon_{i+1} = (\epsilon_j \cdots \epsilon_1) \eta_0 \cdots \eta_{j-1} \epsilon_i \eta_j \\
&= (\epsilon_j \cdots \epsilon_1) \epsilon_{i-j} \eta_0 \cdots \eta_j = \epsilon_j \cdots \epsilon_2 \epsilon_{i-j+1} \epsilon_1 \eta_0 \cdots \epsilon_j \\
&= \epsilon_i \epsilon_j \cdots \epsilon_1 \eta_0 \cdots \eta_j = T(\epsilon_{i-1}) h^{n-1,j}.
\end{aligned}$$

for $i > j+1$. Finally, it is to verify that the morphisms $h^{n,j}$ intertwine the coface morphisms in the right way. For $i \le j$ the a repeated application of the simplicial identity $\eta_j \eta_i = \eta_i \eta_{j+1}$ yields the equalities

$$\begin{aligned}
h^{n,j} T(\eta_i) &= \epsilon_j \cdots \epsilon_1 \eta_0 \cdots \eta_j \eta_{i+1} = \epsilon_j \cdots \epsilon_1 \eta_0 \cdots \eta_{j-1} \eta_{i+1} \eta_{j+1} \\
&= \epsilon_j \cdots \epsilon_1 \eta_0 \cdots \eta_i \eta_{i+1} \cdots \eta_{j+1} = \epsilon_j \cdots \epsilon_1 \eta_0 \cdots \eta_{j+1}
\end{aligned}$$

Observe that the simplicial identity $\epsilon_j \epsilon_i = \epsilon_i \epsilon_{j-1}$ for $i < j$ implies that the morphism $\eta_j \cdots \epsilon_1 : [n+1-j] \to [n+1]$ can also be written as ϵ_1^j or $\epsilon_i^{j-i} \epsilon_i \cdots \epsilon_1$ for each $i \le j$. So the above expression for $h^{n,j} T(\eta_i)$ can be rewritten as

$$h^{n,j}T(\eta_i) = \mathrm{id}\epsilon_j \cdots \epsilon_1 \eta_0 \cdots \eta_{j+1} = \eta_{i+1}\epsilon_{i+1}\epsilon_j \cdots \epsilon_1 \eta_0 \cdots \eta_{j+1}$$
$$= \eta_{i+1}\epsilon_{i+1}^{j+1-i}\epsilon_i \cdots \epsilon_1 \eta_0 \cdots \eta_{j+1} = \eta_{i+1}\epsilon_{j+1} \cdots \epsilon_1 \eta_0 \cdots \eta_{j+1}$$
$$= T(\eta_i)h^{n+1,j+1}.$$

Last, for $i > j$ the simplicial identities $\eta_j \eta_i = \eta_{i-1}\eta_j$ and $\epsilon_j \eta_i = \eta_{i+1}\epsilon_j$ imply the equalities

$$h^{n,j}T(\eta_i) = \epsilon_j \cdots \epsilon_1 \eta_0 \cdots \eta_j \eta_{i+1} = \epsilon_j \cdots \epsilon_1 \eta_0 \cdots \eta_{j-1}\eta_i \eta_j$$
$$= \epsilon_j \cdots \epsilon_1 \eta_{i-j}\eta_0 \cdots \eta_j = \epsilon_j \cdots \epsilon_2 \eta_{i-j+1}\epsilon_1 \eta_0 \cdots \eta_j$$
$$= \eta_i \epsilon_j \cdots \epsilon_1 \eta_0 \cdots \eta_j = T(\eta_{i-1})h^{n+1,j}.$$

Therefore the collection of morphisms $h^{n,j}$ forms a cosimplicial homotopy from ip to the identity of T. □

For each simplicial object C in a category $\mathbf{C}$, the composition of C with the constant functor $[0]$ in $\mathcal{O}$ is the constant simplicial object $C([0])$, which we also denote by C_0. The embedding $i : [0] \to T$ induces a morphism $C(i) : CT \to C_0$ from CT onto the constant simplicial object C_0 in $\mathbf{C}$. The morphism $p : T \to [0]$ induces an embedding $C(p) : C_0 \to CT$ of the constant simplicial object C_0 in CT, which is a right inverse to $C(i)$.

Theorem F.6.4. *For every simplicial object C in a category $\mathbf{C}$, the simplicial object CT is homotopic to the constant simplicial object C_0.*

Proof. The homotopy $ip \sim T$ in $\mathcal{O}$ provided in Theorem F.6.3 induces a simplicial homotopy from id_{CT} to the endomorphism $C(p)C(i)$ of CT and the composition $C(i)C(p) = C(pi) = C(\mathrm{id}_{[0]}) = \mathrm{id}_{C_0}$ is the identity of C_0. □

Dually, for each cosimplicial object C in a category $\mathbf{C}$, the composition of C with the constant functor $[0]$ in $\mathcal{O}$ is the constant cosimplicial object $C([0])$, which we also denote by C^0. The embedding $i : [0] \to T$ induces an embedding $C(i) : C^0 \to CT$ and the morphism $p : T \to [0]$ induces a morphism $C(p) : CT \to C^0$, which is a left inverse to the embedding $C(i) : C^0 \to CT$. So we observe:

Theorem F.6.5. *For every cosimplicial object C in a category $\mathbf{C}$, the cosimplicial object CT is homotopic to the constant cosimplicial object C^0.*

Proof. The homotopy $ip \sim T$ in $\mathcal{O}$ provided in Theorem F.6.3 induces a cosimplicial homotopy from the endomorphism $C(p)C(i)$ of CT to the identity id_{CT} and the composition $C(p)C(i) = C(ip) = C(\mathrm{id}_{[0]}) = \mathrm{id}_{C^0}$ is the identity of C^0. □

Lemma F.6.6. *Simplicially homotopic morphisms $f \simeq g : A \to B$ between simplicial objects A and B in an Ab-category induce homotopic morphisms of the associated chain complexes. Similarly homotopic morphisms $f \simeq g : A \to B$ between cosimplicial objects A and B in an Ab-category induce homotopic morphisms of the associated cochain complexes.*

Proof. The proof of [May92, Proposition 5.3] carries over to this more general case: Let f, g be morphisms of simplicial objects and $h_{n,i}$ be a homotopy from f to g and define morphisms $s_n : A([n]) \to B([n+1])$ via $s_n := \sum(-1)^i h_{n,i}$, then $sd + ds = f - g$, i.e. f_* is chain homotopic to g_*. The proof for cosimplicial morphisms is dual. □

There exist an alternate characterisation of simplicial homotopies in sufficiently nice categories. Suppose that $\mathbf{A}$ is a subcategory of **Top** or **Set** containing the simplicial discrete space (or set) $\hom(-, [1])$ and binary products with the components

$\hom([n], [1])$ thereof. The constant morphisms 1 in $\hom(-, [1])$ form a simplicial singleton subspace of $\hom(-, [1])$. We identify this simplicial subspace with the simplicial space $\hom(-, [0])$ via the embedding

$$\hom(-, \epsilon_0) : \hom(-, [0]) \to \hom(-, [1]).$$

For any simplicial object A in $\mathbf{A}$ we denote denote the slice $A \hookrightarrow A \times \hom(-, [1])$ through this singleton subspace by $j_{A,0}$. Analogously, the zero morphisms in $\hom(-, [1])$ form a singleton simplicial subspace of $\hom(-, [1])$. We identify this simplicial subspace with the simplicial space $\hom(-, [0])$ via the embedding $\hom(-, \epsilon_1)$ and denote the slice $A \hookrightarrow A \times \hom(-, [1])$ through this singleton subspace by $j_{A,1}$.

Theorem F.6.7. *If $\mathbf{A}$ is either a subcategory of the category $\mathbf{Top}$ of topological spaces containing the discrete simplicial space $\hom(-, [1])$ and binary products with this space or the category $\mathbf{Set}$ of sets, then there exists a one-to-one correspondence between simplicial homotopies $f \simeq g : A \to B$ in $\mathbf{A}$ and simplicial morphisms $h : A \times \hom(-, [1]) \to B$ making the following diagram commutative:*

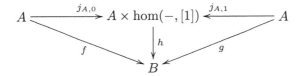

Proof. The case of simplicial sets is covered in [May92, Proposition 6.2]. All the maps used in the proof thereof are continuous if f and g are morphisms of simplicial topological spaces, so the proof is also valid for the category of simplicial topological spaces. $\square$

For Ab-categories $\mathbf{A}$ with finite direct sums there exists a similar characterisation of simplicial homotopies between morphisms $f, g : A \to B$. Here one uses the convention that the expression $A \times \hom(-, [1])$ stands for a certain simplicial object composed of direct sums. The n-th component of this simplicial object the direct sum

$$\bigoplus_{\alpha \in \hom([n], [1])} A([n])$$

For each $\alpha \in \hom([n], [1])$ let i_α denote the inclusion of the α-th summand and p_α denote the projection onto this summand. The morphism of $A \times \hom(-, [1])$ induced by a morphism $\beta : [m] \to [n]$ in $\mathcal{O}$ is given by

$$\bigoplus_{\alpha \in \hom([n], [1])} A([n]) \xrightarrow{\sum i_{\alpha\beta} A(\beta) p_\alpha} \bigoplus_{\alpha \in \hom([m], [1])} A([m]).$$

The expressions $A \times \hom(-, [k])$ for different $[k]$ are understood in an analogous way. The embedding $\hom(-, \epsilon_0) : \hom(-, [0]) \to \hom(-, [1])$ of the singleton simplicial subspace of $\hom(-, [1])$ consisting of the constant morphisms 1 induces an inclusion

$$j_{A,0} : A \cong A \times \hom(-, [0]) \to A \times \hom(-, [1])$$

of A into $A \times \hom(-, [1])$, which embeds the object $A([n])$ as the last summand of the direct sum $\oplus_{\alpha \in \hom([n], [1])} A([n])$. Analogously the embedding of the zero morphisms in $\hom(-, [1])$ induces an embedding

$$j_{A,1} : A \cong A \times \hom(-, [0]) \to A \times \hom(-, [1])$$

of A into $A \times \hom(-, [1])$, which embeds the object $A([n])$ as the first summand of the direct sum $\oplus_{\alpha \in \hom([n], [1])} A([n])$. Using this convention, one obtains a result similar to the result for the categories $\mathbf{Top}$ and $\mathbf{Set}$:

Theorem F.6.8. *If* **A** *is an additive category, then there exists a one-to-one correspondence between simplicial homotopies* $f \simeq g : A \to B$ *in* **A** *and simplicial morphisms* $h : A^{\hom(-,[1])} \to B$ *making the following diagram commutative:*

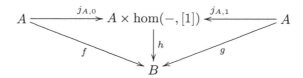

Proof. A proof for simplicial homotopies in abelian categories can be found in [Wei94, Theorem 8.312]. It works equally well for Ab-categories.   □

The analogous statement for cosimplicial homotopies are $f, g : A \to B$ between cosimplicial objects A and B are obtained similarly. Here one only demands the existence of finite direct products and uses the the simplicial object $B^{\hom(-,[1])}$. (Recall that the " exponential functor" $\mathbf{A} \times \mathbf{Set} \to \mathbf{A}$, which sends a pair (A, S) to the finite product A^S is contravariant in the second argument, so $A^{\hom(-,[1])}$ is a cosimplicial object.) The n-th component of this cosimplicial object the finite product

$$B([n])^{\hom([n],[1])}$$

by definition. The embedding $\hom(-, \epsilon_0) : \hom(-, [0]) \to \hom(-, [1])$ of the singleton simplicial subspace of $\hom(-, [1])$ consisting of the constant morphisms 1 induces a projection

$$p_{D,0} : A^{\hom(-,[1])} \to B^{\hom(-,[0])} \simeq B$$

of $B^{\hom(-,[1])}$ onto B, which projects the object $B^{\hom(-,[1])}$ onto the last factor of the product $B([n])^{\hom([n],[1])}$. Analogously the embedding of the zero morphisms in $\hom(-, [1])$ induces a projection

$$p_{B,1} : B^{\hom(-,[1])} \to B^{\hom(-,[0])} \cong B$$

of $B^{\hom(-,[1])}$ onto B, which projects the object $B^{\hom(-,[1])}$ onto the first factor of the product $B([n])^{\hom([n],[1])}$. Using this notation, one obtain a result similar to the result for the categories **Top** and **Set**:

Theorem F.6.9. *If* **A** *is a category with finite products, then there exists a one-to-one correspondence between simplicial homotopies* $f \simeq g : A \to B$ *in* **A** *and simplicial morphisms* $h : A \to B^{\hom(-,[1])}$ *making the following diagram commutative:*

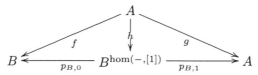

Proof. A proof can be found in [Mey90, Proposition 2.3].   □

In case the category **A** considered is not an Ab-category, we replace the direct sums in $A \times (-, [1])$ by finite products.

Definition F.6.10. *Two morphisms* $f, g : A \to B$ *of simplicial topological algebras in a category* **A** *of topological algebras are called* loop homotopic, *if there exists a morphism* $h : A \times \hom(-, [1]) \to B$ *of simplicial topological spaces satisfying the following conditions:*

1. The morphism h makes the following diagram commutative:

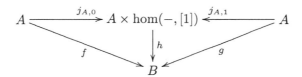

2. For every morphism $\alpha \in \hom([n],[1])$ the restriction of h to $A([n]) \times \{\alpha\}$ is a morphism in $\mathbf{A}$.

Remark F.6.11. Condition 1 in the definition of loop homotopy implies that loop homotopic morphisms are homotopic.

F.7 The Dold-Kan Correspondence

Let $\mathbf{A}$ be an abelian category. There exists an equivalence between the category $\mathbf{A}^{\mathcal{O}^{op}}$ of simplicial objects in $\mathbf{A}$ and the category $\mathbf{Ch}_{\geq 0}$ of positive chain complexes in $\mathbf{A}$. One direction is given by the normalised (Moore) chain complex functor $(N-)_* : \mathbf{A}^{\mathcal{O}^{op}} \to \mathbf{Ch}_{geq0}(\mathbf{A})$ which assigns to each simplicial object A in $\mathbf{A}$ its normalised chain complex NA_*. This functor $(N-)_*$ has a left adjoint $K : \mathbf{Ch}_{geq}(\mathbf{A}) \to \mathbf{A}^{\mathcal{O}^{op}}$. This left adjoint can be constructed as follows: Given a chain complex $C \in \mathbf{Ch}(\mathbf{A})$ we define $K(C)([n])$ to be the finite direct sum

$$K(C)([n]) = \bigoplus_{p \leq n} \bigoplus_{\eta : [n] \twoheadrightarrow [p]} C_\eta,$$

where η ranges over all surjections from $[n]$ onto $[p]$ and $C_{p,\eta}$ denotes a copy of C_p. If $\alpha : [m] \to [n]$ is a morphism in $\mathcal{O}$ then $K(\alpha) : KC([n]) \to KC([m])$ is defined by giving its restrictions to the summands C_η. For each surjection $\eta : [n] \twoheadrightarrow [p]$ the morphism $\eta\alpha$ has a unique epi-monic factorisation:

$$
\begin{CD}
[m] @>\alpha>> [n] \\
@V\eta'VV @VV\eta V \\
[q] @>\epsilon>> [p]
\end{CD}
$$

In the case $p = q$ (and thus $\eta\alpha = \eta'$) we define $K(\alpha, \eta)$ to be the natural identification of C_η with $C_{\eta'}$. For $p = q + 1$ and $\epsilon = \epsilon_p$ we define $K(\alpha, \eta)$ to be the composition

$$C_\eta \cong C_p \xrightarrow{d} C_{p-1} \cong C_{\eta'} \leq KC([m]).$$

In all other cases we define $K(\alpha, \eta)$ to be zero. The morphism $K(\alpha)$ is now defined to be the unique morphism whose restriction to a summand C_η is the morphism $K(\alpha, \eta)$.

Lemma F.7.1. *The normalised chain complex functor $(N-)_*$ reflects isomorphisms, i.e. a morphism $f : A \to B$ is an isomorphism if its image $Nf_* : NA_* \to NB_*$ is an isomorphism.*

Proof. See [Wei94, Lemma 8.4.5] for a proof.   □

Theorem F.7.2. *The functor $K : \mathbf{Ch}_{\geq 0}(\mathbf{A}) \to \mathbf{A}^{\mathcal{O}^{op}}$ is left adjoint to the normalisation functor $(N-)_*$.*

524   F Simplicial Objects

Proof. A proof can be found in [Wei94, Chapter 8.4]. □

Theorem F.7.3. *The functors* $(N-)_*$ *and* K *establish an equivalence between the categories* $\mathbf{A}^{\mathcal{O}^{op}}$ *and* $\mathbf{Ch}_{\geq 0}(\mathbf{A})$. *Under this correspondence, simplicial homotopy corresponds to homology (i.e.* $\pi_*(A) \cong H_*(NA)$*).*

Proof. A proof can be found in [Wei94, Chapter 8.4]. □

Theorem F.7.4. *The normalised chain complex functor* $(N-)_*$ *preserves and reflects homotopy.*

Proof. See [DP61, Satz 3.31] for a proof. □

F.8 Bisimplicial Objects

Definition F.8.1. *A bisimplicial object* A *in a category* $\mathbf{A}$ *is a contravariant functor from* $\mathcal{O} \times \mathcal{O}$ *to* $\mathbf{C}$.

If $\mathbf{A}$ is an additive category then one can associate a first quadrant double complex $A_{*,*}$ to each bisimplicial object $A \in \mathbf{A}$. The object parts of this double complex are given by $A_{p,q} = A([p],[q])$ and the horizontal and vertical differentials are given by

$$d_{p,q}^h : A_{p,q} \to A_{p-1,q}, \qquad d_{p,q}^h = \sum_{i=0}^{p-1}(-1)^i A(\eta_i, \mathrm{id}_{[q]})$$

and

$$d_{p,q}^v : A_{p,q} \to A_{p,q-1}, \qquad d_{p,q}^v = (-1)^p \sum_{i=0}^{q-1}(-1)^i A(\mathrm{id}_{[p]}, \eta_i).$$

Definition F.8.2. *The double complex* $A_{*,*}$ *constructed above is called the* (first quadrant) *double complex associated to* A.

Any morphism $f : A \to B$ of bisimplicial objects in an additive category $\mathbf{A}$ induces a morphism $f_{*,*} : A_{*,*} \to B_{*,*}$ between the associated double complexes. Thus the assignment $A \mapsto A_{*,*}$ (for bisimplicial objects in an additive category $\mathbf{A}$) is the object part of a functor $(-)_{*,*} : \mathbf{A}^{\mathcal{O}^{op} \times \mathcal{O}^{op}} \to \mathbf{Dc}(\mathbf{A})$ from the category $\mathbf{A}^{\mathcal{O}^{op} \times \mathcal{O}^{op}}$ of bisimplicial objects in $\mathbf{A}$ to the category $\mathbf{Dc}(\mathbf{A})$ of double complexes in $\mathbf{A}$.

Definition F.8.3. *The diagonal* $\mathrm{diag}A$ *of a bisimplicial object in a category* $\mathbf{A}$ *is the composition of* A *with the diagonal functor* $\mathcal{O} \to \mathcal{O} \times \mathcal{O}$.

Remark F.8.4. If A is a bisimplicial object in $\mathbf{A}$ then its diagonal $\mathrm{diag}A$ is a simplicial object in A.

In the following we work exclusively in an additive category $\mathbf{A}$. If A is a bisimplicial object in $\mathbf{A}$ then the total complex $\mathrm{Tot}A_{*,*}$ of its associated double complex is a chain complex in A. We are going to define a morphism from the chain complex $(\mathrm{diag}A)_*$ associated to the diagonal $\mathrm{diag}A$ to the total complex $\mathrm{Tot}A_{*,*}$ of the double complex $A_{*,*}$ associated to the bisimplicial object A. We start by defining morphisms

$$AW_{p,q} : (\mathrm{diag}A)_{p+q} = A([p+q],[p+q]) \to A_{p,q}, \quad AW_{p,q} = d_{p+1}^h \cdots d_{p+q}^h (d_0^v)^p$$

$$AW_n : (\mathrm{diag}A)_n = A([n],[n]) \to (\mathrm{Tot}A_{*,*})_n, \quad AW_n := \bigoplus_{p+q=n} AW_{p,q}$$

Lemma F.8.5. *The morphisms AW_n assemble to a morphism $(\mathrm{diag}A)_* \to \mathrm{Tot}A_{*,*}$ of chain complexes.*

Proof. A proof for modules over a commutative ring can be found in [ML63, Theorem 8.5]. The proof carries over in verbatim. Alternatively one may use the Freyd-Mitchell Embedding Theorem.   □

Definition F.8.6. *The morphism $AW_* : (\mathrm{diag}A)_* \to \mathrm{Tot}A_{*,*}$ of chain complexes is called the* Alexander-Whitney morphism *for A.*

Let A and B bisimplicial objects A and B in **A**. A morphism $f : A \to B$ of bisimplicial objects induces morphisms $(\mathrm{diag}f)_* : (\mathrm{diag}A)_* \to (\mathrm{diag}B)_*$ and $\mathrm{Tot}A_{*,*} \to \mathrm{Tot}B_{*,*}$ of chain complexes. As is clear from the definition of the Alexander-Whitney morphisms for A and B, they intertwine the induced morphisms of chain complexes. Thus the Alexander Whitney morphisms form a natural transformation

$$AW_* : (\mathrm{diag}-)_* \to \mathrm{Tot}(-)_{*,*}.$$

Definition F.8.7. *The natural transformation $AW_* : (\mathrm{diag}-)_* \to \mathrm{Tot}(-)_{*,*}$ is called the* Alexander-Whitney transformation.

As a counterpart to the Alexander Whitney transformation one can also define a natural transformation $EZ_* : \mathrm{Tot}(-)_{*,*} \to (\mathrm{diag}-)_*$ in the reverse direction. Here we use the notion of a shuffle

Definition F.8.8. *A (p,q)-shuffle $(p, q \geq 0)$ is a permutation μ of the set $\{1, \ldots, p+q\}$ of integers such that the restrictions $\mu_{|[1,p]}$ and $\mu_{|[p+1,p+q]}$ are order preserving.*

Using shuffles one can now define morphisms $EZ_{p,q} : A_{p,q} \to \mathrm{diag}A_{p+q}$ via

$$EZ_{p,q} = \sum_{\mu}(-1)^{\mathrm{sign}\mu} A\big(\epsilon_{\mu(n)} \cdots \epsilon_{\mu(p+1)}, \epsilon_{\mu(p)} \cdots \epsilon_{\mu(1)}\big)$$

where the sum is taken over all (p,q) shuffles.

Lemma F.8.9. *The morphisms $EZ_n := \bigoplus_{p+q=n} EZ_{p,q}$ assemble to a morphism $EZ_* : \mathrm{Tot}A_{*,*} \to (\mathrm{diag}A)_*$ of chain complexes.*

Proof. A proof for modules over a commutative ring can be found in [ML63, Theorem VIII.8.8]. The proof carries over in to our more general setting.   □

Definition F.8.10. *The morphism $EZ_* : \mathrm{Tot}A_{*,*} \to (\mathrm{diag}A)_*$ is called the* Eilenberg-Zilber morphism *for A.*

Let A and B bisimplicial objects A and B in **A**. As is clear from the definition of the Eilenberg-Zilber morphisms for A and B, they intertwine the induced morphisms of chain complexes. Thus the Eilenberg-Zilber morphisms form a natural transformation (cf. [ML63, Theorem 8.8])

$$EZ_* : \mathrm{Tot}(-)_{*,*} \to (\mathrm{diag}-)_*.$$

Definition F.8.11. *The natural transformation $EZ_* : (\mathrm{diag}-)_* \to \mathrm{Tot}(-)_{*,*}$ is called the* Eilenberg-Zilber transformation.

Theorem F.8.12 (Eilenberg-Zilber). *The compositions AW_*EZ_* and EZ_*AW_* are both chain homotopic to the identity.*

Proof. See [DP61, Satz 2.5] for a proof.   □

Corollary F.8.13. *The induced morphisms $H_*(AW_*)$ and $H_*(EZ_*)$ in homology are inverse to each other; so the Alexander-Whitney and Eilenberg-Zilber transformations induce isomorphisms in homology.*

Assume that the category $\mathbf{A}$ is monoidal with a product $\otimes : \mathbf{A} \times \mathbf{A} \to \mathbf{A}$. In the follwing we shall identify the products $(A \otimes B) \otimes C$ and $A \otimes (B \otimes C)$ for all objects A, B and C in $\mathbf{A}$.

Let A, B and C be simplicial objects in $\mathbf{A}$. The tensor product $A \otimes B$ of A and B is given pointwise. It is a simplicial object whose associated chain complex $(A \otimes B)_*$ is the pointwise tensor product $A_* \otimes B_*$ of the chain complexes A_* and B_* associated to the simplicial objects A resp. B in $\mathbf{A}$. The simplicial object $A \otimes B$ is the diagonal of the bisimplicial object

$$\otimes \circ (A \times B) : \mathcal{O} \times \mathcal{O} \to \mathbf{A}, \quad ([n],[m]) \mapsto A([n]) \otimes B([m]), \quad (\alpha,\beta) \mapsto A(\alpha) \otimes B(\beta)$$

in $\mathbf{A}$. Thus the Alexander-Whitney morphism for $\otimes(A \times B)$ induces a morphism of chain complexes $A_* \otimes B_* \to \mathrm{Tot}(\otimes(A \times B))_{*,*}$. If C is another simplicial object in $\mathbf{A}$ then analogous considerations can be made for the simplicial objects $\mathrm{Tot}(\otimes \circ (A \times B))$ and C. The Alexander-Whitney morphisms for the resulting chain complexes may be composed:

$$(A \otimes B)_* \otimes C_* \xrightarrow{AW_* \otimes \mathrm{id}_{C_*}} (\mathrm{Tot}(\otimes(A \times B))_{*,*}) \otimes C_* \xrightarrow{AW_*} \mathrm{Tot}\left(\mathrm{Tot}(\otimes(A \times B))_{*,*} \otimes C_*\right)$$

Here the chain complex on the right is the total complex of the trisimplicial object $\otimes \circ (\otimes \times \mathrm{id}_{\mathcal{O}})(A \times B \times C)$ in $\mathbf{A}$. Its parts in dimension n are the objects

$$(\otimes \circ (\otimes \times \mathrm{id}_{\mathcal{O}})(A \times B \times C))_n = \sum_{k+l+m=n} A([k]) \otimes B([l]) \otimes C([m]).$$

The composition of the Alexander-Whitney maps does not depend on the order of the components on which they are applied:

Lemma F.8.14. *The Alexander-Whitney morphism is associative.*

Proof. A proof can be found in [ML63, Proposition VIII.8.7]. □

This statemant may also be expressed by passing to the differential graded objects $\bigoplus_n A([n])$, $\bigoplus_n B([n])$, and $\bigoplus_n C([n])$ associated to the simplicial objects A, B, and C in the additive category $\mathbf{A}$. The differential graded object associated to the diagonal of the bisimplicial object $\otimes(A \times B)$ in $\mathbf{A}$ is given by

$$\bigoplus_{n \in \mathbb{N}} A([n]) \otimes B([n]), \quad d = d_A \otimes d_B.$$

The differential graded object associated to the total complex $\mathrm{Tot} \otimes (A \times B)$ of the bisimplicial object $\otimes(A \times B)$ in $\mathbf{A}$ is the tensor product of the differential graded objects $\bigoplus_n A([n])$ and $\bigoplus_n B([n])$, which is by definition given by

$$\left(\bigoplus_p A([p])\right) \otimes \left(\bigoplus_q B([q])\right) = \bigoplus_{n \in \mathbb{N}} \left(\bigoplus_{p+q=n} A([p]) \otimes B([q])\right),$$

$$d_{|A([p]) \otimes B([q])} = d_A \otimes \mathrm{id}_{B([q])} + (-1)^p \mathrm{id}_{A([p])} \otimes d_B.$$

The Alexander-Whitney morphism AW_* between the chain complexes $A_* \otimes B_*$ and $\mathrm{Tot} \otimes (A \times B)$ induces a morphism

$$AW = \bigoplus_{n \in \mathbb{N}} AW_n : \bigoplus_{n \in \mathbb{N}} A([n]) \otimes B([n]) \to \left(\bigoplus_p A([p])\right) \otimes \left(\bigoplus_q B([q])\right) \qquad \text{(F.4)}$$

of the associated differential graded objects, which is associative by Lemma F.8.14.

Definition F.8.15. *The induced morphism AW of differential graded objects in Equation F.4 is also called the* Alexander-Whitney morphism.

Lemma F.8.16. *The Alexander-Whitney morphism AW between differential graded objects is associative.*

Proof. This is a consequence of Lemma F.8.14.   □

Similar considerations can be made for the Eilenberg-Zilber morphisms. For three simplicial objects A, B and C in **A** the Eilenberg-Zilber morphisms EZ for $A \times B$ and $[\mathrm{Tot} \otimes (A \times B)] \times C$ may be composed to obtain a morphism

$$\mathrm{Tot}\left(\mathrm{Tot}(\otimes(A \times B))_{*,*} \otimes C_*\right) \xrightarrow{EZ_*} (\mathrm{Tot}(\otimes(A \times B))_{*,*}) \otimes C_* \xrightarrow{EZ_* \otimes \mathrm{id}_{C_*}} (A \otimes B)_* \otimes C_*$$

of chain complexes.

Proposition F.8.17. *The Eilenberg-Zilber morphisms for the bisimplicial objects A, B and C are associative, i.e. $EZ_*(EZ_* \times \mathrm{id}_{C_*}) = EZ_*(\mathrm{id}_{A_*} \times EZ_*)$.*

Proof. The proof of [EML53, Theorem 5.3,(5.10)] carries over.   □

This also can be expressed using the differential graded objects associated to the simplicial objects $\mathrm{Tot} \otimes (A \times B)$ and $A_* \times B_*$. Here the Eilenberg-Zilber morphism $EZ_* : \mathrm{Tot} \otimes (A \times B) \to A_* \otimes B_*$ induces a morphism

$$EZ = \bigoplus_n EZ_n : \left(\bigoplus_p A([p])\right) \otimes \left(\bigoplus_q B([q])\right) \to \bigoplus_n A([n]) \otimes B([n])$$

of differential graded objects. By Proposition F.8.17 this morphism EZ of differential graded objects is associative up top homotopy.

Lemma F.8.18. *The Eilenber-Zilber morphism EZ between differential graded objects is associative.*

Proof. This is a consequence of Lemma F.8.17.   □

G

Smooth Alexander-Spanier Cohomology and de Rham Cohomology

In this appendix we discuss a smooth version of the Alexander-Spanier cohomology of differential manifolds. We examine the relation of the smooth Alexander-Spanier cohomology $H^*_{AS,s}(M;V)$ with coefficients in a topological vector space V of a differential manifold M with the de Rham cohomology $H^*_{dR}(M;V)$ of M. We introduce a "derivation homomorphism" $\tau^* : \Omega^*(M;V) \to A^*_{AS}(M;V)$ of chain complexes which explicitly assigns a differential n-form $\tau^n(f)$ to each smooth Alexander-Spanier cochain f on M. This homomorphism goes back to an observation of van Est and Korthagen in the appendix of [vEK64]. All topological vector spaces and manifolds in this appendix are considered over a fixed topological field $\mathbb{K}$ (e.g. $\mathbb{K} = \mathbb{R}$ or $\mathbb{K} = \mathbb{C}$).

G.1 Smooth Alexander-Spanier Cohomology

Let (G, M) be a smooth transformation group, i.e. G is a Lie group acting smoothly on a differential manifold M. The (equivariant) smooth Alexander-Spanier cohomology of a differential manifold is defined similar to the classical Alexander-Spanier cohomology; the only difference is that one only uses (equivariant) smooth functions on neighbourhoods of the diagonals in the spaces M^{n+1}.

Definition G.1.1. *Let $\mathfrak{U}$ be a family of subsets of a topological space M. An n-simplex $(m_0, \ldots, m_n)$ in $\mathrm{SS}(M)([n])$ is called $\mathfrak{U}$-small if there exists a set $U \in \mathfrak{U}$ such that the product space U^{n+1} contains $(m_0, \ldots, m_n)$. A subset W of M is called $\mathfrak{U}$-small if there exists a set $U \in \mathfrak{U}$ containing W.*

Definition G.1.2. *The subspace of $\mathrm{SS}(M)([n])$ consisting only of $\mathfrak{U}$-small simplices is denoted by $\mathrm{SS}(M; \mathfrak{U})([n])$.*

In case there is no restriction on the n-simplices (e.g. if the family $\mathfrak{U}$ contains the space M,) we simply write $\mathrm{SS}(M)$ for the space of $\mathfrak{U}$-small n-simplices. Since boundaries and degeneracies of $\mathfrak{U}$-small simplices are $\mathfrak{U}$-small as well, the spaces $\mathrm{SS}(M, \mathfrak{U})([n])$ form a simplicial subspace of the standard simplicial space $\mathrm{SS}(M)$ of the topological space M.

Definition G.1.3. *The simplicial space $\mathrm{SS}(M, \mathfrak{U})$ is called the $\mathfrak{U}$-small standard simplicial space of M.*

The topological spaces we are interested in this appendix are smooth manifolds. The category of smooth manifolds and smooth maps will be denoted by **Mf**. Observe that for each family $\mathfrak{U}$ of open subsets of M the spaces $\mathrm{SS}(M, \mathfrak{U})([n])$ of $\mathfrak{U}$-small n-simplices in M are open subspaces of M^{n+1} and therefore differential manifolds themselves. In particular $\mathrm{SS}(M, \mathfrak{U})$ is a simplicial differential manifold in this case. This justifies the following definition:

Definition G.1.4. *For each family $\mathfrak{U}$ of open subsets of M the simplicial manifold* $SS(M,\mathfrak{U})$ *is called the $\mathfrak{U}$-small standard simplicial manifold of M.*

Definition G.1.5. *For each smooth manifold M, every family $\mathfrak{U}$ of open subsets of M and Hausdorff topological vector space V the cosimplicial Abelian group* $\hom_{mf}(SS(M,\mathfrak{U}),V)$ *is denoted by $A_s(X,\mathfrak{U};V)$. It is called the $\mathfrak{U}$-small smooth standard cosimplicial Abelian group of X.*

Let $\mathfrak{U}$ and $\mathfrak{V}$ be families of open subsets of a differential manifold M. We write $\mathfrak{V} \prec \mathfrak{U}$ if $\mathfrak{V}$ is a refinement of $\mathfrak{U}$. In this case every $\mathfrak{V}$-small simplex also is $\mathfrak{U}$-small, so the simplicial manifold $SS(M,\mathfrak{V})$ is a simplicial submanifold of $SS(M,\mathfrak{U})$. The inclusion of $SS(M,\mathfrak{V})$ as a simplicial submanifold induces a restriction morphism $i_{\mathfrak{V},\mathfrak{U}} : A_s(M,\mathfrak{U};V) \to A_s(M,\mathfrak{V};V)$ of cosimplicial Abelian groups. Thus the $\mathfrak{U}$-small smooth standard cosimplicial Abelian groups $A(X,\mathfrak{U};V)$ for different families $\mathfrak{U}$ of subsets of X (with the relation "is refined by") form a directed set of cosimplicial Abelian groups. The same is true if one restricts oneself to open coverings of M.

Definition G.1.6. *The colimit cosimplicial Abelian group* $\operatorname{colim}_{\mathfrak{U}} A(M,\mathfrak{U};V)$ *where $\mathfrak{U}$ ranges over all open coverings of M is denoted by $A_{AS,s}(M;V)$. It is called the smooth Alexander-Spanier cosimplicial group with coefficients V of X. The elements of $A^n_{AS,s}(X;V)([n])$ are called smooth Alexander-Spanier n-cochains.*

This is the cosimplicial topological group of germs of smooth functions defined on a neighbourhood of the diagonal. In case the differential manifold M is smoothly paracompact, the smooth Alexander-Spanier cosimplicial Abelian group $A_{AS,s}(X;V)$ of M can also be obtained from the smooth standard cosimplicial Abelian group $A_s(X;V)$ by factoring out the cosimplicial subgroup formed by the subgroups

$$A_0(M;V)([n]) := \{f \in A_s(M;V)([n]) \mid \exists \text{ open covering } \mathfrak{U} \text{ of } M : f_{|SS(M,\mathfrak{U})([n])} = 0\}$$

of smooth functions that vanish on a neighbourhood of the diagonal. The morphisms forming the colimit cocone under the directed set of cosimplicial groups $A_s(M,\mathfrak{U};V)$ will be denoted by $\varrho_{\mathfrak{U}} : A_s(M,\mathfrak{U};V) \to A_{AS,s}(M;V)$.

Definition G.1.7. *The functor $H_{AS,s}(-;V) := H \circ A_{AS,s}(-;V)$ is called the smooth Alexander-Spanier topological cohomology functor with coefficients V. The graded Abelian topological group $H_{AS,s}(M;V)$ associated to a manifold M is called its smooth Alexander-Spanier cohomology with coefficients V.*

G.2 The Derivation Morphism

Let M be a smooth manifold, V a Hausdorff topological vector space and the function $f : M^{n+1} \to V$ be smooth. We denote the points of M^{n+1} by $\mathbf{m} = (m_0,\ldots,m_n)$, where the first index is zero. If X is a smooth vector field on M we define a smooth function $\partial_i(X)f$ on M^{n+1} by the partial derivative

$$\partial_i(X)f : M^{n+1} \to V, \quad [\partial_i(X)f](\mathbf{m}) = df(\mathbf{m})(0,\ldots,0,X_i,0,\ldots,0)$$

of f. For vector fields $X_1,\ldots,X_n$ on M one can iterate this process to obtain a smooth function

$$\partial_1(X_1)\cdots\partial_n(X_n)f : M^{n+1} \to V.$$

Applying this procedure to open subsets U of M and antisymmetrising in the arguments $X_1,\ldots,X_n$ leads to a "derivation homomorphism" $A^n_s(M;V) \to \Omega^n(M;V)$ which assigns to each smooth Alexander-Spanier n-cochain a differential n-form.

Recall that a smooth Alexander-Spanier n-cochain is the germ of a smooth function f on a neighbourhood of the diagonal in M^{n+1}. So, if $[f] \in A_s^n(M;V)$ is an Alexander-Spanier n-cochain, then there exists an open covering $\mathfrak{U}$ of M such that $[f]$ is the germ of a smooth function

$$f : \mathrm{SS}(M,\mathfrak{U})([n]) \to V$$

on the open neighbourhood $\mathrm{SS}(M,\mathfrak{U})([n])$ of the diagonal in M^{n+1}. If $X_1,\ldots,X_n$ are tangent vectors at a point $m \in M$, then there exists a $\mathfrak{U}$-small open neighbourhood W of m such that the tangent vectors $X_1,\ldots,X_n \in T_m M$ can be extended to smooth vector fields $X_1,\ldots,X_n$ on W.

Lemma G.2.1. *For each smooth Alexander-Spanier n-cochain $[f] \in A^n(M;V)$ on M and tangent vectors $X_1,\ldots,X_n$ at a point $m \in M$ and local extensions $\tilde{X}_1,\ldots,\tilde{X}_n$ to smooth vector fields on a neighbourhood of m the value of the smooth function*

$$\partial_1(\tilde{X}_1)\cdots\partial_n(\tilde{X}_n)f : M^{n+1} \to V$$

on $T_m^{\oplus n}M$ does not depend on the extensions $\tilde{X}_1,\ldots\tilde{X}_n$ of $X_1,\ldots,X_n$ chosen. Moreover the assignment

$$\tau^n(f) : T^{\oplus n}M \to V,$$
$$\tau^n(f)(X_1,\ldots,X_n) = \sum_{\sigma \in S_n} \mathrm{sign}(\sigma) \cdot [\partial_1(\tilde{X}_{\sigma(1)})\cdots\partial_n(\tilde{X}_{\sigma(n)})f](m,\ldots,m)$$

for any local extension $\tilde{X}_1,\ldots,\tilde{X}_n$ of the tangent vectors $X_1,\ldots,X_n$ to smooth vector fields is a differential n-form on M.

Proof. For the proof we refer to the appendix A in [Nee04]. $\square$

Since the maps $\tau^n : A_s^n(M;V) \to \Omega^n(M;V)$ are linear in the Alexander-Spanier cochains by construction, they form homomorphisms of (non-topological) $\mathbb{K}$-vector spaces.

Definition G.2.2. *The homomorphisms $\tau^n : A_s^n(M;V) \to \Omega^n(M;V)$ of abstract vector spaces are called* derivation homomorphisms.

Examining the derivation homomorphisms further one finds that they intertwine the differentials on the chain complexes $A_s^*(M;V)$ and $\Omega^*(M;V)$:

Lemma G.2.3. *The maps τ^n form a morphism $\tau^* : A_s^*(M;V) \to \Omega^*(M;V)$ of chain complexes.*

Proof. The proof of [Nee04, Proposition A.6] carries over to our slightly more general case. $\square$

Definition G.2.4. *The morphism $\tau^* : A_s^*(M;V) \to \Omega^*(M;V)$ of chain complexes id called the* derivation morphism.

Now consider a action $\varrho : G \times M \to M$ of a group G by diffeomorphisms on the manifold M. Then the group G also acts by diffeomorphisms on the tangent bundle TM. If we also equip the coefficients V with a G-action by continuous linear maps then we find that the assignment

$$C^\infty(M;V) \times TM \to V, \quad (f,X) \mapsto df(X)$$

intertwines the actions of G on $C^\infty(M;V)$, TM and V. As a consequence we note:

Lemma G.2.5. *The derivation morphism τ^* is equivariant.*

Proof. This follows from the fact that for each action of G by diffeomorphisms on M and every action of G on V by continuous linear maps taking (partial) derivatives in prescribed directions is an equivariant map $C^\infty(M;V) \times TM \to V$. $\square$

References

[Are52]   Richard Arens. Extension of functions on fully normal spaces. *Pacific J. Math.*, 2:11–22, 1952.

[Are53]   Richard Arens. Extension of coverings, of pseudometrics, and of linear-space-valued mappings. *Canadian J. Math.*, 5:211–215, 1953.

[Ark62]   Martin Arkowitz. The generalized Whitehead product. *Pacific J. Math.*, 12:7–23, 1962.

[Ark71]   Martin Arkowitz. Whitehead products as images of Pontrjagin products. *Trans. Amer. Math. Soc.*, 158:453–463, 1971.

[AS68]   R. A. Alò and H. L. Shapiro. Extensions of totally bounded pseudometrics. *Proc. Amer. Math. Soc.*, 19:877–884, 1968.

[BB04]   Francis Borceux and Dominique Bourn. *Mal'cev, protomodular, homological and semi-abelian categories*, volume 566 of *Mathematics and its Applications*. Kluwer Academic Publishers, Dordrecht, 2004.

[BC05]   F. Borceux and Maria Manuel Clementino. Topological semi-abelian algebras. *Adv. Math.*, 190(2):425–453, 2005.

[BC06]   F. Borceux and Maria Manuel Clementino. Topological protomodular algebras. *Topology Appl.*, 153(16):3085–3100, 2006.

[Beg87]   Edwin J. Beggs. The de Rham complex on infinite-dimensional manifolds. *Quart. J. Math. Oxford Ser. (2)*, 38(150):131–154, 1987.

[BGN04]   W. Bertram, H. Glöckner, and K.-H. Neeb. Differential calculus over general base fields and rings. *Expo. Math.*, 22(3):213–282, 2004.

[Bis02]   Daniel K. Biss. The topological fundamental group and generalized covering spaces. *Topology Appl.*, 124(3):355–371, 2002.

[BM78]   Ronald Brown and Sidney A. Morris. Embeddings in contractible or compact objects. *Colloq. Math.*, 38(2):213–222, 1977/78.

[Bou71]   N. Bourbaki. *Éléments de mathématique. Topologie générale. Chapitres 1 à 4*. Hermann, Paris, 1971.

[Bou03]   Dominique Bourn. The denormalized 3×3 lemma. *J. Pure Appl. Algebra*, 177(2):113–129, 2003.

[Bre97a]   Glen E. Bredon. *Sheaf theory*, volume 170 of *Graduate Texts in Mathematics*. Springer-Verlag, New York, second edition, 1997.

[Bre97b]   Glen E. Bredon. *Topology and geometry*, volume 139 of *Graduate Texts in Mathematics*. Springer-Verlag, New York, 1997. Corrected third printing of the 1993 original.

[BT82]   Raoul Bott and Loring W. Tu. *Differential forms in algebraic topology*, volume 82 of *Graduate Texts in Mathematics*. Springer-Verlag, New York, 1982.

[CS52a]   Henri Cartan and Jean-Pierre Serre. Espaces fibrés et groupes d'homotopie. I. Constructions générales. *C. R. Acad. Sci. Paris*, 234:288–290, 1952.

[CS52b]   Henri Cartan and Jean-Pierre Serre. Espaces fibrés et groupes d'homotopie. II. Applications. *C. R. Acad. Sci. Paris*, 234:393–395, 1952.

[DG86]   Jerzy Dydak and Ross Geoghegan. The singular cohomology of the inverse limit of a Postnikov tower is representable. *Proc. Amer. Math. Soc.*, 98(4):649–654, 1986.

534   References

[DG88]   Jerzy Dydak and Ross Geoghegan. Correction to: "The singular cohomology of
          the inverse limit of a Postnikov tower is representable" [Proc. Amer. Math. Soc.
          98 (1986), no. 4, 649–654; MR0861769 (87m:55024)]. *Proc. Amer. Math. Soc.*,
          103(1):334, 1988.

[DK70]   B. J. Day and G. M. Kelly. On topologically quotient maps preserved by pull-
          backs or products. *Proc. Cambridge Philos. Soc.*, 67:553–558, 1970.

[DL59]   Albrecht Dold and Richard Lashof. Principal quasi-fibrations and fibre homo-
          topy equivalence of bundles. *Illinois J. Math.*, 3:285–305, 1959.

[DP61]   Albrecht Dold and Dieter Puppe. Homologie nicht-additiver Funktoren. Anwen-
          dungen. *Ann. Inst. Fourier Grenoble*, 11:201–312, 1961.

[DT58]   Albrecht Dold and René Thom. Quasifaserungen und unendliche symmetrische
          Produkte. *Ann. of Math. (2)*, 67:239–281, 1958.

[Dug50]   J. Dugundji. A topologized fundamental group. *Proc. Nat. Acad. Sci. U. S. A.*,
          36:141–143, 1950.

[Dug89]   James Dugundji. *Topology*. Wm. C. Brown Publishers, Dubuque, Iowa, 2 edition,
          1989. Reprinting of the 1966 original, Allyn and Bacon Series in Advanced
          Mathematics.

[ELS04]   Martín Escardó, Jimmie Lawson, and Alex Simpson. Comparing Cartesian
          closed categories of (core) compactly generated spaces. *Topology Appl.*, 143(1-
          3):105–145, 2004.

[EML53]   Samuel Eilenberg and Saunders Mac Lane. On the groups of $H(\Pi, n)$. I. *Ann.
          of Math. (2)*, 58:55–106, 1953.

[Eps66]   D. B. A. Epstein. Semisimplicial objects and the Eilenberg-Zilber theorem.
          Invent. Math., 1:209–220, 1966.

[EVdL04] T. Everaert and T. Van der Linden. Baer invariants in semi-abelian categories.
          II. Homology. *Theory Appl. Categ.*, 12:No. 4, 195–224 (electronic), 2004.

[EZ50]   Samuel Eilenberg and J. A. Zilber. Semi-simplicial complexes and singular ho-
          mology. *Ann. of Math. (2)*, 51:499–513, 1950.

[EZ53]   Samuel Eilenberg and J. A. Zilber. On products of complexes. *Amer. J. Math.*,
          75:200–204, 1953.

[FHT01a] Yves Félix, Stephen Halperin, and Jean-Claude Thomas. *Rational homotopy
          theory*, volume 205 of *Graduate Texts in Mathematics*. Springer-Verlag, New
          York, 2001.

[FHT01b] Yves Félix, Stephen Halperin, and Jean-Claude Thomas. The Serre spectral
          sequence of a multiplicative fibration. *Trans. Amer. Math. Soc.*, 353(9):3803
          3831 (electronic), 2001.

[Fuc03]   Martin Fuchssteiner. Kohomologien von Lie-Gruppen. Diplomarbeit, Technsiche
          Universität Darmstadt, 2003.

[Gan68]   T. E. Gantner. Extensions of uniformly continuous pseudometrics. *Trans. Amer.
          Math. Soc.*, 132:147–157, 1968.

[GGH06]   H. Glöckner, R. Gramlich, and T. Hartnick. Final group topologies, Kac-Moody
          groups and Pontryagin duality. *ArXiv Mathematics e-prints*, March 2006.

[Glö04]   H. Glöckner. Extending the Scope of Lie Theory to Larger Classes of Groups.
          Habilitationsschrift, Technische Universität Darmstadt, 2004.

[Gra51]   M. I. Graev. Free topological groups. *Amer. Math. Soc. Translation*,
          1951(35):61, 1951.

[GZ67]   P. Gabriel and M. Zisman. *Calculus of fractions and homotopy theory.* Ergeb-
          nisse der Mathematik und ihrer Grenzgebiete, Band 35. Springer-Verlag New
          York, Inc., New York, 1967.

[Ham82]   Richard S. Hamilton. The inverse function theorem of Nash and Moser. *Bull.
          Amer. Math. Soc. (N.S.)*, 7(1):65–222, 1982.

[HM98]   Karl H. Hofmann and Sidney A. Morris. *The structure of compact groups*,
          volume 25 of *de Gruyter Studies in Mathematics*. Walter de Gruyter & Co.,
          Berlin, 1998. A primer for the student—a handbook for the expert.

[Hof79]   Rudolf-E. Hoffmann. On weak Hausdorff spaces. *Arch. Math. (Basel)*,
          32(5):487–504, 1979.

[HS53a]   G. Hochschild and J.-P. Serre. Cohomology of group extensions. *Trans. Amer.
          Math. Soc.*, 74:110–134, 1953.

[HS53b] G. Hochschild and J.-P. Serre. Cohomology of Lie algebras. *Ann. of Math. (2)*, 57:591–603, 1953.

[Isb64] J. R. Isbell. *Uniform spaces.* Mathematical Surveys, No. 12. American Mathematical Society, Providence, R.I., 1964.

[Iva85] N. V. Ivanov. Foundations of the theory of bounded cohomology. *Zap. Nauchn. Sem. Leningrad. Otdel. Mat. Inst. Steklov. (LOMI)*, 143:69–109, 177–178, 1985. Studies in topology, V.

[Jac89] Nathan Jacobson. *Basic algebra. II.* W. H. Freeman and Company, New York, second edition, 1989.

[Kan58a] Daniel M. Kan. A combinatorial definition of homotopy groups. *Ann. of Math. (2)*, 67:282–312, 1958.

[Kan58b] Daniel M. Kan. On homotopy theory and c.s.s. groups. *Ann. of Math. (2)*, 68:38–53, 1958.

[KM97] Andreas Kriegl and Peter W. Michor. *The convenient setting of global analysis*, volume 53 of *Mathematical Surveys and Monographs*. American Mathematical Society, Providence, RI, 1997.

[KS02] Keith Kearnes and Luís Sequeira. Hausdorff properties of topological algebras. *Algebra Universalis*, 47(4):343–366, 2002.

[KS06] Masaki Kashiwara and Pierre Schapira. *Categories and sheaves*, volume 332 of *Grundlehren der Mathematischen Wissenschaften [Fundamental Principles of Mathematical Sciences]*. Springer-Verlag, Berlin, 2006.

[Lam76] W. F. Lamartin. Epics in the category of T_2 k-groups need not have dense range. *Colloq. Math.*, 36(1):37–41, 1976.

[Lam77] W. F. Lamartin. On the foundations of k-group theory. *Dissertationes Math. (Rozprawy Mat.)*, 146:32, 1977.

[May92] J. Peter May. *Simplicial objects in algebraic topology.* Chicago Lectures in Mathematics. University of Chicago Press, Chicago, IL, 1992. Reprint of the 1967 original.

[McC69] M. C. McCord. Classifying spaces and infinite symmetric products. *Trans. Amer. Math. Soc.*, 146:273–298, 1969.

[McC01] John McCleary. *A user's guide to spectral sequences*, volume 58 of *Cambridge Studies in Advanced Mathematics*. Cambridge University Press, Cambridge, second edition, 2001.

[Mey90] Jean-Pierre Meyer. Cosimplicial homotopies. *Proc. Amer. Math. Soc.*, 108(1):9–17, 1990.

[Mic66] Ernest Michael. $\aleph_0$-spaces. *J. Math. Mech.*, 15:983–1002, 1966.

[Mic68] Ernest Michael. Bi-quotient maps and Cartesian products of quotient maps. *Ann. Inst. Fourier (Grenoble)*, 18(fasc. 2):287–302 vii (1969), 1968.

[Mil57] John Milnor. The geometric realization of a semi-simplicial complex. *Ann. of Math. (2)*, 65:357–362, 1957.

[Mil67] R. James Milgram. The bar construction and abelian H-spaces. *Illinois J. Math.*, 11:242–250, 1967.

[ML63] Saunders Mac Lane. *Homology.* Die Grundlehren der mathematischen Wissenschaften, Bd. 114. Springer-Verlag, Berlin-Göttingen-Heidelberg, 1963.

[ML70] Saunders Mac Lane. The Milgram bar construction as a tensor product of functors. In *The Steenrod Algebra and its Applications (Proc. Conf. to Celebrate N. E. Steenrod's Sixtieth Birthday, Battelle Memorial Inst., Columbus, Ohio, 1970)*, Lecture Notes in Mathematics, Vol. 168, pages 135–152. Springer, Berlin, 1970.

[ML98] Saunders Mac Lane. *Categories for the working mathematician*, volume 5 of *Graduate Texts in Mathematics*. Springer-Verlag, New York, second edition, 1998.

[MLM94] Saunders Mac Lane and Ieke Moerdijk. *Sheaves in geometry and logic.* Universitext. Springer-Verlag, New York, 1994. A first introduction to topos theory, Corrected reprint of the 1992 edition.

[MMO73] John Mack, Sidney A. Morris, and Edward T. Ordman. Free topological groups and the projective dimension of a locally compact abelian group. *Proc. Amer. Math. Soc.*, 40:303–308, 1973.

[Mor56] Kiiti Morita. On decomposition spaces of locally compact spaces. *Proc. Japan Acad.*, 32:544–548, 1956.

[Mor69]   Sidney A. Morris. Varieties of topological groups. *Bull. Austral. Math. Soc.*, 1:145–160, 1969.

[Mor70a]   Sidney A. Morris. Varieties of topological groups. II. *Bull. Austral. Math. Soc.*, 2:1–13, 1970.

[Mor70b]   Sidney A. Morris. Varieties of topological groups. III. *Bull. Austral. Math. Soc.*, 2:165–178, 1970.

[Mor73]   Sidney A. Morris. Varieties of topological groups and left adjoint functors. *J. Austral. Math. Soc.*, 16:220–227, 1973. Collection of articles dedicated to the memory of Hanna Neumann, II.

[Mor82]   Sidney A. Morris. Varieties of topological groups: a survey. *Colloq. Math.*, 46(2):147–165, 1982.

[Nee02a]   Karl-Hermann Neeb. Central extensions of infinite-dimensional Lie groups. *Ann. Inst. Fourier (Grenoble)*, 52(5):1365–1442, 2002.

[Nee02b]   Karl-Hermann Neeb. Nancy lectures on infinite-gimensional lie groups. Preprint, March 2002.

[Nee04]   Karl-Hermann Neeb. Abelian extensions of infinite-dimensional Lie groups. In *Travaux mathématiques. Fasc. XV*, Trav. Math., XV, pages 69–194. Univ. Luxemb., Luxembourg, 2004.

[Nee06]   Karl-Hermann Neeb. Lie algebra extensions and higher order cocycles. *J. Geom. Symmetry Phys.*, 5:48–74, 2006.

[NV03]   K.-H. Neeb and C. Vizman. Flux homomorphisms and principal bundles over infinite dimensional manifolds. *Monatsh. Math.*, 139(4):309–333, 2003.

[O'M71]   Paul O'Meara. On paracompactness in function spaces with the compact-open topology. *Proc. Amer. Math. Soc.*, 29:183–189, 1971.

[Pes82]   V. G. Pestov. Some properties of free topological groups. *Vestnik Moskov. Univ. Ser. I Mat. Mekh.*, (1):35–37, 77, 1982.

[Pog70]   Detlev Poguntke. Epimorphisms of compact groups are onto. *Proc. Amer. Math. Soc.*, 26:503–504, 1970

[Rob96]   Derek J. S. Robinson. *A course in the theory of groups*, volume 80 of *Graduate Texts in Mathematics*. Springer-Verlag, New York, second edition, 1996.

[Sam54]   H. Samelson. Groups and spaces of loops. *Comment. Math. Helv.*, 28:278–287, 1954.

[Seg68]   Graeme Segal. Classifying spaces and spectral sequences. *Inst. Hautes Études Sci. Publ. Math.*, (34):105–112, 1968.

[Seg70]   Graeme Segal. Cohomology of topological groups. In *Symposia Mathematica, Vol. IV (INDAM, Rome, 1968/69)*, pages 377–387. Academic Press, London, 1970.

[Sha66]   H. L. Shapiro. Extensions of pseudometrics. *Canad. J. Math.*, 18:981–998, 1966.

[Sip86]   O. V. Sipacheva. Description of the topology of free topological groups without using universal uniform structures. In *General topology (Russian)*, pages 122–130, 168. Moskov. Gos. Univ., Moscow, 1986.

[Sip00]   Ol′ga V. Sipacheva. Free topological groups of spaces and their subspaces. *Topology Appl.*, 101(3):181–212, 2000.

[Sip03]   O. V. Sipachëva. The topology of a free topological group. *Fundam. Prikl. Mat.*, 9(2):99–204, 2003.

[Ste67]   N. E. Steenrod. A convenient category of topological spaces. *Michigan Math. J.*, 14:133–152, 1967.

[Ste68]   N. E. Steenrod. Milgram's classifying space of a topological group. *Topology*, 7:349–368, 1968.

[Str06]   Markus Stroppel. *Locally compact groups*. EMS Textbooks in Mathematics. European Mathematical Society (EMS), Zürich, 2006.

[Sug55]   Masahiro Sugawara. *H*-spaces and spaces of loops. *Math. J. Okayama Univ.*, 5:5–11, 1955.

[Tan05]   Yoshio Tanaka. On products of *k*-spaces. *Topology Appl.*, 146/147:593–602, 2005.

[Tay77]   Walter Taylor. Varieties of topological algebras. *J. Austral. Math. Soc. Ser. A*, 23(2):207–241, 1977.

[tD87]   Tammo tom Dieck. *Transformation groups*, volume 8 of *de Gruyter Studies in Mathematics*. Walter de Gruyter & Co., Berlin, 1987.

[tD91]   Tammo tom Dieck. *Topologie.* de Gruyter Lehrbuch. [de Gruyter Textbook].
           Walter de Gruyter & Co., Berlin, 1991.

[Tka83]   M. G. Tkachenko. On the completeness of free abelian topological groups. *Soviet
           Math. Dokl.*, 27(2):314–318, 1983.

[Usp94]   V. V. Uspenskij. The epimorphism problem for Hausdorff topological groups.
           Topology Appl., 57(2-3):287–294, 1994.

[vE53]   W. T. van Est. Group cohomology and Lie algebra cohomology in Lie groups. I,
           II. *Nederl. Akad. Wetensch. Proc. Ser. A.* **56** = *Indagationes Math.*, 15:484–492,
           493–504, 1953.

[vE55a]   W. T. van Est. On the algebraic cohomology concepts in Lie groups I,II. *Nederl.
           Akad. Wetensch. Proc. Ser. A.* **58** = *Indag. Math.*, 17:225–233, 286–294, 1955.

[vE55b]   W. T. van Est. Une application d'une méthode de Cartan-Leray. *Nederl. Akad.
           Wetensch. Proc. Ser. A.* **58** = *Indag. Math.*, 17:542–544, 1955.

[vE58]   W. T. van Est. A generalization of the Cartan-Leray spectral sequence. I, II.
           Nederl. Akad. Wetensch. Proc. Ser. A 61 = *Indag. Math.*, 20:399–413, 1958.

[vE62a]   W. T. van Est. Local and global groups. I. *Nederl. Akad. Wetensch. Proc. Ser.
           A 65* = *Indag. Math.*, 24:391–408, 1962.

[vE62b]   W. T. van Est. Local and global groups. II. *Nederl. Akad. Wetensch. Proc. Ser.
           A 65* = *Indag. Math.*, 24:409–425, 1962.

[vEK64]   W. T. van Est and Th. J. Korthagen. Non-enlargible Lie algebras. *Nederl. Akad.
           Wetensch. Proc. Ser. A 67=Indag. Math.*, 26:15–31, 1964.

[War83]   Frank W. Warner. *Foundations of differentiable manifolds and Lie groups*, vol-
           ume 94 of *Graduate Texts in Mathematics*. Springer-Verlag, New York, 1983.
           Corrected reprint of the 1971 edition.

[Wei94]   Charles A. Weibel. *An introduction to homological algebra*, volume 38 of *Cam-
           bridge Studies in Advanced Mathematics*. Cambridge University Press, Cam-
           bridge, 1994.

[Whi78]   George W. Whitehead. *Elements of homotopy theory*, volume 61 of *Graduate
           Texts in Mathematics*. Springer-Verlag, New York, 1978.

Wissenschaftlicher Werdegang

	Schulischer Werdegang
1979 – 1983	Christian-Morgenstern Schule, Darmstadt
1983 – 1992	Ludwig-Georgs-Gymnasium, Darmstadt
25.05.1992	Allgemeine Hochschulreife
01.07.1992 - 30.06.1993	Wehrdienst
	Beruflicher Werdegang
01.10.1993	Beginn des Studiums der Mathematik (Diplom) mit Nebenfach Physik an der TU Darmstadt
01.10.1993	Beginn des Studiums der Physik (Diplom) mit Nebenfach Physikalische Chemie an der TU Darmstadt
30.09.1995	Vordiplom im Studiengang Mathematik mit der Note "Schr gut"
30.09.1995	Vordiplom im Studiengang Physik mit der Note "Gut"
Oktober 1996	Diplomprüfung Physik "Astronomie und Astrophysik" mit der Note "Sehr gut"
Oktober 1997	Diplomprüfung Physik "Gruppen- und Darstellungstheorie" mit der Note "Sehr gut"
14. Oktober 2003	Diplom Mathematik ("mit Auszeichnung bestanden")
01.10.2004 – 30.09.2009	Wissenschaftlicher Mitarbeiter am Fachbereich Mathematik der TU Darmstadt
Februar 2010	Promotion in Mathematik ("mit Auszeichnung bestanden")